Second Edition

Advanced Digital
Optical Communications

Optics and Photonics

Series Editor

Le Nguyen Binh

Huawei Technologies, European Research Center, Munich, Germany

Second Edition

Advanced Digital
Optical Communications

Le Nguyen Binh

EUROPEAN RESEARCH CENTER,
HUAWEI TECHNOLOGIES, MUNICH, GERMANY

CRC Press
Taylor & Francis Group
Boca Raton London New York

CRC Press is an imprint of the
Taylor & Francis Group, an **informa** business

CRC Press
Taylor & Francis Group
6000 Broken Sound Parkway NW, Suite 300
Boca Raton, FL 33487-2742

First issued in paperback 2017

© 2015 by Taylor & Francis Group, LLC
CRC Press is an imprint of Taylor & Francis Group, an Informa business

No claim to original U.S. Government works

ISBN-13: 978-1-4822-2652-2 (hbk)
ISBN-13: 978-1-138-74954-2 (pbk)

Library of Congress Cataloging-in-Publication Data

Binh, Le Nguyen.
  [Digital optical communications]
  Advanced digital optical communications / author, Le Nguyen Binh. -- Second edition.
     pages cm. -- (Optics and photonics)
  Includes bibliographical references and index.
  ISBN 978-1-4822-2652-2 (alk. paper)
  1. Optical communications. 2. Digital communications. 3. Computer networks. I. Title.

TK5103.59.B52 2014
621.382--dc23
                                                                          2014035373

Visit the Taylor & Francis Web site at
http://www.taylorandfrancis.com

and the CRC Press Web site at
http://www.crcpress.com

To my parents
To my wife, Phuong, and my son, Lam.

Contents

Preface

The trends in long-haul terrestrial and intercontinental telecommunications systems and networks are toward longer transparent distance and increased information capacity with cost-effective implementation of the communications systems.

One of the possibilities is the use of modulation formats and coherent reception incorporating digital signal processors (DSPs), analog-to-digital converters (ADCs), and digital-to-analog converters (DACs) providing a sampling rate of 120 GSa/s rather than the on–off keying intensity modulation and direct detection traditionally employed in current optical communications networks; that is, the use of the amplitude, phase, and frequency of the lightwave carrier and formats of return-to-zero and non-return-to-zero with and without suppression of the carrier to represent the information for effective transmission. These modulation formats must be robust to noise, distortion, and nonlinear impairments and have high spectral efficiency. The DSP operating at an extremely high speed allows system engineers to overcome several problems in the reception subsystems of the coherent transmission technology initiated in the 1980s.

Thus, we have been witnessing the merging of the fields of digital communications and optical communications in advanced transmission techniques for long-haul optically amplified communications in global systems networks. Furthermore, the DSP-coherent reception techniques with polarization multiplexing and complex modulation schemes have pushed the current optical transmission technology to a new generation, with bit rates of 100G and beyond deployed extensively in several terrestrial and undersea telecommunications networks as well as in metropolitan and access networking environments. Therefore, in this book, digital modulation techniques are applied to optical communications systems, including binary and quadrature amplitude shift keying, phase shift keying, differential shift keying, continuous phase shift keying, frequency shift keying, and multilevel modulation as well as the use of multi-subcarriers such as orthogonal frequency division multiplexing. Both coherent and self-coherent reception techniques are described in this second edition, whereas only the latter technique was treated in the previous edition.

Further, the formats of the modulation with return-to-zero or non-return-to-zero and suppression of carrier can be employed so as to minimize the contribution of the carrier energy, to combat the nonlinear impairment with the enhancement of the energy contained within the signal spectra. Furthermore, with the advances in digital signal processing with a high sampling rate, lightwave-modulated signals can be detected coherently to recover the phase of the signals and then processed via DSPs. Thus, we have now witnessed the revival of coherent communications, which was developed during the 1980s to improve the optical signal to noise ratio for extending the repeaterless distance, before optical amplifiers made their appearance.

The modulation and transmission are treated theoretically, experimentally, and by simulation. MATLAB® and Simulink® is extensively described in the chapters as this modeling platform is now universally available in research laboratories in industries and universities around the world. So it is expected that the models provided in this book will assist researchers and engineers in further developing specific optical communications systems employing new modulation formats in MATLAB and Simulink is preferred as the software package within MATLAB because it is based on block sets, thus resulting in ease of use and shortening of learning and development time. The use of MATLAB and Simulink would require users to understand the principles of digital communications with various communications and mathematical blocks. There are no such optical communications block sets in MATLAB and Simulink, so one of the main objectives of this book is to provide the operational principles of optical communications blocks as examples for users who wish to model their systems.

The objectives of the book are (1) to describe the principles of digital communications within the context of optical and photonic transmission technology; (2) to present the advanced optical transmission of various digital modulation formats over optically amplified fiber links within the contexts of experiment, theory, and modeling; and (3) to identify the roles of the frequency spectra of digital modulation formats for effective transmission of signals, to combat transmission impairments, and to increase the communications capacity.

This book, *Advanced Digital Optical Communications*, is the second edition of the book *Digital Optical Communications* that was published in 2008 [1]. It has extensive chapters treating modern aspects of coherent homodyne reception techniques using algorithms incorporated in DSP systems and DSP-based transmitters to overcome several transmission linear and nonlinear impairments and frequency mismatching between the local oscillator and the carrier as well as clock recovery and cycle slips. The differences between the two editions of the book are given in Chapter 1.

The author can be contacted via le.nguyen.binh@huawei.com.

REFERENCE

1. L. N. Binh, *Digital Optical Communications*, Boca Raton, FL: CRC Press, 2008.

MATLAB® is a registered trademark of The MathWorks, Inc. For product information, please contact:

The MathWorks, Inc.
3 Apple Hill Drive
Natick, MA 01760-2098 USA
Tel: 508 647 7000
Fax: 508-647-7001
E-mail: info@mathworks.com
Web: www.mathworks.com

For information on samples of MATLAB and Simulink models and problem solving techniques, especially problems given at the end of a number of chapters refer to www.crcpress/products/ISBN9781482226522.com

Acknowledgments

I am grateful to Huawei Technologies for the availability of laboratories and facilities to carry out the studies. The chapters of this edition were finalized while I was technical director of the company's European Research Centre in Munich, while the first version was structured at the Christian Albretchs University of Kiel in Germany, where I spent a sabbatical year as a professorial fellow in the Chair of Communications. Thus, I thank Prof. Dr. W. Rosenkranz for the many fruitful discussions with him.

I am grateful to Dr. Thomas Lee of SHF AG, Berlin, Germany, and formerly of Nortel Networks, for advice and the loan of various optical transmitters and receivers and purchase of 40 Gbps BERT for the experiments on ASK and DPSK using various formats that I have conducted since 2004. I am also indebted to my undergraduate and postgraduate students for discussions over the years on development and research in optical communications, including Dr. S. V. Chung of Corning Optical Systems (Australia); Prof. John Ngo of Nanyang Technological University of Singapore; Dr. K. F. Chang, Dr. T. L. Huynh, Dr. W. J. Lai, Dr. N. Nguyen, C. Li, H. S. Chong, H. C. Chong, and H. Q. Lam of the Department of Electrical and Computer Systems Engineering of Monash University of Melbourne, Australia. While attached to the Chair of Communications of the Christian Albretchs University of Kiel, as a professorial fellow in 2008 to write the chapters of the first edition of the book, I am grateful for the interesting discussions with Dr. Ing. Jochen Lebrich, Dr. Ing. Abdulamir Ali, Dr. Ing. Chumin Xia, and Stefan Schoemann, and the availability of the doctoral theses of Dr. C. Wree and Dr. Lebrich as well as the lecture notes on high-speed systems and networks of Prof. Dr. Rosenkranz.

I thank my colleagues at Huawei Technologies, especially Xu Xiaogeng, Yang Ning, Bruce Liu, Chen Ming, Dr. Mao Bangning, Dr. Xie Changsong, and Prof. Nebojsa Stojanovic for fruitful exchanges of ideas and information.

Author

Le Binh earned the degrees of BE (Hons) and PhD in electronic engineering and integrated photonics in 1975 and 1978, respectively, both from the University of Western Australia. He is currently working with Huawei Technologies European Research Centre as technical director in Munich, Germany.

He has authored and co-authored more than 300 papers in leading journals and eight books (besides refereeing conferences) in the fields of photonic signal processing, digital optical communications, and integrated optics, all published by CRC Press, Taylor & Francis Group of Boca Raton, Florida, USA (www.crcpress.com). His current research interests are in advanced modulation formats for long-haul optical transmission, electronic equalization techniques for optical transmission systems, ultra-short-pulse lasers and photonic signal processing, optical transmission systems and network engineering, and Si-on-SiO_2-integrated photonics, especially the 5G optical switched flexible grid wireless-optical networks. He was Chair of Commission D (Electronics and Photonics) of the National Committee for Radio Sciences of the Australian Academy of Sciences (1995–2005). Dr. Binh was professorial fellow at Nanyang Technological University of Singapore and the Christian Albretchs University of Kiel, Germany, and several Australian universities. He is an alumnus of the Phan Chu Trinh and Phan Boi Chau high schools of Phan Thiet, Vietnam.

Currently, he is the series editor of "Photonics and Optics" for CRC Press.

Acronyms

Acronym	Meaning
ADC	Analog-to-digital converter
AM	Amplitude modulation/modulator
ASE	Amplified spontaneous emission
ASK	Amplitude shift keying
BDPSK	Binary differential phase shift keying
CD	Chromatic dispersion
CMA	Constant modulus algorithm
CPE	Constant phase estimation
CPFSK	Continuous phase frequency shift keying
CSRZ	Carrier-suppressed return-to-zero format
DAC	Digital-to-analog converter
DB	Duobinary
DBM	Duobinary modulation
DCF	Dispersion-compensating fiber
DCM	Dispersion-compensating module
DD	Direct detection
demux	Demultiplexer
DFB	Distributed feedback (laser)
DI	Delay interferometer
DP	Dual polarization
DPSK	Differential phase shift keying
DQPSK	Differential quadrature phase shift keying
DSF	Dispersion shifted fiber
DSP	Digital signal processing (processor)
EDFA	Erbium-doped fiber amplifier
FEC	Forward error coding
FMF	Few-mode fiber
FWM	Four-wave mixing
GVD	Group velocity dispersion
IF	Intermediate frequency
IM	Intensity modulation/modulator
IM/DD	Intensity modulation/direct detection
I–Q	In-phase and quadrature
ITU	International Telecommunications Union
LDPC	Low-density parity check
LO	Local oscillator
MADPSK	Multilevel (M-ary) amplitude differential phase shift keying
MIMO	Multiple-input multiple-output
MMF	Multimode optical fibers
MSK	Minimum shift keying
mux	Multiplexers

MZI	Mach–Zehnder interferometer
MZDI	Mach–Zehnder delay interferometer
MZIM	Mach–Zehnder interferometer modulator or Mach–Zehnder interferometric intensity modulator
NF	Noise figure
NLPN	Nonlinear phase noise
NLSE	Nonlinear Schrödinger equation
NRZ	Non-return-to-zero
NZDSF	Nonzero dispersion-shifted fiber (ITU-655)
OA	Optical amplifier
ODPSK	Offset differential phase shift keying
OFDM	Orthogonal frequency division multiplexing
oOFDM	Optical OFDM
OOK	On–off keying or amplitude shift keying (ASK)
OPLL	Optical phase-locked loop
OSNR	Optical signal-to-noise ratio
PDF	Probability density function
PDP	Photodetector pair
PLC	Planar lightwave circuits
PLL	Phase-locked loop
PM	Phase modulator
PMD	Polarization-mode dispersion
PMF	Polarization maintaining fiber or 4QAM
QAM	Quadrature amplitude modulation
QPSK	Quadrature phase shift keying
ROA	Raman optical amplifier
RZ	Return-to-zero
RZ33	RZ pulse of width of 33% of bit period format
RZ50	RZ pulse of width of 50% of bit period format
RZ67	RZ pulse of width of 67% of bit period format (normally CSRZ)
SMF	Single-mode fiber
SPM	Self-phase modulation
SSMF	Standard single-mode fiber (ITU-652)
STAR-QAM	Star constellation quadrature amplitude modulation
TE	Transverse electric
TM	Transverse magnetic
TOD	Third-order dispersion
XPM	Cross-phase modulation

1 Introduction

This chapter outlines briefly the historical development, emergence, and merging of the fundamental digital communication and optical communication techniques to fully exploit and respond to the challenges of the availability of ultra-high frequency and ultra-wideband in the optical spectra of optical fiber communications technology. The organization of rest of the chapters of the book is outlined in this chapter.

1.1 DIGITAL OPTICAL COMMUNICATIONS AND TRANSMISSION SYSTEMS: CHALLENGING ISSUES

Starting from the proposed dielectric waveguides by Kao and Hockham [1,2] in 1966, the first research phase attracted intensive interest around the early 1970s in the demonstration of fiber optics, and optical communications has greatly progressed over the past three decades. The first-generation lightwave systems were commercially deployed in 1983 and operated in the first wavelength window of 800 nm over multimode optical fiber (MMF) at transmission bit rates of up to 45 Mbps [1–3]. After the introduction of ITU-G652 standard single-mode fiber (SSMF) in the late 1970s [3,4], the second generation of lightwave transmission systems became available in the early 1980s [5,6]. The operating wavelengths were shifted to the second window of 1300 nm, which offers much lower attenuation for silica-based optical fiber as compared to the previous 800 nm region. In particular the chromatic dispersion (CD) factor is close zero. This spectral window is current attracting lots of interests for optical interconnections for data centers. These second-generation systems could operate at bit rates of up to 1.7 Gbps and have a repeaterless transmission distance of about 50 km [7]. Further research and engineering efforts were also devoted to the improvement of the receiver sensitivity by coherent detection techniques, and the repeaterless distance reached 60 km in installed systems with a bit rate of 2.5 Gbps. Optical fiber communications then evolved to third-generation transmission systems that utilized the lowest-attenuation 1550 nm wavelength window and operated up to a bit rate of 2.5 Gbps [7,8]. These systems were commercially available in 1990 with a repeater spacing of around 60–70 km [7,9]. At this stage, the generation of optical signals was based on direct modulation of the semiconductor laser source and either direct detection. Since the invention of erbium-doped fiber amplifiers (EDFAs) in the early 1990s [10–12], lightwave systems have rapidly evolved to wavelength division multiplexing (WDM) and shortly after that to dense WDM (DWDM) optically amplified transmission systems that are capable of transmitting multiple 10 Gbps channels. This is because the loss is no longer a major issue for external optical modulators, which normally suffer an insertion loss of at least 3 dB. These modulators allow the preservation of the narrow linewidth of distributed feedback lasers (DFBs). These high-speed and high-capacity systems extensively exploited the external modulation in their optical transmitters. The present optical transmission systems are considered as the fifth generation, having a transmission capacity of a few terabits per second [7].

Coherent detection, homodyne or heterodyne, was the focus of extensive research and development during the 1980s and early 1990s [13–18] and was the main detection technique used in the first three generations of lightwave transmission systems. At that time, the main motivation for the development of coherent optical systems was to improve the receiver sensitivity, commonly by 3–6 dB [14,17]. Thus, the repeaterless transmission distance could be extended to more than 60 km of SSMF (with a 0.2 dB/km attenuation factor). However, coherent optical systems suffer severe performance degradation due to fiber dispersion impairments. In addition, the phase coherence for lightwave carriers of the laser source and the local laser oscillator were very difficult to maintain. On the contrary, the incoherent detection technique minimizes the linewidth obstacles of the laser source as well as the local laser oscillator, and thus relaxes the requirement of the phase coherence. Moreover, incoherent detection mitigates the problem of polarization control in the mixing of transmitted lightwaves and the local laser oscillator in the multiterahertz optical frequency range. The invention of EDFAs, which are capable of producing optical gains of 20 dB and above, has also greatly contributed to the progress of incoherent digital photonic transmission systems up to now.

Recent years have witnessed a huge increase in demand for broadband communications driven mainly by the rapid growth of multimedia services, peer-to-peer networks, and IP streaming services, in particular IP TV. It is most likely that such tremendous growth will continue in the coming years. This is the main driving force for local and global telecommunications service carriers to develop high-performance and high-capacity next-generation optical networks. The overall capacity of WDM or DWDM optical systems can be boosted either by increasing the base transmission bit rate of each optical channel, multiplexing more channels in a DWDM system or, preferably, by combining both of these schemes. However, while implementing these schemes, optical transmission systems encounter a number of challenging issues, which are outlined in the following paragraphs Figure 1.1.

Current 10 Gbps transmission systems employ intensity modulation (IM), also known as on–off keying (OOK), and utilize non-return-to-zero (NRZ) pulse shapes. The term OOK can also be used interchangeably with amplitude shift keying (ASK) [1,2].* For high-bit-rate transmission such as 40 Gbps, the performance of OOK photonic transmission systems is severely degraded owing to fiber impairments, including fiber dispersion and fiber nonlinearities. The fiber dispersion is classified as CD and polarization-mode dispersion (PMD), causing the intersymbol interference (ISI) problem. On the contrary, severe deterioration in the system performance due to fiber nonlinearities result from high-power spectral components at the carrier and signal frequencies of OOK-modulated optical signals. It is also of concern that existing transmission networks comprise millions of kilometers of SSMF, which have been installed for approximately two decades. These fibers do not have as advanced properties as the state-of-the-art fibers used in recent laboratory "hero" experiments, and they have degraded after many years of use.

The total transmission capacity can be enhanced by increasing the number of multiplexed DWDM optical channels. This can be carried out by reducing the frequency spacing between these optical channels, for example, from 100 GHz down to 50 GHz, or even 25 GHz and 12.5 GHz [19,20]. The reduction in the channel spacing also results in narrower bandwidths for the optical multiplexers (mux) and demultiplexers (demux). On passing through these narrowband

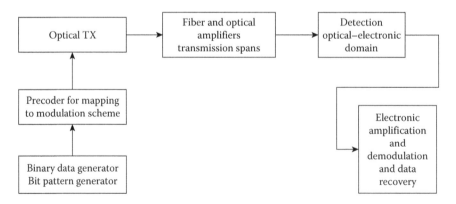

FIGURE 1.1 Schematic diagram of the modulation and the electronic detection and demodulation of an advanced modulation format optical communications system.

---

* The OOK format simply implies the on–off states of the lightwaves where only the optical intensity is considered. On the contrary, the ASK format is a digital modulation technique representing the signals in the constellation diagram by both the amplitude and phase components.

optical filters, signal waveforms are distorted and optical channels suffer the problem of interchannel crosstalk. The narrowband filtering problems are becoming more severe at high data bit rates, for example, 40 Gbps, thus degrading the system performance significantly.

Together with the demand for boosting the total system capacity, another challenge for the service carriers is to find cost-effective solutions for the upgrading process. These cost-effective solutions should require minimum renovation of the existing photonic and electronic subsystems; that is, the upgrading should only take place at the transmitter and receiver ends of an optical transmission link. Another possible cost-effective solution is to extend significantly the uncompensated reach of optical transmission links, that is, without using dispersion compensation fibers (DCFs), thus considerably reducing the number of required in-line EDFAs. This network configuration has recently attracted the interest of both the photonic research community as well as service carriers.

Over the past few years, extensive research and development has proved that coherent reception incorporating digital signal processing (DSP) can push the bit rates per wavelength channel to 100 Gbps by employing 25 GBaud polarization multiplexing and QPSK (two bits/symbol) and/or M-ary quadrature amplitude modulation (M-QAM) to aggregate to 200 Gbps or 400 Gbps. Furthermore, the channels can be made pulse shaped by using digital-to-analog converter (DAC) to pack the channels into superchannels to generate terabits per second (Tbps) per channel with subcarriers. The advances of ultra-high sampling rate analog-to-digital converters (ADCs) and DACs at 64 GSa/s allow the DSP to recover the clock, and hence the sampling rate and time, combating the linear and nonlinear impairments due to CD, PMD, self-phase modulation (SPM), cross-phase modulation (XPM), and other effects. The transmission for 100 Gbps would reach 3500 km in field trials and 1750 km for 200 Gbps over optically amplified and non-DCF fiber span transmission distances.

Therefore, the principal motivations of this book are to describe the employment of digital communications in modern optical communications. The fundamental principles of digital communications, both coherent and incoherent transmission and detection techniques, are described with a focus on the technological developments and limitations of the optical domain. The enabling technologies, research results, and demonstrations on laboratory experimental platforms for the development of high-performance and high-capacity next-generation optical transmission systems impose significant challenges for the engineering of optical transmission systems in the near future and techniques for network monitoring.

1.2   ENABLING TECHNOLOGIES

1.2.1   Modulation Formats and Optical Signal Generation

Modulation is the process of facilitating the transfer of information over a medium. In optical communications, the process of converting information so that it can be successfully sent through the optical fiber is called optical modulation.

There are three basic types of digital modulation techniques: amplitude shift keying (ASK), frequency shift keying (FSK), and phase shift keying (PSK), in which the parameter that is varied is the amplitude, frequency, or phase, respectively, of the carrier to represent the information to be sent. Digital modulation is the process of mapping such that the digital data of "1" and "0" or symbols of "1" and "0" are converted into some aspects of the carrier such as the amplitude, the phase, or both amplitude and phase, and then transmitting the modulated carrier, which is the lightwaves in the context of this book. The modulated and transmitted lightwave carrier is then remapped at the reception systems back to a near copy of the information data.

1.2.1.1   Binary Level

Modulation is a process that facilitates the transport of information over the medium; in this book, our medium is the optical-guided fiber and associate photonic components. In digital communications, there are three basic types of digital modulation techniques: ASK, PSK, and FSK.

Under these modulation techniques, the phases or amplitudes or frequencies of the lightwaves are varied to represent the information bits "1" and "0".

In ASK, the amplitude of the lightwave carrier, normally generated by a narrow-linewidth laser source, is changed in response to the digital data, keeping everything else fixed. That is, bit 1 is transmitted by the lightwave carrier of a particular amplitude. To transmit 0, the amplitude is changed keeping the frequency unchanged, as shown in Figure 1.2. NRZ or RZ can be assigned depending on the occupation of the state 1 during the time length of a bit period. For RZ, normally only half of the bit period is occupied by the digital data.

In addition to NRZ and RZ formats, in optical communications, the carrier can be suppressed under these formats so as to achieve non-return-to-zero carrier suppression (NRZ-CS) and return-to-zero carrier suppression (RZ-CS). This is normally generated by biasing the optical modulator in such a way that the carriers passing through the two parallel paths of an interferometric modulator have a π phase shift difference with each other. Thus, the carrier at the center frequency is suppressed, but the sidebands of the modulated signals remain unchanged.

In PSK, the phase of the lightwave carrier is changed to represent the information. The phase in this context is the shift of the angle at the phasor vector initial position at which the sinusoidal carrier starts. To transmit a 0, the phase would be shifted by π and a 1 with no change of phase. The phase angle can be changed and take a value of a set of phases corresponding to the mapping of the symbols as shown in Figure 1.3.

In FSK, the frequency of the carrier represents the digital information. One particular frequency is assigned to 1, and another frequency is assigned to 0, as shown in Figure 1.4. FSK can be considered as continuous phase modulation, for example, a continuous phase modulation MSK (minimum shift keying) whose frequency separation f_d is selected such that the signals carried by these frequencies are orthogonal.

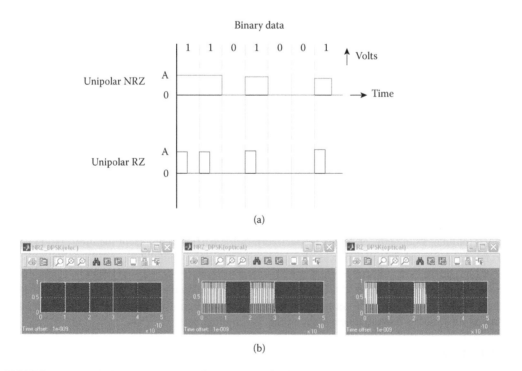

FIGURE 1.2 (a) NRZ and RZ pulse amplitude-modulation formats for a sequence of {1 0 1 0 1 0 1 0 1 0 1 0} and (b) generated ASK signals with carrier (not to scale and high-density area) data and carrier-modulated NRZ and NZ formats.

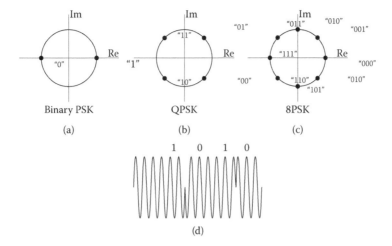

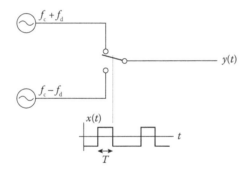

FIGURE 1.3 Signal-space constellation of discrete PM: (a) binary PSK, (b) quadrature binary PSK, (c) eight PSK and (d) phase of the carrier under modulation with π phase shift of the BPSK at the edge of the pulse period.

FIGURE 1.4 Schematic diagram of a FSK transmitter. f_d is the deviation frequency from the carrier frequency f_c.

ASK can be combined with PSK to create a hybrid modulation scheme such as QAM, where both the phase and amplitude of the carrier are changed at the same time. The carriers are expected to follow a similar pattern as that of differential phase shift keying (DPSK) in Figure 1.3d but with different frequencies of the carrier under the envelope of the bits 0 and 1. For MSK signals, the carrier frequency is chirped up or down depending on the 0 or 1, that is, the phase of the carrier is continuously varied during the bit period, and the carrier frequencies of the bits are such that there is an orthogonality of the carriers and the signal envelope.

1.2.1.2 Binary and Multilevel

An additional degree of freedom for detection can be used to effectively enhance the capacity owing to the effective equivalence of the multilevel and symbol rates, and hence the detection of the received optical signals. A widely used and mature detection scheme for optical signals is the direct detection scheme in which the optical power $P = [E]^2$, the square of a complex optical field amplitude. The photodetector would not be able to distinguish between a 0 or π phase shift of the carrier lightwave embedded within the pulse. The carrier phase can only be possibly extracted if and only if photonic processing to extract the phase at the front end of the receiver is performed. Thus, a + or − field complexity would be seen as identical in the photodetector.

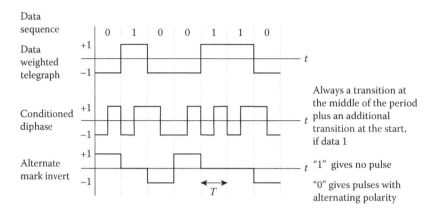

FIGURE 1.5 Illustration of pseudo-multilevel or polybinary baseband signals. Binary data sequence, weighted signals, diphase RZ, and alternate mark inversion formats.

This ambiguity of the phase detection process would allow one to shape the optical spectra of optical signals, thus creating a modulation format more resilient to the distortion effects accumulated during the transmission process.

Formats making use of the tri-level are illustrated in Figure 1.5 and could be termed as pseudo-multilevel, tri-level, or polybinary signals. These tri-level signals can be represented in terms of the phase or frequency of the lightwave carrier.

The use of more than two symbols to encode a single bit of information is to increase the information bits carried by a symbol. However, the transmission is still at the symbol rate B_s. Under optical transmission, the -1 and $+1$ can be coded in terms of the variation in the phase of the carrier as there is no negative intensity representation unless the field of the lightwaves is used. The tri-level uses $\{+|E|, -|E|, \text{and } 0\}$, and its equivalent phase representation can be mapped to $\{0, |E|^2\}$ at the optical receiver, for example, the duobinary format, which will be described in a later chapter. A phase difference of π and 0 between the three levels would minimize the pulse dispersion as they are propagating along the fiber with a relative phase difference of π, thus resulting in destructive interference of any pulse spreading due to dispersion.

This tri-level must not be mixed up with the truly multilevel signaling in which $\log_2 M$ bits are encoded on N symbols, and then transmitted at a reduced rate $B_s/\log_2 N$. Both multilevel amplitude or APSK and DQPSK are multilevel optical modulation techniques. The multilevel modulation formats will be described in Chapter 7.

1.2.1.3 In-Phase and Quadrature-Phase Channels

Another form of modulation that would enhance the capacity of the transmission is the use of the orthogonal channels, in which the information can be coded into the in-phase and quadrature (I–Q) components in polar or Cartesian coordinates, as shown in Figure 1.6.

QPSK is most commonly used in differential and nondifferential phase modulation, in which I and Q components are used extensively owing to its bit error rate (BER); its corresponding energy per bit is similar to that of PSK, doubling the capacity of that of PSK. QPSK is an extension of the binary PSK signals but with a phase change of only $\pi/2$ instead of π. Mathematically, the signal $s(t)$ can be expressed as

$$s(t) = A_c p_s(t) \cos\left(2\pi f_c t + \frac{2\pi_i}{M}\right) \tag{1.1}$$

where $p_s(t)$ is the pulse shaping of the data, M is the quantized level or the total number of phase states of the modulation, and π_i is the phase modulation index. QPSK can be combined with ASK

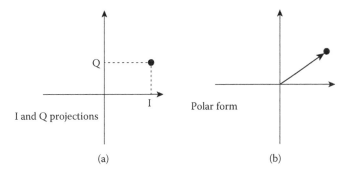

FIGURE 1.6 Signal vectors plotted in signal space: (a) Cartesian coordinates and (b) polar form.

to generate QAM, where the phase and amplitude can be used to map a symbol of data information into one of the points on the signal space.

1.2.1.4 External Optical Modulation

External modulation is the essential technique for modulating the lightwaves so that its linewidth preserves its narrowness, and only the sidebands of the modulation scheme dominate the spectral property of the generated passband characteristics. Figures 1.7 and 1.8 show the typical structure of optical transmitters for the generation of NRZ and RZ optical signals.

The laser is always switched on, and its lightwaves are modulated via the electro-optic modulator using the principles of interferometric constructive and destructive interference to represent the ON and OFF states of the lightwaves. The RZ can then be similarly generated but with an additional optical modulator that would generate periodic optical pulses whose width is half of that of the bit period. The phase and frequency modulation can also be generated using these electro-optic modulators by biasing conditions and controlling the amplitudes of the electrical pulses. These optical transmitters are described in Chapter 2. We note here that the fiber that connects the two modulators of Figure 1.8 must be of the polarization maintaining (PM) type. Otherwise, there would be polarization fluctuation and hence reduction in the coupling of the lightwave power to each other.

The laser source would normally be a narrow-linewidth laser that is turned on at all times to preserve its narrowness characteristics. The lightwaves are generated and coupled to the optical modulator via the pigtails of both devices. The modulator is driven with data pulse sequence which is the output of a bit pattern generator (BPG) whose voltage level is conditioned to the appropriate driving level required by the $V\pi$ value of the modulator. The phase variation of the carrier can be similarly implemented for phase modulation. When a modulation format is necessary, then an electronic precoder is required to code the serial sequence for appropriate coding. The precoder can be a differential coding, multilevel coding, or IFFT to generate orthogonal data subchannels in case of orthogonal frequency division multiplexing (OFDM).

The pulses of all modulation formats can take the form of NRZ or RZ. For the RZ format, an additional optical modulator is required to generate or condition the 1 NRZ to RZ, as shown in Figure 1.8. The second modulator can exchange its position with that of the other modulator without affecting the generation of the modulation formats. This second modulator is usually called the *pulse carver*.

Note that if the RZ modulator is biased such that the phase difference at the biasing condition is π, then we would have carrier suppression located at the central location of the spectra. The sidebands of the optical signals remain unchanged. The bandwidth of the modulator determines the rise time and fall time of the edges of the pulse sequence shown in Figures 1.7 and 1.8. The details of the optical transmitters for different modulation formats are given in Chapter 3.

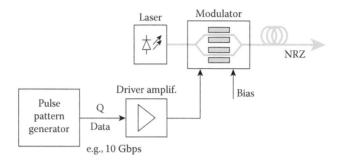

FIGURE 1.7   Generation of optical signals of format NRZ using an external modulator.

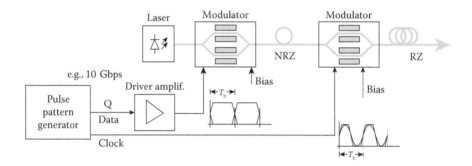

FIGURE 1.8   Generation of optical signals of format RZ.

1.2.2 Advanced Modulation Formats

The aforementioned problems facing contemporary optical fiber communications can be effectively overcome by utilizing spectrally efficient transmission schemes via the implementation of advanced modulation formats. A number of modulation formats have recently been reported as alternatives for the OOK format, including RZ pulses in OOK/ASK systems [21–23], DPSK [19,21–26], and more recently, MSK [27–34]. These formats are adopted into photonic communications from a knowledge of wireline and wireless communications.

DPSK has received much attention over the last few years, particularly when it is combined with RZ pulses. The main advantages of RZ-DPSK are (1) a 3 dB improvement in the receiver sensitivity over the OOK format by using an optical balanced receiver [22,35,36] and (2) high resilience to fiber nonlinearities [7,21,35,37] such as intrachannel SPM and interchannel XPM. Several experimental demonstrations of DPSK long-haul DWDM transmission systems for 10 Gbps, 40 Gbps, and higher bit rates have been reported recently [19,24,38–40]. However, there are few practical experiments addressing the performance of cost-effective 40 Gbps DPSK–10 Gbps OOK hybrid systems for gradually upgrading the existing installed SSMF transmission infrastructure [41,42]. In addition, the performance of 40 Gbps DPSK for use in this hybrid transmission scheme has not been thoroughly studied.

The MSK format offers a spectrally efficient modulation scheme compared to the DPSK counterpart at the same bit rate. As a subset of continuous phase frequency shift keying (CPFSK), MSK possesses spectrally efficient attributes of the CPFSK family. The frequency deviation of MSK is equal to a quarter of the bit rate, and this frequency deviation is also the minimum spacing to maintain the orthogonality between two FSK-modulated frequencies. On the contrary, MSK can also be considered as a particular case of offset differential quadrature phase shift keying (ODQPSK) [43–47], which enables MSK to be represented by I and Q components on the signal constellation. The advantageous characteristics of the optical MSK format can be summarized as follows:

1. Compact spectrum, which is of particular interest for spectrally efficient and high-speed transmission systems. This also provides robustness to tight optical filtering.
2. High suppression of spectral side lobes in the optical power spectrum compared to DPSK. The roll-off factor follows f^{-4} rather than f^{-2} as in the case of DPSK. This also reduces the effects of interchannel crosstalk.
3. No high-power spectral spikes in the power spectrum, thus reducing fiber nonlinear effects compared to OOK.
4. As a subset of either CPFSK or ODQPSK, MSK can be detected either incoherently based on the phase or the frequency of the lightwave carrier or coherently based on the popular I–Q detection structure.
5. Constant envelope property, which eases the measure of the average optical power.

Several studies have been conducted recently on the generation and direct detection of externally modulated optical MSK signals [27–30]. However, there are few studies investigating the performance of the externally modulated MSK format for digital photonic transmission systems, particularly at high bit rates such as 40 Gbps [27,48,49]. Furthermore, if MSK can be combined with a multilevel modulation scheme, the transmission baud rate would be reduced in addition to the spectral efficiency of the MSK formats. This is of great interest for long-haul and metropolitan optical networks and thus provides the main motivation for proposing the dual-level MSK modulation format. In addition, the potential of optical dual-level MSK format transmission is yet to be explored.

1.2.3 Incoherent Optical Receivers

The modulation formats studied in this research, optical DPSK and MSK-based formats, can be demodulated incoherently by using an optical balanced receiver that employs a Mach–Zehnder delay interferometer (MZDI). In the case of the optical DPSK format, MZDI is used to detect differentially coded phase information between every two consecutive symbols [7,35,36,50]. This detection is carried out in the photonic domain as the speed of the electrical domain is insufficient, especially at very high bit rates of 40 Gbps or above. The MZDI-balanced receiver is also used for incoherent detection of optical MSK signals [27,29,30] by also detecting the differential phase of MSK-modulated optical pulses. However, with the MZDI-based detection scheme, it is found that optical MSK provides a slight improvement for the CD tolerance over its DPSK and OOK counterparts [29,31].

As a subset of the CPFSK family, MSK-modulated lightwaves can also be incoherently detected based on the principles of optical frequency discrimination. Thus, an optical frequency discrimination receiver (OFDR) employing dual narrowband optical filters and an optical delay line (ODL) is proposed in this research. This receiver scheme effectively mitigates CD-induced ISI effects and enables breakthrough CD tolerances for optical MSK transmission, as reported in my first-authored papers [31,32]. In addition, the feasibility of this novel receiver is based on recent advances in the design of optical filters, in particular, the micro ring resonator filter. Such optical filters have very narrow bandwidths, that is, less than 2 GHz (3 dB bandwidth), and they have been realized commercially by Little Optics [51–54]. This research thus provides a comprehensive study of this OFDR scheme, from the operational principles to the analysis of the receiver design, and to the performance of OFDR-based MSK optical transmission systems.

1.2.4 DSP-Coherent Optical Receivers

Coherent detection and transmission techniques were extensively exploited in the mid-1980s to extend the repeaterless distance a further 20–40 km of SSMF with an expected improvement in the receiver sensitivity of 10–20 dB, depending on the modulation format and receiver structure using phase or polarization diversity.

In general, a coherent receiver would operate on the beating the received optical signals and that of the field of a local laser oscillator. The beating optical signals are then detected by the photodetector with the phase of the carrier preserved, which permits the detection of the phase of the carriers. Hence, the phase modulation and continuous phase or frequency modulation signals can be processed in the electronic domain. With the advancement of the digital electronic processor, the processing of the received signals either in the IF or base band of the heterodyne and homodyne detection, respectively, can be processed to determine the phase of the modulated and transmitted signals. Coherent receivers for different modulation formats are described in this book, especially in Chapter 4.

The three type of coherent receivers, namely homodyne, heterodyne, and intradyne detection techniques, are possible and dependent on the frequency difference of zero, intermediate frequency greater or smaller than the passband of the signals between the local oscillator laser and that of the signal carrier. With modern advanced optically amplified fiber communications, broadband amplified spontaneous emission (ASE) noise always exists, and under coherent detection, the beating between local laser source and ASE dominates the electronic noise of the receiver at the front end. These noise considerations are described in Chapter 4.

The advanced aspects of coherent reception systems are currently advanced by a generation by incorporating DSP subsystems and naturally the ultra-high sampling rate ADCs and DACs, permitting the bit rate per channel to reach 100 Gbps and higher by polarization division multiplexing (PDM) QPSK or an M-QAM modulation scheme with M of 16 or 32 or even higher. The DSP in association with advanced algorithms have overcome a number of the difficulties faced by "analog" coherent reception techniques extensively reported in the 1980s, such as frequency matching of the local oscillator and that of the signal carrier; clock recovery and generation of sampling for retiming; and compensation of dispersion due to CD and PMD impairments and nonlinear distortions. These DSP techniques are comprehensively covered in this second edition with four chapters and another chapter on optical hardware and digitalization circuitry for these systems.

1.2.5 Transmission of Ultra-Short Pulse Sequence

It has been proposed to employ bit rates of up to 100 and 160 Gbps for all optical transmission systems and networks. At this bit rate, the processing speed of electronics would face severe challenges. The generation of data pulse sequences at this speed can be derived from ultra-short pulses of mode-locked fiber lasers of subpicoseconds width. In order to deal with the dispersiveness of these ultra-short pulses, equalization can be done using the principles of temporal imaging or effectively focusing and defocusing time-domain pulses using optical quadratic phase modulation.

1.2.6 Electronic Equalization

Electronic equalizers have recently become one of the most potential solutions for future high-performance optical transmission systems. The Si-Ge technological development has enhanced electron mobility and hence has decreased the rise and fall time of pulse propagation, thus increasing the processing speed. The sampling rate can now reach several giga-samples/second, which enables the processing of 10 Gbps bit rate data channel without any difficulty. Hence, it is very likely that electronic processing and equalization will be implemented in real systems in the very near future Figure 1.9.

The channel is a single-mode optical fiber whose dispersion factor can be either negative or positive leading to some residual dispersion even if compensation of dispersion is used, such as the fiber dispersion-compensating module. Thus, the distortion of the pulse is purely phase distortion prior to the detection by the photodiode, which follows the square law rule for the direct detection case. On the contrary, for coherent detection, the beating between the local oscillator and the signal in the photodetection device would lead to the preservation of the phase, and the distortion would be considered a pure phase distortion. Again, in the case of direct detection, after the square law detection, the phase distortion is transferred to the amplitude distortion. In order to conduct the equalization process, it is important for us to know the impulse and step responses of the fiber channel $h_F(t)$ and $s_F(t)$. We will then present the fundamental aspects of equalization using feed-forward equalization, decision feedback equalization with maximum mean square error (MMSE), or maximum-likelihood sequence estimation of Viterbi algorithm (MLSE).

1.2.6.1 Feed-Forward Equalizer

The feed-forward equalization (FFE) is a linear equalizer that has been the most widely studied; a transversal filter structure would offer a linear processing of the signal prior to the decision. The structure of an FFE transversal filter consists of a cascade delay of the input sample, and at each delay the signal is tapped and multiplied with a coefficient. These tapped signals, whose delay tap time is the bit period, are then summed to give the output sample. The coefficients of the transversal filter must take values that match the channel so that the convolution of the channel impulse response and that of the filter result in unity, so as to achieve a complete equalized pulse sequence at the output. Figure 1.10a and b show the linear equalization scheme that uses either feed-forward or feedback equalization; the difference between these two schemes is the tapped signals either at the output of the transversal filter or at the output of the feedback that minimize the input sequence.

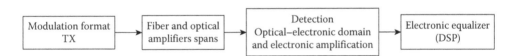

FIGURE 1.9 Schematic diagram of the location of the electronic equalizer at the receiver of an advanced optical communications system (DSP denotes digital signal processor).

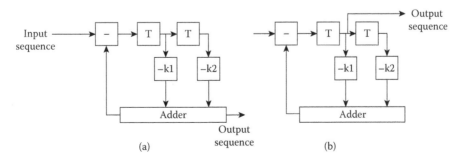

FIGURE 1.10 Linear (a) feed-forward equalization (transversal equalization) and (b) feedback equalization scheme.

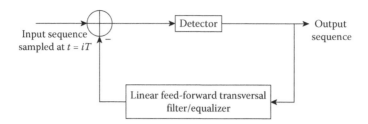

FIGURE 1.11 Schematic diagram of receiver using nonlinear equalization by decision-directed cancellation of ISI.

On the other hand, a decision feedback equalizer differs from a linear equalizer in that it has a decision detector that determines the signal amplitude required for feedback of the difference error at the input, as shown in Figure 1.11.

1.2.6.2 Decision Feedback Equalization

The FFE method is based on the use of a linear filter with adjustable coefficients. The equalization method that exploits the use of previous detected symbols to suppress the ISI in the present symbol being detected is termed as decision feedback equalization.

The decision feedback equalizer is shown in Figure 1.11. It consists of m coefficients and m delay taps. Each tap spacing equals the bit duration T. From Figure 1.11, we can see that the received signal sequence goes through the forward filter first. After making decisions on previously detected symbols, the feedback filter provides information from the previously detected symbols for the current estimation.

1.2.6.3 Minimum Mean Square Error Equalization

Consider next the general case where the linear equalizer is adjusted to minimize the mean square error due to both ISI and noise. This is called minimum mean square error equalization (MMSE).

1.2.6.4 Placement of Equalizers

Linear and nonlinear equalization are possible as they are well known in the field of signal processing. The principles of equalization with equalizers placed at the transmitter, or the receiver, or sharing between the transmitter and receiver are critical for practical networks.

1.2.6.5 MLSE Electronic Equalizers

Among the electronic equalization techniques, maximum-likelihood sequence estimation (MLSE), which can be implemented effectively with the Viterbi algorithm, has attracted considerable research interest [55–60]. However, most of the studies on MLSE equalizers have focused on either ASK or DPSK formats [55,56,59,61]. Apart from a co-authored paper reported recently [33], there has not been any study on the performance of MLSE equalizers for optical MSK transmission, especially when OFDR is used as the detection scheme. The performance of OFDR-based MSK optical transmission systems is significantly enhanced with the incorporation of postdetection MLSE electronic equalizers because the ISI problems caused by either fiber dispersion impairments or tight optical filtering effects are effectively mitigated. Therefore, this decision process considered as an equalizer is included as a case study in Chapter 14 to comprehensively investigate the performance of MLSE equalizers for OFDR-based MSK optical transmission systems.

1.2.7 Ultra-Short Pulse Transmission

The need for ultra-high-speed transmission beyond the processing speed of electronic devices drives the interest in transmission techniques in the photonic domain using ultra-short pulse with sufficient energy to interact with nonlinear SPM effects for balancing the linear dispersion, that is, the solitons.

The operation of the first installed soliton long-haul transmission system between South and Western Australia has proved the superiority of solitons. However, challenges due to environmental effects still remain to be resolved. The fundamental issues of soliton transmission are described. Temporal imaging for the refocusing of dispersive pulse sequences is also outlined. This temporal imaging is also used to minimize the jittering of soliton pulses due to noise generated from cascade optical amplifiers.

1.3 ORGANIZATION OF THE BOOK CHAPTERS

The presentation of this book follows the progression of the digital communications and integration of modulation techniques in optical communications under self-coherent and coherent reception techniques (see also Figure 1.12) as follows.

In this edition, Chapter 2 introduces optical fibers as the transmission medium for long-haul, metro, and access optical networks. Both geometricaland profile structures as well as the propagation of the modulation envelope under the influence of interference due to the different propagation velocities owing to the guided mode confinement of the fiber core and the wavelength-dependent

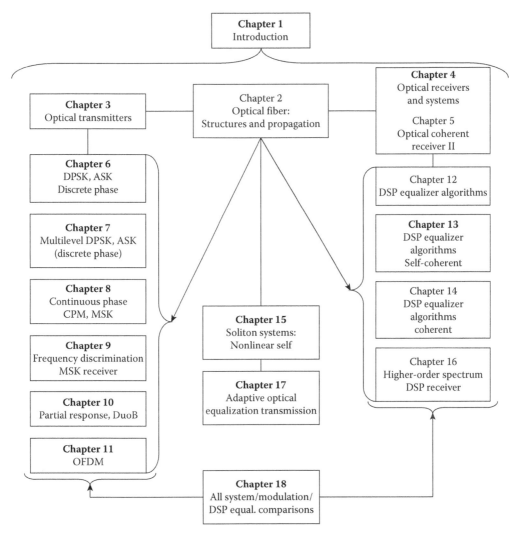

FIGURE 1.12 Flowchart of Chapters 1–18. Bold letters indicate those chapters contained in the 1st edition, with updates in the 2nd edition. Likewise, unbold letters are for new chapters in this 2nd edition.

refractive index are treated to give insights into the propagation of lightwaves and distortion on the modulated signals so that the equalization and compensation in optical and DSP electronic domains can be developed in later chapters of the book.

In Chapter 3, optical transmitter configurations based on the principles of operation of interferometric effects for the generation of phase and frequency modulation, either the CPFSK format or the I–Q structure of the ODQPSK format, are proposed for generating optical MSK signals. On the contrary, the generation scheme of the optical dual-level MSK format comprises two MSK optical transmitters connected in parallel. In addition, the spectral characteristics of these formats are also discussed in detail. In an optical transmitter, data modulation is implemented by using either external electro-optic phase modulators (EOPMs) or Mach–Zehnder intensity modulators (MZIMs). Phasor principles are extensively applied in this chapter to derive the modulation of the carrier phase and amplitude.

Chapter 3 also provides an insight into a number of investigated advanced optical modulation formats, including RZ, carrier-suppressed RZ (CSRZ) pulse shaping, DPSK, and DQPSK (quadrature), with discrete phase shift keying. Modulation with a continuity of the phase at the transition of the logical bits is also demonstrated, such as CPFSK, MSK, and the multilevel version of the latter, dual-level MSK. The focus of these chapters is on the generation and detection schemes of these formats. Two optical transmitter configurations that are based on the principles of either the CPFSK format or the I–Q structure of the ODQPSK format are proposed for generating optical MSK signals. On the contrary, the generation scheme of the optical dual-level MSK format comprises two MSK optical transmitters connected in parallel. In addition, the spectral characteristics of these formats are also discussed in detail. Further details of the transmitters and receivers can be found in later chapters that address optical transmission systems with advanced modulation formats.

Chapter 4 provides an insight into a number of advanced optical modulation formats, including RZ and NRZ pulse shaping, and MSK and its multilevel version, dual-level MSK. The focus of this chapter is on the generation and detection schemes of these formats. Chapter 4 also describes the modeling of critical optical subsystems of a digital photonic transmission system. The system modeling is sequentially described end to end, from the optical transmitter to the fiber channel to the optical receiver. Lightwaves generated from the optical transmitter and propagating along the optical fiber are degraded by fiber dispersion impairments and fiber nonlinearities. Detailed descriptions of these impairments, particularly fiber CD, are presented. These fiber dynamics are embedded in the nonlinear Schrödinger equation (NLSE), which is solved numerically by the symmetric split-step Fourier method (SSFM). This chapter also describes the modeling of ASE noise and receiver noise sources as well as the modeling of optical and electrical filters. Key signal quality metrics and methods used to evaluate the performance of optical transmission systems are described in detail. The final part of Chapter 4 discusses the advantages of the MATLAB® and Simulink® modeling platform developed for advanced digital photonic transmission systems.

A fast method for the evaluation of the statistical properties of the distribution of the received eye diagrams is described, enabling the measurement of the BER from the receive eye diagram rather than resorting to the Monte Carlo method, which would consume a considerable amount of time for computing the errors.

Why should Simulink be developed when there are a number of commercial simulation packages available, such as VPI Systems, and so on Simulink is a modular toolbox attached to MATLAB that is now used as a universal computing tool in the global university. Its communications blockset contains many toolboxes for digital communications in the baseband as well as the passband. Thus, these toolboxes can be used as a modular set for the generation and detection of lightwaves signals. As the frequency of the lightwaves is much higher than that of the baseband signals, it is not likely that the carrier is contained under the envelope of the signals, because an extremely high sampling rate would be required that is currently not possible in computing systems. Thus, complex envelope signals are used to represent the phase of the carrier. Therefore, the simulation models for digital optical communications can be modeled very accurately and with ease owing to the modularity of Simulink. Simulink models are given in this book, and Annex 5 is dedicated to the modeling techniques of digital optical communications. The Simulink techniques and samples of models may be made available as a toolbox in MATLAB in the future to facilitate teaching and research activities in global laboratories. Thus, the Simulink models described in this book would be easily integrated into the standard MATLAB software package in university, research, and engineering laboratories as well as in the field, especially because the toolset "Digital Optical Communications" is most universal and affordable for advanced-level students and postgraduate students of telecommunications engineering.

Chapters 4 and 5 describe fundamental techniques of self-coherent and coherent optical communications, the optical receivers, and associated noise in such receiving systems. The principal motivation behind the introduction of Chapters 4 and 5 is the emerging technological developments in photonic, optoelectronic components, and digital signal processors. The limitations and obstacles due to the linewidth of the laser source are no longer a major hurdle. They are now used in both the transmitters and as the local oscillator at the receiver. The high sampling speed of electronic digital processors enables the estimation of the phase of the lightwave carrier at the coherent receiver in which the signals are beating with the local oscillator to produce the baseband-product signals. Thus, it is necessary for coherent techniques to be described for its applications in modern digital optical communication systems. These high-speed digital signal processors are also employed as electrical equalization systems to compensate for disturbance or residual dispersion in optical transmission systems. The equalization techniques in the electrical domain of digital optical communications are described in Chapter 10. In coherent receivers, the noise sources due to the beating between the oscillator and the ASE noise sources would dominate the electronic noise at the front end of the receiver. Analysis and simulation are described in this chapter. A typical schematic of the coherent reception for PDM-QPSK 100G system is shown in Figure 1.13, in which the optical signals are polarization demultiplexed in the optical domain as well as in the real and complex parts of the QPSK channels, and then detected using balanced detection. They are next sampled by high-speed ADCs and processed by a DSP in the digital domain to recover and compensate the distortion impairment due to fibers and the frequency difference between the local oscillator and signal carriers.

Chapter 6 describes discrete phaseshift keying modulation and reception under self-coherent optical communications systems. The differential mode of detection is most appropriate for such

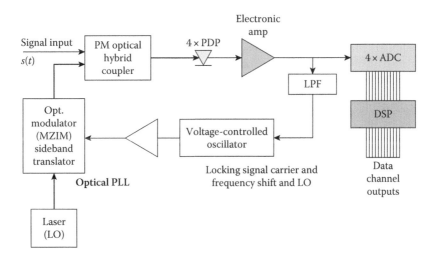

FIGURE 1.13 Schematic diagram of optical homodyne detection incorporating ADC-DSP systems. Electrical line (thin line) and optical line (thick line) using an optical phase-locked loop.

systems in which the phases contained in the two consecutive bit periods are assigned discrete values. An experimental demonstration of DPSK modulation formats is also described. This chapter presents the theoretical and experimental results of studies on the performance characteristics of 40 Gbps optical DPSK, with the focus on its robustness and tight optical filtering. This chapter also introduces phase discrete modulation, especially differential phase coding. The purpose is to show the feasibility of implementing cost-effective hybrid optical transmission systems in which 40 Gbps DPSK channels can be co-transmitted simultaneously with 10 Gbps OOK channels over the existing 10 Gbps network infrastructure. These experiments were conducted in collaboration with commercial and industrial engineering partners, SHF Communications Technology AG (Germany) and Research Laboratories-Telstra Corporation (Australia).

Chapter 7 gives an introduction to multilevel amplitude and phase modulation formats. The driving conditions for a dual-drive MZIM for the generation of multilevel amplitude DPSK are derived, as well as the models, using the MATLAB and Simulink platform. Dual-level MSK signals are also described. A 16-ADPSK and 16-Star QAM are studied as two typical case studies that would allow the reduction of the symbol rate from 100 Gbps to 25 Gbps, at which speed electronic processing would be able to assist in the detection using either direct or coherent detection.

Chapter 8 continues with the modulation of the phase of the carrier, but the subject is continuous modulation rather than discrete modulation (as in Chapter 6), in which MSK is described extensively as it is the most efficient continuous phase modulation owing to the orthogonal properties of the spectra of the modulated signals at the two distinct carriers.

A novel self-coherent detection scheme, the OFDR, is proposed in Chapter 9 for optical MSK systems aiming to extend significantly uncompensated optical links. The receiver design is optimized by analyzing the significance of key subsystem optical components for the receiver performance. These receiver design guidelines are verified with analytical results as well as simulation results. In addition, the performance of OFDR-based optical MSK systems are evaluated numerically for the receiver sensitivities, CD and PMD tolerances, resilience to SPM nonlinear effects, and transmission limits due to the high PMD coefficient of legacy fiber.

Chapter 10 describes optical transmission systems employing partial response techniques such as duobinary optical modulation techniques, single-sideband modulation, and self-homodyne reception for the modulation systems.

Chapter 11 introduces multi-subcarrier modulation with its carriers orthogonal to each other, and OFDM, in which the FFT and IFFT digital signal processors are extensively exploited so as to combat the dispersion compensation in long-haul optically amplifier transmission systems. The roles of DAC and ADC are very important in these OFDM transmission systems. Thus, there is a possibility that dispersion-compensating fiber sections can be discarded, reducing the number of optical amplifiers in spans and hence reducing ASE noise and lengthening transmission reach. This chapter deals with multi-subcarrier transporting information data with orthogonal property of adjacent subchannels, that is, OFDM. The lowering of the data rates of the subchannels assists the mitigation of linear and nonlinear impairments of the fiber channel. The principles and performances of the scheme are described.

Chapter 12 introduces the processes in the electronic domain that equalize the transmitted pulse sequence degraded by impairments of the transmission media, the quadratic phase optical fibers. Both linear and nonlinear equalization processes are described. Noise contributions in the equalization processes are stated and derived for the BER. Transmission examples are given for duobinary modulation and MSE with Viterbi trellis tracing. The chapter briefly reviews the principles of the MLSE equalization technique, Viterbi algorithm, and state trellis structure. Detailed explanations of state-based Viterbi-MLSE equalizers for optical communications are given. The chapter then investigates the performance of MLSE equalizers for 40 Gbps OFDR-based optical MSK systems. In this receiver scheme, OFDR serves as the optical front end and is integrated with a postdetection MLSE equalizer. The CD tolerance performance of both Viterbi–MLSE and template-matching MLSE equalizers is studied. The performance limit of Viterbi–MLSE equalizers with 2^4–2^{10} states is then investigated based on maximum uncompensated transmission distances. The performance of 16-state Viterbi-MLSE equalizers for PMD equalization is also investigated. This number of states reflects the feasibility of high-speed electronic signal processing in the near future. The significance of multisample sampling schemes (two and four samples per one-bit period) over the conventional single-sample sampling technique is also highlighted in this chapter [62,63].

Chapter 13 outlines the principle of self-homodyne reception and associated DSP systems and gives an introduction to algorithms for DSPs applied in self-coherent reception in which only the intensity detection with high noise level and is considered. These algorithms can also be applied in coherent reception systems, as described in Chapter 14.

Chapter 14 addresses the DSP algorithms for self-homodyne reception systems operating under the different modulation formats described in Chapters 6–11. Chapter 14 also describes advances in coherent reception techniques incorporating DSP and ultra-high sampling rates and processing algorithms for real-time data processing at the receiver side.

Chapter 15 completes the picture of ultra-short pulse transmission by describing the transmission of solitons over optically amplified transmission spans. Although soliton transmission was developed in the last decade of the twentieth century, the first soliton, at 10 Gbps, was installed between Adelaide and Perth in Australia in the late 1990s by Marconi Co. Ltd. of England, and proved to be robust in practice. Thus, this chapter reactivates optical soliton transmission.

Chapter 16 describes an advanced technique in digital processing applicable in both the DSP electronic and photonic domains in which multidimensional spectra are used to separate the linear and nonlinear impairments upon transmitting an optical sequence, hence compensating distortion in both regions.

Chapter 17 introduces the impulse and step responses of the quadratic phase guided optical media which are the single-mode optical fiber, then leading to the development of an all-optical equalizer by photonic modulation. This chapter describes an equivalent temporal imaging in the time domain for pulse multiplication and compensation of ultra-short pulse transmission. Simulation results are given to demonstrate the photonic equalization of ultra-short pulse propagation.

Chapter 18 summarizes the key achievements of digital optical communications in the transmission of several terabits per second capacity over single-mode optical fibers employing both self-coherent and coherent reception incorporating DSP. A brief comparison of the modulation formats is given incorporating experimentally measured data. The roles of digital communications in the emerging advanced photonic transmission systems and networks are identified. Furthermore, emerging ultra-high-speed electronic processors have made it possible to achieve real-time equalization of the distortion and other disturbance effects in long-haul ultra-high-speed optical communications. Thus, additional chapters on coherent reception and DSP algorithms for such coherent transmission technology are added in this edition. Concluding remarks are provided to give an overview of the chapters of the book. We expect an explosion in research and development in Peta-bps in the near future.

Practice problems for some chapters are given wherever appropriate at the end of each chapter. A number of appendices are provided to supplement technical information.

REFERENCES

1. C. Kao and G. Hockham, Dielectric-fibre surface waveguides for optical frequencies, *Proc. IEEE*, Vol. 113, No. 7, pp. 1151–1158, 1966.
2. I. P. Kaminow and T. Li, *Optical Fiber Communications,* Vol. IVB, New York: Elsevier Science, 2002.
3. R. S. Sanferrare, Terrestrial lightwave systems, *AT&T Technol. J.*, Vol. 66, pp. 95–107, 1987.
4. C. Lin, H. Kogelnik, and L. G. Cohen, Optical pulse equalization and low dispersion transmission in single-mode fibers in the 1.3–1.7 mm spectral region, *Opt. Lett.*, Vol. 5, pp. 476–478, 1980.

5. A. H. Gnauck, S. K. Korotky, B. L. Kasper, J. C. Campbell, J. R. Talman, J. J. Veselka, and A. R. McCormick, Information bandwidth limited transmission at 8 Gb/s over 68.3 km of single mode optical fiber, in *Proceedings of the OFC'86,* paper PDP6, Atlanta, GA, 1986.

6. H. Kogelnik, High-speed lightwave transmission in optical fibers, *Science*, Vol. 228, pp. 1043–1048, 1985.

7. G. P. Agrawal, *Fiber-Optic Communication Systems*, 3rd ed., New York: Wiley, 2002.

8. A. R. Chraplyvy, A. H. Gnauck, R. W. Tkach, and R. M. Derosier, 8x10 Gb/s transmission through 280 km of dispersion-managed fiber, *IEEE Photon. Technol. Lett.*, Vol. 5, pp. 1233–1235, 1993.

9. H. Kogelnik, High-capacity optical communications: Personal recollections, *IEEE J. Sel. Topics Quant. Electron.*, Vol. 6, No. 6, pp. 1279–1286, 2000.

10. C. R. Giles and E. Desurvire, Propagation of signal and noise in concatenated erbium-doped fiber amplifiers, *IEEE J. Lightwave Technol.*, Vol. 9, No. 2, pp. 147–154, 1991.

11. P. C. Becker, N. A. Olsson, and J. R. Simpson, *Erbium-Doped Fiber Amplifiers, Fundamentals and Technology*, San Diego: Academic Press, 1999.

12. M. C. Farries, P. R. Morkel, R. I. Laming, T. A. Birks, D. N. Payne, and E. J. Tarbox, Operation of erbium-doped fiber amplifiers and lasers pumped with frequency-doubled Nd:YAG lasers, *IEEE J. Lightwave Technol.*, Vol. 7, No. 10, pp. 1473–1477, 1989.

13. T. Okoshi, Heterodyne and coherent optical fiber communications: Recent progress, *IEEE Trans. Microwave Theory Tech.*, Vol. 82, No. 8, pp. 1138–1149, 1982.

14. T. Okoshi, Recent advances in coherent optical fiber communication systems, *IEEE J. Lightwave Technol.*, Vol. 5, No. 1, pp. 44–52, 1987.

15. J. Salz, Modulation and detection for coherent lightwave communications, *IEEE Commun. Mag.*, Vol. 24, No. 6, p. 38, 1986.

16. T. Okoshi, Ultimate performance of heterodyne/coherent optical fiber communications, *IEEE J. Lightwave Technol.*, Vol. 4, No. 10, pp. 1556–1562, 1986.

17. P. S. Henry, *Coherent Lightwave Communications*, New York: IEEE Press, 1990.

18. A. F. Elrefaie, R. E. Wagner, D. A. Atlas, and A. D. Daut, Chromatic dispersion limitation in coherent lightwave systems, *IEEE J. Lightwave Technol.*, Vol. 6, No. 5, pp. 704–710, 1988.

19. G. Charlet, E. Corbel, J. Lazaro, A. Klekamp, W. Idler, R. Dischler, S. Bigo, et al., Comparison of system performance at 50, 62.5 and 100 GHz channel spacing over transoceanic distances at 40 Gbit/s channel rate using RZ-DPSK, *Electron. Lett.*, Vol. 41, No. 3, pp. 145–146, 2005.

20. P. S. Cho, V. S. Grigoryan, Y. A. Godin, A. Salamon, and Y. Achiam, Transmission of 25-Gb/s RZ-DQPSK signals with 25-GHz channel spacing over 1000 km of SMF-28 fiber, *IEEE Photon. Technol. Lett.*, Vol. 15, No. 3, pp. 473–475, 2003.

21. K. Ishida, T. Kobayashi, J. Abe, K. Kinjo, S. Kuroda, and T. Mizuochi, A comparative study of 10 Gb/s RZ-DPSK and RZ-ASK WDM transmission over transoceanic distances, in *Proceedings of the OFC'03*, Vol. 2, pp. 451–453, 2003.

22. W. A. Atia and R. S. Bondurant, Demonstration of return-to-zero signaling in both OOK and DPSK formats to improve receiver sensitivity in an optically preamplified receiver, in *Proceedings of the IEEE LEOS'99*, Vol. 1, pp. 226–227, 1999.

23. G. Bosco, A. Carena, V. Curri, R. Gaudino, and P. Poggiolini, On the use of NRZ, RZ, and CSRZ modulation at 40 Gb/s with narrow DWDM channel spacing, *IEEE J. Lightwave Technol.*, Vol. 20, No. 9, pp. 1694–1704, 2002.

24. A. H. Gnauck, G. Raybon, P. G. Bernasconi, J. Leuthold, C. R. Doerr, and L. W. Stulz, 1-Tb/s (6/spl times/170.6 Gb/s) transmission over 2000-km NZDF using OTDM and RZ-DPSK format, *IEEE Photon. Technol. Lett.*, Vol. 15, No. 11, pp. 1618–1620, 2003.

25. B. Zhu, L. E. Nelson, S. Stulz, A. H. Gnauck, C. Doerr, J. Leuthold, L. Gruner-Nielsen, M. O. Pedersen, J. Kim, and R. L. Lingle, Jr., High spectral density long-haul 40-Gb/s transmission using CSRZ-DPSK format, *IEEE J. Lightwave Technol.*, Vol. 22, No. 1, pp. 208–214, 2004.

26. A. Hirano, Y. Miyamoto, and S. Kuwahara, Performances of CSRZ-DPSK and RZ-DPSK in 43-Gbit/s/ch DWDM G.652 single-mode-fiber transmission, in *Proceedings of the OFC'03*, Vol. 2, Anaheim, CA, pp. 454–456, 2003.

27. L. N. Binh and T. L. Huynh, Linear and nonlinear distortion effects in direct detection 40Gb/s MSK modulation formats multi-span optically amplified transmission, *Opt. Commun.*, Vol. 237, No. 2, pp. 352–361, 2007.

28. J. Mo, D. Yi, Y. Wen, S. Takahashi, Y. Wang, and C. Lu, Optical minimum-shift keying modulator for high spectral efficiency WDM systems, in *Proceedings of the ECOC'05*, Vol. 4, pp. 781–782, 2005.

29. J. Mo, Y. J. Wen, Y. Dong, Y. Wang, and C. Lu, Optical minimum-shift keying format and its dispersion tolerance, in *Proceedings of the OFC'05,* paper JThB12, Anaheim, CA, 2005.

30. M. Ohm and J. Speidel, Optical minimum-shift keying with direct detection (MSK/DD), *Proc. SPIE Opt. Transm. Switch. Syst.*, Vol. 5281, pp. 150–161, 2004.

31. T. L. Huynh, T. Sivahumaran, L. N. Binh, and K. K. Pang, Narrowband frequency discrimination receiver for high dispersion tolerance optical MSK systems, in *Proceedings of the Coin-Acoft'07,* paper TuA1-3, Melbourne, Australia, 2007.

32. T. L. Huynh, T. Sivahumaran, L. N. Binh, and K. K. Pang, Sensitivity improvement with offset filtering in optical MSK narrowband frequency discrimination receiver, in *Proceedings of the Coin-Acoft'07,* paper TuA1-5, Melbourne, 2007.

33. T. Sivahumaran, T. L. Huynh, K. K. Pang, and L. N. Binh, Non-linear equalizers in narrowband filter receiver achieving 950 ps/nm residual dispersion tolerance for 40Gb/s optical MSK transmission systems, in *Proceedings of the OFC'07*, paper OThK3, Anaheim, CA, 2007.

34. T. Sakamoto, T. Kawanishi, and M. Izutsu, Optical minimum-shift keying with external modulation scheme, *Opt. Express*, Vol. 13, pp. 7741–7747, 2005.

35. A. H. Gnauck and P. J. Winzer, Optical phase-shift-keyed transmission, *IEEE J. Lightwave Technol.*, Vol. 23, No. 1, pp. 115–130, 2005.

36. J. A. Lazaro, W. Idler, R. Dischler, and A. Klekamp, BER depending tolerances of DPSK balanced receiver at 43Gb/s, in *Proceedings of the IEEE/LEOS Workshop on Advanced Modulation Formats 2004*, pp. 15–16, 2004.

37. H. Kim and A. H. Gnauck, Experimental investigation of the performance limitation of DPSK systems due to nonlinear phase noise, *IEEE Photon. Technol. Lett.*, Vol. 15, No. 2, pp. 320–322, 2003.

38. S. Bhandare, D. Sandel, A. F. Abas, B. Milivojevic, A. Hidayat, R. Noe, M. Guy, and M. Lapointe, 2/spl times/40 Gbit/s RZ-DQPSK transmission with tunable chromatic dispersion compensation in 263 km fibre link, *Electron. Lett.*, Vol. 40, No. 13, pp. 821–822, 2004.

39. T. Mizuochi, K. Ishida, T. Kobayashi, J. Abe, K. Kinjo, K. Motoshima, and K. Kasahara, A comparative study of DPSK and OOK WDM transmission over transoceanic distances and their performance degradations due to nonlinear phase noise, *IEEE J. Lightwave Technol.*, Vol. 21, No. 9, pp. 1933–1943, 2003.

40. C. Xu, X. Liu, L. F. Mollenauer, and X. Wei, Comparison of return-to-zero differential phase-shift keying and ON-OFF keying in long-haul dispersion managed transmission, *IEEE Photon. Technol. Lett.*, Vol. 15, No. 4, pp. 617–619, 2003.

41. L. N. Binh and T. L. Huynh, Phase-modulated hybrid 40Gb/s and 10Gb/s DPSK DWDM long-haul optical transmission, in *Proceedings of the OFC'07,* paper JWA94, Anaheim, CA, 2007.

42. T. Ito, K. Sekiya, and T. Ono, Study of 10 G/40 G hybrid ultra long haul transmission systems with reconfigurable OADM's for efficient wavelength usage, in *Proceedings of the ECOC'02,* paper 1.1.4, Copenhagen, Denmark, 2002.

43. K. Iwashita and N. Takachio, Experimental evaluation of chromatic dispersion distortion in optical CPFSK transmission systems, *IEEE J. Lightwave Technol.*, Vol. 7, No. 10, pp. 1484–1487, 1989.

44. J. G. Proakis, *Digital Communications*, 4th ed., New York: McGraw-Hill, 2001.

45. J. G. Proakis and M. Salehi, *Communication Systems Engineering*, 2nd ed., NJ: Upper Saddle River, Prentice Hall, pp. 522–524, 2002.

46. K. K. Pang, *Digital Transmission*, Melbourne, Australia: Mi-Tec Media, 2005.

47. K. Iwashita and T. Matsumoto, Modulation and detection characteristics of optical continuous phase FSK transmission system, *IEEE J. Lightwave Technol.*, Vol. 5, No. 4, pp. 452–460, 1987.

48. J. Mo, Y. J. Wen, and Y. Wang, Performance evaluation of externally modulated optical minimum shift keyed data, *Opt. Eng.*, Vol. 46, No. 3, pp. 035001–035008, 2007.

49. T. L. Huynh, L. N. Binh, and K. K. Pang, Optical MSK long-haul transmission systems, in *Proceedings of the SPIE APOC'06,* paper 6353-86, Thu9a, 2006.

50. A. F. Elrefaie and R. E. Wagner, Chromatic dispersion limitations for FSK and DPSK systems with direct detection receivers, *IEEE Photon. Technol. Lett.*, Vol. 3, No. 1, pp. 71–73, 1991.

51. B. E. Little, Advances in microring resonator, in *Integrated Photonics Research,* Washington, D.C., June 15, 2003, session Semiconductor Micro-Ring Resonators (ITuE), invited paper.

52. V. Van, B. E. Little, S. T. Chu, and J. V. Hryniewicz, Micro-ring resonator filters, in *Proceedings of the LEOS'04*, Vol. 2, pp. 571–572, 2004.

53. P. P. Absil, S. T. Chu, D. Gill, J. V. Hryniewicz, F. Johnson, O. King, B. E. Little, F. Seiferth, and V. Van, Very high order integrated optical filters, in *Proceedings of the OFC'04*, Vol. 1, Anaheim, CA, 2004.

54. L. Brent, C. Sai, C. Wei, C. Wenlu, H. John, G. Dave, K. Oliver, et al., Advanced ring resonator based PLCs, *IEEE Lasers Electro-Optics Soc.*, pp. 751–752, 2006.

55. N. Alic, G. C. Papen, R. E. Saperstein, L. B. Milstein, and Y. Fainman, Signal statistics and maximum likelihood sequence estimation in intensity modulated fiber optic links containing a single optical pre-amplifier, *Opt. Express*, Vol. 13, No. 12, pp. 4568–4579, 2005.

56. N. Alic, G. C. Papen, and Y. Fainman, Performance of maximum likelihood sequence estimation with different modulation formats, in *Proceedings of the IEEE/LEOS Workshop on Advanced modulation Formats*, pp. 49–50, 2004.

57. V. Curri, R. Gaudino, A. Napoli, and P. Poggiolini, Electronic equalization for advanced modulation formats in dispersion-limited systems, *IEEE Photon. Technol. Lett.*, Vol. 16, No. 11, pp. 2556–2558, 2004.

58. J. D. Downie, M. Sauer, and J. Hurley, Flexible 10.7 Gb/s DWDM transmission over up to 1200 km without optical in-line or post-compensation of dispersion using MLSE-EDC, in *Proceedings of the OFC'06*, paper THB5, Anaheim, CA, 2006.

59. O. E. Agazzi, M. R. Hueda, H. S. Carrer, and D. E. Crivelli, Maximum-likelihood sequence estimation in dispersive optical channels, *IEEE J. Lightwave Technol.*, Vol. 23, No. 2, pp. 749–762, 2005.

60. A. Napoli, Limits of maximum-likelihood sequence estimation in chromatic dispersion limited systems, in *Proceedings of the OFC'06*, paper JThB36, Anaheim, CA, 2006.

61. H. Haunstein, PMD and chromatic dispersion control for 10 and 40Gb/s systems, in *Proceedings of the OFC'04*, invited paper, ThU3, Anaheim, CA, 2004.

62. J. Qi, B. Mao, N. Gonzalez, L. N. Binh, and N. Stojanovic, Generation of 28GBaud and 32GBaud PDM-Nyquist-QPSK by a DAC with 11.3GHz analog bandwidth, in *Proceedings of the OFC 2013*, San Francisco, CA, 2013.

63. N. Stojanović, C. Xie, Y. Zhao, B. Mao, N. Guerrero Gonzalez, J. Qi, and L. N. Binh, Modified Gardner phase detector for Nyquist coherent optical transmission systems, in *Proceedings of the OFC 2013*, paper JTh2A.50.pdf, San Francisco, 2013.

2 Optical Fibers

2.1 OVERVIEW

Planar optical waveguides comprise a guiding region and a slab embedded between a substrate and a superstrate having identical or different refractive indices. The lightwaves are confined by the boundary, and guided under the condition that oscillation is achieved, hence the wave exhibits evanescent penetration in the cladding region. The number of oscillating solutions that satisfy the boundary constraints is the number of modes which can be guided. The guiding of lightwaves in an optical fiber is similar to that of the planar waveguide except that lightwaves are guided through a circular core embedded in a circular cladding layer.

Within the context of this book, optical fibers are treated as circular optical waveguides that can support a single mode with two polarized modes or a few modes with different polarizations. We must point out the progressive development stages in optical fiber communications systems in the following, so that we can focus on modern optical systems:

1. The step-index and graded-index multimode optical fibers find very limited applications in systems and networks for long-haul applications. This type of fiber was extensively developed in the early 1970s. These fibers offered only a limited bandwidth–distance product of about 900 MHz km owing to the interference of different modes, which are estimated at around a few thousand for a step index with a 0.1 index difference and a profile core of 50 μm.

2. Single-mode optical fibers (SMFs) can be designed and realized with a very small difference in the refractive indices between the core and cladding regions (about 0.03%) and a core diameter of about 8 μm. Thus, the guiding in modern optical fiber for telecommunications is called "weak" guiding. This development was intensively debated and agreed upon by the optical fiber communications technology community during the late 1970s. The SMF led to extensive development of coherent reception optical systems and long-haul intensity modulation and direct detection (IM/DD) transmission systems in the 1980s.

3. The invention of optical amplification in rare-earth doped SMFs in the late 1980s has transformed the design and deployment of optical fiber communications systems and networks in the last decade and the coming decades of the twenty-first century. The optical losses of the fiber and those of the optical components of the optical networks can be compensated by inserting the fiber in-line optical amplifiers (OAs).

4. Therefore, the pulse broadening of optical signals during transmission and distribution in the networks become much more important for system design engineers.

5. Recent advances in ultra-high sampling rate for analog-to-digital converters (ADCs) and digital-to-analog converters (DACs), currently at 64 GSa/s have allowed DSP processors incorporating these devices within the coherent reception subsystems. These transmission systems would now consist of only optically amplified standard single-mode optical fiber (SSMF) spans and no DCF (dispersion-compensating fiber) are no longer necessary. Hence, only one OA is required per span. This will minimize the accumulated amplification stimulated emission (ASE) noises at the receiver front end. Thus, under DSP-based reception coherent systems with polarization division multiplexing, the aggregate bit rate per channel can reach 100G or 200G by employing QPSK or 16QAM modulation formats.

6. Furthermore, recently, owing to several demonstrations of the use of digital signal processing of coherently received modulated lightwaves, multiple-input multiple-output (MIMO) techniques can be applied to significantly enhance the sensitivity of optical receivers and hence the transmission distance and the capacity of optical communication systems [1]. MIMO techniques would offer some possibilities for uses of different guided modes through a single fiber, for example, few-mode fibers (FMFs), which can support more than one mode but not too many as in the case of multimode types. Each mode would carry a channel, and at the output they would be spatially demultiplexed, detected in the electrical domain, and then digitally processed using MIMO techniques. Thus, the conditions under which circular optical waveguides can operate as an FMF are also described in this chapter.

Because of the above development, we shall focus on the theoretical approach on understanding optical fibers from the standpoint of the practical aspects of the design of optical fibers with minimum dispersion or for a specified dispersion factor. This can be confirmed by practical measurements, that the optical field distribution follows a Gaussian distribution. Knowing the field distribution, one can obtain the propagation constant of the single guided mode, and hence the spot size of this mode and the energy concentration inside the core of the optical fiber. The basic concept of optical dispersion is that by using the definition of group velocity and group delay, we would be able to derive the chromatic dispersion in SMFs. After arming ourselves with the basic equations for dispersion, we would be able to embark on the design of optical fibers with a specified dispersion factor.

Annex 1 gives the technical parameters of two typical optical fibers (Corning types): (1) G652, commonly known as the SSMFs and (2) G655, known as large effective area fibers.

Apart from understanding the physical phenomena of guiding optical waves and the interference of the modulated wave spectral components, the fiber transfer function in the linear domain and the distortion due to nonlinear effects are also very important for system design. Hence, Sections 2.3 and 2.6 deal with these topics, while Sections 2.2, 2.4, and 2.5 explain the linear physical phenomena of the attenuation and distortion of modulated lightwaves propagating through single-mode fibers.

2.2 OPTICAL FIBER: GENERAL PROPERTIES

2.2.1 GEOMETRICAL STRUCTURES AND INDEX PROFILE

An optical fiber consists of two concentric dielectric cylinders. The inner cylinder, or core, has a refractive index of $n(r)$ and radius a. The outer cylinder, or cladding, has index n_2 with $n_2 < n(r)$ for all positions in the core region. A core size of about 4–9 μm and a cladding diameter of 125 μm are typical values for silica-based SMF. A schematic diagram of the structure of a circular optical fiber is shown in Figure 2.1. Figure 2.1a shows the core and cladding region of the circular fiber, while Figure 2.1b and c show the figure of the etched cross sections of a multimode and single-mode, respectively. The silica fibers are etched in a hydrogen peroxide solution so that the core region doped with impurity would be etched faster than that of pure silica, and thus the core region would be exposed as observed.

Figure 2.2 shows the index profile and the structure of the circular fibers. The refractive index profile can be step or graded.

The refractive index $n(r)$ of a circular optical waveguide usually varies with radius r from the fiber axis ($r = 0$) and is expressed by

$$n^2(r) = n_2^2 + \mathrm{NA}^2 s\left(\frac{r}{a}\right) \qquad (2.1)$$

where NA is the numerical aperture at the core axis, while $s(r/a)$ represents the profile function that characterizes any profile shape ($s = 1$ at maximum) with a scaling parameter (usually the core radius).

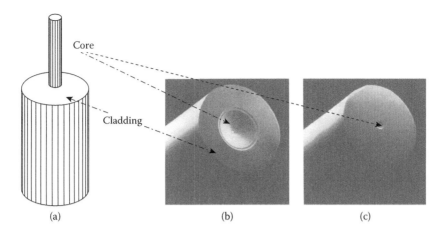

(a)   (b)   (c)

FIGURE 2.1 (a) Schematic diagram of the step-index fiber: coordinate system structure. The refractive index of the core is uniform and slightly larger than that of the cladding. For silica glass, the refractive index of the core is about 1.478 and that of the cladding is about 1.47 in the 1550 nm wavelength region. (b) Cross section of an etched fiber—multimode type—50 µm diameter. (c) Single-mode optical fiber etched cross section.

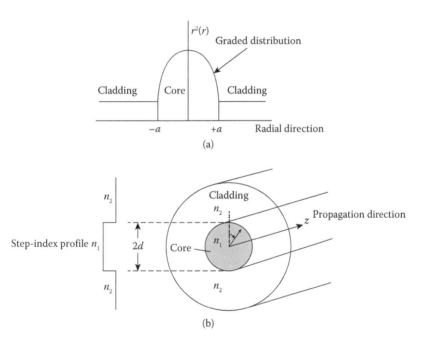

FIGURE 2.2 (a) Refractive index profile of a graded-index profile and (b) fiber cross section and step-index profile with a as the radius of the fiber.

For a step-index profile, the refractive index remains constant in the core region, and thus

$$s\left(\frac{r}{a}\right) = \begin{cases} 1 & r \leq a \\ 0 & r > a \end{cases} \xrightarrow[\text{ref\_index}]{\text{hence}} n^2(r) = \begin{cases} n_1^2 & r \leq a \\ n_2^2 & r > a \end{cases} \tag{2.2}$$

For a graded-index profile, we can consider the two most common types of graded-index profiles: power-law index and the Gaussian profile.

For a power-law profile, the core refractive index of optical fiber usually follows a graded-index profile. In this case, the refractive index rises gradually from the value n_2 of the cladding glass to the value n_1 at the fiber axis. Therefore, $s(r/a)$ can be expressed as

$$s\left(\frac{r}{a}\right) = \left\{1 - \left(\frac{r}{a}\right)^{\alpha} \quad \text{for } r \leq a \text{ and } = 0 \quad \text{for } r < a \right. \tag{2.3}$$

with α as the power exponent. Thus, the index profile distribution $n(r)$ can be expressed in the usual way (by using Equations 2.2 and 2.3), by substituting $\text{NA}^2 = n_1^2 - n_2^2$, as

$$n^2(r) = \left\{ \begin{array}{ll} n_1^2 \left[1 - 2\Delta\left(\dfrac{r}{a}\right)^{\alpha}\right] & \text{for } r \leq a \\[3mm] n_2^2 & \text{for } r > a \end{array} \right. \tag{2.4}$$

where $\Delta = \text{NA}^2/n_1^2$ is the relative refractive difference with a small difference between that of the cladding and the core regions. The profile shape given in Equation 2.4 offers three special distributions: (1) $\alpha = 1$: the profile function $s(r/a)$ is linear, and the profile is called a triangular profile; (2) $\alpha = 2$: the profile is a quadratic function with respect to the radial distance, and the profile is called the parabolic profile; and (3) $\alpha = \infty$: the profile is a step type.

For a Gaussian profile, the refractive index changes gradually from the core center to a distance very far away from it, and $s(r)$ can be expressed as

$$s\left(\frac{r}{a}\right) = e^{-\left(\frac{r}{a}\right)^2} \tag{2.5}$$

2.2.2 FUNDAMENTAL MODE OF WEAKLY GUIDING FIBERS

The electric and magnetic fields $E(r,\phi,z)$ and $H(r,\phi,z)$ of the optical fibers in cylindrical coordinates can be found by solving Maxwell's equations. Only the lower-order modes of ideal step-index fibers are important for digital optical communication systems. The fact is that for $\Delta < 1\%$, the optical waves are confined very weakly and are thus gently guided. Thus, the electric and magnetic fields E and H can then take approximate solutions of the scalar wave equation in a cylindrical coordinate system (x,θ,ϕ).

$$\left[\frac{\delta^2}{\delta r^2} + \frac{1}{r}\frac{\delta}{\delta r} + k^2 n_j^2\right]\varphi(r) = \beta^2 \varphi(r) \tag{2.6}$$

where $n_j = n_1, n_2$, and $\phi(r)$ is the spatial field distribution of the nearly transverse EM waves.

$$E_x = \psi(r)e^{-i\beta z}$$
$$H_y = \left(\frac{\varepsilon}{\mu}\right)^{1/2} E_x = \frac{n_2}{Z_0} E_x \tag{2.7}$$

where E_y, E_z, H_x, H_z are negligible, $\varepsilon = \varepsilon_0 n_2^2$, and $Z_0 = (\varepsilon_0\mu_0)^{1/2}$ is the vacuum impedance. We can assume that the waves can be seen as a plane wave traveling down along the fiber tube. This plane wave is reflected between the dielectric interfaces; in other words, it is trapped and guided along the core of the optical fiber. Note that the electric and magnetic components are spatially orthogonal to each other. Thus, for a single mode, there are always two polarized components that are then the polarized modes of single-mode fiber. It is further noted that Maxell's equations, not Snell's law of reflection, would be applicable for single-mode propagation. However, we will see in the next section that the field distribution of SMFs closely follows a Gaussian shape. Hence, the solution to the wave equation (Equation 2.6) can be assumed, and hence the eigenvalue or the propagation constant of the guided wave can be found or optimized to achieve the best fiber structure. However, currently, because of the potential of digital signal processing, multimode fiber modes can be put to beneficial use, so few-mode optical fibers are intensively investigated.

Thus, in the next section, we give a brief analysis of the wave equations subject to the boundary conditions so that the eigenvalue equation can be found; hence, the propagation or wave number of the guided modes can be found and thence the propagation delay of these group of lightwaves along the fiber transmission line. Next, we will revisit the single-mode fiber with a Gaussian mode field profile to gain an insight into the weakly guiding phenomenon that is so important for the understanding of the guiding of lightwaves over very long distances with minimum loss and optimum dispersion: the group delay difference.

2.2.2.1   Solutions of the Wave Equation for Step-Index Fiber

The field spatial function $\phi(r)$ would have the form of Bessel functions given by Equation 2.6 as

$$\varphi(r) = \begin{cases} A\dfrac{J_0\left(ur/a\right)}{J_0(u)} & 0 < r < a - \text{core} \\[2em] A\dfrac{K_0\left(vr/a\right)}{K_0(v)} & r > a - \text{cladding} \end{cases} \tag{2.8}$$

where J_0, K_0 are the Bessel functions of the first kind and the modified second kind, respectively, and u, v are defined as

$$\frac{u^2}{a^2} = k^2 n_1^2 - \beta^2 \tag{2.9a}$$

$$\frac{v^2}{a^2} = -k^2 n_2^2 + \beta^2 \tag{2.9b}$$

Thus, following the Maxwell's equations relation, we can find that E_z can take two possible solutions that are orthogonal as

$$E_z = -\frac{A}{kan_2}\left(\begin{array}{c} \sin\phi \\ \cos\phi \end{array}\right)\begin{cases} \dfrac{uJ_1\left(u\dfrac{r}{a}\right)}{J_0(u)} & \text{for } 0 \leq r < a \\[2em] \dfrac{vK_1\left(\dfrac{vr}{a}\right)}{K_0(v)} & \text{for } r > a \end{cases} \tag{2.10}$$

The terms u and v must simultaneously satisfy two equations

$$u^2 + v^2 = V^2 = ka\left(n_1^2 - n_2^2\right)^{1/2} = kan_2\left(2\Delta\right)^{1/2} \tag{2.11}$$

$$u\frac{J_1(u)}{J_0(u)} = v\frac{K_1(v)}{K_0(v)} \tag{2.12}$$

where Equation 2.12 is obtained by applying the boundary conditions at the interface $r = a(E_z)$ is the tangential component and must be continuous at this dielectric interface. Equation 2.12 is commonly known as the eigenvalue equation of the wave equation bounded by the continuity at the boundary of the two dielectric media, and is thus the condition for guiding the transverse plane such that the maximum or fastest propagation velocity is obtained in the axial direction. The solution of this equation would give specific discrete values of β, the propagation constants of the guided lightwaves.

2.2.2.2 Single-Mode and Few-Mode Conditions

Over the years, since the demonstration of the guiding in circular optical waveguides, the eigenvalue equation (Equation 2.12) has been employed to find the number of modes supported by the waveguide and their specific propagation constants. Next, the Gaussian mode spatial distribution can be approximated for the fundamental mode based on experimental measurement of the mode fields, and the eigenvalue equation is no longer needed when single-mode fiber is extensively used. However, with the current extensive research interest in spatial multiplexing in DSP-based coherent optical communication systems, FMFs have attracted quite a lot of interest owing to their potential support of many channels with their modes and their related field polarizations. This section is thus extended to consider both the fundamental mode and higher-order modes.

Equation 2.11 shows that the longitudinal field is in the order of $u/(kan_2)$ with respect to the transverse component. In practice, $\Delta \ll 1$, and by using Equation 2.11, we observe that this longitudinal component is negligible compared to the transverse component. Thus, the guided mode is *transversely polarized*. The fundamental mode is then usually denoted as LP_{01} mode (LP = linearly polarized), for which the field distribution is shown later in the chapter in Figure 2.4a and b. The graphical representation of the eigenvalue equation (Equation 2.12) can be estimated with the variation of $b = \beta/k$ *defined* as the normalized propagation constant, and the V parameter is shown later in the chapter in Figure 2.5d. There are two possible polarized modes, the horizontal and vertical polarizations, which are orthogonal to each other. These two polarized modes can be employed for transmission of different information channels. They are currently exploited, in the first two decades of the twenty-first century, in optical transmission systems employing polarization division multiplexing so as to offer transmission bit rates of 100 Gbps and beyond.

Furthermore, when the number of guided modes is higher than two polarized modes, they form a set of modes over which information channels can be simultaneously carried and spatially demultiplexed at the receiving end so as to increase the transmission capacity as illustrated later in the chapter in Figure 2.4a and b [2,3]. Such FMFs are employed in the most modern optical transmission system, whose schematic is shown later in the chapter in Figure 2.6. The mode multiplexer performs as mode spatial mixing. Similarly at the receiver side, the demultiplexer splits the modes LP_{11}, LP_{01} into individual modes and thence injecting into the coherent receiver. Obviously, there must be mode spatial demultiplexing and then modulation and then multiplexing back into the transmission fiber for transmission. Similar structures would be available at the receiver to separate and detect the channels. Note the two possible polarizations of the mode LP_{11}. Note that there are four polarized modes of the LP_{11} mode, but only two polarized modes are shown in this diagram. The delay due to the propagation velocity, from the difference in the propagation constant, can be easily compensated in the DSP processing algorithm, similar to that due to polarization-mode dispersion (PMD). The main problem to resolve in this spatial mode multiplexing optical transmission system is the optical amplification for all modes so that long-haul transmission can be achieved, that is, the amplification in the multimode fiber structure.

The number of guided modes is determined by the number of intersecting points of the circle of radius V and the curves representing the eigenvalue solutions (Equation 2.12). Thus, for a single-mode fiber, the V parameter must be less than 2.405, and for an FMF, this value is higher; for example, if $V = 2.8$, we have three intersecting points between the circle of radius V and three curves, and then the number of modes would be LP_{01}, LP_{11}, and their corresponding alternative polarized modes would be as shown in Figure 2.3 [4] and then Figure 2.4b and c. For single mode, there are two polarized modes whose polarizations can be vertical or horizontal. Thus, a single-mode fiber is not a monomode but supports two polarized modes! The main issues are also related to the optical amplification gain for the transmission of modulated signals in such FMFs. These remain the principal obstacles.

We can illustrate the propagation of the fundamental mode and higher-order modes as in Figure 2.5a and b. The rays of these modes can be axially straight or skewed and twisted around the principal axis of the fiber. Thus, there are different propagation times between these modes. This property can be employed to compensate for the chromatic dispersion effect [5].

Figure 2.5 shows a spectrum of the graphical solution of the modes of optical fibers. In Figure 2.5d, the regions of single operation and then a higher-order, second-order mode region as determined by the value of the V parameter are indicated. Naturally, owing to the manufacturing process, the mode regions would be variable from fiber to fiber.

Figure 2.6 shows a setup for ultra-high-capacity transmission using an FMF with spatial division multiplexing (SDM) and demultiplexing devices. For example, suppose that a lightwave source at wavelength 850 nm is launched into the standard single-mode fiber (SSMF) which is designed to be

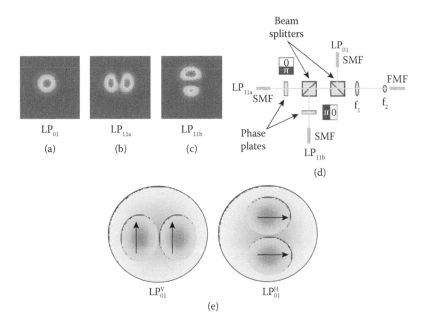

FIGURE 2.3 (a–c) Intensity profiles distributed in the core of first few order modes of a few-mode optical fiber employed for 5 × 65 Gbps optical transmission system and optical system arrangement for spatially demux and mux of modal channels. (d) Schematic of the optical system setup and (e) Horizontal (H) vertical (V) polarized modes $LP_{01}^{V,H}$, polarization directions indicated by arrows. (From S. Randel et al., *Opt. Express.*, Vol. 19, No. 17, p. 16697, 2011.)

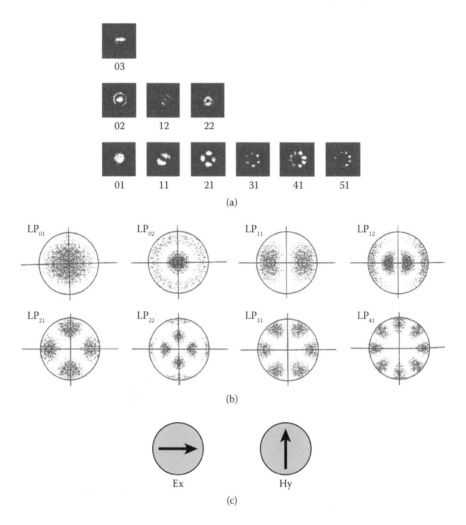

FIGURE 2.4 (a) Spectrum of guided modes in an FMF (numbers indicate the order of modes), (b) calculated intensity distribution of LP-guided modes in a step-index optical fiber with $V = 7$, and (c) electric and magnetic field distribution of an LP_{01} mode polarized along Ox (H-mode) and Oy (V-mode) of the fundamental mode of a single-mode fiber.

in the single mode region at wavelength 1550 nm. The SSMF is no longer single mode at 850 nm but becomes few mode and can support several modes up to the order of LP41. Thus, 16 partial modes can be launched into the SSMF in which the bit rate can reach 100 Gbps per spatial channel, resulting in 1.6 Tbps for the transmission over the FMF. The transmission would be less than 10 km, though. Another advantage of this SDM transmission is that at the receiver, multiple DSP systems can be combined and joint processing can be conducted under the multiple-input-multiple-output (MIMO) algorithms to optimize and reduce significantly the bit error rate (BER; Figure 2.7).

2.2.2.3 Gaussian Approximation: Fundamental Mode Revisited

We note again that the E and H are approximate solutions of the scalar wave equation, and the main properties of the fundamental mode of weakly guiding fibers that can be observed are as follows:

1. The propagation constant β (in the z-direction) of the fundamental mode must lie between the core and cladding wave numbers. This means the effective refractive index of the guided mode lies within the range of the cladding and core refractive indices.

2. Accordingly, the fundamental mode must be nearly a transverse electromagnetic wave as described in Equation 2.9.

$$\frac{2\pi n_2}{\lambda} < \beta < \frac{2\pi n_1}{\lambda} \tag{2.13}$$

3. The spatial dependence $\psi(r)$ is a solution of the scalar wave equation (Equation 2.6).

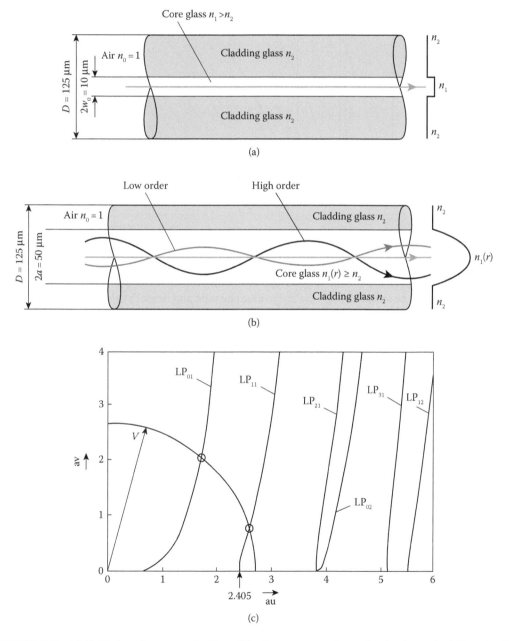

FIGURE 2.5   (a) Guided modes as seen by "a ray" in the transverse plane of a circular optical fiber; "ray" model of lightwave propagating in single-mode fiber, (b) ray model of propagation of different modes guided in a few-/multimode graded-index fiber, and (c) graphical illustration of solutions for eigenvalues (propagation constant–wave number of optical fibers).   *(Continued)*

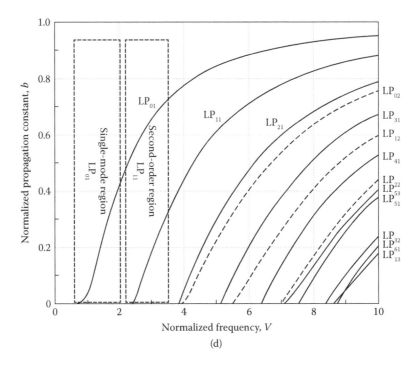

FIGURE 2.5 (Continued) (d) b–V characteristics of guided fibers.

The main objectives are to find a good approximation for the field $\psi(r)$ and the propagation constant β. This can be found through the eigenvalue equation and Bessel's solutions as shown in the previous section. It is desirable if we can approximate the field with a good accuracy to obtain simple expressions to have a clearer understanding of light transmission on the SMF without employing graphical or numerical methods. Furthermore, experimental measurements and numerical solution for step and power-law profiles show that $\psi(r)$ is approximately Gaussian in appearance. We thus approximate the field of the fundamental mode as

$$\phi(r) \cong A e^{-\frac{1}{2}\left(\frac{r}{r_0}\right)^2} \tag{2.14}$$

where r_0 is defined as the spot size, that is, the size at which the intensity equals to e−1 of the maximum. Thus, if the wave equation (Equation 2.6) is multiplied by $r\psi(r)$ and the identity

$$r\phi\frac{\delta^2\phi}{\delta r^2} + \phi\frac{\delta\phi}{\delta r} = \frac{\delta}{\delta r}\left(r\phi\frac{\delta\phi}{\delta r}\right) - r\left(\frac{\delta\phi}{\delta r}\right)^2 \tag{2.15}$$

is used, then, by integrating from 0 to infinity and using $\left[r\psi\frac{d\phi}{dr}\right]_0^\infty = 0$, we have

$$\beta^2 = \frac{\int_0^\infty \left[-\left(\frac{\delta\phi}{\delta r}\right)^2 + k^2 n^2(r)\phi^2\right] r\,\delta r}{\int_0^\infty r\phi^2\,\delta r} \tag{2.16}$$

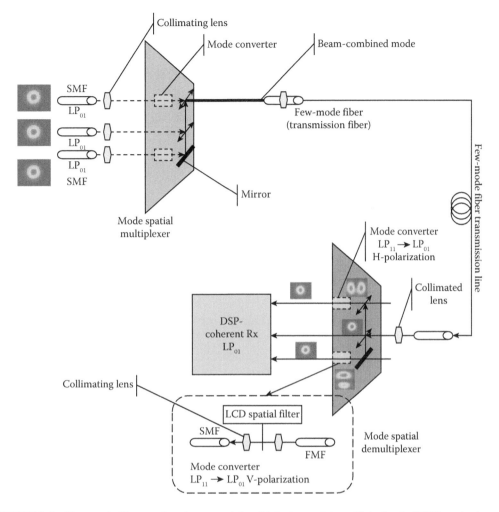

FIGURE 2.6 Few-mode fiber employed as a spatial multiplexing and demultiplexing in DSP-based coherent optical transmission systems operating at 100 Gbps and higher bit rate.

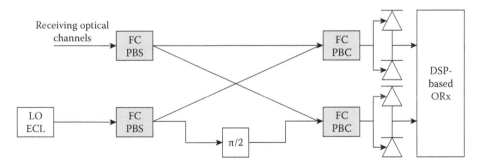

FIGURE 2.7 $\pi/2$ hybrid coupler for polarization demultiplexing and mixing with local oscillator in a coherent receiver of modern DSP-based optical receiver for detection of phase-modulated schemes.

The procedure to find the spot size is then followed by substituting $\psi(r)$ (Gaussian) in Equation 2.14 into Equation 2.16, differentiating, and setting $\delta^2\beta/\delta r$ evaluated at r_0 to zero; that is, the propagation constant β of the fundamental mode *must* give the largest value of r_0. Therefore, knowing r_0 and β, the fields E_x and H_y (Equation 2.7) are fully specified.

2.2.2.3.1 Step-Index Profile

Substituting the step-index profile given by Equation 2.7 and $\psi(r)$ in Equation 2.14 and then Equation 2.16 leads to an expression for β in term of r_0 given by

$$V = \text{NA}.\text{k}.a = \text{NA}\frac{2\pi}{\lambda}a \tag{2.17}$$

The spot size is thus evaluated by setting

$$\frac{\delta^2\beta}{\delta r_0} = 0 \tag{2.18}$$

and r_0 is then given by

$$r_0^2 = \frac{a^2}{\ln V^2} \tag{2.19}$$

Substituting Equation 2.19 in Equation 2.17, we have

$$(a\beta)^2 = (akn_1)^2 - \ln V^2 - 1 \tag{2.20}$$

This expression is physically meaningful only when $V > 1$, that is, when r_0 is positive, which is naturally feasible.

2.2.2.3.2 Gaussian Index Profile Fiber

Similarly, for the case of a Gaussian index profile, by following the procedures for a step-index profile fiber, we can obtain

$$(a\beta)^2 = (an_1k)^2 - \left(\frac{a}{r_0}\right)^2 + \frac{V^2}{\left(\dfrac{a}{r_0}+1\right)} \tag{2.21}$$

and

$$r_0^2 = \frac{a^2}{V-1} \text{ by using } \frac{\delta^2\beta}{\delta^2 r_0} = 0 \tag{2.22}$$

That is, the propagation constant of the guided waves is maximized. The propagation constant is maximum when the "light ray" is very close to the horizontal direction. Substituting Equation 2.22 in Equation 2.21, we have

$$(a\beta)^2 = (akn_1)^2 - 2V + 1 \tag{2.23}$$

Thus, Equations 2.22 and 2.23 are physically meaningful only when

$$V > 1 \ (r_0 > 0). \tag{2.24}$$

It is obvious from Equation 2.24 that the spot size of the optical fiber with a V parameter of 1 is extremely large. It is very important that the optical fiber must not be designed with a near-unity value of the V parameter, because under this scenario all the optical field is distributed in the cladding region. In practice, we observe that the spot size is large but finite (observable). In fact, if V smaller than 1.5, the spot size becomes large. The next chapter investigates this in detail.

2.2.2.4 Cutoff Properties

Similar to the case of planar dielectric waveguides, from Figure 2.5, we observe that when $V < 2.405$, only the fundamental LP_{01} exists. It is noted that for single-mode operation, the V parameter must be ≤ 2.405. However, in practice, $V < 3$ may be acceptable for single-mode operation. Indeed, the value 2.405 is the first zero of the Bessel function $J_0(u)$. In practice, one cannot really distinguish between the V values of 2.3 and 3.0; experimental observation also shows that optical fiber can still support only one mode. Thus, designers do usually take the value of V as 3.0 or less to design an SMF.

The V parameter is inversely proportional to the optical wavelength, which is directly related to the operating frequency. Thus, if an optical fiber is launched with lightwaves whose optical wavelength is smaller than the operating wavelength at which the optical fiber is single mode, then the optical fiber is supporting more than one mode. The optical fiber is said to be operating in a few-mode region and then the multimode region when the total number of modes reaches a few hundreds.

Thus, one can define the cutoff wavelength for optical fibers as follows: the wavelength (λ_c) *above which* only the fundamental mode is guided in the fiber is called the *cutoff wavelength* λ_c. This cutoff wavelength can be found by using the V parameter as follows: $V_C = V|_{\text{cutoff}} = 2.405$, and thus

$$\lambda_c = \frac{2\pi a \text{NA}}{V_c} \tag{2.25}$$

In practice, the fibers tend to be effectively single mode for larger values of V, say, $V < 3$, for the step profile, because the higher-order modes suffer radiation losses due to fiber imperfections. Thus, if $V = 3$, from Equation 2.11 we have $a < (3\lambda/2) \cdot \text{NA}$; when $\lambda = 1$ μm, the numerical aperture NA must be very small ($\ll 1$) for the radius a to have a reasonable dimension. Usually, Δ is about 1% or less for SSMF employed in long-haul optical transmission systems so as to minimize the loss factor and the dispersion.

2.2.2.5 Power Distribution

The axial power density or intensity profile $S(r)$, the z-component of the Poynting vector, is given by

$$S(r) = \frac{1}{2} E_x H_y^* \tag{2.26}$$

Substituting Equation 2.7 in Equation 2.26, we have

$$S(r) = \frac{1}{2}\left(\frac{\varepsilon}{\mu}\right)^{\frac{1}{2}} e^{-\left(\frac{r}{r_0}\right)^2} \tag{2.27}$$

The total power is then given by

$$P = 2\pi \int_0^\infty r S(r) dr = \frac{1}{2}\left(\frac{\varepsilon}{\mu}\right)^{\frac{1}{2}} r_0^2 \tag{2.28}$$

and hence the fraction of the power $\eta(r)$ within $0 \rightarrow r$ across the fiber cross section is given by

$$\eta(r) = \frac{\int_0^r r S(r) dr}{\int_0^\infty r S(r) dr} = 1 - e^{-\left(\frac{r^2}{r_0^2}\right)} \tag{2.29}$$

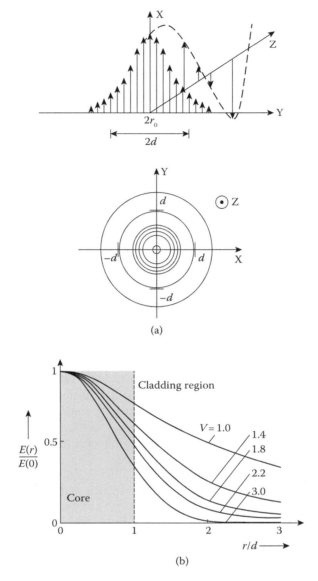

FIGURE 2.8 (a) Intensity distribution of the LP_{01} mode and (b) variation of the spot size–field distribution with radial distance r with V as a parameter.

Thus, given a step profile or approximated profile, one can either analytically or numerically estimate the power confined in the core region and the propagation constant, and hence the phase velocity of the guided mode. Experimentally or in production, the mode spot size can be obtained from the digital image from an infrared camera, and then Equations 2.28 and 2.29 can be used to obtain the power of the mode confined inside the core.

Figure 2.8 depicts the intensity distribution of the LP_{01} mode and the variation in the spot size field distribution with the radial distance r, with V as a parameter, using the equations developed in this subsection.

2.2.2.6 Approximation of Spot Size r_0 of Step-Index Fiber

As stated earlier, the spot size r_0 would play a major role in determining the performance of single-mode fiber. It is useful if we can approximate the spot size as long as the fiber is operating over

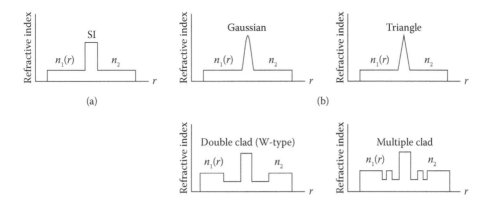

FIGURE 2.9 Index profiles of a number of modern fibers, for example, dispersion-shifted single-mode fibers. (a) SSMF and (b) dispersion-shifted single-mode fiber.

a certain wavelength. When a single-mode fiber is operating above the cutoff wavelength, a good approximation (greater than 96% accuracy) for r_0 is given by

$$\frac{r_0}{a} = 0.65 + 1.619V^{-3/2} + 2.879V^{-6} = 0.65 + 0.434\left(\frac{\lambda}{\lambda_c}\right)^{+3/2} + 0.0419\left(\frac{\lambda}{\lambda_c}\right)^{+6}$$

(2.30)

$$\text{for} \quad 0.8 \le \frac{\lambda}{\lambda_c} \le 2.0 \quad \text{single\_mode\_region}$$

2.2.3 EQUIVALENT STEP-INDEX DESCRIPTION

We have seen that there are two possible orthogonally polarized modes (E_x, H_y) and (E_y, H_x) that can be propagating *simultaneously*. The superposition of these modes can be usually approximated by a single linearly polarized (LP) mode. These modes' properties are well known and well understood for step-index optical fibers and analytical solutions are also readily available. Unfortunately, practical SMFs never have a perfect step-index profile owing to the variation in the dopant diffusion and polarization. These non-step-index fibers can be approximated, under some special conditions, by the *equivalent step-index* (ESI) profile technique.

A number of index profiles of modern single-mode fibers, for example, nonzero dispersion-shifted fibers (NZ-DSFs), are shown in Figure 2.9. The ESI profile is determined by approximating the fundamental mode electric field spatial distribution $\psi(r)$ by a Gaussian function as described earlier. The electric field can thus be totally specified by the e^{-1} width of this function or the mode spot size (r_0). Alternatively, the term *mode field diameter* is also used and equivalent to twice the size of the mode spot size r_0. The ESI method is important in practice because the measured refractive index profile of manufactured fibers would never follow the ideal geometrical profile; so the ESI is applied to reconstruct it to an equivalent ideal distribution so that the analytical estimation can be used to confirm that obtained by experimental measurement.

2.3 NONLINEAR EFFECTS

In this section, the nonlinear effects on the guided lightwaves propagating through a long length of optical fibers, the single-mode type, are described. These effects play important roles in the transmission of optical pulses along SMFs as distortion owing to the modification of the phase of the lightwaves. The nonlinear effects can be classified into three types: (1) the effects that change the refractive index of the guided medium owing to the intensity of the pulse, the self-phase modulation; (2) the scattering of the lightwave to other frequency-shifted optical waves when the intensity exceeds a certain threshold, the Brillouin and Raman scattering phenomena; and (3) the mixing of optical waves to generate fourth waves, the degenerate four-wave mixing. Besides these nonlinear effects, there is also the photorefractive effect, which is due to the change in the refractive index of silica due to the intensity of ultraviolet optical waves. This phenomenon is used to fabricate grating whose spacing between dark and bright region satisfies the Bragg diffraction condition. These are fiber Bragg gratings and would be used as optical filters and dispersion compensators when the spacing varies or is chirped.

In modern coherent optical communication systems incorporating digital signal processors at the receiver, the compensation can be done in the electronic domain, and back propagation of the lightwaves can be implemented to reverse the nonlinear effects imposing on the phase of the guided mode using the frequency-domain transfer function [6–8].

2.3.1 Nonlinear Self-Phase Modulation Effects

All optical transparent materials are subject to a change in the refractive index with the intensity of the optical waves, the optical Kerr effect. This physical phenomenon is due to the harmonic responses of electrons of optical fields, leading to a change in the material susceptibility. The modified refractive index $n_{1,2}^K$ of the core and cladding regions of the silica-based material can be written as

$$n_{1,2}^K = n_{1,2} + \overline{n}_2 \frac{P}{A_{\text{eff}}} \tag{2.31}$$

where n_2 is the nonlinear index coefficient of the guided medium, the average typical value of which is about 2.6×10^{-20} m²/W. P is the average optical power of the pulse, and A_{eff} is the effective area of the guided mode. The nonlinear index changes with the doping materials in the core. Although the nonlinear index coefficient is very small, the effective area is also very small, about 50–70 μm², the length of the fiber under the propagation of the optical signals is very long, and the accumulated phase change is quite substantial. This leads to SPM and cross-phase modulation (XPM) effects in the optical channels.

2.3.2 Self-Phase Modulation

The SPM is due to the phase variation of the guided lightwaves caused by the intensity of its own power or field accumulated along the propagation path, which is quite long, possibly a few hundreds to thousands of kilometers. Under a linear approximation, we can write the modified propagation constant of the guided LP mode in an SMF as

$$\beta^K = \beta + k_0 \overline{n}_2 \frac{P}{A_{\text{eff}}} = \beta + \gamma P \tag{2.32}$$

where

$$\gamma = \frac{2\pi \overline{n}_2}{\lambda A_{\text{eff}}}$$

where n_2 and γ are the nonlinear coefficient and parameter of the guided medium taking, respectively, effective values of 2.3×10^{-23} m⁻² and from 1 to 5 (kmW)⁻¹, depending on the effective area of the guided mode and the operating wavelength. Thus, the smaller the mode spot size or mode field diameter, the larger the nonlinear SPM effect. For dispersion-compensating fiber, the effective area is about 15 μm², while for SSMF and NZ-DSFs, the effective area ranges from

50 to 80 μm^2. Thus, the nonlinear threshold power of DCF is much lower than that of SSMF and NZ-DSF. The maximum launched power into DCF would be limited to about 0 dBm or 1.0 mW in order to avoid the nonlinear distortion effect, while it is about 5 dBm for SSMF.

The accumulated nonlinear phase changes due to the nonlinear Kerr effect over the propagation length L is given by

$$\phi_{NL} = \int_0^L \left(\beta^K - \beta\right) dz = \int_0^L \gamma P(z) dz = \gamma P_{in} L_{eff} \tag{2.33}$$

with $P(z) = P_{in} e^{-\alpha z}$

This equation represents the phase change under nonlinear effects over a length L along the propagation direction z. When considering that the nonlinear SPM effect is small compared with the linear chromatic dispersion effect, one can set $\phi_{NL} \ll 1$ or $\phi_{NL} = 0.1$ rad and the effective length of the propagating fiber is set at $L_{eff} = 1/\alpha$ with optical losses equalized by cascaded optical amplification subsystems. Then the maximum input power to be launched into the fiber can be set at

$$P_{in} < \frac{0.1\alpha}{\gamma N_A} \tag{2.34}$$

If $\gamma = 2$ (W·km)$^{-1}$ and $N_A = 10$, $\alpha = 0.2$ dB/km (or 0.0434×0.2 km^{-1}), then $P_{in} < 2.2$ mW or about 3 dBm. Similarly, this threshold level is about 1.0 mW for DCF with an effective area of 15 μm^2. In practice, owing to the randomness of the arrival of "1" and "0," this nonlinear threshold input power can be set at about 10 dBm as the total average power of all wavelength multiplexed channels of wavelength division multiplexed (WDM) transmission systems launched into the fiber link.

2.3.3 Cross-Phase Modulation

The change in the refractive index of the guided medium as a function of the intensity of the optical signals can also lead to the phase changes of optical channels in different spectral regions, hence distortion imposes on the propagating channels. This is XPM, which is critical in WDM channels, and even more critical in dense WDM when the frequency spacing between channels is 50 GHz or even narrower, the cross interference between channels generating unwanted noises in the optical domain and thence to the detected electronic signals at the receiver. In such systems, the nonlinear phase shift of a particular channel depends not only on its power but also on the powers of other multiplexed channels. The phase shift of the ith channel can be written as [9]

$$\phi_{NL}^i = \gamma L_{eff} \left(P_{in}^i + 2 \sum_{j \neq i}^M P_j \right) \tag{2.35}$$

with M number of multiplexed channels.

The factor of 2 in Equation 2.35 is due to the bipolar effects of the susceptibility of silica materials, and the total phase noises are integrated over both sides of the channel spectrum. The XPM thus depends on the bit pattern and the randomness of the synchronous arrival of the 1. It is difficult to estimate analytically, so numerical simulations would normally be employed to obtain the XPM distortion effects using the nonlinear Schrödinger wave propagation equation involving

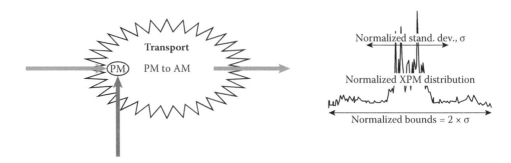

FIGURE 2.10 Illustration of XPM effects: phase modulation conversion to amplitude modulation and hence interference between adjacent channels.

the signal envelopes of all channels. The evolution of slow-varying complex envelopes $A(z,t)$ of optical pulses along an SMF is governed by the nonlinear Schrödinger equation (NLSE) [7]

$$\frac{\partial A(z,t)}{\partial z} + \frac{\alpha}{2} A(z,t) + \beta_1 \frac{\partial A(z,t)}{\partial t} + \frac{j}{2}\beta_2 \frac{\partial^2 A(z,t)}{\partial t^2} - \frac{1}{6}\beta_3 \frac{\partial^3 A(z,t)}{\partial t^3} = -j\gamma |A(z,t)|^2 A(z,t) \quad (2.36)$$

where z is the spatial longitudinal coordinate, α accounts for fiber attenuation, β_1 indicates differential group delay (DGD), β_2 and β_3 represent the second- and third-order factors of fiber CD, and γ is the nonlinear coefficient. This equation would be derived from Maxwell's equations under external perturbation.

The phase change imposes on the phases of the guided channel due to nonlinear phase effects is then converted to amplitude changes and thence the crosstalk onto adjacent channels. This is shown in Figure 2.10.

2.3.4 STIMULATED SCATTERING EFFECTS

Scattering of lightwaves by impurities can occur owing to the absorption and vibration of electrons and dislocation of molecules in silica-based materials. The backscattering and absorption are commonly known as Raleigh scattering losses in fiber propagation, in which the frequency of the optical carrier does not change. Other scattering processes in which the frequency of the lightwave carrier is shifted to another frequency region are commonly known as inelastic scattering and commonly known as Raman scattering and Brillouin scattering. In both cases, the scattering of photons to a lower-energy level photon with the energy difference between these levels is fallen with the energy of the phonons. Optical phonons result from the electronic vibration for Raman scattering, while acoustic phonons or mechanical vibration of the linkage between molecules lead to Brillouin scattering. At high power, when the intensity exceeds a certain threshold, the number of scattered photons exponentially grows, and then the phenomenon is a simulated process. Thus, the phenomenon can be called stimulated Brillouin scattering (SBS) and stimulated Raman scattering (SRS). SRS and SBS were first observed in the 1970s [10–12].

2.3.4.1 Stimulated Brillouin Scattering

Brillouin scattering comes from the compression of the silica materials in the presence of an electric field, the electrostriction effect. Under the pumping of an oscillating electric field of frequency f_p, an acoustic wave of frequency F_a is generated. Spontaneous scattering is an energy transfer from the pump wave to the optical wave followed by a phase matching to transfer a frequency shifted optical wave of frequency as a sum of the optical signal waves and the acoustic wave. This acoustic wave frequency shift is around 11 GHz with a bandwidth of around 50–100 MHz (due to the gain coefficient of the SBS), and a beating envelope would be modulating the optical signals.

Thus, jittering of the received signals at the receiver would be formed, and hence the closure of the eye diagram in the time domain.

Once the acoustics wave is generated, it beats with the signal waves to generate the sideband components. This beating beam acts as a source and further transfers the signal beam energy into the acoustic wave energy and further amplifies this wave to generate further jittering effects. Brillouin scattering process can be expressed by the following coupled equations [13]

$$\frac{dI_p}{dz} = -g_B I_p I_s - \alpha_p I_p$$

$$-\frac{dI_s}{dz} = +g_B I_p I_s - \alpha_s I_s$$

(2.37)

The SBS gain g_B is frequency dependent with a gain bandwidth of around 50–100 MHz for the pump wavelength at around 1550 nm. For silica fiber, g_B is about 5e−11 mW^{-1}. The threshold power for the generation of SBS can be estimated (using Equation 2.37) as

$$g_B P_{\text{th\_SBS}} \frac{L_{\text{eff}}}{A_{\text{eff}}} \approx 21$$

(2.38)

$$\text{with the\_effective\_length } L_{\text{eff}} = \frac{1 - e^{-\alpha L}}{\alpha}$$

where
   I_p = intensity of pump beam
   I_s = intensity of signal beam
   g_B = Brillouin scattering gain coefficient
   α_s, α_p = losses of signal and pump waves

For SSMF, this SBS power threshold is about 1.0 mW. Once the launched power exceeds this power threshold level, the beam energy is reflected back. Thus, the average launched power is usually limited to a few decibel-milliwatts due to this low threshold power level.

2.3.4.2 Stimulated Raman Scattering

SRS occurs in silica-based fiber when a pump laser source is launched into the guided medium. The scattering light from the molecules and dopants in the core region would be shifted to a higher energy level and then jump down to a lower energy level, and hence amplification of photons occurs in this level. Thus, a transfer of energy from photons with different frequency and energy levels occurs. The stimulated emission occurs when the pump energy level exceeds the threshold level. The pump intensity and signal beam intensity are related via the coupled equations

$$\frac{dI_p}{dz} = -g_R I_p I_s - \alpha_p I_p$$

$$-\frac{dI_s}{dz} = +g_R I_p I_s - \alpha_s I_s$$

(2.39)

where
   I_p = intensity of pump beam
   I_s = intensity of signal beam
   g_R = Raman scattering gain coefficient
   α_s, α_p = losses of signal and pump waves

The spectrum of the Raman gain depends on the decay lifetime of the excited electronic vibration state. The decay time is in the range of 1 ns, and the Raman-gain bandwidth is about 1 GHz. In SMFs, the bandwidth of the Raman gain is about 10 THz. The pump beam wavelength is usually about 100 nm below the amplification wavelength region. Thus, in order to extend the gain spectra, a number of pump sources of different wavelengths is used. Polarization multiplexing of these beams is also used to reduce the effective power launched in the fiber so as to avoid damage to the fiber. The threshold for stimulated Raman gain is given by

$$g_R P_{th\_SRS} \frac{L_{eff}}{A_{eff}} \approx 16$$

$$(2.40)$$

$$\text{with the\_effective\_length } L_{eff} = \frac{1 - e^{-\alpha L}}{\alpha} \text{ or } \approx 1/\alpha \text{ for long\_length}$$

For SSMF with an effective area of 50 μm², g_R ~ 1e–13 m/W, and the threshold power is about 570 mW near the C-band spectral region. This would require at least two pump laser sources that should be polarization multiplexed. SRS is used frequently in modern optical communications systems, especially when no undersea optical amplification is required, and the distributed amplification of SRS offers significant advantages as compared with lumped amplifiers such as EDFA. The broadband gain and low gain ripple of SRS are other advantages of DWDM transmission.

2.3.4.3 Four-Wave Mixing Effects

Four-wave mixing (FWM) is considered to be a scattering process in which three photons are mixed to generate the fourth wave. This happens when the momenta of the four waves satisfy a phase-matching condition, that is, the condition of maximum power transfer.

Figure 2.11 illustrates the mixing of different wavelength channels to generate interchannel crosstalk. The phase matching can be represented by a relationship between the propagation constant along the z-direction in an SMF as

$$\beta(\omega_1) + \beta(\omega_2) - \beta(\omega_3) - \beta(\omega_4) = \Delta(\omega)$$

$$(2.41)$$

where β is the propagation constant; ω_1, ω_2, ω_3, ω_4 are the frequencies of the first to fourth waves; and Δ is the phase-mismatching parameter. When the channels are equally spaced with a frequency spacing of Ω as in DWDM optical transmission, $\omega_1 = \omega_2$; $\omega_3 = \omega_1 + \Omega$; $\omega_4 = \omega_1 - \Omega$. One can use

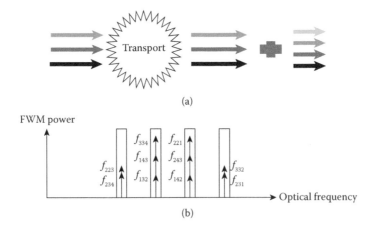

FIGURE 2.11 Illustration of FWM of optical channels. (a) Momentum vectors of channels and (b) frequencies resulting from mixing of different channels.

the Taylor series expansion around the propagation constant at the center frequency of the guide carrier β_0 to obtain [14]

$$\Delta(\omega) = \beta_2 \Omega^2 \qquad (2.42)$$

The phase matching is thus optimized when the dispersion parameter β_2 is nullified, indicating that in the region where there is no dispersion, the FWM is largest, and hence has maximum inter-channel crosstalk. This is the reason why DSF is not commonly used when the zero-dispersion wavelength is in the spectral region of operation of the channel. Instead, nonzero DSFs are used whose dispersion-zero wavelength is shifted away from the spectral region of the active channels so that there exists some dispersion value that would not satisfy the FWM condition, and hence there will be minimum generation of the fourth waves. In modern transmission fiber, the zero-dispersion wavelength (λ_{ZD}) is shifted to just outside the C-band, say 1510 nm, so that the dispersion is small, and hence a mismatch in the phase velocities of adjacent waves at 1550 nm and the C-band ranging from 2 to 6 ps/(nm · km). Therefore, no FWM occurs, for example, Corning LEAF or NZ-DSF. This small amount of dispersion is sufficient to avoid the FWM with a channel spacing of 100 GHz or 50 GHz.

The XPM signal is proportional to instantaneous signal power. Its distribution is bounded by <5 channels and is otherwise effectively unbounded. Thus, the link budgets include XPM evaluated at the maximum outer bounds.

2.4 SIGNAL ATTENUATION IN OPTICAL FIBERS

The optical loss in optical fibers is one of the two main fundamental limiting factors as it reduces the average optical power reaching the receiver. The optical loss is the sum of three major components: intrinsic loss, microbending loss, and splicing loss.

2.4.1 INTRINSIC OR MATERIAL ABSORPTION LOSSES

Intrinsic loss consists mainly of absorption loss due to OH impurities and Rayleigh scattering loss. The intrinsic is a function of λ^{-4}. Thus, the longer the operating wavelength, the lower the loss. However, it also depends on the transparency of the optical materials used to form the optical fibers. For silica fiber, the optical material loss is low over the wavelength range 0.8 to 1.8 μm. Over this wavelength range, there are three optical windows that optical communications utilizes. The first window over the central wavelength of 810 nm is an approximately 20.0 nm spectral window over the central wavelength. The second and third windows are most commonly used in present optical communications and are over 1300 and 1550 nm with a range of 80 and 40 nm, respectively. The intrinsic losses are about 0.3 and 0.15 dB/km at 1310 and 1550 nm regions, respectively.

This is a few hundred thousand times improvement over the original transmission of signal over 5.0 m with a loss of about 60 dB/km. Most communication fibers systems are operating at 1300 nm owing to its minimum dispersion at this range. For "power-hungry" optical transmission system, especially the extra-long reach systems, the lightwave sources should be in the 1550 nm spectral region as the silica fiber loss is minimum.

The absorption loss in silica glass is composed mainly of ultraviolet (UV) and infrared (IR) absorption region of pure silica. The IR absorption tale of pure silica has been shown to be due to the vibration of the basic tetrahedron, and thus strong resonances occur at around 8 to 13 μm with a loss about 10^{-10} dB/km. This loss is shown in Curve IR of Figure 2.1. The overtones and combinations of these vibrations lead to various absorption peaks in the low-wavelength range, as shown by Curve UV.

Various impurities that also lead to spurious absorption effects in the wavelength range of interest (1.2–1.6 μm) are transition metal ions and water in the form of OH ions. These sources of absorption have been practically reduced in recent years.

The Raleigh scattering loss, L_R, which is due to microscopic nonhomogeneities of the material, shows a $\lambda-4$ dependence and is given by

$$L_R = (0.75 + 4.5\Delta)\lambda^{-4} \text{ dB/Km} \qquad (2.43)$$

where Δ is the relative index difference as defined earlier, and λ is the wavelength in micrometers. Thus, to minimize the loss, Δ should be made as low as possible.

2.4.2 WAVEGUIDE LOSSES

The losses due to waveguide structure arise from power leakage, bending, microbending of the fiber axis, and defects and joints between fibers. Power leakage is significant only for depressed cladding fibers.

When a fiber is bent, the plane wave fronts associated with the guided mode are pivoted at the center of curvature, and their longitudinal velocity along the fiber axis increases with the distance from the center of curvature. As the fiber is bent further over a critical curve, the phase velocity would exceed that of the plane wave in the cladding, and radiation occurs.

The bending loss L_B for a radius R, the radius of curvature, is given by

$$L_B = -10\log_{10}\left(1 - 890\frac{r_0^6}{\lambda^4 R^2}\right) \text{for silica} \qquad (2.44)$$

Microbending loss results from power coupling from the guided fundamental mode of the fiber to radiation modes. This coupling takes place when the fiber axis is bent randomly at a high spatial frequency. Such bending can occur during packing of the fiber during the cabling process. The microbending loss of an SM fiber is a function of the fundamental mode spot size r_0. Fibers with a large spot size are extremely sensitive to microbending. It is therefore desirable to design the fiber to have as small a spot size as possible to minimize the bending loss. The microbending loss can be expressed by the relation (Figure 2.12)

$$L_m = 2.15 \times 10^{-4} r_0^6 \lambda^{-4} L_{mm} \text{ dB/km} \qquad (2.45)$$

where L_{mm} is the microbending loss of a 50 μm core multimode fiber having an NA of 0.2.

Ultimately, the fibers will have to be spliced together to form the final transmission link. With fiber cable that averages 0.4–0.6 dB/km, a splice loss in excess of 0.2 dB/splice drastically reduces the repeaterless distance that can be achieved. It is therefore extremely important that the fiber be designed such that the splicing loss is minimized.

Splicing loss is mainly due to axial misalignment of the fiber core, as shown in Figure 2.13. Splicing techniques, which rely on aligning the outside surface of the fibers, require extremely tight tolerances on core to outside surface concentricity. Offsets on the order of 1 μm can produce significant splice loss. This loss is given by

$$L_s = \frac{10}{\ln 10}\left(\frac{d}{r_0}\right)^2 \text{ dB} \qquad (2.46)$$

where d is the *axial* misalignment of the fiber cores. It is obvious that minimizing the optical loss involves making trade-offs between the different sources of loss. It is advantageous to have a large spot size to minimize both Raleigh and splicing losses, whereas minimizing bending and microbending losses requires a small spot size. In addition, as will be described in the next section, the spot size plays a significant role in the chromatic dispersion properties of single-mode fibers.

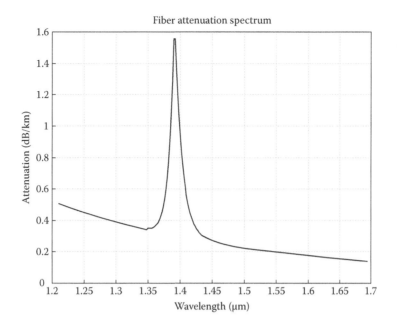

FIGURE 2.12 Attenuation of optical signals as a function of wavelength. The minimum loss at wavelength: at $\lambda = 1.3$ μm, about 0.3 dB/km, and at $\lambda = 1.5$ μm, a loss of about 0.13 dB/km. For cabled fibers, the attenuation factor at 1550 nm is 0.25 dB/km.

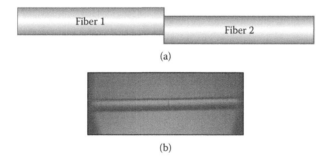

FIGURE 2.13 (a) Misalignment in splicing two optical fibers generating losses and (b) aligned spliced fibers.

2.4.3 ATTENUATION COEFFICIENT

Under general conditions of power attenuation inside an optical fiber, the attenuation coefficient of the optical power P can be expressed as

$$\frac{dP}{dz} = -\alpha P \qquad (2.47)$$

where α is the attenuation factor in a linear scale. This attenuation coefficient can include all the effects of power loss when signals are transmitted though the optical fibers.

Assume that optical signals with an average optical power of P_{in} enter at the input of the fiber of length L, and P_{out} is the output optical power; then P_{in} and P_{out} are related to the attenuation coefficient α as follows

$$P_{out} = P_{in}e^{(-aL)} \qquad (2.48)$$

It is customary to express α in dB/km by using the relation

$$\alpha\left(\text{dB/km}\right) = -\frac{10}{L}\log_{10}\left(\frac{P_{\text{out}}}{P_{\text{in}}}\right) = 4.343\alpha \tag{2.49}$$

Standard optical fibers with a small Δ would exhibit a loss of about 0.2 dB/km; that is, the purity of the silica is very high. Such purity of a bar of silica would allow us to see though a 1 km glass bar without distortion the person standing at the other end! The attenuation curve for silica glass is shown in Figure 2.1.

2.5   SIGNAL DISTORTION THROUGH OPTICAL FIBERS

Consider a *monochromatic* field given by

$$E_x = A \cos\left(\omega t - \beta z\right) \tag{2.50}$$

where A is the wave amplitude, ω is the radial frequency, and β is the propagation constant along the z-direction. If $(\omega t - \beta z)$ is set constant, then the wave phase velocity is given by

$$v_p = \frac{dz}{dt} = \frac{\omega}{\beta} \tag{2.51}$$

Now, consider that the propagating wave consists of two monochromatic fields of frequencies $\omega + \delta\omega$ and $\omega - \delta\omega$

$$E_{x1} = A \cos\left[(\omega + \delta\omega)t - (\beta + \delta\beta)z\right] \tag{2.52}$$

$$E_{x2} = A \cos\left[(\omega - \delta\omega)t - (\beta - \delta\beta)z\right] \tag{2.53}$$

The total field is then given by

$$E_x = E_{x1} + E_{x2} = 2A \cos\left[(\omega t - \beta z)\cos(\delta\omega t + \delta\beta z)\right] \tag{2.54}$$

If $\omega \gg \delta\omega$, then $\cos(\omega t - \beta z)$ varies much faster than $\cos(\delta\omega t - \delta\beta z)$, and hence, by setting $(\delta\omega t - \delta\beta z)$ invariant, we can define the group velocity as

$$v_g = \frac{d\omega}{d\beta} \rightarrow v_g^{-1} = \frac{d\beta}{d\omega} \tag{2.55}$$

The group delay t_g per unit length (setting L as 1.0 km) is thus given as

$$t_g = \frac{L\left(\text{of 1 km}\right)}{v_g} = \frac{d\beta}{d\omega} \tag{2.56}$$

The pulse spread $\Delta\tau$ per unit length due to the group delay of light sources of spectral width α_λ, that is, the full-width-half-mark (FWHM) of the optical spectrum of the light source is

$$\Delta\tau = \frac{dt_g}{d\lambda}\sigma_\lambda \tag{2.57}$$

The spread of the group delay due to the spread of the source wavelength can be in ps/km. Thus, the linewidth of the light source contributes significantly to the distortion of the optical signal transmitted through the optical fiber because of the delay differences between the guided modes carried by the spectral components of the lightwaves. Hence, the narrower the source linewidth, the less dispersed are the optical pulses. The typical linewidth of Fabry–Perot semiconductor lasers is about 1–2 nm, while the DFB (distributed feedback) lasers would exhibit a linewidth of 100 MHz (What is the equivalent spectral window of 100 GHz in frequency-domain band?). Later, we will see that when the source linewidth is very narrow, for example, the external cavity laser (ECL), then the band width of the channel would play the principal role in the distortion.

Optical signal traveling along a fiber becomes increasingly distorted. This distortion is a consequence of *intermodal* delay effects and *intramodal* dispersion. Intermodal delay effects are significant in multimode optical fibers because each mode has a different value of group velocity at a specific frequency; intermodal dispersion, on the contrary, is pulse spreading that occurs within a single mode. It is the result of the group velocity being a function of the wavelength λ and is therefore referred as chromatic dispersion.

The two main causes of intermodal dispersion are (1) material dispersion, which arises from the variation of the refractive index $n(\lambda)$ as a function of wavelengths, causing a wavelength dependence of the group velocity of any given mode and (2) waveguide dispersion, which occurs because the mode propagation constant $\beta(\lambda)$ is a function of the wavelength λ, core radius a, and the refractive index difference.

The group velocity associated with the fundamental mode is frequency dependent because of chromatic dispersion. As a result, different spectral components of the light pulse travel at different group velocities, a phenomenon referred to as *group velocity dispersion*, intramodal dispersion, or as material dispersion and waveguide dispersion.

2.5.1 MATERIAL DISPERSION

The refractive index of silica as a function of wavelength is shown in Figure 2.14. The refractive index is plotted over the wavelength region of 1–2 µm, which is the most important range for silica base optical communications systems as the loss is lowest at 1300 and 1550 nm windows.

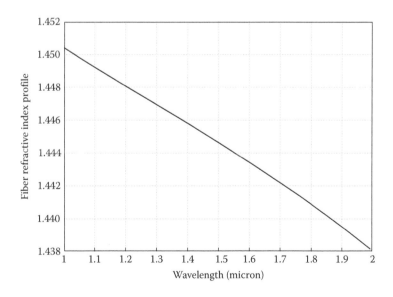

FIGURE 2.14 Variation in the refractive index as a function of the optical wavelength of silica.

The propagation constant β of the fundamental mode guided in the optical fiber can be written as

$$\beta(\lambda) = \frac{2\pi n(\lambda)}{\lambda} \tag{2.58}$$

The group delay t_{gm} per unit length then (Equation 2.58) can be obtained as

$$t_{gm} = \frac{d\beta}{d\omega} \tag{2.59}$$

where we can replace

$$d\omega = d\left(\frac{2\pi c}{\lambda}\right) = -\frac{2\pi c}{\lambda^2} d\lambda \tag{2.60}$$

which leads to

$$t_{gm} = -\frac{\lambda^2}{2\pi c} \frac{d\beta}{d\lambda} \tag{2.61}$$

Substituting Equation 2.58 in Equation 2.61, we have

$$t_{gm} = \frac{1}{c}\left[n(\lambda) - \frac{\lambda dn(\lambda)}{d\lambda}\right] \tag{2.62}$$

Thus, the pulse dispersion per unit length $\Delta\tau_m/\Delta\lambda$ due to material (using Equation 2.62) for a source having an RMS spectral width σ_λ

$$\Delta\tau_m = -\frac{\lambda}{c} \frac{d^2 n}{d\lambda^2} \sigma_\lambda \tag{2.63}$$

If $\Delta\tau_m = M(\lambda)\sigma_\lambda$, then

$$M(\lambda) = -\frac{\lambda}{c} \frac{d^2 n}{d\lambda^2} \tag{2.64}$$

$M(\lambda)$ is called the *material dispersion factor or material dispersion parameter*, and its unit is commonly expressed in ps/(nm·km). Thus, if the refractive index can be expressed as a function of the optical wavelength, then the material dispersion can be calculated. In fact, in practice, optical material engineers have to characterize all optical properties of new materials.

The refractive index $n(\lambda)$ can usually be expressed in Sellmeier's dispersion formula as

$$n^2(\lambda) = 1 + \sum_k \frac{G_k \lambda^2}{(\lambda^2 - \lambda_k^2)} \tag{2.65}$$

where G_k are Sellmeier's constants, and k is an integer normally taken in the range of 1–3. In the late 1970s, several silica-based glass materials were manufactured and their properties measured. The refractive indices are usually expressed using Sellmeier's coefficients. These coefficients for several optical fiber materials are given in Table 2.1.

By using curve fitting, the refractive index of pure silica $n(\lambda)$ can be expressed as

$$n(\lambda) = c_1 + c_2\lambda^2 + c_3\lambda^{-2} \tag{2.66}$$

where $c_1 = 1.45084$, $c_2 = -0.00343 \ \mu m^{-2}$, and $c_3 = 0.00292 \ \mu m^2$. Thus, from Table 2.1 and Equation 2.66, we can use Equation 2.64 to determine the material dispersion factor for a certain wavelength range.

TABLE 2.1

Sellmeier's Coefficients for Several Optical Fiber Silica-Based Materials with Germanium Doped in the Core Region

| Sellmeier's Constants | Germanium Concentration, C (mole%) | | | |
	0 (pure silica)	3.1	5.8	7.9
G1	0.6961663	0.7028554	0.7088876	0.7136824
G2	0.4079426	0.4146307	0.4206803	0.4254807
G3	0.8974794	0.8974540	0.8956551	0.8964226
$\lambda 1$	0.0684043	0.0727723	0.0609053	0.0617167
$\lambda 2$	0.1162414	0.1143085	0.1254514	0.1270814
$\lambda 3$	9.896161	9.896161	9.896162	9.896161

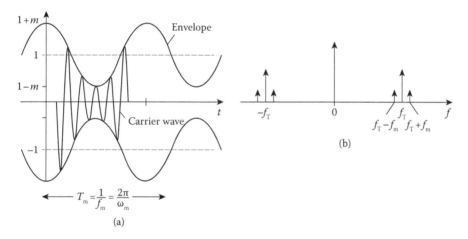

FIGURE 2.15   (a) Time signal and (b) spectrum.

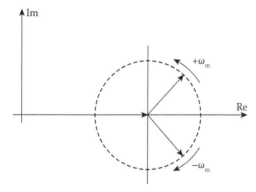

FIGURE 2.16   Vector phasor diagram of the complex envelope.

Figure 2.15 illustrates the time-domain signal and its corresponding spectrum when the carrier is modulated by this signal. Figure 2.16 illustrates the effects on the complex signal using a vector phasor diagram representing the complex envelope. Figure 2.17 then illustrates the temporal signal by the magnitude of the signal complex envelope when it is not sinusoidal or the envelope is subject to nonlinear distortions.

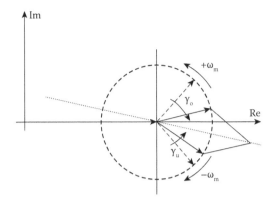

FIGURE 2.17 Magnitude of complex envelope when not sinusoidal, the envelope being subject to nonlinear distortions.

For the doped core of the optical fiber, Sellmeier's expression (Equation 2.65) can be approximated by using a curve fitting technique to approximate it to the form in Equation 2.66. The material dispersion factor $M(\lambda)$ becomes zero at wavelengths of around 1350 nm and about −10 ps/(nm·km) at 1550 nm. However, the attenuation at 1350 nm is about 0.4 dB/km compared with 0.2 dB/km at 1550 nm, as shown in Table 2.1.

Plots of the material dispersion factors and their total due to material and waveguide group velocity as a function of wavelength are shown in Figure 2.18.

Plotted curves representing the material dispersion factor as a function of optical wavelength for silica base optical fiber (yellow curve) with a zero-dispersion wavelength at 1290 nm. This curve is generated as an example. For SSMFs that are currently installed throughout the world, the total dispersion is around +17 ps/(nm·km) at 1550 nm and almost zero at 1310 nm. Can we estimate the waveguide dispersion curve for the SSMF at around 1300 and 1550 nm windows?

2.5.2 WAVEGUIDE DISPERSION

The effect of waveguide dispersion can be approximated by assuming that the refractive index of the material is independent of wavelength. Let us now consider the group delay, that is, the time required for a mode to travel along a fiber of length L. This kind of dispersion depends strongly on the Δ and V parameters. To obtain the results of fiber parameters, we define a *normalized propagation constant b* as

$$b = \frac{\dfrac{\beta^2}{k^2} - n_2^2}{n_1^2 - n_2^2} \tag{2.67}$$

for small Δ. We note that the β/k is in fact the *effective refractive index* of the guided optical mode propagating along the optical fiber; that is, the guided waves traveling the axial direction of the fiber "sees" it as a medium with a refractive index of an equivalent "effective" index.

If fiber is a weakly guiding waveguide and the effective refractive index takes a value that is significantly close to that of the core or cladding index, Equation 2.67 can then be approximated by

$$b \cong \frac{\dfrac{\beta}{k} - n_2}{n_1 - n_2} \tag{2.68}$$

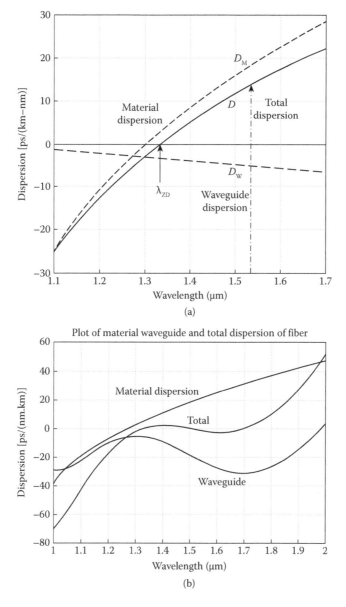

FIGURE 2.18 Chromatic dispersion factor of (a) SSMF and (b) dispersion flatten fiber.

Solving Equation 2.68 for β, we have $\beta = n_2 k(b\Delta + 1)$; the group delay for waveguide dispersion is then given by (per unit length)

$$t_{wg} = \frac{d\beta}{d\omega} = \frac{1}{c}\frac{d\beta}{dk} \tag{2.69}$$

$$t_{wg} = \frac{1}{c}\left[n_1 + n_2\Delta\frac{d(bk)}{dk}\right] = \frac{1}{c}\left[n_1 + n_2\Delta\frac{d(bk)}{dk}\right] = \frac{1}{c}\left[n_1 + n_2\Delta\frac{d(bV)}{dV}\right] \tag{2.70}$$

Equation 2.70 can be obtained from Equation 2.69 by using the expression for V. Thus, the pulse spreading $\Delta\tau_\omega$ due to the waveguide dispersion per unit length by a source having an optical bandwidth (or linewidth σ_λ) is given by

$$\Delta\tau_\omega = \frac{dt_{gw}}{d\lambda}\sigma_\lambda = -\frac{n_2\Delta}{c\lambda}V\frac{d^2(Vb)}{dV^2}\sigma_\lambda \tag{2.71}$$

and similar to the definition of the material dispersion factor, the *waveguide dispersion factor* or *waveguide dispersion parameter* can be defined as

$$D(\lambda) = -\frac{n_2(\lambda)\Delta}{c\lambda}V\frac{d^2(Vb)}{dV^2} \tag{2.72}$$

This waveguide factor can take units of ps/(nm · km). In the range of $0.9 < \lambda/\lambda_c < 2.6$, the factor $V\frac{d^2(Vb)}{dV^2}$ can be approximated (to <5% error) by

$$V\frac{d^2(Vb)}{dV^2} \cong 0.080 + 0.549(2.834 - V)^2 \tag{2.73}$$

or alternatively using the definition of the cutoff wavelength and the expression of the V parameters, we obtain

$$V\frac{d^2(Vb)}{dV^2} \cong 0.080 + 3.175\left(1.178 - \frac{\lambda_c}{\lambda}\right)^2 \tag{2.74}$$

It is not so difficult to prove the equivalent equation of Equations 2.69 through 2.74. Further, the sign assignment of the material and waveguide dispersion factors must be the same. Otherwise, negative and positive dispersion factors would create confusion. Can you explain what would happen to the pulse if it is transmitted through an optical fiber having a total negative dispersion factor?

Thus, from Equations 2.73 and 2.74, we can calculate the waveguide dispersion factor and hence the pulse dispersion factor for a particular source spectral width σ_λ. Note that the dispersion considered in this chapter is for step-index fiber only. For graded-index fiber, ESI parameters must be found, and the chromatic dispersion can then be calculated. Figure 2.18 shows a design of single-mode optical fibers with the total dispersion factor contributed by material and waveguide effects.

2.5.2.1 Alternative Expression for Waveguide Dispersion Parameter

Alternatively, the waveguide dispersion parameter can be expressed as a function of the propagation constant β by using $\omega = 2\pi c/\lambda_g$ and Equation 2.74, when the waveguide dispersion factor can be written as

$$D(\lambda) = -\frac{2\pi c}{\lambda^2}\beta_2 = -\frac{2\pi c}{\lambda^2}\frac{d\beta^2}{d\omega^2} \tag{2.75}$$

Thus, the waveguide dispersion factor is directly related to the second-order derivative of the propagation constant with respect to the optical radial frequency. An example of a design of an optical fiber operating in the single-mode region is given in Figure 2.19. The cladding material is pure silica. Shown in this figure are the curves of the material dispersion factor, waveguide dispersion factor, and total dispersion for an SMF with a nonuniform refractive index profile in the core.

A typical measurement setup for determination of the dispersion is the measurement of the differential group delay as shown in Figure 2.19, in which a narrow-linewidth source is modulated

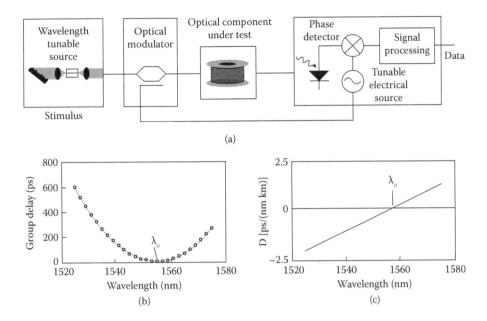

FIGURE 2.19 (a) Chromatic dispersion measurement of two-port optical device, (b) relative group delay versus wavelength, and (c) dispersion parameter versus wavelength.

by an optical modulator, the Mach–Zehnder interferometric modulator (MZIM), whose RF signal comes from the detected signal at the end of the device under test (DUT). This setup is thus very similar to an RF network analyzer but with the RF signal transferred to the optical domain by modulating the optical modulator, recovered at the output of the optical device, and then tuned to vary the RF signals as feedback to the modulator to scan the excitation frequency. The group delay can then be estimated without much difficulty, and from it the dispersion factor.

2.5.2.2 Higher-Order Dispersion

We observe also from Figure 2.19 that the bandwidth–length product of the optical fiber can be extended to infinity if the system is operating at the wavelength at which the total dispersion factor is zero. However, the dispersive effects do not disappear completely at this zero-dispersion wavelength. Optical pulses still experience broadening because of higher-order dispersion effects. It is easy to imagine that the total dispersion factor cannot be made zero to "flatten" it over the optical spectrum. This is higher-order dispersion, which is governed by the slope of the total dispersion curve, called the dispersion slope; $S = d[D(\lambda) + M(\lambda)]/d\lambda$; $S(\lambda)$ can thus be expressed as

$$S(\lambda) = \left(\frac{2\pi c}{\lambda^2}\right)^2 \frac{d^3\beta}{d\lambda^3} + \left(\frac{4\pi c}{\lambda^3}\right)\frac{d^2\beta}{d\lambda^2} \qquad (2.76)$$

$S(\lambda)$ is also known as the differential-dispersion parameter.

2.5.3 Polarization-Mode Dispersion

The delay between two PSPs is normally negligibly small at bit rates less than 10 Gbps (Figure 2.20). However, at high bit rates and in ultra-long-haul transmission, PMD severely degrades the system performance [15–18]. The differential delay between the polarized modes creates the PMD effect, which is illustrated in Figure 2.21. The instantaneous value of DGD ($\Delta\tau$) varies along the fiber and follows a Maxwellian distribution [19,20] (see Figures 2.21 and 2.22).

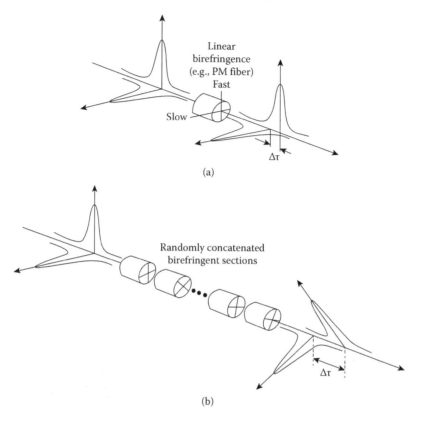

FIGURE 2.20 Conceptual model of polarization-mode dispersion (PMD). (a) Simple birefringence device, and (b) randomly concatenated birefringence.

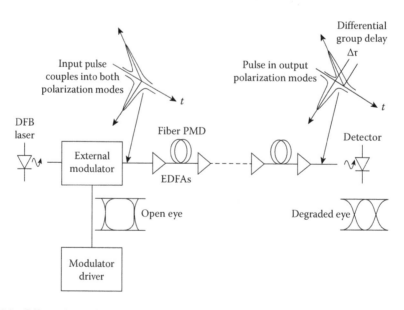

FIGURE 2.21 Effect of polarization-mode dispersion (PMD) in a digital optical communication system, degradation of the received eye diagram.

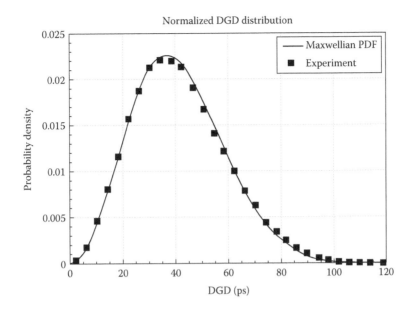

FIGURE 2.22 Maxwellian distribution of polarization-mode dispersion (PMD) random process.

The Maxwellian distribution is governed by the following expression

$$f(\Delta\tau) = \frac{32(\Delta\tau)^2}{\pi^2 \langle\Delta\tau\rangle^3} \exp\left\{-\frac{4(\Delta\tau)^2}{\pi\langle\Delta\tau\rangle^2}\right\} \quad \Delta\tau \geq 0 \tag{2.77}$$

The mean DGD value $\langle\Delta\tau\rangle$ is commonly termed *fiber PMD* and included in the fiber specifications. The following expression gives an estimate of the maximum transmission limit L_{max} due to the PMD effect as

$$L_{max} = \frac{0.02}{\langle\Delta\tau\rangle^2 \cdot R^2} \tag{2.78}$$

where R is the bit rate. On the basis of Equation 2.78, L_{max} for both old fiber vintage and contemporary fibers are obtained as follows: (1) $\langle\Delta\tau\rangle = 1$ ps/km (old fiber vintage): for a bit rate of $R = 40$ Gbps, the maximum distance $L_{max} = 12.5$ km; for $R = 10$ Gbps, $L_{max} = 200$ km; (2) $\langle\Delta\tau\rangle = 0.1$ ps/km (contemporary fiber for modern optical systems): if the bit rate $R = 40$ Gbps, the maximum transmission distance is $L_{max} = 1,250$ km; for $R = 10$ Gbps; $L_{max} = 20,000$ km.

2.6 TRANSFER FUNCTION OF SINGLE-MODE FIBERS

2.6.1 LINEAR TRANSFER FUNCTION

The treatment of the propagation of modulated lightwaves through single-mode fiber in the linear and nonlinear regimes has been well documented [21–26]. For completeness of the transfer function of SMFs, in this section, we restrict our study to the frequency transfer function and impulse responses of the fiber to the linear region of the media. Furthermore, the delay term in the NLSE can be ignored, as it has no bearing on the size and shape of the pulses. From the NLSE, we can thus model the fiber simply as a quadratic phase function. This is because the nonlinear term of the NLSE can be removed, and a Taylors series approximation around the operating frequency (central wavelength) can be obtained as well as the frequency and impulse responses of the single-mode fiber. The input–output relationship of the pulse can therefore be depicted. Equation 2.79 expresses the time-domain impulse response $h(t)$ and the frequency-domain transfer function $H(\omega)$ as a Fourier transform pair

$$h(t) = \sqrt{\frac{1}{j4\pi\beta_2}} \exp\left(\frac{jt^2}{4\beta_2}\right) \leftrightarrow H(\omega) = \exp\left(-j\beta_2\omega^2\right) \tag{2.79}$$

where β_2 is well known as the group velocity dispersion parameter. The input function $f(t)$ is typically a rectangular pulse sequence, and β_2 is proportional to the length of the fiber. The output function $g(t)$ is the dispersed waveform of the pulse sequence. The propagation transfer function in Equation 2.79 is an exact analogy of diffraction in optical systems (see item 1, Table 2.1, p. 14, A. Papoulis Ref. [27]). Thus, the quadratic phase function also describes the diffraction mechanism in one-dimensional optical systems, where the distance x is analogous to the time t. This analogy affords us to borrow much of the imagery and analytical results that have been developed in diffraction theory. Thus, we may express the step response $s(t)$ of the system $H(\omega)$ in terms of Fresnel cosine and sine integrals as follows

$$s(t) = \int_0^t \sqrt{\frac{1}{j4\pi\beta_2}} \exp\left(\frac{jt^2}{4\beta_2}\right) dt = \sqrt{\frac{1}{j4\pi\beta_2}}\left[C\left(\sqrt{\frac{1}{4}\beta_2}t\right) + jS\left(\sqrt{\frac{1}{4}\beta_2}t\right)\right] \tag{2.80}$$

with

$$C(t) = \int_0^t \cos\left(\frac{\pi}{2}\tau^2\right) d\tau$$

$$S(t) = \int_0^t \sin\left(\frac{\pi}{2}\tau^2\right) d\tau \tag{2.81}$$

where $C(t)$ and $S(t)$ are the Fresnel cosine and sine integrals.

Using this analogy, one may argue that it is always possible to restore the original pattern $f(x)$ by refocusing the blurry image $g(x)$ (e.g., image formation, item 5, Table 2.1, Ref. [15]). In the electrical analogy, the implication is that it is possible to compensate for the quadratic phase media perfectly. This is not surprising. The quadratic phase function $H(\omega)$ in Equation 2.79 is an all-pass transfer function, and thus it is always possible to find an inverse function to recover $f(t)$. One can express this differently in information theory terminology: the quadratic phase channel has a theoretical bandwidth of infinity, and hence its information capacity is infinite. Shannon's channel capacity theorem states that there is no limit on the reliable rate of transmission through the quadratic phase channel. Figure 2.23 shows the pulse and impulse responses of the fiber. Note that only the envelope of the pulse is shown, and the phase of the lightwave carrier is included as the complex values of the amplitudes. It can also be observed that the chirp of the carrier is significant at the edges of the pulse. At the center of the pulse, the chirp is almost negligible for some limited fiber length, and thus the frequency of the carrier remains nearly the same as at its original starting value. One could obtain the impulse response quite easily, but we are of the opinion that the pulse response is much more relevant for the investigation of the uncertainty in the pulse sequence detection. The impulse response is much more important for the process of equalization.

The uncertainty of the detection depends on the modulation formats and the detection process. The modulation can be implemented by manipulation of the amplitude, the phase, or the frequency

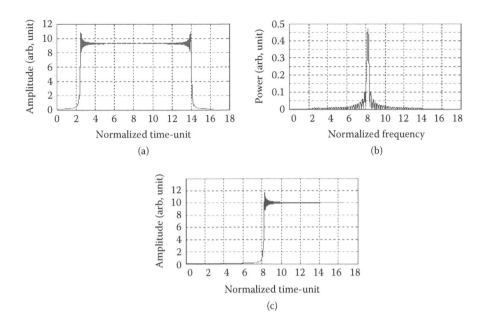

FIGURE 2.23 Rectangular pulse transmission through a SMF. (a) Pulse response, (b) frequency spectrum, and (c) step response of the quadratic phase transmittance function. Note horizontal scale in normalized unit of time.

of the carrier; or both the amplitude and phase; or multiple subcarriers, such as orthogonal frequency division multiplexing (OFDM) [16]. The amplitude detection would be mostly affected by the ripples of the amplitudes of the edges of the pulse. The phase of the carrier is mostly affected near the edge owing to the chirp effects. However, if differential phase detection is used, then the phase change at the transition instant is the most important and is the opening of the detected eye diagram. For frequency modulation, the uncertainty in the detection is not very critical provided the chirping does not enter the region of the neighborhood of the center of the pulse, in which the frequency of the carrier remains almost constant.

The picture changes completely if the detector/decoder is allowed only a finite time window to decode each symbol. In the convolution coding scheme, for example, it is the decoder's constraint length that manifests due to the finite time window. In the adaptive equalization scheme, it is the number of equalizer coefficients that determines the decoder window length. Because the transmitted symbols have already been broadened by the quadratic phase channel, if they are next gated by a finite time window, the information received could be severely reduced. The longer the fiber, the broader the pulses become, and hence the more uncertain the decoding. It is the interaction of the pulse broadening on the one hand and the restrictive detection time window on the other that gives rise to the finite channel capacity.

It is observed that the chirp occurs mainly near the edge of the pulses when it is in the near-field region, about a few kilometers for standard single-mode fibers. In this near-field distance, the accumulation of nonlinear effects is still very weak, and thus this chirp effectively dominates the behavior of the single-mode fiber. The nonlinear Volterra transfer function presented in the next section would thus have a minimal influence. This point is important for understanding the behavior of lightwaves circulating in short-length fiber devices in which both linear and nonlinear effects are to be balanced such as active mode locked soliton and multibound soliton lasers [28,29]. In the far field, the output of the fiber is Gaussian like for a square pulse launched at the input. In this region, the nonlinear effects would dominate over the linear dispersion effect as they have been accumulated over a long distance.

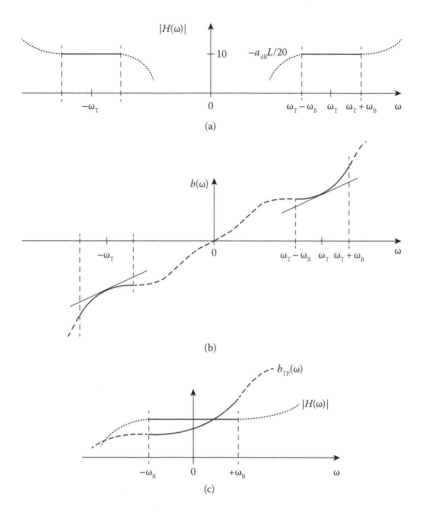

FIGURE 2.24 Frequency response of a single-mode optical fiber. (a) Magnitude, (b) phase response in bandpass regime, and (c) baseband equivalence.

A linear time-variant system such as the single-mode fiber would have a transfer function of

$$H(f) = |H(f)| e^{-j\alpha(f)} \tag{2.82}$$

where $\alpha = \pi^2 \beta_2 L = -\pi D L \lambda^2 / 2c$ is proportional to the length L and the dispersion factor $D(\lambda)$ (ps/nm/km). The phase of the frequency transfer response is a quadratic function of the frequency, and thus the group delay would follow a linear relationship with respect to the frequency as observed in Figure 2.24. The frequency response in the amplitude term is infinite and is a constant, while the phase response is a quadratic function with respect to the frequency of the baseband signals. The carrier is chirped accordingly as observed in Figures 2.25 and 2.26.

The chirping effect is very significant near the edge of the rectangular pulse and almost nil at the center of the pulse, in the near-field region of less than 1 km of standard single-mode fiber. In the far-field region, the pulse becomes Gaussian like. Thus, the response of the fiber in the linear region can be seen as shown in Figure 2.27 for a Gaussian pulse input to the fiber. The output pulse can also be shown to be Gaussian by taking the Fourier transform of the input pulse and multiplying it by the fiber transfer function. An inverse Fourier transform would indicate that the output pulse shape follows a Gaussian profile.

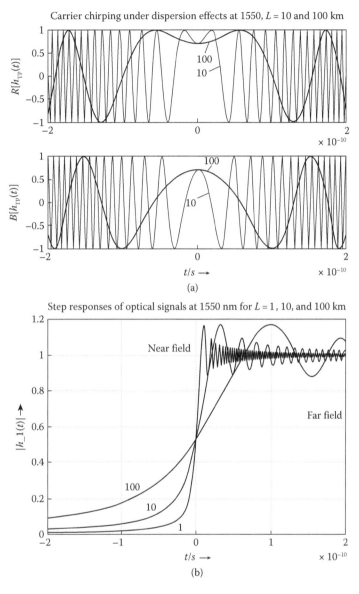

FIGURE 2.25 (a) Carrier chirping effects and (b) step response of a single-mode optical fiber of $L = 1$, 10 and 100 km.

This leads to the following rule of thumb for consideration of the scaling of the bit rate and transmission distance: *Given that a modulated lightwave with a bit rate B can be transmitted over a maximum distance L of single-mode optical fiber with a BER of error-free level, then if the bit rate is halved, the transmission distance can be increased by four times.* For example, for the 10 Gbps amplitude shift keying modulation format, if signals can be transmitted over 80 km of standard single-mode optical fiber, then at 40 Gbps, only 5 km can be transmitted for a BER of 10^{-9}. Figure 2.24 shows a typical frequency response in magnitude and phase and their corresponding baseband equivalence when the optical signals are recovered to the electrical domain. Ideally, one can see that the amplitude response of the fiber is constant throughout all frequencies for no attenuation or constant attenuation throughout the entire frequency range. Only the phase of the lightwaves is altered, that is, the chirping of the carrier. It is this chirping of the carrier that would then limit the response frequency range.

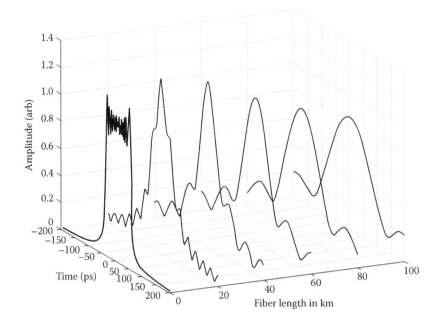

FIGURE 2.26 Pulse response from near field (~<2 km) to far field (>80 km).

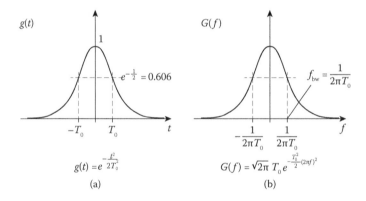

FIGURE 2.27 (a) Fiber response to Gaussian pulse input (time domain) and (b) Gaussian shape in frequency domain after propagation through chromatic dispersive fiber.

In the case of phase modulation, this chirp would rotate the constellation of signals modulated by QAM given in Chapter 1. The digital signal processing could then be used to determine the exact dispersion and hence the rotation of the constellation by an angle such that it is recovers back to its original position. The chirp of the carrier can be seen in Figure 2.25, in which the chirp is much less at the center but significant near the edge of the pulse. The step responses shown in Figures 2.26 and 2.27 indicate the damping oscillation of the step pulse, which is due to the chirp of the carrier as we also observe from the calculated impulse and step responses given in Figure 2.23. Figure 2.26 shows the Gaussian-like impulse response when the transmission distance is large. This is the typical pulse shape in long-haul nondispersion-compensating transmission. These dispersive pulse sequences are then coherently detected and processed by the digital signal processors. The number of samples must be long enough to cover the dispersive sequence, and the number of taps of the finite impulse response (FIR) filter must be high enough to ensure that the whole dispersive pulse is covered and falls within the filter length. So, the longer the transmission length, the higher the number of taps of the FIR. This is the first step in the DSP-based optical receiver: using the constant modulus algorithm (CMA) to compensate for the dispersion effects before the compensating and recovering the phase constellation of the modulated and transmitted channels can be implemented. Note that the chirp of the pulse envelope is much higher when the pulse is in the near-field region than in the far field, as can be observed in the step responses given in Figures 2.25a and 2.26. Thus, if a Gaussian pulse is launched and propagates in the single-mode fiber, we would obtain a Gaussian pulse shape at the output (see Figure 2.26), provided the fiber length is sufficient long, typically more than 20 km, which is commonly found in real-world transmission systems.

2.6.2 Nonlinear Fiber Transfer Function

The weakness of most of the recursive methods in solving the NLSE is that they do not provide much useful information to help the characterization of nonlinear effects. The Volterra series model provides an elegant way of describing a system's nonlinearities and enables the designers to see clearly where and how the nonlinearity affects the system performance. Although Refs. [6,30] have given an outline of the kernels of the transfer function using the Volterra series, for clarity and physical representation of these functions, brief derivations are given here on the nonlinear transfer functions of an optical fiber operating under nonlinear conditions.

The Volterra series transfer function (VSTF) of a particular optical channel can be obtained in the frequency domain as a relationship between the input spectrum $X(\omega)$ and the output spectrum $Y(\omega)$, as

$$Y(\omega) = \sum_{n=1}^{\infty} \int_{-\infty}^{\infty} \cdots \int_{-\infty}^{\infty} H_n(\omega_1, \cdots, \omega_{n-1}, \omega - \omega_1 - \cdots - \omega_{n-1}) \times X(\omega_1) \cdots X(\omega_{n-1})$$

$$X(\omega - \omega_1 - \cdots - \omega_{n-1}) d\omega_1 \cdots d\omega_{n-1} \tag{2.83}$$

where $H_n(\omega_1, \ldots, \omega_n)$ is the nth-order frequency-domain Volterra kernel including all signal frequencies of orders 1 to n. The wave propagation inside a single-mode fiber can be governed by a simplified version of the NLSE given earlier in this chapter with only the SPM effect included as

$$\frac{\partial A}{\partial z} = -\frac{\alpha_0}{2} A - \beta_1 \frac{\partial A}{\partial t} - j\frac{\beta_2}{2} \frac{\partial^2 A}{\partial t^2} - \frac{\beta_3}{6} \frac{\partial^3 A}{\partial t^3} + j\gamma |A|^2 A \tag{2.84}$$

where $A = A(t, z)$. The proposed solution of the NLSE can be written with respect to the VSTF model of up to the fifth order as

$$A(\omega, z) = H_1(\omega, z) A(\omega) + \int_{-\infty}^{\infty} \int_{-\infty}^{\infty} H_3(\omega_1, \omega_2, \omega - \omega_1 + \omega_2, z)$$

$$\times A(\omega_1) A^*(\omega_2) A(\omega - \omega_1 + \omega_2) d\omega_1 d\omega_2$$

$$+ \int_{-\infty}^{\infty} \int_{-\infty}^{\infty} \int_{-\infty}^{\infty} \int_{-\infty}^{\infty} H_5(\omega_1, \omega_2, \omega_3, \omega_4, \omega - \omega_1 + \omega_2 - \omega_3 + \omega_4, z) \tag{2.85}$$

$$\times A(\omega_1) A^*(\omega_2) A(\omega_3) A^*(\omega_4) \times A(\omega - \omega_1 + \omega_2 - \omega_3 + \omega_4) d\omega_1 d\omega_2 d\omega_3 d\omega_4$$

where $A(\omega) = A(\omega, 0)$, that is, the amplitude envelope of the optical pulses at the input of the fiber. Taking the Fourier transform of (2.85) and assuming $A(t, z)$ is of sinusoidal form, we have

$$\frac{\partial A(\omega, z)}{\partial z} = G_1(\omega) A(\omega, z) \int_{-\infty}^{\infty} \int_{-\infty}^{\infty} G_3(\omega_1, \omega_2, \omega - \omega_1 + \omega_2) A(\omega_1, z) A^*(\omega_2, z)$$

$$\times A(\omega - \omega_1 + \omega_2, z) d\omega_1 d\omega_2 \tag{2.86}$$

where $G_1(\omega) = \dfrac{-\alpha_0}{2} + j\beta_1\omega + j\dfrac{\beta_2}{2}\omega^2 - j\beta_3/6\omega_3$ and $G_3(\omega_1, \omega_2, \omega_3) = j\gamma$. ω is taking the values over the signal bandwidth and beyond in overlapping the signal spectra of other optically modulated carriers, while $\omega_1 \ldots \omega_3$ are also taking values over a similar range as that of ω. For the general expression, the limit of integration is indicated over the entire range to infinity.

Substituting Equation 2.85 in Equation 2.86 and equating both sides, the kernels can be obtained after some algebraic manipulations

$$
\frac{\partial}{\partial z}\Bigg[H_1(\omega, z)A(\omega) + \int_{-\infty}^{\infty}\int_{-\infty_i}^{\infty} H_3(\omega_1, \omega_2, \omega - \omega_1 + \omega_2, z)A(\omega_1)A^*(\omega_2)A(\omega - \omega_1 + \omega_2)d\omega_1 d\omega_2
$$

$$
+ \int_{-\infty}^{\infty}\int_{-\infty}^{\infty}\int_{-\infty}^{\infty}\int_{-\infty}^{\infty} H_5(\omega_1, \omega_2, \omega_3, \omega_4, \omega - \omega_1 + \omega_2 - \omega_3 + \omega_4, z)
$$

$$
\times A(\omega_1)A^*(\omega_2)A(\omega_3)A^*(\omega_4)A(\omega - \omega_1 + \omega_2 - \omega_3 + \omega_4)d\omega_1 d\omega_2 d\omega_3 d\omega_4 \Bigg]
$$

$$
= G_1(\omega)\Bigg[H_1(\omega, z)A(\omega) + \int_{-\infty}^{\infty}\int_{-\infty_i}^{\infty} H_3(\omega_1, \omega_2, \omega - \omega_1 + \omega_2, z)
$$

$$
\times A(\omega_1)A^*(\omega_2)A(\omega - \omega_1 + \omega_2)d\omega_1 d\omega_2
$$

$$
+ \int_{-\infty}^{\infty}\int_{-\infty}^{\infty}\int_{-\infty}^{\infty}\int_{-\infty}^{\infty} H_5(\omega_1, \omega_2, \omega_3, \omega_4, \omega - \omega_1 + \omega_2 - \omega_3 + \omega_4, z)A(\omega_1)A^*(\omega_2)A(\omega_3)A^*(\omega_4)
$$

$$
\times A(\omega - \omega_1 + \omega_2 - \omega_3 + \omega_4)d\omega_1 d\omega_2 d\omega_3 d\omega_4 \Bigg] + \int_{-\infty}^{\infty}\int_{-\infty}^{\infty} G_3(\omega_1, \omega_2, \omega - \omega_1 + \omega_2)
$$

$$
\times \Bigg[H_1(\omega_1, z)A(\omega_1) + \int_{-\infty}^{\infty}\int_{-\infty}^{\infty} H_3(\omega_{11}, \omega_{12}, \omega_1 - \omega_{11} + \omega_{12}, z)
$$

$$
\times A(\omega_{11})A^*(\omega_{12})A(\omega_1 - \omega_{11} + \omega_{12})d\omega_{11}d\omega_{12}
$$

$$
+ \int_{-\infty}^{\infty}\int_{-\infty}^{\infty}\int_{-\infty}^{\infty}\int_{-\infty}^{\infty} H_5(\omega_{11}, \omega_{12}, \omega_{13}, \omega_{14}, \omega_1 - \omega_{11} + \omega_{12} - \omega_{13} + \omega_{14}, z)
$$

$$
\times A(\omega_{11})A^*(\omega_{12})A(\omega_{13})A^*(\omega_{14}) \times A(\omega_1 - \omega_{11} + \omega_{12} - \omega_{13} + \omega_{14})d\omega_{11}d\omega_{12}d\omega_{13}d\omega_{14} \Bigg]
$$

$$
\quad (2.87)
$$

$$
\times \Bigg[H_1(\omega_1, z)A(\omega_1) + \int_{-\infty}^{\infty}\int_{-\infty}^{\infty} H_3(\omega_{11}, \omega_{12}, \omega_1 - \omega_{11} + \omega_{12}, z)
$$

$$
\times A(\omega_{11})A^*(\omega_{12})A(\omega_1 - \omega_{11} + \omega_{12})d\omega_{11}d\omega_{12}
$$

$$
+ \int_{-\infty}^{\infty}\int_{-\infty}^{\infty}\int_{-\infty}^{\infty}\int_{-\infty}^{\infty} H_5(\omega_{21}, \omega_{22}, \omega_{23}, \omega_{24}, \omega_2 - \omega_{21} + \omega_{22} - \omega_{23} + \omega_{24}, z)
$$

$$
\times A(\omega_{21})A^*(\omega_{22})A(\omega_{23})A^*(\omega_{24})
$$

$$
A(\omega_2 - \omega_{21} + \omega_{22} - \omega_{23} + \omega_{24})d\omega_{21}d\omega_{22}d\omega_{23}d\omega_{24} \Bigg]^* \times \Bigg[H_1(\omega - \omega_1 + \omega_2, z)A(\omega - \omega_1 + \omega_2)
$$

$$
+ \int_{-\infty}^{\infty}\int_{-\infty}^{\infty} H_3(\omega_{31}, \omega_{32}, \omega - \omega_1 + \omega_2 - \omega_{31} + \omega_{32}, z) \times A(\omega_{31})
$$

$$
\times A^*(\omega_{32})A(\omega - \omega_1 + \omega_2 - \omega_{31} + \omega_{32})d\omega_{31}d\omega_{32}
$$

$$
+ \int_{-\infty}^{\infty}\int_{-\infty}^{\infty}\int_{-\infty}^{\infty}\int_{-\infty}^{\infty} H_5(\omega_{31}, \omega_{32}, \omega_{33}, \omega_{34}, \omega - \omega_1 + \omega_2 - \omega_{31} + \omega_{32} - \omega_{33} + \omega_{34}, z)
$$

$$
\times A(\omega_{31})A^*(\omega_{32})A(\omega_{33})A^*(\omega_{34})
$$

$$
\times A(\omega - \omega_1 + \omega_2 - \omega_{31} + \omega_{32} - \omega_{33} + \omega_{34}) \times d\omega_{31}d\omega_{32}d\omega_{33}d\omega_{34} \Bigg]
$$

Equating the first-order terms on both sides, we obtain

$$\frac{\partial}{\partial z} H_1(\omega, z) = G_1(\omega) H_1(\omega, z) \tag{2.88}$$

Thus, the solution for the first-order transfer function (Equation 2.88) is then given by

$$H_1(\omega, z) = e^{G_1(\omega) z} = e^{\left(-\frac{\alpha_0}{2} + j\beta_{1\omega} + j\frac{\beta_2}{2}\omega^2 - j\frac{\beta_3}{6}\omega^3\right) z} \tag{2.89}$$

This is in fact the linear transfer function of an SMF with the dispersion factors β_2 and β_3 as already shown in the previous section.

Similarly, for the third-order terms, we have

$$\frac{\partial}{\partial z} \int_{-\infty}^{\infty} \int_{-\infty}^{\infty} H_3(\omega_1, \omega_2, \omega - \omega_1 + \omega_2, z) \times A(\omega_1) A^*(\omega_2) A(\omega - \omega_1 + \omega_2) d\omega_1 d\omega_2$$

$$= \int_{-\infty}^{\infty} \int_{-\infty}^{\infty} G_3(\omega_1, \omega_2, \omega - \omega_1 + \omega_2) H_1(\omega_1, z) A(\omega_1) H_2^*(\omega_2, z) \tag{2.90}$$

$$\times A(\omega_2) H_1(\omega - \omega_1 + \omega_2) A(\omega - \omega_1 + \omega_2) d\omega_1 d\omega_2$$

Now, letting $\omega_3 = \omega - \omega_1 + \omega_2$, it follows that

$$\frac{\partial H_3(\omega_1, \omega_2, \omega_3, z)}{\partial z} = G_1(\omega_1 - \omega_2 + \omega_3) H_3(\omega_1, \omega_2, \omega_3, z)$$

$$+ G_3(\omega_1, \omega_2, \omega_3) H_1(\omega_1, z) H_1^*(\omega_2, z) H_1(\omega_3, z) \tag{2.91}$$

The third order kernel transfer function can be obtained as

$$H_3(\omega_1, \omega_2, \omega_3, z) = G_3(\omega_1, \omega_2, \omega_3) \times \frac{e^{\left(G_1(\omega_1) + G_1^*(\omega_2) + G_1(\omega_3)\right) z} - e^{G_1(\omega_1 - \omega_2 + \omega_3) z}}{G_1(\omega_1) + G_1^*(\omega_2) + G_1(\omega_3) - G_1(\omega_1 - \omega_2 + \omega_3)} \tag{2.92}$$

The fifth-order kernel can similarly be obtained as

$$H_5(\omega_1, \omega_2, \omega_3, \omega_4, \omega_5, z)$$

$$= \frac{H_1(\omega_1, z) H_1^*(\omega_2, z) H_1(\omega_3, z) H_1^*(\omega_4, z) H_1(\omega_5, z) - H(\omega_1 - \omega_2 + \omega_3 - \omega_4 + \omega_5, z)}{G_1(\omega_1) + G_1^*(\omega_2) + G_1(\omega_3) + G_1^*(\omega_4) + G_1(\omega_5) - G_1(\omega_1 - \omega_1 + \omega_3 - \omega_4 + \omega_5)}$$

$$\times \left[\frac{G_3(\omega_1, \omega_2, \omega_3 - \omega_4 + \omega_5) G_3(\omega_3, \omega_4, \omega_5)}{G_1(\omega_3) + G_1^*(\omega_4) + G_1(\omega_5) - G_1(\omega_3 - \omega_4 + \omega_5)} + \frac{G_3(\omega_1, \omega_2 - \omega_3 + \omega_4, \omega_5) G_3^*(\omega_2, \omega_3, \omega_4)}{G_1^*(\omega_2) + G_1(\omega_3) + G_1^*(\omega_4) - G_1^*(\omega_2 - \omega_3 + \omega_4)} \right.$$

$$\left. + \frac{G_3(\omega_1 - \omega_2 + \omega_3, \omega_4, \omega_5) G_3(\omega_1, \omega_2, \omega_3)}{G_1(\omega_1) + G_1^*(\omega_2) + G_1(\omega_3) - G_1(\omega_1 - \omega_2 + \omega_3)} \right] - \frac{G_3(\omega_1, \omega_2, \omega_3 - \omega_4 + \omega_5) G_3(\omega_3, \omega_4, \omega_5)}{G_1(\omega_3) + G_1^*(\omega_4) + G_1(\omega_5) - G_1(\omega_3 - \omega_4 + \omega_5)}$$

$$\times \frac{H_1(\omega_1, z) H_1^*(\omega_2, z) H_1(\omega_1 - \omega_2 + \omega_3, z) - H_1(\omega_1 - \omega_2 + \omega_3 - \omega_4 + \omega_5, z)}{G_1(\omega_1) + G_1^*(\omega_2) + G_1(\omega_3 - \omega_4 + \omega_5) - G_1(\omega_1 - \omega_2 + \omega_3 - \omega_4 + \omega_5)} \tag{2.93}$$

$$- \frac{G_3(\omega_1 - \omega_2 + \omega_3, \omega_4, \omega_5) G_3^*(\omega_2, \omega_3, \omega_4)}{G_1^*(\omega_2) + G_1(\omega_3) + G_1^*(\omega_4) G_1^*(\omega_2 - \omega_3 + \omega_4)}$$

$$\times \frac{H_1(\omega_1, z) H_1^*(\omega_2 - \omega_3 + \omega_4, z) H_1(\omega_5, z) - H_1(\omega_1 - \omega_2 + \omega_3 - \omega_4 + \omega_5, z)}{G_1(\omega_1) + G_1^*(\omega_2 - \omega_3 + \omega_4) + G_1(\omega_5) + G_1(\omega_1 - \omega_2 + \omega_3 - \omega_4 + \omega_5)}$$

$$- \frac{G_3(\omega_1 - \omega_2 + \omega_3, \omega_4, \omega_5) G_3(\omega_1, \omega_2, \omega_3)}{G_1(\omega_1) + G_1^*(\omega_2) + G_1(\omega_3) - G_1(\omega_1 - \omega_2 + \omega_3)}$$

$$\times \frac{H_1(\omega_1 - \omega_2 + \omega_3, z) H_1^*(\omega_4, z) H_1(\omega_5, z) - H_1(\omega_1 - \omega_2 + \omega_3 - \omega_4 + \omega_5, z)}{G_1(\omega_1 - \omega_2 + \omega_3) + G_1^*(\omega_4) + G_1(\omega_5) - G_1(\omega_1 - \omega_2 + \omega_3 - \omega_4 + \omega_5)}$$

Higher-order terms can be derived with ease if higher accuracy is required. However, in practice, such higher-order terms would not exceed the fifth rank. We can understand that for a length of a uniform optical fiber, the first- to nth-order frequency spectrum transfer can be evaluated, indicating the linear to nonlinear effects of the optical signals transmitted through it. Indeed, the third- and fifth-order kernel transfer functions based on the Volterra series indicate the optical field amplitude of the frequency components that contribute to the distortion of the propagated pulses. The inverse of these higher-order functions would give the signal distortion in the time domain. Thus, the VSTFs allow us to conduct distortion analysis of optical pulses and hence an evaluation of the BER of optical fiber communications systems.

The superiority of such Volterra transfer function expressions allow us to evaluate each effect individually, especially the nonlinear effects, so that we can design and manage the optical communications systems under linear or nonlinear operations. Currently, this linear–nonlinear boundary of operations is critical for system implementation, especially for optical systems operating at 40 Gbps where linear operation and a carrier-suppressed return-to-zero format are employed. As a norm in series expansion, the series needs to converge to a final solution. It is this convergence that would allow us to evaluate the limit of nonlinearity in a system.

2.6.3 TRANSMISSION BIT RATE AND THE DISPERSION FACTOR

The effect of dispersion on the system bit rate B_r is obvious and can be estimated by using the criterion

$$B_r \cdot \Delta\tau < 1 \tag{2.94}$$

where $\Delta\tau$ is the total pulse broadening. When the fiber length is L, the total dispersion $D_T = M(\lambda) + D(\lambda)$, and the source linewidth σ_λ, the criterion becomes

$$B_r \cdot L \cdot |D_T| \sigma_\lambda \leq 1 \tag{2.95}$$

For a total dispersion factor of 1 ps/(nm.km) and a semiconductor laser of linewidth 2–4 nm, the bit rate–length product cannot exceed 100 Gbps-km. That is, if a 100 km transmission distance is used, then the bit rate cannot be higher than 1.0 Gbps. However, with the digital signal processing algorithms (DSP algorithms) in coherent reception, this pulse broadening can be compensated in the electronic digital domain, and the dispersive transmission distance can reach a few thousands of kilometers if the modulation format is QPSK and intradyne reception is employed.

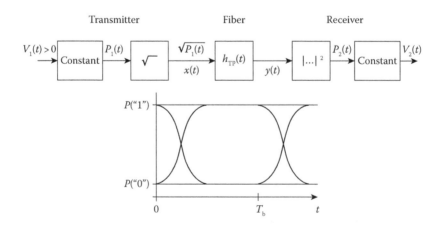

FIGURE 2.28 Schematic of an optical transmission system and its equivalent transfer functions.

Figure 2.28 shows the schematic of a transmission system and the power amplitude level for the one and zero bits under the effects of rise and fall time of all subsystems forming the link.

Figure 2.29a through c show typical simulated waveforms for such systems at the transmitter end, after 10 km transmission distance at medium range, and then the distorted sequence at 80 km of SSMF with two pulse periods (left side) and with several pulse periods (right side).

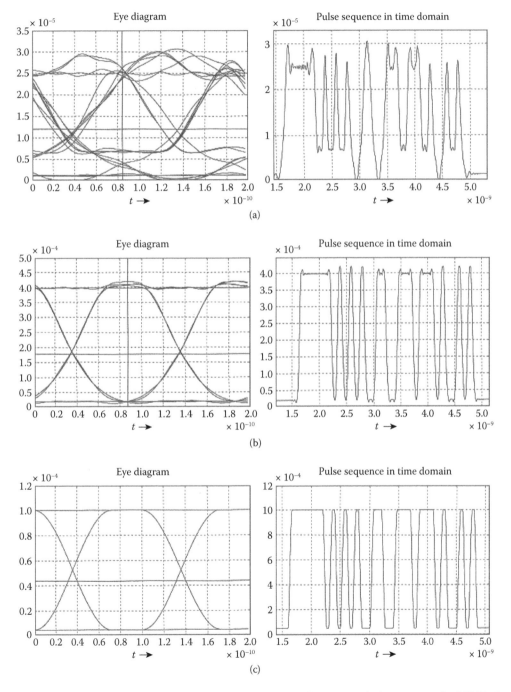

FIGURE 2.29 Eye diagram (left) of time signals (right) by 10 Gbps transmission over standard SMF after 0, 20, and 80 km length.

2.7 FIBER NONLINEARITY REVISITED

The nonlinear effects in optical fibers were described in Section 2.3. This section revisits these effects and their influence on the propagation of optical signals over long fiber lengths. The nonlinearity and linear effects in optical fibers can be classified as shown in Figure 2.30.

The elastic nonlinear effects are due to the change in the refractive index of the core region caused by the intensity of the lightwaves, and hence the corresponding changes in their phases and interference effects, or due to the different propagation times of the components of the modulated spectra. The fiber RI is dependent on both the operating wavelengths and the lightwave intensity. This intensity-dependent phenomenon is known as the Kerr effect and is the cause of fiber nonlinear effects.

On the contrary the nonelastic nonlinear effects are due to the electronic vibration of the atoms of the impurities doped in the fiber core. The electronic vibrations can then transfer the energy of the lightwaves to another component whose frequency would be some distance away, for example, 100 nm away or a few terahertz by Raman scattering or only 100 MHz amplitude modulation by Brillouin scattering.

2.7.1 SPM AND XPM EFFECTS

The power dependence of RI is expressed as

$$n' = n + \overline{n_2}\left(\frac{P}{A_{\text{eff}}}\right) \tag{2.96}$$

where P is the average optical power of the guided mode, $\overline{n_2}$ is the fiber nonlinear coefficient, and A_{eff} is the effective area of the fiber.

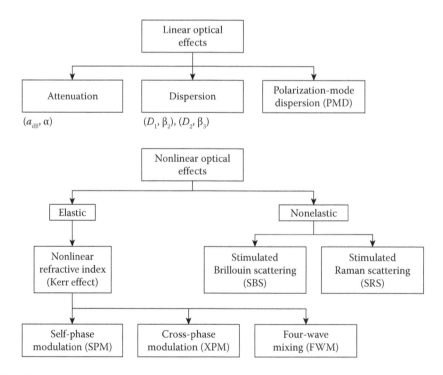

FIGURE 2.30 Linear and nonlinear fiber properties in single-mode optical fibers.

Fiber nonlinear effects include intrachannel SPM, interchannel XPM, FWM, SRS, and SBS. SRS and SBS are not the main degrading factors as their effects only become noticeably large with very high optical power. On the contrary, FWM severely degrades the performance of an optical system with the generation of ghost pulses only if the phases of the optical signals are matched with each other. However, with high local dispersions such as in SSMF, the effects of FWM become negligible. In terms of XPM, its effects can be considered to be negligible in a DWDM system in the following scenarios: (1) highly locally dispersive system and (2) large channel spacing. However, XPM should be taken into account for optical transmission systems deploying NZ-DSF fiber where local dispersion values are small. Thus, SPM is usually the dominant nonlinear effect for systems employing transmission fiber with high local dispersions, for example, SSMF and DCF. The effect of SPM is normally coupled with the nonlinear phase shift ϕ_{NL} defined as

$$\phi_{NL} = \int_0^L \gamma P(z)\, dz = \gamma L_{eff} P$$

$$\gamma = \frac{\omega_c \overline{n_2}}{A_{eff} c} \tag{2.97}$$

$$L_{eff} = \frac{\left(1 - e^{-\alpha L}\right)}{\alpha}$$

where ω_c is the lightwave carrier, L_{eff} is the effective transmission length, and α is the fiber attenuation factor, which normally has a value of 0.17–0.2 dB/km in the 1550 nm spectral window. The temporal variation of the nonlinear phase ϕ_{NL} results in the generation of new spectral components far apart from the lightwave carrier ω_c, indicating the broadening of the signal spectrum. This spectral broadening $\delta\omega$ can be obtained from the time dependence of the nonlinear phase shift as follows

$$\delta\omega = -\frac{\partial \phi_{NL}}{\partial T} = -\gamma \frac{\partial P}{\partial T} L_{eff} \tag{2.98}$$

Equation 2.98 indicates that $\delta\omega$ is proportional to the time derivative of the average signal power P. In addition, the generation of new spectral components occurs mainly at the rising and falling edges of optical pulses; that is, the amount of generated chirps is substantially larger for an increased steepness of the pulse edges.

The wave propagation equation can be represented as

$$\frac{\partial A(z,t)}{\partial z} + \frac{\alpha}{2} A(z,t) + \beta_1 \frac{\partial A(z,t)}{\partial t} + \frac{j}{2}\beta_2 \frac{\partial^2 A(z,t)}{\partial t^2} - \frac{1}{6}\beta_3 \frac{\partial^3 A(z,t)}{\partial t^3}$$

$$= -j\gamma |A(z,t)|^2 A(z,t) - \frac{1}{\omega_0}\frac{\delta}{\delta t}\left(|A|^2 A\right) - T_R A \frac{\delta\left(|A|^2\right)}{\delta t} \tag{2.99}$$

in which we have ignored the pure delay factor involving β_1. The last term in the RHS represents the Raman scattering effects.

2.7.2 SPM AND MODULATION INSTABILITY

Nonlinear effects such as the Kerr effect, where the refractive index of the fiber medium strongly depends on the intensity of the optical signal, can severely impair the transmitted signals. In a

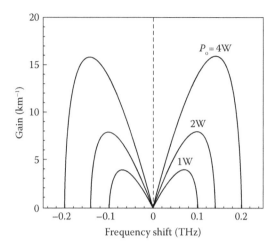

FIGURE 2.31 Spectrum of the optical gain due to modulation instability at three different average power levels in an optical fiber with $\beta_2 = 20$ ps²/km and $\gamma = 2$ W/km.

nonsoliton system, the Kerr effect broadens the signal optical spectrum through SPM; this broadened spectrum is mediated by fiber dispersion and causes performance degradation. In addition, FWM between the signal and the amplified spontaneous emission (ASE) noise generated from the in-line OAs has been reported to cause performance degradation in transmission systems employing in-line OAs. This latter effect is commonly referred to as modulation instability. Thus, the instability results from the conversion of the phase noises to intensity noises. The gain spectrum of the modulation instability is shown in Figure 2.31 [31].

2.7.3 Effects of Mode Hopping

Up to now, we have assumed that the source center emission wavelength is unaffected by the modulation. In fact, when a short current pulse is applied to a semiconductor laser, its center emission wavelength may hop from one mode to its neighbor. When a multilongitudinal-mode laser is used, this hopping effect is negligible; however, it is very significant for a single longitudinal laser.

2.7.4 SPM and Intrachannel Nonlinear Effects

Under considerations only the SPM of all the nonlinear effects on the optical signals transmitting through a dispersive transmission link. Thus can drop the cross-coupling terms but $\Delta\Omega$, and the nonlinear effect is thus contributed by additional intrachannel effects with ω_1, ω_2 taking the values fallen within the spectra of the optical signal and not crossing over the spectra of other adjacent channels. With the substituting of the fundamental order transfer function, we arrive at

$$H_3\left(\omega_1,\omega_2,\omega\right)_{\text{SPM,inter}} = -\frac{j\gamma_L}{4\pi^2}\left(e^{-j\omega^2 N_s \beta_2 L_s/2}\right)\left(1-e^{-(\alpha+j\beta_2\Delta\Omega)L_s}\right)$$

$$x\left(L_s\sqrt{\alpha^2+\beta_2^2\Delta\Omega^2}\,e^{j\tan^{-1}\frac{\alpha}{\beta_2\Delta\Omega}}\right)\sum_{k=0}^{N_s-1}e^{-jk\beta_2 L_s\Delta\Omega}$$

(2.100)

The nonlinear distortion noises contributed to the signals when operating under the two regimes of large and negligible dispersion are given in Ref. [23], depending on the dispersion factor of the fiber spans. The nonlinear transfer function H_3 indicates the power penalty due to nonlinear distortion, and can be approximated as [32]

$$H_3^i(\omega_1,\omega_2,\omega) \approx \begin{cases} j\gamma_L e^{-[\alpha/2 - j\beta_2(\omega_1-\omega_2)(\omega-\omega_2)]} \\ \times\left[L_s^{\text{eff}} - j\beta_2(\omega_1-\omega_2)(\omega-\omega_2)\right]\left(\dfrac{L_s - L_s^{\text{eff}}}{\alpha} - L_s L_s^{\text{eff}}\right) \end{cases} \tag{2.101}$$

Thus, if the ASE noise of the in-line OA is weak compared with signal power, we can obtain the nonlinear distortion noise for *highly dispersive* fiber spans (e.g., G.652 SSMF) as

$$K_{N,\beta}(\omega_0) = N_s\left[Q(\omega_0) + 2\left(\frac{\gamma_L}{2\pi}\right)^2\frac{\Omega^2}{\alpha^2}\frac{P^3}{\Delta\omega_c^3}\partial\left(\frac{\omega_0}{\Delta\omega_c},\frac{\beta\Omega^2}{\alpha}\right)\right] \tag{2.102}$$

and for *mildly dispersive* fiber spans (e.g., G.655 fiber spans)

$$K_{N,\beta\ll 20}(\omega_0) \approx N_s\left[Q(\omega_0) + 2\left(\frac{\gamma_L}{2\pi}\right)^2\frac{\Omega^2}{\alpha^2}\frac{P^3}{\Delta\omega_c^3}\partial\left(\frac{\omega_0}{\Delta\omega_c},0\right)\right] \tag{2.103}$$

with

$$\partial_\chi(x,\xi) = \int_{-\infty}^{\infty}dx_1\int_{-\infty}^{\infty}dx_2^*$$

$$\begin{pmatrix} \dfrac{(1+e^{-\alpha L}) - 2e^{-\alpha L}\cos[\alpha L\xi(x-x_1)(x_2-x_2)]}{1+\xi^2(x-x_1)^2(x_1-x_2)^2} \\ *\eta(x_1)\eta(x_2)\eta(x-x_1+x_2) \end{pmatrix} \tag{2.104}$$

and

$$\eta(x) = \begin{cases} 1, & \text{for} \quad x = [-1/2, 1/2] \\ 0 & \text{else\_where} \end{cases} \tag{2.105}$$

The nonlinear power penalty thus consists of the linear OA noise; the second is the SPM noise from the input signal and nonlinear interference between the input and the OA noise, which may be ignored when the ASE is weak. Equations 2.102 and 2.103 show the variation of the penalty and hence the channel capacity of dispersive fibers of transmission systems operating under the influence of nonlinear effects with optically amplified multispan transmission line whose fiber dispersion parameter varies from 0 to −20 ps²/km. With nondispersive fiber, the spectral efficiency is limited to about 3–4 b/s/Hz and 9–6 b/s/Hz with 4 and 32 spans, respectively, for a dispersion factor of −20 ps²/km with 100 DWDM channels of 50 GHz spacing between the channels with an optical spectral noise density of 1 µW/GHz. The fiber length of each span is 80 km.

By definition, the nonlinear threshold is determined at the 1 dB penalty deviation level from the linear OSNR, the contribution of the nonlinear noise term, and from Equation 2.102, we can obtain the maximal launched power at which the degradation of the channel capacity begins as

$$\text{max\_}P = \sqrt{0.1\frac{\omega_c^3}{2N_s\left(\dfrac{\gamma_L}{2\pi}\right)^2\dfrac{\Omega^2}{\alpha^2}}} \tag{2.106}$$

An example of the estimation of the maximum level of power per channel allowable to be launched to the fiber before reaching the nonlinear threshold 1-dB penalty level for an overall 100 channels of 150 GHz spacing with $\Omega_T \approx 200$ nm, $P_{th} \approx 58$ µW/GHz. Thus, for a bandwidth of 25 GHz, we have the threshold power level at $P_{th\ blow} = 0.15$ mW per channel. For highly dispersive and 8-wavelength channels, we have $P_{th\ bhigh} = 7$–$10\ P_{th\ blow}$ or the threshold level may reach 1.5 mW/channel. The estimations given here, as an example, are consistent with the analytical expression obtained in Equation 2.106. Thus, this clearly shows the following: (1) Dispersive multispan long-distance transmission under a coherent ideal receiver would lead to a better channel capacity than a low-dispersive transmission line. (2) If a combination of low- and high-dispersive fiber spans is used, then we can expect that, from our analytical Volterra approach, the penalty would reach the same level of threshold power, so that a 1-dB penalty on the OSNR is reached. Note that this approach relies on the average level of optical power of the light-wave-modulated sequence. It may not be easy to estimate if a simulation model is employed. (3) However, under simulation, the estimation of average power consumes a lot of time, and thus commonly the instantaneous power is estimated at the sampled time interval of a symbol. This sampled amplitude, and hence the instantaneous power, can be deduced. Next, the nonlinear phase is estimated and superimposed on the sampled complex envelope for further propagation along the fiber length. The sequence high–low dispersive spans would offer slightly better performance than the low–high combination.

The argument in step (3) can be further strengthened by representing a fiber span by the VSTFs as shown in Figure 2.32. Any swapping of the sequence of low- and high-dispersion fiber spans would result in the same power penalty owing to nonlinear phase distortion except the accumulated noise contributed from the ASE noise of the in-line OAs of all spans. Thus, we can see that the noise figure (NF) of both configurations can be approximated in the same way. This is in contrast to the simulation results reported in Ref. [33]. We believe that the difference in the power penalty for different order of arrangements of low- and high-dispersion fiber spans reported in Ref. [15] is due to numerical error, because possibly the split-step Fourier method (SSFM) was employed, and the instantaneous amplitude of the complex envelope was used. This does not indicate the total average signal power of all channels. Therefore, we can conclude that the simulated nonlinear threshold power level would be suffering an additional artificial OSNR penalty due to the instantaneous power of the sampled complex amplitude of the propagating amplitude.

Tang [23] reported a variation in the channel capacity against the input power/channel with dispersion as a parameter $-2 \rightarrow -20$ ps²/km with a noise power spectral density of 10 µW/GHz over 4 spans; for 4 and 32 spans of dispersive fibers of 0 & -20 ps²/km with a channel spacing of 50 GHz and 100 channels, the noise spectral density was 10 µW/GHz. The deviation in the capacity is observed at the onset of the power per channel of 0.1, 2, and 5 mW. Further observations can be made here. The noise responses indicate that the nonlinear frequency transfer function of a highly dispersive fiber link is related directly to the fundamental linear transfer function of the fiber link. When the transmission is highly dispersive, the linear transfer function of the fiber acts as a low-pass filter. Thus all the energy concentrates in the passband of this filter which may be narrower than that of the signal at the transmitting end. Thus distortion occurs. This may thus lower the nonlinear effects as given in Equation 2.100. For a lower-dispersive fiber, this transfer function would represent a low-pass filter with a 3 dB roll-off frequency, which is much higher than that of a dispersive fiber. For example, the G.655 fiber would have a dispersion factor that is three times lower than that of the G.652 fiber. This wideband low-pass filter will allow nonlinear effects of intrachannels and interchannel interactions. The dispersive accumulation term $\sum_{k=0}^{N_s-1} e^{-jk\beta_2 L_s \Delta\Omega}$ dominates when the number of spans is high.

In the simulation results given in Ref. [34], the VSTFs were applied to dispersive fiber spans. It is expected from Equation 2.100 that the arrangement of alternating positions between G.655

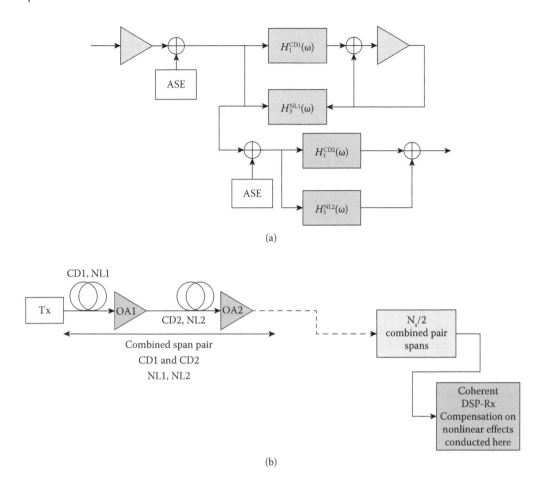

(a)

(b)

FIGURE 2.32 (a) System of concatenation of fiber spans consisting of pairs of different CDs and NLs, models for simulation and (b) optically amplified N_s-span fiber link without DCF and DSP-based optical receiver with compensation of nonlinear effects by back propagation or Volterra series transfer function conducted by digital signal processing. This model represents real-time processing in practical optical transmission systems.

and G.652 would not exert any penalty. The simulation reported in Ref. [6] indicates a 1.5 dB difference at 10×2 spans (SSMF + TWC) and no difference at 20×2 spans. The contribution by the ASE noise of the OAs at the end of each span would influence the phase noise and hence the effects on the error vector magnitude (EVM) of the sampled signal detected constellation.

From the transfer functions including both linear and nonlinear kernels of the dispersive fibers, we could see that if the noise is the same, the nonlinear effects would not be different regardless whether high- or low-dispersive fiber spans are placed at the front or back. However, if the nonlinear noises are accounted for, especially the intrachannel effects, then if less dispersive fibers are placed in the front, higher noise is expected, and thus a lower nonlinear threshold (at which a 1-dB penalty is incurred on the OSNR). This is in contrast to the simulation results presented in Ref. [26]. However, the accumulated noise is much smaller than the average signal power. Under simulation, depending on the numerical approach to solve the NLSE, the estimation of signal power at the sampled instant is normally obtained from the sampled amplitude at this instant and is thus different from the average launched power into the fiber span. This creates discrepancies on the order of the high- or low-dispersion fiber spans. Thus, there are possibilities that the peak above average amplitude of the very dispersive pulse sequence at some locations along the propagation path where superposition of several pulses would occur. This amplitude may reach a level much higher than the nonlinear threshold and thus create a different distortion penalty due to nonlinear effects.

VSTFs offer better accuracy and cover a number of SPM and parametric scattering, but at the expense of costs of computing resources due to two-dimensional fast Fourier transform (FFT) for the SPM and XPM. This model should be employed when such extra nonlinear phase noise is required such as in the case of superchannel transmission.

2.7.5 Nonlinear Phase Noises in Cascaded Multispan Optical Link

Gordon and Mollenauer [35] showed that when OAs are used to compensate for fiber loss, the interaction of amplifier noise and the Kerr effect causes phase noise, even in systems using constant-intensity modulation. This nonlinear phase noise, often called the Gordon–Mollenauer effect or, more precisely, SPM-induced nonlinear phase noise, or simply nonlinear phase noise (NLPN), corrupts the received phase and limits the transmission distance in systems using M-ary QAM. The NLPN in turn would create random variation in the intensity, and thus a transfer or conversion of the NLPN into intensity noise, or modulation intensity.

In modern long-haul transmission systems, cascading of optically amplified nondispersion-compensating fiber spans is commonly used to form a multispan transmission link. The highly dispersive signal sequences emerged at the end of the link are coherently detected and then sampled and digitalized then further processed in the digital domain by digital signal processor (DSP). Ho and Kahn [36] have studied and derived the variances and co-variances of DWDM optical transmission systems. For an electric field E_0 of the optical waves launched at the input of the first span, the field at the input of the kth span would be superimposed by the noises accumulated over k spans as $E_k = E_0 + n_1 + n_2 + \ldots + n_{k-1}$; then the variance $\sigma_{\phi_{NL}}$ of the nonlinear phase shift is given as

$$\sigma_{NLPN}\left(\alpha_{NL}\right) = \left(\gamma L_{eff}\right)^2 \left[\sigma_{NL}^2\left(N-1\right) + \left(\alpha_{NL}-1\right)^2 f\left(N\sigma_1^2\right) - 2\left(\alpha_{NL}-1\right)\sum_{k=1}^{N} f\left(N\alpha_1^2\right)\right] \qquad (2.107)$$

where α is the scaling factor, $f(N\sigma^2)$ is the expected value of the optical electric field between two consecutive spans, and σ is the variance of the field under superposition with the noises. N is the total number of optically amplified spans greater than the optimal factor and can be found by differentiating Equation 2.107 with respect to this factor

$$a_{NL} \approx -gL_{eff}\frac{N+1}{2} \qquad (2.108)$$

At high OSNR $\gg 1$, this variance can be found to be

$$\sigma_{NLNP}^2 \gg \frac{4}{3}N^3\left(\gamma L_{eff}\sigma|E_0|^2\right)^2 \qquad (2.109)$$

with $\sigma \equiv \sigma_{|E_0+n_1+n_2+\ldots n_k|}$ as the variance of the field superimposed by noises after kth span of k cascade spans. The expected value of the nonlinear phase shift can be approximated as

$$\langle\phi_{NL}\rangle \approx N\gamma L_{eff}|E_0|^2 \qquad (2.110)$$

Then, the NLPN variance of N cascaded spans can be given as

$$\sigma_{NLNP}^2 \gg \frac{4}{3}N^3\left(\gamma L_{eff}\sigma|E_0|^2\right) = \frac{4}{3}N\frac{\langle\phi_{NL}\rangle^2}{OSNR_L^2} \qquad (2.111)$$

where $OSNR_L$ is the optical signal-to-noise ratio in the linear scale, and the mean phase noise is given by Equation 2.110. Thus, we can see that the nonlinear phase rotation due to SPM in the N-cascade-span link is the total phase rotation accumulated over the spans.

The variances of the residual NLPN are also given as

$$\sigma_{NLNP,res}^2 \approx \frac{1}{6} \frac{\langle \phi_{NL} \rangle^2}{OSNR_L^2} \tag{2.112}$$

The NLPN power variance is also proportional to the square of the accumulated phase rotation. This allows a compensation algorithm for nonlinear impairments by rotating the phase of the digital samples of the in-phase and quadrature-phase components at the end of each span. This is indeed a linear operation based on the derived and observed rotation of the constellation due to SPM. This linear phase rotation simplifies the numerical computation and hence the computing resources of the DSP. The phase-to-intensity conversion and the instability problem will create some degradation of the OSNR due to this increase of the noise intensity over the 0.4 nm band which is commonly measured of the noise in practice. Thus, we expect a logarithmic reduction of the OSNR with respect to the number of cascaded optically amplified fiber spans due to the SPM-induced and modulation instability.

2.8   SPECIAL DISPERSION OPTICAL FIBERS

At the beginning of the 1980s, there was great interest in reducing the total dispersion $[M(\lambda) + D(\lambda)]$ of the SMF at 1550 nm, where the loss is lowest for silica fiber. There were two significant trends: one was to reduce the linewidth and stabilize the laser center wavelength and the other was to reduce the dispersion at this wavelength. The fibers designed for long-haul transmission systems usually exhibit a near-zero-dispersion at a certain spectral window. These are called DSFs, and their total dispersion approaches zero: $[M(\lambda) + D(\lambda)] \sim 0$. The material dispersion factor $M(\lambda)$ is natural and slightly affected by variation in the doping material and concentration. The waveguide dispersion factor, $D(\lambda)$, can be tailored by designing appropriate refractive index profiles and a geometrical structure to balance the material dispersion effects so that the total dispersion factor reaches a null value at a specific wavelength or spectral window. Note that the dispersion factors due to the material and the waveguide take algebraic values, and so they can be designed to take opposite values to cancel each other out.

The problems facing DSFs are that FWM can occur easily owing to the phase-matching condition that can be satisfied at the zero dispersion from the equally spaced wavelength channels. Thus, the zero-dispersion wavelength is usually shifted to outside the C-band to avoid FWM. These types of fibers are NZ-DSFs, whose zero-dispersion wavelength is commonly placed around 1510 nm so that only a small dispersion amount occurs in the C-band to avoid the FWM problems.

Advanced optical fiber design techniques can design dispersion to flatten fibers where the dispersion factor is flat over the wavelength range from 1300 to 1600 nm, by tailoring the refractive index profile of the core of optical fibers in such a distribution as the W-profile, segmented profile, multilayer core structure, and so on.

Another type of optical fiber would be required to compensate for the dispersion effect of the optical signal after transmission over a length of optical fiber; this is the dispersion-compensated fiber, whose dispersion factor is many times larger than that of the standard communication fiber with the opposite sign. This can be designed by setting the total dispersion to the required compensated dispersion, and thus the waveguide dispersion can be found over the required operating range. Optical fiber structures can then be designed to obtain the core radius and the refractive index profile with the optimum mode spot size.

2.9 SMF TRANSFER FUNCTION: SIMPLIFIED LINEAR AND NONLINEAR OPERATING REGION

In this section, a closed expression of the frequency transfer function of dispersive and nonlinear SMFs for broadband operation can be derived, similar to the case under microwave photonics. The expression takes into account both chromatic dispersion and SPM effects and is valid for optical double-sideband modulation, optical single-sideband (SSB) modulation, and chirped optical transmitters.

The evolution along the propagation path z of the "small-signal" IM, or complex power $\overline{p}(\omega,z)$ and phase rotation (PM) $\overline{\phi}(\omega,z)$ during the propagation of the guided mode through the SMF, taking into account both the chromatic dispersion and the nonlinearity (SPM) effects, is governed by the following set of differential equations [37–40]

$$\frac{\delta \overline{p}(\omega,z)}{\delta z} = \beta_2 \omega^2 \overline{p_0} \overline{\phi}(\omega,z); \; \overline{A}(\omega,z) = \sqrt{\overline{p}(\omega,z)} \tag{2.113}$$

$$\frac{\delta \overline{\phi}(\omega,z)}{\delta z} = -\left[\frac{\beta_2 \omega^2}{4\overline{p_0}} + \gamma e^{-\alpha z}\right]\overline{p}(\omega,z) \tag{2.114}$$

where $\overline{A}(\omega,z)$ and $\overline{\phi}(\omega,z)$ are defined as the normalized complex amplitude and phase, respectively, of the optical field in the Fourier domain, ω is the radial frequency of the RF or broadband signal, z is the distance along the propagation axis of the fiber, α is the attenuation coefficient in the linear scale of SMF, and β_2 is the first-order dispersion coefficient, that is, the group delay factor expressed as a function of the optical wavelength. β_2 is given by

$$\beta_2 = -\frac{\lambda^2 D(\lambda)}{2\pi c} \tag{2.115}$$

where c is the velocity of light in vacuum, and $D(\lambda)$ is the dispersion factor of the fiber, typically taking a value of 17 ps/nm/km for silica SMF at the operating wavelength $\lambda = 1550$ nm.

γ is the nonlinear SPM coefficient defined by

$$\gamma = \frac{2\pi n_2}{\lambda A_{\text{eff}}}; \quad \text{with } A_{\text{eff}} = \pi r_0^2 \tag{2.116}$$

where n_2 is the nonlinear index coefficient of the fiber, where typically $n_2 = 1.3 \times 10^{-23}$ m^2/W for SMF-28, and $A_{\text{eff}} = \pi r_0^2$ is the effective area of the fiber, which is the area of the Gaussian mode spot of the guided mode in an SMF under the weakly guiding condition [41].

These equations are derived from the observer positioned on the moving frame of the phase velocity of the waves, which is normally expressed by the NLSE

$$\frac{\partial A(t,z)}{\partial z} = -\left[\alpha A(t,z) + \frac{j}{2}\beta_2 \frac{\partial A^2(t,z)}{\partial t^2}\right] + j\gamma |A(t,z)|^2 A(t,z) \tag{2.117}$$

Presently, digital signal processing subsystems are integrated at the optical transmitter to compensate for the short fall in bandwidth of the modulator and also at the receiver following the opto-electronic detection and electronic amplification so as to recover the clock pulse sequence of the data channel, to recover and decode the modulated sequence and finally to compensate for any distortion due to the propagation medium. A schematic of the transmission is shown in Figure 2.33,

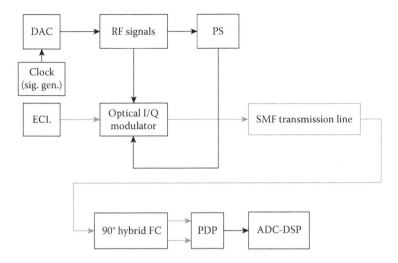

FIGURE 2.33 Digital-based optical transmitter and coherent reception with real-time sampling and digital signal processing. DAC = digital-to-analog converter, ADC = analog-to-digital converter, DSP = digital signal processing, PDP = photodetector pair, FC = fiber coupler, I/Q = in-phase/quadrature phase.

in which the transmitter can generate an optical sequence or near-single-frequency sinusoidal waves at a frequency reaching 30 GHz using a Fujitsu DAC sampling rate of 65 GSa/s.

The optical modulator is a typical Fujitsu I–Q modulator modulated by electrical signals output from the DAC and phase shifted in the RF domain by an electrical phase shifter PS. The RF phase can be set such that when the signals are $\pi/2$-shifted with respect to each other, the suppression of one of the single sidebands can be achieved at the output spectrum. The main carrier can be suppressed by biasing the "children" Mach–Zehnder intensity modulators (MZIMs) at the minimum transmission point (Figure 2.34).

By differentiating Equation 2.113 and substituting in Equation 2.37, we obtain

$$\frac{\delta^2 \overline{p}(\omega,z)}{\delta z^2} = -\left[\frac{\beta_2^2 \omega^4}{4} - \beta_2 \omega^2 + \gamma \overline{P_0} e^{-\alpha z}\right] \overline{p}(\omega,z) \tag{2.118}$$

Subject to the initial conditions of

$$\overline{p}(\omega,0) = \overline{p}_{in}(\omega) \text{ and } \frac{\delta \overline{p}(\omega,0)}{\delta z} = \beta_2 \omega^2 \overline{P_0} e^{-\alpha z} \phi(\omega,0) = \beta_2 \omega^2 \overline{P_0} e^{-\alpha z} \phi_{in}(\omega) \tag{2.119}$$

where the subscript *in* indicates the input location, where the propagation of the modulated optical waves through the SMF commences.

Now, by changing some variables by setting

$$x = 2\sqrt{B} e^{-\alpha z/2}; \ B = -\frac{\beta_2 \omega^2 \gamma P_0}{\alpha^2} \tag{2.120}$$

Equation 2.118 can be rewritten as

$$\left[x \frac{\partial^2}{\partial x^2} + x \frac{\partial}{\partial x} - \left(x^2 - \upsilon^2\right)\right] \overline{p}(\omega,z) = 0 \tag{2.121}$$

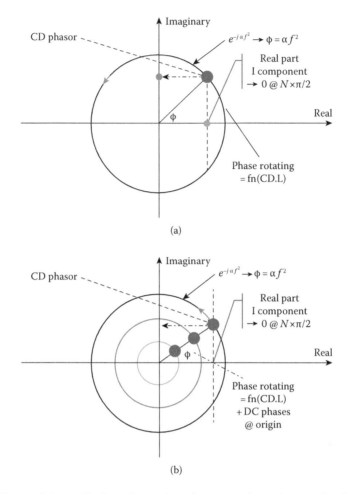

FIGURE 2.34 Phasor of the amplitude evolution along the propagation path, assuming the amplitude is not affected by attenuation. (a) Phasor and (b) phase or the phase constellation of M-QAM ($M = 16$ square QAM), three amplitude level.

The solution of this equation is a combination of purely imaginary Bessel functions L and K and is subject to the initial conditions of Equation 2.119. Thus, the evolution of the complex amplitude of the intensity-modulated optical waves along the propagation path is given by [39,42]

$$\bar{p}(\omega,z) = \frac{2\sinh(\pi\upsilon)}{\pi}\left\{ \begin{array}{l} \sqrt{B}\,\bar{p}_{\mathrm{in}}(\omega)\left[\dfrac{\partial L_{i\upsilon}}{\partial x}\left(2\sqrt{B}\right)K_{i\upsilon}(x) - \dfrac{\partial K_{i\upsilon}}{\partial x}\left(2\sqrt{B}\right)I_{i\upsilon}(x)\right] \\[4mm] + \dfrac{\alpha B}{\gamma}\phi_{\mathrm{in}}(\omega)\left[K_{i\upsilon}\left(2\sqrt{B}\right)L_{i\upsilon}(x) - L_{i\upsilon}\left(2\sqrt{B}\right)K_{i\upsilon}(x)\right] \end{array} \right\} \tag{2.122}$$

with $\upsilon = -\dfrac{\beta_2\omega^2}{\alpha}$. The first term on the right-hand side of Equation 2.122 is the magnitude part, and the second term is the phase part, that is, the in-phase and quadrature components of the QAM signal. Thus, the in-phase and quadrature parts of the complex magnitude can be expressed as

$$\bar{p}_I(\omega,z) = \frac{2\sinh(\pi\upsilon)}{\pi}\left\{\sqrt{B}\,p_{\mathrm{in}}(\omega)\left[\dfrac{\partial L_{i\upsilon}}{\partial x}\left(2\sqrt{B}\right)K_{i\upsilon}(x) - \dfrac{\partial K_{i\upsilon}}{\partial x}\left(2\sqrt{B}\right)I_{i\upsilon}(x)\right]\right\} \tag{2.123}$$

$$\overline{p}_Q(\omega,z) = \frac{\alpha\beta}{\gamma}\phi_{in}(\omega)\frac{2\sinh(\pi\upsilon)}{\pi}\Big[K_{i\upsilon}\big(2\sqrt{B}\big)L_{i\upsilon}(x) - L_{i\upsilon}\big(2\sqrt{B}\big)K_{i\upsilon}(x)\Big] \qquad (2.124)$$

The variations of the in-phase (real part) and quadrature-phase (imaginary part) of the complex power can be estimated by referring to Figures 2.35 and 2.36 [39], respectively.

The in-phase and quadrature-phase components move along the horizontal and vertical axes within the normalized ±1 limits, meaning that as the phase rotates around the unit circle, these

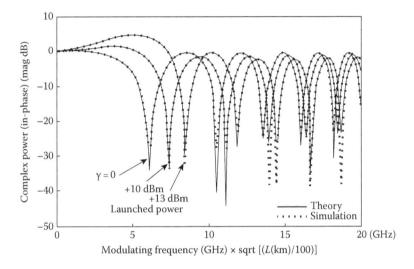

FIGURE 2.35 Variation of the magnitude of the optical field intensity with frequency, the intensity frequency response of standard SMF of $L = 100$ km (at $z = 100$ km) under coherent detection with normalized amplitude of the intensity of the optical waves under linear ($\gamma = 0$) and nonlinear operating conditions. SSMF parameters: $\alpha = 0.2$ dB/km, $D = 17$ ps^2/(nm.km); $n_2 = 3.2.10^{-23}$ m^2/W, $L = 100$ km.

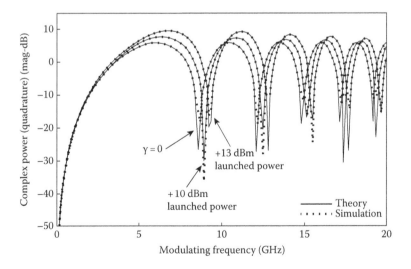

FIGURE 2.36 Variation of the magnitude of phase component with frequency (PM–IM conversion) of standard SMF of $L = 100$ km (at $z = 100$ km) under coherent detection with normalized amplitude of the intensity of the optical waves under linear ($\gamma = 0$) and nonlinear operating conditions. SSMF parameters: $\alpha = 0.2$ dB/km, $D = 17$ ps^2/(nm.km); $n_2 = 3.2.10^{-23}$ m^2/W. $L = 100$ km. (Extracted from F. Ramos and J. Martí, *IEEE Photon. Technol. Lett.*, Vol. 12, No. 5, pp. 549–551, 2000.)

components oscillate in a manner such that when the phase is $(2M + 1)(M = 0, 1, 2,...)$ or odd number of $\pi/2$, then the in-phase component becomes nullified, and so on, and likewise, the quadrature phase at $N\pi(N = 0, 1, 2,...)$. In the case of the QAM scheme, for example, 16QAM, there would be three amplitude levels in its phase constellation which are rotated relatively with respect to each other as shown in Figure 2.36. The initial phase is set by the initial position of the constellation point of square 16 QAM, but the oscillation and nullified locations would be very much similar to that of Figure 2.35.

Under the linear operating regime $\gamma = 0$, we can obtain the expressions for the complex power amplitude and phase and the overall fiber transfer function under a modulation transfer as

$$\bar{p}_I(\omega, z) = \cos\frac{\beta_2\omega^2 z}{2}$$

$$p_\phi(\omega, z) = 2\sin\frac{\beta_2\omega^2 z}{2}$$

$$(2.125)$$

$$H_F(\omega, z) = \bar{p}_I(\omega, z) + \frac{H_{PM}(\omega, z)}{2} p_Q(\omega, z)$$

$$H_{PM}(\omega, z) = \text{frequency\_response\_modulated\_signals}$$

For SSB-modulated signals $H_{PM}(\omega, z) = j$, which are purely complex, we can obtain the transfer characteristics as shown in Figure 2.37.

Note that for SSB signals, the frequency response is flat over a very wideband, and the notches of the linear and nonlinear responses are significantly reduced. This is because SSB signals have at least half the band of the DSB's band, and thus the nonlinear effects are also reduced. The nonlinearity is estimated with the average power launched into 100 km fiber at 10 dBm.

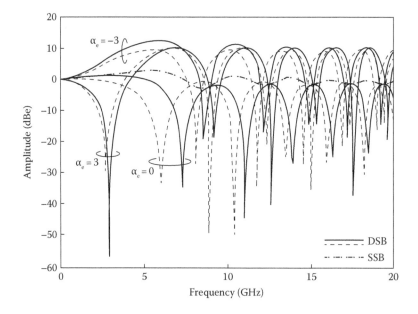

FIGURE 2.37 Frequency response of 100 km SSMF for DSB and SSB spectral signals under linear and nonlinear (+10 dBm). - - - - = linear; continuous line nonlinear DSB and -·· ··· ··· SSB with chirp parameters α of a directly modulated laser diode as a parameter. Note the flat response of SSB signals.

2.10 NUMERICAL SOLUTION: SPLIT-STEP FOURIER METHOD

In practice, with the extremely high-speed operation of the transmission systems, it is very costly to simulate by experiment, especially when the fiber transmission line is very long, for example, a few thousands of kilometers. It is then preferable to conduct computer simulations to guide the experimental setup. In such simulations, the propagation of modulated lightwave channels plays a crucial role in achieving a transmission performance of the systems that is close to that of practical systems. The main challenge in the simulation of the propagation of lightwave channels employing the NLSE, which can be derived from Maxwell equations, is whether the signal presented in the time domain be propagating though the fiber and its equivalent in the frequency domain when the nonlinearity is effective. The propagation techniques for such modulated channels are described in a later subsection.

2.10.1 SYMMETRICAL SSFM

The evolution of slow-varying complex envelopes $A(z,t)$ of optical pulses along an SMF is governed by the NLSE

$$\frac{\partial A(z,t)}{\partial z} + \frac{\alpha}{2} A(z,t) + \beta_1 \frac{\partial A(z,t)}{\partial t} + \frac{j}{2}\beta_2 \frac{\partial^2 A(z,t)}{\partial t^2} - \frac{1}{6}\beta_3 \frac{\partial^3 A(z,t)}{\partial t^3} = -j\gamma |A(z,t)|^2 A(z,t) \quad (2.126)$$

where z is the spatial longitudinal coordinate, α accounts for the fiber attenuation, β_1 indicates DGD, β_2 and β_3 represent second- and third-order dispersion factors of fiber CD, and γ is the nonlinear coefficient as also defined earlier. In a single-channel transmission, Equation 2.126 includes the following effects: fiber attenuation, fiber CD and PMD, dispersion slope, and SPM nonlinearity. The fluctuation in optical intensity caused by the Gordon–Mollenauer effect is also included in this equation. We can observe that the term involving β_2 and β_3 relates to the phase evolution of optical carriers under the pulse envelope. The term β_1 relates to the delay of the pulse when propagating through a length of the fiber. So, if the observer is situated on top of the pulse envelope, then this delay term can be eliminated.

The solution of the NLSE and hence the modeling of pulse propagation along an SMF is solved numerically by using the SSFM so as to facilitate the solution of the nonlinear equation when nonlinearity is involved. In the SSFM, the fiber length is divided into a large number of small segments δz. In practice, fiber dispersion and nonlinearity are mutually interactive at any distance along the fiber. However, these mutual effects are small within δz, and thus the effects of fiber dispersion and fiber nonlinearity over δz are assumed to be statistically independent of each other. As a result, the SSFM can separately define two operators: (1) the linear operator that involves fiber attenuation and fiber dispersion effects and (2) the nonlinearity operator that takes into account fiber nonlinearities. These linear and nonlinear operators are formulated as follows

$$\hat{D} = -\frac{j\beta_2}{2} \frac{\partial^2}{\partial T^2} + \frac{\beta_3}{6} \frac{\partial^3}{\partial T^3} - \frac{\alpha}{2}$$

$$\hat{N} = j\gamma |A|^2$$

$$\quad (2.127)$$

where $j = \sqrt{-1}$; A replaces $A(z,t)$ to simplify the notation, and $T = t - z/v_g$ is the reference time frame moving at the group velocity, meaning that the observer is situated on top of the pulse envelope. Equation 2.127 can be rewritten in a shorter form

$$\frac{\partial A}{\partial z} = \left(\hat{D} + \hat{N}\right) A \quad (2.128)$$

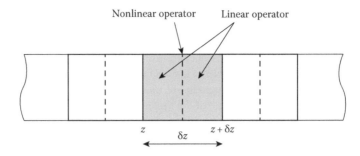

FIGURE 2.38   Schematic illustration of symmetric SSFM.

and the complex amplitudes of optical pulses propagating from z to z + δz are calculated using the following approximation

$$A(z+h,T) \approx \exp\left(h\hat{D}\right)\exp\left(h\hat{N}\right)A(z,T) \qquad (2.129)$$

Equation 2.129 is accurate to the second order of the step size δz. The accuracy of the SSFM can be improved by including the effect of fiber nonlinearity in the middle of the segment rather than at the segment boundary. This modified SSFM is known as the symmetric SSFM, and is illustrated in Figure 2.38.

Equation 2.129 can now be modified as

$$A(z+\delta z,T) \approx \exp\left(\frac{\delta z}{2}\hat{D}\right)\exp\left(\int_{z}^{z+\delta z}\hat{N}(z')dz'\right)\exp\left(\frac{\delta z}{z}\hat{D}\right)A(z,T) \qquad (2.130)$$

This method is accurate to the third order of the step size δz. In the symmetric SSFM, the optical pulse propagates along a fiber segment δz in two stages. First, the optical pulse propagates through the linear operator, which has a step of δz/2 in which the fiber attenuation and dispersion effects are taken into account. An FFT is used here in the propagation step so that the output of this half-size step is in the frequency domain. Note that the carrier is removed here and the phase of the carrier is represented by the complex part, the phase evolution. Hence, the term *complex amplitude* is coined.

Then, the fiber nonlinearity is superimposed on the frequency-domain pulse spectrum at the middle of the segment. After that, the pulse propagates through the second half of the linear operator via an inverse FFT to get back the pulse envelope in the time domain.

The process continues repetitively over consecutive segments of size δz until the end of the fiber length. It should again be noted that the linear operator is computed in the time domain, whereas the nonlinear operator is calculated in the frequency domain.

2.10.1.1 Modeling of PMD

As described earlier, PMD results from the delay difference between the propagation of each polarized mode of the LP modes LP_{01}^H and LP_{01}^V of the horizontal and vertical directions, respectively, as illustrated in Figure 2.3. The parameter DGD, the differential group delay, determines the first-order PMD, which can be implemented by modeling the optical fiber as two separate paths representing the propagation of two principal states of polarization (PSPs). The symmetrical SSFM can be implemented in each step on each polarized transmission path, and then their outputs are superimposed to form the output optical field of the propagated signals. The transfer function to represent the first-order PMD is given by

$$H(f) = H^+(f) + H^-(f) \qquad (2.131)$$

where

$$H^+ (f) = \sqrt{k} \exp\left[j2\pi f\left(-\frac{\Delta\tau}{2} \right) \right] \qquad (2.132)$$

and

$$H^- (f) = \sqrt{k} \exp\left[j2\pi f\left(-\frac{\Delta\tau}{2} \right) \right] \qquad (2.133)$$

In which k is the power-splitting ratio ($k = 0.5$ when a 3 dB or 50:50 optical coupler/splitter is used), $\Delta\tau$ is the instantaneous DGD value is the average value of a statistical distribution that follows a Maxwell distribution (refer to Equation 2.77) [43,44]. This randomness is due to the random variations of the core geometry, the fiber stress, and hence anisotropy due to the drawing process, variation of temperature, and so on in the installed fibers.

2.10.1.2 Optimization of Symmetrical SSFM

2.10.1.2.1 Optimization of Computational Time

A huge amount of time can be spent for the symmetric SSFM via the uses of FFT and IFFT operations, in particular, when fiber nonlinear effects are involved. In practice, when optical pulses propagate toward the end of a fiber span, the pulse intensity has been greatly attenuated owing to the fiber attenuation. As a result, fiber nonlinear effects become negligible for the rest of that fiber span, and hence the transmission operates in a linear domain in this range. Symmetric SSFM can be used to reduce the computational time. If the peak power of an optical pulse is lower than the nonlinear threshold of the transmission fiber, for example, around −4 dBm, symmetrical SSFM is switched to a linear mode operation. This linear mode involves only fiber dispersions and fiber attenuation, and its low-pass equivalent transfer function for the optical fiber is

$$H(\pi) = \exp\left\{ -j\left[\left(\frac{1}{2} \right)\beta_2\pi^2 + \left(\frac{1}{6} \right)\beta_3\pi^3 \right] \right\} \qquad (2.134)$$

If β_3 is not considered in this fiber transfer function, which is normally the case owing to its negligible effects on 40 Gbps and lower-bit-rate transmission systems, the foregoing transfer function has a parabolic phase profile [43,44].

2.10.1.2.2 Mitigation of Windowing Effect and Waveform Discontinuity

In symmetric SSFM, the mathematical operations of FFT and IFFT play very significant roles. However, owing to the finite window length required for FFT and IFFT operations, these operations normally introduce overshooting at two boundary regions of the FFT window, commonly known as the windowing effect of FFT. In addition, because the FFT operation is a block-based process, there exists the issue of waveform discontinuity; that is, the right-most sample of the current output block does not start at the same position as the left-most sample of the previous output block. The windowing effect and the waveform discontinuity problems are resolved with the following technique; see also Figure 2.39.

The actual window length for FFT/IFFT operations consists of two blocks of samples, and hence the sample length is $2N$. The output, however, is a truncated version with the length of one block (N samples), and output samples are taken in the middle of the two input blocks. The next FFT window overlaps the previous one by one block of N samples.

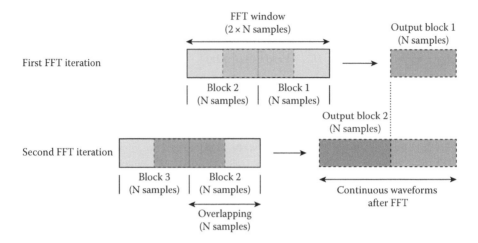

FIGURE 2.39 Proposed technique for mitigating windowing effect and waveform discontinuity caused by FFT/IFFT operations.

2.11 CONCLUDING REMARKS

Although the treatment of optical, in particular single-mode, fibers has been extensively presented by several research papers and textbooks, currently, single-mode and few-mode fibers have been extensively revisited for the design and considerations of compensation of linear and nonlinear distortion effects in the digital domain of the coherent reception end and pre-equalization at the transmitting end. This chapter has thus presented the fiber characteristics in static parameters and dynamic characteristics in terms of dispersion properties and nonlinear phase rotation effects.

A brief introduction to the compensation of nonlinear effects via the use of nonlinear transfer functions and back propagation techniques will be described in Chapter 17 for compensation of such nonlinear impairments in long-haul transmission systems. Further treatment of these digital processing techniques will be given in later chapters of this book.

The applications of digital signal processing for spatial multiplexed channels warrants the revisiting of the FMFs and the guiding as well as splitting and combining of these modes and their polarized partners; such an investigation is especially appropriate now, with the development of high-capacity extremely long-haul optical fiber transmission systems. However, the availability of in-line optical amplification for all these modal channels remains an important topic for systems engineering.

REFERENCES

1. C. Xia, N. Bai, I. Ozdur, X. Zhou, and G. Li, Supermodes for optical transmission, *Opt. Express*, Vol. 19, No. 17, p. 16653, 2011.
2. A. Safaai-Jazi and J. C. McKeeman, Synthesis of intensity patterns in few-mode optical fibers, *IEEE J. Lightwave Technol.*, Vol. 9, No, 9, p. 1047, 1991.
3. M. Salsi, C. Koebele, D. Sperti, P. Tran, P. Brindel, H. Mardoyan, S. Bigo, et al., Transmission at 2100Gb/s, over two modes of 40km-long prototype few-mode fiber, using LCOS based mode multiplexer and demultiplexer, in *Optical Fiber Conference, OFC 2011, and National Fiber Optic Engineers Conference (NFOEC)*, Los Angeles, CA, March 6, 2011, postdeadline session B (PDPB).
4. S. Randel, R. Ryf, A. Sierra, P. J. Winzer, A. H. Gnauck, C. A. Bolle, R.-J. Essiambre, D. W. Peckham, A. McCurdy, and R. Lingle, Jr., 6×56-Gb/s mode-division multiplexed transmission over 33-km few-mode fiber enabled by 6×6 MIMO equalization, *Opt. Express*, Vol. 19, No. 17, p. 16697, 2011.
5. C. D. Poole, J. M. Wiesenfeld, D. J. DiGiovanni, and A. M. Vengsarkar, Optical fiber-based dispersion compensation using higher order modes near cutoff, *IEEE J. Lightwave Technol.*, Vol. 12, No. 10, pp. 1746–1758, 1994.

6. L. N. Binh, L. Liu, and L. C. Li, Volterra series transfer function in optical transmission and nonlinear compensation, in *Nonlinear Optical Systems*, Ed. L. N. Binh and D. V. Liet, Boca Raton, FL: CRC Press, pp. 373–410, 2012.

7. A. Mecozzi, C. B. Clausen, and M. Shtaif, Analysis of intrachannel nonlinear effects in highly dispersed optical pulse transmission, *IEEE Photon. Technol. Lett.*, Vol. 12, pp. 392–394, 2000.

8. K. Kikuchi, Coherent detection of phase-shift keying signals using digital carrier-phase estimation, in *Proceedings of the IEEE Conference on Optical Fiber Communications*, paper OTuI4, Institute of Electrical and Electronics Engineers, Anaheim, 2006.

9. G. P. Agrawal, *Fiber Optic Communications Systems*, 3rd ed., New York: John Wiley, 2002.

10. R. H. Stolen, E. P. Ippen, and A. R. Tynes, Raman oscillation in glass optical waveguide, *Appl. Phys. Lett.*, Vol. 20, p. 62, 1972.

11. E. P. Ippen and R. H. Stolen, Stimulated Brillouin scattering in optical fibers, *Appl. Phys. Lett.*, Vol. 21, p. 539, 1972.

12. R. G. Smith, Optical power handling capacity of low optical fibers as determined by stimulated Raman and Brillouin scattering, *Appl. Opt.*, Vol. 11, pp. 2489–2494, 1972.

13. G. P. Agrawal, *Fiber Optic Communications Systems*, 3rd ed., New York: John Wiley, p. 60, 2002.

14. G. P. Agrawal, *Fiber Optic Communications Systems*, 3rd ed., New York: John Wiley, p. 67, 2002.

15. J. P. Gordon and H. Kogelnik, PMD fundamentals: Polarization mode dispersion in optical fibers, *Proc. Natl. Acad. Sci. U. S. A.*, Vol. 97, No. 9, pp. 4541–4550, 2000.

16. Corning Inc., *An Introduction to the Fundamentals of PMD in Fibers*, White Paper, Corning, NY: Corning Inc., 2006.

17. A. Galtarossa and L. Palmieri, Relationship between pulse broadening due to polarisation mode dispersion and differential group delay in long single-mode fiber, *Electron. Lett.*, Vol. 34, No. 5, pp. 492–493, 1998.

18. J. M. Fini and H. A. Haus, Accumulation of polarization-mode dispersion in cascades of compensated optical fibers, *IEEE Photon. Technol. Lett.*, Vol. 13, No. 2, pp. 124–126, 2001.

19. A. Carena, V. Curri, R. Gaudino, P. Poggiolini, and S. Benedetto, A time-domain optical transmission system simulation package accounting for nonlinear and polarization-related effects in fiber, *IEEE J. Sel. Areas Commun.*, Vol. 15, No. 4, pp. 751–765, 1997.

20. S. A. Jacobs, J. J. Refi, and R. E. Fangmann, Statistical estimation of PMD coefficients for system design, *Electron. Lett.*, Vol. 33, No. 7, pp. 619–621, 1997.

21. G. P. Agrawal, *Fiber Optic Communication Systems*, New York: Academic Press, 2002.

22. A. F. Elrefaie, R. E. Wagner, D. A. Atlas, and D. G. Daut, Chromatic dispersion limitations in coherent lightwave transmission systems, *IEEE J. Lightwave Technol.*, Vol. 6, No. 6, pp. 704–709, 1998.

23. J. Tang, The channel capacity of a multispan DWDM system employing dispersive nonlinear optical fibers and an ideal coherent optical receiver, *IEEE J. Lightwave Technol.*, Vol. 20, No. 7, pp. 1095–1101, 2002.

24. B. Xu and M. Brandt-Pearce, Comparison of FWM- and XPM-induced crosstalk using the volterra series transfer function method, *IEEE J. Lightwave Technol.*, Vol. 21, No. 1, pp. 40–54, 2003.

25. J. Tang, The Shannon channel capacity of dispersion-free nonlinear optical fiber transmission, *IEEE J. Lightwave Technol.*, Vol. 19, No. 8, pp. 1104–1109, 2001.

26. J. Tang, A comparison study of the Shannon channel capacity of various nonlinear optical fibers, *IEEE J. Lightwave Technol.*, Vol. 24, No. 5, pp. 2070–2075, 2006.

27. A. Papoulis, *Systems and Transforms With Applications in Optics*, Robert Krieger Publishing Company, Malabar, Florida,1968.

28. L. N. Binh, *Digital Optical Communications*, Boca Raton, FL: CRC Press, 2009.

29. L. N. Binh and N. Nguyen, Generation of high-order multi-bound-solitons and propagation in optical fibers, *Opt. Commun.*, Vol. 282, pp. 2394–2406, 2009.

30. K.V. Peddanarappagari and M. Brandt-Pearce, Volterra series transfer function of single-mode fibers, *J. Lightwave Technol.*, Vol. 15, No. 12, pp. 2232–2241, 1997.

31. G. P. Agrawal, *Nonlinear Fiber Optics*, 3rd ed., San Diego, CA: Academic Press, 2001.

32. K. V. Peddanarappagari and M. Brandt-Pearce, Volterra series approach for optimizing fiber-optic communications systems designs, *IEEE J. Lightwave Technol.*, Vol. 16, No. 11, pp. 2046–2055, 1998.

33. J. Pina, C. Xia, A. G. Strieger, and D. V. D. Borne, Nonlinear tolerance of polarization-multiplexed QPSK transmission over fiber links, in *ECOC2011*, Geneva, 2011.

34. L. N. Binh, Linear and nonlinear transfer functions of single mode fiber for optical transmission systems, *J. Opt. Soc. Am. A*, Vol. 26, No. 7, pp. 1564–1575, 2009.

35. J. P. Gordon and L. F. Mollenauer, Phase noise in photonic communications systems using linear amplifiers, *Opt. Lett.*, Vol. 15, pp. 1351–1353, 1990.

36. K.-P. Ho and J. M. Kahn, Electronic compensation technique to mitigate nonlinear phase noise, *IEEE J. Lightwave Technol.*, Vol. 22, No. 3, p. 779, 2004.

37. A. V. T. Cartaxo, B. Wedding, and W. Idler, New measurement technique of nonlinearity coefficient of optical fibre using fibre transfer function, in *Proceedings of the ECOC98*, Madrid, Spain, pp. 169–170, 1998.

38. A. V. T. Cartaxo, B. Wedding, and W. Idler, Influence of fiber nonlinearity on the phase noise to intensity noise conversion in fiber transmission: Theoretical and experimental analysis, *IEEE J. Lightwave Technol.*, Vol. 16, No. 7, p. 1187, 1998.

39. F. Ramos and J. Martí, Frequency transfer function of dispersive and nonlinear single-mode optical fibers in microwave optical systems, *IEEE Photon. Technol. Lett.*, Vol. 12, No. 5, p. 549, 2000.

40. G. Agrawal, *Nonlinear Fiber Optics*, 2nd ed., San Diego, CA: Academic Press, 1995.

41. L. N. Binh, *Guided Wave Photonics*, Boston: CRC Press, 2012.

42. T. M. Dunster, Bessel functions of purely imaginary order, with an application to second-order linear differential equations having a large parameter, *SIAM J. Math. Anal.*, Vol. 21, No. 4, pp. 995–1018, 1990.

43. A. F. Elrefaie and R. E. Wagner, Chromatic dispersion limitations for FSK and DPSK systems with direct detection receivers, *IEEE Photon. Technol. Lett.*, Vol. 3, No. 1, pp. 71–73, 1991.

44. A. F. Elrefaie, R. E. Wagner, D. A. Atlas, and A. D. Daut, Chromatic dispersion limitation in coherent lightwave systems, *IEEE J. Lightwave Technol.*, Vol. 6, No. 5, pp. 704–710, 1988.

3 Optical Transmitters

A directly modulated photonic transmitter can consist of a single or multiple lightwave sources that can be modulated directly by manipulating the driving current of the laser diode. Alternatively, the laser source can be turned on all the time, and its output lightwaves are externally modulated via an integrated optical modulator, which is then known as an externally modulated optical transmitter. This externally modulated transmitter preserves the linewidth of the laser and hence its coherence.

The idea of external modulation was first proposed by P.K. Tien in 1969 in an article on integrated optics and reviewed in 1977 [1].

This chapter presents the techniques for the modulation of lightwaves, externally—not directly manipulating the stimulated emission from inside the laser cavity—via the use of electro-optic and electroabsorption effects. Advanced modulation formats have attracted much attention for the enhancement of the transmission efficiency since the mid-1980s for coherent optical communications. Hence, the preservation of the narrow linewidth of the laser source is critical for operation bit rates in the range of several tens of gigabits per second. Thus, external modulation is essential.

Three typical types of optical modulators are presented in this chapter, including the lithium niobate ($LiNbO_3$) electro-optic modulators, the electroabsorption (EA) modulators, and polymeric integrated modulators. Their operating principles, device physical structures, device parameters, and their applications and driving condition for the generation of different modulation formats as well as their impact on system performance are discussed.

3.1 OPTICAL MODULATORS

The modulation of lightwaves via an external optical modulator can be classified into three types depending on the special effects that alter the lightwaves' property, especially the intensity or the phase of the lightwave carrier. In an external modulator, the intensity is usually manipulated by manipulating the phase of the carrier lightwaves guided in one path of an interferometer. The Mach–Zehnder interferometric structure is the most common type, especially the lithium niobate ($LiNbO_3$) type [2–6].

The EA modulator employs the Franz–Keldysh effect, which is observed as the lengthening of the wavelength of the absorption edge of a semiconductor medium under the influence of an electric field [7,8]. In quantum structures such as the multiquantum well structure, this effect is called the Stark effect, or the EA effect. The EA modulator can be integrated with a laser structure on the same integrated circuit chip. For the $LiNbO_3$ modulator, the device is externally connected to a laser source via an optical fiber.

The total insertion loss of a semiconductor intensity modulator is about 8–10 dB, including the fiber–waveguide coupling loss, which is rather high. However, this loss can be compensated by a semiconductor optical amplifier (SOA) that can be integrated on the same circuit. Compared with the $LiNbO_3$ type, its total insertion loss is about 3–4 dB, which could be affordable as the Er-doped fiber amplifier (EDFA) is now readily available.

The driving voltage for the EA modulator is usually lower than that required for $LiNbO_3$. However, the extension ratio is not as high as that of the $LiNbO_3$ type, which is about 25 dB as compared to 10 dB for the EA modulator. This feature is in contrast to the operating characteristics of the $LiNbO_3$ and EA modulators. Although the driving voltage for optical modulator is about 2–3 V for EA type and 5–7 V for $LiNbO_3$, the former type would be preferred for intensity or phase modulation formats owing to this extinction ratio, which offers a much lower "zero" noise level and hence a high-quality factor.

3.1.1 Phase Modulators

The phase modulator is a device that manipulates the "phase" of optical carrier signals under the influence of an electric field created by an applied voltage. When a voltage is not applied to the RF electrode, the number of periods of the lightwaves, n, exists in a certain path length. When voltage is applied to the RF electrode, one or a fraction of one period of the wave is added, which now means $(n + 1)$ waves exist in the same length. In this case, the phase has been changed by 2π and the half voltage of this is called the driving voltage. In the case of long-distance optical transmission, waveform is susceptible to degradation due to nonlinear effects such as self-phase modulation, and so on. A phase modulator can be used to alter the phase of the carrier to compensate for this degradation. The magnitude of the phase change depends on the change in the refractive index due to the electro-optic effect, which in turn depends on the orientation of the crystal axis with respect to the direction of the electric field established by the applied signal voltage.

An integrated optic phase modulator operates in a similar manner except that the lightwave carrier is guided through an optical waveguide created by diffusion or ion-exchange of impurity to increase the refractive index of the region. Such diffused waveguide structures can be fabricated in $LiNbO_3$, and rib type structures in semiconductor materials such as InGaAsP for 1550 nm wavelength operations. Two electrodes are deposited so that an electric field can be established across the waveguiding cross section so that refractive index changes owing to the electro-optic or EA effect, as shown in Figure 3.1. For ultra-fast operation, one of the electrodes is a traveling wave type or hot electrode and the other is a ground electrode. The traveling wave electrode must be terminated with a matching impedance at the end to avoid wave reflection. Usually, a quarter wavelength impedance is used to match the impedance of the traveling wave electrode to that of the 50 Ω transmission line.

A phasor representation of a phase-modulated lightwave can be circular rotation at a radial speed of ω_c. Thus, the vector with an angle φ represents the magnitude and phase of the lightwave.

3.1.2 Intensity Modulators

The basic structured litium niobate (LN) modulator comprises (1) two waveguides, (2) two Y-junctions, and (3) RF/DC traveling wave electrodes (Figure 3.2). The optical signals coming from the lightwave source are launched into the LN modulator through the polarization maintaining fiber and then equally split into two branches at the first Y-junction on the substrate. When no voltage is applied to the RF electrodes, the two signals are recombined constructively at the second Y-junction and coupled into a single output. In this case, output signals from the LN modulator are recognized as ONE. When a voltage is applied to the RF electrode, due to the electro-optic effects of the LN crystal substrate, the waveguide refractive index is changed, and hence the carrier phase in one arm is advanced, though it is retarded in the other arm. The two signals are then recombined destructively at the second Y-junction, transformed

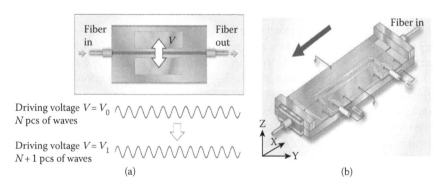

FIGURE 3.1 Electro-optic phase modulation in an integrated modulator using $LiNbO_3$. Electrode impedance matching is not shown. (a) Schematic diagram and (b) integrated optic structure.

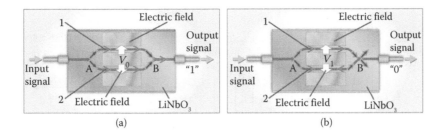

(a)   (b)

FIGURE 3.2 Intensity modulation using interferometric principles in guide wave structures in LiNbO$_3$. (a) ON: constructive interference mode and (b) destructive interference mode. OFF: Optical-guided wave paths 1 and 2. Electric field is established across the optical waveguide.

into a higher-order mode, and radiated at the junction. If the phase retarding is in multiple odd factors of π, the two signals are completely out of phase, the combined signals are radiated into the substrate, and the output signal from the LN modulator is recognized as a ZERO. The voltage difference that induces this ZERO and ONE is called the driving voltage of the modulator and is one of the important parameters in determining the modulator's performance.

3.1.2.1 Phasor Representation and Transfer Characteristics

Consider an interferometric intensity modulator consisting of an input waveguide, then split into two branches and recombined to a single output waveguide. If the two electrodes are initially biased with voltages V_{b1} and V_{b2}, then the initial phases exerted on the lightwaves would be $\varphi_1 = \pi V_{b1}/V_\pi = -\varphi_2$, which are indicated by the bias vectors shown in Figure 3.3b. From these positions, the phasors swing according to the magnitude and sign of the pulse voltages applied to the electrodes. They can be switched to the two positions, which can be constructive or destructive. The output field of the lightwave carrier can be represented by

$$E_0 = \frac{1}{2} E_{iRMS} e^{j\omega_c t} \left(e^{j\phi_1(t)} + e^{j\phi_2(t)} \right) \tag{3.1}$$

where ω_c is the carrier radial frequency, E_{iRMS} is the root mean square value of the magnitude of the carrier, and $\varphi_1(t)$ and $\varphi_2(t)$ are the temporal phases generated by the two time-dependent pulse sequences applied to the two electrodes. With the voltage levels varying according to the magnitude of the pulse sequence, one can obtain the transfer curve as shown in Figure 3.3a. This phasor representation can be used to determine exactly the biasing conditions and magnitude of the RF or digital signals required for driving the optical modulators to achieve 50%, 33%, or 67% bit period pulse shapes.

The power transfer function of Mach–Zehnder modulator is expressed as[*]

$$P_0(t) = \alpha P_i \cos^2 \frac{\pi V(t)}{V_\pi} \tag{3.2}$$

where $P_0(t)$ is the output transmitted power, α is the modulator total insertion loss, P_i is the input power (usually from the laser diode), $V(t)$ is the time-dependent signal applied voltage, and V_π is the driving voltage so that a π phase shift is exerted on the lightwave carrier. It is necessary to set the static bias on the transmission curve through a bias electrode. It is common practice to set the bias point at the 50% transmission point or a $\pi/2$ phase difference between the two optical waveguide branches, the quadrature bias point. As shown in Figure 3.3, the electrical digital signals can be transferred to the optical domain by a swinging voltage with a bias point V_B at quadrature $V_\pi/2$ and oscillating within a peak-to-peak magnitude of one V_π.

---

[*] Note that this equation represents a single-drive MZIM—it is the same for a dual-drive MZIM, provided that the bias voltages applied to the two electrodes are equal and opposite in sign. The transfer curve of the field representation would have half the periodic frequency of the transmission curve shown in Figure 3.3.

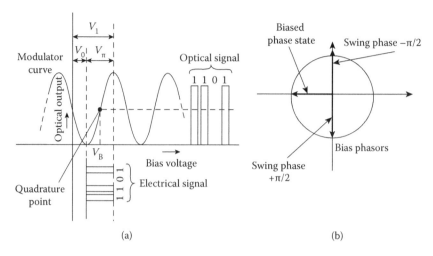

FIGURE 3.3 Electrical to optical transfer curve of an interferometric intensity modulator. (a) Optical power versus applied voltage-transfer characteristics and (b) phasor representations of biased state and the swinging phase states due to signal amplitude.

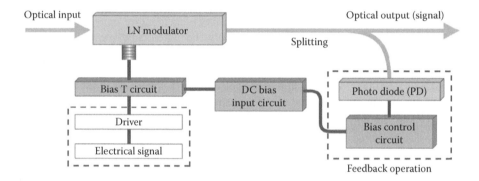

FIGURE 3.4 Arrangement of bias control of integrated optical modulators.

One factor that affects the modulator performance is the drift of the bias voltage. For the Mach–Zehnder interferometric modulator (MZIM) type, it is critical when it is required to bias at the quadrature point or at the minimum or maximum locations on the transfer curve. DC drift occurs in LiNbO$_3$ owing to the buildup of charges on the surface of the crystal substrate. Under this drift, the transmission curve gradually shifts in the long term [9,10]. In the case of the LiNbO$_3$ modulator, the bias point control is vital as the bias point will shift in the long term. To compensate for the drift, it is necessary to monitor the output signals and feed them back into the bias control circuits to adjust the DC voltage so that the operating points stay at the same point, as shown in Figure 3.4, for example, the quadrature point. It is the manufacturer's responsibility to reduce the DC drift so that the DC voltage does not exceed the limit throughout the lifetime of the device.

3.1.2.2 Chirp-Free Optical Modulators

Owing to the symmetry of the crystal refractive index of the uniaxial anisotropy of LiNbO$_3$ crystal structure as class 3 m, the crystal cut and the propagation-direction of the optical and the electric fields affecting both the modulation efficiency, the magnitude of the driving voltage and modulator chirp. The uniaxial property of LiNbO$_3$ is shown in Figure 3.5.

As shown in Figure 3.6, in the case of the Z-cut structure, as a hot electrode is placed on top of the waveguide, the RF field flux is more concentrated, and this improves the overlap between

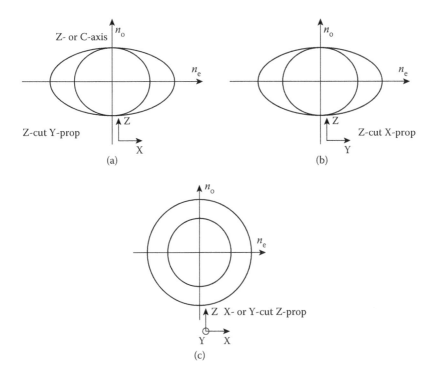

FIGURE 3.5 Refractive index contours of LiNbO₃ uniaxial crystal with Z- or C-denoting the principal axis. (a) Lightwave propagation in the Z-axis polarized along Y-cut LiNbO₃, (b) propagation-direction Z-axis and X-cut crystal, and (c) Y-prop and Y-cut crystal.

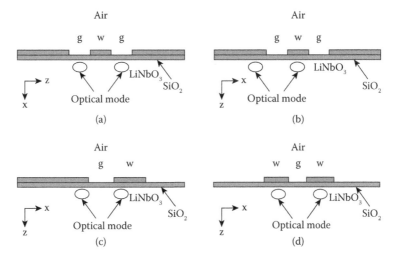

FIGURE 3.6 Commonly used electrode structure and crystal orientation for interferometric modulation to maximize the use of the overlap integral between the optical-guided mode and the electric field distribution for largest electro-optic coefficients. (a) x-cut CPW, (b) z-cut CPW, (c) z-cut ACPS, and (d) z-cut CPS.

the RF and the optical field. However, the overlap between the RF in the ground electrode and the waveguide is reduced in the Z-cut structure, so that the overall improvement in the driving voltage for a Z-cut structure compared to an X-cut structure is approximately 20%. The different overlapping area for the Z-cut structure results in a chirp parameter of 0.7, whereas X-cut and Z-propagation have almost zero chirp due to their symmetric structure. A number of commonly arranged electrode and waveguide structures are shown in Figure 3.6 to maximize the interaction between the traveling electric field and the optical-guided waves. Furthermore, a buffer layer, normally SiO_2, is used to match the velocities between these waves so as to optimize to optical modulation bandwidth.

3.1.3 Structures of Photonic Modulators

Figure 3.7a and b show the structure of an MZ intensity modulator using single and dual-electrode configurations, respectively. The thin-line electrode is called the "hot" electrode or traveling wave electrode. RF connectors are required for launching RF data signals to establish the electric field required for electro-optic effects. Impedance termination is also required. Optical fiber pigtails are also attached to the end faces of the diffused waveguide. The mode spot size of the diffused waveguide is not symmetric, and hence some diffusion parameters are controlled so that the coupling between the fiber and the diffused or rib waveguide can be maximized. Owing to this mismatching between the mode spot sizes of the circular and diffused optical waveguides, coupling loss occurs. Furthermore, the difference between the refractive indices of fiber and $LiNbO_3$ is quite substantial, and thus Fresnel reflection loss would also occur.

Figure 3.7c shows the structure of a polarization modulator, which is essential for multiplexing of two polarized data sequences so as to double the transmission capacity, for example, 40 Gbps to 80 Gbps. Furthermore, this type of polarization modulator can be used as a polarization rotator in a polarization dispersion-compensating subsystem [11,12]. Currently, these types of modulators with I–Q modulation have been reported [13].

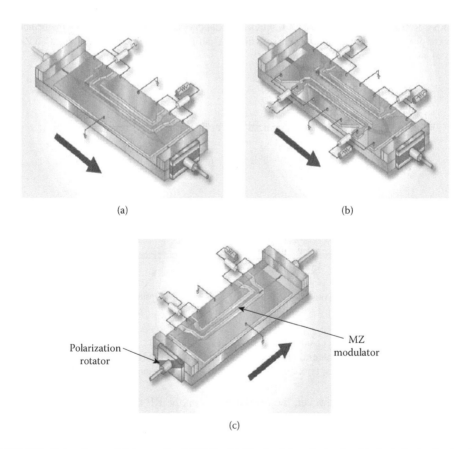

(a)

(b)

(c)

FIGURE 3.7 Intensity modulators using $LiNbO_3$. (a) Single-drive electrode, (b) dual-electrode structure, and (c) electro-optic polarization scrambler using $LiNbO_3$.

3.1.4 OPERATING PARAMETERS OF OPTICAL MODULATORS (TABLE 3.1)

TABLE 3.1
Typical Operational Parameters of Optical Intensity Modulators

Parameters	Typical Values	Definition/Comments
Modulation speed	10 Gbps	Capability to transmit digital signals
Insertion loss	Max 5 dB	Defined as the optical power loss within the modulator
Driving voltage	Max 4 V	The RF voltage required to have a full modulation
Optical bandwidth	Min 8 GHz	3dB roll-off in efficiency at the highest frequency in the modulated signal spectrum
ON/OFF extinction ratio	Min 20 dB	The ratio of maximum optical power (ON) and minimum optical power (OFF)
Polarization extinction ratio	Min 20 dB	The ratio of two polarization states (TM and TE guided modes) at the output

3.2 RETURN-TO-ZERO OPTICAL PULSES

3.2.1 GENERATION

Figure 3.8 shows the conventional structure of a return-to-zero amplitude-shift keying (RZ ASK) transmitter in which two external LiNbO$_3$ MZIMs can be used. The MZIM shown in this transmitter can be either a single- or dual-drive (push–pull) type. The operational principles of the MZIM were presented in Chapter 2, Section 2.2. The optical OOK transmitter would normally consist of a narrow-linewidth laser source to generate lightwaves whose wavelength satisfies the ITU grid standard.

The first MZIM, commonly known as the pulse carver, is used to generate the periodic pulse trains with a required return-to-zero (RZ) format. The suppression of the lightwave carrier can also be carried out at this stage if necessary, commonly known as the carrier-suppressed RZ (CSRZ). Compared to other RZ types, the CSRZ pulse shape is found to have attractive attributes for long-haul WDM transmissions, including the π phase difference of adjacent modulated bits, suppression of the optical carrier component in the optical spectrum, and narrower spectral width.

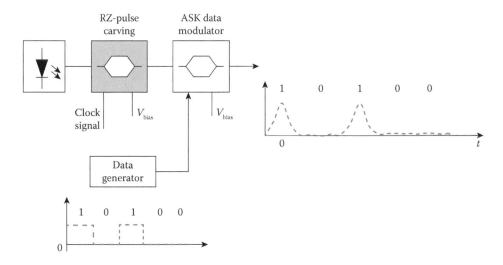

FIGURE 3.8 Conventional structure of an OOK optical transmitter utilizing two MZIMs.

Different types of RZ pulses can be generated depending on the driving amplitude of the RF voltage and the biasing schemes of the MZIM. The equations governing the RZ pulses electric field waveforms are

$$E(t) = \begin{cases} \sqrt{\dfrac{E_b}{T}} \sin\left[\dfrac{\pi}{2}\cos\left(\dfrac{\pi t}{T}\right)\right] & \text{67\% duty-ratio RZ pulses or CSRZ} \\[3mm] \sqrt{\dfrac{E_b}{T}} \sin\left[\dfrac{\pi}{2}\left(1+\sin\left(\dfrac{\pi t}{T}\right)\right)\right] & \text{33\% duty-ratio RZ pulses or RZ33} \end{cases} \tag{3.3}$$

where E_b is the pulse energy per transmitted bit, and T is a bit period.

The 33% duty-ratio RZ pulse is denoted as RZ33 pulse, whereas the 67% duty cycle RZ pulse is known as the carrier-suppressed return-to-zero (CSRZ) type. The art in generating these two RZ pulse types stays at the difference of their biasing points located on the transfer curve of an MZIM as given in Figure 3.3 and Figure 3.9.

Under the bias voltage conditions and the pulse shape of these two RZ types, the carrier suppression and nonsuppression of maximum carrier can be implemented with the biasing points at the minimum and maximum transmission points of the transmittance characteristics of the MZIM, respectively. The peak-to-peak amplitude of the RF driving voltage is $2V_\pi$, where V_π is the required driving voltage to obtain a π phase shift of the lightwave carrier. Another important point is that the RF signal is operating at only half the transmission bit rate. Hence, pulse carving is actually implementing the frequency doubling. The generations of RZ33 and CSRZ pulse trains are demonstrated in Figure 3.9a and b.

The pulse carver can also utilize a dual-drive MZIM, which is driven by two complementary sinusoidal RF signals. This pulse carver is biased at $-V_{\pi/2}$ and $+V_{\pi/2}$ with the peak-to-peak amplitude of $V_{\pi/2}$. Thus, a π phase shift is created between the states 1 and 0 of the pulse sequence, and hence the RZ with alternating phases 0 and π. If carrier suppression is required, then the two electrodes are supplied with voltages V_π and a swing voltage amplitude of V_π.

Although RZ modulation offers improved performance, RZ optical systems usually require more complex transmitters than those in the non-return-to-zero (NRZ) ones. Compared to only one stage for modulating data on the NRZ optical signals, two modulation stages are required for generation of RZ optical pulses [14–17].

3.2.2 Phasor Representation

Recalling Equation 3.1, we have

$$E_o = \frac{E_i}{2}\left[e^{j\varphi_1(t)} + e^{j\varphi_2(t)}\right] = \frac{E_i}{2}\left[e^{j\pi v_1(t)/V_\pi} + e^{j\pi v_2(t)/V_\pi}\right] \tag{3.4}$$

It can be seen that the modulating process for the generation of RZ pulses can be represented by a phasor diagram as shown in Figure 3.10, where the optical frequency component $e^{j\omega t}$ has been removed to indicate that the wave vector is rotating at this angular frequency and is considered to be stationary. This technique gives a clear understanding of the superposition of the fields at the coupling output of two arms of the MZIM. Here, a dual-drive MZIM is used, that is, the data driving signals [$V_1(t)$] and inverse data [$\overline{\text{data}}$: $V_2(t) = -V_1(t)$] are applied into each arm of the MZIM, respectively, and the RF voltages swing in inverse directions. Applying the phasor representation, vector addition, and simple trigonometric calculus, the process of generation of RZ33 and CSRZ is explained in detail and verified.

The width of these pulses are commonly measured at the position of full-width half maximum (FWHM). Note that the measured pulses are intensity pulses, whereas we are considering the

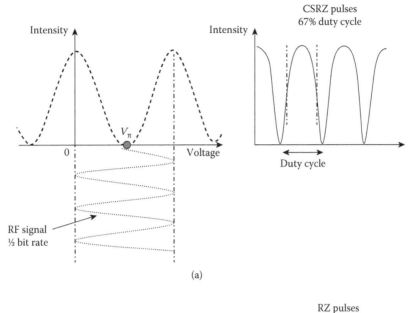

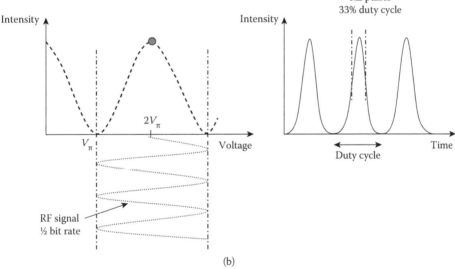

FIGURE 3.9 Bias point and RF driving signals for generation of (a) CSRZ and (b) RZ33 pulses.

addition of the fields in the MZIM. Thus, the normalized E_o field vector has a value of $\pm 1/\sqrt{2}$ at the FWHM intensity pulse positions, and the time interval between these points gives the FWHM values.

3.2.2.1 Phasor Representation of CSRZ Pulses

Key parameters including V_{bias} and the amplitude of the RF driving signal are shown in Figure 3.11a. Accordingly, its initialized phasor representation is shown in Figure 3.11b.

The values of the key parameters are outlined as follows: (1) V_{bias} is $\pm V_\pi/2$; (2) the swing voltage of the driving RF signal on each arm has an amplitude of $V_\pi/2$ (i.e., $V_{p-p} = V_\pi$); (3) the RF signal operates at half the bit rate ($B_R/2$); (4) at the FWHM position of the optical pulse, $E_{out} = \pm 1/\sqrt{2}$ and the component vectors V_1 and V_2 form with the vertical axis a phase of $\pi/4$, as shown in Figure 3.12.

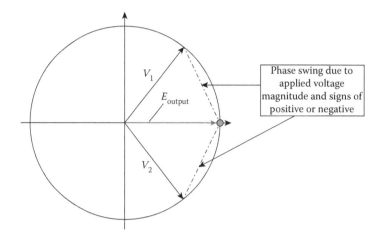

FIGURE 3.10 Phasor representation for generation of output field in dual-drive MZIM.

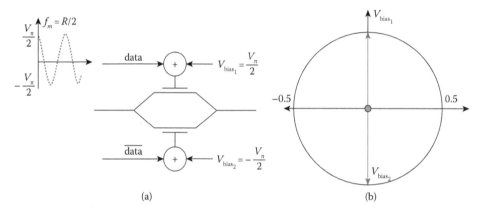

(a)   (b)

FIGURE 3.11 Initial stage for generation of CSRZ pulse: (a) RF driving signal and the bias voltages and (b) initial phasor representation.

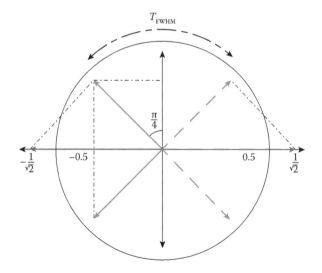

FIGURE 3.12 Phasor representation of CSRZ pulse generation using dual-drive MZIM.

Considering the scenario for the generation of a 40 Gbps CSRZ optical signal, the modulating frequency is f_m ($f_m = 20$ GHz $= B_R/2$). At the FWHM positions of the optical pulse, the phase is given by the following expressions

$$\frac{\pi}{2}\sin(2\pi f_m) = \frac{\pi}{4} \Rightarrow \sin 2\pi f_m = \frac{1}{2} \Rightarrow 2\pi f_m = \left(\frac{\pi}{6}, \frac{5\pi}{6}\right) + 2n\pi \tag{3.5}$$

Thus, the calculation of T_{FWHM} can be carried out, and hence the duty cycle of the RZ optical pulse can be obtained as given in the following expressions

$$T_{FWHM} = \left(\frac{5\pi}{6} - \frac{\pi}{6}\right)\frac{1}{R2\pi} = \frac{1}{3}\pi \times \frac{1}{R} \Rightarrow \frac{T_{FWHM}}{T_{BIT}} = \frac{1.66 \times 10^{-4}}{2.5 \times 10^{-11}} = 66.67\% \tag{3.6}$$

The result obtained in Equation 3.6 clearly verifies the generation of CSRZ optical pulses from the phasor representation.

3.2.2.2 Phasor Representation of RZ33 Pulses

Key parameters, including V_{bias}, the amplitude of the driving voltage, and its corresponding initialized phasor representation are shown in Figure 3.13a and b, respectively.

The values of the key parameters are (1) V_{bias} is V_π for both arms; (2) swing voltage of the driving RF signal on each arm has an amplitude of $V_\pi/2$ (i.e., $V_{p-p} = V_\pi$); (3) the RF signal operates at half the bit rate ($B_R/2$).

At the FWHM position of the optical pulse, $E_{output} = \pm 1/\sqrt{2}$ and the component vectors V_1 and V_2 form a phase of $\pi/4$ with the horizontal axis, as shown in Figure 3.14.

Considering the scenario for the generation of 40 Gbps CSRZ optical signals, the modulating frequency is f_m ($f_m = 20$ GHz $= B_R/2$). At the FWHM positions of the optical pulse, the phase is given by the following expressions

$$\frac{\pi}{2}\cos(2\pi f_m t) = \frac{\pi}{4} \Rightarrow t_1 = \frac{1}{6f_m} \tag{3.7}$$

$$\frac{\pi}{2}\cos(2\pi f_m t) = -\frac{\pi}{4} \Rightarrow t_2 = \frac{1}{3f_m} \tag{3.8}$$

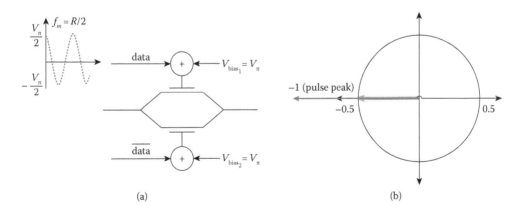

(a)   (b)

FIGURE 3.13 Initialized stage for generation of RZ33 pulse: (a) RF driving signal and the bias voltage, and (b) initial phasor representation.

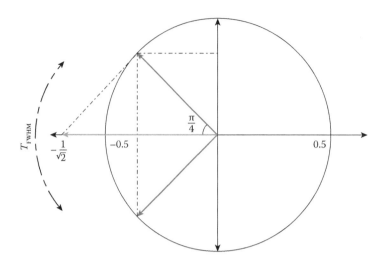

FIGURE 3.14 Phasor representation of RZ33 pulse generation using dual-drive MZIM.

Thus, the calculation of T_{FWHM} can be carried out, and hence the duty cycle of the RZ optical pulse can be obtained as given in the following expressions

$$T_{\text{FWHM}} = \frac{1}{3f_m} - \frac{1}{6f_m} = \frac{1}{6f_m} \quad \therefore \quad \frac{T_{\text{FWHM}}}{T_b} = \frac{1/6f_m}{1/2f_m} = 33\% \tag{3.9}$$

The result obtained in Equation 3.9 clearly verifies the generation of RZ33 optical pulses from the phasor representation.

3.3 DIFFERENTIAL PHASE SHIFT KEYING

3.3.1 BACKGROUND

Digital encoding of data information by modulating the phase of the lightwave carrier is referred to as optical phase shift keying (PSK). In the early days, optical PSK was studied extensively for coherent photonic transmission systems. This technique requires the manipulation of the absolute phase of the lightwave carrier. Thus, precise alignment of the transmitter and demodulator center frequencies for the coherent detection is required. These coherent optical PSK systems face severe obstacles such as broad linewidth and chirping problems of the laser source. Meanwhile, the differential phase shift keying (DPSK) scheme overcomes those problems, because the DPSK optically modulated signals can be detected incoherently. This technique only requires the coherence of the lightwave carriers over one-bit period for the comparison of the differentially coded phases of the consecutive optical pulses.

A binary 1 is encoded if the present input bit and the past encoded bit have opposite logic, whereas a binary 0 is encoded if the logic is similar. This operation is equivalent to an XOR logic operation. Hence, an XOR gate is employed as a differential encoder. NOR can also be used to replace the XOR operation in differential encoding as shown in Figure 3.15a. In DPSK, the electrical data 1 indicates a π phase change between the consecutive data bits in the optical carrier, whereas binary 0 is encoded if there is no phase change between the consecutive data bits. Hence, this encoding scheme gives rise to two points located exactly at π phase difference with respect to each other in the signal constellation diagram. For continuous PSK such as the minimum shift keying, the phase evolves continuously over a quarter of the section, and thus there is a phase change of $\pi/2$ between one phase state to the other. This is indicated by the inner bold circle shown in Figure 3.15b.

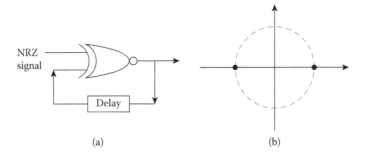

(a)   (b)

FIGURE 3.15 (a) DPSK precoder, (b) signal constellation diagram of DPSK. (From P. K. Tien, *Rev. Mod. Phys.*, Vol. 49, pp. 361–420, 1977.)

3.3.2 OPTICAL DPSK TRANSMITTER

Figure 3.16 shows the structure of a 40 Gbps DPSK transmitter in which two external LiNbO$_3$ MZIMs are used. The operational principles of an MZIM were presented earlier. The MZIMs shown in Figure 3.16 can be either of single or dual-drive type. The optical DPSK transmitter also consists of a narrow-linewidth laser to generate a lightwave whose wavelength conforms to the ITU grid.

The RZ optical pulses are then fed into the second MZIM, through which the RZ pulses are modulated by the precoded binary data to generate RZ-DPSK optical signals. Electrical data pulses are differentially precoded in a precoder using the XOR coding scheme. Without a pulse carver, the structure shown in Figure 3.16 is an optical NRZ-DPSK transmitter. In data modulation for the DPSK format, the second MZIM is biased at the minimum transmission point. The precoded electrical data have a peak-to-peak amplitude of $2V_\pi$ and operates at the transmission bit rate. The modulation principles for the generation of optical DPSK signals are shown in Figure 3.17.

The electro-optic phase modulator can also be used for the generation of DPSK signals instead of MZIM. Using an optical phase modulator, the transmitted optical signal is chirped, whereas using MZIM, especially the X-cut type with Z-propagation, chirp-free signals can be produced. However, in practice, a small amount of chirp might be useful for transmission [18].

3.4 GENERATION OF MODULATION FORMATS

Modulation is the process facilitating the transfer of information over a medium, for example, a wireless or optical environment. Three basic types of modulation techniques are based on the manipulation of a parameter of the optical carrier to represent the information in the digital data. These are amplitude shift keying (ASK), phase shift keying (PSK), and frequency shift keying (FSK). In addition to the manipulation of the carrier, the occupation of the data pulse over a single period would also determine the amount of energy and the speed of the system required for transmission. The pulse can remain constant over a bit period or return to the zero level within a portion of the period. These formats would be named non-return-to-zero (NRZ) or return-to-zero (RZ). They are combined with the modulation of the carrier to form various modulation formats, which are presented in this section. Figure 3.18 shows the baseband signals of the NRZ and RZ formats, and its corresponding block diagram of a photonic transmitter.

3.4.1 AMPLITUDE–MODULATION ASK-NRZ AND ASK-RZ

3.4.1.1 Amplitude–Modulation OOK-RZ Formats

There are a number of advanced formats used in advanced optical communications, based on the intensity of the pulse, that include NRZ, RZ, and duobinary. These ASK formats can also be

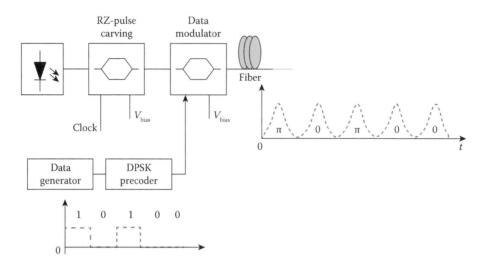

FIGURE 3.16 DPSK optical transmitter with RZ pulse carver.

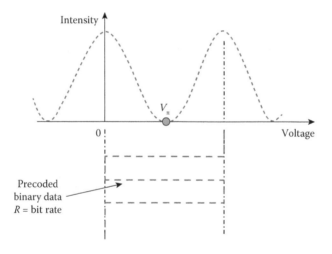

FIGURE 3.17 Bias point and RF driving signals for generation of optical DPSK format.

integrated with phase modulation to generate discrete or continuous phase NRZ or RZ formats. Currently, the majority of 10 Gbps installed optical communications system have been developed with NRZ owing to its simple transmitter design and bandwidth-efficient characteristic. However, RZ format is more robust to fiber nonlinearity and polarization-mode dispersion. In this section, the RZ pulse is generated by an MZIM commonly known as a *pulse carver* as arranged in Figure 3.19.

There are a number of variations in the RZ format based on the biasing point in the transmission curve shown in Table 3.2. The phasor representation of the biasing and driving signals can be observed in Table 3.2.

CSRZ has been found to have more attractive attributes in long-haul WDM transmission compared to the conventional RZ format, because of the possibility of reducing the upper level of the power contained in the carrier that serves no purpose in the transmission but only increases the total energy level, and so the nonlinear threshold level is approached faster. The CSRZ pulse has an optical phase difference of π in adjacent bits, removing the optical carrier component in the optical spectrum and reducing the spectral width. This offers an advantage in compact WDM channel spacing.

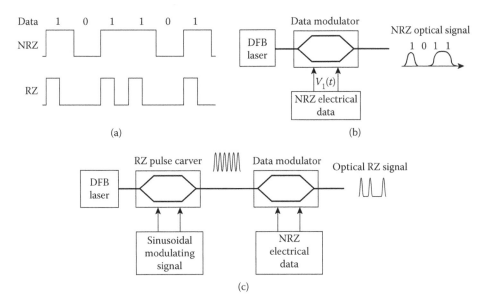

FIGURE 3.18 (a) Baseband NRZ and RZ line coding for 101101 data sequence, (b) block diagram of NRZ photonics transmitter, and (c) RZ photonics transmitter incorporating a pulse carver.

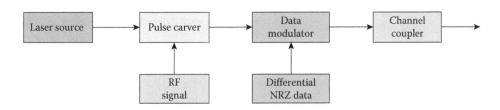

FIGURE 3.19 Block diagrams of RZ-DPSK transmitter.

3.4.1.2 Amplitude–Modulation CSRZ Formats

The suppression of the carrier can be implemented by biasing the MZ interferometer in such a way that there is a π phase shift between the two arms of the interferometer. The magnitude of the sinusoidal signals applied to an arm or both arms would determine the width of the optical output pulse sequence. The driving conditions and phasor representation are shown in Table 3.2.

3.4.2 Discrete Phase Modulation NRZ Formats

The term *discrete phase modulation* refers to DPSK, whether DPSK or differential quadrature phase shift keying (DQPSK), to indicate the states of the phases of the lightwave carrier are switched from one distinct location on the phasor diagram to the other state, for example, from 0 to π or $+\pi/2$ to $-\pi/2$ for binary PSK (BPSK), or even more evenly spaced PSK levels as in the case of M-ary PSK.

3.4.2.1 Differential Phase Shift Keying

Information encoded in the phase of an optical carrier is commonly referred to as optical PSK. In the early days, PSK required precise alignment of the transmitter and demodulator center frequencies [14]. Hence, the PSK system was not widely deployed. When the DPSK scheme was introduced, coherent detection is not critical because DPSK detection only requires source coherence over one-bit period by a comparison of two consecutive pulses.

TABLE 3.2

Summary of RZ Format Generation and Characteristics of Single-Drive MZIM Based on Biasing Point, Drive Signal Amplitude, and Frequency

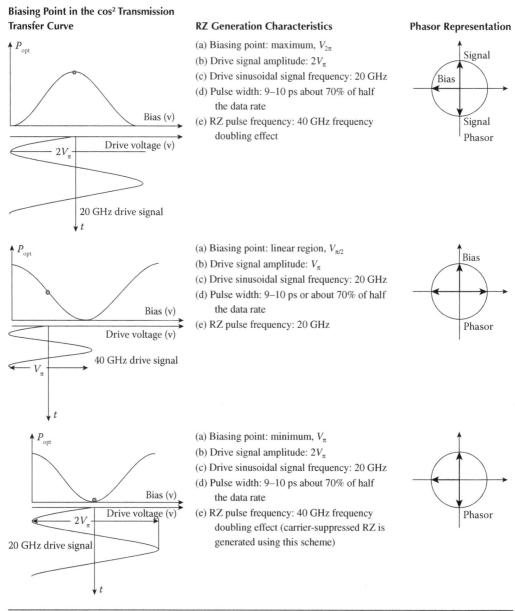

Biasing Point in the cos² Transmission Transfer Curve	RZ Generation Characteristics	Phasor Representation
	(a) Biasing point: maximum, $V_{2\pi}$	
	(b) Drive signal amplitude: $2V_\pi$	
	(c) Drive sinusoidal signal frequency: 20 GHz	
	(d) Pulse width: 9–10 ps about 70% of half the data rate	
	(e) RZ pulse frequency: 40 GHz frequency doubling effect	
	(a) Biasing point: linear region, $V_{\pi/2}$	
	(b) Drive signal amplitude: V_π	
	(c) Drive sinusoidal signal frequency: 20 GHz	
	(d) Pulse width: 9–10 ps or about 70% of half the data rate	
	(e) RZ pulse frequency: 20 GHz	
	(a) Biasing point: minimum, V_π	
	(b) Drive signal amplitude: $2V_\pi$	
	(c) Drive sinusoidal signal frequency: 20 GHz	
	(d) Pulse width: 9–10 ps about 70% of half the data rate	
	(e) RZ pulse frequency: 40 GHz frequency doubling effect (carrier-suppressed RZ is generated using this scheme)	

A binary 1 is encoded if the present input bit and the past encoded bit are of opposite logic, and a binary 0 is encoded if the logic is similar. This operation is equivalent to the XOR logic operation. Hence, an XOR gate is usually employed in the differential encoder. NOR can also be used to replace the XOR operation in differential encoding, as shown in Figure 3.20.

In optical application, electrical data 1 is represented by a π phase change between the consecutive data bits in the optical carrier, while state 0 is encoded with no phase change between the consecutive data bits. Hence, this encoding scheme gives rise to two points located exactly at

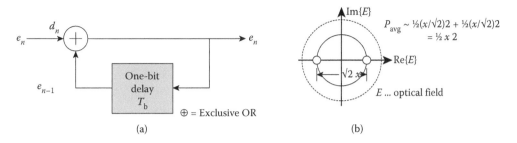

FIGURE 3.20 (a) The encoded differential data are generated by $e_n = d_n \oplus e_{n-1}$ and (b) signal constellation diagram of DPSK.

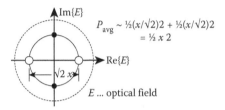

FIGURE 3.21 Signal constellation diagram of DQPSK. Two bold dots are orthogonal to the DPSK constellation.

a π phase difference with respect to each other in the signal constellation diagram, as indicated in Figure 3.20b.

An RZ-DPSK transmitter consists of an optical source, pulse carver, data modulator, differential data encoder, and a channel coupler. The channel coupler block is not included in the simulation model because no coupling losses are assumed when an optical RZ-DPSK modulated signal is launched directly into the optical fiber transmission line. The modulation scheme can combine the functionalities of the dual-drive MZI modulator to generate pulse carving and phase modulation as described in the Subsection 3.4.1.

The pulse carver, usually an MZ interferometric intensity modulator, is driven by a sinusoidal RF signal for a single-drive MZIM and two complementary electrical RF signals for a dual-drive MZIM to carve pulses out from the optical carrier signal forming the RZ pulses. These optical RZ pulses are fed into the second MZ intensity modulator, where the RZ pulses are modulated by differential NRZ electrical data to generate RZ-DPSK. This data phase modulation can be performed using a straight line phase modulator, but the MZ waveguide structure has several advantages over the phase modulator owing to its chirpless property. Electrical data pulses are differentially precoded in a differential precoder as shown in Figure 3.20a. Without a pulse carver and a sinusoidal RF signal, the output pulse sequence follows the NRZ-DPSK format; that is, the pulse would occupy 100% of the pulse period, and there is no transition between the consecutive 1s.

3.4.2.2 Differential Quadrature Phase Shift Keying

This differential coding is similar to DPSK except that each symbol consists of two bits, which are represented by the two orthogonal axial discrete phase at $(0,\pi)$ and $(-\pi/2,+\pi/2)$ as shown in Figure 3.21, or two additional orthogonal phase positions are located on the imaginary axis of Figure 3.20b but in opposite sides of the origin.

Figure 3.22 shows the block diagram of a typical NRZ-DPSK transmitter. The differential precoder of electrical data is implemented using the logic explained in the previous section. In phase modulating the optical carrier, the MZ modulator known as the data phase modulator is biased at the minimum point and driven by a data swing of $2V_\pi$. The modulator showed excellent behavior: the phase of the optical carrier will be altered by π exactly when the signal transits the minimum point of the transfer characteristic.

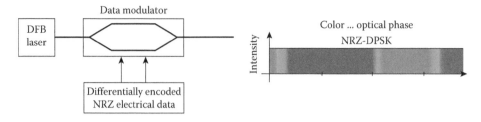

FIGURE 3.22   Block diagram of NRZ-DPSK photonics transmitter.

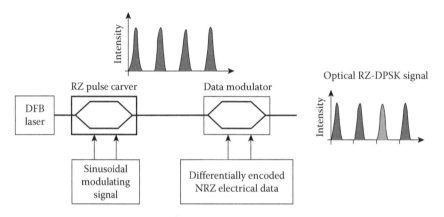

FIGURE 3.23   Block diagram of RZ-DPSK photonics transmitter.

The arrangement of the RZ-DPSK transmitter is essentially similar to the RZ-ASK, as shown in Figure 3.23, with the data intensity modulator replaced by the data phase modulator. The difference between them is the biasing point and the electrical voltage swing. Different RZ formats can also be generated.

3.4.2.3   Generation of M-ary Amplitude Differential Phase Shift Keying Using One MZIM

As an example, a 16-ary M-ary amplitude differential phase shift keying (M-ary ADPSK) signal can be represented by the constellation shown in Figure 3.24. It is, indeed, a combination of a 4-ary ASK and a DQPSK scheme. At the transmitting end, each group of four bits $[D_3D_2D_1D_0]$ of user data are encoded into a symbol; among them, the two least significant bits $[D_1D_0]$ are encoded into four phase states $[0, \pi/2, \pi, 3\pi/2]$, and the other two most significant bits, $[D_3D_2]$, are encoded into four amplitude levels. At the receiving end, as MZ delay interferometers (DIs) are used for phase comparison and detection, a diagonal arrangement of the signal constellation shown in Figure 3.24a is preferred. This simplifies the design of transmitter and receiver and minimizes the number of phase detectors, leading to high receiver sensitivity.

In order to balance the bit error rate (BER) between the ASK and DQPSK components, the signal levels corresponding to the four circles of the signal space should be adjusted to a reasonable ratio that depends on the noise power at the receiver. As an example, if this ratio is set to $[I_0/I_1/I_2/I_3] =$ [1/1.4/2/2.5], where I_0, I_1, I_2, and I_3 are the intensities of the optical signals corresponding to circle 0, circle 1, circle 2, and circle 3, respectively, then by selecting E_i equal to the amplitude of the circle 3 and V_π equal to 1, the driving voltages should have the values given in Table 3.3. Conversely, one can set the outermost level such that its peak value is below the nonlinear SPM threshold, and the voltage level of the outermost circle would be determined. The innermost circle is limited to the condition that the largest signal-to-noise ratio (SNR) should be achieved. This is related to the optical SNR

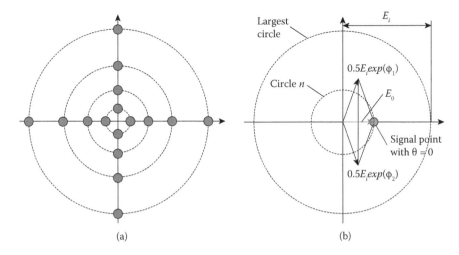

(a)   (b)

FIGURE 3.24 (a) Signal constellation of 4-ary ADPSK format and (b) phasor representation of a point on the constellation point for driving voltages applied to dual-dive MZIM.

TABLE 3.3
Driving Voltages for 16-Ary MADPSK Signal Constellation

	Circle 0			Circle 1			Circle 2			Circle 3	
Phase	$V_1(t)$ volt	$V_2(t)$ volt	Phase	$V_1(t)$ volt	$V_2(t)$ volt	Phase	$V_1(t)$ volt	$V_2(t)$ volt	Phase	$V_1(t)$ volt	$V_2(t)$ volt
0	0.38	−0.38	0	0.30	−0.30	0	0.21	−0.21	0	0.0	0.0
$\pi/2$	0.88	0.12	$\pi/2$	0.80	0.20	$\pi/2$	0.71	0.29	$\pi/2$	0.5	0.5
π	−0.62	0.62	π	−0.7	0.70	π	−0.79	0.79	π	−1.0	1.0
$3\pi/2$	−0.12	−0.88	$3\pi/2$	−0.20	−0.8	$3\pi/2$	−0.29	−0.71	$3\pi/2$	−0.5	−0.5

required for a certain BER. Thus, from the largest amplitude level and the smallest amplitude level, we can design the other points of the constellation.

Furthermore, to minimize the effect of intersymbol interference, the 66%-RZ and 50%-RZ pulse formats are also used as alternatives to the traditional NRZ counterpart. These RZ pulse formats can be created by a pulse carver that precedes or follows the dual-drive MZIM modulator. Mathematically, the waveforms of the NRZ and RZ pulses can be represented by the following equations, where E_{on}, $n = 0, 1, 2, 3$ are the peak amplitudes of the signals in circle 0, circle 1, circle 2, and circle 3 of the constellation, respectively

$$p(t) = \begin{cases} E_{on} & \text{for NRZ} \\ E_{on}\cos\left(\dfrac{\pi}{2}\cos^2\left(\dfrac{1.5\pi t}{T_s}\right)\right) & \text{for 66\%-RZ} \\ E_{on}\cos\left(\dfrac{\pi}{2}\cos^2\left(\dfrac{2\pi t}{T_s}\right)\right) & \text{for 50\%-RZ} \end{cases} \quad (3.10)$$

A typical arrangement of the signals of the precoder and driving signals for the MZIM is shown in Figure 3.25.

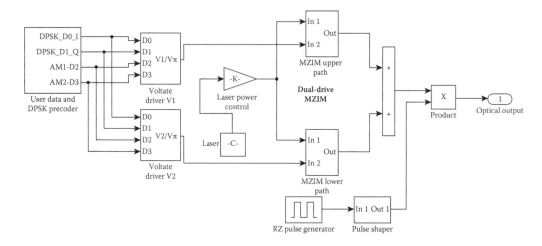

FIGURE 3.25 MATLAB® and Simulink® model of an MADPSK photonic transmitter. The MZIM is represented by two phase shifter blocks.

3.4.3 CONTINUOUS PHASE MODULATION PM-NRZ FORMATS

In the previous section, the optical transmitters for discrete PSK modulation formats have been described. Obviously, the phase of the carrier has been used to indicate the digital states of the bits or symbols. These phases are allocated in a noncontinuous manner around a circle corresponding to the magnitude of the wave. Alternatively, the phase of the carrier can be continuously modulated and the total phase changes at the transition instants, usually at the end of the bit period, would be the same as those for discrete cases. Because the phase of the carrier is continuously varied during the bit period, this can be considered as an FSK modulation technique, except that the transition of the phase at the end of one bit and the beginning of the next bit would be continuous. The continuity of the carrier phase at these transitions would reduce the signal bandwidth and hence make it more tolerable to dispersion effects and higher energy concentration for effective transmission over the optical-guided medium. One of the examples of a reduction in the phase at the transition is the offset DQPSK, which is a minor but important variation of the QPSK or DQPSK. In OQPSK, the Q channel is shifted by half a symbol period so that the transition instants of I and Q channel signals do not occur at the same time. The result of this simple change is that the phase shifts at any one time are limited. and hence the offset QPSK has more constant envelope that the normal QPSK.

The enhancement of the efficiency of the bandwidth of the signals can be further improved if the phase changes at these transitions are continuous. In this case, the change of the phase during the symbol period is continuously changed by using a half-cycle sinusoidal driving signal with the total phase transition over a symbol period being a fraction of π, depending on the levels of this PSK modulation. If the change is $\pi/2$, then we have a minimum shift keying (MSK) scheme. The orthogonality of the I and Q channels will also further reduce the bandwidth of the carrier-modulated signals. In this section, we describe the basic principles of optical MSK and the photonic transmitters for these modulation formats. Ideally, the driving signal to the phase modulator should be a triangular wave so that a linear phase variation of the carrier in the symbol period is linear. However, when a sinusoidal function is used then some nonlinear distortion would happen, especially near the ends of the sine wave period; we thus term this type of MSK a nonlinear MSK format. This nonlinearity contributes to some penalty in the optical SNR (OSNR), which will be explained in a later chapter. Furthermore, the MSK as a special form of differential quadrature phase shift keying is also described for optical systems.

3.4.3.1   Linear and Nonlinear MSK

MSK is a special form of continuous phase FSK (CPFSK) signal in which the two frequencies are spaced in such way that they are orthogonal and hence have a minimum spacing between them, defined by

$$s(t) = \sqrt{\frac{2E_b}{T_b}} \cos\left[2\pi f_1 t + \theta(0)\right] \quad \text{for symbol 1} \tag{3.11}$$

$$s(t) = \sqrt{\frac{2E_b}{T_b}} \cos\left[2\pi f_2 t + \theta(0)\right] \quad \text{for symbol 0} \tag{3.12}$$

As shown by the foregoing equations, the signal frequency change corresponds to a higher frequency for data 1 and lower frequency for data 0. Both frequencies, f_1 and f_2, are defined by

$$f_1 = f_c + \frac{1}{4T_b} \tag{3.13}$$

$$f_2 = f_c - \frac{1}{4T_b} \tag{3.14}$$

Depending on the binary data, the phase of the signal changes; data 1 increases the phase by $\pi/2$, while data 0 decreases the phase by $\pi/2$. The variation in the phase follows paths of sequences of straight lines in phase trellis (Figure 3.26) [19], in which the slopes represent frequency changes. The change in carrier frequency from data 0 to data 1, or vice versa, is equal to half the bit rate of the incoming data [6]. This is the minimum frequency spacing that allows the two FSK signals representing symbols 1 and 0 to be coherently orthogonal, in the sense that they do not interfere with one another in the process of detection.

An MSK signal consists of both I and Q components, which can be written as

$$s(t) = \sqrt{\frac{2E_b}{T_b}} \cos\left[\theta(t)\right] \cos\left[2\pi f_c t\right] - \sqrt{\frac{2E_b}{T_b}} \sin\left[\theta(t)\right] \sin\left[2\pi f_c t\right] \tag{3.15}$$

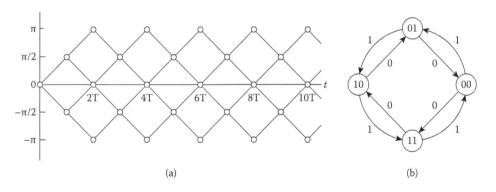

(a)   (b)

FIGURE 3.26   (a) Phase trellis for MSK and (b) state diagram for MSK. (From K. K. Pang, *Digital Transmission*, Melbourne: Mi-Tec Publishing, 2002, p. 58.)

The in-phase component consists of a half-cycle cosine pulse defined by

$$s_I(t) = \pm\sqrt{\frac{2E_b}{T_b}}\cos\left(\frac{\pi t}{2T_b}\right), \quad -T_b \le t \le T_b \tag{3.16}$$

while the quadrature component would take the form

$$s_Q(t) = \pm\sqrt{\frac{2E_b}{T_b}}\sin\left(\frac{\pi t}{2T_b}\right), \quad 0 \le t \le 2T_b \tag{3.17}$$

Refer to Figure 3.26 we can see that, during the even bit intervals, the I component consists of a positive cosine waveform for a phase of 0, and a negative cosine waveform for a phase of π; during the odd bit intervals, the Q component consists of a positive sine waveform for a phase of $\pi/2$, and a negative sine waveform for a phase of $-\pi/2$ (as shown in Figure 3.26). Any of the four states can arise: 0, $\pi/2$, $-\pi/2$, and π. However, only state 0 or π can occur during any even bit interval and only $\pi/2$ or $-\pi/2$ can occur during any odd bit interval. The transmitted signal is the sum of the I and Q components, and its phase is continuous with time.

Two important characteristics of MSK are that (1) each data bit is held for a two-bit period, meaning the symbol period is equal to two-bit periods (h = $^1/_2$) and (2) the I and Q components are interleaved. The I and Q components are delayed by one-bit period with respect to each other. Therefore, only the I or the Q component can change at a time (when one is at the zero-crossing, the other is at its maximum peak). The precoder can be a combinational logic, as shown in Figure 3.27.

A truth table can be constructed based on the logic state diagram and the foregoing combinational logic diagram (Table 3.4). For a positive half-cycle cosine wave and a positive half-cycle sine wave, the output is 1; for a negative half-cycle cosine wave and a negative half-cycle sine wave, the output is 0. Then, a K-map can be constructed to derive the logic gates of the precoder, based on the truth table. The following three precoding logic equations are derived

$$S_0 = \overline{b_n}\,\overline{S_0'}\,\overline{S_1'} + b_n\,\overline{S_0'}S_1' + \overline{b_n}S_0'S_1' + b_nS_0'\,\overline{S_1'} \tag{3.18}$$

$$S_1 = \overline{S_1'} = \overline{b_n}\,\overline{S_1'} + b_n\,\overline{S_1'} \tag{3.19}$$

$$\text{Output} = \overline{S_0} \tag{3.20}$$

The resultant logic gates' construction for the precoder is as shown in Figure 3.27.

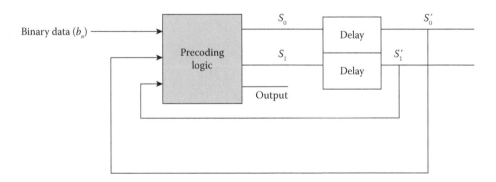

FIGURE 3.27 Combinational logic, the basis of the logic for constructing the precoder.

TABLE 3.4
Truth Table Based on MSK State Diagram

$b_n S_0' S_1'$	$S_0' S_1'$	Output
100	01	1
001	00	1
100	01	1
101	10	0
010	01	1
101	10	0
110	11	0
111	00	1
000	11	0
011	10	0

3.4.3.2 MSK as a Special Case of CPFSK

CPFSK signals are modulated in the upper and lower sideband frequency carriers f_1 and f_2 as expressed in Equations 3.18 and 3.19.

$$s(t) = \sqrt{\frac{2E_b}{T_b}} \cos\left[2\pi f_1 t + \theta(0)\right] \qquad \text{for '1'} \tag{3.21}$$

$$s(t) = \sqrt{\frac{2E_b}{T_b}} \cos\left[2\pi f_2 t + \theta(0)\right] \qquad \text{for '0'} \tag{3.22}$$

where $f_1 = f_c + \dfrac{1}{4T_b}$ and $f_2 = f_c - \dfrac{1}{4T_b}$ with T_b denoting the bit period.

The phase slope of the lightwave carrier changes linearly or nonlinearly with the modulating binary data. In the case of linear MSK, the carrier phase linearly changes by $\pi/2$ at the end of the bit slot with data 1, while it linearly decreases by $\pi/2$ with data 0. The variation of phase follows paths of well-defined phase trellis in which the slopes represent frequency changes. The change in the carrier frequency from data 0 to data 1, or vice versa, equals half the bit rate of incoming data [13]. This is the minimum frequency spacing that allows the two FSK signals representing symbols 1 and 0 to be coherently orthogonal, in the sense that they do not interfere with one another in the process of detection. The MSK carrier phase is always continuous at bit transitions. The MSK signal in Equations 3.21 and 3.22 can be simplified as

$$s(t) = \sqrt{\frac{2E_b}{T_b}} \cos[2\pi f_c t + d_k \frac{\pi t}{2T_b} + \Phi_k], \quad kT_b \le t \le (k+1)T_b \tag{3.23}$$

and the baseband equivalent optical MSK signal is represented as

$$\tilde{s}(t) = \sqrt{\frac{2E_b}{T_b}} \exp\left\{j\left[d_k \frac{\pi t}{2T} + \Phi(t,k)\right]\right\}, \quad kT \le t \le (k+1)T$$

$$= \sqrt{\frac{2E_b}{T_b}} \exp\left\{j\left[d_k 2\pi f_d t + \Phi(t,k)\right]\right\} \tag{3.24}$$

where $d_k = \pm 1$ are the logic levels, f_d is the frequency deviation from the optical carrier frequency, and $h = 2f_d T$ is defined as the frequency modulation index. In the case of optical MSK, $h = 1/2$ or $f_d = 1/(4T_b)$.

3.4.3.3 MSK as ODQPSK

Equation 3.18 can be rewritten to express MSK signals in the form of I–Q components as

$$s(t) = \pm \sqrt{\frac{2E_b}{T_b}} \cos\left(\frac{\pi t}{2T_b}\right)\cos\left[2\pi f_c t\right] \pm \sqrt{\frac{2E_b}{T_b}} \sin\left(\frac{\pi t}{2T_b}\right)\sin\left[2\pi f_c t\right] \qquad (3.25)$$

The I and Q components consist of half-cycle sine and cosine pulses defined by

$$s_I(t) = \pm \sqrt{\frac{2E_b}{T_b}} \cos\left(\frac{\pi t}{2T_b}\right) \qquad -T_b < t < T_b \qquad (3.26)$$

$$s_Q(t) = \pm \sqrt{\frac{2E_b}{T_b}} \sin\left(\frac{\pi t}{2T_b}\right) \qquad 0 < t < 2T_b \qquad (3.27)$$

3.4.3.4 Configuration of Photonic MSK Transmitter Using Two Cascaded Electro-Optic Phase Modulators

Electro-optic phase modulators and interferometers operating at a high frequency using resonant-type electrodes have been studied and proposed in Refs. [20–22]. In addition, high-speed electronic driving circuits evolved with the ASIC technology using 0.1 μm GaAs P-HEMT or InP HEMTs [23] enable the realization of the optical MSK transmitter structure. The baseband equivalent of the optical MSK signal is represented in Equation 3.25.

The first electro-optic phase modulator (E-OPM) enables the frequency modulation of data logic into upper sidebands (USBs) and lower sidebands (LSBs) of the optical carrier with a frequency deviation of f_d. Differential phase precoding is not necessary in this configuration owing to the nature of the continuity of the differential phase trellis. By alternating the driving sources $V_d(t)$ to sinusoidal waveforms for simple implementation or a combination of sinusoidal and periodic ramp signals, which was first proposed by Amoroso in 1976 [16], different schemes of linear and nonlinear phase-shaping MSK-transmitted sequences can be generated whose spectra are shown later in the chapter in Figure 3.37.

The second E-OPM enforces the phase continuity of the lightwave carrier at every bit transition. The delay control between the E-OPMs is usually implemented by the phase shifter shown in Figure 3.28. The driving voltage of the second E-OPM is precoded to fully compensate the transitional phase jump at the output $E_{01}(t)$ of the first E-OPM. The phase continuity characteristic of the optical MSK signals is determined by the algorithm given in Equation 3.25.

$$\Phi(t,k) = \frac{\pi}{2}\left(\sum_{j=0}^{k-1} a_j - a_k I_k \sum_{j=0}^{k-1} I_j\right) \qquad (3.28)$$

where $a_k = \pm 1$ are the logic levels; $I_k = \pm 1$ is a clock pulse whose duty cycle is equal to the period of the driving signal $V_d(t)$; f_d is the frequency deviation from the optical carrier frequency; and $h = 2f_d T$ was previously defined as the frequency modulation index. In the case of optical MSK, $h = 1/2$ or $f_d = 1/(4T)$. The phase evolution of the continuous phase optical MSK signals is explained in Figure 3.28. In order to mitigate the effects of unstable stages of the rising and falling edges of the electronic circuits, the clock pulse $V_c(t)$ is offset with the driving voltages $V_d(t)$ (Figure 3.29).

3.4.3.5 Configuration of Optical MSK Transmitter Using Mach–Zehnder Intensity Modulators: I–Q Approach

The conceptual block diagram of the optical MSK transmitter is shown in Figure 3.30. The transmitter consists of two dual-drive electro-optic MZMs generating chirpless I and Q components of MSK-modulated signals, which is considered as a special case of staggered

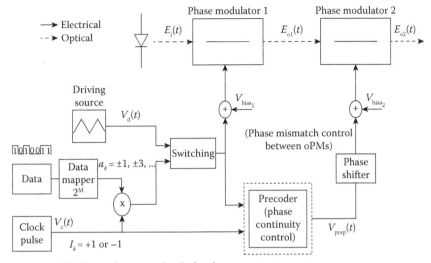

FIGURE 3.28 Block diagram of optical MSK transmitter employing two cascaded optical phase modulators.

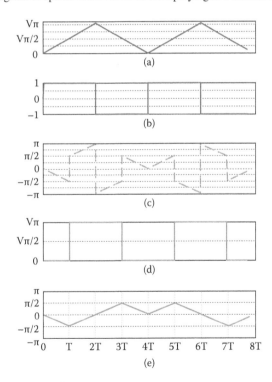

FIGURE 3.29 Evolution of time-domain phase trellis of optical MSK signal sequence [−1 1 1 −1 1 −1 1 1] as inputs and outputs at different stages of the optical MSK transmitter. The notation is denoted in Figure 3.28 accordingly: (a) $V_d(t)$, periodic triangular driving signal for optical MSK signals with a duty cycle of four-bit periods; (b) $V_c(t)$, the clock pulse with a duty cycle of $4T$; (c) $E_{o1}(t)$, phase output of oPM1; (d) $V_{prep}(t)$, pre-computed phase compensation driving voltage of oPM2; and (e) $E_{o2}(t)$, phase trellis of a transmitted optical MSK sequences at output of oPM2.

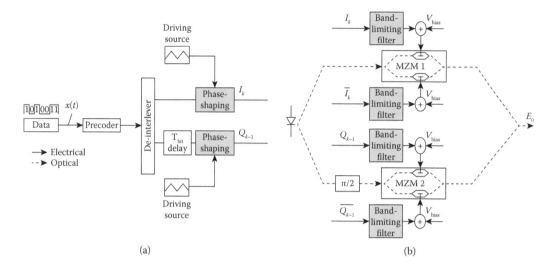

FIGURE 3.30   Block diagram configuration of (a) band-limited (b) phase-shaped optical MSK.

FIGURE 3.31   Phase trellis of linear and nonlinear MSK-transmitted signals. (a) linear and (b) nonlinear.

or offset QPSK. The binary logic data are precoded and de-interleaved into even and odd bit streams that are interleaved with each other by one-bit duration offset.

Two arms of the dual-drive MZM are biased at $V_\pi/2$ and $-V_\pi/2$ and driven with *data* and $\overline{data}$. Phase-shaping driving sources can be a periodic triangular voltage source in the case of linear MSK generation or simply a sinusoidal source for generating a nonlinear MSK-like signal, which also achieves the linear phase trellis property but with small ripples introduced in the magnitude. The magnitude fluctuation level depends on the magnitude of the phase-shaping driving source. A high spectral efficiency can be achieved with tight filtering of the driving signals before modulating the electro-optic MZMs. Three types of pulse-shaping filters are investigated including Gaussian, raised-cosine, and squared-root raised-cosine filters. The optical carrier phase trellis of linear and nonlinear optical MSK signals are shown in Figure 3.31.

3.4.4 SINGLE-SIDEBAND OPTICAL MODULATORS

3.4.4.1 Operating Principles

A single-sideband (SSB) modulator can be formed using a primary interferometer with two secondary MZM structures, the optical Ti-diffused waveguide paths that form a nested primary MZ structure as shown in Figure 3.32 [14]. Each of the two primary arms contains an MZ structure. Two RF ports are for RF modulation, and three DC ports are for biasing the two secondary MZMs and one primary MZM. The modulator consists of an X-cut Y-propagation LiNbO$_3$ crystal, where SSB

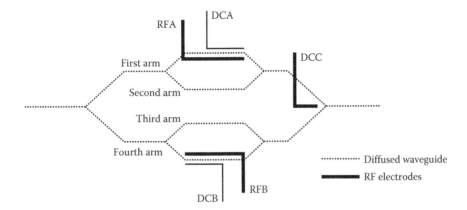

FIGURE 3.32 Schematic diagram (not to scale) of a single-sideband (SSB) FSK optical modulator formed by nested MZ modulators.

modulation can be produced just by driving each MZ. DC voltage is supplied to produce the π phase shift between upper and lower arms. DC bias voltages are also supplied from DC_B to produce the phase shift between third and fourth arms. A DC bias voltage is supplied from DC_C to achieve a $\pi/2$ phase shift between $MZIM_A$ and $MZIM_B$. The RF voltages applied $\Phi_1(t) = \Phi \cos\omega_m t$ and $\Phi_2(t) = \Phi \sin\omega_m t$ are inserted from RF_A and RF_B, respectively, by using a wideband $\pi/2$ phase shifter. Φ is the modulation level, and ω_m is the RF angular frequency.

3.4.4.2 Optical RZ MSK

The RZ format of the optical MSK modulation scheme can also be generated. A PDM I–Q modulator is used to generate the ASK-like feature of the bit sequence in the optical domain. A Simulink® structure of such a transmitter is shown in Figure 3.33. The encoder as shown in the far left of the model provides two outputs, one for MSK signal generation and the other for amplitude modulation for generation of the RZ or NRZ format. The amplitude and phase of the RZ signals at the receivers are shown in Figure 3.34 after a 100 km span transmission and are fully compensated. The phase of the RZ MSK must be nonzero so as to satisfy the continuity of the phase from one state to the other.

3.4.5 MULTICARRIER MULTIPLEXING OPTICAL MODULATORS

Another modulation format that can offer much higher single-channel capacity and flexibility in dispersion and nonlinear impairment mitigation is multicarrier multiplexing (MCM). When these subcarrier channels are orthogonal, the term orthogonal frequency division multiplexing (OFDM) is used [24–26].

Our motivation for introducing OFDM is its potential as a ultra-high-capacity channel for the next-generation Ethernet, the optical Internet. The network interface cards for 1- and 10-Gbps Ethernet are readily commercially available. Traditionally, Ethernet data rates have grown in 10 Gbps increments, so the data rate of 100 Gbps can be expected as the speed of the next generation of Ethernet. The 100 Gbps all-electrically time-division-multiplexed (ETDM) transponders are becoming increasingly important because they are viewed as a promising technology that may be able to meet speed requirements of the new generation of Ethernet. Despite the recent progress in high-speed electronics, ETDM [27] modulators and photodetectors are still not widely available, and so alternative approaches to achieving a 100 Gbps transmission using commercially available components and QPSK are very attractive. However, the use of polarization division

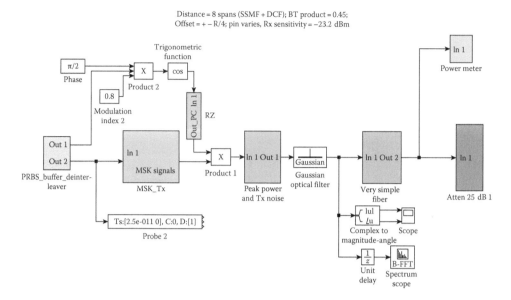

Distance = 8 spans (SSMF + DCF); BT product = 0.45;
Offset = + – R/4; pin varies, Rx sensitivity = –23.2 dBm

FIGURE 3.33 MATLAB® and Simulink® model of a RZ MSK optical transmission system.

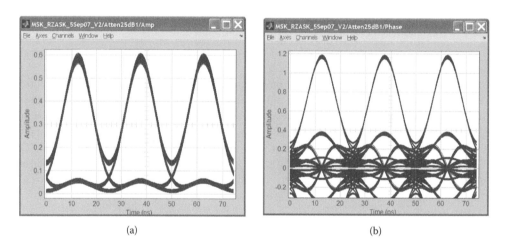

(a)   (b)

FIGURE 3.34 RZ eye diagram at output of (a) the amplitude receiver and (b) phase detection.

multiplexing (PDM) and 25 GBaud QPSK would offer 100 Gbps transmission with superiority under coherent reception incorporating digital signal processing (DSP) [17]. This PDM-QPSK technique will be described in later chapters.

OFDM is a combination of multiplexing and modulation. The data signal is first split into independent subsets and then modulated with independent subcarriers. These subchannels are then multiplexed to OFDM signals. OFDM is thus a special case of FDM, but instead of one stream, several small streams are combined into a bundle.

A schematic signal flow diagram of an MCM is shown in Figure 3.35. The basic OFDM transmitter and receiver configurations are given in Figure 3.36a and b, respectively [28]. Data streams (e.g., 1 Gbps) are mapped into a two-dimensional signal point from a point signal constellation such as QAM. The complex-valued signal points from all subchannels are considered as

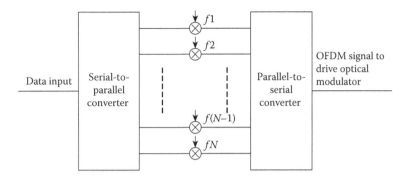

FIGURE 3.35 Multicarrier FDM signal arrangement. The middle section is the RF converter as shown in Figure 3.26.

the values of the discrete Fourier transform (DFT) of a multicarrier OFDM signal. The serial-to-parallel converter arranges the sequences into equivalent discrete frequency domains. By selecting the number of subchannels to be sufficiently large, the OFDM symbol interval can be made much larger than the dispersed pulse width in a single-carrier system, resulting in an arbitrary small intersymbol interference (ISI). The OFDM symbol, shown in Figure 3.36c and d, is generated with software processing (for example, by using an arbitrary waveform generator) as follows: input QAM symbols are zero-padded to obtain input samples for the inverse fast Fourier transform (IFFT), the samples are inserted to create the guard band, and the OFDM symbol is multiplied by the window function, which can be represented by a raised-cosine function. The purpose of the cyclic extension is to preserve the orthogonality among subcarriers even when the neighboring OFDM symbols partially overlap owing to dispersion.

3.4.6 SPECTRA OF MODULATION FORMATS

Utilizing this double-phase modulation configuration, different types of linear and nonlinear CPM phase-shaping signals, including MSK, weakly nonlinear MSK, and linear-sinusoidal MSK can be generated. The third scheme was introduced by Amoroso [29], and its side lobes decay with a factor of f-8 compared to f-4 of MSK. The simulated optical spectra of DBPSK and MSK schemes at 40 Gbps are contrasted in Figure 3.37.

Table 3.5 outlines the characteristics and spectra of all the different modulation schemes.

3.5 SPECTRAL CHARACTERISTICS OF DIGITAL MODULATION FORMATS

Figure 3.38 shows the power spectra of the DPSK-modulated optical signals with various pulse shapes, including NRZ, RZ33, and CSRZ types.

For the convenience of comparison, the optical power spectra of the return-to-zero OOK counterparts are also shown in Figure 3.37.

Several key features are observed from Figures 3.38 and 3.39:

1. The optical power spectrum of the OOK format has high-power spikes at the carrier frequency or at signal modulation frequencies, which contribute significantly to the severe penalties caused by the nonlinear effects. Meanwhile, the DPSK optical power spectra do not contain these high-power frequency components.
2. RZ pulses are more sensitive to fiber dispersion owing to their broader spectra. In particular, the RZ33 pulse type has the broadest spectrum at the point of −20 dB down from the peak.

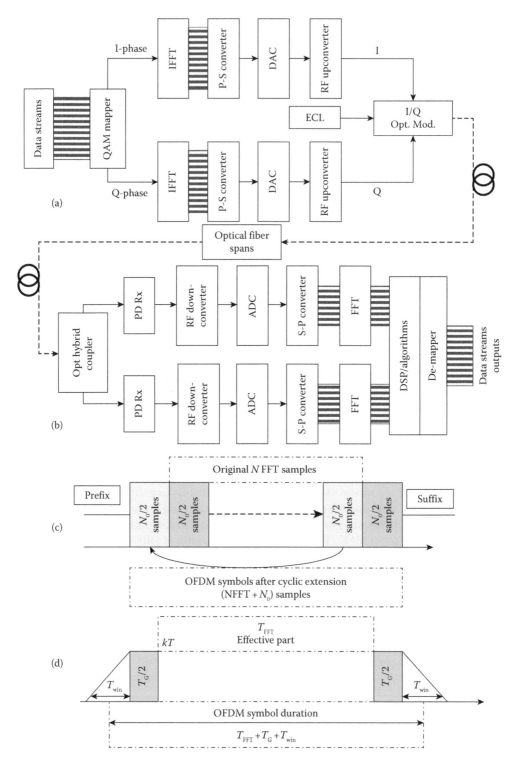

FIGURE 3.36 Schematic diagram of the principles of generation and recovery of OFDM signals. (a) Electronic processing and optical transmitter, (b) optoelectronic and DSP-based receiver configurations, (c) OFDM symbol cyclic extension, and (d) OFDM symbol after windowing (T_{win}).

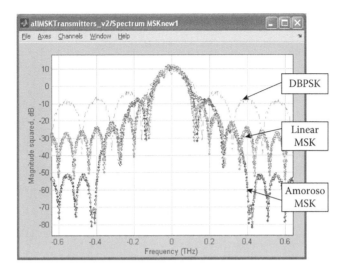

FIGURE 3.37 Spectra of 40 Gbps DBPSK, and linear and nonlinear MSK.

TABLE 3.5
Typical Parameters of Optical Intensity Modulators for Generation of Modulation Formats

Modulation Techniques	Spectra	Formats	Definition/Comments
Amplitude modulation – ASK-NRZ	DSB + carrier	ASK-NRZ	Biased at quadrature point or offset for prechirp.
AM–ASK-RZ	DSB + carrier	ASK-RZ	Two MZIMs are required: one for RZ pulse sequence and the other for data switching.
ASK–RZ-Carrier-suppressed	DSB – CSRZ	ASK-RZCS	Carrier-suppressed, biased at π phase difference for the two electrodes.
Single sideband	SSB + carrier	SSB NRZ	Signals applied to MZIM are in phase quadrature to suppress one sideband. Alternatively, an optical filter can be used.
CSRZ-DSB	DSB – carrier	CSRZ-ASK	RZ pulse carver is biased to give a π phase shift between the two arms of the MZM to suppress the carrier and then switch on and off phase modulation via a data modulator.
DPSK-NRZ, DPSK-RZ, CSRZ-DPSK		Differential BPSK-RZ or NRZ/RZ-carrier-suppressed	
DQPSK		DQPSK-RZ or NRZ	Two bits per symbol.
MSK	SSB equivalent	Continuous phase modulation with orthogonality	Two bits per symbol and efficient bandwidth with high side-lobe suppression.
Offset DQPSK			
MCM (multicarrier multiplexing, e.g., OFDM)	Multiplexed bandwidth—base rate per subcarrier		
Duobinary	Effective SSB		Electrical low-pass filter required at the driving signal to the MZM.
FSK			
Continuous phase FSK			
Phase modulation (PM)	Chirped carrier phase		Chirpless MZM should be used to avoid inherent crystal effects, hence carrier chirp.

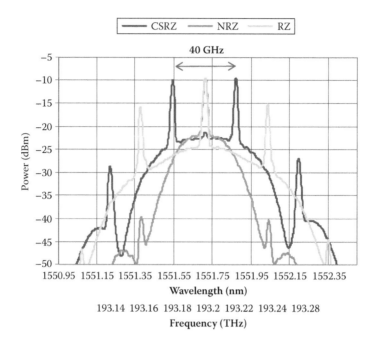

FIGURE 3.38   Spectra of CSRZ/RZ33/NRZ-DPSK modulated optical signals.

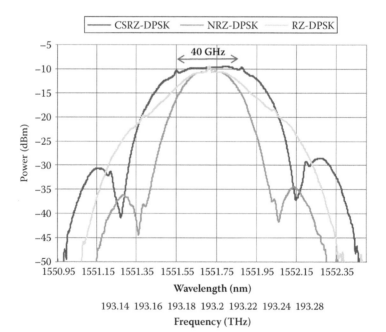

FIGURE 3.39   Spectra of CSRZ/RZ/NRZ-OOK modulated optical signals.

This property of the RZ pulses thus leads to faster spreading of the pulse when propagating along the fiber. Thus, the peak values of the optical power of these CSRZ or RZ33 pulses decrease much more quickly than the NRZ counterparts. As the result, the peak optical power quickly turns to be lower than the nonlinear threshold of the fiber, which means the effects of fiber nonlinearity are significantly reduced.

3. However, NRZ optical pulses have the narrowest spectrum. Thus, they are expected to be most robust to the fiber dispersion. As a result, there is a trade-off between RZ and NRZ pulse types. RZ pulses are much more robust to nonlinearity but less tolerant to fiber dispersion. The RZ33/CSRZ-DPSK optical pulses have proved to be more robust against impairments, especially self-phase modulation and PMD compared to the NRZ-DPSK and the CSRZ/RZ33-OOK counterparts.

The optical power spectra of three I–Q optical MSK modulation formats—linear, weakly nonlinear, and strongly nonlinear—are shown in Figure 3.40. It can be observed that there are no significant distinctions in the spectral characteristics among these three schemes. However, the strongly nonlinear optical MSK format does not highly suppress the side lobe as compared to the linear MSK type. All three formats offer better spectral efficiency compared to the DPSK counterpart as shown in Figure 3.41. This figure compares the power spectra of three modulation formats: 80 Gbps dual-level MSK, 40 Gbps MSK, and NRZ-DPSK optical signals. The normalized amplitude levels into the two optical MSK transmitters comply with the ratio of 0.75/0.25.

Several key features can be observed from Figure 3.41:

1. The power spectrum of the optical dual-level MSK format has identical characteristics to that of the MSK format. The spectral width of the main lobe is narrower than that of the DPSK. The base width takes the value of approximately ±32 GHz on either side compared to ±40 GHz in the case of the DPSK format. Hence, the tolerance to fiber dispersion effects is improved.
2. High suppression of the side lobes occurs with a value of approximately greater than 20 dB in the case of 80 Gbps dual-level MSK and 40 Gbps optical MSK power spectra; thus, there is more robustness to interchannel crosstalk between DWDM channels.
3. The confinement of signal energy in the main lobe of spectrum leads to a better SNR. Thus, the sensitivity to optical filtering can be significantly reduced [2–7]. A summary of the spectra of different modulation formats is given in Table 3.5.

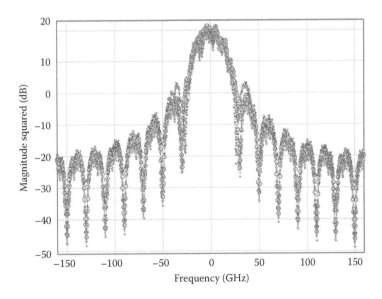

FIGURE 3.40 Optical power spectra of three types of I–Q optical MSK formats: linear (very light), weakly nonlinear (light), and strongly nonlinear (dark).

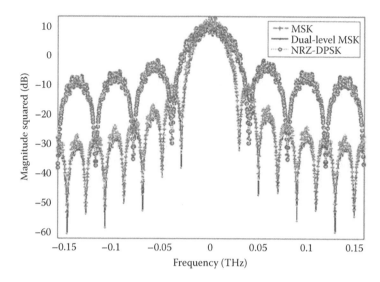

FIGURE 3.41 Spectral properties of three modulation formats: MSK (very light, dash), dual-level MSK (dark, solid), and DPSK (light, dot).

3.6 I–Q INTEGRATED MODULATORS

3.6.1 IN-PHASE AND QUADRATURE-PHASE OPTICAL MODULATORS

We have described in the earlier sections the modulation of QPSK and M-ary quadrature amplitude modulation (QAM) schemes using single-drive or dual-drive MZIM devices. However, there is another alternative technique to generate the states of constellation of QAM: using I–Q modulators. I–Q modulators are devices in which the amplitude of the in-phase and the quadrature components are modulated in synchronization, as illustrated in Figure 3.42b. These components are $\pi/2$ out of phase, and thus we can achieve the in-phase and quadrature-phase components of QAM. In optics, this phase quadrature at the optical frequency can be implemented by a low-frequency electrode with an appropriate applied voltage as observed by the $\pi/2$ block in the lower optical path of the structure given in Figure 3.42a. This type of modulation can offer significant advantages when multilevel QAM schemes are employed, for example, 16-QAM (see Figure 3.42c) or 64-QAM. Integrated optical modulators have been developed in recent years, especially in electro-optic structures such as $LiNbO_3$, for coherent QPSK and even with polarization division multiplexed optical channels. In summary, the I–Q modulator consists of two MZIMs performing ASK modulation incorporating a quadrature-phase shift.

Multilevel or multicarrier modulation formats such as QAM and OFDM have been demonstrated as a promising technology to support high capacity and high spectral efficiency in ultra-high-speed optical transmission systems. Several QAM transmitter schemes have been experimentally demonstrated using commercial modulators [30–32] and integrated optical modules [23,33] with binary or multilevel driving electronics. The integration techniques could offer a stable performance and effectively reduce the complexity of the electronics in QAM transmitters with binary driving electronics. The integration schemes based on parallel structures usually require hybrid integration between $LiNbO_3$ modulators and silica-based planar lightwave circuits (PLCs). Except the DC electrodes for the bias control of each sub-MZM (child MZM), several additional electrodes are required to adjust the relative phase offsets among embedded sub-MZMs. Shown in Figure 3.42a is a 16-QAM transmitter using a monolithically integrated quad Mach–Zehnder in-phase/quadrature (QMZ-I–Q) modulator. Unlike parallel integration, four sub-MZMs are integrated and arranged in a single I–Q superstructure, where two of them are cascaded in each of the arms (I and Q arms).

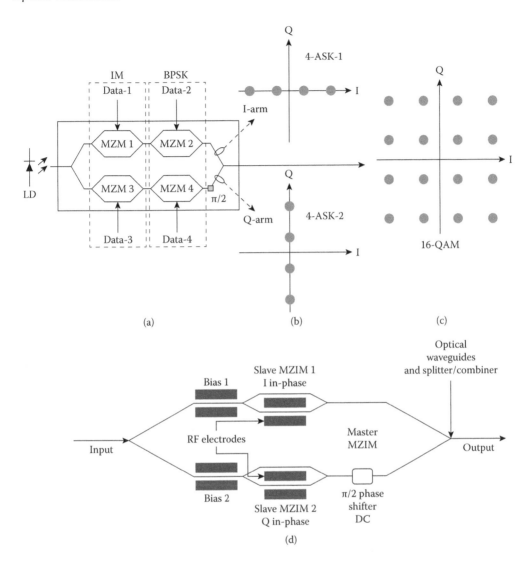

FIGURE 3.42 Schematic of (a) integrated I–Q optical modulator and (b) amplitude modulation of lightwave path in the in-phase and quadrature components, (c) constellation of a 16 QAM modulation scheme, and (d) alternate structure of an I–Q modulator using two slave MZIMs and one master MZIM.

These two pairs of child MZMs are combined to form a master or parent MZ interferometer with a π/2 phase shift incorporated in one arm to generate the quadrature-phase shift between them, and thus we have the in-phase arm and the quadrature-phase optical components. In principle, only one electrode is required to obtain the orthogonal phase offset in these I–Q superstructures, which makes the bias control much easier to handle, and thus ensures a stable performance. A 16-QAM signal can be generated using the monolithically integrated quadrature Mach–Zehnder in-phase and quadrature (QMZ-I–Q) modulator with binary driving electronics, as shown in Figure 3.42c, by modulating the amplitude of the lightwaves guided through both I (in-phase) and Q (quadrature-phase) paths as indicated in Figure 3.42b. Hence, we can see that QAM can be implemented by modulating the amplitude of these two orthogonal I and Q components so that any constellation of M-ary QAM, e.g., 16QAM or 256QAM, can be generated. Alternatively, the structure of Figure 3.42d gives an arrangement of two slave MZIMs and one master MZIM for the I–Q modulator, which is commonly designed by Fujitsu of Japan.

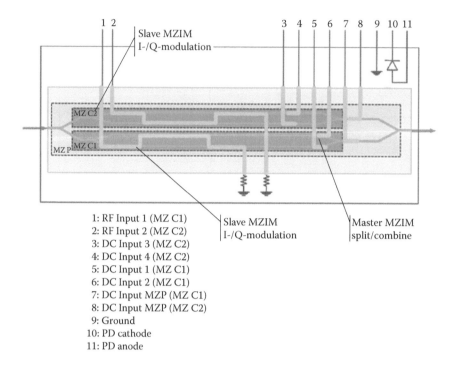

1: RF Input 1 (MZ C1)
2: RF Input 2 (MZ C2)
3: DC Input 3 (MZ C2)
4: DC Input 4 (MZ C2)
5: DC Input 1 (MZ C1)
6: DC Input 2 (MZ C1)
7: DC Input MZP (MZ C1)
8: DC Input MZP (MZ C2)
9: Ground
10: PD cathode
11: PD anode

FIGURE 3.43 Schematic diagram of Fujitsu PDM I–Q modulator with assigned electrodes.

In addition, a number of electrodes would be incorporated so that biases can be applied to ensure that the amplitude modulation operates at the right point of the transfer curve characteristics of the output optical field versus the applied voltage of the MZIM in each interferometric branch of the master MZ interferometer. This can be commonly observed and simplified as in the I–Q modulator manufactured by Fujitsu [34], shown in Figure 3.43 (top view only).

The arrangement of the high-speed I–Q Mach–Zehnder modulator using Ti-diffused lithium niobate ($LiNbO_3$) waveguide technology in which the DC bias can be adjusted at the separate DC port of the modulator. This type of modulator can be employed for various modulation formats such as NRZ, DPSK, optical duobinary, DQPSK, DP-BPSK, DP-QPSK, M-ary QAM, and so on. A built-in PD monitor and coupler function for auto bias control of 100 Gbps optical transmission equipment (NRZ, DPSK, optical duobinary, DQPSK, DP-BPSK, PDM-QPSK) can be generated by this I–Q modulator in which four wavelength carriers are used with 28–32 Gbps per channel to form 100 Gbps, including extra error coding bits and a payload of 25 Gbps for each channel.

3.6.2 I–Q Modulator and Electronic Digital Multiplexing for Ultra-High Bit Rates

Recently, we [20,21,35–37] have seen reports on the development of an electrical multiplexer whose speed can reach 165 Gbps. This multiplexer will allow the interleaving of high-speed sequences so that we can generate a higher bit rate with the symbol rate as shown in Figure 3.44. This type of multiplexing in the electrical domain has been employed for the generation of the superchannel in Ref. [22]. Thus, we can see that the data sequence can be generated from the DACs, and then the analog signals at the output of the DACs can be conditioned to the right digital level of the digital multiplexer, and with the assistance of a clock generator, the multiplexing occurs.

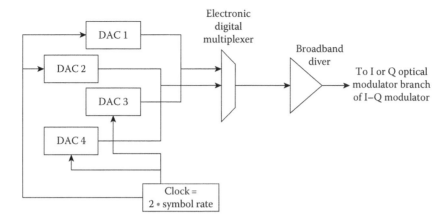

FIGURE 3.44   Time division multiplexing using high-speed sequences from DACs.

Note that the multiplexer operates in digital mode, so any pre distortion of the sequence for dispersion precompensation will not possible unless some predistortion can be done at the output of the digital multiplexer.

This time-domain interleaving can be combined with the I–Q optical modulator to generate an M-level QAM signal, thus further increasing the bit rate of the channels. Several of these high-bit-rate channels can be combined with pulse shaping, for example, the Nyquist shape, to generate superchannels, which will be illustrated in Chapter 7. With a digital multiplexer operating higher than 165 Gbps, 128 Gbps bit rate can be generated with a 32 GSy/s data sequence. Thus, with polarization multiplexing and 16QAM, we can generate 8 × 128 Gbps for one channel or 1.32 Tbps per channel. If 8 of these 1.32 Gbps channels form a superchannel, then the bit rate capacity exceeds 10 Tbps. This capacity is very probable with the baud rate reaching 50 GBaud as the bandwidth of relevant electronic components and photonic devices can reach 75 GHz and higher for implementing the optical transmitter and opto-electronic digital receiver.

3.7   DIGITAL-TO-ANALOG CONVERTER FOR DSP-BASED MODULATION AND TRANSMITTER

Recently, we have witnessed the development of ultra-high sampling rate DACs and ADCs by Fujitsu and NTT of Japan. Figure 3.45a shows an IC layout of the InP-based DAC produced by NTT [38].

3.7.1   Fujitsu DAC

A differential input stage incorporating D-type flip flops and adders to produce analog signals representing the digital samples are shown in Figure 3.45a and b. These DACs make possible the programmable sampling and generation of digitalized signals which then converted to analog signals to feed to the optical I–Q modulator to modulate the lightwaves as described in Section 3.5. Thus they allow shaping of the pulse sequence for predistortion to combat dispersion effects when necessary to add another dimension of compensation in combination with the function implemented at the receiver. Test signals used are ramp waveforms for ensuring the linearity of the DAC at 27 GS/s and are shown in Figure 3.46a, and eye diagrams of generated 16QAM signals after modulation at 22 and 27 GSy/s are shown in Figure 3.46b.

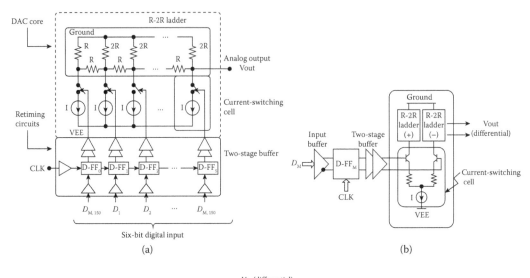

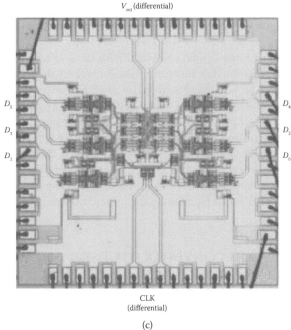

FIGURE 3.45 NTT InP-based DAC with 6-bit: (a) schematic, (b) differential input stage, and (c) IC layout. (Extracted with permission from T. Sakamoto et al., 50-Gbps 16-QAM by a quad-parallel Mach–Zehnder modulator, *ECOC 2007*, paper PDP2.8, 2007.)

3.7.2 STRUCTURE

A DSP-based optical transmitter can incorporate a DAC for pulse shaping, pre-equalization, and pattern generation as well as digitally multiplexing for higher symbol rates. A schematic structure of the DAC and functional blocks fabricated by Si-Ge technology is shown in Figure 3.47a and b, respectively. An external sinusoidal signal is fed into the DAC so that multiple clock sources can be generated for sampling at 56–64 GSa/s. Thus, the noise and clock accuracy depend on the stability and noise of this synthesizer/signal generator. Four DAC submodules are integrated in one IC with four pairs of eight outputs of $(V_I^+, V_Q^+)(H_I^+, H_Q^+)\_$ and $\_(V_I^-, V_Q^-)(H_I^-, H_Q^-)$.

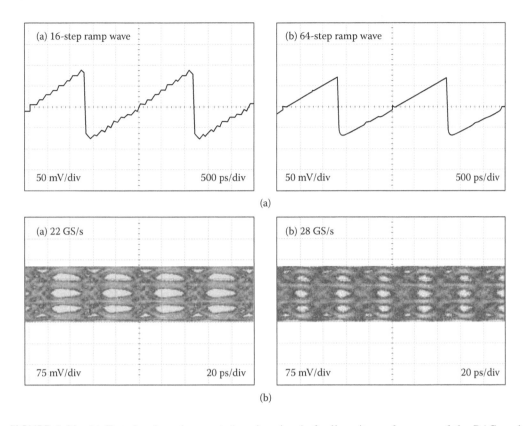

FIGURE 3.46 (a) Test signals and converted analog signals for linearity performance of the DAC, and (b) 16QAM four-level-generated signals at 22 and 27 GSy/s.

3.7.3 GENERATION OF I AND Q COMPONENTS

The electrical outputs from the quad DACs are in mutual complementary pairs of positive and negative signs. Thus, it would be able to form two sets of four output ports from the DAC development board. Each output can be independently generated with off-line uploading of pattern scripts. The arrangement of the DAC and PDM I–Q optical modulator is depicted in Figure 3.48. Note that we require two PDM I–Q modulators for the generation of odd and even optical channels.

As Nyquist pulse-shaped sequences are required, a number of processing steps can be implemented:

1. Characterization of the DAC transfer functions
2. Pre-equalization in the RF domain to achieve equalized spectrum in the optical domain, that is, at the output of the PDM I–Q modulator

The characterization of the DAC is conducted by launching to the DAC sinusoidal wave at different frequencies and measuring the waveforms at all eight output ports. As can be observed in the insets of Figure 3.48, the electrical spectrum of the DAC is quite flat provided that pre-equalization is done in the digital domain launching to the DAC. The spectrum of the DAC output without equalization is shown in Figure 3.49a and b. The amplitude spectrum is not flat owing to the transfer function of the DAC as given in Figure 3.50, which is obtained by driving the DAC with sinusoidal waves of different frequencies. This shows that the DAC acts as a low-pass filter with the amplitude of its passband gradually decreasing when the frequency is increased.

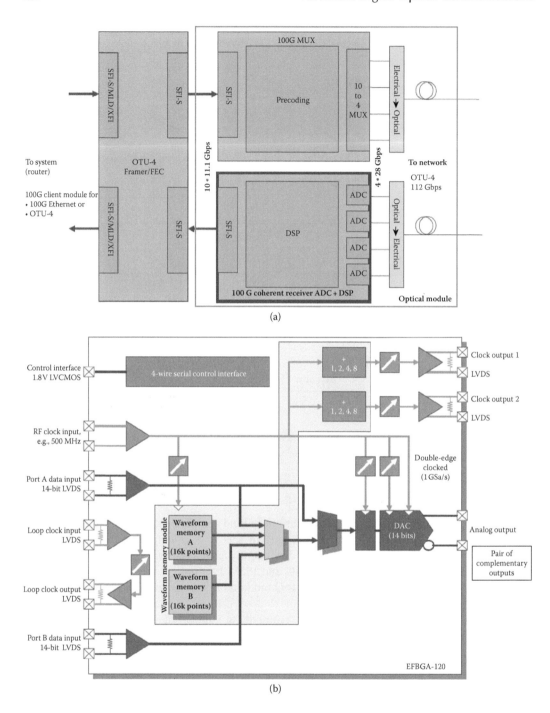

FIGURE 3.47 (a) Structure of the Fujitsu DAC (note four DACs are structured in one integrated chip) and (b) functional block diagram.

This effect can occur because the number of samples is reduced when the frequency is increased as the sampling rate can only be set in the range of 56–64 GSa/s. The equalized RF spectra are depicted in Figure 3.49c and d. The time-domain waveforms corresponding to the RF spectra are shown in Figure 3.49e and f, and Figure 3.49g and h show the waveforms for coherent detection after conversion back to electrical domain from the optical modulator via the real-time sampling

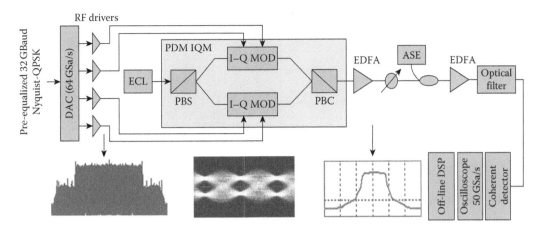

FIGURE 3.48   Experimental setup of 128 Gbps Nyquist PDM-QPSK transmitter and B2B performance evaluation.

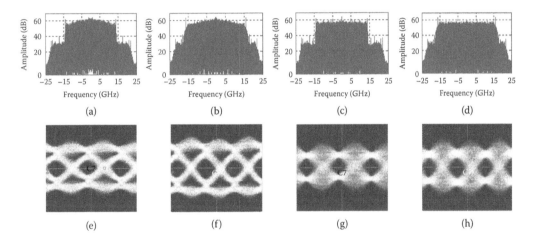

FIGURE 3.49   (a) Spectrum and (e) eye diagram of 28 GBaud RF signals after DAC without pre-equalization, (b, f) for 32 GBaud; (c) spectrum and (g) eye diagram of 28 GBaud RF signals after DAC with pre-equalization, (d, h) for 32 GBaud.

oscilloscope Tektronix DPO 73304A or DSA 720004B. Furthermore, the noise distribution of the DAC is shown in Figure 3.50b, indicating that the sideband spectra of Figure 3.49 come from these noise sources.

It noted that the driving conditions for the DAC are very sensitive to the supply current and voltage levels, which are given in Annex 2 with a resolution of even down to 1 mV. With this sensitivity, care must be taken when new patterns are fed to the DAC to drive the optical modulator (Figure 3.51). Optimal procedures must be conducted with the evaluation of the constellation diagram and BER derived from such a constellation. However, we believe that the new version of the DAC supplied from Fujitsu Semiconductor Pty Ltd. of Maidenhead, Berkshire, England Europe have somehow overcome these sensitivity problems. However, we still recommend that care be taken and an inspection of the constellation after the coherent receiver must be made to ensure an error-free B2B connection. Various time-domain signal patterns obtained in the electrical time domain generated by DAC at the output ports can be observed. Obviously, the variations of the in-phase and quadrature signals give rise to the noise, and hence blurry constellations.

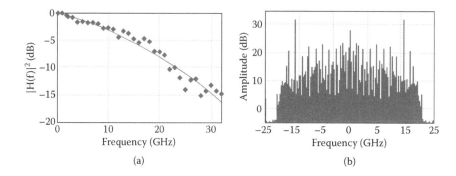

FIGURE 3.50 Frequency transfer characteristics of the DAC. Note the near linear variation of the magnitude as a function of the frequency. (a) Magnitude square as a function of frequency and (b) Noise spectral characteristics of the DAC.

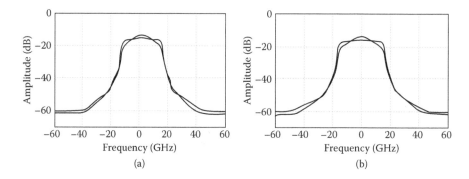

FIGURE 3.51 Optical spectrum after PDM IQM. The light gray line is for without pre-equalization, dark gray line for with pre-equalization: (a) 28 GSym/s and (b) 32 GSym/s.

3.8 CONCLUDING REMARKS

Currently, photonic transmitters play a principal part in the extension of the modulation speed to the range of several gigahertz and enable the modulation of the amplitude, phase, and frequency of the optical carriers and their multiplexing. The performance of photonic transmitters using $LiNbO_3$ has been proved in laboratory and installed systems. The principal optical modulator is the MZIM, which can be a single unit or a combination of these modulators for binary or multilevel amplitude or phase modulation and even more effective for discrete or continuous phase shift keying techniques. The effects of the modulation on the transmission performance will be described in later chapters.

The spectral properties of the optical 80 Gbps dual-level MSK, 40 Gbps MSK, and 40 Gbps DPSK with various RZ pulse shapes are compared. The spectral properties of the first two formats are similar. Compared to the optical DPSK, the power spectra of optical MSK and dual-level MSK modulation formats have more attractive characteristics. These include high spectral efficiency for transmission, higher energy concentration in the main spectral lobe, and more robustness to interchannel crosstalk in DWDM due to greater suppression of the side lobes. In addition, the optical MSK offers the orthogonal property, which may offer a great potential in coherent detection, in which the phase information is reserved via I and Q components of the transmitted optical signals. In addition, the multilevel formats would permit the lowering of the bit rate and hence a substantial reduction in the signal effective bandwidth and the possibility of reaching the highest speed limit of electronic signal processing and digital signal processing for equalization and compensation of distortion effects. The demonstration of the ETDM receiver at 80 Gbps and higher speeds would increase the potential for applications of these modulation formatted schemes in ultra-high-speed transmission.

In recent years, research on new types of optical modulators using silicon waveguides has attracted several groups. In particular, a graphene thin layer deposited [39] on silicon waveguides enables the improvement of the EA effects and enables the modulator structure to be incredibly compact and potentially perform at speeds up to ten times faster than current technology allows, reaching higher than 100 GHz and even 500 GHz. This new technology will significantly enhance our capabilities in ultra-fast optical communication and computing. This may be the world's smallest optical modulator, and the modulator in data communications is the heart of speed control. Furthermore, these graphene silicon modulators can be integrated with Si- or SiGe-based micro-electronic circuits such as the DAC and ADC, DSP. The sampling rate of these digital electronic systems can, at present, reach 56–64 GSamples/sec.

3.9 PROBLEMS ON TRANSMITTER (TX) FOR ADVANCED MODULATION FORMATS FOR LONG-HAUL TRANSMISSION SYSTEMS

Problem 3.1

A bit pattern "1 1 0 1 0 0 1 1 0 1" with a bit rate of 10 Gbps is input into two separate modulators to generate ASK and BDPSK modulation formats.

    a. Sketch the modulated bit pattern, including the carrier, that can be drawn with two or three periods within the duration of a bit period. Hence, sketch the phase distribution or the scatter diagram of the modulated sequence.
    b. A Mach–Zehnder intensity modulator (MZIM) is used as an optical modulator; its V_π is 5 V. Show how the data sequence can be conditioned for feeding into the electrode port of the modulator so that ASK or BDPSK signals can be generated. Make sure that you show the biasing voltage.
    c. Repeat (b) for the case that the modulator is a dual-drive type of optical modulator.
    d. Give the structure of a precoder for DPSK modulation. That is, the precoder would generate the differential codes that are then used to drive the optical modulator with signal conditioning to an appropriate level for driving the electrodes.

Problem 3.2

Sketch the schematic diagram of an intensity optical modulator that employs an optical waveguide as the input light guide, which is then split into two parallel paths whose refractive indices would be modulated by an electrode, hence resulting in phase modulation of the lightwaves that pass through these waveguides. This type of optical modulator can be termed as an interferometric intensity modulator.

For a single-drive modulator, only one path of the lightwave is modulated. What is the total phase change exerted on the lightwaves if the following parameters are employed for the optical modulation: electro-optic coefficient: $r = 10^{-11}$ m/V......., electrode length = 10 mm. Separation between active electrode and earth electrode = 5 μm.

First, estimate the change in the refractive index due to the applied voltage via the electro-optic effect and then:

The change of the velocity of the lightwave, the total phase change over the electrode length; note that a phase change of 2π is equivalent to the slowing down of one wavelength. You may confirm that the phase change can be given by $\Delta\phi = \dfrac{c\Delta n}{n^2} L \dfrac{2\pi}{\lambda}$.

Estimate the voltage that should be applied to the electrode so that a π phase change occurs in the lightwave carrier passing through it.

Write down an expression that represents the lightwave at the input and for the split waves propagating in the two parallel lightpaths of the interferometer. Represent their phasors in a plane.

Find the sum of the two phasors, and project this total phasor vector on the real axis. Obtain the equation of the lightwave at the output of the modulator as a function of the applied voltage V.

Problem 3.3

Figure Problem 3.3 shows the pulse sequence and its optical modulated spectrum for NRZ and RZ formats.

What is the bit rate of the sequence?

Examine the spectra, and identify their principal features such as the lightwave carrier and its power, the 3 dB bandwidth of the passband spectrum. Compare the 3 dB of RZ and NRZ modulation formats. What modulation schemes can we derive from the spectra?

Problem 3.4

A carrier-suppressed RZ optical transmitter is shown below.

Note: The vertical scale is from 10 dBm and 10 dB per division.

NRZ modulation with cosine edge pulses

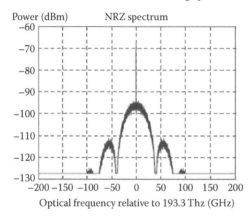
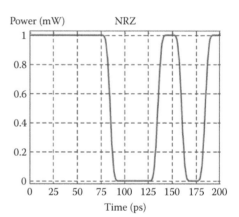

RZ modulation with 50% duty cycle

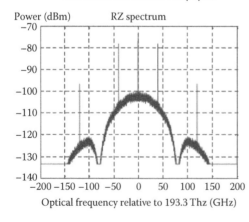
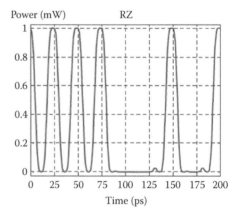

a. Sketch the time-domain pulse sequence over a ten-bit period for a bit rate of 40 Gbps at the output of the pulse pattern generator.

b. Give a brief description of the principles of the suppression of the carrier. Which component of the transmitter would implement the suppression?

c. What are the functions of the laser, the modulator, and the push–pull modulator? If V_π for the two modulators is 5 volts, sketch the transfer characteristics of the modulators—that is, the output power versus the input driving voltage. Make sure that you set appropriate biasing voltages for the modulators. The output power of the laser is 10 dBm, and the total insertion loss for each modulator is 4 dB. For the pulse pattern generator, the output power at the output port data is 10 dBm and that at the clock output port is 2 Vp-p. All line impedances are 50 ohms.

d. Is it necessary to use a booster optical amplifier to increase the total average power launching into an optical fiber for transmission? If it is, then what is the gain and noise of the optical amplifier? Note that the nonlinear limit of an SSMF is around 5 dBm.

e. Sketch the spectra at the outputs of the laser, the modulator, and the push–pull modulator.

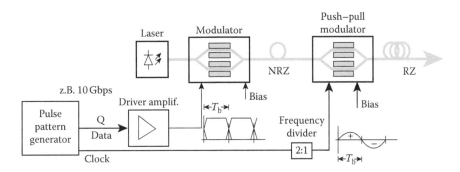

Problem 3.5

This question pertains to an optical modulator and phasor diagram of a dual-drive MZM.

a. Repeat Question 3 with a bias voltage of $V_b = V_\pi/2$ and V_π and a time-varying signal of $v_s(t) = V_{\pi/2}\cos\omega_s t$ with $\omega_s = 2\pi f_s$ and $f_s = 20$ GHz.

b. Now the modulator is a dual-drive MZM; repeat Question 1 with a bias voltage of $V_b = V_\pi/2$ and V_π and a time-varying signal of $v_s(t) = V_{\pi/2}\cos\omega_s t$ with $\omega_s = 2\pi f_s$ and $f_s = 20$ GHz.

Problem 3.6

An optical fiber communication system consists of an optical transmitter using a 1550 nm DFB laser with a linewidth of 10 pm (picometers), an external optical modulator whose bandwidth is 20 GHz, and a total insertion loss of 5 dB.

The modulator is driven with a bit pattern signal generator with a 10 dBm electrical power output into a 50 ohm line. A microwave amplifier is used to boost the electrical data pulse to an appropriate level for driving the optical modulator. The data bit rate is 10 Gbps, and its format is non-return-to-zero (NRZ).

An 80 km SSMF is used for the transmission of the modulated signals.

Sketch the block diagram of the transmission system.

If V_π of the external modulator is 5 V, what is the gain of the microwave amplifier so that an extension ration of 20 dB can be achieved for the output pulses of 1 and 0 at the output of the modulator? Make sure that you sketch the amplitude and power output of the modulator versus the driving voltage into the modulator. What type of connector which you would recommend to connect the microwave amplifier to the modulator operating with the bandwidth and specified bit rate, similarly the connector between the bit pattern generator and the amplifier?

If the DFB laser emits 0 dBm optical power at its pig tail output, then what is the average of optical power contained in the signal spectrum? You may assume that the pulse sequence generated at the output of the bit pattern generator has a perfect rectangular shape.

What is the effective 3 dB bandwidth of the signal power spectrum? Estimate the total pulse broadening of the pulse sequence at the end of the 80 km fiber length. Similarly, estimate the pulse sequence if the bit rate is 40 Gbps.

Now, if a dispersion-compensating fiber (DCF) of 20 km is used to compensate for the signal distortion in the 80 km fiber, what is the required dispersion factor of this fiber so that there would be no distortion? If the loss of the DCF is 1.0 dB/km at 1550 nm, estimate the average optical power of the signal at the output of the DCF.

Based on the dispersion limit given below, plot the dispersion length as a function of the bit rate for NRZ format.

Note: The dispersion limit, under a linear regime operation, can be estimated by the following equation [40].

$$L_D = \frac{c}{\lambda} \frac{\rho}{B_R^2 D}$$

where B_R is the bit rate, D is the dispersion factor (s/m^2), ρ is the duty cycle ration (the ratio between the ON and OFF in a bit period), and L_D is in meters.

Problem 3.7

Repeat Problem 3.1 for return-to-zero (RZ) format and ASK modulation. Sketch the structure of the optical transmitter. Note that an extra optical modulator must be used and coupled with the data modulator of Problem 3.1, the optical pulse carver. Give details of the pulse carver including the driving voltage, driving signal, and synchronization with the data generator.

Problem 3.8

    a. A DWDM optical transmission system can transmit optical channels whose channel spacing is 100 GHz. What is the spectral efficiency if the bit rate of each channel is 40 Gbps and the modulation is NRZ-ASK?
    b. Repeat (a) for the RZ-ASK modulation format.
    c. Repeat (a) and (b) for a channel spacing of 50 GHz.

Problem 3.9

    a. Give the structure of an optical transmitter for the generation of the RZ-ASK modulation format. Make sure that you assign the optical power of lightwaves generated from the light source and ensure that a maximum of 10 dBm is at the output of the optical modulator. This is then launched into the SSMF, so that it is below the nonlinear SPM effect limit.
    b. Describe the operation of the optical modulator, the pulse carver, so that it can generate a periodic pulse sequence before feeding into the data generator. Make sure that you provide the amplitude and intensity versus the driving signal voltage levels that are used to drive the optical modulators.

Problem 3.10

The nonlinear refractive index coefficient of silica-based SSMF is $n_2 = 2.5 \times 10^{-20}$ m^2/W.

What is the effective area of the SSMF? You can refer to the technical specification of the Corning SMF-28 and its mode field diameter to estimate this area.

Estimate the change in the refractive index as a function of the average optical power. Hence, estimate the total phase change due to this nonlinear effect after propagating though a length L (in kilometers) of this fiber.

Hence, estimate the maximum length L of the SSMF that the lightwaves can travel so that not higher than 0.1 rad of the phase change on this lightwave carrier would occur.

Show how you can generate a format that would have an RZ format and suppression of the lightwave carrier. Show that the width of the RZ pulse in this case is 67% of the bit period. **Hint**: You may represent the lightwaves in the path of the optical modulator, an optical interferometer, by using phasors. First, sketch the phasor of the input lightwave, then those of the two paths and the phase applied to these paths. Then sum up at the output to give the resultant output. For the pulse width, you can estimate the width over which the amplitude falls to 1/sqrt(2) of its maximum.

Now show you can generate an RZ pulse sequence with 50% and 33% pulse width of the bit period.

Problem 3.11

Sketch the schematic diagram of an optical balanced receiver, which would consist of a delay interferometer and a back-to-back connected pair of photodetectors with the output connected to the input of an optical preamplifier.

What is the functionality of the delay interferometer? What is the temporal length of the delay unit?

What are the roles of the two optical couplers and their ideal coupling coefficients?

What is the relationship between the two output ports of the delay interferometer?

Suppose that a sequence of four bits of a DPSK 10 Gbps data channel is presented at the input of a balanced receiver. The phases of the lightwave carrier contained within these four bits are π π o π at the transition of the bit period.

Sketch the carrier wave and the pulse envelope. The lightwave has a wavelength of 1550 nm; however, to illustrate the wave you need only sketch a few periods of the waves contained within the bit period at the input of the receiver.

Sketch the electrical signal at the output of the electronic preamplifier not including noise.

Now, assume that an optical amplifier used as an optical preamplifier is placed at the input of the balanced receiver that would give an optical signal power of −10 dBm for the 0 and 1 of the DPSK sequence. The responsibility of the photodetector is 0.9 A/W, and the electronic preamplifier has a transimpedance of 150 ohms and a total equivalent noise current spectral density of 2 pA/(Hz)$^{1/2}$ and a bandwidth of 15 GHz. Sketch the signal waveform at the output of the electronic preamplifier.

Problem 3.12

Design a block diagram of a precoder that would generate tri-level modified duobinary format signals. Make sure that the coefficients of the filters are specified.

If possible, obtain the precoders for AMI and duobinary and their frequency responses. Compare the frequency responses of the three modulation schemes.

Sketch the structure of the tri-level duobinary precoder with its output of −1, 0, +1.

Show how to use the coded signals to drive a dual-drive MZIM to generate optical duobinary signals.

Problem 3.13

A modulation format that would allow the detection of the modulated signals is duobinary, which is a special case of partial response coding.

Give a brief account of the principles of operation of this line code.

A duobinary coder using a delay and add coding structures is shown below. If three-level duobinary coded signals are required, design the precoder for this type of modulation.

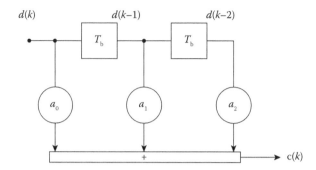

Now, if the delay time Tb is that of a bit period, transform the structure into the z-transform diagram, and hence obtain the transfer function of the coder in the z-domain and thus the frequency response of this coder. Plot the frequency response of the transfer function of the filter in the continuous domain.

Find the impulse response of the coder, and hence the term partial response.

Sketch a block diagram that shows the functionality of precoding, coding, tri-level conversion (offset), and decoding.

A binary sequence d(k) = {0 0 1 1 0 1 0 0 1} is applied to the input of the duobinary coder. Determine the data sequences b(K), c(k), and c'(k) in the electrical domain that can be used to modulate an optical modulator.

Assuming that there is no dispersion in the transmission of the duobinary data sequence, find the output pulse sequence at the output of the decoder. What is the physical realization of the decoder? Sketch the sequence at the output of a decision circuit.

Now, the electrical signals are applied to a microwave amplifier that would condition the signals to appropriate signal levels so as to modulate the optical modulator. The measured spectra are recorded as shown in the diagrams below.

Determine where in the block diagram (as per the attached diagram) each of the spectra belongs to the points of the diagram of the transmission system.

Problem 3.14

DQPSK is a two bits per symbol modulation, that is two bits per symbol, and thus the scheme is spectrally efficient.

Give a brief account of the modulation schemes DPSK and DQPSK.

Give the structure of a precoder for DPSSK – which produces modulation with phase as the codes for 1 and 0.

Now, extend this precoder and the phase quadrature modulation technique for the structure of a DQPSK optical transmitter.

Problem 3.15

Refer to the diagrams shown below for the generation of optical signals with SSB.

State the functionality of the Hilbert transformer. Could you deduce a general principle for suppression of a sideband to generate single-sideband signals?

What is the role of the phase shifter $\pi/2$?

Explain the operation of the optical SSB transmitter, in both the time and the frequency domain. Confirm that the spectrum is correct.

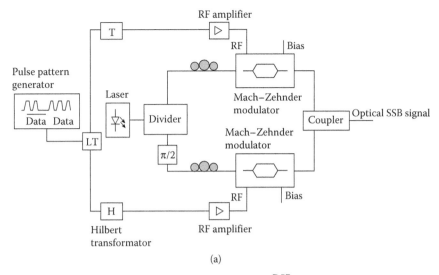

(a)

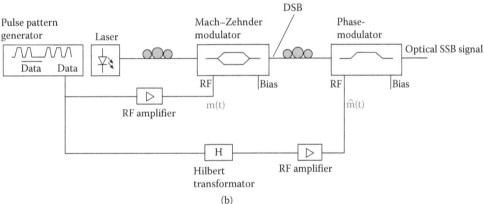

(b)

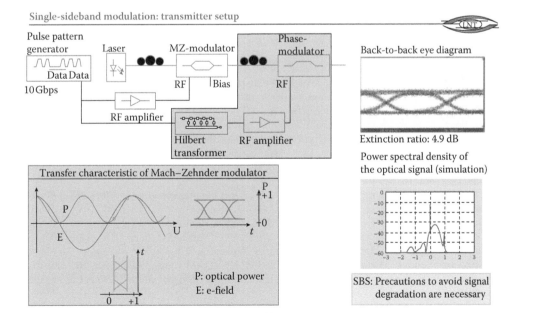

(c)

Problem 3.16

Sketch a structure of an optical coherent receiver. Give a brief description of the roles of each component in your system.

What is the typical modern linewidth of the laser that acts as the local oscillator?

Give a distinction between homodyne and heterodyne coherent systems.

A homodyne optical receiver has the following parameters:

A photodetector with a responsibility of 0.9 is followed by an electronic preamplifier whose total equivalent noise spectral density is 5 pA/(Hz)$^{1/2}$ and an electrical bandwidth of 15 GHz. The transmission bit rate is 10 Gbps.

The local oscillator is a tunable laser source with a linewidth of 100 MHz. The wavelength in vacuum of both the signals and the local oscillator is 1550.92 nm. The average optical power of the local oscillator coupled to the photodetector is 0 dBm.

Sketch the structure of the receiver and then its equivalent small signal circuit, which includes the generated electronic signal current at the output of the photodetector, the total noise currents looking from the input of the electronic preamplifier. What is the dominant noise source in this receiver?

For an optical signal with an average power of −20 dBm, estimate the optical signal-to-noise ratio (OSNR) with 0.1 nm noise power band of the optical signals at the input and the SNR of the signals at the output port of the photodetector.

Recalculate the SNR of the receiver if the frequency of the local oscillator is 20 GHz away from that of the signal carrier frequency.

REFERENCES

1. P. K. Tien, Integrated optics and new wave phenomena in optical waveguides, *Rev. Mod. Phys.*, Vol. 49, pp. 361–420, 1977.
2. R. C. Alferness, Optical guided-wave devices, *Science*, Vol. 234, No. 4778, pp. 825–829, 1986.
3. M. Rizzi and B. Castagnolo, Electro-optic intensity modulator for broadband optical communications, *Fiber Integrated Optics*, Vol. 21, pp. 243–251, 2002.
4. H. Takara, High-speed optical time-division-multiplexed signal generation, *Opt. Quant. Electron.*, Vol. 33, No. 7–10, pp. 795–810, 2001.
5. E. L. Wooten, K. M. Kissa, A. Yi-Yan, E. J. Murphy, D. A. Lafaw, P. F. Hallemeier, D. Maack, et al., A review of lithium niobate modulators for fiber-optic communications systems, *IEEE J. Sel. Topics Quant. Electron.*, Vol. 6, No. 1, pp. 69–80, 2000.
6. K. Noguchi, O. Mitomi, H. Miyazawa, and S. Seki, A broadband Ti: LiNbO$_3$ optical modulator with a ridge structure, *J. Lightwave Technol.*, Vol. 13, No. 6, p. 1164, 1995.
7. J. Noda, Electro-optic modulation method and device using the low-energy oblique transition of a highly coupled super-grid, *IEEE J. Lightwave Technol.*, Vol. LT-4, pp. 1445–1453, 1986.
8. M. Suzuki, Y. Noda, H. Tanaka, S. Akiba, Y. Kuahiro, and H. Isshiki, Monolithic integration of InGaAsP/InP distributed feedback laser and electroabsorption modulator by vapor phase epitaxy, *IEEE J. Lightwave Technol.*, Vol. LT-5, No. 9, p. 127, 1987.
9. H. Nagata, Y. Li, W. R. Bosenberg, and G. L. Reiff, DC drift of X-Cut LiNbO$_3$ modulators, *IEE Photon. Technol. Lett.*, Vol. 16, No. 10, pp. 2233–2335, 2004.
10. H. Nagata, DC drift failure rate estimation on 10 Gb/s X-cut lithium niobate modulators, *IEEE Photon. Technol. Lett.*, Vol. 12, No. 11, pp. 1477–1479, 2000.
11. R. Krahenbuhl, J. H. Cole, R. P. Moeller, and M. M. Howerton, High-speed optical modulator in LiNbO$_3$ with cascaded resonant-type electrodes, *J. Lightwave Technol.*, Vol. 24, No. 5, pp. 2184–2189, 2006.
12. A. Sano, T. Kobayashi, A. Matsuura, S. Yamamoto, S. Yamanaka, Z. Yoshida, Y. Miyamoto, M. Matsui, M. Mizoguchi, and T. Mizuno, 100 x 120-Gb/s PDM 64-QAM transmission over 160 km using linewidth-tolerant pilotless digital coherent detection, in *Proceedings of the OFC 2012*, Los Angeles, CA, 2010.
13. P. Dong, C. Xie, L. Chen, L. L. Buhl, and Y.-K. Chen, 12-Gb/s monolithic PDM-QPSK modulator in silicon, *Opt. Express*, Vol. 20, No. 26, pp. B624–B629, 2012.

14. A. Hirano, Y. Miyamoto, and S. Kuwahara, Performances of CSRZ-DPSK and RZ-DPSK in 43-Gbit/s/ch DWDM G.652 single-mode-fiber transmission, in *Proceedings of the OFC'03*, Vol. 2, pp. 454–456, 2003.
15. A. H. Gnauck, G. Raybon, P. G. Bernasconi, J. Leuthold, C. R. Doerr, and L. W. Stulz, 1-Tb/s (6/spl times/170.6 Gb/s) transmission over 2000-km NZDF using OTDM and RZ-DPSK format, *IEEE Photon. Technol. Lett.*, Vol. 15, No. 11, pp. 1618–1620, 2003.
16. Y. Yamada, H. Taga, and K. Goto, Comparison between VSB, CS-RZ and NRZ format in a conventional DSF based long Haul DWDM system, in *Proceedings of the ECOC'02*, Vol. 4, pp. 1–2, 2002.
17. A. H. Gnauck, X. Liu, X. Wei, D. M. Gill, and E. C. Burrows, Comparison of modulation formats for 42.7-Gb/s single-channel transmission through 1980 km of SSMF, *IEEE Photon. Technol. Lett.*, Vol. 16, No. 3, pp. 909–911, 2004.
18. L. N. Binh, *Digital Processing: Optical Transmission and Coherent Reception Techniques*, New York: CRC Press, 2013.
19. K. K. Pang, *Digital Transmission*, Melbourne: Mi-Tec Publishing, p. 58, 2002.
20. Y. Suzuki, Z. Yamazaki, Y. Amamiya, S. Wada, H. Uchida, C. Kurioka, S. Tanaka, and H. Hida, 120-Gb/s multiplexing and 110-Gb/s demultiplexing ICs, *IEEE J. Solid State Circuits*, Vol. 39, No. 12, pp. 2397–2402, 2004.
21. T. Suzuki, Y. Nakasha, T. Takahashi, K. Makiyama, T. Hirose, and M. Takikawa, 144-Gbit/s selector and 100-Gbit/s 4:1 multiplexer using InP HEMTs, in *IEEE MTT-S International Microwave Symposium Digest*, pp. 117–120, June 2004.
22. X. Liu, S. Chandrasekhar, P. J. Winzer, T. Lotz, J. Carlson, J. Yang, G. Cheren, and S. Zederbaum, 1.5-Tb/s guard-banded superchannel transmission over 56 GSymbols/s 16qam signals with 5.75-b/s/Hz Net spectral efficiency, in *ECOC 2012*, paper Th.3.C.5.pdf ECOC Postdeadline Papers, The Netherlands.
23. H. Yamazaki, T. Yamada, T. Goh, Y. Sakamaki, and A. Kaneko, 64QAM modulator with a hybrid configuration of silica PLCs and LiNbO$_3$ phase modulators for 100-Gb/s applications, in *ECOC 2009*, paper 2.2.1, 2009.
24. I. B. Djordjevic and B. Vasic, 100-Gb/s transmission using orthogonal frequency-division multiplexing, *IEEE Photon. Technol. Lett.*, Vol. 18, No. 15, pp. 1576–1578, 2006.
25. W. Shieh, X. Yi, and Y. Tang, Transmission experiment of multi-gigabit coherent optical OFDM systems over 1000 km SSMF fiber, *Electron. Lett.*, Vol. 43, No. 3, pp. 183–185, 2007.
26. Y. Ma, W. Shieh, and X. Yi, Characterization of nonlinearity performance for coherent optical OFDM signals under the influence of PMD, *Electron. Lett.*, Vol. 43, No. 17, pp. 943–945, 2007.
27. C. Schubert, R. H. Derksen, M. Möller, R. Ludwig, C.-J. Weiske, J. Lutz, S. Ferber, A. Kirstädter, G. Lehmann, and C. Schmidt-Langhorst, Integrated 100-Gb/s ETDM receiver, *IEEE J. Lightwave Technol.*, Vol. 25, No. 1, pp. 122–130, 2007.
28. A. Ali, *Investigations of OFDM Transmission for Direct Detection Optical Systems*, Dr. Ing. Dissertation, Albretchs Christian Universitaet zu Kiel, 2012.
29. F. Amoroso, Pulse and spectrum manipulation in the minimum frequency shift keying (MSK) format, *IEEE Trans. Commun.*, Vol. 24, pp. 381–384, 1976.
30. P. J. Winzer, A. H. Gnauck, C. R. Doerr, M. Magarini, and L. L. Buhl, Spectrally efficient long-haul optical networking using 112-Gb/s polarization-multiplexed 16-QAM, *J. Lightwave Technol.*, Vol. 28, pp. 547–556, 2010.
31. M. Nakazawa, M. Yoshida, K. Kasai, and J. Hongou, 20 Msymbol/s, 64 and 128 QAM coherent optical transmission over 525 km using heterodyne detection with frequency-stabilized laser, *Electron. Lett.*, Vol. 43, pp. 710–712, 2006.
32. X. Zhou and J. Yu, 200-Gb/s PDM-16QAM generation using a new synthesizing method, in *ECOC 2009*, paper 10.3.5, 2009.
33. T. Sakamoto, A. Chiba, and T. Kawanishi, 50-Gb/s 16-QAM by a quad-parallel Mach-Zehnder modulator, in *ECOC 2007*, paper PDP2.8, 2007.
34. Fujitsu Semiconductor Europe GmbH, Building 3, Concorde Park, Concorde Road, Maidenhead, Berkshire SL6 4FJ, UK.
35. J. Hallin, T. Kjellberg, and T. Swahn, A 165-Gb/s 4:1 multiplexer in InP DHBT technology, *IEEE J. Solid State Circuits*, Vol. 41, No. 10, pp. 2209–2214, 2006.
36. K. Murata, K. Sano, H. Kitabayashi, S. Sugitani, H. Sugahara, and T. Enoki, 100-Gb/s multiplexing and demultiplexing IC operations in InP HEMT technology, *IEEE J. Solid State Circuits*, Vol. 39, No. 1, pp. 207–213, 2004.

37. M. Meghelli, 132-Gb/s 4:1 multiplexer in 0.13-μm SiGe-bipolar technology, *IEEE J. Solid State Circuits*, Vol. 39, No. 12, pp. 2403–2407, 2004.

38. M. Nagatani and H. Nosaka, High-speed low-power digital-to- analog converter using InP heterojunction bipolar transistor technology for next-generation optical transmission systems, *NTT Tech. Rev.*, Vol. 9, No. 4, pp. 1–8, 2011.

39. M. Liu, X. Yin, E. Ulin-Avila, B. Geng, T. Zentgraf, L. Ju, F. Wang, and X. Zhang, A graphene-based broadband optical modulator, *Nature*, Vol. 474, pp. 64–67, 2011.

40. F. Forghieri, P. R. Prucnal, R. W. Tkach, and A. R. Chraplyvy, RZ versus NRZ in nonlinear WDM systems, *IEEE Photon. Technol. Lett.*, Vol. 9, No. 7, pp 1035–1037, 1997.

4 Optical Receivers and Transmission Performance

Fundamentals

Signals and noise are important parameters to determine the quality of any engineering system, especially in the telecommunications field. The modulated signals under different modulation formats will be treated in subsequent chapters. This chapter presents an overview of the noise process and mechanism in terms of equivalent noise power and contribution to the optoelectronic process at the optical receiver. Further, the integration of the noise and the statistical process for determination of the receiver sensitivity under transmission of different modulation formats are described together with modeling techniques in the MATLAB® and Simulink® platform. Statistical analyses are also given for the evaluation of the transmission performance, especially when there are no uniform probability distributions.

4.1 INTRODUCTION

Optical detection and the noise interference in such processes are critical to the performance of optical communications systems. A fundamental understanding of the noise and sensitivity of the receiver, the minimum optical power available at the photodetector, requires evaluation of the transmission performance of different modulation formats. Optical receivers have evolved from binary digital direct detection to coherent mixing of signals and local lightwave oscillator and detection with optical preamplification. This chapter presents an overview of the noise process and mechanism in terms of the equivalent noise power and contribution.

A schematic diagram of a single-channel DPSK system is illustrated in Figure 4.1.

As the bit or symbol rate of the optical transmission systems is increased, the demand for modeling is intense, especially for a modeling platform that can simulate the real physical photonic subsystems. In order to enhance the effective transmission capacity with minimum renovation of the photonic and electronic subsystems, there are two possible solutions: (1) increasing the base bit rate or (2) employing a multilevel modulation technique such as the M-ary amplitude and/or phase shift keying [1]. The latter solution would be preferred because for the same as the bit rate, the multilevel offers a much higher transmission capacity (in bits/sec.) but with the same baud rate without increasing the complexity of the transmitters and receivers as well as the dispersion tolerance of the transmission fibers which can be single-mode fiber types ITU G652 or G655. However due to the nonlinear threshold of the fiber the highest level is limited and so an increase in the level of the multilevel modulation system will lower the opening of the eye diagram of sublevels. Thus this may reduce bit error rate.

One possible technique that could offer insights into possible solutions that facilitate the simulation of ultra-high-capacity and ultra-high-bit-rate transmission systems is to develop a comprehensive modeling platform. Further, the modeling platform should take advantage of any user-friendly software platform that is popular, easy to use, and easy to extend. This platform would offer the research community of optical communication engineering a basis for extension and enhance the linkages between research groups.

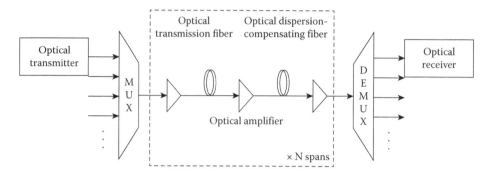

FIGURE 4.1 Schematic diagram of a single-wavelength channel optical transmission system.

Thus, one of the principal objectives of this chapter is to present the development of MATLAB and Simulink 7.0 [2] for computer experiments in optical transmission systems under advanced modulation formats, especially the amplitude and/or phase shift keying modulation of binary or multilevel. Indeed, duobinary in association with an appropriate low-pass filter can offer multiple rates much higher than the basic rate. The advantage of duobinary is that the detection is much simpler than that of DPSK, DQPSK, or M-ary PSK, in which the receiver is a direct detection type receiver. In the photonic domain, the combination of phase and amplitude is taken into account for coding modulation schemes that involve the tri- or higher-order levels. These features can be implemented with ease on the Simulink platform.

We base our system bit rate on a 40 Gbps per channel. The modulation formats of binary RZ, NRZ, and CSRZ-ASK-DPSK are demonstrated with the modeling of the transmission over standard single-mode optical fiber (SSMF), as well as the photonic decoding of this optical signaling. The transmission performance of some optical systems is also given in terms of the bit error rate (BER), the δ-factor (usually known as Q factor), and the receiver sensitivity.

In this chapter, we present, to the best of our knowledge, the first MATLAB and Simulink platform simulation test bed for modeling of the transmission of the amplitude and differential phase modulation incorporating RZ, NRZ, and carrier-suppressed RZ formats for advanced high-capacity and long-haul optical fiber transmission systems. A novel, modified fiber propagation algorithm has been developed and optimized to minimize the simulation processing time and enhance its accuracy. The performance of optical transmission systems can be automatically and accurately evaluated with various methods. Analysis and methodologies for future development of the simulation test bed are also presented.

In recent years, the capacity and reach of optical transmission systems have dramatically increased owing to the accelerating growth of data usage demand (Internet, peer-to-peer network, …). Therefore, the necessity of upgrading of current dense wavelength division multiplexing (DWDM) 10 Gbps systems to DWDM 40 Gbps or even higher bit rates becomes crucial to telecommunications service providers. Accordingly, several modulation formats have been proposed and investigated as the alternatives to the current on–off keying (OOK) intensity modulation, which is severely degraded at high bit rates owing to dispersion and nonlinear effects of the transmission fibers. Apart from the requirements of robustness to the transmission impairments, the cost-effectiveness of system upgrades is also significant. Among the candidates, differential binary and/or quadrature phase shift keying (DPSK/DQPSK) have recently attracted much attention owing to the following advantages: (1) a 3 dB improvement on receiver sensitivity (if the balanced receiving technique is used) [1,2]; (2) high tolerance to fiber nonlinearities, especially to intrachannel nonlinear effects, cross-phase modulation (XPM), and four-wave mixing (FWM) [2,3]; (3) superior spectral efficiency (DQPSK), and hence high tolerance to optical filtering [3]; and finally, (4) advantages in all-optical networks incorporating optical add-drop multiplexers or optical cross-connects [4]. Recently, several experimental demonstrations of DPSK/DQPSK long-haul transmission DWDM systems for 10 and 40 Gbps have been reported [5,6]. Therefore, a simulation test bed is necessary for detailed design, investigation, and verification of the benefits and shortcomings of these advanced modulation formats on fiber-optic transmission systems. In this chapter, we present, to the best of our knowledge, the first MATLAB-and Simulink-based simulation package for photonic DWDM systems. The simulator is still in the first phase of the development, with the focus on single-channel systems, and it is being continuously improved and updated.

This chapter is organized as follows: (i) in Section 4.2, receiver noise sources are outlined; noise analyses and performance of basic binary digital optical receivers are then given in Section 4.3; (ii) Section 4.3 presents the architecture and operational principles of the simulator; (iii) Section 4.4 presents simulation results of the up-to-date simulator, and the corroboration of the developed models in comparison with experimental results are presented in Section 4.5 for both ASK and DPSK modulation formats. Finally, concluding remarks are presented.

4.2 DIGITAL OPTICAL RECEIVERS

This section gives fundamental concepts of receiver structure, sensitivity, and noise. It is mainly focused on direct detection, which would be integrated into more complex receiver structures for different modulation formats.

4.2.1 PHOTONIC AND ELECTRONIC NOISE

Before proceeding to the system calculations to determine the performance of optical receivers, it is necessary to consider the noise generated in the photodetector and the preamplifier front end. One can, in fact, consider investigating the system calculation and returning back to the noise calculation provided that they are taking either the total equivalent noise spectral density at the input of the detector or its noise figure. This section describes all noise mechanisms related to the photodetection process, including the electronic noise associated with the receiver.

4.2.1.1 Electronic Noise of Receiver

At the receiver, noise sources that are superimposed onto the electrical signals detected after photodiodes include electrical shot noise, dark current noise, and the equivalent noise current as seen from the input of the electronic preamplifier. The electronic noise sources, represented as the square of current, can be found by using the noise spectral density as a function of frequency and then integrating over the bandwidth of the system. They can be given as

$$i_{\text{Nshot}}^2 = 2q\Re P_{\text{in}} B_{\text{e}} \tag{4.1}$$

$$i_{\text{ND}}^2 = 2q I_{\text{D}} B_{\text{e}} \tag{4.2}$$

$$i_{\text{EA}}^2 = (I_{\text{eq}})^2 B_{\text{e}} \tag{4.3}$$

4.2.1.2 Shot Noise

Shot noise and thermal noise are the two most significant noises in optical detection systems. Shot noise is generated by either a quantum process or electronic biasing. The noise is specified in noise spectral density, that is, the square of the noise current per unit frequency. Thus, the noise spectral density is to be integrated over the total amplifier bandwidth to obtain the equivalent noise currents.

Electrical shot noise is generated by the random generation of streams of electrons (current). In optical detection, shot noise is generated by (1) biasing currents in electronic devices and (2) photo currents generated by the photodiode.

4.2.1.2.1 Biasing Current Shot Noise

Any biasing current I has a spectral current density S_I given by

$$s_I = \frac{d(i_I^2)}{df} = 2qI \ \text{ in } \ A^2/\text{Hz} \tag{4.4}$$

where q is the electronic charge. The current i_I represents the noise current generated due to the biasing current I.

4.2.1.2.2 Quantum Shot Noise

The average current $<i_s^2>$ generated by the photodetector by an optical signal with an average optical power P_{in} is given by

$$s_Q = \frac{d\langle i_s^2 \rangle}{df} = 2q\langle i_s^2 \rangle \tag{4.5}$$

This is the signal-dependent noise, and it is a unique feature of optical communications. When the avalanche photo-detector (APD) is used, the noise spectral density is given by

$$s_Q = \frac{d\langle i_s^2 \rangle}{df} = 2q\langle i_s^2 \rangle\langle G_n^2 \rangle \tag{4.6}$$

It is noted here that the dark currents are generated in the photodetector is a natural phenomenon in semiconductor material operating at room temperature. This dark current noise must be included in the total equivalent noise current as referred to the input. These currents are generated even in the absence of an optical signal. These dark currents can be eliminated by cooling the photodetector to at least below the temperature of liquid nitrogen ($77°K$).

4.2.1.3 Thermal Noise

At a certain temperature, the conductivity of a conductor varies randomly. The random movement of electrons generates a fluctuating current even in the absence of an applied voltage. The thermal noise of a resistor R is given by

$$s_R = \frac{d\left(i_R^2\right)}{df} = \frac{2k_B T}{R} \tag{4.7}$$

where k_B is the Boltzmann's constant, T is the absolute temperature (in $°K$), R is the resistance in ohms, and i_R denotes the noise current due to R.

4.2.1.4 ASE Noise of Optical Amplifier

The following formulation accounts for all noise terms that can be treated as Gaussian noise due to the optical amplifier.

$$N_{ASE} = mn_{sp}hv(G_{op} - 1)B_o \tag{4.8}$$

where G_{op} = amplifier gain; n_{sp} = spontaneous emission factor; m = number of polarization modes (1 or 2); B_o = mean noise in bandwidth; optical signal-to-noise ratio (OSNR) at the output of EDFA.

4.2.1.5   Optical Amplifier Noise Figure

The amplifier noise figure (NF) is defined at the output of the optical amplifier as the ratio of the output OSNR at the output to that OSNR at the input of the EDFA.

$$F_N = \frac{OSNR_{in}}{OSNR_{out}} \approx 2n_{sp} \text{ for } G_{op} \gg 1 \tag{4.9}$$

Modern optical amplifiers are optimized, and n_{sp} reaches unity; so the NF can reach 3 dB.

4.2.1.6   Electronic Beating Noise

When an optical preamplifier is used in association with an optoelectronic receiver front end, then there would be noise generated due to: the signal-dependent shot noise, beat noise from the beating of the electronic currents generated from the signals, random ASE noise, and the beating noise between the ASE electronic currents. Usually, the signal ASE noise dominates this noise process.

The beating noise generated from the beating between the signal current and the ASE noise current, dominates the detection process in an optical preamplifying optoelectronic receiver and is given by

$$i^2_{sig-ASE} = 2(q\eta G_{op})^2 (2n_{sp}) P_{in} \frac{B_e}{h\upsilon} \tag{4.10}$$

where B_e is the bandwidth of the electronic amplifier system.

Under direct detection, the ASE noise appearing at the input of an electronic preamplifier would follow square detection; that is, the noise vector would be taken with its absolute value, then squared, and multiplied by the responsibility of the detector to obtain the spectral density noise current.

On the contrary, under coherent detection, the ASE noise is superimposed on the electric field of the local oscillator and the signals. The local oscillator amplitude is much higher than that of the signal, and the beating noise between the local oscillator and the ASE noise dominates the noise source presented at the input of the electronic preamplifier. This is the significant difference between the optically amplified and the nonoptically amplified receiver for direct and coherent optical receivers, respectively.

4.2.1.7   Accumulated ASE Noise in Cascaded Optical Amplifiers

Long-haul optical communications would be structured with dispersion managed and loss equalized through tens or hundreds of spans. Error-free transmission over several thousands of kilometers without repeaters has been demonstrated. It is critical to account for the accumulated noise sources that ultimately limit the transmission distance. The amplified spontaneous emission (ASE) noise accumulated is the principal noise source that has been built up over several optical amplifiers in cascade that would be originally from the photons generated randomly under spontaneous emission. The spontaneous noise factor is given as n_{sp}. Practical EDFAs have now reached a mature stage with n_{sp} reaching unity. In this section, an effective noise factor at the end of the transmission spans or at the input of the optical receiver is derived.

Let the linear loss factor of the ith transmission fiber be $\alpha_{L,I}$ and G_i be the linear gain factor of the optical amplifier in the span; then a recursive relation between the noise of the first stage and ith stage can be written as

$$n''_{sp,1} = n'_{sp,1}$$

$$n''_{sp,i} = \frac{n''_{sp,i-1} G_{i-1}}{\alpha_{L,i-1}} + n'_{sp,i} \text{ for } 2 \leq i \leq N \tag{4.11}$$

where $n'_{sp,1}$ is the noise as seen from the input of the amplifier, and $n''_{sp,1}$ is the equivalent noise factor at the output of the amplifier. Then, the accumulated ASE noise over the transmission spans is given by the sum of all the noise sources of the amplifiers. At the output of the final Nth stage, without taking into account the other noise sources, the equivalent ASE accumulated noise is given as

$$n''_{sp,N} = \sum_{i=1}^{N} n'_{sp,i}$$

$$n^e_{sp,i} = \frac{n''_{sp,N}G_N}{\alpha_L} + n_{sp,i}$$

(4.12)

under the assumption that the gain of optical amplifiers has equalized all the losses of transmission and dispersion-compensating fibers (DCFs).

When an optical preamplified receiver and two optical filters are placed before and after the optical preamplifier, the effective ASE noise is given by

$$n^e_{sp} = \frac{n''_{sp,N}G_N\tau_2}{\alpha_L\tau_1} + n_{sp}$$

(4.13)

where τ_1 and τ_2 are the time constants of the two optical filters.

The NF of the equivalent accumulated noise is then given by

$$NF^e = 10Log_{10}n^e_{sp}$$

(4.14)

This is also the amount of degradation of the OSNR due to the accumulated ASE noise at the front of the photodetector.

4.3 PERFORMANCE EVALUATION OF BINARY AMPLITUDE MODULATION FORMAT

As described earlier, the noise of the optical receivers consists of the thermal noise and quantum shot noise due to the bias currents and the photocurrent generated by the photodetector with and without the optical signals. Thus, this quantum shot noise is strongly signal dependent. These noises degrade the sensitivity of the receiver [1] and thus a penalty can be estimated. Another source of interference that would also result in signal penalty is the intersymbol interference (ISI). The goal of this section is to obtain an analytical expression for the receiver sensitivity of the direct detection optical receiver ON/OFF keying (OOK) system.

4.3.1 RECEIVED SIGNALS

Assuming that the received signal power is

$$p(t) = \sum_{j} a_j h_p(t - jT_b)$$

(4.15)

The average output voltage at the output of the electronic preamplifier is thus given by

$$\langle v_o \rangle = \Re\langle G_n \rangle \left[\sum_{j} a_j \frac{1}{T_b} \int_{-T_b/2}^{T_b/2} h_p(t - jT_b)dt \right] R_I A$$

(4.16)

where T_b is the bit period, and $h_p(t-jT_b)$ is the impulse response of the system evaluated at each time interval. R_I is the input resistance of the overall amplifier of the system including both the front end and the linear channel amplifier. It is assumed that the overall amplifier has a flat gain response A over the bandwidth of the system.

$$a_j\big|_{t=0} = \begin{cases} b_0 \approx 0 \\ b_1 \end{cases} \tag{4.17}$$

with b_0 being the energy when a transmitted 0 is received, and b_1 is the energy when a transmitted 1 is received. The sum over a number of periods is necessary to take into account the contribution of adjacent optical pulses. We now have to distinguish between two cases when a 0 or a 1 is transmitted and received.

4.3.1.1 Case 1: OFF or a Transmitted 0 Is Received

Using Equations 4.12 and 4.13, we have

$$\langle v_o \rangle_o = v_{oo} = \Re \frac{b_0}{T_b} GR_I A \cong 0 \tag{4.18}$$

with the total equivalent noise voltage at the output, v_{NTo}^2 is

$$v_{NTo}^2 = v_{NA}^2 = i_{Neq}^2 R_I^2 A^2 \tag{4.19}$$

where i_{Neq} is the total equivalent current at the input of the electronic preamplifier. Appendix 2 (Section 4.12) and Annex 4 gives a method for estimating this noise current for any preamplifier whose equivalent Y-parameters are known.

4.3.1.2 Case 2: ON Transmitted 1 Received

In this case, the average signal voltage at the output is received as

$$\langle v_o \rangle_1 = v_{01} = \Re \frac{b_1}{T_b} \langle G_n \rangle R_I A \tag{4.20}$$

with the total noise equivalent mean voltage at the output given by

$$v_{NT1}^2 = v_{oSN}^2 + v_{NA}^2 \tag{4.21}$$

where v_{oSN}^2 is the signal-dependent shot noise. v_{NA}^2 is the amplifier noise at the output and is given by

$$v_{NA}^2 = i_{Neq}^2 R_I^2 A^2 B \tag{4.22}$$

The signal-dependent noise is, in fact, the quantum shot noise and is given by

$$v_{oSN}^2 = \int_0^B 2q \langle i_s \rangle_1 \langle G_n^2 \rangle R_I^2 A^2 \, df \tag{4.23}$$

where B is the 3 dB bandwidth of the overall amplifier, $\langle i_s \rangle_1$ is the average photocurrent received when a 1 was transmitted. This current can be estimated as follows

$$\langle i_s \rangle_1 = \sum_{-\infty}^{\infty} \Re \frac{b_1}{T_b} \int_{-T_b/2}^{T_b/2} h_p(t - jT_b) dt \tag{4.24}$$

or

$$\langle i_s \rangle_1 = \Re \frac{b_1}{T_b} \int_{-\infty}^{\infty} h_p(t) dt \tag{4.25}$$

with a normalization that is $\int_{-\infty}^{\aleph} h_p(t) dt = 1$.

Equation 4.21 becomes

$$\langle i_s \rangle_1 = \Re \frac{b_1}{T_b} \tag{4.26}$$

4.3.2 Probability Distribution Functions

Though the optical systems under consideration are typical IM/DD systems, they are also suitable for further calculations for modulation formats in digital optical communications. Further development of statistical distributions can be referred to in Section 4.4. In these systems, the optical energy of each pulse period is equivalent to that of at least a few hundred photons. This number is large enough so that a Gaussian distribution of the probability density is warranted. The probability density function (pdf) of a 0 transmitted and received by the optical receiver is thus given by

$$p[v_0|\text{``0''}] = \frac{1}{(2\pi v_{\text{NT0}}^2)^{1/2}} \exp\left[\frac{-(v_0 - v_{00})^2}{2v_{\text{NT0}}^2} \right] \tag{4.27}$$

and similarly for a "1" transmitted

$$p[v_0|\text{``1''}] = \frac{1}{(2\pi v_{\text{NT1}})^{1/2}} \exp\left[\frac{(-v_0 - v_{01})^2}{2v_{\text{NT1}}^2} \right] \tag{4.28}$$

The total probability of error or a BER is defined as

$$\text{BER} = p(1)p(0/1) + p(0)p(1/0) \tag{4.29}$$

where $p(1)$ and $p(0)$ are the probabilities of receiving a 1 and a 0, respectively, and $p(1/0)$ and $p(0/1)$ are the probabilities of deciding 1 when a 0 is transmitted and vice versa. In an OOK bit stream, 1 and 0 are likely to occur equally, that is, $p(1) = p(0) = 0.5$; then Equation 4.25 becomes

$$\text{BER} = \frac{1}{2}\left[p(0/1) + p(1/0) \right] \tag{4.30}$$

Thus, as an equal probability of transmitting a 0 and a 1 is assumed, for a decision voltage level of d, the total probability of error P_E is the sum of the errors of deciding 0 or 1. Integrating the pdf over the overlapping regions of the two pdfs gives

$$P_E = \frac{1}{2} \int_d^\infty p[v_0|\text{``0''}] dv_0 + \frac{1}{2} \int_{-\infty}^d p[v_0|\text{``1''}] dv_0 \tag{4.31}$$

Substituting for the probability distribution using Equations 4.26 and 4.27 leads to

$$\text{BER} = \frac{1}{2\sqrt{\pi}} \int_{\frac{d-v_{00}}{v_{\text{NT}\,0}}}^\infty e^{-\frac{x^2}{2}} dx + \frac{1}{2\sqrt{\pi}} \int_{\frac{v_{01}-d}{v_{\text{NT}1}}}^d e^{-\frac{x^2}{2}} dx \tag{4.32}$$

The functions in Equation 4.28 have the standard form of the complementary error function $Q(\alpha)$ defined as

$$Q(a) = \frac{1}{\sqrt{2\pi}} \int_\alpha^\infty e^{-x^2/2} dx \tag{4.33}$$

and then Equation 4.28 becomes

$$\text{BER} = \frac{1}{2}\left[Q\left(\frac{d}{v_{\text{NT0}}}\right) + Q\left(\frac{v_{01}-d}{v_{\text{NT1}}}\right) \right] \tag{4.34}$$

4.3.3 RECEIVER SENSITIVITY

Again, using the condition $p[v_0/\text{``0''}] = p[v_0/\text{``1''}]$, we have

$$\frac{d}{v_{\text{NT0}}} = \frac{v_{01}-d}{v_{\text{NT1}}} \equiv \partial o \tag{4.35}$$

Now assuming $v_{o0} = 0$, hence

$$\text{BER} = Q(\partial) \tag{4.36}$$

The Marcum $Q(\delta)$ function is a standard function, and this curve is shown in Figure 4.2. Note that for a BER = 10^{-9}, the value of δ is about 6, which is the normal standard for communications at bit rates of 134 Mb/s–40 Gbps.

Thus, by eliminating the decision-level variable d from Equation 4.31, we obtain

$$\partial = \frac{v_{01} - \partial v_{\text{NT0}}}{v_{\text{NT1}}} \tag{4.37}$$

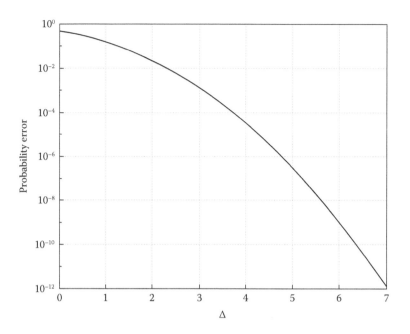

FIGURE 4.2 BER as a function of the δ parameter for NRZ-ASK.

or alternatively

$$v_{01} = \partial(v_{NT0} + v_{NT1}) \tag{4.38}$$

substituting v_{01}, v_{NT1}, v_{NT0}, we have

$$\Re \frac{b_1}{T_b}\langle G_n \rangle R_1 A = \partial\left\{ \left[2\Re q \frac{b_1}{T_b}\langle G_n^2 \rangle BR_1^2 A^2 + v_{NA}^2 \right]^{1/2} + v_{NA} \right\} \tag{4.39}$$

However, the amplifier noise voltage v_{NA} is given by $v_{NA} = i_{Neq} R_1 AB^{1/2}$. Thus, by substituting this noise voltage, eliminating $R_1 A$, and solving for the energy required for the 1 transmitted and received at the photodetector b_1, we obtain

$$b_1 = q\Re\left(\delta^2 G + \frac{2i_{Neq}\delta T_b}{qG} \right) \tag{4.40}$$

where we have used the approximation $\langle G_n \rangle = G$ and $\langle G_n^2 \rangle = G^{2+x}$ with x being the factor dependent on the ionization ratio in an APD. For a pin photodetector, $G = 1$. The optical receiver sensitivity in dBm, denoted as RS in dBm, can thus be obtained as

$$RS = 10\text{Log}_{10}\frac{p_{av}}{p_0} \text{ in dBm} \tag{4.41}$$

where $P_{av} = b_1/T_b$ and $P_o = 1.0$ mW. This is the optical receiver sensitivity defined as the minimum optical power required for the receiver to operate reliably with a BER below a specific value. In Equation 4.36, there are two terms clearly specifying the dependence of the signal dependence δ

and the amplifier noise contribution at a certain data rate and for a certain modulation format. Recall that the term b_1 represents the optical energy of the 1 required for the optical receiver to detect with a certain BER. The term q/R in the right-hand side of Equation 4.36 is equivalent to the optical power required to generate one electron, or the number of photons required to generate one electron. A typical measure for the optical receiver is the number of photons required for it to operate with a specified BER.

Note that the common term of the energy b_1 is q/R, which is equivalent to about one photon energy; we can rewrite Equation 4.36 by using $q/R = h\upsilon/\eta$

$$b_1 = \frac{h\upsilon}{\eta}\left(\delta^2 G + \frac{2i_{\text{Neq}}\delta T_b}{qG}\right) \tag{4.42}$$

Thus, we can observe from equation 4.42 when we have an ideal electronic amplifier, that is, a noiseless amplifier $i_{\text{Neq}} = 0$, then it requires only a number of photon energy of $\delta^2 G$ for error-free detection. The second term in the bracket is thus the number of photons required to overcome the amplifier noise. Note that the determination of the BER using the eye diagram when the pdf is not Gaussian is inaccurate. However, a much more accurate statistical analysis is developed and presented later in Section 4.9.

4.3.4   OSNR and Noise Impact

4.3.4.1   Optical Signal-to-Noise Ratio

For an optically preamplified receiver, the ASE dominates the noise processes, and an OSNR is commonly used that is determined by measuring the noise and optical signal levels in both polarization directions under an optical filtering bandwidth B_0 of 0.1 nm [2].

$$\text{OSNR} = \frac{\text{optical} - \text{signal} - \text{power}}{\text{optical} - \text{noise} - \text{power}} \tag{4.43}$$

For an average output power P_0 of the EDFA, the OSNR can be determined by

$$\text{OSNR} = \frac{P_0}{2n_{\text{sp}}hf_T(G-1)B_0} \tag{4.44}$$

where n_{sp} is the number of spontaneous emissions of the EDFA, h is Plank's constant, and G is the linear power gain coefficient of the EDFA. Then, if the signal ASE beat noise dominates the noises of the detection process, an NF of the EDFA can be used to determine the ratio of the OSNR at the input and output as this is much more practical. Thus, the NF can be approximated by $F_n \sim 2n_{\text{sp}}$. Thus, the OSNR can be rewritten as

$$\text{OSNR} = \frac{P_0}{F_n hf_T(G-1)B_0} \tag{4.45}$$

4.3.4.2   Determination of the Impact of Noise

One way to measure the impact of noise on the received signals of optically amplified systems is by attenuating the signal power at the input of the optical preamplifier and then obtaining the BER. A BER versus received power plot would be obtained. Assuming that only one span is used and the optical gain is much greater than unity, the OSNR can be written for an operating wavelength of 1550 nm as

$$OSNR = 58\,(dBm) - F_n\,(dB) + P_{in}\,(dBm) \tag{4.46}$$

If an eye diagram is used, then the eye opening penalty (EOP) can be determined for the same BER. The EOP can be written as

$$EOP = -10Log_{10}\frac{EO}{EO_n}.....dB \tag{4.47}$$

4.4 QUANTUM LIMIT OF OPTICAL RECEIVERS UNDER DIFFERENT MODULATION FORMATS

Instead of using the equivalent noise current density or the noise-equivalent power (NEP), optical receiver front ends are sometimes also characterized in terms of their receiver sensitivity. While the receiver sensitivity is undoubtedly of great interest in optical receiver design, it comprises not only the degrading effects of noise but also encompasses the essential properties of the received signal, such as the extinction ratio, signal distortions, and ISI, generated either within the transmitter or within the receiver itself. Thus, knowledge of the receiver sensitivity alone does not allow trustworthy predictions on how the receiver will perform for other formats.

Although electronic noise usually dominates shot noise, it can be squeezed to zero. The signal-dependent shot noise, however, is fundamentally present. The limit, when only fundamental noise sources determine receiver sensitivity, is called the quantum limit in optical communications. The existence of quantum limits makes optical receiver design an exciting task, because there is always a fundamental measure against which practically implemented receivers can be compared. Note, however, that each class of receivers in combination with each class of modulation formats has its own quantum limit.

From Equation 4.36, we can observe that when the amplifier is noiseless, the receiver would require an energy equivalent to that of $\delta^2 G$ photon energy for detection. This is the quantum limit of the receiver. For example, for a BER of 1e–9 for a pin detector under ASK modulation, we would need 36 photon energy per bit for detection when both the 1 and 0 are 50–50 randomly received.

Noise plays a major part in the receiving end of any communication system. In optical communications using coherent or noncoherent detection techniques, noise is contributed by (1) the electronic noise of the electronic amplifications following the optoelectronic processes in the photodetector, (2) The quantum shot noise due to the electronic current generated by the optical signals, (3) the quantum shot noise due to the high power of the local oscillator (additional and dominant source of coherent detection), and (4) the beating of the local oscillator and the optical signals.

These noise sources vary from one optical receiver to another depending on their structures and whether they consist of a photodetector and electronic amplifiers with and without optical preamplification under coherent or noncoherent detection. This section thus presents the fundamental issues of the noise processes and their impact on the sensitivity of receiving systems. In particular, we examine the quantum limits of the optical receivers, that is, when the electronic noise is considered to be nullified.

Schematic diagrams of coherent and noncoherent direct detection receivers are shown in Figure 4.3a and b without using an optical amplifier, while Figure 4.3c and d show their counterparts with optical amplifiers. Figure 4.3f and g show the balanced and fiber versions of the detection and receiving systems. The difference between these configurations is the noise generated after the photodetection and at the input of the electronic amplifier. Note that coherent systems are identified with the mixing of the optical signals and local oscillator whose polarization directions are aligned with each other.

4.4.1 Direct Detection

Optical detection for optical fiber communications is in the form of direct modulation and direct detection. Direct detection is the simplest form of detection, which requires only a photodetector followed by an electronic amplifier and the decision circuitry, clock recovery, and data recovery.

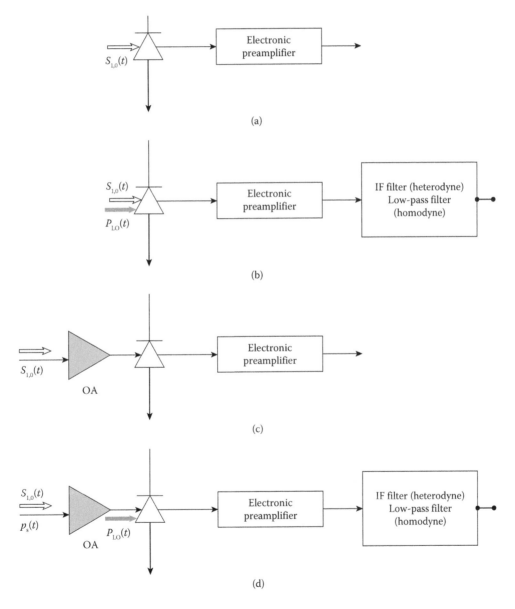

FIGURE 4.3 Schematic diagram of (a) direct detection, (b) coherent detection, (c) direct detection with OA, and (d) coherent detection with OA. OA = optical amplifier, FC = fiber coupler, PD = photodetector, DSP = digital signal processor, ADC = analog-to-digital converter. *(Continued)*

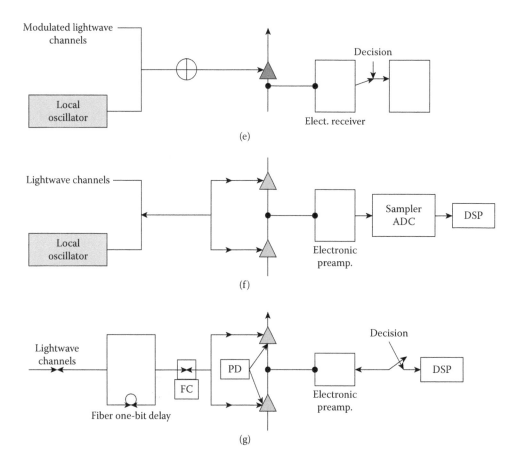

(e)

(f)

(g)

FIGURE 4.3 (Continued) Schematic diagram of (e) fiber version of coherent receiver, (f) coherent receiver using balance detection with two photodetectors connected back to back and a local oscillator (laser), (g) self-homodyne reception balanced receiver with one-bit delay photonic phase comparator. OA = optical amplifier, FC = fiber coupler, PD = photodetector, DSP = digital signal processor, ADC = analog-to-digital converter.

For the ASK system, the signals for the 1 and 0 can be expressed in terms of the amount of photon energy over the entire bit period, contained within a bit period T as

$$s_{1,0}(t) = \begin{cases} \dfrac{n_p}{T} & \text{for "1"} \\[2mm] 0 & \text{for "0"} \end{cases} \qquad 0 \le t \le T \tag{4.48}$$

where n_p is the number of photons, and the energy of the lightwave is normalized with a single photon energy at the operating wavelength. Thus, the total optical energy is

$$E_s = n_p h \upsilon \cdot T \tag{4.49}$$

Thus, the electronic current generated after the photodetector, with R being the responsivity of the photodetection, is

$$\langle i_s \rangle = n_p h \upsilon \cdot T \Re = n_p h \upsilon \cdot T \frac{\eta q}{h \upsilon} \tag{4.50}$$

Thus, we could say that n_p is the number of photons per bit required for the detection of a 1 if there is no noise contributed by the electronic amplifier or detection.

With the probability of 1 and 0 being equal (50%), the probability of error of the detection is

$$P_e = \frac{1}{2} e^{-n_p} \tag{4.51}$$

Thus, for a BER of 1e–9 the argument $n_p = (3\sqrt{2})^2$ or $n_p = 18$ with an allowance of a ½ factor of the single-sided estimation so that $n_p = 20$ for the full detection error. This is the super quantum limit. We also assume a unity responsibility of the photodetection.

4.4.2 COHERENT DETECTION

In the case of coherent homodyne detection with a local oscillator whose optical power P_{LO} is very much larger than the signal average power, the quantum shot noise current dominates the noise process. The detected electronic current is the beating current between the local oscillator lightwave and that of the signal, and thus we have

$$\langle i_{N(LO)}^2 \rangle = 2qP_{LO} \frac{\eta q}{h\upsilon} \frac{1}{T} = 2q^2 P_{LO} \frac{\eta}{h\upsilon} \frac{1}{T} \tag{4.52}$$

and the SNR is given by

$$\text{SNR} = \frac{\langle i_s^2 \rangle}{\langle i_{N(LO)}^2 \rangle} = \frac{4\Re^2 p_s(t) P_{LO}/T}{2q\Re P_{LO}/T} = 2n_p \tag{4.53}$$

$$P_e = \text{erfc}(2n_p) \tag{4.54}$$

Thus, for 10^{-9} BER, we have $2n_p = 3(2)^{1/2}$.

4.4.3 COHERENT DETECTION WITH MATCHED FILTER

Now, it is assumed that a matched filter is inserted after the coherent receiver of Figure 4.3b, and different modulation formats are used for transmission over long-haul optically amplified fiber systems.

In general, and under the assumption that the noise process in the optical detection is Gaussian, the BER is given by

$$\text{BER} = \text{erfc}\left(\frac{d}{N_0}\right) \tag{4.55}$$

where d is the signal power separation between the average level of the 1 and 0 for binary systems and the equidistance between the constellation points of the modulation scheme as shown in Figure 4.4. Let E_1 and E_0 be the field amplitudes of the signals 1 and 0; then, the Euclidean distance d is given by

$$d^2 = E_1^2 + E_0^2 - 2\rho E_1 E_0$$

$$\textit{with } \rho = \frac{2}{T} \int_0^T s_1(t) s_0(t) \, dt \tag{4.56}$$

ρ is the correlation coefficient between the two logic levels or, alternatively, the Euclidean angle between the two vector signals as represented on the scattering plane of the constellation.

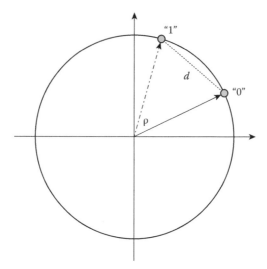

FIGURE 4.4   Signal constellation and energy level and the geometrical distance between 1 and 0.

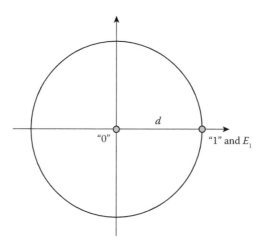

FIGURE 4.5   Signal constellation, energy level, and the geometrical distance between 1 and 0 of ASK system.

4.4.3.1   Coherent ASK Systems

In heterodyne detection with an IF frequency region, the two-bit signals (see also the constellation in Figure 4.5) are given by

$$s_{1,0}(t) = \begin{cases} \sqrt{\dfrac{2n_\mathrm{p}}{T}}\cos\omega_{\mathrm{IF}}t & \text{for} \quad \text{``1''} \\ 0 \quad \text{for} \quad \text{``0''} \end{cases} \Bigg\} 0 \le t \le T \tag{4.57}$$

where the amplitude of the lightwave-modulated signal is expressed in terms of the energy of the photons over the time interval, and thus the square root of this amount is the amplitude of the field of the lightwave. Naturally, the characteristic impedance of medium is set at unity. The distance is then $d = n_p$.

Thus, the BER is given by

$$\text{BER} = \text{erfc}\left(\frac{\sqrt{n_p}}{2N_0}\right) \tag{4.58}$$

By setting $N_0 = 1$, then for a BER of 10^{-9} the required number of photons is $n_p = 4 \times 9 \times 2 = 72$ for the heterodyne receiver, while a 3 dB improvement for the homodyne receiver requires 36 photon energy under the assumption that no electronic noise is contributed by the electronic amplifier. These are the quantum limits of ASK heterodyne and homodyne detection when the power of the local oscillator is much larger than that of the signal.

4.4.3.2 Coherent Phase and Frequency Shift Keying Systems

$$s_{1,0}(t) = \left\{ \begin{array}{ll} \sqrt{\dfrac{2n_p}{T}} \cos\omega_1 t & \text{for "1"} \\[4ex] \sqrt{\dfrac{2n_p}{T}} \cos\omega_0 t & \text{for "0"} \end{array} \right\} 0 \leq t \leq T \tag{4.59}$$

The FSK modulation scheme with two distinct frequencies f_1 and f_2 can be represented with a constant envelope and variation in the carrier frequency, or continuous phase between the two states as shown in Figure 4.6.

The modulation index can be defined as

$$m = \frac{|\omega_1 - \omega_0|}{2\pi} T \tag{4.60}$$

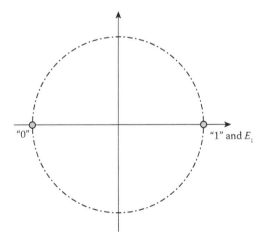

FIGURE 4.6 Signal constellation and energy level evolution of the signal envelope and the continuous phase between the 1 and 0 of FSK system.

If the two frequencies are large enough and the second and higher harmonics are outside the detection region, the signal correlation coefficient is given by

$$
\rho = \frac{2}{T} \int_0^T s_1(t) s_0(t) \, dt = \frac{2}{T} \int^T \cos(\omega_1 t) \cos(\omega_0 t) \, dt
$$

$$
\approx \frac{\sin(2\pi m)}{2\pi m}
$$

(4.61)

Thus, the BER is given as

$$
\mathrm{BER} = \frac{1}{2} \mathrm{erfc}\left(\sqrt{\frac{n_p}{2}\left(1 - \frac{\sin(2\pi m)}{2\pi m}\right)} \right)
$$

(4.62)

Thus, for BER of 1e−9, we have

$$
\mathrm{BER} = 10^{-9} \longrightarrow \sqrt{\frac{n_p}{2}\left(1 - \frac{\sin(2\pi m)}{2\pi m}\right)} = 3\sqrt{2}
$$

$$
\text{then,} \quad n_p = \frac{36}{1 - \dfrac{\sin(2\pi m)}{2\pi m}}
$$

(4.63)

For MSK, the modulation index is 0.25, leading to the required number of photon energy per bit of 60–70, much higher than that of 0.8 at which only 30 photons energy is required per bit. It is, however, shown that the MSK can be optimum for the transmission over a dispersive medium owing to the optimum bandwidth of the modulation scheme and hence minimum dispersive effects on the phase of the carrier.

The FSK can be implemented with continuous phase frequency shift keying (CPFSK); that is, the phase of the carrier is continuously chirped. Assume that the phase is linearly chirped such that the phase variation for the 1 and 0 are given by

$$
s_{1,0}(t) = \begin{cases} \sqrt{\dfrac{2n_p}{T}} \cos\theta_1(t) & \text{for "1"} \\[2mm] \sqrt{\dfrac{2n_p}{T}} \cos\theta_0(t) & \text{for "0"} \end{cases}
$$

where

$$
\theta_{1,0}(t) = \left. \begin{cases} \omega_{IF} t + \dfrac{m\pi}{T} t & \text{for "1"} \\[2mm] \omega_{IF} t - \dfrac{m\pi}{T} t & \text{for "0"} \end{cases} \right\} 0 \le t \le \frac{T}{2m}
$$

(4.64)

$$
\theta_{1,0}(t) = \left. \begin{cases} \omega_{IF} t + \dfrac{\pi}{2} & \text{for "1"} \\[2mm] \omega_{IF} t - \dfrac{\pi}{2} & \text{for "0"} \end{cases} \right\} \frac{T}{2m} \le t \le T
$$

The correlation coefficient is then given by

$$\text{BER} = 10^{-9} = \frac{1}{2}\text{erfc}\left(\sqrt{n_p\left(1-\frac{1}{4m}\right)}\right) \longrightarrow \sqrt{n_p\left(1-\frac{1}{4m}\right)} = 3\sqrt{2}$$

(4.65)

$$\text{then,} \quad n_p = \frac{18}{1-\frac{1}{4m}}$$

The variation in the photon energy with the modulation index for this linear CPFSK is shown in Figure 4.7.

The FSK can be modified with the control of the relative phase of the two carriers between the two bits; thus, the data bits can be written as

$$s_{1,0}(t) = \begin{cases} \sqrt{\dfrac{2n_p}{T}}\,\cos(\omega_1 t + \theta) & \text{for "1"} \\[4mm] \sqrt{\dfrac{2n_p}{T}}\,\cos\omega_0 t & \text{for "0"} \end{cases} \Bigg\} 0 \le t \le T$$

(4.66)

where the phase angle t can be chosen as

$$\theta_{1,0}(t) = \begin{cases} \pi - m\pi & \text{for "}p \le m \le 2p+1\text{" } p = 0,1,2.... \\[2mm] -m\pi & \text{for "}2p+1 \le m \le 2(p+1)\text{"} \end{cases}$$

(4.67)

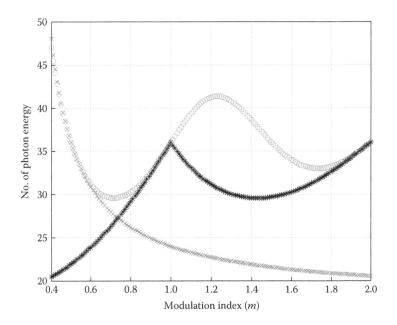

FIGURE 4.7 Fundamental limits of the coherent phase and frequency shift keying detection for (a) mid-gray o– CPFSK, (b) "light gray +" FSK (ω1 and ω2) with optimized correlation coefficient with phase control, and (c) "black *" linear DMPSK.

with the correlation coefficient

$$\rho = \frac{2}{T}\int_0^T s_1(t)s_0(t)dt = \frac{2}{T}\int_0^T \cos(\omega_1 t + \theta)\cos(\omega_0 t)dt$$

$$\approx -\frac{|\sin(\pi m)|}{\pi m}$$

(4.68)

Thus, this gives a BER for a phase-controlled FSK as

$$BER = \frac{1}{2}\operatorname{erfc}\left(\sqrt{\frac{n_p}{2}\left(1 + \frac{|\sin(\pi m)|}{\pi m}\right)}\right)$$

$$\to n_p = \frac{36}{1 + \dfrac{|\sin(\pi m)|}{\pi m}} \quad \text{for} \quad a....BER = 10^{-9}$$

(4.69)

The required number of photon is plotted against the modulation index and shown in Figure 4.7b 'black*'.

4.5   BINARY COHERENT OPTICAL RECEIVER

Another way of amplifying the signal while boosting the accompanying noise above the electronic noise floor is known as coherent detection [3]. A coherent receiver, as depicted in Figure 4.8, combines the signal with a local oscillator (LO) laser by means of an optical coupler. Upon detection, the two fields beat against each other during the electronic generation process within the photodetector, and the average electrical signal results in a mixing product between these time-dependent signals and the local CW. Electronic signals are generated at an intermediate frequency (IF) between the signal and the LO because the IF signal is usually mixed down to the baseband after the photodetection, using standard microwave techniques. The splitting ratio of the optical coupler has to be chosen as high as possible so as not to waste too much signal power and as low as possible to allow sufficient LO power to reach the detector to achieve shot-noise-limited performance (see the following explanation). The heterodyne efficiency η accounts for the degree of spatial overlap as well as for the polarization match between the LO field and the signal field. If both LO and signal are provided co-polarized in single-mode optical fibers, then η approaches unity.

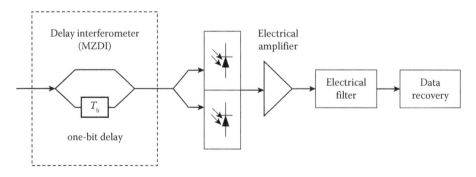

FIGURE 4.8   MZIM optical balanced receiver for optical DPSK detection.

If the frequency of the LO differs from the signal frequency, we speak of a heterodyne receiver. If the LO and the signal have the same frequency, such that IF = 0, then we have a homodyne receiver. Homodyne detection strictly requires optical phase locking between the LO and signal optical fields, which indicates a significant technological development. In the desired range of operation, the LO power is chosen much stronger than the signal power; the beating between the local oscillator and that of the signal would dominate. We are then left with an exact replica of the received optical field's amplitude $(P_s)^{1/2}$ and phase $\phi_s(t)$ (as compared to the optical power accessible in direct detection receivers). Thus, any amplitude or phase modulation scheme can directly be employed in combination with coherent receivers.

Owing to the high LO power reaching the detector, the main noise contribution in a coherent receiver in the absence of optical noise at its input is the shot noise produced by the LO, $\sigma^2_{LO-s} = 2qS(1-\varepsilon)P_{LO}B_e$. If this noise term dominates the electronic noise $\sigma^2_{LO-s} \gg i^2_{Neq}$, the optimum receiver performance should be derived depending on the parameters of the receiver. This limit is known as the *shot-noise limit* in the context of coherent receivers. The highest receiver sensitivity with the potential for practical implementation that is known today can be achieved using homodyne detection of PSK, where the data bits are directly mapped onto the phase ϕ. $s(t)$ of the optical signal, $\{0, 1\} \rightarrow \{0, \pi\}$. Without going into the derivations, the quantum limit for homodyne PSK can be shown to equal only 9 photons/bit, with a reported receiver sensitivity record of 20 photons/bit at 565 Mbit/s. Using OOK instead of PSK, the sensitivity degrades by 3 dB. If heterodyne detection is employed, then an additional loss of 3 dB in terms of receiver sensitivity would occur.

An alternative implementation of coherent balanced receivers is shown in Figure 4.8. It employs balanced detection with a pair of photodetectors connected back to back so that push–pull signal generation can be generated owing to the phase and the antiphase of the combined signals. While a balanced coherent receiver has exactly the same quantum-limited sensitivity as its single-ended equivalent, it offers the advantage of utilizing the complete mixing of the polarization of the fields of the arrived lightwave signal and LO power, and of being more robust to LO relative intensity noise (RIN).

Furthermore, a new type of receiver is the self-heterodyne balanced receiver, as shown in Figure 4.3e through g, whereby the lightwave by itself is split into two equal paths, and one is delayed with respect to the other so as to interferometrically resolve the phases of the lightwaves contained within the two consecutive bit periods. This is then followed by balanced detection.

The main advantages and disadvantages of direct single detection, balanced reception, and self-heterodyne structure of coherent receivers are as follows.

Receiver sensitivities of coherent receivers offer significant enhancement of the sensitivity as compared with those achieved with pin-receivers and APDs, thus allowing for increased transmission distances in un-amplified optical links. However, optically *preamplified* receivers, which exhibit similar receiver sensitivities to coherent receivers, are polarization insensitive and take less serious hits in performance if in-line amplification is presented.

The possibility of correcting for chromatic dispersion in the microwave regime is offered in coherent detection, because both the amplitude and phase of the optical field are converted to an electronic signal. However, efficient and adaptive broadband phase corrections can only be performed on RF bandpass signals, which requires heterodyne detection. Because the IF components have to be chosen about three times, high data rates would be needed, which would in turn require high receiver front-end bandwidths. Conversely, all-optical dispersion compensators and adaptive optical filters are attracting extensive development, allowing for efficient phase corrections in the optical regime.

Coherent receivers allow for the separation of closely spaced WDM channels by means of RF bandpass filters with sharp roll-offs. Optical filter technology has advanced dramatically, thus enabling channel spacing on the order of 10 GHz with sharp optical filter roll-offs that makes possible optical channel filtering even for ultra-dense WDM applications.

Self-heterodyne detection can offer significant improvement in receiver sensitivity and ease the difficulties in detection, as this is a noncoherent detection and requires coherence only within one-bit period.

4.6 NONCOHERENT DETECTION FOR OPTICAL DPSK AND MSK

All the receiver configurations presented here are based on noncoherent detection, which means the detection of the power envelope of the optical lightwaves signals at the input of the photodiodes is direct detection, and there is no mixing of the received signals with a local lightwave source. Noncoherent detection mitigates the issues in the coherent receiver schemes induced from the phase noise and nonzero linewidth of the coherent laser source.

4.6.1 PHOTONIC BALANCED RECEIVER

This receiver employs an electro-optic Mach–Zehnder delay interferometer (MZDI) with the introduction of a delay in one arm that is equal to one-bit period. The MZDI acts as a phase comparator. Unlike in the receiver for wireless communications where the phase comparison is performed after the detection process, the photonic balance receiver performs photonic phase differential demodulation prior to the optoelectronic detection.

The power of the optical signal is detected and converted from the optical domain to the electrical domain by using a photodiode. Because the optoelectronic conversion of a photodiode is based on the optical power, this detection technique is similar to the well-known nonlinear and noncoherent square law detection in wireless communications. This is the superiority of the photonic processing so that the ultra-wideband channel can be demodulated to overcome the electronic limitation.

Balanced receivers are essential to recover the differentially coded phases of the optical signals. This detection scheme offers a 3 dB improvement over direct detection using a single photodetector owing to its push–pull back-to-back connection of the photodetectors. The delay balanced receiver is shown in Figure 4.8, in which the phase information carrier of the received optical field received and its consecutive pulse are compared by an interferometer in which the lightwaves are split into two arms, one delayed by a symbol period and the other without delay, and then recombined via an optical directional coupler.

The recombined optical signals is the sum of the two optical fields whose magnitude are square in the photodetector when converting into electronic current, by applying the square law, thus the beating of the two optical fields in the photodetector and the products in the electrical domain can be written as

$$P_D(t) = \left|E_D(t) + E_D(t - \tau)\right|^2 - \left|E_D(t) - E_D(t - \tau)\right|^2$$

$$= 4\Re\left\{E_D(t)\right\}E_D^*(t - \tau) = \cos(\Delta\phi + \varsigma)$$

(4.70)

where $E_D(t)$ and $E_D(t-\tau)$ are the present and the one-bit delay versions of the optical signals at the delay interferometer, respectively. $\Delta\phi$ and ς represent the differential phase and the phase noise caused by the nonlinear phase noise or the MZ delay interferometer imperfections (MZDI). The latter issue is not a severe degradation factor owing to the high stability of the planar lightwave circuit MZDI using a thin-film heater for tuning any waveguide path mismatch. The MZDI-based optical balanced receiver can be used to detect both DPSK- and MSK-modulated optical signals. However, there is a significant difference in the configuration of the receiver of the two formats: *an additional π/2 phase shift is introduced in the detection of the optical MSK signals*, as shown in Figure 4.9. This constant phase shift can be located at either arm without affecting the performance of the receiver.

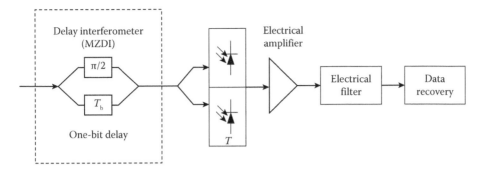

FIGURE 4.9 MZIM optical balanced receiver configuration for optical MSK detection.

4.6.2 OPTICAL FREQUENCY DISCRIMINATION RECEIVER

Detection of optical MSK-modulated signals using frequency discrimination principles was introduced and studied in the early 1990s. The CPFSK and MSK optical lightwave signals could be easily generated with this technique [3]. The detection at the time was mainly based on coherent detection. The dual filters used for discriminating two modulating principle frequencies are electrical filters. This is due to the unavailability of optical filters with the narrow bandwidth necessary for use in noncoherent detection of optical CPFSK and MSK signals.

With the advance of technology in filter design and fabrication, especially the microring resonator filters, the 3 dB bandwidth of contemporary optical filters can be narrowed down to about less than 2 GHz [7]. The availability of such optical filters enables frequency discrimination in the optical domain for noncoherent detection of optical MSK signals. This scheme, to the best knowledge of the author, has not yet been investigated. The proposed scheme has been reported in Ref. [4]. It is expected that the detection technique based on frequency discrimination is less sensitive to the effects of the quadratic phase profile of the single-mode optical fiber. The detailed configuration of the optical frequency discrimination receiver (OFDR) together with the analysis of optimization for the receiver design is presented in Chapter 9.

4.7 TRANSMISSION IMPAIRMENTS

4.7.1 CHROMATIC DISPERSION

This section briefly presents the key theoretical concepts describing the properties of chromatic dispersion (CD) in a single-mode fiber. Another objective of this section is to introduce the key parameters that will be commonly described in the rest of the chapter.

The initial point to note when discussing CD is the expansion of the mode propagation constant or "wave number" parameter, β, which can be approximated around the optical operating frequency using the Taylor series as

$$\beta(\omega) = \frac{\omega n(\omega)}{c} = \beta_0 + \beta_1 \Delta\omega + \frac{1}{2}\beta_2 \Delta\omega^2 + \frac{1}{6}\beta_3 \Delta\omega^3 \qquad (4.71)$$

where β_0 is the propagation constant of the fiber evaluated at the operating frequency, ω is the angular optical frequency, and $n(\omega)$ is the frequency-dependent refractive index of the fiber. The higher-order parameters

$$\beta_n = \left(\frac{d^n\beta}{d\omega^n}\right)\bigg|_{\omega=\omega_0} \qquad (4.72)$$

These parameters have different physical meanings as follows. β_0 is involved in the phase velocity of the optical carrier, which is defined as

$$v_p = \frac{\omega_0}{\beta_0} = \frac{c}{n(\omega_0)} \tag{4.73}$$

β_1 determines the group delay that relates to the velocity v_g and the guided mode propagation constant β of the guided mode by [8,9]

$$v_g = \frac{1}{\beta_1} = \left(\frac{d\beta}{d\omega}\bigg|_{\omega=\omega_0} \right)^{-1} \tag{4.74}$$

β_2 is the derivative of the group velocity with respect to the frequency. This means that different frequency components of an optical pulse travel at different velocities, hence leading to the spreading of the pulse or dispersion. β_2 is therefore known as the group velocity dispersion (GVD). The fiber is said to exhibit normal dispersion for $\beta_2 > 0$ or anomalous dispersion if $\beta_2 < 0$.

A pulse having the spectral width $\Delta\omega$ is broadened by

$$\Delta\tau = \beta_2 L \Delta w \tag{4.75}$$

In practice, a more commonly used factor to represent the CD of a single-mode optical fiber is known as D (ps/(nm·km)). The dispersion factor is closely related to the GVD β_2 and is given by

$$D = -\left(\frac{2\pi c}{\lambda^2} \right) \beta_2 \quad \text{at the operating wavelength } \lambda \tag{4.76}$$

β_3 is defined as $\beta_3 = \dfrac{d\beta_2}{d\omega}$ and contributes to the dispersion slope, $s(\lambda)$, which is an essential dispersion factor for high-speed multichannel DWDM transmission. $S(\lambda)$ can be obtained from the higher-order derivatives of the propagation constant as

$$S = \frac{dD}{d\lambda} = \left(\frac{2\pi c}{\lambda^2} \right) \beta_3 + \left(\frac{4\pi c}{\lambda^3} \right) \beta_2 \tag{4.77}$$

4.7.2 Chromatic Linear Dispersion

From the standpoint of fiber design [5,6], this dispersion factor is actually the sum of the two dispersion components, D_M and D_W, the material and waveguide dispersion parameters, respectively, and is given by

$$D = -\left(\frac{2\pi c}{\lambda^2} \right) \beta_2 \equiv D_M + D_W \tag{4.78}$$

A step-index optical fiber with the core radius notation, a, is considered. The refractive indices of the core and cladding of the SSMF are noted to be n_1 and n_2, respectively. The significant transverse propagation constants of the guided lightwaves u and v in the core and cladding regions are formulated as

$$u = a\sqrt{k^2 n_1^2 - \beta^2}; \; v = a\sqrt{\beta^2 - k^2 n_2^2} \tag{4.79}$$

where $k^2 n_1^2$ and $k^2 n_2^2$ are the plane-wave propagation constants in the core and second cladding layers, respectively. β is the longitudinal propagation constant of the guided waves along the optical fiber, which can be expressed as

$$\beta = \sqrt{k^2 (b(n_1^2 - n_2^2) + n_2^2)} \tag{4.80}$$

where b denotes the critical normalized propagation constant whose value for guided modes falls within the range of [0,1] and is formulated as

$$b = \frac{\dfrac{\beta}{k} - n_2}{n_1 - n_2} \tag{4.81}$$

The normalized frequency is then mathematically expressed as

$$V = ak\sqrt{n_1^2 - n_2^2} \tag{4.82}$$

The material dispersion in an optical fiber is due to the wavelength dependence of the refractive index of the core and cladding. The refractive index $n(\lambda)$ is estimated by the well-known Sellmeier's equation

$$n^2(\lambda) = 1 + \sum_{i=1}^{M} \frac{B_i \lambda^2}{(\lambda^2 - \lambda_i^2)} \tag{4.83}$$

where λ_i indicates the ith resonance wavelength, and B_i is its corresponding oscillator strength. n stands for n_1 or n_2 for the core or cladding regions. These constants of different doping concentrations of GeO_2 in SiO_2 pure silica are tabulated in Table 4A.1 given in the appendix of this chapter [10]. The first three Sellmeier terms, G_1, G_2, and G_3, are normally used. These tabulated coefficients are then used to find the material dispersion factor, D_M, which can be obtained by

$$D_M = -\frac{\lambda}{c}\left(\frac{d^2 n(\lambda)}{d\lambda^2} \right) \tag{4.84}$$

where c is the velocity of light in vacuum. For pure silica over the spectral range 1.25 to 1.66 μm, D_M can also be approximated by the empirical relation [8]

$$D_M = 122\left(1 - \frac{\lambda_{ZD}}{\lambda} \right) \tag{4.85}$$

where λ_{ZD} is the zero material dispersion wavelength.* For instance, $\lambda_{ZD} = 1.276$ μm for pure silica. λ_{ZD} can vary according to various doping concentrations in the core and cladding of different materials such as germanium (Ge) or fluorine (F).

---

* λ_{ZD} is defined as the wavelength at which the material dispersion factor $D_M(\lambda) = 0$.

The waveguide dispersion D_W can be calculated as [8,9]

$$D_W = -\left(\frac{n_1 - n_2}{c\lambda}\right)V\frac{d^2(Vb)}{dV^2} \tag{4.86}$$

where b and V are the normalized propagation constant and the normalized frequency as defined in Equations 4.50 and 4.51, respectively. $Vd^2(Vb)/dV^2$ is defined as the normalized waveguide dispersion parameter.

An effective approximation based on polynomial interpolation has been developed to calculate the waveguide dispersion not only in a simple SI fiber but in a multicladding dispersion-compensating fiber (MC-DCF) as well [11]. See Figure 4.10.

A well-known parameter that governs the effects of CD imposed on the transmission length of an optical system is known as the dispersion length L_D. Conventionally, the dispersion length L_D corresponds to the distance after which a pulse has broadened by one-bit interval [8]. For high-capacity long-haul transmission employing external modulation, the dispersion limit can be estimated with the following equation [8]

$$L_D = \frac{10^5}{D.B^2} \tag{4.87}$$

where B is the bit rate (Gbps), D is the dispersion factor ps/(nm·km), and L_D is in kilometers. Equation 4.87 provides a reasonable approximation, even though the accurate computation of this limit depends on the modulation format, the pulse shaping, and the optical receiver design. It can be seen clearly from this equation that the severity of the effects caused by the fiber CD on externally modulated optical signals is inversely proportional to the square of the bit rate. Thus, for 10 Gbps OC-192 optical transmission on an SSMF medium that has a dispersion of about ±17 ps/(nm·km), the dispersion length L_D has a value of approximately 60 km, which corresponds to a residual dispersion of about ±1000 ps/nm and less than 4 km, or, equivalently, to about ±60 ps/nm in the case of 40 Gbps OC-768 optical systems. These lengths are a great deal smaller than the length limited by

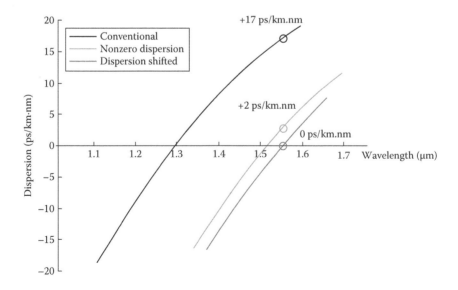

FIGURE 4.10 Dispersion characteristics of different transmission and compensating fibers.

ASE noise accumulation. The CD therefore becomes one of the most critical constraints for modern high-capacity and ultra-long-haul transmission optical systems.

4.7.3 POLARIZATION-MODE DISPERSION

Polarization-mode dispersion (PMD) represents another type of pulse spreading. The PMD is caused by the differential group delay (DGD) between two principle orthogonal states of polarization (PSP) of the propagating optical field (Figure 4.11).

One of the intrinsic causes of PMD is the asymmetry of the fiber core. The other causes are derived from the deformation of the fiber, including stress applied on the fiber, the aging of the fiber, the variation in the temperature over time, or effects from a vibration source. These processes are random, resulting in the dynamics of PMD.

The imperfection of the core or deformation of the fiber may be inherent from the manufacturing process or as a result of mechanical stress on the deployed fiber resulting in the dynamic aspect of PMD. The delay between these two PSPs is normally negligibly small in 10 Gbps optical transmission systems. However, at high transmission bit rates for long-haul and ultra-long-haul optical systems, the PMD effect becomes much more severe and degrades the system performance [8,12–14]. The DGD value varies along the fiber following a stochastic process. It has been proved that these DGD values comply with a Maxwellian distribution [8,15,16] given by the following expression

$$f(\Delta\tau) = \frac{32(\Delta\tau)^2}{\pi^2 \langle\Delta\tau\rangle^3} \exp\left\{-\frac{4(\Delta\tau)^2}{\pi\langle\Delta\tau\rangle^2}\right\} \quad \Delta\tau \geq 0 \tag{4.88}$$

where $\Delta\tau$ is the DGD over a segment of the optical fiber δz. The mean DGD value $\langle\Delta\tau\rangle$ is commonly termed the *fiber* PMD and normally given by the fiber manufacturer. An estimate of the transmission limit due to PMD effect is given as [7]

$$L_{max} = \frac{0.02}{\langle\Delta\tau\rangle^2 \cdot R^2} \tag{4.89}$$

with R being the transmission bit rate. Therefore, we have (1) $\langle\Delta\tau\rangle = 1$ ps/km (older fiber types) and the bit rate = 40 Gbps; $L_{max} = 12.5$ km and bit rate = 10 Gbps; $L_{max} = 200$ km and (2) $\langle\Delta\tau\rangle = 0.1$ ps/km (contemporary fiber for modern optical systems) and thus bit rate = 40 Gbps; $L_{max} = 1250$ km and bit rate = 10 Gbps; $L_{max} = 20.000$ km.

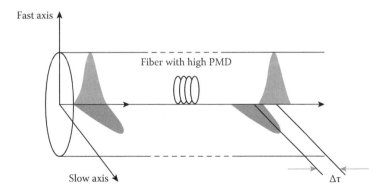

FIGURE 4.11 Illustration of PMD effects.

In short, PMD is an issue for ultra-long distances even at 10 Gbps optical transmission. When upgrading to a higher bit rate and higher capacity, PMD together with CD become the two most critical impairments that limit the performance of optical systems.

4.7.4 FIBER NONLINEARITY

The fiber refractive index is not only dependent on the wavelength but also on the intensity of the lightwave. This well-known phenomenon, the Kerr effect, is normally referred to as the fiber self-phase modulation (SPM) nonlinearity. The power dependence of the refractive index n_r is given as

$$n_r' = n_r + \bar{n}_2(P/A_{\text{eff}}) \tag{4.90}$$

where P is the average optical intensity inside the fiber, $\bar{n}_2$ is the nonlinear-index coefficient, and A_{eff} is the effective area of the fiber.

There are several nonlinear phenomena induced from the Kerr effect, including intrachannel SPM, cross-phase modulation between ichannels (XPM), four-wave mixing (FWM), stimulated Raman scattering (SRS), and stimulated Brillouin scattering (SBS). SRS and SBS are not the main degrading factors compared to the others. The FWM effect degrades the performance of an optical system severely if the local phase of the propagating channels is matched with the introduction of the ghost pulse. However, with a high local dispersion parameter such as in SSMF or even in NZ-DSF, the effect of the FWM becomes negligible. While XPM is strongly dependent on the channel spacing between the channels as well as on the local dispersion factor of the optical fiber [8]. Furthermore XPM on the optical signal is much smaller than that due to the SPM effect.

In this chapter, only SPM nonlinearity is considered. This is the main degradation factor for high-bit-rate transmission systems where the signal spectrum is broadened. The effect of SPM is normally coupled with the nonlinear phase shift, which is defined as

$$\phi_{\text{NL}} = \int_0^L \gamma P(z)\,dz = \gamma L_{\text{eff}} P$$

where

$$\gamma = \omega_c \bar{n}_2 /(A_{\text{eff}} c) \tag{4.91}$$

$$L_{\text{eff}} = (1 - e^{-\alpha L})/\alpha$$

where ω_c is the lightwave carrier, L_{eff} is the effective transmission length, and α is the attenuation factor of an SSMF that normally has a value of 0.17–0.2 dB/km for the currently operating wavelengths within the 1550 nm window.

The temporal variation of the nonlinear phase ϕ_{NL} while the optical pulses propagate along the fiber results in the generation of new spectral components far apart from the lightwave carrier ω_c, implying the broadening of the signal spectrum. The frequency difference $\delta\omega$ is given by

$$\delta\omega = -\frac{\partial\phi_{\text{NL}}}{\partial T} \tag{4.92}$$

The time dependence of $\delta\omega$ is known as frequency chirping.

4.8 MATLAB® AND SIMULINK® SIMULATOR FOR OPTICAL COMMUNICATIONS SYSTEMS

A simulation package based on the MATLAB and Simulink platform for optical transmission systems has been developed. The simulator integrates all the aforementioned transmission impairments over a single optical transmission link. The DWDM optical system can be developed further from this platform without much difficulty.

The simulator provides adequate toolboxes and blocksets for setting up any complicated system configurations under test. Spectral characteristics of new proposed modulation formats and their transmission performance with a focus on tolerance to CD and PMD can be easily achieved with various techniques of system evaluation.

Among nonlinearity impairments, SPM is considered to be the major shortfall in the system. Furthermore, XPM can be considered to be negligible in a DWDM system for the following scenarios: (1) highly locally dispersive system, for example, SSMF-and DCF-deployed systems; (2) large channel spacing, and (3) high spectral efficiency [4,6,17–19]. For transmission systems operating at 40 Gbps, using SSMF 17 ps/(nm·km) at 1550 nm and with 0.8 nm spacing between two adjacent DWDM wavelengths, a walk-off length L_w of about 2.95 km is obtained. However, the XPM should be taken into account for systems deploying nonzero dispersion-shifted fiber (NZ-DSF) where the local dispersion factor is low. The values of the NZ-DSF dispersion factors can be obtained.

The other critical degradation factors in DPSK systems are the nonlinear phase noise due to the fluctuation of the optical intensity caused by ASE noise via the Gordon–Mollenauer effect [20]. The receiver and electrical amplifier noise are also taken into account.

4.8.1 FIBER PROPAGATION MODEL

4.8.1.1 Nonlinear Schrödinger Equation

Evolution of the slow-varying complex envelope $A(z,t)$ of the optical pulses along a single-mode optical fiber is governed by the well-known nonlinear Schrödinger equation (NLSE) [8]

$$\frac{\partial A(z,t)}{\partial z} + \frac{\alpha}{2} A(z,t) + \beta_1 \frac{\partial A(z,t)}{\partial t} + \frac{j}{2}\beta_2 \frac{\partial^2 A(z,t)}{\partial t^2} - \frac{1}{6}\beta_3 \frac{\partial^3 A(z,t)}{\partial t^3}$$

$$= -j\gamma |A(z,t)|^2 A(z,t)$$

$$(4.93)$$

where z is the spatial longitudinal coordinate, α accounts for fiber attenuation, β_1 indicates the DGD, β_2 and β_3 represent the second- and third-order dispersion factors of the GVD, and γ is the nonlinear coefficient. Equation 4.62 involves the following effects in a single-channel transmission fiber: (1) the attenuation, (2) CD, (3) third-order dispersion factor, the dispersion slope, and (4) SPM nonlinearity. Other nonlinear effects such as FWM and SRS can be inserted into the NLSE when necessary.

4.8.1.2 Symmetrical Split-Step Fourier Method

In this chapter, solutions of the NLSE and hence the model of pulse propagation in a single-mode optical fiber are numerically obtained by using the popular approach of the split-step Fourier method (SSFM) [9], in which the fiber length is divided into a large number of segments of a small step size δz.

In practice, dispersion and nonlinearity are mutually interactive while the optical pulses propagate through the fiber. However, the SSFM assumes that over a small length δz, the effects of dispersion and nonlinearity on the propagating optical field are independent. Thus, in SSFM, the linear operator representing the effects of fiber dispersion and attenuation and the nonlinearity operator taking into account fiber nonlinearities are defined separately as

$$\hat{D} = -\frac{i\beta_2}{2}\frac{\partial^2}{\partial T^2} + \frac{\beta_3}{6}\frac{\partial^3}{\partial T^3} - \frac{\alpha}{2}$$

$$\hat{N} = i\gamma \,|\, A \,|^2$$

(4.94)

where A replaces $A(z,t)$ for simpler notation and $T = t - z/v_g$ is the reference time frame moving at the group velocity.

The NLSE in Equation 4.62 can be rewritten as

$$\frac{\partial A}{\partial z} = (\hat{D} + \hat{N})A$$

(4.95)

and the complex amplitudes of optical pulses propagating from z to $z + \delta z$ are calculated using the approximation as

$$A(z+h,T) \approx \exp(h\hat{D})\exp(h\hat{N})A(z,T)$$

(4.96)

Equation 4.33 is accurate to the second order for the step size δz [10]. The accuracy of SSFM can be improved by including the effect of the nonlinearity in the middle of the segment rather than at the segment boundary as illustrated in Figure 4.12.

Equation 4.96 can now be modified as

$$A(z+\delta z,T) \approx \exp\left(\frac{\delta z}{2}\hat{D}\right)\exp\left(\int_z^{z+\delta z}\hat{N}(z')dz'\right)\exp\left(\frac{\delta z}{2}\hat{D}\right)A(z,T)$$

(4.97)

This method is accurate to the third order for the step size δz. The optical pulse is propagated down segment from segment in two stages at each step. First, the optical pulse propagates through the first linear operator (step of $\delta z/2$) with only dispersion effects taken into account. The nonlinearity is calculated in the middle of the segment. It is noted that the nonlinearity effect is considered over the whole segment. Then at $z + \delta z/2$, the pulse propagates through the remaining $\delta z/2$ distance of the linear operator. The process continues repetitively in executive segments δz until the end of the fiber. This method requires an appropriate selection of step sizes δz to ensure the required accuracy.

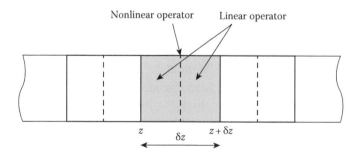

FIGURE 4.12   Schematic illustration of the split-step Fourier method.

4.8.1.3 Modeling of PMD

The first-order PMD effect can be implemented by splitting the optical field into two distinct paths representing two states of polarizations with different propagating delay times $\Delta\tau$, and then implementing SSFM over the segment δz before superimposing the outputs of these two paths for the output optical field. The transfer function for first-order PMD is given by [21].

$$H_f(f) = H_{f+}(f) + H_{f-}(f) \tag{4.98}$$

$$H_{f+}(f) = \sqrt{\gamma}\exp\left[j2\pi f\left(-\frac{\Delta\tau}{2}\right)\right] \text{ and } H_{f-}(f) = \sqrt{\gamma}\exp\left[j2\pi f\left(-\frac{\Delta\tau}{2}\right)\right] \tag{4.99}$$

γ is the splitting ratio, which normally takes a value of 1/2.

4.8.1.4 Optimizing the Symmetrical SSFM

The symmetrical SSFM can be optimized by (1) a split-step length that satisfies the constraint for the convergence and (2) evaluation of incident peak power. If it is lower than the threshold of nonlinearities of the transmission fiber, the SPM effect can be neglected, and SSFM will be switched to the low-pass transfer function (LPTF) method that involves only the fiber dispersion and attenuation effects [12]. The LPTF of the fiber is

$$H_f(f) = \exp\left\{-j\left[(1/2)\beta_2'\varpi^2 + (1/6)\beta_3'\varpi^3\right]L\right\} \tag{4.100}$$

4.8.1.5 Fiber Propagation in Linear Domain

Here, the low-pass equivalent frequency response of the optical fiber, $H(f)$, has a parabolic phase profile and can be modeled by the following equation [22]

$$H_c(f) = e^{-j\alpha_D f^2} \tag{4.101}$$

where $\alpha_D = \pi^2\beta_2 L, \beta_2$ represents the group velocity distortion (GVD) parameter of the fiber, and L is the length of the fiber. The parabolic phase profile is the result of the CD of the optical fiber [23]. The third-order dispersion factor β_3 is not considered in this transfer function of the fiber owing to the negligible effects on 40 Gbps transmission systems. However, if the transmission bit rate is higher than 40 Gbps, and thus the spectrum wider, the effects of β_3 must be taken into account.

In the model of the optical fiber, it is assumed that the signal is propagating in the linear domain, implying that the fiber nonlinearities are not included in the model. These nonlinear effects are investigated numerically. It is also assumed that the optical carrier has a line spectrum. This is a valid assumption considering the state-of-the-art laser sources employed nowadays with a very narrow linewidth and the use of external modulators in signal transmission.

A pure sinusoidal signal of frequency f, propagating through the optical fiber, experiences a delay of $|2\pi f_D\beta_2 L|$. The standard fibers used in optical communications have a negative β_2 and thus, in low-pass equivalent representations, sinusoids with positive frequencies (i.e., frequencies higher than the carrier) have negative delays, which means early arrivals compared to the carrier and the ones with negative frequencies (i.e., frequencies lower than the carrier) experience positive delays.

The DCFs have a positive β_2 and so have reverse effects. The low-pass equivalent channel impulse response of the optical fiber, $h_c(t)$, would also have the following parabolic phase profile

$$h_c(t) = \sqrt{\frac{\pi}{j\alpha_D}} e^{-j\alpha^2 t^2/\alpha_D} \tag{4.102}$$

4.8.2 NONLINEAR EFFECTS VIA FIBER PROPAGATION MODEL

4.8.2.1 SPM Effects

When a single channel has been transmitted and only the SPM effect is considered due to the change in the refractive index of the guided medium as a function of the intensity, then the SPM effects can be represented in the NLSE as the term in the right-hand side of Equation 4.99 of the ith channel

$$\frac{\partial A_i(z,t)}{\partial z} + \frac{\alpha}{2} A_i(z,t) + \beta_1 \frac{\partial A_i(z,t)}{\partial t} + \frac{j}{2}\beta_2 \frac{\partial^2 A_i(z,t)}{\partial t^2} - \frac{1}{6}\beta_3 \frac{\partial^3 A_i(z,t)}{\partial t^3} = -j\gamma |A(z,t)|^2 A(z,t) \tag{4.103}$$

where the nonlinear coefficient, as previously stated, is $\gamma = \dfrac{2\pi n_2}{\lambda A_{\text{eff}}}$; n_2 is the nonlinear refractive index; and A_{eff} is the effective area covered by the area of the fundamental mode. The instantaneous magnitude of the envelope of the optical field plays an important role in the SPM effect, and thus there is a strong interaction between the SPM and the linear second and third-order dispersion effects. The nonlinear dispersion length and the linear and nonlinear dispersion lengths are given by

$$L_D = \frac{T_e^2}{|\beta^2|}............L_{\text{NL}} = \frac{1}{\gamma P_p} \tag{4.104}$$

where P_p is the peak power of the transmitted pulse, and T_e is the e^{-1} pulse width. In the regime where L_{NL} is much longer than the dispersion length, the SPM dominates the pulse broadening and hence is limited by SPM. This normally occurs in DCF, whose effective area is very small and hence the enhancement effects of the SPM. In such a case, the NLSE reduces to

$$\frac{\partial A_i(z,t)}{\partial z} + \frac{\alpha}{2} A_i(z,t) = -j\gamma |A(z,t)|^2 A(z,t) \tag{4.105}$$

That is, all the terms relating to the phase variation due to the propagation constant are neglected, and only the phase due to the nonlinear SPM is included and the attenuation of the fiber. This equation can be solved analytically to give a possible solution of

$$A(L,t) = A(0,t)e^{\frac{-\alpha}{2}L}e^{-j\phi_{\text{SPM}}(L,t)} \tag{4.106}$$

with the nonlinear phase ϕ_{SPM} given by

$$\phi_{\text{SPM}}(L,t) = \gamma |A(0,t)|^2 L_{\text{eff}} \tag{4.107}$$

Equation 4.101 reveals that the nonlinear refractive index varies with the evolution of its own intensity along the effective propagation distance. This phase modulation would then turn into frequency modulation in the form of a broadening of the spectral characteristics of the signals and hence pulse compression or dispersion, depending on the sign of the nonlinear phase variation.

In the regime in which the effective nonlinear length is very close to that of the dispersion, an interplay between the linear and nonlinear dispersion effects occurs, the foregoing equations do not represent the overall dispersive effects, and only numerical solutions could be obtained. This SSFM has already been described in Section 2.10.

If the optical signals propagate in the anomalous regime ($\beta_2 < 0$), then the nonlinear SPM phase change is in the opposite direction of the linear dispersion, and these two effects can compensate each other. Under some specific power of the pulses, the pulse may preserve its shape over a very long distance. This is called solitary wave propagation and is described in Chapter 15. On the contrary, if the change in the linear dispersion is in the same as that in the nonlinear induced dispersion, then the signals broaden further, and a significant broadening of the pulse sequence would occur. Optimization or management of the dispersion effects can be used to minimize the SPM broadening.

4.8.2.2 XPM Effects

If DWM transmission is employed to extend the total transmission capacity of the system and networks, then the nonlinear refractive index generated by the SPM would create intermodulation effects, usually called the cross-phase modulation (XPM). This XPM can be represented in the NLSE by coupled equations between different wavelength channels. In the NLSE, the XPM is shown in the second term of the right-hand side's last term

$$\frac{\partial A_i(z,t)}{\partial z} + \frac{\alpha}{2} A_i(z,t) + \beta_1 \frac{\partial A_i(z,t)}{\partial t} + \frac{j}{2}\beta_2 \frac{\partial^2 A_i(z,t)}{\partial t^2} - \frac{1}{6}\beta_3 \frac{\partial^3 A_i(z,t)}{\partial t^3}$$
$$= -j\gamma \left\{ |A_i(z,t)|^2 + 2\sum_{k,k\neq i} |A_k(z,t)|^2 \right\} A_i(z,t)$$

(4.108)

We now assume only the XPM effect due to two wavelength channels, and then the envelopes of the two wavelength channels are coupled via coupled differential equations

$$\frac{\partial A_1(z,t)}{\partial z} + \frac{\alpha}{2} A_1(z,t) = -j\gamma \left\{ |A_1(z,t)|^2 + 2|A_2(z,t)|^2 \right\} A_1(z,t)$$
$$\frac{\partial A_2(z,t)}{\partial z} + \frac{\alpha}{2} A_2(z,t) + d_{12} \frac{\partial A_2(z,t)}{\partial z} = -j\gamma \left\{ |A_2(z,t)|^2 + 2|A_1(z,t)|^2 \right\} A_2(z,t)$$

(4.109)

The extra term is due to the propagation delay between the two wavelength channels coupled via the propagation constant of the guide waves of different wavelengths given by

$$d_{12}z = D[\lambda_1 - \lambda_2)z]$$

(4.110)

The solution of the coupled differential equations can be written as

$$\phi_1(L,t) = \phi_{1SPM}(L,t) + \phi_{1XPM}(L,t)$$
$$= \gamma |A(0,t)|^2 L_{eff} + 2\gamma_1 \int_0^L |A_2(0,t) + d_{12}z|^2 e^{-\alpha z} dz$$

(4.111)

where the first term of the right-hand side represents the SPM effects on channel 1, and the second term is the intermodulation XPM effect term, which can be expressed in the frequency domain by taking its Fourier transform

$$\left|A_2(0,t)+d_{12}z\right|^2 = P_2(0,t+d_{12}z) = \int_{-\infty}^{+\infty} P_2(0,f)e^{j2\pi f(t+d_{12}z)}df \tag{4.112}$$

Thus, the term $\phi_{1SPM}(L,t)$ can be written in the frequency domain as

$$\phi_{1XPM}(L,f) = 2\gamma_1 \int_{-\infty}^{+\infty} P_2(0,f)e^{j2\pi f d_{12}z}e^{-\alpha z}df = 2\gamma_1 P_2(0,f)\int_{-\infty}^{+\infty} e^{j2\pi f d_{12}z}e^{-\alpha z}df$$
$$= P_2(0,f)H_{12}(f) \tag{4.113}$$

where $H_{12}(f)$ represents the contribution of the XPM effects on the optical passband spectrum of the optical signals

$$H_{12}(f) = 2\gamma_1 \int_{-\infty}^{+\infty} e^{j2\pi f d_{12}z - \alpha z}df = 2\gamma_1 \frac{1 - e^{j2\pi f d_{12}L - \alpha L}}{\alpha - j2\pi f d_{12}} \approx \frac{2\gamma_1}{\alpha - j2\pi f d_{12}} \tag{4.114}$$

Thus, we see that the effect of the XPM is filtering the spectrum of the optical signals with a low-pass filter whose transfer function in the frequency domain is given by Equation 4.110. As an example, consider the transmission fiber SSMF over which the XPM is taking place, with the attenuation coefficient and the dispersion factor D known to be $D = 17$ ps/nm/km and $\alpha = 0.2$ dB/km; and the corner frequency of the XPM low-pass filter is 540 MHz. As a result, the XPM-induced effect would alter the spectrum of the signal channel and would significantly affect the received signals if the bit rate is slow enough, for example, 10 Gbps with a channel spacing on the order of 100 GHz. This low-frequency filtering effect would not be significant for bit rates of 40 and 100 Gbps.

The XPM effect would be significant under coherent detection as the phase is preserved under such scheme, but not for direct detection. Under direct detection, this XPM phase is converted into amplitude modulation, PM to AM intermodulation conversion, and then through square law detection that would generate the distortion and penalty of the eye opening (EO). Further, this XPM creates a propagation delay difference between the channels in DWDM systems and thus walk-off effects between channels.

If intrachannel dispersion is taken into account [8], the XPM increases proportionally with respect to the square of the channel spacing ($\alpha P/\Delta f^2$).

4.8.2.3 FWM Effects

FWM is the phenomenon that involves the mixing of three lightwaves and then generation of the fourth wave, which may fall into the spectrum of one of the transmitted channels. This phenomenon occurs when there is a match between the phases of the lightwave channels. This phase matching between the channels occurs when their group velocities are the same, that is, when the transmission fiber has its zero dispersion in the region of these lightwave channels. Thus, the FWM can be distinguished from other nonlinear effects affected by the strong intensity of the guided waves in that the phase of the optical fields and the satisfaction of the phase-matching condition for the degenerate FWM are much more critical. The generation of unwanted optical signals due to the mixing

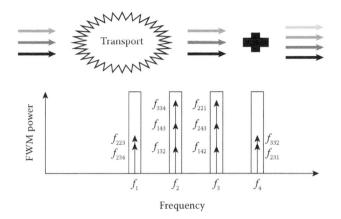

FIGURE 4.13 FWM with the four sources and their mixing to generate other lightwave signals.

of the three lightwave channels is shown in Figure 4.13. The FWM effects can be included in the NLSE, as indicated by the last term of the right-hand side below [9]

$$\frac{\partial A_i(z,t)}{\partial z} + \frac{\alpha}{2} A_i(z,t) + \beta_1 \frac{\partial A_i(z,t)}{\partial t} + \frac{j}{2}\beta_2 \frac{\partial^2 A_i(z,t)}{\partial t^2} - \frac{1}{6}\beta_3 \frac{\partial^3 A_i(z,t)}{\partial t^3}$$

$$= -j\gamma \left\{ |A_i(z,t)|^2 + 2\sum_{k,k\neq i} |A_k(z,t)|^2 \right\} A_i(z,t) + j\gamma \sum_{i=p+q-r,pq\neq r} \frac{d}{3} \left\{ A_p A_q A_r^* e^{j\Delta\beta_{ipqr}z} \right\}$$

(4.115)

where

$$\Delta\beta_{ipqr} = \beta_i - \beta_p - \beta_q + \beta_r = (\omega_p - \omega_r)(\omega_q - \omega_r) + \frac{1}{2}\beta_{3i}(\omega_p + \omega_q - 2\omega_r)$$

$$d = \begin{cases} 3 & \text{for } q = p \\ 6 & \text{for } q \neq p \end{cases}$$

Thus, the number of generated waves is

$$M = \frac{1}{2}(N^3 - N^2)$$

(4.116)

The effect of FWM can best be explained by the nonlinear interaction between the lightwaves of frequencies $f_1, f_2,$ and $f_3,$ which can also be represented by the Volterra series transfer function [14] with $(r \neq p,q)$ that would generate a new lightwave component that would interfere with other DWDM channels that have the same frequency as that of the generated ones.

For continuous wave channels, the efficiency of the degenerate FWM is given by

$$P_{ipqr} = \left(\frac{d}{3}\gamma L_{\text{eff}}\right)^2 P_p(0)P_q(0)P_r(0)e^{-\alpha L}\eta_{ipqr}$$

(4.117)

The efficiency of the FWM is given by

$$\eta_{ipqr} = \frac{\alpha^2}{\alpha^2 + \Delta\beta_{ipqr}^2}\left(1 + \frac{4e^{-\alpha L}\sin^2(\Delta\beta_{ipqr}L/2)}{(1-e^{-\alpha L})^2}\right) \tag{4.118}$$

If the fiber is a dispersion-shifted type, then the phase is matched with $\beta_2 = 0$, $\Delta\beta_{ipqr} = 0$; and thus the efficiency reaches unity. The FWM is maximum under this condition, and this is the reason why the NZ-DSF is designed to reduce the linear dispersion but must avoid the FWM effects.

4.8.2.4 SRS Effects

SRS and SBS are the scattering generated from the generation of the phonons in the fiber under high-energy pumping at some spectral distance away from the signal spectrum. The phonons would then be phase-matched to the pump lightwaves to create a stimulated emission condition in the wavelength regions to enable creation of new photons, SRS and SBS. The difference between SRS and SBS is the vibration of the electronic domain and the elasticity of the crystal structures of silica and doped impurities, respectively.

Various optical fibers already installed in the transmission systems and networks throughout global telecommunication backbone networks are measured for the SRS gain efficiency as shown in Figure 4.14. The spacing of the pump to the Stokes wavelength would be on order of around 100 nm or 13 THz.

The SRS effects can be represented in the NLSE as the last term of the right-hand side of the following equation

$$\frac{\partial A_i(z,t)}{\partial z} + \frac{\alpha}{2}A_i(z,t) + \beta_1\frac{\partial A_i(z,t)}{\partial t} + \frac{j}{2}\beta_2\frac{\partial^2 A_i(z,t)}{\partial t^2} - \frac{1}{6}\beta_3\frac{\partial^3 A_i(z,t)}{\partial t^3}$$

$$= -j\gamma\left\{|A_i(z,t)|^2 + 2\sum_{k,k\neq i}|A_k(z,t)|^2\right\}A_i(z,t) + j\gamma\sum_{i=p+q-r,pq\neq r}\frac{d}{3}\left\{A_pA_qA_r^* e^{j\Delta\beta_{ipqr}z}\right\}A_i(z,t) \tag{4.119}$$

$$+\left\{\sum_{k=1}^{i-1}\frac{\omega_i}{\omega_k}\frac{g_R}{2A_{eff}}|A_k|^2 + \sum_{k=i+1}^{N}\frac{g_R}{2A_{eff}}|A_k|^2\right\}A_i(z,t)$$

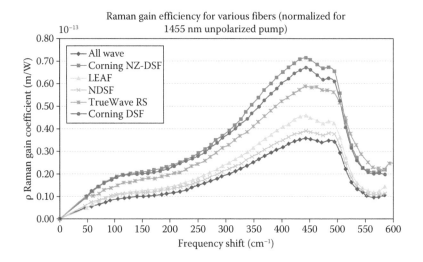

FIGURE 4.14 Raman gain efficiency of various single-mode optical fibers.

If we now consider the effect of the SRS only in the NLSE, we can rewrite the equation as

$$\frac{\partial P_i(z,t)}{\partial z} = -\alpha P_i - \left\{ \sum_{k=1}^{i-1} \frac{f_i - f_k}{A_{\text{eff}}} g_R P_k + \sum_{k=i+1}^{N} \frac{f_k - f_i}{A_{\text{eff}}} g_R P_k \right\} \tag{4.120}$$

where the gain coefficient of the Raman scattering is represented as g_R $(f_i - f_k)$ for a given frequency separation. This equation can be solved analytically to give a solution of

$$P_i(L,t) = P_i(0)\exp\left\{ -\alpha L - \sum_{k=1}^{i-1} \frac{f_i - f_k}{A_{\text{eff}}} g_R P_k + \sum_{k=i+1}^{N} \frac{f_k - f_i}{A_{\text{eff}}} g_R P_k \right\} \tag{4.121}$$

in which the second and third parts of coefficient of the exponential represent the loss and gain of the SRS given as

$$\alpha_{\text{SRS}} = \left\{ \sum_{k=1}^{i-1} \frac{f_i - f_k}{A_{\text{eff}}} g_R P_k + \sum_{k=i+1}^{N} \frac{f_k - f_i}{A_{\text{eff}}} g_R P_k \right\} \tag{4.122}$$

If the gain spectrum of Figure 4.14 can be approximated as a linear distribution, then the gain tilt of the SRS can be written as

$$\alpha_{\text{SRS}} = 10\text{Log}_e\left(\frac{1}{2} \frac{L_{\text{eff}}}{A_{\text{eff}}} \frac{dg_R}{df} P_{\text{channel}}(N-1)N\Delta f \right)\text{dBm} \tag{4.123}$$

The gain tilt would result in different optical signal levels of different wavelength channels, and thus gain equalization should be recommended. This can be implemented by pumping at different pump wavelengths as shown in Figure 4.15.

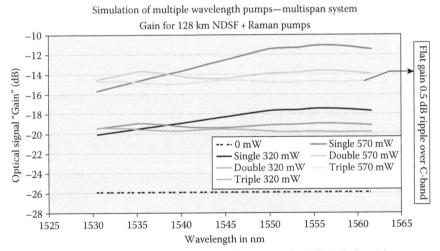

0.8 dB ripple for dual pumps at 1427 and 1455 nm and 0.5 dB ripple for triple pumps at 1423, 1435, and 1459 nm

FIGURE 4.15 Simulation of Raman gain flattening using multiple pump sources and total pump power. Vertical axis = gain in dB.

4.8.2.5 SBS Effects

SBS comes from the nonlinear interaction of the photons and the phonons. SBS is to be considered if the optical power within a narrow bandwidth exceeds a certain threshold. The maximum spectral density of the power that can be launched into a single-mode fiber is given by

$$\text{PSD}_{\text{SBS,th}} = \frac{P_{\text{SBS,th}}}{B_{\text{sig}}} \approx \frac{21 A_{\text{eff}}}{g_{\text{SBS}} L_{\text{eff}} B_{\text{SBS}}} \tag{4.124}$$

where g_{SBS} is the Brillouin gain coefficient, and B_{sig} and B_{SBS} stand for the signal bandwidth and the SBS gain bandwidth, respectively. For SSMF with $A_{\text{eff}} = 50 \ \mu m^2$, $L_{\text{eff}} = 20$ km, $B_{\text{SBS}} = 20$ MHz, and $g_{\text{SBS}} = 4.10^{-11}$, the threshold value of the power spectral density is about $6e^{-11}$ W/Hz.

Under the transmission and detection, the SBS effect can be observed with the low-frequency modulation of the high-frequency signal envelope, or, effectively, the jitter of the eye diagram and hence the reduction of the sampling time interval of the eye diagram or reduction of the time Q factor. We can observe in Chapter 6 that the ASK format exhibits the peak power of the optical carrier and none for phase shift keying formats. Thus, ASK is expected to have a lower SBS threshold and hence a low optical signal power available for transmission.

4.9 PERFORMANCE EVALUATION

The performance evaluation of an optical transmission system via the quality of the electrically detected signals is an essential aspect of both simulation and experiment scenarios. The key metrics reflecting the signal quality include the OSNR and OSNR penalty, EO and EOP, where the BER is the ultimate indicator for the performance of a system.

In an experimental setup and practical optical systems, BER and the quality factor, Q factor, can be obtained directly from the BERT set, and data can be exported to a portable memory for post-processing. However, it is noted that these experimental systems need to be run within at least a few hours for stable and accurate results. For the case of investigation of the performance of an optical transmission system by simulation, several effective statistical methods have been developed and are outlined in this section.

4.9.1 BER FROM MONTE CARLO METHOD

The conventional method of calculating the Q factor, Q dB, and hence the BER is based on the assumption of a Gaussian distribution for the noise. However, new methods based on statistical processes taking into account the distortion dynamics of the optical fibers are necessary to include the common patterning effects and the ISI induced by CD and PMD, especially when the phase or amplitudes of the carrier are under different modulation formats.

The first statistical technique implements the expected maximization theory, in which the pdf of the obtained electrical detected signal is approximated as a mixture of multiple Gaussian distributions.

The second technique is based on the generalized extreme values GEV [15–17] theorem. Although this theorem [18] is well known in other fields such as financial forecasting, meteorology, materials engineering, and so on to predict the probability of the occurrence of extreme values, it has not been much studied for application in optical communications.

Exactly as the BERT set [19] is used in experimental transmissions, the BER in a simulation of a particular system configuration is counted. The BER is the ratio of the occurrence of errors (N_{error}) to the total number of transmitted bits N_{total} and is given as

$$\text{BER} = \frac{N_{\text{error}}}{N_{\text{total}}} \tag{4.125}$$

The Monte Carlo method offers a precise picture via the BER metric for all modulation formats and receiver types. The optical system configuration under a simulation test needs to include all the sources of impairments affecting signal waveforms, including the fiber impairments and ASE (optical)/electronic noise.

It can be seen that a sufficient number of transmitted bits for a certain BER is required, leading to exhaustive computational time. In addition, time-consuming algorithms such as FFT especially carried out in symmetrical SSFM contribute greatly to the long computational time. A BER of 1e−9, which can be considered as error free in most scientific publications, requires at least 1e10 bits to be transmitted.

However, 1e−6 or even 1e−7 is feasible in Monte Carlo simulation [20]. Furthermore, with the use of forward error coding (FEC) schemes in contemporary optical systems, the reference for BERs to be obtained in simulation can be as low as 1e−3 provided there is no sign of an error floor. This is normally known as the FEC limit. The BERs obtained from the Monte Carlo method are a good benchmark for other BER values estimated with other techniques.

4.9.2 BER and Q Factor from Probability Distribution Functions

This method implements a statistical process before calculating the values of the BER and Q factor to determine the normalized probability distribution function (pdf) of received electrical signals (for both 1 and 0 and at a particular sampling instance). The electrical signal is normally in voltage form because the detected current after a photodiode is usually amplified by a transimpedance electrical amplifier. The pdfs can be determined statistically by using the histogram approach. A particular voltage value used as a reference for the distinction between 1 and 0 is known as the threshold voltage (V_{th})

The BER in the case of transmitting bit 1 (received as 0 instead) is calculated from the well-known principle [24]. The integral of the overlap of the normalized pdf of 1 exceeds the threshold. A similar calculation can be applied for bit 0. The actual shape of the pdf is thus very critical to obtaining an accurate BER. If the exact shape of the pdf is known, the BER can be calculated precisely as

$$\text{BER} = P(\text{'1'})P(\text{'0'}|\text{'1'}) + P(\text{'0'})P(\text{'1'}|\text{'0'}) \tag{4.126}$$

where $P(\text{'1'})$ is the probability that a 1 is sent; $P(\text{'0'}|\text{'1'})$ is the probability of error due to receiving a 0 where actually a 1 is sent; $P(\text{'0'})$ is the probability that a 0 is sent; $P(\text{'1'}|\text{'0'})$ is the probability of error due to receiving a 1 where actually a 0 was sent. The probabilities of transmitting a 1 and 0 are equal; that is, $P(\text{'1'}) = P(\text{'0'}) = 1/2$.

A popular approach in both simulation and commercial BERT test-sets is the assumption of a pdf of 1 and 0 following Gaussian/normal distributions, which means the noise sources can be approximated by Gaussian distributions. If the assumption is valid, high accuracy is achieved. This method enables a fast estimation of the BER by using the complementary error functions [24]

$$\text{BER} = \frac{1}{2}\left[\text{erfc}\left(\frac{|\mu_1 - V_{th}|}{\sqrt{2}\sigma_1} \right) + \text{erfc}\left(\frac{|\mu_0 - V_{th}|}{\sqrt{2}\sigma_0} \right) \right] \tag{4.127}$$

where μ_1 and μ_0 are the mean values for the pdf of 1 and 0, respectively, whereas σ_1 and σ_0 are the variance of the pdfs. The quality factor, Q- or δ-factor, which can be either in the linear scale or in the logarithmic scale, can be calculated from the obtained BER through the expression

$$Q \triangleq \delta = \sqrt{2}\,\text{erfc}^{-1}(2\text{BER})$$

$$Q_{dB} = 20\text{Log}_{10}\left(\sqrt{2}\,\text{erfc}^{-1}(2\text{BER}) \right) \tag{4.128}$$

4.9.3 HISTOGRAM APPROXIMATION

The common question is, what are the proper values for the number of bins and the bin width to be used in the approximation of the histogram so that the bias and the variance of the estimator are negligible? According to Ref. [25], with a sufficiently large number of transmitted bits (N_0), a good estimate for the width (W_{bin}) of each equally spaced histogram bin is given by

$$W_{bin} = \sqrt{N_0} \qquad (4.129)$$

4.9.4 OPTICAL SNR

The OSNR is a popular benchmark indicator for assessment of the performance of optical transmission systems, especially those limited by the ASE noise from the optical amplifiers—the EDFAs. The OSNR is defined as the ratio of the optical signal power to the optical noise power. For a single EDFA with output power, P_{out}, the OSNR is given by [21]

$$\text{OSNR} = \frac{P_{out}}{N_{ASE\,(0.1\,nm)}} = \frac{P_{out}}{(\text{NF} \cdot G_{op} - 1)h\nu B_o} \qquad (4.130)$$

where NF is the amplifier NF, G_{op} is the amplifier gain, $h\nu$ is the photon energy, and B_o is the optical bandwidth by measurement, standard setting at 0.1 nm or about 12.5 GHz in the 1550 region. However, the OSNR does not provide a good estimation of the system performance when the main degrading sources involve the dynamic propagation effects such as dispersion (including both CD and PMD) and Kerr nonlinearity effects, for example, the SPM. In these cases, the degradation of the performance is mainly due to waveform distortions rather than the corruption of the ASE or electronic noise.

When addressing the value of an OSNR, it is important to define the optical measurement bandwidth over which it is calculated. The signal power and noise power are obtained by integrating all the frequency components across the bandwidth, leading to the value of the OSNR.

In practice, signal and noise power values are usually measured directly from the optical spectrum analyzer (OSA), which does the mathematics for the users and displays the resultant OSNR versus wavelength or frequency over a fixed resolution bandwidth. A value of $\Delta\lambda = 0.1$ nm or $B_o = 12.5$ GHz in the 1550 nm region is widely used as the typical value for calculation of the OSNR.

The OSNR penalty is determined at a particular BER when varying the value of a system parameter under test. For example, the OSNR penalty at BER = 1e−4 for a particular optical phase modulation format when varying the length of an optical link in a long-haul transmission system configuration.

When measuring the energy contained in the signal band and the noise under optical amplifications, we must note the following: (1) The optical noise under amplification and active signals are quite different as compared to the case in which no signals are present. Figure 4.16 shows the spectra of various DPSK signals and the spectral windows in which Figure 4.17 the signals and noises are measured to obtain the OSNR (0.1 nm).

4.9.5 EYE OPENING PENALTY

The OSNR is a time-averaged indicator of the system performance in which the ratio of the average power of optical carriers to noise is considered. When optical lightwaves propagate through a dispersive and nonlinear optical fiber channel, the fiber impairments, including ISI induced by

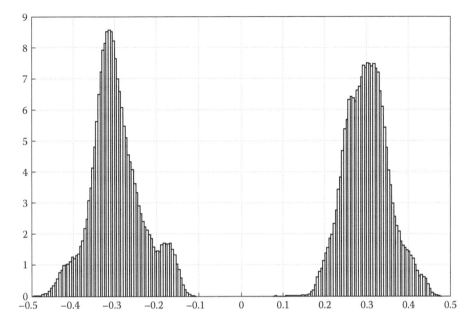

FIGURE 4.16 Demonstration of multipeak/non-Gaussian distribution of the received electrical signal.

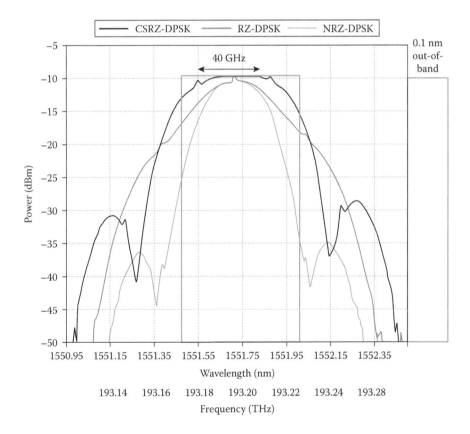

FIGURE 4.17 Spectrum of signals and noise (0.1 nm) bands in which the total power of signals and noise is measured for estimation of the optical signal-to-noise ratio.

CD, PMD, and the spectral effects induced by nonlinearities, cause waveform distortion. Another dynamic cause of waveform distortion is the ISI effects, the result of optical or electrical filtering. In a conventional OOK system, the bandwidth of an optical filter is normally several times larger than the spectral width of the signal.

The EOP is a performance measure defined as the penalty of the "eye" caused by the distortion of the electrically detected waveforms to a referenced EO. The EO is the difference between the amplitudes of the lowest mark and the highest space. The benchmark EO is usually obtained from a back-to-back measurement when the waveform is not distorted at all by any of the foregoing impairments.

The EOP at a particular sampling instance is normally calculated in the log scale (dB unit) and is given by

$$\text{EOP}(t_{\text{samp}}) = \frac{\text{EO}_{\text{ref}}}{\text{EO}_{\text{received}}} \tag{4.131}$$

The EOP is useful for noise-free system evaluations as a good estimate of deterministic pulse distortion effects. The accuracy of the EOP indicator depends on the sampling instance in a bit slot. Usually, the detected pulses are sampled at the instant giving the maximum EO. If noise is present, the calculation of the EOP become less precise because of the ambiguity of the signal levels, which are corrupted by noise.

4.9.6 STATISTICAL EVALUATION TECHNIQUES

The method using a Gaussian-based single distribution involves only the effects of noise corruption on the detected signals and ignores the dynamic distortion effects such as ISI and nonlinearity. These dynamic distortions result in a multipeak pdf as demonstrated in Figure 4.16, which is clearly overlooked by the conventional single distribution technique. As a result, the pdf of the electrical signal cannot be approximated accurately. The addressed issues are resolved with the proposal of two new statistical methods.

Two new techniques proposed to accurately obtain the pdf of the detected electrical signal in optical communications include a mixture of multi-Gaussian distributions (MGD) by employing expectation maximization theory (EM) and the generalized Pareto distribution (GPD) of the GEV theorem. These two techniques are well known in fields of statistics, banking, finance, meteorology, and so on. The required algorithms can be implemented without using MATLAB functions. Thus, these novel statistical methods offer significant flexibility, convenience, and speedy processing while maintaining the errors in obtaining the BER within small and acceptable limits.

4.9.6.1 Multi-Gaussian Distributions via Expectation Maximization Theorem

The mixture density parameter estimation problem is probably one of the most widely used applications of the expectation maximization (EM) algorithm. It comes from the fact that most deterministic distributions can be seen as the result of superposition of different multidistributions. Given a probability distribution function $p(x|\Theta)$ for a set of received data, $p(x|\Theta)$ can be expressed as the mixture of M different distributions

$$p(x|\Theta) = \sum_{i=1}^{M} w_i p_i(x|\theta_i) \tag{4.132}$$

where the parameters are $\Theta = (w_1, ..., w_M, \theta_1, ..., \theta_M)$ such that $\sum_{i=1}^{M} w_i = 1$, each P_i is a pdf by q_i, and each pdf assigned a weight w_i that is the probable event of that pdf.

As a particular case adopted for optical communications, the EM algorithm is implemented with a mixture of MGDs. This method offers great potential solutions for the evaluation of the performance of an optical transmission system for the following reasons: (1) In a linear optical system (low input power into fiber), the conventional single Gaussian distribution fails if taking into account the waveform distortion caused by both the ISI due to fiber CD and the PMD dispersion, and the pattern dependence effects. Hence, the obtained BER is no longer accurate. These issues, however, are overcome by using the MGD method. (2) The computational time for implementing MGD is fast via the EM algorithm, which has become quite popular. It should be noted that the Q factor or single Gaussian distribution is considered a subset of the MGD method.

4.9.6.2   Selection of Number of Gaussian Distributions for MGD Fitting

The critical step affecting the accuracy of the BER calculation is the process of estimating the number of Gaussian distributions applied in the EM algorithm for fitting the received signal's pdf. This number is determined by the estimated number of peaks or valleys in the curves of the first and second derivative of the original data set. The explanation of this procedure employs the well-known "Heming Lake Pike" example as reported in Refs. [24,25]. In this problem, the data of five age groups give the lengths of 523 pikes (*Esox lucius*), sampled in 1965 from Heming Lake, Manitoba, Canada. The components are heavily overlapped, and the resultant pdf is obtained with a mixture of these five Gaussian distributions as shown in Figure 4.18a. The figures are extracted from [25] to demonstrate the procedure. The following flowchart (Figure 4.19) illustrates the implementation process of the MGD method.

We now describe the estimation of the number of Gaussian distributions in the mixed pdf based on the first and second derivatives of the original data set (from Ref. [26]). As seen from Figure 4.19b, the first derivative of the resultant pdf shows clearly four pairs of peaks (dark gray) and valleys (light gray), suggesting that there should be at least four component Gaussian distributions contributing to the original pdf. However, by taking the second derivative, it is clear that there are actually up to five contributed Gaussian distributions, as shown in Figure 4.18.

In summary, the steps for implementing the MGD technique to obtain the BER value is described briefly as follows: (1) Obtaining the pdf from the normalized histogram of the received electrical levels. (2) Estimating the number of Gaussian distributions (N_{Gaus}) to be used for fitting the pdf of the original data set. (c) Applying the EM algorithm with the mixture of N_{Gaus} Gaussian distributions and obtaining the values of mean, variance, and weight for each distribution. (d) Calculating the BER value based on the integrals of the overlaps of the Gaussian distributions when the tails of these distributions cross the threshold.

4.9.7   Generalized Pareto Distribution

The GEV theorem is used to estimate the distribution of a data set of a function in which the possibility of extreme data lengthens the tail of the distribution. Owing to the mechanism of estimation for the pdf of the extreme data set, GEV distributions can be classified into two classes: the GEV distribution and the GPD.

Currently, there are only a few research reports on the application of this theorem in optical communications systems [22–24], which restrict themselves to GEV distributions that involve only the effects of noise and neglect the effects of dynamic distortion factors. Unlike the Gaussian-based techniques, but rather similar to the exponential distribution, the GPD is used to model the tails of

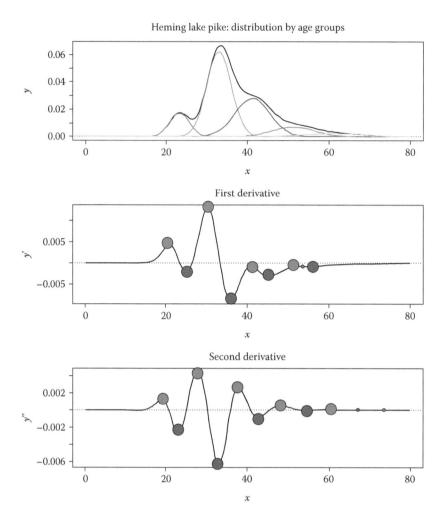

FIGURE 4.18 Estimation of number of Gaussian distributions in the mixed pdf based on first and second derivatives of the original data set.

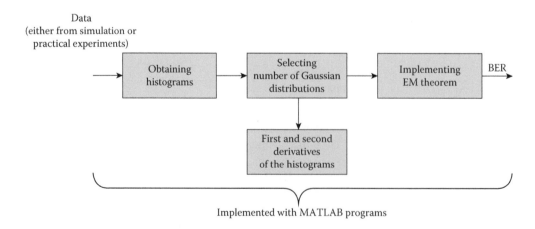

FIGURE 4.19 Flowchart of the implementation process of MGD method.

the distribution [25]. This section provides an overview of the GPD. The pdf for the GPD is defined as follows

$$y = f(x|k,\sigma,\theta) = \left(\frac{1}{\sigma}\right)\left(1 + k\frac{(x-\theta)}{\sigma}\right)^{-1-\frac{1}{k}} \tag{4.133}$$

for $\theta < x$ when $k > 0$ or for $\theta < x < \dfrac{-\sigma}{k}$ $k < 0$ where k is the shape parameter $k \neq 0$, σ is the scale parameter, and the threshold parameter is θ.

Equation 4.80 is subject to some significant constraints: (1) when $k > 0$: $\theta < x$, that is, there is no upper bound for x; (2) when $k < 0$: $\theta < x < -\dfrac{\sigma}{k}$ and zero probability for the case $x > -\dfrac{\sigma}{k}$; (3) when $k = 0$; that is, Equation 4.80 can be turned to $y = f(x|0,\sigma,\theta) = \left(\dfrac{1}{\sigma}\right)e^{-\frac{(x-\theta)}{\sigma}}$ for $\theta < x$; (4) If $k = 0$ and $\theta = 0$, the GPD is equivalent to the exponential distribution; and (5) If $k > 0$ and $\theta = \sigma$, the GPD is equivalent to the Pareto distribution.

The GPD has three basic forms reflecting the different classes of the underlying distributions: (1) distributions whose tails decrease exponentially, such as the normal distribution, lead to a generalized Pareto shape parameter of zero; (2) distributions with tails decreasing as a polynomial, such as Student's t, lead to a positive shape parameter; and (3) distributions having finite tails, such as the beta, lead to a negative shape parameter.

The GPD is widely used in finance, meteorology, material engineering, and so on, for the prediction of extreme or rare events known as the *exceedances*. However, the GPD has not yet been applied in optical communications to obtain the BER. The following reasons suggest that GPD may become a potential and a quick method for evaluation of an optical system, especially when nonlinearity is the dominant degrading factor for the system performance.

The normal distribution has a fast roll-off or short-tail distribution. Thus, it is not a good fit to a set of data involving exceedances, that is, rarely occurring data located in the tails of the distribution. With a certain threshold value, the GPD can be used to provide a good fit to the extremes of this complicated data set.

1. When nonlinearity is the dominating impairment degrading the performance of an optical system, the sampled received signals usually introduce a long-tail distribution. For example, in case of a DPSK optical system, the distribution of nonlinearity phase noise differs from that in the Gaussian counterpart owing to its slow roll-off of the tail. As a result, the conventional BER obtained from assumption of Gaussian-based noise is no longer valid, and it often underestimates the BER.
2. A wide range of analytical techniques have recently been studied and suggested such as importance sampling, multicanonical method, and so on. Although these techniques provide solutions to the problem of obtaining a precise BER, they are usually far too complicated; whereas, calculation of the GPD has become a standard and is available in the recent MATLAB version (since MATLAB 7.1). The GPD therefore may provide a very quick and convenient solution for monitoring and evaluating the system performance. The necessary preliminary steps, which are fast in implementation, need to be carried out to find the proper threshold.
3. Evaluation of contemporary optical systems requires the BER to be as low as 1e−15. Therefore, the GPD can be seen to be quite suitable for optical communications.

4.9.7.1 Selection of Threshold for GPD Fitting

Using statistical methods, the accuracy of the obtained BER strongly depends on the threshold value (V_{thres}) used in the GPD fitting algorithm, that is, the decision where the tail of the GPD curve starts.

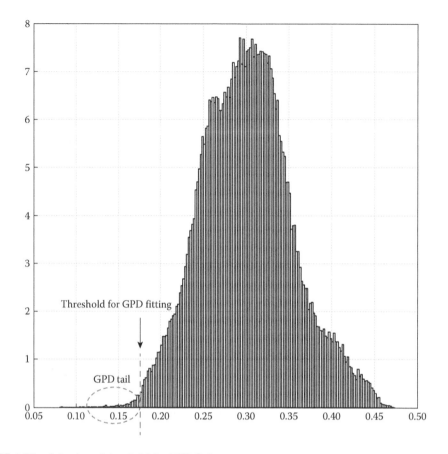

FIGURE 4.20   Selection of threshold for GPD fitting.

Several guidelines have been suggested to help determine the threshold value for the GPD fitting. However, they are not absolute techniques and are quite complicated. In this chapter, a simple technique to determine the threshold value is proposed. The technique is based on the observation that the GPD tail with exceedances normally obeys a slow exponential distribution compared to the faster decaying slope of the distribution close to the peak values. The inflection region between these two slopes gives a good estimation of the threshold value for GPD fitting. This is demonstrated in Figure 4.20.

Whether the selection of the V_{thres} value leads to an adequately accurate BER or not is evaluated by using the cumulative density function (cdf) and the quantile–quantile plot (Q–Q plot). If there is a high correlation between the pdf of the tail of the original data set (with a particular V_{thres}) and the pdf of the GPD, there would be a good fit between empirical cdf of the data set with the GPD-estimated cdf with the focus in the rightmost region of the two curves. In the case of the Q–Q plot, a linear trend would be observed. These guidelines are illustrated in Figure 4.21. In this particular case, the value of 0.163 as shown in Figure 4.21 is selected to be V_{thres}.

Furthermore, as a demonstration of the improper selection of V_{thres}, the value of 0.2 is selected. Figures 4.22 and 4.23 show the noncompliance of the fitted curve with the GPD, which is reflected in the discrepancy in the two cdfs and the nonlinear trend of the Q–Q plot (Figure 4.24).

The following flowchart (Figure 4.25) illustrates the implementation process of the GPD method.

4.9.7.2   Validation of Novel Statistical Methods

A simulation test bed of an optical DPSK transmission system over 880 km SSMF dispersion-managed optical link (eight spans) is set up. Each span consists of 100 km SSMF and 10 km of DCF whose dispersion values are +17 ps/(nm · km) and −170 ps/(nm · km) at 1550 nm wavelength,

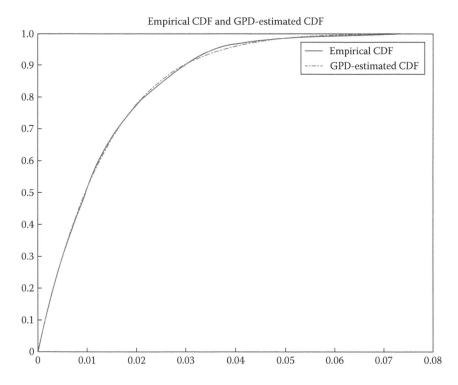

FIGURE 4.21 Comparison between fitted and empirical cumulative distribution functions with the threshold V value of 0.163.

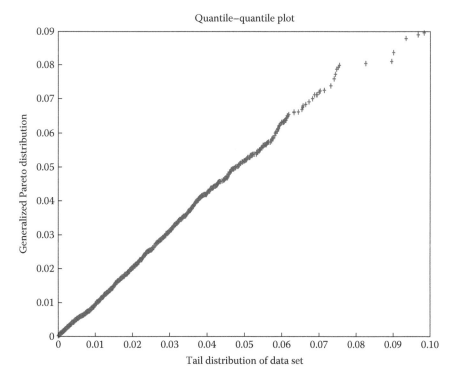

FIGURE 4.22 Quantile–quantile plot corresponding to distribution given in Figure 4.21.

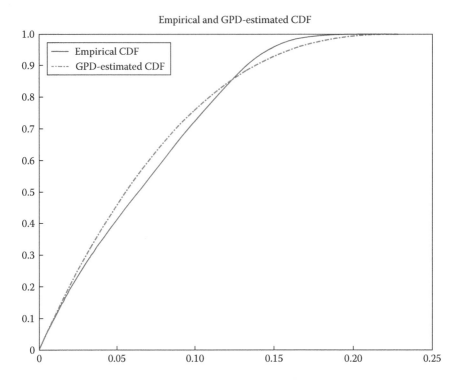

FIGURE 4.23 Comparison between fitted and empirical cumulative distribution functions for a threshold value of 0.2.

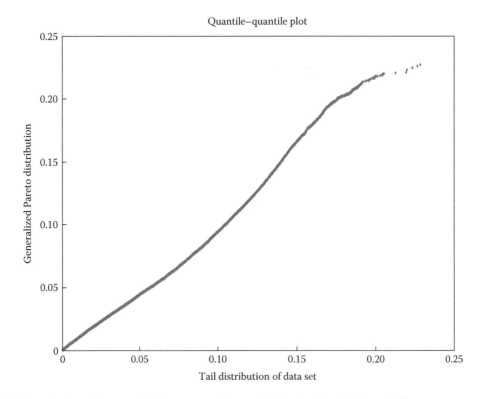

FIGURE 4.24 Quantile–quantile plot corresponding to distribution given in Figure 4.23.

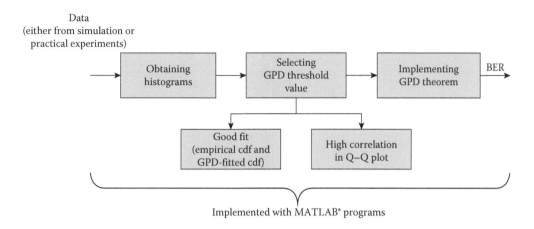

FIGURE 4.25 Flowchart of the implementation process of MGD method.

respectively, and fully compensated, that is, zero residual dispersion. The average optical input power into each span can be set to be higher than the nonlinear threshold of the optical fiber. The degradation of the system performance is therefore dominated by the nonlinear effects, which are of the interest because it is a random process creating nondeterministic errors in the long-tail region of the pdf of the received electrical signals.

The BER results obtained from the novel statistical methods are compared to those from the Monte Carlo simulation as well as from the semi-analytical method. As an example, the well-known analytical expression to obtain the BER of the optical DPSK format is used, which is given in the following equation [26,27]

$$\text{BER} = \frac{1}{2} - \frac{\rho e^{-\rho}}{2} \sum_{k=0}^{\infty} \frac{(-1)^k}{2k+1} \left[I_k\left(\frac{\rho}{2}\right) + I_{k+1}\left(\frac{\rho}{2}\right) \right]^2 e^{-\frac{1}{2}(2k+1)^2 \sigma_{\text{NLP}}^2} \tag{4.134}$$

where ρ is the obtained OSNR, and σ_{NLP}^2 is the variance of nonlinear phase noise.

In this case, in order to calculate the BER of an optical DPSK system involving the effect of nonlinear phase noise, the required parameters including the OSNR and the variance of nonlinear phase noise, and so on, are obtained from the simulation numerical data, which is stored and processed in MATLAB. The fitting curves implemented with the MGD method for the pdf of bit 0 and bit 1 (input power of 10 dBm) as shown in Figures 4.26 and 4.27 are for bit 0 and bit 1, respectively.

The selection of the optimal threshold for GPD fitting follows the guideline as addressed in detail in the previous section. The BERs from various evaluation methods are shown in Table 4.1. The input powers are controlled to be 10 and 11 dBm.

Table 4.1 validates the accuracy of the proposed novel statistical methods with the discrepancies compared to the Monte Carlo and semi-analytical BER to be within one decade. In short, these methods offer a great deal of fast processing while maintaining the accuracy of the obtained BER within the acceptable limits.

Discussions: It is found that the MGD and GPD statistical methods provide sufficiently accurate BERs. The discrepancies in the BER values obtained by the MGD and GPD methods compared to the Monte Carlo and semi-analytical BER are within one decade.

The BER using the Monte Carlo method at a fiber input power of 11 dBm is not provided in the table as it is quite time consuming (it might take weeks) to obtain the BER in the range of 1e−8. In order to obtain a BER in 1e−8 range, there are at least 1e−9 binary bits generated from the random bit pattern generator. Moreover, this is not feasible for simulation to investigate optical

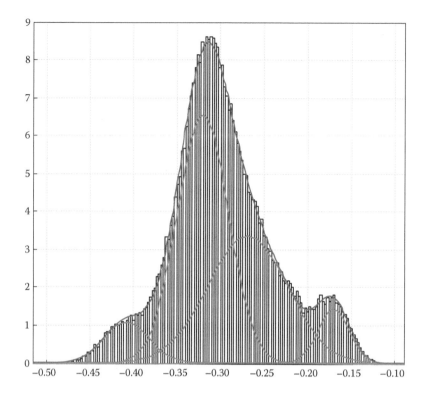

FIGURE 4.26   Demonstration of fitting curves for bit 0 with MGD method for bit 0.

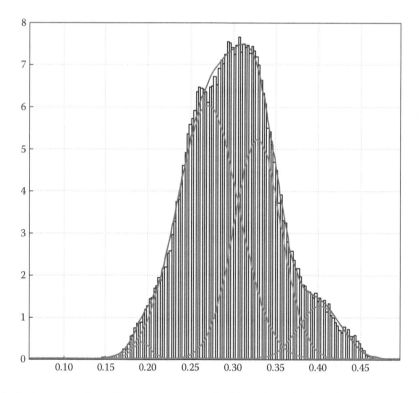

FIGURE 4.27   Demonstration of fitting curves for bit 0 with MGD method for bit 1.

TABLE 4.1

Required Input Launched Power Versus Bit Error Rate under Different Statistical Methods

Evaluation Method → Input Power	Monte Carlo Method	Semi-Analytical Method	MGD Method	GPD Method
10 dBm	$1.7e^{-5}$	$2.58e^{-5}$	$5.3e^{-6}$	$3.56e^{-4}$
11 dBm		$1.7e^{-8}$	$2.58e^{-9}$	$4.28e^{-8}$

transmission systems operating in the nonlinear region (in this case, with the existence of nonlinear phase noise—NLPN). In addition, BERs at an input power of 9 dBm (or lower than 9 dBm) are not provided either, as the NLPN does not significantly affect the system performance.

The MGD method consists of a set of Gaussian distributions that have the short-tail characteristic. Thus, the MGD method tends to underestimate the BER value. As can be observed from the table, the MGD method provides sufficiently accurate BERs within one decade below the Monte Carlo and semi-analytical BERs.

The GPD method provides precise estimation of extreme values (in this case, errors caused by NLPN). These extreme values are represented by the long-tail characteristic of GPD distributions. Thus, the GPD method tends to overestimate the BER value. As can be observed from the table, the GPD method provides sufficiently accurate BERs within one decade above the Monte Carlo and semi-analytical BERs.

As a guideline, the BER value can be the average of the BERs computed from the MGD and GPD methods.

4.9.8 Novel BER Statistical Techniques

The Gaussian-based single distribution involves only the effects of noise corruption on the detected signals and ignores the dynamic distortion effects such as ISI and nonlinearity. These dynamic distortions result in a multipeak pdf, which is clearly overlooked by the conventional single distribution technique. As the result, the pdf of the electrical signal cannot be approximated accurately. The addressed issues can be resolved with two new statistical methods proposed in this section. The two new techniques proposed to accurately obtain the pdf of the detected electrical signal in optical communications include the mixture of MGDs by implementing the EM theory [26–28] and the GPD [26] of the GEV theorem. These two techniques are well known in fields of statistics, banking, finance, meteorology, and so on. They offer a great deal of flexibility, convenience, and fast processing while maintaining the accuracy of the BER obtained.

4.9.8.1 MGDs and EM Theorem

The mixture density parameter estimation problem is probably one of the most widely used applications of the EM algorithm [29]. It comes from the fact that most deterministic distributions can be seen as the result of the superposition of different multidistributions [30,31]. Given a pdf $p(x|\Theta)$ for a set of received data, $p(x|\Theta)$ can be expressed as a superposition of M different distributions given as

$$p(x|\Theta) = \sum_{i=1}^{M} w_i p_i(x|\theta_i) \tag{4.135}$$

where the parameters are $\Theta = (w_1,...,w_M,\theta_1,...,\theta_M)$ such that $\sum_{i=1}^{M} w_i = 1$, each p_i is a pdf by θ_i, and each pdf has a weight w_i that is the probability of that pdf.

As a particular case adopted for optical communications, the EM algorithm is implemented with a mixture of MGDs. This method potentially offers accurate solutions for the evaluation of the performance of optical transmission systems for the following reasons: (1) In a linear optical system (low input power into fiber), the conventional single Gaussian distribution fails to take into account the waveform distortion caused by either the ISI due to fiber CD or PMD, the patterning effects. Hence, the obtained BER is no longer accurate. These issues, however, are overcome by using the MGD method. (2) Computational time for implementing the MGD is fast via the EM algorithm.

4.10   EFFECTS OF SOURCE LINEWIDTH

Direct detection differential quadrature phase shift keying (DD/DQPSK) and other discrete and continuous phase modulation formats such as BDPSK, CPFSK, and MSK have attracted attention and have been analyzed and studied in this book as possible modulation techniques for optical systems. In most published literature, it is generally assumed that the impact of the laser linewidth may be ignored. The aim of this section is to determine when we may neglect the effect of the oscillator phase noise. In a DD/DQPSK system, a pair of Mach–Zehnder interferometers and balanced receivers is used to recover the in-phase and quadrature bits. In an idealized system, the phase difference in the arms of the Mach–Zehnder interferometer is used to recover the phase difference of the two consecutive bits. However, phase noise from the finite linewidth laser will result in this being perturbed. Therefore, this effect is examined to determine when the linewidth of the laser can be ignored.

A 10 GBaud system is assumed in this section. However, the results can be easily scaled to other line rates. This section is a shortened version of the materials presented in Ref. [32] in which the BER performance of DQPSK is analyzed using a quadratic decision variable [3]. In this section, a similar approach may be used for optical DQPSK systems. To simplify the analysis, the electrical low-pass filter can be replaced by an integrate-and-dump receiver, giving the decision variable d to within a constant factor of the received optical field as

$$d = \frac{1}{T_s} \int_0^{T_s} e^{j\theta} E_o^{\mathrm{H}}(t - T_s) E_o^{\mathrm{H}}(t) dt + c.c. \tag{4.136}$$

The superscript H denotes the Hermitian, and θ, the phase offset, is nominally $\pi/4$; however, it may be perturbed by laser phase noise such that $\theta = \pi/4 + \varphi$, φ is the phase perturbation, and *c.c.* represents the complex conjugation. Over the symbol period T_s, the noise and signal may be expressed as a Fourier series, allowing the integral to be evaluated analytically [4]. If the preceding optical filter, which has a bandwidth $B = 1/T_s$, can be assumed, then only the first Fourier component need be considered for each polarization. Consider receiving the symbol $(I_k, Q_k) = (1,1)$ such that within a constant factor, it can be expressed as

$$d = \sum_{k=1}^{2} e^{j\theta} X_k X_k^* + c.c. \tag{4.137}$$

with X_k given as

$$X_1 = \sqrt{P} + n_1 + jn_2$$

$$X_2 = \sqrt{P} + n_3 + jn_4$$

$$X_2 = n_5 + jn_6 \tag{4.138}$$

$$X_1 = n_7 + jn_8$$

where $n_1 - n_4$ are the independent identically distributed components of the filtered ASE in-phase and quadrature components in both polarizations, assumed to be Gaussian with zero mean. If the OSNR can be defined as the ratio of the mean power to the total ASE power measured in an optical bandwidth, then it follows that

$$\langle n_1^2 \rangle = \frac{P}{4 \cdot \text{OSNR}} \tag{4.139}$$

This definition of the OSNR allows the use of published results for the quadratic decision variable [3] with the substitution OSNR = γ; with some algebra detailed in the Ref. [32], the conditional probability can be obtained as

$$\text{Prob}(d < 0|\theta) = \left(\frac{\tan\theta}{2} + \frac{\sec\theta}{2} + \frac{\cot\theta}{2} \right) \frac{e^{-2\gamma(1-\sin\theta)}}{\sqrt{4\pi\gamma\sin\theta}} \tag{4.140}$$

A conditional probability that combines with the PDF of the laser phase noise for both the 1 and 0, allows the determination of the overall BER by evaluating the integral

$$\text{BER} = \frac{1}{2} \int_{-\infty}^{+\infty} [\text{Prob}(d < 0|\phi) + \text{Prob}(d < 0|-\phi)] p(\phi) d\phi \tag{4.141}$$

For a laser with linewidth Δv, the phase noise may be modeled as a Gaussian process, with zero mean and variance. Using an asymptotic technique detailed in the Appendix of Ref. [33], a relatively simple equation for the BER in the presence of the phase noise can be given as

$$\text{BER} = \frac{e^{-2\gamma(\sqrt{2}-2)}}{\sqrt{4\pi\gamma\sqrt{2}}} \left(\frac{3\sqrt{2}}{8} + \frac{1}{2} \right) (1 + \gamma^2 \sigma_\phi^2) \tag{4.142}$$

To validate the approximations made, the analytical solution given by Equation 4.90 can be compared with a numerical evaluation of the integral Equation 4.89 using the conditional probability given by Equation 4.88. The agreement is excellent for error rates lower than 1e−3, where the region is nearly linear (allowing easy determination for a given BER). By expanding as a Taylor series about $\gamma = 20$, a linear fit can be used to obtain the total BER as

$$\log_{10} \text{BER}(\gamma, \sigma_\phi) \approx -6.05 - 0.265(\gamma - 20) \tag{4.143}$$

It is thus possible to consider the scenario under which the effect of the linewidth can be ignored. Two alternating criteria for neglecting the linewidth can be considered, first, when the phase noise causes the BER to double, and second, a 0.1 dB signal SNR penalty. Figure 4.28 illustrates the required linewidth for both of the criteria for 10 Gbps systems.

From Equation 4.91, it is observed that the BER is doubled when $\sigma_\phi^2 = 1/\gamma^2$. For a 10 GBd DQPSK system operating at a BER of 1e9 and in the absence of laser phase noise, the laser linewidth should be less than 16 MHz. It is noted, from Equation 4.93 that a doubling of the value of the BER leads to a SNR penalty of 0.5 dB. To keep the penalty below 0.1 dB, a similar condition can be obtained. For a 10 GBd DQPSK system operating at a BER of 1e−3 under the Monte Carlo simulation environment, the laser linewidth should be less than 3 MHz so that the penalty is less than 0.1 dB. These linewidths are much wider than those of current practical DFB laser sources, and thus direct detection of different modulation formats can be made without the laser linewidth significantly influencing the process.

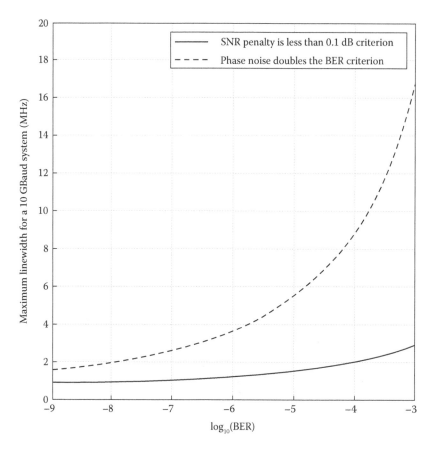

FIGURE 4.28 Criteria for neglecting linewidth in a 10 GBd system. The loose bound is to neglect linewidth if the impact is to double the BER, with the tighter bound being to neglect linewidth if the impact is a 0.1 dB SNR penalty. (From B. B. Brabson and J. P. Palutikof, *J. Appl. Meteorol.*, Vol. 39, pp. 1627–1640, 2000. Reprinted with permission.)

4.11 CONCLUDING REMARKS

An overview of the optical receiver and noise processes in optical signal detection has been presented. We have also demonstrated the Simulink modeling of amplitude and phase modulation formats at 40 Gbps optical fiber transmission. A novel modified fiber propagation algorithm has been used to minimize the simulation processing time and optimize its accuracy. The principles of amplitude and phase modulation, encoding, and photonic-optoelectronic balanced detection and receiving modules have been demonstrated via Simulink modules and can be corroborated with experimental receiver sensitivities. Experimental results on DPSK formats would be illustrated in detail in Chapter 5.

The XPM and other fiber nonlinearities such as Raman scattering and FWM effects are not integrated in the MATLAB and Simulink models. A switching scheme between the linear only and the linear and nonlinear models is developed to enhance the computing aspects of the transmission model.

Other modulations formats such as multilevel M-DPSK and M-ASK that offer a narrower effective bandwidth, simple optical receiver structures, and no chirping effects would also be integrated. These systems are given in Chapter 10. The effects of the optical filtering components in DWDM transmission systems to demonstrate the effectiveness of the DPSK and DQPSK formats have been measured in this paper and will be verified with simulation results in future publications.

Finally, further development stages of the simulator together with simulation results will be reported in future studies.

Furthermore, an illustration of the modeling of various schemes of advanced modulation formats for optical transmission systems is presented. Although transmitters have been described in Chapter 2, the models of transmitter modules integrating lightwaves, precoders, and external modulators can be included to show that they can be modeled with ease under MATLAB and Simulink. As MATLAB is becoming a standard computing language for academic research institutions throughout the world, the models reported here contribute to the wealth of computing tools for modeling optical fiber transmission systems and teaching undergraduates at a senior level and postgraduate research scholars. The models can integrate photonic filters or other photonic components using blocksets available in Simulink. Furthermore, we have used the developed models to assess the effectiveness of the models by evaluating the simulated results and experimental transmission performance of long-haul advanced modulation format transmission systems.

4.12 PROBLEMS

Problem 4.1

a. What are the frequencies of lightwaves with wavelengths of 1300 and 1550 nm in vacuum?
b. What are the *frequencies* and *wavelengths* of these lightwaves if they are propagating in optical fibers with a refractive index of 1.448 at 1330 nm and 1.446 at 1550 nm? You may assume single-mode operation of the lightwaves at these wavelengths.
c. What is the frequency difference between these two lightwaves?
d. What is the frequency difference between lightwaves of 1530 and 1560 nm? Now divide this wavelength range into a grid of 50 GHz spacing. Tabulate your wavelength grid and the corresponding spectral table.
e. What are the colors of the lightwaves of the above parts (a) and (d)?

Note: The velocity of light in vacuum is $c = 299{,}792{,}458$ m/s. Can you give reasons why this velocity must be very accurate?

Problem 4.2

a. What are the attenuations of lightwaves of 1300 and 1550 nm wavelength when propagating through silica-based glass fibers?
b. What is the peak attenuation within this spectral range? Give reasons for this peak attenuation.
c. Could you obtain a relationship between the loss or attenuation as a function of wavelength?
d. Examine Corning single-mode fibers SMF-28 and SMF-28e, which are available on e-mail request, and spot the differences between them. Give reasons why Corning spent time and money to develop SMF-28e for metropolitan optical networks.
e. Write down the frequency responses of an SSMF optical fiber, including the phase and amplitude frequency responses.

Problem 4.3

a. Examine the standard single-mode SMF-28 Corning type fibers, and write down the principal parameters related to
   i. The geometrical aspects of the fiber
   ii. The refractive index distribution
   iii. The operation of the fibers when modulated lightwaves are propagating
   iv. The impairments that may cause signal distortion

b. Repeat (a) for Corning LEAF fibers
c. Repeat (a) for Corning DCFs, which are available on several websites, for example, Avenex.com, JDSU.com, or corning.com (registration required).
d. Examine the dispersion factors given in these fibers. Why are there differences between them? Calculate the temporal broadening of a lightwave pulse sequence of 10 Gbps bit rate with a lightwave carrier wavelength of 1550 nm and similarly at 1560 nm.

Problem 4.4

Assume that the wavelength channels in Problem 4.2(d) are used as optical carriers for optical transmission systems that are modulated with a data rate of 10 Gbps using on–off keying non-return-to-zero format.

a. Sketch the frequency spectrum of a single optical channel whose carrier wavelength is in the region of 1550 nm.
b. Sketch the spectrum of all possible consecutive multiplexed optical channels in the spectral region 1530 to 1565 nm.
c. Repeat (b) for data rate of 100 Gbps. Comment on the spectrum of this part.

Problem 4.5

Refer to the technical data of the following fibers:

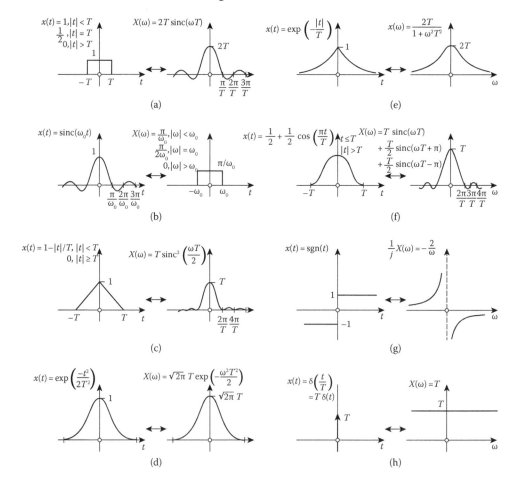

a. For Corning SMF-28, SMF-28e, and LEAF, extract the following information:
   i. Geometrical parameters: Core diameter, cladding diameter, relative refractive index difference, attenuation versus wavelength, core eccentricity, and ITU standard of the fibers.
   ii. Operational parameters: Mode field diameter, effective area, cutoff wavelength or wavelength range (why range?), dispersion factor and its variation as a function of wavelength, dispersion slope and from it the derivation of the values for beta 2 and beta 3, nonlinear coefficient, and polarization-mode dispersion factor.
b. Compare the parameters of these different fibers, and give some applications in optical transmission systems, with reasons. Especially state why a large effective area LEAF fiber is preferred for long-haul transmission systems.
c. Consider the dispersion factor of the LEAF fiber, and explain why the fiber is designed so that its zero dispersion is outside the operational spectrum of the C- or S-band.

Problem 4.6

The wavelength division multiplexing technique is a common method for increasing the total transmission capacity over a single strand of optical fiber. It is now widely employed in long-haul high-speed optical transmission systems throughout the world.

a. Briefly explain the WDM techniques, and state the typical wavelength or frequency spacing between the channels so that they can be called dense wavelength division multiplexing (DWDM) optical transmission systems.
b. What are the *frequencies* and *wavelengths* of these lightwaves so that they can be transmitted with minimum loss? You may consider the spectral regions called S-, C-, and L- bands of a silica-based single-mode optical fiber.
c. Sketch the structure of a long-haul optical transmission system using multiple spans with optical fibers, optical amplifiers, DCFs, optical filters as multiplexers and demultiplexers, and optical transmitters and receivers.
d. Could you search the Internet and find at least one array waveguide multiplexer that would be able to satisfy your DWDM optical transmission system?

Note: The velocity of light in vacuum is $c = 299,792,458$ m/s. Can you state the reasons why this velocity must be very accurate?

Problem 4.7

a. Sketch the basic structure of a symmetric and an asymmetric Mach–Zehnder interferometer.
b. Obtain the optical intensity transmittance between an input port and the other two output ports, for example, T_{11} and T_{12}. Hence, show that a complete power transfer can be achieved with a total phase difference of π. You may state the relationship of the transmittance or derive the equations.
c. You could use the z-transform to obtain the transmittance of the filter and hence the continuous transmittance with an appropriate substitution of the z-variable. Hence, determine the zeroes and poles of the transmittance. Sketch the poles and zeroes on the z-plane. How close to the units circle should the zeroes of the transfer function be so that the filter passband can provide the highest filtering property? Determine the frequency of the cutoff band (-20 dB from the 3 dB roll-off). Sketch the filter passband. Explain the physical meaning of the zeroes and poles of the transfer function.
d. Now, with and without using the equations, can you explain the mechanism of the coupling and filtering of optical waves?

e. Now consider silica on silicon interferometric filters of the above parts. The effective refractive index of the single-mode waveguide of the filter is assumed to be 1.4465 at 1550 nm. Estimate the propagation constant β of the guided wave. Hence, estimate the difference length of an asymmetric interferometer so that a complete filtered channel can be achieved at the cross output port.

f. If there are some path difference fabrication errors, then could you suggest a method so that the filter can be tuned to the exact filtering property?

g. Let the delay difference path be now one-bit long for a 10 Gbps phase modulation system. Suggest a function of this interferometer. Now suppose that two consecutive bits of a data sequence of 10 Gbps whose optical carriers are 0 and π phase or π and π or 0 and 0 phase. Work out the intensity of the outputs of the interferometer under these receiving conditions.

h. Determine the length of the delay path of the interferometer so that a one-bit delay can be achieved for part (d).

Problem 4.8

a. Refer to the ring resonator given in Ref. [34]. State the power transfer function $H_{11}(z)$ or $|H_{11}(z)|^2$ including the frequency responses of its power amplitude, phase, group delay, and dispersion. Use MATLAB to plot these frequency responses. Find the resonant frequency and the 3 dB bandwidth of the filter, and hence its Q factor. If the resonant frequency is to be set at the equivalent vacuum frequency of the wavelength of 1550 nm, what is the length of the ring resonator? The effective refractive index of a guided wave at 1550 nm of the ring resonator is 1.448.

b. By cascading at least two ring resonators, obtain the frequency responses for the power amplitude and phase. This is best achieved by using $H_{11}(z)$, that is, the transfer function would have multiple identical poles. Hence, obtain the corresponding group delay and dispersion at the resonant wavelength.

c. How close to the units circle should the poles of the transfer function be so that the filter passband can provide the highest filtering property? Determine the frequency of the cutoff band (−20 dB from the 3 dB roll-off). Sketch the filter passband. Explain the physical meaning of the poles of the transfer function.

d. If there are losses for lightwaves propagating in the ring, how you would correct them so that the poles of the resonator can be set close to the unit circle?

e. Now that the frequency of the input lightwaves can be shifted to the right or to the left of the resonant peak frequency, obtain the dispersion factor at these edges of the filter passband.

f. If the ring resonator is used as a dispersion compensator of 100 km optical fiber of standard SMF (about +17 ps/nm/km depending on the wavelength), determine the wavelength at which the resonator should be operating so that complete compensation can be achieved.

g. Comment on the stability of the optical ring resonator and that of the interferometric filter of Problem 4.2.

Problem 4.9

The frequency response of the power transmittance of a Fabry–Perot optical filter is given by

$$G_{PF} = \cfrac{1}{1 + \left[\dfrac{2\sqrt{R}}{1-R} \sin\left(\dfrac{2\pi nL}{c} \right) f \right]^2}$$

a. Sketch the structure of the Fabry–Perot filter, and indicate all the parameters of the foregoing equation in this diagram.
b. State the assumptions and approximations for the derivation of this transfer function.
c. Determine the frequencies or wavelengths of the resonant peaks, and hence the free-spectral range as a function of n, L, and R. Sketch the frequency response of the power transfer between the input and output ports of the filter.
d. Determine the 3 dB bandwidth of the filter as a function of n, L, and R, and from it the Q factor of the filter and its finesse. Can you determine the cutoff frequency of the filter passband at which the power is −20dB from the 3 dB roll-off point?
e. The center frequency of the FP filter can be tuned so that it can be matched with the center wavelength of a received data channel. What parameter can be changed so that the center passband of the filter can be tuned?
f. If the FP filter is used for filtering the noise of an optical amplifier, the ASE noise, determine the finesse of the filter for ASE noise reduction at a center wavelength of 1565 nm and a bandwidth of 0.1 nm.

Problem 4.10

a. Based on information provided on the manufacturers' websites (e.g., www.avenex.com, www.jdsu.com), sketch the structure of an integrated optic AWG. State the physical mechanism of the filtering property of the AWG. You may note the path length difference between waveguide channels, and hence the interference effects.
b. Provide the operation parameters such as the center wavelength or frequency spacing, number of channels, insertion loss, passband cutoff wavelength, and so on, for an AWG that can be used either as a multiplexer or demultiplexer in a 50 GHz spacing DWDM optical transmission system.
c. If the average power at the output pigtail of an optical modulator of an optical transmitter is −10 dBm is launched into the AWG multiplexer of (b) and a total average power into the transmission fiber of 10 dBm for 44 optical channels is required, find out whether optical amplification is needed. If so, what gain of the optical amplifier is required?

Problem 4.11

Consider the passband reflectance and group delay characteristics of an FBG as shown in the Figure problem 4.11:

a. What are the center wavelength of the passband and the 3 dB bandwidth (passband) of the filter? Note that the transfer characteristic is plotted in a linear scale.
b. The group delay is measured using a tunable laser source launched into one end of the FBG. Determine the average dispersion factor of the FBG in ps/nm. Note the sign of the dispersion factor. If you want to reverse the sign of the dispersion, what could you do to the FBG to achieve this? Give your reasons.
c. If this FBG is used for compensating a length of standard SMF, determine the length of the fiber.
d. Comment on the ripple of the group delay and its effects on the total dispersion characteristics when it is used for dispersion compensation.
e. Obtain the transmittance of the FBG.
f. Show how you can connect the FBG to an in-line optical system to compensate for any residual dispersion of the fiber transmission line. It is suggested that you use the FBG together with an optical circulator.

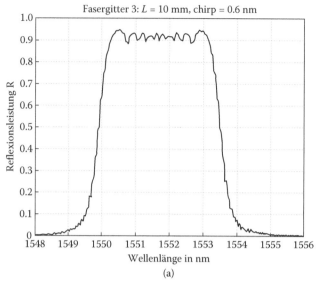

(a)

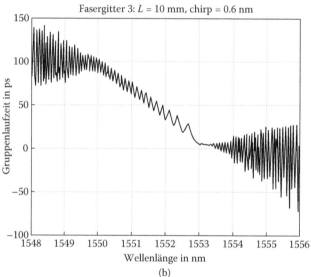

(b)

Problem 4.12

A 2 × 2 coupler has an intensity coupling coefficient ε of 0.5. This is called a 3 dB coupler.

a. Write down the transmittance matrix of the coupler.
b. Write down the relationship between the output optical fields and the input fields for the coupler.
c. Hence, sketch the signal flow model of the coupler.

Problem 4.13

Consider the Mach–Zehnder interferometric filter shown in the figure.

a. Write down the transmittance matrices for the two couplers and the middle section.
b. Hence, write down its relationship with the input and output fields at the input ports 1,2 and the output ports 1,2.

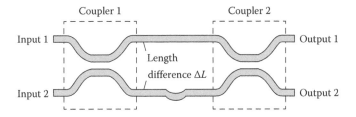

c. Sketch the signal flow graph of the filter, and show that for a delay length ΔL, which is equivalent to the traveling time of a bit period, the output field E_{01} and E_{02} and its input field E_{i1} can be written as

$$E_{01}(z) = H_{11}(z)E_{i1}(z) = (1 + z^{-1})E_{i1}(z) = z^{-\frac{1}{2}}\left[z^{+\frac{1}{2}} + z^{-\frac{1}{2}}\right]E_{i1}(z)$$

where $z = exp(j\beta\Delta L) = exp(j\omega t)$.

d. Plot the zeroes and poles of the transfer function on the z-plane and thereby the transmittance frequency responses of the filter.

Problem 4.14

Wavelength division multiplexing (WDM) and dense WDM networks are now installed throughout global communication backbone networks.

a. Define the wavelength or frequency spacing between channels for dense and coarse WDM systems and networks.
b. Give reasons why Corning fibers SMF-28e and SMF-28 (ITU G.652 standard) are usually selected for CWDM and DWDM in metropolitan networks. What is the main difference between these types of fibers?
c. What are the typical wavelengths that can be used for CWDM? Give reasons. **Hint**: Consider downloading and uploading speed.
d. With reference to the attenuation factor of silica fiber, specify the spectral regions for the S-, C-, and L- bands for silica-based optical fibers.
e. For the trans-Pacific transmission system with a total capacity of 1806 Gbps in 2006 and with a 10 Gbps wavelength channel, how many wavelength carriers are required to carry this capacity? If they are to be transmitted in the C-band, determine the frequency spacing between the channels.
f. Sketch a DWDM optical transmission system with optical transmitters, multiplexers, amplifiers, dispersion compensators, and demultiplexers and receivers. Indicate in your system whether the signals are in the optical or electronic domain.

Problem 4.15

a. What are DXC, ADM, and LTE in an optical network? Explain with a sketch of the structure of an optical network.
b. Sketch the structure of cascade Mach–Zehnder interferometers that would form two-channel and four-channel wavelength demultiplexers. If the wavelength channels are separated by 50 GHz, determine the delay length ΔL.

Note: The reference B. H. Verbeeck et al., Integrated four-channel Mach–Zehnder Multi/Demultiplexer fabricated with Phosphorous Doped SiO_2 Waveguide on silicon, *IEEE J.*

Lightwave Technol. The condition for splitting the wavelength channels 1 and 2 to the two output ports is given as

$$2\pi n_{\mathrm{eff}} \Delta L \left(\frac{1}{\lambda_1} - \frac{1}{\lambda_2} \right) = \pi$$

where n_{eff} is the guided refractive index of the guided mode, and λ_1 and λ_2 are the wavelengths of the two consecutive channels.

Problem 4.16

    a. Sketch a possible structure of the OSI layer model.
    b. Give the structure of the SHF model, and explain how data channel can be integrated from STM-1 to STM-16. Answer the same problems for SONET.
    c. Show clearly how the data formats of SHF or SONET can be integrated into optical transmission systems.

Problem 4.17

    a. Sketch the structure of an 8 × 8 switching matrix for Benes, strict sense nonblocking, and rearrangeably nonblocking architectures. Comment on the advantages and disadvantages of these structures.
    b. What are the typical switching speeds of the optical switches of (a)? Compare their speed with that of electromechanical switches.

Problem 4.18

A wavelength router can be designed using a wavelength converter and an optical demultiplexer. A wavelength converter is usually formed in a semiconductor optical amplifier (SOA).

    a. Sketch the structure of this wavelength router.
    b. Can you use the wavelength router of (a) in the design of a wavelength add/drop?

Problem 4.19

Consider the optical add/drop multiplexer shown in the following figure.

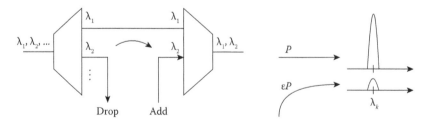

If the crosstalk of −20 dBm power from adjacent channel is imposed on the considered channel then estimate the penalty of the eye diagram due to this crosstalk. The average power of the channel is −10 dBm.

Now, if the same amount of optical power of the optical channel on the left of the operating channel is residual in this channel, re-estimate the eye penalty.

a. A DWDM transmission system has an interchannel spacing of 50 GHz, and the average optical power of each channel is −10 dBm. The C-band has been filled with all channels. Estimate the number of channels that are contained in the C-band from 1530 to 1565 nm. The data bit rate is 40 Gbps, and the modulation is Return-to-Zero ASK.

b. An optical filter is now used to extract the channel of wavelength 1552.93 nm, which has the following characteristics: center wavelength of 1552.93, optical bandwidth of 40 GHz (3 dB), and a roll-off of −12 dB per decade.

    i. Sketch the spectral distribution of all data channels.

    ii. Sketch the filter frequency response.

    iii. Estimate the residual power of the other wavelength channel on to this channel, and hence estimate the eye penalty due to this crosstalk. What kind of crosstalk is this, interchannel or intrachannel? You may assume the pulse shape of the data bits are perfect rectangular or Gaussian.

    iv. What can you propose to reduce the power penalty to the eye opening?

Problem 4.20

The nonlinear coefficient due to the intensity of lightwave propagating in a silica fiber is given by $n_2 = 2.5 \times 10^{-20}$ m²/W. A Gaussian guided mode lightwave is propagating through a length L of standard single-mode optical fiber (SSMF). Its average optical power is 10 dBm.

a. What is the effective area of the SSMF, that is, the ITU Standard G.652?

b. What is the intensity imposed on the cross section of the SSMF?

c. The change in the refractive index of the silica fiber can be given by $\Delta n_{NL} = n_2 I$. Estimate this change in the refractive index of the fiber due to the propagation of this guided mode.

d. Obtain the change in the phase of the lightwave carrier at 1550 nm after propagating through a length L (km). Hence, estimate the fiber length so that the nonlinear phase change is less than or equal to 0.1 radians.

e. *Referring Chapter 2, Section 2.10, write down the Schrödinger wave equation in which the nonlinear effects such as self-phase modulation are included. Make sure that you state clearly the meaning of all variables and the notation of this equation. By using a diagram, explain the split-step Fourier method that can be used to represent and model the propagation of the complex envelope of the lightwave through an optical-guided fiber/ waveguide.

Problem 4.21

a. Give a brief account of four-wave mixing. Indicate clearly the mixing of the waves and the generated wavelength channel.

b. Give the condition of fiber dispersion so that there is a significant impact of FWM on transmission systems. Sketch the generated waves that interfere with other DWDM channels.

c. Sketch the maximum transmitted power per channel versus fiber distance, and specify the average power per channel so that the transmitted channels would not be affected by FWM.

---

* This part is for high distinction level students.

APPENDIX 4A: SELLMEIER'S COEFFICIENTS FOR DIFFERENT CORE MATERIALS (TABLE 4A.1)

TABLE 4A.1
Sellmeier's Coefficients for Silica-Based Material and Doping Concentration of GeO$_2$ for the Design

Type	Doping Conc.	SiO$_2$	B1	B2	B3	λ1	λ2	λ3
A(1)	0%	100%	0.6961663	0.4079426	0.8974794	0.0684043	0.1162414	9.896161
B(2)	3.1%	96.9%	0.7028554	0.4146307	0.8974540	0.0727723	0.1143085	9.896161
C(3)	5.8%	94.2%	0.7088876	0.4206803	0.8956551	0.0609053	0.1254514	9.896162
D(4)	7.9%	92.1%	0.7136824	0.4254807	0.8964226	0.0617167	0.1270814	9.896161
E(5)	0%	pure	0.696750	0.408218	0.890815	0.069066	0.115662	9.900559
F(6)	13.5%	86.5%	0.711040	0.408218	0.704048	0.064270	0.129408	9.425478
G(7)	9.1%	90.9%	0.695790	0.452497	0.712513	0.061568	0.119921	8.656641
H(8)	13.3%	86.7%	0.690618	0.401996	0.898817	0.061900	0.123662	9.098960
I(9)	1%	99%	0.691116	0.399166	0.890423	0.068227	0.116460	9.993707
J(10)	48.7%	51.3%	0.796468	0.497614	0.358924	0.094359	0.093386	5.999652

APPENDIX 4B: TOTAL EQUIVALENT ELECTRONIC NOISE

Our principal goal in this section is to obtain an analytical expression for the noise spectral density equivalent to a source looking into the electronic amplifier including the quantum shot noise of the photodetector. A general method for deriving the equivalent noise current at the input is by representing the electronic device with a Y-equivalent linear network as shown in Figure 4.6. The two current noise sources $d\left(i_n'\right)^2$ and $d\left(i_n''\right)^2$ represent the sum of all noise currents at the input and at the output of the Y-network. This can be transformed into a Y-circuit with the noise current at the input as follows.

The output voltages V_o for Figure 4B.1a and 4B.1b can be written as

$$V_o = \frac{i_N'(Y_f - Y_m) + i_N''(Y_i + Y_f)}{Y_f(Y_m + Y_i + Y_o + Y_L) + Y_i(Y_o + Y_L)} \tag{4.144}$$

and for Figure 4B.1b as

$$V_o = \frac{(i_N')_{eq}(Y_f - Y_m)}{Y_f(Y_m + Y_i + Y_o + Y_L) + Y_i(Y_o + Y_L)} \tag{4.145}$$

Comparing these two equations, we can deduce that the equivalent noise current at the input of the detector is

$$i_{Neq} = i_N' + i_N'' \frac{Y_i + Y_f}{Y_f - Y_m} \tag{4.146}$$

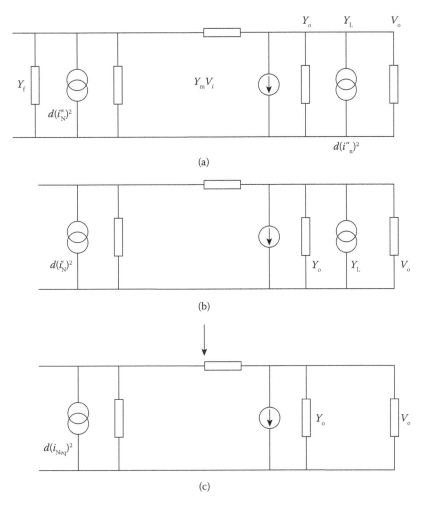

FIGURE 4B.1 Small equivalent circuits including noise sources. (a) Y-parameter model representing the ideal current model and all current noise sources at the input and output ports, (b) with noise sources at the input and output ports, and (c) with a total equivalent noise source at the input.

Reverting to mean square generators for a noise source, we have

$$d(i_{\text{Neq}})^2 = d(i_N')^2 + d(i_N'')^2 \left| \frac{Y_i + Y_f}{Y_f - Y_m} \right|^2 \tag{4.147}$$

It is therefore expected that if the Y-matrix of the front-end low-noise amplifier is known, the equivalent noise at the input of the amplifier can be obtained using Equation 4.144.

REFERENCES

1. G. Jacobsen, *Noise in Digital Optical Transmission Systems*, Artech House, Norwood, MA, USA, 1994.
2. *ITU-Recommendation G.692: Optical Interface for Multichannel Systems with Optical Amplifiers*, Technical report, retrieved http://www.itu.int/rec/T-REC-G.692-199810-I/en, from October 30, 2014.

3. H. Simada, *Coherent Lightwave Communications Technology*, Chapman & Hall, London, 1995.
4. T. L. Huynh, T. Sivahumaran, L. Le Binh, and K. K. Pang, Sensitivity improvement with offset filtering in optical MSK narrowband frequency discrimination receiver, in *ACOFT–COIN 2007*, Melbourne, Australia.
5. G. P. Agrawal, *Fiber Optic Communications Systems*, 2nd ed., New York: John Wiley & Sons, 2002.
6. L. N. Binh and S. V. Chung, Generalized approach to single-mode dispersion-modified optical fiber design, *Opt. Eng.*, Vol. 35, No. 8, pp. 2250–2261, 1996.
7. I. Kaminov and T. Koch, *Optical Fiber Communications*, Vol. IIIa, Academic Press, San Diego, USA.
8. L. N. Binh, *Digital Processing: Optical Transmission and Coherent Reception Techniques*, Boca Raton: CRC Press, 2013.
9. G. P. Agrawal, *Nonlinear Fiber Optics,* 4th Ed., Boston: Academic Press, 2012.
10. G. P. Agrawal, *Nonlinear Fiber Optics*, 3rd Ed., John Wiley & Sons, New York, 2001.
11. L. N. Binh, T. L. H. Huynh, K.-Y. Chin, and D. Sharma, Design of dispersion flattened and compensating fibers for dispersion-managed optical communications systems, *Int. J. Wireless Opt. Commun.*, Vol. 2, No. 1, pp. 63–81, 2004.
12. A. F. Elrefaire, R. E. Wagner, D. A. Atlas, and D. G. Laut, Chromatic dispersion limitations in coherent lightwave transmission systems, *IEEE J. Lightwave Technol.*, Vol. 6. No. 5, pp. 704–709, 1988.
13. J. Leibrich, C. Wree, and W. Rosenkranz, CF-RZ-DPSK for suppression of XPM on dispersion managed long haul optical WDM transmission on standard optical fiber, *IEEE Photon. Technol. Lett.*, Vol. 14, pp. 155–157, 2002.
14. L. N. Binh, Linear and nonlinear transfer functions of single mode fiber for optical transmission systems, *J. Opt. Soc. Am. A, Opt Image Sci Vis.*, Vol. 7, No.7, pp. 1564–1575, 2009.
15. M. Bierlaire, D. Bolduc, and D. McFadden, *Characteristics of Generalized Extreme Value Distributions*, Technical report, http://elsa.berkeley.edu/wp/mcfadden0403.pdf
16. MATLAB Simulink Model, Matworks, The MathWorks, Inc.,3 Apple Hill Drive, Natick, MA, USA, 2008.
17. S. Markose and A. Alentorn, *The Generalized Extreme Value (GEV) Distribution, Implied Tail Index and Option Pricing*, http://www.essex.ac.uk/economics/discussion-papers/papers-text/dp594.pdf
18. S. Kotz and S. Nadarajah, *Extreme Value Distributions: Theory and Applications*, London: ICP Imperial College Press, 2000.
19. SHF Communication Technologies AG, Wihelm-von-Siemens-Str. 23, Berlin, Germany.
20. M. E. J. Newman and G. T. Barkema, *Monte Carlo Methods in Statistical Physics*, Oxford, UK: Oxford University Press, 1999.
21. G. Jacobsen, *Noise in Digital Optical Communications Systems*, Boston, MA: Artech House, 2004.
22. A. C. Davidson, Models for exceedances over high thresholds, *J. Roy. Stat. Soc.*, Vol. B52, pp. 393–442, 1990.
23. R. L. Smith and I. Weissman, Estimating the extremal index, *J. Roy. Stat. Soc.*, Vol. B56, pp. 515–528, 1994.
24. R. L. Smith, Extreme value theory, in *Handbook of Applicable Mathematics*, New York: Wiley, pp. 437–472, 1989.
25. MATLAB Helpdesk, *Statistical Toolbox, Generalized Pareto Distribution,* http://www.mathworks.com/access/helpdesk/help/toolbox/stats
26. S. Borman, *The Expectation Maximization Algorithm: A Short Tutorial*, Retrieved http://www.cs.utah.edu/~piyush/teaching/EM_algorithm.pdf, 2012.
27. U. Fayyad, P. S. Bradley, and C. Reina, Scalable System for Expectation Maximization Clustering of Large Databases, United States Patent 6263337.
28. J. Sakuma and S. Kobayashi, Non-parametric expectation-maximization for Gaussian mixtures, in *Proceedings of the 9th International Conference on Neural Information Processing (ICONIP'OZ)*, Vol. I, pp. 517–522, 2002.
29. M. W. Mak, S. Y. Kung, and S. H. Lin, *Expectation-Maximization Theory*, Prentice Hall, Professional Technical Reference, Prentice-Hall information and system sciences series, New York, 2005.
30. X. Emery and J. M. Ortiz, Histogram and variogram inference in the multi-Gaussian model, *J. Stoch. Environ. Res. Risk Assess. (SERRA)*, Vol. 19, pp. 48–58, 2005.
31. Y. Wang and F. J. Doyle, III, Reachability of particle size distribution in semibatch emulsion, *J. Polymerization Particle Technol. Fluidization, AIChE J.*, Vol. 50, No 12, pp. 3049–3059, 2004.

32. S. Savory and A. Hadjifotiou, Laser linewidth requirements for optical DQPSK systems, *IEEE Photon. Technol. Lett.*, Vol. 16, No. 3, pp. 930–933, 2004.
33. B. B. Brabson and J. P. Palutikof, Test of the generalized Pareto distribution for predicting extreme wind speed, *J. Appl. Meteorol.*, Vol. 39, pp. 1627–1640, 2000.
34. B. E. Little, S. T. Chu, H. A. Haus, J. Foresi, and J.-P. Laine, Microring Resonator Channel Dropping Filters, *IEEE J. Lightwave Tech.*, Vol. 15, No. 6, pp. 998–1005, 1997.

5 Optical Coherent Detection and Processing Systems

Detection of optical signals can be carried out at the optical receiver by direct conversion of optical signal power to electronic current by the photon absorption and generation of electrons in the photodiode. This current is then amplified and converted in to voltage output by a transfer impedance amplifier. This chapter gives a fundamental understanding of coherent detection of optical signals that requires the mixing of the optical fields of the optical signals and that of the local oscillator (LO), a high-power laser, so that their beating product would result in the modulated signals preserving both their phase and amplitude characteristics in the electronic domain. Optical preamplification in coherent detection can also be integrated at the front end of the optical receiver.

5.1 INTRODUCTION

With the exponential increase in data traffic, especially due to the demand for ultra-broad bandwidth driven by multimedia applications, cost-effective ultra-high-speed optical networks have become an urgent need. It is expected that Ethernet technology will not only dominate in access networks but will also become the key transport technology of next-generation metro/ core networks. 100 Gigabit Ethernet (100 GbE) is currently considered to be the next logical evolution step after 10 Gigabit Ethernet (10 GbE). Based on the anticipated 100 GbE requirements, 100 Gbit/s data rate of serial data transmission per wavelength is required. To achieve this data rate while complying with current system design specifications such as channel spacing, chromatic dispersion (CD), and polarization-mode dispersion (PMD) tolerance, coherent optical communication systems with multilevel modulation formats will be desired, because they can provide high spectral efficiency, high receiver sensitivity, and potentially high tolerance to fiber dispersion effects [1–6]. Compared to conventional direct detection in intensity-modulation/direct detection (IMDD) systems, which only detects the intensity of the light of the signal, coherent detection can retrieve the phase information of the light, and therefore can tremendously improve the receiver sensitivity.

Coherent optical receivers are important components in long-haul optical fiber communication systems and networks in order to improve the receiver sensitivity and thus transmission distances. Coherent techniques were considered for optical transmission systems in the 1980s when the extension of the repeater distance between spans was pushed to 60 km instead of 40 km for single-mode optical fiber at a bit rate of 140 Gbps. However, in the late 1980s, the invention of optical fiber amplifiers has overcome these attempts. Recently, interest in coherent optical communications has attracted significant research activities for ultra-bit-rate DWDM optical systems and networks. The motivation has been possible due to (1) the uses of optical amplifiers in cascade fiber spans, which have added significant noise and thus limit the transmission distance; (2) the advances in digital signal processors whose sampling rate can reach a few tens of giga samples per second, allowing the processing of beating signals to recover the phase or phase estimation; (3) the availability of advanced signal processing algorithms such as Viterbi and Turbo algorithms; (4) the differential coding and modulation; (5) the detection of such signals may not require an optical phase-locked loop, and (6) possible self-coherent detection technique in association with digital signal processing to recover the transmitted signals.

As is well known, a typical arrangement of an optical receiver is that the optical signals are detected by a photodiode (a pin diode or APD or a photon-counting device); electrons generated in the photodetector are then electronically amplified through a front-end electronic amplifier. The electronic signals are then decoded for recovery of the original format. However, when the fields of incoming optical signals are mixed with those of an LO whose frequency can be identical or different to that of the carrier, the phase and frequency property of the resultant signals reflect those of the original signals. Coherent optical communication systems have also been reviving dramatically owing to electronic processing and the availability of stable narrow-linewidth lasers.

This chapter deals with the analysis and design of coherent receivers with optical phased lock loop and the mixing of optical signals and that of the LO in the optical domain and subsequently by the optoelectronic receivers following this mixing. Thus, both optical mixing and photodetection devices act as the fundamental elements of a coherent optical receiver. Depending on the frequency difference between the lightwave carrier of the optical signals and that of the LO, coherent detection can be termed as heterodyne or homodyne detection. For heterodyne detection, there is a difference in the frequency, and thus the beating signal falls in the passband region in the electronic domain. Thus, all the electronic processing at the front end must be in this passband region. In homodyne detection, on the contrary, there is no frequency difference, and thus the detection is in the baseband of the electronic signal. Both cases would require a locking of the LO and carrier of the signals. An optical phase-locked loop is thus treated in this chapter.

This chapter is organized as follows: Section 5.2 gives an account of the components of coherent receivers; Section 5.3 outlines the principles of optical coherent detection under heterodyne, homodyne, or intradyne techniques; and Section 5.4 gives details of the optical phase-locked loop, which is a very important development for modern optical coherent detection.

5.2 COHERENT RECEIVER COMPONENTS

The design of an optical receiver depends on the modulation format of the signals. The modulation of the optical carrier can be in the form of amplitude, phase, and frequency. Furthermore, phase shaping also plays a critical role in the detection and the bit error rate of the receiver, and thence of transmission systems. In particular, pulse shaping is dependent on the modulation in analog or digital, Gaussian or exponential pulse shape, on–off keying or multiple levels, etc.

Figure 5.1 shows the schematic diagram of a digital coherent optical receiver, which is similar to the direct detection receiver but with an optical mixer at the front end. Figure 5.2 shows the small signal equivalent circuits of such a receiver's front end. However, the phase of the signals at the baseband or passband of the detected signals in the electrical domain would remain in the generated

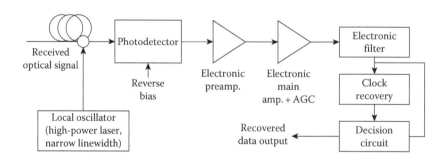

FIGURE 5.1 Schematic diagram of a digital optical coherent receiver with an additional local oscillator mixing with the received optical signals before being detected by an optical receiver.

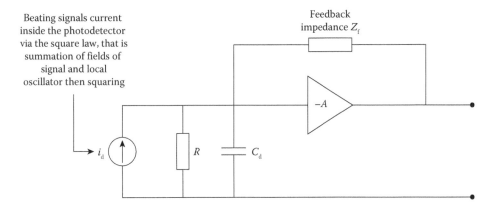

FIGURE 5.2 Schematic diagram of electronic preamplifier in an optical receiver of transimpedance electronic amplifier at the front end. The current source represents the electronic current generated in the photodetector due to the beating of the local oscillator and the optical signals. C_d = photodiode capacitance.

electronic current and voltages at the output of the electronic preamplifier. An optical front end is an optical mixer combining the fields of the optical waves of the local laser and the optical signals so that the envelope of the optical signals can be beating with each other to a product with the sum of the frequencies and the difference of the frequencies of the lightwaves. Only the lower-frequency term, which falls within the absorption range of the photodetector, is converted into the electronic current preserving both the phase and amplitude of the modulated signals.

Thus, an optical receiver front end, is connected following the optical processing front end consisting of a photodetector for converting lightwave energy into electronic currents; an electronic preamplifier for further amplification of the generated electronic current; followed by an electronic equalizer for bandwidth extension, usually in voltage form; a main amplifier for further voltage amplification; a clock recovery circuitry for regenerating the timing sequence; and a voltage-level decision circuit for sampling the waveform for the final recovery of the transmitted and received digital sequence. Therefore, the optoelectronic preamplifier is followed by a main amplifier with an automatic control to regulate the electronic signal voltage to be filtered and then sampled by a decision circuit with synchronization by a clock recovery circuitry.

An in-line fiber optical amplifier can be incorporated in front of the photodetector to form an optical receiver with an optical amplifier front end to improve its receiving sensitivity.

The structure of the receiver thus consists of four parts: the optical mixing front, the front-end section, the linear channel of the main amplifier and automatic gain control (AGC) if necessary, and the Data Recovery section. The optical mixing front end sums the optical fields of the LO and that of the optical signals. Polarization orientation between these lightwaves is very critical in order to maximize the beating of the additive field in the photodiode. Depending on whether the frequency difference is finite or null between these fields of the resulting electronic signals derived from the detector, the electronic signals can be in the baseband or the pass band, and the detection technique is termed as a heterodyne or homodyne technique, respectively.

5.3   COHERENT DETECTION

Optical coherent detection can be distinguished by the "demodulation" scheme in communications techniques in association with the following definitions: (1) Coherent detection is the mixing between two lightwaves or optical carriers, one of which is the information-bearing lightwave and the other a local oscillator with an average energy much larger than that of the signals; (2) demodulation refers to the recovery of baseband signals from the electrical signals.

A typical schematic diagram of a coherent optical communication employing a guided wave medium and components is shown in Figure 5.1, in which a narrow band laser incorporating an optical isolator cascaded with an external modulator is usually the optical transmitter. Information is fed via a microwave power amplifier to an integrated optic modulator, usually the LiNbO$_3$ or EA types. Coherent detection is a principal feature of coherent optical communications, which can be further divided into heterodyne and homodyne techniques depending whether or not there is a difference between the frequencies of the LO and that of the carrier of the signals. A LO is a laser source whose frequency can be tuned and is approximately equivalent to a monochromatic source; a polarization controller would also be used to match its polarization with that of the information-bearing carrier. The LO and the transmitted signal are mixed via a polarization maintaining coupler and then detected by a coherent optical receiver. Most previous coherent detection schemes are implemented in a mixture of photonic and electronic/microwave domains.

Coherent optical transmission has become the focus of research. One significant advantage is the preservation of all the information of the optical field during detection, leading to enhanced possibilities for optical multilevel modulation. This section investigates the generation of optical multilevel modulation signals. Several possible structures of optical M-ary phase shift keying (M-ary-PSK) and M-ary quadrature amplitude modulation (M-ary QAM) transmitters are shown and theoretically analyzed. Differences in the optical transmitter configuration and the electrical driving lead to different properties of the optical multilevel modulation signals. This is shown by deriving general expressions applicable to every M-ary-PSK and M-ary-QAM modulation format, and with special emphasis on Square-16-QAM modulation.

Coherent receivers are divided into synchronous and asynchronous receivers. Synchronous detection requires an optical phase-locked loop (OPLL) that recovers the phase and frequency of the received signals to lock the LO to that of the signal so as to measure the absolute phase and frequency of the signals relative to that of the LO. Thus, synchronous receivers allow direct mixing of the bandpass signals and the baseband, and so this technique is termed as homodyne reception. For asynchronous receivers, the frequency of the LO is approximately the same as that of the receiving signals, and no OPLL is required. In general, the optical signals are first mixed with an intermediate frequency (IF) oscillator that is about two to three times the 3 dB passband. The electronic signals can then be recovered using electrical PLL at a lower carrier frequency in the electrical domain. The mixing of the signals and an LO of an IF frequency is referred to as heterodyne detection.

If no LO is used for demodulating of the digital optical signals, then differential or self-homodyne reception may be utilized, which is classically termed as the autocorrelation reception process or self-heterodyne detection.

Coherent communications were an important technique in the 1980s and the early 1990s, but then optical amplifiers were developed in the late 1990s that offered up to 20-dB gain without difficulty. Nowadays, however, coherent systems have once again attracted interest, owing to the availability of digital signal processing and low-priced components, the partly relaxed receiver requirements at high data rates, and several other advantages that coherent detection provides. The preservation of the temporal phase of coherent detection enables new methods for adaptive electronic compensation of CD. With regard to WDM systems, coherent receivers offer tunability and allow channel separation via steep electrical filtering. Furthermore, only the use of coherent detection permits convergence to the ultimate limits of spectral efficiency. To reach higher spectral efficiencies, the use of multilevel modulation is required. For this, too, coherent systems are also beneficial, because all the information of the optical field is available in the electrical domain. This way complex optical demodulation with interferometric detection—which has to be used in direct-detection systems—can be avoided, and the complexity is transferred from the optical to the electrical domain. Several different modulation formats based on the modulation of all four quadratures of the optical field were proposed in the early 1990s, describing the possible transmitter and receiver structures and calculating the theoretical BER performance. However, a more detailed and practical investigation of multilevel modulation coherent optical systems for today's networks and data rates has not been performed so far.

Currently, coherent reception has attracted significant interest for the following reasons:

1. The received signals of the coherent optical receivers are in the electrical domain, whose amplitude and phase remain proportional to those of the optical domain. This, in contrast to the direct detection receivers, allows exact electrical equalization or exact phase estimation of the optical signals.
2. Using heterodyne receivers, DWDM channels can be separated in the electrical domain by using electrical filters with sharp roll-off of the passband to the cutoff band. Presently, the availability of ultra-high sampling rate digital signal processors (DSPs) allows users to conduct filtering in the DSP in which the filtering can be changed with ease.

However, there are disadvantages that coherent receivers would suffer from: (1) Coherent receivers are polarization sensitive, which requires polarization tracking at the front end of the receiver; (2) homodyne receivers require OPLL and electrical PLL for heterodyne that would need control and feedback circuitry, optical or electrical, which may be complicated; and (3) for differential detection, the compensation may be complicated owing to the differentiation receiving nature.

In Chapters 6 to 9 when some advanced modulation formats are presented for optically amplified transmission systems, photonic components are extensively exploited to take advantage of the advanced technology of integrated optics, planar lightwave circuits. The modulation formats of signals depend on whether the amplitude, the phase, or the frequency of the carrier is manipulated as mentioned in Chapter 2. In this chapter, the detection is coherently converted to the intermediate frequency range in the electrical domain and the signal envelope. The down-converted carrier signals are detected and then recovered. Both binary-level and multilevel modulation schemes employing amplitude, phase, and frequency shift keying modulation are described in this chapter.

Thus, coherent detection can be distinguished by the difference between the central frequency of the optical channel and that of the LO. Three types can be classified: (1) heterodyne, when the difference is higher than the 3 dB bandwidth of the baseband signal; (2) homodyne, when the difference is nil; and (3) intradyne, when the frequency difference falls within the baseband of the signal.

It is noted that in order to maximize the beating signals at the output of the photodetector, the polarizations of the LO and the signals must be aligned. In practice, this can be best implemented by the polarization diversity technique (Figure 5.3) [2,3].

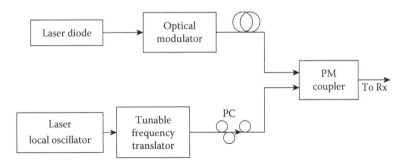

FIGURE 5.3 Typical arrangement of coherent optical communications systems. LD/LC is a very-narrow-linewidth laser diode that acts as a local oscillator without any phase locking to the signal optical carrier. PM coupler is a polarization maintaining fiber-coupled device. PC = polarization controller.

5.3.1 Optical Heterodyne Detection

The basic configuration of optical heterodyne detection is shown in Figure 5.4. The local oscillator, whose frequency can be higher or lower than that of the carrier, is mixed with the information-bearing carrier, thus allowing down- or up-conversion of the information signals to the intermediate frequency range. The down-converted electrical carrier and signal envelope is received by the photodetector. This combined lightwave is converted by the PD into electronic current signals, which are then filtered by an electrical bandpass filter (BPF) and then demodulated by a demodulator. An LPF is also used to remove higher-order harmonics of the nonlinear detection photodetection process, the square law detection. With an envelope detector, the process is asynchronous, and hence the term *asynchronous detection*. If the down-converted carrier is recovered and then mixed with IF signals, this is synchronous detection. Note that the detection is conducted at the intermediate frequency range in the electrical domain, and hence the stability of the frequency spacing between the signal carrier and the local oscillator has to be controlled. This means the mixing of these carriers would result in an IF carrier in the electrical domain prior to the mixing process or envelope detection to recover the signals.

The coherent detection thus relies on the electric field component of the signal and the local oscillator. The polarization alignment of these fields is critical for optimum detection. The electric field of the optical signals and the local oscillator can be expressed as

$$E_s(t) = \sqrt{2P_s(t)} \cos\{\omega_s t + \phi_s + \phi(t)\} \tag{5.1}$$

$$E_{LO} = \sqrt{2P_L} \cos\{\omega_{LO}t + \phi_{LO}\} \tag{5.2}$$

where $P_s(t)$ and P_{LO} are the instantaneous signal power and average power of the signals and local oscillator, respectively; $\omega_s(t)$ and ω_{LO} are the signal and local oscillator angular frequencies; ϕ_s and ϕ_{LO} are the phases, including any phase noise of the signal and the local oscillator; and $\psi(t)$ is the modulation phase. The modulation can be amplitude modulation, with the switching on and off (amplitude shift keying—ASK) of the optical power or phase or frequency with the discrete or

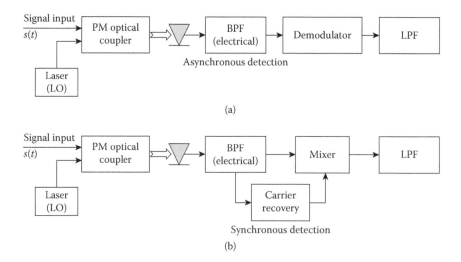

FIGURE 5.4 Schematic diagram of optical heterodyne detection. (a) Asynchronous and (b) synchronous receiver structures. LPF is a low-pass filter. BPF = bandpass filter, PD = photodiode.

continuous variation of the time-dependent phase term. For the discrete phase, it can be phase shift keying (PSK), differential PSK (DPSK), or differential quadrature PSK (DQPSK), and when the variation of the phase is continuous, we have frequency shift keying if the rate of variation is different for the bit 1 and 0 as given in Chapter 4.

Under an ideal alignment of the two fields, the photodetection current can be expressed by

$$i(t) = \frac{\eta q}{h\upsilon}\left[P_s + P_{LO} + 2\sqrt{P_s P_{LO}} \cos\left\{ (\omega_s - \omega_{LO})t + \phi_s - \phi_{LO} + \phi(t) \right\} \right] \tag{5.3}$$

where the higher-frequency term (the sum) is eliminated by the photodetector frequency response, η is the quantum efficiency, q is the electronic charge, h is Plank's constant, and v is the optical frequency.

Thus, the power of the local oscillator dominates the shot-noise process and at the same time boosts the signal level, hence enhancing the signal-to-noise ratio (SNR). The oscillating term is the beating between the local oscillator and the signal and is signal proportional with the amplitude is the square root of the product of the amplitudes of the local oscillator and the signal.

The electronic signal power S and shot noise N_s can be expressed as

$$S = 2\Re^2 P_s P_{LO}$$
$$N_s = 2q\Re(P_s + P_{LO})B \tag{5.4}$$
$$\Re = \frac{\eta q}{hv} = \text{responsivity}$$

where B is the 3 dB bandwidth of the electronic receiver. Thus, the optical signal-to-noise ratio (OSNR) can be written as

$$\text{OSNR} = \frac{2\Re^2 P_s P_{LO}}{2q\Re(P_s + P_{LO})B + N_{eq}} \tag{5.5}$$

where N_{eq} is the total electronic noise equivalent power at the input to the electronic preamplifier of the receiver. From this equation, we can observe that if the power of the local oscillator is significantly increased so that the shot noise dominates the equivalent noise, at the same time increasing the SNR, the sensitivity of the coherent receiver can only be limited by the quantum noise inherent in the photodetection process. Under this quantum limit, the OSNR_{QL} is given by

$$\text{OSNR}_{QL} = \frac{\Re P_s}{qB} \tag{5.6}$$

5.3.1.1 ASK Coherent System

Under the ASK modulation scheme, the demodulator of Figure 5.4 is an envelope detector (in lieu of the demodulator) followed by a decision circuit. That is, the eye diagram is obtained, and a sampling instant is established with a clock recovery circuit. While synchronous detection would require the phase locking between the carrier frequencies, the beating product can be obtained. The frequency of the local oscillator can be tuned to the central frequency of the signal spectrum to give the maximum output current. The amplitude-demodulated envelope can be expressed as

$$r(t) = 2\Re\sqrt{P_s P_{LO}} \cos(\omega_{IF})t + n_x \cos(\omega_{IF})t + n_y \sin(\omega_{IF})t$$
$$\omega_{IF} = \omega_s - \omega_{LO} \tag{5.7}$$

The IF frequency ω_{IF} is the difference between the frequencies of the LO and the signal carrier, and n_x and n_y are the expected values of the orthogonal noise power components, which are random variables.

$$r(t) = \sqrt{[2\Re P_s P_{LO} + n_x]^2 + n_y^2} \cos(\omega_{IF}t + \Phi)t \qquad \text{with} ...\Phi = \tan^{-1}\frac{n_y}{2\Re P_s P_{LO} + n_x} \qquad (5.8)$$

5.3.1.1.1 Envelope Detection

The noise power terms can be assumed to follow a Gaussian probability distribution and are independent of each other with a zero mean and a variance σ, and the probability density function (PDF) can thus be given as

$$p(n_x, n_y) = \frac{1}{2\pi\sigma^2} e^{\frac{-(n_x^2 + n_y^2)}{2\sigma^2}} \qquad (5.9)$$

With respect to the phase and amplitude, this equation can be written as [3]

$$p(\rho, \phi) = \frac{\rho}{2\pi\sigma^2} e^{\frac{-(\rho^2 + A^2 - 2A\rho\cos\phi)}{2\sigma^2}} \qquad (5.10)$$

where

$$\rho = \sqrt{[2\Re\sqrt{P_s(t)P_{LO}} + n_s(t)]^2 + n_y^2(t)}$$
$$A = 2\Re\sqrt{P_s(t)P_{LO}} \qquad (5.11)$$

The pdf of the amplitude can be obtained by integrating the phase amplitude pdf over the range of 0 to 2π and given as

$$p(\rho) = \frac{\rho}{\sigma^2} e^{\frac{-(\rho^2 + A^2)}{2\sigma^2}} I_0\left\{\frac{A\rho}{\sigma^2}\right\} \qquad (5.12)$$

where I_0 is the modified Bessel function. If a decision level is set to determine the 1 and 0 levels, then the probability of error and the bit error rate BER can be obtained assuming an equal probability of error between the 1s and 0s

$$\text{BER} = \frac{1}{2}P_e^1 + \frac{1}{2}P_e^0 = \frac{1}{2}\left[1 - Q\left(\sqrt{2\delta}, d\right) + e^{-\frac{d^2}{2}}\right] \qquad (5.13)$$

where Q is the Magnum function given in Chapter 4, and δ is given by

$$\delta = \frac{A^2}{2\sigma^2} = \frac{2\Re^2 P_s P_{LO}}{2q\Re(P_s + P_{LO})B + i_{N_{eq}}^2} \qquad (5.14)$$

When the power of the local oscillator is much larger than that of the signal and the equivalent noise current power, then this SNR becomes

$$\delta = \frac{\Re P_s}{qB} \qquad (5.15)$$

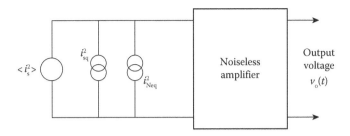

FIGURE 5.5 Equivalent current model at the input of the optical receiver, average signal current, and equivalent noise current of the electronic preamplifier as seen from its input port.

The physical representation of the detected current, the noise current due to the quantum shot-noise, and then the input-referred total noise equivalent of the electronic preamplifier can be represented by the current sources as shown Figure 5.5, in which the signal current can be general and derived from the output of the detection scheme, that from a photodetector or a back-to-back (B2B) pair of photodetectors of a balanced receiver for detecting the phase difference of DPSK or DQPSK or CPFSK signals and converting to amplitudes.

The BER is optimum when setting its differentiation with respect to the decision level δ, and an approximate value of the decision level can be obtained as

$$d_{\text{opt}} \cong \sqrt{2 + \frac{\delta}{2}} \Rightarrow \text{BER}_{\text{ASK}-e} \cong \frac{1}{2} e^{-\frac{\delta}{4}} \tag{5.16}$$

5.3.1.1.2 Synchronous Detection

ASK can be detected using synchronous detection,[*] and the BER is given by

$$\text{BER}_{\text{ASK}-S} \cong \frac{1}{2} \text{erfc} \frac{\sqrt{\delta}}{2} \tag{5.17}$$

5.3.1.2 PSK Coherent System

Under the PSK modulation format, the detection is similar to that of Figure 5.4 for heterodyne detection (see Figure 5.6), but after the BPF, an electrical mixer is used to track the phase of the detected signal.

The received signal is given by

$$r(t) = 2\Re\sqrt{P_s P_{\text{LO}}} \cos[(\omega_{\text{IF}})t + \phi(t)] + n_x \cos(\omega_{\text{IF}})t + n_y \sin(\omega_{\text{IF}})t \tag{5.18}$$

The information is contained in the time-dependent phase term $\phi(t)$.

When the phase and frequency of the voltage control oscillator (VCO) are matched with those of the signal carrier, the received electrical signal can be simplified as

$$r(t) = 2\Re\sqrt{P_s P_{\text{LO}}} a_n(t) + n_x$$
$$a_n(t) = \pm 1 \tag{5.19}$$

Under the Gaussian statistical assumption, the probability of the received signal of a 1 is given by

$$p(r) = \frac{1}{\sqrt{2\pi\sigma^2}} e^{-\frac{(r-u)^2}{2\sigma^2}} \tag{5.20}$$

---

[*] Synchronous detection is implemented by mixing the signals and a strong local oscillator in association with the phase locking of the local oscillator to that of the carrier.

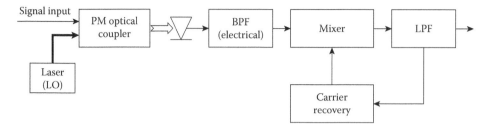

FIGURE 5.6   Schematic diagram of optical heterodyne detection for PSK format.

Furthermore, the probability of the 0 and 1 are assumed to be equal. We can obtain the BER as the total probability of the received 1 and 0 as

$$\text{BER}_{\text{PSK}} = \frac{1}{2}\text{erfc}(\delta) \qquad (5.21)$$

5.3.1.3   Differential Detection

As has been observed, in synchronous detection, a carrier recovery circuit is necessary and is usually implemented using a PLL, which complicates the overall receiver structure. It is possible to detect the signal by a self-homodyne process by beating the carrier of one-bit period to that of the next consecutive bit; this is called differential detection. The detection process can be modified as shown in Figure 5.7, in which the phase of the IF carrier of one bit is compared with that of the next bit, and a difference is recovered to represent the bit 1 or 0. This requires differential coding at the transmitter and an additional phase comparator for the recovery process.

In Chapter 6 on differential PSK, differential decoding is implemented in the photonic domain via a photonic phase comparator in the form of an MZ delay interferometer (MZDI) with a thermal section for tuning the delay time of the optical delay line. The BER can be expressed as

$$\text{BER}_{\text{DPSK}-\text{e}} \cong \frac{1}{2}e^{-\delta} \qquad (5.22)$$

$$r(t) = 2\Re\sqrt{P_{\text{s}}P_{\text{LO}}} \, \cos[\pi A_k s(t)] \qquad (5.23)$$

where $s(t)$ is the modulating waveform, and A_k represents the bit 1 or 0. This is equivalent to the baseband signal, and the ultimate limit is the BER of the baseband signal.

The noise is dominated by the quantum shot noise of the local oscillator, with its square noise current given by

$$i_{\text{N}-\text{sh}}^2 = 2q\Re(P_{\text{s}} + P_{\text{LO}}) \int_0^\infty |H(j\omega)|^2 d\omega \qquad (5.24)$$

where $H(j\omega)$ is the transfer function of the receiver system, normally a transimpedance of the electronic preamp and that of a matched filter. As the power of the local oscillator is much larger than the signal, integrating over the dB bandwidth of the transfer function, this current can be approximated by

$$i_{\text{N}-\text{sh}}^2 \simeq 2q\Re P_{\text{LO}}B \qquad (5.25)$$

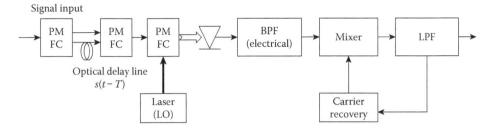

FIGURE 5.7   Schematic diagram of optical heterodyne and differential detection for PSK format.

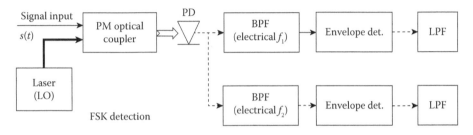

FIGURE 5.8   Schematic diagram of optical homodyne detection of FSK format.

Hence, the (power) SNR is given by

$$\text{SNR} \equiv \delta \approx \frac{2\Re P_s}{qB} \tag{5.26}$$

The BER is the same as that of a synchronous detection and is given by

$$\text{BER}_{\text{homodyne}} \cong \frac{1}{2}\,\text{erfc}\sqrt{\delta} \tag{5.27}$$

The sensitivity of the homodyne process is at least 3 dB better than that of the heterodyne, and the bandwidth of the detection is half of its counterpart owing to the double-sideband nature of the heterodyne detection.

5.3.1.4   FSK Coherent System

The nature of FSK is based on the two frequency components that determine the bits 1 and 0. There are a number of formats related to FSK depending on whether the change of the frequencies representing the bits is continuous or noncontinuous, the FSK or CPFSK modulation formats. For noncontinuous FSK, the detection is usually performed by a structure of dual frequency discrimination as shown in Figure 5.8, in which two narrowband filters are used to extract the signals. For CPFSK, both the frequency discriminator and balanced receiver for PSK detection can be used. The frequency discrimination is indeed preferred as compared to the balanced receiving structures because it would eliminate the phase contribution by the local oscillator or optical amplifiers, which may be used as an optical preamp.

When the frequency difference between the 1 and 0 equals a quarter of the bit rate, the FSK can be termed as the minimum shift keying (MSK) modulation scheme. At this frequency spacing, the phase is continuous between these states.

5.3.2 Optical Homodyne Detection

Optical homodyne detection matches the transmitted signal phases to that of the local oscillator phase signal. A schematic of the optical receiver is shown in Figure 5.9.

The field of the incoming optical signals is mixed with the local oscillator, whose frequency and phase are locked with that of the signal carrier waves via a PLL. The resultant electrical signal is then filtered. The pulse sequence is then recovered back to the original data by a decision subsystem.

5.3.2.1 Detection and Optical PLL

Optical homodyne detection requires the phase matching of the frequency of the signal carrier and that of the local oscillator. This type of detection would give a very sensitive detection, in principle, of 9 photons/bit. Implementation of such a system would normally require an optical PLL, described in [4] as shown in Figure 5.10. The local oscillator frequency is locked into the carrier frequency of the signals by shifting it to the modulated sideband component via the use of the optical modulator. A single-sideband optical modulator is preferred. However, a double sideband may also be used. This modulator is excited by the output signal of a voltage-controlled oscillator whose frequency is determined by the voltage level of the output of an electronic BPF condition to meet the required voltage level for driving the electrode of the modulator. The frequency of the local oscillator is normally tuned to the region such that the frequency difference with respect to the signal carrier falls within the passband of the electronic filter. When the frequency difference is zero, there

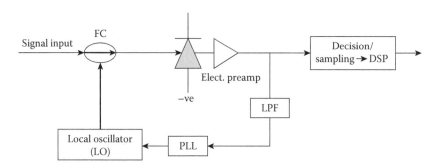

FIGURE 5.9 General structure of an optical homodyne detection system. FC = fiber coupler, LPF = low-pass filter, PLL = phase-locked loop.

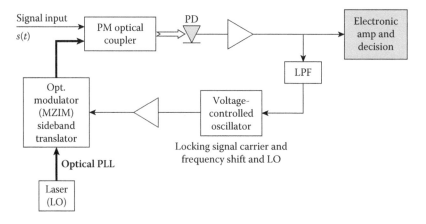

FIGURE 5.10 Schematic diagram of optical homodyne detection—electrical line (thick line) and optical line (thin line) using an optical phase lock loop.

is no voltage level at the output of the filter, and thus the optical PLL has reached the final stage of locking. The bandwidth of the optical modulator is important so that it can extend the locking range between the two optical carriers.

Any difference in frequency between the LO and the carrier is detected, and noise is filtered by the LPF. This voltage level is then fed to a VOC to generate a sinusoidal wave that is then used to modulate an intensity modulator modulating the lightwaves of the LO. The output spectrum of the modulator would exhibit two sidebands and the LO lightwave. One of these components would then be locked to the carrier. A closed loop would ensure stable locking. If the intensity modulator is biased at the minimum transmission point and the voltage level at the output of the VCO is adjusted to $2V_\pi$ with driven signals of $\pi/2$ phase shift with each other, then we would have carrier suppression and sideband suppression.

Under perfect phase matching, the received signal is given by

$$i_s(t) = 2\Re\sqrt{p_s P_{LO}}\,\cos\{\pi a_k s(t)\} \tag{5.28}$$

where a_k takes the value ± 1, and $s(t)$ is the modulating waveform. This is a baseband signal, and thus the error rate is the same as that of the baseband system.

The shot-noise power induced by the local oscillator and the signal power can be expressed as

$$i_{NS}^2 = 2q\Re\left(p_s + P_{LO}\right)\int_0^\infty |H(j\omega)|\,d\omega \tag{5.29}$$

where $|H(j\omega)|$ is the transfer function of the receiver, whose expression, if under a matched filtering, can be

$$|H(j\omega)|^2 = \left[\frac{\sin(\omega T/2)}{\omega T/2}\right]^2 \tag{5.30}$$

where T is the bit period. Then the noise power becomes

$$i_{NS}^2 = q\Re\left(p_s + P_{LO}\right)\frac{1}{T} \simeq \frac{q\Re P_{LO}}{T} \tag{5.31}$$

when $p_s \ll P_{LO}$.

Thus, the SNR is

$$\text{SNR} = \frac{2\Re p_s P_{LO}}{q\Re P_{LO}/T} = \frac{2p_s T}{q} \tag{5.32}$$

and the bit error rate is

$$P_E = \frac{1}{2}\text{erfc}\left(\sqrt{\text{SNR}}\right) \to \text{BER} = \text{erfc}\left(\sqrt{\text{SNR}}\right) \tag{5.33}$$

5.3.2.2 Quantum Limit Detection

For homodyne detection, a super quantum limit can be achieved. In this case, the local oscillator is used in a very special way that matches the incoming signal field in polarization, amplitude, and frequency and is assumed to be phase-locked to the signal. Assuming that the phase signal is perfectly modulated such that it acts in-phase or counter-phase with the LO, the homodyne detection would give a normalized signal current of

$$i_{sC} = \frac{1}{2T} \left[\mp \sqrt{2n_p} + \sqrt{2n_{LO}} \right]^2 \dots \text{for} \dots 0 \leq t \leq T \tag{5.34}$$

Assuming further that $n_p = n_{LO}$, the number of photons for the LO for generation of detected signals, then the current can be replaced with $4n_p$ for the detection of a 1 and nothing for a 0 symbol.

5.3.2.3   Linewidth Influences

5.3.2.3.1   Heterodyne Phase Detection

When the linewidth of the light source is significant, the intermediate frequency deviates owing to a phase fluctuation, and the PDF is related to this linewidth conditioned on the deviation $\delta\omega$ of the intermediate frequency. For a signal power of p_s, the total probability of error is given as

$$P_E = \int_{-\infty}^{\infty} P_C(p_s, \partial\omega) p_{IF}(\partial\omega) \partial\omega \tag{5.35}$$

The PDF of the intermediate frequency under a frequency deviation can be written as [5]

$$p_{IF}(\partial\omega) = \frac{1}{\sqrt{\Delta \upsilon B T}} e^{-\frac{\partial\omega^2}{4\pi\Delta\upsilon B}} \tag{5.36}$$

where Δv is the full IF linewidth at FWHM of the power spectral density, and T is the bit period.

5.3.2.3.2   Differential Phase Detection with Local Oscillator

5.3.2.3.2.1   DPSK Systems   DPSK detection requires an MZDI and a balanced receiver either in the optical domain or in the electrical domain. If in the electrical domain, then the beating signals in the PD between the incoming signals and the LO would give the beating electronic current, which is then split. One branch is delayed by one-bit period and then summed up. The heterodyne signal current can be expressed as [6]

$$i_s(t) = 2\Re\sqrt{P_{LO} p_s} \cos(\omega_{IF} t + \phi_s(t)) + n_x(t)\cos\omega_{IF} t - n_y(t)\sin\omega_{IF} t \tag{5.37}$$

The phase $\phi_s(t)$ is expressed by

$$\phi_s(t) = \phi_s(t) + \left\{ \phi_N(t) - \phi_N(t+T) \right\} - \left\{ \phi_{pS}(t) - \phi_{pS}(t+T) \right\} - \left\{ \phi_{pL}(t) - \phi_{pL}(t+T) \right\} \tag{5.38}$$

The first term is the phase of the data and takes the value 0 or π. The second term represents the phase noise due to shot noise of the generated current, and the third and fourth terms are the quantum shot noise due to the LO and the signals. The probability of error is given by

$$P_E = \int_{-\pi/2}^{\pi/2} \int_{-\infty}^{\infty} p_n(\phi_1 - \phi_2) p_q(\phi_1) \partial\phi_1 \, \partial\phi_1 \tag{5.39}$$

where $p_n(.)$ is the PDF of the phase noise due to the shot noise, and $p_q(.)$ is for the quantum phase noise generated from the transmitter and the LO [7].

The probability or error can be written as

$$p_N(\phi_1 - \phi_2) = \frac{1}{2\pi} + \frac{\rho e^{-\rho}}{\pi} \sum_{m=1}^{\infty} a_m \cos(m(\phi_1 - \phi_2))$$

$$a_m \sim \left\{ \left[\frac{2^{m-1} \Gamma\left[\frac{m+1}{2}\right] \Gamma\left[\frac{m}{2}+1\right]}{\Gamma[m+1]} \right] \left[I_{m-1/2} \frac{\rho}{2} + I_{(m+1)/2} \frac{\rho}{2} \right] \right\}^2 \qquad (5.40)$$

where $\Gamma(.)$ is the gamma function which is the modified Bessel function of the first kind. The pdf of the quantum phase noise can be given as [8]

$$p_q(\phi_1) = \frac{1}{\sqrt{2\pi D\tau}} e^{\frac{\phi_1^2}{2D\tau}} \qquad (5.41)$$

where D is the phase diffusion constant, and the standard deviation from the central frequency

$$\Delta\upsilon = \Delta\upsilon_R + \Delta\upsilon_L = \frac{D}{2\pi} \qquad (5.42)$$

is the sum of the transmitter and the LO FWHM linewidth. Substituting Equations 5.41 and 5.40 in Equation 5.39, we obtain

$$P_E = \frac{1}{2} + \frac{\rho e^{-\rho}}{2} \sum_{n=0}^{\infty} \frac{(-1)^n}{2n+1} e^{-(2n+1)^2 \pi\Delta\upsilon T} \left\{ I_{n-1/2} \frac{\rho}{2} + I_{(n+1)/2} \frac{\rho}{2} \right\}^2 \qquad (5.43)$$

This equation gives the probability of error as a function of the received power. The probability of error is plotted against the receiver sensitivity, and the product of the linewidth and the bit rate (or the relative bandwidth of the laser linewidth and the bit rate) is shown in Figure 5.11 for the DPSK modulation format at 140 Mbps bit rate and the variation of the laser linewidth from 0 to 2 MHz.

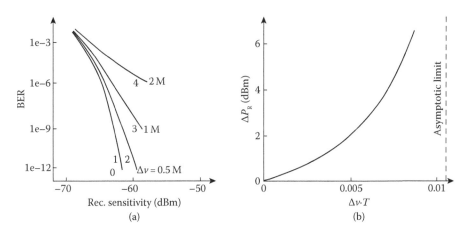

FIGURE 5.11 (a) Probability of error versus receiver sensitivity with linewidth as a parameter in megahertz. (b) Degradation of optical receiver sensitivity at BER = 1e–9 for DPSK systems as a function of the linewidth and bit period; bit rate = 140 Mbps. (Extracted from G. Nicholson, *Electron. Lett.*, Vol. 20/24, pp. 1005–1007, 1984.)

5.3.2.3.3 Differential Phase Detection under Self-Coherence

Recently, the laser linewidth requirement for DQPSK modulation and differential detection for DQPSK have also been studied. No LO is used, which means self-coherent detection. It has been shown that a linewidth of up to 3 MHz of the transmitter laser would not significantly influence the probability of error as shown in Figure 5.12 [8,9]. Figure 5.13 [8] shows the maximum linewidth of a laser source or in a 10 GSymbol/s system. The loose bound is to neglect linewidth if the impact is to double the BER, and the tighter bound is to neglect linewidth if the impact is a 0.1-dB SNR penalty.

5.3.2.3.4 Differential Phase Coherent Detection of Continuous Phase FSK Modulation Format

The probability of error of CPFSK can be derived by taking into consideration the delay line of the differential detector, the frequency deviation, and phase noise [10]. Similar to Figure 5.8, the differential detector configuration is shown in Figure 5.14a, and the conversion of the frequency to voltage relationship is shown in Figure 5.14b [10]. If heterodyne detection is employed, then a BPF is used to bring the signals back to the electrical domain.

The detected signal phase at the shot-noise limit at the output of the LPF can be expressed as

$$\phi(t) = \omega_c t + a_n \frac{\Delta\omega}{2}\tau + \phi(t) + \phi_n(t)$$

$$\text{with} \ldots \omega_c = 2\pi f_c = (2n+1)\frac{\pi}{2\tau}$$

(5.44)

where τ is the differential detection delay time, $\Delta\omega$ is the deviation of the angular frequency of the carrier for the 1 or 0 symbol, $\phi(t)$ is the phase noise due to the shot noise, and $n(t)$ is the phase noise

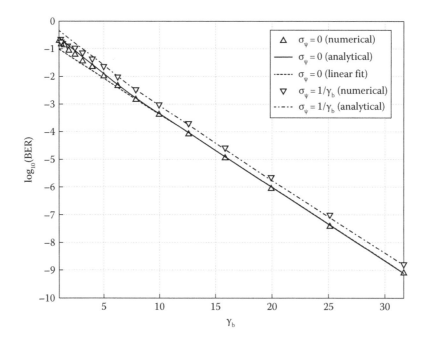

FIGURE 5.12 Analytical approximation (solid line) and numerical evaluation (triangles) of the BER for the cases of zero linewidth and that required to double the BER. The dashed line is the linear fit for zero linewidth. Bit rate 10 Gbps per channel. (Extracted from K. Iwashita and T. Masumoto, *IEEE J. Lightwave Technol.*, Vol. LT-5/4, pp. 452–462, 1987.)

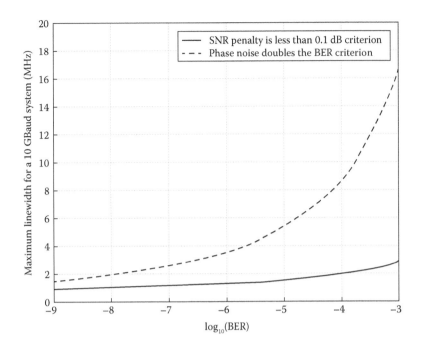

FIGURE 5.13 Criteria for neglecting linewidth in a 10 GSymbols/s system. The loose bound is to neglect linewidth if the impact is to double the BER, and the tighter bound is to neglect linewidth if the impact is a 0.1 dB SNR penalty. Bit rate 10 GSymbols/s. (Extracted from Y. Yamamoto and T. Kimura, *IEEE J. Quant. Electron.*, Vol. QE-17, pp. 919–934, 1981.)

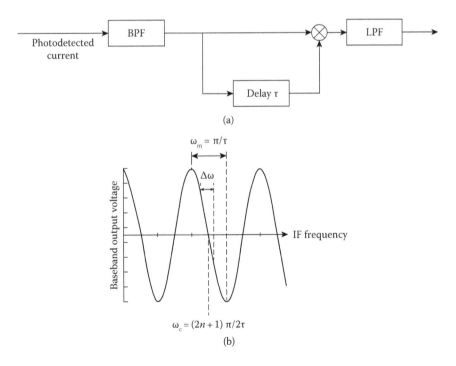

FIGURE 5.14 (a) Configuration of a CPFSK differential detection and (b) frequency to voltage conversion relationship of FSK differential detection. (Extract from K. Iwashita and T. Masumoto, *IEEE J. Lightwave Technol.*, Vol. LT-5/4, pp. 452–462, 1987 (Figure 1).)

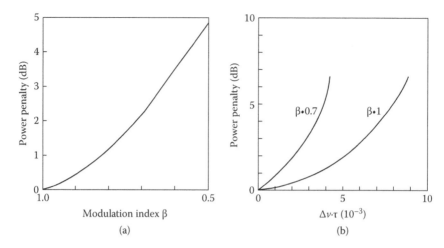

FIGURE 5.15 (a) Dependence of receiver power penalty at BER of 1e−9 on modulation index β (ratio of frequency deviation to maximum frequency spacing between f1 and f2) and (b) receiver power penalty at BER 1e−9 as a function of product of beat bandwidth and bit delay time—effects excluding LD phase noise. (Extracted from K. Iwashita and T. Masumoto, *IEEE J. Lightwave Technol.*, Vol. LT-5/4, 1987, pp. 452–462. (Figures 2 and 3); Y. Yamamoto and T. Kimura, *IEEE J. Quant. Electron.*, Vol. QE-17, pp. 919–934, 1981.)

due to the transmitter and the LO quantum shot noise and takes the values of ±1, the binary data symbol.

Thus, by integrating the detected phase from $-\dfrac{\Delta\omega}{2}\tau \longrightarrow \pi - \dfrac{\Delta\omega}{2}\tau$, we obtain the probability of error as

$$P_E = \int_{-\frac{\Delta\omega}{2}\tau}^{\pi - \frac{\Delta\omega}{2}\tau/2} \int_{-\infty}^{\infty} p_n(\phi_1 - \phi_2)p_q(\phi_1)\,\partial\phi_1\,\partial\phi_1 \tag{5.45}$$

Similar to the case of DPSK systems, substituting Equations 5.40 and 5.41 in Equation 5.45, we obtain

$$P_E = \frac{1}{2}\frac{\rho e^{-\rho}}{2}\sum_{n=0}^{\infty}\frac{(-1)^n}{2n=1}e^{-(2n+1)^2\pi\Delta\upsilon\tau}\left\{I_{n-1/2}\frac{\rho}{2}+I_{(n+1)/2}\frac{\rho}{2}\right\}^2 e^{-(2n+1)^2\pi\Delta\upsilon\tau}\cos\left\{(2n+1)\alpha\right\} \tag{5.46}$$

$$\alpha = \frac{\pi(1-\beta)}{2}\ldots \text{and}\ldots \beta = \frac{\Delta\omega}{\omega_m} = 2\pi\tau/T_0$$

where ω_m is the deviation of the angular frequency with m being the modulation index, and T_0 is the pulse period or bit period. The modulation index parameter b is defined as the ratio of the actual frequency deviation to the maximum frequency deviation. Figure 5.15 shows the dependence of degradation of the power penalty to achieve the same BER as a function of the linewidth factor $\Delta\upsilon\tau$ and the modulation index β [8].

5.3.3 Optical Intradyne Detection

Optical phase diversity receivers combine the advantages of homodyne receivers, minimum signal processing bandwidth, and heterodyne reception, with no optical phase locking required. The term *diversity* is well known in radio transmission links; it describes transmission over more than one path. In optical receivers, the optical path is considered as being due to different polarization

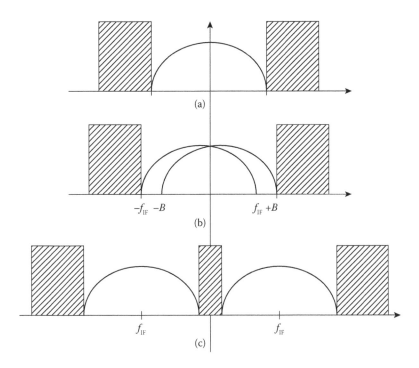

FIGURE 5.16   Spectrum of coherent detection: (a) homodyne, (b) intradyne, and (c) heterodyne.

and phase paths. In intradyne detection, the frequency difference, the intermediate frequency or the LOFO (local oscillator frequency offset) between the LO and the central carrier is nonzero and lies within the signal bandwidth of the baseband signal, as illustrated in Figure 5.16 [11].

Naturally, the control and locking of the carrier and the LO cannot be exact, sometimes owing to jittering of the source. Most of the time, the laser frequency is locked stably by oscillating the reflection mirror, and hence the central frequency is varied by a few hundreds of kilohertz. Thus, intradyne coherent detection is more realistic. Furthermore, the digital signal processor in modern coherent reception systems would be able to extract this difference without much difficulty in the digital domain [12]. Obviously, heterodyne detection would require a large frequency range of operation of electronic devices, whereas homodyne and intradyne reception require simpler electronics. Either the differential or the nondifferential format can be used in DSP-based coherent reception. For differential-based reception, differential decoding would gain an advantage when there are slips in the cycles of bits due to walk-off of the pulse sequences over very long transmission noncompensating fiber lines.

The diversity in phase and polarization can be achieved by using a $\pi/2$ hybrid coupler that splits the polarization of the LO and the received channels and mixing with a $\pi/2$ optical phase shift, after which the mixed signals are detected by a balanced photodetector. This diversity detection is described in the next few sections (see also Figure 5.21 later in the chapter).

5.4 SELF-COHERENT DETECTION AND ELECTRONIC DSP

The foregoing coherent techniques would offer a significant improvement, but facing difficulty owing to the availability of a stable LO and an OPLL for locking the frequency of the LO and that of the signal carrier.

DSPs have been widely used in wireless communications and play key roles in the implementation of DSP-based coherent optical communication systems. DSP techniques have been applied to coherent optical communication systems to overcome the difficulties of OPLL, and also to improve the performance of the transmission systems in the presence of fiber-degrading effects, including CD, PMD, and fiber nonlinearities.

Coherent optical receivers have the following advantages: (1) the shot-noise-limited receiver sensitivity can be achieved with a sufficient LO power; (2) closely spaced WDM channels can be separated with electrical filters having sharp roll-off characteristics; and (3) the phase detection capability can improve the receiver sensitivity compared with the IMDD system [13]. In addition, any kind of multilevel phase modulation format can be introduced by using the coherent receiver. While the spectral efficiency of binary modulation formats is limited to 1 bit/s/Hz/polarization (which is called the Nyquist limit), multilevel modulation formats with N bits of information per symbol can achieve a spectral efficiency of up to N bit/s/Hz/polarization. Recent research has focused on M-ary PSK and even QAM with coherent detection, which can increase the spectral efficiency by a factor of $\log_2 M$ [14–16]. Moreover, for the same bit rate, because the symbol rate is reduced, the system can have a higher tolerance to CD and PMD.

However, one of the major challenges in coherent detection is to overcome the carrier phase noise when using an LO to beat with the received signals to retrieve the modulated phase information. Phase noise, which can result from lasers, will impose a power penalty on the receiver sensitivity. Self-coherent multisymbol detection of optical differential M-ary PSK is introduced to improve the system performance; however, higher analog-to-digital conversion resolution and more digital signal processing power are required as compared to a digital coherent receiver [17]. Further, differential encoding is also necessary in this scheme. As for the coherent receiver, initially, an optical PLL is an option to track the carrier phase with respect to the LO carrier in homodyne detection. However, an optical PLL operating at optical wavelengths in combination with distributed feedback (DFB) lasers may be quite difficult to implement because the product of the laser linewidth and loop delay is too large [18]. Another option is to use electrical PLL to track the carrier phase after down-converting the optical signal to an intermediate frequency (IF) electrical signal in a heterodyne detection receiver as mentioned earlier. Compared with heterodyne detection, homodyne detection offers better sensitivity and requires a smaller receiver bandwidth [19]. On the contrary, coherent receivers employing high-speed analog-to-digital converters (ADCs) and high-speed baseband DSP units are becoming increasingly attractive rather than using an optical PLL for demodulation. A conventional block Mth power phase estimation scheme is proposed in Refs. [13,18] to raise the received M-ary PSK signals to the Mth power to estimate the phase reference in conjunction with a coherent optical receiver. However, this scheme requires nonlinear operations, such as taking the Mth power and the $\tan^{-1}(\cdot)$, and resolving the $\pm 2\pi/M$ phase ambiguity, which imposes a large latency on the system. Such nonlinear operations would limit further potential for real-time processing of the scheme. In addition, nonlinear phase noise always exists in long-haul systems owing to the Gordon–Mollenauer effect [20], which severely affects the performance of a phase-modulated optical system [21]. The results in Ref. [22] show that such Mth power phase estimation techniques may not be able to effectively deal with nonlinear phase noise.

The maximum-likelihood (ML) carrier phase estimator derived in Ref. [23] can be used to approximate ideal synchronous coherent detection in optical PSK systems. The ML phase estimator requires only linear computations, and thus it is more feasible for online processing for real systems. Intuitively, one can show that the ML estimation receiver outperforms the Mth power block phase estimator and conventional differential detection, especially when the nonlinear phase noise is dominant, thus significantly improving the receiver sensitivity and tolerance to the nonlinear phase noise. The algorithm of the ML phase estimator is expected to improve the performance of coherent optical communication systems using different M-ary PSK and QAM formats. The improvement by DSP at the receiver end can be significant for transmission systems in the presence of fiber-degrading effects, including CD, PMD, and nonlinearities for both single-channel and also DWDM systems.

5.5 ELECTRONIC AMPLIFIERS: RESPONSES AND NOISE

5.5.1 INTRODUCTION

The electronic amplifier as a preamplification stage of an optical receiver plays a major role in the detection of optical signals so that the optimum SNR and therefore the OSNR can be derived based on the photodetector responsivity. Under coherent detection, the amplifier noise must be much less than that of the quantum shot noise contributed by the high-power level of the LO, which is normally about 10-dB above that of the signal average power.

Thus, this section gives an introduction to electronic amplifiers for wideband signals applicable to ultra-high-speed, high-gain, and low-noise transimpedance amplifiers (TIAs). We concentrate on differential input TIAs, but address the detailed design of a single input single output with the noise suppression technique of Annex 2 with the design strategy of achieving stability in the feedback amplifier as well as low noise and wide bandwidth. We define the electronic noise of the preamplifier stage as the total equivalent input noise spectral density, that is, all the noise sources (current and voltage sources) of all elements of the amplifier are referred to the input port of the amplifier and thus an equivalent current source is found, from which the current density is derived. Once this current density is found, the total equivalent at the input can be found when the overall bandwidth of the receiver is determined. When this current is known, and with the average signal power, we can obtain without difficulty the SNR at the input stage of the optical receiver, and then the OSNR. On the contrary, if the OSNR required at the receiver is determined for any specific modulation format, then with the assumed optical power of the signal available at the front of the optical receiver and the responsivity of the photodetector, we can determine the maximum electronic noise spectral density allowable by the preamplification stage and hence the design of the amplifier electronic circuit.

The principal function of an optoelectronic receiver is to convert the received optical signals into equivalent electronic signals, followed by amplification and sampling and processing to recover properties of the original shapes and sequence. So, at first, the optical-domain signals must be converted to electronic current in the photodetection device, a photodetector of either p-I-n or an avalanche photodiode (APD), in which the optical power is absorbed in the active region and both electrons and holes generated are attracted to the positive- and negative-biased electrodes, respectively. Thus, the generated current is proportional to the power of the optical signals, and hence the name *square law detection*. The p-I-n detector is structured with p+ and n+ doped regions sandwiched by the intrinsic layer in which the absorption of optical signal occurs. A high electric field is established in this region by reverse-biasing the diode, and electrons and holes are attracted to either sides of the diode, resulting in generation of current. Similarly, an APD works with the reverse-biasing level close to the reverse breakdown level of the *pn* junction (no intrinsic layer) so that electronic carriers can be multiplied in the avalanche flow when the optical signals are absorbed.

This photo-generated current is then fed into an electronic amplifier whose transimpedance must be sufficiently high and noise low so that a sufficient voltage signal can be obtained and then further amplified by a main amplifier, a voltage-gain type. For high-speed and wideband signals, the transimpedance amplification type is preferred as they offer wideband, much wider than the high-impedance type though the noise level might be higher. With TIAs, there are two types, the single input single output port and two differential inputs and single output. The output ports can be differential with a complementary port. The differential input TIA offers a much higher transimpedance gain (Z_T) and wider bandwidth as well. This contributes to the use of a long-tail pair at the input and hence a reasonably high input impedance that would facilitate feedback stability [24–26].

In Section 5.5.3, a case study of a coherent optical receiver is described from the design to implementation, including feedback control and noise reduction. Although the corner frequency is only a few hundreds of megahertz but with a limited transistor transition frequency, this bandwidth is remarkable. The design is scalable to ultra-wideband reception subsystems.

5.5.2 WIDEBAND TIAS

Two types of TIAs are described. They are distinguished by whether a single input or differential inputs are used, which depends on whether a differential pair, long-tail pair, or a single transistor stage is used at the input of the TIA.

5.5.2.1 Single Input, Single Output

We prefer to treat this section as a design example and describe the experimental demonstration of a wideband and low-noise amplifier in Annex 2. In the next section, the differential input TIA is described. This type of modern TIA offers high value of the transimpedance and reasonably low noise.

5.5.2.2 Differential Inputs, Single/Differential Output

An example circuit of the differential input transimpedance amplifier is shown in Figure 5.17 [27], in which a long-tail pair or differential pair is employed at the input stage. Two matched transistors are used to ensure the minimum common mode rejection and maximum differential mode operation. This pair has a very high input impedance and thus the feedback from the output stage can be stable.

Thus, the feedback resistance can be increased up to the limit of the stability locus of the network pole. This thus offers a high transimpedance Z_T and a wide bandwidth. A typical Z_T of 3000–6000 Ω can be achieved with 30-GHz 3-dB bandwidth, as shown in Figures 5.18 [28] and 5.19. In addition, the chip image of the TIA can be seen in Figure 5.18a. Such a TIA can be implemented in either InP or SiGe material. The advantage of SiGe is that the circuit can be integrated with a high-speed Ge-APD detector and ADC and DSP. On the contrary, if implemented in InP, the high-speed p-I-n or APD can be integrated and then RF interconnected with ADC and DSP. The differential group delay may be significant and must be compensated in the digital processing domain.

5.5.3 AMPLIFIER NOISE REFERRED TO INPUT

There are several noise sources in any electronic system, which include thermal noise, shot-noise, and quantum shot noise, especially in optoelectronic detection. Thermal noise is due to the operating temperature being well above the absolute temperature at which no random

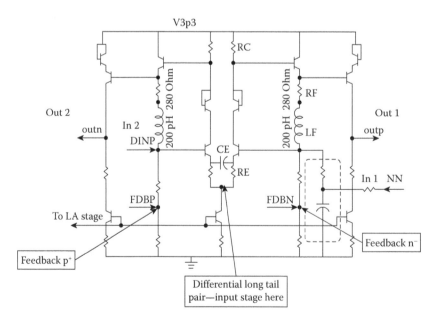

FIGURE 5.17 A typical structure of a differential TIA with differential feedback paths.

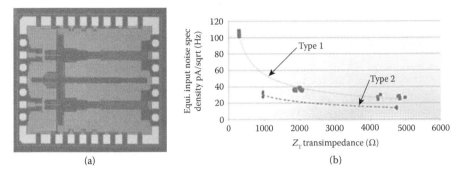

(a)   (b)

FIGURE 5.18 Differential amplifiers: (a) chip-level image and (b) referred input noise equivalent spectral noise density {Inphi TIA 3205 (type 1) and 2850 (type 2)}. (Courtesy of Inphi Inc., Technical information on 3205 and 2850, TIA device, 2012.)

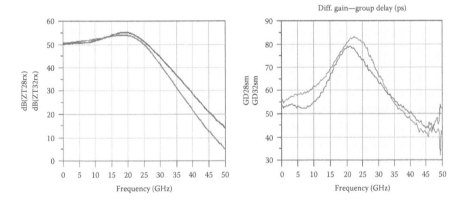

FIGURE 5.19 Differential amplifier: frequency response and differential group delay.

movement of electrons occur. This type of noise depends on the ion temperature. Shot noise is due to the current flowing and random scattering of electrons, and thus this type of noise depends on the strength of the flowing current such as the biasing current in electronic devices. Quantum shot noise is generated owing to the current emitted from optoelectronic detection processes, which are dependent on the strength of the intensity of the optical signals or sources imposed on the detectors. Thus, this type of noise is dependent on signals. In the case of coherent detection, the mixing of the LO laser and signals normally occur with the strength of the LO being much larger than that of the average signal power. Thus, the quantum shot noise is dominated by that from the LO.

In practice, an equivalent electronic noise source is the total noise referred to the input of electronic amplifiers which can be measured by measuring the total spectral density of the noise distribution over the whole bandwidth of the amplification devices. Then, the total noise spectral density can be evaluated referred to the input port. For example, if the amplifier is a transimpedance-type, then the transimpedance of the device is measured first, and then the measured voltage spectral density at the output port can be referred to the input. In this case, it is the total equivalent noise spectral density. The common term employed and specified for TIAs is the total equivalent spectral noise density over the midband region of the amplifying device. The midband range of any amplifier is defined as the flat gain region from DC to the corner 3-dB point of the frequency response of the electronic device.

Figure 5.20 illustrates the meaning of the total equivalent noise sources as referred to the input port of a two-port electronic amplifying device. A noisy amplifier with an input excitation

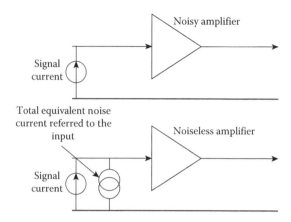

FIGURE 5.20 Equivalent noise spectral density current sources.

current source, typically a signal current generated from the PD after the optical to electrical conversion, can be represented with a noiseless amplifier and a current source in parallel with a noise source whose strength is equal to the total equivalent noise current referred to the input. Thus, the total equivalent current can be found by taking the product of this total equivalent current noise spectral density and the 3 dB bandwidth of the amplifying device. Thus, one can find the SNR at the input of the electronic amplifier

$$SNR = \frac{square\_of\_current\_generated}{square\_of\_current\_generated + total\_equivalent\_noise\_current\_power} \tag{5.47}$$

From this SNR referred to the input of the electronic front end, one can estimate the eye opening of the voltage signals at the output of the amplifying stage that is normally required by the ADC for sampling and conversion to digital signals for processing. Thus, one can then estimate the required OSNR at the input of the photodetector and hence the launched power required at the transmitter over several span links with certain attenuation factors.

Detailed analyses of amplifier noise and the equivalent noise sources as referred to input ports are given in Annex 2. It is noted that noise has no direction of flow as they always add and do not subtract, and thus noise is measured as noise power and not as current. Thus, electrical spectrum analyzers are commonly used to measure the total noise spectral density, or the distribution of noise voltage over the spectral range under consideration, which is thus defined as the noise power spectral density distribution.

5.6 DIGITAL SIGNAL PROCESSING SYSTEMS AND COHERENT OPTICAL RECEPTION

5.6.1 DSP-Assisted Coherent Detection

Over the years, since the introduction of optical coherent communications in the mid-1980s, the invention of optical amplifiers has left coherent reception behind until recently, when long-haul transmission suffered from the nonlinearity of dispersion-compensating fibers (DCF) and SSMF transmission line due to its small effective area. Furthermore, the advancement of DSP in wireless communication also contributed to the application of DSP in modern coherent communication systems. Thus, the name *DSP-assisted coherent detection*, that is, when a real-time DSP is incorporated after the optoelectronic conversion of the total field of the local oscillator and that of the signals, the analog received signals are sampled by a high-speed ADC and then the digitalized signals are processed in a DSP. Currently, real-time DSP processors are intensively researched for practical implementation. The main difference between real-time and off-line processing is that the real-time processing algorithm must be effective owing to the limited time available for processing.

When polarization division multiplexed (PDM) QAM channels are transmitted and received, then polarization and phase diversity receivers are employed. The schematics of such receivers are shown in Figure 5.21a. Further, the structures of such reception systems incorporating DSP with the diversity hybrid coupler in the optical domain are shown in Figure 5.21b through d. The polarization diversity section with the polarization beam splitters at the signal and LO inputs facilitate the demultiplexing of polarized modes in the optical waveguides. The phase diversity using a 90° optical phase shifter here allows the separation of the in-phase (I-) and quadrature-phase (Q-) components of the QAM channels. Using a 2 × 2 coupler also enables balanced reception using photodetector pair (PDP) connected back-to-back and hence results in a 3 dB gain in the sensitivity. Section 3.6, Chapter 3, has described the modulation scheme QAM using I–Q modulators for single-polarization or dual-polarization multiplexed channels.

5.6.1.1 DSP-Based Reception Systems

The schematic of a synchronous coherent receiver based on DSP is shown in Figure 5.22. Once the polarization and the I- and Q-optical components are separated by the hybrid coupler, the positive and negative parts of the I- and Q- are coupled into a balanced optoelectronic receiver as shown in

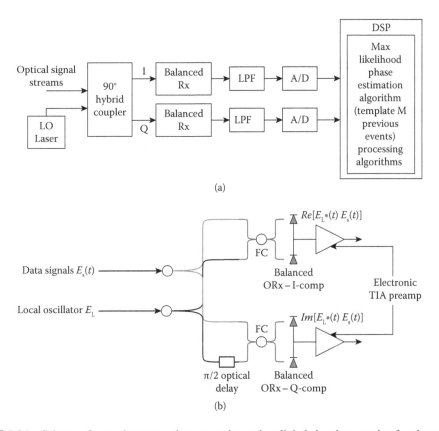

(a)

(b)

FIGURE 5.21 Scheme of a synchronous coherent receiver using digital signal processing for phase estimation of coherent optical communications. (a) Generic scheme and (b) detailed optical receiver using only one polarization phase diversity coupler. (Adapted from S. Zhang et al., in *Conference on Lasers and Electro-Optics/Quantum Electronics and Laser Science and Photonic Applications Systems Technologies*, Technical Digest, paper CThJJ2.) *(Continued)*

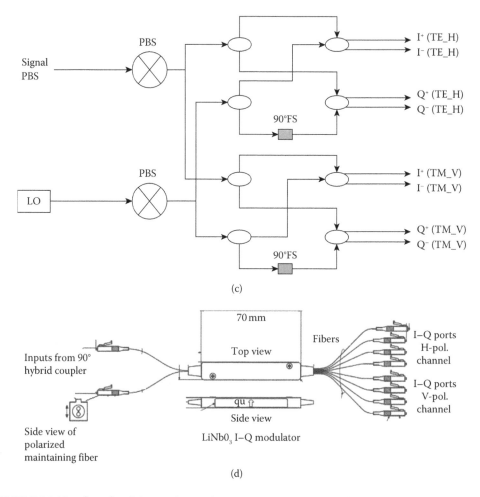

(c)

(d)

FIGURE 5.21 (Continued)   Scheme of a synchronous coherent receiver using digital signal processing for phase estimation of coherent optical communications. (c) hybrid 90° coupler for polarization and phase diversity; and (d) typical view of a hybrid coupler with two input ports and 8 output ports of the structure in (c). TE_V, TE_H = transverse electric mode with vertical (V) or horizontal (H) polarized mode, TM = transverse magnetic mode with polarization orthogonal to that of the TE mode, FS = phase shifter, PBS = polarization beam splitter. (Adapted from S. Zhang et al., in *Conference on Lasers and Electro-Optics/Quantum Electronics and Laser Science and Photonic Applications Systems Technologies*, Technical Digest, paper CThJJ2.)

Figure 5.21b. Two photodiodes (PDs) are connected back-to-back so that the push–pull operation can be achieved, and hence there is a 3 dB improvement as compared to a single PD detection. The current generated from the B2B connected PDs is fed into a transimpedance amplifier so that a voltage signal can be derived at the output. A further voltage-gain amplifier is used to boost these signals to the right level of the ADC so that sampling can be conducted and conversion of the analog signals to the digital domain can happen. These digitized signals are then fetched into DSPs, and processing in the "soft domain" can be conducted. Thus, a number of processing algorithms can be employed in this stage. A number of processing algorithms are employed to compensate for linear and nonlinear distortion effects due to optical signal propagation through the optical-guided medium, to recover the carrier phase and the clock rate for resampling of the data sequence, and so on. Chapter 13 will describe in detail the fundamental aspects of these processing algorithms. Figure 5.22 shows a schematic of possible processing phases in the DSP incorporated in a DSP-based coherent receiver.

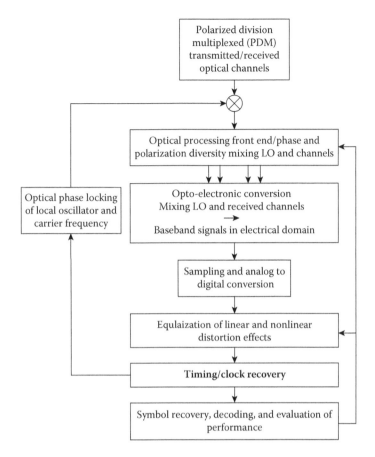

FIGURE 5.22   Flow of functionalities of DSP processing in a QAM coherent optical receiver with possible feedback control.

Besides, all the soft processing, the optical phase locking as described in Chapter 5 is necessary in order to lock the frequencies of the LO and that of the signal carrier within a certain limit within which the algorithms for clock recovery can function, for example, within ±2 GHz.

5.6.2   COHERENT RECEPTION ANALYSIS

5.6.2.1   Sensitivity

At ultra-high bit rates, the laser must be externally modulated, so that the phase of the lightwave is conserved along the fiber transmission line. The detection can be direct detection, self-coherent, or homodyne and heterodyne. The sensitivity of a coherent receiver is also important for the transmission system, especially the PSK scheme under both homodyne and heterodyne transmission techniques. This section gives the analysis of a receiver for synchronous coherent optical fiber transmission systems. Consider that the optical fields of the signals and LO are coupled via a fiber coupler with two output ports, namely 1 and 2. The output fields are then launched into two photodetectors connected back-to-back, and then the electronic current is amplified using a transimpedance-type amplifier and further equalized to extend the bandwidth of the receiver. Our objective is to obtain the receiver penalty and its degradation due to imperfect polarization mixing and unbalancing effects in the balanced receiver. A case study of the design, implementation, and measurements of an optical balanced receiver electronic circuit and noise suppression techniques is given in Annex 2.

The following parameters are commonly used in analysis (Figure 5.23):

E_s	Amplitude of signal optical field at the receiver
E_L	Amplitude of local oscillator optical field
P_s, P_L	Optical power of signal and local oscillator at the input of the photodetector
$s(t)$	The modulated pulse
$\langle i_{NS}^2(t) \rangle$	Mean square noise current (power) produced by the total optical intensity on the photodetector
$\langle i_S^2(t) \rangle$	Mean square current produced by the photodetector by $s(t)$
$S_{NS}(t)$	Shot-noise spectral density of $\langle i_S^2(t) \rangle$ and local oscillator power
$i_{Neq}^2(t)$	Equivalent noise current of the electronic preamplifier at its input
$Z_T(\omega)$	Transfer impedance of the electronic preamplifier
$H_E(\omega)$	Voltage transfer characteristic of the electronic equalizer followed the electronic preamplifier

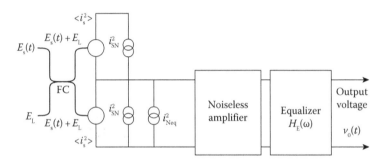

FIGURE 5.23 Equivalent current model at the input of the optical balanced receiver under coherent detection, average signal current, and equivalent noise current of the electronic preamplifier as seen from its input port and equalizer. FC = fiber coupler.

The combined field of the signal and LO via a directional coupler can be written with their separate polarized field components as

$$E_{sX} = \sqrt{K_{sX}}\, E_S \cos(\omega_s t - \phi_{m(t)})$$

$$E_{sY} = \sqrt{K_{sY}}\, E_S \cos(\omega_s t - \phi_{m(t)} + \delta_s)$$

$$E_{LX} = \sqrt{K_{LX}}\, E_L \cos(\omega_L t) \qquad (5.48)$$

$$E_{LY} = \sqrt{K_{LY}}\, E_L \cos(\omega_L t + \delta_L)$$

$$\phi_{m(t)} = \frac{\pi}{2} K_m s(t)$$

with $\phi_{m(t)}$ representing the phase modulation and K_m the modulation depth; $K_{sX}\, K_{sY}\, K_{LX}\, K_{LY}$ are the intensity fraction coefficients in the X- and Y-directions of the signal and LO fields, respectively.

Thus, the output fields at ports 1 and 2 of the FC in the X-plane can be obtained using the transfer matrix as

$$\begin{bmatrix} E_{R1X} \\ E_{R2X} \end{bmatrix} = \begin{bmatrix} \sqrt{K_{sX}(1-\alpha)}\cos(\omega_s t - \phi_{m(t)}) & \sqrt{K_{LX}\alpha}\,\sin(\omega_L t) \\ \sqrt{K_{sX}\alpha}\,\sin(\omega_s t - \phi_{m(t)}) & \sqrt{K_{LX}(1-\alpha)}\cos(\omega_L t) \end{bmatrix} \begin{bmatrix} E_s \\ E_L \end{bmatrix} \qquad (5.49)$$

$$\begin{bmatrix} E_{R1Y} \\ E_{R2Y} \end{bmatrix} = \begin{bmatrix} \sqrt{K_{sY}(1-\alpha)}\cos(\omega_s t - \phi_{m(t)}) & \sqrt{K_{LY}\alpha}\sin(\omega_L t + \delta_L) \\ \sqrt{K_{sY}\alpha}\sin(\omega_s t - \phi_{m(t)}) & \sqrt{K_{LY}(1-\alpha)}\cos(\omega_L t) + \delta_L \end{bmatrix} \begin{bmatrix} E_s \\ E_L \end{bmatrix} \tag{5.50}$$

with α defined as the intensity coupling ratio of the coupler. Thus, the field components at ports 1 and 2 can be derived by combining the X- and Y-components from Equations 5.49 and 5.50, and hence the total power at ports 1 and 2 is given as

$$P_{R1} = P_s(1-\alpha) + P_L\alpha + 2\sqrt{P_s P_L \alpha(1-\alpha)K_p}\,\sin(\omega_{IF}t + \phi_{m(t)} + \phi_p - \phi_e)$$

$$P_{R2} = P_s\alpha + P_L(1-\alpha) + 2\sqrt{P_s P_L \alpha(1-\alpha)K_p}\,\sin(\omega_{IF}t + \phi_{m(t)} + \phi_p - \phi_e + \pi)$$

with $\qquad K_p = K_{sX}K_{LX} + K_{sY}K_{LY} + 2\sqrt{K_{sX}K_{LX}K_{sY}K_{LY}}\cos(\delta_L - \delta_s) \tag{5.51}$

$$\phi_p = \tan^{-1}\left[\frac{\sqrt{K_{sX}K_{LY}}\sin(\delta_L - \delta_s)}{\sqrt{K_{sX}K_{LX}} + \sqrt{K_{sY}K_{LY}}\cos(\delta_L - \delta_s)} \right]$$

ω_{IF} is the intermediate angular frequency and equals the difference between the frequencies of the LO and the carrier of the signals. ϕ_e is the phase offset, and $\phi_p - \phi_e$ is the demodulation reference phase error.

In Equation 5.51, the total fields of the signal and the LO are added and then the product of the field vector and its conjugate gives the power. Only the term with a frequency that falls within the range of the sensitivity of the photodetector would produce the electronic current. Thus, the term with the sum of the frequencies of the wavelength of the signal and the LO would not be detected, and only the product of the two terms would be detected as given.

Now, assuming a binary PSK (BPSK) modulation scheme, the pulse has a square shape with amplitude +1 or −1, the PD is a p-I-n type, and the PD bandwidth is wider than the signal 3-dB bandwidth followed by an equalized electronic preamplifier. The signal at the output of the electronic equalizer or the input signal to the decision circuit is given by

$$\hat{v}_D(t) = 2K_H K_p \sqrt{P_s P_L \alpha(1-\alpha)K_p} \int_{-\infty}^{\infty} H_E(f)\,df \int_{-\infty}^{\infty} (t)\,dt \sin\left(\frac{\pi}{2}K_m\right)\cos(\phi_p - \phi_e)$$

$$\rightarrow \hat{v}_D(t) = 2K_H K_p \sqrt{P_s P_L \alpha(1-\alpha)K_p}\,\sin\left(\frac{\pi}{2}K_m\right)\cos(\phi_p - \phi_e) \tag{5.52}$$

$$K_H = 1 \text{ for homodyne}; K_H = 1/\sqrt{2} \text{ for .. heterodyne}$$

For a perfectly balanced receiver $K_B = 2$ and $\alpha = 0.5$, otherwise $K_B = 1$. The integrals of the first line in Equation 5.52 are given by

$$\int_{-\infty}^{\infty} H_E(f)\,df = \frac{1}{T_B} \qquad \because H_E(f) = \sin c(\pi T_B f)$$

$$\int_{-\infty}^{\infty} s(t)\,dt = 2T_B \tag{5.53}$$

$V_D(f)$ is the transfer function of the matched filter for equalization, and T_B is the bit period. The total noise voltage as a sum of the quantum shot noise generated by the signal and the local oscillator and the total equivalent noise of the electronic preamplifier at the input of the preamplifier, at the output of the equalizer, is given by

$$\langle v_N^2(t) \rangle = \frac{[K_B \alpha S'_{IS} + (2 - K_B)S_{Ix} + S_{IE}] \int\limits_{-\infty}^{\infty} |H_4(f)|^2 \, df}{K_{IS}^2}$$

or

(5.54)

$$\langle v_N^2(t) \rangle = \frac{[K_B \alpha S'_{IS} + (2 - K_B)S_{Ix} + S_{IE}]}{K_{IS}^2 T_B}$$

For homodyne and heterodyne detection, we have

$$\langle v_N^2(t) \rangle = \Re q \frac{P_L}{\lambda T_B}[K_B \alpha S'_{IS} + (2 - K_B)S_{Ix} + S_{IE}]$$

(5.55)

where the spectral densities S'_{IX}, S'_{IE} are given by

$$S'_{IX} = \frac{S_{IX}}{S'_{IS}}$$

(5.56)

$$S'_{IE} = \frac{S_{IE}}{S'_{IS}}$$

Thus, the receiver sensitivity for binary PSK and equiprobable detection with a Gaussian probability density distribution, we have

$$P_e = \frac{1}{2}\text{erfc}\left(\frac{\delta}{\sqrt{2}}\right)$$

(5.57)

with δ given by

$$P_e = \frac{1}{2}\text{erfc}\left(\frac{\delta}{\sqrt{2}}\right) \quad \text{with}....\delta = \frac{\hat{v}_D}{2\sqrt{\langle v_N^2 \rangle}}$$

(5.58)

Thus, using Equations 5.52, 5.55, and 5.58, the receiver sensitivity in the linear power scale is

$$P_s = \langle P_s(t) \rangle = \frac{\Re q \delta^2}{4\lambda T_B K_H^2} \frac{[K_B \alpha S'_{IS} + (2 - K_B)S_{Ix} + S_{IE}]}{\eta K_p(1-\alpha)\alpha K_B^2 \sin^2\left(\frac{\pi}{2}K_m\right)\cos^2(\phi_p - \phi_e)}$$

(5.59)

5.6.2.2 Shot-Noise-Limited Receiver Sensitivity
When the power of the LO dominates the noise of the electronic preamplifier and the equalizer, the receiver sensitivity (in the linear scale) is given as

$$P_s = \langle P_{sL} \rangle = \frac{\Re q \delta^2}{4\lambda T_B K_H^2}$$

(5.60)

This shot-noise-limited receiver sensitivity can be plotted as shown in Figure 5.24 [29].

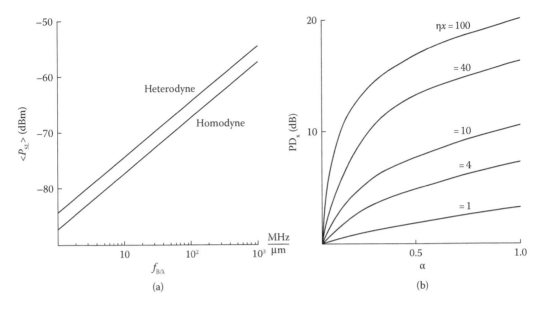

FIGURE 5.24 (a) Receiver sensitivity of coherent homodyne and heterodyne detection, signal power versus bandwidth over the wavelength. (b) Power penalty of the receiver sensitivity from the shot-noise-limited level as a function of the excess noise of the local oscillator. (Extracted from I. Hodgkinson, *IEEE J. Lightwave Technol.*, Vol. 5, No. 4, 1987, pp. 573–587, Figures 1 and 2).

5.6.2.3 Receiver Sensitivity under Nonideal Conditions

Under nonideal conditions, the receiver sensitivity deviates from the shot-noise-limited sensitivity and is characterized by the receiver sensitivity penalty PD_T as

$$PD_T = 10 \text{Log} \frac{\langle P_s \rangle}{\langle P_{sL} \rangle} \text{dB} \tag{5.61}$$

$$
\begin{aligned}
PD_T = {} & 10\text{Log}_{10} \left[\frac{K_B \alpha S'_{IS} + (2 - K_B) S_{Ix} + S_{IE}}{K_B \alpha} \right] \\
& - 10\text{Log}_{10} \left[K_B (1 - \alpha) \right] \\
& - 10\text{Log}_{10} \left([\eta][K_p] \sin^2 \left(\frac{\pi}{2} K_m \right) \cos^2(\phi_p - \phi_e) \right)
\end{aligned}
\tag{5.62}
$$

where η is the LO excess noise factor.

The receiver sensitivity is plotted against the ratio f_B/λ for the case of homodyne and heterodyne detection and is shown in Figure 5.24a, and the power penalty of the receiver sensitivity against the excess noise factor of the LO is shown in Figure 5.24b. The receiver power penalty can be deduced as a function of the total electronic equivalent noise spectral density, and as a function of the rotation of the polarization of the LO (this can be found in Ref. [29]).

Furthermore, in Ref. [30], the receiver power penalty and the normalized heterodyne center frequency can vary as a function of the modulation parameter and as well as the optical power ratio at the same polarized channels.

5.6.3 Digital Processing Systems

A generic structure of the coherent reception and digital signal processing system is shown in Figure 5.25 in which the digital signal processing system is placed after the sampling and conversion from analog state to the digital form. Obviously, the optical signal fields are beating with the LO laser, whose frequency would be approximately identical with the signal-channel carrier. The beating occurs in the square law photodetectors; that is, the sum of the two fields is squared, and the product term is decomposed into the difference and summation terms, and thus only the difference term falls back into the baseband region and is amplified by the electronic preamplifier, which is a balanced differential transimpedance-type.

If the signals are complex, then there are the real and imaginary components, which form a pair. The other pair comes from the other polarization-mode channel. The digitized signals of both the real and the imaginary parts are processed in real time or off-line. The processors contain algorithms to combat a number of transmission impairments such as the imbalance between the in-phase and the quadrature components created at the transmitter, the recovery of the clock rate and timing for resampling, the carrier phase estimation for estimation of the signal phases, adaptive equalization for compensation of propagation dispersion effects using MLSE, and so on. These algorithms are built into the hardware processors or memory and loaded to processing subsystems.

The sampling rate must normally be twice that of the signal bandwidth to ensure that the Nyquist criterion is satisfied. Although this rate is very high for 25 to 32 GBaud optical channels, the Fujitsu ADC has reached this requirement with a sampling rate of 56 to 64 GSamples/s (GSa/s) as depicted later in the chapter in Figure 5.37.

The linewidth resolution of the processing for semiconductor device fabrication has progressed greatly over the year in an exponential trend as shown in Table 5.1. This progress was possible owing to the successes in lithographic techniques using optical at short wavelengths such as UV light source, the electron beam, and X-ray beam in association with sensitive photoresist such as the SU-80. This would allow the line resolution can reach 5 nm in 2020 and hence even higher sampling rate can be possible. So, if we plot the trend in a log-linear scale as shown in Figure 5.26, a linear line is obtained, meaning that the resolution is reduced exponentially. When the gate width is reduced, the electronic speed would increase tremendously; at 5 mm, the speed of an electronic

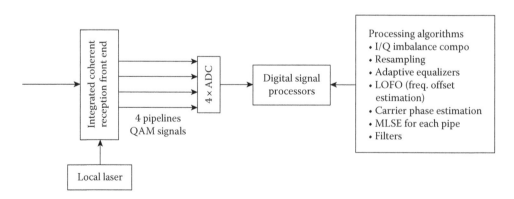

FIGURE 5.25 Coherent reception and digital signal processing system.

TABLE 5.1

Milestones of Progress in Linewidth Resolution

Semiconductor Manufacturing Processes and Spatial
Resolution (Gate Width)

10 μm—1971
3 μm—1975
1.5 μm—1982
1 μm—1985
800 nm (0.80 μm)—1989: UV lithography
600 nm (0.60 μm)—1994
350 nm (0.35 μm)—1995
250 nm (0.25 μm)—1998
180 nm (0.18 μm)—1999
130 nm (0.13 μm)—2000
90 nm—2002: Electron lithography
65 nm—2006
45 nm—2008
32 nm—2010
22 nm—2012
14 nm—approx. 2014: X-ray lithography
10 nm—approx. 2016: X-ray lithography
7 nm—approx. 2018: X-ray lithography
5 nm—approx. 2020: X-ray lithography

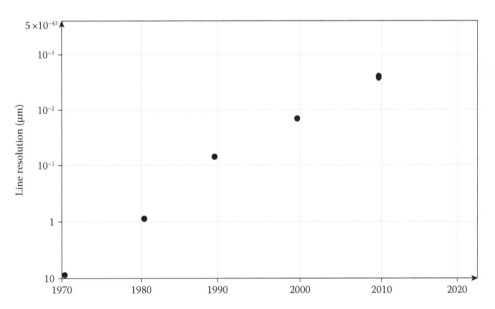

FIGURE 5.26 Progress of linewidth resolution over last three decades of semiconductor manufacturing technology.

CMOS device in SiGe would reach several tens of gigahertz. Regarding the high-speed ADC and DAC, the clock speed is increased by interleaving techniques and summation of all the digitized digital lines to form a very high-speed operation. For example, for a Fujitsu 64 GSa/s DAC or ADC, the applied clock sinusoidal waveform is only 2 GHz. Figure 5.27 shows the progress in the speed development of the Fujitsu ADC and DAC.

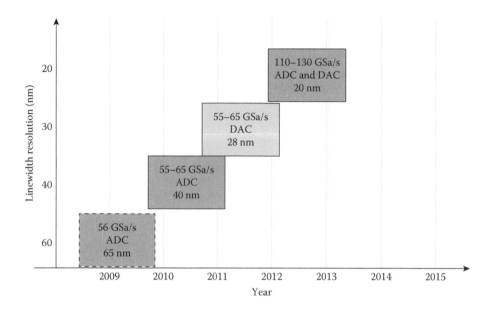

FIGURE 5.27   Evolution of ADC and DAC operating speed with corresponding linewidth resolution.

5.6.3.1   Effective Number of Bits

5.6.3.1.1   Definition

The effective number of bits (ENOB) is a measure of the quality of a digitized signal. The resolution of a DAC or ADC is commonly specified by the number of bits used to represent the analog value, in principle, giving 2^N signal levels for an N-bit signal. However, all real signals contain a certain amount of noise. If the converter is able to represent signal levels below the system noise floor, the lower bits of the digitized signal only represent system noise and do not contain useful information. The ENOB specifies the number of bits in the digitized signal above the noise floor. Often, the ENOB is used also as a quality measure for other blocks such as sample-and-hold amplifiers. In this way, analog blocks can be easily included in signal-chain calculations because the total ENOB of a chain of blocks is usually below the ENOB of the worst block.

Thus, we can represent the ENOB of a digitized system by writing

$$\text{ENOB} = \frac{\text{SINAD} - 1.76}{6.02} \tag{5.63}$$

where all values are given in dB, and signal-to-noise and distortion ratio (SINAD) is the ratio of the total signal including distortion and noise to the wanted signal; the 6.02 term in the divisor converts decibels (a $\log_{10}$ representation) to bits (a $\log_2$ representation); the 1.76 term comes from quantization error in an ideal ADC [31].

This definition compares the SINAD of an ideal ADC or DAC with a word length of ENOB bits with the SINAD of the ADC or DAC being tested.

Indeed, SINAD is a measure of the quality of a signal from a communications device, often defined as

$$\text{SINAD} = \frac{P_{\text{sig}} + P_{\text{noise}} + P_{\text{distortion}}}{P_{\text{noise}} + P_{\text{distortion}}} \tag{5.64}$$

where P is the average power of the signal, noise, and distortion components. SINAD is usually expressed in decibels and is quoted alongside the receiver sensitivity, to give a quantitative evaluation of the receiver sensitivity. Note that with this definition, unlike SNR, a SINAD reading can never be less than 1 (i.e., it is always positive when quoted in decibels).

When calculating the distortion, it is common to exclude the DC components. Owing to its widespread use, SINAD has a few different definitions: (1) the ratio of (a) total received power, that is, the signal to (b) the noise-plus-distortion power. This is modeled by Equation 5.64. (2) The ratio of (a) the power of original modulating audio signal, that is, from a modulated RF carrier to (b) the residual audio power, that is, noise-plus-distortion powers remaining after the original modulating audio signal is removed. With this definition, it is now possible for SINAD to be less than 1. This definition is used when SINAD is used in the calculation of the ENOB for an ADC.

Example: Consider the following as measurements of a three-bit unipolar D/A converter with reference voltage $V_{ref} = 8$ V:

Digital input	000	001	010	011	100	101	110	111
Analog output (V)	−0.01	1.03	2.02	2.96	3.95	5.02	6.00	7.08

The offset error in this case is −0.01 V or −0.01 LSB as 1 V = 1 LSB in this example. The gain error is

$$\frac{7.08 + 0.01}{1} - \frac{7}{1} = 0.09 \text{ LSB}$$

LSB stands for least significant bits. Correcting the offset and gain errors, we obtain the following list of measurements: (0, 1.03, 2.00, 2.93, 3.91, 4.96, 5.93, 7) LSB. This allows the integral nonlinearity (INL) and differential nonlinearity (DNL) to be calculated: INL = (0, 0.03, 0, −0.07, −0.09, −0.04, −0.07, 0) LSB, and DNL = (0.03, −0.03, −0.07, −0.02, 0.05, −0.03, 0.07, 0) LSB.

DNL: For an ideal ADC, the output is divided into $2N$ uniform steps, each with Δ width as shown in Figure 5.28. Any deviation from the ideal step width is the DNL and is measured in number of counts (LSBs). For an ideal ADC, the DNL is 0 LSB. In a practical ADC, DNL error comes from its architecture. For example, in a SAR ADC, DNL error may be caused near the mid-range owing to mismatching of its DAC.

The INL is a measure of how closely the ADC output matches its ideal response. The INL can be defined as the deviation in LSB of the actual transfer function of the ADC from the ideal transfer curve. The INL can be estimated using the DNL at each step by calculating the cumulative sum of DNL errors up to that point. In reality, the INL is measured by plotting the ADC transfer characteristics.

The INL is popularly measured using either (1) the best-fit (best straight line) method or (2) the end-point method.

Best-Fit INL: The best-fit method of INL measurement considers the offset and gain error. One can see in Figure 5.29 that the ideal transfer curve considered for calculating the best-fit INL does not go through the origin. The ideal transfer curve here is drawn such that it depicts the nearest first-order approximation to the actual transfer curve of the ADC.

The intercept and slope of this ideal curve can yield the values of the offset and gain error of the ADC. Quite intuitively, the best-fit method yields better results for the INL. For this reason, often this is the number present on ADC datasheets.

The only real use of the best-fit INL number is to predict distortion in time-variant signal applications. This number would be equivalent to the maximum deviation for an AC application. However, it is always better to use the distortion numbers than INL numbers. To calculate the error budget, end-point INL numbers provide a better estimation. Also, this is the specification that is generally provided in datasheets. So, one has to use this instead of the end-point INL.

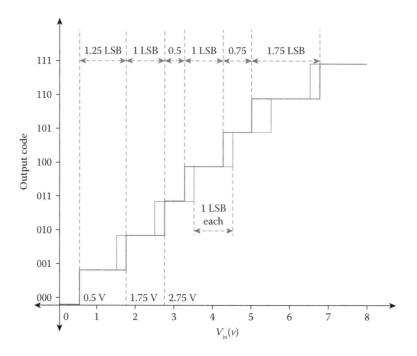

FIGURE 5.28 Representation of DNL in a transfer curve of an ADC.

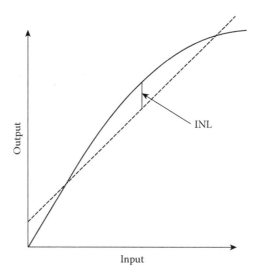

FIGURE 5.29 Best-fit INL.

End-point INL: The end-point method provides the worst-case INL. This measurement passes the straight line through the origin and the maximum output code of the ADC (refer to Figure 5.6). As this method provides the worst-case INL, it is more useful to use this number as compared to the one measured using the best fit for DC applications. This INL number would be typically useful for error budget calculation. This parameter must be considered for applications involving precision measurements and control (Figure 5.30).

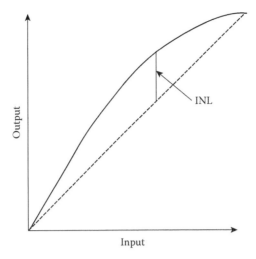

FIGURE 5.30 End-point integral nonlinearity.

The absolute and relative accuracies can now be calculated. In this case, the ENOB absolute accuracy is calculated using the largest absolute deviation D, in this case 0.08 V.

$$D = \frac{V_{ref}}{2^{ENOB}} \rightarrow ENOB = 6.64 \text{ bits} \tag{5.65}$$

The ENOB relative accuracy is calculated using the largest relative (INL) deviation d, in this case 0.09 V.

$$d = \frac{V_{ref}}{2^{ENOB}} \rightarrow ENOB_{rel} = 6.47 \text{ bits} \tag{5.66}$$

For this kind of ENOB calculation, note that the ENOB can be larger or smaller than the actual number of bits. When the ENOB is smaller than the ANOB, this means that some of the least significant bits of the result are inaccurate. However, one can also argue that the ENOB can never be larger than the ANOB, because you always have to add the quantization error of an ideal converter, which is ±0.5 LSB. Different designers may use different definitions of the ENOB.

5.6.3.1.2 High-Speed ADC and DAC Evaluation Incorporating Statistical Property

The ENOB of an ADC is considered as the number of bits that an analog signal can be converted to its digital equivalent by the number of levels represented by the modulo-2 levels, which are reduced due to the noise contributed by electronic components in such a convertor. Thus, only an effective number of equivalent bits can be accounted for. Hence, the term ENOB is proposed.

As shown in Figure 5.31b, a real ADC can be modeled as a cascade of two ideal ADCs, additive noise sources, and an AGC amplifier [32]. The quantized levels are thus equivalent to a specific ENOB as far as the ADC is operating in the linear nonsaturated region. If the normalized signal amplitude/power surpasses unity, the saturated region, then the signals are clipped. The decision level of the quantization in an ADC normally varies following a normalized Gaussian PDF, and thus we can estimate the RMS noise introduced by the ADC as

$$\text{RMS\_noise} = \sqrt{\int_{\frac{LSB}{2}}^{\frac{LSB}{2}} \int_{-\infty}^{\infty} \frac{x^2 \frac{1}{\sqrt{2\pi\sigma^2}} \exp\left(-\frac{(x-y)^2}{2\sigma^2}\right)}{LSB} dx\,dy} = \sqrt{LSB^2/12 + \sigma^2} \tag{5.67}$$

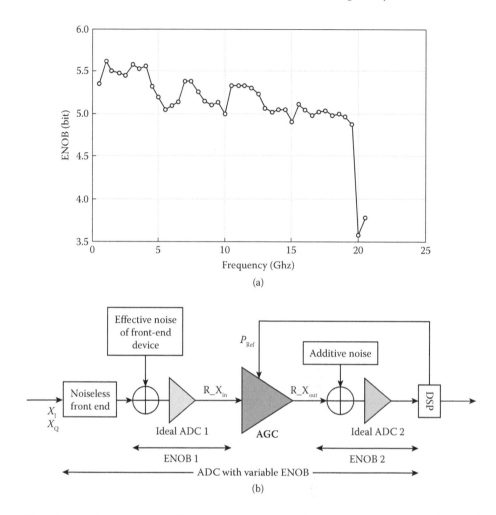

FIGURE 5.31 (a) The measured ENOB frequency response of a commercial real-time DSA of 20 GHz bandwidth and sampling rate of 50 GSa/s and (b) the deduced ADC model of variable ENOB based on the experimental frequency response of (a) and the spectrum of broadband signals.

where σ is the variance; x is the variable related to the integration of decision voltage; and similarly y for integration inside one LSB. Given the known quantity $LSB^2/12$ by the introduction of the ideal quantization error, σ^2 can be determined via the Gaussian noise distribution. We can thus deduce the ENOB values corresponding to the levels of Gaussian noise as

$$\text{ENOB} = N - \log_2 \left(\frac{\sqrt{LSB^2/12 + (A\sigma)^2}}{LSB/\sqrt{12}} \right) \tag{5.68}$$

where A is the RMS amplitude derived from the noise power. According to the ENOB model, the frequency response of the ENOB of the DSA is shown in Figure 5.31a with the excitation of the DSA by sinusoidal waves of different frequencies. As observed, the ENOB varies with respect to the excitation frequency in the range 5–5.5. With a known frequency response of the sampling device, what is the ENOB of the device when excited with broadband signals? This indicates the different resolutions of the ADC of the receiver with the transmission operating under different noisy and dispersive conditions, and thus an equivalent model of the ENOB for performance evaluation is essential. We note that the amplitudes of the optical fields arriving at the receiver vary depending on the conditions of the optical transmission line. The AGC has a nonlinear gain characteristic in which the input sampled signal power level is normalized with respect to the saturated (clipping) level. The gain is significantly high in the linear region and saturated at a high level. The received signal R_X_{in} is scaled with a gain coefficient according to $R\_X_{out} = R\_X_{in}/\sqrt{P_{in\_av}/P_{Ref}}$, where the signal averaged power P_{in_av} is estimated, and the gain is scaled relative to the reference power level P_{Ref} of the AGC, and then a linear scaling factor is used to obtain the output sampled value R_X_{out}. The gain of the AGC is also adjusted according to the signal energy, via the feedback control path from the DSP (see Figure 5.31b). Thus, new values of the ENOB can be evaluated with the noise distributed across the frequency spectrum of the signals, by an averaging process. This signal-dependent ENOB is now denoted as $ENOB_S$.

5.6.3.1.3 Impact of ENOB on Transmission Performance

Shown in Figure 5.32a is the BER variation with respect to the OSNR under B2B transmission using the simulated samples at the output of the eight bit ADC with $ENOB_S$ and full ADC resolution as parameters. The difference is due to the noise distribution (Gaussian or uniform) [30]. Figure 5.32b depicts the variation of BER versus OSNR with $ENOB_S$ as the variation parameter in the case of off-line data with the ENOB of the DSA shown in Figure 5.1a. Several more tests were conducted to ensure the effectiveness of our ENOB model. When the sampled signals of different amplitudes are presented to the ADC, controlled and gain nonlinearly adjusted by the AGC, different degrees of the clipping effect would be introduced. Thus, the clipping effect can be examined for the ADCs of different quantization levels but with identical $ENOB_S$ as shown in Figure 5.33a for the B2B experiment. Figure 5.33b through e show, with the BER as a parameter, the contour plots of the variation of the adjusted reference power level of the AGC and $ENOB_S$ for the cases of 1500 km long-haul transmission of full CD compensation and non-CD compensation operating in the linear (0-dBm launch power in both links) and nonlinear regimes of the fibers with a launch power of 4 and 5 dBm, respectively. When the link is fully CD compensated, the nonlinear effects further contribute to the ineffectiveness of the ADC resolution, and hence moderate AGC freedom in the performance is achieved. In the case of the non-CD compensation link (Figure 5.33d and e), the dispersive pulse sampled amplitudes are lower with less noise, allowing the resolution of the ADC to be higher via the nonlinear gain of the AGC, and thus effective phase estimation and equalization can be achieved. We note that the offline data sets employed prior to

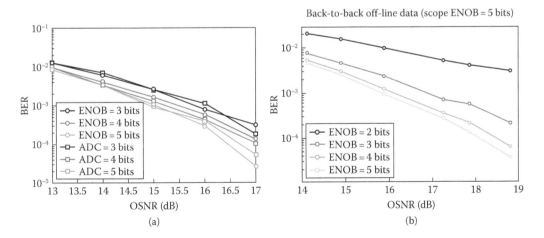

FIGURE 5.32 (a) B2B performance with different $ENOB_S$ values of the ADC model with simulated data (eight bit ADC) and (b) OSNR versus BER under different simulated $ENOB_S$ of off-line data obtained from experimental digital coherent receiver. (From B. N. Mao et al., Investigation on the ENOB and clipping effect of real ADC and AGC in coherent QAM transmission system, *Proc. ECOC 2011*, Geneva, 2011.)

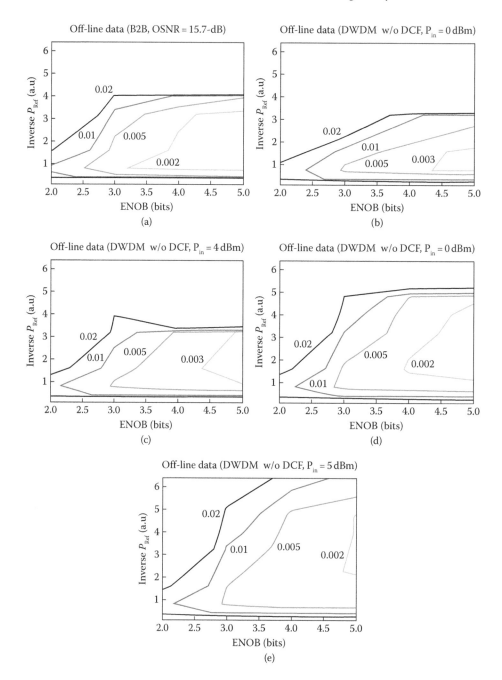

FIGURE 5.33 Comprehensive effects of AGC clipping (inverse of P_{Ref}) and $ENOB_S$ of coherent receiver, experimental transmission under (a) B2B, (b) DWDM linear operation with full chromatic dispersion (CD) compensation and (b) linear, (c) nonlinear regions with 4 dBm launch power; and (d) non-CD compensation and (d) linear, (e) nonlinear region with launch power of 5 dBm.

the processing using $ENOB_S$ to obtain the contours of Figure 5.33 produce the same BER contours of 2e−3 for all cases. Hence, a fair comparison is made when the $ENOB_S$ model is used. The opening distance of the BER contours indicates the dynamic range of the $ENOB_S$ model, especially the AGC. It is obvious from Figure 5.33a through e that the dynamic range of the model is higher for noncompensating than for full CD compensated transmission and even in the case of B2B.

However, for the nonlinear scenario for both cases, the requirement for $ENOB_S$ is higher for the dispersive channel (Figure 5.33c through e). This may be due to the cross-phase modulation effects of adjacent channels, which produce more noise.

5.6.3.2 Digital Processors

The structures of the DAC and ADC are shown in Figures 5.34 and 5.35, respectively. Normally there would be four DACs in an IC in which each DAC section is clocked with a clock sequence that is derived from a lower-frequency sinusoidal wave injected externally into the DAC. Four units are required for the in-phase and quadrature-phase components of QAM-modulated polarized channels, and thus the notation of I_{DAC}, Q_{DAC} shown in the diagram. Similarly, the optical received signals of PDM-QAM (polarization division multiplexed–quadrature amplitude modulation) would be sampled by a four-phase sampler and then converted to digital form into four groups of I and Q lanes for processing in the DSP subsystem. Due to the interleaving of the sampling clock waveform, the digitized bits appear simultaneously at the end of a clock period. The period is long enough so that all sampled values can appear at the outputs. For example, in Figure 5.35, 1024 samples appear at the outputs after a period corresponding to a 500 MHz cycle clock for an eight bit ADC. Thus, the clock has been slowed down by a factor of 128, or alternatively, the sampling interval is $1/(500 \cdot 128) = 1/64$ GHz s. The sampling is implemented using a CHAIS (CHArged mode Interleaved Sampler).

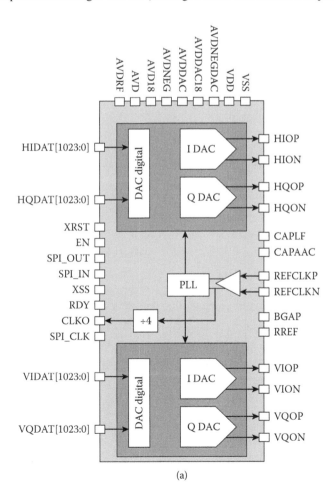

(a)

FIGURE 5.34 Fujitsu DAC structures for four-channel PDM-QPSK signals: (a) schematic diagram.

(Continued)

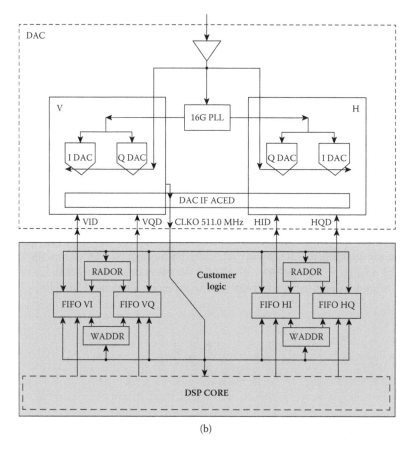

(b)

FIGURE 5.34 (Continued) Fujitsu DAC structures for four-channel PDM-QPSK signals: (b) processing function.

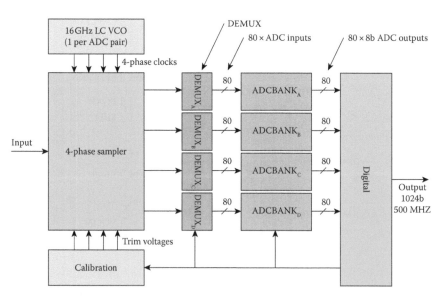

FIGURE 5.35 ADC principles of operations (CHAIS).

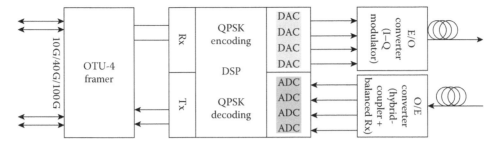

FIGURE 5.36 Schematic of typical structure of ADC and ADC transceiver subsystems for PDM-QPSK modulation channels.

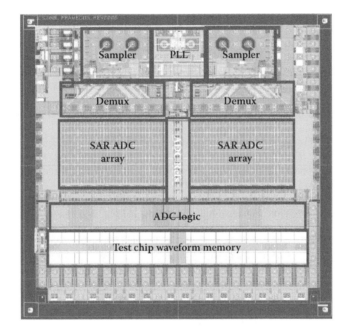

FIGURE 5.37 Fujitsu ADC subsystems with dual converter structure.

Figure 5.36 shows a generic diagram of an optical DSP-based transceiver employing both DAC and ADC under QPSK-modulated or QAM signals. The current maximum sampling rate of 64 GSa/s is available commercially. An IC image of the ADC chip is shown in Figure 5.37.

5.7 CONCLUDING REMARKS

This chapter has described the principles of coherent reception and associated techniques with noise considerations and main functions of the DSPs whose algorithms will be described in separate chapters.

Furthermore, the matching of the LO laser and that of the carrier of the transmitted channel is very important for effective coherent detection, if not degradation of the sensitivity of results. The ITU standard requires that for a digital-processing-based coherent receiver, the frequency offset between the LO and the carrier must be within the limit of ±2.5 GHz. Furthermore, in practice, it is expected that in network and system management, the tuning of the LO is to be done remotely as also automatic locking of the LO with some prior knowledge of the frequency region to set the LO initial frequency. Thus, this section briefly describes the optical phase locking of the local oscillator source for an intradyne coherent reception subsystem.

REFERENCES

1. A. H. Gnauck and P. J. Winzer, Optical phase-shift-keyed transmission, *J. Lightwave Technol.*, Vol. 23, pp. 115–130, 2005.
2. R. C. Alferness, Guided wave devices for optical communication, *IEEE J. Quant. Electron.*, Vol. QE-17, pp. 946–959, 1981.
3. W. A. Stallard, A. R. Beaumont, and R. C. Booth, Integrated optic devices for coherent transmission, *IEEE J. Lightwave Technol.*, Vol. LT-4, No. 7, pp. 852–857, 1986.
4. V. Ferrero and S. Camatel, Optical phase locking techniques: An overview and a novel method based on single sideband sub-carrier modulation, *Opt. Express*, Vol. 16, No. 2, pp. 818–828, 2008.
5. I. Garrett and G. Jacobsen, Theoretical analysis of heterodyne optical receivers for transmission systems using (semiconductor) lasers with nonneglible linewidth, *IEEE J. Lightwave Technol.*, Vol. LT-3/4, pp. 323–334, 1986.
6. G. Nicholson, Probability of error for optical heterodyne DPSK systems with quantum phase noise, *Electron. Lett.*, Vol. 20/24, pp. 1005–1007, 1984.
7. S. Shimada, *Coherent Lightwave Communications Technology*, London: Chapman and Hall, p. 27, 1995.
8. Y. Yamamoto and T. Kimura, Coherent optical fiber transmission system, *IEEE J. Quant. Electron.*, Vol. QE-17, pp. 919–934, 1981.
9. S. Savory and T. Hadjifotiou, Laser linewidth requirements for optical DQPSK optical systems, *IEEE Photon. Technol. Lett.*, Vol. 16, No. 3, pp. 930–932, 2004.
10. K. Iwashita and T. Masumoto, Modulation and detection characteristics of optical continuous phase FSK transmission system, *IEEE J. Lightwave Technol.*, Vol. LT-5/4, pp. 452–462, 1987.
11. F. Derr, Coherent optical QPSK intradyne system: Concept and digital receiver realization, *IEEE J. Lightwave Technol*, Vol. 10, No. 9, pp. 1290–1296, 1992.
12. G. Bosco, I. N. Cano, P. Poggiolini, L. Li, and M. Chen, MLSE-based DQPSK transmission in 43Gb/s DWDM long-haul dispersion managed optical systems, *IEEE J. Lightwave Technol.*, Vol. 28, No. 10, pp. 1573–1581, 2010.
13. D.-S. Ly-Gagnon, S. Tsukamoto, K. Katoh, and K. Kikuchi, Coherent detection of optical quadrature phase-shift keying signals with carrier phase estimation, *J. Lightwave Technol.*, Vol. 24, pp. 12–21, 2006.
14. E. Ip and J. M. Kahn, Feedforward carrier recovery for coherent optical communications, *J. Lightwave Technol.*, Vol. 25, pp. 2675–2692, 2007.
15. L. N. Binh, Dual-ring 16-Star QAM direct and coherent detection in 100 Gb/s optically amplified fiber transmission: Simulation, *Opt. Quant. Electron.*, Vol. 40, pp. 707–732, 2008.
16. L. N. Binh, Generation of multi-level amplitude-differential phase shift keying modulation formats using only one dual-drive Mach-Zehnder interferometric optical modulator, *Opt. Eng.*, Vol. 48, No. 4, pp. 635337–635342, 2009.
17. M. Nazarathy, X. Liu, L. Christen, Y. K. Lize, and A. Willner, Self-coherent multisymbol detection of optical differential phase-shift-keying, *J. Lightwave Technol.*, Vol. 26, pp. 1921–1934, 2008.
18. R. Noe, PLL-free synchronous QPSK polarization multiplex/diversity receiver concept with digital I&Q baseband processing, *IEEE Photon. Technol. Lett.*, Vol. 17, pp. 887–889, 2005.
19. L. G. Kazovsky, G. Kalogerakis, and W.-T. Shaw, Homodyne phase-shift-keying systems: Past challenges and future opportunities, *J. Lightwave. Technol.*, Vol. 24, pp. 4876–4884, 2006.
20. J. P. Gordon and L. F. Mollenauer, Phase noise in photonic communications systems using linear amplifiers, *Opt. Lett.*, Vol. 15, pp. 1351–1353, 1990.
21. H. Kim and A. H. Gnauck, Experimental investigation of the performance limitation of DPSK systems due to nonlinear phase noise, *IEEE Photon. Technol. Lett.*, Vol. 15, pp. 320–322, 2003.
22. S. Zhang, P. Y. Kam, J. Chen, and C. Yu, Receiver sensitivity improvement using decision-aided maximum likelihood phase estimation in coherent optical DQPSK system, in *Conference on Lasers and Electro-Optics/Quantum Electronics and Laser Science and Photonic Applications Systems Technologies*, Technical Digest (CD) (Optical Society of America, 2008), paper CThJJ2.
23. P. Y. Kam, Maximum-likelihood carrier phase recovery for linear suppressed-carrier digital data modulations, *IEEE Trans. Commun.*, Vol. COM-34, pp. 522–527, 1986.
24. E. M. Cherry and D. A. Hooper, *Amplifying Devices and Amplifiers*, New York: John Wiley & Sons, 1965.

25. E. Cherry and D. Hooper, The design of wide-band transistor feedback amplifiers, *Proc. IEEE*, Vol. 110, No. 2, pp. 375–389, 1963.

26. N. M. S. Costa and A. V. T. Cartaxo, Optical DQPSK system performance evaluation using equivalent differential phase in presence of receiver imperfections, *IEEE J. Lightwave Technol.*, Vol. 28, No. 12, pp. 1735–1744, 2010.

27. H. Tran, F. Pera, D. S. McPherson, D. Viorel, and S. P. Voinigescu, 6-kΩ, 43-Gb/s differential transimpedance-limiting amplifier with auto-zero feedback and high dynamic range, *IEEE J. Solid State Circuits*, Vol. 39, No. 10, pp. 1680–1689, 2004.

28. Inphi Inc., Technical information on 3205 and 2850, TIA device, 2012.

29. I. Hodgkinson, Receiver analysis for optical fiber communications systems, *IEEE J. Lightwave Technol.*, Vol. 5, No. 4, pp. 573–587, 1987.

30. B. Mao, N. Stojanovic, C. Xie, L. N. Binh, and M. Chen, Investigation on the ENOB and clipping effect of real ADC and AGC in coherent QAM transmission system, in *Proceedings of the ECOC 2011*, Geneva, 2011.

31. Maxim Inc., Maxim Integrated, San Jose, CA, USA, Glossary of Frequently Used High-Speed Data Converter Terms, Application Note, Maxim, p. 740, December 17, 2001.

32. N. Stojanovic, An algorithm for AGC optimization in MLSE dispersion compensation optical receivers, *IEEE Trans. Circuits Syst.*, Vol. I, No. 55, pp. 2841–2847, 2008.

6 Differential Phase Shift Keying Photonic Systems

This chapter investigates differential phase shift keying (DPSK) modulation for noncoherent transmission and detection optical communications systems. The differential mode of detection is most appropriate for such systems in which the phases contained in the two consecutive bit periods are assigned discrete values, for example, 0, π or 0, $\pi/2$, π and $3\pi/2$. An experimental demonstration of DPSK modulation formats is also described.

6.1 INTRODUCTION

Owing to the tremendous growing demand for high-capacity transmission over the Internet, a high data rate of 40 Gbps per channel is an attractive feature of the next generation of lightwave communications system. Under the current 10 Gbps DWDM optical system, overlaying 40 Gbps on the existing network can be considered to be the most cost-effective method for upgrading purposes. However, there are a number of technical difficulties faced by communications engineers involving interoperability that requires a 40 Gbps line system to have signal optical bandwidth, tolerance to chromatic dispersion, resistance to nonlinear crosstalk, and susceptibility to accumulated noise over multispan of optical amplifier to be similar to a 10 Gbps system.

In view of this, advanced modulation formats have been demonstrated as an effective scheme to overcome 40 Gbps system impairments. The DPSK modulation format has attracted extensive studies owing to its benefit over conventional on–off keying (OOK) signaling format, including a 3 dB lower optical signal-to-noise ratio (OSNR) [1] at a given bit error rate (BER), more robust to narrow band optical filtering, and even more robust against some nonlinear effects such as cross-phase modulation and self-phase modulation. Moreover, spectral efficiencies can be improved by using multilevel signaling. On top of that, coherent detection is not critical as DPSK detection requires comparison of two consecutive pulses, and hence the source coherence is required only over one-bit period.

Nevertheless, the DPSK format involves rapid phase change, which causes intensity ripples due to chromatic dispersion that induce a pattern-dependent SPM-GVD (group velocity dispersion) effect [2]. Therefore, a return-to-zero (RZ) pulse can be employed in conjunction with DPSK to generate more tolerance to the data-pattern-dependent SPM-GVD effect. In addition, RZ improves dispersion tolerance and nonlinear effects, particularly in long-haul networks at high data rates. Specific RZ formats such as carrier-suppressed RZ (CSRZ) help to reduce the inherent wide spectral bandwidth.

At 40 Gbps, generation of RZ pulses is not feasible as it is at 10 Gbps owing to the large bandwidth requirement. Thus, 40 Gbps RZ signals are produced optically in a "pulse carver" by driving the modulator with a 20 GHz RF signal. With the remarkable advances in external modulator, especially the Mach–Zehnder interferometric (intensity) modulator (MZIM), this is easily achieved by utilizing the microwave optical transfer characteristic of the MZIM.

Despite the telecom boom and subsequent bust, Internet traffic has been steadily growing. This growth requires new investment in telecommunications infrastructure to provide long-haul communications capacity between major population centers in the world. The transmission route between Melbourne to Sydney and vice versa, for example, is one of the most intensive, demanding upgradation to higher capacity, especially from the current 10–40 Gbps. One of the most cost-effective ways to provide such upgrades is to use the existing fiber infrastructure, and upgrade the transmitters and receivers at either end of the long-haul-links. That implies a transmission rate of 40 Gbps over the existing 10G RZ format dense multiwavelength optical communications systems.

Because of the properties of the installed fiber (which is older and also more degraded than the state-of-the-art fibers used in laboratory "hero" experiments), the transmission methods must be highly tolerant to chromatic dispersion (CD) and polarization-mode dispersion (PMD). This favors the use of advanced modulation formats (such as variants of phase shift keying) rather than ultra-high-rate time division-multiplexed (TDM) schemes, because the effects of CD and second-order PMD are proportional to the square of the bit rate. However, there are various optical filters installed throughout DWM optical transmission systems such as optical multiplexers (muxes), demultiplexers (demuxes), and add-drop muxes that would affect the spectral properties of the multiplexed channels, particularly for a hybrid transmission of 10 Gbps and 40 Gbps channels.

Recently, advanced modulation formats are considered to play a significant role in enhancing the effectiveness of bandwidth reduction and effective transmission over long distances [1,2]. These advanced modulation formats are new to optical communications but are very well known in wire and wireless communications systems. Accordingly, the modulation can manipulate the amplitude, the frequency, or the phase of the carriers corresponding to the coding of the input data sequence. Either coherent or incoherent transmission and detection techniques have been employed over the years for optical systems. However, incoherent detection is preferred so as to minimize the linewidth obstacles of the lasers at the transmitter and the local oscillators required for homodyne or heterodyne coherent detection systems. Furthermore, the phase comparison in differential detection can be easily implemented using delay interferometric photonic components. If such phase comparison is implemented in the electrical domain operating at high bit rate, it is quite difficult. Therefore, differential, discrete, or continuous phase modulation formats can be the preferred techniques.

This chapter investigates the transmission of 40 Gbps channels over 10 Gbps DWDM optical system in which "standard" single optical filters (SSMFs) normally employed for 10 Gbps systems are used. DPSK is transmitted and compared with those employing RZ- and NRZ-amplitude shift keying (ASK). We demonstrate that with the passband of optical filters on the order of 0.5 nm, the transmitted 40 Gbps channels are not penalized and vice versa for 10 Gbps DWDM channels. Transmission BER and receiver sensitivities are reported for different transmission scenarios.

Owing to the high demand for data transmission capacity, with the advent of wavelength division multiplexing (WDM) technology, data transmission has reached a revolutionary stage for increasing higher system capacity. Higher bit rates up to 40 Gbps and growing numbers of WDM channels are increasingly challenging the limit of the ASK format.

6.2 OPTICAL DPSK MODULATION AND FORMATS

Figure 6.1 shows the structure of a 40 Gbps DPSK transmitter in which two external LiNbO$_3$ optical interferometric modulators (MZIMs) are used [3].

The operational principles of MZIMs were presented in Chapter 2. The MZIM used in Figure 6.1 can be either a single- or a dual-drive type. As previously described, the optical DPSK transmitter consists of (i) a narrow-linewidth laser source to generate a lightwave of a wavelength that conforms to the ITU grid, (ii) two cascade MZIMs are employed to modulate the lightwaves, one to generate a periodic (clock-like) return zero or CSRZ formats and the differential phase of the lightwave and one data modulator switching on and off the optical clock sequence. This shift can be continuous between the states as in the case of continuous phase FSK or minimum shift keying modulation formats, which will be presented in later chapters.

6.2.1 GENERATION OF RZ PULSES

The first MZIM, commonly known as the pulse carver, is used to generate the periodic pulse trains with the required RZ format. The suppression of the lightwave carrier can also be carried out at this

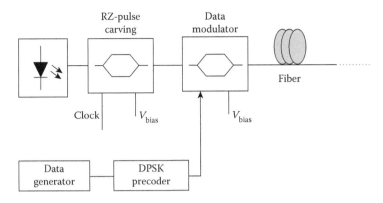

FIGURE 6.1   Schematic diagram of a discrete PSK optical transmitter using cascaded MZI modulators.

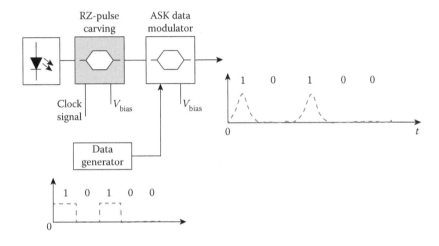

FIGURE 6.2   Cascaded MZIM for generation of RZ or CSRZ-DPSK lightwave signals.

stage if necessary and is known as carrier-suppressed RZ (CSRZ). The CSRZ pulse shape has been found to have attractive attributes in long-haul WDM transmissions compared to other RZ types, including an optical phase difference of π in adjacent bits, suppression of the optical carrier component in the optical spectrum, and smaller spectral width. Figure 6.2 shows the structure of a 40 Gbps ASK transmitter in which two external LiNbO$_3$ optical interferometric modulators (MZIMs) [4] are used. The operational principles of MZIMs were presented in Chapter 2. The MZIM used in Figure 6.2 can be either a single- or dual-drive type. The optical ASK transmitter would consist of a narrow-linewidth laser source to generate a lightwave of wavelength conforming to the ITU grid.

Different types of RZ pulses can be generated depending on the driving amplitude of the RF voltage and the biasing schemes of the MZIM. The equations governing the RZ pulses electric field waveforms are [1]

$$E(t) = \begin{cases} \dfrac{1}{\sqrt{E_b}}\sin\left[\dfrac{\pi}{2}\cos\left(\dfrac{\pi t}{T_b}\right)\right] & \text{67\% duty-ratio RZ pulses} \\[4mm] \dfrac{1}{\sqrt{E_b}}\sin\left[\dfrac{\pi}{2}\left(1+\sin\left(\dfrac{\pi t}{T_b}\right)\right)\right] & \text{33\% duty-ratio RZ pulses} \end{cases} \tag{6.1}$$

where E_b is the pulse energy per transmitted bit.

The first type of pulse, the 33% duty-ratio RZ pulse, is commonly known as the "carrier max or conventional" (RZ33) pulse, whereas the 67% duty cycle RZ pulse is the CSRZ pulse type. The difference in the generation between these two RZ pulse types is due to different biasing point placed on the transfer curve of the MZIM. The bias voltage conditions and the pulse shape of these two RZ types, the carrier suppression and nonsuppression of the maximum carrier, can be implemented with the biasing points at the minimum and maximum transmission points of the transmittance characteristics of the MZIM, respectively. The peak-to-peak amplitude of the RF driving voltage is $2V_\pi$, where V_π is the required driving voltage to achieve a π phase shift of the lightwave carrier. Another important point is that the RF signal is operating at only half of the transmission bit rate. Hence, pulse carving is actually implementing the frequency doubling. The generation of RZ33 and CSRZ pulse trains is shown in Figure 6.3a and b.

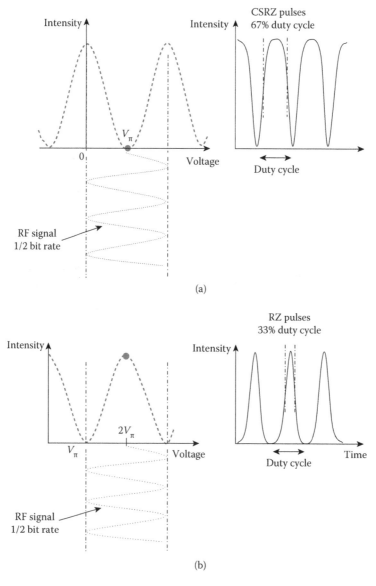

(a)

(b)

FIGURE 6.3 Biasing and driving sinusoidal electrical waveform to generate (a) CSRZ and (b) RZ33 (pulse width occupies 33% pulse period).

The pulse carver can also utilize a dual-drive MZIM, which is driven by two complementary sinusoidal RF signals. This pulse carver is biased at $-V_{\pi/2}$ and $+V_{\pi/2}$ with a peak-to-peak amplitude of $V_{\pi/2}$. Thus, a π phase shift is created between the states 1 and 0 of the pulse sequence and hence the RZ with alternating phases 0 and π. If carrier suppression is required, then the two electrodes are supplied voltages V_π with a swing voltage amplitude of V_π.

RZ optical systems are proved to be more robust against impairments, especially self-phase modulation and PMD. Although RZ modulation offers improved performance, RZ optical systems usually require more complex transmitters than those in the NRZ ones. Compared to only one stage for modulating data on the NRZ optical signals, two modulation stages are required for the generation of RZ optical pulses.

6.2.2 PHASOR REPRESENTATION

Recalling the derivation of a phase-modulated interferometer given in Chapter 2, we see that the output field of the MZIM is given by

$$E_o = \frac{E_i}{2}\left[e^{j\varphi_1(t)} + e^{j\varphi_2(t)}\right] = \frac{E_i}{2}\left[e^{j\pi v_1(t)/V_\pi} + e^{j\pi v_2(t)/V_\pi}\right] \tag{6.2}$$

It can be seen that the modulating process for the generation of RZ pulses can be represented by a phasor diagram as shown in Figure 6.4. This technique gives a clear understanding of the superposition of the fields at the coupling output of the two arms of the MZIM. Here, a dual-drive MZIM is used, that is, the data $[V_1(t)]$ and the complementary data [data: $V_2(t) = -V_1(t)$] are applied to each arm of the MZIM, and the RF voltages swing in inverse directions. Applying the phasor representation, vector addition, and simple trigonometric calculus, the process of generation of RZ33 and CSRZ is explained in detail and verified.

The width of these pulses is commonly measured at the position of full-width half maximum (FWHM). Note that the measured pulses are intensity pulses, whereas we are considering the addition of the fields in the MZIM. Thus, the normalized E_0 field vector has the value of $\pm 1/\sqrt{2}$ at the FWHM intensity pulse positions, and the time interval between these FWHM points gives the FWHM values.

6.2.3 PHASOR REPRESENTATION OF CSRZ PULSES

Key parameters, including the V_{bias}, the amplitude of driving voltage, and its corresponding initialized phasor representation, are shown in Figure 6.5a and b, respectively.

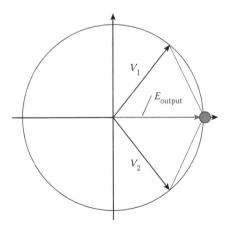

FIGURE 6.4 Phasor representation for generation of output field in dual-drive MZIM.

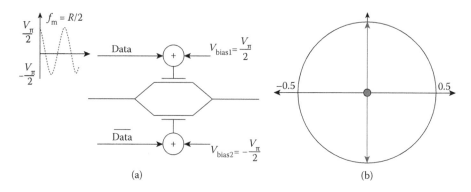

FIGURE 6.5   Biasing voltages of (a) MZIM structure and (b) phasor representation for RZ pulse generation. V_{bias} is $\pm V_\pi/2$.

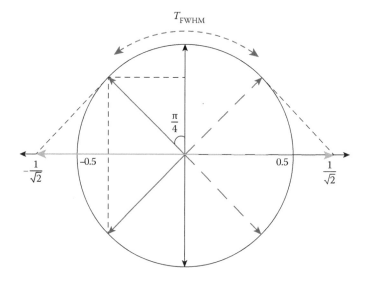

FIGURE 6.6   Phasor representation of evolution of CSRZ pulse generation using dual-drive MZIM.

The swing voltage of the driving RF signal on each arm has the amplitude of $V_\pi/2$ (i.e., $V_{\text{p-p}} = V_\pi$). The RF signal operates at half of the bit rate ($B_R/2$). At the FWHM position of the optical pulse, $E_{\text{output}} = \pm 1/\sqrt{2}$ and the component vectors V_1 and V_2 form with the vertical axis a phase of $\pi/4$ as shown in Figure 6.6.

Considering the scenario for the generation of 40 Gbps CSRZ optical signals (25 ps pulse width), the modulating frequency is $f_m = 20$ GHz. $= B_R/2$. At the FWHM positions of the optical pulse, the phase is given by the following expressions

$$\frac{\pi}{2}\sin(2\pi f_m) = \frac{\pi}{4} \Rightarrow \sin 2\pi f_m = \frac{1}{2} \Rightarrow 2\pi f_m = \left(\frac{\pi}{6}, \frac{5\pi}{6}\right) + 2n\pi \tag{6.3}$$

Thus, the calculation of the T_{FWHM} can be carried out, and hence the duty cycle of the RZ optical pulse can be obtained as given in the following expressions

$$T_{\text{FWHM}} = \left(\frac{5\pi}{6} - \frac{\pi}{6}\right)\frac{1}{B_R\,2\pi} = \frac{1}{3}\pi \times \frac{1}{B_R} \Rightarrow \frac{T_{\text{FWHM}}}{T_{\text{BIT}}} = \frac{1.66\times10^{-11}}{2.5\times10^{-11}} = 66.67\% \tag{6.4}$$

The result clearly verifies the generation of CSRZ optical pulses.

6.2.4 Phasor Representation of RZ33 Pulses

Key parameters, including V_{bias}, the amplitude of the driving voltage, and its corresponding initialized phasor representation, are shown in Figure 6.7a and b, respectively.

At the FWHM position of the optical pulse, $E_o = \pm 1/\sqrt{2}$ and the component vectors V_1 and V_2 form with the horizontal axis a phase of $\pi/4$ as shown in Figure 6.8.

Considering the scenario for generation of a 40 Gbps CSRZ optical signal (25 ps pulses), the modulating frequency $f_m = 20$ GHz $= B_R/2$. At the FWHM positions of the optical pulse, the phase is given by the following expressions

$$\frac{\pi}{2}\cos(2\pi f_m t) = \frac{\pi}{4} \Rightarrow t_1 = \frac{1}{6f_m} \tag{6.5}$$

$$\frac{\pi}{2}\cos(2\pi f_m t) = -\frac{\pi}{4} \Rightarrow t_2 = \frac{1}{3f_m} \tag{6.6}$$

Thus, the calculation of TFWHM can be carried out, and hence the duty cycle of the RZ optical pulse can be obtained as given in the following expressions

$$T_{FWHM} = \frac{1}{3f_m} - \frac{1}{6f_m} = \frac{1}{6f_m} \quad \therefore \quad \frac{T_{FWHM}}{T_b} = \frac{1/6f_m}{1/2f_m} = 33\% \tag{6.7}$$

Therefore, the result clearly verifies the generation of RZ33 optical pulses.

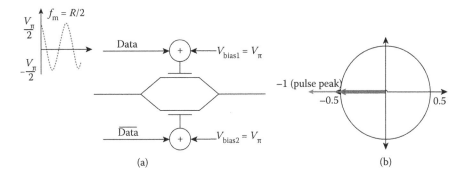

(a) (b)

FIGURE 6.7 (a) Optical scheme for generation of 33% RZ lightwave pulse sequence. (b) Phasor representation of the generation. V_{bias} is V_π for both arms, Swing voltage of driving RF signal on each arm has the amplitude of $V_\pi/2$ (i.e $V_{p-p} = V_\pi$) and RF signal operates at half of bit rate ($R/2$).

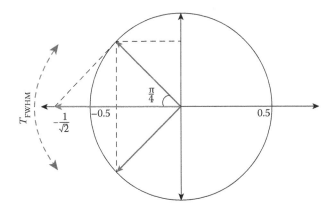

FIGURE 6.8 Phasor representation of RZ33 pulse generation using dual-drive MZIM.

6.2.5 Discrete Phase Modulation—DPSK

6.2.5.1 Principles of DPSK and Theoretical Treatment of DPSK and DQPSK Transmission

For PSK signals, the data information is contained in the discrete phase shifts relative to the phase of the unmodulated carrier. The required absolute phase at the receiver must be supplied by the recovered carrier such as the OPLL or relatively the electrical PLL, which would be complicated. Thus, it is more effective to process the phase difference of the two consecutive symbols.

Information encoded in the phase of an optical carrier is commonly referred to as optical phase shift keying. In the early days, PSK required precise alignment of the transmitter and demodulator center frequencies. Hence, the PSK system was not widely deployed. With the introduction of the DPSK scheme, coherent detection is not critical because DPSK detection only requires source coherence over one-bit period by a comparison of two consecutive pulses.

A binary 1 is encoded if the present input bit and the past encoded bit are of opposite logic, and a binary 0 is encoded if the logic is similar. This operation is equivalent to the XOR logic operation. Hence, an XOR gate is usually employed in the differential encoder. A NOR operation can also be used to replace the XOR operation in differential encoding, as shown in Figure 6.9a. The decoding logic circuit is also shown in this figure.

In optical applications, electrical data 1 is represented by a π phase change between the consecutive data bits in the optical carrier, whereas state 0 is encoded with no phase change between the consecutive data bits. Hence, this encoding scheme gives rise to two points located exactly at a π phase difference with respect to each other in the signal constellation diagram. For continuous phase shift keying such as minimum shift keying, the phase evolves continuously over a quarter of the section, and thus a phase change of $\pi/2$ between one state and another [5] as indicated by the bold circles shown in Figure 6.9b.

For differential quadrature phase shift keying (DQPSK), another quadrature component or subsystem would be superimposed on the DPSK (the in-phase) such that they are orthogonal to each other. In other words, the phase constellation of Figure 6.9b would have another imaginary ($\pi/2$) axis with the states $+\pi/2$ and $-\pi/2$.

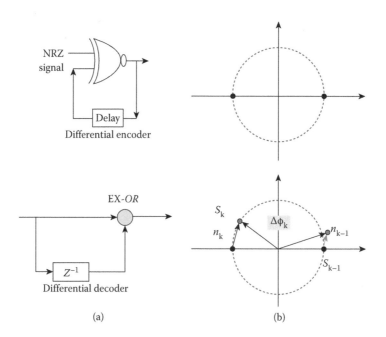

(a)   (b)

FIGURE 6.9 (a) DPSK precoder and decoder for encoding and decoding of differential data and (b) signal constellation diagram of phasor representation.

6.2.5.2 Optical DPSK Transmitter

Figure 6.10 shows the structure of a 40 Gbps DPSK transmitter in which two external LiNbO$_3$ optical interferometric modulators [3] (MZIMs) are used. The operational principles of MZIMs were presented in Chapter 2. The MZIM used in Figure 6.10 can be either a single- or dual-drive type. The optical DPSK transmitter would consist of a narrow-linewidth laser source to generate a lightwave of wavelength conforming to the ITU grid.

As shown in Figure 6.11, the optical RZ pulses are then fed into the second MZIM through which the RZ pulses are modulated by the precoded NRZ binary data to generate RZ-DPSK optical signals.

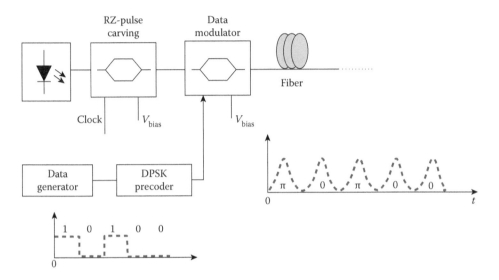

FIGURE 6.10 Generation of RZ formats DPSK modulation.

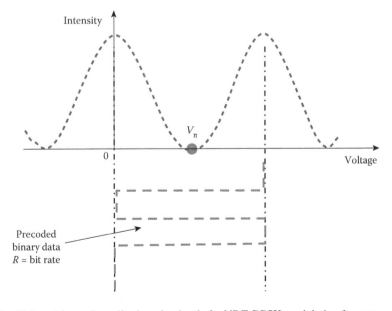

FIGURE 6.11 Voltage bias and amplitude swing levels for NRZ-DPSK modulation formats.

Electrical data pulses are differentially precoded in the precoder using the XOR coding scheme. Without the pulse carver and the sinusoidal RF signal, the system becomes an NRZ-DPSK transmitter.

In DPSK modulation, the MZIM is biased at the minimum transmission point. The precoded electrical data have a peak-to-peak amplitude equal to $2V_\pi$ and operate at the transmission bit rate. The modulation principles are demonstrated in Figure 6.11.

An electro-optic phase modulator (PM) might also be used for the generation of DPSK signals, but the Mach–Zehnder wave guide structure has several advantages over the PM described in Ref. [1]. Using an optical PM, the transmitted optical signal is chirped, whereas using an electro-optic intensity modulator (MZIM, especially X-cut versions), chirp-free signals are produced. In practice, a small amount of chirp might be useful for transmission [6].

6.2.6   DPSK-BALANCED RECEIVER

Consider two consecutive symbols whose signals can be represented by

$$s_{k-1} = A\sin(2\pi f_c t + \phi_{k-1}) \quad \text{for } (k-1)T_b \leq t \leq kT_b$$

$$s_k = A\sin(2\pi f_c t + \phi_k) \quad \text{for } kT_b \leq t \leq (k+1)T_b$$

(6.8)

where A is an arbitrary amplitude and f_c the carrier frequency. At the receiver side, the phase difference $\Delta\phi_k = \phi_k - \phi_{k-1}$ has to be determined. If there is no distortion, then the phase difference is that of the original phase difference generated at the transmitter. However, in general, the signals are distorted when reaching the receiver, and thus we have

$$x_{k-1}(t) = s_{k-1}(t) + n_{k-1}(t)$$

$$x_k(t) = s_k(t) + n_k(t)$$

(6.9)

Figure 6.9b shows the position of the data signals at the receiver side, including the additional contribution of the noise vectors of the two consecutive intervals. The phase angle between the two vectors can then be written as

$$\cos\Delta\phi_k = \frac{x_k x_{k-1}}{|x_k||x_{k-1}|} \quad \text{and} \quad \sin\Delta\phi_k = \frac{x_k x_{k-1}^*}{|x_k||x_{k-1}^*|}$$

(6.10)

$$\text{with} \quad x_k x_{k-1} = \int_0^T x_k(t)x_{k-1}(t)\,dt \quad \text{and} \quad x_k x_{k-1}^* = \int_0^T x_k(t)x_{k-1}^*(t)\,dt$$

representing the scalar products between the two received signal vectors x_k and x_{k-1}. For the nondistortion case of the binary DPSK modulation format, the phase difference would be 0 and π; thus, $\Delta\pi_k = \pi$. Thus, the received signals are in push–pull states, hence the phase difference is bipolar. Equation 6.8 can then be reinterpreted as

$$b_k = \text{sgn}(x_k x_{k-1})$$

$$\rightarrow b_k = \text{sgn}(\cos\Delta\phi_k)$$

(6.11)

The receiver for determination of the differential phase can be implemented using a balanced receiver operating in the push–pull mode. The delay balanced receiver is shown at the left side in Figure 6.12, in which the received optical field is split into two arms, one delayed in the optical

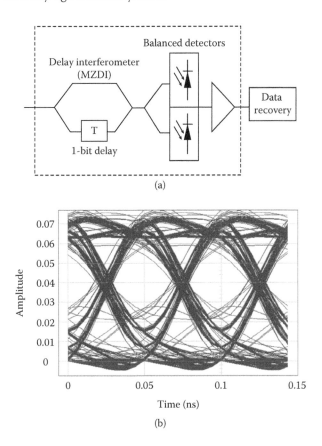

(a)

(b)

FIGURE 6.12 Schematic diagram of the balanced receiver for detection of the phase difference of two consecutive bits: (a) structure and (b) a sample of phase detected eye diagram. Bit rate = 20 Gbps format RZ.

domain by a symbol period and the other with no delay. The delay time must be matched to the propagation time of the lightwave carrier over the guided region of the optical fiber or integrated waveguide. A thermal tuning section is normally used to adjust to any mismatched time delay section. Two photodetectors are connected in a balanced configuration, producing the push–pull eye diagram (Equation 6.10) as shown in Figure 6.12b.

Figure 6.12 shows the structure of an MZ delay interferometer (MZDI)-based DPSK-balanced receiver with an insert of a typical eye diagram at the output of the MZDI-balanced receiver for 50 ps, that is, 20 Gbps DPSK optical pulses. Note that the diagram shown is of a RZ format so there is no continuous baseline. The received signal at the output of the balanced receiver is given as [7]

$$P_D(t) = |E_D(t) + E_D(t-\tau)|^2 - |E_D(t) - E_D(t-\tau)|^2$$
$$= 4\Re\{E_D(t)\} E_D^*(t-\tau) = \cos(\Delta\phi + \varsigma) \tag{6.12}$$

where $E_D(t)$ and $E_D(t-\tau)$ are the current and the one-bit delay version of the optical DPSK pulses, respectively. $\Delta\phi$ and ς represent the differential phase and the phase noise caused by the MZDI imperfections. The latter issue is not a serious degradation factor in modern MZDI-based DPSK receivers owing to the use of a thin-film heater for tuning any waveguide path mismatch. The tuning has high stability with the implementation of the electronic feedback control circuit.

6.3 DPSK TRANSMISSION EXPERIMENT

6.3.1 COMPONENTS AND OPERATIONAL CHARACTERISTICS

The components and features of the optical transmission system shown in Figure 6.13 for ASK and DPSK modulation and related formats of RZ, NRZ, and CSRZ are given in Table 6.1.

6.3.2 SPECTRA OF MODULATION FORMATS

The center wavelength is set at 1551.72 nm. The experimental spectra of NRZ, RZ, and CSRZ of the ASK and DPSK formats are shown in Figure 6.14. Note that for DPSK spectra, the carrier at the center is at the same level as that of the signal. The RZ spectra are wider than that of the NRZ as expected owing to the shorter pulse width. The influence of the spectra of the optical signals on the dispersion and nonlinear impairments will be discussed in Section 6.5.2. For CSRZ, the carrier is clearly suppressed. Owing to the resolution of the spectrum analyzer, deep suppression is not observed.

6.3.3 DISPERSION TOLERANCE OF OPTICAL DPSK FORMATS

A typical experimental test bed is set up as shown in Figure 6.15.

There are significant points for consideration in the setup of the test bed: (1) 40 Gbps Tx is the SHF 5003 DPSK transmitter, which can generate both ASK and DPSK data; (2) in the case of ASK, the Lab Buddy optical receiver (45 GHz with built-in amplifier) is used; (3) in the case of DPSK, the SHF 5008 DPSK receiver is used (MZ—one-bit delay); also, the electrical amp is

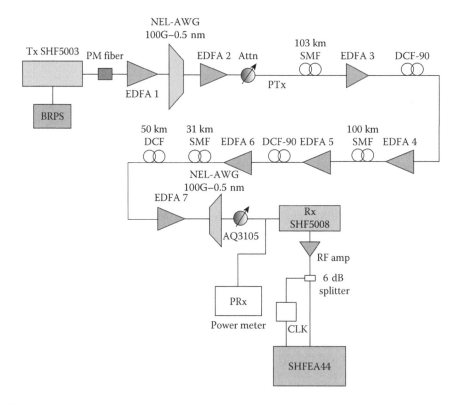

FIGURE 6.13 Experiment setup for evaluation of modulation formats of ASK and DPSK modulation, 230 km three optically amplified spans transmission, 234 km SMF and 230 km DCF.

TABLE 6.1

Photonic Components and Operating Characteristics

Photonic Subsystems	Description	Remarks
Laser source	Anritsu tunable laser set at 1551.72 nm for center wavelength ($\lambda 5$); 1548.51 nm for $\lambda 1$ and 1560 nm for $\lambda 16$.	max output power of 8 dBm
Multiplexer and demultiplexer	NEL-AWG = frequency spacing 100G 3 dB BW of 0.5 nm. Circulating property so spectrum would appear cyclic (note the spectrum). Note the channel input and output, e.g., input at port 5 of $\lambda 5$, then output at port 8; input $\lambda 1$ at port 1, then output at port 8 ($\lambda 1$). This demux is a NEL array waveguide product. The AWG has a circulant property and thus can be used as cascade filters.	
Photonic transmitters (Tx) for different modulation format generation	The Tx can be set at CSRZ-DPSK or RZ-DPSK; no DQPSK format facility is available.	A pair of external modulators of MZIM single drive. SHF model 5003
Clock recovery Module	Clock recovery using 6 dB splitter.	SHF module—SHF 1120A
Optical receiver	Balanced receiver used when DPSK format is used. Otherwise Discovery Semiconductor Lab Buddy DS-10H is employed. An MZD interferometer with thermal tuning of the delay path included.	Tuning of MZ phase decoder at the Rx is necessary when wavelength of the channels are changed
In-line optical amplifiers	EDFA 1 driven at 178 mW pump power (saturation mode). EDFA2 driven at 197 mW (saturation mode).	All EDFAs are driven in the saturation mode
Optical filters	NEL-AWG mux/demux 0.45nm bandwidth* AWG JDS 8 channel mux/demux 0.35 nm bandwidth JDS—1.2 nm bandwidth Santec—0.5 nm bandwidth with slow roll-off.	
PRBS	Bit pattern generator SHF-BPG 44.	
Error analyzer	SHF model EA-44.	
Directly derived clock	Direct synchronization of the sync signals to the error analyzer. If no directly derived clock is indicated, then a clock recovery module is used for synchronization and sampling of the received data sequence.	
Signal monitoring	Oscilloscope: Tektronix oscilloscope with 70 GHz optical plug-in and Agilent oscilloscope with 50 GHz optical plug-in spectrum analyzer—Photonetics OSA.	
Other components	Optical attenuator is used to vary the optical power input at the receiver for setting the operating condition in the linear or nonlinear region. About 5 dBm is required to set the onset level of nonlinear operation. Optical attenuator is inserted in front of the receiver to ensure no damage of the photodetector.	

* The bandwidth here is considered as the 3 dB bandwidth.

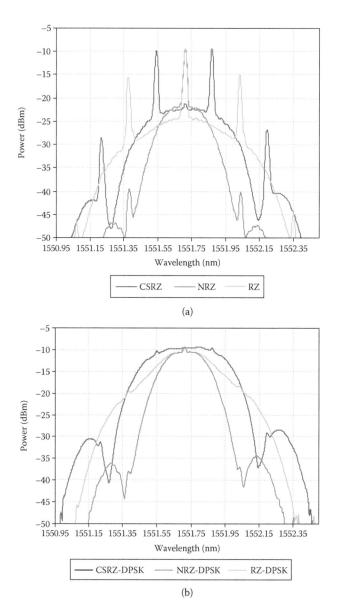

FIGURE 6.14 Optical spectra of transmitted optical signals of formats NRZ, RZ, and CSRZ for (a) ASK and (b) DPSK.

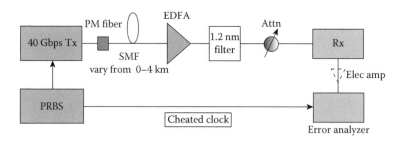

FIGURE 6.15 Experimental test bed for investigation of dispersion tolerance of transmission system.

utilized to drive the error analyzer; (4) a JDS 1.2 nm filter was utilized after EDFA to decrease the ASE noise level; and (v) power launched into the SMF (right after the transmitter Tx) is low, and hence nonlinearities do not have any impact. The launched power is recorded as shown in Table 6.2.

Using a directly derived clock, that is, directly from the bit pattern generator and hence without the clock recovery unit, the error analyzer directly uses the clock pulse of the PRBS. The laser source center wavelength is set at 1551.72 nm, and the pulse pattern of PRBS is $2^{31} - 1$. Figures 6.16 and 6.17 depict the eye diagrams at the receiver under back-to-back connection and transmission over a distance of SSMF without dispersion compensating under modulation formats DPSK and ASK using self-homodyne reception. Figure 6.18 shows the performance of CSRZ/RZ/NRZ-DPSK formats over lengths of 0–4 km (from top down to bottom) of SSMF using the eye diagrams of 40 Gbps transmission. Hence, dispersion tolerance to the transmission fiber SSMF with a dispersion factor of $D = +17$ ps/(nm · km) for the DPSK format versus the conventional ASK format can be obtained.

TABLE 6.2

Maximum Power Launched into Input of the Optical Fiber Transmission for Different Modulations ASK and DPSK with RXZ, NRZ, and CSRZ Formats

	NRZ-ASK	RZ-ASK	CSRZ-ASK	NRZ-DPSK	RZ-DPSK	CSRZ-DPSK
Launched power (dBm)	−4.2	−7.7	−6.2	−3.1	−6.2	−4.8

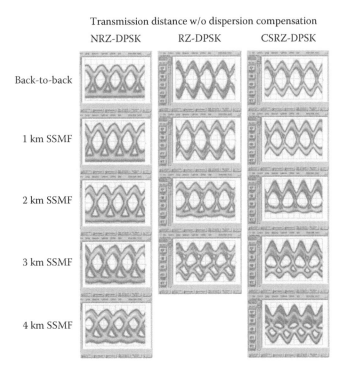

FIGURE 6.16 Eye diagrams of DPSK transmission after 0, 1, 2, 3, and 4 km SSMF for evaluation of the dispersion tolerance.

Noncompensation
transmission distance

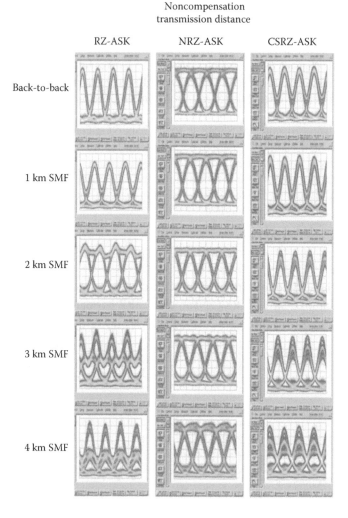

FIGURE 6.17 Eye diagrams of ASK transmission after 0, 1, 2, 3, and 4 km SSMF for evaluation of the dispersion tolerance of the transmission system.

It can be seen clearly that the DPSK system offers an improvement of approximately 3 dB in receiver sensitivity compared to the same ASK system. The obtained results confirm the theoretical analysis (Figure 6.19).

6.3.4 OPTICAL FILTERING EFFECTS

Initially, the impacts of the optical filtering characteristics of the mux are evaluated using a back-to-back setup as shown in Figure 6.20. We set the sampling clock directly from the auxiliary clock output of the $2^{31} - 1$ pattern generator. Although two AWGs can be used as muxes and demuxes, we use only one AWG at either the transmitter or receiver sites. The other filter can be substituted with a multilayer thin-film optical filter. Two optical filters acting as mux and demux at the transmitting and receiving ends are then used to evaluate their impacts on 40 Gbps channels. We observe no significant degradation of the BER as shown in Figure 6.21.

Figure 6.22 shows the BER versus the receiver sensitivity curves obtained for ASK, and DPSK and DQPSK with RZ, NRZ, or CSRZ formats are shown in Figure 6.21. The sensitivities do not

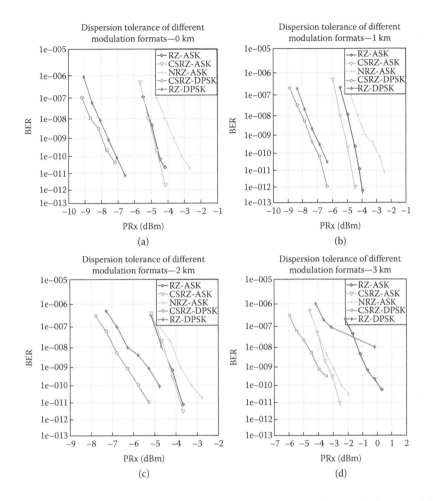

FIGURE 6.18 Performance of dispersion tolerance of advanced modulation formats in the case of (a) back to back (b) 1 km (c) 2 km; and (d) 3 km of SSMF residual length.

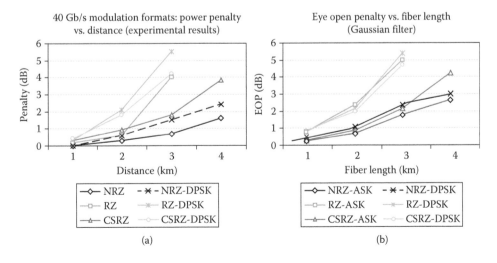

FIGURE 6.19 (a) Experimental results and (b) simulation results on power penalties induced chromatic dispersion limits of ASK and DPSK formats.

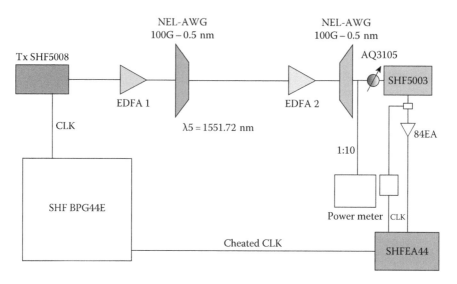

FIGURE 6.20 Back-to-back experimental system for investigation of the optical AWG filtering impacts on modulation formats.

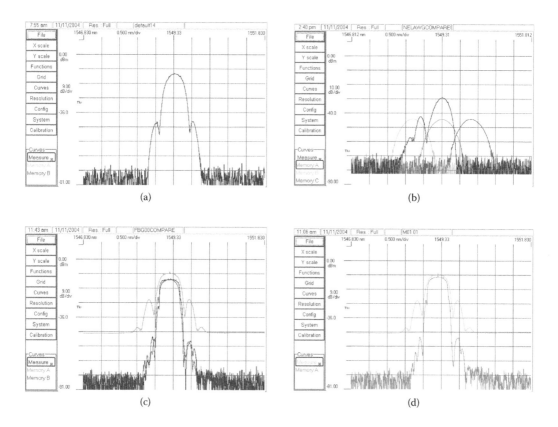

FIGURE 6.21 Optical passbands of AWG filters. (a) Top-left corner: signal spectrum of a channel. (b) Top-right corner: optical passbands of the multiplexed output of the AWG. Note the parabolic passband characteristics and the "black" curve of the output spectra of a wavelength channel. (c) Signal spectrum and its output of the AWG. (d) Same as (c) but a different wavelength region.

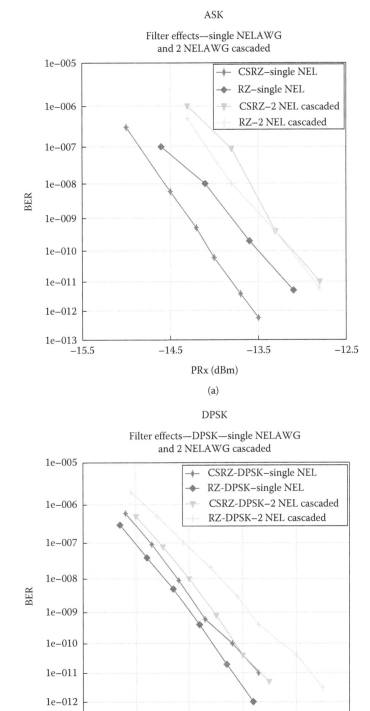

FIGURE 6.22 BER versus input power level of (a) ASK and (b) DPSK modulation with various formats.

change significantly under the influence of the 0.5 nm optical filter on 40 Gbps channels operating under different modulation formats. We can now observe the optical spectra of ASK and PSK formats as shown in Figure 6.1a and b. The Gaussian-like or $\cos^2$ profile of the pulses generated at the output of both optical modulators and the parabolic passband properties of the AWG can tolerate a wide signal spectrum. We do not observe any degradation of the BER versus the sensitivity for the cases of wideband optical filters (1.2 nm) and 0.5 nm optical mux filtering. The detected eye diagrams under RZ- and CSRZ-DPSK transmission are shown in Figure 6.16. The MZDI acting as a phase comparator is thermally tuned so as to obtain a maximum eye opening, and thus an optimum BER.

The characteristics of multiplexers/demultiplexers that are currently utilized for 10 Gbps systems still offer sufficiently good performance for 40 Gbps transmission systems. The experimental results suggest the feasibility of upgrading the currently deployed system to a higher transmission rate: 40 Gbps. A higher bit rate such as 80 or 160 Gbps needs to be further investigated.

6.3.5 Performance of CSRZ-DPSK over a Dispersion-Managed Optical Transmission Link

A dispersion-managed optical transmission link over 328 km length is set up as shown in Figure 6.13 with three optically amplified spans. The dispersion factor of the transmission fiber is matched with that of a length of dispersion compensation fiber. The bit pattern generator (BPBG) is used to drive the modulators of the transmitter. RZ and CSRZ can also be generated by a signal synthesizer. The nonlinear effect is explored by varying the optical power launched into the SMF from 0 to 15 dBm as shown in Figure 6.23. The power launched into the dispersion-compensating fiber (DCF) is kept unchanged at 0 dBm (nonlinearities are generated in SMF, not DCF).

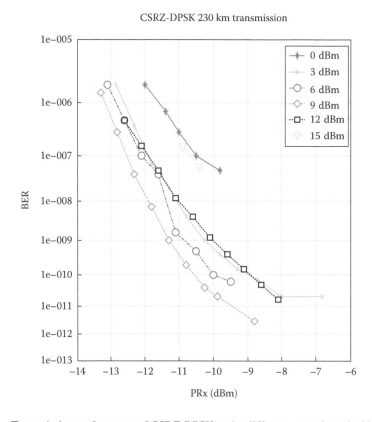

FIGURE 6.23   Transmission performance of CSRZ-DPSK under different power launched levels.

Note that mopping up of the residual dispersion using a tunable dispersion compensator is used and set at the wavelength 1555.75 nm, which can be shown to be optimum compared with the BER performance curve obtained at $\lambda_s = 1551.72$ nm. Hence, the laser source is tuned to this wavelength.

6.3.6 MUTUAL IMPACT OF ADJACENT 10G AND 40G DWDM CHANNELS

A 320 km transmission is also conducted to assess the performance of 10G NRZ-ASK and a CSRZ-DPSK 40G channel. The transmission of adjacent and nonadjacent channels is demonstrated and evaluated with a 100 GHz AWG mux and a 1.2 nm tunable filter at the input of the Rx. An insignificant power penalty is observed when the adjacent 40G channel is co-transmitted. The setup of the transmission system is shown in Figure 6.24 which consists of (1) a 320 km SSMF and a dispersion-compensating module with two optical filters and splitters, (2) the optical transmitters, (3) 100 GHz AWG as optical multiplexers (at transmitter side) and demultiplexers (at the receiving side), and then (3) the optical reception subsystems. An optical filter is inserted at the front end to measure the impact of adjacent CSRZ-DPSK 40G channels on the 10G NRZ-ASK performance. No significant impact was noted when 4a 0G channel, adjacent or nonadjacent, was co-transmitted, as evaluated in Figure 6.25.

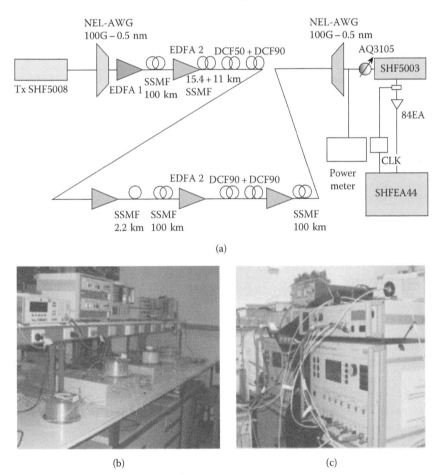

(a)

(b)   (c)

FIGURE 6.24 (a) Setup of the optically amplified long-haul optical transmission system: the span length can be varied for 50–320 km. (b) Optically amplified and dispersion-compensated fiber transmission line—transmitter and receiver plus (c) 40 Gbps bit pattern generator and error analyzer.

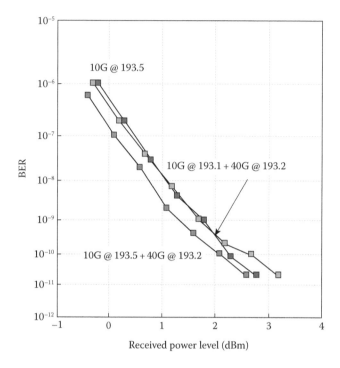

FIGURE 6.25 320 km transmission 40 impact on 10G channel: BER versus receiver sensitivity (dBm)—effects of 40G (CSRZ-DPSK) with 10G (NRZ-ASK) channel simultaneously transmitted for NRZ-ASK and CSRZ-DPSK formats—light gray dots for 1.2-nm-thin film filter and dark gray dots for 0.5-nm-thin AWG filter (demux with 100 GHz spacing).

6.4 DQPSK MODULATION FORMAT

When digitizing data for transmission across many hundreds of kilometers, many digital modulation formats have been proposed and investigated. This section investigates the DQPSK modulation format. This modulation scheme, although having been in existence for quite some time, has only been implemented in the electrical domain. Its application to optical systems proved difficult in the past as constant phase shifts in optical carriers (OCs) were required to be maintained. However, with the improvement of optical technology and alternative transmitter design setups, these difficulties have been eliminated.

6.4.1 DQPSK

One of the main attractive features of the DQPSK modulation format is that it offers both twice as much bandwidth and increased spectral efficiency compared to OOK. As an example, comparisons between 8 × 80 Gbps DQPSK systems and 8 × 40 Gbps OOK systems show that DQPSK modulation offers more superior performance, that is, spectral efficiency. The very nature of the signaling process also makes noncoherent detection at a receiver feasible, thus reducing the overall cost of the system design.

As the name suggests, the idea behind the DQPSK modulation format is to apply a differential form of phase shift modulation to the optical carrier which encodes the data. DQPSK is an extension to the simpler DPSK format. Rather than having two possible symbol phase states (0 or π phase shift) between adjacent symbols, DQPSK employs a four-symbol equivalent {0, $\pi/2$, π, or $3\pi/2$}. Depending on the desired dibit combination to be encoded, the difference in phase between the two adjacent symbols (optical carrier pulses) is varied systematically. Table 6.3 outlines this behavior.

The parameters φ_1 and φ_2 as indicated in Table 6.3 are the phases of adjacent symbols. The table can also be represented in the form commonly known as a *constellation diagram* (see Figure 6.26), which graphically explains the signal's state in both amplitude and phase.

The DQPSK transmitter design is based on a design that was previously experimentally proposed [8]. First, we implement an RZ pulse carving MZM that generates the desired RZ pulse shape as described in Chapter 2. Next, an MZIM (generating 0 or π phase shift of the OC) is to be coupled along with a PM that induces a 0 or $\pi/2$ phase shift of the OC. When placed in this configuration, the four phase states of the OC required by the DQPSK modulation format, {0, $\pi/2$, π, or $3\pi/2$}, can be achieved. Both the MZM and PM are to be driven by random binary generators operating at 40 Gbps. Figure 6.27 is a schematic of the system transmitters implementing RZ-DQPSK modulation. We originally started with the NRZ-DQPSK design; however, the RZ format is used more readily in practice and has proved to be more robust to system nonlinearities. The pulse sequence generated after the pulse carver is shown in Figure 6.28.

For DQPSK, the signal constellation can be considered to be two DPSKs orthogonal to each other. That is, there is a phase shift of $\pi/2$ between the constellation points. Thus, the transmitter and receiver can be implemented with the following phase difference

$$\Delta\phi_k = \left\{0, \frac{\pi}{2}, \pi, -\frac{\pi}{2}\right\} \quad \text{or} \quad \Delta\phi_k = \left\{\frac{\pi}{4}, \frac{3\pi}{4}, -\frac{\pi}{4}, -\frac{3\pi}{4}\right\} \tag{6.13}$$

TABLE 6.3
DQPSK Modulation Phase Shifts

Dibit	Phase Difference $\Delta\phi = \phi_2 - \phi_1$ (degrees)
00	0
01	90
10	180
11	270

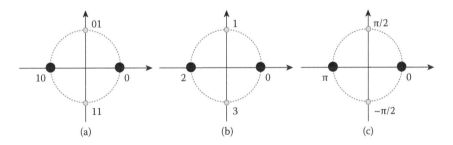

(a) (b) (c)

FIGURE 6.26 DQPSK signal constellation and assignment of phase symbols: (a) quarternary digits, (b) two-digit binary word, and (c) phase complex plane.

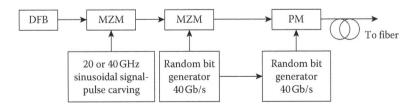

FIGURE 6.27 Schematic of a channel implementing RZ-DQPSK modulation.

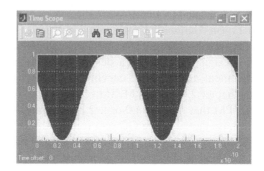

FIGURE 6.28 The optical carrier after RZ pulse carving using an MZM driven by a 10 GHz electrical signal.

In the latter case, the receiver simply outputs the binary bits of the imaginary and real parts of the DQPSK signals.

$$b_{k\_I} = \cos\Delta\phi_k = \frac{x_k x_{k-1}}{|x_k||x_{k-1}|} \quad \text{and} \quad b_{k\_Q} = \sin\Delta\phi_k = \frac{x_k x_{k-1}^*}{|x_k||x_{k-1}^*|}$$

$$\text{with } x_k x_{k-1} = \int_0^T x_k(t) x_{k-1}(t)\,dt \quad \text{and} \quad x_k x_{k-1}^* = \int_0^T x_k(t) x_{k-1}^*(t)\,dt$$

(6.14)

This can be explained as follows. The data information transmitted during the time interval $[(k-1)T_s, kT_s]$ with T_s as the symbol period is carried by the in-phase and quadrature components of the signal and thus given by the sign of the terms *sin* and *cos* as given in Equation 6.13. An offset of the signal constellation by $\pi/4$ would assist the simplification of the transmitter and receiver due to the fact that

$$\Delta\phi_k = \left\{0, \frac{\pi}{2}, \pi, -\frac{\pi}{2}\right\} \rightarrow \sin\left\{0, \frac{\pi}{2}, \pi, -\frac{\pi}{2}\right\}. \rightarrow \{0, \pm1\}$$

(6.15)

Or effectively, there would be three output levels while the $\pi/4$ shift would give binary levels. For the noiseless case, the correlation between the two bits is given as

$$s(t)s(t-T_b) = A\sin\left(2\pi f_c t + \phi_k\right) \cdot A\sin\left(2\pi f_c t - 2\pi f_c T_b + \phi_{k-1}\right)$$

$$= \frac{A^2}{2}\left\{\cos\left(\phi_k - \phi_{k-1} + \alpha\right) - \cos\left(4\pi f_c + \phi_k + \phi_{k-1} - \alpha\right)\right\}$$

(6.16)

detected signals   filtered by low-pass filter

The phase difference $\Delta\phi_k = \phi_k - \phi_{k-1} \in \{0, \pi\}$ leads to $\cos(\Delta\phi_k + K\pi) \in \{1, -1\}$, ensuring the maximum distance between the two difference symbols.

In the rest of this section, some of the key principles allowing for the successful decoding of the DQPSK-modulated signal are described. We describe an exact receiver model before fiber propagation to allow comparisons between pre- and postfiber effects of the received eye diagram. The comparison of the phase of the lightwave carrier contained within the two consecutive bits is the optical delay interferometer shown in Figure 6.29. Note that due to the differential nature of the modulation process, the demodulation and detection stage can be considered as a noncoherent or direct detection; or effectively,

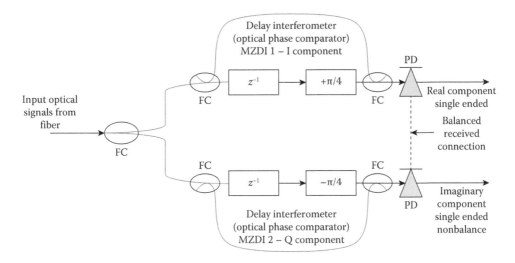

FIGURE 6.29 DQPSK receiver configuration using noncoherent direct detection. FC = fiber coupler, MZDI = Mach–Zehnder delay interferometer.

this can be known as a delay autocorrelation or a self-heterodyne detection scheme. The absence of a local oscillator (LO) and other extra hardware that is required in conventional detectors makes this demodulation technique attractive. The receiver configuration shown in Figure 6.28 is capable of demodulating the signal transmitted along the 96 km total fiber span. Because the modulation format used in this section typically encodes two bits of data per symbol, it is necessary to extract both bits (termed *real* and *imaginary* bits [8]) from the one received symbol.

We now outline the purpose of the $\pm\pi/4$ additional phase shift of the optical carrier implemented in the MZDI of the receiver. Because the two bits to be encoded at the Tx are implemented using an MZIM (0 or π phase shift) and the second bit is encoded via the PM (0 or $\pi/2$ phase shift), the two devices are said to be in *quadrature* to one another. This implies that there is a $\pi/2$ phase difference between all signaling phase states (Figure 6.26). The additional $+\pi/4$ and $-\pi/4$ give a total $\pi/2$ phase difference between the upper and lower receiver branches shown in Figure 6.22. Thus, recovery can be implemented by comparing the real and imaginary bits received to those transmitted. However, one can consider only the "real component" of the received signal and assess the overall performance of the system via eye diagram analysis (using the Q factor method and the BER). In practice, the demodulation of the DQPSK signal using the two-branch configuration in Figure 6.29 can be successfully demonstrated [9]. A balanced detection structure has been proven to be more sensitive (3 dB better) than the case of only a single photodetector. The eye diagram of the balanced detection of the transmission over noncompensated SSMF distance of 10 km and 5 km, are shown in Figure 6.30.

6.4.2 Offset DQPSK Modulation Format

In this section, we introduce offset DQPSK (O-DQPSK) transmission. ODQPSK has been applied in transmission over nonlinear satellite channels that offer a smoother transition between the phase states and hence avoids the π phase jumps as shown in Figure 6.31. In contrast to DQPSK, which requires detection at the symbol rate, the demodulation of ODQPSK can, however, be achieved at the bit rate, thus allowing the detection of the ODQPSK with only one set of balanced receivers and one set of MZDIs. Optical ODQPSK can be generated by two binary RZ-DPSK signals operated at half the bit rate that can be merged optically by a 3 dB coupler with a one-bit delay in one path as shown in Figure 6.32.

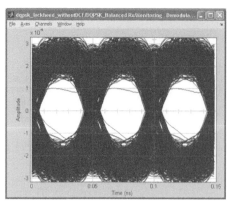

Eye diagram of balanced detection of
I component message (10 km SSMF mismatch)

(a)

Eye diagram of balanced detection of
Q component message (10 km SSMF mismatch)

(b)

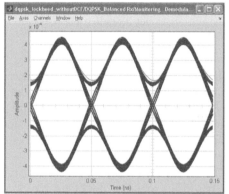

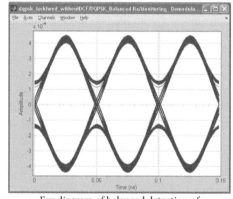

Eye diagram of balanced detection of
I component message (5 km SSMF mismatch)

(c)

Eye diagram of balanced detection of
Q component message (5 km SSMF mismatch)

(d)

FIGURE 6.30 Eye diagrams achieved for I component: (a) 10 km SSMF, (b) 10 km SSMF, (c) 5 km SSMF; and for Q components (d) 5 km SSMF mismatch of dispersion over two 200 km dispersion compensated and optically amplified spans—RZ-DQPSK 50% RZ pulse shape.

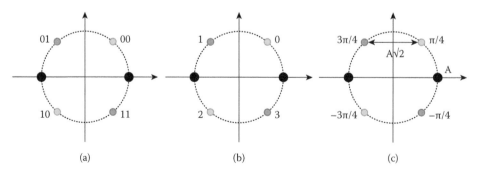

FIGURE 6.31 ODQPSK signal constellation and assignment of phase symbols: (a) quaternary digits, (b) two-digit binary word, and (c) phase complex plane with two state (horizontal line) circles are for DPSK constellation points.

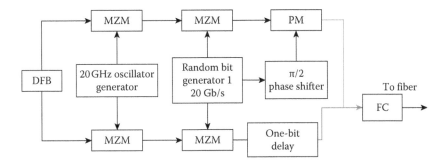

FIGURE 6.32 Schematic diagram of the optical offset DQPSK transmitter.

Similar to the case of DQPSK, a carrier phase difference of $\pi/2$ has to be guaranteed between two parallel signals to ensure the $\pi/2$ phase shift between the states of the constellation points similar to the phase plane shown in Figure 6.32. The receiver for the DQPSK modulation format can be implemented using only one MZDI and a balanced receiver, but only one MZDI is required and a $\pi/2$ phase shift must be inserted into one path of the MZDI to ensure the demodulation of the $\pi/2$ phase difference between consecutive bits, and hence the in-phase and quadrature components. If there is a cosine continuity of the phase shift between the constellation points, then the scheme ODQPSK becomes an MSK scheme, which was described in Chapter 5.

The receiver sensitivities of DQPSK, ODPSK, and DPSK have also been measured for a transmission over a 400 km span with optical amplifiers and DCF as shown in Figure 6.32. The bit rate for DPSK is 40 Gbps and 20 Gbps for DQPSK, and the input power to the fiber is set at 0 dBm to ensure that the systems operate under the linear regime. There are no differences between DQPSK and ODQPSK. A 6 dB difference between DQPSK and DPSK receiver sensitivities using a balanced receiver and a 3.5 dB difference between a single detector receiver and a balanced receiver for ODQPSK systems that is expected as the distance between the 1 and 0 is double that of a single detector as shown by the detected eye diagrams shown in Figure 6.16.

Two principal differences between the receiver sensitivities of RZ-DPSK and RZ-DQPSK are as follows: (1) Using the same total average power (i.e., the same radius on the phasor diagram; Figure 6.32), the binary level to the quaternary level would require an increase by a factor of $\sqrt{2}$ as we can observe from the signal constellations of DPSK and DQPSK. Thus, a 3 dB increase in the power is required for DPSK as compared with DQPSK. (2) For MZDI self-heterodyne detection, the detection seems to add an additional power penalty in the splitting and combining of the received optical fields. This is the complexity of the self-homodyne detection as compared to the coherent detection, in which the polarization can be diversified, while in the MZDI integrated lightwave circuitry, the polarization of the input lightwave coupled from the fiber would be reduced owing to the strong polarization dependence of the rib waveguide of the MZDI.

These original mechanisms of the reduction of the receiver sensitivity can be further explained as follows.

6.4.2.1 Influence of the Minimum Symbol Distance on Receiver Sensitivity

Assuming the noise pdf distribution at the receiver is Gaussian, coherent detection with matched filter would follow the well-known rule of the bit error probability, which depends on the Euclidean distance between the symbols and the variances at the levels of the symbols. In general, the noise power would have a mean of zero and a variance of σ^2. With a minimum distance d between the symbols, the bit error probability P_e is given by

$$P_e = \text{erfc}\left(\frac{d/2}{\sigma}\right) \tag{6.17}$$

If the total average power of the signals is A^2, then a binary DPSK has a P_e of

$$P_e = \text{erfc}\left(\frac{A}{\sigma}\right) \qquad (6.18)$$

while the DQPSK would have a P_e of

$$P_e = \text{erfc}\left(\frac{A/\sqrt{2}}{\sigma}\right) \qquad (6.19)$$

The DQPSK allows the receiving of four bits in a symbol period while only two for the binary DPSK, thus the capacity of the DQPSK is double of that of the DPSK without increasing the noise contribution at the receiver output.

6.4.2.2   Influence of Self-Homodyne Detection on Receiver Sensitivity

The complexity of coherent detection with the use of a local laser and the synchronization of the phase of both the signals and the LO has been described in Chapter 3. With the advances in integrated lightwave technology, the MZDI can be implemented without difficulty and self-heterodyne detection can be assisted with the phase comparison; thus, self-heterodyne detection has attracted attention in current DPSK and DQPSK receivers or any receiver that would require the detection of the I and Q components using a balanced receiver. Under a matched filter condition of binary DPSK, the BER is approximately given by

$$P_e = \frac{1}{2}\exp\left(-\frac{A^2}{\sigma^2}\right) \qquad (6.20)$$

For QPSK signals using an MZDI-balanced receiver for self-homodyne detection, the BER is given by Ref. [10]

$$P_e = Q(\vartheta,\mu) - \frac{1}{2}\exp\left(-\frac{\vartheta^2}{\sigma^2}\right)I_0\left(\frac{\vartheta^2\sqrt{2}}{\sigma^2}\right)$$

$$\text{with}..\mu = A'\sqrt{2}\cos\frac{\pi}{8}.....\vartheta = A'\sqrt{2}\sin\frac{\pi}{8} \qquad (6.21)$$

where Q represents the Marcum function and I_0 the modified Bessel function of the zeroth order, and A' is the amplitude of the signal at the input of the PD. The BER versus the SNR can be estimated from Equations 6.18 through 6.21 which can also be modified for the case of coherent detection using balanced detectors [11]. However, a 2 dB improvement for DQPSK for the case of coherent over self-heterodyne detection is obtained. These results were obtained without considering the noise contribution in both the positive and negative electrical signal levels.

6.4.3   MATLAB® AND SIMULINK® MODEL

6.4.3.1   Simulink® Model

MATLAB® and Simulink® models of the transmitter and receivers for the DQPSK modulation format are shown in Figures 6.33 and 6.34, respectively, with a $\pi/4$ phase offset of the signal constellation. A RZ-DQPSK optical transmission system is modeled with a pulse carver placed in

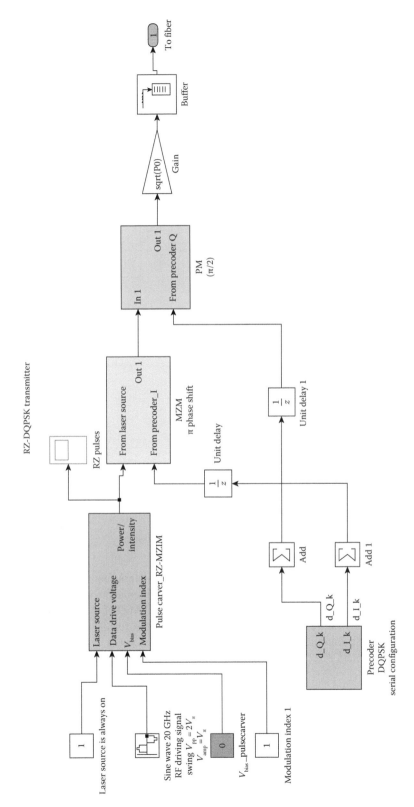

FIGURE 6.33 Transmitter MATLAB® and Simulink® model for DQPSK optical transmission.

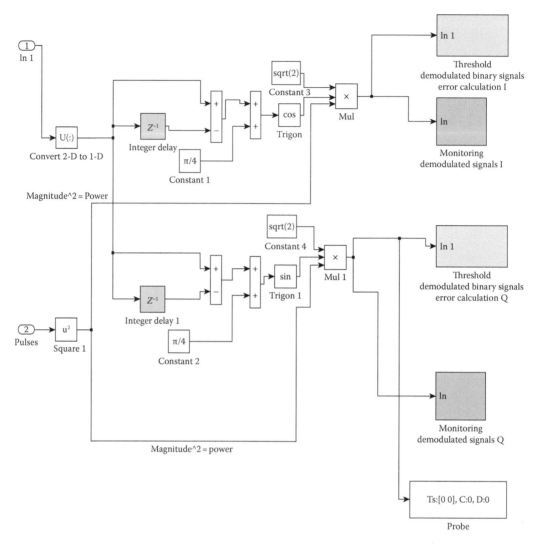

FIGURE 6.34 Receiver Simulink® model for DQPSK optical transmission.

cascade with the data optical modulator MZIM. A DQPSK precoder of two-bit per symbol provides the two output signals, the in-phase and the quadrature-phase components in the electrical domain which are amplified and used to drive the two arms of the optical MZIM. A $\pi/2$ PM is also used to assign the in-phase or quadrature-phase components of the DQPSK format.

In the receiver side, differential detection or self-homodyne receiving structures are used with a phase offset of $\pi/4$ to obtain directly the amplitude and phase of the receiving signals as discussed earlier.

6.4.3.2 Eye Diagrams

The transmission of the DQPSK with a bit rate of 40 Gbps or 20 Giga symbol/s conducted over SSMF lengths of 10 and 5 km are shown in Figure 6.35.

The length of SSMF is set to 10 km, and the length of DCF is set to be 2 km with the dispersion factor of the DCF set to be five times more negative than that of SSMF (full compensation) as shown in Figure 6.36.

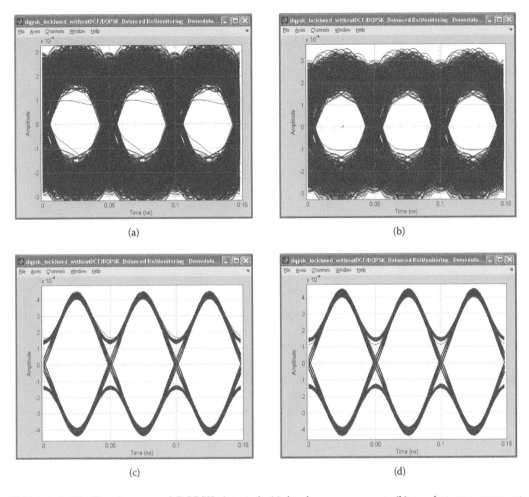

FIGURE 6.35 Eye diagram of DQPSK detected: (a) in-phase component, (b) quadrature component 10 km SSMF, and eye pattern of (c) the in-phase and (d) the quadrature components after 5 km SSMF transmission.

FIGURE 6.36 Arrangement of dispersion-compensated spans for transmission of DQPSK signals.

The Simulink models shown in Figures 6.37 and 6.38 illustrate how to set in the DCF block in the model. The length of DCF is set at 2 km and the dispersion factor to be five times more negative than that of SSMF for full dispersion compensation.

The measured eye diagrams of DQPSK are shown in Figure 6.39 at a total bit rate of 108 and 110 Gbps (or around 50 Gsymbol/s) when the I and Q components are detected separately and when both channels are turned on.

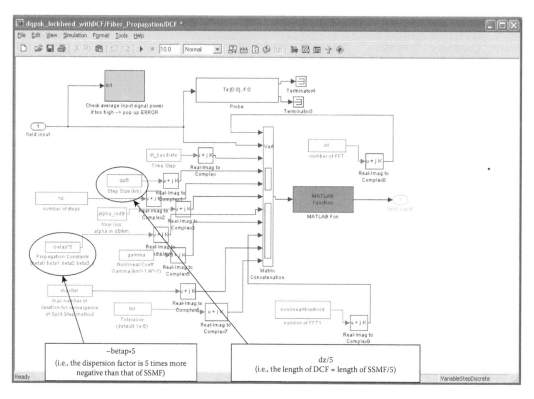

FIGURE 6.37 Simulink® model of the fiber sections of dispersion-compensating fiber (DCF) for fully compensated span.

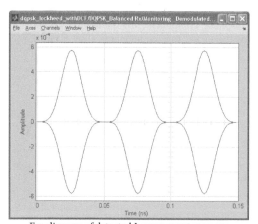

Eye diagram of detected I-component message
(10 km SSMF and 2 km DCF with DCF
dispersion factor about −85 ps/[nm · km])

(a)

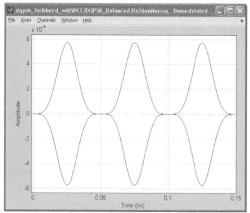

Eye diagram of detected Q component message
(10 km SSMF and 2 km DCF with DCF
dispersion factor about −85 ps/[nm · km])

(b)

FIGURE 6.38 Simulated eye diagram of DQPSK detected: (a) in-phase component and (b) quadrature component, of a fully dispersion-compensated SSMF and DCF.

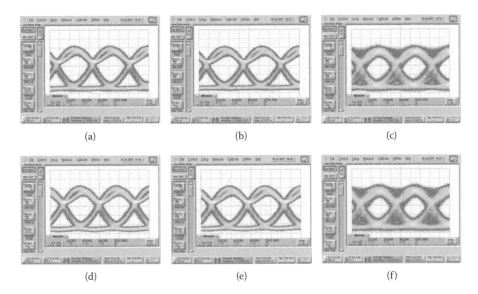

(a)   (b)   (c)

(d)   (e)   (f)

FIGURE 6.39 Measured eye diagrams of 108 Gbps RZ-DQPSK detected (50 Gsymbols/s): (a) in-phase component, (b) quadrature component, (c) both in phase and quadrature channels, (d) in-phase component, (e) quadrature component, and (f) both channels transmitted and one component is selected by tuning the MZDI. (Courtesy of SHF AG Berlin Germany).

6.5   COMPARISONS OF DIFFERENT FORMATS AND ASK AND DPSK

6.5.1   BER AND RECEIVER SENSITIVITY

6.5.1.1   RZ-ASK and NRZ-ASK

It is obvious that if the pulse sequences of the NRZ and RZ (50% duty) ASK whose average powers are the same then because the FWHM of the RZ is half of that of NRZ, the peak amplitude of the RZ sequence would be higher than that of the NRZ. The higher the peak power of the RZ pulses, the greater the distance between the levels 0 and 1 and hence a possibility of an enhancement of the eye opening and thus BER.

Assuming now an equal average power of the two random NRZ and RZ data sequences, theoretically speaking, a higher peak power might not lead to better receiver sensitivity. However, according to the matched filtering concept [12,13], for any signal corrupted by additive white Gaussian noises in transmission systems with impulse responses of transmitter and receiver $h_T(t)$ and $h_R(t)$, the maximum SNR at the output of the matched filter is achieved and $h_R(t) = h_T(t-T)$ with T being the sampling delay time. Or alternatively, in the frequency domain, we have $H_R(f)$. $H_T(f)$ $exp(-j2\pi fT) = 1$.

In the optical domain, such matched filters can be implemented using chirp fiber Bragg grating (FBG), thin-film filter, or microfilters for RZ pulses. Experimentally, it has been observed, as described earlier, that an improvement of about 2 dB can be achieved for the receiver sensitivity for RZ ASK as compared with NRZ-ASK, whereas the theoretical value is 3 dB.

6.5.1.2   RZ-DPSK and NRZ-DQPSK

Figure 6.40 shows the difference between the receiver sensitivity of the modulation formats RZ-DPSK and RZ-DQPSK under the detection structures of single and balanced detectors. A 3 dB improvement of the balanced receiver over that of the single detector is as expected from the push–pull mechanism of the balanced detection. This is known to be the maximum performance achievable with a balanced receiver with narrow band optical filtering at the input of the MZDI [14].

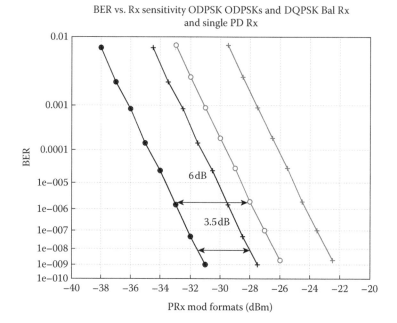

FIGURE 6.40 Optical DPSK and ODPSK transmission over 4 × 100 km SSMF of DWDM 8 × 40 Gbps transmission. Simulated results of BER versus receiver sensitivity of modulation formats: (o) Offset DQPSK–Balanced Rx, (left +) Offset DQPSK–single detector, (far right *) DQPSK–Balanced Rx, (far right +) DQPSK–single PD. * – offset DQPSK.

For RZ-DPSK and RZ-DQPSK under-balanced detection, there is a difference of 6–7 dB between these schemes. This is due to two possibilities: (1) Using the same averaged signal power, a $\sqrt{2}$ factor is reduced between the distance of the 1 and 0 for DQPSK as is obvious from the signal constellations. (2) Using an MZDI at the input of the balanced detector pair, the optical power is split between the two MZDIs, so there is no gain in power, and even a loss occurred sometimes due to unused optical ports and scattering loss in the power split coupler.

6.5.1.3 RZ-ASK and NRZ-DQPSK

The comparison among ASK, DPSK, and DQPSK is important when differential phase with balanced receiving is used in the upgradation of 10 to 40 G rates as mentioned in Section 6.3.6. Theoretically and experimentally demonstrated as shown in Figure 6.22, the modulation format RZ-DPSK is expected to offer a 3 dB enhancement over the RZ-ASK and about 6 dB improvement as compared with NRZ-ASK. Thus, the performance of RZ-DQPSK would nearly be the same as RZ-ASK but with a symbol rate twice of that of NRZ or RZ-ASK. The Simulink and MATLAB model for RZ-DQPSK is given in the appendix at the end of this chapter.

This can be explained as follows: (1) The distance between the 1 and 0 of the ASK is nearly the same as that for DQPSK signals of the same average power as seen from the signal constellation shown in Figure 6.41; (2) the RZ format outperforms the NRZ format by 3 dB (maximum), and thus, RZ-DQPSK is expected to improve over NRZ-ASK by the same amount or less; and (3) the use of MZDI as an optical phase comparator at the input of the balanced detector pair would introduce an optical power loss owing to the splitting and unused ports, which can be improved using planar lightwave circuit (PLC) technology.

In summary, the use of multilevel modulation formats may result in optical loss and splitting when the MZDI-balanced receiving technique is used. These multilevel schemes would offer an effective high information capacity with the symbol rates much lower than that of the binary

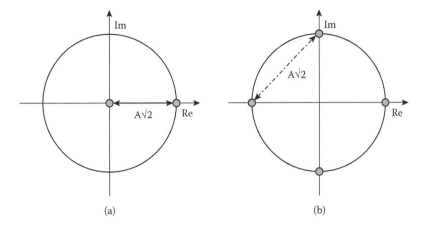

FIGURE 6.41   Signal constellation of (a) ASK and (b) DQPSK for the same average power.

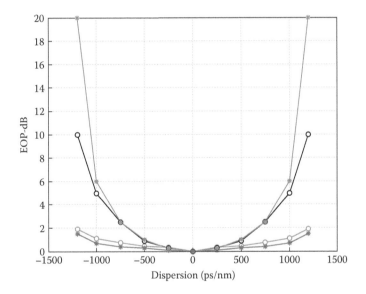

FIGURE 6.42   Measured and simulated dispersion tolerance of ASK and DQPSK (* gray – RZ-DQPSK, o dark gray RZ-ASK)—both RZ format of 50% duty cycle and NRZ-ASK (light gray *), NRZ-DPSK (light gray circle).

schemes, for example, 100 Gbps can be reduced to 25 Gsymbols/s if a two-level star-QAM is used. These multilevel modulation schemes will be described in Chapter 7.

6.5.2   Dispersion Tolerance

Dispersion tolerance is the measure of the penalty of the eye opening at the receiver after the transmission whose dispersion over the number of spans is completely compensated with some extra residual dispersion. This is very important in practice because it would specify the maximum length of uncompensated fiber allowable during the installation of the transmission link.

Figure 6.42 shows the simulated and measured dispersion tolerances of 10 Gbps RZ-ASK and RZ-DPSK at 20 Gbps, which are effectively equivalent in the symbol rate. The tolerances of the modulation formats seem very close. This can be explained by observing the spectral distributions of these two modulation formats (Figure 6.14). The spectra of RZ-ASK and RZ-DPSK are very much the same except the peak power at the center of the passband of the RZ-ASK, which is understood as the contribution of the amplitude rising from 1 to 0 or vice versa. The 3 dB bandwidth is thus the same, and this contributes to the interference between the sidebands of the modulated lightwave. The RZ-DQPSK at 20 Gbps is equivalent to that of 10 Gbps DPSK, and thus we could state that the dispersion tolerance of the RZ-DQPSK would be equivalent to that of the 10 Gbps DPSK modulation format. A 1 dB penalty of the eye diagram or the eye opening penalty (EOP) is at a dispersion of 650 ps/nm for the two modulation formats, equivalent to 650/17 km of SSMF. The EOPs of the NRZ-ASK and NRZ-DPSK formats are also plotted against the dispersion to give a very similar pattern, as expected from their spectra. Note that for NRZ-DPSK, the data modulator is a dual modulator, and no chirp is introduced.

6.5.3 PMD Tolerance

Refs. [15–19] have investigated the impact of PMD on DPSK, DQPSK direct detection formats. In systems with impairments dominated by first-order PMD, smaller duty cycle RZ formats would improve the resilience to the PMD by the ASK and DPSK systems as the ISI is smaller for narrower pulses. RZ-DQPSK is much more resilient to first-order PMD. DQPSK allows a DGD that is at least two times higher for the same EOP due to PMD as compared to that suffered by ASK and DPSK.

In systems in which the PMD is compensated, the second-order PMD would be minimized for modulation formats that offer narrower spectra because the second order is wavelength dependent, and hence a narrower spectrum would offer more resilience owing to second-order PMD effects.

6.5.4 Robustness toward Nonlinear Effects

6.5.4.1 Robustness toward SPM

The setup for investigation, both experimental and simulation, consists of four spans with 100 km SSMF and dispersion-compensating modules (DCMs) and associated EDF amplifiers at the end of SSMF and at the output of DCM with a booster EDFA and attenuator to adjust the input power launched to the transmission link. Four wavelength channels in the middle of the C-band are used. The EOP of a wavelength channel at 1552.95 nm is monitored and plotted against the launched power as shown in Figure 6.43 for the RZ-ASK and RZ-DQPSK modulation formats.

It is observed that the nonlinear SPM threshold for 1 dB EOP is 3 dBm, which is consistent with a 0.1π phase shift due to the nonlinear phase effect of a fiber with a nonlinear coefficient $n_2 = 2.3 \times 10^{-20}$ m^{-1}. The nonlinear threshold for RZ-DQPSK is observed at around 7 dBm. By inspecting the spectra of ASK and DPSK, we note that because the amplitude is switched on and off, there is transitional time, or, equivalently, the sampling time of the sequence and the Fourier transform of the sequence would have spikes at the sampling frequency. This peak power level and hence the average power of the ASK sequence is contained mostly in the carrier peaks. Unlike in DPSK or DQPSK, the envelope is constant and the energy is contained mostly in the signal as we could observe no peaks in their spectra. Thus, RZ-DQPSK and RZ-DPSK data sequence would be more tolerant to the SPM effects.

6.5.4.2 Robustness toward Cross-Phase Modulation

In 40 Gbps systems, the nonlinear SPM effects dominate the impairments due to fiber nonlinearity. However, in 10 Gbps DWDM systems, the situation is different, especially for 100 GHz spacing [20]. The cross-phase modulation (XPM) is caused by the nonlinear phase effects of a channel in the DWDM system, especially when its intensity is fluctuating, which would cause a phase disturbance

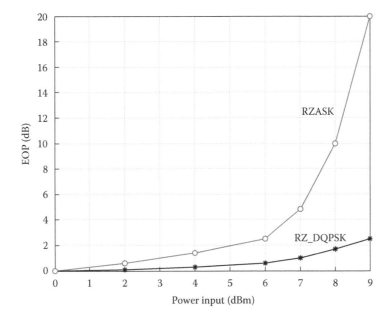

FIGURE 6.43 EOP induced by SPM versus input power at the beginning of the fiber transmission links of 4 × 100 km fully compensated SSMF spans at a bit rate of 40 Gbps. RZ-ASK (circle) RZ-DQPSK (star).

in other channels. This PM disturbance would then be transferred to other channels with amplitude modulation. The XPM effects can be considered to contribute to the frequency spectrum via a transfer function [21] given by

$$H_{12} \approx \frac{2\gamma_1}{\alpha - j2\pi f d_{12}} \qquad (6.22)$$

where γ_1 is the nonlinear factor due to the intensity fluctuation of wavelength λ_1 channel, α is the attenuation factor, and d_{12} ($d_{12}z = D(\lambda_1-\lambda_2)z$) is the group delay difference between the two channels λ_1 and λ_2 over a distance z. Thus, for 100 GHz spacing and over SSMF of $D = 17$ ps/nm/km, the 3 dB passband of this low-pass filter transfer function is at about 540 MHz, which would fall into the signal bands of a 10 Gbps bit rate system.

Shown in Figure 6.44 is the EOP due to an XPM of 10 Gbps 100 GHz spacing DWDM transmission of RZ-ASK and RZ-DPSK over multispans link in which both dispersion and attenuation effects are equalized with DCM and EDFAs. Clearly, the DPSK format is much more tolerable to the XPM effects as compared to that of RZ-ASK. The XPM effect could be reduced by using an electrical filter of a corner frequency about the 540 MHz with a sharp roll-off factor placed at the output of the electronic preamplifier.

6.5.4.3 Robustness toward Four-Wave Mixing

Unlike the impairments due to SPM and XPM mainly caused by the optical power of the channels, four-wave mixing (FWM) effects come from the relative phase difference between the channels. We can summarize the influence of FWM on DPSK and DQPSK signals as follows.

For dispersion-shifted fibers, the phase velocity of different channels around the zero-dispersion wavelength region is the same, and thus phase matching would occur and enhance the generation of the fourth wave that would fall within the active band of an equally spaced DWDM system.

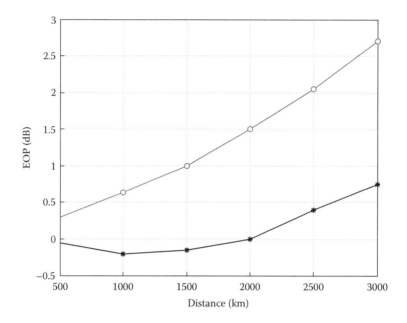

FIGURE 6.44 EOP induced by XPM versus total transmission distance (dispersion-compensated and attenuation-equalized spans).

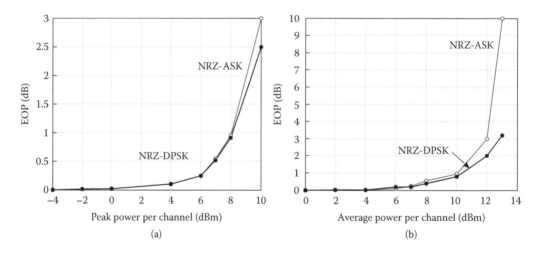

FIGURE 6.45 EOP induced by FWM versus (a) launched power (peak) and (b) average power of three NRZ-ASK and NRZ-DPSK channels after 100 km DSF with zero-dispersion wavelength at 1552.93 nm.

Figure 6.45a and b shows simulated results of the EOP of the fourth channel versus the peak and average power of three NRZ-ASK and NRZ-DPSK channels with 100 GHz spacing in the region of zero-dispersion wavelength of DSF.

6.5.4.4 Robustness toward Stimulated Raman Scattering

The impact of stimulated Raman scattering (SRS) on different modulation formats has been reported [22], including the formats DPSK and DQPSK, which have been shown to improve significantly the resilience to SRS crosstalk, especially when the RZ formats are employed,

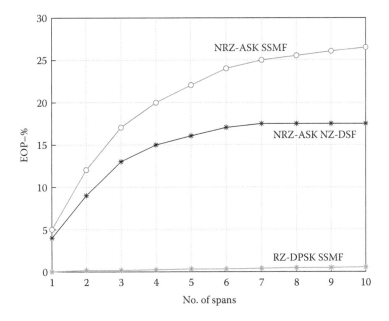

FIGURE 6.46   Percentage broadening of the 1 level (% of pulse period) due to SRS crosstalk for 11 channels with 4 nm spacing and 10 Gbps bit rate under SSMF and NZ-DSF fiber spans. Modulation format NRZ-ASK. Span length = 100 km SSMF fully dispersion compensated.

the SRS effects are negligible. The principal reasons are the periodic variation of the optical power or field propagating through the fiber, and the consequent absence of low-frequency components to enhance the SRS effects. The effect is very much like that of XPM but at a lower frequency. Thus, RZ-DPSK would not exhibit low frequency due to SRS. Simulated results of the effects of SRS on the broadening of NRZ pulse ASK modulation over the number of spans of SSMF and NZ-DSF, completely optically amplified and dispersion compensated, are shown in Figure 6.46. The SRS effects on the pulse broadening of RZ-DPSK are also shown to be significantly undisturbed by the periodic variation of the phase of the carrier, which explains the strong resilience of the modulation format to SRS. The standard deviation of the pulse amplitude over the average signal level is measured as a percentage.

6.5.4.5   Robustness toward Stimulated Brillouin Scattering

All DPSK and DQPSK do not exhibit the carrier component at the center of the passband of the spectrum, and thus it can be stated that stimulated Brillouin scattering would play no role in phase shift keying modulation formats.

6.6   CONCLUDING REMARKS

This chapter gives the fundamental aspects of the optical transmission of discrete phase modulation formats (DPSK, DQPSK, and ASK) as well as the structures of the transmitters and receivers. Experimental setups for the transmission of modulation formatted signals, NRZ, RZ, CSRZ, NRZ-DPSK, RZ-DPSK, CSRZ DPSK, and so on over optically amplified dispersion-compensating systems have been described. Experiments have investigated the filtering effects on performance of the transmission systems, the dispersion tolerance, and receiver sensitivity performance of RZ-ASK and RZ/CSRZ-DPSK. The filtering properties of the muxes and demuxes do not significantly affect the transmission performance in terms of the BER and receiver sensitivity of the 40 Gbps RZ or CSRZ-DPSK modulation formats.

FIGURE 6.47 Siemens TranXpress multiwavelength transport systems.

We have also reported and simulated the nonlinear effects, including SRS, SPM, XPM, and SBS, on the eye opening at the receiver of phase shift keying modulation, in particular, DPSK and DQPSK with RZ formats. Phase shift keying modulation is much more resilient to the nonlinear effects as compared to ASK.

The negligible mutual effects of adjacent 40 Gbps DPSK channels on 10 Gbps ASK-modulated transmission channels enable the feasibility of the upgradeability of the 10 Gbps current system to 40 Gbps optical DPSK-modulated transmission. We have also measured the transmission quality of both 40 and 10 Gbps channels and observed no significant degradation of either channel by the other. Our next step is to launch the 40 Gbps modulation format over a commercial system, the Siemens multiwavelength transport TranXpress system, as shown in Figure 6.47, and then over an installed multiwavelength 10 Gbps transmission link between cities, for example, Melbourne and Sydney of Australia.

APPENDIX 6A: MATLAB® AND SIMULINK® MODEL FOR DQPSK OPTICAL SYSTEM

A number of models in MATLAB and Simulink are given in Figures 6A.1 through 6A.4. Figure 6A.1 gives the block diagram of the entire transmission system, generally an optical transmitter, the optically amplified fiber spans, and then the receiver, including subsystems for evaluation of the BER versus parameters such as receiver sensitivity, EOP, and so on. Figure 6A.2 gives the model of

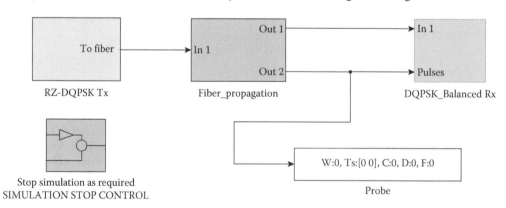

FIGURE 6A.1 General schematic diagram of the DQPSK system.

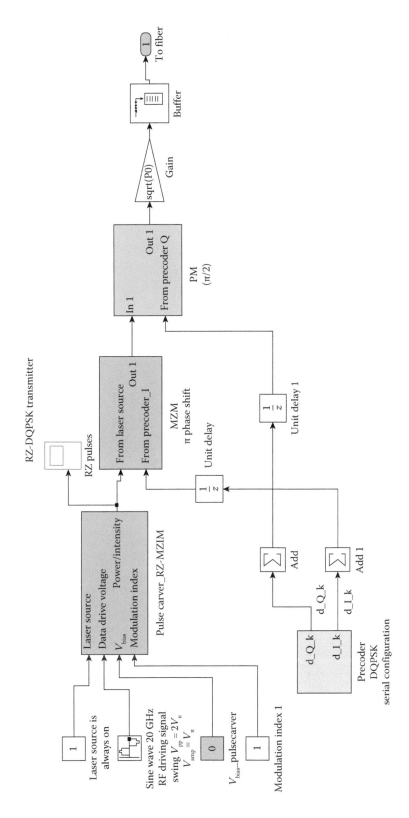

FIGURE 6A.2 Simulink® model of the optical transmitter for RZ-DQPSK.

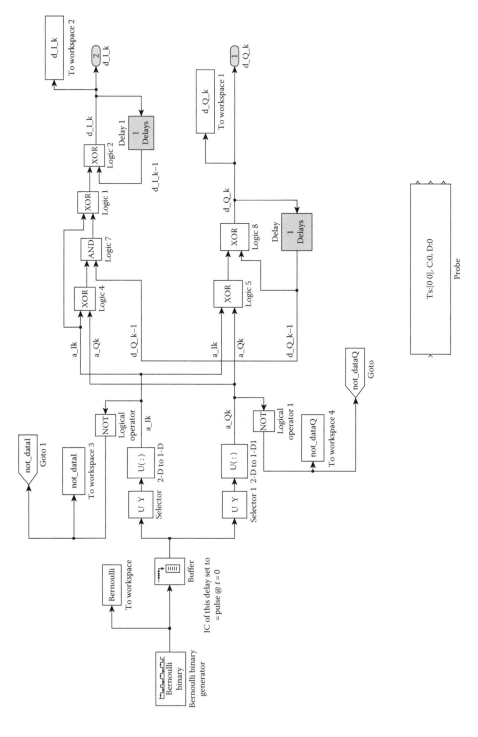

FIGURE 6A.3   Simulink® model of the electrical precoder for optical transmitter for RZ-DQPSK.

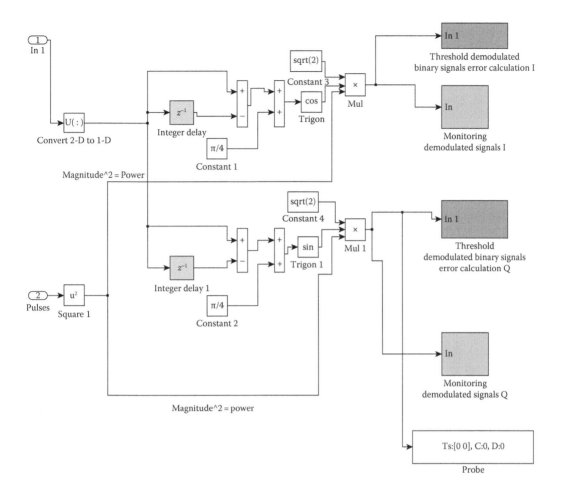

FIGURE 6A.4 Simulink® model of the MZDI optical balanced receiver for RZ-DQPSK.

optical transmitters together with Figure 6A.3 as a precoder in the electrical domain to generate signals for driving the electrodes of the MZIM. Figure 6A.4 gives the model of a balanced receiver incorporating the MZDI as a phase comparator for DPSK.

REFERENCES

1. P. J. Winzer, Optical transmitters, receivers, and noise, in *Wiley Encyclopedia of Telecommunications*, J. G. Proakis, Ed., New York: Wiley, 2002, pp. 1824–1840.
2. P. J. Winzer and R.-J. Essiambre, Advanced optical modulation formats, in *Proceedings of the ECOC 2003*, Rimini, Italy, 2003, invited paper Th2.6.1, pp. 1002–1003.
3. L. N. Binh, Lithium niobate optical modulators, in *International Conference on Material and Technology Symposium M: Material and Devices*, Singapore, July 2005.
4. R. A. Linke and A. H. Gnauck, High-capacity coherent lightwave systems, *J. Lightwave Technol.*, Vol. 6, No. 11, pp. 1750–1769, 1988.
5. J.-X. Cai, D. G. Foursa, C. R. Davidson, Y. Cai, G. Domagala, H. Li, L. Liu, et al., A DWDM demonstration of 3.73 Tb/s over 11 000 km using 373 RZ-DPSK channels at 10 Gb/s, in *Proceedings of the OFC 2003*, postdeadline paper PD22, Atlanta, GA, 2003.
6. J. C. Livas, E. A. Swanson, S. R. Chinn, E. S. Kintzer, R. S. Bondurant, and D. J. DiGiovanni, A one-watt, 10-Gbps high-sensitivity optical communication system, *IEEE Photon. Technol. Lett.*, Vol. 7, pp. 579–581, 1995.

7. A. H. Gnauck, X. Liu, X. Wei, D. M. Gill, and E. C. Burrows, Comparison of modulation formats for 42.7-Gb/s single-channel transmission through 1980 km of SSMF, *IEEE Photon. Technol. Lett.*, Vol. 16, pp. 909–911, 2004.

8. C. Wree, *RZ-DQPSK Format with High Spectral Efficiency and High Robustness towards Fiber Nonlinearities*, Technical report, University of Kiel, Kiel, Germany.

9. C. Wree, Experimental investigation of receiver sensitivity of RZ-DQPSK modulation format using balanced detection, in *Proceedings of the OFC2003*, paper ThE5, Vol. 2, p. 456, 2003.

10. Y. Okunev, *Phase and Phase Difference in Digital Communications*, Artec House, Boston, 1997.

11. C. Wree, *Differential Phase Shift Keying for Long Haul Fiber-Optic Transmission Based on Direct Detection*, Dr. Ing. Dissertation, CAU University zu Kiel, Kiel, Germany, p. 75, 2004.

12. S. Benedetto and E. Biglieri, *Principles of Digital Communications with Wireless Applications*, New York: Kluwer Academics, 1999.

13. K. K. Pang, *Digital Communications*, Lecture notes, Monash University, Melbourne, Australia, 2002.

14. A. H. Gnauck and P. J. Winzer, Optical phase-shift-keyed transmission, *IEEE J. Lightwave Technol.*, Vol. 23, No. 1, pp. 115–130, 2005.

15. C. Xie, L. Moeller, H Haustein, and S. Hunsche, Comparison of system tolerance to polarization mode dispersion between different modulation formats, *IEEE Photon. Technol. Lett.*, Vol. 15, No. 8, pp. 1168–1170, 2003.

16. J. Wang and J. M. Kahn, Impact of chromatic and polarization mode dispersion on DPSK systems using interferometric demodulation and direct detection, *IEEE J. Lightwave Technol.*, Vol. 10, pp. 96–100, 2004.

17. H. Kim, C. R. Doerr, R. Pafchek, L. W. Stulz, and P. Bernasconi, Polarization-mode dispersion impairments in direct-detection differential phase-shift-keying systems, *Electron. Lett.*, Vol. 38, No. 18, pp. 1047–1048, 2002.

18. R. A. Griffin, R. I. Johnstone, R. G. Walker, J. Hall, S. D. Wadsworth, K. Berry, A. C. Carter, et al., 10 Gb/s optical differential quadrature phase shift key (DQPSK) transmission using GaAs/AlGaAs integration, in *Proceedings of the OFC 2002*, postdeadline paper FD6, Anaheim, CA, 2002.

19. R. A. Griffin and A. C. Carter, Optical differential quadrature phase shift keying (oDPSK) for high capacity optical transmission, in *Proceedings of the Optical Fiber Communications Conference OFC 2002*, paper WX-6, 2002.

20. P Mitra and J Stark, Nonlinear limits to the information capacity of optical fiber communications, *Nature*, Vol. 411, pp. 1027–1070, 2001.

21. C. Wree, *Differential Phase Shift Keying for Long Haul Transmission Based on Direct Detection*, Dr. Ing Dissertation, CAU University zu Kiel, Kiel, Germany, pp. 24, 2002.

22. S. Schoemann, J. Leibrich, C. Wree, and W Rosenkranz, Impact of SRS-induced crosstalk for different modulation formats, in *Proceedings of the Optical Fiber Conference., OFC 2004*, paper FA5, 2004.

7 Multilevel Amplitude and Phase Shift Keying Optical Transmission

This chapter presents the modulation formats that combine the amplitude modulation and differential phase modulation schemes and the multilevel amplitude and phase shift keying (PSK). Comparisons between multilevel and binary modulation are made. Critical issues for transmission performance for multilevel modulation are identified. A simulation platform based on MATLAB® and Simulink® is described. Furthermore, for reaching 100 Gbps Ethernet, a number of multilevel modulations such as PSK and orthogonal frequency division multiplexing (OFDM) are proposed and described in the last section of the chapter.

7.1 INTRODUCTION

Under the conventional on–off keying (ASK) modulation format, the transmission bit rate beyond 40 Gbps per optical channel is very costly because the electronic signal processing technology may have reached its fundamental speed limit. It is expected that advanced photonic modulation formats such as M-ary amplitude and differential phase shift keying would replace ASK in the near future. These advanced formats would offer efficient spectral properties, thus making it possible to increase the transmission rate without placing stringent requirements on high-speed electronics and to use the same existing photonic communication infrastructure.

Coherent communications developed in the mid-1980s and has extensively exploited different modulation techniques to improve the optical signal-to-noise ratio (OSNR) [1]. However, coherence detection has faced considerable difficulties owing to the stability of the source spectrum and the laser linewidth for gain in the receiver sensitivity of a mere 3 dB for heterodyne detection and 6 dB for homodyne detection in order to extend the repeaterless distance of 60–80 km of standard single-mode fiber (SSMF).

The invention of the optical amplifier (OA) in the early 1990s has overcome the fiber attenuation limit and thus offers a significant improvement in optical transmission technology. Because of this, ultra-long-haul and ultra-high-capacity optical transmission systems have been deployed widely around the world in the last decade. The technology has matured with ASK modulation reaching 10 Gbps per optical channel, a total channel count of hundreds, and with 100/50 GHz channel spacing [2].

Based on proven efficient spectra and transmission technology, especially the controllable total dispersion of the transmission and compensating fibers, it is much more advantageous that these spectral regions be efficiently used. Therefore, the contribution of advanced modulation techniques and formats would offer higher spectral efficiency for photonic transmission.

Further, digital modulation techniques have been well established over the last half century with amplitude, frequency, or phase modulations [3]. These techniques, especially phase modulation, rely principally on the detection schemes, that is, on whether it is coherent or pseudo-coherence differential detection, and have been heavily exploited in wireless communication networks. In the photonic domain, for a long time the technological difficulties associated with manufacturing narrow-linewidth lasers have prevented the use of coherent and differential phase modulation. Only over the last several years, due to the maturity of the laser technology, particularly the successful development of distributed feedback (DFB) laser, has the laser linewidth reached a level that is much smaller than the modulation bandwidth. The coherence of the sources is now sufficient for differential phase modulating and detecting applications that require the phase of the sources to remain stable over at least two consecutive symbol periods [4].

Recently, advanced modulation techniques have attracted significant interest from the photonic transmission research and systems engineering community. Several modulation schemes and formats such as binary differential phase shift keying (BDPSK), differential quadrature phase shift keying (DQPSK), duobinary ASK associated with non-return-to-zero (NRZ), return-to-zero (RZ), and carrier-suppress return-to-zero (CSRZ) formats have been widely studied [5–8]. However, what have not been widely explored are optical multilevel modulation schemes. Although multilevel schemes have been intensively exploited in wireless communications [6,9,10], there are only a small number of studies to date that incorporate the in-coherent multilevel optical amplitude phase shift keying modulation schemes that offer the following advantages:

1. Lower symbol rate; hence for the same available spectral region, a multilevel modulation scheme would offer a transmission capacity higher than its binary modulation counterparts.
2. Efficient bandwidth utilization; photonic transmission of these multilevel signals could be implemented over the existing optical fiber communications infrastructure without significant alteration of the system architecture, thus saving the cost of capital investment and easing the system management.
3. The complexity of the coder and demodulation subsystems falls within the technological capabilities of current microwave and photonic technologies.

The principal objectives of this chapter are

1. To evaluate different modulation and coding techniques and signal pulse formats for long-haul ultra-high-capacity transmission and thus determine novel modulation schemes, such as multilevel amplitude phase shift keying, and others to be determined, for research studies.
2. To develop analytical, simulation, and experimental test beds to demonstrate the uniqueness and superiority of our novel schemes.
3. A comparative study of the modulation formats so as to unveil the principal directions for photonic modulation and transmission technologies for the next transmission generation.
4. A novel photonic communication system based on advanced multilevel optical modulation formats and implementation of the system on the Simulink platform to demonstrate its effectiveness and superiority to its counterparts and to demonstrate its feasibility as a useful platform for desktop computer simulation.

Thus, a conceptual photonic transmission system is proposed based on a hybrid technique that combines phase and amplitude modulation, the multilevel amplitude differential phase shift keying (MADPSK) format. This technique combines two modulation formats: the well-known M-ary ASK and the M-ary DPSK to take advantage of high receiver sensitivity and dispersion tolerance (DPSK), and the enhancement of total transmission capacity (M-ASK) as compared to the traditional ASK format.

The models of the MADPSK transmitter and receiver have been structured for MADPSK signaling. A simulation model based on the MATLAB and Simulink platform has been developed for the proof of concept. The system performance is evaluated for back-to-back and long-haul transmission. The analytical and simulation results of the transmission configurations are demonstrated. The following are presented:

1. Noise mechanisms, for example, quantum shot noise, quantum phase noise, optically amplified noise, noise statistics, nonlinear phase noise; hence, the design of an optimum detection and decision-level schemes for MADPSK
2. Linear and nonlinear and polarization dispersion impairments and their impact on MADPSK system performance
3. Matched filter design for optimum MADPSK signal detection
4. Offset MADPSK (O-MADPSK) modulation schemes
5. MAMSK modulation
6. MADPSK modulation for applications in subcarrier transmission systems, especially for metropolitan wide area multi-add/drop networks
7. Other issues or additional modulation formats suitable for MADPSK

This chapter is thus organized as follows: Section 7.1 gives a brief review of a number of advanced photonic modulation formats. Section 7.2 reviews and compares different modulator structures used for generating advanced photonic modulation signals and emphasizes the advantages of a dual-drive Mach–Zehnder intensity modulator (MZIM) as a modulator for generating an MADPSK signal, the main objective of the study. Section 7.3 identifies a number of critical issues and alternative multilevel signaling for optical systems. In Section 7.4, a novel photonic transmission system with the MADPSK modulation format is proposed. Section 7.5 summarizes the preliminary studies and results.

7.2 AMPLITUDE AND DIFFERENTIAL PHASE MODULATION

7.2.1 ASK MODULATION

7.2.1.1 NRZ-ASK Modulation

ASK has been the dominant modulation technique from the early days of optical communications. The main advantage of this modulation is that the ASK signal is not sensitive to the phase noise. ASK modulation can take two principal formats: the first one is called NRZ-ASK, in which the one optical bit occupies the entire bit period; in the second one, RZ-ASK, the one bit is present in only the first half of the bit period.

Figure 7.1 shows the spectrum of a 40 Gbps NRZ-ASK signal, with the carrier seen at the highest peak and the 3 dB bandwidth reaching the bit rate. The main advantage of the NRZ-ASK

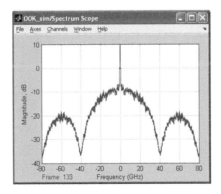

FIGURE 7.1   Spectrum of 40 Gbps NRZ-ASK signal.

signal is that its spectrum is generally the most compact compared with that of other formats such as RZ-ASK and CSRZ-ASK. On the contrary, the NRZ-ASK signal is affected by fiber chromatic dispersion (CD) and is more sensitive to fiber nonlinear effects as compared to its RZ- and CSRZ-ASK counterparts.

7.2.1.2 RZ-ASK Modulation

The RZ-ASK signal, shown in Figure 7.2 [7], is similar to the NRZ-ASK signal, except that the one bit occupies only the first half of the bit period. This signal can be generated by a transmitter shown in the same figure in which an NRZ-ASK transmitter is followed by a pulse carver driven by a pulse train synchronized with the data source. The pulse train has a frequency equal to the data rate. The RZ-ASK pulse width can take the form of 33%, 50%, and 66% duty ratio. Because of its narrower pulse width, the spectrum of the RZ-ASK signal, shown in Figure 7.3, is larger than that of the NRZ-ASK signal, which decreases the spectrum efficiency. In this spectrum, the carrier is seen as the highest peak, and the two side peaks are RF modulating signals positioned 80 GHz apart.

7.2.1.3 CSRZ-ASK Modulation

The CSRZ-ASK modulation format [11] is similar to the standard RZ-ASK format, except that the neighboring optical pulses have a π phase difference. The carrier in neighboring time slots is thus cancelled out and effectively excluded from the signal spectrum. The CSRZ-ASK signal can be generated by a transmitter with the scheme shown in Figure 7.4 [11]. In this scheme, the first MZIM modulates the intensity of the optical signal coming from a laser source, while the second MZIM,

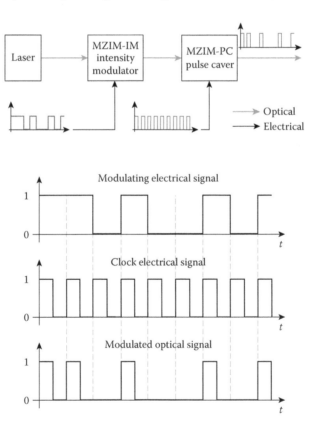

FIGURE 7.2   RZ-ASK transmitter and signal. (Adapted from T. Mizuochi et al., *IEEE J. Lightwave Technol.*, Vol. 21, No. 9, pp. 1933–1943, 2003.)

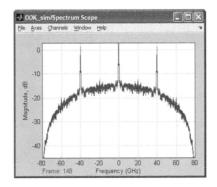

FIGURE 7.3   Spectrum of 40 Gbps 50% RZ-ASK signal.

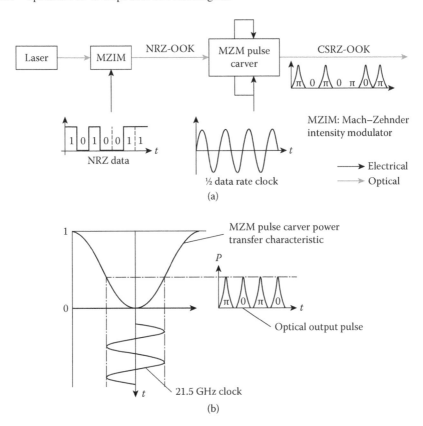

FIGURE 7.4   (a) Block diagrams of CSRZ-ASK transmitter and (b) generation of optical pulse with alternative phase using biasing control and amplitude. (Adapted from Y. Miyamoto et al., *IEEE Electron. Lett.*, Vol. 35, No. 23, pp. 2041–2042, 1999.)

driven by a clock signal whose rate is equal to half of that of the data rate, caves the NRZ pulses into RZ ones. Because the second MZIM is biased at the minimum-intensity point, it provides an RZ pulse train at the data rate with alternating phases 0 and π for neighboring time slots. The CSRZ-ASK signal can also be detected by a direct detection receiver as it would not be phase sensitive.

The main advantages of CSRZ-ASK include a narrower spectrum, higher tolerance to dispersion, and stronger robustness against fiber nonlinear effects as compared with standard RZ-ASK. Because its peak optical power is much lower than that of other formats, it is less affected by both self-phase modulation (SPM) and cross-phase modulation (XPM) [7]. Figure 7.5 shows the spectrum of 40 Gbps CSRZ-ASK signal with a very low carrier power level [12].

ASK is a modulation technique that generates a signal $s(t)$ by multiplying a digital signal $m(t)$ by a carrier f_c [13]

$$s(t) = A \cdot m(t) \cos 2\pi f_c t, \ 0 < t < T \qquad (7.1)$$

where A is the amplitude envelope; the digital signal $m(t)$ may take one of M levels $[b_0, b_1, ..., b_M]$. When $M = 2$, $s(t)$ is a binary ASK signal with ASK as a special case. ASK is also implemented in NRZ, RZ, and CSRZ formats, whose spectra are shown in Figure 7.6 in the same graph for

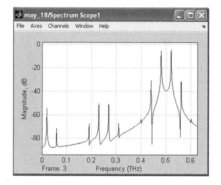

FIGURE 7.5   Spectrum of 40 Gbps CSRZ-ASK. (Extracted from L. N. Binh et al., DPSK RZ modulation formats generated from dual-drive interferometric optical modulators. Unpublished works.)

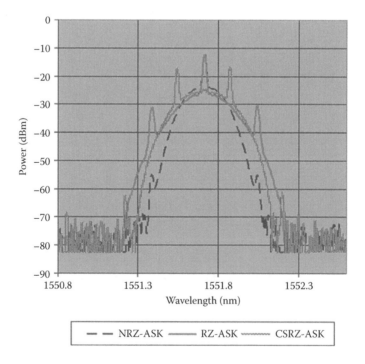

FIGURE 7.6   Spectrum of NRZ-ASK, RZ-ASK, and CSRZ-ASK signals.

the purpose of comparison. Like their ASK analogs, NRZ-ASK has the most compact spectrum, whereas RZ-ASK has the broadest. In terms of energy, CSRZ-ASK has the lowest peak power because the carrier signal has been effectively removed.

7.2.2 Differential Phase Modulation

Under ASK/ASK modulation schemes with the associated NRZ, RZ, and CSRZ formats, the amplitude of the optical carrier varies accordingly. Phase modulation, on the contrary, modulates the carrier phase and thus facilitates the use of bipolar signals "±1." This distinguishing feature means that phase modulation offers a significant improvement in receiver sensitivity as compared with ASK modulation. With the recent advances in photonic lightwave technology, especially the integrated optic delay interferometer, differential phase modulation and demodulation and the balanced receiver have become realizable. This section gives a brief overview of the differential modulation techniques and their implementations in the photonic domain, especially the MADPSK.

The term NRZ-BPSK, or traditionally NRZ-DPSK, is commonly used for denoting a modulation technique in which the optical carrier is always present with a constant power, with its phase alternating between 0 and π. The modulation rule is as follows

1. At the transmitter: Initially a reference 0 bit is entered as the present encoded bit. Then the next data bit is compared with the present encoded bit. If they are different, then the next encoded bit is 1, for which a phase change of π occurs, else the next encoded bit is 0, which causes no (or 0) phase change.
2. At the receiver, the phase of the carrier at the present bit slot is compared with that of the previous one. If the phase difference is π, then the data is decoded as 1, otherwise the data is 0 when the phase difference is 0.

One of the NRZ-DPSK transmitter structures is shown in Figure 7.7 [14]. User data are first encoded by a differential encoder into the driving voltage which then alternates the phase of the carrier signal between 0 and π. In detecting a NRZ-DPSK signal, a delay Mach–Zehnder interferometer (MZI) in combination with a balanced optoelectronic receiver can be used. The interferometer acts as the phase comparator with constructively and destructively interfered outputs.

As shown in Figure 7.7, the received optical signal is split into two arms of an MZI, one of which has a one-bit optical delay. The MZI compares the phase of each bit with the phase of the previous bit, and the photodetector converts the phase difference to the intensity. When there is no phase shift between two bits, they are added constructively hence giving maximum resultant amplitude to the output signal; otherwise, they cancel out when the phase shift equals π. If the differential phase shift is $\Delta\phi$, then the differential current at the output of the balanced photodetector can be written as

$$i = A^2 \cos \Delta\phi \qquad (7.2)$$

Because the balance receiver uses both the constructive and destructive ports of the MZI, the detected signal level can swing from 1 to −1. Compared with ASK or with the use of the unbalanced receiver where the signal amplitude is limited between 1 and 0, DPSK can offer a 3 dB improvement in receiver sensitivity.

Due to its constant envelope, the NRZ-DPSK signal is less sensitive to power modulation-related nonlinear effects, such as SPM and XPM, than its NRZ-ASK counterpart [15,16]. On the contrary, long-haul DPSK systems, including both NRZ and RZ, with OA are affected by nonlinear phase noise. The amplified spontaneous emission (ASE) noise of OAs is converted into phase noise, leading to waveform distortion and, consequently, signal degradation. The spectrum of the NRZ-DPSK signal is shown in Figure 7.8 [17], together with other DPSK formats. It can be seen that the NRZ-DPSK signal has the most compact spectrum compared with that of other DPSK formats.

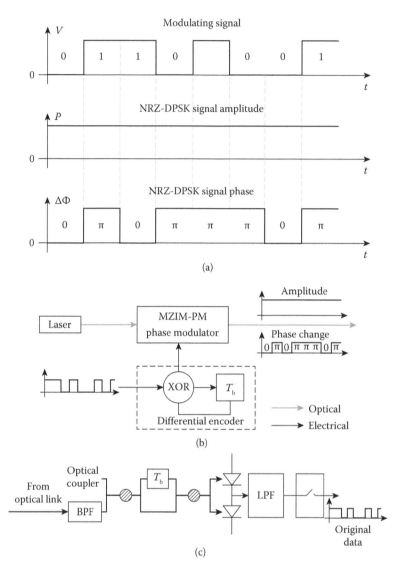

FIGURE 7.7 (a) NRZ-DPSK signal, (b) transmitter, and (c) receiver. (Adapted from R. Hui et al., *Advanced optical modulation formats and their comparison in fiber-optic systems*, Technical Report, Information and Telecommunication Technology Center, University of Kansas, 2004.)

This can be explained by the fact that the NRZ-DPSK signal amplitude remains constant regardless whether bit 1 or bit 0 is transmitted, and thus the energy is distributed more equally compared to RZ- and CSRZ-DPSK signals.

The RZ-DPSK format is similar to the NRZ-DPSK format, the only difference being that instead of constant optical power, a pulse narrower than the bit period appears in each bit slot as shown in Figure 7.9. The RZ-DPSK transmitter, however, resembles an RZ-ASK transmitter in which the phase modulator (PM) replaces the intensity modulator (IM). The RZ-DPSK signal can also be detected by the same receiver used for the NRZ-DPSK signal. Owing to its narrow pulse, the RZ-DPSK format is expected to minimize the effects of intersymbol interference and is thus capable of achieving a longer transmission distance [7]. A narrow pulse, however, spreads the spectrum of the RZ-DPSK signal wider than that of the NRZ-DPSK signal, making RZ-DPSK systems more susceptible to CD. To reduce the effect of this impairment, CD compensation devices are used.

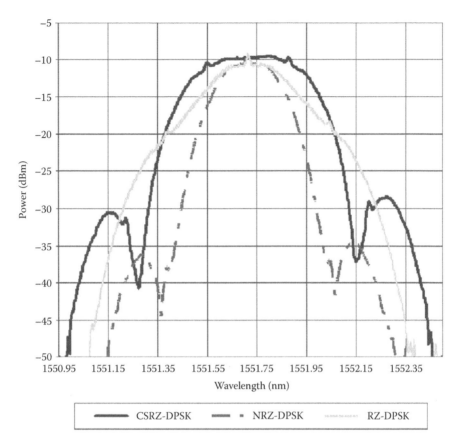

FIGURE 7.8 Experimentally measured spectra of NRZ-DPSK, RZ-DPSK, and CSRZ-DPSK signals. (Extracted from T. L. Huynh et al., Long-haul ASK and DPSK optical fiber transmission systems: Simulink modeling and experimental demonstration test beds, *Proc. IEEE Tencon'05*, Melbourne, Australia, November 2005.)

RZ-DPSK signal energy is not distributed equally as in the case of NRZ-DPSK. Most of it is concentrated in only a fraction of the bit duration, while it reduces to nearly zero for the rest of the time. This large energy fluctuation makes the signal more susceptible to fiber nonlinearity and makes signal detection more difficult.

The carrier suppression technique can also be used in conjunction with RZ-DPSK modulation to produce a CSRZ-DPSK signal, which has been demonstrated as one of the most attractive modulation formats in high-spectral-efficiency wavelength division multiplexing (WDM) and dense WDM (DWDM) systems [15].

It is due to the suppression of the carrier, the CSRZ-DPSK modulation format offers higher energy and spectral efficiency, and hence more resilience to impairments due to the fiber nonlinearity, CD and polarization-mode dispersion (PMD) as compared its RZ-DPSK counterpart. The spectra of CSRZ-DPSK, RZ-DPSK, and NRZ-DPSK are shown together in Figure 7.8 for comparison.

The CSRZ-DPSK signal can be generated by a transmitter whose scheme, shown in Figure 7.10, for the ASK parts is similar to that of CSRZ-ASK. The main difference is that in the CSRZ-DPSK transmitter a PM replaces the IM used in the CSRZ-ASK transmitter. The receiver for CSRZ-DPSK has the same structure as that of the NRZ-DPSK scheme.

To increase transmission bit rate without an increased bandwidth requirement, one can code more than one bit into a data symbol. DQPSK modulation is the first step in the realization of this idea [18–20].

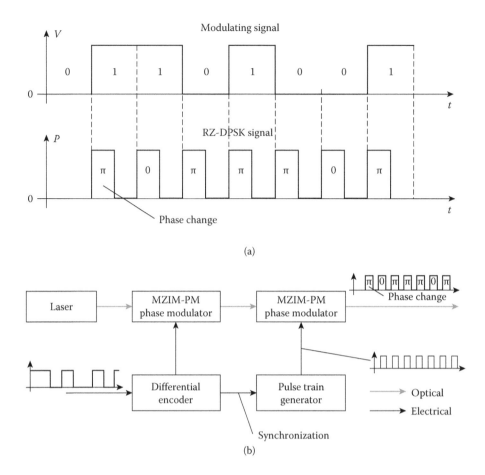

FIGURE 7.9   (a) RZ-DPSK signal and (b) and transmitter structure.

A signal constellation or signal space is the best way to represent a DQPSK signal, in which the points representing phase-modulated signals are located on two orthogonal axes called I and Q (for in-phase and quadrature components, respectively). Each of the two data bits $[D_1, D_0]$ is first precoded into a symbol, and then the symbol is encoded into a phase shift that may take one of four values $[0, \pi/2, \pi, 3\pi/2]$ depending on the bit combination it represents. The DQPSK symbol rate is thus equal to only half the bit rate. The constellation of QPSK is shown in Figure 7.11. Intuitively, one can say that with the same bandwidth available, DQPSK offers twice the transmission capacity compared with its ASK and binary DPSK counterparts.

The DQPSK signal can be generated by a transmitter as shown in Figure 7.12. This structure consists of two MZIMs connected in parallel. A $+\pi/2$ phase shift is introduced in one of these MZIMs, making the optical signals in the two paths orthogonal to each other. A precoder encodes user data in accordance with the differential rule to generate the I and Q driving voltages, which then modulate the carrier's phase in two optical paths. The modulated carrier components are then combined at the output of the MZI. If the two normalized driving signals are denoted by I and Q, respectively, then the output signal is [21]

$$E_{\text{output}} = I \cos 2\pi f_c t + Q \sin 2\pi f_c t \tag{7.3}$$

where f_c is the frequency of the optical carrier. The coding and mapping bits $[D_1, D_0]$ into I and Q and the signal constellation points follow the rule shown in Table 7.1.

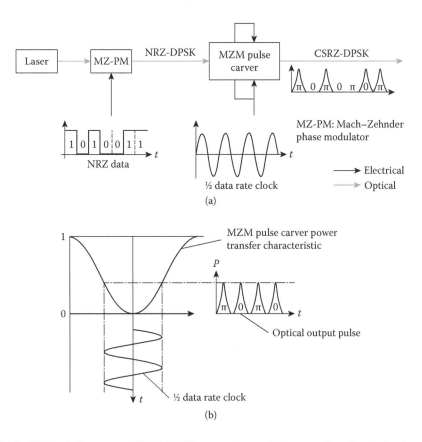

FIGURE 7.10 (a) Block diagrams of CSRZ-DPSK transmitter and (b) generation of optical pulse with alternative phase by driving the dual-drive MZIM with a 2 V_π voltage swing.

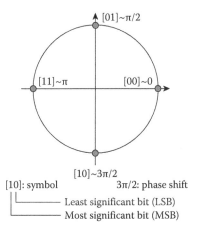

FIGURE 7.11 DQPSK signal constellation.

The DQPSK receiver uses two sets of MZ delay interferometers (DIs) and balance receivers to detect in-phase (I) and quadrature-phase (Q) components of the received signal. Each set is similar to the one used in the NRZ-DPSK receiver. There are, however, two main differences: first, the delay introduced in the first branches of interferometers is now replaced by the symbol duration T_s; second, the phases of the signal in the second branches are shifted by $+\pi/4$ and $-\pi/4$ for I and Q

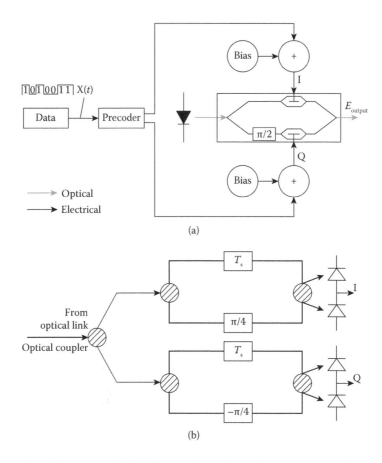

(a)

(b)

FIGURE 7.12   (a) Parallel structure of DQPSK transmitter using I–Q modulator and (b) self-homodyne optical receiver for DQPSK signal sequence. T_s = symbol duration.

TABLE 7.1
DQPSK Signal Bit-Phase Mapping

D_1	D_0	I	Q	Phase Shift
0	0	0	0	0
0	1	0	1	$\pi/2$
1	1	1	1	π
1	0	1	0	$3\pi/2$

components, respectively. These additional phase shifts are needed to separate the two orthogonal phase components I and Q.

The spectra of typical NRZ-DQPSK and RZ-DQPSK are shown in Figure 7.13. Likewise, the spectra for the modulation formats of 67% RZ-DQPSK, CSRZ-DQPSK, and 16 MAPSK are shown in Figure 7.14. Figure 7.13a shows the spectrum of the 40 Gbps NRZ-DQPSK signal with the single-sided bandwidth of the main lobe equal to 20 GHz, which is only half of the transmitted bit rate. The spectrum of the RZ-DQPSK signal, Figure 7.13b, is much broader with strong harmonics beside the main lobe.

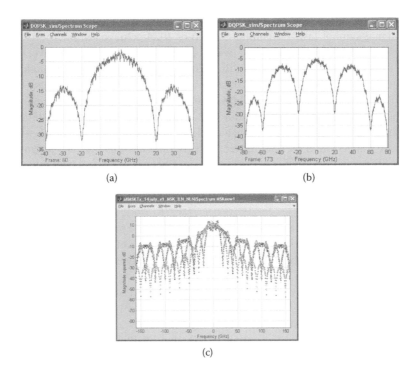

FIGURE 7.13 Optical spectra of 40 Gbps (a) NRZ-DQPSK, (b) 50% RZ-DQPSK, and (c) DPSK as compared with MSK (dotted curve).

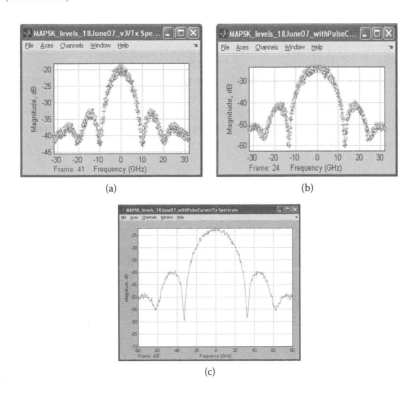

FIGURE 7.14 Optical spectra of 40 Gbps (a) NRZ-DQPSK, (b) 67% CSRZ-DQPSK, and (c) 100 Gbps CSRZ 16-ADPSK.

Despite numerous advancements in optical modulation techniques, the number of levels encoded in a signal symbol falls far behind the 256 or 1024 achieved in microwave modulation schemes [6]. The phase noise associated with optical sources and OAs have hindered the use of phase-related modulation schemes to current fluctuations in the photodetection, and hence the degradation of the BER. The differential phase demodulation process based on the phase comparison of two consecutive symbols requires that the phase should remain stable over two symbol periods. Thus, narrow-linewidth lasers are critical for phase-modulated systems. It has been shown that to achieve a power penalty less than 1 dB [22], $\Delta v/B < 1\%$ with Δv and B, the laser linewidth and system bit rate, respectively. In optical transmission systems where OAs are used, the ASE noise intermingles with the fiber nonlinear phase effect, and thus enhances the nonlinear phase noise. While SPM-induced nonlinear phase noise is the dominant phase noise in single optical channel systems, XPM-induced phase noise is the main phase noise for multi-channel (WDM) systems.

Significant phase noise caused by optical sources and OAs have prevented optical DPSK schemes from having many levels in each symbol. Increasing the number of levels in the signal space, and hence the number of bits per symbol higher than one, the most popular solution is a combination of the DQPSK and ASK modulation formats.

Figure 7.15 shows a typical eye diagram for multilevel MAPSK with an amplitude detection section of 40 Gbps after transmission of 5 km SSMF and under only quadrature-phase detection.

Recently, Hayase et al. [9] have demonstrated experimentally a 30 Gbps eight-states per symbol optical modulation system using a combined ASK and DQPSK modulation scheme as shown in Figure 7.16 [6,9]. It maps three bits into a symbol, and thus creates a transmission bit rate that is three times higher than the symbol rate. The transmitter consists of two cascaded PMs and an amplitude modulator (AM). The first PM, driven by data bit D_0, creates 0 and π phase shifts, while the second, driven by D_1, forces two further phase shifts 0 and $\pi/2$, the quadrature phase to generate four distinct phases of the DQPSK signal. The AM, driven by D_2 bit, shifts the four phases between two amplitudes to create totally eight signal points.

At the receiver side, optical signals are detected in both amplitude and differential phase. An ASK demodulator detects the D_2 bit. The other is a DQPSK demodulator and detected to recover D_1 and D_0 bits. Sekine et al. [6] reported experimentally a similar scheme, but with four bits $[D_3,D_2,D_1,D_0]$ mapped into a symbol: $[D_1,D_0]$ bits are used to generate a "normalized" DQPSK signal, while $[D_3,D_2]$ bits manipulate the amplitude of this DQPSK signal between four concentric circles. Thus, a 16-ary MADPSK signal can be generated. This would offer 40 Gbps bit rate with a symbol rate of only 10 GBauds.

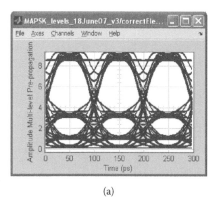

(a)

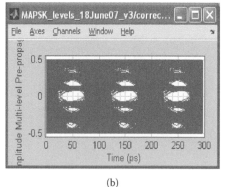

(b)

FIGURE 7.15 Eye diagram showing amplitude detection section of 40 Gbps. (a) 5 km SSMF transmission and (b) quadrature-phase detection.

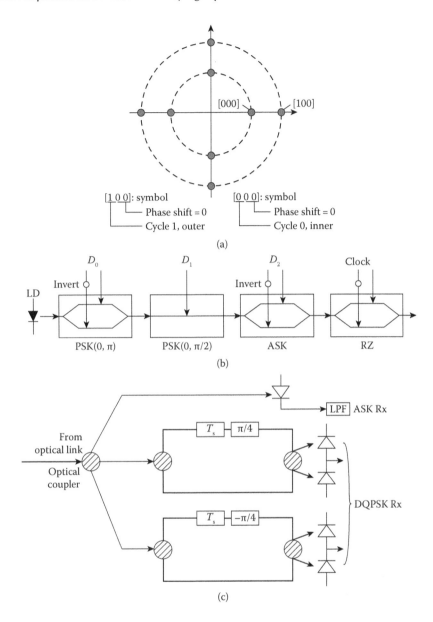

FIGURE 7.16 8-ary APSK modulation experimental configuration (Extracted from K. Sekine et al., *IEEE Electron. Lett.*, Vol. 41, No. 7, 2005.), 10 GHz clock assignment synchronization of symbol rate, data modulator, and quadrature-phase shift in optical domain using the PM, two balanced receivers for differential phase shift detection and direct detection for amplitude detection. (a) 8-ary ASK-DPSK signal, (b) transmitter configuration, and (c) receiver configuration. (Adapted from S. Hayase et al., Proposal of 8-state per symbol (binary ASK and QPSK) 30-Gb/optical modulation demodulation scheme, *Proc. European Conf. Opt. Commun.*, paper Th2.6.4, ECOC 2003, Rimini, Italy, pp. 1008–1009, September 2003.)

7.2.3 COMPARISON OF DIFFERENT AMPLITUDE AND PHASE OPTICAL MODULATION FORMATS

Different amplitude and phase optical modulation formats are summarized in Table 7.2. In most cases, NRZ-ASK parameters are used as references. From the comparison, it can be concluded that MADPSK is advantageous compared to other modulation formats in terms of spectral efficiency and the ability to significantly increase the transmission bit rate, which are very, if not the most,

TABLE 7.2

Comparison of Different Optical Modulation Formats

Mod. Format	Spectral Width	Receiver Sensitivity	Resilience to Dispersion	Resilience to SPM	Resilience to XPM	Current Transmission Bit Rate Limits
NRZ-ASK	Narrowest	Lowest	Worst	High	High	40 Gbps
RZ-ASK	2 × RZ-ASK (at 50% duty ratio)	Higher NRZ-ASK	Higher NRZ-ASK	High	High	40 Gbps
CSRZ-ASK	Same as RZ-ASK	Higher NRZ-ASK	Higher NRZ-ASK	High	High	40 Gbps
NRZ-DPSK	Same as NRZ-ASK	3 dB better than NRZ-ASK	Higher NRZ-ASK	Worse than ASK	Worse than ASK	40 Gbps
RZ-DPSK	Same as RZ-ASK	3 dB better than RZ-ASK	Higher NRZ-ASK	Worse than ASK	Worse than ASK	40 Gbps
CSRZ-DPSK	Same as CSRZ-ASK	3 dB better than CSRZ-ASK	Higher CSRZ-ASK	Higher than ASK	Higher than ASK	40 Gbps
DQPSK	1/2 DPSK	1.5 dB better than ASK	UR*	Worse than ASK	Worse than ASK	2 × DPSK
MADPSK	1/M DPSK					$(\log_2 M) \times$ DPSK expected

important parameters for an optical transmission system. It is also expected that MADPSK inherits good properties (and of course the bad ones, if any) from two basic ASK and DPSK modulation formats.

7.2.4 MULTILEVEL OPTICAL TRANSMITTER USING SINGLE DUAL-DRIVE MZIM TRANSMITTER

In this section, several optical transmitter structures used for generating the DQPSK signal are described. This is necessary because a novel optical transmission system will be developed based on the DQPSK modulation format. All these structures have MZIM as their base component, which can be a single- or dual-electrode structure. Figure 7.11 displays a constellation of QPSK modulation.

Unlike a single-drive MZIM, a dual-drive electrode structure with two traveling wave RF electrodes can modulate the phase of optical signals in both branches, and hence push–pull operation. Interference at the output of a dual-drive MZIM will produce a phase-modulated signal. However, when the effects of phase modulation in the two branches are exactly equal but opposite in sign, the output signal becomes intensity modulated. In this manner, a dual drive can be used for both phase and intensity modulation. The relationship between the input and output signals of a dual-drive MZIM can be described by [12,23]

$$E_{\text{output}} = \frac{E_{\text{input}}}{2}\left[\exp\left(j\pi\frac{V_1(t)}{V_\pi}\right) + \exp\left(j\pi\frac{V_2(t)}{V_\pi}\right)\right] \qquad (7.4)$$

where $V_1(t)$ and $V_2(t)$ are the driving voltages applied to the modulator, and V_π is the voltage required to provide a π phase shift of the carrier in each branch of MZIM. Note that unlike in single-drive MZIMs, the chirp effect does not exist in dual-drive MZIMs.

The transmitter structure shown in Figure 7.12 is called the parallel type. It is only one of several structures that can be used for generating the DQPSK signal, namely, parallel structure, serial structure, single PM structure, and dual-drive MZIM structure. These terms are used to indicate the structuring of MZIMs whether they are connected in tandem, parallel, or just a pure PM with a single electrical drive port.

In an electro-optic transmitter of the serial type shown in Figure 7.17, an MZIM generating an in-phase component and a PM generating a quadrature component are connected in tandem. Pre-encoded data generate two signals: one is used for driving the MZIM and the other for driving the PM. Usually, the square shape of the precoded waveforms is replaced by the raised-cosine one before being fed to the modulators [21]. Furthermore, the biasing conditions and the amplitude of the modulators can be used to generate 33% to 67% pulse width RZ formats. It is also noted that the pulse shape would also follow a $\cos^2$ profile owing to the property of the IM. This transmitter would suffer from the chirping effects owing to the rise time of the electrical driving signals and hence would contribute to the distortion of the lightwave signals, in particular when switching between the lowest level to the highest level.

The single PM structure shown in Figure 7.18 uses only one MZIM as the PM. Precoded data are added up to create a single driving voltage. One of the two precoded data is amplified and together with the other signal represents four positions of the DQPSK signal [21].

The dual-drive MZIM structure in Figure 7.19 uses two driving voltages for modulating the optical carrier phase in two branches of an MZIM. Data are first precoded following the differential rule and then used to create driving voltages $V_1(t)$ and $V_2(t)$ corresponding to the signal constellation points [21].

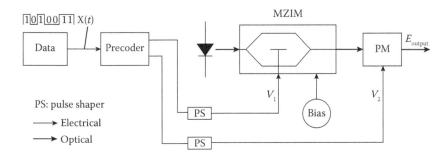

FIGURE 7.17 Cascade PM and MZIM for DQPSK signal generation.

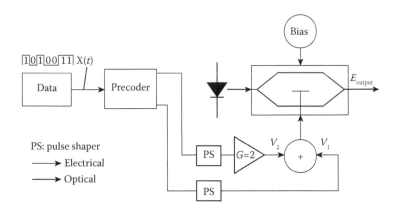

FIGURE 7.18 Single PM structure.

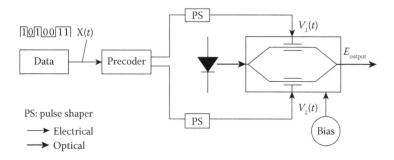

FIGURE 7.19   Dual-drive MZIM structure.

In the four transmitter structures described earlier, the parallel and serial structures are the most complex and difficult to implement because they have discrete devices connected together. The dual-drive MZIM and single PM structures, on the contrary, are much simpler because they require fewer discrete devices. Furthermore, as will be shown in the next section, dual-drive MZIMs can be configured to work as both PM and AM at the same time, so they can easily generate not only DQPSK but also MADPSK signals. Thus, a dual-drive MZIM is the principal part of the MAPSK transmission system. Table 7.3 gives a comparison between the different transmitter structures; the dual-drive modulator is outstanding for combined amplitude and phase switching between the states of a multicircular constellation.

The main reason why the dual-drive MZIM structure has attracted our attention in this chapter is that it can play the role of both AM and PM simultaneously, which is impossible with other transmitter structures. This means that to generate an MADPSK signal, there is no need to employ separate PM and AM, as has been implemented by Sekine et al. [6] and Hayase et al. [9]. This section describes a method for generating a 16-ary MADPSK signal using this dual-drive MZIM structure.

The 16-ary MADPSK signal constellation of interest is shown in Figure 7.20. It is actually a combination of a 4-ary ASK and a DQPSK signal, with four bits $[D_3, D_2, D_1, D_0]$ mapped into a symbol. Among them, two bits $[D_1 D_0]$ are coded into four phases $[0, \pi/2, \pi, 3\pi/2]$, and two bits $[D_3 D_2]$ are coded into four amplitude levels $[I_3, I_2, I_1, I_0]$. As has been shown in Ref. [6], with the use of a balanced receiver and a DI.

The MADPSK signal sequence can produce clear DQPSK eye patterns whose decision level is located at the zero-voltage level.

Recall that the signal at the output of the dual-drive MZIM can be represented as [12,23]

$$E_o = \frac{E_i}{2} e^{j\phi_1} + \frac{E_i}{2} e^{j\phi_2} \qquad (7.5)$$

with $\phi_1 = \pi \dfrac{V_1(t)}{V_\pi}$, $\phi_2 = \pi \dfrac{V_2(t)}{V_\pi}$, where E_i and E_o are electrical fields of the input and output optical signal, respectively; $V_1(t)$, $V_2(t)$ are driving voltages applied to the modulator; and V_π is the voltage required to provide a π phase shift for the carrier in each MZIM branch.

Equation 7.5 suggests that with a properly chosen input signal E_{input} and driving voltages $V_1(t)$, $V_2(t)$, all signal points of the constellation in Figure 7.20 can be constructed from two phasor signals $\dfrac{E_i}{2} e^{j\phi_1}$ and $\dfrac{E_i}{2} e^{j\phi_2}$. Indeed, if E_i is chosen to equal the electrical field corresponding to the signal points in the largest circle of the constellation, then a constellation signal point E_{output} with the phase shift θ_i in the circle n can be found as a sum of two vectors $\dfrac{E_i}{2} e^{j\phi_{ni1}}$ and $\dfrac{E_i}{2} e^{j\phi_{ni2}}$,

TABLE 7.3
Comparison of DQPSK Transmitter Structures

Parameters for Comparison	Parallel MZIM	Serial MZIM & PM	Single PM	Dual-Drive MZIM
Complexity of circuit design	Complicated in matching of ultra-high-frequency electrical paths; high insertion loss. Flexible in biasing.	Complicated in matching of ultra-high-frequency electrical paths; high insertion loss. Flexible in biasing.	Simple in photonics but complicated in realization of ultra-high-frequency signal connections.	Simplest but requires multilevel voltage switching at symbol rate (microwave speed).
Ability to create MADPSK signal	Not possible. A separate ASK modulator required.	Not possible. A separate ASK modulator required.	Impossible. A separate ASK modulator required.	Dual-drive MZIM acts as ASK and DPSK simultaneously.

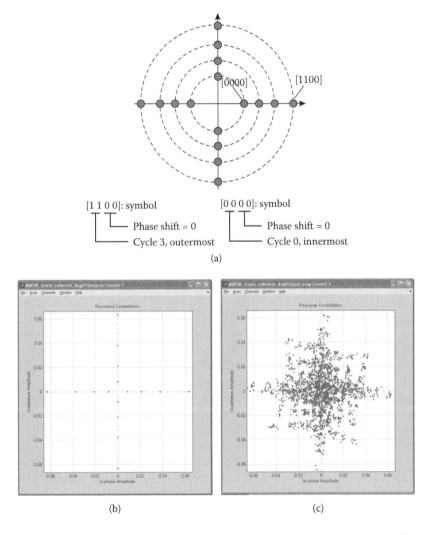

FIGURE 7.20 16-ary MADPSK signal bit-phase mapping: (a) design, (b) Simulink® scattering plot before transmission, and (c) after 200 km transmission with 2 km mismatch in dispersion.

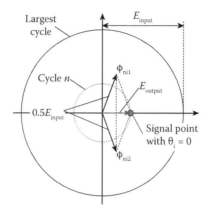

FIGURE 7.21   Relationship between E_i, E_0, ϕ_{ni1}, and ϕ_{ni2} using phasor representation.

where $\phi_{ni1} = \theta_i + \arccos\left(E_n/E_{input}\right)$, $\phi_{ni2} = \theta_i - \cos^{-1}\left(E_n/E_i\right)$. The subscriptions i and n are used to denote the phase position and the order of the circle of interest. Figure 7.21 illustrates the relationship between E_i, E_0, ϕ_{ni1}, and ϕ_{ni2}. For simplicity, the signal point is chosen with $\theta_i = 0$.

By substituting ϕ_1 and ϕ_2 in Equation 7.5, the driving voltages for this point can be obtained

$$V_{ni1}(t) = \frac{V_\pi}{\pi}\left[\theta_i + \cos^{-1}\left(E_n/E_i\right)\right],\ V_{ni2}(t) = \frac{V_\pi}{\pi}\left[\theta_i - \cos^{-1}\left(E_n/E_i\right)\right] \qquad (7.6)$$

7.3   MADPSK OPTICAL TRANSMISSION

In general, the structures of the MADPSK can be given as shown in Figure 7.22. A model has been constructed for investigating the performance of systems based on the MADPSK modulation format. It consists of a signal coding model, transmitter model, receiver model, and transmission and dispersion compensation fiber models.

The 16-ary MADPSK signal model described in Section 7.2 will be used. To balance the ASK and DQPSK sensitivities, ASK signal levels are preliminary adjusted to the ratio $I_3/I_2/I_1/I_0 = 3/2/1.5/1$ [6]. as shown in Figure 7.23. These level ratios can be determined from the signal-to-noise ratio at each separation distance of the eye diagram or Q factor. The noise is assumed to be dominated by the beat noise between the signal level and that of the ASE noise.

The transmitter model shown in Figure 7.24 is used to produce the 16-ary MADPSK signal. It consists of a DFB laser source generating continuous wave (CW) light (carrier), which is then modulated in both phase and amplitude by a dual-drive MZIM. Each of the four bits of user data $[D_3D_2D_1D_0]$ is first grouped into a symbol and then encoded to generate two electrical driving signals $V_1(t)$ and $V_2(t)$ under which the amplitude and phase of the carrier in two optical paths of the dual-drive MZIM will be modulated to produce the NRZ 16-ary MADPSK signal. The RZM-PC then converts the NRZ pulse train into RZ one in order to minimize the effects of intersymbol interference.

The receiver model shown in Figure 7.25 consists of two phase demodulators, an amplitude demodulator and a data multiplexer MUX. Two phase demodulators are used for extracting the $[D_1D_0]$ bits, and they work exactly in the same way as the ones in the DQPSK receiver described in the foregoing section. The amplitude demodulator (AD) is used for detecting four amplitude levels of the MADPSK signal. It is a well-known direct detection scheme consisting of a photodiode followed by an electronic receiver. The amplitude modulated signal is then threshold detected in association with a clock recovery circuit to recover two bits $[D_3D_2]$. The two bits $[D_3D_2]$ are interleaved with the two bits $[D_1D_0]$ by the MUX to reconstruct the original binary data stream.

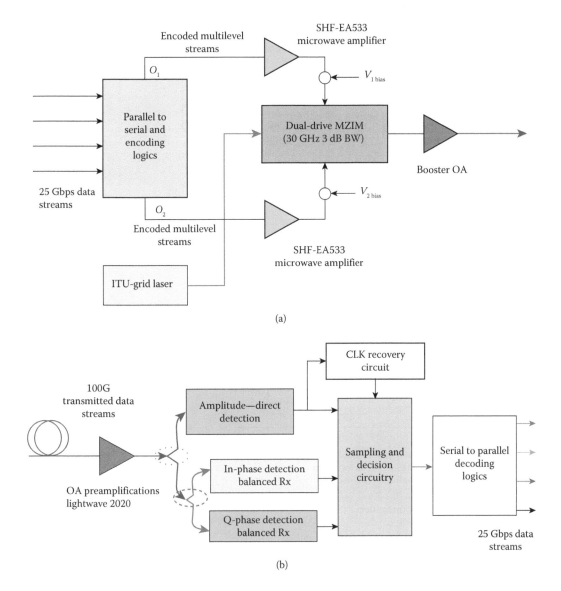

FIGURE 7.22 Schematic diagram of the photonic transmitters and receivers for the 16ADQPSK transmission scheme: (a) transmitter and (b) receivers with branches for detection of amplitude, in-phase and quadrature-phase components.

7.3.1 PERFORMANCE EVALUATION

Under performance evaluation, the following main parameters are investigated:

1. The system bit error rate (BER) versus SNR: A solution for the system BER will be found analytically, and the BER will be computed against different SNR values and bit rates. The system BER versus the SNR will also be obtained by system simulation and cross-checked with the BER against that obtained analytically. Graphs of BER versus SNR will be plotted.
2. The system BER versus receiver sensitivity: The BER versus receiver sensitivity will be obtained analytically and by simulation, and the results will be cross checked. Graphs of BER versus receiver sensitivity will be plotted.

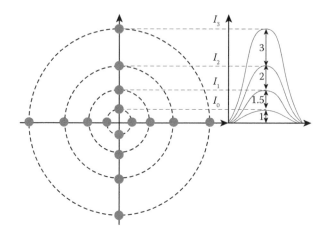

FIGURE 7.23   ASK interlevel spacing.

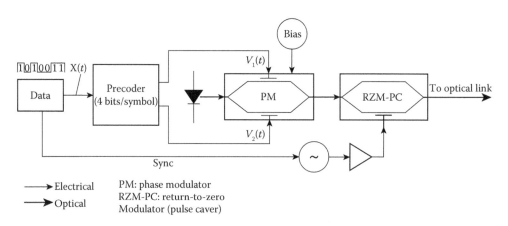

FIGURE 7.24   MADPSK transmitter.

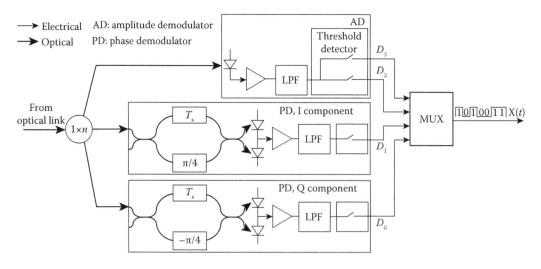

FIGURE 7.25   Amplitude direct detection and photonic phase comparator with balanced receiver for MADPSK demodulation.

3. Dispersion tolerance: Transmission over fibers of types ITU-G652, ITU-G.655, and LEAF with corresponding dispersion factors will be considered. Graphs of the power penalty due to the dispersion as compared with back-to-back transmission will be plotted against the dispersion factor in ps/nm.
4. Tolerance to other system impairments: For example, with dispersion tolerance (DT), the power penalty due to other impairments such as laser and OA phase noise, and receiver phase error will be investigated. The corresponding graphs will be plotted.

Performance evaluation is conducted under the effects of the following conditions or contexts:

1. Different pulse shapes: raised cosine, rectangular, and Gaussian
2. Modulation formats: NRZ, RZ, and CSRZ
3. ASE noise of OAs
4. Transmitter impairments: laser noise
5. Receiver impairments: phase error of DI-based phase demodulators
6. Change of ASK interlevel spacing
7. Optical and electrical filtering
8. Multichannel environment: system performance in combination with DWDM technology would be reported in future

7.3.2 IMPLEMENTATION OF MADPSK TRANSMISSION MODELS

The following simulation models have been built on the MATLAB and Simulink platform for proving the working principles and for investigating the performance of systems using optical MADPSK modulation. A transmitter is simulated to generate a 16-ary MADPSK signal, and a receiver is to reconstruct the original binary signal. These models run over a simulated single-mode optical fiber. Laser chirp, OA phase noise, nonlinearities, CD, PMD, and other impairments will be involved in later stages to evaluate the different performance characteristics of the modulation format: system BER, receiver sensitivity, and power penalties due to different impairments.

The phases and the driving voltages for creating the signal points of the 16-ary MADPSK constellation are computed and tabulated in Table 7.4.

7.3.3 TRANSMITTER MODEL

The MATLAB and Simulink model of the system is shown in Figure 7.26.

The transmitter model using the dual-drive MZIM structure is shown in Figure 7.27. The purpose of the blocks are as follows:

1. The User Data and ADPSK precoder block generates a pseudo-random data sequence to simulate user data stream and encodes each group of four data bits into a symbol.
2. The Voltage driver 1 and Voltage driver 2 blocks map precoded data into driving voltages for modulating the amplitude and phase of the carrier in the dual-drive MZIM.
3. Two Complex Phase Shift blocks simulate two optical paths of the dual-drive MZIM.
4. The Sum block simulates the combiner at the output of the MZIM.
5. The Gaussian Noise Generator block simulates a noise source.
6. The Amplifier block simulates an OA.

7.3.4 RECEIVER MODEL

The receiver structure is shown in Figure 7.28.

TABLE 7.4

Phase and Driving Voltages for 16-Ary MADPSK Constellation

Positions	θ_i	V_{i1}	V_{i2}
		Circle 3	
1100	0°	$0.0V_\pi$	$0.0V_\pi$
1101	90°	$0.5V_\pi$	$0.5V_\pi$
1111	180°	$1.0V_\pi$	$1.0V_\pi$
1110	270°	$1.5V_\pi$	$1.5V_\pi$
		Circle 2	
1000	0°	$0.2952V_\pi$	$-0.2952V_\pi$
1001	90°	$0.7949V_\pi$	$0.2046V_\pi$
1011	180°	$1.2947V_\pi$	$0.7043V_\pi$
1010	270°	$1.7944V_\pi$	$1.2041V_\pi$
		Circle 1	
0100	0°	$0.3919V_\pi$	$-0.3919V_\pi$
0101	90°	$0.8917V_\pi$	$0.1078V_\pi$
0111	180°	$1.3914V_\pi$	$0.6076V_\pi$
0110	270°	$1.8912V_\pi$	$1.1073V_\pi$
		Circle 0	
0000	0°	$0.4575V_\pi$	$-0.4575V_\pi$
0001	90°	$0.9573V_\pi$	$0.0422V_\pi$
0011	180°	$1.4570V_\pi$	$0.5420V_\pi$
0010	270°	$1.9568V_\pi$	$1.0417V_\pi$

The functions of the blocks are as follows:

1. Each DI is simulated by a set of two Magnitude-Angle blocks, a Delay block and a Sum block. The Delay block stores the phase of the previous symbol, and the Magnitude-Angle blocks extract the phase and amplitude of the present and previous symbols, which will be used in the followed different phase demodulation and detection operations.
2. The Constant $\pi/4$ and Constant $\pi/4$ and the next two Sum blocks simulate an extra phase delay in each branch of the DI.
3. Two Cos blocks and two Product blocks simulate two balanced receivers.
4. The Amplitude Detectors, D2 and D3 blocks, simulate the ASK detector for D_2 and D_3 bits.
5. Three Analog Filter Design blocks simulate electrical low-pass filters.
6. The Phase Detector D0_I and Phase Detector D1_Q blocks simulate the threshold detectors for D_0 and D_1 bits (I and Q components of a DQPSK signal), respectively.

7.3.5 TRANSMISSION FIBER AND DISPERSION COMPENSATION FIBER MODEL

The propagation of an optical signal in a fiber medium that is dispersive and nonlinear is best described by the nonlinear Schrödinger equation (NLSE) [22] as described in Chapter 2. Other parameters are explained in the following text. The transmission fiber model shown in Figure 7.29 is used to simulate the propagation of an optical signal. This fiber model simulates the impairments that impact the system performance.

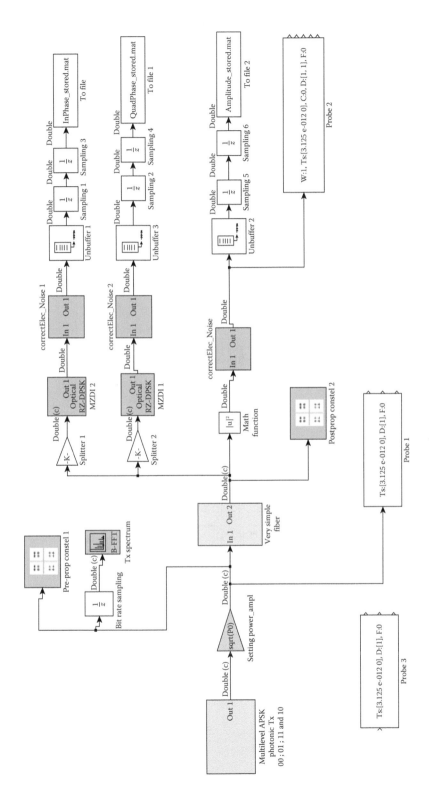

FIGURE 7.26 MATLAB® simulated system model.

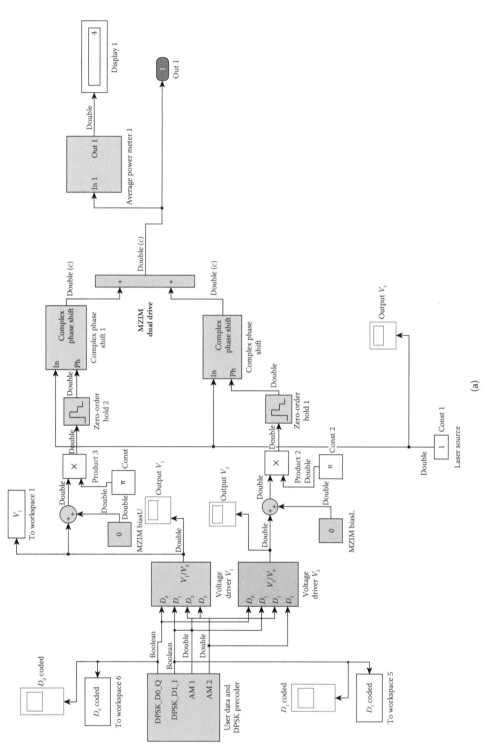

FIGURE 7.27 MATLAB® simulated MADPSK: (a) transmitter.

(a)

(Continued)

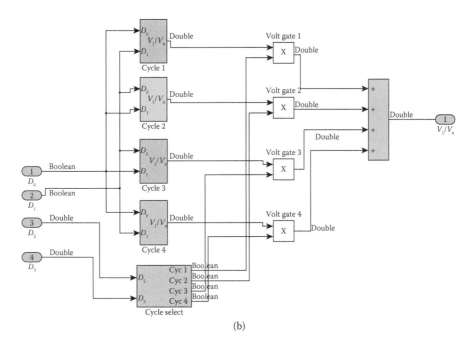

(b)

FIGURE 7.27 (Continued) MATLAB® simulated MADPSK: (b) logic precoder.

All characteristic parameters of the fiber medium together with the optical input signal are taken by the Matrix Concatenation block and then processed by a MATLAB function that solves the NLSE using the split-step Fourier method [24].

The dispersion compensation fiber model has the same structure as the propagation fiber model, except that the signs of the propagation constant beta2 in the two models are opposite.

7.3.6 Transmission Performance

7.3.6.1 Signal Spectrum, Signal Constellation, and Eye Diagram

The spectrum of a 40 Gbps 16-ary MDAPSK signal obtained by the running transmitter model is given in Figure 7.30. As is seen clearly in the graph, the single-sided bandwidth of the main lobe equals 10 GHz. Numerically, this amounts to only one-fourth of the transmission bit rate, and from that it can be concluded that MADPSK is a high bandwidth-efficient modulation format.

Figure 7.31 shows the signal constellation recovered at the receiving end. The noise and nonlinear property of the fiber cause amplitude and phase fluctuations and scatter signal points around some mean value. The MADPSK eye diagram is shown in Figure 7.32 for the I component (the Q component should have a similar diagram). This eye diagram clearly shows four amplitude levels associated with the two phase shifts 0 and π.

7.3.6.2 BER Evaluation

The MADPSK system can be considered as consisting of two subsystems, ASK and DQPSK, and its error probability can be evaluated as a joint error probability of the two

$$P_{\text{ADPSK}} = \left[\frac{1}{2} P_{\text{ASK}} + \frac{1}{2} P_{\text{DPSK}} - \frac{1}{2} P_{\text{ASK}} \cdot \frac{1}{2} P_{\text{DPSK}} \right] = \frac{1}{2} \left[P_{\text{ASK}} + P_{\text{DPSK}} - P_{\text{ASK}} \cdot P_{\text{DPSK}} \right] \quad (7.7)$$

where P_{ASK} and P_{DPSK} are the error probabilities of the ASK and DQPSK subsystems, respectively.

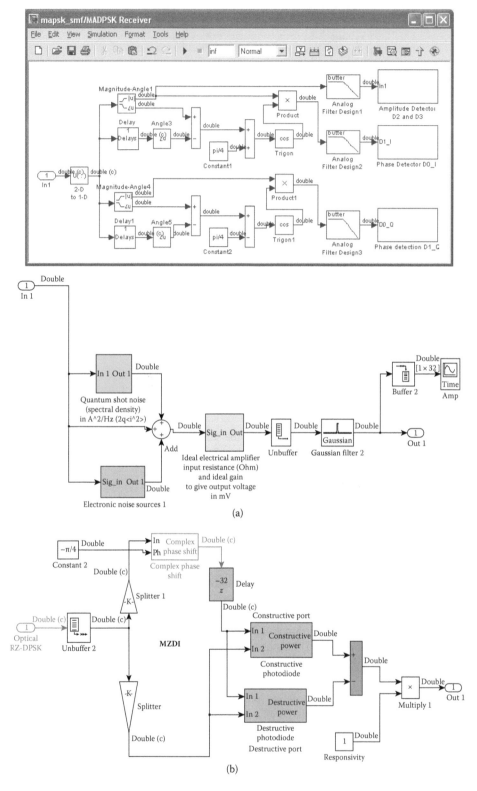

FIGURE 7.28 MATLAB® simulated MADPSK receiver: (a) amplitude direct detection and (b) balanced receiver detection—in-phase and quadrature.

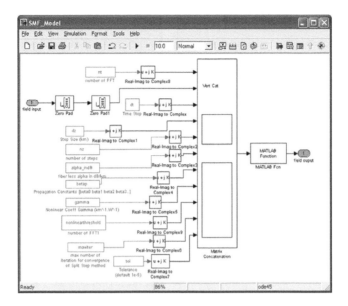

FIGURE 7.29   Single-mode fiber model.

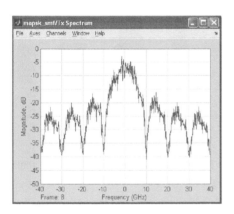

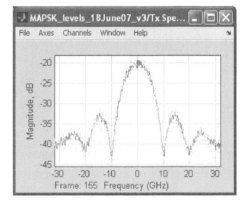

FIGURE 7.30   40 Gbps MADPSK spectrum.

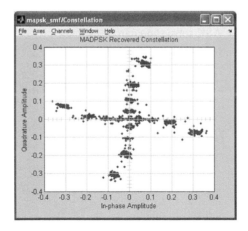

FIGURE 7.31   40 Gbps MADPSK constellation recovered at the receiver.

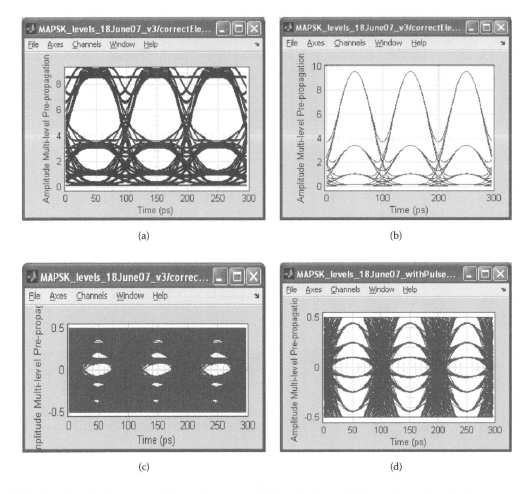

FIGURE 7.32 40 Gbps MADPSK eye diagram at OSNR = 20 dB: (a) NRZ amplitude, (b) CSRZ amplitude, (c) NRZ in phase, and (d) CSRZ in phase.

7.3.6.3 ASK Subsystem Error Probability

Figure 7.33 shows four ASK signal levels b_0, b_1, b_2, b_3, three decision levels d_1, d_2, d_3, and the standard deviation of noise at different signal levels σ_0, σ_1, σ_2, σ_3.

The error probability of the ASK subsystem can be evaluated by [25]

$$P_{ASK} = \frac{2}{M+1} \sum_{1}^{M} Q\left(\frac{b_i - d_i}{\sigma_i}\right) = \frac{2}{3+1}\left[Q\left(\frac{b_1 - d_1}{\sigma_1}\right) + Q\left(\frac{b_2 - d_2}{\sigma_2}\right) + Q\left(\frac{b_3 - d_3}{\sigma_3}\right) \right] \tag{7.8}$$

For example, in our system: (1) $b_1 = 8.08e - 2$, $b_2 = 1.45e - 1$, $b_3 = 2.42e - 1$, (2) $d_1 = 5.11e - 2$, $d_2 = 1.08e - 1$, $d_3 = 1.88e - 1$, (3) $\sigma_1 = 5.00e - 3$, $\sigma_2 = 6.70e - 3$, $\sigma_3 = 8.85e - 3$ at an OSNR = 20 dB.

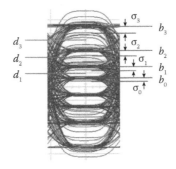

FIGURE 7.33 MADPSK eye diagram: signal levels, decision levels, and standard deviation of noise.

The error probability of the ASK subsystem thus equals

$$P_{\text{ASK}} = \frac{1}{2}\left[Q\left(\frac{(8.08e-2)-(5.11e-2)}{5.00e-3}\right) + Q\left(\frac{(1.45e-1)-(1.08e-1)}{6.70e-3}\right) \\ + Q\left(\frac{(2.42e-1)-(1.88e-1)}{8.65e-3}\right) \right]$$

$$P_{\text{ASK}} = \frac{1}{2}\left[Q(5.94) + Q(5.52) + Q(6.24) \right]$$

$$= \frac{1}{2}\left[(1.47e-9) + (1.73e-8) + Q(2.26e-10) \right]$$

$$= 9.49e-9$$

The error probability of the ASK subsystem over a range of OSNR from 6 to 24 dB is evaluated and shown in Figure 7.34.

7.3.6.4 DQPSK Subsystem Error Probability Evaluation

In terms of differential phase shift keying modulation, the system can be broken up into four independent DQPSK subsystems corresponding to circle 0, circle 1, circle 2, and circle 3 of the signal constellation. The error probability of each subsystem is evaluated first, and then they are averaged to obtain the error probability of the DQPSK subsystem.

Each DQPSK subsystem in turn can be thought of as being made up of two 2-ary DPSK subsystems. The error probability of each 2-ary DPSK subsystem is evaluated, and then they are averaged to get the error probability of the DQPSK subsystem

$$P_{\text{DQPSK}} = 1 - (1 - P_{\text{DPSK\_I}})(1 - P_{\text{DPSK\_Q}}) = P_{\text{DPSK\_I}} + P_{\text{DPSK\_Q}} - P_{\text{DPSK\_I}} \cdot P_{\text{DPSK\_Q}} \qquad (7.9)$$

where $P_{\text{DPSK\_I}}$ and $P_{\text{DPSK\_Q}}$ is the error probability of the in-phase (I) and quadrature-phase (Q) components of each DPSK subsystem (circle). Because I is coded by bit D_0, Q is coded by bit D_1, I and Q are detected in the same way, and D_0 and D_1 are supposed to be equally probable, then (20) becomes

$$P_{\text{DQPSK}} = 2P_{\text{DPSK\_I}} - P_{\text{DPSK\_I}}^2 = 2P_{\text{DPSK\_Q}} - P_{\text{DPSK\_Q}}^2 \qquad (7.10)$$

$P_{\text{DPSK\_I}}$ is evaluated based on the δ-factor [22]

$$P_{\text{DPSK\_I}} = \frac{1}{2}\left(\frac{\delta}{\sqrt{2}}\right) \approx \frac{\exp\left(-\delta^2/2\right)}{\delta\sqrt{2\pi}} \qquad (7.11)$$

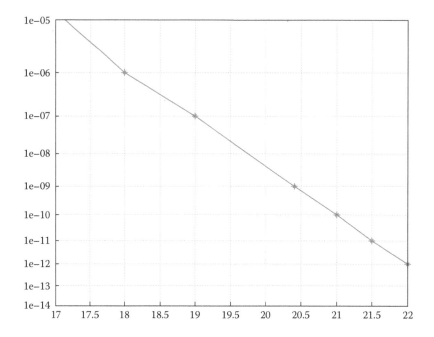

FIGURE 7.34   Error probability of ASK subsystem versus OSNR.

where $Q = \dfrac{i_H - i_L}{\sigma_H + \sigma_L}$, i_H, i_L and σ_H, σ_L are the mean value and standard deviation of signal currents at high and low levels at the input of the receiver, respectively. For example, the transmission parameters can be set as follows: $i_H = 3.23e - 02$, $i_L = (-3.23e - 02)$, at OSNR = 20 dB $\sigma_H = \sigma_H = 3.16e - 3$. The δ-factor for a single DQPSK subsystem of circle 0 thus equals, and the corresponding error probability is $P_{DPSK\_1\_CYCLE0} = \dfrac{1}{2}\operatorname{erfc}\left(\dfrac{10}{\sqrt{2}}\right) \approx 7.7e - 24$.

The error probability of circle 0 (the innermost circle) is $P_{DQPSK\_CYCLE0} = 2*(7.7e - 24) - (7.7e - 24)^2 = 1.54*10e - 23$. Thus, the error probability of all four circles is

$$P_{DQPSK} = \frac{1}{4}\left[P_{DQPSK\_CYCLE0} + P_{DQPSK\_CYCLE1} + P_{DQPSK\_CYCLE2} + P_{DQPSK\_CYCLE3} \right] \qquad (7.12)$$

P_{DQPSK} over a range of OSNRs from 6 to 24 dB is evaluated and shown in Figure 7.35.

7.3.6.5   MADPSK System BER Evaluation

The MADPSK system error probability is evaluated based on Equation 7.17. Figure 7.36 shows the graphs of the error probability for the ASK subsystem, DQPSK subsystem, and MADPSK system in the same coordinates for comparison purpose. As can be observed from Figure 7.36, at OSNR = 24 dB, the MADPSK. It is also clear that for the same value of the OSNR, especially when it is high, the DQPSK subsystem outperforms its ASK counterpart, and the overall performance of the MADPSK system is dominated by the ASK subsystem performance. Thus, the spaces between the ASK levels could be adjusted for a better balance between the BER ASK and the BER DQPSK to achieve a better overall MADPSK BER performance. This probably is caused mainly by the intersymbol interference during the transition of different levels.

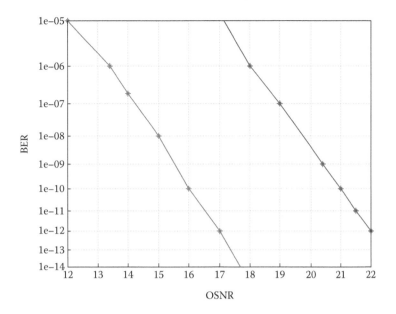

FIGURE 7.35 Error probability of DQPSK subsystem versus OSNR.

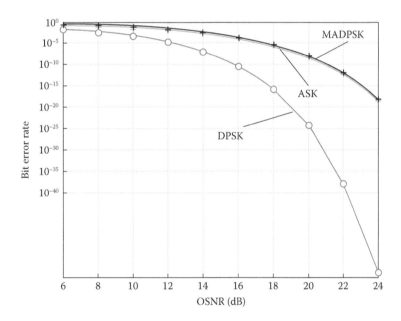

FIGURE 7.36 Error probability of MADPSK (black) system versus OSNR; logarithm scale. Error probability of ASK (light gray) and MADPSK almost coincide.

Figure 7.36 shows the simulation results of 16ADPSK at 100 Gbps transmission (extreme left graph) in comparison with other modulation formats such as Duobinary 50 and Duobinary 67 and experimental results of CSRZ-DPSK. The bit rates of these other transmission results are at 40 Gbps. It is observed that for 16MADPSK, the receiver sensitivity is close to the −28 dBm performance standard used in 10 Gbps NRZ transmission, and performs better at 100 Gbps than the other modulations operating at the lower rate of 40 Gbps. However, this superior performance at 100 Gbps is still with a penalty of approximately 3 dB compared with

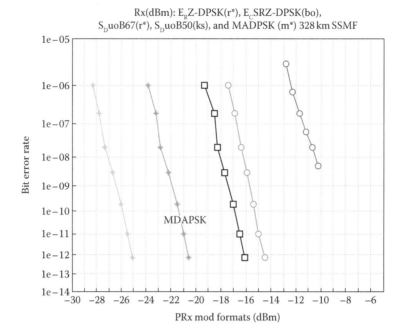

FIGURE 7.37 BER versus receiver sensitivity for MADPSK format and other duobinary and ASK (simulation) and CSZ and CSRZ-DPSK (experimental). Legend: mid gray * is the MDAPSK.

10 Gbps transmission systems. Fortunately, this penalty can easily be compensated for by using a low-noise optical preamplifier at the receiver end. For example, a 15 dB gain optical preamplifier with a 3 dB noise figure would satisfactorily resolve the issue. The BER versus the receiver sensitivity of 16ADPSK and duobinary formats and ASK are shown in Figure 7.37. It indicates a 2–3 dB improvement of the MADPSK.

The detection of the lowest level may have been affected by the noise level of the optical preamp when only the amplitude information is used. This can be improved significantly if both phase detection and amplitude detection are used, as is observable from Figure 7.32c and d.

7.3.6.6 Chromatic DT

The residual CD of the optical link is characterized by the DL product, which is defined as the product of the dispersion coefficient D and the total fiber length L. Figure 7.38 shows the signal phase evolution under the effect of CD. It can be seen that with a predetermined DL = 50 ps/nm, all signal points are rotated around the [0,0] origin by the same angle of approximately 0.125 rad. This confirms the parabolic phase shift due to the CD. This phenomenon is called linear phase distortion, in contrast to the nonlinear phase distortion caused by the fiber nonlinearity.

Figure 7.39 shows the BER penalty versus different values of the DL product. It can be seen very clearly that the BER performance of the NRZ format is severely affected by the fiber dispersion. When the DL increases from 0 ps/nm (fully CD compensated) to 35 ps/nm, its BER performance is improved by 1.5 dB, but sharply degraded by a 28 dB penalty at DL = 50 ps/nm, and should be worse for a higher value of dispersion. This leads to the conclusion that it is undesirable to use the NRZ format in MADPSK systems because the optical link residual dispersion usually cannot be compensated to a small amount, and an ineffective dispersion management and control plan could lead to a very high BER.

The 66%-RZ format, on the contrary, can tolerate a much higher degree of CD. Its BER performance is even slightly improved at DL ≈ 50 ps/nm, and the BER penalty is less than 1 dB at DL = 100 ps/nm. This is equivalent to the transmission over 6 km of uncompensated standard SMF fiber without significantly compromising the BER performance.

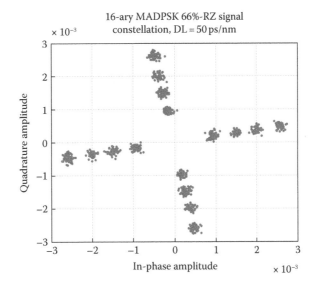

FIGURE 7.38 Evolution of the phase scattering of the MADPSK signal constellation under chromatic dispersion effects.

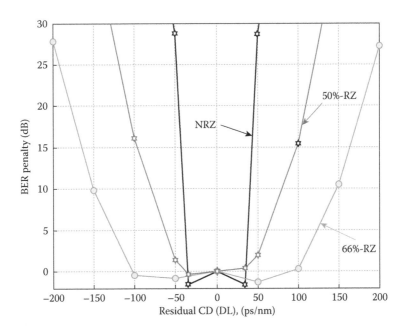

FIGURE 7.39 Error probability of MADPSK system versus the dispersion-length DL product.

The MAPSK offers a lower symbol rate and hence a higher channel capacity that would allow the upgradation to a higher rate over a low-bit-rate optical fiber transmission system without modifying the photonic infrastructure of the optical networks.

7.3.6.7 Critical Issues

This section outlines the critical issues involved in the evaluation of the performance of MADPSK systems.

7.3.6.7.1   Noise Mechanism and Noise Effect on MADPSK

Although receiver noise in multilevel amplitude modulation was investigated intensively in the 1980s, little has been reported for multilevel phase and differential phase modulation. One of the principal goals in the system design, especially for long-haul transmission systems, is to achieve high receiver sensitivity. At a given optical power, the error probability depends on the noise power, and hence the receiver sensitivity.

Quantum shot noise is the fundamental noise mechanism in photodiodes, which leads to a fluctuation in the detected electrical current even when the incident optical signal has a constant or variable power. Thus, it is signal dependent. Furthermore, the beating of the currents of the signal and the optical phase noise would generate an amplitude-dependent noise at different-level signals of the MADPSK. It is caused by random generation of electrons contributing to the photoelectric current, which is a random variable. All photodiodes generate some current even in the absence of an optical signal because of the stray light and/or thermal generation of electron–hole pairs, the dark current.

In MADPSK, the amplitude of the signal of the outermost circle of the constellation would be affected by the quantum shot noise, which is strongly signal amplitude dependent, especially when there is an optical preamplifier. On the contrary, it is desirable that the innermost constellation would have the largest magnitude to maximize the optical signal energy for long-haul transmission. Therefore, an optimum receiving scheme must be developed both analytically and by modeling and eventually by experimental demonstration.

However, the amplitude of the outermost constellations is limited by the nonlinear SPM effects, which will be further explained in the next few sections. Thus, the lower and upper limits of the amplitude of the MADPSK would be extensively investigated in the next phases of the research.

The electronic equivalent noise as seen from the input of the electronic preamplifier following the photodetector can be measured and taken into account for the total noise process caused by the thermal noise of the input impedance, the biasing current shot noise, and the noise at the output of the electronic preamplifier. These noises are combined with the signal-dependent quantum shot noise so as to gauge their contribution to the MADPSK receiver. Thus, we may consider new structures of electronic amplifiers or a matched filter at the input of the receiver to achieve the optimum MADPSK receiver structure.

For long-haul transmission systems, the ASE of an OA is probably the most important noise mechanism. In OAs, even in the absence of an input optical signal, spontaneous emission always occurs stochastically when electron–hole pairs recombine and release energy in the form of light. This spontaneous emission is noise, and it is amplified by the OAs together with the useful optical signal and accumulated along the optical transmission link [26].

Noise reduces the SNR, and hence the system BER and receiver sensitivity. Noise models also affect the design of optimum detection schemes such as decision thresholds. To the best of my knowledge, a thorough investigation of the noise mechanism and its impact on multilevel signaling has never been reported except some preliminary results for 10 Gbps 4-ary ASK schemes [10]. Thus, all noise sources and the mechanism by which they affect the system performance must be thoroughly investigated. These noises are used to estimate the optimum decision level of the detection of the amplitude of the multilevel eye diagram (Figure 7.40).

7.3.6.7.2   Transmission Fiber Impairments

For optical signals, the transmission medium is an optical fiber with associated OAs and dispersion compensation devices, or a leased wavelength running on top of a DWDM system. Impairments are always part of the transmission medium; among them, CD, PMD, and nonlinearity are critical.

When an optical pulse propagates along a fiber, its spectral components disperse owing to the differential group delay (DGD), and the output pulse will be broadened. CD is proportional to the fiber length and the laser linewidth, especially the spectrum of the lightwave modulate signals.

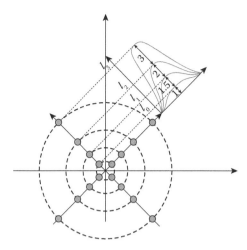

FIGURE 7.40 ASK interlevel spacing and offset modulation and detection line.

CD may cause optical pulses to overlap each other, thus leading to intersymbol interference, and increase the system BER, especially for ASK systems. DPSK systems, on the contrary, are more CD tolerant. For MADPSK systems, the eye diagrams and phase constellation are as shown in Figures 7.41 and 7.42, respectively. The phase constellation is rotating when the MADPSK is under the linear CD effect. Another idea that is well known—and is developed in our model—is that this CD can be compensated by dispersion-compensating fiber modules. However, the mismatching of the dispersion slopes of the transmission and compensating fibers is very critical for multichannel multilevel modulation schemes.

The optical pulse is also broadened by PMD, which is actually the time mismatch between two orthogonal polarizations of the optical pulse when they traverse along a fiber. In the ideal optical fiber having a truly homogeneous glass and a truly coaxial geometry of the core, the two optical polarizations would propagate with the same velocity. However, this is not the case for a real fiber, so the two polarizations have different speeds and will reach the fiber end at different times.

Similar to the CD effects, PMD can cause pulse overlapping and thus increase the system BER. However, unlike CD, which is practically constant over time and can be in a large scale compensated, PMD is a stochastic process and cannot be managed easily. It is well known that PMD has the Maxwellian probability density function with a mean value $\langle PMD \rangle = K_{PMD} \cdot \sqrt{L}$, where K_{PMD} is defined as a PMD coefficient whose measured values vary from fiber to fiber in the range $[0.01 - 1\,\mathrm{ps}/\sqrt{\mathrm{km}}]$, and L is the fiber length. Under the MADPSK, the signal space of the constellations would be affected either in the magnitude or phase by PMD, but is expected to be dominated by the phase distortion. It is well known that the PMD first and second effects are critical for ASK modulation. For DPSK, it is expected that the principal axes of the polarization modes propagating through the fiber would be minimally affected. Thus, under the hybrid amplitude–phase modulation scheme, several issues remain to be resolved. Under the MADPSK scheme, the delay of the polarization modes would generate the phase difference or phase distortion on the I and Q components, and hence an enhancement of the distortion effects of the ISI. The amplitude distortion would then be increased but is considered to be a secondary effect.

7.3.6.7.3 Nonlinear Effects on MADPSK

Nonlinear effects occur owing to the nonlinear response of the fiber glass to the applied optical power. Fiber nonlinearity can be classified into stimulated scattering and the Kerr effect. Among several stimulated scattering effects, stimulated Raman scattering, caused by the interaction between light and the acoustical vibration modes in the fiber glass, is the most critical. Under this

0 km mismatch SSMF

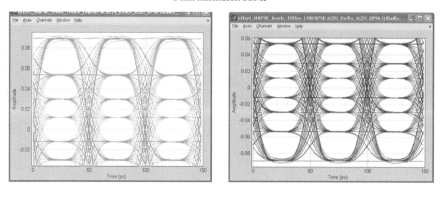

2 km mismatch dispersion SSMF

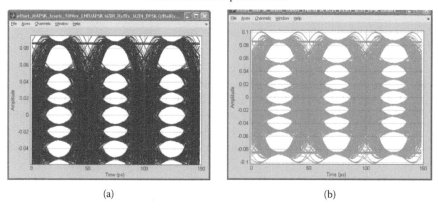

(a)   (b)

FIGURE 7.41 40 Gbps MADPSK eye diagrams of the (a) I and (b) Q components (A) 0 km—back-to-back, (B) 2 km SSMF mismatch over three 100 km SSMF transmission spans (dispersion compensated).

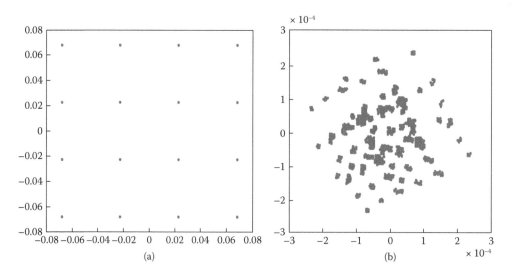

(a)   (b)

FIGURE 7.42 Constellation of 16-square QAM after two optically amplified spans and 2 km SSMF dispersion mismatch: (a) prepropagation and (b) postpropagation.

mechanism, the optical signal is reflected back to the transmitter, and in WDM systems its power is also transferred from shorter to longer wavelengths, thus attenuating the signal and causing crosstalk. The Kerr effect is the cause of the intensity-dependent phase shift of the optical field. It manifests in three forms: self-phase modulation (SPM), cross-phase modulation (XPM), and four-wave mixing (FWM) provided the phase matching is satisfied.

SPM is usually the dominant effect in a single-channel DPSK system. The changes in instantaneous power of the optical pulses together with the ASE from associated OAs lead to intensity-dependent changes, the Kerr effect, in the guided medium refractive index, and hence the effective index of the guided mode. These changes are converted to the phase shifts or phase noise of the lightwave carriers. At the receiver, the phase noise is transferred back to intensity noise, which degrades the BER [27]. As mentioned earlier, the contribution of noise to different levels of the MADPSK scheme is very critical to determine the optimum decision thresholds. This is further complicated by these additional nonlinear effects, especially the nonlinear phase noise (NLPN) usually contributed by the SPM due to the outermost constellation. These NLPN effects from the outermost constellation to other inner circle signal spaces have never been investigated.

XPM becomes the most critical nonlinearity in WDM systems where the phase shifts (noise) in one channel comes from refractive index fluctuations caused by power changes in other channels. XPM becomes more pronounced when neighboring channels have equal bit rates [27]. FWM is basically a crosstalk phenomenon in WDM systems. When three wavelengths with frequencies, ω_1, ω_2, and ω_3 propagate in a nonlinear fiber medium at which the dispersion is zero, they combine and create a degenerate fourth wavelength that would fall in an active wavelength channel. If these parametric wavelengths fall in other channels, they cause crosstalk and degrade the performance of the system. Although FWM is expected to reduce the receiver sensitivity, in the proposed system, to minimize the effects of fiber nonlinearity, the maximum power of the optical signal should not be set higher than a certain threshold. This maximum power dictates the amplitude of signal points in the outermost circle (circle 3), and hence other circles, of the signal space. Thus, optimization of the signal amplitude levels for MADPSK is critical.

7.3.6.8 Offset Detection

The 16-ary MADPSK signal model described in Section 7.3.2 can be modified. To balance the ASK and DQPSK sensitivities, the ASK signal levels are preliminary adjusted to the ratio $I_3/I_2/I_1/I_0 = 3/2/1.5/1$ and rotated by $\pi/4$ [6] as shown in Figure 7.40. These level ratios can be determined from the signal-to-noise ratio at each separation distance of the eye diagram or Q factor. The noise is assumed to be dominated by the beat noise between the signal level and that of the ASE noise. The eye opening is expected to improve significantly as shown in Figure 7.41.

7.4 STAR 16-QAM OPTICAL TRANSMISSION

This section gives a description of the simulation of the transmission performances of optical transmission systems over ten spans of dispersion-compensated and optically amplified fiber transmission systems. The modulation format is focused on the Star 16-QAM with two level and eight phase state constellation. Optical transmitters and coherent receivers are the main transmission terminal equipment; other constellations of the 16-QAM are described very briefly. Simulation results have shown that it is possible to transmit and detect the data symbols for 43 Gbps with the possibility of scaling to 107 Gbps without much difficulty. The OSNR with 0.1 nm optical filters is achieved with 18 and 23 dB for back-to-back and long-haul transmission cases with a DT of 300 ps/nm.

7.4.1 Introduction

To increase the channel capacity and bandwidth efficiency in optical transmission, the multilevel modulation formats such as QAM formats are of interest [43–50]. In digital transmission with multilevel (M levels) modulation, m bits are collected and mapped onto a complex symbol from an alphabet with M = 2m possibilities at the transmitter side.

The symbol duration is $T_s = mT_B$ with T_B as the bit duration, and the symbol rate is $f_s = f_B/m$ with $f_B = 1/T_B$ as the bit rate. This shows that for a given bit rate, the symbol rate decreases if the modulation level increases. That means a higher bandwidth efficiency can be achieved by a higher-order modulation format. For the 16-QAM format, $m = 4$ bits are collected and mapped to one symbol from an alphabet with $M = 16$ possibilities. In comparison to the case of the binary modulation format, only $m = 1$ bit is mapped to one symbol from an alphabet with $M = 2$ possibilities. With 16-QAM format and a data source with a bit rate of $f_B = 40$ Gbps, only a symbol rate of $f_s = 10$ GBaud/s is necessary. From the commercial point of view, this means a 40 Gbps data rate can be transmitted with 10 Gbps transmission devices. In the case of binary transmission, the transmitter needs a symbol rate of $f_s = 40$ GBaud/s. This means 16-QAM transmission requires four times slower transmission devices than that for the binary transmission. It is noted here that 10.7 Gsymbol/s is used as the symbol rate so as to compare the simulation results with the well-known 10.7 Gbps modulation schemes such as DPSK, CSRZ-DPSK, and so on. For a 107 Gbps bit rate, the transmission performance, that is, the sensitivity and the OSNR, can be scaled accordingly without any difficulty.

This section gives a general approach to the design and simulation of Star 16-QAM with two amplitude levels and eight phase states forming two star circles. We term this Star 16-QAM as 2A-8P Star 16-QAM, two amplitude level and eight phase states. The transmission format is discussed with theoretical estimates and simulation results to determine the transmission performance. The optimum Euclidean distance is defined for the design of star 16-QAM. Then in the second Section 7.4.4, the two detection schemes, namely direct detection and coherent detection, for Star QAM constellations are discussed.

7.4.2 Design of 16-QAM Signal Constellation

There are many ways to design a 16-QAM signal constellation. The three most popular constellations for 16-QAM modulation schemes are (1) Star 16-QAM, (2) Square 16-QAM, and (3) Shifted-square 16-QAM. The first two of these constellations are implemented. However, only the Star 16-QAM with two amplitudes and eight phases per amplitude level are employed in this section.

7.4.3 Signal Constellation

The signal constellation for Star 16-QAM with Gray coding is shown in Figure 7.43. The binary presentation of the symbols in the figure is shown in the symbol to bit presentation mapping of Table 7.5.

As can be seen from the figure, the symbols are evenly distributed on two rings, and the phase differences between the neighboring symbols on the same ring are equal ($\pi/4$). In order to detect a received symbol, its phase and amplitude must be determined. In other words, between two amplitude levels of the rings and among eight phase possibilities, there are a number of ways to build this constellation.

The ring ratio (RR) for this constellation is defined as RR = b/a, where a and b are the ring radii as shown in Figure 7.43. The RR can be set to different values to optimize the transmission performance.

7.4.4 Optimum Ring Ratio for Star Constellation

From Figure 7.46 (later in the chapter), it can be seen that there are many ways to choose the RR for the star 16-QAM constellation. Here, the theoretically best RR is defined to minimize the error probability in an AWGN channel by maximizing the minimum distance d_{min} between the neighboring symbols. The results for the AWGN channel can be used approximately for optical transmission. For Star 16-QAM, the minimum distance d_{min} is maximized when

$$d_1 = d_2 = b - a = d_{min} \qquad (7.13)$$

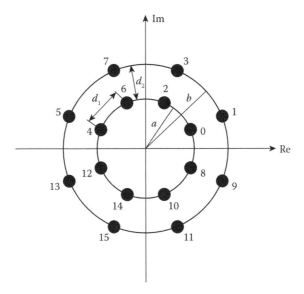

FIGURE 7.43   Theoretical arrangement of the modulation constellation for Star 16-QAM.

TABLE 7.5

Symbol Mapping and Coding for Star 16-QAM

Symbol to Bit Presentation

$0 \rightarrow 0000$	$4 \rightarrow 0100$	$8 \rightarrow 1000$	$12 \rightarrow 1100$
$1 \rightarrow 0001$	$5 \rightarrow 0101$	$9 \rightarrow 1001$	$13 \rightarrow 1101$
$2 \rightarrow 0010$	$6 \rightarrow 0110$	$10 \rightarrow 1010$	$14 \rightarrow 1110$
$3 \rightarrow 0011$	$7 \rightarrow 0111$	$11 \rightarrow 1011$	$15 \rightarrow 1111$

Gray Coding for Star 16-QAM

$0 \rightarrow 1$	$4 \rightarrow 7$	$8 \rightarrow 15$	$12 \rightarrow 9$
$1 \rightarrow 0$	$5 \rightarrow 6$	$9 \rightarrow 14$	$13 \rightarrow 8$
$2 \rightarrow 3$	$6 \rightarrow 5$	$10 \rightarrow 13$	$14 \rightarrow 11$
$3 \rightarrow 2$	$7 \rightarrow 4$	$11 \rightarrow 12$	$15 \rightarrow 10$

Mapping for Star 16-QAM

$0 \rightarrow 1$	$4 \rightarrow 9$	$8 \rightarrow 13$	$12 \rightarrow 5$
$1 \rightarrow 0$	$5 \rightarrow 8$	$9 \rightarrow 12$	$13 \rightarrow 4$
$2 \rightarrow 3$	$6 \rightarrow 11$	$10 \rightarrow 15$	$14 \rightarrow 7$
$3 \rightarrow 2$	$7 \rightarrow 10$	$11 \rightarrow 14$	$15 \rightarrow 6$

Gray Coding for Square 16-QAM

$0 \rightarrow 12$	$4 \rightarrow 11$	$8 \rightarrow 13$	$12 \rightarrow 4$
$1 \rightarrow 10$	$5 \rightarrow 2$	$9 \rightarrow 5$	$13 \rightarrow 3$
$2 \rightarrow 15$	$6 \rightarrow 8$	$10 \rightarrow 14$	$14 \rightarrow 7$
$3 \rightarrow 9$	$7 \rightarrow 1$	$11 \rightarrow 6$	$15 \rightarrow 0$

Mapping for Square 16-QAM

$0 \rightarrow 10$	$4 \rightarrow 11$	$8 \rightarrow 4$	$12 \rightarrow 15$
$1 \rightarrow 6$	$5 \rightarrow 1$	$9 \rightarrow 14$	$13 \rightarrow 3$
$2 \rightarrow 5$	$6 \rightarrow 2$	$10 \rightarrow 13$	$14 \rightarrow 0$
$3 \rightarrow 9$	$7 \rightarrow 8$	$11 \rightarrow 7$	$15 \rightarrow 12$

With some geometrical calculations, it can be obtained that

$$d_{min} = 2a \cdot \sin(22.5°) \qquad (7.14)$$

which leads to an optimal RR of

$$RR_{opt} = b/a = (d_{min} + a)/a = (2a \cdot \sin 22.5 + a)/a \approx 1.77 \qquad (7.15)$$

The average power of the star 16-QAM constellation can be determined as

$$P_0 = (8a^2 + 8b^2)/16 = (a^2 + b^2)/2 \qquad (7.16)$$

Thus, we have the relationship between the average optical power and the minimum distance between the two rings of the two amplitude levels as

$$d_{min} \approx 0.53(P_0)^{1/2} \qquad (7.17)$$

The obtained $RR_{opt} = 1.77$ does not depend on P_0 and is constant for each P_0 value. For an average power of 5 dBm (3.16 mW), $d_{min} = 2.98 \cdot 10^{-2} \sqrt{W}$, $a = 3.89 \cdot 10^{-2} \sqrt{W}$, and $b = 6.87 \cdot 10^{-2} \sqrt{W}$ are obtained.

7.4.4.1 Square 16-QAM

The signal constellation of the square 16-QAM with Gray coding is shown in Figures 7.44. A generic digram of the signal constellation of the 16-QAM modulation scheme is shown in Figure 7.45.

The binary presentation of the symbols in the figure is shown in the symbol to bit presentation mapping in Table 7.5. In the constellation of the square 16-QAM, the 16 symbols are equally separated from their direct neighbors and have a total of 12 different phases, that is, three phases per quarter, distributed on three rings. The phase differences between neighboring symbols on the inner and outer rings are equal ($\pi/2$), but the phase differences between neighboring symbols on the middle ring are different (37° or 53°). If the distance between direct neighbors in the square 16-QAM is rotated as $2d$, the average symbol power (P_0) of the constellation is

$$P_0 = 10 \cdot d_2 \qquad (7.18)$$

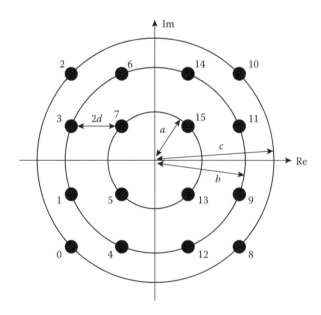

FIGURE 7.44 Square 16-QAM signal constellation.

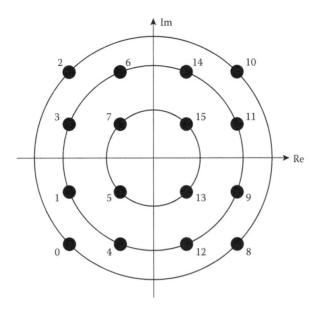

FIGURE 7.45   Generic square 16-QAM signal constellation.

For an average power of 5 dBm (3.16 mW), it can be computed that $d = 1.77 \cdot 10^{-2} \sqrt{W}$ and from it: $a = 2.5 \cdot 10^{-2} \sqrt{W}$, $b = 5.6 \cdot 10^{-2} \sqrt{W}$, and $c = 7.5 \cdot 10^{-2} \sqrt{W}$. In comparison with star 16-QAM, here the distances between the middle ring and the outer ring are much smaller. This means that to achieve the same BER, square 16-QAM needs a higher average power than star 16-QAM. The decision method for square 16-QAM is more complicated than that for Star 16-QAM. First, the decision between the three amplitude possibilities of each ring should be made; then, depending on the ring level, the decision is made between four or eight phase possibilities.

7.4.4.2   Offset-Square 16-QAM

To optimize the phase detection of the middle ring, it is envisaged that the phase differences between neighboring symbols on the middle ring in square 16-QAM should be equal. Thus, the shifted-square 16-QAM is introduced by shifting (rotation) symbols on the middle ring to obtain equal phase differences between all neighboring symbols as shown in Figure 7.7. After shifting the symbols on the middle ring, the distances between all direct neighbors are not necessarily equal. In comparison with square 16-QAM, this constellation may offer more robust detection against phase distortions according to our amplitude and phase detection method introduced in Section 7.4.7.

7.4.5   Detection Methods

In the case of differential encoding for the 16-QAM format, as described in Chapters 4 and 5, two different detection methods can be employed to demodulate and recover the data in the receiver: (1) direct detection and (2) coherent detection. Further details on the receivers will be given in Section 7.4.7.

In this section, direct detection means detection with Mach–Zehnder delay interferometric (MZDI) or (2 × 4) 90° hybrid, and coherent is similar except that a local oscillator (LO), a very narrow-linewidth laser, is used to mix the signal and its lightwaves to generate the IF or baseband signals with preservation of the modulated phase states. Each of these two receiving methods has different implementations, which can be introduced as follows.

7.4.5.1 Direct Detection

Unlike coherent detection, differential decoding is done for direct detection in the optical domain. Indeed, this is equivalent to self-homodyne coherent detection. This has the disadvantage of the transmitted absolute phase being lost after differential decoding. However, the relative phase (the phase of differential decoded signal) remains in the electrical domain, which makes electrical equalization still possible. The equalization with relative phases is more difficult, and the results can be worse than that with absolute phases. The advantage of direct detection, compared to coherent detection, is that the synchronization of a local laser with that of the signal lightwave is omitted. There are two methods to implement direct detection: one is with MZDI, and the other is with a (2×4) $90°$ hybrid coupler.

7.4.5.2 Coherent Detection

In a coherent receiver, an LO is used to mix its signal with the incoming signal lightwave for demodulation. As a result, the phase can be preserved in the electrical domain. This makes the electrical equalization very effective in coherent detectors. For coherent detectors, differential decoding is done in the electrical domain. On the basis of the intermediate frequency (f_{IF}) defined as $f_{IF} = f_s - f_{LO}$, three different coherent methods can be distinguished: (1) homodyne receiver, (2) heterodyne receiver, and (3) intradyne receiver. Only the homodyne receiver is included in this section, and the other two are only briefly mentioned.

7.4.5.2.1 Homodyne Receiver

A receiver is called homodyne when the carrier frequency (f_s) and the LO frequency (f_{LO}) are the same

$$f_{IF} = f_s - f_{LO} = 0 \tag{7.19}$$

In practice, because of the laser linewidth, a carrier synchronization must be implemented to set the center frequency and the phase of the LO to the same values as those in the incoming signal. For homodyne receivers, carrier synchronization can be implemented in the optical domain via an optical phase-locked loop (OPLL). Carrier synchronization failure causes degradation in the receiver's performance, but in this chapter, this effect is not considered, and perfect synchronization in the receiver (a perfect single spectrum line) is assumed. Alternatively, as mentioned later, a heterodyne receiver using only one $\pi/2$ hybrid coupler with the associated electronic demodulation circuitry can be used to simplify the receiver configuration for coherent detection. Polarization control is another critical difficulty in all coherent receivers, which too is not included in this book. The implementation of homodyne receivers for the star 16-QAM is described in several textbooks.

7.4.5.2.2 Heterodyne Receiver

For this kind of receiver, the following applies

$$f_{IF} = f_s - f_{LO} \neq 0 > B_{opt} \tag{7.20}$$

B_{opt} is the optical bandwidth of the transmitted signal. The IF will be mixed in the electrical domain with a synchronous or asynchronous method in the low-pass domain. In the case of synchronous demodulation, the phase synchronization can be done in the electrical domain. The implementation complexity of heterodyne receivers in the optical domain is less than that of homodyne receivers.

7.4.5.2.3 Intradyne Receiver

The intradyne receiver requires

$$f_{IF} = f_s - f_{LO} \neq 0 < B_{opt} \tag{7.21}$$

The phase synchronization in the intradyne receiver can be done in the digital domain. That makes the intradyne receiver less complex in the optical domain than the homodyne receiver.

The intradyne receiver compared to the heterodyne receiver has the advantage that its processing bandwidth is smaller. The disadvantage of the intradyne receiver is that it has a higher laser line-width requirement than the heterodyne receiver.

7.4.6 TRANSMITTER DESIGN

There are many ways to implement the transmitter for star 16-QAM described in the previous section. For the simulations in this book, the parallel transmitter shown in Figure 7.46 is implemented. The bit stream enters the differential encoder module after serial-to-parallel conversion.

The differential encoder implements the following processes: (1) The four parallel bits that have arrived at the module are mapped (Gray coding) into symbols according to the gray coding for star 16-QAM mapping; (2) the precoded symbols are differentially encoded (differential coding); and (3) the differentially encoded symbols are mapped again to other symbols to drive the Mach–Zehnder modulators (MZMs) according to the star 16-QAM mapping of Table 7.5.

Each symbol at the output of the differential encoder module is represented by four bits. The bits are sent to pulse formers. The first two bits drive the first MZMs, with lightwaves generated from the CW laser. If the input bit is equal to 1, then the output of the MZM is −1, and in the other case the output of the MZM is a 1 (after sampling). After combining the output signals of these two MZMs and considering the 90° phase delay in one arm, we obtain the QPSK signal shown in Figure 7.46.

The third bit from the differential encoder output drives a PM to obtain the 8-PSK signal constellation from the QPSK signal [50–57]. If this bit is equal to 1, then the QPSK symbol will rotate by $\pi/4$. The 8-PSK signal constellation is shown in Figure 7.47. To achieve the two-level star 16-QAM signal constellation, another MZM is used to generate the second amplitude. If the fourth bit of the differential encoder output is 1, then this output symbol is set on the outer ring of the constellation, otherwise on the inner ring. This MZM sets the RR of the constellation. The signal constellation after MZM3 is shown in Figures 7.48 and 7.49.

The signal constellation in Figure 7.48 can be constructed from the whole constellation in Figure 7.43 with a rotation of $\pi/6°$. The advantage of this rotation is that on the real and imaginary axis of the constellation, only eight different amplitude levels instead of nine levels exist. So, another PM can be used between MZM3 and MZM-RZ to rotate the constellation by $\pi/6°$. This additional PM is not shown in Figure 7.46. To increase the receiver sensitivity and reduce the signal chirp, a RZ pulse carving with a duty cycle of 50% should be implemented at the end of the transmitter with an MZM driven by a sinus signal generator (SG). In our simulations, the MZMs in Figure 7.46 worked in push–pull operation, and the PMs were MZMs working as phase modulators.

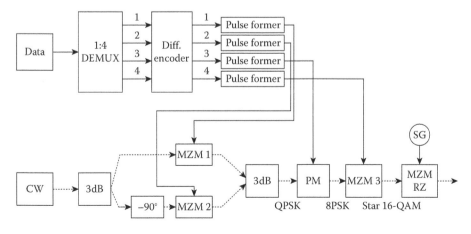

FIGURE 7.46 Schematic diagram of the optical transmitter for Star 16-QAM.

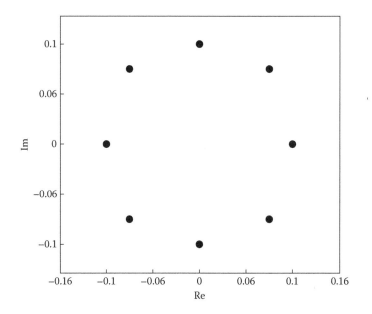

FIGURE 7.47   Constellation of the first amplitude level generated from the optical transmitter for Star 16-QAM.

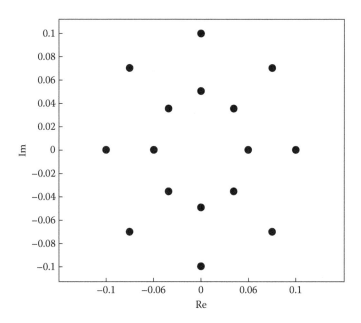

FIGURE 7.48   Constellation of the first and second amplitude levels generated from the optical transmitter for Star 16-QAM.

7.4.7   RECEIVER FOR 16-STAR QAM

In this section, the implementation of a direct detection receiver and coherent receiver for star 16-QAM is explained. For coherent detection, there are many possibilities in the digital domain of the receiver to recover the data. The two methods implemented in this study detect the symbol before realizing differential decoding. The difference is that one detects the symbol directly using

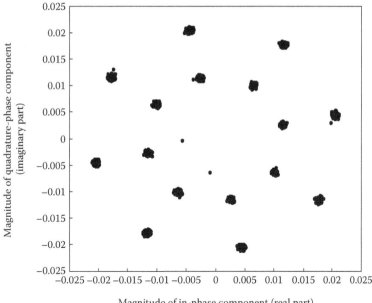

FIGURE 7.49   Constellation of the first and second amplitude level at the receiver of Star 16-QAM after ten spans of dispersion-compensated standard SMF links.

the method described in Section 7.4.2.1, while the other method employs a phase estimation algorithm, described in Section 7.4.2.2, before the symbol detection in order to cancel out the phase distortions (phase synchronization between the LO and the received signal). Another possibility, which is not implemented in this book, is to first do the differential decoding of the incoming signal and then the symbol detection.

7.4.7.1   Coherent Detection Receiver without Phase Estimation

The structure of a coherent detection receiver is shown in Figure 7.50a. After transmission over fiber, the signal is amplified by an EDFA. The input power of the EDFA can be changed via an attenuator to set the OSNR. The output signal of the EDFA is sent to a bandpass (BP) filter in order to reduce the noise bandwidth.

An attenuator used to set the OSNR is required. The output signal of the EDFA is sent to a BP filter in order to reduce the noise bandwidth. The signal from the LO and the output of the BP filter are sent to a (2 × 4) π/2-hybrid and after it to two balanced detectors. The (2 × 4) 90°-hybrid and the balanced detectors demodulate and separate the received signal into in-phase (I) and quadrature (Q) components. The structure of the (2 × 4)π/2-hybrid coupler, the balanced detectors, and their mathematical description can be found in many published works on coherent optical communication technology. This coherent detector can be simplified further if heterodyne detection is used, as shown in Figure 7.50b, and is commercially available from Discovery Semiconductor [2].

Furthermore, an amplitude direct detection of in the electronic digital processor must be able to process the magnitude of the vector formed by I and Q components to determine the amplitude and phase of the received signals and hence their corresponding positions on the constellation and the decoding of the data symbols.

After LP filtering and sampling of the I and Q components, the samples are sent to the symbol detection and differential decoding module. The sampling is done in the center of the eye diagram. In the symbol detection and differential decoding module, we first recover the symbols

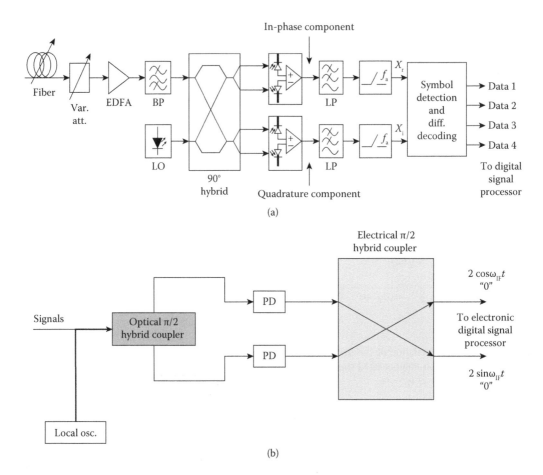

FIGURE 7.50 Coherent receiver for Star 16-QAM: (a) homodyne and heterodyne I and Q detection model and (b) heterodyne model with optical and electrical $\pi/2$ hybrid couplers-electrical detection of I and Q components.

from the incoming samples, then perform the differential decoding of symbols. In order to recover the symbols, the I and Q components are added together to a complex signal. Now, according to the original signal constellation, a decision must be made regarding to which symbol our complex sample must be mapped. This decision has two parts. First, an amplitude decision is made to determine to which ring the sample belongs. After this, a phase decision is made to determine to which of the eight possible symbols on a ring our sample belongs.

For the amplitude decision, a known bit sequence for the receiver (training sequence) is used, and an amplitude threshold a_{th} is defined according to

$$a_{th} = \{\max 1 \le k \le n \mid s_1(k)\mid + \min 1 \le k \le m \mid s_2(k)\mid\}/2 \qquad (7.22)$$

where $s_1(k)$ is the kth complex received samples of symbols on the inner ring, and n is the total number of symbols on the inner ring. $s_2(k)$ is the kth complex received sample of symbols on the outer ring and m is the total number of symbols on the outer ring. If the amplitude of one sample is larger than the threshold, the symbol is decided to be on the upper ring, otherwise on the inner ring. For the phase decision, the complex plane is divided into eight equiphase intervals. An index from 0 to 7 is assigned to this sample according to the interval the phase of the sample falls in. The next steps are differential decoding and mapping. The amplitude differential decoding and phase differential decoding are done separately. From their results, the symbol detection and after it the symbol-to-bit mapping is done in an inverse manner to that of the encoding in the transmitter.

Alternatively, the detection can be conducted with a heterodyne receiver that uses only a single π/2 hybrid optical coupler. It is then detected by the two photodiodes and coupled through a π/2 electrical hybrid coupler to detect the I and Q components for the phase and amplitude reconstruction of the received signals as shown in Figure 7.50b. The phase estimation can then be estimated by processing the I and Q signals in the electronic domain as described in the next section.

Owing to the two levels of Star 16-QAM, there must be an amplitude detection subsystem that can be implemented using a single photodetector followed by an electronic preamplifier as shown in later in the chapter in Figure 7.52. An electronic processor would be able to determine the position of the received signals on the constellation, and hence decoding can be implemented without any problem. The transmission performance presented here would not be affected. Only the technological implementation would be affected, and hence the electronic noise or optical noises' contribution to the receiver can thus be taken into account.

7.4.7.2 Coherent Detection Receiver with Phase Estimation

The method and structure of this receiver is almost the same as for the previous receiver shown in Figure 7.50. The difference is that here a phase estimation is done before the phase decision in the symbol detection and differential decoding module. The dispersion of the single-mode optical fiber is purely phase effects and thus causes phase rotation, resulting in a phase decision error. The effect of dispersion on the Star 16-QAM format is shown in Figure 7.51.

The left plot (a) in the figure shows the sampled input signal into the fiber, and the plot on the right-hand side shows the sampled output of the fiber with a CD of 300 ps/nm. The fiber is considered to be linear in this simulation. A comparison of point A in both figures shows that this point is spread and rotated owing to dispersion. The spreading causes both phase and amplitude distortion, while the rotation causes only phase distortion. To solve the problem of phase rotation, the following phase estimation method is implemented. In general, this phase estimation method is for phase synchronization between the LO with the linewidth and signal to replace the OPLL.

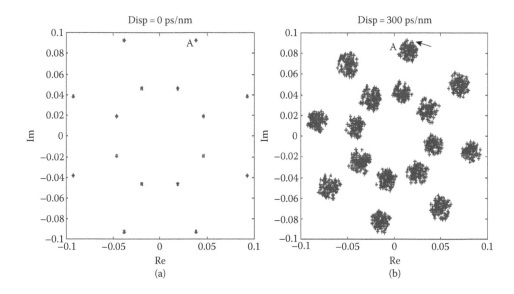

FIGURE 7.51   Signal constellation (a) before and (b) propagation through the optical fiber link-phase rotation.

The incoming signal after sampling can be described as

$$c(k) = A e^{j[\phi'_{tot}(k) + \phi_{mod}(k)]} \tag{7.23}$$

where ϕ'_{tot} is the phase distortion due to the dispersion and noise, and ϕ_{mod} is the signal phase that must be recovered. Now, ϕ_{tot} must be eliminated from c(k). ϕ_{mod} values are $\pi/8$, $3\pi/8$, $5\pi/8$, $\pi/8$, $9\pi/8$, $11\pi/8$, $13\pi/8$, and $15\pi/8$. If these phase values are multiplied by 8, then

$$c_8(k) = c^8(k) = A^8 e^{j(8\phi'_{tot}(k) + 8\phi_{mod}(k))} = A^8 e^{j(8\phi'_{tot}(k) + \phi_{mod}(k))} = -A^8 e^{j(8\phi'_{tot}(k))} \tag{7.24}$$

and from this

$$\phi'_{tot}(k) = 1/8 \arg(-c_8(k)) \tag{7.25}$$

$\phi'_{tot}(k)$ is the estimated phase for $\phi'_{tot}(k)$. In our simulations, $\arg(c_8(k))$ is filtered (the filter takes the average of 20 neighbor symbols) to avoid the phase jumps from symbol to symbol. The filter order of 20 is not optimized for each CD.

Now, the signal phase $\phi_{mod}(k)$ can be estimated as

$$\phi_{mod}(k) = \arg(c(k)) - \phi'_{tot}(k) = \phi_{mod}(k) + \phi'_{tot}(k) - \phi'_{tot}(k). \tag{7.26}$$

After this phase estimation, the signal decision takes place by employing the same method as for the amplitude decision case.

7.4.7.3 Direct Detection Receiver

The block diagram of the direct detection receiver is shown in Figure 7.52. After the optical filter, the signal is split into two branches via a 3 dB coupler. We name these two branches the intensity branch and the phase branch. In the phase branch, the phase differential demodulation is done in the optical domain. The signal and the delayed signal at T_s (symbol duration) are sent into the (2 × 4) 90°-Hybrid and after that again into balanced detectors. At the output of the balanced detectors, the in-phase and quadrature components of the demodulated and differential decoded and received signal can be derived. After electrical filtering, the signal is sampled and then sent into the symbol detection module. In the amplitude branch, the amplitude is determined and differentially decoded. After the photodiode, the signal is low-pass-filtered and sampled and then fed into the amplitude

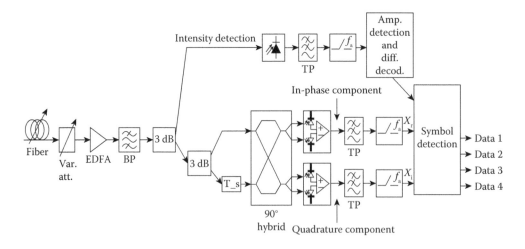

FIGURE 7.52   Direct detection receiver for Star 16-QAM.

detection and differential decoding module, a well-known optical coherent structure can be used to accomplish the amplitude decision and differential decoding. At the end, the in-phase and quadrature components and the amplitude branch are sent to the symbol detection module for further processes such as symbol detection and symbol-to-bit mapping.

7.4.7.4 Coherent Receiver without Phase Estimation

In this section, the required OSNR at 10^{-4} BER (using Monte Carlo simulation) is determined for the coherent and incoherent direct detection receivers. For each detection method, the optimum RR is obtained to minimize the OSNR at BER = 10^{-4} (the BER is determined via Monte Carlo simulations). After that, the DT at 2 dB OSNR penalty at BER = 10^{-4} is determined. The OSNR penalty is only 2 dB in our work and can have other values. The OSNR penalty is defined as the OSNR difference in decibels between the OSNR of the back-to-back case and the OSNR of other CD values. The DT is the CD interval that can be achieved with a certain OSNR penalty. In practice, the DT describes how much dispersion (residual dispersion) a system can tolerate with an OSNR penalty smaller than 2 dB. The simulations in this book are done with the simulation tool for both the linear and nonlinear channels. The simulation parameters are given in Table 7.6. The average input power of nonlinear fiber in our work is always 5 dBm.

7.4.7.4.1 Linear Channel

In Figure 7.53b, the optimum RR (RR_{opt}) can be seen for each CD. RR_{opt} is the RR that minimizes the OSNR for the given CD.

The optimum RR changes here nearly linear with CD and can be expressed as

$$RR_{opt} = -0.002\,|CD| + 1.92 \quad \text{for} \quad 50\,\text{ps/nm} \le |CD| \le 300\,\text{ps/nm} \tag{7.27}$$

RR_{opt} increases with CD because an increase in CD means an increase in the phase rotation due to dispersion. This causes more phase detection errors and thus a higher OSNR requirement. An increase in the RR reduces the phase error probability but increases the amplitude error probability. RR_{opt} is the best trade-off between phase errors and amplitude errors for each CD.

In the back-to-back case, RR_{opt} is around 1.87. The theoretical value of RR_{opt} is obtained as 1.77. The difference is because in the introduced coherent receiver, at first the symbol detection is done and then the differential decoding. In the case of differential decoding before symbol detection, the optimum RR is around 1.77 as excepted. To determine the RR_{opt} for each CD, the RR is changed for each CD. The RR value that yields the smallest OSNR is RR_{opt}. The RR step (it determines the RR accuracy) in simulations is 0.05. The characteristic for other CDs is similar to Figure 7.53a. To compare the OSNRs, the reference in this work is the OSNR from the back-to-back case. In the case of residual dispersion, it is of interest how the system performance changes with the change of RR. The simulation results of three different RRs can be seen in Figure 7.52.

TABLE 7.6

Simulation Parameters for 16-Star QAM

$f_s = 10.7$ GHz	$\lambda_c = 1550$ nm
$P_{laser} = 7$ dBm	$\alpha_{att} = 0.21$ dB/km
$\gamma = 0.00137$ 1/W/m	$F_n = 5$ dB
$G_{EDFA} = 30$ dB	fil-opt. = Gaussian 1. order
$B_{opt} = 44$ GHz	$\lambda_{LO} = 1550$ nm
$P_{LO} = 0$ dBm	fil-el. = Butterworth 3. degree
$B_c = 11$ GHz	$R_{photodiode} = 1$ A/W

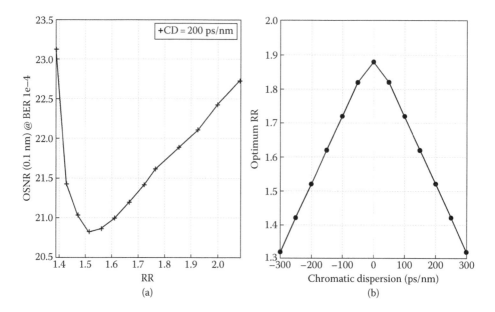

FIGURE 7.53 (a) OSNR versus RR (left) and (b) optimum RR versus CD (right) for coherent receiver without phase estimation.

As shown in Figure 7.52a, the OSNR performance for the back-to-back case is decreased if (in the simulated interval) the RR from 1.87 decreases. A degradation of 6.5 dB is determined if the RR decreases from 1.87 to 1.32. From the other side, the DT at 2 dB OSNR penalty increases (Figure 7.9b). The DT for RR = 1.87 is 220 ps/nm and for RR = 1.32 is 460 ps/nm. To understand the reason for this behavior, Figure 7.53a should be considered again. For each CD, if the RR decreases from the RR_{opt} value, the OSNR increases rapidly. That means the RR value of 1.32 for the back-to-back case has increased the OSNR, but CD = 300 ps/nm has the minimum OSNR for this RR. The result is that the OSNR for the back-to-back case increases and decreases for the case without phase estimation in the linear channel. CD = 300 ps/nm. This effect causes a larger DT at a certain OSNR penalty. In a practical system, according to the higher requirement for the OSNR or DT, the RR can be chosen.

In Figure 7.54a (left), it can be seen as well that the required OSNR for |CD| = 300 ps/nm makes a jump compared to other CDs. With |CD| = 350 ps/nm, it is not possible to achieve a BER of 10^{-4}. The reason is that |CD| = 350 ps/nm is the limit of the system. For this CD, it is not possible to transmit error free even without noise owing to phase rotations caused by dispersion (phase detection error). The signal constellation of the received signal after sampling with CD = 350 ps/nm can be seen in Figure 7.55; for example, some of the received symbols of A are over the phase threshold line, and they generate the detection errors.

A typical eye diagram at the output of the coherent receiver at the limit of the distortion is shown in Figure 7.56. Phase estimation can be implemented in the digital signal processor. Note that the signal constellation is rotated uniformly owing to the phase evolution of the spectral components of the modulated signals when propagating through the single-mode fiber as described in Chapter 2, and thus in the processing of the constellation. It is best if the reference frame of the phasor diagram is rotated to align with the constellation, which may simplify the phase estimation process at $\pi/8$ and its multiple values.

7.4.7.4.2 Nonlinear Effects

The optimum RR in the case of a nonlinear channel for different CDs is shown in Figure 7.57. As mentioned earlier, for linear channels, RR_{opt} decreases with increase of CD too (Figure 7.58). The difference from Figure 7.53 is that the diagram is not symmetric. The reason for this is the

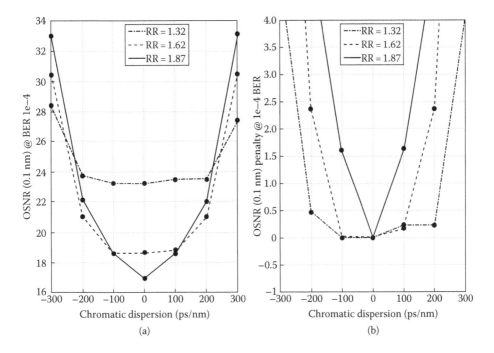

FIGURE 7.54 (a) OSNR versus CD and (b) OSNR penalty versus CD for coherent receiver.

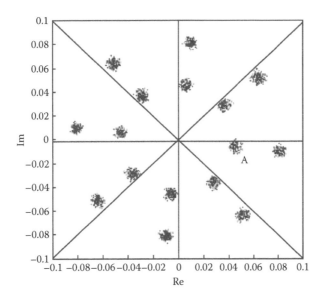

FIGURE 7.55 Received signal constellation with CD = 350 ps/nm without noise for coherent receiver without phase estimation in linear channel.

interaction between the dispersion and the SPM. For a positive CD, the dispersion and the SPM have a constructive interaction, which results in a better OSNR performance. For a negative CD, the dispersion and the SPM have a destructive interaction that results in a much worse OSNR performance.

This effect can be seen if Figure 7.53b (right) is compared with Figure 7.57 as shown in Figure 7.58. The curve slope for negative CD is higher in the nonlinear channel, which means the

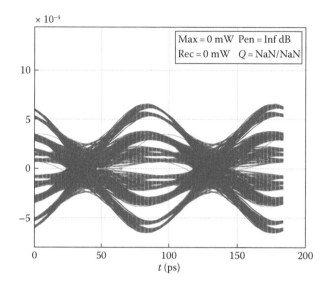

FIGURE 7.56 Typical eye diagram at the output of the coherent receiver under significant distortion limit of the 2A-8P Star 16-QAM.

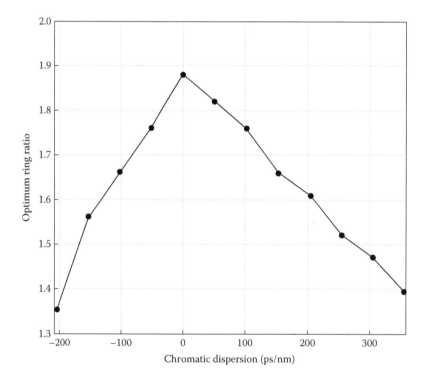

FIGURE 7.57 RR_{opt} versus CD in nonlinear channel for coherent receiver without phase estimation.

phase distortion is higher in the nonlinear channel. For positive CD, the slope in the nonlinear channel is lower, which means the phase distortion is less. The simulation results for different RRs are shown in Figure 7.59. Here again, it can be seen that the required OSNR for the back-to-back case increases with a decrease in the RR. Similarly, for the same reason as for the linear case, the DT at 2 dB OSNR penalty increases as well.

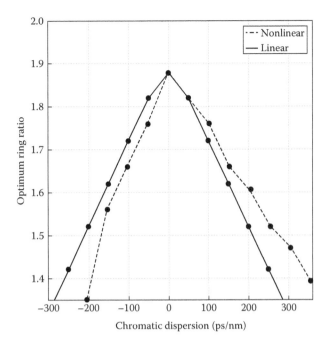

FIGURE 7.58 RR_{opt} comparison in linear and nonlinear channels for coherent receiver without phase estimation.

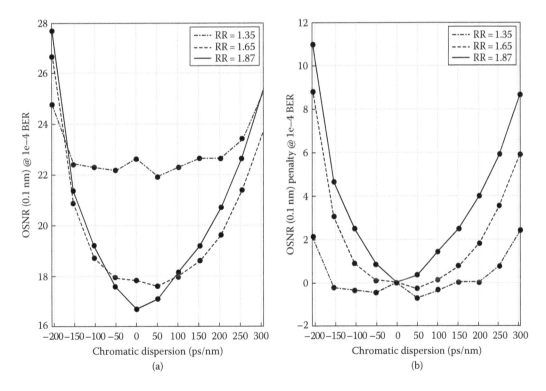

FIGURE 7.59 (a) OSNR versus CD and (b) OSNR penalty versus CD for coherent receiver without phase estimation in nonlinear channel. Note: No equalization, only phase estimation processing of I and Q received components in the electrical domain.

7.4.7.5 Remarks

The design of a Star 16-QAM modulation scheme is proposed for coherent detection for ultra-high-speed ultra-high-capacity optical fiber communications schemes. Two amplitude levels and eight phases (2A-8P 16-QAM) are considered to offer simple transmitter and receiver configurations and at the same time the best receiver sensitivity at the receiver. An optical SNR of about 22 dB is required for the transmission of Star 16-QAM over an optically amplified transmission dispersion-compensated link. A DT of 300 ps/nm is possible with a 1 dB penalty of the eye opening at 40 Gbps bit rate or 10 Giga-symbols/s with an OSNR of 18 dB. The OSNR could be about 22 dB for a 107 Gbps bit rate and a symbol rate of 26.75 G symbols/s. The transmission link consists of several spans of a total of 1000 km dispersion-compensated optically amplified transmission link. The optical gain of the in-line OAs is set to compensate for the attenuation of the transmission and compensating fibers with a noise figure of 3 dB.

The optical transmitter and receivers incorporating commercially available coherent receiver are structured and are sufficient for engineering the optical transmission terminal equipment for a bit rate of 107 Gbps and a symbol rate of 26.3 Gsymbol/s.

Furthermore, electronic equalization of the receiver PSK signals can be done using blind equalization, which would further improve the DT. For a symbol rate of 10.7 Gbps, this DT for a 1 dB penalty would reach 300 km of standard SMF. This electronic equalization can be implemented without any difficulty at 10.7 Gsymbol/s. For a 107 Gbps bit rate, a similar improvement of the DT can be obtained at 26.5 Gsymbol/s provided the electronic sampler can offer more than 50 GSamples/sec sampling rate.

7.4.8 OTHER MULTILEVEL AND MULTI-SUBCARRIER MODULATION FORMATS FOR 100 GBPS ETHERNET TRANSMISSION

Numerous technologies have been introduced in recent years to cope with the ever-growing demand for transmission capacity in optical communications. Although the optical single-mode fiber offers enormous bandwidth in the order of magnitude of 10 THz, efficient exploitation of bandwidth started to become an issue a couple of years ago. Moreover, the limited speed of electronic and electro-optic devices such as modulators and photo receivers are considered a bottleneck for further increase of the data rate based on binary modulation.

For all these reasons, optical modulation formats offering a high ratio of bits per symbol are an essential technology for next-generation high-speed data transmission. In this way, data throughput can be increased while the required bandwidth in the optical domain as well as for electronic devices is kept at a lower level.

Because of the demand for transmission technologies offering high ratios of bits per symbol, two promising candidates to achieve a data rate of 100 Gbps per optical carrier are discussed, namely optical orthogonal frequency division multiplexing and 16-ary multilevel modulation. Their performance is analyzed by means of numerical simulation and by experiment.

7.4.8.1 Multilevel Modulation

Optical modulation formats incorporating four or eight bits per symbol were investigated in numerous contributions in the last couple of years (e.g., DQPSK [28] and 8-DPSK [29]). However, to carry out the step from 10 to 40 Gbps data rate using devices designed for 10 Gsymbol/s, the main challenge is to find the optimal combination of ASK and DPSK formats. Several approaches, which are reviewed in Ref. [30], are possible.

The simplest structure can be an extension of a 30 Gbps 8-DPSK by an additional PM resulting in 40 Gbps 16-DPSK. That is, in the complex plane, 16 symbols are placed onto a unit circle as shown in Figure 7.60a. Depending on the current bit at the data input of the additional PM, the 8-DPSK symbol is shifted by $\pi/8$ in case of a 1, while in case of a 0, the incoming phase of the symbol is preserved. Although it seems simple, experimental implementations have shown that

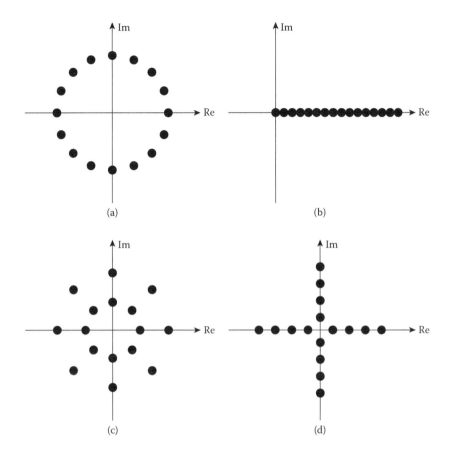

FIGURE 7.60 Constellation of symbols in the complex plane for (a) 16-DPSK, (b) 16-ASK, (c) Star 16-QAM, and (d) 16ADPSK.

the phase stability of the modulators and corresponding demodulators is very critical. Thus, the experimental setup must be stabilized. Moreover, 16-DPSK is suboptimal regarding exploitation of the full area that the complex plane offers. That is, the ratio of the symbol distance to signal power is low, resulting in poor receiver sensitivity.

Similar behavior can be seen for the other extreme case of 16-ASK, as shown in Figure 7.60b. Here, the 16 symbols are placed on the positive real axis. The symbol distance is extremely narrow, resulting in poor sensitivity as well.

A much improved performance can be achieved by combining the amplitude and phase shift keying modulations of ASK and DPSK, the M-ary ADPSK as described earlier. There are a number of combinations of M-ary ADPSK. One approach can be the extension of a 8-DPSK by two rings of ASK levels. Thus, the 16 symbols appear as two rings in the complex plane with eight symbols per ring as shown in Figure 7.60c. Alternatively, this topology is known as Star-16 QAM. A second structure is given by combining DQPSK with four-level ASK (or equivalently, M-ary ADPSK), resulting in four rings with four symbols each as analyzed in earlier sections and shown in Figure 7.60d. Both structures effectively utilize the complex plane. However, both structures require that the sensitivity or the magnitude (diameters) of the rings have to be optimized to compromise the sensitivity performance of the DPSK and the ASK geometrical distribution. Especially, the inner ring has to be of a sufficient size to enable the different phases of the symbols on this ring to be distinguished. They are limited by the nonlinear effects of the transmission fiber and the noise contributed by the receiver and the in-line OAs. Hence, the distance between the constellations is limited by these two limits.

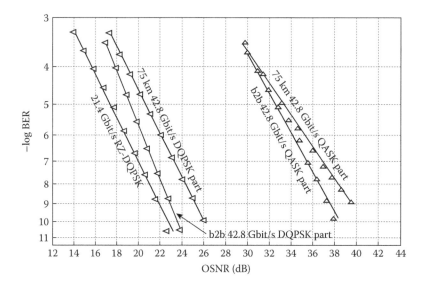

FIGURE 7.61   Measurement BER results for 16-ary inverse RZ-ADPSK modulation. Legend: b2b = back-to-back.

A strategy to mitigate this trade-off was introduced in Ref. 48 by using a special pulse shaping called inverted RZ. For binary ASK in conjunction with inverse RZ, a 0 is encoded as a temporary breakdown of the optical power, while for a 1 the optical power remains at a high level. Using this pulse shaping, for the M-ary ADPSK format, the four levels of the QASK part are transmitted by means of four different values for temporary decay of optical power. The DQPSK part, however, is transmitted by modulating the phase of the signal in the time slot in between, which implies that in the transmitter, the phase of the signal can be detected while the signal has maximum power.

Measurement results for this modulation format are depicted in Figure 7.61, where the BER is plotted as function of the OSNR measured within a 0.1 nm optical filter bandwidth. The main outcome is the fact that the DQPSK part is insignificantly disturbed by an additional QASK part. Moreover, even after transmission over 75 km of SSMF, the DPSK part shows a very low penalty. However, the QASK component inherently shows low performance due to the low symbol Euclidean distance. An improvement might be achieved by optimizing the duty cycle of inverse RZ and the RR. A simulated eye diagram of the 16-square QAM is shown in Figure 7.62 with the constellation before and after the optical transmission link of two optically amplified fiber spans with 2 km SSMF mismatch are shown in Figure 7.42.

7.4.8.2   Optical Orthogonal Frequency Division Multiplexing

Orthogonal frequency division multiplexing (OFDM) is a transmission technology that is primarily known from wireless communications and wired transmission over copper cables [31,32]. It is a special case of the widely known frequency division multiplexing (FDM) technique in which digital or analog data are modulated onto a certain number of carriers and transmitted in parallel over the same transmission medium. The main motivation for using FDM is the fact that due to parallel data transmission in the frequency domain, each channel occupies only a small frequency band. Signal distortions originating from frequency-selective transmission channels, the fiber CD, can be minimized. The special property of OFDM is its very high spectral efficiency. While for conventional FDM, the spectral efficiency is limited by the selectivity of the bandpass filters required for demodulation, OFDM is designed such that the different carriers are pairwise orthogonal. In this way, for the sampling point, the intercarrier interference (ICI) is suppressed, although the channels are allowed to overlap spectrally.

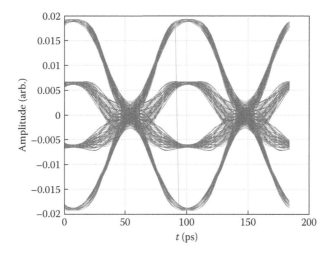

FIGURE 7.62 Eye diagram of 16-QAM detected at the output of a balanced receiver. Bit rate of 40 Gbps and baud rate of 10 Gbps.

Orthogonality is achieved by placing the different RF carriers onto a fixed frequency grid and assuming rectangular pulse shaping. It can be shown that in this special case, the OFDM signal can be described as the output of a discrete inverse Fourier transform with the parallel complex data symbols as input. This property has been one of the main driving aspects for OFDM in the past because modulation and demodulation of a large number of carriers can be realized by simple digital signal processing (DSP) instead of using many LOs in the transmitter and receiver. Recently, OFDM has become an attractive topic for digital optical communications [28–30,33,34]. It is just another example of the current tendency in optical communications to consider technologies that are originally known from classical digital communications. Using OFDM appears to be very attractive because the low bandwidth occupied by a single OFDM channel increases the robustness toward fiber dispersion, allowing the transmission of high data rates of 40 Gbps and more over hundreds of kilometers without the need for dispersion compensation [35]. In the same way as for modulation formats such as DPSK or DQPSK that were introduced in recent years, in OFDM also the challenge for optical system engineers is to adapt a classical technology to the special properties of the optical channel and the requirements of optical transmitters and receivers.

Thus, two approaches have been reported recently. An intuitive approach introduced by Llorente et al. [29] makes use of the fact that the WDM technique itself already realizes data transmission over a certain number of different carriers. By means of special pulse shaping and carrier wavelength selection, the orthogonality between the different wavelength channels can be achieved, resulting in the so-called orthogonal WDM technique (OWDM). However, with this technique, the option of simple modulation and demodulation by means of the discrete Fourier transforms (DFT) cannot be utilized as this kind of DSP is not available in the optical domain.

An alternative popular method is the generation of an electrical OFDM signal by means of electrical signal processing followed by modulation onto a single optical carrier [30,33,34]. This approach is known as optical OFDM (oOFDM). Here, the modulation is a two-step process: first, the electrical OFDM in Ref. [58] signal already is a broadband bandpass signal, which is then modulated onto the optical carrier. Second, to increase data throughput, oOFDM can be combined with WDM resulting in multi-Tbps transmission system as shown in Figure 7.63. Nevertheless, oOFDM itself offers different options for implementation. An important issue is optical demodulation, which can be realized either by means of direct detection (DD) or coherent detection (CoD) using an LO. DD is preferable owing to its simplicity. However, for DD the optical intensity has to

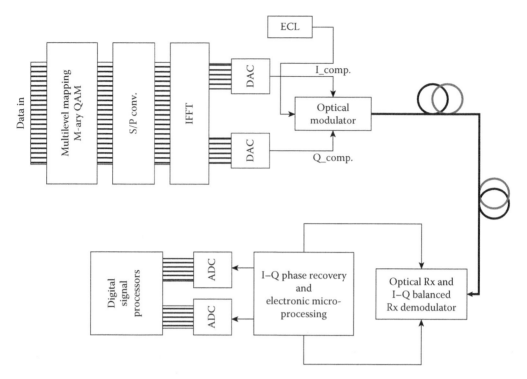

FIGURE 7.63 Schematic diagram of optical OFDM transmitter of a long-haul optical transmission system using multilevel modulation formats. ECL = external cavity laser.

be modulated. Because the electrical OFDM signal is quasi-analog with a zero mean and a high peak-to-average ratio, the majority of the optical power has to be wasted for the optical carrier (i.e., an additional DC value of the complex baseband signal), resulting in low receiver sensitivity. For CD, in addition, the bandwidth efficiency is twice as high as for DD because for pure intensity modulation inherently a double-sideband signal is generated. For CD, a complex optical I–Q modulator composed of two real modulators in parallel followed by superposition with a $\pi/2$ phase shift allows for transmission of twice as much data within the same bandwidth. For intensity modulation, the bandwidth efficiency may be increased by suppressing one of the redundant sidebands, resulting in optical OFDM with single-sideband (SSB) transmission. First, the serial data can be converted to parallel streams. These parallel data sequences are then mapped to QAM constellation in the frequency domain, and then converted back by IFFT to the time domain. The time-domain signals are in the I and Q components, which are then fed into an I and Q optical modulator. This optical modulation can be DQPSK or any other multilevel modulation subsystem. At the end of the optical fiber transmission, the I and Q components are detected either by DD or CD. For CD, a (2×4) $90°$ hybrid coupler is used to mix the polarizations of the LO and that of the received signals. The outputs of the couplers are then fed into the balanced optical receivers. The mixing of the local laser source and that of the signals preserves the phase of the signals, which are then processed by a high-speed electronic processor. For direct detection, the I and Q components are detected differentially, and the amplitude and phase detection are then compared and processed similarly as for the coherent case.

In order to show the robustness of oOFDM toward fiber dispersion and also fiber nonlinearity, numerical simulations are carried out for a data stream at a 42.7 Gbps data rate. The number of OFDM channels can be varied between $N_{min} = 256$ and $N_{max} = 2048$. A guard interval of 12 ns can be inserted, a strategy belonging inherently to OFDM technology that ensures the orthogonality of the different channels in case of a transmission channel with memory. For the optical modulation,

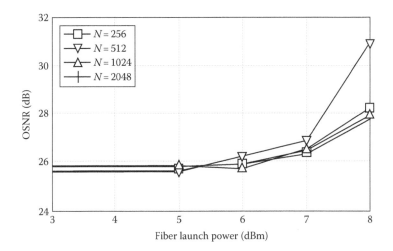

FIGURE 7.64 Simulation result for 42.7 Gbps OFDM transmission over 640 km of standard single-mode fiber; OSNR required for BER = 10^{-3} (Monte Carlo simulation) as function of fiber launch power.

intensity modulation using a single MZM in conjunction with SSB filtering and direct detection was implemented. The nonlinear optical transmission channel consisted of eight 80 km non-DCF spans of SSMF. As a criterion for performance, the required OSNR for a BER of 10^{-3} (Monte Carlo) is measured. Using forward error correct (FEC), after decoding this is transferred into a BER below 10^{-9} depending on the specific code.

Figure 7.64 shows the required OSNR as function of the fiber launch power for different values of N. The most important result is that transmission is possible over 640 km over SSMF without any dispersion compensation. This can be explained by the fact that even for the lowest value of N_{min} = 256, each subchannel occupies a bandwidth of approximately 42.7 GHz/256 = 177 MHz, resulting in high robustness toward fiber dispersion.

The principal difficulties of optical OFDM are that the pure delay, owing to the variation of the refractive index of the fiber with respect to the optical frequency, leads to bunching of the subchannels and hence an increase of the optical power; thus, an unexpected SPM may occur in a random manner. Chapter 11 describes this OFDM modulation format.

7.4.8.3   100 Gbps 8-DPSK–2-ASK 16-Star QAM

7.4.8.3.1   Introduction

A multilevel modulation scheme enables the transmission baud rate to be reduced, thus obtaining spectral efficiency. Another significant advantage of this modulation scheme is the reduction of the requirement of high-speed processing electronics. This is of particular interest for high-speed optical transmission systems.

This part of the chapter investigates a multilevel modulation scheme that has eight phases and two amplitude levels. This scheme, which is called 8-DPSK–2-ASK, effectively utilizes four bits per symbol for transmission, in which is the first three bits are for coding phase information while the coding of the amplitude levels is implemented with the fourth bit. As a result, the transmission baud rate is equivalently a quarter of the bit rate from the bit pattern generator.

This section is organized as follows: Section 7.4.8.3.2 presents a detailed description of the optical transmitter for generating 8-DPSK–2-ASK signals. In Section 7.4.8.3.3, the detailed configuration of the receiver is provided. The configuration of the optical transmitter and receiver is based on that reported by Djordjevic and Vasic [36]. Section 7.4.8.3.4 describes a study on DT and the transmission performance of the 8-DPSK–2-ASK scheme. Finally, a short summary of the report is provided.

7.4.8.3.2 Configuration of 8-DPSK–2-ASK Optical Transmitter

There have been several different configurations of an optical transmitter for generating multiphase/level optical signals with the use of AMs or PMs arranged in either serial or parallel configurations [32,36–39]. However, the optical transmitters reported in Refs. [32,36–39] require a precoder with high complexity. However, the configuration reported by Ivan et al. [36] utilizes the Gray mapping technique to differentially encode the phase information, and this significantly reduces the complexity of the optical transmitter. In addition, as elaborated in more detail in Section 7.3, this precoding technique enables a detection scheme using the I–Q demodulation techniques, equivalent to those employed in coherent transmission systems.

The optical transmitter of the 8-DPSK–2-ASK scheme employs the I–Q modulation technique with two MZIMs in parallel and a $\pi/2$ optical PM, as shown in Figure 7.65. At each kth instance, the absolute phase of transmitted lightwaves θ_k is expressed as $\theta_k = \theta_{k-1} + \Delta\theta_k$, where θ_{k-1} is the phase at $(k-1)$th instance, and $\Delta\theta_k$ is the differentially coded phase information. The encoding of this $\Delta\theta_k$ for generating 8-DPSK–2-ASK modulated optical signals (four bits per transmitted symbol) follows the well-known Gray mapping rules. This Gray mapping phasor diagram is shown in Figure 7.66. The phasor is normalized with the maximum energy on each branch, that is, $E_1/2$.

The amplitude levels are optimized so that the Euclidean distances d_1, d_2, and d_3 are equal, that is, $d_1 = d_2 = d_3$. After derivation, we obtain: $r_1 = 0.5664$. The I and Q field vectors corresponding to the Gray mapping rules from the M-ADPSK precoder (see Figure 7.65) are provided in Table 7.7.

The foregoing transmitter configuration can be replaced with a dual-drive MZIM. The explanation and derivation for generating 8-DPSK–2-ASK optical signals are also based on the phasor diagram of Figure 7.66. In this case, the output field vector is the sum of two component field vectors, each of which is not only determined by the amplitude but also by initially biased phases [51–56].

7.4.8.3.3 Configuration of 8-DPSK–2-ASK Detection Scheme

The detection of 8-DPSK–2-ASK optical signals is implemented with the use of two Mach–Zehnder delay interferometric (MZDI) = balanced receivers (see Figure 7.67).

Several key features of this detection structure are as follows:

1. The MZDI introduces a delay corresponding to the baud rate.
2. One arm of the MZDI has a $\pi/4$ optical phase shifter, while the other arm has an optical phase shift of $-\pi/4$.

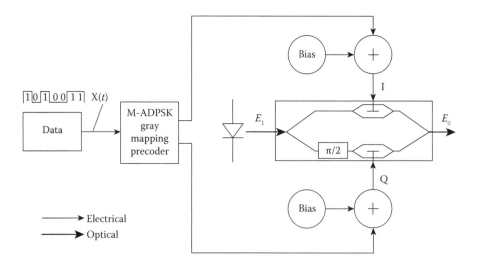

FIGURE 7.65 Optical transmitter configuration of the 8-DPSK–2-ASK modulation scheme.

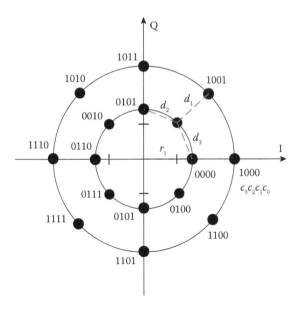

FIGURE 7.66 Gray mapping for optimal 8-DPSK–2-ASK modulation scheme.

TABLE 7.7
I and Q Field Vectors in 8-DPSK–2-ASK Modulation Scheme Using Two MZIMs in Parallel

Binary Sequence	$(\Delta\theta_k,$ Amplitude)	I_k	Q_k
1000	(0, 1)	1	0
1001	($\pi/4$, 1)	$\sqrt{2}/2$	$\sqrt{2}/2$
1011	($\pi/2$, 1)	0	1
1010	($3\pi/4$, 1)	$-\sqrt{2}/2$	$\sqrt{2}/2$
1110	(π, 1)	-1	0
1111	($-3\pi/4$, 1)	$-\sqrt{2}/2$	$-\sqrt{2}/2$
1101	($-\pi/2$, 1)	0	-1
1100	($-\pi/4$, 1)	$\sqrt{2}/2$	$-\sqrt{2}/2$
0000	(0, 0.5664)	$1*0.5664$	0
0001	($\pi/4$, 0.5664)	$\sqrt{2}/2*0.5664$	$\sqrt{2}/2*0.5664$
0011	($\pi/2$, 0.5664)	0	$1*0.5664$
0010	($3\pi/4$, 0.5664)	$-\sqrt{2}/2*0.5664$	$\sqrt{2}/2*0.5664$
0110	(π, 0.5664)	$-1*0.5664$	0
0111	($-3\pi/4$, 0.5664)	$-\sqrt{2}/2*0.5664$	$-\sqrt{2}/2*0.5664$
0101	($-\pi/2$, 0.5664)	0	$-1*0.5664$
0100	($-\pi/4$, 0.5664)	$\sqrt{2}/2*0.5664$	$-\sqrt{2}/2*0.5664$

3. The outputs from two balanced receivers are superimposed positively and negatively, which leads to I and Q detected signals, respectively. The I and Q detected components are expressed as $I = \mathrm{Re}\left\{E_k E_{k-1}^*\right\}$ and $Q = \mathrm{Im}\left\{E_k E_{k-1}^*\right\}$.
4. The I–Q detected components are demodulated using the popular I–Q demodulator in the electrical domain. These detected signals are then sampled and represented as shown in the signal constellation.

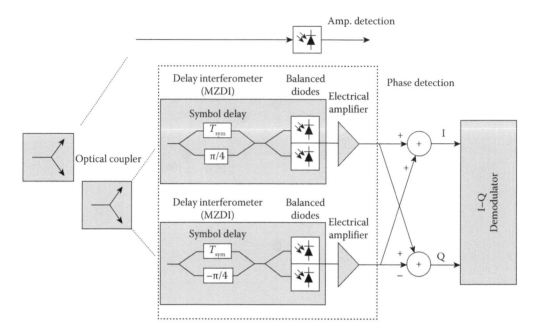

FIGURE 7.67 Detection configuration for the 8-DPSK–2-ASK modulation scheme.

7.4.8.3.4 Transmission Performance of 100 Gbps 8-DPSK–2-ASK Scheme

Performance characteristics of the 8-DPSK–2-ASK scheme operating at 100 Gbps bit rate are studied in terms of receiver sensitivity, DT, and the feasibility for long-haul transmission. The BERs are the pre-FEC BERs, and the pre-FEC limit is conventionally referenced at 2e−3. In addition, the BERs are evaluated by the Monte Carlo method.

7.4.8.3.5 Power Spectrum

The power spectrum of 8-DPSK–2-ASK optical signals is shown in Figure 7.68. It can be observed that the main lobe spectral width is about 25 GHz, as the symbol baud rate of this modulation scheme is equal to a quarter of the bit rate from the bit pattern generator. The harmonics are not highly suppressed, thus requiring bandwidth of the optical filter to be necessarily large in order not to severely distort signals.

7.4.8.3.6 Receiver Sensitivity and DT

The receiver sensitivity is studied by connecting the optical transmitter of the 8-DPSK–2-ASK scheme directly to the receiver for a back-to-back setup (see Figure 7.69). On the contrary, the DT is studied by varying the length of SSMF from 0 to 5 km (|D| = 17 ps/(nm · km). The received power is varied by using an optical attenuator. The optical Gaussian filter has BT = 3 (B is approximately 75 GHz). Modeling of receiver noise sources includes shot noise, an equivalent noise current density of 20 pA/$\sqrt{\text{Hz}}$ at the input of the trans impedance electrical amplifier, and a dark current of 10 nA for each of the two photodiodes in the balanced structure. A fifth-order Bessel electrical filter with a bandwidth of BT = 0.8 is used.

The numerical BER curves of the receiver sensitivity for cases of 0–5 km SSMF are shown in Figure 7.70. The receiver sensitivity of the 8-DPSK–2-ASK scheme is approximately −18.5 dBm at BER = 1e−4. The receiver sensitivity at BER = 1e−9 can be obtained by extrapolating the BER curve of the 0 km case. The power penalty versus residual dispersion results are then obtained

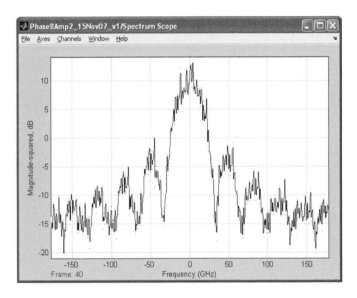

FIGURE 7.68 Power spectrum of 8-DPSK–2-ASK signals.

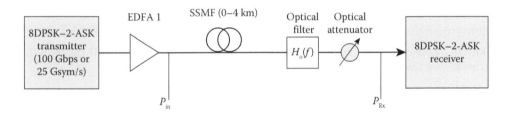

FIGURE 7.69 Setup for the study of receiver sensitivity (back-to-back) and dispersion tolerance (0–4 km SSMF) for the 8-DPSK–2-ASK modulation scheme.

and plotted in Figure 7.71. It is realized that the 2 dB penalty occurs for the residual dispersion of approximately 60 ps/nm, or, equivalently, to 3.5 km SSMF.

7.4.8.3.7 Long-Haul Transmission

The long-haul transmission performance of this modulation format is conducted over ten optically amplified and fully compensated spans, and each span is composed of 100 km SSMF and 10 km DCF100 (Sumitomo fiber). As a result, the length of the transmission fiber link is 1100 km. This long-haul range is selected to reflect the distance between Melbourne and Sydney. The wavelength of interest is 1550 nm, and the dispersion at the end of the transmission link is fully compensated. The simulation setup is shown in Figure 7.69. In addition, the fiber attenuation due to SSMF and DCF is also fully compensated by using two EDFAs with optical gains as depicted in Figure 7.72. These EDFAs have a noise figure set at 5 dB.

Numerical transmission BERs are plotted against the received powers in Figure 7.73 and compared to the back-to-back BER curve. It can be observed that the BER curve of 1100 km follows a linear trend and feasibly reaches 1e−9 if extrapolated as shown in Figure 7.73. It should be noted that this transmission performance can be significantly improved with the use of a high-performance FEC scheme.

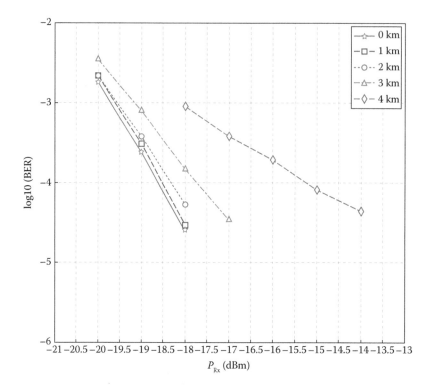

FIGURE 7.70 Receiver sensitivity (back-to-back) and dispersion tolerance (0–4 km SSMF) for the 8-DPSK–2-ASK modulation scheme.

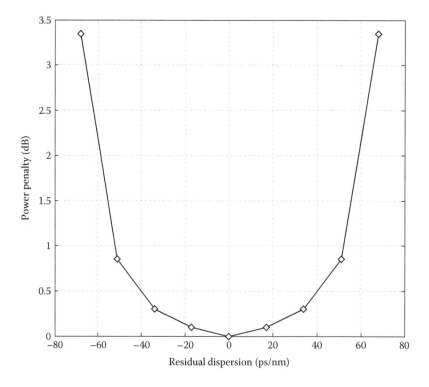

FIGURE 7.71 Power penalty due to residual dispersions for the 8-DPSK–2-ASK modulation scheme.

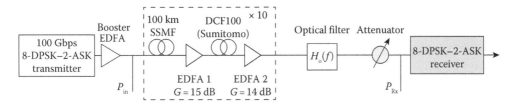

FIGURE 7.72   Transmission setup of 1100 km optically amplified and fully compensated fiber link.

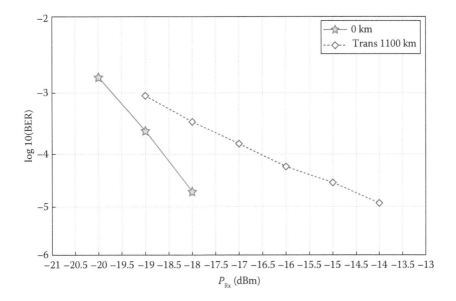

FIGURE 7.73   BER versus receiver sensitivity for 8-DPSK–2-ASK modulation format transmission.

7.4.9 CONCLUDING REMARKS

The ever-increasing bandwidth hunger in telecommunication networks based mainly on optical fiber communication technology indicates that low bandwidth-efficient modulation formats such as ASK would no longer satisfy the transmission capacity demands, and new advanced optical modulation schemes should replace ASK in the near future. Advanced optical modulation schemes, especially the multilevel amplitude and phase schemes presented in this chapter, are able to (1) provide long reach, error-free, and high transmission capacity; (2) provide high bandwidth efficiency (no. bits/Hz parameter); (3) push the bit rate well above what could be offered by electronic technology, for example, 100 Gbps with the detection at the symbol rate; (4) tolerate dispersion and nonlinearity; and (5) to maximally utilize the existing optical network infrastructure.

Current developments of photonic technology have enabled the use of differential phase modulation and demodulation in the optical domain. Presently, the BDPSK and DQPSK formats have received great attention owing to their improvement in the receiver sensitivity as compared with ASK. Furthermore, the RZ and CSRZ formats would assist the battle against nonlinear effects. However, as long as transmission capacity and bandwidth efficiency are concerned, MADPSK modulation formats would offer a better performance in the trading off for its complexity in the receiver structures.

Alternatively, there are other multilevel modulation schemes that offer further improvement of the optical transmission performance [10].

7.4.9.1 Offset MADPSK Modulation

Binary DPSK and quaternary DPSK (DQPSK) are just special cases of a more general class of differential phase modulation formats that map data bits into the phase difference between neighboring symbols. This phase difference $\Delta\phi_i$ can be described as

$$\Delta\phi_i = \theta + \frac{2\pi(i-1)}{M}, \quad i = 1,2,...M-1 \tag{7.28}$$

where θ is the initial phase, and M is the total number of phase levels. This class of formats is called offset DPSK (ODPSK) and denoted as θ-M-DPSK. Specifically, with $\theta = 0$, $M = 2$ or M = 4, θ-M-DPSK becomes 0-2-DPSK or 0-4-DPSK, the conventional binary DPSK and DQPSK mentioned earlier.

ODPSK has been used to transmit over satellite nonlinear channels because its phase transition is smooth and can avoid a 180° phase jump [40]. As the fiber medium also exhibits nonlinearity, this modulation format attracts our attention as a candidate, together with ASK, for creating a new multilevel modulation format, possibly termed Offset MADPSK.

7.4.9.2 MAMSK Modulation

Minimum shift keying (MSK) is a form of OQPSK with sinusoidal pulse weighting [41]. In MSK, data bits are first coded into bipolar signals ±1, which are then separated into $V_I(t)$ and $V_Q(t)$ streams consisting of even and odd bits, respectively. In the next stage, $V_I(t)$ and $V_Q(t)$ are used to modulate a carrier f_c to create a MSK signal $s(t)$, which can be presented as [41]

$$s(t) = V_I(t)\cos\left(\frac{\pi t}{2T}\right)\cos\left(2\pi f_c t\right) + V_Q(t)\cos\left(\frac{\pi t}{2T}\right)\sin\left(2\pi f_c t\right) \tag{7.29}$$

MSK is a well-known modulation format in wireless communications with efficient spectral characteristics owing to the high compactness of its main lobe as compared with DPSK, and high suppression of the side lobes [42]. These characteristics indicate that the MSK signal is highly dispersion tolerant. A combination of MSK and ASK into MAMSK modulation would even improve the transmission performance without increasing the complexity of the detection scheme [43].

Multilevel techniques for 100 Gbps are given; in particular, OFDM with multicarrier and multilevel amplitude modulation with orthogonality between adjacent channels are proved to be cost-effective and appropriate for current electronic technologies. Two interesting approaches to achieve data transmission of 40 Gbps and beyond (e.g., 100 Gbps Ethernet) based on low symbol rate are discussed. On the one hand, oOFDM can combine a large number of parallel data streams into one broadband data stream with high spectral efficiency. Simulation results are shown for different values of the number of parallel data streams in a nonlinear environment [58]. On the other hand, 16-ary modulation formats enable 40 Gbps transmission with 10 GSym/s (i.e., 100 Gbps with 25 GSymbols/s). For a special case, namely inverse RZ 16ADQPSK, measurement results are demonstrated.

7.4.9.3 Star QAM Coherent Detection

The last two sections of the chapter describe two optical transmission schemes using both coherent and incoherent transmission and detection techniques, the 2A-8P Star QAM and 8DPSK 2ASK 16-Star QAM for 100 Gbps or 25 GSymbols/s [44–50].

First, the design of a Star 16-QAM modulation scheme is proposed for coherent detection for ultra-high-speed ultra-high-capacity optical fiber communications schemes. Two amplitude levels and eight phases (2A-8P 16-QAM) are considered to offer significant simple transmitter and receiver configurations and at the same time the best receiver sensitivity at the receiver. An optical

SNR of about 22 dB is required for the transmission of Star 16-QAM over an optically amplified transmission dispersion-compensated link. A DT of 300 ps/nm is possible with a 1 dB penalty of the eye opening at 40 Gbps bit rate or 10 Giga-symbols/s with an OSNR of 18 dB. The OSNR could be about 22 dB for a 107 Gbps bit rate and a symbol rate of 26.75 G symbols/s. The transmission link consists of several spans of total 1000 km dispersion-compensated optically amplified transmission link. The optical gain of the in-line OAs are set to compensate for the attenuation of the transmission and compensating fibers with a noise figure of 3 dB. The optical transmitter and receivers incorporating commercially available coherent receivers are structured and are sufficient for engineering of the optical transmission terminal equipment for a bit rate of 107 Gbps and a symbol rate of 26.3 GBaud/s.

Furthermore, electronic equalization of the receiver phase shift keying signals can be done using blind equalization, which would improve the DT much further. For a symbol rate of 10.7 Gbps, this DT for a 1 dB penalty would reach 300 km of standard SMF. This electronic equalization can be implemented without any difficulty at 10.7 GBaud/s. For a 107 Gbps bit rate, a similar improvement in the DT can be at 26.5 GBaud/s provided that the electronic sampler can offer a >50 Gig sampling rate.

Second, the transmitter and receiver configurations for generating 8-DPSK–2-ASK optical signals and direct detection are described as well as the transmission performance. In addition, the performance characteristics of this modulation format at 100 Gbps (equivalently, 25 GBaud/s) has also been investigated in terms of the receiver sensitivity, DT, and the long-haul transmission performance. The simulation results show that 8-DPSK–2-ASK is a promising modulation for very high-speed (100 Gbps) and long-haul optical communications.

Furthermore, another multidimension system using multiple carriers such as the OFDM scheme is introduced, and more details will be given in Chapter 11.

REFERENCES

1. R. A. Linke and A. H. Gnauck, High-capacity coherent lightwave systems, *IEEE J. Lightwave Technol.*, Vol. 6, No. 11, pp. 1750–1769, 1988.
2. L. N. Binh, *Monash Optical Communication System Simulator. Part I: Ultra-long Ultra-High Speed Optical Fiber Communication Systems*, Manual for Technical Development, Melbourne, Australia: ECSE Monash University, 2004.
3. G. Proakis, *Digital Communication*, New York: McGraw-Hill, 2001.
4. L. N. Binh and B. Laville, *Simulink Models for Advanced Optical Communications: Part IV-DQPSK Modulation Format*, Technical Report, Melbourne, Australia: ECSE Monash University, 2005.
5. A. H. Gnauck, X. Liu, X. Wei, D. M. Gill, and E. C. Burrows, Comparison of modulation formats for 42.7Gbs single-channel transmission through 1980 km of SSMF, *IEEE Photon. Technol. Lett.*, Vol. 16, No. 3, pp. 909–911, 2004.
6. K. Sekine, N. Kikuchi, S. Sasaki, and S. Hayase, 40 Gb/s 16-ary (4 bit/symbol) optical modulation/ demodulation scheme, *IEEE Electron. Lett.*, Vol. 41, No. 7, pp. 430–432, 2005.
7. T. Mizuochi, K. Ishida, T. Kobayashi, J. Abe, K. Kinjo, K. Motoshima, and K. Kasahara, A comparative study of DPSK and ASK WDM transmission over transoceanic distances and their performance degradations due to nonlinear phase noise, *IEEE J. Lightwave Technol.*, Vol. 21, No. 9, pp. 1933–1943, 2003.
8. T. Hoshida, O. Vassilieva, K. Yamada, S. Choudhary, R. Pecqueur, and H. Kuwahara, Optimal 40 Gb/s modulation formats for spectrally efficient long-haul DWDM systems, *IEEE J. Lightwave Technol.*, Vol. 20, No. 12, pp. 1989–1996, 2002.
9. S. Hayase, N. Kikuchi, K. Sekein, and S. Sasaki, Proposal of 8-state per symbol (binary ASK and QPSK) 30-Gb/optical modulation demodulation scheme, in *Proceedings of European Conference on Optical Communication*, paper Th2.6.4, ECOC 2003, Rimini, Italy, pp. 1008–1009, September 2003.
10. S. Walklin and J. Conradi, Multilevel signaling for increasing the reach of 10 Gb/s lightwave systems, *IEEE J. Lightwave Technol.*, Vol. 17, No. 11, pp. 2235–2248, 1999.
11. Y. Miyamoto, A. Hirano, K. Yonenaga, A. Sano, H. Toba, K. Murata, and O. Mitomi, 320 Gb/s (8 × 40 Gb/s) WDM transmission over 367-km zero-dispersion-flattened line with 120-km repeater spacing using carrier-suppressed return-to-zero pulse format, *IEEE Electron. Lett.*, Vol. 35, No. 23, pp. 2041–2042, 1999.

12. L. N. Binh, H. S. Tiong, T. L. Huynh, and D. D. Tran, DPSK RZ modulation formats generated from dual drive interferometric optical modulators. Unpublished works.

13. L. N Binh and Y. L. Cheung, *DWDM Advanced Optical Communication–Simulink Models: Part I–Optical Spectra*, Technical Report, Melbourne, Australia: ECSE Monash University, 2005.

14. R. Hui, S. Zhang, B. Zhu, R. Huang, C. Allen, and D. Demarest, *Advanced Optical Modulation Formats and their Comparison in Fiber-Optic Systems*, Technical Report, Information and Telecommunication Technology Center, University of Kansas, 2004.

15. D. S. Lee, S. K. Daejeon, M. S. Lee, Y. J. Wen, and A. Nirmalathas, Electrically band-limited CSRZ-DPSK signal with a simple transmitter configuration and reduced linear crosstalk in high spectral efficiency DWDM systems, *IEEE Photon. Technol. Lett.*, Vol. 16, No. 9, pp. 2135–2137, 2004.

16. C. Wree, *RZ-DQPSK Format with High Spectral Efficiency and High Robustness Towards Fiber Nonlinearities*, University of Kiel, 2002.

17. T. L. Huynh, L. N. Binh, D. D. Tran, and Q. H. Lam, Long-haul ASK and DPSK optical fiber transmission systems: Simulink modeling and experimental demonstration test beds, in *Proceedings of IEEE Tencon'05*, Melbourne, Australia, November 2005.

18. H. Kim and P. J. Winzer, Robustness to laser frequency offset in direct detection DPSK and DQPSK systems, *IEEE J. Lightwave Technol.*, Vol. 21, No. 9, pp. 1887–1891, 2003.

19. R. A. Griffin et al., 10 Gb/s optical differential quadrature phase shift key (DQPSK) transmission using GaAs/AlGaAs integration, in *Proceedings of OFC 2002*, postdeadline paper FD6, Anaheim, CA, USA, 2002.

20. N. Yoshikane and I. Morita, 1.14 b/s/Hz spectrally-efficient 50x85.4 Gb/s transmission over 300 km using co-polarized CS-RZ DQPSK signals, in *Proceedings of OFC 2004*, postdeadline paper PDP38, Los Angeles, CA, 2004.

21. D. D. Tran, L. N. Binh, T. L. Huynh, and H. S. Tiong, Geometrical and phasor representation of multilevel amplitude-phase modulation formats and photonic transmitter structures, in *Proceedings of the IEEE Tencon'05*, Melbourne, Australia, November 2005.

22. G. P. Agrawal, *Fiber-Optic Communication Systems*, New York: Wiley, 1992.

23. K.-P. Ho, Generation of arbitrary quadrature signals using one dual-drive modulator, *IEEE J. Lightwave Technol.*, Vol. 23, No. 2, pp. 764–770, 2005.

24. G. P. Agrawal, *Nonlinear Fiber Optics*, 2nd ed., Boston, MA: Academic Press, 1995.

25. T. V. Muoi, Stepped-index optical fiber systems, Ph.D. Thesis, University of Western of Australia, 1978.

26. L. N. Binh et al., *Monash Optical Communication System Simulator. Wavelength Division Multiplexed Optical Communication Systems and Networks*, Manual for Technical Development, Melbourne, Australia: Monash University, 1997.

27. A. Gumaste and T. Antony, *DWDM Network Designs and Engineering Solutions*, Cisco Press, Milton Keynes, UK, 2002.

28. S. L. Jansen, I. Morita, N. Takeda, and H. Tanaka, 20-Gb/s OFDM transmission over 4160-km SSMF enabled by RF-pilot tone phase noise compensation, in *Proceedings of OFC 2007*, paper PDP15, Anaheim, CA, USA, March 2007.

29. R. Llorente, J. H. Lee, R. Clavero, M. Ibsen, and J. Martí, Orthogonal wavelength-division-multiplexing technique feasibility evaluation, *J. Lightwave Technol.*, Vol. 23, pp. 1145–1151, 2005.

30. A. J. Lowery, L. Du, and J. Armstrong, Orthogonal frequency division multiplexing for adaptive dispersion compensation in long haul WDM systems, in *Proceedings of OFC 2006*, paper PDP39, Anaheim, CA, USA, March 2006.

31. J. Leibrich, A. Ali, and W. Rosenkranz, Optical OFDM as a promising technique for bandwidth-efficient high-speed data transmission over optical fiber, in *Proceedings of the 12th International OFDM-Workshop 2007, InOWo 2007*, 29, 30, August 2007, Hamburg, Germany.

32. H. Yoon, D. Lee, and N. Park, Performance comparison of optical 8-ary differential phase-shift keying systems with different electrical decision schemes, *Opt. Express*, Vol. 13, No. 2, pp. 371–376, 2005.

33. L. Hanzo, M. Münster, B. J. Choi, and T. Keller, *OFDM and MC-CDMA for Broadband Multi-User Communications, WLANs and Broadcasting*, New York: Wiley, 2003.

34. W. Shieh and C. Athaudage, Coherent optical orthogonal frequency division multiplexing, *Electron. Lett.*, Vol. 42, pp. 587–589, 2006.

35. L. N. Binh, *Monash Optical Communication System Simulator. Part II: Ultra-long Ultra-High Speed Optical Fiber Communication Systems*, Manual for Technical Development, Melbourne, Australia: ECSE Monash University, 2004.

36. I. B. Djordjevic and B. Vasic, Multilevel coding in M-ary DPSK/differential QAM high-speed optical transmission with direct detection, *IEEE J. Lightwave Technol.*, Vol. 24, pp. 420–428, 2006.

37. M. Serbay, C. Wree, and W. Rosenkranz, Implementation of different precoder for high-speed optical DQPSK transmission, *Electron. Lett.*, Vol. 40, No. 20, pp. 1–2, 2004.
38. M. Serbay, C. Wree, and W. Rosenkranz, Experimental investigation of RZ-8DPSK at 3×10.7 Gb/s, in *Proceedings of LEOS 2005*, paper WE3, Sydney, Australia, October 2005.
39. M. Seimetz, M. Noelle, and E. Patzak, Optical systems with high-order DPSK and star QAM modulation based on interferometric direct detection, *IEEE J. Lightwave Technol.*, Vol. 25, pp. 1515–1530, 2007.
40. C. Wree et al., Offset-DQPSK modulation format for 40 Gb/s and comparison to RZ-DQPSK in WDM environment, in *Proceedings of OFC 2004*, paper MF62, Los Angeles, CA, USA, February 2004.
41. S. Pasupathy, Minimum shift keying: A spectrally efficient modulation, *IEEE Commun. Mag.*, Vol. 17, pp. 14–22, 1979.
42. T. Sakamoto et al., Optical minimum-shift keying with external modulation scheme, *Opt. Express*, Vol. 13, pp. 7741–7747, 2005.
43. M. Ohm and J. Speidel, Quaternary optical ASK-DPSK and receivers with direct detection, *IEEE Photon. Technol. Lett.*, Vol. 15, No. 1, pp. 159–161, 2003.
44. O. Boyraz, T. Indukuri, and B. Jalali, Self-phase modulation induced spectral broadening in silicon waveguides, in *Proceedings of CLEO2004*, Vol. 2, paper CThj2, May 2004.
45. G. Goeger, M. Wrage, and W. Fischler, Cross-phase modulation in multispan WDM systems with arbitrary modulation formats, *IEEE Photon. Technol. Lett.*, Vol. 16, No. 8, pp. 1858–1860, 2004.
46. G. Turin, An introduction to matched filters, *IEEE Trans. Information Theory.*, Vol. 6, No. 3, pp. 311–329, 1960.
47. B. Zhu, 3.08 Tbit/s (77x42.7 Gb/s) WDM transmission over 1200 km fiber with 100 km repeater spacing using dual C- and L-band hybrid Raman/erbium-doped inline amplifiers, *IEEE Electron. Lett.*, Vol. 37, No. 13, p. 844, 2001.
48. H. Rongqing, Z. Benyuan, H. Renxiang, T. A. Christopher, R. D. Kenneth, and D. Richards, Subcarrier multiplexing for high-speed optical transmission, *IEEE J. Lightwave Technol.*, Vol. 20, No. 3, pp. 417–427, 2005.
49. A. H. Gnauck, and P. J. Winzer, Optical phase-shift-keyed transmission, *IEEE J. Lightwave Technol.*, Vol. 23, No. 1, pp. 115–130, 2005.
50. K.-P. Ho, Error probability of DPSK signals with cross-phase modulation induced nonlinear phase noise, *IEEE J. Sel. Top. Quant. Electron.*, Vol. 10, No. 2, pp. 421–427, 2004.
51. G. P. Agrawal, Self-phase modulation and spectral broadening of optical pulses in semiconductor laser amplifiers, *IEEE J. Quant. Electron.*, Vol. 25, No. 11, pp. 2297–2306, 1989.
52. J. P. Gordon and L. F. Mollenauer, Phase noise in photonic communications systems using linear amplifiers, *Opt. Lett.*, Vol. 15, No. 23, pp. 1351–1353, 1990.
53. C. Michel, Techniques for estimating the bit error rate in the simulation of digital communication systems, *IEEE J. Sel. Areas Commun.*, Vol. SAC-2, No. 1, pp. 153–170, 1984.
54. H. Kim and A. H. Gnauck, Experimental investigation of the performance limitation of DPSK systems due to nonlinear phase noise, *IEEE J. Lightwave Technol.*, Vol. 15, No. 2, pp. 320–322, 2005.
55. C. Wree, J. Leibrich, and W. Rosenkranz, RZ-DQPSK format with high spectral efficiency and high robustness towards fiber nonlinearities, in *Proceedings of ECOC 2002*, paper 9.6.6, Copenhagen, Denmark, September 2002.
56. N. Kikuchi, Amplitude and phase modulated 8-ary and 16-ary multilevel signaling technologies for high-speed optical fiber communication, in *Proceedings of APOC 2005*, paper 602127, Shanghai, China, October 2005.
57. M. Serbay, T. Tokle, P. Jeppesen, and W. Rosenkranz, 42.8 Gbit/s, 4 bits per symbol 16-ary inverse-RZ-QASK-DQPSK transmission experiment without polmux, in *Proceedings of OFC 2007*, paper OThL2, Anaheim, CA, USA, March 2007.
58. A. Ali, Investigation of OFDM-transmission for direct optical systems, Dr. Ing. Dissertation, Lehrstuhl für Nachrichten- und Übertragungstechnik, Christian ALbrechts Unviversitaet zu Kiel, Germany, 2012.

8 Continuous Phase Modulation Format Optical Systems

This chapter describes the modulation scheme that uses the modulation of the phase lightwave carrier, in which the change of the phase during a one-bit state would be continuous, that is, two frequencies, and hence the term continuous phase frequency shift keying (CPFSK). When the two-frequency signal components are orthogonal, the CPFSK is considered as minimum shift keying (MSK). This scheme is also presented. The generation of modulated signals and propagation and detection schemes are then presented, after which the transmission performance is described.

8.1 INTRODUCTION

The generation of MSK requires a linear variation of the phase, and hence a constant frequency of the optical carrier. However, the generation of the optical phase may be preferred by driving an optical modulator using a sinusoidal signal for practical implementation. Thus, a nonlinear variation of the carrier phase results, and hence some distortion effects are produced. In this chapter, we investigate the use of linear and nonlinear phase-shaping filtering and their impact on MSK-modulated optical signals transmission over optically amplified long-haul communications systems. The evolution of the phasor of the in-phase and quadrature components is illustrated for lightwave-modulated signal transmission. The distinct features of three different MSK modulation formats—linear MSK, weakly nonlinear MSK, and strongly nonlinear MSK—and their transmissions are simulated. The transmission performance obtained indicates the resilience of the MSK signals in transmission over optically amplified multispans.

Recent years have witnessed intensive interest in the employment of advanced modulation formats to explore their advantages and performance in high-density and long-haul transmission systems. Return-to-zero (RZ) and non-return-to-zero (NRZ) without or with carrier suppression (CSRZ) formats are associated with shift keying (SK) modulation schemes such as amplitude (ASK), phase (PSK), and differential phase (DPSK, DQPSK) [1,2]. Differential detection offers the best technological implementation owing to the nonrequirement of coherent sources and avoidance of polarization control of the mixing of the signals and a local oscillator in the multi-THz frequency range. Continuous phase modulation (CPM) is also another form of phase shift keying (PSK) in which the phase of the optical carrier evolves continuously from one phase state to the other. For MSK, the phase change is limited to $\pi/2$. Although MSK is a well-known modulation format in radio frequency digital communications, it has only been attracting interest in optical system research in the past few years [3–5]. The phase continuous evolution of MSK has many interesting features: the main lobe of the power spectrum is wider than that of quadrature phase shift keying (QPSK) and differential phase shift keying (DPSK) and the side lobes of the MSK signal spectrum are much lower, allowing ease of optical filtering and hence less distortion due to dispersion effects. In addition, higher signal energy is concentrated in the main lobe of MSK spectrum than its side lobes, leading to a better signal-to-noise ratio at the receiver.

Advanced modulation formats for 40 Gbps and higher-bit-rate long-haul optical transmission systems have recently attracted intensive research, including various amplitude and discrete differential phase modulation and pulse shape formats (ASK-NRZ/RZ/CSRZ, DPSK, and DQPSK-NRZ/RZ/CSRZ). However, there are only a few reports on optical CPM schemes using external electro-optical modulators [1–4]. Compared to discrete phase modulation, CPM signals have very interesting and attractive characteristics including high spectral efficiency, higher energy concentration in the signal bands, and more robustness to interchannel crosstalk in dense wavelength division multiplexed (DWDM) owing to greater suppression of the side lobes, which has recently arisen as a critical issue in DWDM optical systems [5,6].

In bandwidth-limited digital communications, including wireless and satellite digital transmission, MSK has proved to be a very efficient modulation format owing to its prominent spectral efficiency, high sideband suppression, constant envelope, and high energy concentration in the main spectral lobe. In modern high-capacity and high-performance optical systems, we have witnessed an acceleration of the transmission bit rate approaching 40 Gbps and even up to 160 Gbps. At these ultra-high bit transmission speeds, the essence of efficient modulation formats for long-haul transmission is critical. There are several technical published works on advanced modulation techniques for optical transmission, which mostly focus on discrete phase modulation including binary DPSK and DQPSK [1–3] with various pulse carving formats including NRZ and CSRZ [2,4]. However, there is only a limited number of studies on the MSK modulation format [5–8]. The features of MSK compared to other modulation formats have prompted researchers to investigate its suitability for high-capacity long-haul transmission. The side lobes of the MSK power spectrum are greatly suppressed, giving it good dispersion tolerance and avoiding interchannel crosstalk. Thus, this modulation is of interest for further investigation.

Two different proposals for transmitter configurations for the generation of linear and nonlinear phase optical MSK modulation are reported in Ref. [9]. Brief operational descriptions of these two transmitter configurations are presented in Section 8.2, whereas the direct detection techniques for both linear and nonlinear optical MSK signals are covered in Section 8.3. Our simulation models for the modulation and system transmission are based on the MATLAB® and Simulink® platform [10] with detailed discussions in Section 8.4. In this chapter, we have also proved that the optical MSK signal is capable of propagating over an optically amplified and fully dispersion-compensated system. The performance of the conventional MSK format and also that of the weakly and strongly nonlinear optical MSK sequences are evaluated.

The MSK format exhibits dual alternating frequencies and offers the orthogonality in phase of the envelopes of two consecutive bit periods. More interestingly, MSK can be considered as either a special case of CPFSK or an staggered/offset QPSK in which I and Q components are interleaved with each other [7]. These characteristics are greatly advantageous for optically amplified long-haul transmission because such a compact spectrum potentially gives a good dispersion tolerance, making MSK a suitable candidate for DWDM systems.

This chapter thus investigates a number of novel structures of photonic transmitters for the generation of optical MSK signals. The theoretical background of MSK modulation formats discussing employing two different approaches is presented in Section 8.2. Section 8.3 proposes two configurations of optical MSK transmitters that employ (1) two cascaded electro-optic phase modulators (E-OPMs) and (2) parallel dual-drive Mach–Zehnder modulators (MZMs). Different types of linear and nonlinear phase-shaped optical MSK sequences can be generated. The precoder for the I–Q optical MSK structure is also derived in this section. We present in Section 8.4 a simple noncoherent configuration for detection of MSK and nonlinear MSK-modulated sequences. The optical detection of MSK and nonlinear MSK signals employs a $\pi/2$ phase shift in one arm of the Mach–Zehnder interferometric delay (MZIM) balanced receiver. New techniques in the measurement of the BERs with the probability density function (pdf) as described in Chapter 4 of the received signals after decision sampling can be computed with superposition of a number of weighted Gaussian pdfs. The following performance results and observations are obtained in Section 8.5: (1) spectral characteristics of linear and nonlinear MSK-modulated signals and (2) improvement in dispersion tolerance of MSK and nonlinear MSK over ASK and DPSK counterparts.

Recently, advanced modulation formats have attracted intensive research for long-haul optical transmission systems including various amplitude and discrete differential phase modulation and pulse shape formats (ASK-NRZ/RZ/CSRZ, DPSK, and DQPSK-NRZ/RZ/CSRZ). For the case of phase modulation, the phases of the optical carrier are discretely coded with "0", "π" (DPSK) or "0", "π" and "$\pi/2$, $-\pi/2$" (DQPSK). Although, the differential phase modulation techniques offer better spectral properties, a higher energy concentration in the signal bands, and more robustness to combat the nonlinearity impairments as compared to the amplitude modulation formats, phase continuity would offer even better spectral efficiency and at least 20 dB better suppression of the side lobes. MSK exhibits a dual alternating frequency between the two consecutive bit periods and is considered to offer the best scheme as this offers the orthogonal property of the two-frequency modulation of the carrier lightwaves embedded within the consecutive bits. Furthermore, it offers maximum simplicity in the implementation of the modulation in the photonic domain.

This chapter thus investigates a number of novel structures of photonic transmitters and differential noncoherent balanced receivers for the generation and detection of MSK optical signals as follows.

Cascaded electro-optic phase modulators MSK transmitter: This structure employs two cascaded optical phase modulators (OPMs). The first OPM plays the role of modulating the binary data logic into two carrier frequencies deviating from the optical carrier of the laser source by a quarter of the bit rate. The second OPM enforces the phase continuity of the lightwave carrier at every bit transition. The driving voltage of this second OPM is precoded in such as a way that the phase discrepancy due to frequency modulation of the first OPM will be compensated, hence preserving the phase continuity characteristic of the MSK signal. Utilizing this double phase modulation configuration, different types of linear and nonlinear CPM phase-shaping signals, including MSK, weakly MSK, and linear-sinusoidal MSK, can be generated. The optical spectra of the modulation scheme obtained confirm the bandwidth efficiency of this novel optical MSK transmitter.

Parallel dual-drive MZM optical MSK transmitter: In this second configuration, MSK signals with small ripple of approximately 5% signal amplitude level can be generated. The optical spectrum is demonstrated to be equivalent to the configuration depicted in Figure 8.1. This configuration can be implemented using commercially available dual-drive intensity interferometric electro-optic modulators.

We also present a simple noncoherent configuration for detection of the MSK and nonlinear MSK-modulated sequences. The optical detection of MSK and nonlinear MSK signals employs a $\pi/2$ OPM followed by the MZIM balanced receiver.

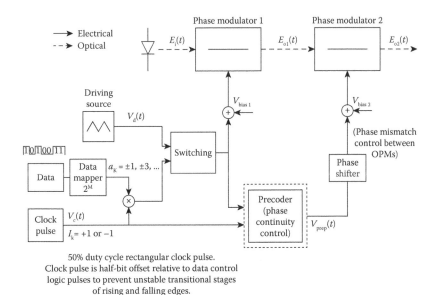

FIGURE 8.1 Block diagram of optical MSK transmitter employing two cascaded optical phase modulators.

The following performance results and observations are obtained:

1. The modeling platform based on MATLAB and Simulink is developed for the transmission systems.
2. Spectral characteristics of linear and nonlinear MSK-modulated signals are obtained.
3. Improvement of the dispersion tolerance of MSK and nonlinear MSK over ASK and DPSK is achieved.
4. Novel techniques in the measurement of BERs with the pdf of received signals after decision sampling to be computed with the superposition of a number of weighted Gaussian pdfs. This technique has been proved to be effective, convenient, and accurate for estimating the non-Gaussian pdf for the case of CPM and its transmission over long-haul optically amplified systems. The MSK format has been demonstrated via modeling to be a promising candidate in the selection of advanced modulation formats for DWDM long-haul optical transmission system.

8.2 GENERATION OF OPTICAL MSK-MODULATED SIGNALS

The two optical MSK transmitter configurations can be referred to in more detail in Ref. [6]. The descriptions of these configurations are briefly addressed in Section 7.2.

The optical MSK transmitter can be formed by using two cascaded E-OPMs as shown in Figure 8.1. The evolution of the time-domain phase trellis of the optical MSK signal sequence [–1 1 1 –1 1 –1 1 1] as inputs and outputs at different stages of the optical MSK transmitter is tabulated in Figure 8.2.

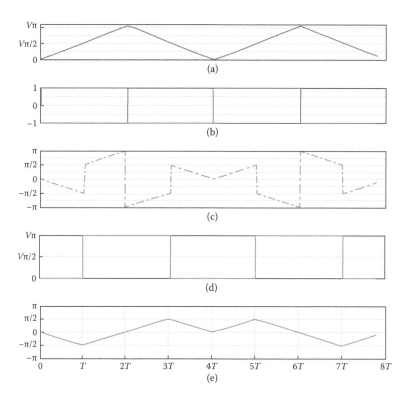

FIGURE 8.2 Evolution of time-domain phase trellis of optical MSK signal sequence [–1 1 1 –1 1 –1 1 1] as inputs and outputs at different stages of the optical MSK transmitter. The notation is denoted in Figure 8.1 accordingly. (a) $V_d(t)$: periodic triangular driving signal for optical MSK signals with a duty cycle of a four-bit period; (b) $V_c(t)$: the clock pulse with a duty cycle of $4T$; (c) $E_{01}(t)$: phase output of OPM1; (d) $V_{prep}(t)$: pre-computed phase compensation driving voltage of OPM2; and (e) $E_{02}(t)$: phase trellis of a transmitted optical MSK sequences at an output of OPM2.

The demonstration of phase compensation for enforcement of phase continuity at bit transitions is given in Figure 8.3. E-OPMs and interferometers operating at a high frequency using resonant-type electrodes have been studied and proposed in Refs. [7,8]. In addition, high-speed electronic driving circuits evolved with the ASIC technology using 0.1 μm GaAs P-HEMT or InP HEMTs [9] enables the realization of the proposed optical MSK transmitter structure. The baseband equivalent optical MSK signal is represented in Figure 8.2.

The first E-OPM enables the frequency modulation of data logic into upper sidebands (USBs) and lower sidebands (LSBs) of the optical carrier with a frequency deviation of f_d. Differential phase precoding is not necessary in this configuration owing to the nature of the continuity of the differential phase trellis. By alternating the driving sources $V_d(t)$ to sinusoidal waveforms for simple implementation or a combination of sinusoidal and periodic ramp signals that was first proposed by Amoroso in 1976 [10], different schemes of linear and nonlinear phase-shaping MSK-transmitted sequences can be generated.

The second E-OPM enforces the phase continuity of the lightwave carrier at every bit transition. The delay control between the E-OPMs is usually implemented by the phase shifter shown in Figure 8.1. The driving voltage of the second E-OPM is precoded to fully compensate for the

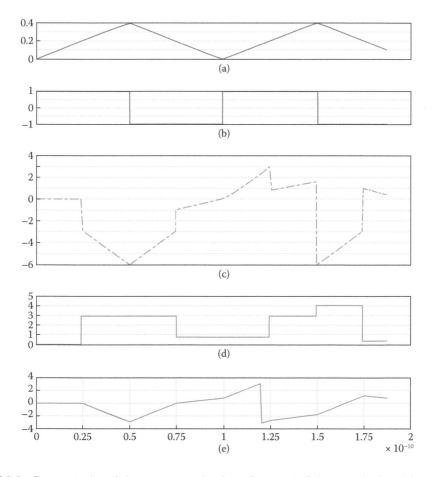

FIGURE 8.3 Demonstration of phase compensation for enforcement of phase continuity at bit transitions: (a) shows the periodic triangular driving signal whose peak voltage is $V\pi/8$ and a duty cycle of four-bit period, (b) shows the clock pulse corresponding to the driving signal, (c) shows frequency switching with $h = \pm 1/8, \pm 7/8$, (d) shows the computed phase compensation, and (e) shows the phase trellis of an optical quaternary CPFSK.

transitional phase jump at the output $E_{01}(t)$ of the first E-OPM. The phase continuity characteristic of the optical MSK signals is determined by the algorithm as

$$\Phi(t,k) = \frac{\pi}{2}\left(\sum_{j=0}^{k-1} a_j - a_k I_k \sum_{j=0}^{k-1} I_j\right) \tag{8.1}$$

where $a_k = \pm 1$ are the logic levels; $I_k = \pm 1$ is a clock pulse whose duty cycle is equal to the period of the driving signal $V_d(t)$, f_d is the frequency deviation from the optical carrier frequency, and $h = 2f_d T$ is previously defined as the frequency modulation index. In the case of optical MSK, $h = 1/2$ or $f_d = 1/(4T)$. The phase evolution of the continuous phase optical MSK signals are shown with detailed discussion in Figure 8.2. In order to mitigate the effects of the unstable stages of rising and falling edges of the electronic circuits, the clock pulse $V_c(t)$ is offset with the driving voltages $V_d(t)$.

The generated optical signal envelope can be written as

$$\tilde{s}(t) = A\exp\left\{j\left[a_k I_k f_d \frac{2\pi t}{T} + \Phi(t,k)\right]\right\}, \quad kT \le t \le (k+1)T \tag{8.2}$$

$$\text{with } a_k = \pm 1, \pm 3, \ldots$$

8.3   DETECTION OF M-ARY CPFSK-MODULATED OPTICAL SIGNAL

The detection of linear and nonlinear optical M-ary CPFSK utilizes the well-known MZDI-balanced receiver. The addition of a $\pi/2$ phase on one arm of the MZDI is also introduced. The time delay being a fraction of a bit period enables the phase trellis detection of optical M-ary CPFSK. The detected phase trellis using the proposed technique is shown in Figure 8.3. Figure 8.4 shows the eye phase trellis detection of an optical M-ary CPFSK-modulated signal.

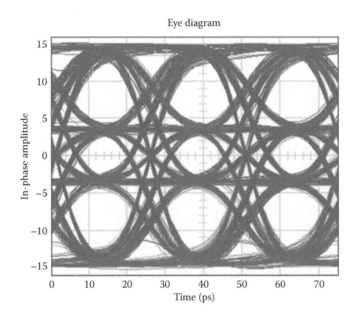

FIGURE 8.4   Eye phase trellis detection of optical M-ary CPFSK-modulated signal.

An optimized ratio of switching frequencies results in the maximum eye open. The detection of optical M-ary CPFSK with delay of $T_b/2$ at $t = (k + 1)T_b/2$ is expressed in Equation 8.3

$$\sin(\Delta\Phi) = \sin\left(a_{k+1} \frac{2\pi f_d(k+1)\frac{T_b}{2}}{T_b} - a_k \frac{2\pi f_d k \frac{T_b}{2}}{T_b} \right) \quad (8.3)$$

On the same slope of phase in the phase trellis, the differential phase, and hence the modulated frequency levels, can be mapped to detected amplitude levels via Equation 8.4.

$$\sin(\Delta\Phi) = \sin(a_{k+1}\pi f_d) \quad (8.4)$$

8.3.1 Optical MSK Transmitter Using Parallel I–Q MZIMs

The conceptual block diagram of the optical MSK transmitter is shown in Figure 3.30, Chapter 3. The transmitter consists of two dual-drive electro-optic MZM modulators generating chirpless I and Q components of MSK-modulated signals, which is considered as a special case of staggered or offset QPSK. The binary logic data are precoded and de-interleaved into even and odd bit streams, which are interleaved with each other by one-bit duration offset. Figure 3.30, Chapter 3 shows the block diagram configuration of band-limited phase-shaped optical MSK. Two arms of the dual-drive MZM modulator are biased at $V\pi/2$ and $-V\pi/2$ and driven with *data* and *data complement*. Phase-shaping driving sources can be a periodic triangular voltage source in the case of linear MSK generation or simply a sinusoidal source for generating a nonlinear MSK-like signal, which also obtains the linear phase trellis property but with small ripples introduced in the magnitude.

The magnitude fluctuation level depends on the magnitude of the phase-shaping driving source. High spectral efficiency can be achieved with tight filtering of the driving signals before modulating the electro-optic MZMs. Three types of pulse-shaping filters are investigated including Gaussian, raised-cosine, and squared-root raised-cosine filters. The optical carrier phase trellis of linear and nonlinear optical MSK signals are shown in Figure 8.5.

The generation of linear and nonlinear optical MSK sequences are now briefly discussed.

8.3.1.1 Linear MSK

The pulse-shaping waveform for linear MSK is a triangular waveform with a duty cycle of 4 Tb. The triangular waveform for the quadrature path is delayed by one-bit period with respect to the in-phase path, and hence they are interleaved with each other. The optical MSK signal is the superposition of both even and odd waveforms from the MZIMs. The amplitude of the signal is a constant, clearly displaying the constant amplitude characteristic of CPM. The phase trellis is linear, and the phase transition is continuous as shown in Figure 8.5a. The signal constellation is a perfect circle.

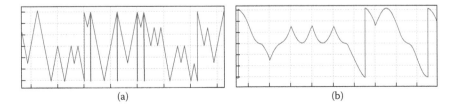

(a)   (b)

FIGURE 8.5 Phase trellis of (a) linear and (b) nonlinear MSK-transmitted signals.

8.3.1.2 Weakly Nonlinear MSK

The pulse-shaping waveform for weakly nonlinear MSK is a sinusoidal waveform with an amplitude of $1/4V\pi$, and its symbol period is equal to a two-bit period (symbol rate of 20 Gbits/s). The in-phase pulse shaper is a cosine waveform, whereas the quadrature pulse shaper is a sine waveform. There is a ripple of approximately 5%. Owing to the sinusoidal pulse shaper, the variation of the phase with time, which is represented by the phase trellis, is nonlinear. Therefore, the rate of frequency change is not constant. This causes a mismatch of the MZIM when the modulated waveforms are added up, resulting in the ripple, as shown in Figure 8.5b.

8.3.1.3 Strongly Nonlinear MSK

The pulse-shaping waveform for strongly nonlinear MSK is a sinusoidal waveform, the same as for the weakly nonlinear case. However, the amplitude of the pulse shaper is $V\pi/2$. The waveforms are interleaved with each other. The optical MSK signal has a ripple of approximately 26%. This ripple is also caused by the mismatch of the MZIM as the modulating waveform is strongly nonlinear. The effect of nonlinearity is obvious in the phase trellis in Figure 8.4b.

The signal state constellations and eye diagrams of the optical MSK sequences are shown in Figure 8.6a through c for both linear and nonlinear schemes. Note that the magnitude ripples. In the case of the nonlinear configuration, MSK signals with a small ripple of approximately 5% signal amplitude level can be generated. This configuration can be implemented using commercially available dual-drive intensity interferometric electro-optic modulators.

The conceptual block diagram of the optical MSK transmitter is shown in Figures 8.1. The transmitter consists of two dual-drive electro-optical MZM modulators generating chirpless I and Q components of MSK-modulated signals, which is considered as a special case of staggered or

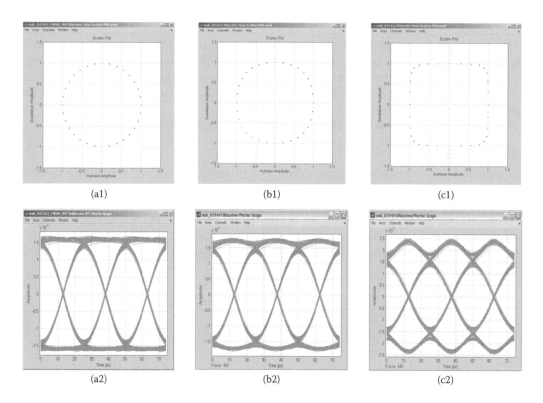

FIGURE 8.6 Constellation diagrams and eye diagrams of optical MSK-transmitted signals: (a) a1–a2, linear; (b) b1–b2, weakly nonlinear; and (c) c1–c2, strongly nonlinear transmission.

offset QPSK. The binary logic data are precoded and de-interleaved into even and odd bit streams that are interleaved with each other by one-bit duration offset.

Two arms of the dual-drive MZM modulator are biased at $V\pi/2$ and $-V\pi/2$ and driven with *data* and *inverted data*. Phase-shaping driving sources can be a periodic triangular voltage source in the case of linear MSK generation or simply a sinusoidal source for generating a nonlinear MSK-like signal, which also obtains the linear phase trellis property but with small ripples introduced in the magnitude. The magnitude fluctuation level depends on the magnitude of the phase-shaping driving source. A high spectral efficiency can be achieved with tight filtering of the driving signals before modulating the electro-optic MZMs. Three types of pulse-shaping filters are investigated including Gaussian, raised-cosine, and square-root raised-cosine filters. A narrow spectral width and high suppression of the side lobes can be achieved.

The logic gates in the precoder are constructed based on the state diagram as this approach eases the implementation of the precoder. As seen from the state diagram, the current state of the signal is dependent on the previous state because the state of the signal advances corresponding to the binary data from the previous state. Therefore, memory is needed to store the previous state. The state diagram in Figure 8.7a is developed into a logic state diagram in Figure 8.7b to enable the construction of the truth table. $S_0 S_1 = 00$ or $S_0' S_1' = 00$ corresponds to state 0, $S_0 S_1 = 01$ or $S_0' S_1' = 01$ corresponds to state $\pi/2$, $S_0 S_1 = 10$ or $S_0' S_1' = 10$ corresponds to state π, while $S_0 S_1 = 11$ or $S_0' S_1' = 11$ corresponds to state $-\pi/2$, with $S_0 S_1$ as the current state and $S_0' S_1'$ as the previous state.

Two delay units in Figure 8.7b function as memory by delaying the current state and feedback into the precoding logic block as the previous state. The precoding logic block, which consists of logic gates, would compute the current state and output based on the feedback state (previous state) and binary data from the Bernoulli binary generator.

The truth table is constructed based on the logic state diagram and combinational logic diagram given earlier. For a positive half-cycle cosine wave and a positive half-cycle sine wave, the output is 1; for a negative half-cycle cosine wave and a negative half-cycle sine wave, the output is 0. Then, Karnaugh maps are constructed to derive the logic gates within the precoding logic block, based on the truth table. The following three precoding logic equations are derived as

$$S_0 = \overline{S_0' S_1'} + b_n \overline{S_0' S_1'} + \overline{b_n} S_0' S_1' + b_n S_0' \overline{S_1'} \tag{8.5}$$

$$S_1 = \overline{S_1'} = \overline{b_n} \overline{S_1'} + b_n \overline{S_1'} \tag{8.6}$$

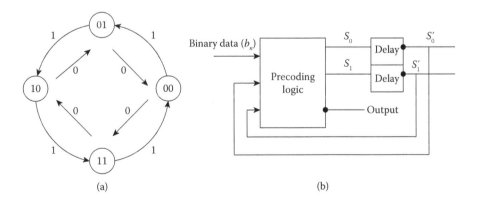

(a) (b)

FIGURE 8.7 (a) State diagram for MSK. The arrows indicate continuous increment or decrement of the phase of the carrier and (b) combinational logic, the basis of the logic for constructing the truth table of the precoder.

TABLE 8.1
Truth Table Based on MSK State Diagram

$b_n S_0' S_1'$	$S_0 S_1$	Output
100	01	1
001	00	1
010	01	1
101	10	0
110	11	0
111	00	1
000	11	0
011	10	0

$$\text{Output} = \overline{S_0} \tag{8.7}$$

The final logic gate construction for the precoder is as shown in Table 8.1.

8.3.2 OPTICAL MSK RECEIVERS

The optical detection of MSK and nonlinear MSK signals employs a $\pi/2$ OPM followed by an optical phase comparator, an MZDI, and then a balanced receiver (BalRx). This detection for optical linear and nonlinear MSK signals is similar to that of the well-known structure MZDI as introduced in Chapter 7 to detect a $\pm\pi/2$ phase difference of two adjacent bits of the MSK optical sequence.

A new technique for the evaluation of the BERs is implemented. The probability density functions (pdfs) of noise-corrupted received signals after decision sampling are computed with the superposition of a number of weighted Gaussian pdfs. The technique implements the expected maximization theorem and has shown its effectiveness in determining arbitrary distributions [11,12].

8.4 OPTICAL BINARY AMPLITUDE MSK FORMAT

8.4.1 GENERATION

The optical MSK transmitters from [4,13–14] can be integrated in the proposed generation of optical MAMSK signals. However, in this chapter, we propose a new simple-in-implementation optical MSK transmitter configuration employing high-speed cascaded E-OPMs as shown in Figure 8.2. E-OPMs and interferometers operating at high frequency using resonant-type electrodes have been studied and proposed in Refs. [7,8]. In addition, high-speed electronic driving circuits evolved with the ASIC technology using 0.1 μm GaAs P-HEMT or InP HEMTs [9] enables the feasibility in realization of the proposed optical MSK transmitter structure. The baseband equivalent optical MSK signal is represented in Equation 8.8.

$$\tilde{s}(t) = A \exp\left\{ j\left[a_k I_k 2\pi f_d t + \Phi(t,k) \right] \right\}, \quad kT \le t \le (k+1)T$$

$$= A \exp\left\{ j\left[a_k I_k \frac{\pi t}{2T} + \Phi(t,k) \right] \right\} \tag{8.8}$$

where $a_k = \pm 1$ are the logic levels; $I_k = \pm 1$ is a clock pulse whose duty cycle is equal to the period of the driving signal $V_d(t)$; f_d is the frequency deviation from the optical carrier frequency; and $h = 2f_d T$ is defined in Equations 8.2 and 8.3 as the frequency modulation index. In the case of optical MSK, $h = 1/2$ or $f_d = 1/(4T)$.

The first E-OPM enables the frequency modulation of data logic into the USBs and LSBs of the optical carrier with a frequency deviation of f_d. Differential phase precoding is not necessary in this configuration owing to the nature of the continuity of the differential phase trellis. By alternating the driving sources $V_d(t)$ with sinusoidal waveforms or a combination of sinusoidal and periodic ramp signals, different schemes of linear and nonlinear phase-shaping MSK-transmitted sequences can be generated [15]. The second E-OPM enforces the phase continuity of the lightwave carrier at every bit transition. The delay control between the E-OPMs is usually implemented by the phase shifter, as shown in Figure 8.2. The driving voltage of the second E-OPM is precoded to fully compensate for the transitional phase jump at the output $E_{01}(t)$ of the first E-OPM. The phase continuity characteristic of the optical MSK signals is determined by the algorithm in Equation 8.2. In order to mitigate the effects of the unstable stages of the rising and falling edges of the electronic circuits, the clock pulse $V_c(t)$ is offset with the driving voltages $V_d(t)$.

$$\Phi(t,k) = \frac{\pi}{2}\left(\sum_{j=0}^{k-1} a_j - a_k I_k \sum_{j=0}^{k-1} I_j \right) \tag{8.9}$$

The binary amplitude MSK (BAMSK) modulation format is proposed for optical communications. We report numerical results of a 80 Gbps two-bit-per-symbol BAMSK optical system on spectral characteristics and residual dispersion tolerance to different types of fibers. BER of 1e–23 is obtained for 80 Gbps optical BAMSK transmission over 900 km Vascade fiber systems, enabling the feasibility of long-haul transmission for the proposed format.

8.4.2 OPTICAL MSK

MSK, which is well-known in radio frequency and wireless communications, has recently been adapted into optical communications. A few optical MSK transmitter configurations have recently been reported [15,16]. For optically amplified communications systems, if multilevel concepts can be incorporated in those reported schemes, the symbol rate would be reduced, and hence bandwidth efficiency can be achieved. This is the principal motivation for the proposed modulation scheme.

BAMSK is a special case of the M-ary CPM format that enables a binary-level (PAM- or QAM-like) transmission scheme, while the bandwidth efficiency due to transitional phase continuity properties between two consecutive symbols (CPM-like signals) are preserved. The generation of M-ary CPM sequences can be expressed in Equation 8.10 [17]

$$s(t) = A_n \cos(\omega_c t + \phi_n(t,a)) + \sum_{m=1}^{N-1} B_m \cos\left(\omega_c t + \phi_m(t,b_m)\right) \tag{8.10}$$

where

$$\phi_N(t,a) = \pi h a_n q(t - nT) + \pi h \sum_{k=-\infty}^{n-1} a_k \quad nT \le t \le (n+1)T \tag{8.11}$$

$$\phi_m(t,b_m) = \pi a_n\left(h + \frac{b_{mn}+1}{2}\right)q(t-nT) + \pi \sum_{k=-\infty}^{n-1} a_k\left(h + \frac{b_{mk}+1}{2}\right) \quad mT \le t \le (m+1)T \tag{8.12}$$

In a generalized M-ary CPM transmitter, values of a_n and b_{mn} are statistically independent and taken from the set of $\{\pm 1, \pm 3,...\}$. A_n and B_m are the signal state amplitude levels, which are either in-phase or π phase shift, with the largest level component at the end of the nth symbol interval; $q(t)$ is the pulse-shaping function, and h is the frequency modulation index. In the case of BAMSK, Equations 8.2 and 8.3, which show the constraints of ϕ_m to maintain the phase continuity characteristic of CPM sequences, are simplified to Equations 8.4 and 8.5, respectively, where $h = \frac{1}{2}$ and the phase-shaping function $q(t - nT)$ is a periodic ramp signal with a duty cycle of $4T$.

$$\phi_n(t,a) = \pi h a_n \frac{t - nT}{T} + \pi h \sum_{k=-\infty}^{n-1} a_k \quad nT \le t \le (n+1)T \tag{8.13}$$

$$\phi_m(t,b_m) = \pi I_n \left(h + \frac{b_{mn}+1}{2} \right) \frac{t - nT}{T} + \pi \sum_{k=-\infty}^{n-1} b_k \left(h + \frac{b_{mk}+1}{2} \right) \quad mT \le t \le (m+1)T \tag{8.14}$$

Any configuration of the reported optical MSK transmitters in Refs. [4,14–16] can be utilized in the proposed generation scheme of optical BAMSK signals. Figure 8.1a shows the block structure of the optical BAMSK transmitter in which two optical MSK transmitters are integrated in a parallel configuration. The amplitude levels are determined from Equations 8.1, 8.4, and 8.5 by the splitting ratio at the output of a high-precision power splitter. The logic sequences $\{\pm 1,..\}$ of a_n and b_n are precoded from the binary logic $\{0,1\}$ of d_n as $a_n = d_n - 1$ and $b_n = a_n(1 - d_n - 1/h)$ [17]. The signal-space trajectories of BAMSK signals are shown in Figure 8.1b.

A simple noncoherent configuration for detection of the optical BAMSK sequences consists of phase and amplitude detections. In the case of BAMSK, that is, $n = 2$, the system effectively implements two bits per symbol with two amplitude levels. Phase detection is enabled with the employment of the well-known integrated optic phase comparator MZDI-balanced receiver with one-bit time delay on one arm of the MZDI [18]. An additional $\pi/2$ phase shift is introduced. Figure 8.8a and b show the block diagram and the eye diagrams of the amplitudes and phases of the optical BAMSK signals, respectively. In phase detection, the decision threshold, which is plotted in broken-line style, is at a zero level, whereas the amplitude levels are determined by different thresholds. A new technique in BER calculation for dispersive and noise-corrupted received signals that exploits the expected maximization (EM) theorem is implemented with superposition of a number of weighted Gaussian probability distribution functions [11,19].

A simple noncoherent configuration for detection of linear and nonlinear optical M-ary MSK sequences consists of phase and amplitude detections that are very well known in the discrete phase shift keying schemes such as DPSK or quadrature DPSK [20]. Phase detection is enabled with the employment of the well-known integrated optic phase comparator MZDI-balanced receiver with a one-bit time delay on one arm of the MZDI. An additional $\pi/2$ phase shift is introduced to detect the differential $\pi/2$ phase shift difference of two adjacent optical MSK pulses. Figure 8.9a and b show the eye diagrams of the amplitudes and phases of the optical BAMSK-detected signals, respectively. In the case of $N = 2$ and $N = 3$, the system effectively implements the two-bits-per-symbol scheme and the three-bits-per-symbol scheme with two and four amplitude levels, respectively. In phase detection, the decision threshold, which is plotted in broken-line style, is at zero level because only in-phase and π differential phases are of interest.

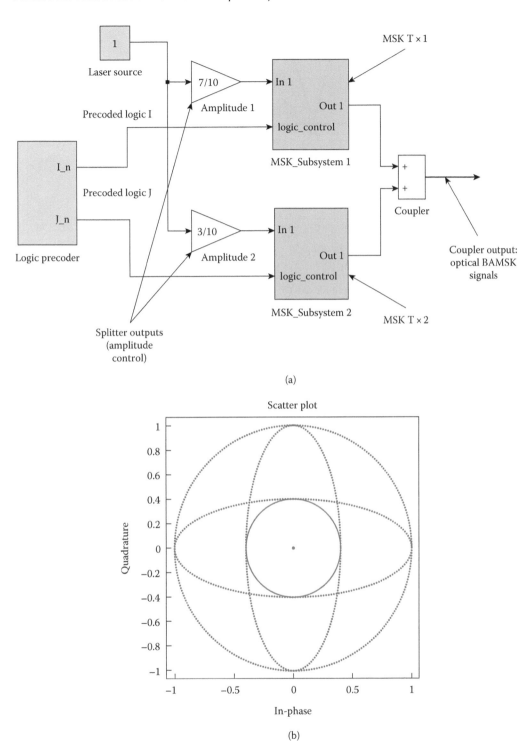

(a)

(b)

FIGURE 8.8 (a) Block diagram of optical BAMSK transmitter and (b) signal trajectories of optical BAMSK-transmitted signals of multiamplitude MSK.

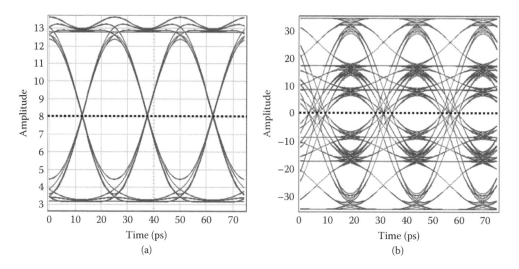

FIGURE 8.9 Eye diagrams of (a) amplitude and (b) phase detection of optical BAMSK received signals with a normalized amplitude ratio of 0.285/0.715. The decision threshold is shown in broken-line style.

8.5 NUMERICAL RESULTS AND DISCUSSION

The state diagram for MSK is shown in Figure 8.10. The continuous wave carrier source is modulated by the two cascaded MZIMs that are driven by a signal sequence which is the output of a broadband amplifier whose input is from the output of the precoder. The arrows indicate continuous increment or decrement of the phase of the carrier. These information-bearing lightwave signals are then propagated along the fiber spans and detected via a photonic phase comparator, the MZDI, and then detected via a balanced receiver. The obtained eye diagram is then statistically analyzed. A more efficient detection scheme using frequency discrimination will be presented in Chapter 8.

8.5.1 TRANSMISSION PERFORMANCE OF LINEAR AND NONLINEAR OPTICAL MSK SYSTEMS

The block diagram of the simulation setup is shown in Figure 8.10. The dispersion tolerance of linear, weakly nonlinear, and strongly nonlinear optical MSK signals is numerically investigated, and the results are shown in Figure 8.11. Among the three types, linear MSK is most tolerant to residual dispersion with a 1 dB eye open penalty at 72 ps/nm/km. Strongly nonlinear MSK suffers a severe penalty when the residual dispersion exceeds 85 ps/nm/km, or equivalently, 5 km standard single-mode fiber (SSMF). Figure 8.12 shows the performance of three types of optical MSK-modulated signals versus the optical signal-to-noise ratio (OSNR) in transmission over 540 km Vascade fibers of optically amplified and fully compensated multispan links (6 spans × 90 km/span). The receiver sensitivity of the differential phase comparison balanced receiver is −24.6 dBm. In Vascade fibers, the dispersion factors of the dispersion-compensating fiber is negatively opposite to that of the transmission fiber, an SSMF, of +17.5 ps/nm/km at 1550 nm wavelength. In addition, the dispersion slopes of these fibers are also matched. An optical amplifier of the EDFA type is placed at the end of the transmission fiber, and another is placed after the DCF so that it would boost the optical power to the right level, which is equal to that of the launched power. An EDFA optical gain of 19 dB and 5 dB ASE noise figure is used. The noise margin reduces greatly after the propagation over 6 × 90 km spans. The effects of positive and negative dispersion mismatch and mid-link nonlinearity on phase evolution are shown in Figure 8.11.

The tolerance of these MSK modulations to nonlinear effects is also studied using transmission models, and the simulation results are shown in Figures 8.11 and 8.12. The input power into the fiber span is varied, whereas the length of transmission link is constant at 180 km. At BER = 1e−9,

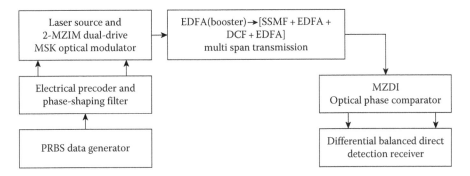

FIGURE 8.10 Schematic diagram of an optically amplified optical transmission system.

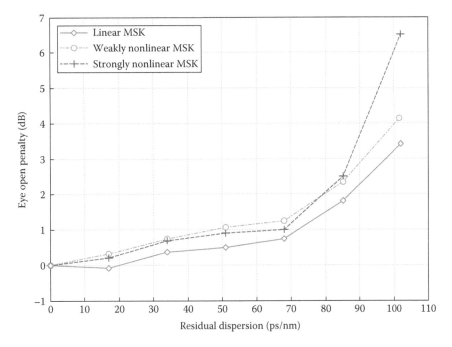

FIGURE 8.11 Dispersion tolerance of 40 Gbps linear MSK, weakly nonlinear MSK, and strongly nonlinear MSK optical signals.

linear MSK could tolerate an increase of the input launched power up to 10.5 dBm, and weakly nonlinear could tolerate up to 10.2 dBm compared to 9.2 dBm in the case of strongly nonlinear MSK. The nonlinear phase shift is proportional to the input power; therefore, increasing the input power would increase nonlinear phase shift as well. This nonlinear phase shift is observed through the asymmetries in the eye diagram and through the scatter plot, which shows the phases of the in-phase and quadrature components have shifted from the x- and y-axes. The maximum phase shift that could be tolerated is approximately 15° of arc. Although increasing the input power increases the noise margin of the eye diagram, it is paid off by the large distortion at the sampling time, causing the SNR to decrease. Typical eye diagrams under full compensation and after transmission over 540 km Vascade fibers of optically amplified and fully compensated multispan links (6 spans × 90 km/span) are shown in Figure 8.13.

We note that if the sampling is conducted at the center of the bit period, then the error is minimum as the ripples of the eyes fall in this position. This is the principal reason why the MSK signals can suffer minimal pulse spreading due to residual and nonlinear phase dispersion. Linear, weakly,

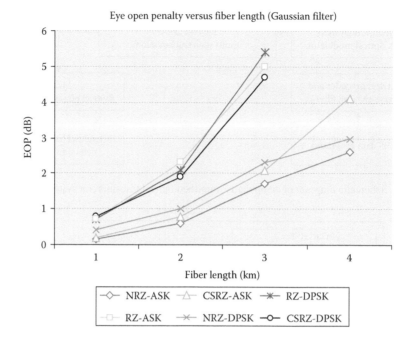

FIGURE 8.12   Simulation of EOP versus transmission distance 1–4 km of SSMF.

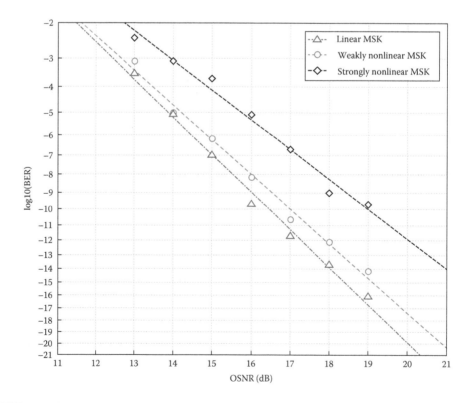

FIGURE 8.13   BER versus optical SNR for transmission of three types of modulated optical MSK signals over 540 km Vascade fibers of optically amplified and fully compensated multispan links (6 spans × 90 km/span).

and strongly nonlinear MSK phase-shaping functions are investigated. It has been proved that optical MSK is a very efficient modulation scheme that offers excellent performance. With an OSNR of about 17 dB, the BER is obtained to be 1e–9 and reaches 1e–17 for an optical SNR of 19 dB under linear MSK modulation. The modulation formats of linear and nonlinear phase-shaping MSK are also highly resilient to nonlinear effects. Nonlinear distortion appears when the total average power reaches about 9 dBm, that is, about 3–4 dB above that of the NRZ ASK format over a 50-μm-diameter SSMF fiber.

The nonlinear phase-shaping filters offer better implementation structures in the electronic domain for driving the dual-drive MZIMs than the linear type but suffer some power penalty; however, they are still better than those candidates of other amplitude or phase or differential phase modulation formats.

Weakly nonlinear MSK offers a much lower power penalty and ease of implementation for precoders and phase-shaping filters; thus, it would be the preferred MSK format for long-haul transmission over optically amplified multispan systems. At a BER of 1e–12, linear MSK is 0.3 dB and 1.2 dB more resilient to nonlinear phase effects than weakly nonlinear MSK and strongly nonlinear MSK, respectively. Figure 8.14 depicts the BER versus input power, and the robustness to nonlinearity of linear, weakly nonlinear, and strongly nonlinear optical MSK signals with transmission over 180 km Vascade fibers of two optically amplified and fully compensated span links.

Compared to the DPSK counterpart in 40 Gbps transmission, various types of filtered MSK-modulated signals are more tolerant to residual dispersion. The eye open penalty of 3 dB is obtained

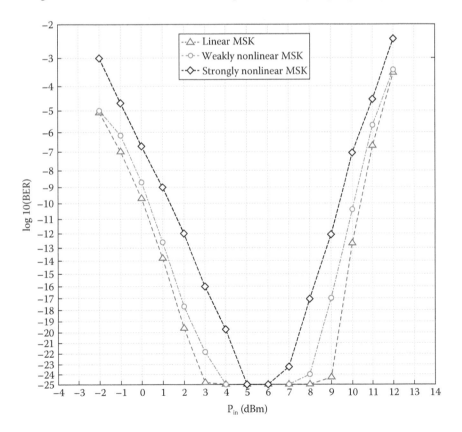

FIGURE 8.14 BER versus input power showing robustness to nonlinearity of linear, weakly nonlinear, and strongly nonlinear optical MSK signals with transmission over 180 km Vascade fibers of two optically amplified and fully compensated span links.

at 4 km of SSMF in the case of NRZ-DPSK/ASK or 68 ps/nm, whereas linear MSK can tolerate up to 98 ps/nm or 6 km accordingly.

It is noted that the pulse shaping using the raised-cosine filter offers a better dispersion tolerance and a lower penalty for RZ-DPSK and CSRZ-DPSK after transmitting 3 km. An approximately 2 dB improvement is observed. Thus, pulse shaping compromises the deficits of RZ-pulses owing to a broader spectrum compared to NRZ pulse shapes. We observe no significant improvement in NRZ pulses for both DPSK and ASK signals.

8.5.2 Transmission Performance of Binary Amplitude Optical MSK Systems

Figure 8.15 numerically compares the power spectra of 80 Gbps optical BAMSK and 40 Gbps optical MSK and DPSK signals. The normalized amplitude levels of two optical MSK transmitters take the ratio of 0.715/0.285. In general, the power spectrum of the optical BAMSK format has identical characteristics to that of the MSK format, including a narrow spectral width and highly suppressed side lobe, and it outperforms the DPSK counterpart [21].

Figure 8.16 shows numerical results of residual dispersion tolerance in both amplitude and phase of 80 Gbps optical BAMSK systems with a normalized amplitude ratio of 0.285/0.715. Standard single-mode fibers (SSMF) and Corning LEAF fibers with a dispersion factor of ±17 ps/nm/km and ±4.5 ps/nm/km, respectively, are used. As expected, the severe penalty due to fiber dispersion derives from the distortion of the waveform amplitudes whose values dramatically jumps over a 20 dB penalty compared to 3 dB in the case of phase distortion at 50 ps/nm SMF residual dispersion. The LEAF fiber enables the system tolerance to a residual dispersion of approximately 150 ps/nm for a 3 dB penalty. It should be kept in mind that the optical 80 Gbps system under test has an effective transmission rate of approximately 40 Gbps due to the two-bit-per-symbol BAMSK scheme.

Figure 8.15 numerically reports the transmission feasibility of 80 Gbps optical BAMSK signals over 900 km Vascade fibers with possible BER values less than 1e−12 for both normalized input amplitude ratios of 0.285/0.715 and 0.25/0.75, respectively. In Figure 8.17, the performance curve with diamond markers and dashed line represents the first ratio, whereas round markers and solid line curves are used for the latter ratio. The single-channel transmission system consists of an 80 Gbps pseudo random generator with a 128-bit sequence, 10 spans of 90 km Vascade fibers

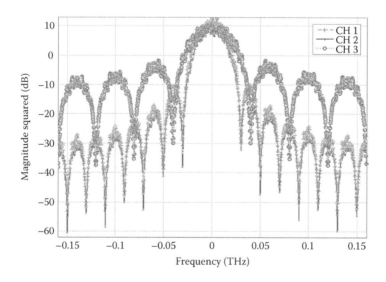

FIGURE 8.15 Comparison of spectra of 80 Gbps optical BAMSK, 40 Gbps optical MSK, and 40 Gbps optical binary DPSK signals.

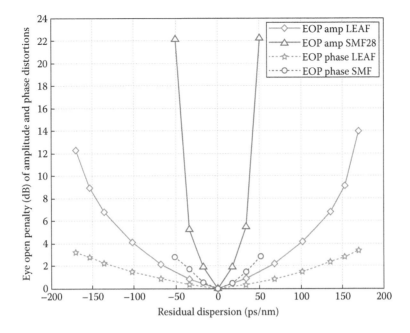

FIGURE 8.16 Numerical results on residual dispersion tolerance of 80 Gbps optical BAMSK systems (effectively 40 Gbps symbol rate) with a normalized amplitude ratio of 0.285/0.715 in both amplitude and phase detection.

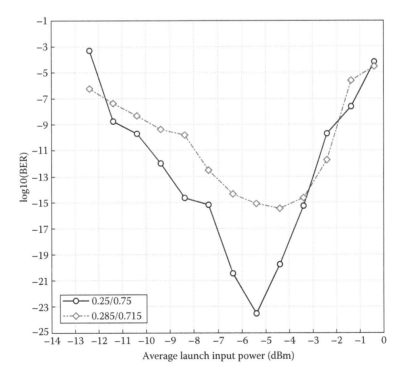

FIGURE 8.17 Transmission performance of 80 Gbps optical BAMSK over 900 km Vascade fiber systems in two cases of normalized amplitude ratios: 0.25/0.75 (round markers and solid line) and 0.285/0.715 (diamond markers and dashed line), respectively.

(60 km of +17 ps/nm/km and 30 km of −34 ps/nm/km and fully compensated for both chromatic dispersion and dispersion slope), and an optical filter with a bandwidth of 80 GHz. The electronic noise of the receiver is modeled with an equivalent noise current density of the electrical amplifier of 20 pA/(Hz)$^{1/2}$ and a dark current of 2 × 10 nA (two photodiodes in a balanced receiver structure). This configuration yields the back-to-back receiver sensitivity at BER = 1e−9 to be approximately −23 dBm. The eye diagrams are obtained after fifth-order Bessel electrical filters with a bandwidth of 36 GHz. The launched peak input power is varied from −10 dBm to +3 dBm, whose corresponding average powers range from −12.5 dBm to −0.5 dBm. The phase noise is dominant in the total BER in low average launched input powers (less than −3 dBm). The results raise the necessity of optimizing the amplitude levels for the optimum BER for optical BAMSK signal transmission.

We have proposed, with detailed operational principles, a new configuration of optical MSK transmitter using two cascaded E-OPMs that reduces the complexity of photonic components. These transmitters can be integrated in a parallel configuration for the inaugural proposed generation of optical M-ary MSK signals. The number of signal states can increase with the number of transmitters. The spectral properties of optical M-ary MSK are similar to those of optical MSK and better than the DPSK counterparts. The main source of penalty of fiber residual dispersion is caused by the waveform amplitude distortions, which can be overcome with advanced dispersion equalization techniques. The numerical results of the transmission performance over 900 km Vascade fibers have been presented in two different cases of normalized amplitude ratios of the BAMSK modulation format. A BER of 1e−23 enables the long-haul transmission feasibility of the proposed format. The simulation test bed is successfully developed based on the MATLAB and Simulink platform.

8.6 CONCLUDING REMARKS

We have proposed two new schemes of optical MSK generation and detection Ref. [21]. These two optical transmitter configurations can generate linear and various types of nonlinear optical MSK modulation formats. The precoder for the I–Q optical MSK structure is shown. The direct detection of optical lightwave is utilized with implementation of the well-known differential noncoherent balanced receivers. Simulated spectral characteristics and dispersion tolerance to 40 Gbps transmission are presented and compared to those of the ASK and DPSK counterparts.

We have proposed the BAMSK modulation format for optical communications. We have reported the configuration for generation and noncoherent detection of the optical BAMSK sequences. The number of signal states can be easily increased with additional optical MSK transmitters. Spectral properties of the 80 Gbps optical BAMSK are similar to those of the 40 Gbps optical MSK and better than the 40 Gbps optical DPSK counterparts. The main source of penalty in dispersion tolerance is the waveform amplitude distortions, which can be overcome with advanced dispersion equalization techniques. The numerical results of transmission performance for an 80 Gbps optical BAMSK system over 900 km Vascade fibers have been reported. A BER of 1e−23 enables the long-haul transmission feasibility of the proposed format. The simulation test bed is developed on the MATLAB and Simulink platform.

It has been proved that MSK transmitter models for these modulation formats can be implemented using the parallel structure of dual-drive MZIM data modulators. The differences in the implementations of these transmitter models are the pulse-shaping waveforms, in which linear MSK follows a triangular periodic waveform, and weakly nonlinear and strongly nonlinear MSKs have sinusoidal waveforms but differ in their amplitudes. The models are simulated under different conditions to investigate the effects of fiber characteristics such as fiber loss, nonlinear effects, and dispersion on the performance of the models. The rotation of the scatter plots confirms the behavior of MSK-modulated signals due to linear chromatic dispersion and nonlinear phase-shaping effects. The simulated optically amplified distance is 540 km fully compensated SSMF. The linear and nonlinear optical MSK modulation formats can thus be offered as an alternative advanced modulation format for long-haul optically amplified transmission.

The visualization of the evolutions of the MSK signal phasor under self-phase modulation is not fully described in this chapter. An electronic compensation technique can be implemented by design of the pre distorted MSK signals at the input of the shaping filters.

REFERENCES

1. T. Hoshida, Optimal 40 Gb/s modulation formats for spectrally efficient long haul DWDM systems, *IEEE J. Lightwave Technol.*, Vol. 20, pp. 1989–1996, 2002.
2. Y. Zhu, K. Cordina, N. Jolley, R. Feced, H. Kee, R. Rickard, and A. Hadjifotiou, 1.6 bit/s/Hz orthogonally polarized CSRZ–DQPSK transmission of 8x40 Gbit/s over 320 km NDSF, in *OFC'04*, Tu-F1, 2004.
3. T. Sakamoto, T. Kawanishi, and M. Izutsu, Initial phase control method for high-speed external modulation in optical minimum-shift keying format, in *ECOC'05*, Vol. 4, pp. 853–854, 2005.
4. M. Ohm and J. Spiedel, Optical minimum shift keying with direct detection, in *Proceedings of the SPIE on Optical Transmission, Switching and Systems*, Vol. 5281, pp. 150–161, 2004.
5. T. L. Huynh, L. N. Binh, and K. K. Pang, Optical MSK long-haul transmission systems, in *SPIE Proceedings of the APOC'06*, paper 6353–86, Thu9a, 2006.
6. T. L. Huynh, L. N. Binh, and K. K. Pang, Linear and weakly nonlinear optical continuous phase modulation formats for high performance DWDM long-haul transmission, in *ECOC'06*, Cannes, France, 2006.
7. T. Kawanishi, S. Shinada, T. Sakamoto, S. Oikawa, K. Yoshiara, and M. Izutsu, Reciprocating optical modulator with resonant modulating electrode, *Electron. Lett.*, Vol. 41, No. 5, pp. 271–272, 2005.
8. R. Krahenbuhl, J. H. Cole, R. P. Moeller, and M. M. Howerton, High-speed optical modulator in LiNbO$_3$ with cascaded resonant-type electrodes, *J. Lightwave Technol.*, Vol. 24, No. 5, pp. 2184–2189, 2006.
9. I. P. Kaminow and T. Li, *Optical Fiber Communications,* Vol. IVA, chap. 16, Elsevier Science, San Diego, 2002.
10. F. Amoroso, Pulse and spectrum manipulation in the minimum frequency shift keying (MSK) format, *IEEE Trans. Commun.*, Vol. 24, pp. 381–384, 1976.
11. L. Ding, W. D. Zhong, C. Lu, and Y. Wang, New bit-error-rate monitoring technique based on histograms and curve fitting, *Opt. Express.*, Vol. 12, No. 11, pp. 2507–2511, 2004.
12. I. Shake, H. Takara, and S. Kawanishi, Simple Q factor monitoring for BER estimation using opened eye diagrams captured by high-speed asynchronous, *Photon. Technol. Lett.*, Vol. 15, pp. 620–622, 2003.
13. J. Mo, D. Yi, Y. Wen, S. Takahashi, Y. Wang, and C. Lu, Novel modulation scheme for optical continuous-phase frequency-shift keying, in *OFC'03*, paper OFG2, 2004.
14. J. Mo, D. Yi, Y. Wen, S. Takahashi, Y. Wang, and C. Lu, Optical minimum-shift keying modulator for high spectral efficiency WDM systems, in *ECOC'05*, Vol. 4, pp. 781–782, 2005.
15. T. L. Huynh, L. N. Binh, K. K. Pang, and L. Chan, Photonic MSK transmitter models using linear and non-linear phase shaping for non-coherent long-haul optical transmission, in *SPIE Proceedings of the APOC'06*, paper 6353–85, September 3–7, 2006.
16. T. Sakamoto, T. Kawanishi, and M. Izutsu, Optical minimum-shift keying with external modulation scheme, *Opt. Express,* Vol. 13, No. 20, pp. 7741–7747, 2005.
17. J. G. Proakis, *Digital Communications*, 4th ed., New York: McGraw-Hill, pp. 199–202, chap. 4, 2001.
18. C. Wree, J. Leibrich, J. Eick, W. Rosenkranz, and D. Mohr, Experimental investigation of receiver sensitivity of RZ-DQPSK modulation using balanced detection, in *OFC'03*, Vol. 2, pp. 456–457, 2003.
19. R. Redner and H. Walker, Mixture densities, maximum likelihood and the EM algorithm, *SIAM Rev.,* Vol. 26, No. 2, pp. 195–239, 1984.
20. K. Sekine, N. Kikuchi, S. Sasaki, S. Hayase, C. Hasegawa, and T. Sugawara, 40 Gbit/s, 16-ary (4bit/symbol) optical modulation/demodulation scheme, *Electron. Lett.*, Vol. 41, No. 7, pp. 430–432, 2005.
21. J. G. Proakis, *Digital Communications*, 4th ed., New York: McGraw-Hill, pp. 185–213, 2001.

9 Frequency Discrimination Reception for Optical Minimum Shift Keying

9.1 INTRODUCTION

Improving the fiber dispersion limit has become an unavoidable challenge for future high-performance and high-capacity dense wavelength division multiplexed (DWDM) long-haul and metropolitan optical systems. The last few years have experienced a fast-growing demand to considerably extend the reach of uncompensated optical links, especially for metropolitan optical networks where the transmission distance is approximately in the range of 300–500 km of standard single-mode fiber (SSMF).

As presented in previous chapters, advanced modulation formats with narrow spectrum and high dispersion tolerance limits have recently attracted extensive research effort. Schemes for the generation and detection of optical minimum shift keying (MSK) signals have been presented in Chapter 8. Other optical MSK modulation transmitters have been reported by Refs. [1–4]. As presented in Chapter 2, or from the Refs. [1–4], the conventional way for self-heterodyne coherent detection of optical MSK signals is to utilize the Mach–Zehnder delay interferometer (MZDI)-balanced receiver, as described in Chapter 5. Here, the differential phase information between two adjacent transmitted symbols is detected. However, the performances of these receivers are severely degraded by the intersymbol interference (ISI) caused by the parabolic phase characteristic of the optical fiber [5].

The dispersion limit for externally modulated signals is approximately inversely proportional to the square of the bit rate. As addressed in Ref. [6], a residual dispersion limit of about ±1000 and ±60 ps/nm corresponding to about 60 and 4 km SSMF are reported, respectively, for 10 and 40 Gbps optical signals, that is, OC-192 and OC-768 signals, respectively. Also, recently reported in the case of 40 Gbps optical MSK transmission with MZDI-based detection, a residual dispersion tolerance of approximately ±100 ps/nm, equivalent to 6 km SSMF, is achieved at 2 dB power penalty [7,8]. Thus, using the MZDI-based receiver, the optical MSK modulation does not offer considerable improvement on the dispersion robustness compared to the other modulation formats.

This chapter provides insights into a novel but simple noncoherent detection scheme for optical-MSK-transmitted signals, a dual-frequency discrimination scheme. The optical narrowband frequency discrimination receiver (ONFDR) employs dual narrowband optical filters and an optical delay line (ODL). The operation of the detection is based on the principles of a matched filter with narrowband frequency discrimination, rather than relying on the phase of the optical carrier. Thus, the performance of the receiver is expected to be less sensitive to the fiber phase characteristics.

ONFDR offers the following advantages over the conventional detection:

1. ONFDR mitigates the effects of ISI much more effectively and, thus, significantly improves the transmission limits due to the fiber chromatic dispersion. Indeed, in Section 9.5.2, it is shown that a dispersion tolerance of up to ±340 ps/nm, equivalent to length 20 km of SSMF, can be achieved at 1 dB power penalty.
2. Narrowband optical filtering offers a great deal of suppression of accumulated ASE noise, thereby improving the OSNR and receiver sensitivity significantly.

3. Imperfection of MZD results in severe degradations in performance. The imperfection can be caused by thermal instability, the phase mismatch in one-bit delay. These problems are overcome in the ONFDR scheme.
4. Combined features of narrowband frequency discrimination of OFNDR and the spectral characteristics of the optical MSK modulation format with narrow main-lobe spectral width and high suppression of the side lobes will enable the possibility of using 50 GHz channel spacing in 40 Gbps optical MSK transmission systems.
5. Frequency discrimination receiver is robust to polarization-mode dispersion (PMD) with the use of narrowband optical filters.

It should be noted that the receiver presented in this chapter can be used with any optical MSK transmitters, as reported in Refs. [1–4,7]. The availability of narrowband filters [9] facilitates the practical implementation of the receive configuration. This chapter also shows that optimizing the center frequencies of the optical discrimination filters from its nominal center frequencies offers significant improvement in optical receiver sensitivity and eases the requirement of an optical filter with a narrow bandwidth. It is also worth noting that ONFDR can be used for the detection of both the discrete and continuous phase frequency shift keying (FSK/CPFSK) optical signals.

In this chapter, the following are presented:

1. Novel but simple receiver configuration with significant improvement of chromatic dispersion (CD) tolerance but robust to polarization-mode dispersion (PMD) and resilient to SPM nonlinearity
2. Overview of the receiver configuration and operational principles of ONFDR
3. Optimization of the receiver design
4. Numerical results showing the dispersion tolerance (CD and PMD), the resilience to nonlinearity (SPM), and the transmission limits of an optical MSK system with noncoherent detection based on ONFDR

This chapter is organized as follows: the operational principles of the proposed detector are presented in Section 9.2. In Section 9.3, modeling of the receiver is addressed systematically. Section 9.4 addresses the importance of optimizing the receiver design and shows the selection for the optimum bandwidth and center frequency of the optical discrimination filters as well as the optimum delay for the ODL. In the final section, numerical results are presented on the followings (1) dispersion tolerance (both CD and PMD), (2) the resilience of the optical MSK detection to nonlinearity (SPM), and (3) the transmission limits of an ultra-long-haul 40 Gbps optical transmission system with use of the ONFDR. Finally, a summary of the chapter is given.

9.2 ONFDR OPERATIONAL PRINCIPLES

In MSK modulation format, binary information is modulated in upper sideband (USB) frequency (f_1) and lower sideband (LSB) frequency (f_2), which are defined as $f_1 = f_c + f_d$ and $f_2 = f_c - f_d$, where f_c is the carrier frequency of the lightwave, and f_d is the frequency deviation from f_c. In the case of optical-MSK-modulated signals, f_d is equal to a quarter of the transmission bit rate (B_R), that is, $f_d = B_R/4$. The block diagram of the proposed narrowband frequency discrimination receiver is shown in Figure 9.1.

The operation of the receiver is based on the principles of frequency discrimination. Optical MSK lightwaves at the output of the optical fiber are split into two paths by a 3 dB optical power splitter after being amplified with a low-noise optical amplifier as shown in Figure 9.2. Two optical narrowband filters F1 and F2 are used to discriminate the USB and the LSB frequencies, respectively. The center frequencies and the bandwidths of the two filters F1 and F2 are selected so that F1 would capture most of the MSK signal only when a "+1" is transmitted and F2 would capture most of the MSK signal only when a "0" is transmitted.

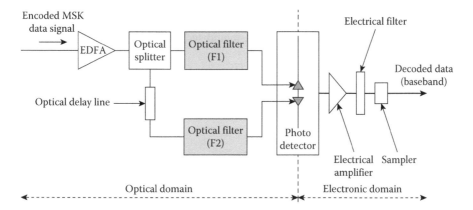

FIGURE 9.1 Narrowband optical filter detector for optical MSK.

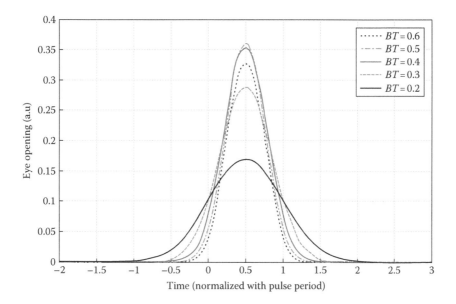

FIGURE 9.2 Eye opening for different *BT* products to demonstrate effects of power leakage when using dual-optical-frequency discrimination filters.

A constant ODL, which is easily implemented in integrated optics, is introduced in one path to compensate for the differential group delay $2\pi f_D \beta_2 L$ between f_1 and f_2, where $f_D = |f_1 - f_2| = R/2$; β_2 represents the group velocity delay (GVD) parameter of the fiber; and L is the fiber length. If the differential group delay is compensated completely, the optical MSK pulses at the output of the two optical discrimination filters arrive at the photodiodes simultaneously. An optical delay path ODL is selected in such a way that its GVD has the opposite sign with respect to that of the propagation medium. In the normal case of SMF optical fiber, that is, $\beta_2 < 0$, ODL is used in the path of the USB frequency.

It is noted that the receiver sensitivity of the scheme can be controlled by adjusting the gain of the optical amplifier. By using dual narrowband optical filters, ASE noise induced by the optical amplifier is mostly filtered and hence highly suppressed. As shown in Figure 9.2, the output of the filters is then converted into the electrical domain through a pair of photodiodes arranged in a balanced receiver configuration.

9.3   RECEIVER MODELING

The baseband-equivalent description of an MSK-modulated signal $x(t)$ is given by Ref. [10]

$$x(t) = \sqrt{\frac{E_b}{T}} e^{j\Phi(t;I)} \tag{9.1}$$

where, $\Phi(t;I)$ is given by

$$\Phi(t;I) = \left[\pi/2 \sum_{k=-\infty}^{n-1} I_k \right] + 2\pi f_d(t + T/2 - nT)I_n \tag{9.2}$$

$$= \theta_{n-1} + 2\pi f_d(t + T/2 - nT)I_n, \quad nT - T/2 \le t \le nT + T/2$$

In Equation 9.2, $I_n \in \{+1,-1\}$ is the data symbol transmitted at symbol interval n; T is the symbol period; and $f_d = 1/4T$ is the frequency deviation. Thus, when $I_n = +1$, the carrier's frequency in the nth symbol period is shifted by $+f_d$ and, when $I_n = -1$, the carrier's frequency is shifted by $-f_d$. The switching of frequencies at the rate $1/T$, however, results in the spectrum of MSK spilling over a range of frequencies centered around f_d and $-f_d$. For the purpose of analysis, we express the baseband-equivalent MSK signal as a sum of MSK pulses in each symbol period, that is

$$x(t) = \sum_n x_n(t) \tag{9.3}$$

where $x_n(t)$ is defined as

$$x_n(t) = \sqrt{\frac{E_b}{T}} e^{j(2\pi f_d(t+T/2-nT)I_n + \theta_{n-1})} s(t - nT) \tag{9.4}$$

In Equation 9.4, $s(t)$ is a square pulse of duration T, that is, $s(t) = u(t - T/2) - u(t + T/2)$, where $u(t)$ is the unit step function. Thus, $x_n(t)$ describes the MSK signal in the time interval $nT - T/2 \le t \le nT + T/2$ and has zero amplitude elsewhere.

The low-pass equivalent frequency response of the optical fiber, $H(f)$, has a parabolic phase profile and can be modeled [11] by the following equation as

$$H_c(f) = e^{-j\alpha_D f^2} \tag{9.5}$$

where $\alpha_D = \pi^2 \beta_2 L$, β_2 represents the GVD parameter of the fiber and L is the length of the fiber. The parabolic phase profile is the result of the chromatic dispersion of the optical fiber [2]. As in Ref. [2], we have omitted the constant phase term and the phase terms linear in f as they do not introduce distortion.* Thus, a pure sinusoidal signal of frequency f, propagating through the optical fiber, experiences a delay of $2\pi f \beta_2 L$. The standard fibers used in optical communications have a negative β_2 and thus, in low-pass-equivalent representation, sinusoids with positive frequencies (i.e., frequencies higher than the carrier) have negative delays, that is, arrive early compared to the carrier, and the ones with negative frequencies (i.e., frequencies lower than the carrier) have positive delays and arrive delayed. The dispersion-compensating fibers have positive β_2 and so have reverse effects. The low-pass-equivalent channel impulse response of the optical fiber, $h_c(t)$, has also got a parabolic phase profile, and is given as

$$h_c(t) = \sqrt{\frac{\pi}{j\alpha_D}} e^{-j\alpha^2 t^2 / \alpha_D} \tag{9.6}$$

---

* We also do not include any fiber nonlinearities in the model since, as discussed earlier, and it can be easily avoided in MSK transmission.

When an optical MSK signal is passed through the optical channel, the output $y(t)$ is given by the following convolution[*]

$$y(t) = x(t) * h_c(t) = \sum_n x_n(t) * h_c(t) \tag{9.7}$$

The output of each of these filters is then input into the optical MSK narrowband frequency discrimination receiver. The outputs of the delay lines $d_{F1}(t)$ and $d_{F2}(t)$ can be modeled as a convolution of the channel output $y(t)$ with each of the filter impulse responses, $h_{F1}(t)$ and $h_{F2}(t)$, respectively, and correctly delayed delta pulses, $\delta(t \pm t_d/2)$, that is

and

$$d_{F1}(t) = \frac{1}{\sqrt{2}} y(t) * h_{F1}(t) * \delta(t + t_d/2)$$

$$\tag{9.8}$$

$$d_{F2}(t) = \frac{1}{\sqrt{2}} y(t) * h_{F2}(t) * \delta(t + t_d/2)$$

To simplify the preceding convolutions, two template functions $\psi_{F1,I}(t)$, $I \in -1, +1$ for filter F1, and two functions $\psi_{F2,I}(t)$, $I \in -1, +1$ for filter F2, are defined as follows

$$\psi_{F1,I}(t) = \sqrt{\frac{E}{T}} e^{j2\pi f_d(t+T/2)I} s(t) * h_c(t) * h_{F1}(t) * \delta(t + t_d/2) \quad I \in -1, +1 \tag{9.9}$$

$$\psi_{F2,I}(t) = \sqrt{\frac{E}{T}} e^{j2\pi f_d(t+T/2)I} s(t) * h_c(t) * h_{F2}(t) * \delta(t + t_d/2) \quad I \in -1, +1 \tag{9.10}$$

These template functions represent the correctly delayed outputs of filters F1 and F2 for each of the +1 and −1 MSK-modulated pulse inputs. An example of the template function is shown in Figure 9.3 and discussed in the following section.

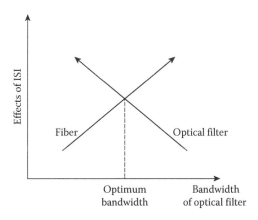

FIGURE 9.3 Trends of ISI effects caused by fiber dispersion and narrowband optical filters.

---

[*] The optical carrier with a line spectrum is assumed. This is a valid assumption, considering the fact that the source is always on and external modulators are used.

Using these template functions, and Equations 9.7 and 9.8, the expression for the output of the delay lines can be expressed concisely as follows

$$d_F(t) = \sum_n \psi_{F,I_n}(t + T/2 - nT)e^{j\theta_{n-1}}, \quad F \in F1, +F2 \tag{9.11}$$

The outputs of the delay lines are then converted into the electrical domain using photodiodes in a balanced receiver configuration [12–14]. The electrical signal is then amplified, sampled, and sent to the decision device. The decision device selects +1 as the transmitted symbol if the input is positive, or −1 otherwise. In order to express the sampled values going to the decision device, let us express the delay line outputs at the sampling instant $t = kT$ by

$$d_F(kT) = \sum_n \psi_{F,I_n(kT-nT)}e^{j\theta_{n-1}} = \left[\underbrace{\psi_{F,I_k}(0)}_{\text{desired}} + \underbrace{\sum_{n \neq k} \psi_{F,I_n}(kT - nT)e^{j\phi_{n,k}}}_{\text{ISI}} \right] e^{j\theta_{k-1}} \tag{9.12}$$

where the phase offset ϕ_n, k is given by

$$\phi_n, k = \pi/2 \left[\sum_{n'=k+1}^{n-1} I_{n'} \right] \text{ if } n > k = -\pi/2 \left[\sum_{n'=n}^{k} I_{n'} \right] \text{ if } n < k \tag{9.13}$$

The last term $e^{j\theta_{k-1}}$, in Equation 9.12, need not be considered further as it is lost when the signal is passed through the photodiode. The output of the balanced receiver at the sampling instant $t = kT$ is then given by $|d_{F1}(kT)|^2 - |d_{F2}(kT)|^2$.

The photodiode introduces quadratic point nonlinearity into the system and complicates the analysis of the receiver design. Therefore, here we present a semi-analytical method to evaluate the receiver performance. This method shows the mutual interactions among the receiver's key parameters and, thus, provides the design guidelines for the selection of their optimum values.

When the system performance is limited by ISI, the eye opening gives a good indication of the system performance. The proposed method relies on obtaining the eye opening for different values of the filter parameter. The parameters that give maximum eye opening will provide the best performance.* To obtain the eye opening, we first consider the transmission of +1 at symbol interval 0, and find out the minimum sample value, $d_{\min}^{+1}$, over all the input train sequences that have +1 as its 0th symbol; that is Ref. [15]

$$d_{\min}^{+1} = \max \left\{ ...,,... |d_{F1}(0)|^2 - |d_{F2}(0)|^2, 0 \right\} \tag{9.14}$$

Similarly, we obtain the minimum negative sample value when −1 is transmitted, that is

$$d_{\min}^{-1} = \max \left\{ ...,,... - |d_{F1}(0)|^2 - |d_{F2}(0)|^2, 0 \right\} \tag{9.15}$$

The eye opening is then given by $d_{\min}^{+1} + d_{\min}^{-1}$. Carrying out the inner minimization over all the possible sequences is prohibitive, and so we limit the length of the sequence to $2N + 1$, that is, N symbols on either side of the 0th symbol. We select the value for N such that the maximum time span over which the template functions have a significant absolute value is less than $(2N + 1)T$. Thus, we limit the minimization operation in Equations 9.14 and 9.15 to be carried over just 2^{2N} sequences

---

* The eye spectrum is necessary to find the optimum parameters. However, considering the fact that most number of errors occur when the eye is at its minimum, considering only the eye opening is accurate enough for most purposes.

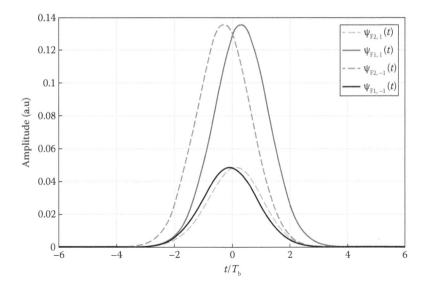

FIGURE 9.4 Eye openings given by the template functions ψ_F of the discrimination filters F1 and F2 for each symbol "+1" and "−1."

without sacrificing the accuracy of the calculated eye opening value. The value of N is usually quite small, for example, 3–6 when the distortion is small, that is, for cases where the eye is open. Usually when the value of N is very high, that is, under severe distortion, the eye will be closed, and thus we do not need to carry out the above minimization over all the sequences. When a sequence with a negative sample value (i.e., representing complete closure of the eye) is obtained, the minimization operation is stopped. Thus, when N is quite large, scanning through the whole 2^{2N} sequences is unnecessary. To obtain the optimum parameters, the eye opening is obtained for different sets of parameters, and the set with the maximum eye opening is selected as the optimum set.

Figure 9.4 shows the sample template functions when the length of the optical fiber ($\beta_2 = -2.17e - 26$) s²/m is 25 km, BT is 0.16, and f_ω is 12.5 GHz. It can be seen that the amplitude spectrum of all the template functions fall within four symbol periods on either side of the center sample, and thus taking $N = 6$ to calculate the eye opening would produce accurate results.

9.4 RECEIVER DESIGN

The proposed optical MSK receiver is based on the principles of narrowband frequency discrimination and relies on the alignment of the signal arrivals at two photodiodes for a receiver-balanced configuration. Therefore, it is significant to optimize the parameters that have effects on the receiver performance. The key parameters include the bandwidth and the center frequency of the optical discrimination filters and the ODL. In this section, an intuitive justification for the effects of key parameters on the receiver performance based on the four template functions $\psi_{F,I}(t)$ is provided. Discussions on the receiver design also provide design guidelines for the selection of optimum filter parameters.

9.4.1 OPTICAL FILTER PASSBAND

The receiver performance critically depends on the bandwidth B of the two optical discrimination filters. They have to be sufficiently narrow compared to the transmission bit rate ($R = 1/T$), in order to effectively discriminate the USB and LSB frequencies, f_d and $-f_d$, respectively (written in low-pass equivalent format). The switching of frequencies at the rate of $1/T$ results in the spectrum of

MSK spilling over a range of frequencies centered around f_d and $-f_d$, corresponding to whether "+1" or "–1" is transmitted. This spectral spilling causes the leakage problem and, in MSK modulation scheme, where the two transmission modulation frequencies are close to each other, the leakage terms are substantial. For example, consider the transmission of a "+1" pulse. The bulk of the energy of a "+1" modulated MSK pulse would be captured by a filter F1, and a small amount of the leakage energy will be captured by a filter F2. For the system to discriminate the "+1" pulse effectively, the leakage term should be comparatively smaller. The effect would be reversed for a "0" MSK-modulated pulse.

The ratio of the signal energy to the leakage energy from F1 and F2 increases with the decrease in the BT product of the filters. However, if the bandwidths of the filters are too narrow, the energy captured by both F1 and F2 goes down, reducing the discrimination property. Thus, there should be an optimal BT product that gives the best performance for the detection. Under this circumstance, the ISI is not taken into account. The preceding justifications are demonstrated in Figure 9.2. It is found that the values of BT within the range of 0.4–0.5 obtain the maximum EOs and, thus, are the optimum values for the optical discrimination filters. Increasing the BT product, that is, increasing the filter bandwidth over 0.5, does not open the eye further.

When propagating through the fiber, different spectral components of the optical pulse experience different delays and thus arrive at the receiver at different times. This leads to pulse spreading in the time domain causing the ISI. When optical signals are passed through the discrimination filter, the filter attenuates frequencies away from its center frequency. The lower the bandwidth of the filter, the more the frequency components of the signal are suppressed, thus reducing the ISI effects caused by fiber CD. As a result, when the length of the fiber increases, that is, the dispersion caused by the fiber is getting more severe, the bandwidth of the optical discrimination filter should be reduced equivalently. However, when the bandwidth of the filter is lowered, its impulse response starts to spread more in the time domain, introducing the ISI caused by the filter's impulse response itself. Therefore, there should be an optimum bandwidth that is the intercept point of these two trends of ISI effects, as illustrated in Figure 9.2.

The performance of the receiver depends critically on the bandwidth B of the two filters. They have to be sufficiently narrow compared to the transmission bit rate ($R = 1/T_b$) in order to discriminate the two frequencies f_1 and f_2 effectively. The bulk of the energy of a "+1" modulated MSK pulse would be captured by the filter F1, and a small amount of leakage energy will be captured by the filter F2. The effect would be reversed for a "0" modulated pulse. The ratio of the energies captured by F1 and F2 increases with the decrease in time–bandwidth product, BT, of the filters. However, if the bandwidth of the filters is too narrow, the energy captured by both F1 and F2 goes down, resulting in a lower ratio. In addition, the narrower the bandwidth, the higher is the ISI. Thus, there is an optimal BT product that gives the best performance for the detection. The two discriminating filters F1 and F2 used in the analysis are modeled as Gaussian filters having the same bandwidth–time product of BT and with center frequencies at $f_1 = f_c + f_d$ and $f_2 = f_c - f_d$, respectively.

When the received signal is passed through a filter, the filter attenuates frequencies away from its center frequency. Thus, the dispersion due to the channel is reduced, and thereby the pulse spreading and ISI are reduced. The lower the bandwidth of the filter, the more the suppression of the frequency components of the signal, leading to lower ISI. However, when the bandwidth of the filter is lowered, its impulse response starts to spread more in the time domain, leading to more ISI. Hence, the optimum bandwidth of the filter is when the pulse spreading after filtering is equal to the spread of the filter's impulse response.

One consequence of this criterion is that, for optimum performance, with increasing fiber lengths, the bandwidth of the filters should be reduced. When the length of the fiber increases, so does the dispersion caused by the channel. Thus, to compensate for this effect, the bandwidth of the filter should be reduced equivalently for optimum performance. Although not shown here, this has been validated by the analysis.

The switching of frequencies at the rate $1/T$ results in the spectrum of MSK spilling over a range of frequencies centered around f_d and $-f_d$, corresponding to whether +1 or −1 is transmitted. When such a signal is passed through the optical channel, the different frequencies travel at different speeds and thus arrive at the receiver at different times. This leads to pulse spreading in the time domain.

9.4.2 Center Frequency of the Optical Filter

The MSK spectrum spilling over a range of frequencies centered around f_d and $-f_d$ causes another problem due to leakage. For example, consider the transmission of a +1 pulse. It has frequency components centered around f_1. The filter F1 with center frequency f_ω will capture most of the signal. However, the frequency components of the pulse falling in the bandwidth of F2 called the "leakage" term will be captured by F2. For the system to discriminate the +1 pulse, the leakage term should be comparatively smaller. In MSK modulation scheme, where the two transmission frequencies are very close to each other, the leakage terms are substantial.

One way to reduce the leakage terms is to select the filters with center frequencies away from the transmission frequencies used by the MSK signal, that is, $f_\omega > f_d$. Although offsetting the center frequency of the filters away from the transmission frequencies reduces the leakage term, it also reduces the signal captured by the primary filter, leading to reduced discrimination properties. Hence, the optimum value for the center frequency is obtained when the reduction in the leakage term offsets the reduction in the amount of signal passing through the primary filter.

One consequence of the preceding design criteria is that, for optimum performance, the center frequency of the filters should be offset more when the bandwidth of the filters increases. When the bandwidth of the filters used is large, the leakage term would be high, and offsetting the center frequency of the filters offers performance improvements.

The selection of the optimum value of ODL, the optimum bandwidth (B), and the offsets from the nominal center frequency of the optical discrimination filters (F1 and F2) is studied.

9.4.3 Optimum ODL

Figure 9.5 shows the analytical results for the eye opening of the receiver for different BT products of F1 and F2 and for different mismatch values of ODL for 25 km SSMF transmission

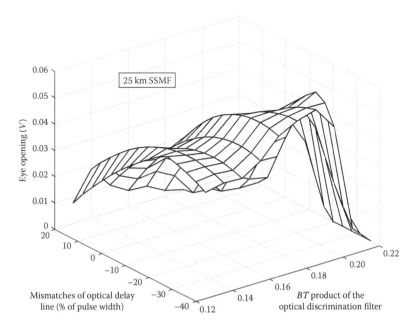

FIGURE 9.5 Analytical results for eye opening of the receiver for different BT products of F1 and F2 and for different mismatch values of ODL for 25 km SSMF.

or, effectively, 425 ps/nm of residual dispersion. Here, and in the rest of the text, length of the fiber refers to a residual distance that is not compensated for the chromatic dispersion. The Gaussian-type optical discrimination filters are used in the analysis. This analysis is based on obtaining the expressions for the eye opening, taking into account the effects of the ISI of both pre-cursor and post-cursor symbols. Zero mismatch represents the value of $2\pi f_D|\beta_2|L$ for the delay, and the mismatches are measured from this value as a percentage of the pulse width T of 25 ps. Larger values for the eye opening signifies better system performance, and a zero value indicates the complete closure of the eye. The mesh shown in Figure 9.5 presents a guideline for optimizing the performance of the proposed receiver at a particular length of fiber, and also indicates whether the system will have an error floor or not for a particular setting.

9.5 ONFDR OPTIMUM BANDWIDTH AND CENTER FREQUENCY

Narrowband filtering has a tradeoff between the discrimination property and the ISI mitigation. Figure 9.6 shows the analytical results for the eye opening against different BT products for various lengths of the fiber.

Here, and in the rest of the text, length of the fiber refers to residual distance that is not compensated for the chromatic dispersion. Larger values of the eye opening signify better system performance, and a zero denotes failure of the receiver. From Figure 9.6, it can be seen that, for long span of SSMF, that is, 25 and 30 km, the optimum BT is obtained at 0.14 (i.e., $B = 5.6$ GHz), while it is 0.28, 0.22, and 0.21 for the length of 5 km, 12 km, and 20 km SSMF, respectively.

Analytical results of offset filtering for different bandwidths (i.e., different BT products where $T = 25$ ps) of the Gaussian-type filters are shown in Figure 9.7 for 25 km SSMF, or effectively 425 ps/nm of residual dispersion. The analysis is based on obtaining the expressions for the eye opening, taking into account the effects of the ISI caused by both pre-cursor and post-cursor symbols. Larger values for the eye opening signifies better system performance, whereas a zero value indicates the failure of the proposed receiver. As shown in Figure 9.8, offset filtering can offer significant performance improvements with appropriate bandwidth for the optical

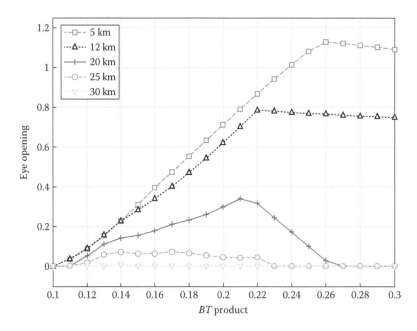

FIGURE 9.6 Eye opening against different BT products under transmission of various lengths of the fiber.

discrimination filters. For higher offset frequencies, the optimum bandwidth of the optical filters is increased. The optimum BT is obtained at about 0.22, and the optimum center frequency of the optical filters is shifted from 10 GHz toward 15 GHz, that is, an offset of 5 GHz.

A mesh for the analysis on transmission of 40 Gbps optical MSK systems over a length of 35 km SSMF is also studied and shown in Figure 9.8. Figure 9.8a shows the eye openings versus BT and center frequencies of optical discrimination filters, while Figure 9.8b shows that there are optimal values of BT and f_c to give the positive value in the eye opening.

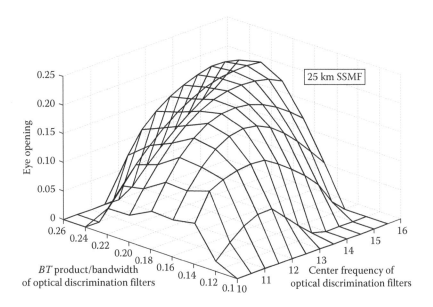

FIGURE 9.7   Mesh diagram for decision of optimum eye opening with respect to the BT product and center frequency of the discrimination filter.

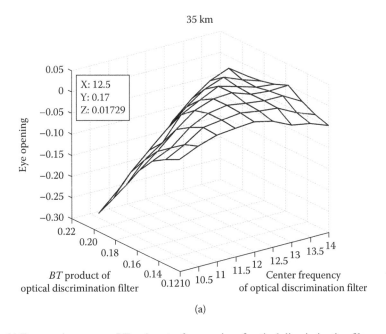

(a)

FIGURE 9.8   (a) Eye openings versus BT and center frequencies of optical discrimination filters.   *(Continued)*

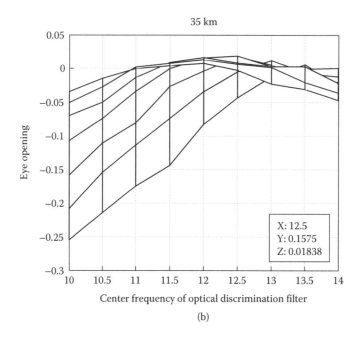

FIGURE 9.8 (Continued)   (b) Optimal values of BT and f_c.

The results show that the optical MSK receiver still has an eye opening and, thus, the narrowband frequency discrimination receiver may offer error-free demodulation of data for an optical link of up to 35 km SSMF non compensated optical link for 40 Gbps optical MSK transmission.

9.6   RECEIVER PERFORMANCE: NUMERICAL VALIDATION

The analysis shown in Figure 9.9 is validated with simulation results. Figure 9.9 shows the simulation results of the receiver sensitivity penalty against different center frequency offsets for various BT products of the optical discrimination filters at an SSMF length of 25 km. For relatively narrow bandwidths ($BT = 0.13 - 0.17$), the receiver achieves optimum receiver sensitivity with center frequencies in the range of 11–12.5 GHz. Within this optimum range, the change in the receiver performance is gradual, whereas the performance severely degrades when the offset frequencies fall outside this range. However, when considering relatively large bandwidths of the optical discriminating filters ($BT = 0.2$ and 0.25), the optimum center frequency shifts toward 14.5 GHz rather than 12.5 GHz. The simulation results in Figure 9.9 closely agree with the analytical results.

Figure 9.10 shows the numerical results for the receiver sensitivity penalty for different optical delay mismatches. Zero mismatch represents the value of $2\pi f_D \beta_2 L$ for the delay, and mismatches are measured from this value as a percentage of the pulse width T ($T = 25$ ps). As shown in Figure 9.10, when the distance increases, the optimum delays shift to smaller values. For 5 km SSMF, the optimum value is at zero, whereas for 25 km SSMF, it is at −12.5%. Also, the receiver is less sensitive to delay mismatches at short fiber lengths or small residual dispersions. In the case of 5 km SSMF, the tolerance range is about 45% (−20% to +25%) at 1 dB power penalty, and about 22% in the case of 25 km SSMF.

9.7   ONFDR ROBUSTNESS TO CHROMATIC DISPERSION

Figure 9.11 shows the simulation system configuration considered for the investigation of the performance and dispersion tolerance of noncoherent 40 Gbps optical MSK systems using the proposed narrowband filter receiver.

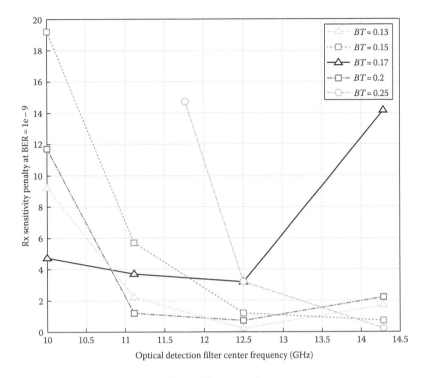

FIGURE 9.9   Receiver penalty versus the offset of the center frequency.

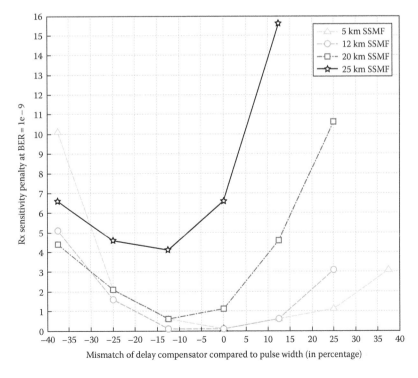

FIGURE 9.10   Receiver penalty versus the mismatch of the delay comparator.

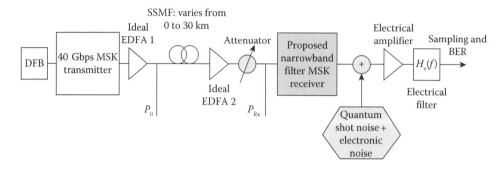

FIGURE 9.11 Simulation setup for investigation on dispersion tolerance of the 40 Gbps MSK system using narrowband filter receiver.

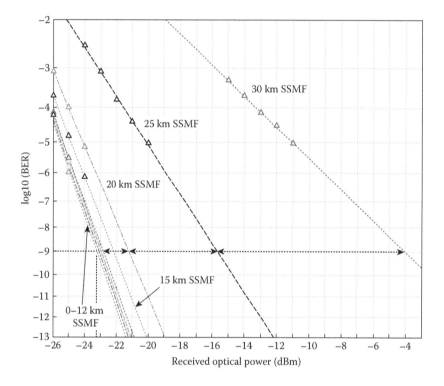

FIGURE 9.12 BER versus optical received powers for transmissions over 0–30 km SSMF without dispersion compensation.

The input power into fiber (P_0) was kept lower than the nonlinear threshold power. The ideal EDFA2 provides the gain to completely compensate for the attenuation of the signals propagating through SSMF. As shown in Figure 9.12, the optical received power (P_{Rx}) is measured at the input of the proposed MSK receiver. In order to obtain the BER curve at a particular length of fiber, P_{Rx} is varied by adjusting the attenuator. The key simulation parameters used in the experiment are given in Table 9.1. Various residual dispersions are obtained by varying the lengths of SSMF from 0 to 30 km. The receiver sensitivity of −23.2 dBm at BER = 1e−9 is achieved in back-to-back experiments in case of nominal filtering. The ODL value is based on the differential group delay between F1 and F2 calculated as $2\pi f_D \beta_2 L$, where $f_D = |f_1 - f_2| = R/2$, β_2 represents the GVD parameter of the fiber, and L is the fiber length. The numerical results are obtained via Monte Carlo simulation.

TABLE 9.1

Key Parameters Used in the Simulation

Input power: $P_0 = -3$ dBm	Narrowband Gaussian filter: $B = 5.2$ GHz or $BT = 0.13$
Operating wavelength: $\lambda = 1550$ nm	Constant delay: $t_d = |2\pi f_d \beta_2 L|$
Bit rate: $R = 40$ Gbps	EDFA: $G = 15$ dB and noise figure $= 5$ dB
SSMF fiber: $|\beta_2| = 2.68e-26$ or $|D| = 17$ ps/nm/km	$i_d = 10$ nA
Attenuation: $\alpha = 0.2$ dB/km	$N_{eq} = 20$ pA/(Hz)$^{1/2}$
Input power: $P_0 = -3$ dBm	

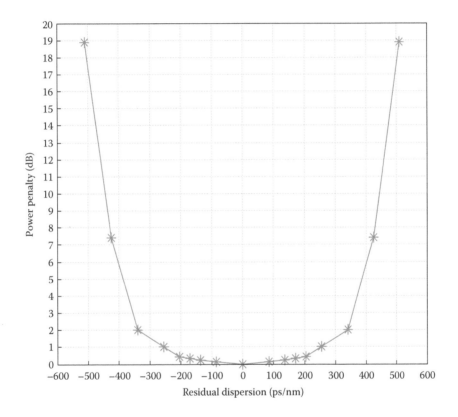

FIGURE 9.13   Dispersion tolerance with optimum values of key parameters.

9.7.1 DISPERSION TOLERANCE

Figure 9.12 shows the BER versus optical received power curves corresponding to different lengths of transmission fiber. The power penalty caused by residual dispersion is shown in Figure 9.13. The numerical results are obtained via Monte Carlo simulation (triangular markers as shown in Figure 9.14) with the low BER tail of the curve linearly extrapolated. There is a negligible effect on the BER performance when SSMF transmission length is from 0 to 12 km.

As shown in Figure 9.14, the 1 and 2 dB power penalties of the 40 Gbps optical MSK systems are obtained when the transmission length is 15 and 20 km, respectively, which corresponds to residual dispersions of ±225 and ±340 ps/nm, respectively. Another important point shown in Figure 9.12 is the ability of optical MSK transmission systems to reach 1e−9 at the received optical power of −4 dBm for 30 km SSMF transmission length.

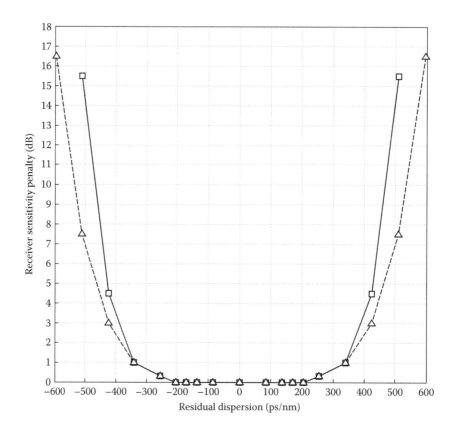

FIGURE 9.14 Dispersion tolerance of 40 Gbps optical MSK systems with ONFDR in two cases: nominal values (square markers and solid line) and optimum values (triangle markers and dashed line) of key parameters of ONFDR.

Comparison of dispersion tolerance of the proposed 40 Gbps optical MSK narrowband filter receiver in both cases of optimum offset filtering (dashed line and triangular markers) and nominal filtering with center frequencies at ±10 GHz (solid line and square markers) are shown in Figure 9.14. The filter bandwidth in this case is optimum at $BT = 0.14$.

It can be seen that there is no advantage in using offset filtering at low residual dispersion less than ±340 ps/nm, or equivalent to 20 km SSMF, in 40 Gbps transmission. However, offset filtering offers significant improvement in receiver sensitivity at high residual dispersion.

Approximately 2–9 dB gain of receiver sensitivity is achieved at residual dispersions of ±425 and ±510 ps/nm, or equivalently to 25 and 30 km of SSMF, respectively. Also, the system can operate up to ±595 ps/nm, or effectively 35 km SSMF, whereas nominal filtering fails when residual dispersion exceeds ±510 ps/nm. It is noted that, in terms of 10 Gbps transmission systems, the preceding figures may effectively correspond to ±10,115 ps/nm or 560 km SSMF, respectively.

9.7.2 10 Gbps Transmission

Figure 9.15 shows the simulation test bed for the 10 Gbps optical MSK transmission system over a 560 km SSMF link. Figure 9.16 shows the BER versus optical received power corresponding to different lengths of transmission fiber. The numerical results are obtained via Monte Carlo simulation. As shown in Figure 9.16, the 1 and 2 dB power penalties of the 10 Gbps optical MSK systems are obtained when the transmission length is approximately 320 and 390 km SSMF, respectively, which corresponds to residual dispersions of ±5440 and ±6630 ps/nm, respectively. It is noted that, in the

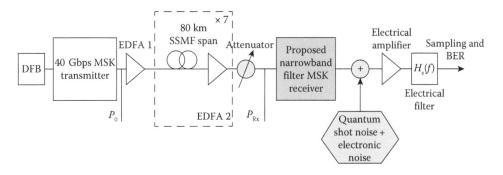

FIGURE 9.15 Simulation test bed for transmission of 10 Gbps MSK system using the proposed narrowband filter receiver over distances of up to 560 km SSMF optical link.

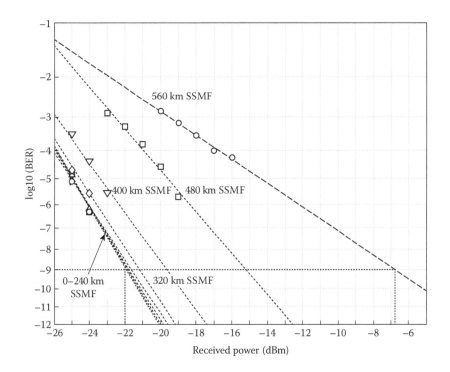

FIGURE 9.16 BER versus received optical powers for 10 Gbps optical MSK transmission over uncompensated optical links from back-to-back 560 km SSMF with detection based on optical narrowband frequency discrimination receiver.

case of transmission over 560 km SSMF uncompensated optical link, the performance of 10 Gbps optical MSK narrowband frequency discrimination receiver is able to reach the BER of 1e−9 at $P_{Rx} = -7$ dBm, even though it suffers about 15 dB penalty in receiver sensitivity. Equivalently, a residual dispersion tolerance of up to ±9520 ps/nm is achieved.

9.7.3 ROBUSTNESS TO PMD OF ONFDR

Figure 9.17 shows the tolerance to PMD of 40 Gbps optical MSK signals with use of the narrowband optical filter receiver. At a normalized ratio of $<\Delta\tau>/T_o = 0.35$ or $<\Delta\tau> = 9.75$ ps, the receiver performance degrades by 1 dB power penalty.

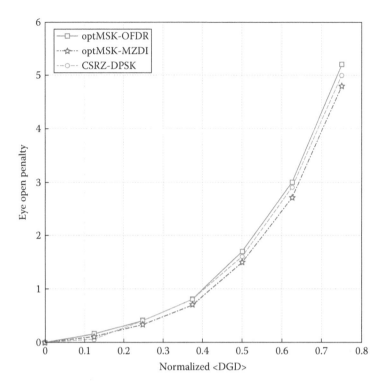

FIGURE 9.17   PMD tolerance of 40 Gbps optical MSK signals with optical narrowband frequency receiver and comparison to the MZDI-based receiver and CSRZ-DPSK.

9.7.4 Resilience to Nonlinearity (SPM) of ONFDR

The BER versus average optical input power P_o are shown in Figure 9.18. The results are obtained via Monte Carlo simulation. The optical MSK is more sensitive to SPM than CSRZ-DPSK and RZ-DPSK with the power penalties to be approximately 3 dB and 4 dB, respectively, at 1e−4. However, optical MSK signal is more robust to SPM than the CSRZ-ASK counterpart.

9.7.5 Transmission Limits of OFDR-Based Optical MSK Systems

One common question, and also a requirement for a new modulation format or new detection scheme in optical communications, is on the transmission limit of a non regenerated optical link compared to the well-known ones. Thus, in this section, the transmission limit of optical MSK transmission systems with the OFDR for detection is numerically investigated. The simulation results also compare the OFDR detection performance to the MZDI counterpart and also to the CSRZ-DPSK modulation format. The setting of the input power level and PMD are $P^{in} = -3$ dBm, $\langle \Delta\tau \rangle = 0.5$ ps/$\sqrt{\text{km}}$ for contemporary and new SMF fiber (Figures 9.19 and 9.20).

9.8   DUAL-LEVEL OPTICAL MSK

The migration of externally modulated MSK format into optics has recently been reported in Refs. [1–6], and is also discussed in Chapters 9 and 10. If a multilevel scheme can be incorporated into the MSK format, the symbol rate would be reduced, which is of particular interest in optically amplified long-haul communications systems. This is the principal feature of the dual-level MSK modulation scheme, whose generation and detection configurations are proposed for the first time,

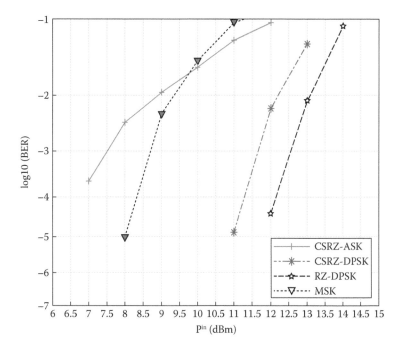

FIGURE 9.18 Comparison of robustness to SPM nonlinearity of 40 Gbps optical MSK signals with narrow-band optical filter receiver to CSRZ/RZ-DPSK and ASK.

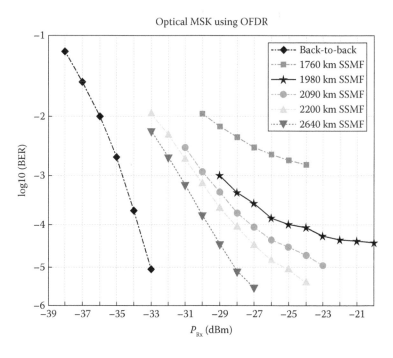

FIGURE 9.19 Transmission performance over multispan optical link of length 1760 and 2640 km SSMF under MSK-OFDR and MZDI receivers.

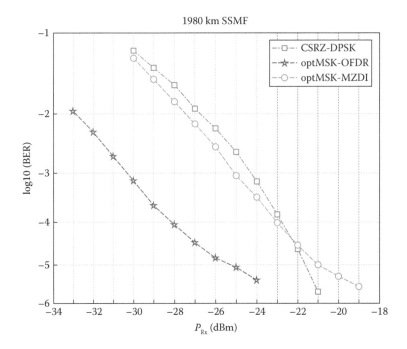

FIGURE 9.20 Comparison of BER against receiver sensitivity for transmission over 1980 km SSMF under MSK-OFDR detection and MZDI receiver and compared with CSRZ-DPSK.

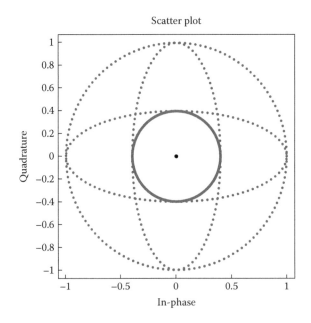

FIGURE 9.21 Signal trajectory of optical dual-level MSK.

to the best of the author's knowledge. Dual-level MSK can be seen as a superposition of two optical MSK signals of different amplitudes [7]. This modulation scheme enables the transmission bandwidth efficiency due to the utilization of two bits per modulated symbol, at the same time exploiting the advantage of narrow spectral characteristics of the MSK format. The constellation of the dual-level MSK-modulated signals is shown in Figure 9.21.

TABLE 9.2

Mapping of Binary Data for Dual-Level MSK Format

a	b	Remarks on Signal Constellation
1	1	Amplitude unchanged, phase increased
1	−1	Amplitude unchanged, phase decreased
−1	1	Amplitude changed, phase increased
−1	−1	Amplitude changed, phase decreased

The mapping scheme of the data information for the dual-level MSK format is shown in Table 9.2.

9.8.1 GENERATION SCHEME

Any configuration of the reported optical MSK transmitters can be utilized for the proposed generation of the optical dual-level MSK signals. Figure 9.22 shows the block structure of the optical dual-level MSK transmitter in which two optical MSK transmitters are integrated in a parallel configuration.

The ratio of the lightwave intensity levels input into the optical MSK transmitters are critical since this is used to manipulate the amplitude level of the transmitter signal. This ratio can be obtained with the utilization of a high-precision power splitter or simply by using a 3 dB coupler followed by the optical attenuator on each arm for the intensity manipulation. At an nth instance, the logic sequence of a_n and b_n of {±1} are precoded from the binary logics d_n of {0,1}, such that $a_n = 2d_n - 1$ and $b_n = a_n (1 - d_n - 1/h)$, where $h = 1/2$ in the case of MSK.

9.8.2 INCOHERENT DETECTION TECHNIQUE

As a marriage between optical MSK and multilevel modulation formats, the demodulation of dual-level MSK optical signals requires both amplitude and phase detections. Similar to the incoherent detection technique for optical MSK, an MZDI-balanced receiver is implemented for phase detection. It is worth mentioning again that an additional $\pi/2$ phase shift is introduced in one arm of the MZDI. The amplitude component of the lightwaves is detected by a single photodetector. A differential decoder is used, so that the amplitude changes between two consecutive bits can be demodulated. Figure 9.23 shows the detected eye diagrams of the dual-level MSK optical signals.

9.8.3 OPTICAL POWER SPECTRUM

Figure 9.24 compares the optical power spectra of three modulation formats: 80 Gbps dual-level MSK, 40 Gbps MSK, and 40 Gbps NRZ-DPSK. The intensity-splitting ratio for the optical dual-level MSK format is set at "0.8/0.2."

The power spectrum of the optical dual-level MSK format has identical characteristics to that of the MSK format. The main lobe of the spectral width is narrower than that of the NRZ-DPSK. The base-width takes a value of approximately ±32 GHz on either side compared to ±40 GHz in the case of the DPSK format.

9.8.4 RECEIVER SENSITIVITY

Three values of the intensity-splitting ratio are studied and compared: "0.7/0.3," "0.8/0.2," and "0.9/0.1." The amplitude and phase receiver sensitivities are shown in Figure 9.25. The average received optical powers are measured after the 3 dB optical power splitter. It is found that,

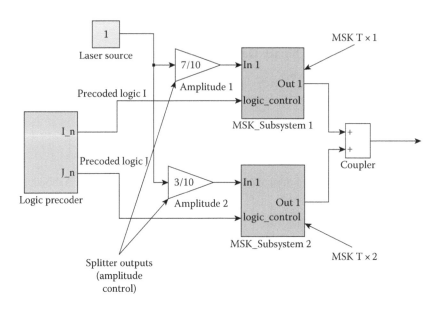

FIGURE 9.22 Block diagram of optical dual-level MSK transmitter.

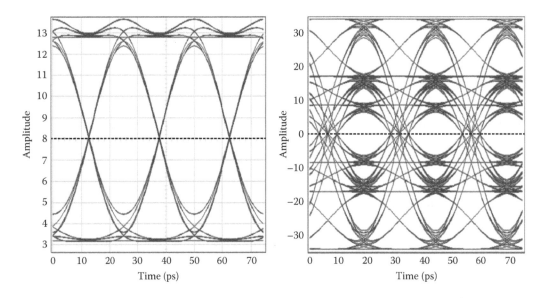

FIGURE 9.23 Eye diagrams of amplitude and phase detection of optical dual-level MSK signals.

at BER = 1e−9, the receiver sensitivities for amplitude and phase detection of "0.8/0.2" is −21.5 and −17.5 dBm, whereas they are −20.2 and −21 dBm for "0.9/0.1." The performance of 80 Gbps optical dual-level MSK systems is then evaluated over 540 km Vascade fiber comprising of six optically amplified spans (90 km per span). The Vascade fiber has a complete compensation for both chromatic dispersion and dispersion slope. The simulation results are shown in Figure 9.26. The BERs are obtained with the Monte Carlo method and linearly interpolated. "0.8/0.2" has a slower roll-off linear trend compared to the "0.9/0.1." At BER = 1e−9, the required OSNRs for ratios of "0.8/0.2" and "0.9/0.1" are about 24 dB. This is much lower than the required OSNR of approximately 32 dB for the case of "0.7/0.3" ratio, thus offering a 6–8 dB gain. The NRZ, CSRZ, and RZ

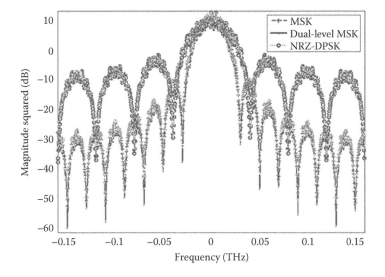

FIGURE 9.24 Spectral properties of 40 Gbps MSK (dash), 80 Gbps dual-level MSK (solid), and 40 Gbps NRZ-DPSK (dot).

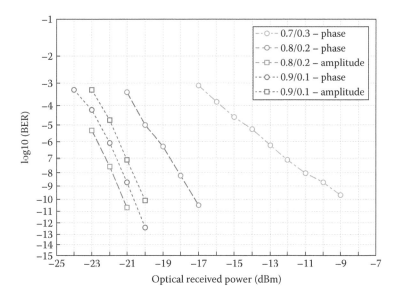

FIGURE 9.25 Receiver sensitivities of both amplitude and phase detections for dual-level MSK format with splitting power ratios of "0.8/0.2," "0.9/0.1," and "0.7/0.3."

modulation formats of DQPSK modulation are also investigated, as shown in Figure 9.26. The slope of the BER curves for DQPSK formats is steeper than that of the dual-level DPSK. However, the dual-MSK format still performs better than that of the CSRZ-DQPSK [16].

9.8.5 REMARKS

Spectral properties of the 80 Gbps optical dual-MSK are similar to those of the 40 Gbps optical MSK and better than the 40 Gbps optical DPSK counterpart. This clearly indicates the highly spectral efficiency for transmission of the proposed format. The obtained BER performances show

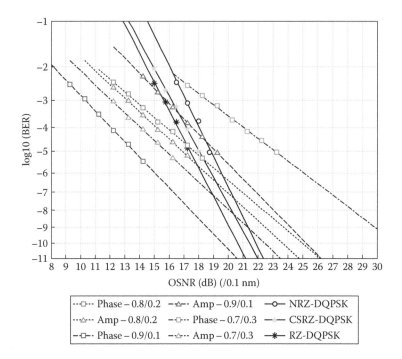

FIGURE 9.26 Performance for both amplitude and phase detections of 80 Gbps optical dual-level MSK format with splitting ratios of "0.8/0.2," "0.9/0.1," and "0.7/0.3" and DQPSK modulation of NRZ, CSRZ, and RZ formats.

the feasible receiver sensitivities and prove the potential for long-haul transmission of the optical 80 Gbps dual-level MSK format. Finally, the generation scheme can be extended to generate 4-ary MSK signals which employs 3-bit-per-symbol scheme.

9.9 CONCLUDING REMARKS

This chapter presents a simple receiver configuration for noncoherent detection of 40 Gbps optical MSK signals with the utilization of the ONFDR. The operational principles of the receiver are explained in detail. Optimum values for key parameters in the receiver design, including the bandwidth and the center frequency of the optical discrimination filters and the ODL, are obtained. The proposed 40 Gbps optical MSK receiver offers significant residual dispersion tolerance of up to ±340 ps/nm for 1 dB penalty, and it is a promising candidate for future high-capacity optical transmission systems.

The noncoherent detection of 10 Gbps optical MSK signals over uncompensated 560 km SSMF optical link is shown to be possible. This transmission distance covers the typical ring or mesh configuration of metropolitan optical networks.

The optical MSK narrowband frequency discrimination receiver significantly improves the residual chromatic dispersion tolerance up to ±10,115 ps/nm. The receiver is also shown to be robust to PMD. Numerical studies on the resilience of 40 Gbps optical MSK transmission to SPM nonlinearity have been conducted over 880 km SSMF dispersion-managed links and compared to the DPSK and ASK counterparts. However, there is a trade-off between significant improvement in dispersion tolerance and the nonlinearity resilience. With careful power distribution consideration, the optical MSK modulation format has been proven to be promising for long-haul and metropolitan transmission.

A dual-level optical MSK transmission system is also introduced as a multilevel MSK scheme. Spectral properties of the 80 Gbps optical dual-MSK are similar to those of the 40 Gbps optical MSK and narrower than the 40 Gbps optical DPSK counterparts. This clearly indicates the highly spectral efficiency for transmission of the multilevel format. The obtained BER performances show the feasible receiver sensitivities and prove the potential for long-haul transmission of the optical 80 Gbps dual-level MSK format. The generation scheme can be extended to generate 4-ary MSK signals in which 3-bits-per-symbol can be used.

REFERENCES

1. T. Sakamoto, T. Kawanashi, and M. Izutsu, Optical minimum-shift keying with external modulation scheme, *Opt. Express*, Vol. 13, No. 20, pp. 7741–7747, 2005.
2. M. Ohm and J. Speidel, Optical minimum-shift keying with direct detection (MSK/DD), in *Proceedings of the SPIE*, Vol. 5281, pp. 150–161, 2004.
3. T. L. Huynh, L. N. Binh, K. K. Pang, and L. Chan, Photonic MSK transmitter models using linear and non-linear phase shaping for non-coherent long-haul optical transmission, in *SPIE Proceedings of the APOC'06,* paper 6353-85, 2006.
4. J. Mo, Y. J. Wen, Y. Dong, Y. Wang, and C. Lu, Optical minimum-shift keying format and its dispersion tolerance, in *OFC'05*, paper JThB12, 2005.
5. A. F. Elrefaie, R. E. Wagner, D. A. Atlas, and A. D. Daut, Chromatic dispersion limitation in coherent lightwave systems, *IEEE J. Lightwave Technol.*, Vol. 6, No. 5, pp. 704–710, 1988.
6. I. P. Kaminow and T. Li, *Optical Fiber Communications*, Vol. IVB, chap. 5, San Diego, CA: Elsevier Science, 2002.
7. L. N. Binh and T. L. Huynh, Linear and nonlinear distortion effects in direct detection 40 Gbps MSK modulation formats multi-span optically amplified transmission, *Opt. Commun.*, Vol. 237, No. 2, pp. 352–361, 2007.
8. T. L. Huynh, L. N. Binh, and K. K. Pang, Optical MSK long-haul transmission systems, in *SPIE Proceedings of the APOC'06,* paper 6353-86, Thu9a, 2006.
9. B. E. Little, Advances in microring resonator, in *Integrated Photonics Research Conference 2003*, invited paper, 2003.
10. J. G. Proakis, *Digital Communications*, 4th ed., New York: McGraw-Hill, pp. 185–213, 2001.
11. A. F. Elrefaie and R. E. Wagner, Chromatic dispersion limitations for FSK and DPSK systems with direct detection receivers, *IEEE Photon. Technol. Lett.*, Vol. 3, No. 1, pp. 71–73, 1991.
12. N. Alic, G. C. Papen, R. E. Saperstein, L. B. Milstein, and Y. Fainman, Signal statistics and maximum likelihood sequence estimation in intensity modulated fiber optic links containing a single optical preamplifier, *Opt. Express*, Vol. 13, No. 12, pp. 4568–4579, 2005.
13. H. Haunstein, PMD and chromatic dispersion control for 10 and 40 Gbps systems, in *OFC'04,* invited paper, ThU3, 2004.
14. O. E. Agazzi, M. R. Hueda, H. S. Carrer, and D. E. Crivelli, Maximum-likelihood sequence estimation in dispersive optical channels, *J. Lightwave Technol.*, Vol. 23, No. 2, pp. 749–762, 2005.
15. J. G. Proakis and M. Salehi, *Communication Systems Engineering*, 2nd ed., Upper Saddle River, NJ: Prentice Hall, 2002.
16. T. L. Huynh, T. Sivahumaran, K. K. Pang, and L. N. Binh, A narrowband filter receiver achieving 225 ps/nm residual dispersion tolerance for 40 Gbps optical MSK transmission systems, in *Proceedings of the OFC'07, Optical Fiber Conference,* Los Angeles, CA, 2007.

10 Partial Responses and Single-Sideband Optical Modulation

Optical fiber communication systems have continuously evolved over the years. The increasing demand for a higher transmission capacity has driven the development of communication systems at ultra-high capacities and ultra-high bit rates. The fact that a 40 Gbps optical fiber communication system can have an extended reach and improved capacity makes it an attractive alternative to the 10 Gbps optical fiber communication system. System performance is further enhanced by employing various advanced modulation formats, such as duobinary (DB), return-to-zero differential phase shift keying (RZ-DPSK), and non-return-to-zero differential phase shift keying (NRZ-DPSK) [1–10]. Research and investigations have been carried out to determine the most appropriate and efficient formats that meet the current as well as future demand.

DB format, in the photonic domain, offers three-level coding, but, unlike electronic or wireless communication systems, the "−1" and "+1" are coded using the phase of the lightwave carrier, that is, either "0" or "π." This coding would overcome the dispersion due to its single-sideband property, and the detection scheme is much simpler as direct detection technique can be employed. The single sideband can also be implemented using vestigial sideband modulation technique, that is, an optical filter (OF) can be inserted after the optical modulator to filter half of the band of the spectrum. Alternatively, the modulators can be conditioned with two Hilbert transform signals and hence the suppression of half of the band [11–13].

The first part of this chapter presents a comprehensive modeling platform for duobinary modulation (DBM) format optical fiber transmission system. The modeling of the system is developed on the MATLAB® and Simulink® 7.0 or higher. Simulink has been chosen due to the availability of blocks, such as the communication blocksets and signal processing blocksets, which eases the process of implementation.

Further, it is demonstrated that the transmission of 40 Gbps alternating phase 0 and π DBM format with 33% and 50% pulse width is demonstrated with a CD tolerance of 850 ps/nm and at least 2 dB receiver sensitivity improvement as compared with that employing carrier-suppressed DPSK.

The second part of this chapter describes the transmission of optical multiplexed channels of 40 Gbps using vestigial single-sideband modulation format over a long-reach optical fiber transmission system. Thus, it is essential that an OF be designed to follow the optical modulator to filter half of the signal band. The effects on the Q factor of fibers dispersion, the passband and roll-off frequency of the OFs, and the channel spacing are described. The performance of the optical transmission using low and nonzero-dispersion fibers and/or dispersion compensation is then given. It has been demonstrated that a BER of 10^{-12} (without FEC) or better can be achieved across all channels, and that minimum degradation of the channels can be obtained under this modulation format. OFs are designed with asymmetric roll-off bands. Simulations of the transmission system are also given and compared for channel spacings of 20, 30, and 40 GHz. It is shown that the passband of 28 GHz and 20 dB cutoff band perform best for 40 GHz channel spacing.

All receptions employed in the systems described here are self-homodyne coherent types, as explained in Chapter 5.

10.1 PARTIAL RESPONSES: DBM FORMATS

10.1.1 INTRODUCTION

The demand for high-capacity long-haul telecommunication systems has increased over the recent years. To achieve high throughput of signals with minimum errors, different advanced modulation formats, such as ASK, PSK (coherent and differential in-coherent), and FSK, have been proposed, and comparisons are made to determine which modulation format would offer the best transmission performance. In countering performance degradation, modulation formats aim to narrow down the optical spectrum to enable close channel spacing in the network. They increase symbol duration, so that more uncompensated dispersion accumulates before intersymbol interference (ISI) becomes significant. Furthermore, this format is more resilient to fiber nonlinearities and optical signal distortion.

Modulation format is important in determining the performance of 40 Gbps optical fiber communication systems. DB and continuous phase modulation (CPM) are shown to offer high spectral efficiency [14,15]. DB modulation formats minimize ISI impairments in a controlled way instead of trying to eliminate it. It is possible to achieve a signaling rate of equal to the Nyquist rate of $2W$ symbols per second in a channel of bandwidth W Hz. Optical DB technique has received much attention due to its high dispersion tolerance and high-frequency utilization efficiency by means of spectral narrowing. DBM format is similar to the non-return-to-zero (NRZ) format, with inclusion of phase coding. The phase characteristics of DBM signals compensate for the group velocity dispersion by reducing the spectral component in conventional NRZ modulation. ISI is reduced, since bit patterns such as 101 are transmitted with the ones carrying the opposite phase. Therefore, if pulses spread out into the zero time slot, due to dispersion, they tend to cancel each other out. The recovering of signals at the receiver is relatively simple. Furthermore, direct detection receiver can be used to simplify and can be offered as a low cost solution. There are two types of DBM schemes, which are constant phase and alternating phase in blocks of logics "1s."

This chapter presents the models for photonic transmission with optical channels operating under DBM format. This includes the development and implementation of the photonic transmitter, the optical fiber propagation, and the optoelectronic receiver. DBM encodes two-level electrical signal to three-level electrical signal before modulating the lightwave carrier. The transmitter of the Simulink model will consist of a DB encoder and a dual-drive Mach–Zehnder interferometric modulator (MZIM). A baseband modulation is first implemented in the DB encoder, which encodes the binary signal into three levels of signals—"1," "0," and "−1." MZIM is an electro-optic modulator that converts the electrical signal to optical signal.

The DB or phase-shape binary modulation formats can be generated by modulating a dual-drive MZIM with the bias voltage located at the V_π and the amplitude of the signals must be at 0 and $2V_\pi$. Recent works have shown that the driving voltages for the modulator can be reduced to generate variable-pulse-width DB optical signals. However, the pulse width of the DB-DPSK has not been thoroughly investigated under the alternating phase of the "1" coded bits. That means the "0," "π," "0," "π" phases of consecutive "1" in contrary of conventional DB formats. We also present modeling performance of alternating-phase DB modulation with a full-width half mark (FWHM) ratio with respect to the bit period of 100%, 50%, and 33%, and compare with experimental transmission of carrier-suppressed DPSK over 50 km of SSMF and dispersion compensation. For the DB case, the transmission without dispersion compensation over the same SSMF length offers better performance for 50% FWHM DB modulation and slightly poorer for 33%. The transmission performance, the bit error rate versus receiver sensitivity, of these DB modulation formats are compared with the carrier-suppressed DPSK experimental transmission.

Section 10.2 of this chapter describes the fundamental aspects of the DBM format. It is followed by a description of each component of the 40 Gbps DBM photonic transmission system in Section 10.3. Section 10.4 describes the implementation of the Simulink model of the communication system. Next, Section 10.5 gives the simulated results. Finally, comparisons with theoretical analyses and other modulation formats are given.

10.1.2 DBM Formatter

Modulation format aims to modulate one or more field properties to suit system needs. There are four types of field properties, which are intensity, phase, polarization, and frequency. Symbols are constellated in one or more dimensions, in order to carry more information and to travel a further distance. Data modulation format is the information-carrying property of the optical field.

DBM format has become an attractive modulation format over the recent years, compared to other formats, such as NRZ and return-zero (RZ). This is due to the fact that it can overcome the fiber chromatic dispersion in high-capacity transmissions. The characteristics of the DBM format make it the preferred format.

DBM schemes can be described as correlative-level coding or partial-response-signaling schemes. Correlative coding schemes can be implemented by inserting ISI to the transmitted signal in a controlled manner.

A signaling rate equal to the Nyquist rate of $2W$ symbols per second in a channel of bandwidth W Hz can then be achieved. "Duo" in the word duobinary indicates the doubling of transmission capacity of a conventional binary system. DBM format is, in fact, NRZ modulation with an inclusion of phase coding. The one bits in the data input are phase modulated. For instance, for a bit pattern of 101, this data will be transmitted with the ones carrying opposite phase, 0 and π. If the pulse of the one bits spreads out to the zero time slot in between, they will cancel each other. This effect increases the dispersion tolerance, and allows the signal to be transmitted over a longer distance.

DB coding converts a two-level binary signal of 0's and 1's into a three-level signal of "−1," "0," and "+1." This is done by, first, applying the binary sequence to a pulse-amplitude modulator to produce two-level short pulses of amplitudes −1 and +1, with −1 corresponding to 0 and +1 corresponding to 1. This sequence is, then, applied to a DB encoder to produce a three-level output of "−2," "0," and "2."

As shown in Figure 10.1 input sequence, $\{a_k\}$ of uncorrelated two-level pulses is transformed into $\{c_k\}$, which is a sequence of correlated three-level pulses. The correlation between adjacent pulses is equivalent to introducing ISI into the transmitted signal in an artificial manner. The DB encoder is simply a filter involving a single delay element and a summer, as shown in Figure 10.2. However, once errors are made, they tend to propagate through the output. This is because a decision made on the current input a_k depends on the decision made on the previous input a_{k-1}. Therefore, precoding is needed to avoid this error propagation phenomenon. Binary sequence, $\{b_k\}$, is converted into another binary sequence, $\{d_k\}$, by modulo-2 addition, exclusive OR of b_k and d_{k-1}, as shown in Figure 10.3.

$$c_k = a_k + a_{k-1} \tag{10.1}$$

$$d_k = b_k \oplus d_{k-1} \tag{10.2}$$

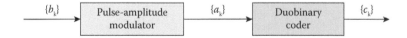

FIGURE 10.1 Brief overview of duobinary signaling.

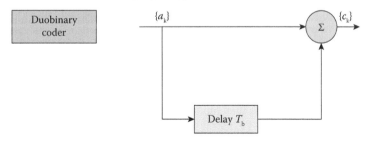

FIGURE 10.2 Duobinary encoder—the block at the left is represented by the signal flow diagram shown in the right.

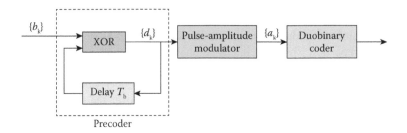

FIGURE 10.3 Duobinary modulation scheme with precoder.

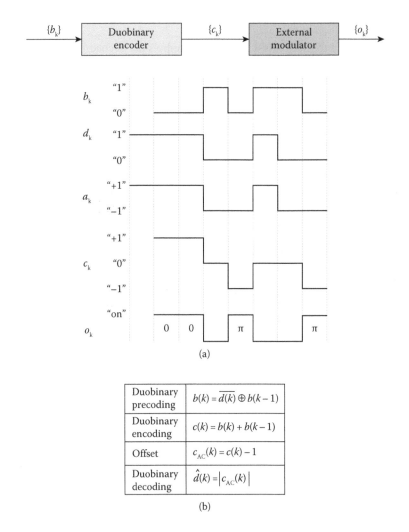

(a)

Duobinary precoding	$b(k) = \overline{d(k)} \oplus b(k-1)$
Duobinary encoding	$c(k) = b(k) + b(k-1)$
Offset	$c_{AC}(k) = c(k) - 1$
Duobinary decoding	$\hat{d}(k) = \left| c_{AC}(k) \right|$

(b)

FIGURE 10.4 (a) Example of original binary signal (b_k), precoded signal (d_k), duobinary-encoded signal (c_k), and optical duobinary signal (o_k). (b) Summary of coding rule.

The three-level DB output, $\{c_k\}$, is then modulated into a two-level optical signal by an external modulator. The most commonly used external modulator is the MZIM. The optical DB signal has two intensity levels, "on" and "off." The "on" state signal can have one of two optical phases, 0 and π. The two "on" states correspond to the logic states "1" and "–1" of the DB-encoded signal, $\{a_k\}$, and the "off" state corresponds to the logic state "0." Figure 10.4a shows an example of the original binary signal, the DB-encoded signal, and optical DB signal, and then Figure 10.4b shows a summary of the coding rule.

10.1.3 40 Gbps DB Optical Fiber Transmission Systems

Ultra-long terrestrial networks, transmitting signals at a bit rate of 40 Gbps, has matured over recent years. Various advanced modulation schemes, such as RZ, NRZ, NRZ-DPSK, and RZ-DPSK, have been proposed to achieve extended reach and improved capacity of communication systems.

Figure 10.5 shows the typical DWDM optical fiber communication system. Signals are modulated at the transmitters and are multiplexed together at the wavelength multiplexer before transmitting them into the fiber. The fiber link is divided into a number of spans. Each span consists of an SSMF and a dispersion-compensating fiber (DCF). The EDFA is used to compensate for the optical power loss of the transmission span. At the end of the fiber, the signals are demultiplexed and detected at the receivers.

DBM format has become popular compared to other modulation formats because it extends the transmission distance as limited by fiber loss, without regenerative repeaters. It extends the dispersion limit without additional optical components, such as the DCF. Chromatic dispersion has become a main effect that limits the transmission distance. The optical three-level transmission can overcome this limitation since narrowband signals have higher tolerance to chromatic dispersion compared to broadband signals. Furthermore, DB optical fiber communication system can suppress stimulated Brillouin scattering (SBS).

The main modules of the communication system are the transmitter, optical fiber, and the optoelectronic receiver, as shown in Figure 10.6. The transmitter consists of the DB encoder and the MZIM. A series of 0's and 1's is modulated under the DBM scheme. These three-level electrical data are then used to modulate the laser source, producing a two-level optical signal. This modulated signal are transmitted along an optical fiber transmission link toward the electro-optic receiver. The signal will be detected using a photodetector, which converts the two-level optical signal back into an electrical signal. Optical amplification can be done at some point along this transmission link to minimize the effect of fiber loss.

10.1.4 Electro-Optic Duobinary Transmitter

A transmitter modulates and converts incoming electrical signals into the optical domain. Depending on the nature of the signal, the resulting modulated light may be turned on and off, or may vary linearly in intensity between the two levels. The output of the DB transmitter is the modulated lightwaves switched on and off at transitional instances of the input electrical signal. In general,

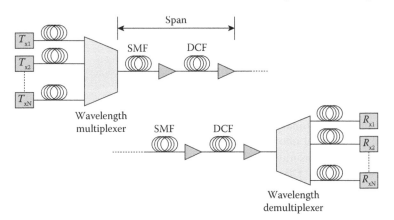

FIGURE 10.5 Ultra-long-haul fiber transmission.

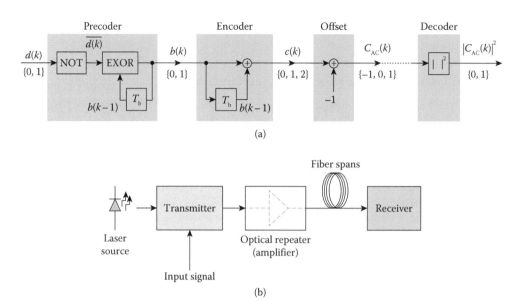

(a)

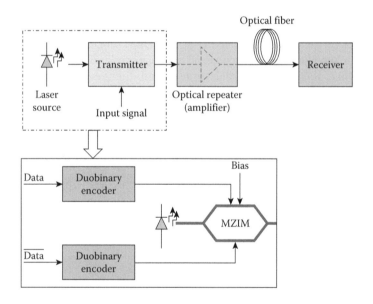

(b)

FIGURE 10.6 Main modules of a duobinary modulation optical communication system: (a) coder and decoder and (b) generic transmission system.

FIGURE 10.7 The transmitter of the 40 Gbps DBM photonic transmission system.

a DBM transmitter is shown in Figure 10.7, consisting of a monochromatic laser source, a coder, and a photonic modulator. Binary data are encoded by a DB encoder. This resulting three-level electrical signal is converted into a two-level optical signal using the folding characteristic of an optical MZIM. It is then transmitted into the optical fiber.

There are two types of DB transmitters. The conventional DB transmitter, as previously mentioned, includes a dual-drive MZIM driven by three-level electrical signals in a push–pull configuration. It is proposed that three-level signals may experience significant distortion in electrical amplifiers operating in saturation, leading to penalties for long word lengths. It may also cause the degradation of receiver sensitivity. For these reasons, a second type of DB transmitter has been proposed. This type of transmitter has the MZ modulator driven by only two-level electrical signals. The optical DB signal generated is the same as the DB transmitter type one, that is, constant phase in blocks of 1's.

10.1.5 DB Encoder

The DB encoder encodes the binary signals, which is a sequence of 0's to a three-level electrical signal. The DB signal is a fundamental correlative coding in partial response signaling. A DB encoder consists of a precoder and a DB coder. A precoder is used before DB coding to allow for easier recovery of binary data at the receiver, and to avoid error propagation. The precoder is a simple binary digital circuit that consists of an exclusive OR (XOR) and a one-bit delay feedback. The DB coder is a filter consisting of a single delay element and a summer.

Binary data input is precoded, with initialization of the one-bit delay to 0. The output of the DB precoder is, then, modulated by a pulse-amplitude modulator, to produce a two-level electrical signal with amplitude of −1 and 1. The DB signal is produced by adding the data delayed by one-bit period to the present data. This DB signal is a three-level electrical signal of −2, 0, and 2. Finally, it is converted to a level of −1, 0, and 1. The three-level is mapped into the optical domain by modulating both amplitude and phase. The "+1" and "−1" levels have the same optical intensity but opposite optical phase. The DB encoder structure is shown in Figure 10.8.

10.1.6 External Modulator

Although the electro-optic modulator has been described in Chapter 2, it is essential here to revisit the operation of the MZIM for DB operation. In an MZIM, the input optical carrier is split into two paths via a Y junction. This Y junction splits the input signal into $E_i/\sqrt{2}$ each. The resultant signal is

$$E_o = \frac{E_i}{2}\left[1 + \exp\left(j\pi\frac{V(t)}{V_\pi}\right)\right] \quad (10.3)$$

where V_π is the voltage to provide a π phase shift of each phase modulator, and $V(t)$ is the driving voltage. Note that, in this equation, the term $e^{j\omega t}$ (ω = optical angular frequency) indicating

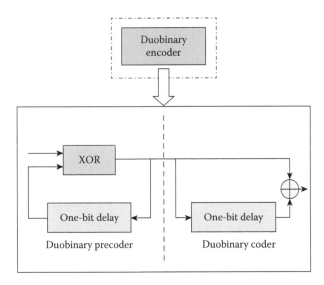

FIGURE 10.8 The DBM encoder of the 40 Gbps DBM photonic transmission system.

the optical wave phase is removed. The transfer relationship between the input and output of this MZIM is as shown in Figure 10.9. It is accompanied by a phase modulation of $\exp(j\varphi(t))$ with $\varphi(t) = \pi V(t)$. For $V(t)$ from 0 to V_π, E_o and E_i have the same phase, and as for $V(t)$ from V_π to $2V_\pi$, E_o and E_i have different phases.

MZIM can be single drive or dual drive. Single-drive x-cut LiNbO$_3$ MZM has no phase modulation along with the amplitude modulation. It follows the transfer characteristics of Figure 10.9. Dual-drive x-cut LiNbO$_3$ MZIM, on the contrary, has two paths phase modulated with opposite phase shifts in a push–pull operation. The V_π in Figure 10.9 is reduced by half in this case. For dual-drive y-cut LiNbO$_3$ MZIM, two paths are driven by complementary signal with V_1 equals to $-V_2$. The output electric of a dual-drive MZIM is

$$E_o = \frac{E_i}{2}\left[\exp\left(j\pi\frac{V_1}{V_\pi}\right) + \exp\left(j\pi\frac{V_2}{V_\pi}\right)\right] \tag{10.4}$$

DB optical signal is generated by driving a dual-drive MZIM with push–pull operation, as shown in Figure 10.10.

One arm is driven by the DB signal and the second arm is driven by the inverted DB signal. Figure 10.11 [6] shows the operation of the MZIM. The output electric field $E_o(t)$ can be expressed as

$$E_o(t) = E_i \cos\frac{\Delta\phi(t)}{2} \cdot \exp\left(-j \cdot \frac{\phi_0}{2}\right) \tag{10.5}$$

where E_i is the input electric field, $\Delta\phi(t)$ is the phase difference between the lightwaves propagating in two optical waveguides, and ϕ_0 is a constant when the MZIM is driven in a push–pull operation.

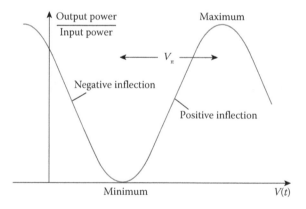

FIGURE 10.9 Input–output transfer characteristics of the MZIM.

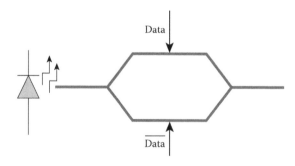

FIGURE 10.10 Dual-drive MZIM.

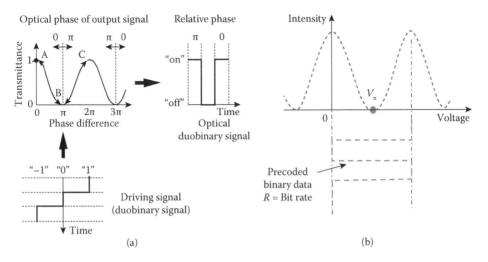

FIGURE 10.11 Driving operation of dual-drive MZM (a) duobinary signaling and (b) DPSK. (From T. Franck et al., *IEEE Photon. Technol. Lett.*, Vol. 10, No. 4, pp. 597–599, 1998.)

At point B of Figure 10.11, the phase of the output optical signal is inverted. The optical DB signal is dependent on the biasing point of the driving signal, which is the electrical DB signal. By biasing at point B in Figure 10.11, "−1" and "+1" level of the electrical DB signal will correspond to the "on" state of the optical signal, whereas the "0" level will correspond to the "off" state. To achieve the effect of carrier suppressed, there must be a π phase different between the two arms.

10.1.7 DB Transmitters and Precoder

The transmitter model, generally, consists of the DB coder and the MZIM. The DB coder encodes the incoming binary sequence of 0's and 1's to DB electrical signal. This signal is, then, used to drive the arms of the dual-drive MZIM. One arm is driven by the DB signal, and the other arm is driven by the inverted DB signal. The Bernoulli binary generator generates a random sequence of binary electrical signals. It is set to generate the data at a rate of 40 Gbps. This signal is encoded by the DB encoder, which consists of a DB precoder and a DB coder. The first output of the encoder is shifted up by 1 to produce levels of "0," "1," and "2." Figure 10.12 shows a conventional Simulink model of the transmitter for DB. This electrical DB signal is sent to the phase shift block, as shown in Figure 10.13, to represent these levels with a certain phase. This, in fact, represents the biasing point on the transmittance curve. For dual-drive MZIM, the driving signal is biased at $V_{\pi/2}$. The second output of the DB encoder is the inversion of output 1. This output is used to modulate the second arm of MZIM. The output 2 signal is shifted down by −1, to bias at the point $-V_{\pi/2}$ of the transmittance curve. MZIM is an amplitude modulator, accompanied by a shift of phase. This modulation is also called AM-PSK. This lightwave, which is the sine wave produced by the sine wave function, is modulated by the DB signal through the complex phase shift block. The input sine wave is shifted by the amount specified at the Ph input.

The DB coder consists of a precoder and a coder, as shown in Figure 10.14. The precoder is a differential coder, with an exclusive OR (XOR) gate and a one-bit delay feedback path. The addition of −0.5 and division by 2 function as the amplitude modulator, which shifts levels of the signal from "0" and "1" to "−1" and "+1." The signal is, then, added to its one-bit delay to produce a three-level DB signal of level "−2," "0," and "+2," followed by a conversion to a level of "−1," "0," and "+1." The summation of the signal with its one-bit delay is the DB coder. For the second output, the output of the differential coder is inverted, before going through the same operation as Out1. Zero-order

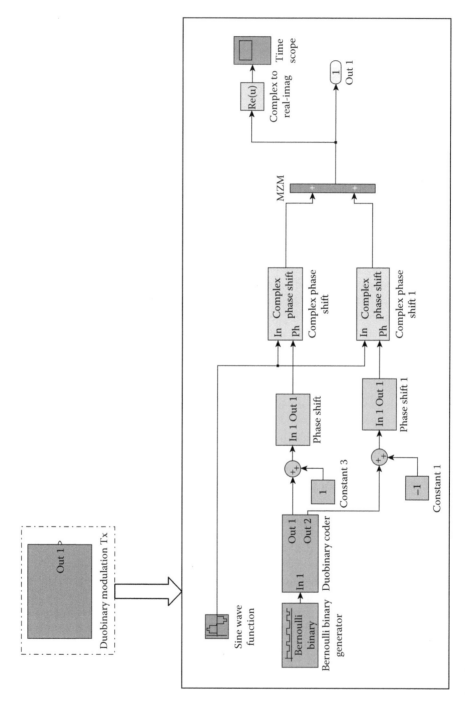

FIGURE 10.12   The conventional transmitter model of the 40 Gbps duobinary optical fiber communication system.

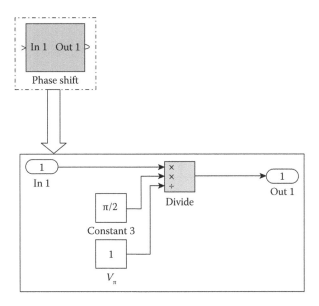

FIGURE 10.13 Phase shift block of the transmitter model.

hold is placed before the output of the DB encoder functions to discretize the signals to have a fast-to-slow transition of signals. It holds and samples the signal before transmitting it out. If the signals are transmitted out without the zero-order hold, the transition to "0" level will be overseen. The signal will only have two levels, which are "−1" and "+1."

10.1.8 ALTERNATIVE PHASE DB TRANSMITTER

Two types of DB transmitter model are proposed. The conventional DB transmitter, as mentioned previously, uses a dual-drive Mach–Zehnder modulator driven by three-level DB electrical signals. The MZIM shown in Figures 10.15 and 10.16 is the DB transmitter type 2, and is usually driven by a two-level electrical signal. In some cases, this three-level driving signal may experience significant distortion in electrical amplifiers operating at saturation. This may lead to penalties for long word lengths. It may also occur that there will be a degradation of receiver sensitivity. Due to these uncertainties, an alternative DB transmitter, as shown in Figure 10.15 and 10.17, is proposed. This second type of DB transmitter has the MZIM driven by two-level electrical signals. It consists of a differential encoder, a one-bit period electrical time delay, and an MZ modulator. One arm of the dual-drive MZ modulator is driven with the signal from the original signal, whereas the other arm is driven by the inverted signal, delayed by a one-bit period. Both DB transmitters produce the same result, which is a constant phase in blocks of 1's.

10.1.9 FIBER PROPAGATION

As described in Chapter 3, the fiber propagation model models the linear and nonlinear dispersion effects that exist over the entire length of the optical fiber can be represented by the NLSE including the SPM, other nonlinear effects can also be included, given by

$$\frac{\partial A}{\partial z} = +\beta_1 \frac{\partial A}{\partial t} + \frac{j}{2}\beta_2 \frac{\partial^2 A}{\partial t^2} - \frac{1}{6}\beta_3 \frac{\partial^3 A}{\partial t^3} = j\gamma |A|^2 A \tag{10.6}$$

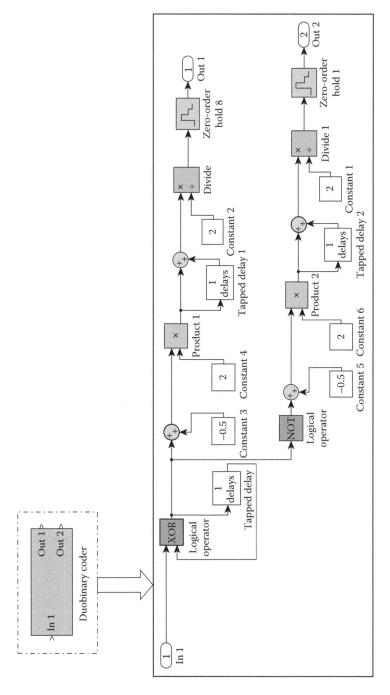

FIGURE 10.14 The duobinary coder of the transmitter model.

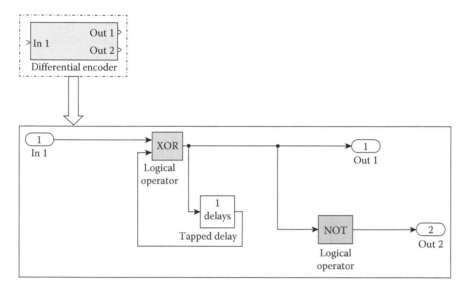

FIGURE 10.15 Duobinary modulation transmitter (type 2).

With β_1 the lightwave group delay, β_2 and β_3 are the first-order and second-order group velocity dispersions, γ is the nonlinear coefficient, and A is the pulse envelope. NLSE can be solved by the split-step Fourier method (SSFM) that integrates two main steps of the split-step model, the nonlinear effects act alone in frequency domain and then the linear effects in the linear Fourier transform domain.

$$\frac{\partial A}{\partial t} = (L + N) A \tag{10.7}$$

where L and N represent linear and nonlinear operators, respectively, which can be extracted from the NLSE.

The fiber propagation block, as shown in Figure 10.18, consists of a Gaussian filter, a gain factor, and the single-mode fiber (SMF) model. The SMF model is shown in Figure 10.19. This model is based on the SSF method. It splits the fiber into a number of small sections. All parameters needed for the SSF operations are concatenated into a matrix, before passing it to the MATLAB function. The MATLAB function block solves the NLSE using the SSFM. Linear operation is implemented in all steps. When the peak power is greater than the nonlinear (SPM) threshold, the nonlinear operator is activated. In this fiber propagation model, a buffer is incorporated at various locations along the propagation path. The buffer is used to redistribute the input samples to a new frame size—in this case, a larger frame size than that of the input. Buffering to a larger frame size yields an output with a slower frame rate than the input. The "Unbuffer" block unbuffers the frame-based input into a sample-based output. The buffer used in this model determines the number of bits sent into the fiber, which is the SMF block.

Probes are connected at two different points of the fiber propagation model. These probes determine the sampling time at these points. The sampling time of $2.5e^{-11}$ indicates down-sampling to baseband. At the baseband, the complex envelope of the signal is extracted and transmitted through the fiber, which is represented by the SMF block shown in Figure 10.18. However, the phase contents of the signal are maintained and are transmitted through the fiber. These phase components of the DB signal are important, because it increases the dispersion tolerance and, thus, allow the signal to travel a longer distance.

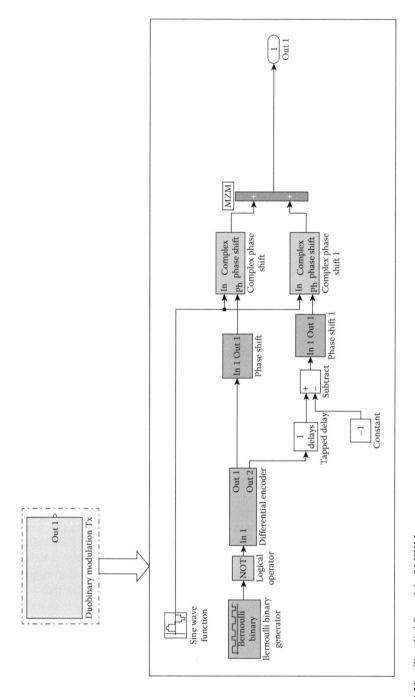

FIGURE 10.16 Simulink® model of MZIM.

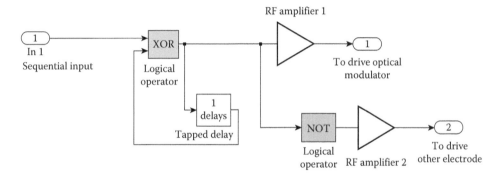

FIGURE 10.17 The differential encoder of Type 2 duobinary transmitter incorporating electrical RF amplifier for driving the MZIM optical modulator.

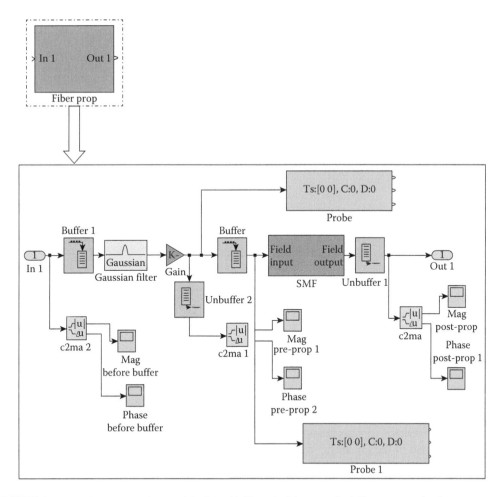

FIGURE 10.18 Fiber propagation model of the 40 Gbps duobinary optical fiber communication system.

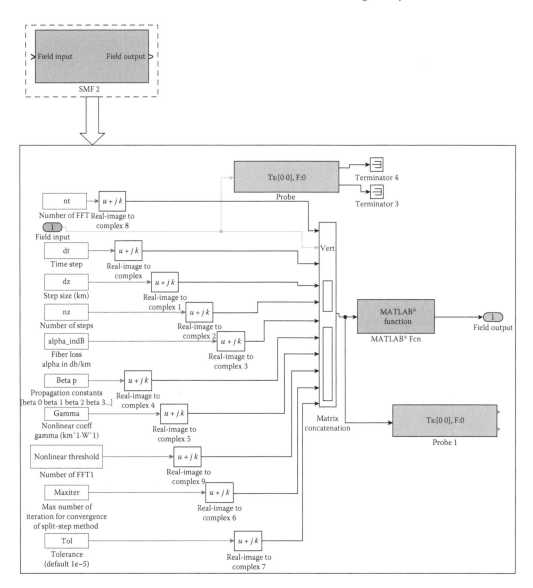

FIGURE 10.19 The SMF models using the SSFM.

10.2 DB DIRECT DETECTION RECEIVER

The receiver model of 40 Gbps DB optical fiber communication system consists of a Gaussian filter and scopes to observe the performance of the system (Figure 10.20). The receiver is a conventional direct detection type. Therefore, a demodulator or decoder is not needed. Power spectrum and eye diagram can be observed directly from this point. The Gaussian filter functions as a baseband filter. The absolute value of the incoming signal is taken because the DB receiver detects the intensity of the signal. The probe is used to determine the sampling time at that point. The "Demodsignals" block shown collects all the data at this point and saves it in the workspace. This data are used to plot the histogram, which is used to determine the Q factor and BER of the system.

Demodulation is needed at the receiver, depending on the modulation format used. For instance, the DPSK receiver consists of a Mach–Zehnder delay interferometer (MZDI), which demodulates the incoming signal before detecting it using a photodetector. MZDI lets two adjacent bits interfere

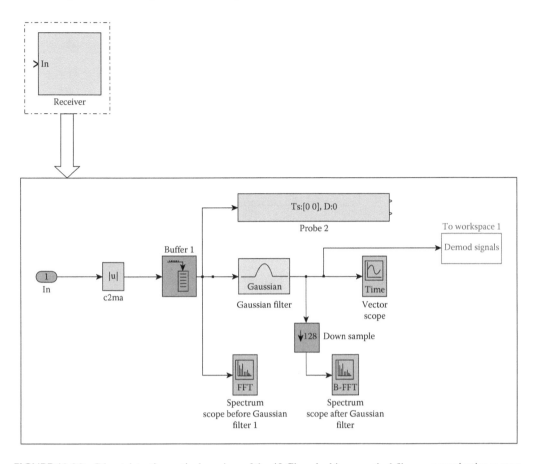

FIGURE 10.20 Direct detection optical receiver of the 40 Gbps duobinary optical fiber communication system.

with each other at its output port. This interference creates the presence, or absence, of power at the output port, depending on whether the interference is constructive or destructive with each other. The preceding bit in the DPSK signal acts as a phase reference for demodulating the current bit. The DI output ports are detected by the balanced detectors. The optical DB signals can be demodulated into a binary signal with an optical receiver of conventional direct detection type, as shown in Figure 10.21. A decoder is not required in this case. The received signal is directly detected by a photodiode operating as a square law detector. The optical DB signal consists of two states, "on" and "off." The photodiode works by detecting the incoming intensity of the signal. The recovery of the original electrical signal can be done by simply inverting the signal detected at the photodiode. This inversion is done within the circuit shown as the decision circuit in Figure 10.20.

The signal is observed at this point of the system. The most commonly used parameters to test and observe the performance of a system are the power spectrum, eye diagram, Q factor, BER, and the received optical power. The Q factor is the quality factor of the system, under the Gaussian noise distribution. It is determined by the mean voltage level and the standard deviation of the noise.

$$\delta = \frac{\mu_1 - \mu_0}{\sigma_1 + \sigma_0} \tag{10.8}$$

where μ_1 is the mean voltage level of the "1" received, μ_0 is the mean voltage level of "0" received, σ_1 is the standard deviation of noise of the "1" received, and σ_0 is the standard deviation of the noise of the "0" received.

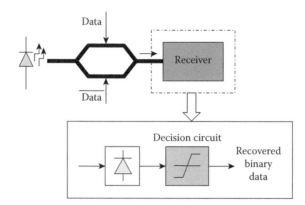

FIGURE 10.21   The receiver of the 40 Gbps DB photonic transmission system. 40 Gbps DB transmission Simulink® model.

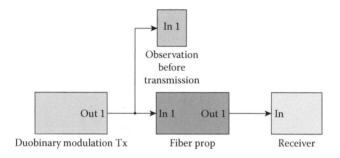

FIGURE 10.22   40 Gbps duobinary optical fiber communication system (Simulink® model).

$$\text{BER} = \frac{1}{2}\text{erfc}\left(\frac{\delta}{\sqrt{2}}\right) \qquad (10.9)$$

The 40 Gbps DB optical fiber communication system can be implemented on the Simulink platform. Simulink has been chosen as the computer software for this development of the model because it consists of a variety of communication blocks that can assist in simplifying the process of implementing and improving this model. Figure 10.22 shows the overall system of the 40 Gbps DB optical fiber communication system. The main modules of this communication system are the DB transmitter, named as the DBM Tx, the fiber propagation, and the receiver. The DB transmitter, in general, consists of the DB encoder and the MZIM. The fiber propagation block introduces the linear and nonlinear dispersion effect that exists over the entire length of the optical fiber to the signal. The receiver is of direct detection type. Therefore, the output at the receiver is directly observed using scope, in the form of eye diagrams and power spectrum.

10.3   SYSTEM TRANSMISSION AND PERFORMANCE

The overall performance of this 40 Gbps DB optical fiber communication system can be observed at the receiver of the system. The eye diagram and power spectrum are the parameters that are used to observe the performance of the system. Multiple debugging and testing processes have to be carried out to prove that the system is functioning as expected. Simulation using Simulink has vastly reduced the time and difficulties involved in these processes.

The first step of testing involved the testing of the transmitter. It is important to ensure that the DB optical signal is generated at the output of the transmitter. The DB encoder is critical in this case. Its output is checked using the time scope to ensure that the three-level electrical signal is produced. Power spectrum and eye diagram are connected at the output of the transmitter. These obtained results are compared with those obtained experimentally and theoretically to verify that the DB transmitter is generating the correct signals. A fiber propagation model is then connected to the output of the transmitter. The fiber propagation model will introduce the linear and non-linear effects of fiber, depending on the distance the signal travels. The DB receiver, which is of conventional direct detection type, is connected after the fiber propagation. The signal is observed directly at the receiver. By observing the eye diagram at this point, the distance that the signal can propagate without severe distortion can be estimated. The testing is started with a fiber length of 1 km. The distance is increased until the point when significant distortion to the eye diagram can be observed, and the "eye" of the eye diagram has closed.

10.3.1 DB ENCODER

The DB encoder is the first implemented model in this 40 Gbps DB optical fiber communication system for the generation of three-level DB electrical signals. Scopes are probing at various points of the DB encoder, as shown in Figure 10.14. All time scopes are set to the same range to allow for comparisons of the bits within the same range. The temporal sequences at the outputs of the encoder for driving the MZIM are shown in Figure 10.23. Scope 1 shows the data generated by the Bernoulli binary generator. Scopes 2–4 show the output at each arm. A bit "0" is encoded as "+2" or "−2," while the bit "1" is encoded as 0. This agrees with the DB coding scheme. They also show a three-level electrical signal. Scope 4 is the inverted version of scope, as expected, due to the NOT gate applied to the second arm.

10.3.2 TRANSMITTER

The transmitter includes the DB encoder and MZIM. After the output of the DB encoder is veri-fied and checked to produce the correct signal, the levels of the signal are represented with a phase. This phase is used to shift the laser source, produced by the sine wave function. The testing of the implemented DB transmitter includes observing the eye diagram and power spectrum at the output. Time scopes are attached at various points of the transmitter to observe the signal at these points. Spectrum scopes are connected at the output of the transmitter, which is after the summation of both arms of MZIM and to each arm of the MZIM to observe the power spectrum at these arms. Some adjustments need to be made to the transmitter model in order to observe the power spec-trum. The sine wave function is set to produce a wave of 200 GHz, rather than 193 THz. This is to

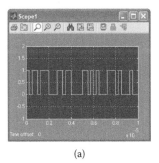
(a)

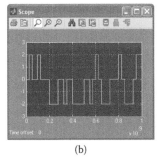
(b)

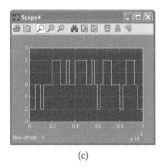
(c)

FIGURE 10.23 DB encoder outputs: (a) Bernoulli binary sequence, and (b, c) DB electrical sequences for driving the electrodes of the MZIM.

enable the observation of the spectrum centered at a lower frequency, and better spectral resolution. The zero-order hold that functions to hold and sample the incoming signal is set to 1e–12, so that the x-axis of the spectrum scope is in the THz range. The obtained results, as shown in Figure 10.24, reflect the results expected from experiments and theories. The power spectrum for a 5 Gbps DB optical fiber communication system is shown in Figure 10.25a [6]. This is used to verify the power spectrum obtained from the Simulink model. Spectrum (c) of Figure 10.26 corresponds to the Spectrum Scope in Figure 10.18. The obtained spectrum reflects on that in Figure 10.24a, with the shape approximately the same. The bandwidth of the obtained spectrum corresponds to the data rate, as in Figure 10.25a. The spectrums obtained are centered at 200 GHz, which is the carrier frequency of the model. Spectrum (c) is carrier suppressed, while the other two spectrums are not. This is as expected from the DBM format. Due to the π-phase difference in the two arms of MZIM, the output is expected to have its carrier suppressed.

Further verification of the DB transmitter Simulink model is carried out by monitoring the eye diagram at the output of the transmitter. The observation prior to the transmission block in the overall system, shown in Figure 10.27, is used to observe the eye diagram before transmission into the fiber. The eye diagram obtained, as shown in Figure 10.28, has an "open eye" with an amplitude of 0.6. It reflects the one in Figure 10.25. For a range of 100 ps, four "eyes" are obtained, giving one "eye" 25 ps. This proves that the eye diagram is correct, since the data rate is 40 Gbps, which equals 1 bit every 25 ps.

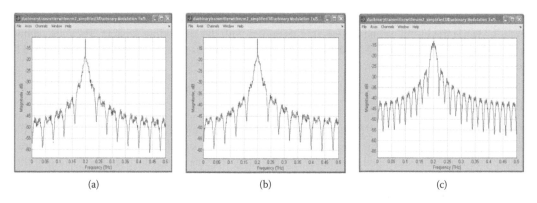

(a)   (b)   (c)

FIGURE 10.24 (a, b) Power spectrum obtained from each arm of the MZIM and (c) the output of duobinary transmitter.

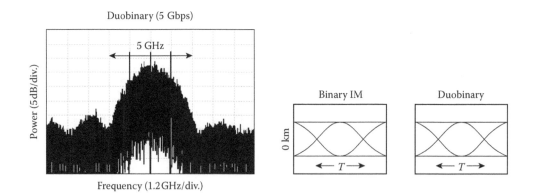

FIGURE 10.25 DB power spectrum and eye diagram for 5 Gbps duobinary optical fiber communication system. (Extracted from T. Franck et al., *IEEE Photon. Technol. Lett.*, Vol. 10, No. 4, pp. 597–599, 1998.)

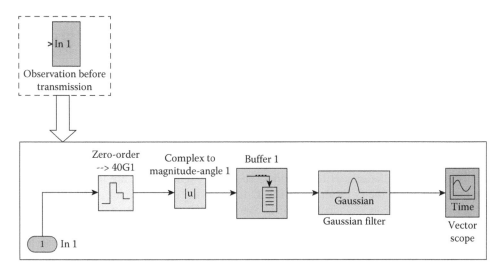

FIGURE 10.26 Signal and eye diagram monitored before the transmission block of the 40 Gbps duobinary optical fiber communication system.

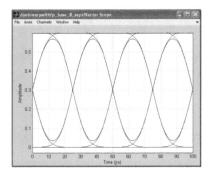

FIGURE 10.27 Eye diagram obtained before propagation block.

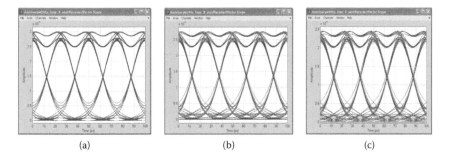

FIGURE 10.28 Duobinary eye diagram for (a) 1 km, (b) 5 km, and (c) 10 km.

10.3.3 Transmission Performance

The overall performance is observed at the receiver of the 40 Gbps DB optical fiber communication system. The eye diagram and power spectrum are observed by the spectrum scope and vector scope attached inside the receiver. Eye diagrams show the distortion and attenuation of signals at various lengths of the fiber. When the "eye" of the eye diagram closes, the signal is severely distorted and dispersed, and, thus, recovering of the signal will be impossible. The eye diagram can also be used to calculate the received optical power. The Q factor and BER can be obtained by plotting the histogram of the received signal, followed by some calculations to obtain the mean voltage level and the standard deviation of noise.

The eye diagrams of various lengths of the fiber are obtained. It can be observed from Figures 10.28 through 10.30 that, as the length of the fiber increases, so do the dispersion and noise. The "eye" of the eye diagram closes eventually. DB signals are expected to be able to travel up to 200 km without the "eye" closing completely. This has been proved by the model. It can be observed that, at the length of 250 km, as shown in Figure 10.30b, the "eye" of the eye diagram is yet to be fully closed. Dispersion effects can be observed, and they progressively increase with the distance of the fiber.

Data obtained from the *demod*signals block, shown in Figure 10.31, are used to plot the histogram that determines the Q factor and BER. Q factor is the quality factor of the system. From the Q factor, the BER can be calculated. It is expected that, for a BER of 10^{-9}, the Q factor is approximately 6. The histograms for 1 and 10 km are shown in Figure 10.32. The two points on the histogram are compared with the histogram at the receiver to obtain the mean value, μ_0 and μ_1,

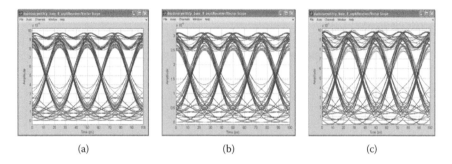

(a)　　　　　　　　　　(b)　　　　　　　　　　(c)

FIGURE 10.29　Duobinary eye diagram for (a) 50 km, (b) 100 km, and (c) 150 km.

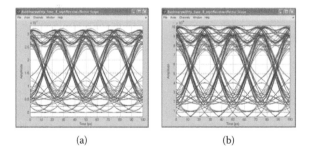

(a)　　　　　　　　　　(b)

FIGURE 10.30　Duobinary eye diagram for (a) 100 km and (b) 150 km.

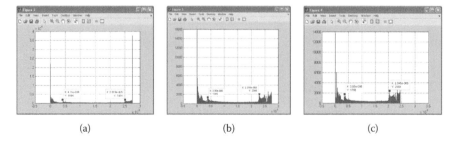

(a)　　　　　　　　　　(b)　　　　　　　　　　(c)

FIGURE 10.31　The histogram obtained from data *demod signals* for (a) 1 km, (b) 5 km and (c) 10 km.

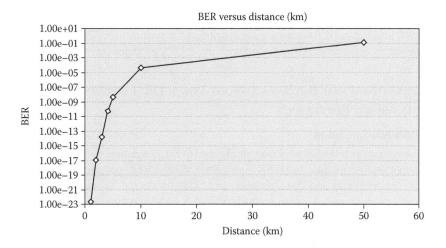

FIGURE 10.32 Plot of BER versus distance (km) of SSMF.

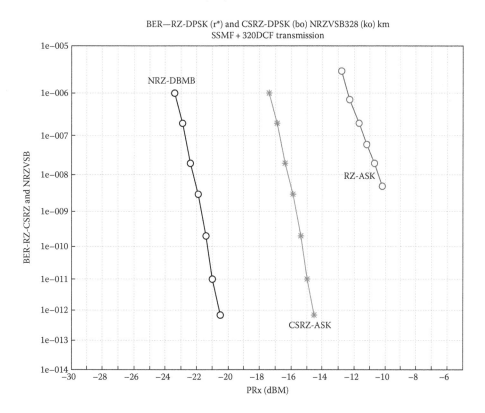

FIGURE 10.33 BER versus receiver sensitivity (dBm) of simulated NRZ-DBM (black o), and experimental CSRZ-ASK, RZ-ASK for transmission system of 328 km of dispersion-compensated optically amplified SSMF spans.

and the standard deviation of noise, σ_0, and the BER, or bit error rate, can be calculated from the Q factor found. Figure 10.32 shows the plot of BER versus distance. It can be observed that, as the distance of propagation of the signal increases, so does the bit error rate. This is as expected, since the linear and nonlinear effects of the fiber introduce noise and errors to the transmitting signal. Figure 10.33 is the plot of BER versus receiver sensitivity. The BER is plotted against the receiver sensitivity for the range from −23 dBm to −20 dBm. Experimental results of the transmissions of the CSRZ-ASK and RZ-ASK formats are also included for system over 328 km SSMF and DCM modules, completely dispersion compensated with five EDFA modules integrated.

10.3.4 ALTERNATING-PHASE AND VARIABLE-PULSE-WIDTH DB: EXPERIMENTAL SETUP AND TRANSMISSION PERFORMANCE

10.3.4.1 Transmission Setup

A transmission system is arranged as shown in Figure 10.34a, consisting of 50 km of SSMF and the associated compensating module; a DPSK transmitter (SHF-5300); optical booster and preamplifiers; wavelength multiplexer and demultiplexer; an MZDI; an optical balanced receiver SHF-5008; a clock recovery module; and an error analyzer SHF-EA 44A. A typical eye diagram obtained for 40 Gbps CSRZ-DPSK modulation format recovered after transmission is shown in Figure 10.34b. This test bed is also used to investigate the transmission of DWDM channels of various modulation formats. Hence, the wavelength mux and demux are included here. The BER versus the receiver sensitivity for the CSRZ-DPSK is shown in Figure 10.35. The curves for DB modulation with alternating phase of the "1" intensity coded pulses are also obtained by simulation, as given in the next section.

DB optical transmission can be experimentally determined via a simulation whose platform can be implemented on MATLAB and Simulink for several modulation formats, especially the phase-modulated optical transmission. A typical system arrangement of optical transmitter using a dual-drive MZIM for generation of DPSK and alternating-phase DB modulation for simulation is shown in Figure 10.36. The fiber propagation model follows the well-known split-step Fourier method, with provisions for switching between linear and nonlinear power level propagation, so as to minimize the computing time.

The FWHM of the DB modulation format can be generated by setting the amplitude of the swing voltage levels applied to the dual electrodes of the MZIM. The biasing condition of the MZIM can be varied between the minimum and maximum transmission and the quadrature points of the transfer characteristics of the modulator to obtain carrier suppression and alternating-phase properties of the modulated lightwaves. RZ formats and suppression of the carrier can also be generated by using another dual-drive MZIM biased at minimum transmission and a half-bit-rate frequency synthesizer. Our Simulink models have been extensively tested, and the system performance agrees well between experiments and modeling. Figure 10.35 shows the agreement between the BER versus the receiver sensitivity for CSRZ-DPSK modulation format, obtained both in experiment and by simulation.

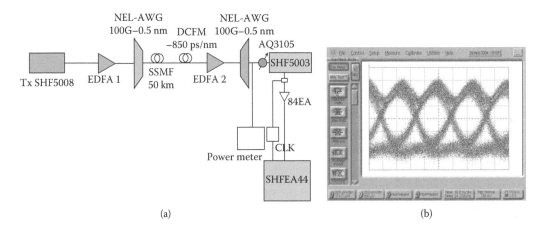

(a)   (b)

FIGURE 10.34 System test bed for CSRZ-DPSK transmission over 50 km SSMF, balanced receiver (a) schematic diagram of the test bed and (b) dispersion-compensated received eye diagram of CSRZ-DPSK.

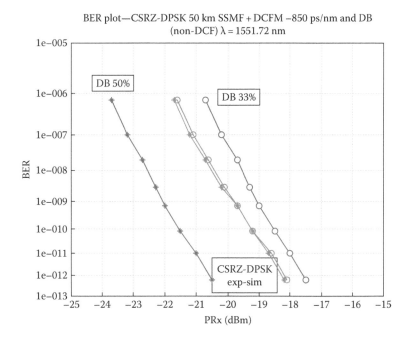

FIGURE 10.35 BER versus receiver sensitivity for 50 km transmission: CSRZ-DPSK transmission with complete compensation (dark gray curve) and 50 and 33% FWHM alternating phase DB modulation (light gray curves).

Noises of optical amplifiers have also been taken into account in both cases. We select the CSRZ-DPSK format to compare with alternating-phase DB modulation because it has been proven in practice to offer superior performance as compared with RZ-DPSK and NRZ-DPSK. The MZIM is modulated and biased, such that the width of the DB-modulated pulses can be altered with ease.

Simulation results are obtained for alternating-phase DB with an FWHM of 50% and 33% of the bit period for 40 Gbps, as shown by the light gray curves in Figure 10.35. An optical Gaussian filter type is also used at the output of the transmitter prior to transmission. We observe almost 2 dB better receiver sensitivity of the 50% FWHM alternating-phase DB formats as compared to the CSRZ-DPSK. However, the 33% FWHM DB case proves to be 1 dB less sensitive. This indicates the effectiveness of the DBM. It is noted again that the 50 km SSMF without compensation is used for 40 Gbps DB transmission. A typical eye diagram obtained at the output of the balanced receiver for 33% FWHM DB is shown in Figure 10.36b. The simulation is conducted with 256 random bit patterns and several frames sufficient for measurement of the Q factor without resorting to the Monte Carlo technique. We also assume a Gaussian distribution of the ISI and phase noises. This assumption may have suffered 0.5–1 dB penalty as compared with the chi-square distribution for phase error in DPSK transmission [16,17].

The DBM with alternating phase and control of the FWHM of the pulse sequence offer 2 dB improvement as compared to that of CSRZ-DPSK. The 50% FWHM with a Gaussian profile allows the possibility of lower bandwidth demand on the optical modulators [2]. Still, this format offers better performance due to the reduction of the signal bandwidth.

10.3.4.2 Test Bed for Variable-Pulse-Width Alternating-Phase DBM Optical Transmission

In this section, we investigate the transmission performance of alternating-phase DBM of an FWHM ratio with respect to the bit period of 100%, 50%, and 33%, and compare with experimental transmission of carrier-suppressed DPSK over 50 km of standard single-mode fiber (SSMF)

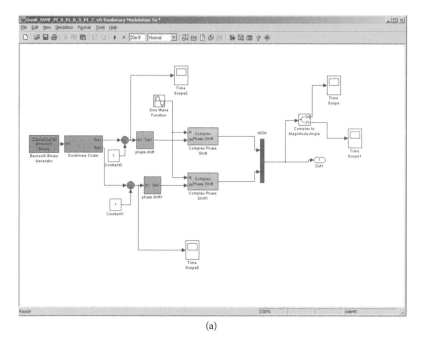

(a)

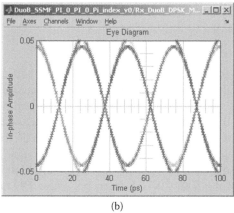

(b)

FIGURE 10.36 (a) Optical transmitter Simulink® model for generation of alternating-phase duobinary modulation format and (b) nondispersion-compensating balanced received eye diagram for DB 33% FWHM.

and ispersion compensation. For the DB case, the transmission without dispersion compensation over the same SSMF length offers better performance for 50% FWHM DB modulation and slightly poorer performance for 33%. The transmission performance, the BER versus receiver sensitivity, of these DB modulation formats are compared with the carrier-suppressed DPSK experimental transmission.

These transmission performances are compared with the noncompensating 50 km SSMF transmission of DBM with pulse widths of 33% and 50%. The later format offers at least 2 dB improvement in terms of the BER and receiver sensitivity over that of CSRZ-DPSK. The 50% and 33% FWHM DB modulation schemes offer simpler driving circuitry for the optical modulators due to its lower swing voltage levels applied to the electrodes of the dual-drive interferometric modulator. This is very important when the bit rate is in the multi-GHz region. Furthermore, the 50% and 33% pulse width and Gaussian profile will further reduce the demand on the bandwidth of optical modulators—that is, a 30 GHz or lower transmittance bandwidth can be used. A Simulink model has also been developed for simulation of the transmission performance of the DBM formats. Simple driving conditions can be achieved, and its effectiveness demonstrated.

We describe a generic simulation platform on MATLAB and Simulink for several modulation formats, especially the phase-modulated and partial-response optical transmission as shown in Figure 10.37. It consists of two DB transmitters, a fiber transmission model, and a direct detection optoelectronic receiver. A typical system arrangement of optical transmitter using a dual-drive Mach–Zehnder interferometric modulator for the generation of DPSK and alternating-phase DB modulation for simulation is shown in Figures 10.38 and 10.39. The fiber propagation model follows the well-known split-step Fourier method with provision for switching between linear and nonlinear power level propagation so as to minimize the computing time.

The pulse FWHM of the DB modulation format can be generated by setting the amplitude of the swing voltage levels applied to the dual electrodes of the MZIM. The biasing condition of the MZIM can be varied between minimum and maximum transmission and the quadrature points of the transfer characteristics of the modulator to obtain carrier suppression and alternating-phase properties of the modulated lightwaves. RZ formats and suppression of the carrier can also be generated by using another dual-drive MZIM biased at minimum transmission points of both sides of the voltage-optical intensity transfer curve of MZIM. This is the pulse carver that would be required to generate variable-pulse-width optical "clock pulses" before feeding into the data modulator (Figures 10.40 through 10.42).

We note also that half-bit-rate frequency synthesizer is used as a driving source applied to the two electrodes for generation of RZ periodic pulse sequence. The optical spectrum of a lightwave of 50% RZ DB modulated signals, as shown in Figure 10.43, confirms the estimation of the modulation technique.

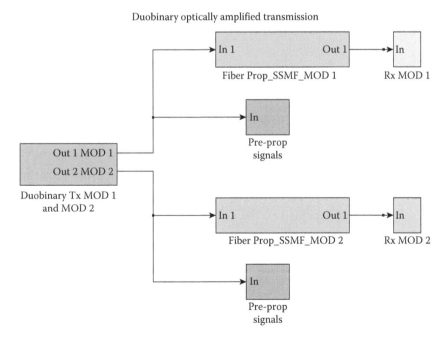

Duobinary optically amplified transmission

FIGURE 10.37 Generic simulation model for duobinary variable-pulse-width transmission. The transmitter is shown in the far left, consisting of two subsystem transmitters—one is the NRZ duobinary, and the other a different variable-pulse-width duobinary format for comparison. Each span consists of a transmission fiber (SSMF) in association with a dispersion-compensating fiber and two in-line amplifiers (see Figure 10.41). Cascaded spans are identical with 100 km SSMF and 100 km DCF (negative dispersion factor and matched slope) and 20 dB gain plus 5 dB NF EDFAs.

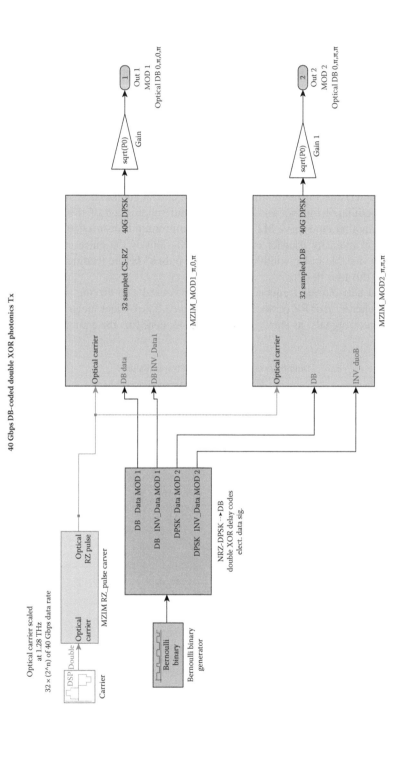

FIGURE 10.38 MATLAB® and Simulink® model showing the pulse carver or optical "clock sources" and the data dual-drive MZIM for generating of variable-pulse-width RZ-DB optical transmitter.

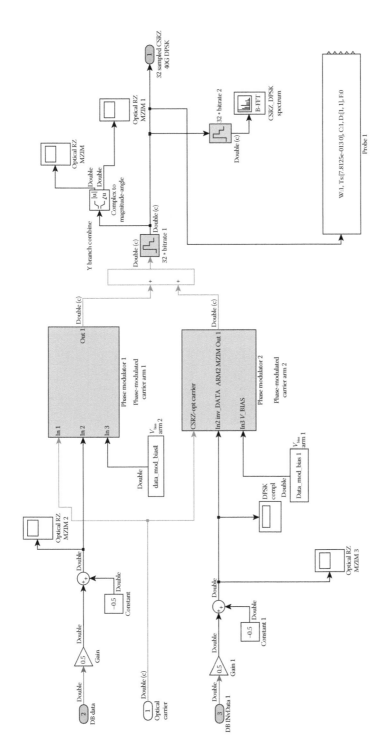

FIGURE 10.39 Electrical filter inserted before modulating the data optical modulator. Electrical filter can be Gaussian or raised-cosine type, or a Bessel filter of the fifth order can be inserted to reduce the required transmitting signal bandwidth.

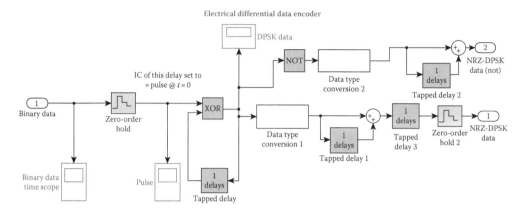

FIGURE 10.40 Simulink® simulation model of the data encoder for duobinary format using differential encoding scheme of DPSK. The differential signals can be delayed or nondelayed and then added with its inverted version to generate three-level electrical signals to two-level signals for driving the data modulator electrodes.

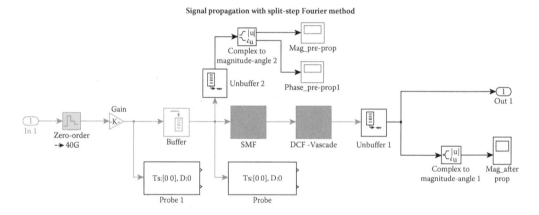

FIGURE 10.41 Simulink® simulation model for the propagation of duobinary modulated carrier over dispersion-compensated transmission link.

The Simulink models have been extensively tested, and the system performance agrees well between experiments and modeling. Later in the chapter, Figure 10.47 shows the agreement between the BER versus the receiver sensitivity for CSRZ-DPSK modulation format obtained both in experiment and by simulation. Noises of optical amplifiers have also been taken into account in both cases. We select CSRZ-DPSK format to compare with alternating-phase DB modulation because it has been proven in practice to offer superior performance as compared with RZ-DPSK and NRZ-DPSK. The MZIM is modulated and biased, such that the width of the DB-modulated pulses can be altered.

Simulation results are obtained for alternating-phase DB with an FWHM of 50% and 33% of the bit period for 40 Gbps, as shown by the light gray curves indicated in Figure 10.47 (see later in the chapter). An optical Gaussian filter type is also used at the output of the transmitter prior to transmission. We observe almost 2 dB better receiver sensitivity of the 50% FWHM alternating-phase DB formats as compared to CSRZ-DPSK [7]. However, the 33% FWHM DB case offers 1 dB less sensitive. A single photodetector can be used to recover DB modulated lightwave signals (Figure 10.44). The simulation is conducted with 256 random bit patterns and several frames sufficient for the measurement of the Q factor without resorting to the Monte Carlo technique. We also assume a Gaussian distribution of the ISI and phase noises. This assumption may have suffered 0.5–1 dB penalty as compared with the chi-square distribution for phase error in DPSK transmission [5].

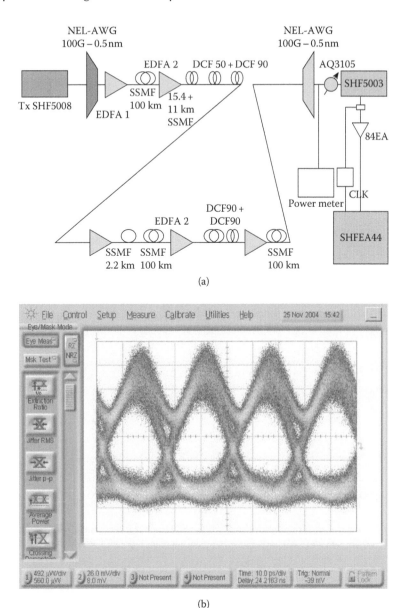

FIGURE 10.42 Experimental and simulated transmission systems: (a) system setup consisting of multispans of total 328 km SSMF plus 320 km DCF and associated optical amplifiers EDFAs, optical receivers employing self-homodyne balanced receiver; modulation formats including CSRZ-DPSK, NRZ-DPSK, and RZ-DPSK and duobinary and (b) received eye diagram of CSRZ-DPSK with 34 ps/nm residual dispersion.

10.3.4.3 CSRZ-DPSK Experimental Transmission Platform and Transmission Performance

A transmission system is arranged as shown in Figure 10.43, consisting of 50 km of SSMF and the associated compensating module; a DPSK transmitter (SHF-5300); optical booster and preamplifiers; wavelength multiplexer and demultiplexer; an MZDI; an optical balanced receiver SHF-5008; a clock recovery module; and an error analyzer SHF-EA 44A. A typical eye diagram obtained for 40 Gbps CSRZ-DPSK modulation format recovered after transmission is shown in Figure 10.44. This test bed is also used to investigate the transmission of DWDM channels of various modulation formats. Hence, the wavelength mux and demux are included here.

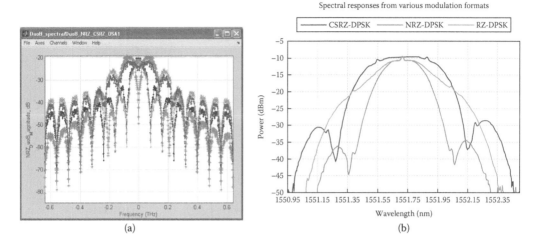

FIGURE 10.43 (a) Optical spectrum as monitored at the output of the 40 Gbps duobinary transmitter with formats NRZ (dotted-black), CSRZ 67% pulse width (gray +), 50% RZ (light gray dot), and RZ 33% (mid gray*). All normalized with respect to the optical carrier, which is set at 192.52 THz; and (b) spectra of CSRZ-DPSK, NRZ-DPSK, and RZ-DPSK modulation formats obtained by experiment.

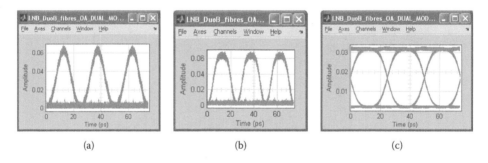

FIGURE 10.44 Received eyes after (three spans) 320 km SSMF fiber fully compensated including two in-line optical amplifiers (20 dB gain and 5 dB noise figure) transmission: (a) DB 33% FWHM and (b, c) 50% duobinary.

The BER versus the receiver sensitivity for the CSRZ-DPSK and DB of different RZ pulse width is shown later in the chapter in Figure 10.47. The curves for DB modulation with alternating phase of the "1" intensity coded pulses are also obtained by simulation.

The DB simulation setup is structured similar to the experimental setup for DB formats in contrast to CSRZ-DPSK transmission over 320 km SSMF optical fiber transmission multispan links. Although the modulators employed in the SHF-5008 optical transmitter are of single drive, the simulated modulators are of dual-driven type, and the signaling and coding blocks are described in the previous section. The receiver model is simple as direct detection receiver circuitry is used, including a wideband photodetector and microwave amplifier. A Besssel fifth-order or Gaussian (80% or DB signal bandwidth) electrical filter can also be incorporated to minimize the noise contribution.

The optical spectra of DPSK signals under different pulse widths and pulse carver operations are shown in Figure 10.43b, while Figure 10.43a shows the simulation spectra of their counterparts under DB-coded scheme. The spectra are centered at the optical carrier frequency. The gray curve (+) shows clearly the suppression of the carrier, while the other DB schemes show minimum carrier suppression but substantial signal power levels that indicate the specific characteristics of the di-phase modulation, especially when the data modulator is biased such that the phase difference between the two arms of the MZIM is π.

Figure 10.36 shows the simulated eye diagrams at the receiver (after the photodetector including electronic amplifier noises) of the DB-coded sequence of NRZ (c), 50% RZ (b), and 33% RZ (a). Gaussian electrical filtering has been applied to the NRZ DB, and none applied to the other RZ formats of the scheme. No electrical filtering is used at the transmitter. The bit rate is 40 Gbps.

The effects of electrical filtering at the transmitter on the received eye diagrams are shown in Figure 10.45a through d. A 30% bit rate (BT = 0.3) electrical Gaussian filter bandwidth would still sustain the eye opening after 320 km transmission with and without filtering effects (Figure 10.45a and b)—that is, the decision sampling time must be very accurate to be error free even under fully compensated transmission. A 50% BT would be, as observed under simulation, much more tolerable to the sampling error. Figure 10.45c and d shows that such filtering effects would not exist if a raised-cosine electrical filter with a 0.5 roll-off factor for 40 Gbps CSRZ 67% DB format transmission over 320 km consisting of three-spans of optically amplified and fully DC-compensated -SSMF 100 km length.

Fully compensated eye diagrams, after transmission over the 320 km optically amplified multispan SSMF transmission link, are shown later in the chapter in Figure 10.47a, and the dispersion tolerance versus the BER is shown in Figure 10.47b. It is observed that the dispersion tolerance

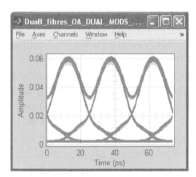

CSRZ 67% with 30%
bandwidth electrical filter.
5th order Bessel filter at receiver.
(a)

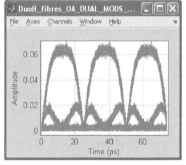

CSRZ 67% with 30%
bandwidth Gaussian electrical filter.
No filter at receiver.
(b)

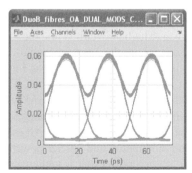

CSRZ 67% without electrical
filter at transmitter and
with filter at receiver.
(c)

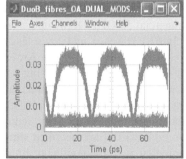

CSRZ 67% with 30%
bandwidth raise cosine electrical filter.
No filter at receiver.
(d)

FIGURE 10.45 Received eyes after (~three spans) 320 km SSMF fiber fully compensated including two in-line optical amplifiers (20 dB gain and 5 dB noise figure) 40 Gbps transmission. (a) CSRZ 67% duobinary with the integration of Gaussian electrical filter with 30% bit rate; (b) same as (a) but without filtering at the receiver; and (c) CSRZ 67% duobinary with raised-cosine filtering at transmitter and (d) no filter at the receiver.

is similar for variable-pulse-width DB signals. The 33% and 50% RZ-modified DB formats shown indicate error-free transmission with a residual distance of about 6 km DB format, while we could experimentally obtain error-free status only after 3 km SSMF residual dispersion for CSRZ 67% DPSK format.

Experimentally and under simulation, the launched power is adjusted, and the BER versus receiver sensitivity is obtained later in the chapter in Figure 10.47. Also included in this figure is the performance curve of the CSRZ-DPSK transmission. The BER curves indicate that 50% RZ and 67% NRZ DB formats outperform the CSRZ-DPSK by at least 2 dB. Although these DB performance curves are from simulation, we expect that the comparison with those of experimental CSRZ-DPSK would not be much different, as we have taken into account the electronic noise and the signal-dependent shot noises in the models of the photodetector. In both experimental and modeling cases, it is reasonable to assume that the polarization-mode dispersion effects are negligible. We expect that the 33% RZ DB performance would fall between these two formats. We have not had sufficient time to conduct detailed transmission results for this format, and these results will be reported in the near future.

It is noted that the model exactly represents the photonic behavior of lightwave-modulated modulation formats, that is, the square law direct detection incorporating the equivalent electronic noises of a 40 GHz receiver. Both signal-dependent shot noise and equivalent electronic noise current are modeled. The bandwidth of the receiver is adjusted according to the effective bandwidth of the modulation scheme. This allows us to justify the comparison of practical implementation of the direct detection and balanced receivers. The pulse sequence at the receiver is integrated, and an average power is obtained. Then its equivalent quantum shot noise is calculated over the electronic amplifier bandwidth. The equivalent electronic noise current at the input of the electronic amplifier is then added to these signal-dependent noises, and a signal-to-noise ratio is obtained. Thus, the received optical power can be derived and plotted with respect to the BER obtained from processing the eye diagram obtained after each transmission. The simulation duration is set long enough, so that the number of pulses transmitted and received is sufficiently high. Although we understand that the noise statistics of the phase-modulated optical transmission is asymmetrical and may follow a Maxwellian distribution, it is assumed in this chapter that the noise statistics follow a Gaussian distribution. This assumption would suffer only about 0.5–1.0 dB difference from the true distribution.

The DBM with alternating phase and control of the FWHM of the pulse sequence offer a 5 dB improvement as compared to that of the CSRZ-DPSK obtained experimentally. For the simulated DB 50% pulse format, a 1.2 dB improvement on receiver sensitivity is achieved as compared with CSRZ-DPSK. Equivalent noise current at the receiver of 2.5 pA/(Hz)$^{1/2}$ is used, which is compatible to that of a commercial 50 GHz DPSK optical receiver. The bandwidth of the receiver is adjusted to be narrower for 67% DB case, and hence less noises for both the quantum-signal-dependent shot noise and total equivalent electronic noise power. The 40 Gbps DB signals in the time domain and its spectrum are shown in Figure 10.46 using a filter of 17 GHz in the transmitter. The 3 dB bandwidth shows clearly the single-sideband and embedded carrier. The 50% FWHM with a Gaussian profile allows the possibility of lower bandwidth demand on the optical modulators [2]. Still, this format offers better performance due to the reduction of the signal bandwidth. All the performance curves shown in Figure 10.47 are obtained when the modulation and carrier power are operating under the linear region, with the nonlinear threshold set at a power that would create a total phase change of 0.1π. A 4.2 dB improvement of 67%, as compared with 50% DB, is due to the reduction of the electronic and quantum shot noise at the receiver.

10.3.5 Remarks

The effectiveness of DBM formats has been demonstrated over the transmission by simulation. MATLAB and Simulink models are employed for this simulation platform. The system performance is error free, with receiver sensitivity of −60 dBm without receiver electronic noise. The nondispersion

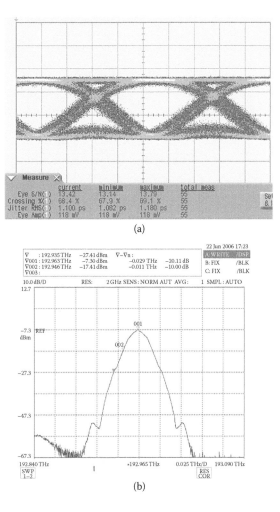

FIGURE 10.46 Duobinary signals: (a) eye diagram and (b) measured optical spectrum. (Courtesy of SHF Ltd., Berlin, Germany.)

compensation transmission can reach 9–10 km for 40 Gbps, equivalent to 150 km for 10 Gbps. The Simulink model is successfully developed. We have also demonstrated both experimentally and in modeling the CSRZ-DPSK transmission. These transmission performances are compared with the noncompensating 50 km SSMF transmission of DBM with a pulse width of 33% and 50%. The later format offers at least 2 dB improvement in terms of the BER and receiver sensitivity over that of CSRZ-DPSK. The 50% and 33% FWHM DB modulation schemes offer simpler driving circuitry for the optical modulators due to its lower swing voltage levels applied to the electrodes of the dual-drive MZIM. This is very important when the bit rate is in the multi-GHz region. Furthermore, the 50% and 33% pulse width and Gaussian profile will further reduce the demand on the bandwidth of optical modulators—that is, a 30 GHz or lower MZIM can be used. A Simulink model has also been developed for simulation of the transmission performance of the DBM formats. Simple driving conditions can be achieved, and its effectiveness demonstrated.

We have demonstrated, both experimentally and in modeling, the CSRZ-DPSK and DB transmissions. These transmission performances are compared with noncompensating 50 km SSMF transmission of DBM with pulse width of 50% and 67% (NRZ). The later formats with 50% and 67% pulse width, offer at least 5 dB and 1.2 dB improvement, respectively, in terms of the BER and receiver sensitivity over that of CSRZ-DPSK. The 67%, 50%, and 33% FWHM DB modulation

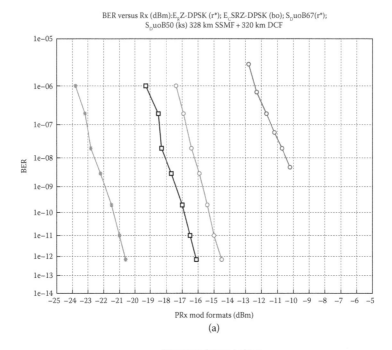

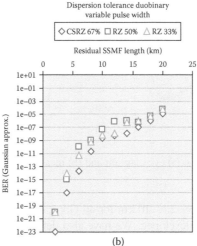

FIGURE 10.47 (a) BER versus receiver power for 320 km SSMF and 320 km (effective negative 320 km SSMF) DCF transmission: "o" middle curve: experimental CSRZ-DPSK transmission with complete compensation; "o" far right: experimental RZ-DPSK; "*" curve: simulated duobinary 67%; "□" square curve: simulated duobinary 50% FWHM alternating phase modulation. Note: 67% improves over 50% duobinary by 4.2 dB. (b) Simulated dispersion tolerance for 40 Gbps duobinary 67% (diamond) and 50% (square) and 33% (triangular) FWHM-BER versus residual equivalent-SSMF length. Launched power 0 dBm at the output of transmitter and input to the fiber transmission line.

schemes offer simpler driving circuitry for the optical modulators due to its lower swing voltage levels applied to the electrodes of the dual-drive MZIM. Simpler and narrower bandwidth of the electronic amplifier for 67% DB case would also improve the receiver sensitivity. We thus expect that 33% DB would worst than the case of CSRZ-DPSK, and hence no simulation is conducted for this scenario. This is very important when the bit rate is in the multi-GHz region. Furthermore, the 67%

and 50% pulse widths and Gaussian profile will further reduce the demand on the bandwidth of optical modulators—that is, a 30 GHz or lower MZIM can be used. Simulink models have also been developed for simulation of the transmission performance of the DBM formats. The balanced receivers are modeled exactly as a real practical subsystem. This allows a fair comparison with the transmission performance of the CSRZ-DPSK formats. Simple driving conditions can be achieved, and its effectiveness demonstrated.

Although we have not conducted the transmission of DB format pulse sequences of variable pulse width over the multispan optically amplified distance in the power threshold region at which the self-phase modulation effects may occur. It is expected that the CSRZ 67%, RZ 33%, and RZ 50% DB format would outperform their DPSK counterparts. It is also expected that the RZ 33% DB would suffer much less nonlinear distortion effects than its 50% and 67% RZ counterparts. These performances will be reported in the future. Similarly, the effects of the electrical filtering at the transmitter will also be investigated.

10.4 DWDM VSB MODULATION-FORMAT OPTICAL TRANSMISSION

This section presents the transmission of optical multiplexed channels of 40 Gbps using vestigial single-sideband modulation format over a long-reach optical fiber transmission system. The effects on the Q factor of fiber dispersion, the passband and roll-ff frequency of the OFs, and the channel spacing are described. The performance of the optical transmission using low and nonzero-dispersion fibers or/and dispersion compensation is presented. It has been demonstrated that a BER of 10^{-12} or better can be achieved across all channels, and that minimum degradation of the channels can be obtained under this modulation format. OFs are designed with asymmetric roll-off bands. Simulations of the transmission system are also given and compared for, channel spacing of 20, 30, and 40 GHz. It is shown that the passband of 28 GHz and 20 dB cutoff band performs best for 40 GHz channel spacing.

Even though optical communications have been extensively developed for ultra-high-capacity transport networks, the demand for high-speed communication systems over ultra-long-reach and ultra-long-haul offering greater capacity is expected to offer challenges for further technical development for bandwidth-efficient networking. The global Internet traffic has been growing rapidly, typically doubling the backbone traffic each year. This drives a requirement for higher channel speeds per Internet port. As Internet backbone routers are currently moving to 40 Gbps, 100 Gbps and even higher in the Tbps with flexible spectrum as the core long-haul networks. It is foreseen that if the information economy continues its growth unabated, then efficient transmission techniques for increasing the bit rates may be required in the near future.

The most common modulation formats for 40 Gbps optical systems are RZ, NRZ, and CSRZ, which can also be integrated with differential PSK coding [18–22]. The vestigial sideband (VSB) RZ modulation format can also be considered as the appropriate choice for Tbps long-haul optical transmission systems due to its half-band property, and hence are highly tolerant to dispersion and nonlinearity. On the contrary, VSB with NRZ could provide higher spectral efficiency than the VSB-RZ format since NRZ occupies only half of the RZ bandwidth and requires a lower peak transmit power in order to maintain the same energy per bit. This would offer the same BER as the RZ format.

This chapter/section gives a numerical simulation of the VSB-NRZ modulation format for long-haul optical fiber transmission systems and the effects of the OFs on its transmission performance. The OFs used for eliminating the unwanted optical sideband are critical in the generation of the desired format. The design of such filters is given so as to alter the passbands and roll-off bands of the filters to investigate their effects on the transmission system performance. Furthermore, the channel spacing of the multiplexed channels is also important and plays a major factor in the specification of OFs. Simulation is presented for eight channels DWDM of 40 GHz channel spacing optical fiber transmission systems. The channels are transmitted at 40 Gbps over a dispersion-managed 100 km span. The effects of these filters on back-to-back transmission and dispersion tolerance are the principal objectives of this work.

This section is thus structured as follows. Section 10.4.2 gives an overview of the simulation design and the transmission system. Sections 10.2.3 and 10.2.4 describe the optical transmitter and modulation format coding. The properties of the OF passband and the signal spectra of the VSB channels are also given. Section 10.4.5 presents the details of the design of the VSB filters and the DWDM, multiplexing, and demultiplexing. Section 10.2.6 discusses the roles of the group velocity dispersion (GVD) factor of the transmission and DCFs. Section 10.2.7 gives the evaluation of the transmission performance using the received eye diagram and the Q factor and the effects of several factors on the BER. Finally, concluding remarks are given.

10.4.1 TRANSMISSION SYSTEM

The schematic diagram of the optical transmission employing the VSB modulation format is shown in Figure 10.48. An optical source operating in the continuous mode is launched into an external optical modulator that is modulating the lightwave carrier via a random bit pattern generator and associate microwave power amplifiers. The external modulator, an X-cut LiNbO$_3$ MZ intensity modulator, is used to offer chirp-free modulated output. For 40 Gbps, the stabilization and linearity of the external modulator is critical. Normally, two modulators would be used, one for generating the required NRZ format, and the other for either carrier suppression or phase modulation, if required, depending on the transmission format.

NRZ data format are used in this simulation. The VSB modulation technique is used and generated by the use of an OF that filters the unwanted sideband of the information channel. After the filtering, a number of optical VSB channels can be multiplexed and then transmitted through a number of optically amplified dispersion-managed fiber spans. Each optically amplified dispersion-compensating in-line unit would consist of an in-line optical amplifier followed by a dispersion-compensating module, and then another optical amplification booster that would then enhance the total average optical power for transmission to the next span. Eye diagrams are obtained, and the BER can be deduced. For example, the back-to-back eye diagram is observed when dispersion compensation fibers are used with appropriate dispersion slope for equalization. Naturally, at the end of the transmission line, the multiplexed channels are separated via a demultiplexer. The principal objective of this chapter is to study the effects of the OF on the VSB system performance, and thus we do not include optical amplification noises in our modeling in the preamplifiers and booster amplifiers located at each transmission span. Both RZ and NRZ formats can be used. The main advantage of the NRZ format is that it occupies half of the RZ format's bandwidth. VSB modulation allows a small amount of the unwanted sideband existing at the output of the modulator. This depends on the roll of passband of the filter. Instead of eliminating the entire second sideband,

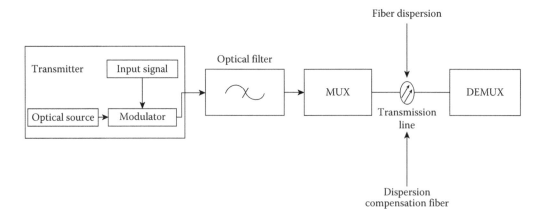

FIGURE 10.48 Schematic of the modeled VSB transmission system.

such as in the case of SSB, the VSB modulation suppresses most of, but not completely, the second sideband. Using this technique, the difficulty in generating a very sharp cutoff can be overcome. The VSB modulation can be implemented with OFs to eliminate most of the second sideband. The spectrum of a VSB signal can be obtained as [23]

$$x_c(t) = \frac{1}{2} AE \cos(\omega_c - \omega_1)t + \frac{1}{2} A(1 - E)\cos(\omega_c + \omega_1)t + \frac{1}{2} B\cos(\omega_c + \omega_2)t \qquad (10.10)$$

where E is the magnitude of the optical field envelope, ω_c is the optical carrier radial frequency, and ω_1 and ω_2 are the arbitrary radial frequencies of the signals. The signal can be demodulated by multiplying with $4\cos(\omega_c t)$ and applying the low-pass OF, leading to

$$E(t) = AE\cos\omega_1 t + A(1 - E)\cos\omega_1 t + B\cos\omega_2 t$$
$$e(t) = A\cos\omega_1 t + B\cos\omega_2 t$$
$$(10.11)$$

The basic characteristics of a VSB OF is illustrated in Figure 10.49. We note that the roll-off bands and passbands are asymmetric.

10.4.2 VSB Filtering and DWDM Channels

To achieve the VSB-modulation-formatted signals, OF is implemented. In this chapter, a number of low-pass elliptic filters (LPEF) are chosen because this filter type offers the steepest transition region between the passband and stopband without stability problems. The elliptic filter is a combination of the Chebyshev Type I and Chebyshev Type II, and exhibit some amplitude response ripples in the passband and stopband. The main advantage of the elliptic filter is that the width of the transition band is minimized for a finite ripple limit in the passband and a minimum attenuation

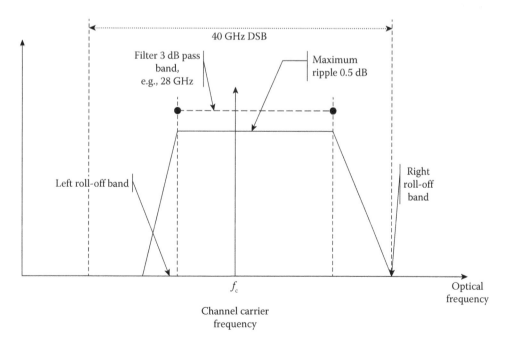

FIGURE 10.49 Spectral property of the VSB modulation format as compared to that of DSB (double sideband): and details of optical VSB filter passbands and roll-off bands.

in the stopband. Furthermore, these filters can be implemented using planar circuit technology [24]. The spectral response of the optical LPEF of the order N is given by [24]

$$|H_{\text{LP}}(j\omega)|^2 = \frac{1}{1 + \varepsilon^2 E_N{}^2(\omega)} \tag{10.12}$$

where $E_N{}^2(\omega)$ is the Chebyshev rational function and can be determined from the specified ripple characteristics. Similarly, the s-domain transfer function of the LPEF of order N can be obtained as

$$H_{\text{LP}}(s) = \frac{H_0}{D(s)} \prod_{i=1}^{r} \frac{s^2 + A_{0i}}{s^2 + B_{1i}s + B_{0i}} \tag{10.13}$$

where $s = j\omega$, $r = \dfrac{N-1}{2}$ for odd N, and $r = \dfrac{N}{2}$ for even N. The definitions of all the bands of the filters can be referred to in Figure 10.49. Figure 10.50 illustrates a typical response of the LPEF. The carrier wavelength of 1550 nm is used as the center wavelength corresponding to the optical frequency of 193.41 THz. The characteristic of the elliptic filter is designed with the following properties: passband ripple = 0.5 dB or less; minimum stopband attenuation = 10 dB; passband region = 28 GHz; stopband = 2 GHz; and 20 dB cutoff band = 10 GHz and 20 GHz for the left and right sides, respectively. The filter spectral characteristics are illustrated in Figures 10.50 and 10.51. The eight multiplexed signals at the output of an LPEF filter are plotted in Figure 10.52. This type of OFs can be implemented by using fibers Bragg gratings or multistage silica-on-silicon planar integrated MZDI. We note that the MZDI can also be considered as an al-pass (all-zero) filter that would offer no resonant peaks, and hence its group delay is almost constant and thus no dispersion would be introduced by these OFs. In this work, we would take into account the dispersion characteristics deduced from the group delay of the design filter, and this would be compensated for by the dispersion-compensating module.

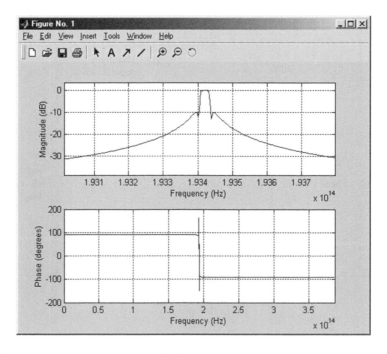

FIGURE 10.50 Frequency responses of the elliptic filter including the amplitude (upper curve) and phase (lower graph).

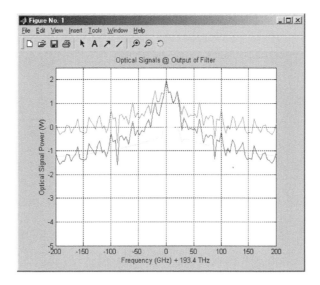

FIGURE 10.51 Filter characteristics (bottom) and signal spectra before filtering (most upper) and after filtering (middle curve). The filter frequency response is also included and indicated as the bottom graph.

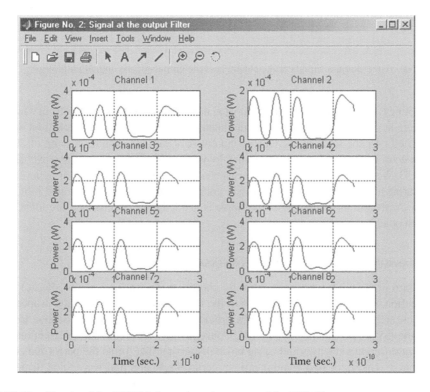

FIGURE 10.52 Signals of the DWDM channels at the output of the VSB filters.

The optical transmission system is simulated for eight wavelength channels in which the 1550 nm wavelength is taken as the center wavelength and the frequency spacing of 20, 30, and 40 GHz wavelength grid, and so on. For simplicity, we use 1550 nm as the center wavelength rather than the exact ITU spectral grid. After filtering out the unwanted sideband, the multichannels are then multiplexed and then propagated through the single-mode low nonzero dispersion-shifted fibers (NZ-DSFs).

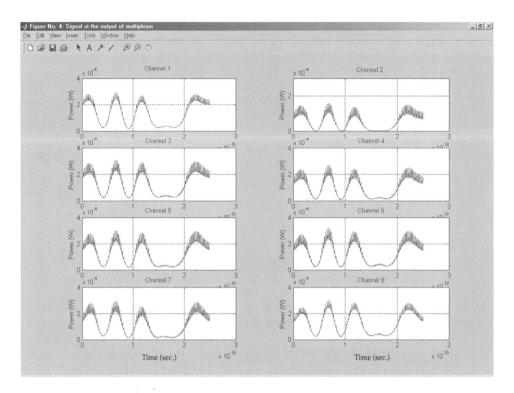

FIGURE 10.53 Time-domain signals of the channels at the outputs of the optical multiplexer.

The multiplexing is based on the wavelength division multiplexing (WDM) technique in which multiple optical carriers at different wavelengths are modulated using independent electrical bit streams and then transmitted over the same fibers. Our design employs variable channel spacing of 20–40 GHz. The signal spectra in the frequency and the time domain at the output of the multiplexer are depicted in Figure 10.53. We observe that there are some ripples in the temporal signals. It is believed that these ripples appearing at the output of the multiplexer are due to the crosstalks generated in multichannel transmission. At the output of the LPEF, the ripple is negligible as the channels are filtered independently to each other.

10.4.3 Transmission Dispersion and Compensation Fibers

SMFs, standard or NZ-DSF types, are naturally chosen as the transmission media in this simulation. This section briefly describes the design of the fibers so that their dispersion properties can be specified accurately instead of using the data provided by fiber manufacturers. Due to chromatic dispersion, the GVD is frequency dependent. Consequently, different spectral components of the pulse travel at different velocities and cause pulse dispersion that limits the performance of SMFs. Fiber dispersion consists of two components: material dispersion and waveguide dispersion. The transmission fibers and the matched DCFs as described in Ref. [25] with matched dispersion factors and dispersion slopes for complete compensation of the channels across the signal spectral band. A brief summary of the design can be explained as follows.

For the transmission fiber, the material $M(\lambda)$ and $D_w(\lambda)$ waveguide dispersion factors can be calculated by using

$$M(\lambda) = -\frac{\lambda}{c}\frac{d^2n}{d\lambda^2} \text{ and } D_w(\lambda) = -\frac{n_2(\lambda)\Delta}{c\lambda}V\frac{d^2(Vb)}{dV^2} \tag{10.14}$$

The waveguide-dependent factor can be approximated by when the V parameter limited in range of 1.3–2.6 micro-meters for SSMF whose waveguide and dispersion factors have been given in Chapter 2 as

$$V\frac{d^2(Vb)}{dV^2} = 0.080 + 0.549(2.834 - V)^2 \tag{10.15}$$

Hence, the total dispersion fibers is given by

$$D_T = M(\lambda) + D_w(\lambda) \tag{10.16}$$

In order to obtain nonzero low total dispersion of the SMMF and the DCF over a wide wavelength range, techniques for design of dispersion-flattened fibers are gain considered here, thus the repeat of a number of equations given earlier in Chapter 2. An optimized large-effective-area fiber with pure silica material for the cladding is designed for the transmission medium since it gives the best dispersion slope for matching with those of the dispersion compensation module. The Sellmeier constants for pure silica fibers are: $G_1 = 0.696750$ and $\lambda_1 = 0.069066e-6$; $G_2 = 0.408218$ and $\lambda_2 = 0.115662e-6$; and $G_3 = 0.890815$ and $\lambda_3 = 9.9900559e-6$. The refractive index of pure silica can be expressed as

$$n(\lambda) = c_1 + c_2\lambda^2 + c_3\lambda^{-2} \tag{10.17}$$

with $c_1 = 1.45084$, $c_2 = -0.00343$ μm^{-2}, and $c_3 = 0.00292$ μm^2, and hence the refractive index [$n(\lambda)$] for pure silica fiber is 1.45. For DCF these coefficients would be changed accordingly, even the index profile so that the wave guide and dispersion factors can meet the criteria for compensation over the whole spectral window of the multichannels. We have designed the transmission and dispersion-compensating optical fibers with a specific dispersion factor of 17 ps/nm/km and a dispersion slope of 0.3 ps/(nm²·km) for transmitting several channels. The dispersion and dispersion slope values of the dispersion-compensating fiber are designed with a factor of five times those of the transmission fiber, with a residual dispersion of about 1.5 ps/nm/km for the highest and lowest wavelength channels. These fiber parameters are used in this simulation: fibers radius $a = 1.6$ μm; relative refractive index difference $\Delta = 0.0339$. The dispersion factors of the centered channel (only channel 5 is illustrated) are indicated in Figure 10.54.

The pulse broadening after transmitting through a fiber of length L can be expressed as

$$\Delta\tau = D(\lambda) \cdot \sigma_\lambda \cdot L \tag{10.18}$$

where $D(\lambda)$ is the fiber's spectral dispersion (ps/(nm·km)); σ_λ is the signal bandwidth when the laser linewidth is much smaller than that of the VSB signals; and L is the fiber's transmission distance. Figure 10.54 shows the signals at the output of the demultiplexer for each channel (channels 1 through 8) without dispersion compensation.

From Figure 10.54, the simulation results show that over a 100 km span transmission line, the eye of the transmitted data closes completely, as expected, due to fiber dispersion. It is observed that the eye is still open at a BER of about 10^{-9} after 16 km designed fiber transmission. This is a significant improvement over the NRZ format.

We now turn to the DCFs. Since the GVD limits the system performance, it is essential to implement a DCF. The condition of dispersion compensation can be expressed as

$$D_1L_1 + D_2L_2 + \Gamma = 0 \tag{10.19}$$

where D_1 is dispersion factor of the transmission section (ps/(nm·km)); L_1 is the transmission distance (km); D_2 is the dispersion compensation distance (ps/(nm·km)); L_2 is the dispersion

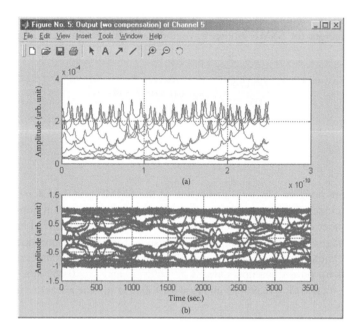

FIGURE 10.54 Signal outputs at the end of transmission fibers without dispersion compensation of the centered channel 5 at 1550 nm, with $D_T = 2.7455\text{e}{-}6$ ps/(nm · km): (a) time sequence and (b) eye diagram.

compensation distance (km); and Γ is the dispersion factor contributed by the VSB OF. Obviously, Equation 10.10 shows that the DCF must have its GVD negative at 1550 nm to compensate for the positive dispersion of the transmission fibers and the OF. The parameters of the DCF can be optimized for minimum nonlinear self-phase-modulation effect with: fiber radius $a = 1.7$ µm; relative refractive index difference $\Delta = 0.0243$. Figure 10.55 shows the transmitted data at the end of the dispersion-managed system with a dispersion compensation for channel 5 (of the eight channels). This shows the effectiveness of dispersion compensation fibers in the improvement of the BER because the received eye diagram is quite close to its original pattern as obtained at the output of the transmitter. The residual ripple of the compensated pulses clearly shows the effects of the filter passband on the time-domain pulses. These ripples contribute the penalty on the eye closure, and hence the Q factor. An optical multiplexer then combines the modulated lightwaves to the transmitting fibers. At the receiver, the channels are separated into different signals by an optical demultiplexer. Figure 10.54 gives the simulation results of transmitted signals at the output of the multiplexer without dispersion compensation, while Figure 10.55 shows the transmitted signals at the output of multiplexers with DCF in cascade with the transmission fibers. We note that the optical amplifiers at the input and booster amplifiers at the output are not included in this simulation.

10.4.4 Transmission Performance

The eye diagram is used to deduce the Q factor and hence the system BER. The random digital optical pulse sequence suffers distortion by noise, pulse broadening and timing jitter errors introduced in the optically amplified fiber link. Assuming that the probability density function of "1"and "0" are equal and Gaussian, the eye diagram can be used to estimate the Q factor by

$$Q = \frac{\mu_1 - \mu_0}{\sigma_1 - \sigma_0} \qquad (10.20)$$

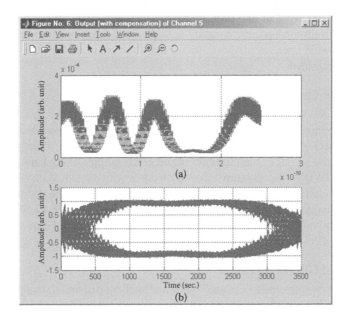

FIGURE 10.55 Signal outputs at the end of the transmission line with dispersion compensation of the center channel—channel 5 at 1550 nm with DCF dispersion factor of −2.7918e−6 ps/(nm·km): (a) temporal sequence and (b) eye diagram.

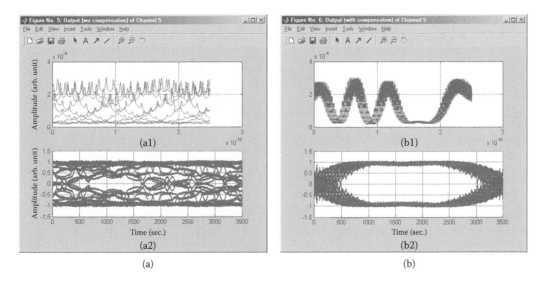

FIGURE 10.56 Signals received after transmission: (a) without dispersion compensation and (b) with dispersion compensation. (a1) and (b1) are the temporal sequence, and (a2) and (b2) are the corresponding eye diagrams.

where μ_1 and μ_0 are the means of the current at the output of the photodetector of the decision circuitry of the receiver at the sampling instant, respectively, for symbols "1" and "0." While σ_1 and σ_0 are the standard deviations of the current at the decision circuit input at the sampling instant, for the symbols "1" and "0", respectively.

Figure 10.56a and b shows the eye diagram of the transmitted pulse random sequence at 1550 nm (channel 5) with 40 GHz channel spacing, which depicts the eye diagram at the end of the transmission line without and with dispersion compensation fibers, respectively. The eye diagram of data with dispersion compensation is wide open compared to that without dispersion compensation. The wide-open eye diagram shows that the system performance is good, since the error has been compensated.

10.4.4.1 Effects of Channel Spacing on Q Factor

The capacity of the WDM system depends on how close channels can be packed into the wavelength domain. The minimum channel spacing is limited by interchannel crosstalk. The typical value of channel spacing should exceed four times that of the bit rate. The Q factor is obtained for each channel at the output of the multiplexer and at the output of the demultiplexer, at the end of the 16 channels DWDM over the transmission distance with a residual dispersion equivalent to 10km SSMF as depicted in Figure 10.57. The Q factor at the output of the demultiplexer decreases significantly for a channel spacing of 20–30 GHz. While the Q factor for 40 GHz channel spacing decreases only slightly. This indicates the resilience of the VSB modulation format to the chromatic dispersion.

10.4.4.2 Effects of GVD on Q Factor

The effects of GVD on transmitted channels with and without dispersion compensation fibers are evaluated with the Q factor as the reference performance parameter. Dispersion compensation fibers play an important role in optical transmission, since the GVD limits the transmission performance.

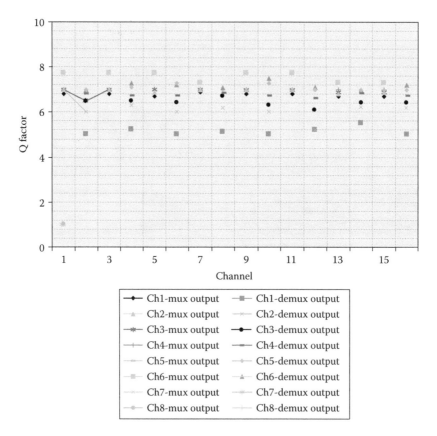

FIGURE 10.57 Q factor for the multiplexed channels 1–16 at the output of the optical mux (before transmission) as a function of the Channel number over the C band. The Q factor (or eye diagram-eye opening quality) is monitored at the output of demux (after transmission distance). Horizontal axis—channel number.

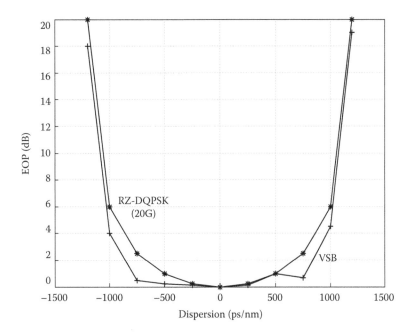

FIGURE 10.58 Dispersion tolerance of VSB format at 40 Gbps (+) transmission as compared with 20 GBaud/s RZ-DQPSK (*). Eye-opening penalty version dispersion factor at BER of 1e–9 of channel 5.

By adding dispersion compensation fibers that have the opposite sign of the GVD, the effect of GVD can be considerably reduced. Figure 10.58 shows the dispersion tolerance of the transmission system—the DCF has significantly improved the Q factor at the receiver to an equivalent BER close to an error-free 10^{-9} at a bit rate of 40 Gbps. Shown also in this figure is the dispersion tolerance of RZ-DQPSK of 20 GBaud/s. The performance of the two modulation formats is very closely similar. However, the VSB format may offer the advantage of a simplified transmitter with additional OF rather than another parallel transmitter arm for generation of the quadrature constellation in addition to the in-phase components [20].

10.4.4.3 Effects of Filter Passband on the Q Factor

The effect of VSB filtering on the Q factor and its dispersion tolerance can be examined by varying the passband characteristic of the VSB filter and measuring the Q factor for each lightwave channel at the output of the demultiplexer, as shown in Figure 10.59. It is clear that, for the 40 Gbps NRZ data format, the system penalty is one unit of the Q factor, which is equivalent to one decade of BER or about 10 dBQ when the passband of the OF is extended from 20 to 24 GHz. This penalty could be due to the cutoff of the signal band by the roll-off band of the OF with more than half of the bandwidth eliminated. For 28 GHz passband, the Q factor increases considerably to 7.8, or BER = 10^{-15}—that is, error-free transmission when the sampling is at the center of the eye. This shows that the performance of the VSB modulation format is better than doublesideband (DSB) modulation, since the VSB format eliminates most but not all of the redundant sideband.

When DCFs and pure gain optical amplifiers are incorporated in front and after the DCF is used, the transmission distance can be extended to 15 spans with a BER of 10^{-12} for all 16 wavelength channels.

Figure 10.60 shows the BER versus the receiver sensitivity of the VSB experiment simulation and measured experimental values of transmission using CSRZ-DPSK [25]. We observe that the OFs

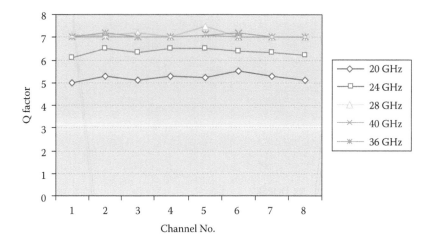

FIGURE 10.59 Variation of the Q factor as a function of the VSB filter frequency passband.

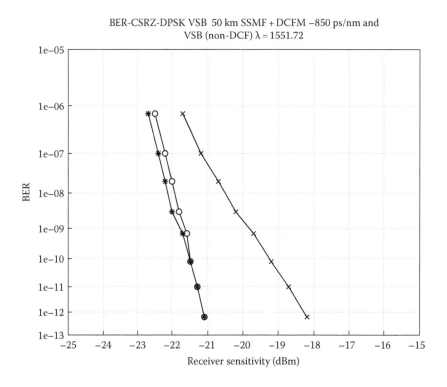

FIGURE 10.60 VSB experiment simulation and CSRZ-DPSK experimental values. (o) VSB with Chebyshev-type optical filter; (*) VSB raised-cosine optical filter having the same cutoff passband as that of the Chebyshev filter; (+) CSRZ-DPSK (experiment).

with a raised-cosine shape and that of the Chebyshev type would offer nearly the same transmission performance of BER versus receiver sensitivity and about 2 dB at a BER of 10^{-9} poorer than the CSRZ modulation format. This is reasonable due to the slow roll-off of the cutoff band of the filter. The advantage of the VSB modulation format of the simpler transmitter structure can compensate for this penalty.

10.5 SINGLE-SIDEBAND MODULATION

10.5.1 Hilbert Transform SSB MZ Modulator Simulation

The Hilbert transform of signal $m(t)$ is defined to be the RF signal whose frequency components are all phase shifted by $\pi/2$ radians (Figures 10.61 and 10.62).* Therefore

$$\hat{m}(t) = H\{m(t)\} \rightarrow m(t) = A\cos 2\pi f_o t \text{ then } \hat{m}(t) = A\cos\left(2\pi f_o t - \frac{\pi}{2}\right) \quad (10.21)$$

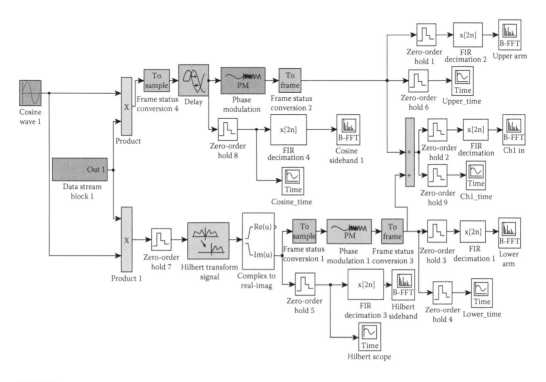

FIGURE 10.61 SSB Hilbert transform phase shift modulator.

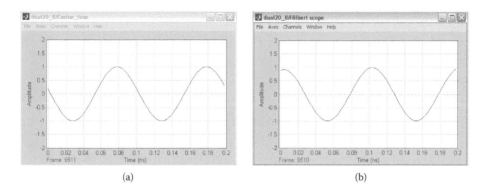

(a)   (b)

FIGURE 10.62 (a) Input RF signal time scope (cosine signal) and (b) Hilbert transform the input RF signal $H\{x(t)\} = A\cos(2\pi f_o t - \pi/2) = A\sin(2\pi f_o t)$. *(Continued)*

---

* Hilbert transform single sideband, *Digital and Analog Communication Systems*, 6th edition, p. 312.

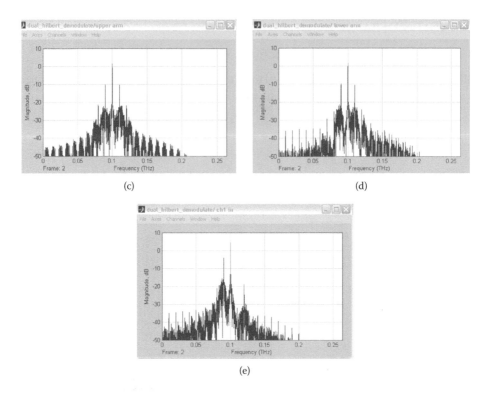

FIGURE 10.62 (Continued) (c) upper arm of dual-electrode MZ modulator with delay 10 ns in Hilbert transform phase shift system; (d) lower arm of dual-electrode MZ modulator with Hilbert transform phase shift $\theta = \pi/2$; and (e) dual-electrode MZ modulator single-sideband system spectrum result with lower SB cancels out.

Taking the Fourier transform, we have

$$M(f) = -j\,\mathrm{sgn}(f)\frac{A}{2}\big[\delta(f+f_{\mathrm{o}}) + \delta(f-f_{\mathrm{o}})\big] = \frac{A}{j2}\big[-\delta(f+f_{\mathrm{o}}) + \delta(f-f_{\mathrm{o}})\big] \qquad (10.22)$$

$$\text{Thus, } \hat{m}(t) = A\cos\left(2\pi f_{\mathrm{o}}t - \frac{\pi}{2}\right) = A\sin 2\pi f_{\mathrm{o}}t \qquad (10.23)$$

10.5.2 SSB Demodulator Simulation

The modulated signal is

$$u(t) = A_{\mathrm{c}}/2\left\{\cos\big(\omega_{\mathrm{c}}t + \gamma\pi + \alpha\pi\cos\omega_{\mathrm{rf}}t\big) + \cos\big(\omega_{\mathrm{c}}t + \alpha\pi\cos(\omega_{\mathrm{rf}}t+\theta)\big)\right\} \qquad (10.24)$$

The demodulated signal is (Figures 10.63 and 10.64)

$$s(t) = A_{\mathrm{c}}\cos 2\pi f_{\mathrm{c}}t * u(t) \qquad (10.25)$$

The low-pass filter is used to filter high-frequency components in the signal. The only component left is the modulating signal (10 Gbps [binary * cosine signal]) after demodulation. It is required to

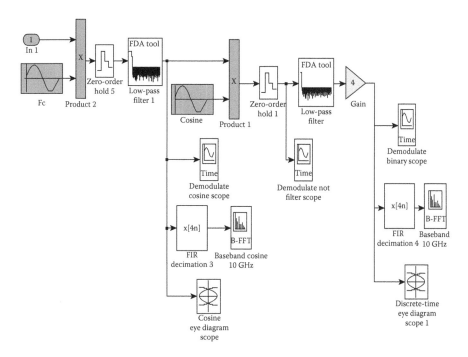

FIGURE 10.63 Single-sideband demodulator platform.

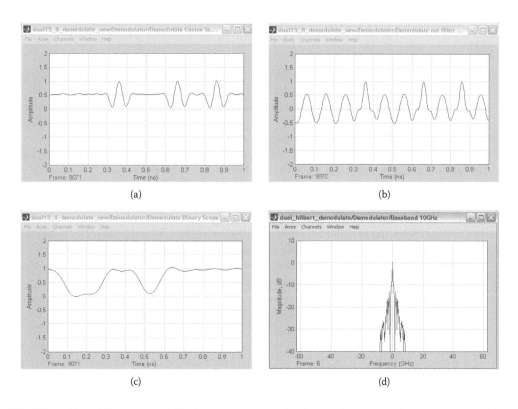

FIGURE 10.64 (a) Demodulated Fc signal time scope; (b) demodulate cosine signal time scope before low-pass filter (c) demodulated signal time scope after low-pass filter, and (d) demodulated signal spectrum.

(Continued)

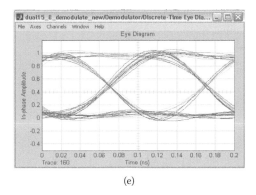

(e)

FIGURE 10.64 (Continued) (e) demodulated signal eye diagram NRZ-SSB with Q = 8 and BER = 10^{-15}.

multiply the demodulated signal by in-phase cosine signal and also there is high-frequency signal, so that LPF is required to filter the high-frequency components in the signal.

10.6   CONCLUDING REMARKS

We have presented the simulation results of 40 Gbps DWDM optical transmission systems with NRZ-VSB modulation format. The effects of filtering on the Q factor with respect to fiber dispersion and compensation, channel spectral spacing, and the effects of symmetry and asymmetry of the filter pass and cutoff bands are examined. We can draw the following conclusions:

1. The Q factor at the output of the demultiplexer decreases significantly for channel spacing of 20–30 GHz due to noise and crosstalk interference. However, the Q factor decreases slightly for 40 GHz channel spacing. A BER of 10–12 can be achieved over all channels with this channel spacing. For dense and superdense WDM optical transmission, the demands on the roll-off, cutoff, and passband of VSB OFs are high.
2. Fiber dispersion can reduce the system performance dramatically. Using dispersion compensation fibers along the transmission line will improve the system performance significantly. Specific designed fibers with deterministic dispersion are used.
3. The 20 and 24 GHz passband OFs give low performance. The low performance could be caused by the unstable system since about half of the bandwidth is eliminated. While, for 28 GHz passband, the Q factor is considerably improved to 7.5. Thus, the performance of VSB modulation format is comparable and marginally enhanced as compared to that of the double-sideband NRZ-amplitude shift keying modulation format since the VSB format eliminates most, but not all, of the redundant sideband.
4. Several types of OFs are examined, and the sharp roll-off band is expected to contribute to the improvement of the VSB transmission. However, the roll-off band must be at least 20 dB below the power level of the center of the spectrum of the signal and carrier frequency to avoid the pulse ripples.
5. The all-zero all-pass OFs can be designed and implemented in planar lightwave circuit technology to substitute for the two-poles elliptic filters to eliminate the filter contribution to the total dispersion of the transmission system.

Nonlinear effects such as the self-phase modulation, cross-phase modulation, Raman scattering, and patterning effects are not considered in this chapter.

REFERENCES

1. S. K. Kim, J. Lee, and J. Jeong, Transmission performance of 10-Gbps optical DB transmission systems considering adjustable chirp of non-ideal LiNbO$_3$ MZIMs due to applied voltage ratio and filter bandwidth, *IEEE J. Lightwave Technol.*, Vol. 19, No. 4, pp. 465–470, 2004.

2. H. Gnauk and P. J. Winzer, Optical phase-shift-keyed transmission, *IEEE J. Lightwave Technol.*, Vol. 23, No. 1, pp. 115–118, 2004.

3. S. Zhang, *Advanced Optical Modulation Formats in High Speed Lightwave System*, Lawrence, KS: University of Kansas, pp. 17–32, 2004.

4. K. Yonenaga, S. Kuwano, S. Norimatsu, and N. Shibata, Optical DB transmission system with no receiver sensitivity degradation, *Electron. Lett.*, Vol. 31, No. 4, pp. 302–304, 1995.

5. K. Yonenage and S. Kuwano, Dispersion-tolerant optical transmission system using DB transmitter and binary receiver, *IEEE J. Lightwave Technol.*, Vol. 15, No. 8, pp. 1530–1537, 1997.

6. T. Franck, P. B. Hansen, T. N. Nielsem, and L. Eskildson, DB transmitter with lower intersymbol interference, *IEEE Photon. Technol. Lett.*, Vol. 10, No. 4, pp. 597–599, 1998.

7. F. Elrefaie, R. E Wagner, D. A. Atlas, and D. G. Daut, Chromatic dispersion limitations in coherent lightwave transmission system, *IEEE J. Lightwave Technol.*, Vol. 6, No. 5, pp. 704–709, 1988.

8. T. Ono, Y. Yuno, K. Fukuchi, T. Ito, H. Yamazaki, M. Yamaguchi, and K. Emura, Characteristics of optical DB signals in terabit/s capacity, high-spectral efficiency WDM systems, *IEEE J. Lightwave Technol.*, Vol. 16, No. 5, p. 7, 1998.

9. Y. Miyamoto, M. Yoneyama, T. Otsuji, K. Yonenaga, and N. Shimizu, 40 G-bits/s TDM transmission technologies based on ultra-high-speed IC's, *IEEE J. Solid-State Circuits*, Vol. 34, No. 9, pp. 1246–1253, 1999.

10. O. Sinkin, J. Zweck, and C. R. Menyuk, Comparative study of pulse interactions in optical fiber transmission systems with different modulation formats, *Opt. Express*, Vol. 9, No. 7, pp. 339–352, 2001.

11. M. Yoneyama, K. Yonenaga, Y. Kisaka, and Y. Miyamoto, Differential precoder IC modules for 20- and 40-Gbits/s optical DB transmission system, *IEEE Trans. Microwave Theory Tech.*, Vol. 47, No. 12, pp. 2263–2270, 1999.

12. K. P. Ho and J. M. Kahn, Spectrum of externally modulated optical signals, *IEEE J. Lightwave Technol.*, Vol. 22, No. 2, pp. 658–663, 2004.

13. S. Haykin, *Communication Systems*, 4th ed., John Wiley & Sons, New York, USA, 2001.

14. S. Ramachandran, B. Mikkelsen, L. C. Cowsar, M. F. Yan, G. Raybon, L. Boivin, and M. Fishteyn, All-fiber grating-based higher order mode dispersion compensator for broad-band compensation and 1000-km transmission at 40 Gb/s, *IEEE Photon. Technol. Lett.*, Vol. 13, No. 6, pp. 632–634, 2001.

15. Y. Kim, J. Lee, Y. Kim, and J. Jeong, Evaluation of transmission performance in cost-effective optical duobinary transmission utilizing modulator's bandwidth or low-pass filter implemented by a single capacitor, *Opt. Fib. Technol.*, Vol. 10, No. 4, pp. 312–324, 2004.

16. D. Boivin, M. Hanna, and J. R. Barry, Reduced-bandwidth duobinary differential continuous-phase Modulation format for optical communications, Photonics Technology Letters, *IEEE Photon. Technol. Lett.*, Vol. 17, No. 6, pp. 1331–1333, 2005.

17. K. P. Ho, *Phase-Modulated Optical Communications Systems*, Berlin, Germany: Springer Verlag, p. 212, 2005.

18. N. S. Bergano et al., Chirped return to zero formats for ultra-long-haul fibers communications, in *2004 IEEE Workshop on Advanced Modulation Formats*, San Francisco, CA, 2004.

19. T. Ohm, Comparison of different DQPSK transmitters with NRZ and RZ impulse shaping, in *IEEE Workshop on Advanced Modulation Formats*, San Francisco, CA, 2004.

20. M. Serbay et al., Comparison of six different RZ-DQPSK transmitter set-ups regarding their tolerance towards fibers impairment in 8x40 Gbps WDM-systems, in *IEEE Workshop on Advanced Modulation Formats*, San Francisco, CA, 2004.

21. Y. Zhu et al., Highly spectral efficient transmission with CSRZ-DQPSK, in *IEEE Workshop on Advanced Modulation Formats*, San Francisco, CA, 2004.

22. T. Tokle et al., Transmission of RZ-DQPSK over 6500 km with 0.66 bits/Hz spectral efficiency, in *IEEE Workshop on Advanced Modulation Formats*, San Francisco, CA, 2004.

23. R. E. Ziemer, W. H. Tranter, and D. R. Fannin, *Signal and Systems: Continuous and Discrete*, New York: Macmillan, 1989.

24. L. N. Binh et al., *Photonic Signal Processing—Part II.2: Tuneable Photonic Filters using Cascaded All-Pole Micro-Rings and All-Zero Interferometers*, Technical Report code MECSE-12-2005, Clayton: Monash University, 2005.

25. L. N. Binh, T. L. Huynh, K.-Y. Chin, and D. Sharma, Design of dispersion flattened and compensating fibers for dispersion-managed optical communications systems, *Int. J. Wireless Opt. Commun.*, Vol. 1, pp. 1–21, 2004.

11 OFDM Optical Transmission Systems

This chapter deals with optical transmission employing orthogonal frequency division multiplexing (OFDM) techniques in which several subcarriers are employed to carry partitioned data for transmission over the fiber channel. The channels are frequency multiplexed and orthogonal. The orthogonality comes easily from the generation of the fast Fourier transform (FFT) and inverse fast Fourier transform of the data sequence in the electronic domain. These subcarriers in the baseband are then modulating the optical carrier under some modulation formats that are most appropriate for optical fiber channels. The narrow band of the data of subchannels ensure the low broadening of the pulses, hence combating effectively the impairments expected from the fiber. This advanced modulation and optical transmission technique (oOFDM) is emerging over the last few years as one of the most potential candidates for long-haul optically amplified communication systems and networks and for low-cost metro and access optical networks.

The main reasons are that IFFT and FFT can be implemented in the electronic digital domain at low cost due to availability of ultra-high sampling rate Analog-to-digital converter (ADC) and digital signal processing (DSP), hence the frequency shifting of subcarriers of the OFDM signals.

11.1 INTRODUCTION

OFDM is a transmission technology that is primarily known for wireless communications and wired transmission over copper cables. It is a special case of the widely known frequency division multiplexing (FDM) technique for which digital or analog data are modulated onto a certain number of carriers and transmitted in parallel over the same transmission medium. The main motivation for using FDM is the fact that due to parallel data transmission in frequency domain, each channel occupies only a "small" frequency band. Signal distortions originating from frequency-selective transmission channels, the fiber chromatic dispersion, can be minimized. The special property of OFDM is characterized by its very high spectral efficiency. While for conventional FDM, the spectral efficiency is limited by the selectivity of the bandpass filters required for demodulation, OFDM is designed such that the different carriers are pairwise orthogonal. This way, for the sampling point the intercarrier interference (ICI) is suppressed although the channels are allowed to overlap spectrally.

11.1.1 PRINCIPLES OF oOFDM: OFDM AS A MULTICARRIER MODULATION FORMAT

11.1.1.1 Spectra

OFDM is a multicarrier transmission technique based on the use of multiple frequencies to simultaneously transmit multiple signals in parallel form. Each subchannel is assigned with a subcarrier which is within the range allowable for an optical channel within the DWDM Optical transmission system. Unlike the normal FDM technique in which the spectra of the subchannels are separated by a guard band such that there are no overlapping between them, in OFDM the term orthogonality comes from the property that adjacent channels are orthogonal, that is, they are perpendicular or the dot product of the channel is zero as shown in Figure 11.1a and b, respectively. This allows the overlapping of the spectra of adjacent channels without creating any cross talks between them [1].

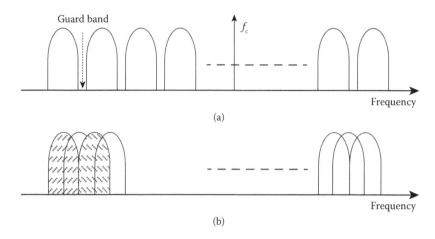

FIGURE 11.1 Multicarrier modulation technique: (a) spectrum of FDM subcarriers and (b) spectrum of OFDM subcarriers.

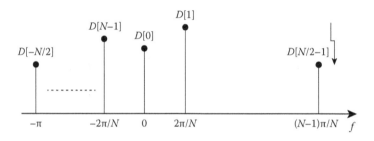

FIGURE 11.2 Frequency-domain representation of an OFDM symbol.

The frequency-domain representation of an OFDM symbol in the sampled discrete plane is shown in Figure 11.2. Thus, a schematic arrangement of the generation/mapping of OFDM symbols and reception/demapping can be shown in Figure 11.3.

The entire bandwidth allocated for a wavelength channel may be occupied. The data source is distributed over all subcarriers. Thus, each subcarrier transports a small amount of the information. Thus, by this lowering of the bit rate per subcarrier channel, the ISI due to the distortion of the channel can be significantly reduced.

11.1.1.2 Orthogonality

The complex baseband OFDM signal $s(t)$ can be written as

$$s(t) = \sum_{k} \sum_{n=0}^{N-1} a_n(k) g_n(t - kT) \tag{11.1}$$

where k is the time index, N is the total number of subcarriers, $T = NT_s$ is the stretched OFDM symbol period as a result of the conversion from serial to parallel width, and T_s is the sampling period. $a_n(k)$ is the kth data symbol of the nth subcarrier and $g_n(t)$ is the baseband data pulse given by

$$g_n(t) = e^{j2\pi f_n} g(t)$$
$$f_n = f_c \pm n\Delta f \tag{11.2}$$

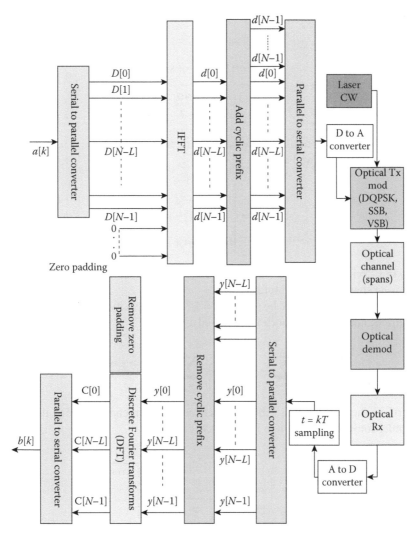

FIGURE 11.3 Block diagram of the sampled discrete time model of OFDM system using N-point FFT and IFFT.

$g(t)$ is the pulse-shaping function, which normally is very close to Gaussian after propagating through the optical MZIM.

The orthogonality of the channels requires

$$\int_{-\infty}^{\infty} f_n(t) f_m^*(t)\, dt = \frac{1}{T} \int_{-T/2}^{T/2} f_n(t) f_m^*(t)\, dt \begin{cases} 1 & n = m \\ 0 & n \neq m \end{cases} = \delta_0[n - m] \tag{11.3}$$

with the asterisk * indicating the complex conjugate. Thus, it requires that the selection of the carrier frequency and the pulse-shaping function in such a way that the orthogonality can be achieved.

11.1.1.3 Subcarriers and Pulse Shaping

For orthogonality, each subcarrier should take an integer number of cycles over a symbol period T, and the number of cycles between adjacent subcarriers differs by exactly unity. That means that the frequencies of the subcarriers are multiple of each other.

The pulse shapes of the signals can take the form of rectangular or sinc function and can be written as

$$g(t) = \frac{\dfrac{4\alpha t}{T}\left\{\cos(1+\alpha)\dfrac{\pi t}{T} + \sin(1+\alpha)\dfrac{\pi t}{T}\right\}}{\dfrac{\pi}{T}\left(1 - \left[\dfrac{4\alpha t}{T}\right]^2\right)} \tag{11.4}$$

where α is the roll-off factor, similar to the raise cosine function.

For the rectangular pulse, we have

$$g(t) = \begin{cases} 1 \; |t| \le \dfrac{T}{2} \\[2ex] 0 \; \text{otherwise} \end{cases} \tag{11.5}$$

However, this "brick wall" pulse shape would be very difficult to be realized in practice, and the raised-cosine shape would be preferred.

The orthogonality can be achieved by setting $\Delta f = 1/T$. For simplicity, we can assume that the signals for all sub-carriers, take the shape function expressed as

$$s(t) = \sum_{n=0}^{N} a_n g_n(t - kT) = \sum_{n=0}^{N} a_n g(t) e^{j2\pi(f_c + n\Delta f)} \tag{11.6}$$

where "$a_n g(t)$" is the data stream fed into each sub carrier channel from the output of the serial-to-parallel converter as shown in Figure 11.4.

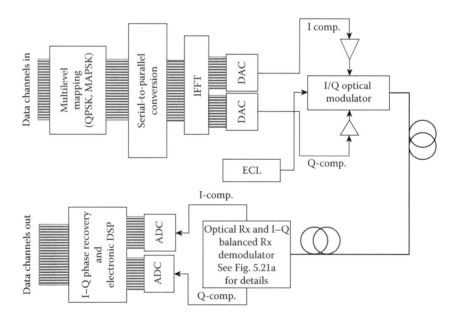

FIGURE 11.4 Schematic diagram of optical OFDM transmitter of a long-haul optical transmission system using OFDM and I–Q multilevel modulation formats. Separate polarized channels PDM are not shown. I = in-phase, Q = quadrature phase, ECL = external cavity lasers, DSP = digital signal processing, IFFT = inverse fast Fourier transform.

Taking the Fourier transform of (6) we have

$$S(f) = \sum_{n=0}^{N} a_n G(f - (f_c - n\Delta f)) \qquad (11.7)$$

which is the summation of all sinc functions if the pulse is rectangular shaped. These spectral sinc functions would cross over at the null of each other and their peaks. On the other hand if the temporal function of the data pulses for the sub-carrier channels follow a $\sin(t)/t$ profile, then the frequency spectrum of the channel would be rectangular and thus they can be separated with minimum crosstalk. Thus, we can have

$$g(t) = \frac{1}{T} \mathrm{sinc}\left(\pi \frac{t}{T} \right) = \Delta f \, \mathrm{sinc}(\pi \Delta ft) \to G(f) = \mathrm{rect}\left(\frac{f}{\Delta f} \right) \qquad (11.8)$$

with $\Delta f = 1/T$.

Therefore, the spectrum of OFDM signal takes the form

$$S(f) = \sum_{n=0}^{N-1} a_n \mathrm{rect}\left(\frac{f - f_c + n\Delta f}{\Delta f} \right) \qquad (11.9)$$

which is made up by a number of rectangular spectra separated by Δf.

In practice, the availability of rectangular time pulse shape would not be possible but with certain raised-cosine features, while for sinc shape pulses can be generated without difficulty by using a transmission filter in the electrical domain. The spectrum of OFDM signal is thus the summation of the sinc-shaped spectra of all the subchannels. If the number of subcarrier channels is sufficiently high, then one would obtain a flat rectangular spectrum [2].

The orthogonal subchannels can be generated using IFFT or FFT normally available from digital signal processors.

11.1.1.4   OFDM Receiver

In a conventional FDM system, the subchannels are normally separated by filtering using banks of filters and demodulators. For orthogonal signals, the subchannels can be separated by correlation techniques [3,4].

11.1.2   FFT- AND IFFT-BASED OFDM PRINCIPLES

To recover or demodulate the subcarrier channels, it is possible to use a bank of filters or alternatively use a local oscillator at the receiver to tune to the sub-channels in order to separate the subchannels. The use of a bank of filters is very costly and processing by DSP under FFT would ease the decoding. Recently, the speed of DSP has increased significantly. Certainly, the FFT and IFFT are the main features of these processors. The main feature of the FFT and IFFT is that the sampling and their sampled transformed subchannels are orthogonal. Thus, they can be effectively used in the transmitter and the receiver of OFDM systems. The OFDM discrete model is shown in Figure 11.4. Thus, the signal is defined in the frequency domain at the transmitter. It is a sampled digital signal and is defined such that the discrete Fourier spectrum exists only at the discrete frequencies. Each OFDM subchannel corresponds to one element of this discrete Fourier spectrum. The magnitude and phase of the carrier depends on the data transmitted. The serial-stream input data sequence is converted into parallel that would then be represented as the components of the frequency spectra, $D[n]$, with equally spaced frequencies $f_n = 2\pi n/N$ for $n = 0,1,\ldots,N$. Thus, $D[0]$ represents the DC component of the frequency spectrum of OFDM signals. The frequency representation of the signal are shown in Figures 11.1 and 11.2.

An N-point inverse FFT block (the DSP Processor) then generates the time-domain components whose prefix indexes are inserted at the transmitter side and then removed at the receiving side, as shown in Figure 11.3. Thus, the signal at this stage is in analog form, as shown in Figure 11.4, padded with zeros for a cyclic discrete Fourier transform. They are then converted from parallel form into serial form and used for modulating the optical modulator. The passband optical signals are then transmitted over the channel.

The definition of the N-point inverse discrete Fourier transform (IDFT) is

$$d[k] = \text{IDFT}\{D[n]\} = \frac{1}{N}\sum_{n=0}^{N-1} D[n]e^{j2\pi k/N} \qquad n,k=0,1...N-1 \tag{11.10}$$

where N is the symbol length. After the pulse shaping of a rectangular function of

$$\text{rect}\left[\frac{k}{N}\right] = \begin{cases} 1 & \text{for } k=0,1...N-1 \\ 0 & \text{otherwise} \end{cases} \tag{11.11}$$

The sequence $d[n]$ in the time domain is given by

$$d[k] = \frac{1}{N}\sum_{n=0}^{N-1} D[n]e^{j2\pi k/N}\, \text{rect}\left[\frac{k}{N}\right] \qquad n,k=0,1...N-1$$
$$= \frac{1}{N}\sum_{n=0}^{N-1} D[n]g_n[k] \tag{11.12}$$

with $g_n[k] = e^{j2\pi k/N}\text{rect}\left[\dfrac{k}{N}\right]$

$g_n[k]$ can be considered as the baseband pulse, which is considered to be orthogonal and satisfying the condition.

$$g_n[k] = e^{j2\pi k/N}\text{rect}\left[\frac{k}{N}\right] \tag{11.13}$$

Thus, a rectangular or sinc pulse shaper can be inserted between the serial-to-parallel and the IFFT blocks, and then multiplied by the subcarrier frequency and then superimposed and multiplied by a factor $1/N$.

At the receiver, after the photodetection and electronic preamplification, the noisy time-domain OFDM samples are converted to its corresponding frequency-domain symbols by an N-point FF. The coefficients of an N-point discrete Fourier transforms (DFT) are given by

$$C[n] = \text{DFT}\{y[k]\} = \sum_{n=0}^{N-1} y[k]e^{j2\pi k/N} \qquad n,k=0,1...N-1$$
$$= \sum_{-\infty}^{+\infty} y[k]e^{j2\pi k/N}\text{rect}\left[\frac{k}{N}\right] \tag{11.14}$$

An N-point DFT is equivalent to a bank of N-orthogonal "matched filters" matched to the corresponding "baseband pulses" in IDFT, followed by samplers that sample once per N symbols. The impulse of the matched filter can be written as

$$g_n^*[-k] = e^{j2\pi k/N}\text{rect}\left[\frac{-k}{N}\right] \tag{11.15}$$

This follows the logical sequence of

$$C[n] = y[k] * g_n^*[-k] = y[n] * e^{j2\pi k/N}\text{rect}\left[\frac{k}{N}\right] \qquad n,k = 0,1...N-1$$

$$= \sum_{-\infty}^{+\infty} y[k]e^{j2\pi k/N}\text{rect}\left[\frac{-(k-u)}{N}\right] \tag{11.16}$$

After sampling at the instant $k = 0$, the frequency spectral component $C[n]$ becomes

$$C[n] = \sum_{-\infty}^{+\infty} y[u]e^{j2\pi u/N}\text{rect}\left[\frac{u}{N}\right]$$

$$\text{with } u \to k \to C[n] = \sum_{-\infty}^{+\infty} y[u]e^{j2\pi u/N}\text{rect}\left[\frac{k}{N}\right] = \sum_{k=0}^{N-1} y[k]e^{j2\pi k/N} \tag{11.17}$$

which is the DFT of $y[k]$.

11.2 OPTICAL OFDM TRANSMISSION SYSTEMS

Orthogonality is achieved by placing the different RF carriers onto a fixed frequency grid and assuming rectangular pulse shaping. For OFDM, the signal can be described as the output of a discrete inverse Fourier transform block using the parallel complex data symbols as input. This property has been one of the main driving aspects for OFDM in the past, since modulation and demodulation of a high number of carriers can be realized by simple DSP instead of using many local oscillators in the transmitter and receiver. Recently, OFDM has become attractive for digital optical communications [5,6]. Using OFDM appears to be very attractive since the low bandwidth occupied by a single OFDM channel increases the robustness toward fiber dispersion drastically, allowing the transmission of high data rates of 40 Gbps and more over hundreds of kilometers without the need for dispersion compensation [2]. In the same way as for modulation formats such as DPSK or DQPSK that were introduced in recent years, for OFDM too, the challenge for optical system engineers is to adapt a classical technology to the special properties of the optical channel and the requirements of optical transmitters and receivers.

However, Ref. [7] shows that an optical SSB modulation can assist in the combat of fiber dispersion. SSB can be achieved by driving the MZIM with two microwave signals $\pi/2$ phase shift with each other or by optical filtering (VSB), as shown in Chapters 2 and 9. However, the truly SSB transmitter using dual-drive and $\pi/2$ phase shift is preferred so as to preserve the energy contained within the bands of the signals. In optical SSB, the phase information can be preserved after the square law detection of the photodetector and the chromatic dispersion (CD) is limited by reducing the optical spectral bandwidth by a factor of 2 [5].

Two approaches have been reported recently. An intuitive approach introduced by Llorente et al. [6] makes use of the fact that the wavelength-division multiplexing (WDM) technique itself already realizes data transmission over a certain number of different carriers. By means of special pulse shaping and carrier wavelength selection, the orthogonality between the different wavelength channels can be achieved, resulting in the so-called orthogonal WDM technique (OWDM). However, in this way, the option of simple modulation and demodulation by means of DFTs cannot be utilized as this kind of DSP is not available in the optical domain.

The basic OFDM transmitter and receiver configurations are given in Figure 11.5a [8]. Data streams (e.g., 1 Gbps) are mapped into a two-dimensional signal point from a point signal constellation such as QAM. The complex-valued signal points from all subchannels are considered as the values of the DFT of a multicarrier OFDM signal. The serial-to-parallel converter arranges the sequences into equivalent discrete frequency domains. By selecting the number of subchannels, sufficiently large, the OFDM symbol interval can be made much larger than the dispersed pulse width in a single-carrier system, resulting in an arbitrary small intersymbol interference (ISI). The OFDM symbol, shown in Figure 11.5, is generated under software processing, as follows: input QAM symbols are zero-padded to obtain input samples for inverse fast Fourier transform (IFFT); the samples are inserted to create the guard band; and the OFDM symbol is multiplied by the window function, which can be represented by a raised-cosine function. The purpose of cyclic extension is to preserve the orthogonality among subcarriers even when the neighboring OFDM symbols partially overlap due to dispersion.

The principles of FFT and IFFT for OFDM symbol generations are shown in Figure 11.3. Section 3.4.7 of Chapter 3 has also outlined the principles of a OFDM transmission system. A system arrangement of the OFDM for optical transmission in a laboratory demonstration is shown in Figure 11.5. The data sequences are arranged in the sampled domain, and then to the frequency domain, and then to the time domain representing the OFDM waves, which would look like analog waveforms, as shown in Figure 11.5b and c (see also, Section 3.4.8, Figure 3.37). Each individual channel at the input would carry the same data rate sequence. These sequences can be generated from an arbitrary waveform generator. The multiplexed channels are then combined and converted to time domain using the IFFT module, and then converted to the analog version via the two digital-to-analog converters. These orthogonal data sequences are then used to modulate I and Q optical waveguide sections of the electro-optical modulator to generate the orthogonal channels in the optical domain. Similar decoding of I and Q channels are performed in the electronic domain after the optical transmission and optical–electronic conversion via the photodetector and electronic amplifier. Figure 11.6 illustrates similar transmission systems with the demodulation of the I and Q channels in the electrical domain, while these functions are implemented in the hybrid optical coupler in the optical domain in which the $\pi/2$ phase shift for the I and Q can be done by a $\pi/2$ shift in the optical domain by using pyro-optic effects via an electrode heating to change the refractive index of the optical path of either I or Q optical signals. This is commonly done in the PLC (planar lightwave circuit) optical hybrid couple.

In comparison with other optical transmitters, the optical OFDM transmitter will require a DSP and DAC for shaping the pulse spectrum and then the IFFT operation so as to generate the OFDM analog signals to modulate the optical modulator.

The performance of an OFDM coherent and transmission under back-to-back and over 1000 km of SSMF spans without DCM (dispersion compensation modules) as shown in Figure 11.7a and b, respectively, indicates the resilience of OFDM format to dispersion with an OSNR of 8 dB for a BER of 1e−3, which is the error-free rate under FEC integration. Note that polarization multiplexing is employed in the systems where both polarized channels are detected and extracted for BER as a function of the OSNR (measured over 0.1 nm spectral width).

In OFDM, the serial data sequence, with a symbol period of T_s and a symbol rate of $1/T_s$, is split up into N-parallel substreams (subchannels).

An alternative method [9] is the generation of an electrical OFDM signal by means of electrical signal processing followed by modulation onto a single optical carrier [10,11]. This approach is known as *optical OFDM* (oOFDM). Here, the modulation is a two-step process: first, the electrical OFDM signal is already a broadband bandpass signal, which is then modulated onto the optical carrier. Second, to increase data throughput, oOFDM can be combined with WDM, resulting in a multi-Tbps transmission system as shown in Figure 11.4. Nevertheless, oOFDM itself offers different options for implementation. An important issue is optical demodulation that can be realized either by

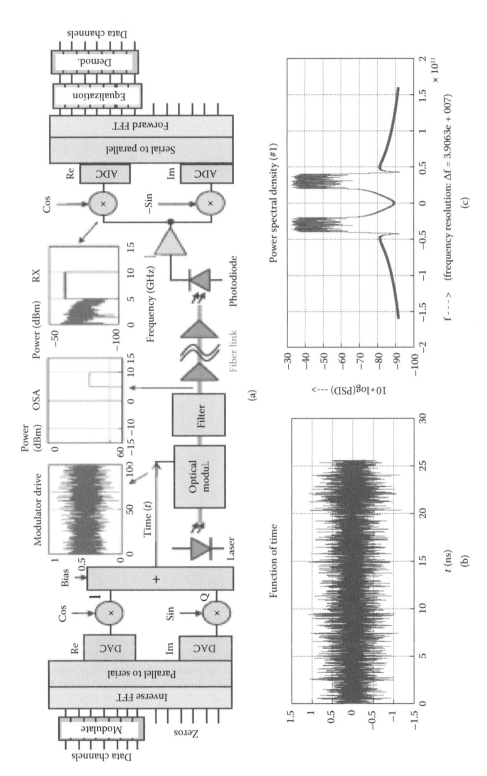

FIGURE 11.5 An optical FFT-IFFT-based (a) OFDM system including representative waveforms and spectra, (b) typical time-domain OFDM signals, and (c) power spectral density normalized in the frequency domain of OFDM signal with 512 subcarriers for 40 Gbps signals and QPSK modulation. (Note: electrical mixers for I and Q components recovery in electrical domain.)

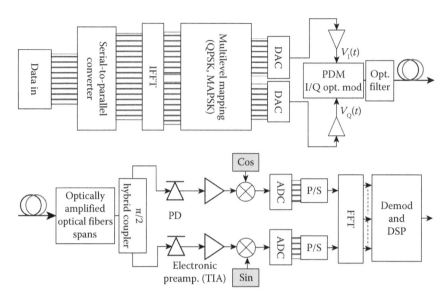

FIGURE 11.6 Schematic of an optical FFT-/IFFT-based OFDM system. S/P and P/S = serial-to-parallel conversion and vice versa. PD = photodetector, TIA = transimpedance amplifier (see also Figure 3.37 of Chapter 3 for more details). (Note electrical demodulation of in-phase and quadrature components.)

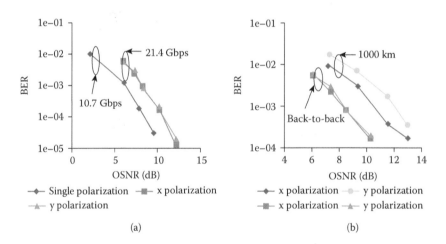

FIGURE 11.7 BER versus OSNR under coherent reception (a) back-to-back and (b) 1000 km non-DCM optically amplified spans. (Extracted with permission from W. Shieh and C. Athaudage, *Coherent Optical Orthogonal Frequency Division Multiplexing*, 2006.)

means of direct detection (DD) or coherent detection CoDet, including a local laser oscillator. DD is preferable due to its simplicity. However, for DD, the optical intensity has to be modulated. Due to the fact that the electrical OFDM signal is quasi-analog with zero mean and high peak-to-average ratio, the majority of the optical power has to be wasted for the optical carrier. That means there is an additional DC value of the complex baseband signal, resulting in low receiver sensitivity. For chromatic dispersion (CD), in addition, the bandwidth efficiency is twice as high as for DD, since for pure intensity modulation, a double-sideband signal is inherently generated. For CoDet, a complex optical I–Q modulator, composed of two real modulators in parallel followed by superposition with π/2 phase shift, allows for transmission of twice as much data within the same bandwidth.

For intensity modulation, the bandwidth efficiency may be increased by suppressing one of the redundant sidebands, resulting in oOFDM with single-sideband (SSB) transmission. First, the serial data sequence in can be converted to parallel streams. These parallel data sequences are then mapped to QAM constellations in the frequency domain, and then by IFFT converted back to the time domain. The time-domain signals are in I and Q components, which are then fed into an I–Q optical modulator. This optical modulation can be DQPSK or any other multilevel modulation subsystem. At the end of the optical fiber transmission, I and Q components are detected either by DD or CD. For coherent detection, a 2×4 $\pi/2$ hybrid coupler is used to mix the polarized optical fields of the local oscillator and that of the received signals. The outputs of the couplers are then fed into balanced optical receivers. The mixing of the local laser source and that of the signals preserves the phase of the signals, which are then processed by a high-speed electronic digital processor. For DD, I and Q components are detected differentially, and the amplitude and phase detection are then compared and processed similarly as for the coherent case.

In order to show the robustness of oOFDM toward fiber dispersion and also fiber nonlinearity, numerical simulations are carried out for a data stream of 42.7 Gbps data rate. The number of OFDM channels can be varied between $N_{min} = 256$ and $N_{max} = 2048$. A guard interval of 12 ns can be inserted, a strategy belonging inherently to OFDM technology that ensures orthogonality of the different channels in case of a transmission channel with memory. For the optical modulation, intensity modulation using a single Mach–Zehnder modulator in conjunction with SSB filtering and direct detection was implemented. The nonlinear optical transmission channel consisted of eight 80 km non-DCF spans of SSMF. MZIM with linearization should be used [12]. As a criterion for performance, the required OSNR for a BER of 10^{-3} (Monte Carlo) is measured. Using FEC subsystem in cascade with the optical receiver, after decoding and error correction the BER is equivalent to an error rate below 10^{-9}. This also depends on the specific code and the extra number of bits of the FEC.

Figure 11.8 [8] shows the required OSNR as a function of fiber launch power for different values of N. The transmission is error free (1e–9) over 640 km of SSMF or a dispersion factor of about 1100 ps/nm without dispersion compensation. It can be explained by the fact that, even for the lowest value of $N_{min} = 256$, each subchannel occupies a bandwidth of approximately 42.7 GHz/256 = 177 MHz, resulting in high robustness toward fiber dispersion.

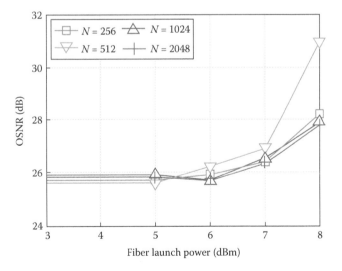

FIGURE 11.8 Simulation result for 42.7 Gbps oOFDM transmission over 640 km of SSMF; OSNR required for BER = 10^{-3} as function of fiber launch power. (Extracted with permission from A. Ali, *Investigations of OFDM Transmission for Direct Detection Optical Systems*, Dr. Ing. Dissertation, Albrechts Christian Universitaet zu Kiel, 2012.)

The principal difficulties of OFDM are that the pure delay is due to the variation of the refractive index of the fiber with respect to the optical frequency, leading to bunching of the subchannels and hence increase in the optical power; thus, unexpected SPM may occur in a random manner.

11.2.1 Impacts of Nonlinear Modulation Effects on Optical OFDM

The MZIM is biased at a quadrature point where the power transfer characteristic is linearizable. By means of low modulation depth, the nonlinear distortions due to the MZM can be considered to be arbitrarily small, hence leading to a low ratio of useful power to carrier power, and thus low sensitivity. Therefore, the modulation depth is a compromise between these constraints.

Figure 11.9 shows [8,13,14] results for the BER versus OSNR under back-to-back transmission, where the BER is obtained by Monte Carlo simulation. The OSNR for BER = 10^{-3} is plotted versus normalized driving voltage, as shown in Figure 11.9. The normalization is performed such that minimum and maximum optical output power can be obtained for instantaneous input voltages of −0.5 and 0.5, respectively. Beyond these values, clipping occurs, and intermodulation distortion also exists, due to MZM characteristics. The OFDM signal is analog, having nearly Gaussian amplitude distribution [4]. In good approximation, the peak voltage is within an interval from plus to minus the triple of the RMS voltage. This relation is used to create the lower from the upper of the two horizontal axes. The clipping threshold obtained for a peak voltage of ±0.5 is also given. Finally, the impact of MZM nonlinearity, which, within the range of acceptable sensitivity, obviously does not depend significantly on N, is identified by means of the dashed line.

Figure 11.9 depicts the nonlinear resilience of the fiber transmission link including 8 × 80 km SSMF without dispersion compensation. In order to investigate the impact of linear factor one by one, the MZM is driven in the quasi-linear range with a normalized effective voltage swing of ≈0.15, resulting in a back-to-back OSNR of ≈25.5 dB. This value is significantly above those reported, for example, in Ref. [15], which is due to incoherent detection and higher bandwidth. For low values of launch power, Figure 11.9 shows the robustness of oOFDM toward fiber dispersion, as the OSNR penalty achieved with ≈11,000 ps/nm accumulated dispersion is negligible. In the nonlinear regime, however, penalty increases rapidly. Beyond 8 dBm launch power at the fiber input, the BER does not fall

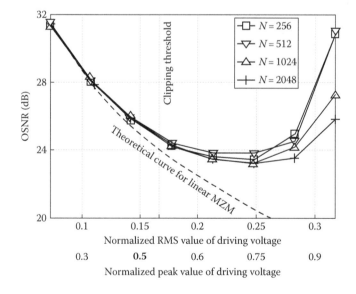

FIGURE 11.9 Required OSNR for BER = 10^{-3} as a function of driving voltage swings fed to MZIM electrodes. (Extracted with permission from A. Ali, *Investigations of OFDM Transmission for Direct Detection Optical Systems*, Dr. Ing. Dissertation, Albrechts Christian Universitaet zu Kiel, 2012.)

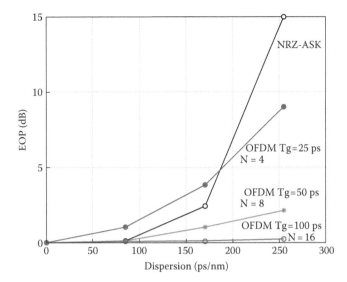

FIGURE 11.10 Eye opening penalty of OFDM as compared with NRZ-ASK with different guard interval times of 25, 50, and 100 ps for 42.7 Gbps bit rate.

below 10^{-3}, due to strong signal distortion. The optical power consists of a strong DC component and a weaker AC component carrying the OFDM signal. Since only the AC component results in signal distortion, the acceptable launch power found in this contribution is higher than for coherent detection.

Figure 11.10 shows the eye opening penalty of modulation formats NRZ-ASK and OFDM with guard intervals of 25, 50, and 100 ps. It shows clearly the superiority of OFDM over ASK at 42.7 Gbps bit rate.[*]

Except for the value for $N = 512$ at a launched power of 8 dBm, attributed to limited simulation accuracy, the result does not depend on N. Increasing N decreases the separation between the subchannels, and fiber nonlinearity is expected to cause strong XPM and FWM. However, with increasing N, the power per subchannel is decreased. Apparently, these two aspects cancelled out each other so that equal performance for all N is achieved.

The impact of nonlinear modulator characteristic and Kerr effect shows that, for an uncompensated oOFDM 8 × 80 km fiber link with a varying number N of subcarriers, both impairments are independent of N. Obviously, oOFDM is quite robust toward the specific nonlinear impairments in fiber-optic transmission systems.

11.2.2 DISPERSION TOLERANCE

The variation of the eye opening penalty as a function of the ratio of the guard interval and the bit period with the fiber SSMF length as a parameter is obtained as shown in Figure 11.11. This indicates that the guard interval is very critical for different lengths of the fibers, and thus also the bit rate and the number of subcarriers.

11.2.3 RESILIENCE TO PMD EFFECTS

Reference [14] reports the first experiment of PMD impact on fiber nonlinearity in coherent optical OFDM (CO-OFDM) systems. The optimal Q value at 10.7 Gbps has been improved by 1 dB after the introduction of DGD of 900 ps with the variation of the Q value against the launched power, as shown

---

[*] Simulation developed under MOVE_IT (MATLAB®), and simulation platform provided by M. Ali of Technische Facultaet, Universit of Kiel.

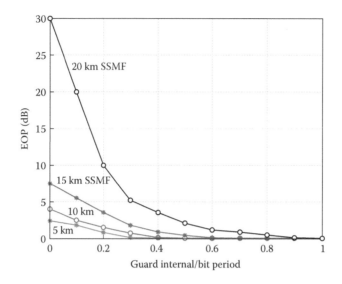

FIGURE 11.11 Eye opening penalty of OFDM against the ratio of the guard interval and bit period with fiber length as a parameter, $N = 8$.

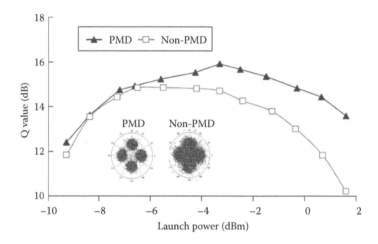

FIGURE 11.12 System Q as a function of launch power for PMD-supported and non-PMD-supported OFDM-QPSK transmission systems. Note the rotation of the signal constellation. (Extracted with permission from Y. Ma et al., *Electron. Lett.*, Vol. 43, pp. 943–945, 2007.)

in Figure 11.12 [16]. 900 ps is equivalent to 22 symbol periods of the baud rate of 25 GB. This is a very substantial resilience rate. Note that, if both polarized channels are employed as such in the Polarization Division Multiplexing 9pDm0 systems, the two polarized channels which are received and demultiplexed, then they can be processed by 2×2 MIMO (multiple input multiple output) processing technique to improve the BER and minimize the required OSNR, as shown in Figure 11.7.

11.3 OFDM AND DQPSK FORMATS FOR 100 Gbps ETHERNET

For 100 Gbps Ethernet, the decomposition of the bit rate to symbol rates under multilevel modulation formats can be referred to Chapter 7. The OFDM offers significantly lower symbol rate due to the decomposition into several subchannels. Reference [8] reported the transmission performance of RZ-DQPSK formats over 75 km of SSMF, as well as RZ-ASK with OSNR of 22 dB and 36 dB, respectively.

On the contrary, the OFDM can combine a large number of parallel data streams into one broadband data stream with high spectral efficiency, and would offer significant reduction of bit rates into symbol rates. This would challenge the use of multilevel modulation formats. However, there are still several issues to be examined by optical OFDM, such as high peak-to-average carrier ratio or FWM effects and other nonlinear effects.

OFDM schemes offer the following advantages to optical long-haul, metro, and access networks:

1. Different users can be assigned with different OFDM subcarriers within one OFDM band of N-subcarriers, as shown in Figure 11.13a. This assignment is very much suitable for the case of a massive MIMO antenna in which each antenna element would be assigned with a specific channel. These MIMO antennae would carry wireless OFDM signals, and this oOFDM transport technique would be most suitable for the convergence between wireless and optical transport layers. Furthermore, each channel can be assigned different slots for the case of TDM (see Figure 11.13b), as well as different wavelengths (see Figure 11.13c).
2. *Regarding speed and distance:* 100 Gbps/λ downstream and same for upstream. 100 km for access, 1000 km for metro, and 5000–10,000 km for long haul.
3. *On flexibility:* OFDM offers adaptive modulation and FEC on subcarrier basis, dynamic bandwidth allocation in time and frequency, transparency to arbitrary services, and optically transparent ONUs.

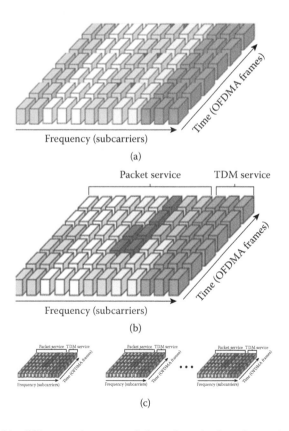

FIGURE 11.13 OFDM in different assignments of channels and subcarriers and wavelength: (a) different subcarriers to different channels, (b) different slots to TDM channels, and (c) different wavelength for OFDM symbols.

4. *On cost efficiency:* OFDM offers colorless architecture; stable, accurate DSP-based operation; and nondisruptiveness to legacy optical data networks (ODNs).

5. *On access passive optical networks (PON):* Orthogonal frequency division multiplexing (OFDM) PON: very well-suited for future PON systems, transparent to emerging heterogeneous applications, highly flexible, dynamic bandwidth allocation, and nondisruptive to legacy ODN.

6. Recent demonstrations of ultra-high-speed OFDMA PON by NEC and Huawei OFDM with aggregate speed of 108 Gbps/downstream and upstream, feasible on class C+ ODN (30+ dB power budget), highly dispersion tolerant (60–100 km transmission).

7. However, OFDM requires advanced DSP, which has been under intense research and development. DSP complexity: IFFT/FFT dominates, ~log(N) scaling, and optimized algorithms.

11.4 CONCLUDING REMARKS

OFDM is strongly suitable as a modulation technique for optical digital communications, especially for lowering the ultra-high-speed 100 Gbps Ethernet to a much more acceptable rate per subchannel. As multicarrier channels with natural orthogonality from the FFT and IFFT, it is much more efficient than the use of a single carrier as seen in other advanced modulation formats described in most chapters of this book. By closely spacing all the subchannels via the orthogonality, one can increase the spectral efficiency.

The pulse-shaping function can certainly influence the orthogonality of the subchannels and minimize the crosstalks between channels. The sinc function in the time domain would be the best choice. The guard bands influence the transmission performance of OFDM transmissions.

OFDM is shown to be more resilient to nonlinearity and PDM, as well as CD. Coherent OFDM offers higher receiver sensitivity or lower eye opening penalty than the direct detection OFDM without much more complexity, as the processing of the received optical sequence can be implemented in the electronic domain.

The trends for oOFDM in optical networking are as follows:

1. DSP-based system cost can be significantly and rapidly reduced by component integration and mass production.
2. Optimized OFDM algorithms reused from wireless and wireline building blocks.
3. Next-generation 100 Gbps long-haul fiber transmission will be heavily DSP-based, and 50+ GSa/s, two-channel ADC chips and 10–30 GS/s, two-channel DAC chips must be commercially available.
4. Intensive ongoing effort in parallelized, real-time DSP architectures.
5. Aggressive 100 Gbps DSP development for fiber transmission expected to have favorable effect on ultra-high-speed OFDM-PON for access and metro networks.

The key challenges to be tackled rest in

1. DSP two-channel ADC/DAC, DSP processor key components.
2. Favorable technology trends in terms of cost profile.
3. 41.25 Gbps real-time, variable-rate receiver for WDM-OFDM and higher bit rates to 100 Gbps must be developed.

REFERENCES

1. R. V. Nee and R. Prasad, *OFDM for Wireless Communications*, Norwood: Artech House, 2000.
2. J. Proakis, *Digital Communications*, New York: McGraw-Hill, 2002.

3. L. Hanzo, S. X. Ng, T. Keller, and W. Webb, *Quadrature Amplitude Modulation: From Basics to Adaptive Trellis Coded, Turbo-Equalized Space-Time Coded OFDM, CDMA and MC-CDMA Systems*, 2nd ed., Chichester, England: IEEE Press, 2004.
4. L. Hanzo, M. Münster, B. J. Choi, and T. Keller, *OFDM and MC-CDMA for Broadband Multi-User Communications, WLANs and Broadcasting*, NY: Wiley, 2003.
5. M. Sieben, J. Conradi, and D. E. Dodds, Optical single-sideband transmission at 10 Gbps using electrical dispersion compensation, *IEEE J. Lightwave Technol.*, Vol. 17, No. 10, pp. 2059–2068, 1999.
6. R. Llorente, J. H. Lee, R. Clavero, M. Ibsen, and J. Martí, Orthogonal wavelength-division-multiplexing technique feasibility evaluation, *J. Lightwave Technol.*, Vol. 23, pp. 1145–1151, 2005.
7. A. J. Lowery and J. Armstrong, Orthogonal-frequency-division multiplexing for dispersion compensation in long haul WDM systems, *Opt. Express*, Vol. 14, No. 6, pp. 2079–2084, 2006.
8. A. Ali, *Investigations of OFDM Transmission for Direct Detection Optical Systems*, Dr. Ing. Dissertation, Albrechts Christian Universitaet zu Kiel, 2012.
9. J. Leibrich, A. Ali, and W. Rosenkranz, Optical OFDM as a promising technique for bandwidth-efficient high-speed data transmission over optical fiber, in *Proceedings of the 12th International OFDM-Workshop 2007* (InOWo 2007), 29–30 August 2007, Hamburg, Germany, 2007.
10. A. J. Lowery, L. Du, and J. Armstrong, Orthogonal frequency division multiplexing for adaptive dispersion compensation in long haul WDM systems, in *Proceedings of the OFC 2006*, Anaheim, CA, USA, paper PDP39, 2006.
11. W. Shieh and C. Athaudage, Coherent optical orthogonal frequency division multiplexing, *Electron. Lett.*, Vol. 42. pp. 587–589, 2006.
12. X. J. Meng, A. Yacoubian, and J. H. Bechtel, Electro-optical pre-distortion technique for linearization of Mach-Zehnder modulators, *Electron. Lett.*, Vol. 37, No. 25, pp. 1545–1547, 2001.
13. A. Ali, J. Leibrich, and W. Rosenkranz, Impact of nonlinearities on optical OFDM with direct detection, in *Proceedings of the ECOC 2007*, Berlin, Germany, 2007.
14. A. Ali, *Orthogonal Frequency Division Multiplexing in Optical Transmission Systems with High Spectral Efficiency*, Master Thesis Dissertation, Technische Facultaet, University of Kiel, Kiel, Germany, 2007.
15. S. L. Jansen, I. Morita, N. Takeda, and H. Tanaka, 20-Gb/s OFDM Transmission over 4,160-km SSMF Enabled by RF-Pilot Tone Phase Noise Compensation, *Proc. Opt. Fiber Conf.*, OFC, Anaheim, CL, 2007.
16. Y. Ma, W. Shieh, and X. Yi, Characterization of nonlinearity performance for coherent optical OFDM signals under the influence of PMD, *Electron. Lett.*, Vol. 43, pp. 943–945, 2007.

12 Digital Signal Processing in Optical Transmission Systems under Self-Homodyne Coherent Reception

Electronic processing is attracting research and development of the enabling technologies for mitigation of impairments at the receiver as equalizer and at transmitters as predistortion. They are thus very important to long-haul transmission systems and networks as well wide-area networks in which the equalizers can be placed at the optical receiver to adaptively equalize any distortion due to different hops in the transmission path in all-optical networks. This chapter introduces the basic concepts of equalization and then gives a case study of equalization of duobinary and minimum shift keying (MSK) modulation formats.

12.1 INTRODUCTION

With the advent of digital signal processors and the decrease in the cost of electronic processing systems, electronic signal processing is becoming very attractive for mitigating various impairments that severely affect optical transmission. Equalization and compensation of the impairment effects of the transmission of optical signals through optically amplified systems have been a principal concern for transmission engineering. Compensation of linear and nonlinear dispersion represents a typical and important application of the processing. Over the years, the compensation has been implemented in long-haul transmission with the insertion of dispersion-compensating fibers (DCF) at the end of a transmission span (e.g., standard single-mode fiber [SSMF] or LEA), or a fiber Bragg grating (FBG) with phase fluctuation. These compensating devices would normally suffer appreciable attenuation, and hence an additional optical amplifier must be used to compensate for these lost factors. Thus, additional optical amplifiers are necessary for long-haul transmission systems. Unavoidable accumulated noises of the ASE noise sources from these amplifiers will shorten the transmission distance. Furthermore, the narrow core of a DCF renders it more susceptible to fiber nonlinearity effects. The DCF is also polarization sensitive.

On the contrary, electrical equalization can offer the advantages of more flexibility and lower cost, and a smaller integrated size for the transceiver electronics. Furthermore the programmability of the digital signal processing provides better system management. Thus, if one could compensate or equalize the pulse sequence in the electrical domain either by predistortion or postcompensation, or a combination of both processes, then the elimination of these amplifying processes can be reduced, thus making possible longer transmission reach.

Furthermore, optical networks with different transmission paths carrying lightpaths of different distances due to the different numbers of hops would suffer dispersion and nonlinear effects due to the nature of ultra-high speeds at 10 Gbps or higher. The 10 Gbps Ethernet is coming into the networks in the near future, and there is an urgent demand that inexpensive techniques be available for network implementation. These techniques must be very adaptive and tunable with control signals. Electronic equalization can offer these solutions for high-speed optical networks. Table 12.1 shows

TABLE 12.1

Eye Opening Penalty—Dispersion Tolerance and Equivalent Length of SSM for Different Bit Rates

Modulation Technique		2 dB Penalty	
Direct Modulation (DM)	External Modulation (EM)	ps/nm	km (SSMF)
DML (2.5 Gbps)	DM	1000–2000	60–120
DML (10 Gbps) (eye opening and distortion)	DM	50–200	3–12
EML (10 Gbps)	EM	400–2000	25–120

the penalty of the eye opening with respect to the bit rate and the length of SSMF when no dispersion compensation is used. This dispersion tolerance becomes very critical when the bit rate is increased. It is important to implement equalization in the electrical domain of digital signal processing, so that it can be tuned or controlled for adaptation to the number of hops of the lightwave channels traveling over the network nodes. These digital signal processing systems for equalization would enhance the received opening eye diagram. Shown in Table 12.1 is the eye opening penalty of some advanced modulation formats due to distortion by chromatic dispersion.

Naturally, the dispersion-limited distance can be improved using optical predistortion techniques such as the control of the chirp of the laser to 200 km SSMF [1] for 10 Gbps, or technique of dispersion-supported transmission to 250 km SSMF [2]. However, these techniques require high complexity of the optical transmitters.

For conventional direct detection receivers, the linear distortion that is induced by chromatic dispersion in the optical domain is transformed into a nonlinear distortion in the electrical signal, which explains why only limited performance improvements can be achieved by using a linear baseband equalizer with only one baseband received signal. This also explains why nonlinear techniques, such as decision feedback equalization (DFE) and maximum-likelihood sequence detection (MLSE), are more effective in combating chromatic dispersion in direct detection receivers. Strictly speaking, the MLSE system attempts neither to compensate the chromatic dispersion effects nor to equalize them, but duly account for these effects in the data detection process.

On the contrary, in a coherent detection system, chromatic dispersion is linear in the electrical signal at the receiver, which is why fractionally spaced equalizers with "complex coefficients" can achieve a performance that is limited only by the number of taps used within the equalizer accounting for the fiber chromatic dispersion only. Recently, with the availability of narrow-linewidth and tunable lasers and ultra-high-speed receivers, coherent lightwave systems have attracted significant attention. However, they have not reached the practical and installation stage so far. This is attributed mainly to the success of wavelength-division multiplexing (WDM) technology with the advent of erbium-doped fiber amplifiers (EDFAs) since the late 1990s. Another reason is the complexity of coherent transmitters and receivers.

The detrimental effect of chromatic dispersion in SSMF can also be reduced by using reduced-bandwidth modulation formats such as optical duobinary signals, single sideband, or vestigial sideband, which can extend the reach of 10 Gbps systems to distances in excess of 200 km, compared with approximately 80 km for conventional NRZ-OOK. In this chapter, optical duobinary signaling is combined with linear electrical pre-equalization schemes [3] to demonstrate the extension of the reach of a 10 Gbps system to several hundred kilometers without any optical chromatic dispersion compensation. In fact, the system reach is influenced by other fiber impairments other than chromatic dispersion. The equalization is based on exploiting the linearity of a coherent lightwave system to move the equalization process to the transmitter, where the data, especially the carrier phase, is still undisturbed. More specifically, the duobinary signal is pre-equalized using two tunable $T/2$-spaced finite-impulse response (FIR) filters. The outputs of the FIR filters then modulate two optical carriers that are in phase quadrature. Thus, the advantages of coherent receiver equalization are still maintained while utilizing a conventional noncoherent direct detection receiver. Although this chapter focuses on duobinary modulation, this choice is principally driven by the reduced-bandwidth advantage, compared with conventional NRZ-OOK modulation, and the ability to use a conventional direct detection receiver. In principle, electrical pre-equalization can also be applied to other modulation formats for potential improvements.

Duobinary offers the most effective mitigation of the linear dispersion due to its single-sideband property and direct detection at the receiver. Duobinary signaling can also be combined with a proposed electrical pre-equalization scheme to extend the reach of 10 Gbps signals that are transmitted over SSMF. The proposed scheme is based on predistorting the duobinary signal using two $T/2$-spaced FIR filters. The outputs of the FIR filters then modulate two optical carriers that are then applied to the two parallel MZIM to generate duobinary optical signals. This equalization and duobinary modulation formats for extending the transmission reach are demonstrated in the first section of this chapter with an adaptation of the algorithm given in Ref. [3].

Other equalization systems that would be implemented in the electrical domain include the MLSE equalizers, which have recently received much interest as an effective solution for overcoming the severe distortion caused by the intersymbol interference (ISI) of the optical channels and for improving the signal-to-noise ratio (SNR) performance, thus extending the reach of the optical transmission systems. Most of the studies on the use of nonliner MLSE equalizers have concentrated on amplitude or differential phase modulation formats [4–6]. However, there are not many reports on the performance of MLSE equalizers for optical transmission systems employing the MSK modulation format.

A narrowband filter receiver with a breakthrough dispersion tolerance limit for the detection of noncoherent 40 Gbps optical MSK transmission system has been proposed and explained in detail in Chapter 8. The performance of this receiver was found to be limited by ISI caused by the differential group delay of the optical channel and the narrow bandwidth of the optical detection filter. In order to combat the ISI, in this chapter, two nonlinear equalizers are proposed and integrated with the narrowband filter receiver for the detection of noncoherent 40 Gbps optical MSK signals. The performances of the proposed schemes are numerically evaluated. The proposed schemes extend the reach of dispersion tolerance of the proposed narrowband optical MSK receiver to ±952 and ±884 ps/nm at BER = 1e−9, or equivalently 52 and 56 km of SSMF with required OSNR of 19 and 23.5 dB, respectively. To the best of the author's knowledge, these dispersion tolerance limits have not been reached before.

Recently, MLSE equalizer for optical transmission systems employing the MSK modulation format has been reported [4]. The uncompensated distance can equivalently reach up to approximately 960 km SSMF for 10 Gbps transmission. The results are comparable to the distance of 1040 km SSMF as reported in Ref. [5] for IMDD systems. In Ref. [5], up to 8192 trellis states are used. Longer reach for IMDD systems require a very high number of states, which is even infeasible in simulation. Due to large-bandwidth optical filters being used in IMDD systems, noise is not greatly suppressed, leading to high values of required optical-signal-to-noise (OSNR) ratios. In addition, noise distribution in IMDD systems no longer follows a Gaussian profile, which increases the complexity of the equalizer in order to estimate the noise distribution for optimal performance of the Viterbi algorithm. These issues can be effectively mitigated with the introduction of narrowband optical filtering in an optical MSK transmission system.

In this section, we numerically described the possibility of transmission over 1472 km SSMF optical link without in-line dispersion compensation for 10 Gbps optical MSK signals. The achievement is enabled by the integration of 1024-state Viterbi–MLSE equalizer with the noncoherent narrowband frequency discrimination receiver. The proposed receiver mitigates the fiber intersymbol interference (ISI) effectively and greatly reduces the noise floor, thus enabling low OSNR for receiver sensitivity. The simulation results also show significant OSNR improvements with two and four samples per bit over the conventional one sample per bit. Finally, the MLSE equalizer in our proposed scheme can operate optimally due to Gaussian noise distribution [6]. Thus, the complexity of the MLSE equalizer is reduced, compared to the IM/DD systems.

This chapter is organized as follows: the next two sections give a general overview and detailed schemes of equalization whereby the equalizers are placed at the receiver, the transmitter, and a share between transmitters and receiver. Essential expressions of the sampled impulse responses of the equalizer subsystems, the FFE, DFE, and combined nonlinear and linear equalizers are summarized from Ref. [7]. The minimization of the square error in the FFE and DFE is also integrated to optimize the equalization process. The doubling sampling or partial response equalization is also included. Two special cases of equalization—the duobinary and MSK modulation formats—are described in detail in Sections 12.5 through 12.8, using double sampling and MLSE techniques, respectively. Section 12.9 then gives some degree of uncertainty in the equalization process, and the gain in the eye opening penalty, the number of taps, or length of the templates in MLSE schemes are obtained. Finally, the concluding remarks are given.

12.2 ELECTRONIC DIGITAL PROCESSING EQUALIZATION

Electronic equalization can be implemented with the possibilities of (1) integration of digital signal processing (DSP) at the receive-side; (2) postequalization, placing the equalization DSP at the transmit-side, or predistortion of the optical signals by modification of the electrical signals driving the external modulator; or (3) sharing the equalization function at both the receive-side and transmit-side, that is, using both predistortion and postequalization.

Figure 12.1 shows the generic arrangement of the equalizers at both the transmit-side and receive-side. The equalization processes can occur by using the predistortion at the transmitter (Tx) only or postcompensation at the receive-side (Rx) only, or sharing between the predistortion and postcompensation at both the Tx and Rx sides. Figure 12.2 shows the equalization processing both in the optical and electronic domains at the receiver and transmitter sides. The optical equalization works with the field of the lightwaves, while the electronic processor works with the electrical signals obtained from the conversion of the optical intensity of the signals, except for the case of coherent reception, the received signals are directly related to the phase in the electrical at the driving input of the optical modulator. A typical layered optical network using IP over WDM is shown in Figure 12.3, in which the equalizer is used to equalize any distortion residual from the optical compensator—for example, phase ripple of the phase group delay of the FBG dispersion compensator, or different traveling hops of the lightpaths. It has been noted that electronic equalizers do not suffer the problems of over-dispersion as in the case of optical compensators, as the electrical type deals with the intensity while the optical type deals with the field and over-compensation or under-compensation, due to the variation of the residual dispersion in different lengths of hops of the lightpaths. Figure 12.4 [4] also shows a typical light of a wavelength channel traveling through the network. The insert in this figure extends the dispersion tolerance with and without equalization. Optical equalizers can be shared by the channels of different wavelengths, while an electronic equalizer is normally integrated into a receiver for a specific channel.

When the equalizer is operating at the receiver, then this postequalization would take the advantage of the knowledge of all distortion effects and deals with the signals depending on the intensity. The distorted signals can then be equalized. If any unexpected distortion occurs, this postequalization can be adaptive to equalize the pulse sequence. On the contrary, an equalizer placed at the transmit-side is processing the lightwave by creating a predistortion on the phase of the carrier. Note that the channel is purely phase distortion, and thus the sequence can be electrically predistorted before driving the external modulator to chirp the phase in such a way that opposes the chirping by the group velocity of the optical fiber.

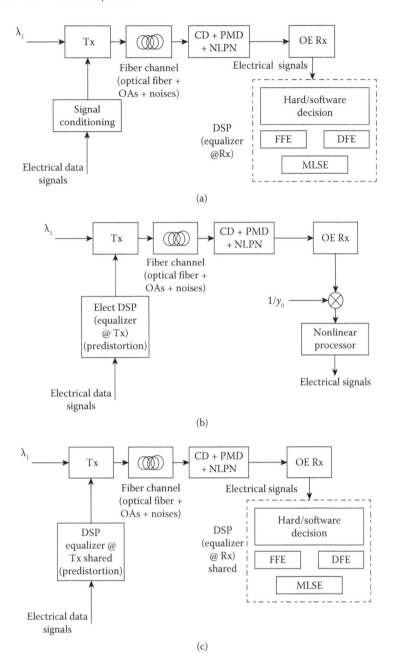

FIGURE 12.1 Arrangement of electronic DSP for equalization in optical transmission systems. Electronic DSP for predistortion, electronic DSP at the Rx is for postequalization and shared predistortion and postequalization at the Tx and Rx: (a) equalization at the transmit-side, (b) equalization at the receive-side, and (c) equalization shared between the Tx and Rx sides.

The postequalization processing usually requires a number of samples to ensure sufficient data for its equalization technique. The higher the number of samples, the better the equalized sequence. However, for long-haul optical transmission systems, the pulse may be very dispersive and the samples may suffer significant distortion. In order to assist the postequalizer, a predistortion compensator may be placed at the transmitter to share the equalization functionality with the postequalizer to extend the reach of the optical transmission system.

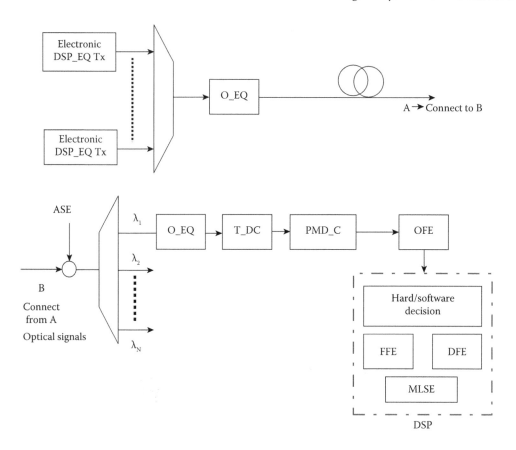

FIGURE 12.2 Typical structure and position of the DSP equalizer of a receive-side equalized optical system. Both optical and electronic equalization processors are integrated. O_EQ = optical equalizer; T_DC-tunable dispersion compensator; PMDC = polarization-mode dispersion compensation; O/E = optical to electrical converter; DFE = decision feedback equalizer; FFE = feed-forward equalizer; MLSE = maximum-likelihood sequence estimator.

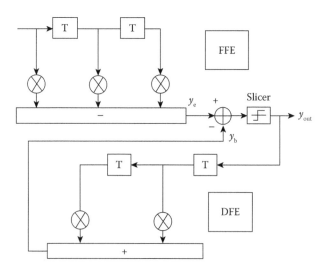

FIGURE 12.3 Typical layered graph model for IP over WDM network with optical receivers and equalizers.

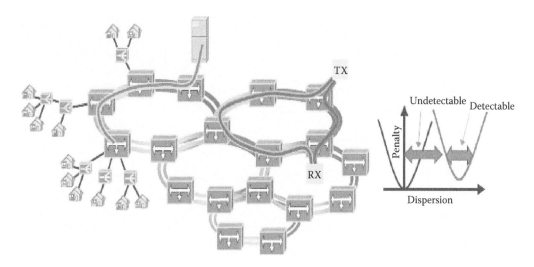

FIGURE 12.4 Typical lightpaths of optical networks and dispersion tolerance without and with electrical equalization. (From N. Alic et al., *Opt. Express*, Vol. 13, No. 12, pp. 4568–4579, 2005.)

TABLE 12.2
Dispersion Tolerance and Equivalent Length of SSM for Different Modulation Formats and Equalization Techniques and Improved Factors

Modulation Technique	Measured/Simulated 10.7 Gbps at 2 dB Penalty BER = 10^{-9}			
	No Electronic Dispersion Compensation (EDC)	Equalization FFE/DFE	MLSE	Improved Factor
Chirp-free NRZ-ASK	50 km SSMF	80	130	× 2.6
Chirp NRZ-ASK	80	130	150	
Duobinary	200	200	260	× 1.3
Chirp managed DM-ASK	200	270	300	
Improved factor	× 2.4		× 2.3	

The benefit of equalization does also depend on the modulation format, as the algorithm must be adapted to the mode of modulation, for example, amplitude, phase, or frequency shift keying. Table 12.2 shows the improvement factors of the equalization techniques on different modulation formats using amplitude shift keying. The duobinary is a tri-level modulation format using both amplitude and phase to represent the levels. However, its detection is intensity based and thus included here.

12.3 SYSTEM REPRESENTATION OF EQUALIZED TRANSFER FUNCTION

12.3.1 GENERIC EQUALIZATION FORMULATION

12.3.1.1 Signal Representation and Channel Pure Phase Distortion

Given that the SMF is a pure phase distortion, the equalization deals with the processing of sampled amplitude or phase of the recovered signals pending on the modulation formats whether its amplitude, phase, or frequency is altered. This section gives a brief strategy of development of the requirement on the transfer function of the equalized system and associate filters. It is essential that the impulse response of the channel (the single-mode optical fiber) be obtained in order to specify the equalization subsystems. Thus, this section gives an analytical derivation of the pure distortion channel. Both the impulse and step responses of the fiber channel are analytically derived.

Let the impulse response of the channel (the single-mode optical fiber) be $y(t)$; then the z-transform of the impulse response with a sampling time equal to the bit period is given by

$$y(nT) = \begin{bmatrix} y_0 & y_1 & .... & y_g & 0 & 0 & .... & 0 \end{bmatrix}$$
$$Y(z) = y_0 + y_1 z^{-1} \quad .... \quad + y_g z^{-g}$$

(12.1)

with $n \geq 2g + 1$ and y_i are complex valued.

Under a purely phase distortion, the channel lost no energy, and hence the conservation of energy requires that

$$\sum_{i=0}^{n} |y_i|^2 = 1$$

(12.2)

with the total energy contained in the impulse assumed to be unity.

Now consider the input signals having complex-valued sampled components, such as the QAM optical signals that comprise two DPSKs in phase quadrature or two AM signals in quadrature, one of which carries the data symbols $\{s_{1,i}\}$. Each AM signal is associated with the corresponding modulator and demodulator, to recover back to the baseband channel. For advanced modulation formats, the detection can be incoherent or coherent. Under coherent detection, the impulse response takes the amplitude and preserves its phase, while for incoherent or direct detection, the impulse response is the result of the beating of different components of the response in the photodetection process, and its response follows intensity-based square law detection. In this case, the response can be estimated by considering the Gaussian distribution with the pulse-spreading factor.

The input signals can now be written as

$$s_i = s_{0,i} + j s_{1,i}$$

(12.3)

with $j = (-1)^{1/2}$, both $s_{0,i}$ and $s_{1,i}$ can independently take on any one of the different m values

$$-(m-1)k, -(m-3)k, ..., (m-1)k$$

(12.4)

An element of the QAM signal carrying the data symbol s_i is itself the sum of two double-sideband-suppressed carrier AM signal elements in phase quadrature, carrying the data symbols $s_{0,i}$ and $s_{1,i}$.

The output signals can be real or complex valued. The sequence of the signals at the output of the channel is, of course, given by the sequence of samples at the output of the samplers. The real and imaginary parts of this sequence are samples of the respective baseband signals at the outputs of the two constituent parallel channels that make up the linear baseband channel. It can now be seen that if the sampled impulse response of the linear baseband channel is real, then there is no coupling between the two parallel channels; the distortion of these two channels are the same and can be determined as for ASK signals. On the contrary, if they are complex valued, the coupling is introduced between the two constituent channels by the imaginary parts of the components of the sampled impulse response. The impulse response of a single-mode optical fiber is purely complex-valued and will be described in the following section.

When the channel introduces pure phase distortion, it introduces a unitary transformation into the transmitted signals [8], such that the received signal elements corresponding to the resultant (complex valued) transmitted signal elements are orthogonal. An orthogonal transformation is a special case of unitary transformation where all signals are real valued.

Under coherent detection, the detected signals are given by

$$r_i = s_i y_0 + w_i \tag{12.5}$$

where we have assumed, without loss of generality, that the impulse response takes only a non-zero value y_0, and w_i is the noise contributed to the signal. The noise is statistically independent Gaussian random variables with zero mean and variance σ^2. Thus, we have

$$r_i = s_{0,i} y_0 + j s_{1,i} y_0 + w_i \tag{12.6}$$

Obviously, the received signals must be detected with both the real and imaginary parts. Thus, in the absence of noise, we have

$$|r_i|^2 = s_{0,i}^2 + s_{1,i}^2 \tag{12.7}$$

Since the impulse response is of pure phase distortion, then the absolute value of components of the sampled time response is unity, and thus when the input signals are convolved with the impulse response, the phase of the components of the received signals are rotated by a fixed amount of phase.

This is equivalent to the reverse of the components of the impulse response pivoting around the central value at $t = 0$ by the complex conjugates; thus, the impulse sequence can be written as

$$\begin{aligned} y_i &= y_i^* \\ y(nT) &= \begin{bmatrix} y_0 & y_1 & .... & 0 & ......y_m^* & y_{m-1}^* & .... & y_1^* \end{bmatrix} \end{aligned} \tag{12.8}$$

Alternatively, we can state that whenever there is a reversal of the complex-conjugate of a sequence pivoted around its components at $t = 0$, then the sequence represents a pure phase distortion. Thus, when convolving the sequence with its original one gives a sequence all of whose components are zero except for the component at time $t = 0$ which is unity.

12.3.1.2   Equalizers at Receiver

12.3.1.2.1   Zero-Forcing Equalization

The equalizer shown in Figure 12.5 is a DFE, which is a nonlinear equalizer in contrast to the linear filter. A linear filter usually consists of a transversal feed-forward equalizer (FFE) that delays the signal by time interval T in which the delayed signals are tapped and multiplied with desired coefficients and then added to form an equalized output. The nonlinear DFE discussed in this chapter is structured with both linear FFE and nonlinear DFE. A zero-forcing equalizer operates as follows.

The linear FFE partially equalizes the signal by setting to zero all components of the channel-sampled impulse response preceding that of the largest magnitude, without changing the relative values of the remaining components. The nonlinear equalizer then completes the equalization process to give accurate equalization of the channel. The equalizer is nonlinear because the detector (or slicer) is a nonlinear device.

Now, let y_l be the largest-magnitude component of the impulse response (Equation 12.1); then, the z-transform of the linear FFE of $(m+1)$-taps is given by

$$C(z) = c_0 + c_1 z^{-1} \quad .... \quad + c_m z^{-m} \tag{12.9}$$

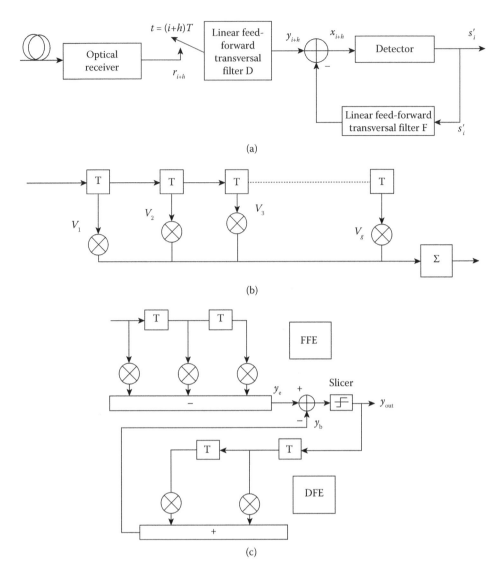

FIGURE 12.5 Schematic of a nonlinear feedback equalizer following an optical receiver with (a) feed-forward transversal filters at the input and at the feedback path (FFE-DFE), (b) linear feed-forward transversal filter (FFE), and (c) combining FFE and DFE and a slicer for equalization.

Such that

$$Y(z)C(z) \approx z^{-h} \tag{12.10}$$

where h is a nonnegative integer in the range $(m + g)$. It is also assumed that the frequency transfer response $Y(z)$ does not have zeros on the unit circle of the z-plane.

Now, let the transfer function of the $(n + 1)$ tap FFE filter D in Figure 12.5 be

$$d(nT) = \begin{bmatrix} d_0 & d_1 & .... & d_n \end{bmatrix}$$
$$D(z) = d_0 + d_1 z^{-1} \quad .... \quad + d_n z^{-n} \tag{12.11}$$

The required sampled impulse response of the combined system consisting of a cascade of the channel and the linear filter D is given by the $(g-l+1)$-component row vector

$$E = \begin{bmatrix} 1 & \dfrac{y_{l+1}}{y_l} & \dfrac{y_{l+2}}{y_l} & \dfrac{y_{l+3}}{y_l} & \cdots & \dfrac{y_g}{y_l} \end{bmatrix} \tag{12.12}$$

$$E(z) = 1 + \frac{y_{l+1}}{y_l} z^{-1} + \frac{y_{l+2}}{y_l} z^{-2} + \frac{y_{l+3}}{y_l} z^{-3} \cdots + \frac{y_g}{y_l} z^{-g+1} \cdots$$

Clearly, E is obtained by removing the first l components and dividing each of the components by y_l. As y_l is the maximum magnitude, we have all the coefficients of the filter E less than unity. $D(z)$ can now be written as

$$D(z) = E(z)C(z)$$

then

$$Y(z)D(z) = Y(z)C(z)E(z) \approx z^{-h}E(z) \tag{12.13}$$

with

$$n = m + g - l$$

The z-transform of the $(i+1)$th transmitted signal element is $s_i z^{-i}$, and thus the z-transform of the $(i+1)$th received signal element at the output of the linear filter D is given by

$$s_i z^{-i-h} E(z) = s_i z^{-i-h} \left(1 + \frac{y_{l+1}}{y_l} z^{-1} + \frac{y_{l+2}}{y_l} z^{-2} + \frac{y_{l+3}}{y_l} z^{-3} \cdots + \frac{y_g}{y_l} z^{-g+1} \right) \tag{12.14}$$

It is thus observed that, whenever there is a nonzero term, there would be intersymbol ISI between the signal elements at the output of the linear filter. This ISI is then removed in the nonlinear equalizer that equalizes the z-transform $z^{-h}E(z)$, as given earlier. Clearly, the nonlinear DFE follows that of Figure 12.2 with the linear FFE with the number of taps of $(g-l)$, with the tap gain coefficients of (instead of the tap coefficients indicated in Figure 12.2b).

$$\frac{y_{l+1}}{y_l}, \quad \frac{y_{l+2}}{y_l}, \quad \frac{y_{l+3}}{y_l} \quad \cdots \quad \frac{y_g}{y_l} \tag{12.15}$$

At the instant $(i+h)T$, the input to the substractor is given as

$$v_{i+h} = s_i + \frac{y_{l+1}}{y_l} s_{i-1} + \frac{y_{l+2}}{y_l} s_{i-2} + \frac{y_{l+3}}{y_l} s_{i-3} + \cdots + \frac{y_g}{y_l} s_{i-g+l} + u_{i+h} \tag{12.16}$$

where s'_{i-j} is the detected value of s_{i-j} at the output of the equalizer. With the correct detection of $s_{i-1}, s_{i-2}, \ldots, s_{i-g+l}$, the signals at the detector input becomes

$$x_{i+h} \simeq s_i + u_{i+h} \tag{12.17}$$

when $x_{i+h} > 0, s'_i = k$, and when $x_{i+h} < 0, s'_i = -k$, especially for the balanced receiver detection of DPSK or DQPSK or QAM constellation. The noise component u_{i+h} is given by

$$u_{i+h} = \sum_{j=0}^{n} w_{i+h-j} d_j \tag{12.18}$$

The noise components are statistically independent Gaussian variables with zero mean and a variance, and thus we have the total variance of the noise component of the detected signal as

$$\eta = \sigma^2 \sum_{j=0}^{n} d_j^2 = \sigma^2 |D|^2 \tag{12.19}$$

where $|D|$ is the Euclidean length of the vector D. Thus, the probability of error of the detection of s_i, given the correct detection of $s_{i-1}, s_{i-2}, \ldots, s_{i-g+l}$, can be approximately given as

$$P_e = \int_{k}^{\infty} \frac{1}{\sqrt{2\pi\eta^2}} e^{-\frac{u^2}{2\eta^2}} du = Q\left(\frac{k}{\eta}\right) = Q\left(\frac{k}{\sigma|D|}\right) \tag{12.20}$$

where k is the average total amplitude of the signal at the output of the receiver. The equalizer presented in this section can be optimized to reduce the probability of error further—for example, the use of two cascaded transfer functions to replace the equivalent transfer function of the fiber, where the linear equalizer takes on one part and the other by the nonlinear equalizer. One could thus reduce the number of taps of the linear FFE. The lesser the number of taps, the better the demands puts on the electronic processor operating very high speeds.

12.3.1.2.2 Equalization that Minimizes Mean Square Error (MSE Equalizer)

Both FFE and DFE can equalize and minimize the error contributed by the noise and distortion that would generally offer a more effective degree of equalization than an equalizer that minimizes peak distortion. Such an equalizer minimizes the main difference between the actual and ideal sampled values at its output for a given of taps for the case of FFE. For DFE, the minimization of the error occurs at the point where the feedback linear FFE and that at the output of the linear filter as shown in Figure 12.5.

Under linear equalization, we recall that the channel-sampled transfer function given in Equation 12.1 of the sampled impulse response of the channel and the equalizer is given by the $(m + g + 1)$th row vector

$$E = [e_0 \, e_1 \, \ldots \, \ldots \, \ldots \, e_{m+g}] \tag{12.21}$$

The ideal value of the sampled impulse response of the channel and the equalizer is given by $(m + g + 1)$-component row vector

$$E_h = \left[\underbrace{0..0...0...0}_{h} \quad 1 \quad .0... \quad 0 \right] \tag{12.22}$$

where h is an integer and is within the range $(m + g)$. The minimization is thus conducted so that the expected mean value

$$E\left[(x_{i+h} - s_i)^2 \right] = k^2 |E - E_h|^2 + \sigma^2 |C|^2 \tag{12.23}$$

where $k^2 |E - E_h|^2$ and $\sigma^2 |C|^2$ are the mean square error (MSE) terms in the received signals x_{i+h} due to ISI and due to the Gaussian noise, respectively, and C is the sampled impulse response of $(m + 1)$-component row of the FFE, denoted by

$$C = [c_0 \, c_1 \, \ldots \, \ldots \, \ldots \, c_m] \tag{12.24}$$

Under nonlinear DFE, the minimization occurs at the error substractor as shown in Figure 12.5; that is, there needs to minimize the MSE of the equalized signals as the expected value of the mean square value which can be expressed as

$$\varepsilon = E\left[\left(x_{i+h} - s_i\right)^2\right] \tag{12.25}$$

The data symbols $s_{i-1}, s_{i-2},\dots,s_{i-\mu}$ can be correctly detected by adjusting the coefficients of the FFE filters and then those of the FFE of the feedback DFE. The sampled impulse responses of these filters can be expressed as

$$D = [d_0\, d_1 \dots \dots \dots d_n] \tag{12.26}$$

$$E = [e_0\, e_1 \dots \dots \dots e_{n+g}] \tag{12.27}$$

$$F = [f_0 f_1 \dots \dots \dots f_\mu] \tag{12.28}$$

$$S = [s_i\, s_{i-1}\, s_{i-2} \dots \dots \dots s_{i-n-g}] \tag{12.29}$$

where E is the sampled impulse of the fiber channel and the linear equalizer FFE D, and S is the sampled signal correctly detected at the output of the equalizer. The signals at the input of the detector in Figure 12.5 at time $t = (i + h)T$ are given by

$$x_{i+h} = \sum_{j=0}^{n+g} s_{i+h}e_j - \sum_{j=1}^{\mu} s_{i-j}f_j + u_{i+h} = S_{i+h}E^T - S_{i+h}F_0^T + u_{i+h}$$

with $\tag{12.30}$

$$F_0 = \left[\begin{array}{cccc} \underbrace{0\dots\dots0}_{h} & f_1 & f_2\dots\dots & f_\mu \end{array}\right]$$

The MSE can thus be written as [9]

$$\varepsilon = k^2|B|^2 + \sigma^2|D|^2$$

with

$$B = E - F_0 - E_h = \left[\begin{array}{ccccc} b_0 & b_1 & .. & .. & b_{n+g} \end{array}\right] \tag{12.31}$$

and

$$E_h = \left[\begin{array}{cccc} \underbrace{0\dots\dots0}_{h} & 1 & 0 & 0 \end{array}\right]$$

The data symbols are assumed to be statistically orthogonal with zero mean and variance k^2. The DFE feedback nonlinear equalizer is assumed to have μ taps in the filter F. This is the lowest number of taps whereby exact decision-directed cancellation can be achieved of all ISI components in x_{i+h}, involving data symbols $s_{i-1}, s_{i-2},\dots,s_{i-\mu}$ that have been detected.

After a number of manipulation steps, the MSE can be obtained as

$$\varepsilon = k^2\left[1 - E_h Z^T \left(ZZ^T + \frac{\sigma^2}{k^2}I\right)^{-1} ZE_h^T\right] \tag{12.32}$$

where $Z = [(n+1) \times (n+g+1)]$, derived from $[Y_c]$ = convolution matrix of the sampled impulse response. The matrix Z can be written as

$$Z = \begin{bmatrix} y_0 & y_1 & y_2\ldots & \ldots 0\ldots 0\ldots 0 \\ 0 & y_0 & y_1\ldots & \ldots 0\ldots 0\ldots 0 \\ 0 & 0 & y_2\ldots & \ldots 0\ldots 0\ldots 0 \\ \ldots & \ldots & \ldots & \ldots 0\ldots 0\ldots 0 \\ 0 & 0 & 0 & \ldots y_0 & y_1 & y_2 & 0\ldots & 0 \\ 0 & 0 & 0 & 0 & y_0 & y_1 & y_2\ldots & 0 \\ 0 & 0 & 0 & 0 & 0 & y_0 & y_1\ldots & 0 \\ 0 & 0 & 0 & 0 & 0 & 0 & y_0\ldots & 0 \end{bmatrix} \tag{12.33}$$

12.3.1.2.3 Nonlinear Equalization that Minimizes SNR

An optimum DFE can be considered as a DFE that further minimizes the BER in the detected signals for a given DFE structure. As previously assumed, the z-transform of the fiber channel in intensity has no zeros or roots on the unit circle of the z-plane. Thus, we have the sampled impulse response of the channel, and the FFE of the DFE is given by

$$Y(z)\,C(z) \simeq z^{-h} \quad \text{or} \quad C(z) \simeq z^{-h} Y^{-1}(z) \tag{12.34}$$

Then the FFE D of the DFE can be composed of the FFE C and a filter E under the constraint that

$$Y(z)\,D(z) \simeq z^{-h}\,E(z) \tag{12.35}$$

where $E(z) = [e_0, e_1, \ldots, e_{n-m}]$. The nonlinear equalizer of Figure 12.5 equalizes $E(z)$ so that the linear FFE of the feedback path F has $(n-m)$ taps with the gains equal, respectively, to the components $e_i, e_2, \ldots, e_{n-m}$ of the vector E.

The signals at the output of the linear filter D, at the instant t, is thus given by $(i+h)T$

$$v_{i+h} \simeq s_i + s_{i-1}e_1 + s_{i-2}e_2 + \ldots\ldots + s_{i-n+m}e_{n-m} \tag{12.36}$$

With the corrected detection of the input signals s_{i-t}, the equalized signals at the output of the DFE can be written as

$$x_{i+h} \simeq s_i + u_{i+h} \tag{12.37}$$

So, when

$$x_{i+h} = \begin{cases} > 0\ldots s_i' = k \\ < 0\ldots s_i' = -k \end{cases}$$

the noise component u_{i+h} is a Gaussian random variable as assumed previously with zero mean and variance of η, given as

$$\eta = \sigma^2 |D|^2 \tag{12.38}$$

Thus, the BER can be obtained as

$$P_e = \int_k^\infty \frac{1}{\sqrt{2\pi\eta^2}} e^{-\frac{u^2}{2\eta^2}} du = Q\left(\frac{k}{\eta}\right) = Q\left(\frac{k}{\sigma|D|}\right) \tag{12.39}$$

So, it is necessary to minimize the Euclidean length of the linear FFE filter or find the minimum number of taps of the filter that still allow the maximizing of the SNR. Thus, the DFE should be adjusted to minimize the vector length $|D|$. This minimization can be implemented as follows. This minimization can be done at the linear filter D of the DFE such that

$$|D| = |G_0 - LM| \tag{12.40}$$

With

$$G_0 = \begin{bmatrix} 0 & c_0 & c_1 & c_2 \text{........} & c_m & 0 \text{..........} 0 \end{bmatrix}$$

$$M = \begin{bmatrix} 0 & c_0 & c_1 & c_2 \text{........} 0 & 0 \\ & 0 & c_0 \text{........} 0 & 0 \\ \text{...} & \text{....} & \text{.....} & \text{.............} & \text{......} \\ & & c_m \text{........} 0 & 0 \\ & & c_{m-1} \text{........} 0 & 0 \\ & & c_{m-2} & c_{m-1} & c_m \end{bmatrix} \tag{12.41}$$

$$L = \begin{bmatrix} -e_1 & -e_2 \text{..........} & -e_{n-2-m} & -e_{n-1-m} & -e_{n-m} \end{bmatrix}$$

The task is now adjusting L so that minimization of the vector length D happens.

12.3.1.2.4   Case Study of Equalization Schemes

This section gives a case study of direct detection with the impulse response of the SMF or the residual dispersion of a dispersion-managed transmission system. The sampled impulse response is a real impulse response after the photodetector and electronic preamplifier. This impulse response follows a Gaussian distribution with the broadening of the impulse at the e^{-1} value of the maximum peak by an amount given by

$$\Delta \tau = D \cdot L \cdot B_{3dB} \tag{12.42}$$

where B_{3dB} is taken as the 3 dB bandwidth of the power spectra of the signals under different modulation formats.

A typical dispersive impulse at the instant $t = iT$ can be given by

$$r_i = 0.3s_i + s_{i-1} = w_i \tag{12.43}$$

where the data symbol $\{s_i\}$ is statistically independent and equally likely to have either value $\pm k$ and $\{w_i\}$ are statistically independent Gaussian random variable with zero mean and a variance σ^2. Effectively, the impulse response of the fiber transmission disperses to 30% of the peak amplitude at the output of the electronic preamplifier. We can compare the tolerance to noises of the following equalization processes: (a) a simple threshold level detector, (b) a linear FFE, (c) a purely NL DFE, (d) linear and nonlinear equalization of two factors of the channel, and (e) optimum DFE.

12.3.1.2.4.1   Nonequalization Direct Detection   At a high SNR, practically all the errors occur when $s_{i+1} = -s_i$, that is, when $r_{i+1} = 0.7s_i + w_i$. Thus, the BER can be derived as

$$P_e = \int_{0.7k}^{\infty} \frac{1}{\sqrt{2\pi\sigma^2}} e^{-\frac{u^2}{2\sigma^2}} du = Q\left(\frac{0.7k}{\sigma}\right) \tag{12.44}$$

(a) Linear Equalization

The sampled impulse response of the fiber channel is given by the two component sequence $Y = [0.3, 1]$, so that the z-transform of the sequence is given by $Y(z) = 0.3 + z^{-1}$. Now, assuming that a six-tap linear FFE is designed for the equalization so as to minimize the peak distortion is $M(z) = 1 + 0.3 z^{-1}$, then the equalizer with a six-tap FFE can be found by taking a long division of $(1 + 0.3 z^{-1})^{-1}$. The long division gives

$$N(z) = 1 - 0.3z^{-1} + 0.09z^{-2} - 0.027z^{-3} + 0.0081z^{-4} - 0.00243z^{-5} \qquad (12.45)$$

Thus, the $C(z)$ of the FFE can be obtained by reversing the order of the $N(z)$ as

$$C(z) = 1.z^{-5} - 0.3z^{-4} + 0.09z^{-3} - 0.027z^{-2} + 0.0081z^{-1} - 0.00243 \qquad (12.46)$$

Hence, the z-transform of the cascaded channel and the FFE is simply given as

$$C(z)Y(z) = 0.000729 + z^{-6} \simeq z^{-6} \qquad (12.47)$$

Thus, the equalized pulse at the instant $t = (i+6)T$ is

$$x_{i+6} = 1w_{i+1} - 0.3w_{i+2} + 0.09w_{i+3} - 0.027w_{i+4} - 0.0081w_{i+5} - 0.00243w_{i+6} \qquad (12.48)$$

The Gaussian noise can be estimated as

$$\eta = 1w_{i+1} - 0.3w_{i+2} + 0.09w_{i+3} - 0.027w_{i+4} + 0.0081w_{i+5} - 0.00243w_{i+6} \qquad (12.49)$$

Thus, the noise variance is obtained as

$$\eta^2 = \sigma^2(1 - 0.3 + 0.09 - 0.027 + 0.0081 - 0.00243) = 1.0989\sigma^2 \qquad (12.50)$$

Thus, the probability of error is given by

$$P_e = \int_k^\infty \frac{1}{\sqrt{2\pi\eta^2}} e^{-\frac{u^2}{2\eta^2}} du = Q\left(\frac{k}{\eta}\right) = Q\left(\frac{k}{1.0989\sigma}\right) = Q\left(\frac{0.9539k}{\sigma}\right) \qquad (12.51)$$

(b) NL DFE

On receiving the signals r_i, the NL DFE determines s'_{i-1} to generate the equalized signals $x_i = r_i - s'_{i-1}$. With the correct detection of s_{i-1}, we have

$$x_i = 0.3s_i + w_i \qquad (12.52)$$

And s_i is now detected from x_i as follows

$$x_{i+h} = \begin{cases} > 0 & s'_i = k \\ < 0 & s'_i = -k \end{cases} \qquad (12.53)$$

An error occurs here in the detection of s_i when w_i has a magnitude greater than $0.3\,k$ and the opposite sign to s_i. Thus, the probability of error of the equalized signals is given by

$$P_e = \int_k^\infty \frac{1}{\sqrt{2\pi\eta^2}} e^{-\frac{u^2}{2\eta^2}} du = Q\left(\frac{k}{\eta}\right) = Q\left(\frac{0.3k}{\sigma}\right) \qquad (12.54)$$

(c) Linear and Nonlinear Equalization

The linear and nonlinear equalization of the two factors here would degenerate into a linear equalizer since the impulse response z-transform of $Y(z)$ does not have zero in the outside region of the unit circle in the z-plane. Thus, the probability of error is the same as in the case (b), that is

$$P_e = Q\left(\frac{0.9539k}{\sigma}\right) \tag{12.55}$$

(d) Optimum DFE

The first FFE of the nonlinear equalizer performs an orthogonal transformation on the receive signals, such that the z-transform of the channel and the filter becomes

$$Y(z)D(z) \simeq z^{-h}(1 + 0.3z^{-1}) \tag{12.56}$$

The filter does not change the SNR and the noise property or any distortion of the amplitude of the received signal. Its output signal at $t = (i + h)T$ is

$$v_{i+h} = s_i + 0.3s_{i-1} + u_{i+h} \tag{12.57}$$

where u_i is a Gaussian random variable of zero mean and a variance of σ^2. Thus, the probability of error is

$$P_e = \int_k^\infty \frac{1}{\sqrt{2\pi\sigma^2}} e^{-\frac{u^2}{2\sigma^2}} du \tag{12.58}$$

On receiving v_{i+h}, the nonlinear filter has formed $0.3s'_{i-1}$, which is subtracted from v_{i+h} to give the equalized signal

$$x_{i+h} = v_{i+h} - 0.3s'_{i-1} \tag{12.59}$$

Then fed to the detector or decision device, the correct detected signal and the equalized signal becomes

$$x_{i+h} = s_i + u_{i+h} \text{ when } x_{i+h} = \begin{cases} > 0....s'_i = k \\ < 0.....s'_i = -k \end{cases} \tag{12.60}$$

An error occurs in the detection of s_i from x_{i+h} when u_{i+h} has a magnitude greater than k and opposite in sign to s_i so that, regardless of the sign of s_i, the probability of detection of s_i can be approximately given as

$$P_e = \int_k^\infty \frac{1}{\sqrt{2\pi\sigma^2}} e^{-\frac{u^2}{2\sigma^2}} du = Q\left(\frac{k}{\sigma}\right) \tag{12.61}$$

Thus, from the preceding analytical estimation of the probability of error as a function of the SNR k/σ, we can see that the resilience to noise of the optimum DFE is best with about 0.4 dB.

The pure nonoptimized nonlinear DFE can suffer -2.7 dB under direct detection without equalization and 0 dB for linear FFE.

12.3.1.2.5 *Nonlinear Equalization under Severe Amplitude Distortion*

The maximum-likelihood sequence estimation (MLSE) is considered as the detection of the transmitted sequence with minimum probability of error, and it depends on the coding and modulation formats of the signals. This method is further explained in detail in Section 12.7.

12.3.1.3 Equalizers at the Transmitter

Consider an arrangement where the equalizer acts as a predistortion and is placed at the transmitter. The nonlinear equalizer in this case converts the sequence of data symbols $\{s_i\}$ into a nonlinear channel $\{f_i\}$ at the output of the transmitter, normally an optical modulator driven by the output of the nonlinear electrical equalizer as shown in Figure 12.1b and remodeled as shown in Figure 12.6.

These symbols are sampled and transmitted as sampled impulses $\{f_i\,\delta(t-iT)\}$. The SNR in this case is given by $E[f_i^2]/\sigma^2$, and the energy is the expected average energy of the predistorted signal sequence, and must be equal to k^2. A nonlinear processor is necessary at the output before the final recovery of the system. This processor performs a modulo-m operation on the received sequence $x[\text{modulo}]-m = x-jm$, with j an appropriate integer. The z-transform $F(z)$ at the output of the nonlinear distortion equalizer is given by

$$F(z) = M\left[S(z) - F(z)(y_0^{-1}Y(z) - 1 \right]\tag{12.62}$$

where $S(z)$ is the z-transform of the sampled input signals, and $Y(z)$ is the z-transform of the channel transfer function. Then, the z-transform at the output of the nonlinear processor at the receiver can be written as

$$X(z) = M\left[F(z)Y(z) + W(z) \right] = M\left[F(z)(y_0^{-1}Y(z) - y_0^{-1}W(z) \right]$$
$$\text{or } X(z) = M\left[S(z) - y_0^{-1}W(z) \right]\tag{12.63}$$

That is, $x_i = M\left[y_0^{-1}r_i \right] = M\left[s_i + y_0^{-1}w_i \right]$, with x_i denoting the signal at the output of the nonlinear processor, and all ISI has clearly been eliminated except the noise contribution.

The nonlinear processor M operates as a modulo-m as

$$M[q] = [(q + 2k)\,\text{modulo} - 4k] - 2k = q - 4jk - 2k \leq M[q] \leq 2k\tag{12.64}$$

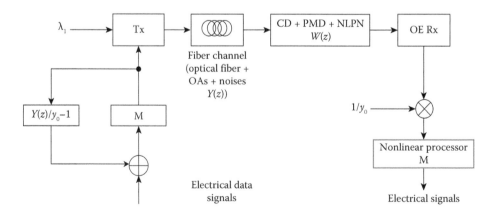

FIGURE 12.6 Schematic of a nonlinear feedback equalizer at the transmitter.

Thus, an error would happen when

$$(4j-3)k \le y_0^{-1}w_i \le (4j-1)k \quad \forall j \tag{12.65}$$

The probability of error is then given as

$$P_e = 2\int_k^\infty \frac{1}{\sqrt{2\pi y_0^{-2}\sigma^2}} e^{-\frac{w^2}{2y_0^2\sigma^2}} dw = 2Q\left(\frac{k|y_0|}{\sigma}\right) \tag{12.66}$$

12.3.1.4 Equalization Shared between Receiver and Transmitter

Long-haul transmission requires equalization so that the extension of the transmission can be as far as possible. Equalization at the receiver can be supplemented with that at the transmitter to further increase the reach.

The equalizer at the transmitter of the preceding section can be optimized by inserting another filter at the receiver. Consider the channel transfer function $Y(z)$ that can be written as a cascade of two linear transfer functions $Y_1(z)$ and $Y_2(z)$ as

$$Y(z) = Y_1(z)Y_2(z)$$

$$\text{with} \begin{cases} Y_1(z) = 1 + p_1 z^{-1} + \ldots\ldots + p_{g-f} z^{-g+f} \\ Y_2(z) = q_0 + q_1 z^{-1} + \ldots\ldots + q_f z^{-f} \end{cases} \tag{12.67}$$

$Y_1(z)$ has all the zeros inside the unit circle and $Y_2(z)$ all outside the unit circle. Then we could find a third system with a transfer function that has all the coefficients in reverse of those of $Y_2(z)$, given as

$$Y_3(z) = q_f + q_{f-1}z^{-1} + \ldots\ldots + q_0 z^{-f} \tag{12.68}$$

with the reverse coefficients, the zeros of the $Y_3(z)$, now lying inside the unit circle.

The linear FFE filter D of the nonlinear filter formed by cascading an FFE and a DFE as shown in the preceding text can minimize the length of the vector of D and minimizing e_0 can be implemented, provided that

$$D(z) = z^{-h}Y_2^{-1}(z)Y_3(z) \tag{12.69}$$

which represents an orthogonal transformation without attenuation or gain. The linear equalizer with a z-transform $Y_2(z)Y_3^{-1}(z)$ is an all-pass pure phase distortion.

This transfer function of the all-pass pure phase distortion is indeed a cascade of optical interferometers, known as half-band filters [10]. This function can also be implemented in the electronic domain. The equalizer can then be followed by an electronic DFE. Hence, the feedback linear filter $F(z)$ in Figure 12.5 can be written as

$$F(z) = Y_1(z)F_3(z) \tag{12.70}$$

The probability of error can be similarly evaluated and given by

$$P_e = \int_k^\infty \frac{1}{\sqrt{2\pi y_0^{-2}\sigma^2}} e^{-\frac{w^2}{2y_0^2\sigma^2}} dw = Q\left(\frac{k|q_f|}{\sigma}\right) \tag{12.71}$$

where q_f is the first coefficient of $Y_3(z)$.

12.3.1.5 Performance of FFE and DFE

The performance of a linear equalizer using an FIR filter with a different order is shown in Figure 12.7 with the penalty of eye opening versus the dispersion factor [11]. The higher the order of the FIR, the better the equalization of the distorted impulse, or the longer the length of the SSMF that the signals can propagate. While the receiver without equalization would suffer an EOP of at last 12 dB, a seventh-order FIR filter can equalize the distorted signals over 150 km with only 1 dB EOP.

Figure 12.8 shows the EOP versus the fiber length (SSMF) for the case of nonlinear equalizer with an FFE and a DFE with the orders of second and third, respectively. With 1 dB EOP, the FFE and DFE combined nonlinear filter can extend the transmission of 10 Gbps NRZ-ASK to 150 km, and to 300 km with nonlinear correction.

Figure 12.8 shows the EOP of modulation formats of NRZ-ASK and NRZ duobinary using FIR of number of taps of 3 and 16 with the distortion of signals. Up to 400 km, SSMF transmission is possible with the equalized system for 1 dB EOP.

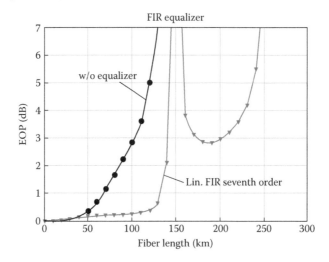

FIGURE 12.7 Eye opening penalty as a function of the dispersion tolerance in kilometers of SSMF without and with linear equalizer FIR.

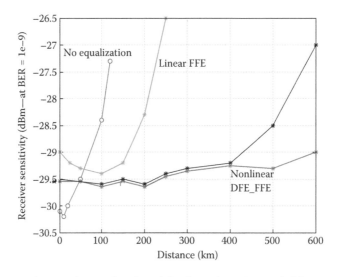

FIGURE 12.8 Eye opening penalty as a function of the dispersion tolerance in kilometers of SSMF without and with linear and nonlinear equalizer FFE-DFE.

In the next section, the impulse response of the fiber is described; especially, it is proven that it is a purely imaginary response as the channel is a pure phase distortion. The impulse response for intensity distortion can be found very easily with Gaussian shape approximation (see Chapter 2) and the broadening of the impulse can be estimated from Equation 12.42.

It is then followed by a detailed description of the equalization using MLSE for continuous phase FSK or optical MSK modulation schemes.

12.3.2 IMPULSE AND STEP RESPONSES OF THE SINGLE-MODE OPTICAL FIBER

The treatment of lightwaves through SMF has been well documented, as given in Chapter 4. In the following, we shall restrict our study to the linear region of the media. Furthermore, the delay term in the transmittance function for the fiber is ignored, as it has no bearing on the size and shape of the pulses. We can thus model the fiber simply as a quadratic phase function. The input–output relationship of the pulse can therefore be depicted as in Figure 12.9.

Equation 12.1 expresses the time-domain impulse response $h(t)$ and the frequency-domain transfer function $H(\omega)$ as a Fourier transform pair

$$h(t) = \sqrt{\frac{1}{j4\pi\beta_2}} \exp\left(\frac{jt^2}{4\beta_2}\right) \leftrightarrow H(\omega) = \exp\left(-j\beta_2\omega^2\right) \tag{12.72}$$

where β_2 is known as the group velocity dispersion (GVD) parameter. The input function $f(t)$ is typically a rectangular pulse sequence; the parameter β_2 is proportional to the length of the fiber. The output function $g(t)$ is the dispersed waveform of the pulse sequence. The electro system of Figure 12.9 is an exact analogy of diffraction in optical systems (see Ref. [12], item 1, Table 1 of Chapter 2, p. 14). Thus, the quadratic phase function also describes the diffraction mechanism in one-dimensional optical systems, where the distance x is analogous to time t. The establishment of this analogy affords us to borrow many of the imageries and analytical results that have been developed in the diffraction theory. Thus, we may express the step response $s(t)$ of the system $H(\omega)$ in terms of Fresnel cosine integral $C(t)$ and Fresnel sine integral $S(t)$ as follows

$$s(t) = \int_0^t \sqrt{\frac{1}{j4\pi\beta_2}} \exp\left(\frac{jt^2}{4\beta_2}\right) dt = \sqrt{\frac{1}{j4\pi\beta_2}} \left[C\left(\sqrt{1/4\beta_2}\,t\right) + jS\left(\sqrt{1/4\beta_2}\,t\right)\right] \tag{12.73}$$

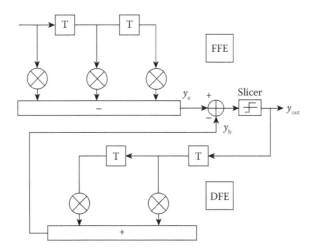

FIGURE 12.9 Representation of the quadratic phase transmittance function in frequency and time domain.

$$C(t) = \int_0^t \cos\left(\frac{\pi}{2}\tau^2\right)d\tau$$

with   (12.74)

$$S(t) = \int_0^t \sin\left(\frac{\pi}{2}\tau^2\right)d\tau$$

Using the electro-optical analogies, one may argue that it is always possible to restore the original pattern $f(x)$ by refocusing the blurry image $g(x)$ (e.g., image formation, item 5, Table 2-1, [12]). In the electrical analogy, it implies that it is possible to compensate the quadratic phase media perfectly. This is not surprising. The quadratic phase function $H(\omega)$ in Figure 12.1 is an all-pass function, it is always possible to find an inverse function to recover $f(t)$ from $g(t)$. Express this differently in information theory terminology: the quadratic phase channel has a theoretical bandwidth of infinity; its information capacity is infinite. Shannon's channel capacity theorem states that there is no limit on the reliable rate of transmission through the quadratic phase channel.

The picture changes completely if the detector/decoder is allowed only a finite time window to decode each symbol—for example, detection using frequency discrimination techniques for continuous phase frequency shift keying (CPFSK), in which a narrow optical filter is used to extract the carrier frequency representing the bits "1" or "0." In the convolutional coding scheme, for example, it is the decoder's constraint length that manifests as the finite time window. In adaptive equalization schemes, it is the number of equalizer's delay elements that determines the decoder window length. Since the transmitted symbols have already been broadened by the quadratic phase channel, if they are next gated by a finite time window, the information received could be severely reduced. The longer the fiber, the more the pulses are broadened, and the more uncertain it becomes in the decoding. It is the interaction of the pulse broadening on one hand, and the restrictive detection time window on the other, that gives rise to the finite channel capacity.

Figure 12.10a shows the impulse response of Gaussian pulse transmission of 100 Gbps pulse sequence through an SMF of length $L = 200$ km, dispersion $= 0e-6$ s/m^2; (a) dispersion slope $S = +0.06e + 5$ s/m^3, while Figure 12.10b shows its response over $L = 2$ km of the same dispersion factor and dispersion slope. It is noted from Equation 12.72 that the impulse response is a pure phase function, and thus phase distortion. The oscillation of the tail of the impulse response for long fiber indicates the phase chirping of the lightwave carrier in the advanced region of time-dependent pulses. Over a short length of fiber the Gaussian pulse remains with minimum change of the phase underneath its envelope. Figure 12.10c shows the impulse tail oscillation on the other side when it propagates through a dispersion-compensated 200 km fiber span with a total residual dispersion of -8.5 ps/nm that is typically found optically amplified fiber transmission system.

The step response (Equation 12.73) consists of the in-phase and quadrature-phase components. Figure 12.11a through c shows the pulse and step responses, its frequency spectrum, and step response after propagating through 1 km SSMF. The overshooting of the lightwave carrier modulated pulse occurs mostly near its edge that indicates that for the detection of the frequency shift keying modulation (FSK) format would be significantly influenced if the frequency filtering is implemented at the pulse center to obtain the best BER and the receiver sensitivity of the optical systems.

12.4 ELECTRICAL LINEAR DOUBLE-SAMPLING EQUALIZERS FOR DUOBINARY MODULATION FORMATS FOR OPTICAL TRANSMISSION

In general, chromatic dispersion is a time-varying impairment mainly because of temperature variations. Implications can be quite critical for 40 Gbps systems that have tight chromatic dispersion tolerances. For 10 Gbps systems, the chromatic dispersion tolerance is far less stringent than

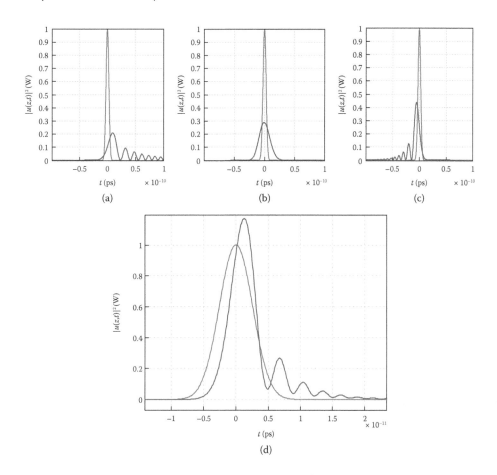

FIGURE 12.10 Gaussian pulse transmission of 100 Gbps pulse sequence through an SMF of length $L = 2$ km, dispersion $= 0e-6$ s/m^2: (a) dispersion slope $S = +0.06e + 5$ s/m^3; fiber length $= 200$ km; (b) $L = 2$ km dispersion $= 17e-6$ s/m^2 and $S = +0.06e + 5$ s/m^3; (c) over 200 km SSMF and completely compensated with a mismatch of -8.5 ps/nm dispersion; and (d) enlarged responses with a dispersion slope of DCF $= 0.3e + 5$ s/m^3. Note the oscillation of the tail of the impulse response.

40 Gbps systems (approximately 1000 ps/nm for a 1 dB power penalty). A variation of approximately 0.15 ps/nm/km over the temperature range from 40°C to 60°C. Compared with a typical value for of 17 ps/nm/km, the variation can be neglected. Therefore, it can be assumed, for our immediate purposes, that chromatic dispersion is a static impairment that can be accurately modeled, and hence static equalizers can be utilized. The ideal zero-forcing equalizer is simply the inverse of the channel transfer function given in Chapter 3. Thus, the required equalizer transfer function for the equalizer $H_{eq}(f)$ is given by

$$H_{eq}(f) = H_f^{-1}(f) = e^{+j\alpha f^2}$$

with

$$\alpha = \pi DL \frac{\lambda}{c^2}$$

(12.75)

However, this transfer function does not maintain conjugate symmetry, that is

$$H_{eq}(f) \neq H_{eq}^{-1}(-f)$$

(12.76)

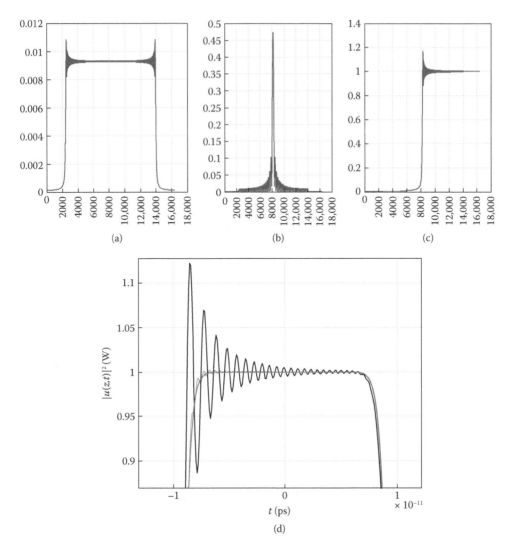

FIGURE 12.11 Rectangular pulse transmission through an SMF, near-field response: (a) pulse response; (b) spectrum of pulse in (a); (c) step response, note horizontal axis—unit normalized with time unit; and (d) enlargement of chirping of pulse at the edge due to dispersion slope.

Thus, the impulse response of the equalizing filter is complex, as shown in the previous section (Equation 12.2). Consequently, this filter cannot be realized by a baseband equalizer using only one baseband received signal, which explains the limited capability of linear equalizers that are used in direct detection receivers to mitigate chromatic dispersion.

On the contrary, it also explains why fractionally spaced linear equalizers, especially the double-sampling type that are used within a coherent receiver, can potentially extend the system reach to distances that are only limited by the number of equalizer taps. However, the nonideal effects such as the laser phase noise and the fiber nonlinearity, which eventually set a limit on the maximum achievable transmission distance, should also be taken into account. The equalizer in a coherent receiver, most of the complexity can be avoided at the receiver by an equalization process at the transmitter. The required filter would thus have complex coefficients, and thus the predistorted data have to modulate two optical carriers that are in phase quadrature.

Although the concept of using two optical carriers that are in phase quadrature is not a conventional one, its practical feasibility was shown in Chapter 4, where two orthogonal optical carriers were used to obtain an optical differential quadrature phase shift keying (ODQPSK) transmission system. To determine the required equalizer coefficients, the simulation setup in Figure 12.12 [4] can be used, where an adaptive algorithm is used to adaptively adjust the FIR filter coefficients to their optimum values. The adaptive algorithm adjusts the FIR filter taps to compensate not only for chromatic dispersion but also other linear dispersion effects such as PMD and any other phase–delay-type filter such as phase ripples of chirp FBGs. Once the filter taps converge to their optimal values, the linearity of the system allows transferring the filter to the transmitter side, where it acts as a predistortion filter. The proposed predistortion scheme is depicted in Figure 12.12. The optical transmitter for duobinary is formed using two parallel MZMs and a $\pi/2$ optical phase shift, as shown in Figure 12.13.

The MZMs are biased for minimum optical transmission and driven by the pre-equalized data from the FIR filters. The duobinary filters can be implemented using analog filters. The analog filters also account for the limited bandwidth of the signal path between the pre-equalizer and the MZM electrodes. The tap spacing of the FIR filter can be as high as a whole bit period, as proposed in Figure 12.14. However, reducing the tap spacing to half the bit period doubles the frequency band over which equalization can be applied, which translates into a substantial improvement in signal quality, but at the expense of a higher clock speed. Alternatively, the FIR coefficients can be computed mathematically using, for example, the minimum MSE criterion. This is accomplished

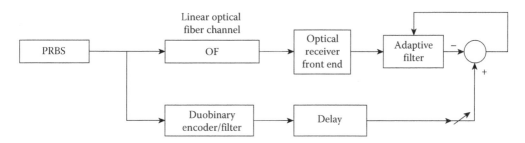

FIGURE 12.12 Schematic of the predistortion equalizer for linear distortion of optical fiber transmission. (Adapted from N. Alic et al., *Opt. Express*, Vol. 13, No. 12, pp. 4568–4579, 2005.)

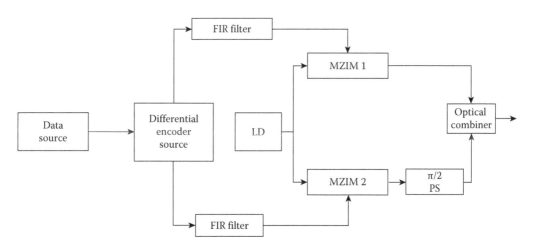

FIGURE 12.13 Schematic of optical modulation structure for generation of NRZ duobinary format. Biasing voltages and microwave amplifiers are not shown for MZIMs.

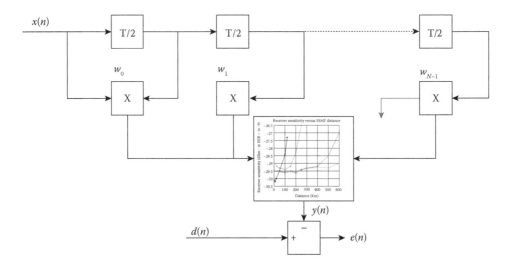

FIGURE 12.14 Structure of FIR filter for predistortion of the electrical driving signal at the optical transmitter.

by solving the Wiener–Hopf equation [3]. The signal input and tap-weight vectors can be defined as shown in Figure 12.14 and written as

$$s^{\mathrm{H}}(n) = \left[s^*(n)..s^*(n-1)..s^*(n-2)...s^*(n-N+1) \right]^{\mathrm{H}}$$
$$s^{\mathrm{T}}(n) = \left[s(n)..s(n-1)..s(n-2)...s(n-N+1) \right]^{\mathrm{T}}$$

(12.77)

and

$$c^{\mathrm{T}} = \left[c_0..c_1..c_2...c_N \right]^{\mathrm{T}}$$
$$c^{\mathrm{H}} = \left[c_0^*..c_1^* c_2^*...c_N^* \right]^{\mathrm{H}}$$

(12.78)

where the superscripts H and T denote the Hermitian and the transpose operation, respectively. The equalizer output can then be expressed in matrix form

$$y(n) = c^{\mathrm{T}} s(n)$$

(12.79)

Note that the samples of the input $s(n)$ and output $y(n)$ of the equalizer are at $T/2$ intervals. However, in the optimization of the equalizer tap weights, only the output samples at T intervals are of concern. Thus, the error signal is defined as

$$e(n) = d(n) - y(2n)$$

(12.80)

Figure 12.14 shows the proposed pre-equalization scheme and the fractionally spaced FIR filter, and thus the performance function to be minimized is written as

$$J = E\left[e^2(n) \right]$$
$$d(n) = \sum_i \gamma_i w(n-i)$$

(12.81)

where γ_i is the impulse response of the digital duobinary filter. This impulse response, if under numerical simulation, would be padded with enough zeros to account for the delay of the transmitter filter and optical fiber, and is the uncorrupted data sequence. The fractionally sampled channel response (includes both the transmitter filter and optical fiber) is expressed as

$$y(nT) = [y_0..y_1..y_2...y_{M-1}]^T \tag{12.82}$$

where h is a sufficiently large integer such that the values of h_i for $i > (M-1)$ are negligible. The optimum tap weights are obtained by solving the Wiener–Hopf equation [3], which yields the optimum tap-weight vector, given by

$$c_0 R^{-1} p$$

with

$$R = E[s(2n)s^H(2n)] \text{ and } p = E[d*(n)s(2n)] \tag{12.83}$$

and

$$s(2n) = Hw(n) + v(2n)$$

$$w(n) = \left[w(n)..w(n-1).....w\left(n - \frac{N}{2} + 1\right) \right]^T \tag{12.84}$$

with

$$H = \begin{bmatrix} y_0 & y_2 & y_4 & y_6 & \cdots & y_{M-2} & 0 & \cdots \\ 0 & y_1 & y_3 & y_5 & \cdots & y_{M-3} & y_{M-1} & \cdots \\ 0 & 0 & y_2 & y_4 & \cdots & y_{M-4} & y_{M-2} & \cdots \\ \cdot & \cdot & \cdot & \cdot & \cdot & \cdot & \cdot & \cdot \\ \cdot & \cdot & \cdot & \cdot & \cdot & \cdot & \cdot & \cdot \end{bmatrix}^T$$

$w(n)$ is a vector of noise samples. Note that the matrix H is a circulant matrix with alternating rows in which a row is formed by shifting one column of the row to two levels above it. The noise, taken to be white noise, is added to avoid excessive filter gains at the frequencies where the magnitude of the channel frequency response is very small. Equation 12.14 may also be rewritten as

$$d(n) = \gamma^T w(n) \tag{12.85}$$

where $d(n)$ is a column vector of the target impulse response. From Equation 12.14, we can obtain

$$R = E[Hss^H H^H] + \sigma^2 I = HIH^H + \sigma^2 I = HH^H + \sigma^2 I \tag{12.86}$$

where σ is the variance of the added noise, and I is an identity matrix. Similarly

$$P = E[Hw^H H^H \gamma] = HI\gamma + HI\gamma \tag{12.87}$$

Thus, the desired optimum $T/2$-spaced FIR equalizer can be written as

$$c_0 = [HH^H + \sigma^2 I]^{-1} H\gamma \tag{12.88}$$

Figure 12.15 [13] shows significant improvement of the sensitivity of the linear equalizer in the 10 Gbps duobinary modulation format under the following conditions as shown in Table 12.3. Recently, the fundamental limits of the direct detection of duobinary modulation signals have been reported [7].

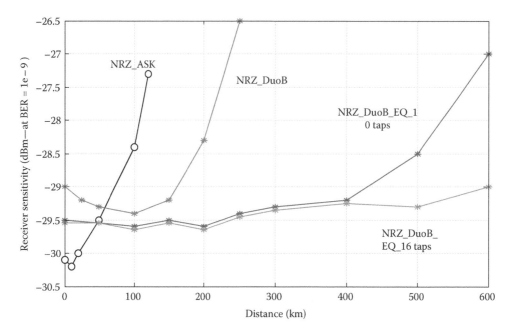

FIGURE 12.15 Receiver sensitivity versus fiber SSMF distance for conventional NRZ-ASK, NRZ duobi-nary, and predistortion equalized duobinary of different number of taps of the linear equalization system. (Adopted and simulation checked from N. Alic et al., Performance of maximum likelihood sequence estimation with different modulation formats, in *Proceedings of LEOS'04*, pp. 49–50, 2004.)

TABLE 12.3

Key Simulation Parameters Used in the LE for Duobinary Modulation Format

Optical amplifier gain 20 dB, ASE noise $n_{sp} = 2$	Narrowband Gaussian filter: $B = 5.2$ GHz or $BT = 0.13$
Operating wavelength: $\lambda = 1{,}550$ nm	Constant delay: $t_d = \left\| 2\pi f_D \beta_2 L \right\|$
Bit rate: $R = 10$ Gbps	Preamp. EDFA of the OFDR: $G = 15$ dB and $NF = 5$ dB
SSMF fiber: $\left\| \beta_2 \right\| = 2.68\mathrm{e}{-}26$ or $\left\| D \right\| = 17$ ps/nm/km, dispersion slope 0.072 ps/(nm²·km), 80 μm² effective diameter	$i_d = 10$ nA
Attenuation: $\alpha = 0.25$ dB/km	$N_{eq} = 20$ pA/(Hz)$^{1/2}$
Laser linewidth	4 MHz
Optical filter bandwidth and duobinary bandwidth	100 and 3.5 GHz (third-order filter Bessel type)

Eye opening penalty as a function of the dispersion tolerance in kms of SSMF under self-homodyne detection without and with linear equalizer FFE-DFE-NLC is shown in Figure 12.16.

12.5   MLSE EQUALIZER FOR OPTICAL MSK SYSTEMS

12.5.1   Configuration of MLSE Equalizer in OFDR

Figure 12.17 shows the block diagram of narrowband filter receiver integrated with nonlinear equalizers for the detection of 40 Gbps optical MSK signals. Two narrowband filters are used to discriminate the USB and LSB frequencies that correspond to logic "1" and "0" transmitted, respectively. A constant optical delay line that is easily implemented in integrated optics is introduced on one branch to compensate for the differential group velocity delay $t_d = 2\pi f_d \beta_2 L$ between F1 and F2, where $f_d = f_1 - f_2 = R/2$, β_2 represents group velocity delay (GVD) parameter of the fiber,

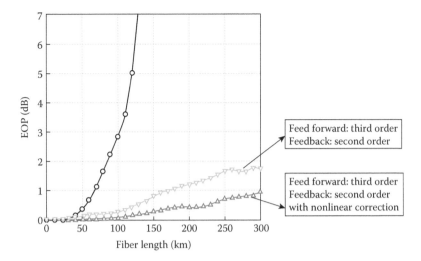

FIGURE 12.16 Eye opening penalty as a function of the dispersion tolerance in kms of SSMF without and with linear equalizer FFE-DFE-NLC.

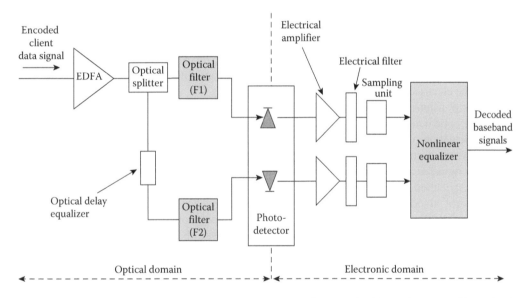

FIGURE 12.17 Block diagram of narrowband filter receiver integrated with nonlinear equalizers for the detection of 40 Gbps optical MSK signals.

and L is the fiber length. If the differential group delay is fully compensated, the optical lightwaves in two paths arrive at the photodiodes simultaneously. The outputs of the filters are then converted into the electrical domain through the photodiodes. These two separately detected electrical signals are sampled before being fed as inputs to the nonlinear equalizer.

12.5.2 MLSE Equalizer with Viterbi Algorithm

At epoch k, it is assumed that the effect of ISI on an output symbol of the finite state machine (FSM) c_k is caused by both executive δ pre-cursor and δ post-cursor symbols on each side. First, a state trellis is constructed with 2δ states for both detection branches of the OFDR. A lookup table per

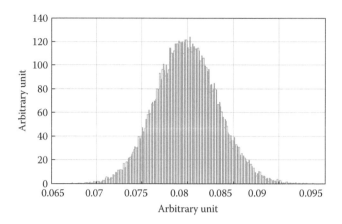

FIGURE 12.18   Noise distribution following Gaussian shape due to narrowband optical filtering.

branch corresponding to symbols "0" and "1" transmitted and containing all the possible $2^{2\delta}$ states of all 11 symbol-length possible sequences is constructed by sending all the training sequences incremented from 1 to $2^{2\delta}$.

The output sequence $c(k) = f(b_k, b_{k-1},...,b_{k-2\delta-1}) = (c_1, c_2,...,c_{2\delta})$ is the nonlinear function representing the ISI caused by the δ adjacent pre-cursor and post-cursor symbols of the optical fiber FSM. This sequence is obtained by selecting the middle symbols c_k of 2δ possible sequences with a length of $2\delta + 1$ symbol intervals.

The samples of the two filter outputs at epoch $k\, y_k^i$, $i = 1, 2$ can be represented as $y^i(k) = c^i(k) + n_{ASE}^i(k) + n_{Elec}^i(k)$, $i = 1, 2$. Here, n_{ASE}^i and n_{Elec}^i represent the amplified spontaneous emission (ASE) noise and the electrical noise, respectively.

In linear transmission of an optical system, the received sequence y_n is corrupted by the ASE noise of the optical amplifiers, n_{ASE}, and the electronic noise of the receiver, n_E. It has been proven that the calculation of branch metric and hence state metric is optimum when the distribution of noise follows the normal/Gaussian distribution—that is, the ASE noise and the electronic noise are collectively modeled as samples from Gaussian distributions. If noise distribution departs from the Gaussian distribution, the minimization process is suboptimum.

The Viterbi algorithm subsystem is implemented on each detection branch of the OFDR. However, the MLSE with Viterbi algorithm may be too computationally complex to be implemented at 40 Gbps with the current integration technology. However, there have been commercial products available for 10 Gbps optical systems. Thus, a second MLSE equalizer using the technique of reduced-state template matching is presented in the next section.

In an optical MSK transmission system, narrowband optical filtering plays the main role in shaping the noise distribution back to the Gaussian profile. The Gaussian-profile noise distribution is verified in Figure 12.18. Thus, branch metric calculation in the Viterbi algorithm, which is based on minimum Euclidean distance over the trellis, can achieve optimum performance. Also, the computational effort is less complex than ASK or DPSK systems due to the issue of non-Gaussian noise distribution.

12.5.3   MLSE Equalizer with Reduced-State Template Matching

The modified MLSE is a single-shot template-matching algorithm. First, a table of $2^{2\delta+1}$ templates, $\mathbf{g}^k$, $k = 1, 2, ..., 2^{2\delta+1}$, corresponding to the $2^{2\delta+1}$ possible information sequences, I^k of length $2\delta + 1$, is constructed. Each template is also a vector of size $2\delta + 1$, which is obtained by transmitting the corresponding information sequence through the optical channel and obtaining the $2\delta + 1$ consecutive received samples. At each symbol period, n the sequence, $\hat{I}_n$ with the minimum metric is selected as $c(k) = \underset{b(k)}{\arg\min}\{m(b(k), y(k))\}$. The middle element of the selected information sequence is then output as the nth decision, $\hat{I}_n$, that is, $\hat{I}_n = \hat{I}_n(\delta+1)$. The minimization is performed over the information sequences, $\mathbf{I}^k$, which satisfy the condition that the $\delta-2$ elements are equal to the previously decoded symbols $\hat{I}_{n-\delta}$, $\hat{I}_{n-(\delta-1)}, \ldots, \hat{I}_{n-2}$ and the metric, $m(\mathbf{I}^k, \mathbf{r}_n)$, is given by

$$m(\mathbf{I}^k, \mathbf{r}_n) = \left\{ \mathbf{w} \cdot \left(\mathbf{g}^k - \mathbf{r}_n \right) \right\}^T \left\{ \mathbf{w} \cdot \left(\mathbf{g}^k - \mathbf{r}_n \right) \right\},$$ where w is a weighting vector that is chosen carefully to improve the reliability of the metric. The weighting vector is selected so that, when the template is compared with the received samples, less weighting is given to the samples further away from the middle sample. For example, we found through numerical results that a weighting vector with elements $\mathbf{w}(i) = 2^{-|i-(\delta+1)|}$, $i = 1, 2, \ldots, 2\delta+1$ gives good results. Here, $\cdot$ represents Hadamard multiplication of two vectors, $(\cdot)^T$ represents transpose of a vector, and $|\cdot|$ is the modulus operation.

12.6 MLSE SCHEME PERFORMANCE

12.6.1 PERFORMANCE OF MLSE SCHEMES IN 40 GBPS TRANSMISSION

Figure 12.19 shows the simulation system configuration used for the investigation of the performance of both the preceding schemes when used with the narrowband optical Gaussian filter receiver for the detection of noncoherent 40 Gbps optical MSK systems. The input power into fiber (P_0) is −3 dBm, which is much lower than the nonlinear threshold power. The EDFA2 provides 23 dB gain to maintain the receiver sensitivity of −23.2 dBm at BER = 1e−9. As shown in Figure 12.19, the optical received power (P_{Rx}) is measured at the input of the narrowband MSK receiver, and the OSNR is monitored to obtain the BER curves for different fiber lengths. The length of SSMF is varied from 48 to 60 km in steps of 4 km to investigate the performance of the equalizers to the degradation caused by fiber cumulative dispersion. The narrowband Gaussian filter with the time bandwidth product of 0.13 is used for the detection filters.

Electronic noise of the receiver can be modeled with equivalent noise current density of electrical amplifier of 20 pA/$\sqrt{\text{Hz}}$ and dark current of 10 nA. The key parameters of the transmission system are given in Table 12.4.

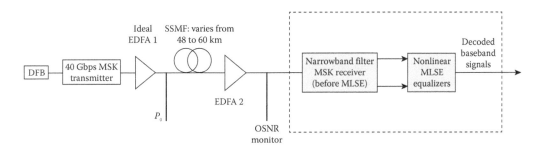

FIGURE 12.19 Simulation setup for performance evaluation of MLSE and modified MLSE schemes for the detection of 40 Gbps optical MSK systems.

TABLE 12.4

Key Simulation Parameters Used in the MLSE-MSK Modulation Format

Input power: $P_0 = -3$ dBm	Narrowband Gaussian filter: $B = 5.2$ GHz or $BT = 0.13$
Operating wavelength: $\lambda = 1550$ nm	Constant delay: $t_d = \left\| 2\pi f_D \beta_2 L \right\|$
Bit rate: $R = 40$ Gbps	Preamp. *EDFA* of the *OFDR*: $G = 15$ dB and $NF = 5$ dB
SSMF fiber: $\left\| \beta_2 \right\| = 2.68\text{e}{-}26$ or $\left\| D \right\| = 17$ ps/nm/km	$i_d = 10$ nA
Attenuation: $\alpha = 0.2$ dB/km	$N_{eq} = 20$ pA/(Hz)$^{1/2}$

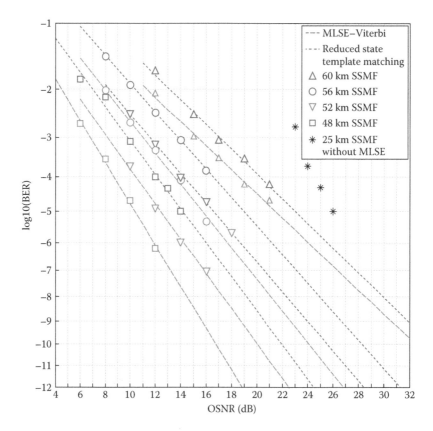

FIGURE 12.20 Performance of modified Viterbi-MLSE and template-matching schemes, BER versus OSNR.

The Viterbi algorithm used with the MLSE equalizer has a constraint length of 6 (i.e., 2^5 number of states) and a trace-back length of 30. Figure 12.20 shows the BER performance of both nonlinear equalizers plotted against the required optical OSNR. The BER performance of the optical MSK receiver without any equalizers for 25 km SSMF transmission is also shown in Figure 12.20 for quantitative comparison. The numerical results are obtained via Monte Carlo simulation (triangular markers as shown in Figure 12.20) with the low BER tail of the curve linearly extrapolated.

The OSNR penalty (at BER = 1e−9) versus residual dispersion corresponding to 48, 52, 56, and 60 km SSMF are presented in Figure 12.21. The MLSE scheme outperforms the modified MLSE schemes, especially at low OSNR. In the case of 60 km SSMF, the improvement at BER = 1e−9 is approximately 1 dB and, in the case of 48 km SSMF, the improvement is about 5 dB. In case of transmission of 52 km and 56 SSMF, MLSE with Viterbi algorithm has 4 dB gain in OSNR, compared to the modified scheme. With residual dispersion at 816, 884, 952, and 1020 ps/nm, at a BER of 1e−9, the MLSE scheme requires 12, 16, 19, and 26 dB OSNR penalty, respectively, while the modified scheme requires 17, 20, 23, and 27 dB OSNR, respectively.

12.6.2 Transmission of 10 Gbps Optical MSK Signals over 1472 km SSMF Uncompensated Optical Link

Figure 12.22 shows the simulation setup for 10 Gbps transmission of optical MSK signals over 1472 km SSMF. The receiver employs an optical narrowband frequency discrimination receiver integrated with a 1024-state Viterbi-MLSE postequalizer. The input power into the fiber (P_0)

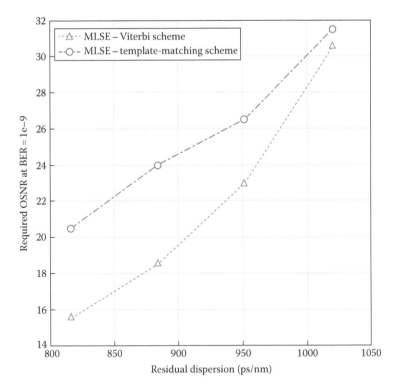

FIGURE 12.21   Required OSNR (at BER = 1e−9) versus residual dispersion in Viterbi-MLSE and template-matching MLSE schemes.

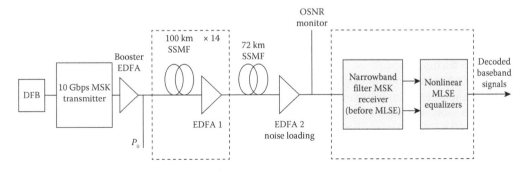

FIGURE 12.22   Simulation setup for transmission of 10 Gbps optical MSK signals over 1472 km SSMF with MLSE-Viterbi equalizer integrated with the narrowband optical filter receiver.

is −3 dBm lower than the fiber nonlinear threshold power. The optical amplifier EDFA1 provides an optical gain to compensate the attenuation of each span completely. EDFA2 is used as a noise-loading source to vary the required OSNR. The receiver electronic noise is modeled with equivalent noise current density of the electrical amplifier of 20 $pA/\sqrt{Hz}$ and dark current of 10 nA for each photodiode. A narrowband optical Gaussian filter with two-sided bandwidth of 2.6 GHz (one-sided $BT = 0.13$) is optimized for detection. A back-to-back OSNR = 8 dB is required for BER at 1e−3 for each branch. The correspondent received power is −25 dBm. This low OSNR is possible due to the large suppression of noise after being filtered by narrowband optical filters. A trace-back length of 70 is used in the Viterbi algorithm. Figure 12.22 shows the simulation results of BER versus the required OSNR for 10 Gbps optical MSK transmission over 1472 and 1520 km SSMF uncompensated optical links with one, two, and four samples per bit, respectively. The 1520 km SSMF with

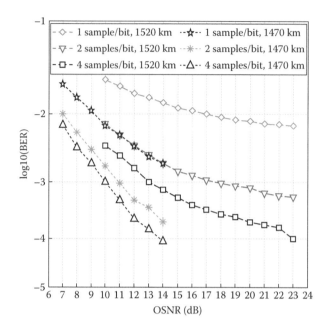

FIGURE 12.23 BER versus required OSNR for transmission of 10 Gbps optical MSK signals over 1472 and 1520 km SSMF uncompensated optical link.

one sample per bit is seen as the limit for 1024-state Viterbi algorithm due to the slow roll-off and the error floor. However, two and four samples per bit can obtain error values lower than the FEC limit of 1e−3.

Thus, 1520 km SSMF transmission of 10 Gbps optical MSK signals can reach error-free detection with the use of a high-performance FEC. In the case of 1472 km SSFM transmission, the error events follow a linear trend without sign of error floor and, therefore, error-free detection can be comfortably accomplished.

Figure 12.23 also shows the significant improvement in OSNR of two and four samples per bit over one sample per bit counterpart with values of approximately 5 and 6 dB, respectively. In terms of OSNR penalty at a BER of 1e−3 from back-to-back setup, four samples per bit for 1472 and 1520 km transmission distance suffers 2 and 5 dB penalty, respectively.

12.6.3 PERFORMANCE LIMITS OF VITERBI-MLSE EQUALIZERS

The performance limits of Viterbi-MLSE equalizer to combat ISI effects are investigated against various SSMF lengths of the optical link. The number of states used in the equalizer is incremented according to this increase and varied from 2^6 to 2^{10}. This range was chosen as reflecting the current feasibility and future advance of electronic technologies that can support high-speed processing of the Viterbi algorithm in the MLSE equalizer. In addition, these numbers of states also provide feasible time for simulation.

In addition, one possible solution to ease the requirement of improving the performance of the equalizer without increasing much the complexity is by multisampling within one-bit period. This technique can be done by interleaving the samplers at different times. Although a greater number of electronic samplers are required, they only need to operate at the same bit rate as the received MSK electrical signals. Moreover, it will be shown later on that there is no noticeable improvement with more than two samples per bit period. Hence, the complexity of the MLSE equalizer can be affordable while improving the performance significantly.

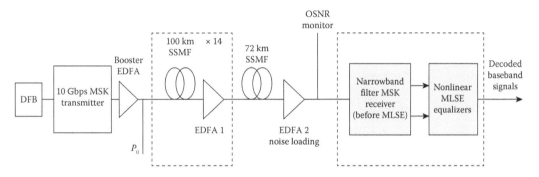

FIGURE 12.24   Simulation setup of OFDR-based 10 Gbps MSK optical transmission for the study of performance limits of the Viterbi-MLSE equalizer.

Figure 12.24 shows the simulation setup for 10 Gbps optical MSK transmission systems with lengths of uncompensated optical links varying up to 1472 km SSMF. In this setup, the input power into fiber (P_0) is set to be −3 dBm, thus avoiding the effects of fiber nonlinearities. The optical amplifier EDFA1 provides an optical gain to compensate for the attenuation of each span completely. The EDFA2 is used as a noise-loading source to vary the required OSNR values. Moreover, a Gaussian filter with two-sided 3-dB bandwidth of 9 GHz (one-sided $BT = 0.45$) is utilized as the optical discrimination filter because this BT product gives the maximized detection's eye openings (refer to Section 9.4). The receiver's electronic noise is modeled with equivalent noise current density of the electrical amplifier of 20 $pA/\sqrt{Hz}$ and a dark current of 10 nA for each photodiode. A back-to-back $OSNR = 15$ dB is required for a BER of 1e−4 on each branch, and the corresponding receiver sensitivity is −25 dBm. A trace-back length of 70 is used for the Viterbi algorithm in the MLSE equalizer. Figures 12.24 through 12.26 show the BER performance curves of 10 Gbps OFDR-based MSK optical transmission systems over 928, 960, 1,280, and 1470 km SSMF uncompensated optical links for different numbers of states used in the Viterbi-MLSE equalizer. In these figures, the performance of Viterbi-MLSE equalizers is given by the plot of the BER versus the required OSNR for several detection configurations: balanced receiver (without the equalizer), the conventional single-sample-per-bit sampling technique, and the multisample-per-bit sampling techniques (two and four samples per bit slot). The simulation results are obtained by the Monte Carlo method.

The significance of multisamples per bit slot in improving the performance for MLSE equalizer in cases of uncompensated long distances is shown. It is found that the tolerance limits to the ISI effects induced from the residual CD of a 10 Gbps MLSE equalizer using 2^6, 2^8, and 2^{10} states are approximately equivalent to lengths of 928, 1280, and 1440 km SSMF, respectively. The equivalent numerical figures in the case of 40 Gbps transmission correspond to lengths of 62, 80, and 90 km SSMF, respectively.

In the case of 64 states over 960 km SSMF uncompensated optical link, the BER curve encounters an error floor that cannot be overcome even by using high-performance FEC schemes. However, at 928 km, the linear BER curve indicates the possibility of recovering the transmitted data with the use of a high-performance FEC. Thus, for 10 Gbps OFDR-based MSK optical systems, a length of 928 km SSMF can be considered as the transmission limit for the 64-state Viterbi-MLSE equalizer. Results shown in Figure 12.26 suggest that the length of 1280 km SSMF uncompensated optical link is the transmission limit for the 256-state Viterbi-MLSE equalizer when incorporating with OFDR optical front end. It should be noted that this is achieved when using the multisample-per-bit sampling schemes.

It is observed from Figure 12.27 that the transmission length of 1520 km SSMF with one sample per bit is seen as the limit for the 1024-state Viterbi algorithm due to the slow roll-off

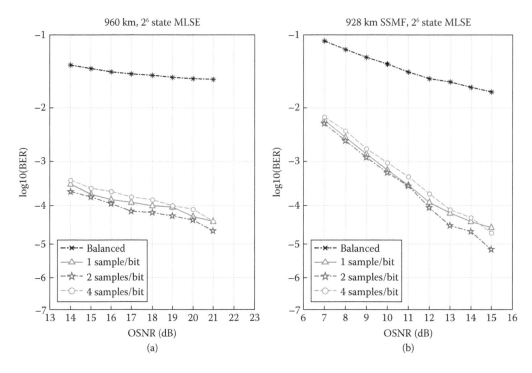

FIGURE 12.25 Performance of 64-state Viterbi-MLSE equalizers for 10 Gbps OFDR-based MSK optical systems over (a) 960 km and (b) 928 km SSMF uncompensated links.

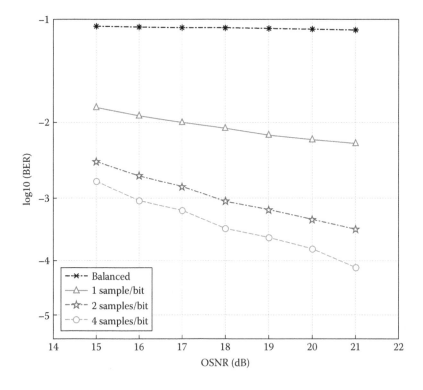

FIGURE 12.26 Performance of 256-state Viterbi-MLSE equalizer for 10 Gbps OFDR-based MSK optical systems over 1280 km SSMF uncompensated link.

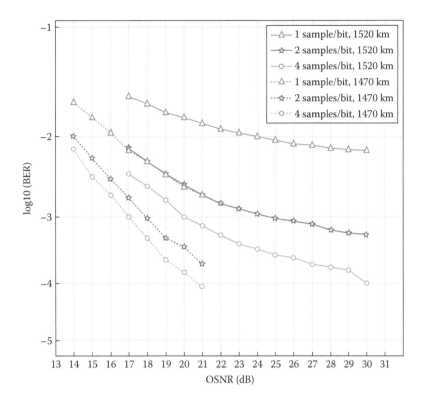

FIGURE 12.27 Performance of 1024-state MLSE equalizer for 10 Gbps optical MSK signals versus required OSNR over 1472 and 1520 km SSMF uncompensated optical link.

and the error floor. However, two and four samples per bit can obtain error values lower than the FEC limit of 1e−3. Thus, 1520 km SSMF transmission of 10 Gbps optical MSK signals can reach error-free detection with the use of a high-performance FEC. In case of 1472 km SSFM transmission, the error events follow a linear trend without the sign of error floor and, therefore, error-free detection can be achieved. Figure 12.27 also shows the significant OSNR improvement of the sampling techniques with two and four samples per bit compared to the single-sample-per-bit counterpart. In terms of OSNR penalty at BER of 1e−3 from back-to-back setup, four samples per bit for 1472 and 1,520 km transmission distance suffers 2 and 5 dB penalty, respectively.

It is found that, for incoherent detection of optical MSK signals based on the OFDR, an MLSE equalizer using 2^4 states does not offer better performance than the balanced configuration of the OFDR itself, that is, the uncompensated distance is not over 35 km SSMF for 40 Gbps or 560 km SSMF for 10 Gbps transmission systems, respectively. It is most likely that, in this case, the severe ISI effect caused by the optical fiber channel has spread beyond the time window of five bit slots (two pre-cursor and two post-cursor bits), the window that 16-state MLSE equalizer can handle.

The trace-back length used in the investigation is chosen to be 70, which guarantees the convergence of the Viterbi algorithm. The longer the trace-back length, the larger the memory that is required. With the state-of-the-art technology for high storage capacity nowadays, memory is no longer a big issue. Very fast processing speed at 40 Gbps operations hinders the implementation of 40 Gbps Viterbi-MLSE equalizers at the mean time. Multisample sampling schemes offer an exciting solution for implementing fast signal processing algorithms. This challenge may also be overcome in the near future together with the advance of the semiconductor industry. At the present, the realization of Viterbi-MLSE equalizers operating at 10 Gbps has been commercially demonstrated [14].

12.6.4   VITERBI-MLSE EQUALIZERS FOR PMD MITIGATION

Figure 12.28 shows the simulation test bed for the investigation of MLSE equalization of the PMD effect. The transmission link consists of a number of spans that comprise 100 km SSMF (with $D = +17$ ps/(nm·km), $\alpha = 0.2$ dB/km) and 10 km DCF (with $D = -170$ ps/(nm·km), $\alpha = 0.9$ dB/km). Input power into each span (P_0) is -3 dBm. The EDFA1 has a gain of 19 dB, hence providing input power into the DCF to be -4 dBm, which is lower than the nonlinear threshold of the DCF. The 10 dB gain of EDFA2 guarantees the input power into next span unchanged of -3 dBm value. An OSNR of 10 dB is required for receiver sensitivity at BER of 1e−4 in case of back-to-back configuration. Considering the practical aspect and complexity a Viterbi-MLSE equalizer for PMD equalization, a small number of four state bits, or effectively 16 states, were chosen for the Viterbi algorithm in the simulation study.

The performance of MLSE against the PMD dynamic of optical fiber is investigated. BER versus required OSNR for different values of normalized average differential group delay (DGD) $<\Delta\tau>$ are shown in Figure 12.29. The mean DGD factor is normalized over one-bit period of 100 and 25 ps for 10 and 40 Gbps bit rate, respectively. The numerical studies are conducted for a range of values from 0 to 1 of normalized DGD $<\Delta\tau>$, which equivalently corresponds to the instant delays of up to 25 or 100 ps in terms of 40 and 10 Gbps transmission bit rate, respectively.

The advantages of using multiple samples per bit over the conventional single sample per bit in MLSE equalizers are also numerically studied, and the Monte Carlo results for normalized DGD values of 0.38 and 0.56 are shown in Figures 12.30 and 12.31. However, the increase from two to four samples per bit does not offer any gain in the performance of the Viterbi-MLSE equalizer. Thus, two samples per bit are preferred to reduce the complexity of the equalizer. Here, and for the rest of this chapter, multisamples per bit implies the implementation of two samples per bit. Performance of MLSE equalizer for different normalized DGD values with two samples per bit is shown in Figure 12.32. From Figure 12.32, the important remark is that 16-state MLSE equalizer implementing two samples per bit enables the optical MSK transmission systems achieving a PMD tolerance of up to one-bit period at BER = 1e−4 with a required OSNR of 8 dB. This delay value starts introducing a BER floor indicating the limit of the 16-state MLSE equalizer. This problem can be overcome with a high-performance FEC. However, the DGD mean value of 0.94 can be considered as the limit for an acceptable performance of the proposed MLSE equalizer without the aid of a high-performance FEC. Figure 12.33 shows the required OSNR for MLSE performance at 1e−4 versus normalized $<\Delta\tau>$ values in configurations of balanced receiver, one sample per bit, and two samples per bit.

A balanced receiver without the use of the MLSE equalizer requires an OSNR = 5 dB to obtain a BER = 1e−4 when there is no effect of the PMD at all compared to the required OSNR of 3 dB and 1 dB in cases of one sample per bit and two samples per bit. The OSNR penalties for various normalized $<\Delta\tau>$ values for the preceding three configurations are shown in Figure 12.34.

It can be observed that the OSNR penalties (back-to-back) of approximately 3 and 1 dB apply to the cases of balanced receiver and one sample per bit, respectively, with reference to the

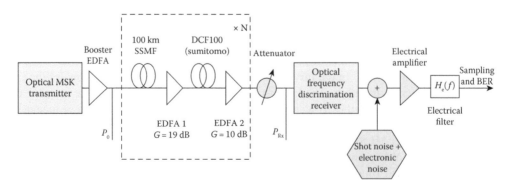

FIGURE 12.28   Simulation test bed for investigation of effectiveness of MLSE equalizer to PMD.

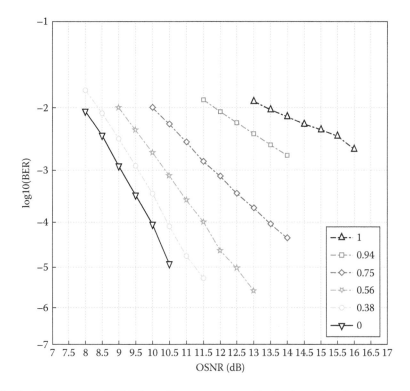

FIGURE 12.29   Performance of MLSE equalizer versus normalized <Δτ> values for one sample per bit.

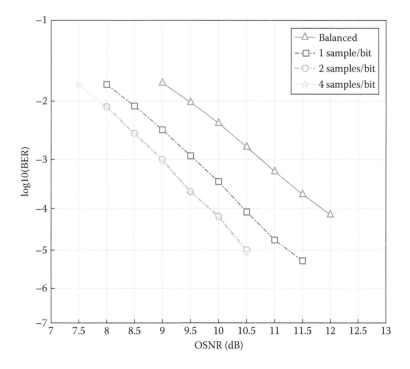

FIGURE 12.30   Comparison of MLSE performance for configurations of one, two, and four samples per bit with normalized <Δτ> value of 0.38.

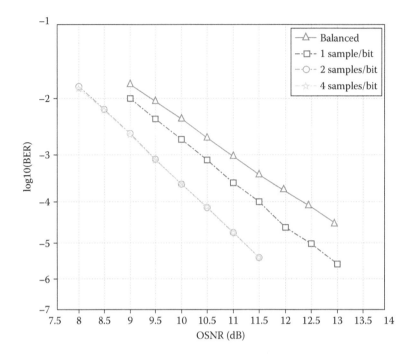

FIGURE 12.31 Performance of Viterbi-MLSE equalizer for configurations of one, two, and four samples per bit with a normalized $<\Delta\tau>$ value of 0.56.

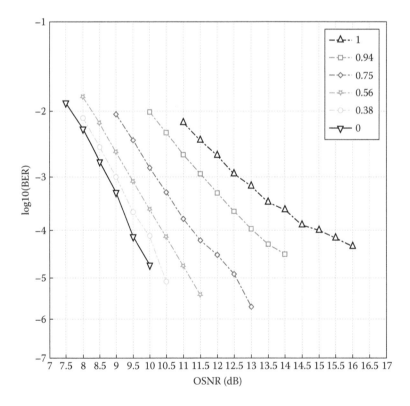

FIGURE 12.32 Performance of Viterbi-MLSE equalizer for different normalized $<\Delta\tau>$ values with two and four samples per bit.

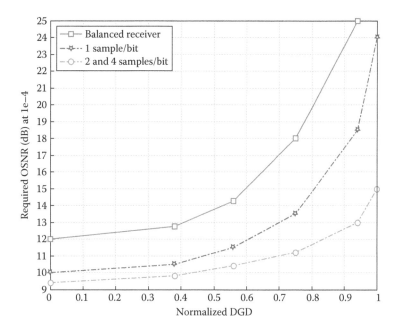

FIGURE 12.33 Required OSNR for MLSE performance at 1e−4 versus normalized <Δτ> values in configurations of balanced receiver and one and two samples per bit.

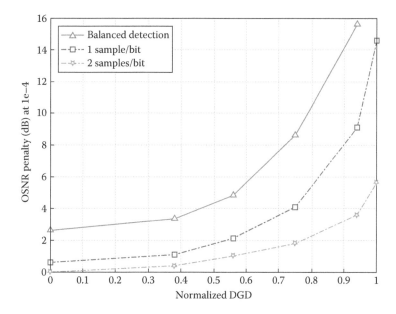

FIGURE 12.34 OSNR penalties of MLSE performance at 1e−4 for various normalized <Δτ> values in configurations of balanced receiver, single-sample, and two-samples-per-bit sampling schemes.

two-samples-per-bit configuration. Another important remark is that the 16-state Viterbi-MLSE equalizer that implements two samples per bit enables the optical MSK transmission systems to achieve a PMD tolerance of up to one-bit period at a BER of 1e−4 with a power penalty of about 6 dB. Moreover, the best 2 dB penalty occurs at 0.75 for the value of normalized <Δτ>. This result shows that the combination of OFDR-based MSK optical systems and Viterbi-MLSE equalizers, particularly with the use of multisample sampling schemes, was found to be highly effective in combating the fiber PMD dynamic impairment and better than the recently reported PMD performance for OOK and DPSK modulation formats [15].

12.6.5 THE UNCERTAINTY AND TRANSMISSION LIMITATION OF THE EQUALIZATION PROCESS

The fundamental limitations of a quadratic phase media on the transmission speed are formulated in signal space for significant applications in signal equalization. These limitations are quantitatively derived for both coherent and incoherent optical systems. This section investigates the dispersive effect of optical fiber on the transmission speed of the media. The main effect of dispersion is pulse broadening (and frequency chirping), as the signal pulses propagate through the quadratic phase media. This causes ISI, thereby limiting the speed of transmission. There are various ways of mitigating the harmful effects of ISI. One way is to use optical fiber with anti-dispersive property to equalize the dispersion. In practice, this optical scheme reduces the signal level, and thus requires amplification. The amplifier introduces noise, which in turn limits the channel speed (Shannon's information capacity theorem). Another way of combating ISI is by digital electronic equalization means such as MLSE, which is applicable for both coherent and incoherent detection schemes [13,15–18]. These reported results show clearly that there is a fundamental limit on the achievable transmission speed for a given fiber length.

This section studies the limitations of the quadratic phase channel, the single-mode optical fiber, in terms of signal space. The signal space is chosen so that it consists of binary signals in an 8-bit block code. As the eight-bit symbols propagate down the quadratic phase channel, the waveform patterns of the symbols would naturally become less and less distinctive, hence making it more and more difficult to discriminate between the symbols. How do we quantify the detrimental effect of the quadratic phase channel? It is obvious that the more information we have at hand, the more accurate one would be able to obtain the BER. We approach this problem entirely from a digital communications perspective, and the results so derived are the fundamental limit imposed by the quadratic phase channel. It is virtually independent of the detection scheme used. Two mechanisms that limit the transmission speed are explained. One is brought about by the finite time window available for detection. In all practical schemes, the decoder must decode each symbol within a finite period of time. The other is brought about by not using the phase information for detection. Depending on the complexity of the detection scheme chosen, the phase information is quite often lost inadvertently when the optical signal is converted into the electrical signal.

12.7 NONLINEAR MLSE EQUALIZERS FOR MSK OPTICAL TRANSMISSION SYSTEMS

12.7.1 NONLINEAR MLSE

MLSE is a well-known technique in communications for the equalization and detection of the transmitted digital signals. The concept of MLSE is discussed in brief here. An MLSE receiver determines a sequence b as the most likely transmitted sequence when the conditional probability $Pr(y|b)$ is maximized, where y is the received output sequence. If the received signal y is corrupted by a noise vector n that is modeled as a Gaussian source (i.e., $y = b + n$), it is shown that the preceding maximization operation can be equivalent to the process of minimization of the Euclidean distance d^4

$$d = \sum_k |\mathbf{y}_k - b_k|^2 \tag{12.89}$$

MLSE can be carried out effectively with the implementation of the Viterbi algorithm based on state trellis structure.

12.7.2 TRELLIS STRUCTURE AND VITERBI ALGORITHM

12.7.2.1 Trellis Structure

Information signal b is mapped to a state trellis structure by an FSM, giving the mapping output signal c as shown in Figure 12.35. The FSM can be a convolutional encoder, a trellis-based detector, or, as presented in more detail later on, the fiber medium for optical communications.

A state trellis structure created by the FSM is illustrated in Figure 12.36. At the epoch nth, the current state B has two possible output branches connecting to states E and F. In this case, these two branches correspond to the two possible transmitted symbols of "0" or "1," respectively. This predefined state trellis applies to the one bit per symbol modulation formats such as binary ASK or DPSK formats. In cases of multibit modulation per symbol, the phase trellis has to be modified. For example, in the case of DQPSK, there are four possible branches leaving the current state B and connecting the next states, corresponding to the possible input symbols of "00," "01," "10," and "11."

Furthermore, for a simplified explanation, several assumptions are made. These assumptions and according notations can be described as follows.

1. Current state B is the starting state of only two branches that connect to states E and F, denoted as b_{BE} and b_{BF}. In general, b_{B*} represents all the possible branches in the trellis that starts from state B. In this case, these two branches correspond to the possible binary transmitted information symbols of "0" or "1," respectively. The notation $c(B, a = 0)$ represents the encoded symbol (at the FSM output) for the branch BE, b_{BE}, which starts from state B and corresponds to a transmitted symbol $a = 0$. Similar conventions are applied for the case of branch BF, b_{BF}, that is, $c(B, a = 1)$, and the all the branches in the trellis.

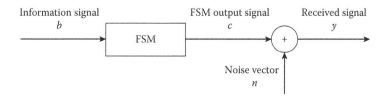

FIGURE 12.35 Schematic of the MLSE equalizer as a finite state machine (FSM).

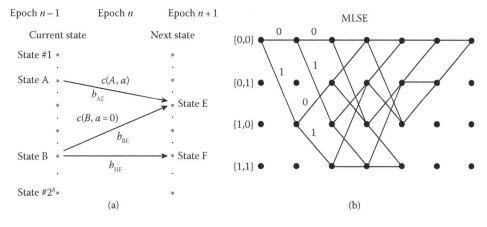

FIGURE 12.36 State trellis of an FSM: (a) trellis path and (b) tracing of possible maximum-likelihood estimation with assigned bit pattern.

2. There are only two possible output branches ending at state E (from state A and state B). They are denoted as b_{AE} and b_{BF}. In general, b_{*E} represents all the possible branches in the trellis that end at state E. The number of states 2^{δ} in the trellis structure is determined by the number of memory bits, δ, also known as the constraint length. Figure 12.36b also shows a possible trace of the trellis with the assigned states of the bit pattern.

12.7.2.2　Viterbi Algorithm

The principles of the Viterbi algorithm follow two principal stages as follows:

12.7.2.2.1　Calculation of State and Branch Metrics

At a sampling index n, epoch n, a *branch metric* is calculated for each of the branches in the trellis. For example, considering the connecting branch between the current state B and the next state E, which corresponds to the case when the "0" symbol is transmitted, the *branch metric* BM_{BE} is calculated as $BM_{BE}(n) = |y(n) - c_{BE}|^2$.

The *state metric* $SM_E(n)$ of the state E is calculated as the sum between the metric of the previous state, that is, state B and the branch metric obtained earlier. The calculation of state metric $SM_E(n)$ of state E is given by

$$SM_E(n) = \min_{\forall B \to E} \left\{ SM_B(n-1) + BM_{BE}(n) \right\} \tag{12.90}$$

The branch giving the minimum state metric $SM_E(n)$ for state E is named as the *preferred path*.

12.7.2.2.2　Trace-Back Process

The process of calculating state and branch metrics continues along the state trellis before terminating at epoch N_{trace}, which is usually referred to as the *trace-back length*. A rule of thumb, as shown in Refs. [19,20], is that the value of N_{trace} is normally taken to be five times the sequence length. This value comes from the results showing that the solution for the Viterbi algorithm converges, giving a unique path from epoch 1 to epoch N_{trace} for the MLSE detection. (1) At N_{trace}, the terminating state with the minimum state metric and its connecting *preferred path* are identified. (2) The previous state is then identified, as well as its previous *preferred path*. (3) The trace-back process continues until reaching epoch 1.

12.7.3　Optical Fiber as an FSM

The FSM shown in Figure 12.36 is now replaced with the optical channel based on optical fibers. It is significant to understand how the state trellis structure is formed for an optical transmission system. In this case, the optical fiber involves all the ISI sources causing the waveform distortion of the optical signals, which includes CD, PMD, and filtering effects. Thus, the optical fiber FSM excludes the signal corruption by noise as well as the random-process nonlinearities. The ISI caused by dispersion phenomena and nonlinear phase and ASE and nonlinear effects has been described in Chapters 3 and 8.

The construction of state trellis structure can be implemented in the following. At epoch n, it is assumed that the effect of ISI on an output symbol of the FSM c_n is caused by both executive δ precursor and δ post-cursor symbols on each side. In this case, c_n is the middle symbol of a sequence that can be represented as $c = (C_{n-\delta}, ..., C_n, ..., C_{n+\delta})$ $b \in \{0,1\}$; $n > 2\delta + 1$. Unlike the conventional FSM used in wireless communications, the output symbol c_n of the optical fiber FSM is selected as the middle symbol of the sequence c. This is due to the fact that the middle symbol gives the most reliable picture about the effects of ISI in optical fiber channels.

A state trellis is defined with a total number of $2^{2\delta}$ possible data sequences. At epoch n, the current state B is determined by the data sequence $(b_{n-2\delta+1}, ..., b_{n-\delta}, ...b_n)$ $b \in \{0,1\}$; $n \geq 2\delta + 1$, where b_n

is the current input symbol into the optical fiber FSM. Therefore, the state trellis structure can be constructed from the optical fiber FSM, and is ready to be integrated with the Viterbi algorithm, especially for the calculation of a branch metric.

The MLSE nonlinear equalizer can effectively combat all the preceding ISI effects, and it is expected that the tolerance to both CD and PMD of optical MSK systems with an OFDR receiver can be improved significantly.

12.8   UNCERTAINTIES IN OPTICAL SIGNAL TRANSMISSION

12.8.1   UNCERTAINTY IN ASK MODULATION OPTICAL RECEIVER WITHOUT EQUALIZATION

There are many ways of coding the data for transmission through the optical fiber. For simplicity, we first choose an NRZ-ASK system for this study, and the transmission rate is operating at 40 Gbps. Later in this section, we investigate the uncertainty of data recovery under MSK modulation and the use of equalization technique, and the MLSE in long-haul optically amplified transmission system.

The NRZ-ASK transmitted signal is organized into eight-bit blocks, and, thus, there are 2^8 (256) possible pattern symbols. As the eight-bit symbols propagate down the quadratic phase channel, the waveform patterns of the symbols would become less and less distinctive. It makes the task of decoding more and more difficult and uncertain. How do we measure this uncertainty? In the field of digital communications, the uncertainty of MLSE detection is measured in symbol error rate (SER), which is a function of the SNR. SNR, in turn, depends on the distances between signal constellation, and it is the shortest distance that dominates the error probability. The SER and the Q factor are expressed as a function of the normalized shortest Euclidean distance metric d_{norm} between all possible received symbol pair combinations as

$$\text{SER} = Q\left(\sqrt{\text{SNR}}\right) = Q\left(\sqrt{d_{norm}^2 \times \text{SNR}_{nofiber}}\right) \qquad 0 < d_{norm} \leq 1 \qquad (12.91)$$

where $\text{SNR}_{nofiber}$ is the optical SNR (OSNR) in the case of a back-to-back setup, that is, without fiber. From Equation 12.91, the logarithmic scale of d_{norm}^2 represents the OSNR penalty as expressed in the following equation

$$d_{norm[dB]}^2 = \text{OSNR penalty} = \text{SNR}_{[dB]} - \text{SNR}_{norfiber[dB]} \qquad (12.92)$$

Figure 12.37 depicts the penalty in the discrimination property, with the OSNR penalty being plotted against the length of the SSMF (the dispersion factor $D = 17$ ps/(nm·km)) for the case of coherent detection schemes, that is, both the magnitude and phase information of the received waveform are used.

The family of curves shown in Figure 12.37 corresponds to a range of different detection window widths; from a window of eight-bit periods (the minimum detection width) to 9-, 10-, and 32-bit periods. As expected, the discrimination property is significantly improved as the detection window length increases. In the case of 32-bit window length, there is insignificant penalty over the fiber lengths of up to 50 km SSMF. This result reaffirms the previous discussion that the channel capacity is infinite, that is, there is no uncertainty if the detection time window is long enough. At the fiber distance of 5 km, a penalty of about 0.7 dB occurs if the eight-symbol detection width is used. Depending on the system noise level, this signal level may or may not be acceptable for a reliable data transmission.

We are now considering incoherent optical transmission systems that utilize a photodiode to convert the optical signals to electrical signals. The use of photodiodes as such could result in the loss

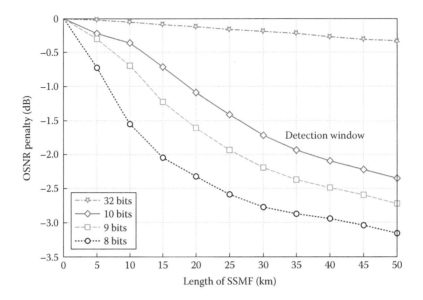

FIGURE 12.37 Discrimination penalty versus transmission distance at 40 Gbps when using both magnitude and phase information.

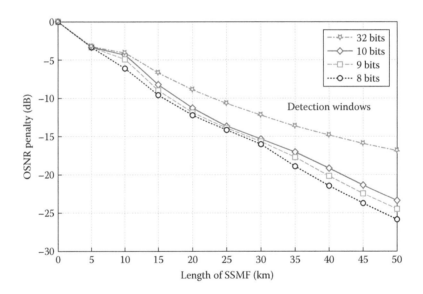

FIGURE 12.38 Discrimination penalty versus transmission distance at 40 Gbps when considering only the magnitude information.

of the phase information of the signal. Furthermore, electrical equalization techniques based on the MLSE principles employ only the magnitude information [17, 12.42–12.44]. To model this incoherent property, the distance metric utilizes the absolute magnitude of the received waveform only. The discrimination penalty versus the length of SSMF is depicted in Figure 12.38.

The family of curves in Figure 12.38 also corresponds to a range of different detection window widths (8-, 9-, 10-, and 32-bit periods). When the length of the fiber is less than about 15 km, there is no appreciable improvement by choosing a larger detection window. However, when the length of the fiber increases further, the deterioration is quite severe, and having a larger detection window helps enormously in reducing this deterioration. At the SSMF length of 5 km, the incoherent detection suffers a penalty of approximately 3.5 dB, compared to the 0.7 dB penalty for the coherent detection, and, thus, leading to an improvement of about 3 dB when using the coherent schemes with the availability of both amplitude and phase information. This agrees with the popularly known advantage on the receiver sensitivity of coherent detections over the incoherent counterparts. In addition, comparing the two sets of results in Figures 12.37 and 12.38 shows clearly the extent of improvements possible if the coherent scheme is used instead of the incoherent scheme. This is achieved at the expense of greater system complexity on the receiver structure of coherent optical systems.

12.8.2 UNCERTAINTY IN MSK OPTICAL RECEIVER WITH EQUALIZATION

The equalization processing system is placed at the output of the optical receiver shown in Figure 12.28, that is, the equalization is completed in the electronic domain, which means the optical impulse and pulses have been converted into the electrical domain and the phase chirping has already converted to amplitude distortion. However, the broadening of the pulse influences the sampled values obtained from the broadened pulses for the processing of the equalization system.

The narrowband filter receiver in Figure 12.35 is now integrated with the MLSE nonlinear equalizers for the detection of 40 Gbps optical MSK signals. Two narrowband filters are used to discriminate the USB and LSB frequencies that correspond to the logical "1" and "0" transmitted, respectively. A constant optical delay line that is easily implemented in integrated optics is introduced on one branch to compensate for the differential group delay $t_d = 2\pi f_d \beta_2 L$ between the "1" and "0" logic pulses of frequencies F1 and F2, where $f_d = f_1 - f_2 = R/2$, β_2 represents the GVD parameter of the fiber, and L is the fiber length. If the differential group delay is fully compensated, the optical lightwaves in two paths arrive at the photodiodes simultaneously. The outputs of the filters are then converted into the electrical domain through the photodiodes. These two separately detected electrical signals are sampled before being fed as the inputs to the nonlinear equalizer. The MLSE equalizer requires values of the samples that occurred in advance as well as in later events. This is the impact of the impulse response of the fiber on this equalization, and the number of events required for the detection and equalization is very critical. Figure 12.27 shows the performance of the optical frequency discrimination integrated with an MLSE equalizer in a long-haul optically amplified fiber transmission system of about 1500 km when the number of one, two, and four samples per bit is used in the equalization processing system. The time distance between the samples is controlled, so that they are within the reasonable time slot under the effects of chirping of the phase of the lightwave.

12.9 ELECTRONIC DISPERSION COMPENSATION OF MODULATION FORMATS

Except the MLSE as the pattern comparison, and the decision made based on the matching, the equalization techniques using FFE, FFE-DFE, nonlinear equalization, and FFE result in different improvements of the eye opening of digital modulation formats of either amplitude shift keying of phase shift keying.

Simulations are conducted for NRZ formats with the modulation of ASK, DuoBinary, DPSK, and SSB. The transmission link setup is implemented with spans of non-CD compensating optically amplified fibers. The bit rate is 10 Gbps, and an optical filter of 50 GHz 3 dB passband is used before the optoelectronic detection. Transmission fibers are SSMF, and 17 ps/nm/km dispersion factor is assumed. The OSNR is plotted against the linear dispersion factor, as shown in Figures 12.39 through 12.42 for NRZ-ASK, NRZ duobinary (50%), NRZ-DPSK, and NRZ-SSB formats. It noted (1) moderate improvement of the OSNR for NRZ-ASK (Figure 12.39), (2) significant improvement of OSNR for the SSB format (see Figure 12.42), (3) mild improvement for the DPSK format, and (4) moderate improvement of the duobinary format.

Thus, effectively, we note that significant improvement of the eye opening is achieved for single-sideband signals. Why this is so? In order to explain these results, see recent clarifications by

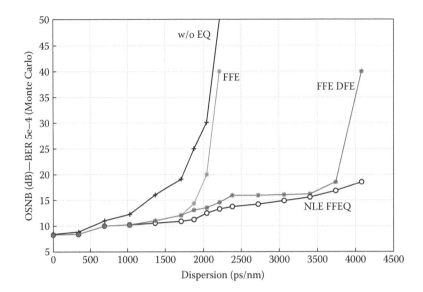

FIGURE 12.39 OSNR required for BER 5e−4 (Monte Carlo simulation) versus linear CD under different equalization schemes for NRZ-ASK modulation format.

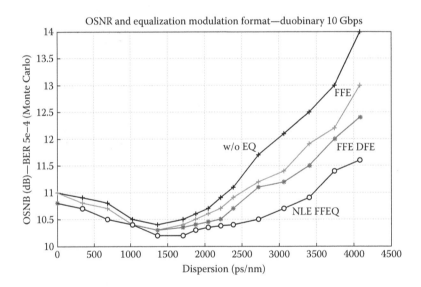

FIGURE 12.40 OSNR required for BER 5e−4 (Monte Carlo simulation) versus linear CD under different equalization schemes for NRZ duobinary modulation format.

Rosenkranz and Xia [21], although attempts have also been made some years ago [22]. The behavior of the signals received after the photodetector, a square law detection device, can be explained using the phasor representation given in Figure 12.43 as follows.

Considering a monochromatic wave of frequency ω_1 of the double sideband of the NRZ-ASK modulated carrier, then, under ASK modulation, the complex field envelope of the received signals entering the photodetector is given by

$$s_{ASK}(t) = 1 + A\frac{m}{2}e^{j(\omega_1 t - \phi_A)} + B\frac{m}{2}e^{j(-\omega_1 t - \phi_B)} \tag{12.93}$$

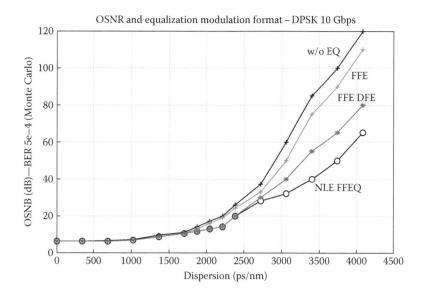

FIGURE 12.41 OSNR required for BER 5e−4 (Monte Carlo simulation) versus linear CD under different equalization schemes for NRZ-DPSK modulation format.

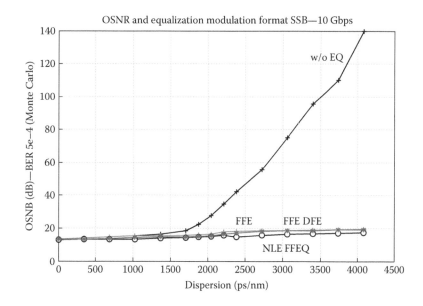

FIGURE 12.42 OSNR required for BER 5e−4 (Monte Carlo simulation) versus linear CD under different equalization schemes for NRZ-SSB modulation format.

The first term represents the carrier term at DC after the detection, the second and third terms come from the upper sideband (USB) and lower sideband (LSB), respectively, and m is the modulation index. The coefficients A and B and the additional phase shifts ϕ_A and ϕ_B are given by

$$A = |H(\omega_1)| = B = |H(-\omega_1)| \tag{12.94}$$

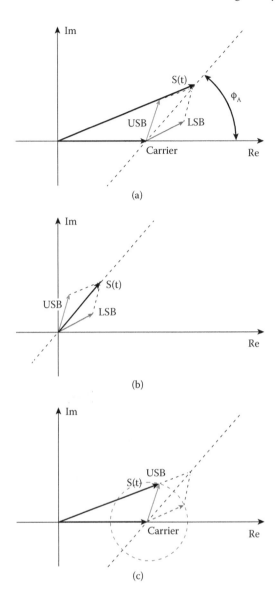

FIGURE 12.43 Phasor representation of optical signals under different modulation formats: (a) ASK—with carrier suppression and lower sideband (LSB) and upper sideband (USB), (b) SSB without carrier suppression, and (c) SSB with carrier suppressed plus USB.

where $H(\omega) = \exp(-j\beta_2\omega^2)$, $\phi_A = \phi_B = \beta_2\omega_1^2L/2$, with $H(\omega)$ being the fiber transfer function in the frequency domain. Thus, the electronic current generated by the photodetector is given by the squaring of the magnitude of the optical field, and can be written as

$$
|s_{\text{ASK}}(t)|^2 = \left|1 + A\frac{m}{2}e^{j(\omega_1 t - \phi_A)} + B\frac{m}{2}e^{j(-\omega_1 t - \phi_B)}\right|^2
$$

$$
= 1 + \frac{m^2}{2} + 2m\cos\phi_A\cos\omega_1 t + \frac{m^2}{2}\cos 2\omega_1 t
$$

(12.95)

We note from this equation that there are three terms: a DC term, a first harmonic term (linear term), and a second harmonic term that represents the distortion due to dispersion.

In the case of the DPSK, duobinary, and SSB modulation formats, we can follow a similar analysis to obtain the electronic currents generated after the photodetector

$$\left|s_{DuoB}(t)\right|^2 = \left|s_{DuoB}(t)\right|^2 = \frac{m^2}{2}(1+2m\cos\omega_1 t)$$

$$\left|s_{SSB}(t)\right|^2 = 1 + \frac{m^2}{4} + m\cos(\omega_1 t - \phi_A)$$

(12.96)

Obviously, the currents are linear, and no nonlinear term existed. A linear equalizer equalizes the signal waveform by minimizing the ISI due to the linear part of Equations 12.95 and 12.96. When the phase term $\cos(\phi_A)$ becomes smaller after a propagation length of about 100 km (or 1700 ps/nm), the linear term diminishes significantly, and the linear equalization would have no effect. The nonlinear distortion term becomes significant and takes over the linear term, as we can observe from Figure 12.39.

It is noted that the linear part comes from the carrier component, whereas for DPSK and duobinary, the carrier is embedded into the signal band and plays no role. As mentioned earlier, this is due to the effective sampling of the waveform by the switching on and off of the amplitude in ASK. Thus, from Equation 12.96, the linear term disappears, and hence the linear equalizer is not effective, as observed from Figures 12.40 and 12.41.

For SSB modulation, we could see that there is a DC term and a linear part. Thus, the linear FFE would perform well and compatible to that by the nonlinear FFE and DFE algorithms as observed in Figure 12.42.

We can visualize the effects of pure phase distortion channel on modulation schemes with DSB and SSB, as shown in Figure 12.44. For DSB modulation, the spectrum shown in Figure 12.44a in

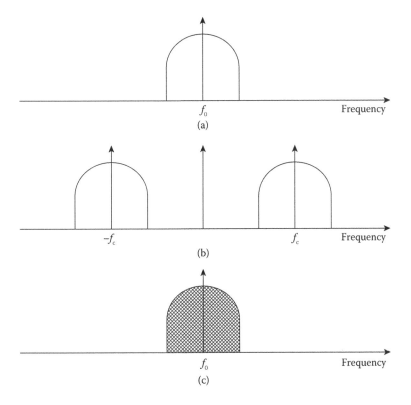

FIGURE 12.44 Spectral illustration of DSB and SSB operation after the lightwave-modulated signals propagating through a fiber length. *(Continued)*

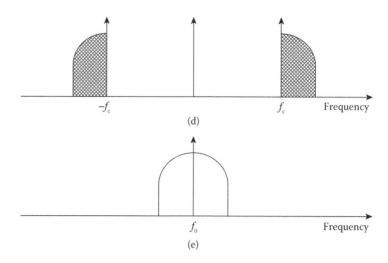

FIGURE 12.44 (Continued) Spectral illustration of DSB and SSB operation after the lightwave-modulated signals propagating through a fiber length.

the baseband is shifted to the passband (Figure 12.44b). After propagating through the fiber, the spectrum amplitude remains the same, but its phase is altered by the fiber phase distortion by an amount $-\beta_2 L f_2$.

Hence, when the signals are recovered in the photodetector, the double sidebands interfere with each other, with the overlapping (hatched area) as shown in Figure 12.44c. Thus, distortion is generated by both the band distortion and the interference between the two sidebands. On the contrary, for SSB, this recovered process would not result in any interference. Hence, distortion could only come from the pure distortion within the bands, and no distortion would be due to spectral overlapping. Thus, linear equalization works well for SSB linear distortion.

12.10 CONCLUDING REMARKS

We have described the equalization of real and imaginary signals using FFE, DFE, and FFE, and linear and nonlinear equalizers that are placed at either the transmitters, the receivers, or shared between the receivers and transmitters. The effects of equalizers on modulation formats are identified.

For phase shift keying and duobinary, the carrier is embedded within the signal band, and hence there is only nonlinear frequency doubling terms that exist in the electronic current generated by the photodetector, and thus nonlinear equalization works well and linear FFE contributes no effect when the linear term diminishes after a certain length of the fiber. The electronic current generated under the SSB modulation format, on the other hand, exhibits only the linear term and no frequency doubling term (no nonlinear term), and thus the linear equalizer FFE works as well as the nonlinear equalizer. Thus, only FFE is required as the implementation of the FFE is much simpler at high speeds. We thus expect that linear FFE also works well for MSK modulation formats.

Supplementing these investigations to prove the effects of the carrier in modulation formats, two case studies have been conducted in detail. First, a study of a linear electrical equalizer for duobinary modulation formats has been conducted, and two nonlinear equalizers are proposed, MLSE using the Viterbi algorithm and the modified MLSE scheme, which are integrated with the narrowband optical filter receiver proposed in Refs. [15,23–25] for noncoherent detection of 40 Gbps optical MSK signals. This chapter has presented the nonlinear MLSE equalizers, especially when integrated with the Viterbi algorithm, for digital photonic communications in general and for OFDR-based MSK optical systems. The backgrounds of MSLE and the Viterbi algorithm are briefly given, and the construction of the state trellis structure for use in optical communications has been explained. It has also been shown that noise after tight optical filtering that follows the Gaussian distribution can be minimized. Thus, the Viterbi algorithm can obtain its optimum performance without requiring more complex algorithms for estimating the exact noise distribution. Two MLSE electronic equalizers based on Viterbi procedures and template-matching algorithms are investigated and formulated for OFDR-based MSK optical transmission systems. In these two schemes, OFDR serves as the optical front end. The proposed equalizers significantly extend the reach of uncompensated links. With a reasonably small number of states (64 states), the uncompensated optical link can equivalently reach up to approximately 960 km SSMF for 10 Gbps transmission or 60 km SSMF for 40 Gbps. The results are comparable to the distance of 1040 km SSMF, as reported previously [26] for OOK/IM systems, in which up to 8192 trellis states are required.

The performance of the Viterbi-MLSE equalizers for OFDR-based optical MSK transmission systems in terms of transmission limits and PMD mitigation was also numerically investigated. Finally, this chapter has proven the improvement in performance of Viterbi-MLSE equalizers by using multisamples per bit compared to the conventional single sample bit. The novel scheme in which Viterbi-MLSE equalizers are incorporated with OFDR-based MSK optical transmission systems provides high robustness to the fiber PMD dynamic and better performance compared to other modulation formats.

Finally, we investigated quantitatively the transmission limits due to the combination of the dispersive effect of the quadratic phase media and restrictiveness of a finite detection and equalization window. The parameter used to quantify these limits is the least Euclidian distance between all symbol combinations, and this distance corresponds to the penalty in the system's OSNR. For the coherent detection, both phase and magnitude information of the received signals is used in the determination of the Euclidian distance, whereas only the magnitude information is used in the incoherent detection. It is found that coherent systems provide a much higher degree of discrimination, which is significant for the signal equalization at the cost of the receiver's complexity.

There is no doubt that the equalizers must be adaptive for automatic control of the adjustment of the equalization due to the nature of the dispersion fluctuation of CD and PMD, as well as the nonlinear effects of signals transmitted over the long haul distance. In practice, the equalizers described in this chapter must be adaptive. One such example is the equalization of channels that have been propagated through different hops of DWDM optical networks.

Electronic equalization emerges as an important technique for the mitigation of impairments of optical signals propagating through optical channels, and will remain a very probable candidate for reshaping the distorted signals. The progresses in the development of high-speed digital electronic processors will assist in the practical implementation of these equalizers in optical transmission systems and networks.

Ref. [22] has reported the electrical equalization of OFDM signals over long-haul transmission. The integration of multicarrier modulation and electrical equalization techniques would enable the extension of transmission beyond the dispersion limit in single-mode optical fiber communications systems. However, the double-sideband modulation of OFDM to the optical passband may face nonlinearity distortion, and the equalizer may have to be nonlinear, causing difficulties in the implementation, unless it is implemented after the recovery of the bit sequence after the PFDM decoder, as described in Chapter 11.

REFERENCES

1. Y. Matsui, D. Mahgerefteh, X. Zheng, C. Liao, Z. F. Fan, K. McCallion, and P. Tayebati, Chirp-managed directly modulated laser (CML), *IEEE Photon. Technol. Lett.*, Vol. 18, No. 2, pp. 385–387, 2006.
2. B. Wedding, B. Franz, and B. Junginger, 10-Gbps optical transmission up to 253 km via standard single-mode fiber using the method of dispersion supported transmission, *J. Lightwave Technol.*, Vol. 12, pp. 1720–1727, 1994.

3. M. M. El Said, J. Stitch, and M. I. Elmasry, An electrically pre-equalized 10 Gbps duobinary transmission system, *IEEE J. Lightwave Technol.*, Vol. 23, No. 1, pp. 388–400, 2005.

4. N. Alic, G. C. Papen, R. E. Saperstein, L. B. Milstein, and Y. Fainman, Signal statistics and maximum likelihood sequence estimation in intensity modulated fiber optic links containing a single optical preamplifier, *Opt. Express*, Vol. 13, No. 12, pp. 4568–4579, 2005.

5. H. Haunstein, PMD and chromatic dispersion control for 10 and 40 Gbps systems, *OFC'04*, invited paper, ThU3, 2004.

6. O. E. Agazzi, M. R. Hueda, H. S. Carrer, and D. E. Crivelli, Maximum-likelihood sequence estimation in dispersive optical channels, *J. Lightwave Technol.*, Vol. 23, No. 2, pp. 749–762, 2005.

7. M. Franceschini, G. Bongioni, G. Ferrari, R. Rahedi, F. Mehli, and A. Castoldi, Fundamental limits of electronic signal processing in direct detection optical communications, *IEEE J. Lightwave Technol.*, Vol. 25, No. 7, pp. 1742–1752, 2007.

8. A. P. Clarke, *Equalizers for High Speed Modems*, London: Pentech Press, 1985.

9. A. P. Clarke, *Equalizers for Digital Modem*, London: Pentech Press, pp. 262–268, 1985.

10. K. Jinguji and M. Oguma, Optical half-band filters, *IEEE J. Lightwave Technol.*, Vol. 18, No. 2, pp. 252–259, 2000.

11. W. Rosenkranz, *Lecture Notes on Optical Transmission II*, Kiel, Germany: Technical Faculty, University of Kiel, 2008.

12. A. Papoulis, *Systems and Transforms with Applications in Optics*, New York: McGraw-Hill, p. 14, 1968 (see item 1, Table 2-1, p. 14).

13. N. Alic, G. C. Papen, and Y. Fainman, Performance of maximum likelihood sequence estimation with different modulation formats, in *Proceedings of LEOS'04*, pp. 49–50, 2004.

14. D. Fritzsche, L. Schurer, A. Ehrhardt, D. Breuer, H. Oeruen, and C. G. Schaffer, *Field trial investigation of 16-states MLSE equalizer for simultaneous compensation of CD, PMD and SPM*, Con. Opti. Fiber Communi, pp. 1–3, San Diego, CA, USA, 2009.

15. T. Sivahumaran, T. L. Huynh, K. K. Pang, and L. N. Binh, Non-linear equalizers in narrowband filter receiver achieving 950 ps/nm residual dispersion tolerance for 40 Gbps optical MSK transmission systems, in *Proceedings of OFC'07*, paper OThK3, Anaheim, CA, USA, 2007.

16. P. Poggiolini, G. Bosco, M. Visintin, S. J. Savory, Y. Benlachtar, P. Bayvel, and R. I. Killey, MLSE-EDC versus optical dispersion compensation in a single-channel SPM-limited 800 km link at 10 Gbit/s, *ECOC 2007*, paper1.3, Berlin, 2007.

17. V. Curri, R. Gaudino, A. Napoli, and P. Poggiolini, Electronic equalization for advanced modulation formats in dispersion-limited systems, *Photon. Technol. Lett.*, Vol. 16, No. 11, pp. 2556–2558, 2004.

18. G. Katz, D. Sadot, and J. Tabrikian, Electrical dispersion compensation equalizers in optical long-haul coherent-detection system, in *Proceedings of ICTON'05*, paper We.C1.5, 2005.

19. J. G. Proakis and M. Salehi, *Communication Systems Engineering*, 2nd ed., NJ: Hoboken, Prentice Hall, 2002.

20. J. G. Proakis, *Digital Communications*, 4th ed., New York: McGraw-Hill, pp. 185–213, 2001.

21. W. Rosenkranz and C. Xia, Electrical equalization for advanced optical communications systems, *Int. J. Electron. Commun.*, Vol. 61, pp. 153–157, 2007.

22. M. Sieben, J. Conradi, and D. E. Dodds, Optical single sideband transmission at 10 Gbps using only electrical dispersion compensation, *IEEE J. Lightwave Technol.*, Vol. 17, No. 10, pp. 1742–1749, 1999.

23. T. L. Huynh, T. Sivahumaran, K. K. Pang, and L. N. Binh, A narrowband filter receiver achieving 225 ps/nm residual dispersion tolerance for 40 Gbps optical MSK transmission systems, in *Proceedings of OFC'07*, Anaheim, CA, 2007.

24. T. L. Huynh, T. Sivahumaran, L. N. Binh, and K. K. Pang, Narrowband frequency discrimination receiver for high dispersion tolerance optical MSK systems, in *Proceedings of COIN-AOFT International Conference*, paper TuA1-3, Melbourne, Australia, June 2007.

25. T. L. Huynh, T. Sivahumaran, L. N. Binh, and K. K. Pang, Sensitivity improvement with offset filtering in optical MSK narrowband frequency discrimination receiver, in *Proceedings of COIN-ACOFT International Conference*, paper TuA1-5, Melbourne, June 2007.

26. S. J. Savory, Y. Benlachtar, R. I. Killey, P. Pavel, G. Bosco, and P. Poggiolini, IMDD transmission over 1,040 km of standard single-mode fiber at 10 Gbps using a one-sample-per-bit reduced complexity MLSE receiver, in *Proceedings of OFC'07*, paper OThK2, Anaheim, CA, 2007.

13 DSP-Based Coherent Optical Transmission Systems

13.1 INTRODUCTION

In Chapter 6, a generic flowchart of the digital signal process (DSP) was introduced (see Figure 6.1). This diagram is now reintroduced here in which the clock/timing recovered signals feeding back into the sampling unit of the analog-to-digital convertor (ADC) so as to obtain the correct timing for sampling the incoming data sequence for processing in the DSP. Any errors made at this stage of the timing will result in high deviation of the bit error rate (BER) in the symbol decoder shown in Figure 13.1. It is also noted that the vertical polarized (V-pol) and horizontal polarized (H-pol) channels are detected, and their in-phase (I-) and quadrature (Q-) components are produced in the electrical domain with signal voltage conditioned for the conversion to digital domain by the ADC.

The processing of the sampled sequence from the received optical data and photodetected electronic signals passing through the ADC relies on the timing recovery from the sampled events of the sequence. The flowing stages of the blocks given in Figure 13.1 may be changed or altered accordingly, depending on the modulation formats and pulse shaping, for example, the Nyquist pulse shapes in Nyquist superchannel transmission systems.

This chapter describes the processing algorithms which are associated within the optical reception subsystems of coherent optical transmission systems over highly dispersive optical transmission lines consisting of multispan optically amplified non-DCM SSMF. First, the quadrature phase shift keying (QPSK) homodyne scheme is examined, and then the 16QAM, incorporating both polarized channels multiplexed in the optical domain, and hence the term PDM-QPSK or PDM-16QAM. We then expand the study for superchannel transmission systems in which several subchannels are closely spaced in the spectral region, so as to increase the spectral efficiency such that the total effective bit rate must reach at least 1 Tbps. Due to the overlapping of adjacent channels, there are possibilities that modifications of the processing algorithms are to be made.

Furthermore, the nonlinearity impairments on transmitted subchannels could degrade the system performance, and the application of back propagation techniques described in Chapter 3 must be combined with linear and nonlinear equalization schemes in order to effectively combat performance degrading.

13.2 QUADRATURE PHASE SHIFT KEYING SYSTEMS

13.2.1 CARRIER PHASE RECOVERY

Homodyne coherent reception requires a perfect match of the frequency of the signal carrier and the local oscillator (LO). Any frequency difference will lead to phase noise of the detected signals. This is the largest hurdle for the first optical coherent system initiated in the mid-1980s. In DSP-based coherent reception systems, the recovery of the carrier phases and hence the frequency is critical to achieving the most sensitive reception with maximum performance in the BER, or evaluation of the probability of error. This section illustrates the recovery of the phase of the carrier of the received QPSK and 16QAM optical transmitted signals. How would the DSP algorithms perform under the physical impairment effects on the recovery of the phase of the carrier?

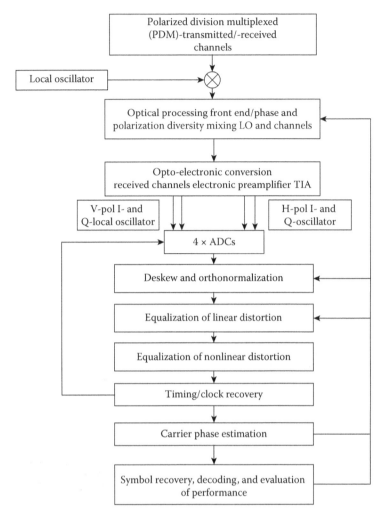

FIGURE 13.1 Flow of functionalities of DSP processing in a coherent optical receiver of a coherent transmission system. A modified diagram from Figure 5.22 (Chapter 5), with feedback path diverted to ADC block.

13.2.2 112G QPSK COHERENT TRANSMISSION SYSTEMS

Currently, several equipment manufacturers are striving to provide commercial advanced optical transmission systems at 100 Gbps, employing coherent detection techniques for long-haul backbone networks and metro networks as well. It has been well known since the 1980s that single-mode optical fibers can support such transmissions due to the preservation of the guided modes and its polarized modes of the weakly guiding linearly polarized (LP) electromagnetic waves [1]. Naturally, both transmitters and receivers must satisfy the coherency conditions of narrow-linewidth sources and coherent mixing with an LO, an external cavity laser (ECL), to recover both the phase and amplitude of the detected lightwaves. Both polarized modes of the LP modes can be stable over long distance so far in order to provide the polarization division multiplexed (PDM) channels, even with polarization-mode dispersion (PMD) effects. All linear distortions due to PMD and chromatic dispersion (CD) can be equalized in the DSP domain employing algorithms in the real-time processors provided in Chapter 6.

It is thus very important to ensure that these subsystems are performing the coherent detection and transmitting functions. This section thus presents a summary of the transmission tests conducted for optical channels under back-to-back (B2B) and long-haul systems employing of QPSK PDM. The symbol rate of the transmission system is 28 GSy/s and the modulation format is PDM-CSRZ-QPSK. It is noted that the differential QPSK (DQPSK) encoder and the bit pattern generator are provided.

The transmission system is arranged as shown in Figure 13.2. The CSRZ-QPSK transmitter consists of a CSRZ optical modulator that is biased at the minimum transmission point of the transfer characteristics of the MZIM driven by sinusoidal signals whose frequency is half of the symbol rate or 14 GHz for 28 GSy/s. A wavelength-division multiplexing (WDM) mux (multiplexer) is employed to multiplex other wavelength channels located within the C-band (1530–1565 nm). An optical amplifier (EDFA type) is employed at the front end of the receiver, so that noises can be superimposed on the optical signals to obtain the OSNR. The DSP signals in the digital domain are carried out off-line, and the BER is obtained.

The transmitter consists of an ECL, an encore type, a polarization splitter coupled with a 45°-aligned ECL beam, two separate CSRZ external LiNbO$_3$ modulators, and two I–Q optical modulators. The linewidth of the ECL is specified at about 100 KHz and, with external modulators, we can see that the spectrum of the output-modulated lightwaves is dominated by the spectrum of the baseband modulation signals. However, we observed that the laser frequency is oscillating at about 300 MHz due to the integration of a vibrating grating so as to achieve stability of the optical frequency.

The receivers employed in this system are of two types. One is a commercialized type, Agilent N 4391, in association with an Agilent external LO, and the other one is a TIA type, including a photodetector pair connected back-to-back (B2B) in a push–pull manner and then a broadband transimpedance amplifier (TIA). In addition to the electronic reception part, a $\pi/2$ hybrid coupler, including a polarization splitter, $\pi/2$ phase shift and polarization combiner that mixed the signal polarized beams and those of the LO (an ECL type identical to the one used in the transmitter), is employed as the optical mixing subsystem at the front end of the receiver. The mixed polarized beams (I–Q signals in the optical domain) are then detected by balanced receivers. I–Q signals in the electrical domain are sampled and stored in a real-time oscilloscope (Tektronix 7200). The sampled I–Q signals are then processed off-line using the algorithms provided in the scope or self-developed algorithms, such as the evaluation of error vector magnitude (EVM) described earlier for Q factor and thence BER.

Both the transmitter and the receiver are functioning with the required OSNR for the B2B of about 15 dB at a BER of 2e−3. It is noted that the estimation of the amplitude and phase of the received constellation is quite close to the received signal power and noise contributed by the balanced receiver with a small difference, owing to the contribution of the quantum shot noise contributed by the power of the LO. The estimation technique was described in Chapter 4. Figure 13.3 shows the BER versus OSNR for B2B QPSK PDM channels. As shown in Figure 13.4, the variation

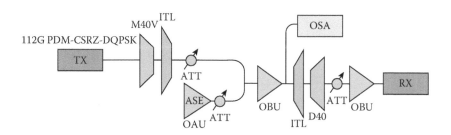

FIGURE 13.2 Setup of the PDM-QPSK optical transmission system. Tx = transmitter, PDM = polarization division multiplexing, ITL = interleaved transmission line, CSRZ = carrier suppressed return-to-zero, OSA = optical spectrum analyzer, Rx = receiver, OBU = optical boosting unit, D40 = optical demux type 40, Att = attenuator, OAU = optical amplifier unit, and ASE = amplification stimulated emission (noise).

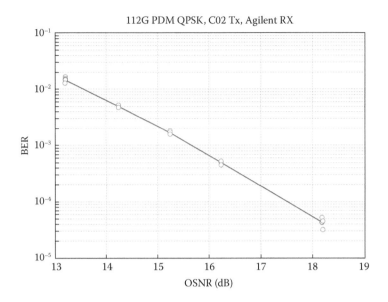

FIGURE 13.3 B2B OSNR versus BER performance with Agilent Rx in 112 Gbps PDM-CSRZ-DQPSK system.

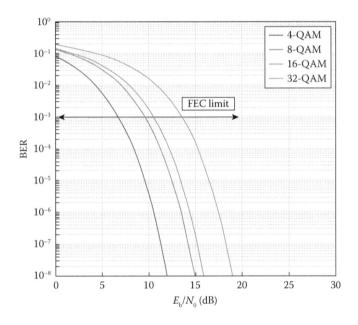

FIGURE 13.4 Theoretical BER versus SNR for different-level QAM schemes obtained by BER tool.m of MATLAB®.

of BER as a function of the signal energy over noise power per bit, for 4QAM or QPSK coherent then the SNR is expected at about 8 dB for BER of 1e−3. Experimental processing of such a scheme in B2B configuration shows an OSNR of about 15.6 dB. This is due to the 3 dB split by the polarized channels, and then additional noises are contributed by the receiver, and hence about 15.6 dB OSNR is required. The forward error coding (FEC) is set at 1e−3. The receiver is the Agilent type, as mentioned earlier.

A brief analysis of the noises at the receiver can be as follows. The noise is dominated by the quantum shot noise generated by the power of the LO, which is about at least ten times greater than that of the signals. Thus, the quantum shot noises generated at the output of the photodetector is

$$i_{N-LO}^2 = 2qI_{LO}B$$

$$I_{LO} = 1.8\,\text{mA}\,@\,0.9 - PD\_\text{quantum}\_\text{efficiency}$$

$$B = 31\,\text{GHz} - BW\_U2t\_\text{Agilennt}\_\text{Rx}$$

$$i_{N-LO}^2 = 2 \times 10^{-19} \times 1.8 \times 10^{-3} \times 30 \times 10^9 = 10.8 \times 10^{-12}\,A$$

$$\longrightarrow i_{N-LO} = 3.286\,\mu A \longrightarrow \text{shot} - \text{noise} - \text{current}$$

The bandwidth of the electronic preamplifier of 31 GHz is taken into account. This shot noise current, due to the LO imposed on the PD pair, is compatible with that of the electronic noise of the electronic receiver, given that the noise spectral density equivalent at the input of the electronic amplifier of the U^2t balanced receiver is specified at $80\,pA/\sqrt{hz}$, that is

$$\left(80 \times 10^{-12}\right)^2 \times 30.10^9 = 19.2 \times 10^{-12}\,A^2 \xrightarrow[\text{noise\_current}]{} i_{Neq.} = 4.38\,\mu A$$

Thus, any variation in the LO would affect this shot noise in the receiver. It is thus noted that, with the transimpedance of the electronic preamplifier estimated at 150 Ω, a dBm difference in the LO would contribute to a change of the voltage noise level of about 0.9 mV in the signal constellation obtained at the output of the ADC. A further note is that the noise contributed by the electronic front end of the ADC has not been taken into account. We note that differential TIA offers at least ten times higher transimpedance of around 3000 Ω over the 30 GHz mid-band. These TIAs offer much higher sensitivities as compared to the single-input TIA type [2,3].

13.2.3 I–Q IMBALANCE ESTIMATION RESULTS

There are imbalances due to the propagation of the polarized channels and the I and Q components. They must be compensated in order to minimize the error. The I–Q imbalance of the Agilent BalRx and U^2t BalRx is less than 2 degrees, which might be negligible for the system, as shown in Figure 13.5. This imbalance must be compensated for in the DSP domain.

13.2.4 SKEW ESTIMATION

In addition to the imbalance of the I and Q due to optical coupling and electronic propagation in high-frequency cables, there are propagation delay time differences between these components that must be compensated. The skew estimation obtained over a number of data sets is shown in Figure 13.6.

Abnormal skew variation from time to time was also observed, which should not happen if there is no modification of the hardware. Considering the skew variation that happened with the Agilent receiver, which only has a very short RF cable and a tight connection, there is a higher probability that the skew happened inside the Tektronix oscilloscope than at the optical or the electrical connection outside.

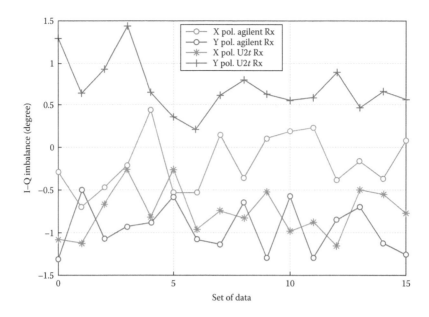

FIGURE 13.5   I–Q imbalance estimation results for both Rx. Note maximum imbalance phase of ±1.5°.

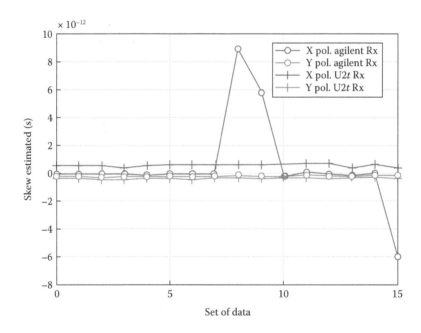

FIGURE 13.6   Skew estimation results for both types of Rx.

Figure 13.7 shows the BER versus OSNR when the skew and imbalance between I and Q components of the QPSK transmitter are compensated. The OSNR of DQPSK is improved by about 0.3~0.4 dB at 1e−3 BER, when compared with the result without I–Q imbalance and skew compensation, as shown in the constellation obtained in Figure 13.8. In the time domain, the required OSNR of DQPSK at a BER of 1e−3 is about 14.7 dB, improved by 0.1 dB. The required OSNR at 1e−3 BER of QPSK is about 14.7 dB, which is about the common performance of the state-of-the-art for QPSK. For a BER = 1e−3, an imbalanced CMRR = −10 dB would create a penalty of 0.2 dB

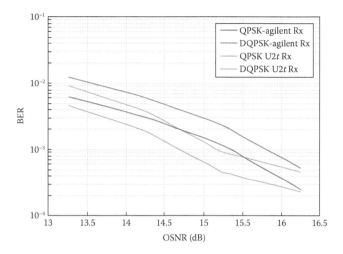

FIGURE 13.7 OSNR versus BER for two types of integrated coherent receiver after compensating I–Q imbalance and skew.

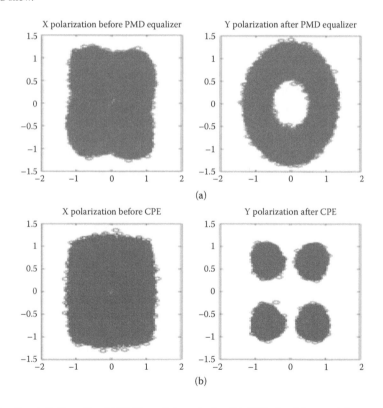

FIGURE 13.8 Constellation after (a) PMD module and after (b) CPE algorithm module.

in the OSNR for the Agilent receiver, and an improvement of 0.7 dB for a commercial balanced receiver employed in the Rx subsystems.

Figure 13.9 shows the structure of the ECL incorporating a reflection mirror that is vibrating at a slow frequency of around 300 MHz. A control circuit would be included to indicate the electronic control of the vibration and cooling of the laser, so as to achieve stability and elimination of Brillouin scattering effects.

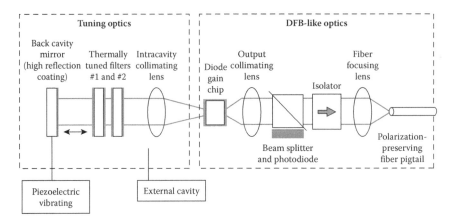

FIGURE 13.9 External cavity structure of the laser with the back mirror vibration in the external cavity structure.

13.2.5 FRACTIONALLY SPACED EQUALIZATION OF CD AND PMD

Ip and Kahn [5] has employed the fractionally spaced equalization scheme with MSE (mean square error) to evaluate its effectiveness of PDM-ASK (amplitude shift keying) with NRZ (non-return-to-zero) or RZ (return-to-zero) pulse-shaping transmission system. Their simulation results are displayed in Figure 13.10 [5] for the maximum allowable CD (normalized in ratio with respect to the dispersion parameter of the single-mode fiber [SMF], as defined in Chapter 2) versus the number of equalizer taps N for a 2 dB power penalty at a launched OSNR of 20 dB per symbol for ASK-RZ and ASK-NRZ pulse shapes, using a Bessel antialiasing filter with a sampling rate of $1/T = M/KT_s$; $T_s = symbol\_period$, with M/K as the fractional ratio, as shown in Figure 13.10 for a fractional ratio of (a) $M/K = 1$, (b) 1.5, (c) 2, and (d), (e), and (f) using a Butterworth antialiasing filter. The sampling antialiasing filter is employed to ensure that artificial fold back to the spectrum is avoided. The filter structures Bessel and Butterworth give much similar performance for fractional spaced equalizers, but less tap for the Bessel filtering case when an equally spaced equalizer is used (see Figure 13.10a and d).

13.2.6 LINEAR AND NONLINEAR EQUALIZATION, AND BACK PROPAGATION COMPENSATION OF LINEAR AND NONLINEAR PHASE DISTORTION

Ip and Kahn [6] first developed and applied back propagation, as given in Chapter 2, to equalize the distortion owing to nonlinear impairment of optical channel transmission through single-mode optical fibers. The back propagation algorithm is simply a reverse-phase rotation at the end of each span of the multispan link. The rotating phase is equivalent to the phase exerted on the signals in the frequency domain with a square of the frequency dependence. Thus, this back propagation is efficient in the aspect that the whole span can be compensated so as to minimize the numerical processes, hence requiring less processing time and central processing unit time of the digital signal processor.

Figure 13.11 [6] shows the equalized constellations of a 21.4 GSy/s QPSK modulation scheme system after transmission through 25 × 80 km non-DCF spans under the equalization using (a) linear compensation only, (b) nonlinear equalization, and (c) using combined back propagation and linear equalization. Obviously, the back propagation contributes to the improvement of the performance of the system.

Figure 13.12 shows the phase errors of the constellation states at the receiver versus a launched power of 25 × 80 km multispan QPSK 21.4 GSy/s transmission system. The results extracted from Ref. [6] show the performance of back propagation phase rotation per span for 21.4 Gbps 50% RZ-QPSK, transmitted over 25 × 80 km spans of SMF with five ROADMs (reconfigurable optical add–drop module), with 10% CD under-compensation. The algorithm is processed off-line of

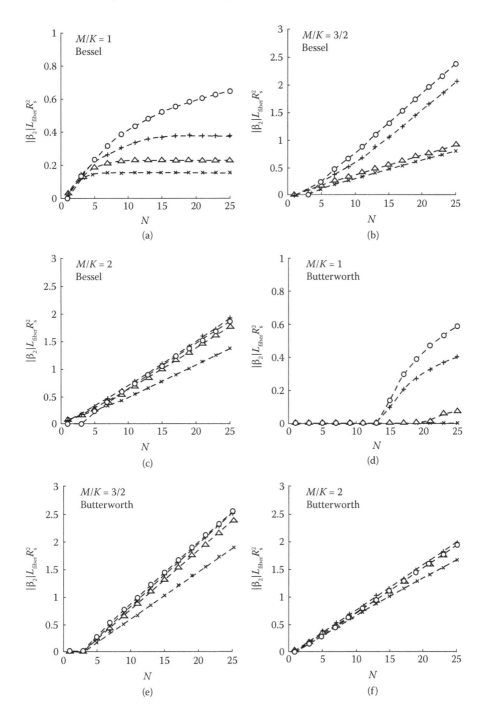

FIGURE 13.10 Maximum allowable CD versus the number of equalizer taps N for a 2 dB power penalty at an input SNR of 20 dB per symbol for RZ and NRZ pulse shapes, using a Bessel antialiasing filter; "o" denotes transmission using NRZ pulses; "x" denotes 33% RZ; "Δ" denotes 50% RZ; and "+" denotes 67% RZ, by fractional spaced equalizer (FSE) and sampling rate $R_s = 1/T = M/KT_s$; $T_s = symbol\_period$, where M/K is the fractional ratio) (a) $M/K = 1$, (b) 1.5, (c) 2, (d), (e), and (f) using a Butterworth antialiasing filter (Adopted from E. Ip and J. M. Kahn, *IEEE J. Lightwave Technol.*, Vol. 25, No. 8, p. 2033, 2007, with permission). The modulation scheme is ASK with NRZ and RZ pulse shaping. N = number of taps of the equalizer.

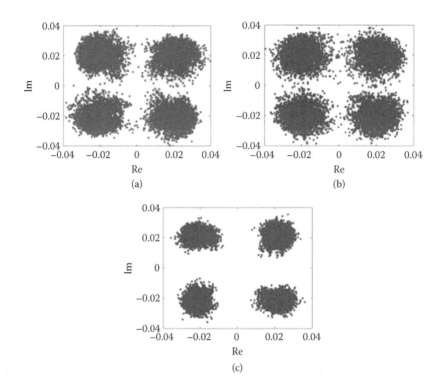

FIGURE 13.11 Constellation of QPSK scheme as monitored at (a) with linear CD equalization only, (b) nonlinear phase noise compensation, and (c) after back propagation processing combined with linear equalization. (Simulation results extracted from E. Ip and J. M. Kahn, *IEEE J. Lightwave Technol.*, Vol. 26, No. 20, p. 3416, 2008.)

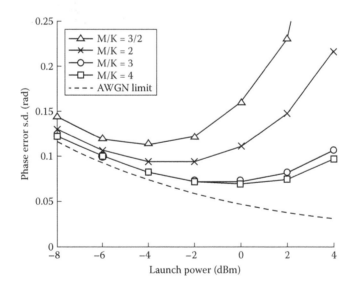

FIGURE 13.12 Phase errors at the receiver versus launched power or 25 × 80 km multispan QPSK at 21.4 GSy/s. Performance of back propagation phase rotation per span for 21.4 Gbps 50% RZ-QPSK transmitted over 25 × 80 km spans of SMF with five ROADMs, with 10% CD under-compensation using fractionally spaced equalizer with *M/K* factor of 1.5 to 4. (Simulation results extracted from E. Ip and J. M. Kahn, *IEEE J. Lightwave Technol.*, Vol. 26, No. 20, p. 3416, 2008. With permission.)

received sampled data after 25 × 80 km SSMF propagation via the use of the nonlinear Schrödinger equation (NLSE) and the coherent reception technique described in Chapter 4. It is desired that the higher the launched power, the better the OSNR that can be employed for longer-distance transmission. Hence, fractional space ratios of 3 and 4 can offer higher launched power and thus they are the preferred equalization scheme as compared with the equal space whose sampling and symbol rate are the same. The ROADM is used to equalize the power of the channel under consideration as compared with other dense wavelength division multiplexing (DWDM) channels.

13.3　16QAM SYSTEMS

Consider the 16QAM received symbol signal whose phase Φ denotes the phase offset. The symbol d_k denotes the magnitude of the QAM symbols, and n_k denotes the noises superimposed on the symbol at the sampled instant. The received symbols can be written as

$$r_k = d_k e^{j\Phi} + n_k; \quad k = 1, 2, \ldots L \tag{13.1}$$

Using the maximum-likelihood sequence estimator (MLSE), the phase of the symbol can be estimated as

$$l(\phi) = \sum_{k=1}^{L} \ln \left\{ \sum_d e^{-\frac{1}{2\sigma^2}|r_k - de^{j\phi}|^2} \right\} \tag{13.2}$$

or effectively one would take the summation of the contribution of all states of the 16QAM on the considered symbol measured as the geometrical distance on a natural logarithmic scale with the noise contribution of a standard deviation σ.

The frequency offset estimation for 16QAM can be conducted by partitioning the 16QAM constellation into a number of basic QPSK constellations, as shown in Figure 13.13 [4]. There are two QPSK constellations in the 16QAM whose symbols can be extracted from the received sampled data set.

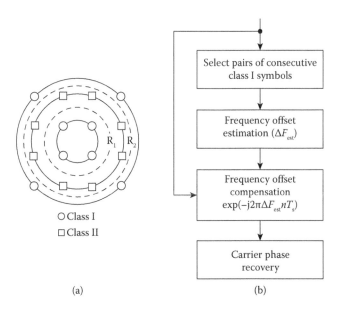

(a)　　　　　　　　　　(b)

FIGURE 13.13 Processing of 16QAM for carrier phase estimation (From I. Fatadin and S. J. Savory, *IEEE Photon. Technol. Lett.*, Vol. 23, No. 17, pp. 1246–1248, 2001.); (a) constellation of 16QAM and (b) processing for carrier phase recovery with classes I and II of circulator subconstellation.

They are then employed to estimate the phase of the carrier, as described in the previous section on carrier phase estimation for QPSK-modulated transmission systems. At first, a selection of the inner-most QPSK constellation is done and this is classified as class I symbols. Then an estimation of the frequency offset of the 16QAM transmitted symbols can be derived from this constellation. Thence, an frequency offset (FO) compensation algorithm is conducted. Thence, the phase recovery of all 16QAM symbols can be derived. Further confirmation of the difference of carrier phase recovery or estimation can be conducted with the constellation of class I, as indicated in Figure 13.13a.

Carrier phase recovery based on the Viterbi-Viterbi algorithm on the class I QPSK subconstel-lation of the 16QAM may not be sufficient, and hence a modified scheme has been reported by Fatadin et al. in Ref. [7]. Further refining of this estimation of the carrier phase for 16QAM by parti-tion and rotating is done, in order to match certain symbol points to those of class I constellation of the 16QAM. The procedures are as shown in the flow diagram of Figure 13.14.

Conduct the partition into different classes of constellation and then rotate class 2 symbols with an angle, either clockwise or anticlockwise of $\pm\theta_{rot} = \pi/4 - \tan^{-1}(1/3)$. In order to avoid opposite rotation with respect to the real direction, the estimation of the error in the rate of changes of the phase variation or frequency estimation can be found by the use of the fourth power of the argument of the angles of two consecutive symbols, given by

$$\Delta F_{\text{est.}} = \frac{1}{8\pi T_s} \arg\left(\sum_{k=0}^{N} S_{k+1} S_k\right)^4 \tag{13.3}$$

to check their quadratic mean. Then select the closer symbol, and then apply the standard Viterbi-Viterbi procedure. Louchet et al. [8] also employed a similar method and have confirmed the effectiveness of such a scheme. The effects on the constellation of the 16QAM due to different

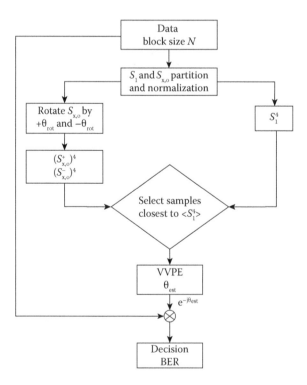

FIGURE 13.14   Flow chart of the refined carrier phase recovery algorithm of 16QAM by rotation of class I and class II subconstellations.

physical phenomena are shown in Figure 13.15 [4]. Clearly, the FO would generate the phase noises in (a), influencing both (i) the I and Q components by the CD of small amounts (so as to see the constellation noises), and (ii) the delay of the polarized components on the I and Q components. These distortions of the constellation allow practical engineers to assess the validity of algorithms that are normally separate and independent and implemented in a serial mode. This is in contrast to the constellations illustrated for QPSK as shown in Figure 13.8. Figure 13.16 also shows the real-time signals that result from the beating of the two sinusoidal waves of FO beating in a real-time oscilloscope.

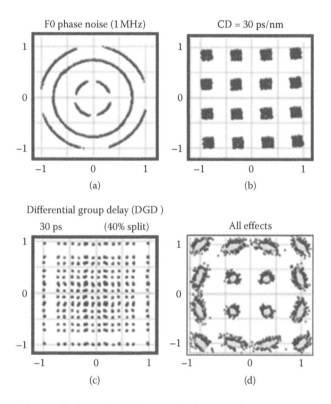

FIGURE 13.15 16QAM constellation under influence; (a) phase rotation due to FO of 1 MHz and no amplitude distortion, (b) residual CD impairment, (c) DGD of PMD effect, and (d) total phase noises effect (From I. Fatadin and S. J. Savory, *IEEE Photon. Technol. Lett.*, Vol. 23, No. 17, pp. 1246–1248, 2001.)

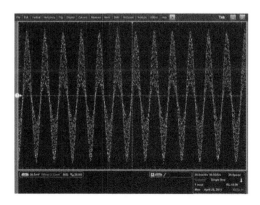

FIGURE 13.16 Beating signals of the two mixed lasers as observed by real-time sampling oscilloscope.

Noe et al. [9] has simulated the carrier phase recovery for QPSK with polarization division-multiplexed (PDM) channels under constant modulus algorithm (CMA) and decision-directed (DD) feedback with and without modification in which the error detected in each stage would be updated. The transmission system under consideration is a back-to-back (BTB) connection in which white Gaussian noises and phase noises are superimposed on the signals, then the BER is measured with the variation of the signal average optical power. For an OSNR of 11 dB, the CMA with modification offers a BER of 1e–3, while it is 4e–2 for a CMA without modification. This indicates that the updating of the matrix coefficients is very critical to recover the original data sequence. The modified CMA was also recognized to be valid for 16QAM.

13.4 TERABITS/SECOND SUPERCHANNEL TRANSMISSION SYSTEMS

13.4.1 OVERVIEW

PDM-QPSK has been exploited as 100 Gbps long-haul-transmission commercial systems, the optimum technologies for 400 Gigabit Ethernet (GE) and/or Terabit Ethernet (TE) transmission for next-generation optical networking, and they have now attracted significant interest for deploying ultra-high-capacity systems over the global Internet backbone networks. Furthermore intense research on Tbps capacity transmission systems has also attracted interests from several research groups, especially when the bit rates reach several 100 Gbps. The development of hardware platforms for 1 to N Tbps is critical for proving the design concept. The Tbps level can be considered as a superchannel that is defined as an optical channel comprising a number of subrate subchannels whose spectra would be the narrowest allowable. Thus, in order to achieve efficient spectral efficiency, phase shaping is required, and one of the most efficient techniques is the Nyquist pulse shaping. Thus, Nyquist-QPSK can be considered as one of the most effective formats for the delivery of high spectral efficiency that is effective in coherent transmission and reception as well as for equalization at both transmitting and reception ends.

Thus, in this section, we describe in detail the design and experimental platform for delivery of Tbps speeds using Nyquist-QPSK at a symbol rate of 28–32 GSa/s and ten subcarriers. The generation of subcarriers has been demonstrated using either recirculating frequency shifting (RFS) or nonlinear driving of an I–Q modulator to create five subcarriers per main carrier, and thus two main carriers are required. Nyquist pulse shaping is used for effectively packing multiplexed channels whose carriers are generated by the comb generation technique. Digital-to-analog converters (DACs) with sampling rates varying from 56 G to 64 GS/s is used for generating the Nyquist pulse shape, including the equalization of the transfer functions of the DAC and optical modulators.

13.4.2 NYQUIST PULSE AND SPECTRA

The raised-cosine filter is an implementation of a low-pass Nyquist filter, that is, one that has the property of vestigial symmetry. This means that its spectrum exhibits odd symmetry about $1/2T_s$, where T_s is the symbol period. Its frequency-domain representation is a "brick-wall-like" function, given by

$$H(f) = \begin{cases} T_s & |f| \leq \dfrac{1-\beta}{2T_s} \\[2mm] \dfrac{T_s}{2}\left[1 + \cos\left(\dfrac{\pi T_s}{\beta}\left\{|f| - \dfrac{1-\beta}{2T_s}\right\}\right)\right] & \dfrac{1-\beta}{2T_s} < |f| \leq \dfrac{1+\beta}{2T_s} \\[2mm] 0 & \text{otherwise} \end{cases} \tag{13.4}$$

with $0 \leq \beta \leq 1$

This frequency response is characterized by two values: β, the *roll-off factor*, and T_s, the reciprocal of the symbol rate in Sym/s—that is, $1/2T_s$ is the half bandwidth of the filter. The impulse response of such a filter can be obtained by analytically taking the inverse Fourier transformation of Equation 13.4, in terms of the normalized sin c function, as

$$h(t) = \sin c\left(\frac{t}{T_s}\right) \frac{\cos\left(\dfrac{\pi\beta t}{T_s}\right)}{1 - \left(2\dfrac{\pi\beta t}{T_s}\right)^2} \tag{13.5}$$

where the roll-off factor, β, is a measure of the *excess bandwidth* of the filter, that is, the bandwidth occupied beyond the Nyquist bandwidth as from the amplitude at $1/2T$. Figure 13.17 depicts the frequency spectra of raised-cosine pulse with various roll-off factors. Their corresponding time-domain pulse shapes are given in Figure 13.18.

When used to filter a symbol stream, a Nyquist filter has the property of eliminating ISI, as its impulse response is zero at all nT (where n is an integer), except when $n = 0$. Therefore, if the transmitted waveform is correctly sampled at the receiver, the original symbol values can

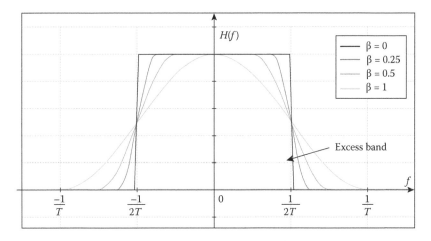

FIGURE 13.17   Frequency response of raised-cosine filter with various values of the roll-off factor β.

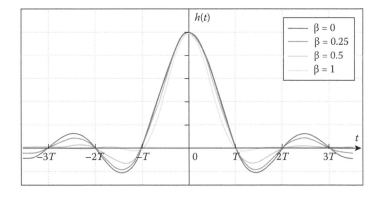

FIGURE 13.18   Impulse response of raised-cosine filter with the roll-off factor β as a parameter.

be recovered completely. However, in many practical communications systems, a matched filter is used at the receiver, so as to minimize the effects of noises. For zero ISI, the net response of the product of the transmitting and receiving filters must equate to $H(f)$, and thus we can write

$$H_R(f)H_T(f) = H(f) \tag{13.6}$$

or alternatively we can rewrite that

$$|H_R(f)| = |H_T(f)| = \sqrt{|H(f)|} \tag{13.7}$$

The filters that can satisfy the conditions of Equation 13.7 are the root-raised-cosine (RRC) filters. The main problem with RRC filters is that they occupy a larger frequency band than that of the Nyquist $\sin c$ pulse sequence. Thus, for the transmission system, we can split the overall raised-cosine filter with RRC filter at both the transmitting and receiving ends, provided the system is linear. This linearity is to be specified accordingly. An optical fiber transmission system can be considered to be linear if the total power of all channels is under the nonlinear SPM threshold limit. When it is over this threshold, a weakly linear approximation can be used.

The design of a Nyquist filter influences the performance of the overall transmission system. Oversampling factor, selection of roll-off factor for different modulation formats, and finite impulse response (FIR) Nyquist filter design are the key parameters to be determined. If taking into account the transfer functions of the overall transmission channel including fiber, wavelength selective switch (WSS), and the cascade of the transfer functions of all optical to electrical (O/E) components, the total channel transfer function is more Gaussian-like. To compensate for this effect, in the Tx-DSP, one would thus need a special Nyquist filter to achieve the overall frequency response equivalent to that of the rectangular or raised cosine with the roll-off factor shown in Figures 13.19 and 13.20.

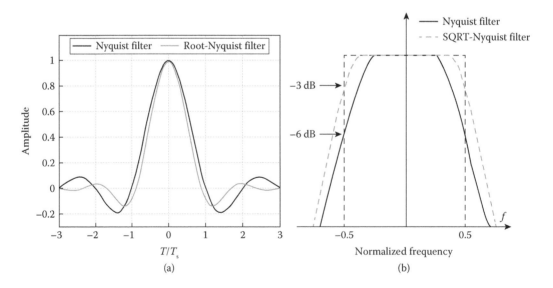

FIGURE 13.19 (a) Impulse and (b) corresponding frequency response of $\sin c$ Nyquist pulse shape or root-raised-cosine (RRC) Nyquist filters.

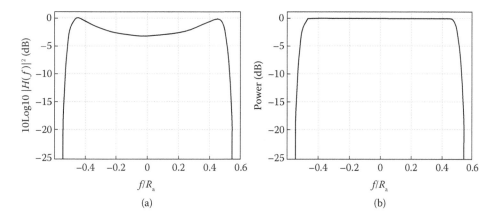

FIGURE 13.20 (a) Desired Nyquist filter for spectral equalization; (b) output spectrum of the Nyquist-filtered QPSK signal.

13.4.3 Superchannel System Requirements

13.4.3.1 Transmission Distance
As the next generation of backbone transport, the transmission distance should be comparable to that of the previous generation, namely the 100 Gbps transmission system. As the most important requirement, we require the 1 Tbps transmission for long-haul to be 1500–2000 km, and for metro applications to be ~300 km.

13.4.3.2 CD Tolerance
Since the SSMF fiber CD factor/coefficient of 16.8 ps/nm is the largest among the currently deployed fibers, CD tolerance should be up to 30,000 ps/nm at the central channel whose wavelength is approximately 1550 nm. At the edge of C-band, this factor is expected to increase by about 0.092 ps/(nm².km), or about 32,760 ps/nm at 1560 nm and 26,400 ps/nm at 1530 nm [10].*

13.4.3.3 PMD Tolerance
The worst case of deployed fiber of 2000 km would have a differential group delay (DGD) of 75 ps, or about three symbol periods for 25 GSy/s per subchannel. Hence, the PMD (mean all-order DGD) tolerance is 25 ps.

13.4.3.4 SOP Rotation Speed
According to the 100 Gbps experiments, state-of-polarization (SOP) rotation can be up to 10 KHz, and we take the same spec as the 100 G system.

13.4.3.5 Modulation Format
PDM-QPSK for long-haul transmission; PDM-16QAM for metro application.

13.4.3.6 Spectral Efficiency
Compared to the 100 G system with an increasing of factor 2, both Nyquist-WDM and coherent OFDM (CO-OFDM) can fulfill this. However, it depends on technological and economical requirements that would determine the suitability of the technology for optical network deployment.

Table 13.1 tabulates the system specifications of various transmission structures with parameters of subsystems, especially when the comb generators are employed using either recirculating or nonlinear generation techniques. The DSP-reception and off-line DSP is integrated in these systems.

---

* Technical specification of Corning fiber G.652 SSMF given in Binh [10].

TABLE 13.1
1 Tbps Off-Line System Specifications

Parameter Technique	Superchannel RCFS Comb Gen.	Superchannel Nonlinear Comb Gen.	Some Specs.	Remarks
Bit rate	1, 2, ..., N Tbps (whole C-band)	1 Tbps, 2, ..., N Tbps	~1.28 Tbps @ 28–32 GB	20% OH for OTN, FEC
Number of ECLs	1	$N \times 2$		
Nyquist roll-off α	0.1 or less	0.1 or less		DAC pre-equalization required
Baud rate (GBauds)	28–32	28–32	28, 30, or 31.5 GBaud	Pending on FEC coding allowance
Transmission distance	2500	2500	1200 (16 span) ~2000 km (25 spans) 2500 km (30 spans)	20% FEC required for long-haul application
			500 km	Metro application
Modulation format	QPSK/16QAM	QPSK/16QAM	Multicarrier Nyquist-WDM PDM-DQPSK/QAM	For long haul
				For long haul
		Multicarrier Nyquist-WDM PDM-16QAM		For metro
Channel spacing			4×50 GHz	For long haul
			2×50 GHz	For long haul
Launch power	<<0 dBm if 20 Tbps is used		~ −3–1 dBm, lower if $N > 2$	Depending on QPSK/16QAM and long haul/metro
B2B ROSNR @2e-2 (BOL) (dB)	14.5	14.5	15 dB for DQPSK 22 dB for 16QAM	1 dB hardware penalty 1 dB narrow filtering penalty
Fiber type	SSMF G.652 (or G.655)	SSMF G.652 (or 655)	G.652 SSMF	
Span loss	22	22	22 dB (80 km)	
Amplifier	EDFA (G>22 dB); NF<5 dB		EDFA (OAU or OBU)	
BER	2e-3	2e-3	Pre-FEC 2e-2 (20%) or 1e-3 classic FEC (7%)	
CD penalty (dB)			0 dB @ +/−3,000 ps/nm <0.3 dB @ +/−30000 ps/nm;	16.8 ps/nm/km and 0.092 ps/(nm².km)
PMD penalty (DGD)			0.5 dB @75 ps, 2.5 symbol periods	
SOP rotation speed	10 kHz	10 kHz	10 kHz	OPLL may require due to oscillation of the LO carrier
Filters cascaded penalty			<1 dB @12 pcs WSS	
Driver linearity	Required	Required	THD <3%	16QAM even more strict

13.4.4 SYSTEM STRUCTURE

13.4.4.1 DSP-Based Coherent Receiver

A possible structure of a superchannel transmission system can be depicted in Figure 13.21. At the transmitter, the data inputs can be inserted into the pulse shaping, and individual data streams can be formed. A DAC can be used for shaping the pulse to a Nyquist-equivalent shape, the raised-cosine form whose spectra also follow a raised cosine with the roll-off factor β varying from 0.1 to 0.5. If this off factor takes the value of 0.1, the spectra would follow an approximate shape of a rectangular brick wall. A comb generator can be used to generate equally spaced subcarriers for the superchannel from a single-carrier laser source, commonly an ECL of very narrow band of linewidth of about 100 KHz. These comb-generated subcarriers (see Figure 13.22) are then demultiplexed into subcarriers and fed into a bank of I–Q optical modulators (as described in Chapter 3), and Nyquist-pulse shape sequence as output of the DAC are then employed to modulate these subcarriers to form the superchannels at the output of an optical multiplexer shown in the block at the left side of Figure 13.21. A more detailed view of the transmitter for superchannels is shown in Figure 13.23. It is noted that the generation of the sub-carriers from a comb source can be implemented by recirculating of the frequency shifted lightwaves around an optical close loop. The frequency shift is the spacing frequency between the subcarriers. Hence, the Nth subcarrier would be the Nth time circulation of the original carrier. There would be superimposing of noises due to the ASE incorporated in the loop, and this would be minimized by inserting into the loop an optical filter whose bandwidth would be the same or wider than that of the superchannel.

The fiber transmission line is an optically amplified optical fiber multispan without incorporating any dispersion-compensating fibers. Thus, the transmission is very dispersive. The broadening of a 40-ps-width pulse would spread across at least 80–100 symbol period after propagating over 3000 km of SSMF. Thus, one can assume that the pulse launched into the fiber of the first span would be considered to be an impulse as compared to that after 3000 km of SSMF propagation.

After propagation over the multispan non-DCF line, the transmitted subchannels are demuxed via a wavelength splitter into individual subchannels, with minimum crosstalk. Each subchannel is then coherently mixed with an LO that is generated from another comb source incorporating an OPLL to lock the comb into that of the subcarriers of the superchannel. Thus, a comb generator is indicated in the right side, the reception system of Figure 13.21. The coherently mixed subchannels are then detected by a balanced receiver (as described in Chapter 4 and Section 2 of this chapter),

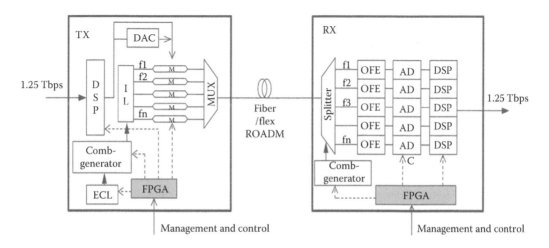

FIGURE 13.21 A possible structure of superchannel Tbps transmission system.

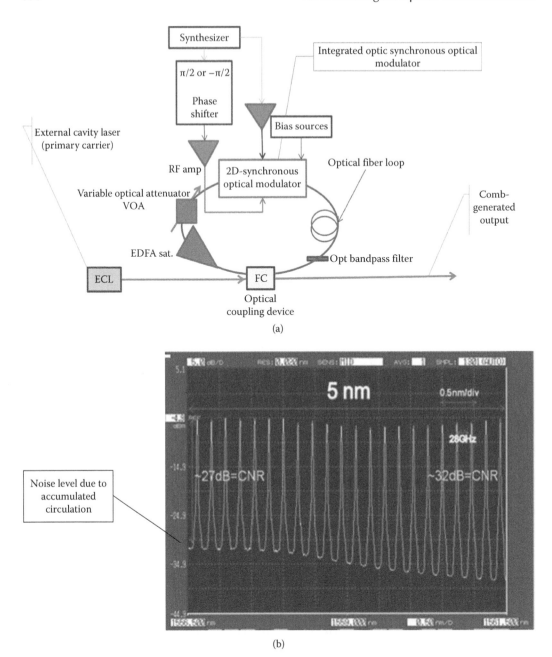

FIGURE 13.22 (a) Block diagram of a recirculating frequency-shifting comb generator and (b) a typical generated spectrum of the comb generator, with 28 GHz spacing between channels over more than 5 nm in the spectral region and about 30 dB carrier-to-noise ratio (CNR).

and then electronically amplified and fed into the sampler and ADC. The digital signals are then processed in the DSP of each subchannel system or parallel and interconnected DSP system. In these DSP systems, the sequence of the processing algorithms is employed to recover the carrier phase, and hence the clock recovery, compensating for the linear and nonlinear dispersion, and the evaluation of the BER versus different parameters such as OSNR, etc. Figure 13.24 shows the modulated spectra of five channels whose subcarriers are selected from the multiple subcarrier source of Figure 13.22. The modulation is QPSK with Nyquist pulse shaping.

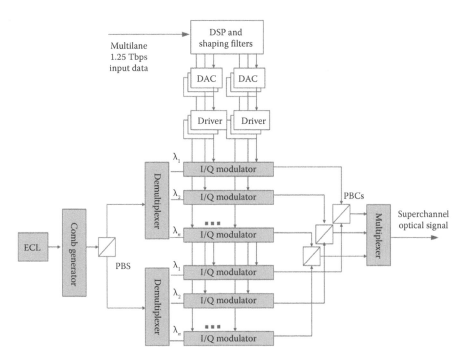

FIGURE 13.23 Generic detailed architecture of a superchannel transmitter. Generic schematic structure of optical recirculating and frequency shifting loop, and spectrum of superchannel. PBC = polarization beam combiner, PBS = polarization beam splitter, I/Q = in-phase–quadrature phase, DAC = digital-to-analog convertor, ECL = external cavity laser.

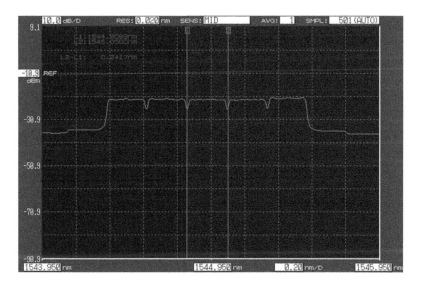

FIGURE 13.24 Selected five subcarriers with modulation.

13.4.4.2 Optical Fourier Transform–Based Structure

Superchannel transmission systems can also be structured using optical fast Fourier transform (OFFT), as demonstrated in Ref. [11,12] and shown in Figure 13.25, in which MZDI components (see Figure 13.25a) act as spectral filter and splitter (see Figure 13.25b), the optical FFT. The outputs of these MZDI are then fed into coherent receivers and processed digitally, as shown in Figure 13.25c,

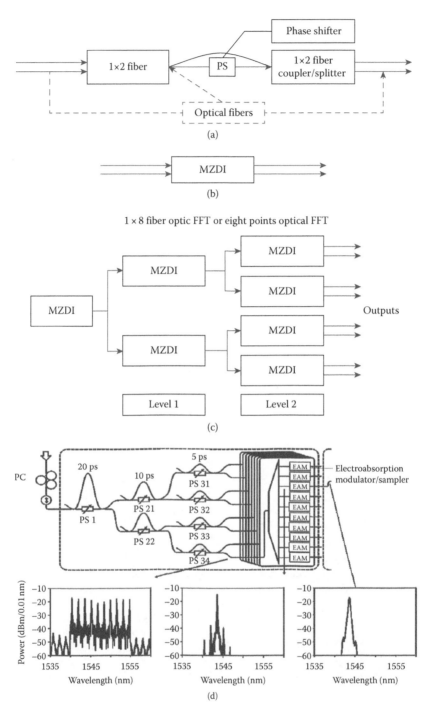

FIGURE 13.25 Operations by guided wave components using fiber optics; (a) guided wave optical path of a Mach–Zehnder delay interferometer or asymmetric interferometer with phase delay tunable with thermal or electro-optic effects, (b) block diagram representation, (c) implementation of optical FFT using cascade stages of fiber optical MZDI structure, and (d) EAM is the electroabsorption modulator used for demultiplexing in the time domain. (Extracted from D. Hillerkuss et al., *Nat. Photonics*, Vol. 5, p. 364, 2011.). Note also the phase shifters employed in MZDIs between stages. Inserts are the spectra of optical signals at different stages, as indicated of the optical FFT (serial type). EAM = electroabsorption modulator used for demultiplexing in the time domain; PC = polarization controller.

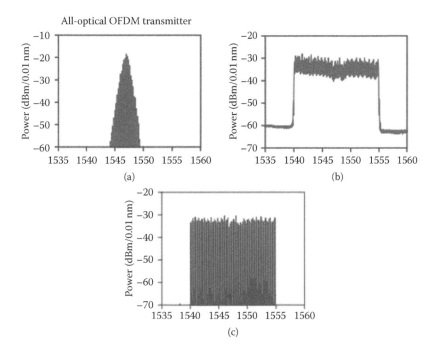

FIGURE 13.26 Spectrum of superchannels and demuxed channels. All subchannels are orthogonal, and thus the name orthogonal frequency division multiplexing (OFDM); (a) at output of transmitter, (b) after fiber propagation, and (b) after polarization demux. (Extracted from I. Fatadin and S. J. Savory, *IEEE Photon. Technol. Lett.*, Vol. 23, No. 17, pp. 1246–1248, 2001. With permission.)

with the electroabsorption modulator (EAM) performing the switching function to time demultiplex the ultra-fast signal speed to a number of lower rate sequences so that the detection system can decode and convert to the digital domain for further processing.

The spectra of superchannels at different positions in the transmission system can be seen in Figure 13.26a through c [4]. It is noted that the pulse shape is Nyquist, and the subchannels are placed close, satisfying the orthogonal condition, and thus the name optical OFDM (orthogonal frequency division multiplexing) is used to indicate this superchannel arrangement.

13.4.4.3 Processing

The processing of superchannels can be considered as similar to the digital processing of individual subchannels, except when there may be crosstalk between subchannels due to overlapping of certain spectral regions between the considered channel and its adjacent channels.

Thus, for the Nyquist-QPSK subchannel, the DSP processing would be much the same as for QPSK-DWDM for 112G described earlier with care taken for overlapping either at the transmitter or at the receiver. The EVM is a parameter that indicates the scattering of the vector formed by I and Q components departing from the center of the constellation point. The variance of this EVM in the constellation plane is used to evaluate the noises of the detected states, thence the Q factor can be evaluated with ease, and thus the BER by using the probability density function and the magnitude of the vector of a state on the constellation plane. The BER of the subchannels of the OFDM superchannel is shown in Figure 13.27b through c [4] for different percentages of overloading due to FEC. The loading factor is important, as this will increase the speed or symbol rate of the subchannel one has to offer. The higher this percentage, the higher is the increase in the symbol rate, thus requiring high-speed devices and components.

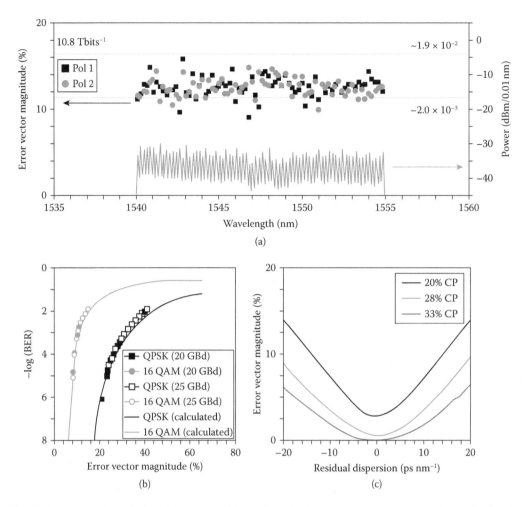

FIGURE 13.27 All-optical 10.8 Tbps OFDM results. (a) Measured error vector magnitude (EVM) for both polarizations (symbols) and for all subcarriers of the OFDM signal decoded with the all-optical FFT. The estimated BER for all subcarriers is below the third-generation FEC limit of 1.9e−2. The optical spectrum far left race, is drawn beneath. (b) Relationship between BER and EVM. Measured points (symbols) and calculated of BER as a function of EVM for QPSK and 16QAM. (c) Tolerance toward residual chromatic dispersion of the implemented system decoded with the eight-point FFT for a cyclic prefix of 20%, 28%, and 33%. (Extracted from I. Fatadin, and S. J. Savory, *IEEE Photon. Technol. Lett.*, Vol. 23, No. 17, pp. 1246–1248, 2001. With permission.)

In the receiver of the optical OFDM superchannel system reported in Ref. [4], the effectiveness of the optical FFT receiver can be evaluated by using three alternative receiver configurations and tested with QPSK signals. A QPSK signal is chosen because it was not possible to receive a 16QAM signal with the alternative receivers owing to their inferior performance [4]. First, a subcarrier with a narrow bandpass filter is used to extract a subchannel. The filter passband is adjusted in for best performance of the received signal (Figures 13.26a and 13.27). The selected filter bandwidth is 25 GHz. The constellation diagram shows severe distortion. When using narrow optical filtering, one has to accept a compromise between crosstalk from neighboring channels (as modulated OFDM subcarriers necessarily overlap) and intersymbol interference (ISI), owing to the increasing length of the impulse response when narrow filters are used. Narrow filters can be used, however, if the ringing from ISI is mitigated by additional time gating. The reception of a subcarrier using a coherent receiver is then performed. In the coherent receiver, the signal is down-converted in a hybrid coupler, as described in Chapter 5, and detected using balanced detectors and sampled in a real-time oscilloscope. Using a combination of error low-pass filtering due to the limited electrical bandwidth of the oscilloscope and DSP, the subcarrier is extracted from the received signal. This receiver performs better than the filtering approach, but a larger electrical bandwidth and sampling rate of the ADC and additional DSP would be needed to eliminate the crosstalk from other subcarriers and then to achieve a performance similar to that of the optical FFT.

Thus, optical OFDM may offer significant advantages for superchannels, but additional processing time would be required, while for Nyquist-QPSK, superchannels allow better performance and less complexity in the receiver DSP subsystem structure.

13.4.5 TIMING RECOVERY IN NYQUIST QAM CHANNEL

Nyquist pulse shaping is one of the most efficient methods to pack adjacent subchannels into a superchannel. The timing recovery of such Nyquist subchannels is critical for sampling the data received and for improving the transmission performance. Timing recovery can be done either before or after the PMD compensator. The phase detector scheme is shown in Figure 13.28, a Godard type [13] that is a first-order linear scheme. After CD compensation (CD^{-1} blocks), the signal is sent to an SOP modifier to improve the clock extraction. The clock performance of NRZ-QPSK signal in the presence of a first-order PMD characterized by a DGD and azimuth is presented in Figure 13.3.

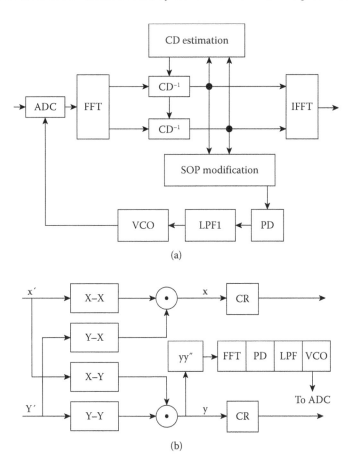

(a)

(b)

FIGURE 13.28 (a) Godard phase detector algorithm and (b) fourth-order 4PPD. VCO = voltage control oscillator, FFT = fast Fourier transform, IFFT = inverse FFT, CD = chromatic dispersion, SOP = state of polarization.

The azimuth of 45° and DGD of a half-symbol/unit interval (UI) completely destroys the clock tone. Therefore, the SOP modifier is required for enabling the clock extraction. In practical systems, a raised-cosine filter is used to generate Nyquist pulses instead of the *Sinc* shape type. A filter pulse response is defined by two parameters, the roll-off factor β and the symbol period T_s, and described by taking the inverse Fourier transform of Equation 13.4. The Godard phase detector cannot recover the carrier phase even with small β, and thus the channel spectra is close to rectangular. A higher-order phase detector must be used to effectively recover the timing clock period, as shown in Ref. [14]. A quadratic-power law PD (4PDP) with prefiltering presented in Ref. [15], as shown in Figure 13.28b, can deal with small β values.

The 4PDP operates by first splitting and forming the combination of these components X- and Y-polarized channels. It then detects the clock frequency by phase recovery and by regenerating through the VCO the frequency shift required for the ADC to assure the correct sampling timing by processing the received samples.

The use of such a phase detector in coherent optical receivers requires a large hardware effort. Due to PMD effects, the direct implementation of this method before PMD compensation is almost impossible. Therefore, the 4PPD implementation in the frequency domain after the PMD compensation is proven to be the most effective and performing well even in the most extreme cases with roll-off factor equal to zero.

13.4.6 128 Gbps 16QAM Superchannel Transmission

An experimental setup by Dong et al. [16] was reported for the generation and transmission of six channels carrying 128 Gbit/s under PDM-16QAM modulation format. The two 16-GBaud electrical 16QAM signals are generated from the two arbitrary waveform generators. Laser sources at 1550.10 nm and the second source with a frequency spacing of 0.384 nm (48 GHz) are generated from two ECLs, each with a linewidth less than 100 kHz and an output power of 14.5 dBm, respectively. Two I/Q MODs are used to modulate the two optical carriers with I and Q components of the 64-Gbps (16-GBaud) electrical 16QAM signals after the power amplification using four broadband electrical amplifiers/drivers, respectively. Two phase shifters with a bandwidth of 5 KHz–22.5 GHz provide two-symbol extra delay to decorrelate the identical patterns. For the operation to generate 16QAM, the two parallel Mach–Zehnder modulators (MZMs) for I/Q modulation are both biased at the null point and driven at full swing to achieve zero-chirp and phase modulation. The phase difference between the upper and the lower branch of I/Q MZIM is also controlled at the null point. The data input is shaped so that about 0.99 roll-off factor of raised-cosine pulse shape can be generated.

The power of the signal is boosted using polarization-maintaining EDFAs. The transmitted optical channels are mixed with an LO and polarization demultiplexed via a π/2 hybrid coupler, and then the I and Q components are detected by four pairs of balanced photodetectors (PDP). They are then transimpedance amplified and resampled.

The sampled data are then processed in the following sequence: CD compensation, clock recovery, then resampling and going through classical CMA, a three-stage CMA, frequency offset compensation, a feed-forward phase equalization, LMS equalizer (LMSE), and then differentially detected to avoid cycle slip effects.

Furthermore, the detailed processing [17] for the electrical polarization recovery is achieved using a three-stage blind equalization scheme: (1) first, the clock is extracted using the "square and filter" method, and then the digital signal is resampled at twice the baud rate based on the recovery clock; (2) second, a $T/2$-spaced time-domain FIR filter is used for the compensation of CD, where the filter coefficients are calculated from the known fiber CD transfer function using the frequency-domain truncation method; (3) third, adaptive filters employing two complex-valued, 13 tap coefficients and partial $T/2$-space are employed to retrieve the modulus of the 16QAM signal.

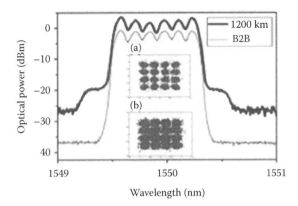

FIGURE 13.29   Spectra of Nyquist channels, 6 × 128 G PDM 16QAM B2B and 1200 km transmission.

The two adaptive FIR filters are based on the classic CMA and followed by a three-stage CMA to realize multimodulus recovery and polarization demultiplexing. The carrier recovery is performed in the subsequent step, where the fourth power is used to estimate the frequency offset between the LO and the received optical signal. The phase recovery is obtained by feed-forward and LMS (least-mean-square) algorithms for local oscillator frequency offset (LOFO) compensation. Finally, differential decoding is used for improving the BER. This is also implemented in the DSP sub-systems incorporated in the receiver.

The spectra of the six Nyquist channels under BTB and 1200 km SSMF transmission are shown in Figure 13.29. It is noted that all even channels are shaped with 0.01 roll-off Nyquist filters, we do not observe the flatness of individual channels in the diagram. The constellation as expected would require significant amounts of equalization and processing.

On average, the achieved BER of $2e^{-3}$ with a launched power of −1 dBm over 1200 km SSMF nondispersion compensation link was demonstrated by the authors of Ref. [17]. The link is determined by a recirculating loop consisting of four spans, with each span length of 80 km SSMF and an in-line EDFA. WSS is employed wherever necessary to equalize the average power of the subchannels. The optimum launched power is about −1 dBm for the six-subchannel superchannel transmission.

13.4.7   450 Gbps 32QAM Nyquist Transmission Systems

Further spectral packing of subchannels in a superchannel can be done with Nyquist pulse shaping and predistortion or pre-equalization at the transmitting side. Zhou et al. [18] has recently demonstrated the generation and transmission of 450 Gbps WDM channels over the standard 50 GHz ITU-T grid optical network at a net spectral efficiency of 8.4 b/s/Hz. This result is accomplished by the use of Nyquist-shaped, polarization-division-multiplexed (PDM) 32QAM, or 5 bits/symbol × 2 (polarized modes) × 45 GSy/s to give 450 Gbps. Both pre- and posttransmission digital equalization techniques are employed to overcome the limitation of the DAC bandwidth. Nearly ideal Nyquist pulse shaping with a roll-off factor of 0.01 allows guard bands of only 200 MHz between subcarriers. To mitigate the narrow optical filtering effects from the 50-GHz-grid reconfigurable optical add-drop multiplexer (ROADM), a broadband optical pulse-shaping method is employed. By combining electrical and optical shaping techniques, the transmission of 5 × 450 Gbps PDM-Nyquist 32QAM on the 50 GHz grid over 800 km and one 50-GHz-grid ROADM was proven with soft DSP equalization and processing. The symbol rate is set at 28 GSy/s.

It is noted that the transmission SSMF length is limited to 800 km due to the reduced Euclidean geometrical distance between constellation points of 32QAM and by avoiding the accumulated

ASE noises contributed by EDFA in each span, Raman optical amplifiers with distributed gain are used in a recirculating loop of 100 km ultra-large-area fibers. The BER performance for all five subchannels is shown in Figure 13.30a, with inserts of the spectra of all subchannels. Note the near-flat spectrum of each subchannel that indicates the near-Nyquist pulse shaping. Figure 13.30b and c show the spectra of a single subchannel before and after WSS, with and without optical filtering that performs as spectra shaping [18]. The original pulse shape can be compared with that displayed in Figure 1.15 in Chapter 1. Further to this published work, Zhou et al. have also time-multiplexed 64QAM and transmitting over 1200 km [19], that is, three circulating around the ring of 400 km of 100 km span incorporating Raman amplification. About 5 dB penalty between 32QAM and

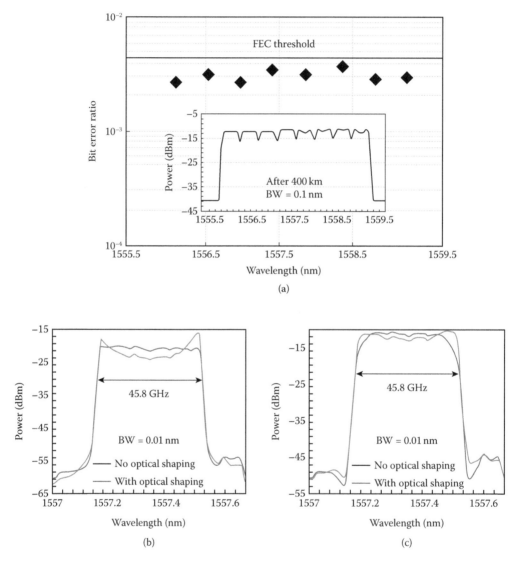

FIGURE 13.30 (a) Spectra of five subchannels of 450 Gbps channel, that is, 5 × 450 Gbps superchannels after 400 km (one loop circulating) of 5 × 80 km plus Raman amplification. BER versus wavelength and FEC threshold at 2.3e−3; (b) spectrum of a single subchannel before; and (c) after WSS with and without optical shaping by optical filters. (Extracted from X. Zhou et al., *IEEE J. Lightwave Technol.*, Vol. 30, No. 4, p. 553, 2012. With permission.) Note: Dark line for original spectra, light line for spectra after spectral shaping.

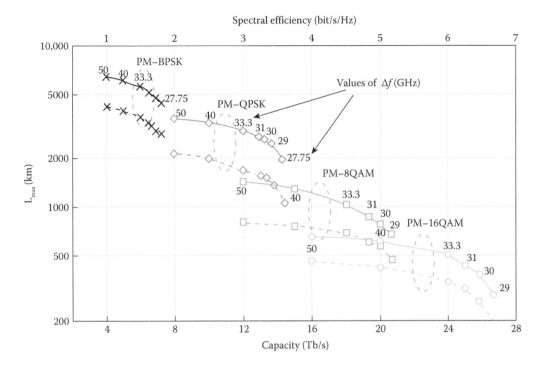

FIGURE 13.31 Transmission reach distance variation with respect to spectral efficiency and total capacity of superchannels. Different QAM schemes are indicated. PM = polarization multiplexing. The numbers are for bandwidth of the subchannel. (Extracted from G. Bosco et al., *IEEE J. Lightwave Technol.*, Vol. 29, No. 1, p. 53, 2011. With permission.)

64QAM in receiver sensitivity at the same FEC BER of 6e−3 is obtained. Note that electrical time division multiplexing of digital sequences from the arbitrary waveform generators was implemented by interleaving, so that higher symbol rates can be achieved. Furthermore, a 20% soft decision FEC using quasi-cyclic LDPC code is employed to achieve a BER threshold of 2.4e−2, and thus the 20% extra-overhead is required on the symbol rate.

Simulated results by Bosco et al. [20] without using soft FEC also shows the variation of the maximum reach distance with a BER of 2e−3 for capacity in the C-band (bottom axis) and spectral efficiency for PM-BPSK, PM-QPSK, PM-8QAM, and PM-16QAM, as shown in Figure 13.31 [20] for SSMF non-DCF optical transmission lines, as well as nonzero dispersion-shifted fibers (NZ-DSF) indicated by dashed lines. The design of NZ-DSF was described in Chapter 2.

13.4.8 DSP-Based Heterodyne Coherent Reception Systems

We have so far described optical transmission systems under homodyne coherent reception, that is, when the frequencies of the lightwave carriers and that of the LO are equal. The original motivation of using homodyne detection is to eliminate the 3 dB degradation as compared with heterodyne technique, and this is possible under DSP-based reception algorithm to avoid the difficulties of locking of the LO and the channel carrier. Under the classical heterodyne reception, the "at least" 3 dB loss comes from the splitting of the received signals in the electrical domain, then multiplied by sinusoidal cosine and sine RF oscillator to extract the in-phase and quadrature components.

So far, we have discussed homodyne reception DSP-based optical transmission systems that are considered for extensive deployment in commercial coherent communication systems for 100 G,

400 G, 1 T, or beyond. However, under the current progresses of wide bandwidth and high-speed electronic ADCs and photodetectors (PDs), once again, the coherent detection with DSP has made highly possible mitigation of impairments of transmitted channels by compensation and equalization in the sampled electrical domain. As we have seen in the previous sections of this chapter and in Chapter 3, for homodyne detection in polarization division multiplexing (PDM) systems, the I/Q components of each polarization state should be separated in optical domain with full information. Thus, four balanced PD pairs incorporating a photonic dual-hybrid structure and four-channel time-delay synchronized ADCs are required.

By up-converting I and Q components to the intermediate frequency (IF) at the same time, not only can heterodyne coherent detection halve the number of the balanced PDs and ADCs of the coherent receiver, but there is also no need to consider the delays between I and Q components in the PDM signal. Therefore, the four output ports of the optical hybrid can be also halved accordingly. However, this heterodyne technique can possibly be restricted by the bandwidth of the photodetectors (PDs). Furthermore, the external down-conversion of the IF signals may enhance the complexity of the reception system. Currently, the tremendous progresses in increasing the sampling rate and bandwidth of ADCs and PDs give a high possibility to exploit a simplified heterodyne detection. With wide bandwidth PDs and ADCs, the functions such as the down-conversion to the IF,the separation of the in-phase and quadrature signals, and the equalization for the PMD and nonlinear distortion effect, can all be realized in the digital domain that can be implemented in the DSP associated within the receiver sub-systems. A heterodyne detection in the transmission system is a limited 5-Gbps 4-ary quadrature amplitude modulation (4QAM) signal over 20 km in Ref. [21] and limited 20-Mbaud 64 and 128QAM over 525 km in Ref. [22], and then reaching Tbps by Dong et al. [23]. High-order modulation formats, such as PDM-16QAM and PMD-64QAM, offer higher capacity, but may suffer some additional penalty.

For the 100 G or beyond coherent system with required transmission distance shorter than 1000 km, the inferiority of the SNR sensitivity in heterodyne detection is not so obvious. Conversely, less number of ADCs and easy implementation of the DSP for IF down-conversion make heterodyne detection a potential candidate for a 100 G or beyond transmission system. Figure 13.32a [23] shows the schematic diagram of the heterodyne reception with digital processing. The quadrature amplitude-modulated signals are transmitted and imposed onto the $\pi/2$ hybrid coupler, which is now simplified, and there is no $\pi/2$ phase shifter as compared to the hybrid coupler discussed in Chapter 4 and the previous sections of this chapter. The coherent mixing between the LO and the modulated channel gives the time-domain beating signals which are the information signal envelope under which the carrier is covered within the symbol period. Thus, both the real and imaginary parts appear in the electronic signals produced after the balanced detection in the PDPs in which the beating happens (refer to Figure 13.32b). The electronic currents produced after the PDPs are then amplified via the TIA to produce voltage-level signals that are conditioned to appropriate levels, so that they can be sampled by the ADCs. Thence, doing the FFT will produce a two-sided spectrum that exhibits the frequency shifting of the baseband to the RF or IF frequency. Both the in-phase and quadrature components are embedded in the two-sided spectrum. One sideband of the spectrum can be used to extract the I- and Q- parts of the QAM signals for further processing. Ideally, the compensation of the CD should be done in the first stage, and then followed by the carrier phase recovery and then resampling with correct timing.

The following sequence of processing in the digital domain was conducted: (1) sampling in ADC and conversion from analog voltage level to digital sampled states; (2) FFT to obtain frequency-domain spectrum; then, (3) extract one-sided spectrum and do a frequency shifting to obtain the baseband samples of the spectrum; (4) resampling with two times the sampling rate; hence (5) compensation of CD and (6) carrier phase and clock recovery; (7) using the recovered clock to resample the data sequence; then (8) conducting normal CMA and three-stage CMA to obtain the initial constellation; then (9) frequency equalization of frequency difference of the LO; (10) feed-forward

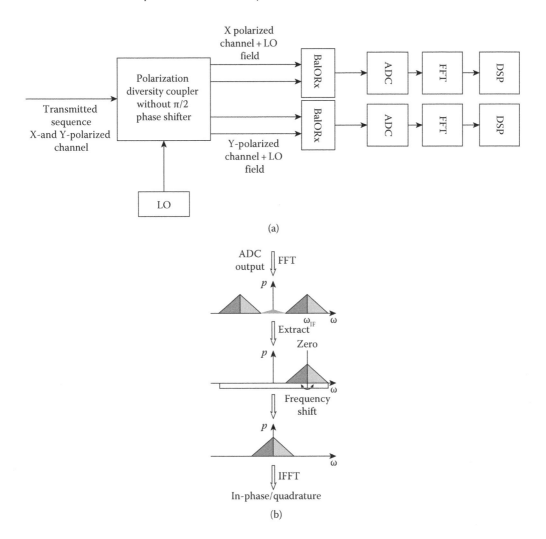

FIGURE 13.32 The principles of heterodyne coherent reception: (a) principal blocks and (b) spectra to recover baseband signals. (Adapted from Z. Dong et al., *Opt. Express*, Vol. 21, No. 2, p. 1773, 2013.)

phase equalization, LMS equalization for PMD compensation; and then, finally, (11) differential decoding of the samples and symbol to determine the transmission performance in term of BER versus the variation of the OSNR which is measured at the position located just before the input of the optical receiver.

It is further noted that, for the equalization based on DSP, a *T*/2-spaced time-domain FIR filter is used for the compensation of CD. The two complex-valued, 13-tap, *T*/2-spaced adaptive FIR filters are based on the classic CMA followed by three-stage CMA to realize multimodulus recovery and polarization demultiplexing.

The B2B BER versus the OSNR of the heterodyne reception optical transmission system is shown in Figure 13.33 [23] for PDM-16QAM 128 Gbps per subchannel with different IF and compared with homodyne detection scheme. It is noted that 3 dB penalty can be suffered by the heterodyne detection as compared with homodyne reception because in such detection system, only signals located in either the upper or lower band spectrum is processed. Thus, at a BER of 4e–3, the required OSNR is about 23 dB. The transmission distance is 720 km of SSMF, incorporating EDFA and non-DCF.

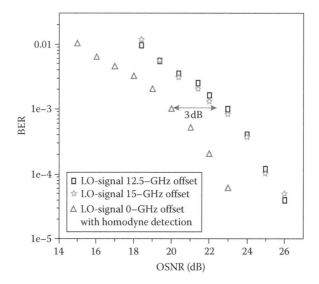

FIGURE 13.33 B2B BER versus OSNR of the heterodyne reception system of PDM-16QAM 112 Gbps × 8 superchannel with different IF frequency. (Extracted from Z. Dong et al., *Opt. Express*, Vol. 21, No. 2, p. 1773, 2013. With permission.)

13.5 CONCLUDING REMARKS

The total information capacity transmitted over a single single-mode-fiber has increased tremendously owing to the spectral packing of subchannels by pulse-shaping techniques and DSP algorithms, as well as soft FEC, so as to allow higher orders of modulation QAM to be realized. Thus, digital processing in real time and digitally based coherent optical transmission systems have been proven to be the most modern transmission systems for the global Internet in the near future. The principal challenges now are the realization of application-specific integrated circuits (ASICs) using microelectronics technology or ultra-fast field-programmable gate array (FPGA)-based systems. The concepts of processing in the digital domain have been proven by off-line systems. In real time processing systems, more efficient algorithms are required. These remain topical research topics and engineering issues to pursue. Higher symbol rates may be possible when wider-bandwidth optical modulators and DAC/ADC systems are available, for example, grapheme plasmonic silicon modulators [24] and their integration with microelectronic DAC and ADC, DSP systems.

The processing algorithms for QAM vary from level to level, but they can essentially be based on the number of circles existing in multilevel QAM as compared to a mono-cycle constellation of the QPSK scheme. The algorithms developed for QPSK can be extended and modified for higher-level circular constellations.

REFERENCES

1. T. E. Bell, Communications: Coherent optical communication shows promise, the FCC continues on its path of deregulation, and satellite communications go high-frequency, *IEEE Spectrum*, Vol. 23, No. 1, pp. 49–52, 1986.
2. Linear Circuits Inc., *From Single Ended to Different Input Trans-Impedance Amplifier*, http://circuits.linear.com/267
3. J. S. Weiner, A. Leven, V. Houtsma, Y. Baeyens, Y.-K. Chen, P. Paschke, Y. Yang, et al., SiGe differential transimpedance amplifier with 50-GHz bandwidth, *IEEE J. Solid-State Circuits*, Vol. 38, No. 9, pp. 1512–1517, 2003.

4. I. Fatadin and S. J. Savory, Compensation of frequency offset for 16-QAM optical coherent systems using QPSK partitioning, *IEEE Photon. Technol. Lett.*, Vol. 23, No. 17, pp. 1246–1248, 2001.

5. E. Ip and J. M. Kahn, Digital equalization of chromatic dispersion and polarization mode dispersion, *IEEE J. Lightwave Technol.*, Vol. 25, No. 8, p. 2033, 2007.

6. E. Ip and J. M. Kahn, Compensation of dispersion and nonlinear impairments using digital back propagation, IEEE *J. Lightwave Technol.*, Vol. 26, No. 20, p. 3416, 2008.

7. I. Fatadin, D. Ives, and S. J. Savory, Laser linewidth tolerance for 16-QAM coherent optical systems using QPSK partitioning, *IEEE Photon. Technol. Lett.*, Vol. 22, No. 9, pp. 631–633, 2010.

8. H. Louchet, K. Kuzmin, and A. Richter, Improved DSP algorithms for coherent 16-QAM transmission, in *Proceedings of the ECOC'08*, paper tu.1.E6, Belgium, September 2008.

9. R. Noe, T. Pfau, M. El-Darawy, and S. Hoffmann, Electronic polarization control algorithms for coherent optical transmission, *IEEE J. Sel. Quant. Electron.*, Vol. 16, No. 5, pp. 1193–1199, 2010.

10. L. N. Binh, *Digital Optical Communications*, Boca Raton, FL: CRC Press, 2010, chap. 3, Appendix.

11. D. Hillerkuss, R. Schmogrow, T. Schellinger, M. Jordan, M. Winter, G. Huber, and T. Vallaitis, 26 Tbit/s line-rate super-channel transmission utilizing all-optical fast Fourier transform processing, *Nat. Photonics*, Vol. 5, p. 364, 2011.

12. D. Hillerkuss, M. Winter, M. Teschke, A. Marculescu, J. Li, G. Sigurdsson, and K. Worms, Simple all-optical FFT scheme enabling Tbit/s real-time signal processing, *Optics Express*, Vol. 18, No. 9, pp. 9324–9340, 2010.

13. N. Godard, Passband timing recovery in an all-digital modem receiver, *IEEE Trans. Commun.*, Vol. 26, pp. 517–523, 1978.

14. N. Stojanovic, N. G. Gonzalez, C. Xie, Y. Zhao, B. Mao, J. Qi, and L. N. Binh, Timing recovery in Nyquist coherent optical systems, in *International Conference on Telecommunications Systems*, Serbia, 2012.

15. T. T. Fang and C. F. Liu, Fourth-power law clock recovery with pre-filtering, in *Proceedings of the ICC*, Vol. 2, pp. 811–815, Geneva, Switzerland, 1993.

16. Z. Dong, X. Li, J. Yu, and N. Chi, 128-Gbps Nyquist-WDM PDM-16QAM generation and transmission over 1200-km SMF-28 with SE of 7.47 b/s/Hz, *IEEE J. Lightwave Technol.*, Vol. 30, No. 24, pp. 4000–4006, 2012.

17. Z. Dong, X. Li, J. Yu, and N. Chi, 128-Gbps Nyquist-WDM PDM-16QAM generation and transmission over 1200-km SMF-28 With SE of 7.47 b/s/Hz, *IEEE J. Lightwave Technol.*, Vol. 30, No. 24, pp. 4000–4006, 2012.

18. X. Zhou, L. E. Nelson, P. Magill, R. Isaac, B. Zhu, D. W. Peckham, P. I. Borel, and K. Carlson, PDM-Nyquist-32QAM for 450-Gbps per-channel WDM transmission on the 50 GHz ITU-T grid, *IEEE J. Lightwave Technol.*, Vol. 30, No. 4, p. 553, 2012.

19. X. Zhou, L. E. Nelson, R. Isaac, P. Magill, B. Zhu, D. W. Peckham, P. Borel, and K. Carlson, 1200 km transmission of 50 GHz spaced, 504-Gbps PDM-32-64 hybrid QAM using electrical and optical spectral shaping, in *OFC'12*, Los Angeles, CA, 2012.

20. G. Bosco, V. Curri, A. Carena, P. Poggiolini, and F. Forghieri, On the performance of Nyquist-WDM terabit superchannels based on PM-BPSK, PM-QPSK, PM-8QAM or PM-16QAM subcarriers, *IEEE J. Lightwave Technol.*, Vol. 29, No.1, p. 53, 2011.

21. R. Zhu, K. Xu, Y. Zhang, Y. Li, J. Wu, X. Hong, and J. Lin, QAM coherent subcarrier multiplexing system based on heterodyne detection using intermediate frequency carrier modulation, in *2008 Microwave Photonics*, pp. 165–168, 2008.

22. M. Nakazawa, M. Yoshida, K. Kasai, and J. Hongou, 20 Msymbol/s, 64 and 128 QAM coherent optical transmission over 525 km using heterodyne detection with frequency-stabilized laser, *Electron. Lett.*, Vol. 42, No. 12, pp. 710–712, 2006.

23. Z. Dong, X. Li, J. Yu, and J. Yu, Generation and transmission of 8 × 112-Gbps WDM PDM-16QAM on a 25-GHz grid with simplified heterodyne detection, *Opt. Express*, Vol. 21, No. 2, p. 1773, 2013.

24. M. Liu and X. Zhang, Graphene-based optical modulators, in *Proceedings of the OFC 2012*, Los Angeles, CA, paper OTu1I.7, 2012.

14 DSP Algorithms and Coherent Transmission Systems

14.1 INTRODUCTION

Digital signal processing (DSP) is the principal functionality of modern optical communication systems employed at both the coherent optical receivers and the transmitter. At the transmitter, DSP is employed for determining the pulse shape at the output of optical modulators, as compensation for the limited bandwidth of the digital-to-analog converter (DAC), as well as that of the transfer characteristics of the electro-optic external modulators, and deskewing of the electrical path differences between the in-phase and quadrature-phase signals that often occur in ultra-high-speed systems. It is noted that the mixing of the received signals and the local oscillator (LO) laser happens in the photodetection stage.

Due to the square law process in the optoelectronic conversion, this mixing gives a number of terms, two square terms related to the intensity of the optical signals and the intensity of the LO laser, and two product terms that result from the beating between the optical signal beam and that of the LO. Only the difference product between these two optical frequencies is detected which is proportional to the signal to be recovered. When the optical frequencies of the signals and the oscillator are the same, the reception process is termed as homodyne coherent detection. If not then the process is a heterodyne type. Thus, provided that perfect homodyne mixing is achieved in the photodetection device, the detected signals, now in electrical domain, are in the baseband, superimposed by noises of the detection process. Thence, after electronic preamplification and then sampled by ADC, the signals are in discrete domain, and thus processed by digital signal processors that can be considered to be similar to processing in wireless communications systems. The main differences are in the physical processes of distortion, dispersion, nonlinear distortion, and clock recovery, broadband properties of optical modulated signals transmitted over long-haul optical fibers or short-reach scenarios.

On the other hand at the receiver end, DSP follows the ADC with the digitalized symbols and processes the noisy and distorted signals with algorithms to compensate for linear distortion impairments such as chromatic dispersion (CD), polarization-mode dispersion (PMD), nonlinear self-phase modulation (SPM) effects, clock recovery, LO frequency offset (FO) with respect to the relieved channel carrier, and so on. A possible flow of sequences in DSP can be illustrated as in Figure 14.1 for modulated channel with the in-phase and quadrature components and polarization multiplexing as recovered in the electronic domain after the optical processing front end as shown at the beginning, the input boxes. Thus, there are two pairs of electrical signal output from the two polarized channels and two for the in-phase and quadrature components. Inter processing of these pairs of signals can be implemented using 2×2 multiple-input multiple-output (MIMO) techniques, which are well known in the digital processing of wireless communications signals [1].

After the optoelectronic detection of the mixed optical signals, the electronic currents of the polarization division multiplexing (PDM) channels are then amplified with a linear transimpedance amplifier and a further amplification main stage if required to boost the electric signals to the level at which the ADC can digitalize the signals into their equivalent discrete states. Once the digitalized signals are obtained, the first task would be to ensure that any delay difference between the channels and the in-phase and quadrature components in the electronic and digitalization processes is eliminated by deskewing. Thence, all the equalization processing can be carried out to compensate for any distortion on the signals due to the propagation through the optical transmission line.

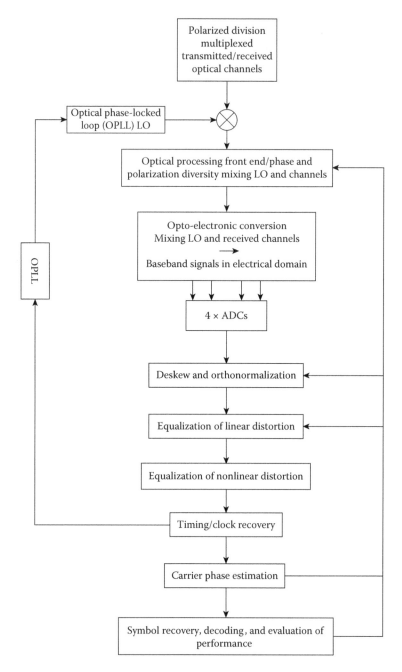

FIGURE 14.1 Flow of functionalities of DSP processing in a QAM coherent optical receiver with possible feedback control. Note that similar chart is also given in Figure 13.1.

Furthermore, nonlinear distortion effects on the signals can also be superimposed at this stage. Then, the clock recovery and carrier phase detection can be implemented. Hence, the processes of symbol recovery, decoding, and evaluation of performance can be implemented.

Regarding noises in the optical coherent detection the noises are mainly dominated by the shot noises generated by the LO power. However, if direct detection, then the quantum shot noises are contributed by the signal power, and hence are signal dependent. These noises will contribute to the convergence of algorithms.

TABLE 14.1

Functionalities of Subsystem Processing in a DSP-Based Coherent Receiver

Function	Description
Deskewing	Alignment of in-phase and quadrature components or polarized channel components temporarily or any different propagation time due to electrical connections
Orthogonalization	Ensuring independence or decorelation between channels
Normalization	Standardization amplitude of components to maximum value of unity
Equalization	Compensation of impairments due to physical effects or imperfections of subsystems
Interpolation	Correction of timing error
Carrier phase estimation	Compensation of phase errors of the carrier
Frequency offset	Correction of the offset frequency between LO and carrier frequencies
OPLL	Optical phase-locked loop to ensure matching between channel lightwave carrier and that of the LO carrier (see locking mechanism given in Chapter 5)

This chapter is thus organized as follows. Section 14.2 gives a basic background of equalization using transversal filtering and zero forcing with and without feedback in the linear sense, the linear equalizer. Then when a decision, in a nonlinear sense, is employed under some criteria to feedback to the input sequence, the equalization is nonlinear, and the equalizer is classified as a nonlinear equalizer (NLE) or direct directing (DD) equalizer. Tolerance to noises is also given, and a simple quality factor of the system performance is deduced.

Clock recovery is important for the recovery of the timing for sampling of the received sequence. This technique is briefly discussed in Section 14.4.4. Determining the reference phase for coherent reception is also critical so as to evaluate several quadrature-modulated transmitted signals. Thus, carrier phase recovery technique is also described using DSP techniques, especially in the case when there is significant offset between the LO and the lightwaves carrying the signals. This is described in Section 14.5.

When the maximum-likelihood sequence estimation (MLSE) used in the NLE is employed, the equalization can be considered optimum, but as consuming substantial memory. We dedicate Section 14.3 to this algorithm an example of this MLSE is given to demonstrate its effectiveness of the equalization algorithm in minimum shift keying (MSK) self-homodyne coherent reception transmission systems under the influence of linear distortion effects with the modulation scheme is an MSK. This has already been described in Chapter 12, and is not repeated. The MLSE algorithm is valid for processing individual I and Q components of each polarized channel as in the case of direct detection. We must note here that, under the I–Q modulation, both the I and Q components are modulated as non-return-to-zero modulation formats with a $\pi/2$ phase shift in the optical carrier. Thus, when detected, they behave similar to the case of direct detection. Their $\pi/2$ phase shift has been taken care of by the hybrid optical coupler. Thus, we can see that four ADC-DSP subsystems are required for coherent PDM-quadrature phase shift keying (QPSK) or PDM-M-ary-quadrature amplitude modulation (QAM) reception systems.

We can thus see that DSP algorithms are most critical in modern DSP-based optical receivers or transponders. This chapter thus introduces the fundamental aspects of these algorithms for optical transmission systems over dispersive channels whether under linear or nonlinear distortion physical effects.

The functionalities of each processing block of Figure 14.1 can be categorized as shown in Table 14.1 [2].

14.2 GENERAL ALGORITHMS FOR OPTICAL COMMUNICATIONS SYSTEMS

Indeed, in coherent optical communication systems incorporating DSP, the reception of the transmitted signals by mixing them with a LO whose frequency is identical or close to equality which is known as the homodyne or intradyne reception techniques. The coherent reception converts the optical signals back to the baseband whose phase and amplitude is complex and distorted due to the transmission medium. The received signals are now transformed back to the digital domain and can thus be processed by a number of digital processing algorithms as previously developed for high-speed modems [3] or in wireless communication systems [4].

Naturally, a number of steps of conversions from optical to electrical and vice versa at the receiver and transmitter have to be conducted with further optical amplifier noises accumulated along the optical transmission line and electronic noises.

In this chapter, we assume that the signal level is quite high as compared to the electronic noise level at the output of the electronic preamplifier that follows the photodetector. Hence, the processing algorithms are conducted in the baseband electrical digital domain after the ADC stage.

14.2.1 Equalization of DAC-Limited Bandwidth for Tbps Transmission

This section considers algorithms developed to compensate for the band-limited DAC, especially the generation of 28 and 32 GBaud PDM-Nyquist-QPSK by a DAC with an analog bandwidth of only 11.3 GHz [5].

14.2.1.1 Motivation

For realizing ultra-high-capacity optical fiber networks, it is extremely important to improve the spectral efficiency in wavelength-division-multiplexed (WDM) transmission systems. High-order modulation formats such as M-ary phase shift keying (PSK) and QAM offer efficient solutions [6,7]. However, further improvement can be achieved by using the well-known "Nyquist pulse shaping" technique [8], applicable to any modulation format. This technique relies on tight signal filtering squeezing the signal bandwidth very close to the symbol rate, meanwhile maintaining zero inter-symbol interference (ISI). In order to minimize the penalty due to crosstalk between closely spaced WDM channels, it is required to reshape the optical spectrum either in optical domain by narrow optical filtering, or, in the electrical domain, by shaping the driving signals applied to optical modulators by combining DSP and DAC.

The main drawback of performing spectrum shaping in the optical domain is the requirement for a very specific transfer function of the optical filter, which may be cost-ineffective and bulky. Alternatively, it is more simple and efficient to generate the Nyquist pulse for any modulation format using transmitter DSP and DAC. Several Nyquist experiments using lower-rate DACs have been reported recently, for example, generation of 9 GBaud PDM-Nyquist-32QAM with a roll-off factor (ROF) of 0.01 using 24 GSa/s DAC [9], Nyquist pulses at 14 GBaud of PDM-16QAM generated by 28 GSa/s DAC [10]. For the generation of Nyquist pulse with high baud rate, one critical limitation comes from the electrical bandwidth at transmitter side, which is dominated by DAC, RF driver, and I–Q modulator. And, in almost all the publications, pre-equalization for this limitation is mentioned.

In this section, we evaluate improvements from bandwidth limitation pre-equalization by considering the generations of 28 and 32 GBaud PDM-Nyquist-QPSK using a DAC working at 2 Sa/s with 11.3 GHz analog bandwidth and 8-bit resolution. As described in Chapter 3, the Nyquist waveform can be generated by the use of off-line DSP and loading to the associate memory for execution in the DAC. A root-raised-cosine (RRC) filter is implemented in frequency domain and combined with a pre-equalizer to compensate for the limited Tx electrical bandwidth. Optical/electrical signal spectrums and eye diagrams together with respectable improvements characterize the limits of high-speed Nyquist signal generation, and the fastest DAC devices reported here will speed up the practical deployment of DAC in ultra-high-bit-rate superchannel transmissions.

14.2.1.2 Experimental Setup and Bandwidth-Limited Equalization

The experimental setup used in our investigation is shown in Figure 14.2. The output of an external cavity laser (ECL) with a linewidth of 100 KHz is coupled to a PDM I–Q modulator driven by a DAC with four independent outputs. Driving signals are generated by off-line DSP and loaded to the DAC

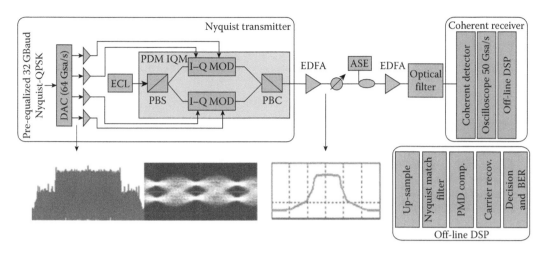

FIGURE 14.2   Experimental setup of 128 Gbps Nyquist PDM-QPSK transmitter and BTB.

evaluation board. To generate the Nyquist waveform, pseudo-random-bit sequences of length $2^{15}-1$ are used in bit-to-symbol mapping to generate the in-phase and quadrature signals, which are then up-sampled to 2 Sa/s and followed by an RRC filter of an ROF of 0.01.

The Nyquist filter has 256 taps and is implemented in the frequency domain. This implementation brings the advantage that bandwidth limitation pre-equalization can be combined with Nyquist pulse shaping, but only one set of FFT/IFFT for each data path is required, and this allows significant reduction of DSP complexity avoiding two separate filters dedicated for Nyquist pulse shaping and bandwidth limitation pre-equalizer, respectively. The DAC is operated at the sample rate of 56 and 64 GSa/s, leading to the symbol rate of 28 and 32 GBaud, respectively (twofold oversampling). The outputs of DACs are amplified by four RF drivers before feeding into the PDM I–Q modulator. Unlike other reports of Nyquist pulse generation at lower speeds, no antialiasing low-pass electrical filter is used to filter out spurious frequency components before drivers. They have already been attenuated after DACs, thanks to the DAC-limited bandwidth being quite below half of the baud rate. Voltages controlling RF driver amplification are carefully adjusted, so that the output amplitudes modulating MZMs are twice less than the $V\pi$ value eliminating nonlinear effect introduced by the MZM nonlinear transfer function. The generated optical Nyquist signal is then combined with ASE noise generated by an erbium-doped fiber amplifier (EDFA) before being amplified and filtered by a 50 GHz optical filter to prevent shooting noise effects.

After optoelectrical conversion using a polarization- and phase-diverse coherent receiver, four electrical signals are captured by a four-channel 50 GSa/s oscilloscope with an analog bandwidth of 20 GHz (million samples per line). The captured data are then processed off-line including up-sampling, Nyquist matched filtering (RRC filter), recovering polarization by an 11-tap 2×2 MIMO adaptive equalizer, carrier frequency and phase recovery (Viterbi & Viterbi), hard limiter, and bit-error-rate (BER) measurement.

To derive the transfer function later used for pre-equalization, two approaches are evaluated: (1) based on DAC's bandwidth characteristics only, and (2) based on the whole electrical and optical filtering effects at the transmitter side. For the former one, we determine the DAC transfer function by driving it with sinusoid waveforms at different frequencies, and the result is shown in Figure 14.3j, in which a 3 dB bandwidth of around 11.3 GHz can be observed. Applying approach (2), the transfer function of the whole transmitter side is obtained by comparing the optical spectrum after PDM I–Q modulator with the ideal Nyquist spectrum. BTB performance comparison between these two approaches is shown in Figure 14.4a. The Q factor difference is calculated by subtracting the Q factor of approach (2) from that of approach (1). And both methods result in similar performance, with

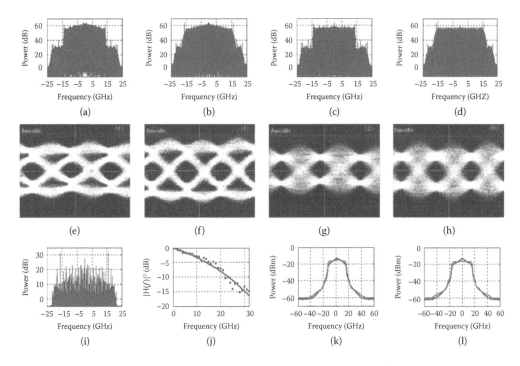

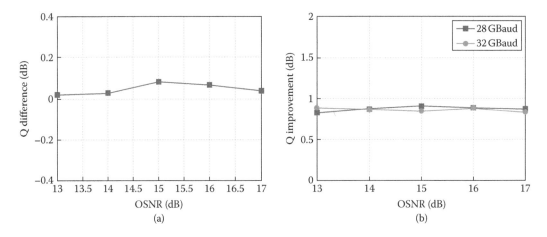

FIGURE 14.3 (a, b) Spectrum and (e, f) eye diagram of RF signals after DAC without pre-equalization at 28/32 GBaud setup; (c, d) spectrum and (g, h) eye diagram of RF signals after DAC with pre-equalization at 28/32 GBaud setup; (i) noise character of DAC; (j) measured DAC bandwidth character; (k, l) optical spectrums of signals after PDM IQM at 28/32 GBaud setup, light/dark gray line for without/with pre-equalization.

FIGURE 14.4 (a) Q factor difference between different transfer functions (all transmitter side imperfection based and only DAC's limited bandwidth based) used for pre-equalization in 32 GBaud system; (b) Q factor improvement from pre-equalization in 28 and 32 GBaud systems. OSNR = optical signal-to-noise ratio.

a bit more emphasized performance for approach (1). Intuitively from this figure, one can conclude that the bandwidth limitation mainly comes from the DAC side. Therefore, all the following pre-compensation results are based on only the DAC's transfer function.

Impacts caused by the limited bandwidth of the transmitter side for a 28 GBaud system can be observed from Figure 14.3a and the light gray line in Figure 14.3k, which depict the electrical and optical spectrums at the outputs of the DAC and the PDM I–Q modulator, respectively.

The Gaussian-like slopes caused by the limited bandwidth can be observed. Furthermore, from the eye diagram shown in Figure 14.2e, the clear transient patterns caused by the non-Nyquist filtering are visible. Similar but stronger influence from the limited bandwidth in case of a 32 GBaud system can be observed from Figure 14.3b and the light gray line in Figure 14.3l.

The effectiveness of pre-equalization for a 28 GBaud system can be observed from both the electrical and optical spectrums depicted in Figure 14.3c and the dark gray line in Figure 14.3k, respectively. In the passband, the spectrums are almost as flat as those of an ideal Nyquist signal, except the side lobes (stopband) caused by DAC noise (will be discussed later).

By examining the eye diagrams of the pre-equalized 28 GBaud RF signal depicted in Figure 14.3g, we can observe that there are no longer any transient patterns, and that the eye opening at the optimum sampling point is only half of that of the nonequalized one. Similar observations come from the electrical/optical spectrum and eye diagram of the 32 GBaud system presented in Figures 14.3d and 14.3l. Improvements from pre-equalization in terms of Q factor (versus optical-signal-to-noise ratio [OSNR]) are presented in Figure 14.4b. More than 0.8 dB is achieved in both scenarios, 28 and 32 GBaud systems. It should be noticed that the side lobes in all the electrical spectrums with or without pre-equalization in Figure 14.3 come from the intrinsic noises of the DAC device. Depicted in Figure 14.3i is the spectrum of output signal from DAC under zero input driving signals. Although the noise is supposed to follow Gaussian distribution with a flat spectrum except for some clock crosstalk, an oscilloscope with 20 GHz bandwidth used to capture the noise gives the steep cutoff edge at around 20 GHz, and a similar spectral shape can be observed as the side lobes of the spectrums shown in Figure 14.3a through d.

Therefore, in this section, the generations of 28 and 32 GBaud PDM-Nyquist-QPSK by a DAC with the sample rates up to 64 GSa/s and 11.3 GHz analog bandwidth is presented to address the band-limited problem of DAC whose sampling rate is a very high 64 Gbps. The impacts from the limited bandwidth of the DAC are evaluated in terms of electrical/optical spectrums and eye diagrams, and the corresponding pre-equalization can bring more than 0.8 dB Q factor gain.

14.2.2 LINEAR EQUALIZATION

Coherent optical communication systems can be considered synchronous serial systems. In transmitting optical signals over uncompensated optical fiber that spans over long-haul multispan links or uncompensated metropolitan networks at very high bit rates (commonly, now in the second decade of the twenty-first century, at 25 or 28–32 GSy/s, including forward error coding overloading), the received signals are distorted due to linear and nonlinear dispersive effects, hence leading to ISI in the baseband after coherent homodyne detection at the receiver. In Chapters 4 and 5 we have described in detail the processes of coherent detection, especially homodyne reception for optical signals including noise processes. Over the nondispersion-compensating multispan fiber links, the most important signal distortion is the linear CD and PMD, especially when PDM is employed and two polarization orthogonal channels are simultaneously transmitted. The ISI naturally occurs due to the band-limiting of the signals over the limited bandwidth of the long fiber link, as described in Chapter 2 on the transfer function of optical fibers. Other nonlinear distortion effects have also been included in the transfer function as indicated by the nonflatness of the passband of the transfer function of the fiber. These ISI effects will degrade the performance of the optical communications and need to be tackled with the equalization processes that are discussed in this chapter.

The optical coherent detection processes and DSP of sampled digital signals may be classified into two separate groups.

The first of these, the received sampled digital signals, are fed through an equalizer that corrects the distortion introduced by the CD and PMD effects and restores the received signals into a copy of the transmitted signals in the electrical domain. The received signals are then detected in the conventional manner as applied to any serial digital signals in the absence of ISI. In other words, the equalizer acts as the inverse of the channel, so that the equalizer and the channel together introduce no signal distortion, and each data symbol is detected as it arrives, independently of each other. The equalizers can be either linear or nonlinear.

In the second group of detection processes, the decision process is modified to take into account the signal distortion that has been introduced by the channel, and no attempt need, in fact, be made here to reduce signal distortion prior to the actual decision process. Although no equalizer is now required, the decision process may be considerably complex, and hence complex algorithms are needed to deal with this type of decision processing.

In addition to these two decision groups, there are additional issues to resolve, involving the synchronization of the sampling at the receivers, the clock recovery, and carrier phase estimation, so that a reference clocking instant can be established, and the referenced phase of the carrier can be used to detect the phase states of the received samples. Concurrently, any skew or delay differences between the I and Q components as well as the polarized multiplexed channels must be detected and deskewed, so that synchronous processes can be preserved in the coherent detection processes. Although they remain in practical implementation under software control, these physical processes are not included in this book's chapters.

Furthermore, any delay difference due to different propagation time of the two different polarized channels or due to different lengths of the electrical connection lines, then a deskewing process can be conducted to minimize these effects, as shown in Figure 14.1.

Thus, in the following section, the LE process is introduced.

14.2.2.1 Basic Assumptions

In general, a simplified block diagram of the coherent optical communication system can be represented in the baseband, that is, after optical field mixing, and detected by optoelectronic receivers and E/O conversion at the transmitter, as shown in Figure 14.5. The input to the transmitter is a sequence of regularly spaced impulses, assuming that the pulse width is much shorter than that of the symbol period, which can be represented by $\sum_i s_i \delta(t - iT)$, with S_i of binary states taking both positive and negative values as the amplitude of the input signal sequence; thus, we can write

$$s_i = \pm k; \qquad k > 0 \qquad (14.1)$$

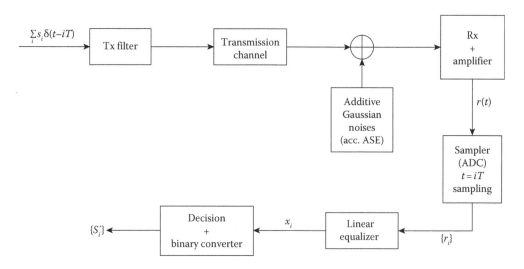

FIGURE 14.5 Simplified schematic of coherent optical communication system. Note equivalent baseband transmission systems due to baseband recovery of amplitude and phase of optically modulated signals.

Thus, the impulse sequence is of a binary polar type. This type of signal representation is advantageous in practice as it can be employed to modulate the optical modulators, especially the Mach–Zehnder interferometric modulator (MZIM), or the I–Q modulator as described in Chapter 3, so that the phase of the carrier under the modulated envelope can take positive or negative phase angles, which can be represented on the phasor diagram. In practice, the impulse can naturally be replaced by a rectangular or rounded or Gaussian waveform by modifying the transmitter filter, which can be a Nyquist raised-cosine type to tailor the transmitting pulse shape.

It is further noted that the phase of the optical waves and the amplitude modulated by the I- and Q-MZIM in the two paths of the I–Q modulator ensure the negative and positive position of the I and Q components on the real and imaginary axes of the constellation, as shown in Figure 3.42. Under coherent reception, these negative and positive amplitudes and phases of the signals are recovered in the electrical domain with superimposed noises. We can see that the synchronization of the modulation signals fed into the two MZIMs is very critical in order to avoid degradation of the sampling instant due to skewing.

14.2.2.2 Zero-Forcing Linear Equalization

Consider a transmission system and receiver subsystem as shown in Figure 14.6, in which the transmitter transfer function can be represented by $H_{Tx}(f)$, the transfer function of the channel in linear domain by $H_C(f)$, and the transfer function of the equalizer or filter by $H_{Eq}(f)$, which must meet the condition for the overall transfer function of all subsystems equating to unity

$$H(f) = H_{Tx}(f)H_C(f)H_{Eq}(f)e^{j\omega t_0} \tag{14.2}$$

Thus, from Equation 14.2, we can obtain, if the desired overall transfer response follows that of a Nyquist raised-cosine filter, the equalizer frequency response as

$$H_{Eq}(f) = \frac{T\,\mathrm{raise}\cos\left(\dfrac{f}{1/T},\rho\right)}{H_{Tx}(f)H_C(f)}e^{j\omega t_0} \tag{14.3}$$

With the raised-cosine function defined with β as the ROF as the frequency response of the equalizer

$$H_{Eq}(f) = \mathrm{raise\_cos}\left(\frac{f}{1/T},\beta\right) = \begin{cases} 1 & 0 \le |x| \le \dfrac{1-\beta}{2} \\ \cos^2\left(\dfrac{\pi}{2}\dfrac{|x|-\dfrac{1-\beta}{2}}{\beta}\right) & \dfrac{1-\beta}{2} \le |x| \le \dfrac{1+\beta}{2} \\ 0 & |x| > \dfrac{1+\beta}{2} \end{cases} \tag{14.4}$$

Implying that the filter is a Nyquist filter of raised-cosine shape; ω is signal baseband frequency; and t_0 is the time at the observable instant. In case two channels are observed simultaneously,

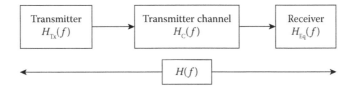

FIGURE 14.6 Schematic of a transmission system including a zero-forcing equalizer/filter.

the right-hand side of Equation 14.2 is a unity matrix and the unitary condition must be satisfied—that is, the conservation of energy.

From Equation 14.3, the magnitude and phase response of the equalizer can be derived. This condition guarantees a suppression of the ISI, and thus the equalizer can be named as a *linear equalizer* (LE) *zero-forcing* (ZF) *filter* (LE-ZF).

14.2.2.3 ZF-LE for Fiber as Transmission Channel

The ideal zero-forcing equalizer is simply the inverse of the channel transfer function given in Equation 14.2. Thus, the required equalizer transfer function $H_{Eq}(f)$ is given by

$$H_{Eq}(f) = H_{SSMF}^{-1}(f) = e^{j\alpha\omega^2} \tag{14.5}$$

where ω is the radial frequency of the signal envelope, and α is the parameter dependent on the dispersion, wavelength, and velocity of light in vacuum as described in Chapter 4. However, this transfer function does not maintain conjugate symmetry, that is

$$H_{Eq}(f) = e^{j\alpha\omega^2} \neq H_{Eq}^{-1}(f) \tag{14.6}$$

This is due to the square law dependence of the dispersion parameter of the fiber on frequency.

Thus, the impulse response of the equalizing filter is complex. Consequently, this filter cannot be realized by a baseband equalizer using only one baseband received signal, which explains the limited capability of LEs employed in direct detection receivers to mitigate the CD. On the other hand, it also explains why fractionally spaced LEs that are used within a coherent receiver can potentially extend the system to reach distances that are only limited by the number of equalizer taps [11]. Under coherent detection, and particularly the QAM, both real and imaginary parts are extracted separately and processed digitally. The schematic of a fractionally spaced finite impulse response (FIR) filter is shown in Figure 14.7, in which the delay time between taps is only a fraction of the bit period. The weighting coefficients $W_0(n),...,W_{N-1}(n)$ are determined, and then multiplied with the

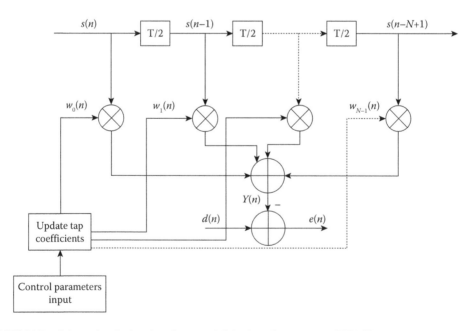

FIGURE 14.7 Schematic of a fractionally spaced finite impulse response (FIR) filter.

signal paths $S(n)$ at the outputs of the taps. These outputs are then summed up and subtracted by the expected filter coefficients $d(n)$ to obtain the errors of the estimation. This process is then iteratively operated until the final output sequence $Y(n)$ is closed to the desired response.

Under ZFE, at the sampling instant, the noise may have been increased due to errors in the delay of the zero crossing point of adjacent pulses, hence inter symbol interference (ISI) effects. Thus, the signal-to-noise ratio (SNR) may not be optimum.

The number of taps must cover the whole length of the dispersive pulses—for example, for a 1000 km standard single-mode optical fiber (SSMF) transmission line, where the BW of the channel is 25 GHz (or ~0.2 nm at 1550 nm), the total pulse spreading is 17 ps/nm/km × 1000 km × 0.2 nm = 3400 ps. Sampling rate would be 50 GSa/s (two samples per period @ 25 Gbps, assuming that modulation results in the BW equal to symbol rate), and one would then have 3400/40 periods or ~85 periods or taps with 170 samples altogether.

14.2.2.4   Feedback Transversal Filter

Transversal equalizers have been considered so far as linear feed-forward (FFE) transversal filters commonly employed in practice over several decades since the invention of digital communications. Equalization may alternatively be achieved by feedback transversal filters, provided certain conditions can be met by the channel responses—in this case, the linear transfer function of the fiber transmission line.

Let the z-transform of the sampled impulse response of the channel, which is the optical fiber under the modulated lightwave propagation connected in series with the receiving sub-systems, as

$$Y(z) = y_0 + y_1 z^{-1} + y_2 z^{-2} + \ldots\ldots + y_g z^{-g} \tag{14.7}$$

where $y_0 \neq 0$, and let $V(z)$ be the z-transform derived from the scaling of $Y(z)$, so that the first term becomes unity. Thus, we have

$$V(z) = \frac{1}{y_0} Y(z) = 1 + \frac{y_1}{y_0} z^{-1} + \frac{y_2}{y_0} z^{-2} + \ldots\ldots + \frac{y_g}{y_0} z^{-g} \tag{14.8}$$

or, written in cascade factorized form,

$$V(z) = \left(1 + \beta_1 z^{-1}\right)\left(1 + \beta_2 z^{-1}\right)\ldots\ldots\left(1 + \beta_g z^{-1}\right) \tag{14.9}$$

where β_i can be real or complex valued. The feedback transversal filter can then be configured as shown in Figure 14.8, which follows the operation just described. Obviously, the received sequence is multiplied with a factor $1/y_0$ in order to normalize the sequence, and delayed by one period with a multiplication ratio then summed up and feedback to subtract with the normalized incoming received sequence. The output is taped from the feedback path as shown in the diagram, at the point where the transversal operation is started.

14.2.2.5   Tolerance to Additive Gaussian Noises

Consider a linear FFE with $(m + 1)$ tap as shown in Figure 14.9, assuming that the equalizer has equalized the input signals in the baseband so that the output of the equalizer is given as

$$x_{i+h} \simeq s_i + u_{i+h} \quad \text{with} \quad \begin{cases} s_i = \pm k; \text{binary\_level\_signal\_amplitude} \\ u_{i+h} = \text{noise} - \text{component} \end{cases} \tag{14.10}$$

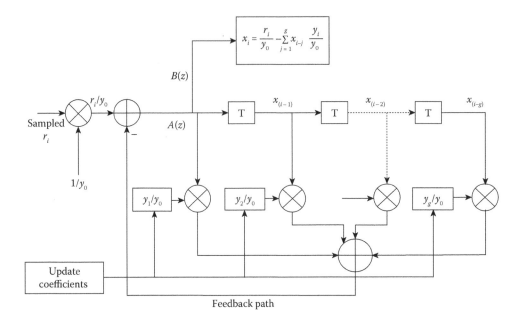

FIGURE 14.8 Schematic structure of a linear feedback transversal equalizer for a channel of a z-transform impulse response $Y(z)$.

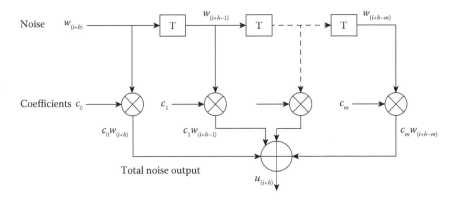

FIGURE 14.9 Noisy signals in an $(m + 1)$ tap linear FFE.

Note that noise always adds to the signal level, so that, for positive k, the noise u_{i+h} follows the same sign (and vice versa for negative value).

The output noise signals of the FFE can thus be written as

$$u_{i+h} = \sum_{j=0}^{m} w_{i+h-j} c_j \tag{14.11}$$

The noise components $\{w_i\}$ are statistically independent Gaussian random variables with zero mean and a variance of σ^2, so that the total accumulated noise output given by Equation 14.11 is also an uncorrelated random Gaussian variable with zero mean and a variance of

$$\eta^2 = \sum_{j=0}^{m} \sigma^2 c_j^2 = \sigma^2 C C^{\mathrm{T}} = \sigma^2 |C|^2 \tag{14.12}$$

where $|C|$ is the Euclidean length of the $(m + 1)$ component vector $[C] = [c_0 \; c_1 \; ..... \; c_m]$, the sampled channel impulse response. Thus, η^2 is the mean square noise of the process and is also the mean square error (MSE) of the output sampled values x_{i+h}.

Thus, the probability noise density function of this Gaussian noise is given by

$$p(u) = \frac{1}{\sqrt{2\pi\eta^2}} e^{-\frac{u^2}{2\eta^2}} \tag{14.13}$$

Thus, with the average value of the magnitude of the received signal in sampled value of k which are equalized with no residual distortion, then the probability of error of the noisy detection process is given by

$$P_e = Q\left(\frac{k}{\eta}\right) = Q\left(\frac{k}{\sigma|C|}\right) \tag{14.14}$$

And, the electrical SNR is given as

$$\text{SNR} = 20\text{Log}_{10}\,(k/\sigma)\text{ dB} \tag{14.15}$$

with $Q(x)$ as the quality function of x, that is, the complementary error function as commonly defined. In practice, this SNR can be measured by measuring the standard deviation of the constellation point and the geometrical distance from the center of the constellation to the central point of the constellation point. One can see in the next section that the equalized constellation would offer much higher SNR as compared with that of the non equalized constellation. Thus, under Gaussian random distribution, the noises that contribute to the equalization process are also Gaussian, and the probability density function and the Euclidean distance of the channel impulse response can be found without much difficulties.

14.2.2.6 Equalization with Minimizing MSE in Equalized Signals

It has been recognized that a linear transversal filter that equalizes and minimizes the MSE in its output signals [12–14] generally gives a more effective degree of equalization than an equalizer that minimizes the peak distortion. Thus, an equalizer that minimizes the mean square difference between the actual and ideal sampled sequence at its output for a given number of taps would be advantageous when noises are present.

Now, revisiting the transmission given in Figure 14.9 in which the sampled impulse response of the channel, the sampled impulse response of $(m + 1)$ tap linear FFE, and the combined sampled impulse response of channel and the equalizer, $[Y]$, $[C]$, and $[E]$, respectively, are given by

$$[Y] = [y_0 \quad y_1 \quad ...... \quad y_g]$$

$$[C] = [c_0 \quad c_1 \quad ...... \quad c_m] \tag{14.16}$$

$$[E] = [e_0 \quad e_1 \quad ...... \quad e_{m+g}]$$

Under the condition that the impulse response of the channel is combined with that of the equalizer so that the recovered output sequence is only a pure delay of h sampling periods, the input sampled sequence we have

$$x_{h+i} \approx s_i + u_{h+i} \tag{14.17}$$

Or it is a pure delay superimposed by the Gaussian noises of the transmission line. For the case of optically amplified multispan transmission link, the system noises are the accumulated ASE noises of the optical amplifiers placed in each span. Thus, the resultant *ideal* vector at the output of the equalizer can be written as

$$[E] = \left[\underbrace{0 \ ...... \ 0}_{h} \quad 1 \ 0 \quad 0 \ 0 \ .... \ 0 \right]; \quad h = \text{integer} \tag{14.18}$$

However, in fact, the equalized output sequence would be (as seen from Equation 14.16)

$$x_{h+i} \approx \sum_{j=0}^{m+g} s_{i+h} - je_j + u_{h+i}; j = 0,1........,(m+g) \tag{14.19}$$

The input sampled values are statistically independent and have equal probability of taking values of $\pm k$, and this indicates that the expected value (denoted by the symbol $\xi[.]$) is

$$\xi\left[s_{i+h-j} s_{i+h-l} \right] = 0; \quad \forall j \neq l \quad \text{and} \quad \xi\left[s_{i+h-j}^2 \right] = k^2 \tag{14.20}$$

The noises superimposed on the sampled values are also statistically independent of the sampled signals with zero mean, and thus the sampled signals and noises are orthogonal, so we have the expected value $\xi\left[s_{i+h-j} u_{i+h} \right] = 0$.

Thus, an LE that minimizes the MSE in its output signals would also minimize the mean square value of $[x_{h+i} - s_i]$, which can be written as

$$\xi\left[(x_{i+h} - s_i)^2 \right] = \left\{ \begin{array}{c} \xi\left[\left(\sum\limits_{j=0}^{m+g} s_{i+h-j} e_j + u_{i+h} - s_i \right)^2 \right] \\ \\ = k^2 \left| E - E_h \right|^2 + \sigma^2 \left| C \right|^2 \end{array} \right. \tag{14.21}$$

with

$$\left| E - E_h \right|^2 = \sum_{\substack{j=0 \\ j \neq h}}^{m+g} e_j^2 + k^2 \left(e_h - 1 \right)^2$$

where $|E - E_h|$ and $|C|$ are the Euclidean distances of the vector $[E - E_h]$ and $C = [C]$, respectively. We can now observe that the first and second terms of Equation 14.21 are the MSEs in x_{i+h} due to ISI and the MSE due to Gaussian noises. Thus, we can see that the MSE process minimizes not only the distortion in terms of the MSE due to ISI but also that of the superimposing noises.

14.2.2.7 Constant Modulus Algorithm for Blind Equalization and Carrier Phase Recovery

14.2.2.7.1 Constant Modulus Algorithm

Suppose that intersymbol interference, additive noise, and carrier FO are considered. Then, a received signal before processing, $x(k)$, can be written as

$$x(k) = \sum_{i=0}^{M-1} h(i)a(k-i)e^{j\phi(k)} + n(k) \tag{14.22}$$

where $h(k)$ is the overall complex baseband-equivalent impulse response of the overall transfer function, including the transmitter, unknown channel, and receiver filter. Note that all E/O and O/E with coherent reception steps are removed. Furthermore it is assumed that there are only noises contribution but no distortion in both phase and amplitude components of the received complex quantities. $n(k)$ are the overall noises which are assumed to be Gaussian, mainly due to ASE noises over cascaded amplifiers with zero mean and standard deviation amount. The input data sequence $a(k-i)$ is assumed to be independent and identically distributed, and $\varphi(k)$ is the carrier phase difference between the signal carrier and LO laser.

Now let the impulse response of the equalizer be $W(k)$; as denoted in the schematic given in Figure 14.10, the output of the equalized signals can be obtained by

$$y\,(k) = X^T\,(k)\,W(k) \tag{14.23}$$

where $W(k) = [w_0, w_1, \ldots\ldots w_k]^T$, the tap weight vector of the equalizer, as described in the preceding sections of this chapter, and $X(k) = [x(k), x(k-11), \ldots\ldots x\,(k-N+1)]^T$ is the vector input to the equalizer, where N is the length of the weight vector.

In fiber transmission, there is a pure phase distortion due to the CD effects in the linear region, and the constellation is rotating in the phase plane. In addition, the carrier phase also creates additional spinning rotation of the constellation, as observed in Equation 14.22. Thus, in order to

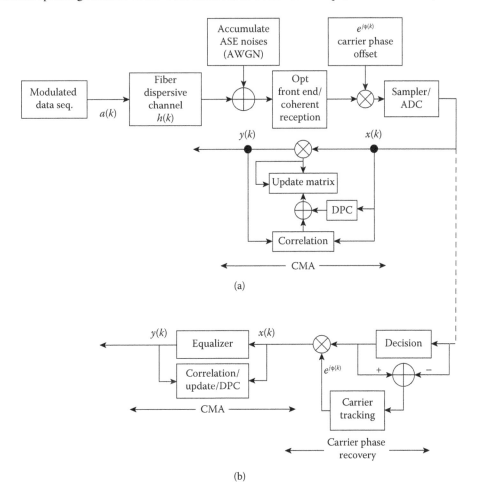

FIGURE 14.10 Equivalent model for baseband equalization: (a) blind constant modulus algorithm (CMA), and (b) modified CMA by cascading CMA with carrier phase recovery. Signal inputs come from the digitalized received samples after coherent reception. DPC = differential phase compensation.

equalize the linear phase rotation effects, one must cancel the spinning effects due to this carrier phase difference created by the mixing between the LO and the signal carrier. Thus, we can see that the constant modulus algorithm must be associated with the carrier phase recovery, so that an additional phase rotation in reverse would be superimposed on the phase rotation due to channel phase distortion to fully equalize the constellation.

The constant modulus (CM) criterion may be expressed by the nonnegative cost function $J_{\text{cma, p, q}}$, parameterized by positive integers p and q, such that

$$J_{\text{cma},p,q} = \frac{1}{pq} E\left\{\left|\left|y_n\right|^p - \gamma\right|^q\right\}; \quad \text{with} \quad \gamma = \text{constant} \tag{14.24}$$

where $E(.)$ indicates the expected statistical value. The CM criterion is usually implemented as constant modulus algorithm (CMA) 2–2, where $p, q = 2$. Using this cost function, the weight vectors can be updated by writing

$$\mathbf{W}(k+1) = \mathbf{W}(k) - \mu \cdot \nabla \mathbf{J}(k)$$

$$= \mathbf{W}(k) - \mu \cdot \left(y(k)|y(k)|^p |y(k)|^{p-2}\left(|y(k)|^p - R_p\right)\right) \cdot \mathbf{X}^*(k) \tag{14.25}$$

where R_p is the constant depending on the type of constellation. Since the final output of the equalizer system would converge to the original input states, one can rewrite Equation 14.25 as

$$\left(a(k)|y(k)|^p |a(k)|^{p-2}\left(|a(k)|^p - R_p\right)\right) = 0 \tag{14.26}$$

Assuming that the in-phase and quadrature-phase components of $a(k)$ to be de correlated with each other and using of the input sequence $a(k)$ and the convolution with the impulse response of the channel $h(k)$ we have $x(k) = h(k) \cdot a(k)$, the constant R_p can be determined as

$$R_{\text{p}} = \frac{E\left[|a(k)|^{2p}\right]}{E\left[|a(k)|^p\right]} \tag{14.27}$$

The flow of the update procedure for the weight coefficients of the filter of the equalizer is shown in Figure 14.11.

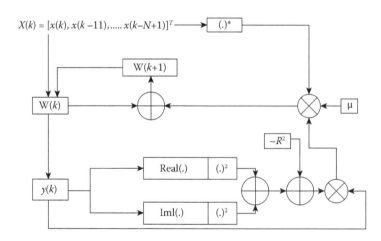

FIGURE 14.11   Block diagram of tap update with CMA cost function.

14.2.2.7.2 Modified CMA: Carrier Phase Recovery Plus CMA

As we know from the cost function (Equation 14.24) used in the CMA, since the cost function is phase-blind, the CMA can converge even in the presence of the phase error. Although it is the merit of the CMA, at convergence, the equalizer output will have a constant phase rotation. This phase rotation is commonly due to the difference of the frequency of the LO and the carrier or LOFO (LO frequency offset). Furthermore, the constellation will be spinning at the carrier FO rate (considered as the phasor rotation) owing to the lack of carrier frequency locking. While this phase-blind nature of the CMA is not a serious problem for constant phase rotation, for some parameters, such as the random PMD and the fluctuation of the phase of the carrier, for example, the 100 MHz oscillation of the feedback mirror in the external cavity of the LO, the performance of the CMA is severely degraded by the randomly rotating phase distortion. Thus, the carrier phase recovery is essential to determine the LOFO, the offset of the LO, and the carrier of the signal channel. The block diagram of the combined CMA blind equalization and carrier tracking and locking is illustrated in Figure 14.10. The carrier recovery (CR) loop uses the error between the output of the equalizer and the corresponding decision. The phase updating rule is given by $\phi(k+1) = \phi(k) - \mu_\phi I[z(k)e*(k)]$, where μ_ϕ is the step-size parameter, $e(k)$ is the error signal, given by $e(k) = z(k) - \hat{a}(k)$, $z(k) = y(k)\,e^{-jf(k)}$ is the equalized output with phase error correction, and $\hat{a}(k)$ is the estimation of $z(k)$ by a decision device. The carrier tracking loop described earlier gives the estimate of the phase error, as shown in Figure 14.10b. Then, a differential phase compensation (DPC) can be implemented.

The modified CMA is implemented by cascading the carrier phase recovery first and then cascaded with the standard CMA, as shown in Figure 14.11. A typical CMA operation with carrier phase recovery is shown in Figure 14.12, for the 16QAM constellation with the optical coherent transmission and reception described in Chapter 4, in which a 100 MHz oscillation of the LO laser is observed. We can observe that the constellation of the nonequalized processing is

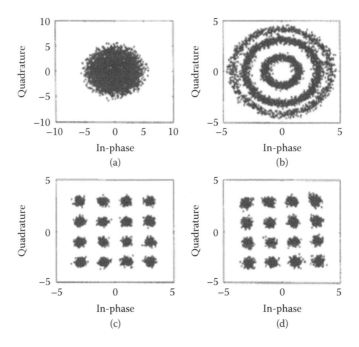

FIGURE 14.12 Constellations of 16QAM modulated signals with a carrier frequency offset with respect to that of the LO laser. (a) Unequalized output, (b) output equalized by CMA, (c) output by joint CMA and carrier phase DD recovery, and (d) output by modified CMA, carrier phase recovery and CMA processes in sequential stages.

noisy and certainly cannot be used to determine the BER of the transmission system. However, as observable in Figure 14.12b, if only the CMA algorithm is employed, then the phase rotation due to the phase offset between the LO and the signal channel carrier can still exist. When carrier phase recovery is applied, the constellation can then be recovered as shown in Figure 14.12c and d, with and without the DD processing stages, which offer similar constellations. The DD processing will be described in the Section 14.2.3.1 dealing with NLEs or decision feedback equalizers (DFEs) in this chapter.

Noe et al. [15] has reported simulation of coherent receivers under polarization multiplexed QPSK employing CMA algorithms under standard and modification in association with the original and modified DD algorithms. Furthermore there are some certain penalties to within 0.5 dB, on the receiver sensitivity.

14.2.3 NLE or DFE

14.2.3.1 DD Cancellation of ISI

It has been demonstrated in both wireless and optical communications systems over recent years that nonlinear (decision feedback) equalizers achieve better performance than their linear counterparts. The method of operation of these DD equalizers for cancellation of ISI is as follows.

Consider the transmission systems given in Figure 14.13a, which is also similar to Figure 14.5, except that the LE is replaced with a nonlinear one. The linear FF equalizer or transversal filter in this NLE is given in Figure 14.13c. The arrow directing from the decision block across the FF transversal filter indicates the updating of coefficients of the filter.

Similar to the LE case, the input sampled data sequence also follows the notations given in Section 14.2. Multilevel data-symbol can be treated in the same way without unduly affecting any of the important results in this analysis.

The pure NLE uses the detected data symbols $\{s_i'\}$ to synthesize the ISI component in a received signal $\{r_i\}$, and then it removes the ISI by subtraction. This is the process of DD cancellation of ISI.

Mathematically setting by following the notation used from Equation 14.10 for the section just before the decision/detection block of Figure 14.13a, or the sampled sequence at the input of Figure 14.13b, the signals entering the detector at the instant $t = iT$ is given as

$$x_i = \frac{r_i}{y_0} - \sum_{j=1}^{g} s_{i-j}' v_j = s_i + \sum_{j=1}^{g} s_{i-j}' v_j + \frac{w_i}{y_0} - \sum_{j=1}^{g} s_{i-j}' v_j \qquad (14.28)$$

So that, with corrected detection of each sampled value s_{i-j} such that

$$s_{i-j}' = s_{i-j} \quad \text{for} \quad j = 1, 2 \ldots g \rightarrow x_i = s_i + \frac{w_i}{y_0} \qquad (14.29)$$

So that, in the correct detection and equalization, we have recovered the sampled sequence $\{x_i\}$. Thus, it is clear that, as long as the data symbols $\{s_i\}$ are correctly detected, their ISI is removed from the equalized signals followed by the NLE, and the channel continues to be accurately equalized.

Under the tolerance of this pure NLE operating in a channel-sampled impulse response whose z-transform given by $\{Y(z)\}$ in Equation 14.5, the equalized sampled signals at the detector input can be expressed by

$$x_i = s_i + \frac{w_i}{y_0}$$

$$s_i = \pm k; \quad \text{binary\_amplitude\_of\_input–seq.} \qquad (14.30)$$

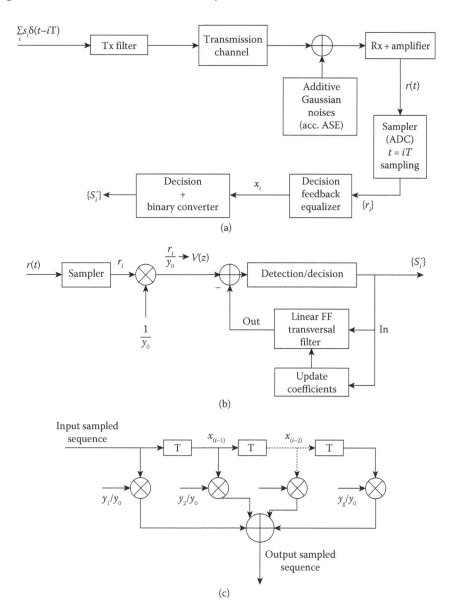

FIGURE 14.13 (a) Simplified schematic of coherent optical communication system incorporating an NLE. Note equivalent baseband transmission systems due to baseband recovery of amplitude and phase of optically modulated signals. (b) Structure of the DFE employing linear FFE and a decision device. (c) Structure of the FF transversal filter.

It is assumed that the sampled sequence is detected as $+k$ or $-k$, depending on whether x_i takes negative or positive values. Errors would occur if the noise levels corrupt the magnitude, that is, if the noise magnitude is greater than k. As in the previous Section 14.2.2.5, with respect to the Gaussian noise contributing to the decision process, we can obtain the probability density function of the random variable noise, w, as

$$p(w) = \frac{1}{\sqrt{2\pi y_0^{-2}\sigma^2}} e^{-\frac{w^2}{2y_0^{-2}\sigma^2}} \qquad (14.31)$$

The probability of error can be found as

$$P_e = \int_k^\infty p(w)\, dw = \int_k^\infty \frac{1}{\sqrt{2\pi y_0^{-2}\sigma^2}} e^{-\frac{w^2}{2y_0^{-2}\sigma^2}}\, dw = \int_{k\frac{|y_0|}{\sigma}}^\infty \frac{1}{\sqrt{2\pi}} e^{-\frac{w^2}{2}}\, dw = Q\left(k\frac{|y_0|}{\sigma}\right) \qquad (14.32)$$

The most important difference between NLEs and LEs is that the NLE can handle the equalization process when there is pulse on the unit circle of the z-plane of the z-transform impulse response of the transmission channel.

A relative comparison of the LE and NLE is as follows:

1. When all zeroes of the channel transfer function in the z-plane lie inside the unit circle in the z-plane, the NLE with samples including the initial instant would gain an advantage to the tolerance of the Gaussian additive noise over that of LE. The NLE would now be preferred in contrast to LE, as the number of taps of the NLE would be smaller than that required for LE.
2. However, when the transmission is purely phase distortion, it means that the poles outside the unit circle of the z-plane and the zeroes are reciprocal conjugates of the poles (inside the unit circles), and then the LE is a matched filter and offers better performance than that by NLE and close to a maximum-likelihood (ML) detector. The complexity of NLE and LE are nearly the same in terms of taps and noise tolerances.
3. Sometimes there would be about 2–3 dB better performance between NLE and LE processing.

14.2.3.2 Zero-Forcing Nonlinear Equalization

When an NLE is used in lieu of the LE of the zero-forcing nonlinear equalization (ZF-NLE) illustrated in Figure 14.6, then the equalizer is called ZF-NLE, as illustrated in the transmission system shown in Figure 14.14 that consists of a linear feed-forward transversal filter and a NLE with a linear feedback transversal filter and a decision/detector operation.

The DFE performs the equalization by zero-forcing, which is similar to the case of ZF-LE with zero-forcing described in Section 14.2.2.2. The operational principles for this type of NLE are as follows:

1. The linear filter of Figure 14.14 partially equalizes the channel by setting to zero all components of the channel-sampled impulse response preceding that of the largest magnitude, without affecting the relative values of the remaining components.
2. The NLE section then completes the equalization process by the operations exactly as described in Section 14.2.3.1.

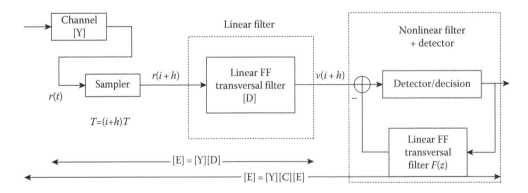

FIGURE 14.14 DFE incorporating a linear and nonlinear filter with overall transfer impulse responses.

Using the notations assigned for the input sequence, the channel impulse response of g-sampled z-transform, $Y(z)$ as given in Equations 14.10–14.12, we can set that the required equalizer impulse response in the z-domain $C(z)$ of m-tap would be

$$Y(z)\,C(z) \simeq z^{-h} \quad \text{with} \quad 0 < h < m + g \qquad (14.33)$$

where h is an integer in the range of 0 to $(m + g)$, with

$$[Y] = [y_0 \quad y_1 \quad ..... \quad y_g]; \text{ Channel}\_\text{impulse}\_\text{response}$$

$$[C] = [c_0 \quad c_1 \quad ..... \quad c_m]: \text{LE}\_\text{impulse}\_\text{response}$$

so that

$$Y(z) = y_0 + y_1 z^{-1} + ... + y_g z^{-g}; \quad \text{and} \quad C(z) = c_0 + c_1 z^{-1} + ... + c_m z^{-m} \qquad (14.34)$$

Let the LE having $[D]$ as an $(n + 1)$-tap gain filter with the impulse response $D(z)$ of $(n+1)$-taps be expressed as

$$[D] = [d_0 \quad d_1 \quad ...... \quad d_n]; \text{ impulse}\_\text{response}\_\text{linear}\_\text{transversal}\_\text{filter} \rightarrow$$

$$D(z) = c_0 + c_1 z^{-1} + ... c_m z^{-m} \qquad (14.35)$$

Thus, the overall transfer function of the channel and the equalizer can be written when setting $[D] = [C][E]$, as

$$Y(z)\,C(z)\,E(z) \simeq z^{-h}\,E(z) \quad \text{with} \quad n = m + g - 1 \qquad (14.36)$$

with $[E]$ as the required sample impulse response, so that the NLE can equalize the input pulse sequence, that is, satisfy the condition that the product of the two transfer function must be unitary, or

$$[E] = \left[1 \quad \frac{y_{l+1}}{y_l} \quad \frac{y_{l+2}}{y_l} \quad \frac{y_g}{y_l} \right] \rightarrow z - \text{transform\_impulse\_response}$$

$$E(z) = 1 + \frac{y_{l+1}}{y_l} z^{-1} + \frac{y_{l+2}}{y_l} z^{-2} + ... + \frac{y_g}{y_l} z^{-g+l} \qquad (14.37)$$

where y_l is the largest component of the channel-sampled impulse response. When this condition is satisfied, then the nonlinear filter ZF-NLE acts the same as ZF-LE but with the tap gain order reduced to $(g-l)$ taps whose coefficients are $\frac{y_{l+1}}{y_l}, \frac{y_{l+2}}{y_l}, ..., \frac{y_g}{y_l}$ instead of g-taps with gains $v_1, v_2, ... v_g$. The probability of error can be evaluated in a similar manner as described earlier for additive noises of Gaussian distribution of the probability density function with zero mean and a standard deviation σ, to give

$$P_e = Q\left(\frac{k}{\sigma |D|} \right); \quad |D| = \text{Euclidean\_length\_of\_impulse\_response}\_[D] \qquad (14.38)$$

14.2.3.3 Linear and Nonlinear Equalization of Factorized Channel Response

In normal case, the z-domain transfer function of the channel $Y(z)$ can be factorized into a cascade of two subchannels, given as

$$Y(z) = Y_1(z)\, Y_2(z) \tag{14.39}$$

So that we can employ the first linear equalizer to equalize the subchannel $Y_1(z)$ and the NLE to tackle the second subchannel $Y_2(z)$. The procedure is a combined cascading of the two processing stages that have been described earlier.

14.2.3.4 Equalization with Minimizing MSE in Equalized Signals

As we have seen, equalization using NLEs is described in the previous section. The decision or detection will enhance the SNR performance of the system. However, under the criteria that the NLE can make the decision block so that an optimum performance can be achieved. The MSE can be employed as described in the preceding text to minimize the errors in the decision stage of the NLE.

14.3 MAXIMUM A POSTERIORI TECHNIQUE FOR PHASE ESTIMATION

14.3.1 Method

Let us assume that we want to estimate an unobserved population parameter θ on the basis of observation variable x. Let f be the sampling distribution of x, so that $f(x|\theta)$ is the probability of x when the underlying population parameter is θ. Then the function that transforms $\theta \rightarrow f(x|\theta)$ is known as the likelihood function, and the estimate $\hat{\theta}_{\mathrm{ML}}(x) = \max_{\theta} f(x|\theta)$ is the ML estimate of θ.

Now assume that a prior distribution g over θ exists. This allows us to treat θ as a random variable as in Bayesian statistics. Then, the posterior distribution of θ can be determined as

$$\theta \rightarrow f(\theta|x) = \frac{f(x|\theta)g(\theta)}{\displaystyle\int_{v \in \Theta} f(x|v)g(v)dv} \tag{14.40}$$

where g density function of is θ, and Θ is the domain of g. This is a straight forward application of Bayes' theorem. The method of maximum a posteriori (MAP) estimation then estimates θ as the mode of the posterior distribution of this random variable

$$\hat{\theta}_{\mathrm{MAP}} = \mathrm{arg\_max} \frac{f(x|\theta)g(\theta)}{\displaystyle\int_{v} f(x|v)g(v)dv} = \mathrm{arg\_max}_{\theta} f(x|\theta)g(\theta) \tag{14.41}$$

The denominator of the posterior distribution (so-called partition function) does not depend on θ and therefore plays no role in the optimization. Observe that the MAP estimate of θ coincides with the ML estimate when the prior g is uniform (i.e., a constant function). The MAP estimate is a limit of Bayes estimators under a sequence of 0–1 loss functions, but generally not a Bayes estimator per se unless θ is discrete.

14.3.2 Estimates

MAP estimates can be computed in several ways:

1. Analytically, the mode(s) of the posterior distribution can be obtained in closed form. This is the case when the conjugate priors are used.

2. Either via numerical optimization such as the conjugate gradient method, or Newton's method, which may usually require first or second derivatives, are to be evaluated analytically or numerically.
3. Via the modification of an expectation-maximization algorithm. This does not require derivatives of the posterior density.

One of the possible processing algorithms can be the ML phase estimation for coherent optical communications [16]. The differential QPSK (DQPSK) optical system can be considered as an example. In this system, as described in Chapter 5, the phase-diversity coherent reception technique, with a mixing of a LO, can be employed to recover the in-phase (I) and quadrature-phase (Q) components of the signals. Such a receiver composes a 90° optical hybrid coupler to mix the incoming signal with the four quadruple states associated with the LO inputs in the complex-field space. A $\pi/2$ phase shifter can be used to extract the quadrature components in the optical domain. The optical hybrid can then provide four lightwave signals to two pairs of balanced photodetectors to reconstruct the I and Q information of the transmitted signal. In case that the LO is the signal itself, then the reception is equivalent to a self-homodyne detection. For this type of a reception subsystem, the balanced receiver would integrate a one-bit delay optical interferometer so as to compare the optical phases of two consecutive bits. We can assume a complete matching of the polarization between the signal and the LO fields, so that only the impact of phase noise should be considered. Indeed, the ML algorithm is the modern MLSE technique [17].

The output signal reconstructed from the photocurrents can be represented by

$$r(k) = E_0 \exp\left[j(\theta_s(k) + \theta_n(k)] + \tilde{n}(k)\right. \tag{14.42}$$

where k denotes the kth sample over time interval $[kT, (k+1)T]$ (T is the symbol period), $E_0 = \Re\sqrt{P_{LO}P_s}$ ($\Re$ is the photodiode responsivity), and P_{LO} and P_s are the powers of LO and received signal, respectively; $\theta_s(k) \in \{0, \pi/2, \pi, -\pi/2\}$ is the modulated phase, the phase difference carrying the data information; $\theta_n(k)$ is the phase noise during the transmission; and $\{\tilde{n}(k)\}$ is complex white Gaussian random variables with $E[\tilde{n}(k)] = 0$ and $E[|\tilde{n}(k)|^2] = N_0$.

In order to retrieve information from the phase modulation $\theta_s(k)$ at time $t = kT$, the carrier phase noise $\theta_n(k)$ is estimated based on the received signal over the immediate past L symbol intervals— that is, based on $\{r(l), k - L \leq l \leq k - 1\}$. In the decision feedback strategy, a complex phase reference vector $v(k)$ is computed by the correlation of L received signal samples $r(l)$ and decisions on L symbols $\{\hat{m}(l), k - L \leq l \leq k - 1\}$, where $\hat{m}(l)$ is the receiver's decision on $\exp(j\theta_s(l))$

$$\mathbf{v}(k) = \sum_{l=k-L}^{k-1} r(l)\hat{m}^*(l) \tag{14.43}$$

Here, * denotes the complex conjugation. An initial L symbol long data sequence is sent to initiate the processor/receiver. Alternatively, the ML decision processor can be trained prior to the transmission of data sequence. Of importance is that, not only can the $\pi/4$-radian phase ambiguity be resolved, the decision-aided method is also now totally linear and efficient to implement based on Equation 14.43. To some extent, the reference $v(k)$ assists the receiver to acquire the channel characteristic. With the assumption that $\theta_n(k)$ varies slowly compared to the symbol rate, the computed complex reference $v(k)$ from the past L symbols forms a good approximation to the phase noise phasor $\exp(j\theta_n(k))$ at time $t = kT$.

Using the phase reference $v(k)$ from Equation 14.43, the decision statistic of the ML receiving processor is given as

$$q_i(k) = \mathbf{r}(k)C_i^* \cdot \mathbf{v}(k), \quad i = 0, 1, 2, 3 \tag{14.44}$$

where $C_i \in \{\pm 1, \pm j\}$ is the DQPSK signal constellation, and $\cdot$ denotes the inner product of two vectors. The detector declares the decision $\hat{m}(k) = C_k$ if $q_k = \max q_i$. This receiver/processor has been shown to achieve coherent detection performance if the carrier phase is a constant and the reference length L is sufficiently long.

The performance of ML, the receiving processor, can be evaluated by simulation using Monte Carlo simulations in two cases: a linear phase noise system and a nonlinear phase noise system. For comparison, the Mth power phase estimator and differential demodulation can also be employed.

Nonlinearity in an optical fiber can be ignored when the launch optical power is below the threshold, so the fiber optic can be modeled as a linear channel. The phase noise difference in two adjacent symbol intervals, that is, $\theta_n(k) - \theta_n(k-1)$, obeys a Gaussian distribution with the variance σ^2 determined by the linewidth of the transmitter laser and the LO, given by the Lorentzian linewidth formula $\sigma^2 = 2\pi\Delta v T$, where Δv is the total linewidth of the transmitter laser and the LO [18]. In the simulations, σ can be set to 0.05, corresponding to the overall linewidth $\Delta v = 8$ MHz when the baud rate is 40 Symbols/s.

As observed in Figure 14.15 [16], the ML processor outperforms the phase estimator and differential demodulation by about 0.25 and 1 dB, respectively. Although the performance gap between an ML processor and the Mth power phase estimation is not very large in the case of an multilevel phase shift keying (MPSK) system, the Mth power phase estimator requires more nonlinear computations, such as an arctan($\cdot$), which incurs a large latency in the system and leads to phase ambiguity in estimating the block phase noise. On the contrary, the ML processor is a linear and efficient algorithm, and there is no need to deal with the $\pi/4$-radian phase ambiguity; thus, it is more feasible for online processing for the real systems. We also extend our ML phase estimation technique to the QAM system. The conventional Mth-power phase estimation scheme suffers from the performance degradation in QAM systems since only a subgroup of symbols with phase modulation $\pi/4 + n \cdot \pi/2$ ($n = 0, 1, 2, 3$) can be used to estimate the phase reference. The maximum tolerance of linewidth per laser in *Square*-16QAM can be improved 10 times compared to the Mth-power phase estimator scheme.

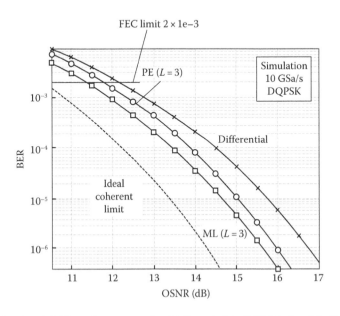

FIGURE 14.15 Simulated BER performances of a 10 Gsymbols/s DQPSK signal in linear optical channel with laser linewidth 2 MHz ($\sigma = 0.05$) and decision-aided length $L = 3$. (Modified from S. Zhang et al., A comparison of phase estimation in coherent optical PSK system, *Photonics Global'08*, paper C3-4A-03. Singapore, 2008.) The FEC limit can be set at $2 \times$ 1e–3 to determine the required OSNR.

Let us now consider the case with nonlinear phase noise. In a multi span optical communication system with EDFAs, the performance of optical DQPSKs is severely limited by the nonlinear phase noise converted by amplified ASE noise through the fiber nonlinearity [19]. The laser linewidth is excluded to only consider the impact of the nonlinear phase noise on the receiver. The transmission system can comprise a single-channel DQPSK-modulated optical signal transmitted at 10–40 GSymbols/s over N 100 km equally spaced amplified whose CD is fully compensated and gain is also equalized as shown in Figure 14.16. The following parameters are assumed for the transmission system: $N = 20$, fiber nonlinear coefficient $\gamma = 2$ W^{-1} km^{-1}, fiber loss $\alpha = 0.2$ dB/km, the amplifier gain $G = 20$ dB with an NF of 6 dB, the optical wavelength $\lambda = 1553$ nm, and the bandwidth of optical filter and electrical filter of $\Delta\lambda = 0.1$ nm and 7 GHz, respectively. The *OSNR* is defined at the location just in front of the optical receiver as the ratio between the signal power and the noise power in two polarization states contained within a 0.1 nm spectral width region. As shown in Figure 14.17a [16], the receiver sensitivity of ML processor has improved by about 0.5 and 3 dB over phase estimation and differential demodulation, respectively, at the BER level of 10^{-4}. Besides, it is noteworthy that differential demodulation has exhibited an error floor due to severe nonlinear phase noise with high launch optical power at the transmitter.

To obtain the impact of the nonlinear phase on the demodulation methods and the noise-loading effects, the number of amplifiers can be increased to $N = 30$. The Q factor is used to show a numerical improvement of the ML processor receiver. As shown in Figure 14.17b, with the increase of launch power (LP), the ML receiver/processor approximates the optimum performance because of the reducing variance of the phase noise. At the optimum point, the ML receiver/processor outperforms the Mth power phase estimator by about 0.8 dB while having about 1.8 dB receiver sensitivity improvement compared to differential detection. It is also found that the optimum performance occurs at an OSNR of 16 dB when the nonlinear phase shift is almost 1 radian. With the LP exceeding the optimum level, nonlinear phase noise becomes severe again. Only the ML phase estimator can offer BER beyond 10^{-4}, while both Mth power phase estimator and differential detection

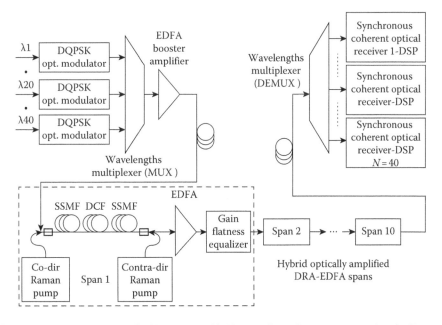

FIGURE 14.16 DWDM optical DQPSK system with N-span dispersion-compensated optic fiber transmission link using synchronous self-homodyne coherent receiver to compensate for nonlinear phase noise under both Raman amplification and EDF amplification.

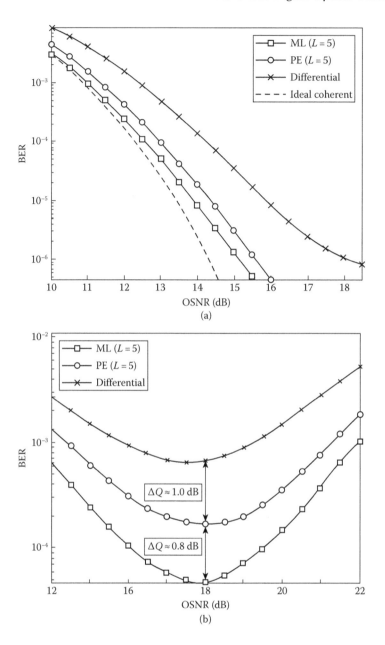

FIGURE 14.17   Monte-Carlo-simulated BER of a 10 Gsymbol/s DQPSK signal in an N-span optical fiber system with template length $L = 5$; (a) $N = 20$; (b) $N = 30$, under different detection schemes of ML estimator, phase estimator, differential demodulation, and ideal coherent detection. ML = maximum likelihood, PE = phase estimation. (Extracted from S. Zhang et al., A Comparison of phase estimation in coherent optical PSK system, *Photonics Global'08*, paper C3-4A-03. Singapore, December 2008.)

exhibit error floor before BER reaches 10^{-4}. From the numerical results, the ML processor carrier phase estimation shows significantly better receiver sensitivity improvement than the Mth power phase estimator and conventional differential demodulation for optical DQPSK signals in systems dominated by nonlinear phase noise. Again, we want to emphasize that ML estimator is a linear and efficient algorithm, and there is no need to deal with the $\pi/4$-radian phase ambiguity; thus, it is feasible for online processing for the real systems.

In summary, the performance of synchronous coherent detection with ML processor carrier phase estimation may offer a linear phase noise system and a nonlinear optical noise system separately. The *ML*-DSP phase estimation can replace the optical PLL, and the receiver sensitivity is improved compared to conventional differential detection and the *M*th power phase estimator. The receiver sensitivity is improved by approximately 1 and 3 dB in these two channels (20 spans for the nonlinear channel), respectively, compared to differential detection. For the 30-span nonlinear channel, only the ML processor can offer BER beyond 10^{-4} (using Monte Carlo simulation), while both *M*th power phase estimator and differential detection exhibit error floor before BER reaches 10^{-4}. At the optimum point of the power, the Q factor of ML receiver/processor outperforms the *M*th power phase estimation by about 0.8 dB. In addition, an important feature is the linear and efficient computation of the ML phase estimation algorithm, which enables the possibility of real-time online DSP processing.

14.4 CARRIER PHASE ESTIMATION

14.4.1 REMARKS

Under the homodyne reception or intra dyne reception, the matching between the LO laser and that of the signal carrier is very critical in order to minimize the deviation of the reception performance, and hence the enhancement of the receiver sensitivity so as to maximize the transmission system performance. This difficulty remains a considerable obstacle for the first generation of coherent systems developed in the mid-1980s, employing analog or complete hardware circuitry.

The first few parts of this chapter deals with the recovery of carrier phase. Then the later parts involve some advanced processing algorithms to deal with the carrier phase recovery under the scenario that the frequency of the LO laser is oscillatory, an operational condition to get the stabilization of an ECL that is essential for homodyne reception with high sensitivity.

In DSP-based optical reception subsystems, this hurdle can be overcome by processing algorithms that are installed in a real-time memory processing system, the application-specific integrated circuit (ASIC), which may consist of an analog-to-digital convertor, a digital signal processor, and high-speed fetch memory.

The FO between the LO and the signal lightwave carrier is due to the oscillation of the LO, commonly used in an external cavity incorporating an external grating that is oscillating by a low frequency of about 300 MHz, so as to stabilize the feedback of the reflected specific frequency line to the laser cavity. This oscillation, however, degrades the sensitivity of the homodyne reception systems. Thus, the later parts of this section illustrate the application of DSP algorithms described in the preceding sections of this chapter to demonstrate the effectiveness of these algorithms in real-time experimental setups.

It is noted that the carrier phase recovery must be realized before any equalization process can be implemented. It is noted here that the optical phase locking (OPL) technique presented in Chapter 5 is another way to reduce the FO between the LO and the signal carrier with operation in the optical domain, while the technique described in this section is at the receiver output in the digital processing domain.

14.4.2 CORRECTION OF PHASE NOISE AND NONLINEAR EFFECTS

Kikuchi and coworkers. [20] at the University of Tokyo used electronic signal processing based on *M*th power phase estimation to estimate the carrier phase. However, DSP circuits for *M*th power phase estimator need nonlinear computations, thus impeding the potential possibility of real-time processing in the future. Furthermore, the *M*th power phase estimation method requires dealing with the $\pi/4$-radian phase ambiguity when estimating the phase noise in adjacent symbol blocks.

While the electronic DSP is based on ML processing for carrier phase estimation to approximate the ideal synchronous coherent detection in optical phase modulation systems, which requires only linear computations, it thus eliminates the optical PLL and is more feasible for online processing for the real systems. Some initial simulation results show that the ML receiver/processor outperforms the Mth power phase estimator, especially when the nonlinear phase noise is dominant, thus significantly improving the receiver sensitivity and tolerance to the nonlinear phase noise.

Liu's coworkers. [21,22] at Alcatel–Lucent Bell Labs uses optical delay differential detection with DSP to detect differential BPSK (DPSK) and DQPSK signals. Since direct detection only detects the intensity of the light, the improvement of DSP is limited after the direct detection. After the synchronous coherent detection of our technique, however, the DSP techniques such as electronic equalization of CD and PMD offer better performance since the phase information is retrieved. And as the level of phase modulation increases, it becomes more and more difficult to apply the optical delay differential detection because of the rather complex implementation of the receiver and the degraded SNR of the demodulated signal, while the ML estimator/processor can still demodulate other advanced modulation formats such as 8-PSK, 16-PSK, 16QAM, and so on.

14.4.3 FORWARD PHASE ESTIMATION QPSK OPTICAL COHERENT RECEIVERS

Recent progresses in DSP [23,24] with the availability of ultra-high sampling rate allow the possibility of DSP-based phase estimation and polarization management make the coherent detection technique in association with digital processors and high sampling ADC and DAC more robust and practically realized. This section is dedicated to the new emerging technology that will significantly influence the optical transmission and detection of optical signals at ultra-high speeds.

Recalling the schematic of a coherent receiver of Figure 14.18, it shows a coherent detection scheme for QPSK optical signals in which they are mixed with the LO field. In the case of modulation format QPSK or DQPSK, a $\pi/2$ optical phase shifter is needed to extract the quadrature component, and thus the real and imaginary parts of phase shift keyed signals can be deduced at the output of a balanced receiver from a pair of identical photodiodes connected back to back (Figure 14.19). Note that, for a balanced receiver, the quantum shot noise contributed by the photodetectors is double, as noise is always represented by the noise power, and no current direction must be applied.

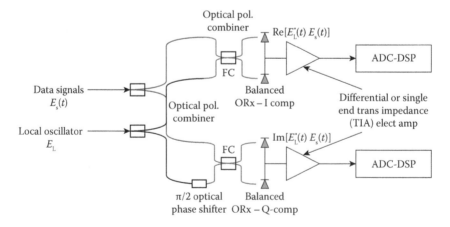

FIGURE 14.18 Schematic of a coherent receiver using balanced detection techniques for I–Q component phase and amplitude recovery, incorporating DSP and ADC.

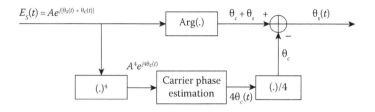

FIGURE 14.19 Schematic of a QPSK coherent receiver.

When the received signal is raised to the quadratic power, the phase of the signal disappears because $e^{j4\theta_s(t)} = 1$ for DQPSK, whose states are 0, $\pi/2$, π, $3\pi/2$, 2π. The data phase is excluded, and the carrier phase can be recovered with its fluctuation with time. This estimation is computed using the DSP algorithm. The estimated carrier phase is then subtracted with the detected phase of the signal to give only the phase states of the signals as indicated in the diagram. This is a feed-forward (FF) phase estimation, and most suitable for DSP implementation either off-line or in real time.

14.4.4 CARRIER RECOVERY IN POLARIZATION DIVISION MULTIPLEXED RECEIVERS: A CASE STUDY

Advanced modulation formats in combination with coherent receivers incorporating DSP subsystems enable high capacity and spectral efficiency [25,26]. PDM, QAM, and coherent detection are dominating the next-generation high-capacity optical networks since they allow information encoding in all the available degrees of freedom [27] with the same requirement of the OSNR. The coherent technique is homodyne or intradyne mixing of the signal carrier and the LO. However, as just mentioned, a major problem in homodyne coherent receiver is the CR, so that matching between the channel carrier and the LO can be achieved to maximize the signal amplitude and phase strength, and hence the SNR in both amplitude and phase planes. This section illustrates an experimental demonstration of the CR of a PDM-QPSK, transmission scheme carrying 100 Gbits/s per wavelength channel, including two polarized modes (H: horizontal; V: vertical) of the linear polarized mode in SSMF—that is, 28 GSy/s × 2 (2 bits/symbol) × 2 (polriazated channels) using some advanced DSP algorithms, especially the Viterbi-Viterbi (V-V) algorithm with MLSE nonlinear decision feedback procedures [28].

Feed-forward carrier phase estimation (FFCPE) [29] has been commonly considered as the solution for this problem. The fundamental operational principle is to assume a time invariant of the phase offset between carrier and local laser during N ($N > 1$) consecutive symbol periods.

In coherent optical systems using tunable lasers, the maximum absolute FO can vary by 5 GHz. To accommodate such a large FO, coarse digital frequency estimation (FDE) and recovery (FDR) techniques [30,31] can be employed to limit the FO to within an allowable range, so that phase recovery can be implemented/managed using fine FFCPE and fine feed-forward carrier phase recovery (FFCPR) techniques [32]. The algorithm presented in Ref. [22], the so-called "differential Viterbi" (DiffV), estimates phase difference between two successive complex symbols. This method enables the estimation of large FO of multiple phase shift keying (M-PSK)-modulated signals up to $\pm f_s/(2M)$, where f_s stands for the symbol rate frequency. It is possible to estimate larger FOs using the technique presented in Ref. [6], which generates insignificant estimation errors that can be handled by FFCPE and FFCPR. However, FFCPE, or "Viterbi and Viterbi" (V-V), would not perform well when the absolute residual FO at the input of the estimation circuit is larger than $\pm f_s/(2MN)$. This problem becomes more serious when the laser frequency oscillates.

An algorithm for coupling carrier phase estimations in polarization division multiplexed (PDM) systems is presented in Ref. [33]. The algorithm requires two loop filters, and the coupling factor should be carefully selected. In the following section, an enhanced CR concept covering large FO and enabling almost zero residual FO for the V-V CPE is discussed. Also, a novel polarization coupling algorithm with reduced complexity is extracted from the work in Ref. [20].

14.4.4.1 FO Oscillations and Q Penalties

Figure 14.20 shows the typical DSP of medium complexity for coherent PDM optical transmission systems. CD and polarization effects are compensated in CD and MIMO blocks, respectively. The DV FE is realized in the microcontroller (MC). Several thousands of data sequences are periodically loaded to the microcontroller that calculates the FO. A small bandwidth of the estimation loop delivers only averaged FO value to the CMOS ASIC part. However, long-time experimental tests with commercially available lasers, for example, the EMCORE laser, show that laser frequency oscillations are sinusoidal with the frequency of 888 Hz in order to stabilize its central frequency by the use of a feedback control, and amplitude of more than 40 MHz (see Figure 14.21 for measured oscillation of the EMCORE external cavity laser as LO). This oscillation is created by the vibration of the external reflector—a grating surface, so that the cavity can be stabilized. The mean offset is close to 300 MHz. Therefore, the DiffV enables the averaging of the FO of 300 MHz, while the residual offset of ±50 MHz is compensated by the V-V algorithm.

Penalties caused by residual offset of 50 MHz have been investigated for case of V-V CPE. Figure 14.22 shows the simulation values of the Q penalties versus V-V averaging window length (WL) for two scenarios under off-line data processing. In first case, we multiplexed 112 G PDM-QPSK channel with 12 10 G on–off keying (OOK) neighbors (10 and 100 G over a 1200 km link), while the second scenario included the transmission of eight 112 G PDM-QPSK channels (100 G WDM—1500 km link). Channel spacing was 50 GHz (ITU-grid). The FO was partly compensated with the residual offset of 50 MHz to check the V-V CPR performance. Three values of LP have been checked: optimum launched power (LP_{opt}), $LP_{opt}-1$ dB (slightly linear regime), and $LP_{opt}+1$ dB (slightly nonlinear regime). Maximum Q penalties of 0.9 dB can be observed, which strongly depend on the averaging window length parameter WL.

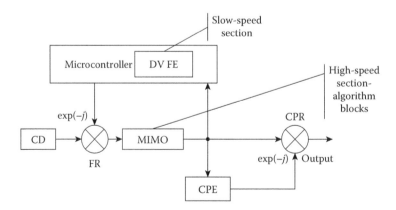

FIGURE 14.20 Typical DSP structure in PDM coherent receivers including a low-speed processor and high-speed DSP system. CPE = carrier phase estimation, MIMO = multiple input multiple output, CPR = carrier phase recovery, FR = frequency recovery, CD = chromatic dispersion.

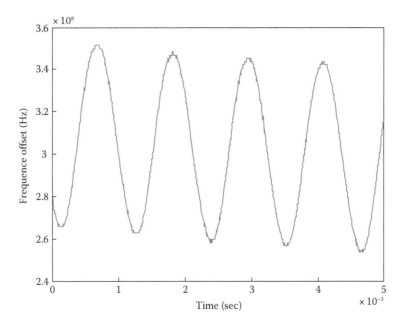

FIGURE 14.21   Laser frequency oscillation of EMCORE external cavity laser as measured from an external cavity laser.

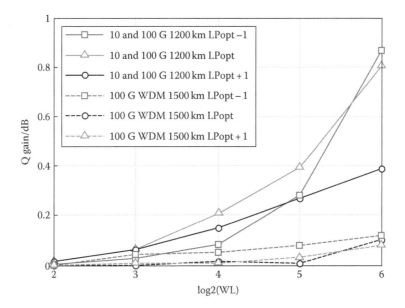

FIGURE 14.22   Penalties of V-V CPR with a residual FO of 50 MHz.

14.4.4.2   Algorithm and Demonstration of Carrier Phase Recovery

A robust FFCPE and FFCPR with a polarization coupling circuit is shown in Figure 14.23. The FFCFE and FFCPR are developed for the recourses of the V-V, and, by simple recourses reallocation, an enhanced CR is obtained. The proposed method is designed to maximize CR performance with a moderate realization complexity. The method consists of four main parts: (1) coarse carrier frequency estimation by processing in slow-speed DSP subsystems (indicated by microcontroller)

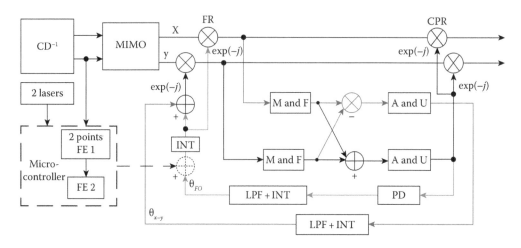

FIGURE 14.23 Modified structure of algorithms for CR. M&F = Mth power operation and averaging filter; A&U = arctan and unwrapping; PD = phase difference calculation; LPF = low-pass filter; INT = integrator. (Extracted from D.-S. Ly-Gagnon et al., *Lightwave Technol.*, Vol. 24, pp. 12–21, 2006.)

and recovery (CCFE-R), (2) feedback carrier frequency estimation and recovery (FBCFE-R, light gray lines in Figure 14.23), (3) X–Y phase offset estimation and compensation (H-V [or X–Y]-POE-C, gray line in Figure 14.23), and (4) joint H-V feed-forward carrier phase estimation and recovery (JFF-CPE-R).

The M-F block conducts the Mth power operation and averaging in case of the MPSK modulation format [6]. For the K-QAM signal, the M constellation points on the ring with a specific radius may be selected and used in Mth power block. More rings can be selected to improve the estimation procedure (e.g., inner and outer rings in 16QAM can be used). Therefore, the use of the method is not limited to PSK schemes.

Large FO [$> \pm f_s/(2MN)$] can be estimated by the method presented in Ref. [31] (denoted by FE1 in Figure 14.23), which can estimate the FO within ±10 GHz (linear curve in case of 112 G PDM-QPSK). A small amount of data is transferred into the microcontroller for calculating the coarse range of the FO. Depending on the FO estimation sign, one can superimpose certain FOs (e.g., 4 GHz) of opposite signs. Using a linear interpolation (FE2), the current FO can be estimated to an error within a few hundreds of MHz. Similarly, depending on available electrical bandwidth and the receiver structure, the compensation of such a large FO can be conducted either in CMOS ASIC or by controlling the frequency of LO, commonly an ECL.

The residual FO after CCFE and CCFER may have values of several hundred MHz plus 50 MHz generated by low processing speeds and laser frequency oscillations. The FBCFE estimates the residual frequency that is compensated for by the CFR feedback loop. The frequency estimation in the range of FBCFE&R is $\pm f_s/(2MN)$ due to its implementation based on the DiffV algorithm.

The FBCFE&R compensates almost the total FO. The residual FO and finite laser linewidth influence can be compensated for by the JFF-CPE&R. This operation can be supported by coupling the carrier estimations from both polarized channels. Prior to the coupling of estimations, both X and Y constellations are aligned using the X–Y-POE&C, and the phase difference between adjacent outputs of the lower unwrapping block can then be used for alignment. It is sufficient to rotate the Y polarization for the constellation angle mismatch.

The algorithm employing MIMO and V-V CPE is compared with the CR as presented in Figure 14.22, in which the uncompensated FO after the MIMO block is set to 50 MHz to simulate a realistic scenario. The measured FO after the MIMO is around 300 MHz. All data processing algorithms are done block-wise with parallellization of 64 symbols per block, and a realistic processing delay is added [20].

Figure 14.24 shows the gain in the quality factor Q for the two measurement cases. Since the FO is recovered by the feedback CR, the penalties shown in Figure 14.22 are completely compensated, with an additional gain coming from the estimations of coupling.

The 10 and 100 G hybrid off-line optical transmissions with homodyne coherent reception experiments are used to demonstrate the convergence of the feedback loops of the modified algorithm shown in Figure 14.23. As shown in Figure 14.25, the FO of 327 MHz can be acquired after 3.4 µs, or 1500 blocks of processing [20]—the steady state of the plateau region. The X–Y constellation offset

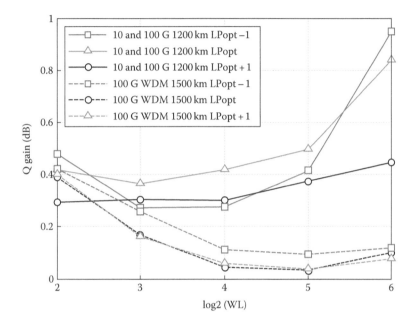

FIGURE 14.24 CR Q gain.

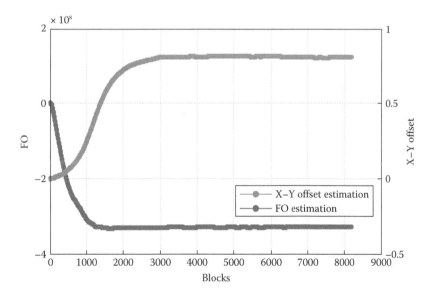

FIGURE 14.25 10 G hybrid off-line experiments LOFO and X–Y phase offset estimation by FO variation with respect to the number of blocks. FO = frequency offset—vertical axis.

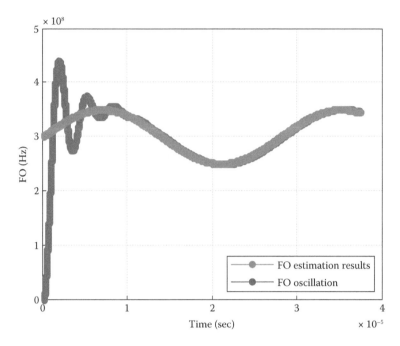

FIGURE 14.26   36 KHz oscillation variation of the laser frequency.

is acquired after 3000 blocks due to the demand for correct FO compensation. The estimated X–Y phase offset is 0.8 rad. The CR can cope with a laser frequency oscillation of 36 MHz. Acquisition and tracking times are shown in Figure 14.26. The CR is able to track the laser frequency changes after 100 ms. Note that a feedback loop can cause delay between real oscillations and estimation (delay between two lines in Figure 14.26, hence resulting in some penalties).

The high precision CR is achieved by the use of the V-V circuits under appropriate allocation and connections. Using feedback and feed-forward CR, significant improvements can be achieved. Laser FO can be compensated by complex circuit design and appropriate parameter selection. The ideal coupling of estimations coming from two polarizations further enhance CR performance with less complexity.

14.4.4.3   Modified Gardner Phase Detector for Nyquist Coherent Optical Transmission Systems

This section describes a timing recovery scheme based on a Gardner phase detector [34] performing excellently in spectrally efficient Nyquist coherent optical systems. Good clock tone performance can be obtained for QAM modulation formats, with the ROF values close to null.

14.4.4.3.1   Motivation

The link traffic can be increased by deploying more fibers, using L band, more efficient IP packaging, dynamic networking, and so on. Also, the spectral efficiency can be improved by applying higher-order modulation formats and efficient spectral shaping. Modulation formats and baud rates should be flexible in future elastic optical networks. In such scenarios, the network management will be able to select the best transmission scenario, depending on the link quality. It requires universal timing recovery, covering most modulation formats of interest. Most commercial timing recovery algorithms are designed for systems using non-return-to-zero and return-to-zero pulse shapes whose spectrums allow the generation of a spectral component at symbol frequency using the simplest nonlinear operations (squaring, rectifying, samples multiplication, etc.) or logical operations (early–late detectors). Several timing recovery candidates for high-speed optical coherent systems are discussed in Ref. [35]. However, in very dense WDM optical systems using Nyquist signaling with very small ROF values, most of the published timing algorithms failed. For example, a very famous square and filter algorithm [36] completely fails for the ROF value equal to zero. A phase detector (PD) described in Ref. [37], the so-called Gardner phase detector (GPD), is very effective in non-Nyquist systems, and can be realized in time domains with acceptable hardware complexity. This detector also fails at small ROF values. The fourth-power law PD (4PPD) presented in Ref. [37] can deal with small ROF values. This solution requires a carefully designed bandpass filter that can be replaced by a filter realized in the frequency domain. Due to PMD effects, the direct implementation of this method before PMD compensation is quite difficult. We propose an improved version of the Gardner PD having lower complexity than the 4PPD. The modified Gardner PD (MGPD) is independent of modulation formats and performs well in the most extreme transmission scenarios, with ROF close and equal to zero.

14.4.4.3.2 Timing Recovery

In coherent optical receivers, linear filters are employed to compensate for deterministic linear channel distortions such as residual CD and PMD. Timing/clock recovery, CR, nonlinear effects compensation, optical performance monitoring, differential decoding, forward error correction, and many other modules are also integral parts of the receiver. A microcontroller is responsible for slow-speed operations such as CD estimation, laser frequency control, controlling analog amplifiers, and so on. The clock tone can be characterized via several parameters [38]: timing error detector characteristics (TEDC), maximum of TEDC (TEDCMAX), peak-to-peak jitter (Jitterpp), and so on. Only the closed loop simulation of the considered transmission scenario can fully characterize timing performance. Timing recovery can be done either before or after PMD compensation. The Godard PD realization in the frequency domain before PMD compensation is shown in Figure 14.27. After CD compensation (CD−1 blocks), the signal is processed by a state-of-polarization (SOP) modifier to improve the clock quality. The SOP modifier is the unavoidable block that solves problems caused by PMD negative effects [39–41].

In practical Nyquist systems, a raised-cosine filter is used to generate Nyquist pulses. Due to the effects of white noise, the RRC filters at the transmitter and receiver sides are often used in simulations and experiments. We simulated the system shown in Figure 14.27 for QPSK (4QAM) case at the energy per bit to noise power spectral density ratio (E_b/N_0), equal to 3 dB (20% redundancy soft FEC threshold). TEDC for ROF values between 0 and 1 are shown in Figure 14.28. The clock tone drops at small ROF values and vanished completely for ROF = 0.

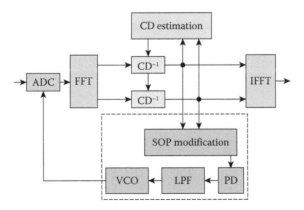

FIGURE 14.27 Implementation of Godard PD.

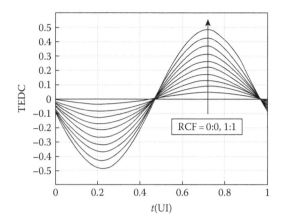

FIGURE 14.28   GPD TEDC; 4QAM; $E_b/N_0 = 3$ dB.

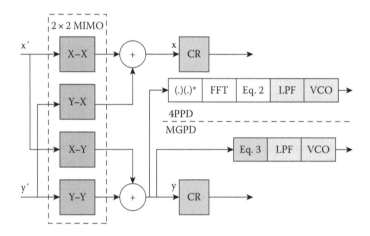

FIGURE 14.29   Implementation 4PPD and MGPD.

14.4.4.3.3   Algorithm Description

Since PDs based on the square and filter method fail for small ROF values, we can investigate the 4PPD and MGPD performance. Their implementations after PMD compensation are shown in Figure 14.29. Four complex filters are usually designed to work with two samples per symbol and to output only one sample. In our schematic, two filters, X–Y and Y–Y, are designed to output two samples per symbol. Therefore, only y output is used for timing extraction (two samples are required). The squared signal $z = yy^*[Z = fft(z)]$ is processed using blocks-generating outputs

$$4PPD: \quad \text{TEDC} = \text{imag} \sum_{k=0}^{N/2-1} Z(k)Z^*(k+N/2)$$

$$(14.45)$$

$$4GPD: \quad \text{TEDC} = y(k)y^*(k)\big(y(k+1)y^*(k+1) - y(k-1)y^*(k-1)\big)$$

The 4PPD requires FFT block (or a time domain FIR filter with an additional squarer). The MGPD consists of the squarer and the real-input Gardner PD. Apparently, the MGPD is less complex than the 4PDD. Since both PDs behave similarly, we continue with MGPD characterization. The implementation of 4PPD and MGPD is by using the experimental setup shown in Figure 14.30.

14.4.4.3.4 Simulation and Experimental Results

In all simulations, we assume a static 2×2 MIMO FIR filter whose coefficients are known in advance and do not vary over time. E_b/N_0 was varied starting from values close to 20% soft FEC thresholds ($E_b/N_{0,FEC}$) for the following modulation formats: 4QAM (3 dB), 16QAM (6 dB), and 64QAM (10 dB). TEDC of 4QAM versus ROF is presented in Figure 14.31. Clock tone is high at small ROF values, while high ROF destroys timing information. Sensitivity to ROF for different modulation formats is shown in Figures 14.32 and 14.33 ($E_b/N_0 = E_b/N_{0,FEC}$). Critical ROF value (low TEDCMAX) is lower for higher-level modulation formats. Anyhow, all of them perform well at small ROF values. TEDCMAX values over E_b/N_0 region of interest are shown in Figure 14.7. The 16QAM and 64QAM formats are almost insensitive to E_b/N_0 variations, while in case of 4QAM the TEDCMAX value is doubled at maximum E_b/N_0 value. Estimating clock tone quality only via TEDCMAX is not valid. Closed loop simulation with carefully adjusted parameters can tell more about the considered scheme. Therefore, we averaged PD output at the equilibrium point by the use of a sliding window filter of length 1024 (symbols). Simulated peak-to-peak jitter over 10^7 symbols is presented in Figure 14.34. Maximum jitter was around 3% of unit interval (UI), which causes negligible BER performance degradation at $E_b/N_{0,FEC}$ values for all modulation formats. It is thus experimentally investigated here, with the timing algorithm using a setup shown in Figure 14.30.

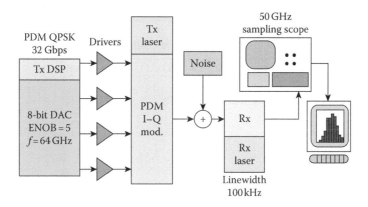

FIGURE 14.30 Experiment setup.

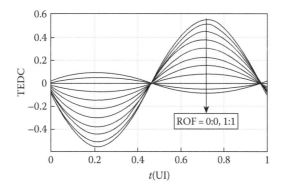

FIGURE 14.31 MGPD TEDC; 4QAM; $E_b/N_0 = 3$ dB.

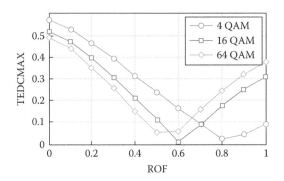

FIGURE 14.32   MGPD TEDCMAX versus ROF.

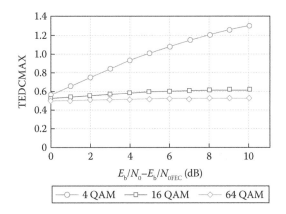

FIGURE 14.33   MGPD. TEDCMAX at ROF = 0 (MGPD).

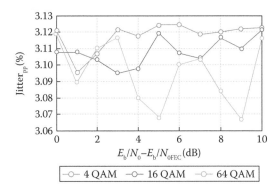

FIGURE 14.34   Jitter peak-to-peak at ROF = 0 (MGPD).

Using DSP, we generated the Nyquist signal including transmitter imperfections compensation. Due to the limited bandwidth of analog devices, DAC crosstalk, and time-varying unpredictable effects, we were not able to generate very accurate ROF values. Results for the GPD and MGPD are presented in Figures 14.35 and 14.36, respectively. One can observe very good agreement between simulations and experiments for both PDs. The proposed PD showed the best performance for small ROF values, while the GPD performance is better at high ROF values.

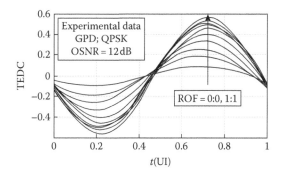

FIGURE 14.35   Experiment-GPD TEDCMAX.

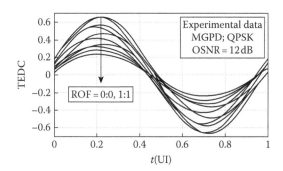

FIGURE 14.36   Experiment-MGPD TEDCMAX.

14.5   SYSTEMS PERFORMANCE OF MLSE EQUALIZER–MSK OPTICAL TRANSMISSION SYSTEMS

14.5.1   MLSE Equalizer for Optical MSK Systems

14.5.1.1   Configuration of MLSE Equalizer in Optical Frequency Discrimination Receiver

Figure 14.37 shows the block diagram of a narrowband filter receiver integrated with NLEs for the detection of 40 Gbps optical MSK signals. The structure and operation of the MSK transmitter have been described in Chapter 3. MSK employs 2 bits/symbol modulation as in the case of QPSK but with continuous phase change at the transition between the bit periods. Continuous phase change can lead to very effective frequency spectrum and hence MSK looks like FDM with the two frequencies are separated by 1/4T. Hence, when the frequency difference between states equals a quarter of the bit rate, the minimum distance between them so that the channel states can be recovered with minimum ISI, then the two frequency states are orthogonal, and then the modulation is termed as *minimum shift keying*, or MSK.

In the reception of MSK channels, two narrowband filters can be used to discriminate the upper sideband (USB) and lower sideband (LSB) channel frequencies that correspond to the logical "1" and "0" transmitted symbols, respectively. A constant optical delay line that is easily implemented in integrated optics is introduced on one branch to compensate for any differential group delay $t_d = 2\pi f_d \beta_2 L$ between f_1 and f_2 where $f_d = f_1 - f_2 = R/2$, β_2, represents group velocity delay (GVD) parameter of the fiber, and L is the fiber length, as described in Chapter 2. Alternatively, deskewing can be implemented in the digital domain to compensate for this delay difference. If the differential group delay is fully compensated, the optical lightwaves in two paths arrive at the photodiodes simultaneously. The outputs of the filters are then converted into

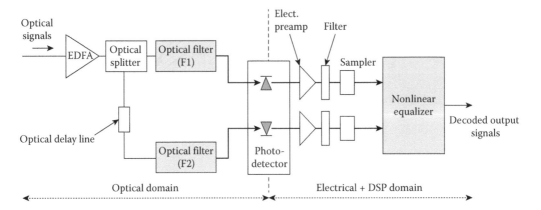

FIGURE 14.37 Block diagram of narrowband filter receiver integrated with nonlinear equalizers for the detection of 40 Gbps optical MSK signals.

electrical domain through the photodiodes. These two separately detected electrical signals are sampled before being fed as the inputs to the NLE.

It is noted here that the MSK reception system described here can be a self-homodyne detection or a coherent detection with the signal channel employed as the LO. Alternatively, the receiver subsystem can employ the coherent detection technique in that an optical LO oscillator is used to mix with the signals so that the resultant power level can be boosted up and hence the increase of the sensitivity of the receiver.

14.5.1.2 MLSE Equalizer with Viterbi Algorithm

At epoch k, it is assumed that the effect of ISI on an output symbol of the FSM c_k is caused by both executive δ pre-cursor and δ post-cursor symbols on each side. First, a state trellis is constructed with 2δ states for both detection branches of the optical frequency discrimination receiver (OFDR). A lookup table per branch corresponding to the symbols "0" and "1" transmitted and containing all the possible $2^{2\delta}$ states of all 11 symbol-length possible sequences is constructed by sending all the training sequences incremented from 1 to $2^{2\delta}$.

The output sequence $c(k) = f(b_k, b_{k-1}, \ldots, b_{k-2\delta-1}) = (c_1, c_2, \ldots, c_{2\delta})$ is the nonlinear function representing the ISI caused by the δ adjacent pre-cursor and post-cursor symbols of the optical fiber FSM. This sequence is obtained by selecting the middle symbols c_k of 2δ possible sequences with length of $2\delta + 1$ symbol intervals.

The samples of the two filter outputs at epoch $k y_k^i$, $i = 1, 2$ can be represented as $y^i(k) = c^i(k) + n_{ASE}^i(k) + n_{Elec}^i(k)$, $i = 1, 2$. Here, n_{ASE}^i and n_{Elec}^i represent the amplified spontaneous emission (ASE) noise and the electrical noise, respectively.

In linear transmission of an optical system, the received sequence y_n is corrupted by ASE noise of the optical amplifiers, n_{ASE}, and the electronic noise of the receiver, n_E. It has been proven that the calculation of branch metric and hence state metric is optimum when the distribution of noise follows the normal/Gaussian distribution—that is, the ASE accumulated noise and the electronic noise of the reception subsystem are collectively modeled as samples from Gaussian distributions. If the noise distribution departs from the Gaussian distribution, the minimization process is suboptimum.

The Viterbi algorithm subsystem is implemented on each detection branch of the OFDR. However, the MLSE with Viterbi algorithm may be too computationally complex to be implemented at 40 Gbps with the current integration technology. However, there have been commercial products available for 10 Gbps optical systems. Thus, a second MLSE equalizer using techniques of reduced-state template matching is presented in the following section.

In an optical MSK transmission system, narrowband optical filtering plays the main role in shaping the noise distribution back to the Gaussian profile. The Gaussian profile noise distribution is verified

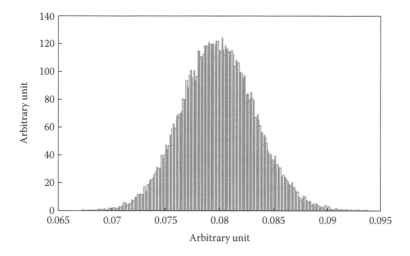

FIGURE 14.38 Noise distribution following Gaussian shape due to narrowband optical filtering.

in Figure 14.38. Thus, branch metric calculation in the Viterbi algorithm, which is based on minimum Euclidean distance over the trellis, can achieve the optimum performance. Also, the computational effort is less complex than ASK or DPSK systems due to the issue of non-Gaussian noise distribution.

14.5.1.3 MLSE Equalizer with Reduced-State Template Matching

The modified MLSE is a single-shot template-matching algorithm. First, a table of $2^{2\delta+1}$ templates, g^k, $k = 1, 2, \ldots, 2^{2\delta+1}$, corresponding to the $2^{2\delta+1}$ possible information sequences, I^k, of length $2\delta+1$ is constructed. Each template is also a vector of size $2\delta+1$, which is obtained by transmitting the corresponding information sequence through the optical channel and obtaining the $2\delta+1$ consecutive received samples. At each symbol period, n, the sequence, $\hat{I}_n$, with the minimum metric is selected as $c(k) = \arg\min_{b(k)}\{m(b(k), y(k))\}$. The middle element of the selected information sequence is then output as the nth decision, $\hat{I}_n$—that is, $\hat{I}_n = I_n(\delta + 1)$. The minimization is performed over the information sequences, I^k, which satisfy the condition that the $\delta-2$ elements are equal to the previously decoded symbols $\hat{I}_{n-\delta}$, $I_{n-(\delta-1)}$, $\ldots\hat{I}_{n-2}$, and the metric, $m(I^k, \mathbf{r}_n)$, is given by $m(\mathbf{I}^k, \mathbf{r}_n) = \{\mathbf{w}\cdot(\mathbf{g}^k - \mathbf{r}_n)\}^T\{\mathbf{w}\cdot(\mathbf{g}^k - \mathbf{r}_n)\}$, where w is a weighting vector that is chosen carefully to improve the reliability of the metric. The weighting vector is selected so that, when the template is compared with the received samples, less weighting is given to the samples further away from the middle sample. For example, we found through numerical results that a weighting vector with elements $w(i) = 2^{-|i-(\delta+1)|}$, $i = 1, 2, \ldots, 2\delta+1$ gives good results. Here, $\cdot$ represents Hadamard multiplication of two vectors, $(.)^T$ represents transpose of a vector, and $|.|$ is the modulus operation.

14.5.2 MLSE SCHEME PERFORMANCE

14.5.2.1 Performance of MLSE Schemes in 40 Gbps Transmission Systems

Figure 14.39 shows the simulation system configuration used for the investigation of the performance of both the preceding schemes when used with the narrowband optical Gaussian filter receiver for the detection of noncoherent 40 Gbps optical MSK systems. The input power into fiber (P_0) is -3 dBm, which is much lower than the nonlinear threshold power. The EDFA2 provides 23 dB gain to maintain the receiver sensitivity of -23.2 dBm at BER = 10^{-9}. As shown in Figure 14.39, the optical received power (P_{Rx}) is measured at the input of the narrowband MSK receiver, and the OSNR is monitored to obtain the BER curves for different fiber lengths. The length of SSMF is varied from 48 to 60 km in steps of 4 km to investigate the performance of the equalizers to the

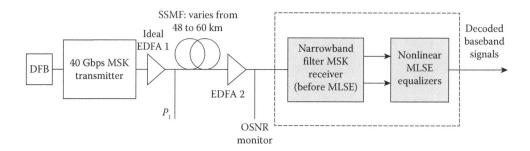

FIGURE 14.39 Simulation setup for performance evaluation of MLSE and modified MLSE schemes for detection of 40 Gbps optical MSK systems.

TABLE 14.2

Key Simulation Parameters Used in the MLSE MSK Modulation Format

Input power: $P_0 = -3$ dBm	Narrowband Gaussian filter: $B = 5.2$ GHz or $BT = 0.13$
Operating wavelength: $\lambda = 1550$ nm	Constant delay: $t_d = |2\pi f_D \beta_2 L|$
Bit rate: $R = 40$ Gbps	Preamp. EDFA of the OFDR: $G = 15$ dB and $NF = 5$ dB
SSMF fiber: $|\beta_2| = 2.68\mathrm{e}{-26}$ or $|D| = 17$ ps/nm/km	$i_d = 10$ nA
Attenuation: $\alpha = 0.2$ dB/km	$N_{eq} = 20$ pA/(Hz)$^{1/2}$

degradation caused by fiber cumulative dispersion. The narrowband Gaussian filter with the time bandwidth product of 0.13 is used for the detection filters.

Electronic noise of the receiver can be modeled with the equivalent noise current density of electrical amplifier of 20 pA/$\sqrt{\mathrm{Hz}}$ and dark current of 10 nA. The key parameters of the transmission system are given in Table 14.2.

The Viterbi algorithm used with the MLSE equalizer has a constraint length of 6 (i.e., 2^5 number of states) and a trace-back length of 30. Figure 14.40 shows the BER performance of both NLEs plotted against the required OSNR. The BER performance of the optical MSK receiver without any equalizers for 25 km SSMF transmission is also shown in Figure 14.40 for quantitative comparison. The numerical results are obtained via Monte Carlo simulation (triangular markers as shown in Figure 14.40) with the low BER tail of the curve linearly extrapolated. The OSNR penalty (at BER = 10^{-9}) versus residual dispersion corresponding to 48, 52, 56, and 60 km SSMF are presented in Figure 14.41. The MLSE scheme outperforms the modified MLSE schemes, especially at low OSNR. In the case of 60 km SSMF, the improvement at BER = 10^{-9} is approximately 1 dB, and in the case of 48 km SSMF, the improvement is about 5 dB. In case of transmission of 52 km and 56 km SSMF, MLSE with Viterbi algorithm has 4 dB gain in OSNR compared to the modified scheme. With residual dispersion at 816, 884, 952, and 1020 ps/nm, at a BER of 10^{-9}, the MLSE scheme requires 12, 16, 19, and 26 dB OSNR penalty, respectively, while the modified scheme requires 17, 20, 23, and 27 OSNR, respectively.

14.5.2.2 Transmission of 10 Gbps Optical MSK Signals over 1472 km SSMF Uncompensated Optical Links

Figure 14.42 shows the simulation setup for 10 Gbps transmission of optical MSK signals over 1472 km SSMF. The receiver employs an optical narrowband frequency discrimination receiver integrated with a 1024-state Viterbi-MLSE postequalizer. The input power into the fiber (P_0) is –3 dBm lower than the fiber nonlinear threshold power. The optical amplifier EDFA1 provides an optical gain to compensate the attenuation of each span completely. The EDFA2 is used as a noise-loading source

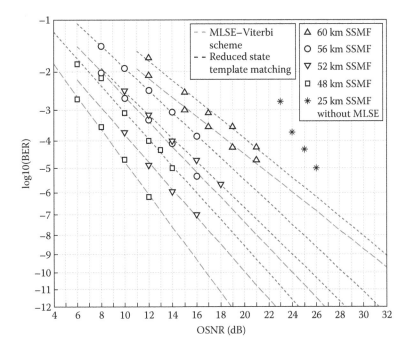

FIGURE 14.40 Performance of modified Viterbi-MLSE and template-matching schemes, BER versus OSNR.

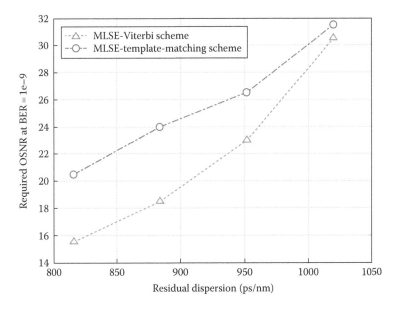

FIGURE 14.41 Required OSNR (at BER = 1e−9) versus residual dispersion in Viterbi-MLSE and template-matching MLSE schemes.

to vary the required OSNR. The receiver electronic noise is modeled with equivalent noise current density of the electrical amplifier of 20 pA/$\sqrt{\text{Hz}}$ and dark current of 10 nA for each photodiode. A narrowband optical Gaussian filter with two-sided bandwidth of 2.6 GHz (one-sided BT = 0.13) is optimized for detection. A back-to-back (BTB) OSNR = 8 dB is required for BER at 10^{-3} for each branch. The correspondent received power is −25 dBm. This low OSNR is possible due to the large suppression of noise after being filtered by narrowband optical filters. A trace-back length of 70 is used in the Viterbi algorithm. Figure 14.42 shows the simulation results of BER versus the required OSNR for 10 Gbps optical MSK transmission over 1472 and 1520 km SSMF uncompensated optical links with one, two, and four samples per bit, respectively. A 1520 km SSMF with one sample per bit is seen as the limit for the 1024-state Viterbi algorithm due to the slow roll-off and the error floor. However, two and four samples per bit can obtain error values lower than the FEC limit of 10^{-3}.

Thus, 1520 km SSMF transmission of 10 Gbps optical MSK signals can reach the error-free detection with the use of a high-performance FEC. In case of 1472 km SSFM transmission, the error events follow a linear trend without sign of error floor and, therefore, error-free detection can be comfortably accomplished.

Figure 14.43 also shows the significant improvement in OSNR of two and four samples per bit over one sample per bit counterpart with values of approximately 5 and 6 dB, respectively. In terms

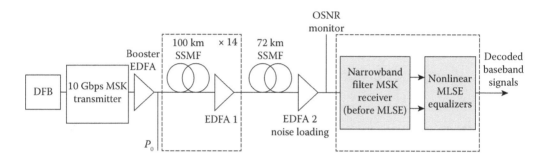

FIGURE 14.42   Simulation setup for transmission of 10 Gbps optical MSK signals over 1472 km SSMF with MLSE-Viterbi equalizer integrated with the narrowband optical filter receiver.

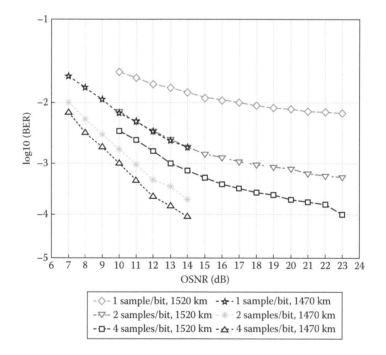

FIGURE 14.43   BER versus required OSNR for transmission of 10 Gbps optical MSK signals over 1472 and 1520 km SSMF uncompensated optical link.

of OSNR penalty at BER of 1e−3 from BTB setup, four samples per bit for 1472 and 1520 km transmission distance suffers 2 and 5 dB penalty, respectively.

14.5.2.3 Performance Limits of Viterbi-MLSE Equalizers

The performance limits of the Viterbi-MLSE equalizer to combat the ISI effects are investigated against various SSMF lengths of the optical link. The number of states used in the equalizer is incremented according to this increase and varied from 2^6 to 2^{10}. This range was chosen as reflecting the current feasibility and future advance of electronic technologies that can support high-speed processing of the Viterbi algorithm in the MLSE equalizer. In addition, these numbers of states also provide feasible time for simulation.

In addition, one possible solution to ease the requirement of improving the performance of the equalizer without increasing much the complexity is by multisampling within one-bit period. This technique can be used by interleaving the samplers at different times. Although a greater number of electronic samplers are required, they only need to operate at the same bit rate as the received MSK electrical signals. Moreover, it will be shown later on that there is no noticeable improvement with more than two samples per bit period. Hence, the complexity of the MLSE equalizer can be affordable while improving the performance significantly.

Figure 14.44 shows the simulation setup for 10 Gbps optical MSK transmission systems with lengths of uncompensated optical links varying up to 1472 km SSMF. In this setup, the input power into fiber (P_0) is set to be −3 dBm, thus avoiding the effects of fiber nonlinearities. The optical amplifier EDFA1 provides an optical gain to compensate for the attenuation of each span completely. The EDFA2 is used as a noise-loading source to vary the required OSNR values. Moreover, a Gaussian filter with two-sided 3 dB bandwidth of 9 GHz (one-sided $BT = 0.45$) is utilized as the optical discrimination filter because this BT product gives the maximized detection eye openings. The receiver electronic noise is modeled with equivalent noise current density of the electrical amplifier of 20 pA/$\sqrt{\text{Hz}}$ and a dark current of 10 nano-Ampere (nA) for each photodiode. A BTB OSNR = 15 dB is required for a BER of 10^{-4} on each branch, and the correspondent receiver sensitivity is −25 dBm. A trace-back length of 70 samples is used for the Viterbi algorithm in the MLSE equalizer. Figures 14.44 through 14.46 show the BER performance curves of 10 Gbps OFDR-based MSK optical transmission systems over 928, 960, 1280, and 1470 km SSMF uncompensated optical links for different number of states used in the Viterbi-MLSE equalizer. In these figures, the performance of Viterbi-MLSE equalizers is given by the plot of the BER versus the required OSNR for several detection configurations: balanced receiver (without incorporating the equalizer), the conventional single-sample-per-bit sampling technique, and the multisample-per-bit sampling techniques (two and four samples per bit slot). The simulation results are obtained by the Monte Carlo method.

The significance of multisamples per bit slot in improving the performance for MLSE equalizer in cases of uncompensated long distances is shown. It is found that the tolerance limits to the ISI

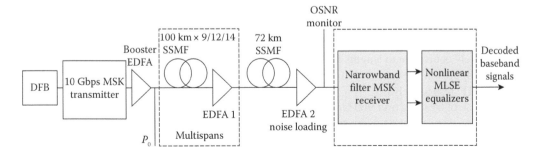

FIGURE 14.44 Simulation setup of OFDR-based 10 Gbps MSK optical transmission for study of performance limits of MLSE-Viterbi equalizer.

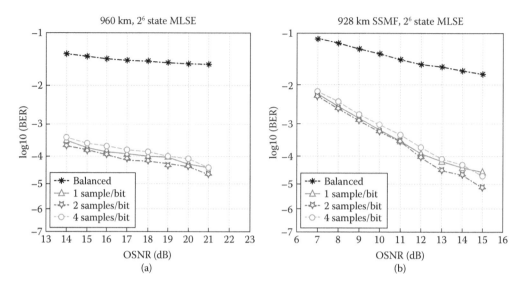

FIGURE 14.45 Performance of 64-state Viterbi-MLSE equalizers for 10 Gbps OFDR-based MSK optical systems over (a) 960 km and (b) 928 km SSMF uncompensated links.

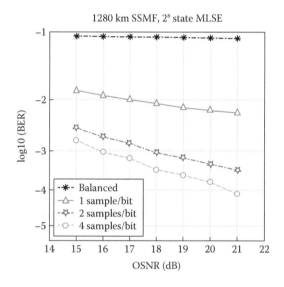

FIGURE 14.46 Performance of 256-state Viterbi-MLSE equalizer for 10 Gbps OFDR-based MSK optical systems over 1280 km SSMF uncompensated links.

effects induced from the residual CD of a 10 Gbps MLSE equalizer using 2^6, 2^8, and 2^{10} states are approximately equivalent to lengths of 928, 1280, and 1440 km SSMF, respectively. The equivalent numerical figures in the case of 40 Gbps transmission are corresponding to lengths of 62, 80, and 90 km SSMF, respectively.

In the case of 64 states over 960 km SSMF uncompensated optical links, the BER curve encounters an error floor that cannot be overcome even by using high-performance FEC schemes. However, at 928 km, the linear BER curve indicates the possibility of recovering the transmitted data with the use of high-performance FEC. Thus, for 10 Gbps OFDR-based MSK optical systems, a length of 928 km SSMF can be considered as the transmission limit for the 64-state Viterbi-MLSE equalizer. Results shown in Figure 14.46 suggest that the length of 1280 km SSMF uncompensated optical link is the transmission limit for the 256-state Viterbi-MLSE equalizer when incorporating with OFDR optical front end. It should be noted that this is achieved when using the multisample-per-bit sampling schemes.

It is observed from Figure 14.47 that the transmission length of 1520 km SSMF with one sample per bit is seen as the limit for the 1024-state Viterbi algorithm due to the slow roll-off and the error floor. However, two and four samples per bit can obtain error values lower than the FEC limit of 10^{-3}. Thus, 1520 km SSMF transmission of 10 Gbps optical MSK signals can reach the error-free detection with use of a high-performance FEC. In case of 1472 km SSFM transmission, the error events follow a linear trend without sign of error floor, and, therefore, error-free detection can be achieved. Figure 14.47 also shows the significant OSNR improvement of the sampling techniques with two and four samples per bit compared to the single-sample-per-bit counterpart. In terms of OSNR penalty at a BER of 1e−3 from BTB setups, four samples per bit for 1472 and 1520 km transmission distance suffers 2 and 5 dB penalty, respectively.

It is found that, for incoherent detection of optical MSK signals based on the OFDR, an MLSE equalizer using 2^4 states does not offer better performance than the balanced configuration of the OFDR itself; that is, the uncompensated distance is not over 35 km SSMF for 40 Gbps or 560 km SSMF for 10 Gbps transmission systems, respectively. It is most likely that, in this case, the severe ISI effect caused by the optical fiber channel has spread beyond the time window of five bit slots (two pre-cursor and two post-cursor bits), the window that a 16-state MLSE equalizer can handle. Thus, it is necessary that the number of taps or states of MLSE cover the full length of the dispersive pulse.

The trace-back length used in the investigation is chosen to be 70, which guarantees the convergence of the Viterbi algorithm. The longer the trace-back length, the larger the memory that is required. With state-of-the-art technology for high storage capacity nowadays, memory is no longer a big issue. Very fast processing speeds at 40 Gbps operations hinder the implementation of 40 Gbps Viterbi-MLSE equalizers at the mean time. Multisample sampling schemes offer an exciting solution for implementing fast signal processing processes. This challenge may also be overcome in the near future together with the advance of the semiconductor industry. At present, the realization of Viterbi-MLSE equalizers operating at 10 Gbps has been commercially demonstrated [42].

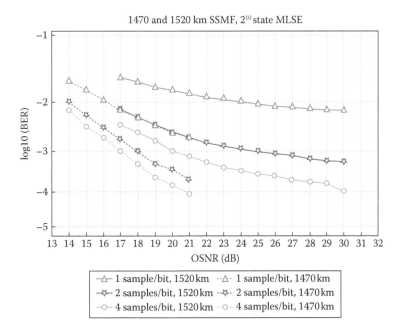

FIGURE 14.47 Performance of 1024-state MLSE equalizer for 10 Gbps optical MSK signals versus the required OSNR over 1472 and 1520 km SSMF uncompensated optical link.

14.5.2.4 Viterbi-MLSE Equalizers for PMD Mitigation

Figure 14.48 shows the simulation test bed for the investigation of MLSE equalization of the PMD effect. The transmission link consists of a number of spans comprising 100 km SSMF (with $D = +17$ ps/(nm km), $\alpha = 0.2$ dB/km) and 10 km DCF (with $D = -170$ ps/(nm km), $\alpha = 0.9$ dB/km). Launched power into each span (P_0) is -3 dBm. The $EDFA_1$ has a gain of 19 dB, hence providing input power into the DCF to be -4 dBm, which is lower than the nonlinear threshold of the DCF. The 10 dB gain of $EDFA_2$ guarantees the input power into next span unchanged with an average power of -3 dBm. An OSNR of 10 dB is required for receiver sensitivity at BER of 10^{-4} in case of back-to-back configuration. Considering the practical aspect and complexity, a Viterbi-MLSE equalizer for PMD equalization, and a small number of four-state bits, or effectively 16 states, were chosen for the Viterbi algorithm in the simulation study.

The performance of MLSE against the PMD dynamic of optical fiber is investigated. BER versus required OSNR for different values of normalized average differential group delay (DGD) $<\Delta t>$ are shown in Figure 14.49. The mean DGD factor is normalized over one-bit period of 100 and 25 ps for 10 and 40 Gbps bit rates, respectively. The numerical studies are conducted for a range of values from 0 to 1 of normalized DGD $<\Delta t>$, which equivalently corresponds to the instant delays of up to 25 or 100 ps in terms of 40 and 10 Gbps transmission bit rate, respectively.

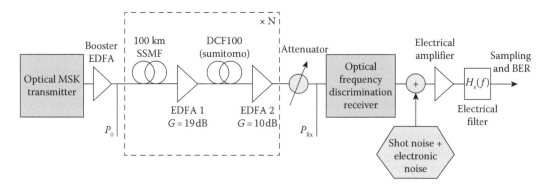

FIGURE 14.48 Simulation test bed for investigation of effectiveness of MLSE equalizer to PMD.

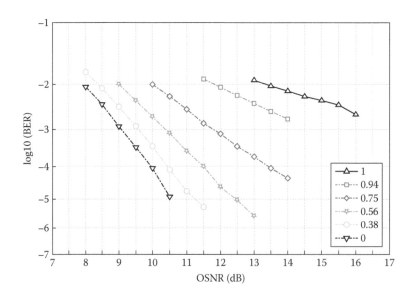

FIGURE 14.49 Performance of MLSE equalizer versus normalized $<\Delta \tau>$ values for one sample per bit.

The advantages of using multiple samples per bit over the conventional single sample per bit in MLSE equalizer are also numerically studied, and the Monte Carlo results for normalized DGD values of 0.38 and 0.56 are shown in Figures 14.50 and 14.51. However, the increase from two to four samples per bit does not offer any gain in the performance of the Viterbi-MLSE equalizer. Thus, two samples per symbol are preferred to reduce the complexity of the equalizer. Here, and for the rest of this chapter, multisamples per bit implies the implementation of two samples per symbol. Performance of MLSE equalizer for different normalized DGD values with two samples per bit is shown in Figure 14.52. From Figure 14.52, the important remark is that 16-state MLSE equalizer

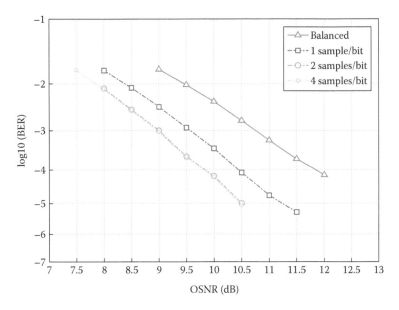

FIGURE 14.50 Comparison of MLSE performance for configurations of one, two, and four samples per bit, with normalized <Δτ> value of 0.38.

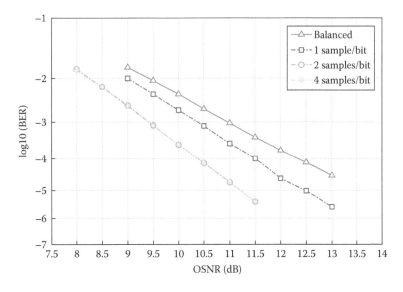

FIGURE 14.51 Performance of Viterbi-MLSE equalizer for configurations of one, two, and four samples per bit, with a normalized <Δτ> value of 0.56.

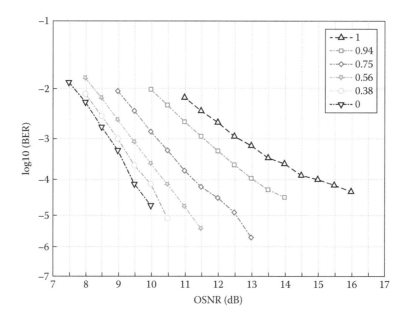

FIGURE 14.52 Performance of Viterbi-MLSE equalizer for different normalized $<\Delta t>$ values with two and four samples per bit.

implementing two samples per symbol enables the optical MSK transmission systems achieving a PMD tolerance of up to one-bit period at BER = 10^{-4}, with a required OSNR of 8 dB. This delay value starts introducing a BER floor, indicating the limit of the 16-state MLSE equalizer. This problem can be overcome with a high-performance FEC. However, the DGD mean value of 0.94 can be considered as the limit for an acceptable performance of the proposed MLSE equalizer without the aid of high-performance FEC. Figure 14.53 shows the required OSNR for MLSE performance at 10^{-4} versus normalized $<\Delta t>$ values in configurations of balanced receiver, one sample per bit, and two samples per bit.

A balanced receiver, with no MLSE equalizer incorporated, requires an OSNR = 5 dB to obtain a BER = 10^{-4} when there are no effects of the PMD compared to the required OSNR of 3 and 1 dB in cases of one sample per bit and two samples per bit, respectively. The OSNR penalties for various normalized $<\Delta t>$ values of the preceding three transmission system configurations are shown in Figure 14.54.

It can be observed that the OSNR penalties (BTB) of approximately 3 and 1 dB apply to the cases of balanced receiver and one sample per bit, respectively, with reference to the two samples per bit configuration. Another important remark is that the 16-state Viterbi-MLSE equalizer that implements two samples per bit enables the optical MSK transmission systems achieving a PMD tolerance of up to one-bit period at a BER of 10^{-4} with a power penalty of about 6 dB. Moreover, the best 2 dB penalty occurs at 0.75 for the value of normalized $<\Delta t>$. This result shows that the combination of OFDR-based MSK optical systems and Viterbi-MLSE equalizers, particularly with the use of multisample sampling schemes, was found to be highly effective in combating the fiber PMD dynamic impairment and better than the recently reported PMD performance for OOK and DPSK modulation formats [43].

14.5.2.5 Uncertainty and Transmission Limitation of the Equalization Process

The fundamental limitations of a quadratic phase media on the transmission speed are formulated in signal space for significant applications in signal equalization. These limitations are quantitatively derived for both coherent and incoherent optical systems. This section investigates the dispersive effect of optical fiber on the transmission speed of the media. The main effect of dispersion is

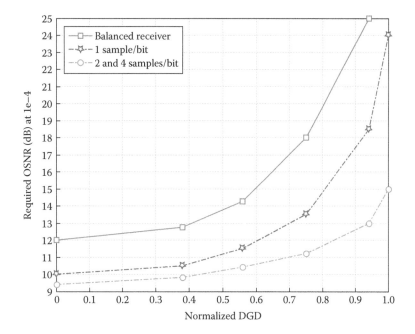

FIGURE 14.53   Required OSNR for MLSE performance at 10^{-4} versus normalized $<\Delta t>$ values in configurations of balanced receiver and one and two samples per bit.

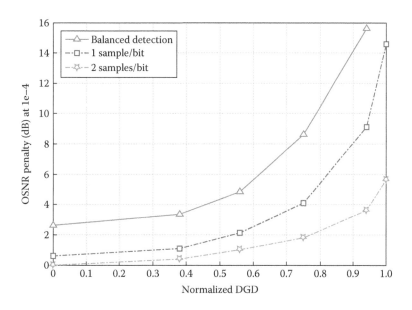

FIGURE 14.54   OSNR penalties of MLSE performance at 10^{-4} for various normalized $<\Delta t>$ values in configurations of balanced receiver, single sample, and two samples per bit sampling schemes.

pulse broadening (and frequency chirping) as the signal pulses propagate through the quadratic phase media. This causes ISI, thereby limiting the speed of transmission. There are various ways of mitigating the harmful effects of ISI. One way is to use optical fiber with antidispersive property to equalize the dispersion. In practice, this optical scheme reduces signal level and thus requires amplification. The amplifier introduces noise, which in turn limits the channel speed

(Shannon's information capacity theorem). Another way of combating ISI is by digital electronic equalization means, such as MLSE, which is applicable for both coherent and incoherent detection schemes [44–48]. These reported results show clearly that there is a fundamental limit on the achievable transmission speed for a given fiber length.

This section studies the limitation of quadratic phase channel, the single-mode optical fiber, in terms of signal space. The signal space is chosen so that it consists of binary signals in eight-bit block code. As the eight-bit symbols propagate down the quadratic phase channel, naturally the waveform patterns of the symbols would become less and less distinctive, and hence more and more difficult to discriminate between symbols. How do we quantify the detrimental effect of the quadratic phase channel? This system performance is approached entirely from a digital communications perspective, and the results so derived are the fundamental limit imposed by the quadratic phase channel (pure phase distortion but quadratic or square dependence of frequency by the linear dispersion). It is virtually independent of the detection scheme used. Two mechanisms that limit the transmission speed are explained. One is brought about by the finite time window available for detection. In all practical schemes, the decoder must decode each symbol within a finite time period. This requires the algorithm to be the least complex as possible. The other is brought about by not using the phase information for detection. Depending on the complexity of the detection scheme chosen, it is quite often that the phase information may be lost inadvertently when the optical signal is converted into the electrical signal. Another issue is the power consumption of the ASICs for implementation of digital algorithms for real-time applications. The more complex the algorithms are, the higher the number of digital circuits, and hence the more the power consumption. It is expected that all ASIC must consume less than 70 W in real-time processing.

14.6 ADAPTIVE JOINT CR AND TURBO DECODING FOR NYQUIST TERABIT OPTICAL TRANSMISSION IN THE PRESENCE OF PHASE NOISE

This section describes an adaptive joint CR and soft low-density parity check (LDPC) code turbo decoding (TD) scheme in the presence of nonlinear phase noise, proposed and experimentally verified with 1.9 dB coding gain in Nyquist Terabit PDM-DQPSK systems [49].

14.6.1 MOTIVATION

Using Nyquist pulse shaping in Tbps transmission systems [50–52] has recently attracted significant research interest, focusing on spectrally efficient ultra-high-capacity optical communications. This is especially important for systems employing fixed or flexible grid wavelength division multiplexing (WDM) grid. The performance of Nyquist coherent systems strongly depends on the capability of DSP algorithms. In combination with DSP, the advanced FEC codes such as LDPCs allow significant increasing transmission distance [53–55].

The success of soft FEC decoding is often related to the quality of log-likelihood ratio (LLR) information, which is mainly related to additive white Gaussian noise (AWGN) channels. However, the noise statistics can be significantly influenced by the imperfect compensation of linear and nonlinear fiber channel impairments, the CR capability, the differential decoding (if employed), and efficiency of DSP algorithms [56,57]. The residual nonlinear phase noise induces substantial performance degradation.

In a previous work, soft differential decoding (SDD) and BCJR decoder performance have been investigated under AWGN channels, including analysis of the performance of quasi-cyclic LDPC (QC-LDPC) codes in differentially encoded 100 G optical transmission systems [44]. This section presents the optimization of CR and soft LDPC TD in practical Nyquist Terabit PDM-DQPSK systems in which the turbo decoder experiences severe performance degradation due to the residual nonlinear phase noise. To overcome this problem, adaptive joint CR and soft LDPC TD algorithm can be proposed. Utilizing probability density function (PDF) of CR output signal derived from a training sequence and post-FEC hard decision bits, the CR module adaptively mitigates more nonlinear phase noise. Furthermore, an associate LLR estimator provides more reliable LLR to turbo decoders. The adaptive scheme is proven to offer a significant system decoding gain in Tbps experimental platform.

14.6.2 Terabit Experiment Setup and Algorithm Principle

In our experiment, we have used 1T Nyquist transmitter (see Figure 14.55), consisting of four high-speed DAC with eight bit resolution (effective number of bits—ENOB equals 5.2—see ADC and DAC described in Chapter 3), a comb generator for subcarrier generation (10 channels; 10×128 G; channel spacing 33.3 GHz), ECL lasers of 100 KHz linewidth, two LiNbO$_3$-based PDM I–Q modulators (MZMs), and optical filters/multiplexers. FEC-encoded Nyquist shaped data are loaded to four DACs. RRC pulses generated at the Tx output closely approached the rectangular spectral shape with an ROF α of 0.01. One of 10 subcarriers from the comb generator is filtered out by a wavelength selective switch (WSS). The other nine are modulated by MZM2 and further combined with the probed channel modulated by MZM1. The optical superchannel is then launched into the transmission system build of several spans. Each span consists of 75 km SSMF, a variable optical attenuator (VOA), and an EDFA. A preamplified Tektronix DP073304D coherent receiver records one of the channels with a sampling frequency of 50 GSa/s. The received signals are then processed off-line. In addition to the BTB case, two transmission scenarios are investigated: 1T superchannel transmission (1200 km link; no DCF; LP 1~4 dBm; the middle subcarrier was chosen as probe channel) and 1T/10 G-OOK hybrid transmission (800 km link with DCF; 50 GHz channel spacing between 10 G and 1T channel; subcarrier channel adjacent to the 10 G channel selected as probe). The DSP module compensates for skew, quadrature error, residual CD, PMD, and frequency and phase offset. The feed-forward carrier recovery (FF-CR) based on the Viterbi-Viterbi algorithm presented in Section 14.5 and also in Chapter 13, is used.

Figure 14.56 illustrates the adaptive joint CR and FEC principle, where PDF and LLR estimators are placed between the CR and FEC decoder. Pre-FEC BER and PDF are interactively estimated by two negative feedback paths, one between the training sequence bits and the post-DSP soft information, and the second between post-FEC hard decision bits and the soft information. Two types of 1T/10 G

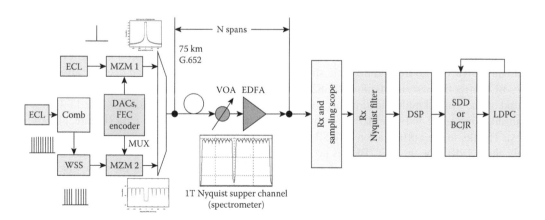

FIGURE 14.55 Experimental setup platform for adaptive joint carrier recovery. Insert is the superchannel spectrum of 1 Tbps with probe channel filtered out.

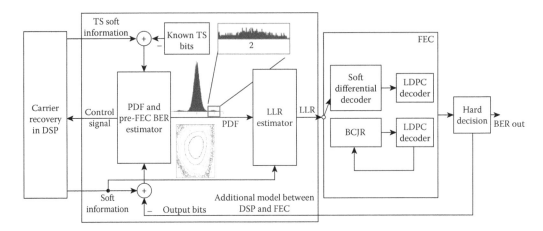

FIGURE 14.56 Adaptive joint carrier recovery and soft FEC decoding, two decoding schemes are applied in FEC block. TS = training sequence; pdf = probability density function.

hybrid transmission PDF estimation results at the output of PDF and pre-FEC BER estimator are also inserted, as shown in Figure 14.56. In the first accumulate distribution approaches a Gaussian profile with the PDF side lobes indicating the cycle slip frequency. The two-dimensional "egg-shaped" PDF contour indicates the CR averaging effects. The estimated PDF profile and the CR soft outputs are then employed in the LLR estimator so as to generate LLR for the proposed FEC decoder. Simultaneously, the PDF estimator analyzes the distortion caused by the phase noise and the probability of cycle slips. Thence, feedback control signals are applied to the CR block that adaptively adjusts the CR averaging window length (dithering mode), so as to decrease the nonlinear phase noise negative effects. The PDF estimator can be realized in a microcontroller without difficulty due to the slow temporal variation of the phase noises.

Two decoding architectures are applied in the FEC block (see Figure 14.56)—SDD plus LDPC (SF) with ten iterations in LDPC decoder, and iterative TD investigated in Ref. [43] (BCJR plus SD-FEC) with ten iterations between LDPC and BJCR decoders. The BER was derived after the CR (pre-FEC BER) and after the FEC block (post-FEC BER).

The experimental data are first processed in the traditional way without the proposed technique. The BER of 1T BTB post-FEC under TD and SF are presented, respectively, as dotted and solid lines connecting, in Figure 14.57. The TD method outperforms the SF one by 2.2 dB at a BER of 1e−5. These improved performances are in much agreement with the simulation presented in Ref. [43]. These results are expected since the EDFA superimposed noise is the principal impairment in the BTB case without the influences of nonlinearity. The CR generates high-quality soft information to the turbo decoder, which reduces to a post-FEC BER of 1.12e−6 from a pre-FEC BER of 0.0161 with an OSNR of 11.5 dB and no cycle slips. However, the BCJR differential decoder almost doubles BER after the first run (BER = 0.0382). The corresponding BER after the SDD is 0.059, as it is based on the symbol multiplication that significantly increases the noise variance. The 1T 1200 km transmission results at 1 dBm LP show similar BER trends as in the BTB case (Figure 14.57, circles). Thanks to the optimization of DSP modules, the transmission penalties are now reduced to only about 1 dB. The pre-FEC BER is no more relevant because of cycle slips. The TD compensation gain is 2.1 dB, almost the same as in the BTB case. This indicates that the noise distribution in the quasi-linear regime at 1 dBm LP is very similar to the BTB one. In these two cases, the joint scheme does not bring any gain as the nonlinear noise was not the dominant negative factor.

As we increased the LP to 4 dBm in the 1200 km 1T experiment, the turbo decoder clearly shows different performances, as depicted in Figure 14.58 (circle lines). The BER curves show large performance degradations resulting from significant nonlinear effects. The TD outperforms the SF by

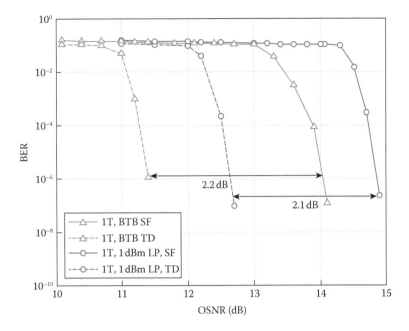

FIGURE 14.57 1T BTB and 1T transmission performances at 1 dBm launch power with measured and processed BER versus OSNR.

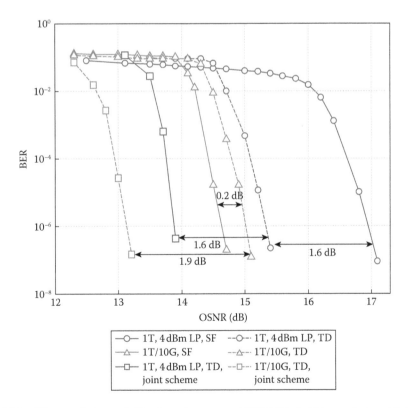

FIGURE 14.58 1T transmission at 4 dBm launch power and 1T/10 G hybrid transmission at 2 dBm launch power. Measured and processed BER versus OSNR.

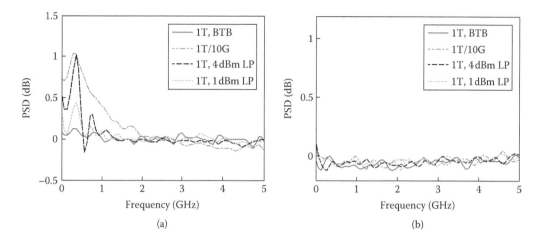

FIGURE 14.59 PSD of signal for four transmission scenarios: (a) without and (b) with joint scheme.

only 1.6 dB, which is 0.5 dB less compared to the moderate LP case. Figure 14.58 also depicts the system performances indicated by triangles for 2 dBm LP under 1T/10 G hybrid transmission. Turbo decoding performs even 0.2 dB worse at post-FEC BER of 1e–5 compared to the much simpler SF method. The experimental data are further processed by the adaptive joint DSP and FEC scheme. The turbo decoder achieves 1.6~1.9 dB OSNR gain (continuous lines connecting in Figure 14.58). The TD outperforms the SF by about 3.2 and 1.7 dB in 1T at high LP and 1T/10 G hybrid transmission, respectively. In order to highlight the effect of adaptive joint CR and TD scheme, a comprehensive analysis of the SD-FEC input data in terms of the power spectral density (PSD) has been performed. Four normalized PSDs without the joint scheme are shown in Figure 14.59a. In the 1T at 4 dBm LP and 1T/10 G hybrid transmission scenarios (strong nonlinear effects), one can observe significant symbol correlation indicating colored phase noise (lower frequencies in Figure 14.59) that influences the decoder performance. In the BTB case and 1T transmission at low LP, the DSP outputs soft values similar to the pure AWGN case with almost uniform PSD. For the 1T transmission at high LP or 1T/10 G hybrid transmission, the colored nonlinear phase noise bandwidth exceeds the CR bandwidth, resulting in BER performance degradation. Utilizing the joint scheme improving significantly the soft information, the residual nonlinear phase noise is effectively mitigated, as shown in Figure 14.59b.

Thus, in this subsection, the performances of soft LDPC TD in terabit Nyquist off-line experiment prove that the colored nonlinear phase noise severely harms TD. The presented adaptive joint CR and soft LDPC TD scheme effectively mitigates the nonlinear colored noise by improving the quality of FEC soft input information.

REFERENCES

1. N. Benuveto and G. Cheubini, *Algorithms for Communications Systems and Their Applications*, Chichester, England: John Wiley & Sons, 2004.
2. S. Savoy, Digital coherent optical receivers, *IEEE J. Sel. Areas Quant. Electron.*, Vol. 16, No. 5, pp. 1164–1178, 2010.
3. A. P. Clark, *Equalizers for High Speed Modem*, London: Pentech Press, 1985.
4. L. Hanzo, S. X. Ng, T. Keller, and W. Webb, *Quadrature Amplitude Modulation: From Basics to Adaptive Trellis Coded, Turbo-Equalized and Space-Time Coded OFDM CDMA and MC-CDMA Systems*, 2nd ed., Chichester, England: John Wiley & Sons, 2004.
5. J. Qi, B. Mao, N. Gonzalez, L. N. Binh, and N. Stojanovic, Generation of 28 GBaud and 32 GBaud PDM-Nyquist-QPSK by a DAC with 11.3 GHz analog bandwidth, in *Proceedings of OFC 2013*, San Francisco, CA, USA, March 2013.

6. P. J. Winzer, A. H. Gnauck, C. R. Doerr, M. Magarini, and L. L. Buhl, Spectrally efficient long-haul optical networking using 112-Gbps polarization-multiplexed 16-QAM, *J. Lightwave Technol.*, Vol. 28, pp. 547–556, 2010.

7. M. Camera, B. Olsson, and G. Bruno, Beyond 100 Gbit/s: System implications towards 400 G and 1T, in *ECOC'10*, Th10G1, 2010.

8. G. Bosco, V. Curri, A. Carena, and P. Poggiolini, On the performance of Nyquist-WDM terabit superchannels based on PM-BPSK, PM-QPSK, PM-QPSK, PM-8QAM or PM-16QAM subcarriers, *J. Lightwave Technol.*, Vol. 29, pp. 53–61, 2011.

9. X. Zhou, L. E. Nelson, P. Magill, R. Isaac, B. Y. Zhu, D. W. Peckham, P. I. Borel, and K. Carlson, PDM-Nyquist-32QAM for 450 Gbps per-channel WDM transmission on the 50 GHz ITU-T grid, *J. Lightwave Technol.*, Vol. 30, pp. 553–559, 2012.

10. R. Schmogrow, M. Meyer, S. Wolf, B. Nebendahl, D. Hillerkuss, B. Baeuerle, and M. Dreschmann, 150 Gbit/s real-time Nyquist pulse transmission over 150 km SSMF enhanced by DSP with dynamic precision, in *OFC'12*, OM2A.6, 2012.

11. J. Winters, Equalization in coherent lightwave systems using a fractionally spaced equalizer, *IEEE J. Lightwave Technol.*, Vol. 8, No. 10, pp. 1487–1491, 1990.

12. R. W. Lucky, J. Salz, and E. J. Weldon, *Principles of Data Communications*, New York: McGraw-Hill, 1968, pp. 93–165.

13. A. Papoulis, *Prophability, Random Variables and Stochastic Processes*, New York: McGraw-Hill, 1965.

14. S. Bennetto and E. Biglieri, *Principles of Digital Transmission with Wireless Applications*, New York: Kluwer Academic, 1999.

15. R. Noe, T. Pfau, M. El-Darawy, and S. Hoffmann, Electronic polarization control algorithms for coherent optical transmission, *IEEE J. Sel. Quant. Electron.*, Vol. 16, No. 5 pp. 1193–1199, 2010.

16. S. Zhang, P. Y. Kam, J. Chen, and C. Yu, A comparison of phase estimation in coherent optical PSK system, in *Photonics Global'08*, paper C3-4A-03, Singapore, December 2008.

17. A. P. Clark, *Equalizers for Digital Modem*, London: Pentec Press, 1985.

18. G. Nicholson, Probability of error for optical heterodyne DPSK system with quantum phase noise, *Electron. Lett.*, Vol. 20, No. 24, pp. 1005–1007, 1984.

19. J. P. Gordon and L. F. Mollenauer, Phase noise in photonic communications systems using linear amplifiers, *Opt. Lett.*, Vol. 15, pp. 1351–1353, 1990.

20. D.-S. Ly-Gagnon, S. Tsukamoto, K. Katoh, and K. Kikuchi, Coherent detection of optical quadrature phase-shift keying signals with carrier phase estimation, *J. Lightwave Technol.*, Vol. 24, pp. 12–21, 2006.

21. X. Liu, X. Wei, R. E. Slusher, and C. J. McKinstrie, Improving transmission performance in differential phase-shift-keyed systems by use of lumped nonlinear phase-shift compensation, *Opt. Lett.*, Vol. 27, No. 18, pp. 1351–1353, 2002.

22. X. Wei, X. Liu, and C. Xu, Numerical simulation of the SPM penalty in a 10-Gbps RZ-DPSK system, *IEEE Photon. Technol. Lett.*, Vol. 15, No. 11, pp. 1636–1638, 2003.

23. E. Ip and J. M. Kahn, Digital equalization of chromatic dispersion and polarization mode dispersion, *IEEE J. Lightwave Technol.*, Vol. 25, pp. 2033–2043, 2007.

24. R. Noe, PLL-free synchronous QPSK polarization multiplex/diversity receiver concept with digital I&Q baseband processing, *IEEE Photon. Technol. Lett.*, Vol. 17, pp. 887–889, 2005.

25. L. N. Binh, *Digital Optical Communications*, Boca Raton, FL: CRC Press, 2009.

26. J. G. Proakis, *Digital Communications*, 4th ed., New York: McGraw-Hill, 2001.

27. Y. Han and G. Li, Coherent optical communication using polarization multiple-input-multiple output, *Opt. Express*, Vol. 13, pp. 7527–7534, 2005.

28. N. Stojanovic, Y. Zhao, B. Mao, C. Xie, F. N. Hauske, and M. Chen, Robust carrier recovery in polarization division multiplexed receivers, in *OFC 2013*, Annaheim, CA, USA, March 2013.

29. H. Meyr et al., *Digital Communication Receivers*, New York: John Wiley & Sons, 1998.

30. A. Leven et al., Frequency estimation in intradyne reception, *IEEE Photon. Technol. Lett.*, Vol. 19, pp. 366–368, 2007.

31. Z. Tao et al., Simple, robust, and wide-range frequency offset monitor for automatic frequency control in digital coherent receivers, in *ECOC 2007*, paper Tu3.5.4, Berlin, Germany, 2007.

32. A. J. Viterbi and A. M. Viterbi, Nonlinear estimation of PSK-modulated carrier phase with application to burst digital transmission, *IEEE Trans. Inf. Theory*, Vol. 29, pp. 543–551, 1983.

33. M. Kuschnerov et al., DSP for coherent single-carrier receivers, *J. Lightwave Technol.*, Vol. 27, pp. 3614–3622, 2009.

34. N. Stojanović, C. Xie, Y. Zhao, B. Mao, N. Guerrero Gonzalez, J. Qi, and L. N. Binh, Modified gardner phase detector for Nyquist coherent optical transmission systems, in *Proceedings of OFC 2013*, paper JTh2A.50.pdf, San Francisco, CA, USA, March 2013.
35. N. Stojanovic, C. Xie, F. N. Hauske, and M. Chen, Clock recovery in coherent optical receivers, in *Proceedings of Photonische Netze*, paper P9, Leipzig, Germany, May 2011.
36. N. Godard, Passband timing recovery in an all-digital modem receiver, *IEEE Trans. Commun.*, Vol. 26, pp. 517–523, 1978.
37. F. M. Gardner, A BPSK/QPSK timing error detector for sampled receivers, *IEEE Trans. Commun.*, Vol. 34, pp. 423–429, 1986.
38. T. T. Fang and C. F. Liu, Fourth-power law clock recovery with prefiltering, in *Proceedings of ICC*, Geneva, Switzerland, vol. 2, pp. 811–815, May 1993.
39. H. Meyr, M. Moeneclaey, and S. A. Fechtel, *Digital Communication Receivers*, New York: John Wiley & Sons, 1998, chap. 2.
40. D. H. Sun and K.-T. Wu, A novel dispersion and PMD tolerant clock phase detector for coherent transmission systems, in *Proceedings of OFC*, paper OMJ.4, Los Angeles, CA, USA, March 2011.
41. N. Stojanović, C. Xie, Y. Zhao, B. Mao, and N. G. Gonzalez, A circuit enabling clock extraction in coherent receivers, in *Proceedings of ECOC*, paper P3.08, Amsterdam, Holland, September 2012.
42. L. N. Binh, T. L. Huynh, K. K. Pang, and T. Sivahumaran, MLSE Equalizers for Frequency Discrimination Receiver of MSK Optical Transmission System, *IEEE J. Lightwave Technol.*, Vol. 26, No. 12, pp. 1586–1595, 2008.
43. T. Sivahumaran, T. L. Huynh, K. K. Pang, and L. N. Binh, Non-linear equalizers in narrowband filter receiver achieving 950 ps/nm residual dispersion tolerance for 40 Gbps optical MSK transmission systems, in *Proceedings of OFC'07*, paper OThK3, 2007.
44. P. Poggiolini, G. Bosco, M. Visintin, S. J. Savory, Y. Benlachtar, P. Bayvel, and R. I. Killey, MLSE-EDC versus optical dispersion compensation in a single-channel SPM-limited 800 km link at 10 Gbit/s, in *ECOC'07*, paper 1.3, Berlin, 2007.
45. N. Alic, G. C. Papen, and Y. Fainman, Performance of maximum likelihood sequence estimation with different modulation formats, in *Proceedings of LEOS'04*, pp. 49–50, 2004.
46. T. Sivahumaran, T. L. Huynh, K. K. Pang, and L. N. Binh, Non-linear equalizers in narrowband filter receiver achieving 950 ps/nm residual dispersion tolerance for 40 Gbps optical MSK transmission systems, in *Proceedings of OFC'07*, paper OThK3, CA, 2007.
47. V. Curri, R. Gaudino, A. Napoli, and P. Poggiolini, Electronic equalization for advanced modulation formats in dispersion-limited systems, *Photon. Technol. Lett.*, Vol. 16, No. 11, pp. 2556–2558, 2004.
48. G. Katz, D. Sadot, and J. Tabrikian, Electrical dispersion compensation equalizers in optical long-haul coherent-detection system, in *Proceedings of ICTON'05*, paper We.C1.5, 2005.
49. Y. Zhao, N. Stojanovic, D. Chang, C. Xie, B. Mao, L. N. Binh, Z. Xiao, and F. Yu, Adaptive joint carrier recovery and turbo decoding for Nyquist Terabit optical transmission in the presence of phase noise, in *Proceedings of OFC 2014*, San Francisco, CA, USA, March 2014.
50. X. Liu et al., 406.6-Gbps PMD-BPSK superchannel transmission over 12,800-km TWRS fiber via nonlinear noise squeezing, in *OFC 2013*, PDP5B.10, 2013.
51. J. Fickers et al., Design rules for pulse shaping in PDM-QPSK and PDM-16QAM Nyquist systems, in *ECOC 2012*, We.1.C.2, 2012.
52. G. Bosco et al., Performance limits of Nyquist-WDM and CO-OFDM in high-speed PM-QPSK systems, *IEEE Photon. Technol. Lett.*, Vol. 22, No. 15, p. 1129–1131, 2010.
53. F. Yu et al., Soft-decision LDPC turbo decoding for DQPSK modulation in coherent optical receivers, in *ECOC 2011*, We.10.P1.70, 2011.
54. Y. Zhao et al., Beyond 100 G optical channel noise modeling for optimized soft-decision FEC performance, in *OFC 2012*, OW1H.3, 2012.
55. I. Djordjevic et al., On the LDPC-coded modulation for ultra-high-speed optical transport in the presence of phase noise, in *OFC 2013*, OM2B.1, 2013.
56. T. Tanimura et al., Co-operation of digital nonlinear equalizers and soft-decision LDPC FEC in nonlinear transmission, in *ECOC 2012*, Mo.3.D.3, 2012.
57. C. Xie et al., Adaptive carrier phase estimation in coherent systems, in *OFC 2012*, OTu2G.5, 2012.

15 Optical Soliton Transmission System

Optical solitons have drawn considerable attention because of its fundamental nature as well as its potential applications for optical fiber communications systems operating at ultra-high bit rates and over extremely long distances. They are analyzed and discussed in this chapter.

The fundamental and higher-order solitons are simulated by employing the numerical approach using the beam propagation method (BPM) and the analytical approach using inverse scattering method (ISM) to solve the basic nonlinear propagation equation, also known as the nonlinear Schrödinger equation (NLSE). Solitons up to fifth order are observed and the "soliton breakdown" effects are discussed. The interaction of solitons of two and three fundamental soliton pulses is simulated using the numerical approach, and its behavior is being observed. A number of methods are suggested to control the soliton interaction.

Furthermore, the optical generators and detectors from mode-locked fiber lasers (MLFLs) are described, and both periodic pulse sequence and multibound solitons are demonstrated. Experimental demonstration of the generation of dual, triple, and quadruple solitons is presented.

15.1 INTRODUCTION

A fascinating manifestation of the fiber nonlinearity occurs in the anomalous dispersion regime where the fiber can support optical solitons through the interplay between the dispersive and nonlinear effects. The term *soliton* refers to special kinds of waves that can propagate undistorted over long distances and remain unaffected after collision with each other. Solitons have been studied extensively in several fields of physics. In the context of optical fibers, solitons are not only of fundamental interest but also has potential application in the field of optical fiber communications. The word *soliton* was coined in 1965 [1] to describe the particle-like properties of pulse envelopes in dispersive nonlinear media.

The basic concepts behind fiber solitons and its basic propagation equation, known as the NLSE, are introduced in Sections 15.3 and 15.4. Numerical approach using BPM to solve the Schrödinger equation is discussed in detail in Section 15.3, and is confirmed in experimental demonstration in Section 15.8.1.4. Analytical approach using the ISM to solve the Schrödinger equation is discussed in detail in Section 15.4. The results obtained through these two approaches are compared and verified.

The results of the numerical approach are discussed in Section 15.5. In this section, the behaviors of fundamental and higher-order solitons are observed. In Section 15.6, soliton interactions are simulated and discussed. Techniques for controlling the soliton interaction effect are then suggested.

Section 15.7 revisits the ISM for the simulation of optical solitons, in particular the interaction between sequential pulses for optical communication systems. Steps for interaction and reconstruction of solitons are described, which are proven to be the most accurate techniques for the design and control of optical solitons' pulses. All the results of our synthesis for optical solitons' propagation and detection techniques for both bright and dark solitons are summarized, and suggestions are made.

Section 15.8 presents the experimental demonstration of the generation of dual-, triple-, and quadruple-bound solitons in an MLFL and their dynamics when circulating around the fiber rings after several thousand times.

We would like to note here that three distinct terms "three solitons," "triple soliton," and "triple-bound solitons" have been used in this chapter, which should not be confused.

The term "three solitons" is used when the solitons are not grouped together or to indicate the general case of the interactions of the three solitons into triple solitons or triple-bound solitons. The term "triple-bound state" indicates the case when the phases of the carriers under the envelope of the three solitons are p and the solitons are in the bound state. Finally, the term "triple soliton" describes the state of three solitons which are about to bind to each other or when they are about to be separated again from the bound state.

15.2 FUNDAMENTALS OF NONLINEAR PROPAGATION THEORY

Similar to all electromagnetic phenomena, the propagation of optical fields either in linear or nonlinear regimes in fibers is governed by Maxwell's equations. From Maxwell's equations, one can obtain [2],

$$\nabla \times \nabla \times E = -\frac{1}{c^2}\frac{\partial^2 E}{\partial t^2} - \mu_0 \frac{\partial^2 P}{\partial t^2} \tag{15.1}$$

where E is the electric field and P is the induced polarization. The nonlinear effects in the optical fibers involve the use of short pulses with widths ranging from 10 ns to 10 fs. When such optical pulses propagate inside the fiber, both dispersive and nonlinear effects influence their shape and spectrum. Thus, Equation 15.1 can be further developed [2] to cater to these effects

$$\nabla^2 E - \frac{1}{c^2}\frac{\partial^2 E}{\partial t^2} = -\mu_0 \frac{\partial^2 P_{\mathrm{L}}}{\partial t^2} - \mu_0 \frac{\partial^2 P_{\mathrm{NL}}}{\partial t^2} \tag{15.2}$$

where PL and PNL are the linear and nonlinear parts of the induced polarization, respectively. Using the method of separation of variables to solve Equation 15.2 and taking perturbation theory into consideration, we could obtain the following relationship

$$\frac{\partial A}{\partial z} + \beta_1 \frac{\partial A}{\partial t} + \frac{j}{2}\beta_2 \frac{\partial^2 A}{\partial t^2} + \frac{\alpha}{2}A = j\gamma|A|^2 A \tag{15.3}$$

where the envelope A is a slowly varying function of z; β_1 and β_2 are the first- and second-order derivatives of β with respect to the optical frequency ω as defined in Chapter 2; γ is the fiber nonlinear coefficient; and α is the intrinsic fiber attenuation express in dB/km. Taking higher-order dispersion effect, stimulated inelastic scattering effect, and Raman gain effect into consideration together with perturbation approach leads to three additional terms in Equation 15.3. The generalized propagation equation takes the form

$$\frac{\partial A}{\partial z} + \frac{\alpha}{2}A + \frac{j}{2}\beta_2 \frac{\partial^2 A}{\partial T^2} - \frac{1}{6}\beta_3 \frac{\partial^3 A}{\partial T^3} = j\gamma\left\{|A|^2 A + \frac{2j}{\omega_0}\frac{\partial}{\partial T}\left(|A|^2 A\right) - T_{\mathrm{R}} A\frac{\partial|A|^2}{\partial T}\right\} \tag{15.4}$$

where ω_0 is the carrier frequency and T_{R} is related to the slope of the Raman scattering gain. Notice that β_1 has disappeared because of the transformation T, that is, the observation frame is situated on the soliton itself, which is given by

$$T = t - \beta_1 z \tag{15.5}$$

Before beginning to simulate the evolution of solitons using either numerical approach or analytical approach, a few important terms of soliton systems need to be clarified. It is useful to write Equation 15.4 in a normalized form by introducing

$$\tau = \frac{t - \beta_1 z}{T_0} \tag{15.6}$$

$$\xi = \frac{z}{L_D} \tag{15.7}$$

$$U = \frac{A}{\sqrt{P_O}} \tag{15.8}$$

where T_o is the soliton pulse width, P_o is the soliton peak power, and the dispersion length LD is defined as

$$L_D = \frac{T_O^2}{|\beta_2|} \tag{15.9}$$

where β_2 is obtained from the total dispersion factor of the optical fiber (see Chapter 2), which is given by

$$\beta_2 = -\left(\frac{D_T \lambda^2}{2\pi c}\right) \tag{15.10}$$

The order of the soliton, N, is defined as

$$N^2 = \frac{\gamma P_O T_O^2}{|\beta_2|} \tag{15.11}$$

The fundamental soliton corresponds to $N = 1$. Thus, in order to maintain $N = 1$, the soliton pulse needs to have its peak power, so that the square root of Equation 15.11 will have an integer value. The fundamental soliton has a secant shape, and its initial amplitude is given by

$$U(0,\tau) = \sec h(\tau) \tag{15.12}$$

This pulse shape is to be launched into the fiber. Its shape remains unchanged during propagation. However, for $N > 1$, such input shape is recovered at each *soliton period* defined as

$$z_O = \frac{\pi}{2} L_D = \frac{\pi}{2} \frac{T_O^2}{|\beta_2|} \tag{15.13}$$

The soliton period z_o and the soliton order N play an important role in the theory of optical solitons.

15.3 NUMERICAL APPROACH

15.3.1 BEAM PROPAGATION METHOD

The propagation equation (Equation 15.4) is a nonlinear partial differential equation that does not generally lend itself to analytic solutions except for some specific cases in which the ISM (Section 15.10.4) can be employed. A numerical approach is therefore often necessary for an understanding of the nonlinear effects in optical fibers. The split-step Fourier method (SSFM), described in Chapter 2, is used to solve the pulse propagation in the nonlinear guided wave medium. Thus, the evolution of solitons can be observed at any time span along the propagation distance. The beam propagation method (BPM) using the split-step method has been presented in detail in Chapter 3. The SSFM is extensively implemented in this chapter for soliton transmission, as demonstrated in Section 15.8.1.4.

15.3.2 ANALYTICAL APPROACH—ISM

The ISM is similar in spirit to BPM, which is commonly used to solve linear partial differential equations. The incident field at $z = 0$ is used to obtain the initial scattering data whose evolution along z is easily obtained by solving the linear scattering problem. The propagated field is reconstructed from the evolved scattering data. The details of the ISM are available in many texts [1,2], and we describe only briefly how this method is used to solve NLSE, given in Equation 15.4.

Using the normalized parameters in Equations 15.1 through 15.13, the parameter N can be eliminated from Equation 15.4 by defining

$$u = NU = \sqrt{\left(\frac{\gamma T_0^2}{|\beta_2|}\right)} A \tag{15.14}$$

and the NLSE becomes

$$j\frac{\partial u}{\partial \xi} + \frac{1}{2}\frac{\partial^2 u}{\partial \tau^2} + |u|^2 u = 0 \tag{15.15}$$

In the ISM, the scattering problem associated with Equation 15.15 is

$$\frac{\partial v_1}{\partial \tau} + j\zeta v_1 = uv_2 \tag{15.16}$$

$$\frac{\partial v_2}{\partial \tau} - j\zeta v_2 = -u^* v_1 \tag{15.17}$$

where v_1 and v_2 are the amplitudes of the waves scattering in a potential $u(\xi, \tau)$, and ζ is the eigenvalue. The soliton order is characterized by the number N of poles or eigenvalues ζ_j ($j = 1$ to N). The general solution is [3]

$$u(\xi, \tau) = -2\sum_{j=1}^{N} \lambda_j^* \psi_{2j}^* \tag{15.18}$$

where

$$\lambda_j = \sqrt{c_j} \exp\left(j\zeta_j \tau + j\zeta_j^2 \xi\right) \tag{15.19}$$

and ψ_{2j} is obtained by solving the linear set of the following equations

$$\psi_{1j} + \sum_{k=1}^{N} \frac{\lambda_j \lambda_k^*}{\zeta_j - \zeta_k^*} \psi_{2k}^* = 0 \tag{15.20}$$

$$\psi_{2j}^* - \sum_{k=1}^{N} \frac{\lambda_k \lambda_j^*}{\zeta_j^* - \zeta_k} \psi_{1k} = \lambda_j^* \tag{15.21}$$

The eigenvalues ζ_j is given by

$$\zeta_j = i\eta_j \tag{15.22}$$

and Equation 15.19 becomes

$$\lambda_j = \sqrt{c_j}\, \exp\!\left(-\eta_j \tau - j\eta_j^2 \xi\right) \tag{15.23}$$

Assuming the soliton is symmetrical about $\tau = 0$, the residues are related to the eigenvalues by the relation

$$c_j = \frac{\displaystyle\prod_{k=1}^{N}(\eta_j + \eta_k)}{\displaystyle\prod_{\substack{k \neq j}}^{N}\left|(\eta_j - \eta_k)\right|} \tag{15.24}$$

The fundamental soliton corresponds to the case of a single eigenvalue η_1, which has a value of 0.5 for $N = 1$. In general, for $N > 1$, the eigenvalues are given by

$$\eta_i = \frac{2i-1}{2} \tag{15.25}$$

where $i = 1, 2, 3,...,N$.

15.3.2.1 Soliton $N = 1$ by Inverse Scattering

The fundamental soliton has only one eigenvalue, which is shown in Equation 15.21. Thus, we obtain $\eta_1 = 0.5$. The residue that is related to the eigenvalue can be found by using Equation 15.20. Then, the eigenvalue and its residue are substituted in Equation 15.19 in order to obtain the eigenfunction λ. The complex eigenvalue can be obtained by using Equation 15.22. The eigenpotential ψ_{21} shown in Equations 15.20 and 15.21 needs to be solved simultaneously, and the process of solving this problem is described in the following. From Equations 15.20 and 15.21, we get

$$\psi_{11} + \frac{\lambda_1 \lambda_1^*}{\zeta_1 - \zeta_1^*}\, \psi_{21}^* = 0 \tag{15.26}$$

$$\psi_{21}^* = \lambda_1^* + \frac{\lambda_1 \lambda_1^*}{\zeta_1^* - \zeta_1}\, \psi_{11} \tag{15.27}$$

Substituting Equation 15.22 in Equation 15.23, we obtain

$$\psi_{21}^* = \frac{\zeta_1 - \zeta_1^*}{\lambda_1 \lambda_1^*}\left(\frac{\lambda_1 \lambda_1^{*2}\left(\zeta_1^* - \zeta_1\right)}{\left(\zeta_1^* - \zeta_1\right)\left(\zeta_1 - \zeta_1^*\right) + \left(\lambda_1 \lambda_1^*\right)^2}\right) \tag{15.28}$$

Thus, by substituting the eigenpotential and eigenfunction in Equation 15.18, we can obtain the fundamental soliton function $U(\xi,\tau)$.

15.3.2.2 Soliton $N = 2$ by Inverse Scattering

All the steps of solving the NLSE are the same, as listed in Section 15.4.1, except that the eigenvalues, eigenfunction, and eigenpotential have to be redefined. Solving the eigenvalues and the corresponding eigenfunction is straightforward. However, solving the eigenpotentials need more effort. Again, from Equations 15.18 and 15.19, we obtain the following expressions

$$\psi_{11} + \left(\frac{\lambda_1 \lambda_1^*}{\zeta_1 - \zeta_1^*}\, \psi_{21}^* + \frac{\lambda_1 \lambda_2^*}{\zeta_1 - \zeta_2^*}\, \psi_{22}^*\right) = 0 \tag{15.29}$$

$$\psi_{21}^* - \left(\frac{\lambda_1 \lambda_1^*}{\zeta_1^* - \zeta_1} \psi_{11} + \frac{\lambda_1 \lambda_2^*}{\zeta_1^* - \zeta_2} \psi_{12} \right) = \lambda_1^* \qquad (15.30)$$

$$\psi_{12} + \left(\frac{\lambda_2 \lambda_1^*}{\zeta_2 - \zeta_1^*} \psi_{21}^* + \frac{\lambda_2 \lambda_2^*}{\zeta_2 - \zeta_2^*} \psi_{22}^* \right) = 0 \qquad (15.31)$$

$$\psi_{22}^* - \left(\frac{\lambda_2 \lambda_1^*}{\zeta_2^* - \zeta_1} \psi_{11} + \frac{\lambda_2 \lambda_2^*}{\zeta_2^* - \zeta_2} \psi_{12} \right) = \lambda_2^* \qquad (15.32)$$

There are four unknown variables and four equations. Thus it is possible to solve them simultaneously. We can represent Equations 15.29 through 15.32 in a matrix form

$$
\begin{pmatrix} \psi_{11} \\ \psi_{12} \\ \psi_{21}^* \\ \psi_{22}^* \end{pmatrix}
=
\begin{pmatrix}
0 & 0 & \frac{\lambda_1 \lambda_1^*}{\zeta_1 - \zeta_1^*} \psi_{21}^* & \frac{\lambda_1 \lambda_2^*}{\zeta_1 - \zeta_2^*} \psi_{22}^* \\
0 & 0 & \frac{\lambda_2 \lambda_1^*}{\zeta_2 - \zeta_1^*} \psi_{21}^* & \frac{\lambda_2 \lambda_2^*}{\zeta_2 - \zeta_2^*} \psi_{22}^* \\
-\frac{\lambda_1 \lambda_1^*}{\zeta_1^* - \zeta_1} \psi_{11} & -\frac{\lambda_1 \lambda_2^*}{\zeta_1^* - \zeta_2} \psi_{12} & 0 & 0 \\
-\frac{\lambda_2 \lambda_1^*}{\zeta_2^* - \zeta_1} \psi_{11} & -\frac{\lambda_2 \lambda_2^*}{\zeta_2^* - \zeta_2} \psi_{12} & 0 & 0
\end{pmatrix}
$$

$$
\cdot \begin{pmatrix} \psi_{11} \\ \psi_{12} \\ \psi_{21}^* \\ \psi_{22}^* \end{pmatrix} + \begin{pmatrix} 0 \\ 0 \\ \lambda_1^* \\ \lambda_2^* \end{pmatrix} \qquad (15.33)
$$

Thus, by substituting the eigenpotential and eigenfunction in Equation 15.22, we can obtain the second order of the soliton function $U(\xi, \tau)$.

The results of both beam propagation (numerical approach) and ISM (analytical approach) are shown in Figures 15.1 and 15.2 for fundamental soliton and second-order soliton, respectively.

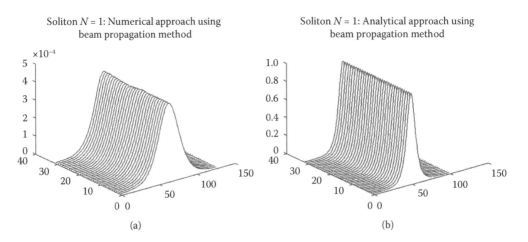

(a)   (b)

FIGURE 15.1 Comparison between (a) numerical approach and (b) analytical approach for soliton $N = 1$. Vertical axes of all diagrams: Amplitude in arbitrary units.

Soliton $N = 2$: Numerical approach using beam propagation method

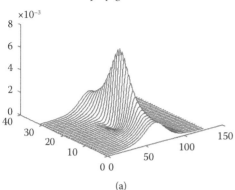

Soliton $N = 2$: Analytic approach using inverse scattering

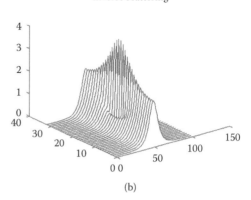

(a)   (b)

FIGURE 15.2 Comparison between (a) numerical approach and (b) analytical approach for soliton $N = 2$.

From the results, we notice that the soliton pulse is recovered at one soliton period. The different pulse widths observed is due to the different setting of the initial soliton pulse width. The results of the numerical approach agree with the analytical results, and hence we conclude that our numerical algorithm in synthesizing solitons' evolution has been a success.

15.4 FUNDAMENTAL AND HIGHER-ORDER SOLITONS

15.4.1 SOLITON EVOLUTION FOR $N = 1, 2, 3, 4$, AND 5

The evolutionary process of optical soliton is shown in Figures 15.1, 15.2, and 15.3 for the fundamental, second, third, fourth, and fifth orders of soliton, respectively. Periodic evolution of soliton is very much depending on the soliton period z_o and the soliton order N. As we can observe in these figures, the soliton pulse begins to evolve and its original shape will be recovered at every soliton period.

For a fundamental soliton, the pulse shape remains the same throughout the propagation distance. However, for soliton $N > 1$, the soliton pulse will be evolved into various shapes periodically, depending on the order of the soliton. Eventually, these distorted pulses will come back to the shape of the fundamental soliton at every soliton period. As for soliton $N = 2$, the pulse begins to contract, and the narrowing effect starts to take place and causes the peak to increase gradually. In the midpoint of the soliton period, the soliton peak has increased by three times (Figure 15.2—numerical results), and there are two minor components on each side of the main component of the soliton pulse.

For soliton $N = 3$, as the pulse propagates along the fiber, it first contracts to a fraction of its initial width, splits into two components, and then merges again to recover its original shape at the end of the soliton period. As for the soliton $N = 4$, the fundamental pulse shape splits into two components, and then further splits into three components. After reaching the midpoint of the soliton period, the three components' pulse begins to merge into two components, and finally recovers its original shape at the end of the soliton period.

Each plot is taken at a certain distance of propagation. Starting at 0 km, we have the fundamental soliton shape. Then it splits into two components at 400 km; further splitting into three components occurs at 600 km; and when it reaches the midpoint of the soliton period, which is at 1210 km, it splits into four components. After the midpoint, the four components begin to merge into three, and into two, and then into the original shape. Having analyzed the first five orders of the soliton, we notice that the maximum number of components split (N_c) at the midpoint is given by

$$N_c = N - 1 \qquad (15.34)$$

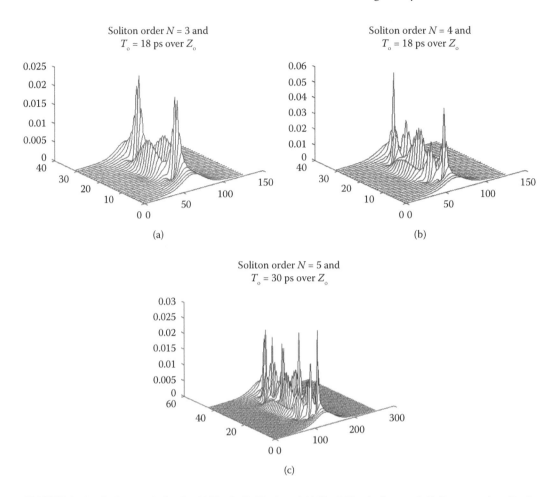

Soliton order $N = 3$ and $T_o = 18$ ps over Z_o

(a)

Soliton order $N = 4$ and $T_o = 18$ ps over Z_o

(b)

Soliton order $N = 5$ and $T_o = 30$ ps over Z_o

(c)

FIGURE 15.3 Soliton evolution for (a) $N = 3$, (b) $N = 4$, and (c) $N = 5$. Vertical axes of all diagrams: Amplitude in arbitrary units.

where N is the order of the soliton. For example, they are five split components at the midpoint of the soliton period for a soliton of order $N = 6$.

15.4.2 SOLITON BREAKDOWN

Another interesting observation of the soliton propagation is the soliton breakdown effect. Referring back to Figure 15.4 for soliton $N = 5$, it is realized that a soliton pulse width of 30 ps is used, which is larger than the pulse width of the lower-order solitons (only 18 ps). In Figure 15.5, we purposely assign a narrow pulse width of 9 ps (instead of 18 ps) to the fourth-order soliton, so that the soliton breakdown may be seen.

This proves that we need a substantial amount of soliton energy to support higher-order soliton in order for it to recover its own shape. Since the pulse width is related to the total power contained in the soliton, we would need to increase the pulse width for higher-order transmission. Otherwise, the soliton breakdown will occur where the soliton pulse will not recover to its original shape. At a certain point in the propagation distance, the pulse will split and may not merge back again. All these effects can be seen in Figure 15.5. Hence, there is a tradeoff between the order of the soliton and the bit rate. For high-bit-rate transmission, we need a narrow soliton pulse width. However, this would limit the order of the soliton that we can use.

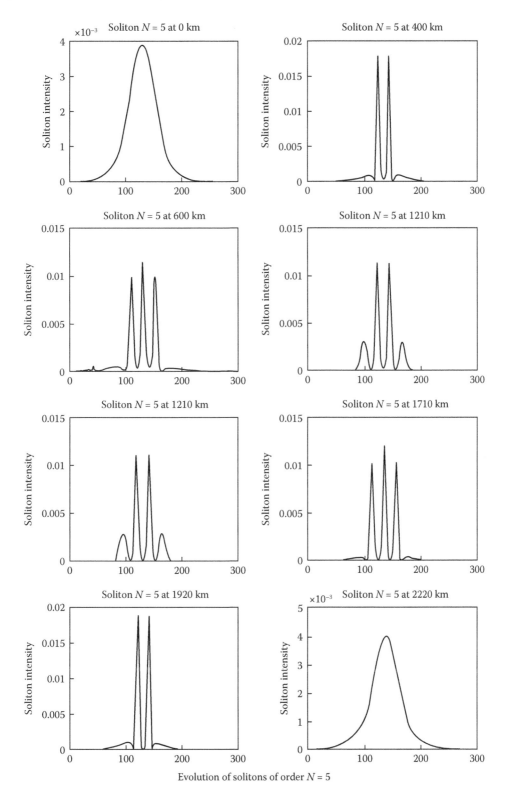

Evolution of solitons of order $N = 5$

FIGURE 15.4 Evolution of soliton of fifth order with respect to different propagation distances as noted in each individual graph. Vertical axes of all diagrams: Amplitude in arbitrary units.

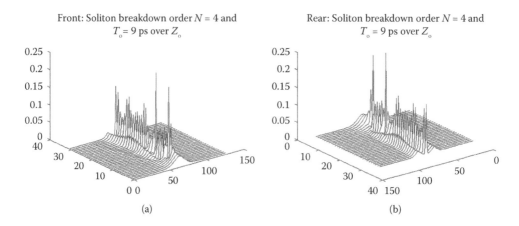

Front: Soliton breakdown order $N = 4$ and $T_o = 9$ ps over Z_o

Rear: Soliton breakdown order $N = 4$ and $T_o = 9$ ps over Z_o

(a)   (b)

FIGURE 15.5 Soliton breakdown phenomena: (a) front region and (b) rear region of the pulse sequence. Vertical axes of all diagrams: Amplitude in arbitrary units.

15.5 INTERACTION OF FUNDAMENTAL SOLITONS

15.5.1 TWO SOLITONS' INTERACTION WITH DIFFERENT PULSE SEPARATION

It is necessary to determine how close two solitons can come to each other without interacting. The nonlinearity that binds a single soliton introduces an interaction force between the neighboring solitons. The amplitude of the soliton pair at the fiber input can be written in the following normalized form

$$u(0,\tau) = \sec h \, (\tau - q_o) + r \sec h \, \{r \, (\tau + q_o)\} \exp \, (j\theta) \qquad (15.35)$$

where q_o is the initial pulse separation and r is the relative amplitude. Thus, soliton interaction can be studied by solving NLSE using the numerical approach, the BPM, discussed in Section 15.3.1. Figure 15.6 displays the evolution pattern showing periodic collapse of a soliton pair under various pulse separations.

The periodic collapse of neighboring solitons is undesirable from the system standpoint. One way to avoid the interaction problem is to increase the separation such that the collapse distance, Z_p, is much larger than the transmission distance LT. From the results in Figure 15.6, we can measure the collapse distance Z_p at each different pulse separation q_o and, thus, draw a rough estimation by interpolating these data. All the data are tabulated in Table 15.1, and the interpolated curve is shown in Figure 15.7.

Data are measured and collected from Figure 15.6. Note that Z_p is defined as the distance where the first collapse of solitons occurs. Thus, the curve shown in Figure 15.7 is useful as a guideline for the selection of optimum pulse separation, given a certain transmission distance. The pulse separation has to be minimized in order to achieve high-bit-rate transmission.

15.5.2 TWO SOLITONS' INTERACTION WITH DIFFERENT RELATIVE AMPLITUDE

Shown in Figure 15.8 is two solitons' interaction with different relative amplitudes. Referring to Equation 15.35, we can set different relative amplitudes, via the parameter r, to the two pulses, and the effect of setting an asymmetrical pair of solitons can be examined in the following.

From Figure 15.8, we observe that setting a slight difference between the amplitudes of the soliton pulses does solve the soliton collapse problem effectively. Another observation we obtain is that, as the relative amplitude increases, the periodical attraction between the two pulses happens more frequently. For example, in case of $r = 1.1$, the two pulses come closer to each other in three incidents, compared to four incidents for a given transmission distance in the case of $r = 1.3$. Thus, by setting a lightly different relative amplitude, we are able to overcome the problem of soliton pair collapse.

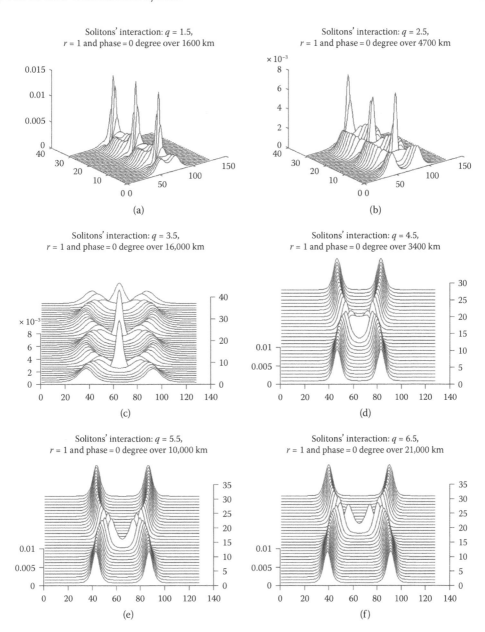

FIGURE 15.6 Dual solitons' interaction at different pulse temporal separation and no phase difference (see notes accompanying each figure): (a) $q = 1.5$, $r = 1$, no phase difference, propagation distance $L = 1600$ km; (b) $q = 2.5$, $r = 1$, no phase difference, $L = 4700$ km; (c) $q = 3.5$, $r = 1$, no phase difference, and $L = 16{,}000$ km; (d) $q = 4.5$, $r = 1$, no phase difference, and $L = 3400$ km; (e) $q = 5.5$, $r = 1$, no phase difference, and $L = 10{,}000$ km; (f) $q = 6.5$, $r = 1$, no phase difference, and $L = 21{,}000$ km. Vertical axes of all diagrams: Amplitude in arbitrary units.

TABLE 15.1

Corresponding Soliton Interaction Distance q_o and Soliton Widths and Propagation Distance Z_p

q_o	1.5	2.5	3.5	4.5	5.5	6.5
Z_p	310 km	910 km	1100 km	1700 km	4500 km	13,000 km

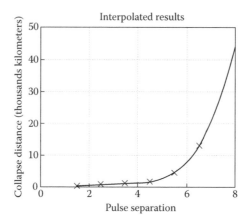

FIGURE 15.7 Interpolated data from Table 15.1.

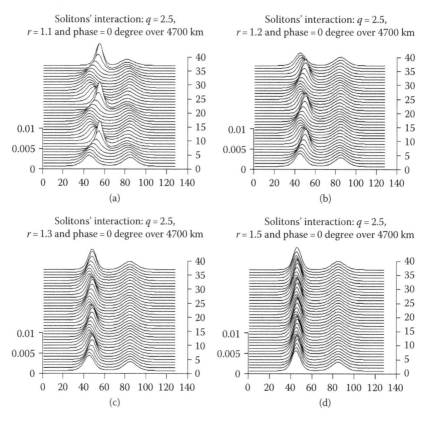

FIGURE 15.8 Two solitons' interaction of no phase difference at $L = 4700$ km with different relative amplitude (ratio r): (a) $r = 1$, (b) $r = 1.1$, (c) $r = 1.2$, and (d) $r = 1.5$. Vertical axes of all diagrams: Amplitude in arbitrary units.

15.5.3   TWO SOLITONS' INTERACTION UNDER DIFFERENT RELATIVE PHASES

As we can observe from the results obtained in Figure 15.9, the strong attractive force turns into a repulsive force for $\theta \neq 0$, such that the soliton pair is separated apart and may not merge back.

Notice that the relative phase of 180° could maintain the oscillation between the soliton pair without merging. The 90° relative phase soliton interaction is in fact the mirror image of the one

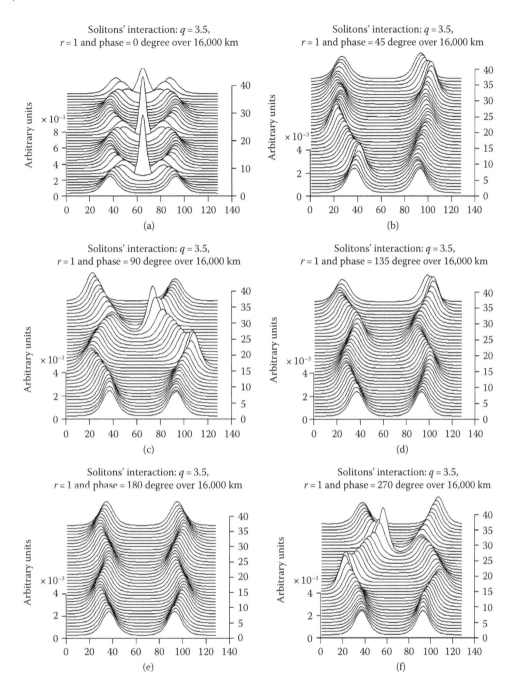

FIGURE 15.9 Two solitons' interaction under different relative phases of the carrier under the soliton envelope: (a) 0, (b) $\pi/4$, (c) $\pi/2$, (d) $3\pi/4$, (e) π, and (f) $3\pi/2$.

with 270°. Apart from those relative phases mentioned earlier, they contribute to soliton pair repulsion. These characteristics are tabulated in Table 15.2.

In conclusion, by adjusting the pulse separation, relative amplitude, and relative phase of the soliton pair, we could increase the bit rate or the transmission capacity. The collapsing problems of the soliton pair can be overcome effectively by varying those three parameters, so that stable soliton transmission system can be achieved.

TABLE 15.2

Characteristic of Soliton Pair Due to the Relative Phase Angle

Relative Phase	Observation and the Effect
0°	Soliton pair is merging periodically
180°	Soliton pair is oscillating periodically without merging
Others	Soliton pair repulsion

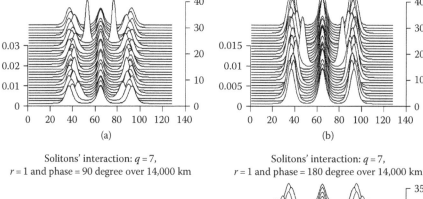

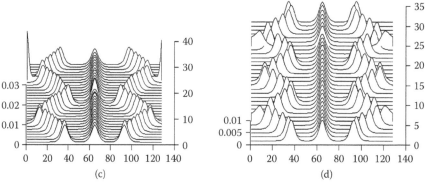

FIGURE 15.10 Three solitons' interaction at different relative phases between the adjacent solitons and the center soliton. (a) 0 rad., (b) π/4, (c) π/2, and (d) π. Vertical axes of all diagrams: Amplitude in arbitrary units.

15.5.4 TRIPLE SOLITONS' INTERACTION UNDER DIFFERENT RELATIVE PHASES

Similar to Equation 15.35, the amplitude of three solitons at the fiber input can be rewritten in the following normalized form

$$u(0,\tau) = \sec h\,(\tau) + r \sec h\,\{r\,(\tau - q_o) + r\,(\tau + q_o)\}\,\exp\,(j\theta) \qquad (15.36)$$

As for the case of three solitons' interaction, the observed characteristics are pretty much the same as those that have been discussed for the soliton pair in terms of the relative phase. In Figure 15.10, we notice that only when the relative phase difference of 180° exists between the centered soliton and its the two-sided solitons, the three solitons would not merge together.

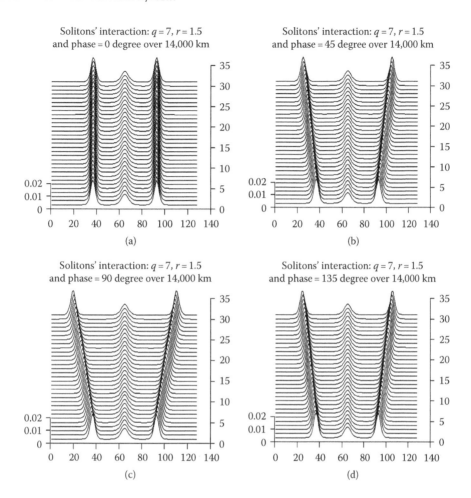

Solitons' interaction: $q = 7$, $r = 1.5$
and phase = 0 degree over 14,000 km

(a)

Solitons' interaction: $q = 7$, $r = 1.5$
and phase = 45 degree over 14,000 km

(b)

Solitons' interaction: $q = 7$, $r = 1.5$
and phase = 90 degree over 14,000 km

(c)

Solitons' interaction: $q = 7$, $r = 1.5$
and phase = 135 degree over 14,000 km

(d)

FIGURE 15.11   Three solitons' interaction at different relative phases between the adjacent and center solitons with different amplitude ratio of $r = 1.5$: (a) 0 rad., (b) $\pi/4$, (c) $\pi/2$, and (d) $3\pi/4$.

The oscillation of the three solitons would create severe problems at the receiving/detection subsystems. Thus, we must stabilize the oscillations by introducing a relative amplitude of $r = 1.5$ for the first and the third pulse described in the next section in Figure 15.11.

15.5.5 TRIPLE SOLITONS' INTERACTION WITH DIFFERENT RELATIVE PHASES AND $r = 1.5$

From Figures 15.11 and 15.12, we can see that, for a relative phase less than 180°, the side solitons are repelling away from the center soliton. The repulsive force reaches its maximum at 90° phase difference. Notice that the repulsive force exerted in the case of 45° and 135° are the same. When the relative phase is adjusted to 180°, there is neither attractive nor repulsive force at all. This is the same for the case when the relative phase = 0.

When the relative phase is greater than 180°, various interesting soliton dynamics can be observed. In Figure 15.12, we see that the two side solitons begin to merge together with the center one at a relative phase of 200°. Indeed, these solitons merge at the propagation distance of approximately 14,500 km (Figure 15.12). When the relative phase = 225°, the three solitons will be scattered apart at a propagation distance of about 10,000 km (Figure 15.12). Higher relative phase would still merge the solitons, and there is no incidence of any repulsive separation.

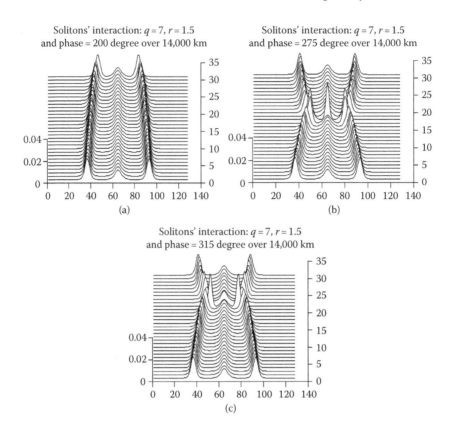

FIGURE 15.12 Three solitons' interaction at different relative phases with relative amplitude difference of $r = 1.5$ observed at $L = 14,000$ km: (a) 200° of arc, (b) 250°, and (c) 315°.

15.6   SOLITON PULSE TRANSMISSION SYSTEMS AND ISM

This section deals with the design of a four-bit soliton pulse train communication system. The dynamics of the optical solitons can be studied via the ISM. Of particular interest is the interaction between the neighboring solitons and its effect on the pulse separation. The design of an optimal four-bit soliton pulse train can then be verified by simulation based on MATLAB®. Obtained results indicate that the system design is functioning well. However, the question as to whether the highest bit rate is achieved, or not, remains unanswered. Furthermore, a re-examination of the phenomenon observed by the simulated results obtained in this section on the soliton breakdown due to "insufficient energy content" can be been proven to be false. The breakdown occurs due to improper adjustment of sampling parameters to accommodate higher-frequency components. A counter-example is shown whereby the breakdown of a fourth-order soliton did not occur even with narrow pulse widths.

We recall here again the wave equation governing optical soliton propagation (Equation 15.3). If this equation can be normalized to a simpler form, and the third-order term neglected, assuming that the operating wavelength is substantially far away from the zero-dispersion wavelength of the optical fiber, then

$$j\frac{\partial U}{\partial \xi} - \text{sgn}(\beta_2)\frac{1}{2}\frac{\partial^2 U}{\partial \tau^2} + N^2|U|^2 U = 0 \qquad (15.37)$$

N denotes the order of the soliton solution. This also indicates how many solitons can be present in the solution.

15.6.1 ISM REVISITED

The ISM was developed to deal with many nonlinear wave equations encountered in the field of physics. One such equation studied was the NLSE (this equation differs slightly from Equation 15.37 in the roles of t and x)

$$iq_t + q_{xx} + \chi\, |q|^2\, q = 0 \tag{15.38}$$

A particular solution of this equation is the soliton wave given as

$$u(x,t) = \sqrt{2\chi\eta}\, \frac{\exp\left[-4j\left(\xi^2 - \eta^2\right)t - 2j\xi x + j\varphi\right]}{\cosh\left[2\eta(x - x_0) + 8\eta\xi t\right]} \tag{15.39}$$

where the constant η characterizes the amplitude as well as the width of the soliton, ξ the velocity of the soliton (with respect to the group velocity), x_0 the initial position of the soliton center, and φ, its initial phase. For a more general solution, we can turn to the ISM.

Given initial boundary constraints, the aim of the ISM is to reconstruct the potential u in the NLSE from the asymptotic data that results from the scattering processes. An outline of the method is given in the following.

The route as presented shows that the ISM is broken up into three subproblems. This seems long and convoluted since later in the chapter there can be a more direct approach. However, one must note that the three steps involved in the inverse method deals with linear processes only, and this is the advantage of the ISM.

15.6.1.1 Step 1: Direct Scattering

The scattering data $S \pm (0)$ corresponding to the initial function $F(x)$ is first obtained by solving the NLSE for the solutions with the asymptotic form e^{-jkx} and e^{+jkx} as $x \to \pm\infty$.

15.6.1.2 Step 2: Evolution of the Scattering Data

Once the scattering data is obtained for $t = 0$, the evolution of the scattering data is determined uniquely for arbitrary time t by

$$C_1\left(\hat{L}^A\right)u_t + C_2\left(\hat{L}^A\right)u_x = 0, \quad x \in \Re, t \geq 0 \tag{15.40}$$

where C_1, C_2 are real analytic functions, and L is basically a reformulation of the NLSE as a linear operator.

15.6.1.3 Step 3: Inverse Spectral Transform

The potential $u(x,t)$ is then reconstructed from the scattering data $S \pm (t)$, usually by solving some system of linear integral equations. Returning to the NLSE (Equation 15.38), we can associate with this equation that the Lax pair

$$\hat{L} = i \begin{bmatrix} 1+p & 0 \\ 0 & 1-p \end{bmatrix} \frac{\partial}{\partial x} + \begin{bmatrix} 0 & u^* \\ u & 0 \end{bmatrix}, \quad \chi = \frac{2}{1-p^2}$$

$$\hat{A} = -p \begin{bmatrix} 1 & 0 \\ 0 & 1 \end{bmatrix} \frac{\partial^2}{\partial x^2} + i \begin{bmatrix} \dfrac{|u|^2}{1+p} & ju_x^* \\ -ju_x & \dfrac{-|u|^2}{1-p} \end{bmatrix} \tag{15.41}$$

and the L operator evolves in time according to the commutation relation

$$\frac{\partial \hat{L}}{\partial t} = \left[\hat{A}, \hat{L} \right] \tag{15.42}$$

Here, L and A are linear differential operators containing the sought function $u(x,t)$ in the form of a coefficient. By appropriate transformations of the eigenfunctions and rescaling of the eigenvalues, the pair could be transformed to resemble the form (see Ref. [4])

$$\hat{L} = j \begin{bmatrix} \dfrac{\partial}{\partial x} & -U(x,t) \\ R(x,t) & -\dfrac{\partial}{\partial x} \end{bmatrix} \tag{15.43}$$

$$\hat{A} = \begin{bmatrix} A(x,t,k) & B(x,t,k) \\ C(x,t,k) & -A(x,t,k) \end{bmatrix} + f(k)\mathbf{I}$$

The key observation of Ablowitz et al., [5] is that a set of nonlinear equations can be obtained from the transformed Lax pair as

$$A_x - UC + RB = 0$$

$$U_t - B_x - 2AU + 2jkB = 0 \tag{15.44}$$

$$R_t - C_x + 2AR + 2jkC = 0$$

This set of equations, in turn, can generate several classes of nonlinear wave equations. A technique for obtaining some of the associated equations is to substitute a polynomial in k for A, B, and C, and solve recursively. This technique reproduces, besides the NLSE, several other important equations. For example, one such equation is the *KdV* equation

$$U_t + 6UU_x + U_{xxx} = 0 \tag{15.45}$$

This equation was generated with the polynomials A, B, and C having the form

$$A = -4ik^3 + 2jU - U_x$$

$$B = -4Uk^2 + 2jU_x k - 2U^2 - U_{xx} \tag{15.46}$$

$$C = -4k^2 + 2U, \quad R = -1$$

15.6.2 ISM SOLUTIONS FOR SOLITONS

Following the scheme as presented in Figure 15.13, we proceed with Step 1, the direct scattering problem.

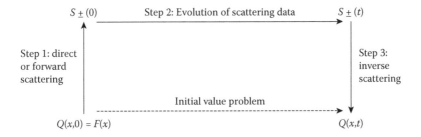

FIGURE 15.13 Steps of the inverse scattering method.

15.6.2.1 Step 1: Direct Scattering Problem

Let us return to the L operator associated with the NLSE. We consider the system of equations

$$\hat{L}\psi = \lambda\psi, \quad \psi = \left\{\begin{array}{c} \psi_1 \\ \psi_2 \end{array}\right\} \tag{15.47}$$

Making the change of variables,

$$\psi_1 = \sqrt{1-p}\,\exp\left(-j\frac{\lambda x}{1-p^2}\right)v_2, \quad \psi_2 = \sqrt{1+p}\,\exp\left(-j\frac{\lambda x}{1-p^2}\right)v_1 \tag{15.48}$$

this equation can be rewritten in the following form

$$\frac{\partial v_1}{\partial t} + j\zeta v_1 = qv_2, \quad \frac{\partial v_2}{\partial t} - j\zeta v_2 = -q^* v_1$$

$$q = \frac{ju}{\sqrt{1-p^2}}, \quad \zeta = \frac{\lambda p}{1-p^2} \tag{15.49}$$

We can define for real $\zeta = \xi$, the functions φ and ψ, as solutions of, with asymptotic values

$$\varphi \to \left\{\begin{array}{c} 1 \\ 0 \end{array}\right\}\exp(-j\xi x), \quad x \to -\infty$$

$$\psi \to \left\{\begin{array}{c} 0 \\ 1 \end{array}\right\}\exp(j\xi x), \quad x \to +\infty \tag{15.50}$$

The solution ψ and its adjoint

$$\bar{\psi} = \left\{\begin{array}{c} \psi_2^* \\ -\psi_1^* \end{array}\right\} \tag{15.51}$$

form a complete system of solutions, and therefore

$$\varphi = a(\xi)\bar{\psi} + b(\xi)\psi \tag{15.52}$$

Note that if ψ is a solution of the system at $\zeta = \xi + i\eta$, then its adjoint satisfies the system at $\zeta^* = \zeta - i\eta$. The points of the upper-half plane, $Im\ \zeta > 0$, $\zeta = \zeta_j$, $j = 1, ..., N$, at which $a(\zeta) = 0$, correspond to the eigenvalues of the problem. Here

$$\varphi(x,\zeta_j) = c_j\psi(x,\zeta_j), \quad j = 1, ..., N \tag{15.53}$$

It can be shown that if q is real, then the zeroes of $a(\zeta)$ lie on the imaginary axis.

15.6.2.2 Step 2: Evolution of the Scattering Data

Revisiting the commutation relation in Equation 15.49, we find that the eigenfunctions ψ obey the equation

$$i\frac{\partial\psi}{\partial t} = \hat{A}\psi \tag{15.54}$$

From this equation, the time dependencies of the coefficients $a(\xi,t)$, $b(\xi,t)$, and $c_j(t)$ can be found as

$$a(\xi,t) = a(\xi), \quad b(\xi,t) = b(\xi,0)\exp(4i\xi^2 t), \quad c_j(t) = c_j(0)\exp(4i\zeta_j^2 t) \tag{15.55}$$

15.6.2.3 Step 3: Inverse Scattering Problem

The task is now to reconstruct $u(x, t)$ from the scattering data $a(\xi)$, $b(\xi,0)$ for $-\infty < \xi < +\infty$ and $c_j(0)$, which we have denoted as $b(\zeta_j,0)$. The eigenfunctions are related to the residues at the poles ζ_j, and the coefficients $a(\xi)$, $b(\xi)$ via the following system of equations. See Ref. [2] for the complete derivation.

$$\phi_1 - c(x,\xi)\frac{1+J}{2}\phi_2^* = -c(x,\xi)\sum_{k=1}^{N}\frac{\exp(-i\zeta_k^* x)}{\xi - \zeta_k^*}c_k^*\psi_2^*(x,\zeta_k)$$

$$c^*(x,\xi)\frac{1-J}{2}\phi_1 + \phi_2^* = c^*(x,\xi) + c^*(x,\xi)\sum_{k=1}^{N}\frac{\exp(i\zeta_k x)}{\xi - \zeta_k}c_k\psi_1(x,\zeta_k)$$

$$\psi_1(x,\zeta_j)\exp(-i\zeta_j x) + \sum_{k=1}^{N}\frac{\exp(-i\zeta_k^* x)}{\zeta_j - \zeta_k^*}c_k^*\psi_2^*(x,\zeta_k) = \frac{1}{2}J\phi_2^* \tag{15.56}$$

$$-\sum_{k=1}^{N}\frac{\exp(i\zeta_k x)}{\zeta_j^* - \zeta_k}c_k\psi_1(x,\zeta_k) + \psi_2^*(x,\zeta_j)\exp(i\zeta_j^* x) = 1 + \frac{1}{2}J\phi_1$$

where $(J\phi)(\xi) = \dfrac{1}{\pi i}\displaystyle\int_{-\infty}^{+\infty}\frac{\phi(\xi')}{\xi' - \xi}d\xi'$, $\phi(\xi) = \dfrac{b(\xi)}{a(\xi)}\exp(i\xi x)\psi(x,\xi)$.

The next step is to find an expression for q because u is related to q by a simple constant, as given in Equation 15.49. This can be done by comparing the asymptotic forms of the preceding system of equations

$$\begin{Bmatrix} \psi_2(x,\zeta) \\ -\psi_1(x,\zeta) \end{Bmatrix}\exp(-i\zeta x) = \begin{Bmatrix} 1 \\ 0 \end{Bmatrix} + \frac{1}{\zeta}\left[\sum_{k=1}^{N}c_k^*\exp(-i\zeta_k^* x)\psi^*(x,\zeta_k) + \frac{1}{2\pi i}\int\phi^*(\xi)d\xi\right]$$

$$+ O\left(\frac{1}{\zeta^2}\right), \quad \zeta \to \infty \tag{15.57}$$

and the differential equation in Equation 15.49 results in

$$\psi(x,\zeta)\exp(-i\zeta x) = \left\{ \begin{array}{c} 0 \\ 1 \end{array} \right\} + \frac{1}{2i\zeta} \left\{ \begin{array}{c} q(x) \\ \displaystyle\int_x^\infty |q(s)|^2\, ds \end{array} \right\} + O\!\left(\frac{1}{\zeta^2}\right)$$ (15.58)

Thus, we get

$$q(x) = -2i\sum c_k^*\exp\!\left(-i\zeta_k^* x\right)\psi_2^*(x,\zeta_k) - \frac{1}{\pi}\int \phi_2^*(\xi)\,d\xi$$

$$\int_x^\infty |q(s)|^2\, ds = -2i\sum c_k\exp(i\zeta_k x)\psi_1(x,\zeta_k) + \frac{1}{\pi}\int \phi_1(\xi)\,d\xi$$ (15.59)

Once a and b are obtained for an initial value, $u(x, t = 0)$, then we can obtain $u(x, t)$ at an arbitrary instant through Equations 15.57 through 15.59. We can now attempt to find an explicit formula for the general solution of the NLSE.

15.6.3 N-Soliton Solution (Explicit Formula)

We consider the inverse scattering problem with the assumption $b(\xi,t) = 0$. Then, $\varphi(\xi) \equiv 0$, and the system of equations (Equation 15.58) reduces to

$$\psi_{1j} + \sum_{k=1}^N \frac{\lambda_j\lambda_k^*}{\zeta_j - \zeta_k^*}\psi_{2k}^* = 0$$

$$-\sum_{k=1}^N \frac{\lambda_k\lambda_j^*}{\zeta_j^* - \zeta_k}\psi_{1k} + \psi_{2j}^* = \lambda_j^*$$

$$\psi_j = \left\{ \begin{array}{c} \psi_{1j} \\ \psi_{2j} \end{array} \right\} = \sqrt{c_j}\,\psi(x,\zeta_j), \quad \lambda_j = \sqrt{c_j}\exp(i\zeta_j x)$$ (15.60)

Subsequently, the potential can also be simplified to

$$q(x) = -2i\sum_{k=1}^N \lambda_k^*\psi_{2k}^*, \quad \int_x^\infty |q(s)|^2\, ds = -2i\sum_{k=1}^N \lambda_k\psi_{1k}$$ (15.61)

To see how a soliton can arise from this system of equations, we examine the case $N = 1$, and $a(\zeta)$ has only one zero in the upper-half plane. The following system of equations can be obtained

$$\psi_1 + \frac{|\lambda|^2}{2i\eta}\psi_2^* = 0, \quad \frac{|\lambda|^2}{2i\eta}\psi_1 + \psi_2^* = \lambda^*$$ (15.62)

This system describes a soliton with amplitude η, and its initial center and phase is given by

$$x_0 = \frac{1}{2\eta}\ln\frac{|\lambda(0)|^2}{2\eta}, \quad \varphi = -2\arg\lambda(0)$$ (15.63)

Returning to the N-soliton case: in order to obtain the explicit solution for q and ultimately u, we need to solve the preceding system of equations (Equation 15.62) for ψ_j. If we represent the preceding equations in the matrix form, then the potential $u(x,t)$ can be shown [2] to be

$$\left|u(x,t)\right|^2 = \sqrt{2\chi}\,\frac{d^2}{dx^2}\ln\det\|\mathbf{A}\| = \sqrt{2\chi}\,\frac{d^2}{dx^2}\ln\det\|\mathbf{BB}^*+1\|$$

(15.64)

$$\mathbf{B}_{jk} = \frac{\sqrt{c_j c_k^*}}{\zeta_j - \zeta_k^*}\exp\left[i\left(\zeta_j - \zeta_k^*\right)x\right]$$

The preceding result is a formula for reconstructing the potential $u(x,t)$. At the present moment, it does not describe a solution to any real physical problem because we have not specified the initial values or boundary conditions. Therefore, let us examine the example from [6], where the initial potential has the form of a soliton, $A\mathrm{sech}(x)$.

Example: $U(x,t) = A\mathrm{sech}(x)$

Let us turn our attention to the NLSE. Note that, in this example, the equation is slightly different from Equation 15.45, given as

$$ju_t = \frac{1}{2}u_{xx} + |u|^2 u$$

(15.65)

We now consider the associated eigenvalue equation

$$jv_x + Uv = \zeta\sigma_3 v$$

$$v = \begin{pmatrix} v_1 \\ v_2 \end{pmatrix},\quad \sigma_3 = \begin{pmatrix} 1 & 0 \\ 0 & -1 \end{pmatrix},\quad U = \begin{pmatrix} 0 & u \\ u^* & 0 \end{pmatrix}$$

(15.66)

Note that, if u is a solution of Equation 15.67, then the eigenvalue ζ is independent of time and v evolves in time according to

$$iv_t = Av$$

$$A = \begin{pmatrix} 1 & 0 \\ 0 & 1 \end{pmatrix}\left(\frac{1}{2}\frac{\rho^2-1}{\rho^2+1}\frac{\partial^2}{\partial x^2} - i\zeta\frac{\partial}{\partial x} + C\right) + \frac{1}{\rho^2+1}\begin{pmatrix} \rho^2|u|^2 & iu_x \\ -j\rho^2 u_x^* & -|u|^2 \end{pmatrix}$$

(15.67)

where C is a constant independent of x. Proceeding with Step 1, we eliminate v_2 in Equation 15.67, and using the initial potential $u(x, t) = A$, sech x, we have

$$s(1-s)\frac{d^2}{ds^2}v_1 + \left(\frac{1}{2}-s\right)\frac{d}{ds}v_1 + \left[A^2 + \frac{\zeta^2 + j\zeta(1-2s)}{4s(1-s)}\right]v_1 = 0,\quad s = \frac{1-\tanh x}{2}$$

(15.68)

Further transformation of the dependent variable v_1 into $s^\alpha(1-s)^\beta\omega_1$ reduces the preceding equation to the hypergeometric function forms,

$$v_{11}(s) = s^{j\zeta/2}(1-s)^{-j\zeta/2}F\left(-A,A,j\zeta+\frac{1}{2};s\right)$$

(15.69)

$$v_{12}(s) = s^{\frac{1}{2}-j\zeta/2}(1-s)^{-j\zeta/2}F\left(\frac{1}{2}-j\zeta+A,\frac{1}{2}-j\zeta-A,\frac{3}{2}-j\zeta;s\right)$$

The solutions for v can be obtained by replacing ζ with $-\zeta$ in the preceding equation,

$$v_{21}(s) = s^{-j\zeta/2}(1-s)^{j\zeta/2} \, F\left(-A, A, -j\zeta + \frac{1}{2}; s\right)$$

$$v_{12}(s) = s^{\frac{1}{2}+j\zeta/2}(1-s)^{j\zeta/2} \, F\left(\frac{1}{2}+j\zeta+A, \frac{1}{2}+j\zeta-A, \frac{3}{2}+j\zeta; s\right)$$

(15.70)

From the asymptotic requirements of Equation 15.50, we can find expressions for the coefficients as

$$\psi = \begin{pmatrix} \dfrac{A}{\xi+j/2}v_{12} \\[2mm] v_{21} \end{pmatrix}, \quad \bar{\psi} = \begin{pmatrix} v_{21}^{*} \\[2mm] -\dfrac{A}{\xi-j/2}v_{12}^{*} \end{pmatrix}$$

$$a(\xi) = \frac{\left[\Gamma\left(-j\xi+\dfrac{1}{2}\right)\right]^{2}}{\Gamma\left(-j\xi+A+\dfrac{1}{2}\right)\Gamma\left(-j\xi-A+\dfrac{1}{2}\right)}$$

(15.71)

$$b(\xi) = \frac{j\left[\Gamma\left(j\xi+\dfrac{1}{2}\right)\right]^{2}}{\Gamma(A)\Gamma(1-A)} = j\frac{\sin(\pi A)}{\cosh(\pi\xi)}$$

The eigenvalues ζ_r are obtained from the zeroes of $a(\zeta)$.

$$\zeta_r = j\left(A - r + \frac{1}{2}\right)$$

(15.72)

r must be positive integers satisfying $A - r + \frac{1}{2} > 0$. Step 2 gives

$$\lambda_k = \sqrt{\frac{b(\zeta_k)}{\partial a(\zeta_k)/\partial t}}\,\exp\left[j\zeta_k x - j\zeta_k^2 t\right], \quad c(x,\xi) = \frac{b(\xi)}{a(\xi)}\exp\left[2j\xi x - 2j\xi^2 t\right]$$

(15.73)

Note the slight difference in the preceding expressions when compared with Equation 15.24. Step 3 is an application of Equations 15.58 through 15.63.

15.6.4 SPECIAL CASE $A = N$

When $A = N$, a positive integer, then, by Equation 15.73, the coefficients $a(\zeta)$ reduces to

$$a(\zeta) = \prod_{r=1}^{N} \frac{(\zeta - \zeta_r)}{\zeta - \zeta_r^{*}}$$

(15.74)

while $b(\xi) = 0$. At the eigenvalues,

$$b(\zeta_k) = j(-1)^{k-1}$$

(15.75)

The coefficients $c(x, \xi)$ are defined by Equation 15.73, and subsequently $\phi_1 = \phi_2^* = 0$. Substituting the results in Equation 15.58, we can solve the equations for ψ_{k1} and ψ_{k2}^*. Explicit solutions for the cases $N = 1$ and $N = 2$ are given in the following

$$u(x,t) = \exp(-jt/2)\operatorname{sech} x, \quad N = 1$$

$$u(x,t) = 4\exp(-jt/2)\frac{\cosh 3x + 3\exp(-4jt)\cosh x}{\cosh 4x + 4\cosh 2x + 3\cos 4t}, \quad N = 2$$

(15.76)

Of interest to us is the long-term dynamic behavior of the soliton, and this will be explored in the following section.

15.6.5 N-Soliton Solution (Asymptotic Form as $\tau \to \pm\infty$)

The long-term behavior of the N-soliton solution is studied here for the special case where there are no two solitons with the same velocity, that is, no ξ_j are the same. It is found that the N-soliton solution breaks up into diverging solitons as $t \to \pm\infty$. To verify this for the case $t \to +\infty$, let us arrange the ξ_j in decreasing order, $\xi_1 > \xi_2 > \ldots > \xi_N$. From Equation 15.57, we have

$$\lambda_j(x,t) = \lambda_j(0)\exp\left[-\eta_j\left(x + 4\xi_j t\right) + j\left(\xi_j x + 2(\xi_j^2 - \eta_j^2)t\right)\right]$$

$$\left|\lambda_j(x,t)\right| = \left|\lambda_j(0)\right|\exp(-\eta_j y_j), \quad y_j = x + 4\xi_j t$$

(15.77)

Let us now consider the reference frame where $y_m = $ constant, as $t \to +\infty$; that is, we are interested in a moving reference frame of one of the solitons. Then

$$y_j \to 0, \quad \left|\lambda_j\right| \to 0 \quad \text{for } j < m$$

$$y_j \to \infty, \quad \left|\lambda_j\right| \to \infty \quad \text{for } j > m$$

(15.78)

It follows from the system of equations (Equation 15.62) that $\psi_{1j}, \psi_{2j} \to 0$ when $j < m$; that is, the effect of these solitons are made negligible. The number of equations is therefore reduced to $2(N-m-1)$, given as

$$\psi_{1m} + \frac{\left|\lambda_m\right|}{2i\eta_m}\psi_{2m}^* = -\lambda_m \sum_{k=m+1}^{N} \frac{1}{\zeta - \zeta_k^*}\phi_{2k}^*$$

$$\frac{\left|\lambda_m\right|^2}{2i\eta_m}\psi_{1m} + \psi_{2m}^* = \lambda_m^* + \lambda_m^* \sum_{k=m+1}^{N} \frac{1}{\zeta^* - \zeta_k}\phi_{1k}$$

$$\sum_{k=m+1}^{N} \frac{1}{\zeta - \zeta_k^*}\phi_{2k}^* = -\frac{\lambda_m^*}{\zeta_j - \zeta_m^*}\psi_{2m}^*$$

$$\sum_{k=m+1}^{N} \frac{1}{\zeta^* - \zeta_k}\phi_{1k} = -1 - \frac{\lambda_m}{\zeta^* - \zeta_m}\psi_{1m}$$

(15.79)

Solving first for ϕ_{1k} *and* $\phi_{2k}{}^{*}$, we obtain

$$\phi_{1k} = a_k + \frac{2i\eta_m}{a_m} \frac{a_k}{\zeta_k - \zeta_m} \psi_{1m}\lambda_m$$

$$\phi_{2k}^{*} = -\frac{2j\eta_m}{a_m^{*}} \frac{a_k^{*}}{\zeta_k^{*} - \zeta_m^{*}} \psi_{2m}^{*}\lambda_m^{*}$$

$$(15.80)$$

$$a_k = \frac{\displaystyle\prod_{p=m+1}^{N} \zeta_k - \zeta_p^{*}}{\displaystyle\prod_{m<p\neq k}^{N} \zeta_k - \zeta_p}, \quad a_m = 2j\eta_m \prod_{p=m+1}^{N} \frac{\zeta_m - \zeta_p^{*}}{\zeta_m - \zeta_p}$$

After some manipulation (see Ref. [4] for details),

$$\psi_{1m} + \frac{\left|\lambda_m^{+}\right|^2}{2i\eta_m} \psi_{2m}^{*} = 0$$

$$(15.81)$$

$$\frac{\left|\lambda_m^{+}\right|^2}{2i\eta_m} \psi_{1m} + \psi_{2m}^{*} = \left(\lambda_m^{+}\right)^{*}, \quad \lambda_m^{+} = \lambda_m \prod_{p=m+1}^{N} \frac{\zeta_m - \zeta_p}{\zeta_m - \zeta_p^{*}}$$

The system coincides with the system (Equation 15.64) and describes a soliton with a displaced position of the center x_0^{+} and phase ϕ^{+},

$$x_{0m}^{+} - x_{0m} = \frac{1}{\eta_m} \sum_{p=m+1}^{N} \ln\left|\frac{\zeta_m - \zeta_p}{\zeta_m - \zeta_p^{*}}\right| < 0$$

$$(15.82)$$

$$\phi_m^{+} - \phi_m = -2 \sum_{p=m+1}^{N} \arg\left(\frac{\zeta_m - \zeta_p}{\zeta_m - \zeta_p^{*}}\right)$$

The calculations are similar for the case $t \to -\infty$. In the reference frames $y = x + \xi t$, where ξ does not coincide with any of the ξ_m as $t \to \pm\infty$, the reduced system tends to zero, hence proving the asymptotic breakdown of the N-soliton solution into solitons. When t changes from $-\infty$ to $+\infty$, the corresponding changes in the soliton centers and phases are

$$\Delta x_{0m} = x_{0m}^{+} - x_{0m}^{-} = \frac{1}{\eta_m}\left(\sum_{k=m+1}^{N} \ln\left|\frac{\zeta_m - \zeta_k}{\zeta_m - \zeta_k^{*}}\right| - \sum_{k=1}^{m-1} \ln\left|\frac{\zeta_m - \zeta_k}{\zeta_m - \zeta_k^{*}}\right|\right)$$

$$(15.83)$$

$$\Delta\phi_m = \phi_m^{+} - \phi_m^{-} = 2\sum_{k=1}^{m-1} \arg\frac{\zeta_m - \zeta_k}{\zeta_m - \zeta_k^{*}} - 2\sum_{k=m+1}^{N} \arg\frac{\zeta_m - \zeta_k}{\zeta_m - \zeta_k^{*}}$$

The formula can be interpreted by assuming that the solitons interact with each other pairwise. In each interaction, the faster soliton moves forward by the first product in the formula, and the slower one shifts backward by the second term in the formula. The total soliton shift is equal to the algebraic sum of its shifts during the paired collisions or interactions. The potential thus takes the form

$$u(x,t) = \sqrt{2\chi} \sum_{j=1}^{N} \eta_j \operatorname{sech}\left[2\eta_j\left(x - x_{0j}\right) + 8\eta_j\xi_j t\right] \exp\left[-4j\left(\xi_j^2 - \eta_j^2\right)t - 2j\xi_j x + j\phi_j\right] \quad (15.84)$$

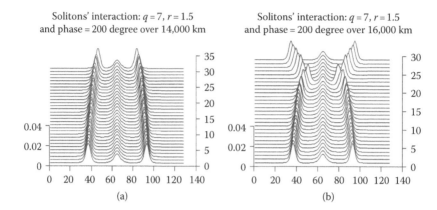

FIGURE 15.14   Triple solitons' interaction at (a) 14,000 km and (b) 16,000 km for 200° phase difference.

which is the sum of all the individual modal solitons. Note that this conclusion was arrived at employing the assumption that $b(\xi)$ vanishes. This may not always be true, but Satsuma and Yajima [6] showed, using methods of complex analysis, that the long-term solution still breaks down into individual solitons.

15.6.6 BOUND STATES AND MULTIPLE EIGENVALUES

From Equation 15.84, one can see that the rate of separation of a pair of solitons is proportional to the difference between the values of the parameters ξ. At equal values of ξ, the solitons do not separate, but form a bound state. Consider the bound state of N solitons with initial potential *Nsechx* as discussed earlier, and suppose that their velocities ξ_j are all zero. Then the state of the soliton varies with time, according to $exp\left(-4j\eta_j^2 t\right)$, from the application of Equation 15.84. Thus, the state is a periodic variation with frequency $4\eta_j^2$. For a bound state of N solitons, there are N frequencies, but we find that the behavior of the solitons is characterized by all the possible frequency differences or beat frequencies. For $N = 2$, the bound state is characterized by the beat frequency $\omega = 4\left(\eta_1^2 - \eta_2^2\right)$. For $N = 3$ or higher, the lowest beat frequency will give the main periodic variation. Refer to Figure 15.14 to see the effect of the periodic variation. The solitons appear to diverge and then coalesce together. This process repeats itself with a frequency determined by the lowest beat frequency.

There needs to study the effect of multiple eigen-solutions. This is particularly important when we consider the interactive effects equal-sized solitons in solitonic transmission systems. Naturally, we would like to use Equations 15.82 or 15.84, which have been developed earlier. Unfortunately, these formulas fail for solitons with equal amplitudes. In Ref. [2], the authors have worked out an approximation for the limiting case when one eigenvalue approaches another. It was also found that the distance between the two solitons with equal amplitudes increases with time as $ln(4\eta^2 t)$. Another recourse would be to turn to perturbation methods as developed by Lisak, Anderson, Karpman, and Solov'ev [7–9].

15.7 INTERACTION BETWEEN TWO SOLITONS IN AN OPTICAL FIBER

We now return to the NLSE describing the nonlinear propagation of lightwaves in optical fibers. The motivation for this study is the eventual design of a soliton communication system. One needs to know how two solitons affect one another. The aim is to find suitable parameters for the soliton separation, widths, phases, or speeds, so that the solitons do not diverge too much or coalesce together. Also, the shape and size of the solitons should change negligibly; otherwise, they may not be recognized by the soliton detector.

Desem and Chu [10] gave an exact derivation of a two-soliton solution with arbitrary initial phase and separation

$$q(x,\tau) = \frac{|\alpha_1|\cosh(a_1 + i\theta_1)e^{j\phi_2} + |\alpha_2|\cosh(a_2 + i\theta_2)e^{j\phi_1}}{\alpha_3 \cosh a_1 \cosh a_2 - \alpha_4\left[\cosh(a_1 + a_2) - \cos(\phi_2 - \phi_1)\right]}$$

$$\phi_{1,2} = \left[\frac{\left(\eta_{1,2}^2 - \xi_{1,2}^2\right)x}{2} - \tau\xi_{1,2}\right] + (\phi_0)_{1,2}, \quad a_{1,2} = \eta_{1,2}(\tau + x\xi_{1,2}) + (a_0)_{1,2}$$

(15.85)

$$|\alpha_{1,2}|e^{i\theta_{1,2}} = \pm\left\{\left[\frac{1}{\eta_{1,2}} - \frac{2\eta}{\Delta\xi^2 + \eta^2}\right] \pm j\frac{2\Delta\xi}{\Delta\xi^2 + \eta^2}\right\}, \quad \alpha_3 = \frac{1}{\eta_1\eta_2}, \quad \alpha_4 = \frac{2}{\eta^2 + \Delta\xi^2}$$

$$\zeta_{1,2} = \frac{\xi_{1,2} + j\eta_{1,2}}{2}, \quad \Delta\xi = \xi_2 - \xi_1, \quad \eta = \eta_1 + \eta_2$$

The authors were interested in the initial condition of the form

$$q(0, \tau) = \text{sech}[\tau - \tau_0] + e^{j\theta} A \, \text{sech}[A(\tau + \tau_0)]$$

(15.86)

15.7.1 Soliton Pair with Initial Identical Phases

Launching the solitons with equal phases, that is, $\theta = 0$, and hence $\xi_1 = \xi_2 = 0$, will result in a bound system. If $\theta \neq 0$, then $\xi_1 \neq \xi_2$, and, from the previous section on the asymptotic behavior of a general soliton solution, the solitons separate eventually. The resulting equation, upon substituting $\theta = 0$, is simplified to

$$q(\tau,x) = Q\left\{\eta_1 \text{ sech } \eta_1(\tau + \gamma_0)e^{j\eta_1^2 x/2} + \eta_2 \text{ sech } \eta_2(\tau - \gamma_0)e^{j\eta_2^2 x/2}\right\}$$

$$Q = \frac{\eta_2^2 - \eta_1^2}{\eta_1^2 + \eta_2^2 - 2\eta_1\eta_2[\tanh a_1 \tanh a_2 - \text{sech } a_1 \text{ sech } a_2 \cos\psi]}$$

(15.87)

$$a_{1,2} = \eta_{1,2}(\tau \pm \gamma_0), \quad \psi = \frac{\left(\eta_2^2 - \eta_1^2\right)x}{2}$$

The two-soliton solution (Equation 15.87) describes the interaction of two solitons with unequal amplitudes ($\eta_1 \neq \eta_2$), bound together by the nonlinear interaction. Here, owing to the slight variation in the coefficients of the NLSE in the optical fiber as compared to the one studied in the inverse scattering techniques, the beat frequencies are now

$$\omega = \left|\frac{\eta_2^2 - \eta_1^2}{2}\right|$$

(15.88)

and the two-soliton pulses undergo a mutual interaction that is periodic with $2\pi/\omega$.

15.7.2 Soliton Pair with Initial Equal Amplitudes

Here, the solitons are launched with equal amplitudes. The eigenvalues are

$$\eta_{1,2} \approx 1 + \frac{2\tau_0}{\sinh 2\tau_0} \pm \text{sech}(\tau_0)$$

(15.89)

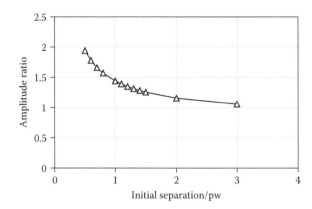

FIGURE 15.15 Amplitude ratio plotted versus initial pulse separation relative to the pulse width (pw).

obtained by equating the first few terms of the Taylor series of Equation 15.85, and applying the first conserved quantity

$$\int_{-\infty}^{\infty} |q|^2 \, d\tau = 2(\eta_1 + \eta_2) \tag{15.90}$$

Referring to Figure 15.15, the period of oscillation in terms of the initial separation τ_0 is

$$\frac{\pi \sinh 2\tau_0 \cosh \tau_0}{2\tau_0 + \sinh 2\tau_0} \tag{15.91}$$

As the initial separation τ_0 increases, the distance of coalescence increases exponentially. Therefore, to minimize interaction, the solitons should be widely separated.

15.7.3 Soliton Pair with Initial Unequal Amplitudes

The eigenvalues for the case of unequal amplitudes are

$$\eta_{1,2} \approx \frac{A+1}{2} + \frac{2\tau_0 \sqrt{A}}{\sinh 2\tau_0 \sqrt{A}} + \left(\sqrt{A} - 1\right) \operatorname{sech} A\tau \pm \left(\frac{A-1}{2} + \operatorname{sech}(A_0)\right) \tag{15.92}$$

and the separation parameter is now

$$\gamma_0 \approx \tau_0 - \left(1 - \frac{2 \operatorname{sech} A\tau_0}{A}\right)\left(\frac{1+A}{2A}\right) \ln\left(\frac{1+A}{A-1}\right) \tag{15.93}$$

One can see from Equation 15.93 that, as the initial separation τ_0 is increased, the eigenvalues η_1, η_2 approach A and 1, respectively. This implies that the oscillation period will no longer be governed by the separation but by the amplitude ratio A. Another key observation, as $|\gamma_0|$ increases by increasing A, is that the value of the coefficient of cosh ψ in Equation 15.91 is decreased. This means that, as the amplitude ratio A is increased, the interaction is minimized. It can be seen that the interaction is still periodic, but at no stage do they coalesce to form one pulse. An approximate expression for the minimum separation τ_{min}, in units of the pulse width of the unity-amplitude solution, is given by

$$\tau_{min} = \tau_0 - \frac{A+1}{4A[1+4A \operatorname{sech} A\tau_0]} \ln\left[1 + 4\left(\frac{A+1}{A-1}\right)^2 \exp\left(\frac{-4A\tau_0}{A+1}\right)\right] \tag{15.94}$$

This finding suggests that, by employing two solitons with unequal amplitudes, we would achieve a much more stable wave formation for long distances, as well as a higher bit rate, since the pulses can be launched closer together. We can admit a higher bit rate because there is a minimum separation for the unequal amplitudes case, whereas, for the case of two solitons with equal amplitudes and zero phase difference, the soliton pulses do coalesce together, resulting in $\tau_{min} = 0$.

A possibly troubling factor in employing unequal amplitudes is the fact that the amplitudes employed are not integers but can have fractions. Satsuma and Yajima [6] studied the behavior of nonintegral amplitudes and concluded that, if $A = N + \alpha$, $|\alpha| < \frac{1}{2}$, then, for large times

$$\frac{\|u\|_{\infty}}{\|u\|_{0}} = 1 - \frac{\alpha^2}{(N + \alpha)^2} \tag{15.95}$$

Numerical simulations show that the solitons oscillate around the limiting value and slowly settle toward the steady or limiting value.

15.7.4 Design Strategy

Using the findings reported by Desem and Chu [10], we consider two designs in which equiphase solitons correspond to the transmission of four digital "1" signals, as this is the worst-case scenario that gives the most interaction. As mentioned earlier, a pair of nonequiphase solitons will result in an unbound system and the solitons will diverge. This is detrimental to the carrier system, as the bit rate should be kept roughly constant. Also, the non-equi-amplitude scheme is implemented, as suggested by Desem and Chu [10], because it promises very little interaction among the soliton pulses.

Working in units of pulse width ($\pi\omega = 2$ for a unity-amplitude soliton), let us impose a design constraint that the minimum separation $2\tau_{min}$ should not be less than $0.7 \cdot 2\tau_0$ (where $2\tau_0$ is the initial separation). The reason is that we do not want the pulses to be too close together. Using Equation 15.94, we obtain, for various values of τ_0, the corresponding A amplitude ratios, as are plotted in Figure 15.15.

Another constraint that needs to be imposed is that the amplitude ratio should not be too large, otherwise the smaller pulse may be mistaken by the detector for a "0" pulse, as the detector may have a certain threshold. Let us assume that the threshold is $0.7 \cdot A^2_{peak}$, where A_{peak} is the largest pulse amplitude (it is assumed that the detector detects the energy or intensity of the wave). This means the largest amplitude ratio possible is 1.2, corresponding to $\tau_0 = 1.75$ pw for the first design and $A = 1.1$, $\tau_0 = 2.5$ pw for the second design. Therefore, the first design seems to be superior to the second one.

Next, one has to check for the variation in the amplitude of the soliton pulses. The constraint is that the amplitude of any soliton pulse should not be less than $\frac{1}{2}$ the amplitude of another pulse. Perturbation theory can be employed to confirm this fact as described in Ref. [8,9]. The other method is to substitute the values of A and τ_0 into Equations 15.94 and 15.95. Then, to find the variation of the soliton peaks, we let $\tau = -\gamma_0$ and $\tau = \gamma_0$ in Equation 15.89 to locate the peaks approximately. To estimate the extremum of the variation of the amplitude, we let $\cos \psi = \pm 1$. Note that this scheme only gives a rough estimate, and that no definite conclusion can be reached from this result. The MATLAB program for the investigation of the four-bit soliton pulse train is developed. It includes an automatic checker that alerts the user (1) if the pulses can no longer be distinguished from one another; that is, if we expect a signal of four "1"s, the pulses should be sufficiently far apart, so that four "1"s can be distinguished; this happens when one pulse merges with another and (2) if pulse separation does not match the specification: 0.7/(bit rate 0 < pulse separation < 1.3/9 bit rate); solitons drift too far apart when drift > 0.35/9 bit rate) and the eye window is less than half the maximum pulse amplitude.

Finally, we are yet to ascertain the maximum range of the four-bit soliton pulse train. From studies on two-soliton interaction, the range can be extended indefinitely for non-equi-amplitude soliton pulses. However, in the transmission scheme that I designed, there are scenarios where there may be interaction between equi-amplitude solitons.

Hence, the maximum range is determined by the interaction between the equi-amplitude solitons. The solitons coalesce at intervals defined by the soliton period. The formula for the soliton period is given by Desem and Chu [10] as follows

$$x_0 = \frac{\pi \sinh(2\tau_0)\cosh(\tau_0)}{2\tau_0 + \sinh(2\tau_0)} \tag{15.96}$$

Substituting the preceding value of $\tau_0 = 3.5$ pw $= 7$, we find that $x_0 = 1722.57$. Assuming that the coalescing process behaves like a cosine function, if we demand that the separation should not be less than 0.7 times the initial separation, then we have the maximum range as equal to $\cos^{-1}(0.7)/(\pi/2) * 1722.57 \sim 872$. This is, of course, in normalized units. In real units and using the normalization constants as used in the MATLAB program, we obtain a value of $872 \times 157 = 58,404$ km. It turns out that, for longer transmission distances, the second design is better, as there is less chance of equi-amplitude soliton interaction. However, of course, the bit rate allowed is lower.

In summary, the design process follows the following steps.

1. Specifies the minimum separation and amplitude threshold for a "1."
2. Uses initial relative phases of zero between neighboring solitons to prevent the spreading apart of the solitons. Then use the equations given by Desem and Chu [10] to obtain a suitable pulse separation and the amplitude ratio A between the solitons.
3. Checks for variation of soliton amplitudes. Determines the maximum range of transmission. Simulates with numerical method to confirm.

15.8   GENERATION OF BOUND SOLITONS

15.8.1   GENERATION OF BOUND SOLITONS IN ACTIVELY PHASE MODULATION MODE-LOCKED FIBER RING RESONATORS

Bound solitons generated in actively mode-locked lasers enable new forms of pulse pairs and multiple pairs or groups of solitons, as described in the preceding sections, in optical transmission or optical logic devices. In this section, we present the generation of stable bound states of multiple solitons in an active MLFL using continuous phase modulation for wideband phase matching. Not only the dual-soliton bound states but also the triple- and quadruple-soliton pulses can be established. Simulation of the generated solitons can be demonstrated in association with experimental demonstration. We can also prove by simulation that the experimental relative phase difference and chirping caused by the phase modulation of an optical modulator in the fiber loop significantly influences the interaction between the solitons. Hence their stability can be obtained as they circulate in the anomalous path-averaged dispersion fiber loop.

15.8.1.1   Introduction

MLFLs are considered as important laser source for generating ultra-short soliton pulses. Recently, soliton fiber lasers have attracted significant research interests with experimental demonstration of bound states of solitons as predicted in some theoretical works [11,12]. These bound soliton states have been observed mostly in passive MLFLs [13–16]. There are, however, a few reports on bound solitons in active MLFLs. Active mode-locking offers significant advantages in the control of the repetition rate that would be critical for optical transmission systems. Observation of bound soliton pairs was first reported in a hybrid frequency modulation (FM) MLFL [17], in which a regime of bound soliton pair harmonic mode-locking at 10 GHz could be generated. There are, however, no reports on multiple bound soliton states. Depending on the strength of soliton interaction, the bound solitons can be classified into two categories: loosely bound solitons and tightly bound solitons, which are determined by the relative phase difference between adjacent solitons. The phase difference may take the value of π or $\pi/2$, or any value depending on the fiber laser structures and mode-locking conditions.

In this section, we demonstrate the generations of bound states of multiple solitons using an active MLFL in which continuous phase modulation or FM mechanism are enforced. By tuning the parameters for the phase matching of the lightwaves circulating in the fiber loop, we could not only observe the dual-soliton bound state, but also the triple- and quadruple-soliton bound states. Relative phase difference and chirping caused by phase modulation of the optical modulator, a $LiNbO_3$ type, in the fiber loop significantly influences the interaction between the solitons, and hence their stability, as they circulate in the anomalous path-averaged dispersion fiber loop.

15.8.1.2 Formation of Bound States in an FM MLFL

Although the formation of the stable bound soliton states, which, determined by the Kerr effect and anomalous-averaged dispersion regime, has been discussed in some configurations of passive MLFLs [14–16], it can be quite distinct in our active MLFL with the contribution of phase modulation of the $LiNbO_3$ modulator in stabilization of bound states. The formation of bound soliton states in an FM mode-locked laser can experience through two stages: (i) a process of pulse splitting and (ii) stabilization of multisoliton states in presence of a phase modulator in the cavity of mode-locked laser.

The first stage is the splitting of a single pulse into multipulses, which occur when the power in the fiber loop increases above a certain mode-locking threshold [17,18]. At higher powers, higher-order solitons can be excited, and, in addition, the accumulated nonlinear phase shift in the loop is so high that a single pulse breaks up into many pulses [19]. The number of split pulses depends on the optical power in the loop, and so there is a specific range of power for each splitting level. The fluctuation of pulses may occur at a region of power where there is a transition from the lower splitting level to the higher. Moreover, the chirping caused by a phase modulator in the loop ensures that the process of pulse conversion from a chirped single pulse into multipulses takes place more easily [20,21].

After splitting into multipulses, the multipulse bound states are stabilized subsequently through the balance of the repulsive and attractive forces between neighboring pulses while circulating in a fiber loop of anomalous-averaged dispersion. The repulsive force comes from direct soliton interaction, depending on the relative phase difference between neighboring pulses [22–25], and the effectively attractive force comes from the variation of a group velocity of soliton pulse caused by the frequency chirping. Thus, in an anomalous average dispersion regime, the locked pulses should be located symmetrically around the extremes of positive phase modulation half-cycle—in other words, the bound soliton pulses acquire an up-chirping when passing the phase modulator. In a specific MLFL setup, beside the optical power level and dispersion of the fiber cavity, the modulator-induced chirp or the phase modulation index determine not only the pulse width but also the time separation of bound soliton pulses at which the interactive effects cancel each other.

The presence of a phase modulator in the cavity to balance the effective interactions among bound soliton pulses is similar to the use of this device in a long-haul soliton transmission system to reduce the timing jitter. For this reason, the simple perturbation theory can be applied to understand the role of phase modulation on mechanisms of bound soliton formation. A multisoliton bound state can be described as following

$$u_{bs} = \sum_{i=1}^{N} u_i(z,t) \tag{15.97}$$

and

$$u_i = A_i \sec h\{A_i [(t - T_i)/T_0)]\} \ \exp (j\theta_i - j\omega_i t) \tag{15.98}$$

where N is the number of solitons in the bound state, T_0 is the pulse width of soliton, and A_i, T_i, θ_i, and ω_i represent the amplitude, position, phase, and frequency of the soliton, respectively. In the simplest case of multisoliton bound state, N equals 2, or we consider the dual-soliton bound state with the identical amplitude of pulse and the phase difference of π value ($\Delta\theta = \theta_{i+1} - \theta_i = \pi$), the ordinary differential equations for the frequency difference and the pulse separation can be derived by using the perturbation method.

$$\frac{d\omega}{dz} = -\frac{4\beta_2}{T_0^3} \exp\left[-\frac{\Delta T}{T_0}\right] - 2\alpha_m \Delta T \tag{15.99}$$

$$\frac{d\Delta T}{dz} = \beta_2 \omega \tag{15.100}$$

where β_2 is the averaged group-velocity dispersion of the fiber loop; ΔT is the pulse separation between two adjacent solitons ($T_{i+1} - T_i = \Delta T$) and $\alpha_m = m\omega_m^2/(2L_{cav})$; L_{cav} is the total length of the loop; and m is the phase modulation index. Equation 15.100 shows the evolution of frequency difference and position of bound solitons in the fiber loop in which the first term on the right-hand side represents the accumulated frequency difference of two adjacent pulses during a round trip of the fiber loop, and the second one represents the relative frequency difference of these pulses when passing through the phase modulator. At steady state, the pulse separation is constant, and the induced frequency differences cancel each other. On the contrary, if setting Equation 15.99 to zero, we have

$$-\frac{4\beta_2}{T_0^3} \exp\left[-\frac{\Delta T}{T_0}\right] - 2\alpha_m \Delta T = 0 \tag{15.101}$$

or

$$\Delta T \exp\left[\frac{\Delta T}{T_0}\right] = -\frac{4\beta_2}{T_0^3} \frac{L_{cav}}{m\omega_m^2} \tag{15.102}$$

We can see the effect of phase modulation to the pulse separation through Equation 15.102— in addition, β_2 and α_m must have opposite signs, which means that, in an anomalous dispersion fiber loop with negative value of β_2, the pulses should be up-chirped. With a specific setup of FM fiber laser, when the magnitude of chirping increases, the bound pulse separation decreases subsequently. The pulse width also reduces according to the increase in the phase modulation index and modulation frequency, so that the variation of the ratio $\Delta T/T_0$ can be high. Thus, the binding of solitons in the FM MLFL is assisted by the phase modulator. Bound solitons in the loop periodically experience the frequency shift, and hence their velocity in response to changes in their temporal positions by the interactive forces in equilibrium state.

15.8.1.3 Experimental Setup and Results

Figure 15.16 shows the experimental setup of the FM MLFL. Two erbium-doped fiber amplifiers (EDFA) pumped at 980 nm are used in the fiber loop to control the optical power in the loop for mode-locking. Both are operating in saturated mode. A phase modulator driven in the region of 1 GHz modulation frequency assumes the role as a mode-locker and controls the states of locking in the fiber ring. At the input of the phase modulator, a polarization controller (PC) consisting of two

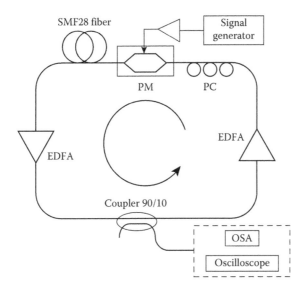

FIGURE 15.16 Experimental setup of the FM mode-locked fiber laser. PM, phase modulator; PC, polarization controller; OSA, optical spectrum analyzer.

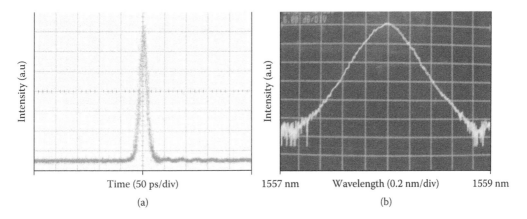

FIGURE 15.17 (a) Oscilloscope trace of generated single soliton and (b) optical spectrum of a single soliton.

quarter-wave plates and one half-wave plate is used to control the polarization of light, which relates to nonlinear polarization evolution and influences multipulse operation in the formation of bound soliton states. A 50 m long Corning SMF-28 fiber is inserted after the phase modulator to ensure that the average dispersion in the loop is anomalous. The fundamental frequency of the fiber loop is 1.7827 MHz, which is equivalent to the 114 m total loop length. The outputs of the mode-locked laser from the 90:10 coupler are monitored by an optical spectrum analyzer (HP 70952B) and an oscilloscope (Agilent DCA-J 86100C) of an optical bandwidth of 65 GHz.

Under normal conditions, the single-pulse mode-locking operation is performed at the average optical power of 5 dBm and modulation frequency of 998.315 MHz (the harmonic mode-locking at the 560th order), as shown in Figure 15.17. The narrow pulses of 8–14 ps width, depending on the RF driving power of the phase modulator, can be observed on the oscilloscope. The measured pulse spectrum has spectral shape of a soliton rather than a Gaussian pulse. By adjusting the polarization states of the PC wave plates at a higher optical power, the dual-bound solitons or bound soliton pairs can be observed at an average optical power circulating inside the fiber loop of about 10 dBm. Figure 15.18a shows the typical time-domain waveform and (b) the corresponding spectrum of the

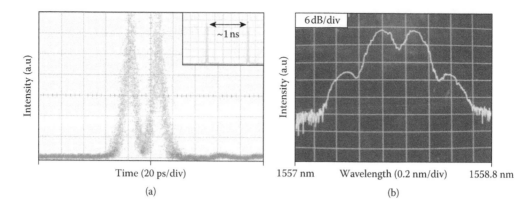

FIGURE 15.18 (a) Oscilloscope trace of periodic bound soliton pairs in time domain and (b) optical spectrum of soliton pair bound state.

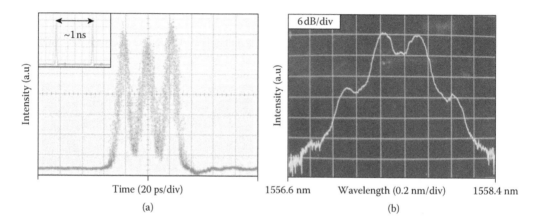

FIGURE 15.19 (a) Oscilloscope trace of periodic groups of triple-bound solitons in time domain and (b) optical spectrum of triple-soliton bound state.

dual-bound soliton state. The estimated full width at half maximum (FWHM) pulse width is about 9.5 ps, and the temporal separation between two bound pulses is 24.5 ps, which are correlated exactly to the distance between two spectral main lobes of 0.32 nm of the observed spectrum. When the average power inside the loop is increased to about 11.3 dBm, and a slight adjustment of the PC is performed, the triple-bound soliton state occurs, as shown in Figure 15.19. We obtain the FWHM pulse width and the temporal separation of two close pulses are 9.5 ps and 22.5 ps, respectively. Insets in Figures 15.18a and 15.19a show the periodic sequence of bound solitons at the repetition rate exactly equal to the modulation frequency. This feature is quite different from that in a passive mode-locked fiber laser, in which the positions of bound solitons is not stable and the direct soliton interaction causes a random movement and phase shift of bound soliton pairs [14]. On the contrary, it is advantageous to perform a stable periodic bound soliton sequence in an FM MLFL.

The symmetrical shapes of optical modulated spectrum in Figures 15.18b and 15.19b indicate clearly that the relative phase difference between two adjacent bound solitons is of π value. At the center of the spectrum, there is a dip due to the suppression of the carrier, with π phase difference in the case of dual-soliton bound state. There is a small hump in case of the triple-soliton bound state, because the three soliton pulses are bound together, with the first and the last pulses in-phase and out-of-phase with the middle pulse. The shape of the spectrum will change, which can be symmetrical or asymmetrical when this phase relationship varies. We believe that the bound state with

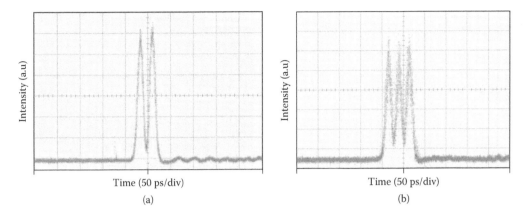

Time (50 ps/div)

(a)

Time (50 ps/div)

(b)

FIGURE 15.20 Oscilloscope traces of (a) the dual-soliton bound state and (b) the triple-soliton bound state after propagating through 1 km SMF.

the relative phase relationship of π between solitons is the most stable in our experimental setup, and this observation also agrees with the theoretical prediction of stability of bound soliton pairs relating to photon-number fluctuation in different regimes of phase difference.

Similar to a single soliton, the phase coherence of bound solitons is still maintained as a unit when propagating through a dispersive medium.

Figure 15.20 shows the dual-soliton bound state and triple-soliton state waveforms on the oscilloscope after they propagate through 1 km SMF fiber. There is also no change in their observed spectral shapes in both cases.

Multisoliton bound operation can be formed in an FM mode-locked fiber by operating the fiber loop of anomalous average dispersion at critical optical power level with the phase modulator-induced chirp. When the average optical power in the loop is increased to a maximum value of 12.6 dBm and the RF driving power applied to the optical modulator is decreased to about 15 dBm, we obtain the bound multiple soliton states. It is really amazing when the quadruple-soliton state is generated, as observed in Figure 15.21. The bound state occurring at the lower phase modulation index is due to maintaining a small enough frequency shifting in a wider temporal duration of bound solitons to hold the balance between interactions of the group of four solitons in the fiber loop. However, the optical power can still not be high enough to stabilize the bound state, and also the noise from EDFAs is larger at higher pump power. This results that the quadruple-soliton states become too noisy, and we can observe that the two main side lobes of the spectrum are inversely proportional to the temporal separation of the pulses. This separation is about 20.5 ps in our experimental conditions. The results show that the pulse separation reduces when the number of pulses in the bound states is larger.

Thus, through the experimental results, both the phase modulation index and the cavity's optical power influence the existence of the multisoliton bound states in the FM MLFL. The optical power determines the number of initially split pulses and maintains the pulse shape in the loop. Observations from the experiment shows that there is a threshold of optical power for each bound level. At threshold value, the bound solitons show strong fluctuations in amplitude. Amplitude oscillations are observed during the transition state between different bound levels. Furthermore collisions of adjacent pulses even occur, as shown in Figure 15.22. The phase difference of adjacent pulses can also change in these unstable states, which mean that the neighboring pulses is not out of phase anymore, but in phase as seen through the measured spectrum in Figure 15.23. Although the decrease in phase modulation index is required to maintain the stability at higher bound soliton level, it increases the pulse width and reduces the peak power. Therefore, the waveform seen is noisier and more sensitive to ambient environment, and its spectrum is not strongly modulated, as shown in Figure 15.24.

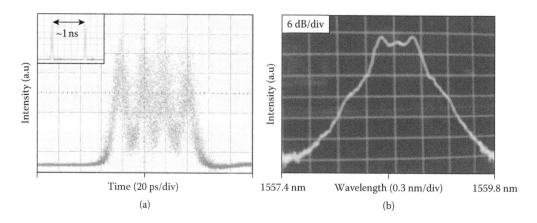

FIGURE 15.21   (a) Oscilloscope trace and (b) optical spectrum of quadruple-soliton bound state.

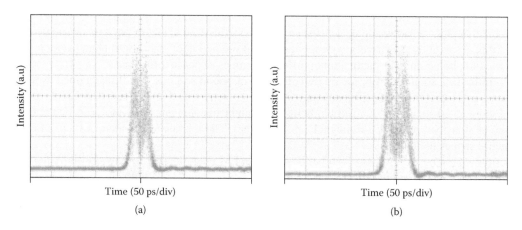

FIGURE 15.22   Oscilloscope traces of (a) the dual-soliton bound state at optical power threshold of 8 dBm and (b) the triple-soliton bound state at 10.7 dBm.

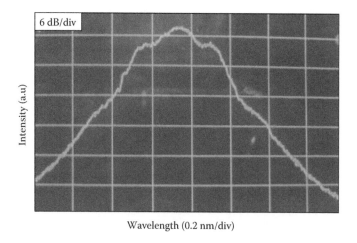

FIGURE 15.23   Spectrum of the bound soliton state with in-phase pulses. Center wavelength located at 1558 nm.

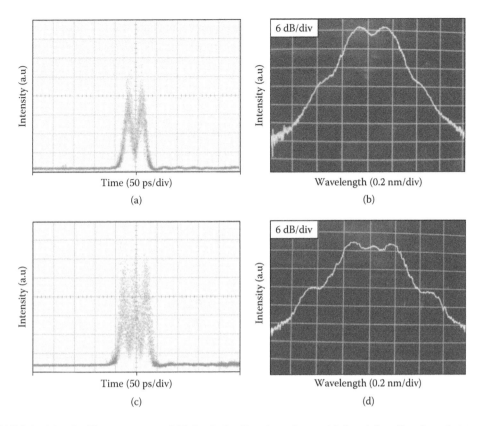

Time (50 ps/div)

(a)

6 dB/div

Wavelength (0.2 nm/div)

(b)

Time (50 ps/div)

(c)

6 dB/div

Wavelength (0.2 nm/div)

(d)

FIGURE 15.24   Oscilloscope traces of (a) the dual-soliton bound state, (c) the triple-soliton bound state, and (b and d) their spectra, respectively, at low RF driving power of 11 dBm.

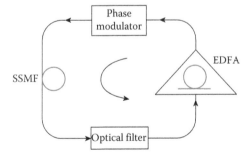

FIGURE 15.25   A circulating model of MLFL for simulating the FM mode-locked fiber ring laser. SSMF: standard single-mode fiber, EDFA: erbium-doped fiber amplifier.

15.8.1.4   Simulation of Dynamics of Bound States in an FM MLFL

15.8.1.4.1   Numerical Model of an FM MLFL

To understand the dynamics of FM bound soliton fiber laser, a simulation technique is used to see the dynamic behavior of the formation and interactions of solitons in the FM mode-locked fiber loop. This is slightly different as compared with the analytical ISM described in Section 15.6. A recirculation model of the fiber loop is used to simulate propagation of the bound solitons in the fiber cavity. The simple model consists of basic components of an FM MLFL, as shown in Figure 15.25; in other words, the cavity of the FM MLFL is modeled as a sequence of different elements.

The optical filter has a Gaussian transfer function with a 2.4 nm bandwidth. The transfer function of the phase modulator is $u_{out} = u_{in} \exp[jm \cos(\omega_m t)]$, where m is the phase modulation index, and $\omega_m = 2\pi f_m$ is angular modulation frequency, assumed to be a harmonic of the fundamental frequency of the fiber loop. The pulse propagation through the optical fibers is governed by the NLSE [22] (reiterated from Equation 15.15).

$$\frac{\partial u}{\partial z} + j\frac{\beta_2}{2}\frac{\partial^2 u}{\partial T^2} - \frac{\beta_3}{6}\frac{\partial^3 u}{\partial T^3} + \frac{\alpha}{2}u = j\gamma|u|^2 u \tag{15.103}$$

where u is the complex envelope of optical pulse sequence; β_2 and β_3 account for the second- and third-order fiber dispersions; and α and γ are the loss and nonlinear parameters of the fiber, respectively. The amplification of signals, including the saturation of the EDFA, can be represented [22] as

$$u_{out} = \sqrt{G}u_{in} \tag{15.104}$$

with

$$G = G_0 \exp\left(-\frac{G-1}{G}\frac{P_{out}}{P_s}\right) \tag{15.105}$$

where G is the optical amplification factor, G_0 is the unsaturated amplifier gain, P_{out} and P_{sat} are the output power and saturation power, respectively.

The difference of dispersion between the SMF fiber and the erbium-doped fiber in the cavity arranges a certain dispersion map, and the fiber loop gets a positive net dispersion or an anomalous average dispersion, which is important in forming the "soliton-like" pulses in an FM fiber laser. Basic parameter values used in our simulations are listed in Table 15.3.

15.8.1.4.2 Simulation of the Formation Process of the Bound Soliton States

First, we simulate the formation process of bound states in the FM MLFL whose parameters are those employed in the experiments described earlier. The lengths of the Er-doped fiber and SMF fiber are chosen to get the cavity's average dispersion $\bar{\beta}_2 = -10.7$ ps^2/km. The Schrödinger equation can be applied under some specific initial conditions with the superposition of random optical noises. The initial amplitudes of soliton waves can then be circulating around the optically amplified ring using this propagation equation so as to observe the formation of bound solitons. The optical power is built up with the phase-matching conditions, and hence the formation of the solitons.

Figure 15.26 shows a simulated dual-soliton bound state building up from initial Gaussian-distributed noise as an input seed over the first 2000 round trips, with a P_{sat} value of 10 dBm, G_0 of 16 dB, and m of 0.6π. The built-up pulse experiences transitions with large fluctuations of intensity, position, and pulse width during the first 1000 round trips before reaching the stable bound

---

TABLE 15.3

Parameter Values Used in Simulations

$\beta_2^{SMF} = -21$ ps^2/km	$\beta_2^{ErF} = 6.43$ ps^2/km	$\Delta\lambda_{filter} = 2.4$ nm
$\gamma^{SMF} = 0.0019$ W^{-1}/m	$\gamma^{ErF} = 0.003$ W^{-1}/m	$f_m \approx 1$ GHz
$\alpha^{SMF} = 0.2$ dB/km	$P_{sat} = 7 \div 13$ dBm	$m = 0.1\pi \div 1\pi$
$L_{cav} = 115$ m	$NF = 6$ dB	$\lambda = 1558$ nm

Notes: SSMF = standard single-mode fiber, ErF = erbium-doped fiber, NF = noise figure of the EDFA.

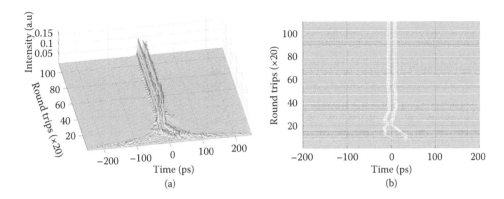

FIGURE 15.26 (a) Numerically simulated dual-soliton bound state formation from noise and (b) the evolution of the formation process in contour plot view.

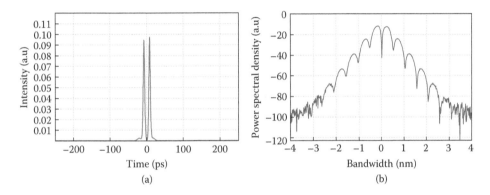

FIGURE 15.27 (a) Waveform and (b) corresponding spectrum of simulated dual-soliton bound state at the 2000th round trip.

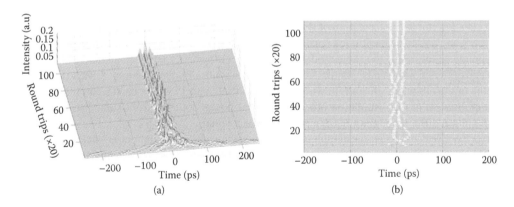

FIGURE 15.28 (a) Numerically simulated triple-soliton bound state formation from noise in 2D view and (b) evolution of the formation process in contour plot view.

steady state. Figure 15.27 shows the time-domain waveform and spectrum of the output signal at the 2000th round trip. The bound states with higher number of pulses can be formed at higher gain of the cavity, and hence when the gain G_0 is increased to 18 dB, which is enhancing the average optical power in the loop, the triple-soliton bound steady state is formed from the noise seed via simulation, as shown in Figure 15.28. In the case of higher optical power, the fluctuation of signal at initial

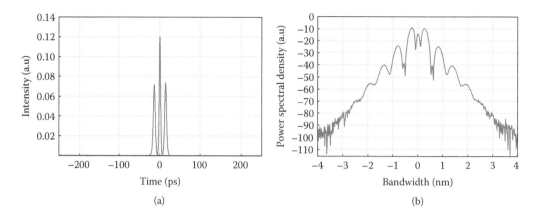

FIGURE 15.29 (a) Waveform and (b) corresponding spectrum of simulated triple-soliton bound state at the end of 2000th round trip.

transitions is stronger, and it needs more round trips to reach a more stable three pulses bound state. The waveform and spectrum of the output signal from the FM MLFL at the 2000th round trip are shown in Figure 15.29a and b, respectively.

Although the amplitude of pulses is not equal, which indicates that the bound state can require a larger number of round trips before the effects in the loop balance, the phase difference of pulses accumulated during circulating in the fiber loop is approximately of π value, which is indicated by strongly modulated spectra. In particular, from the simulation result, the phase difference between adjacent pulses is 0.98π in case of the dual-pulse bound state, and 0.89π in case of the triple-pulse bound state. These simulation results agree with the experimental results (shown in Figures 15.18b and 15.19b) discussed earlier to confirm the existence of multisoliton bound states in an FM MLFL.

15.8.1.4.3 Simulation of the Evolution of the Bound Soliton States in an FM Fiber Loop

The stability of bound states in the FM MLFL strongly depends on the parameters of the fiber loop, which also determine the formation of these states. Besides the phase modulation and group velocity dispersion (GVD), as mentioned earlier, the cavity's optical power also influences the existence of the bound states. Using the same model as in the preceding text, the effects of active phase modulation and optical power can be simulated to see the dynamics of bound solitons. Instead of the noise seed, the multisoliton waveform following Equations 15.97 and 15.98 are used as an initial seed for the simplification of our simulation processes. Initial bound solitons are assumed to be identical, with the phase difference between adjacent pulses of π value.

First, the effect of the phase modulation index on the stability of the bound states is simulated through a typical example of the evolution of dual-soliton bound state over 2000 round trips in the loop at different phase modulation indexes with the same saturation optical power of 9 dBm, as shown in Figure 15.30. The simulation results also indicate that the pulse separation decreases corresponding to the increase in the modulation index. In the first-round trips, there is a periodic oscillation of bound solitons that is considered as a transition of solitons to adjust their own parameters to match the parameters of the cavity before reaching a finally stable state. Simulations in other multisoliton bound states also manifest this similar tendency. The periodic phase modulation in the fiber loop is not only to balance the interactive forces between solitons but also to retain the phase difference of π between them. At too small modulation indices, the phase difference changes or reduces slightly after many round trips, or, in other words, the phase coherence is looser; this leads to amplitude oscillation due to the alternatively periodic exchange of energy between solitons, as shown in Figure 15.31. The higher the number of solitons in the bound state, the more sensitive

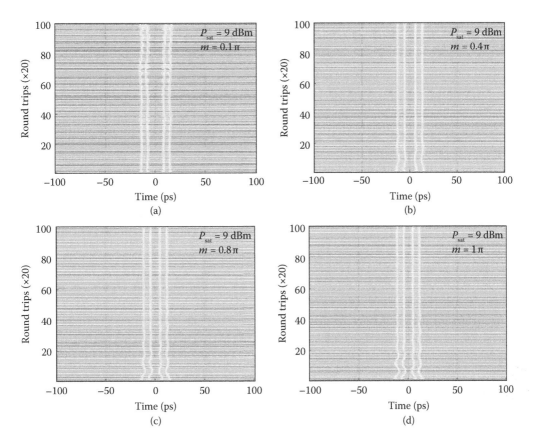

FIGURE 15.30 Simulated evolution of dual-soliton bound state over 2000 round trips in the fiber loop at different phase modulation indexes; (a) $m = 0.1\pi$, (b) $m = 0.4\pi$, (c) $m = 0.8\pi$, and (d) $m = 1\pi$.

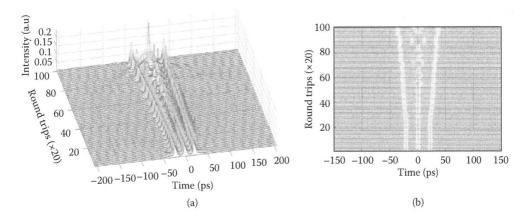

FIGURE 15.31 Simulated evolution of triple-soliton bound state over 2000 round trips at the low phase modulation index $m = 0.1\pi$; (a) 3D view and (b) contour plane view.

it is to changes in phase modulation index. The modulation index determines the time separation between bound pulses, yet, at too high modulation indices, it is more difficult to balance the effectively attractive forces between solitons, especially when the number of solitons in the bound state is larger. The increase in chirping leads to faster variations in the group velocity of pulses when passing the phase modulator, which can create the periodic oscillation of pulse position in the time

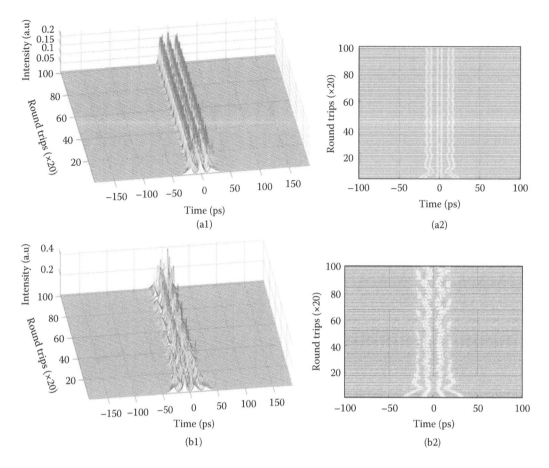

FIGURE 15.32 Simulated evolutions over 2000 round trips in the fiber loop of (a1) triple-soliton bound state at $m = 1\pi$, and (a2) contour plane view of (a1); (b1) quadruple-soliton bound state at $m = 0.7\pi$, and (b2) contour plane views of (b1).

domain evolution. Figure 15.32 shows the evolution of the triple-soliton bound state at the m value of 1π, and the quadruple-soliton bound state at the m value of 0.7π. In the case of quadruple-soliton bound state, solitons oscillate strongly and tend to collide together.

Another factor is the optical power of the fiber loop; it plays an important role not only in the determination of multipulse bound states, as in simulation of the previous section, but also in stabilization of the bound states circulating in the loop. As mentioned in Section 15.8.1.3, each bound state has a specific range of operational optical power. In our simulations, dual-soliton, triple-soliton, and quadruple-soliton bound states are in stable evolution in the loop when the saturated power P_{sat} is about 9 dBm, 11 dBm, and 12 dBm, respectively. When the optical power of the loop is not within these ranges, the bound states become unstable, and they are more sensitive to the change of phase modulation index. At a power lower than the threshold, the bound states are out of bound and switched to a lower level of the bound state. In contrast, at the high-power level region, the generated pulses are broken into random pulse trains or decay into radiation.

Figure 15.33 shows the unstable evolution of the triple-soliton bound states under nonoptimized operating conditions of the loop. Different values of the phase difference between bound solitons can also be simulated. The states of non-π phase difference often leads to unstable behavior of the bound states and can easily be destroyed, as depicted in Figure 15.34. The simulation results agree well with the experimental observations discussed in the preceding text.

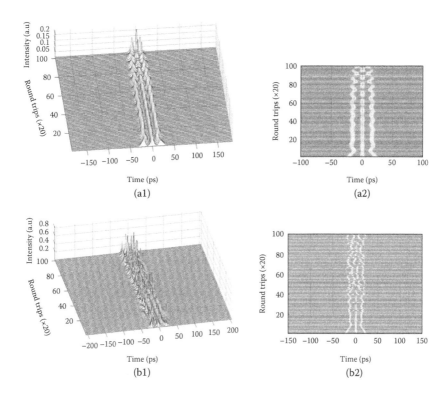

FIGURE 15.33 Simulated dynamics of triple-soliton bound state at unoptimized power (a1) P_{sat} = 9 dBm, m = 0.7π, and (b1) P_{sat} = 13 dBm and m = 0.6π, (a2) and (b2) contour plane views.

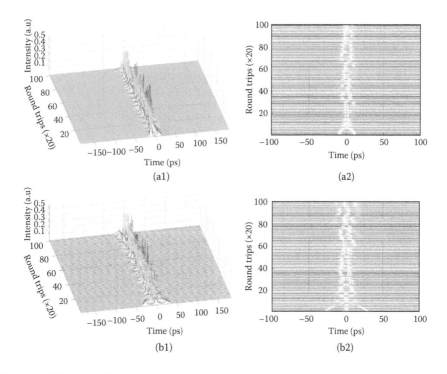

FIGURE 15.34 Time-domain dynamics of dual-soliton and triple-soliton bound states, respectively, with in-phase pulses (a1) and (b1) 3D view and (a2) and (b2) plane contour view.

15.8.2 Active Harmonic MLFL for Soliton Generation

15.8.2.1 Experiment Setup

Figure 15.35 shows the schematic of the active MLFL. The fiber ring configuration incorporates an isolator to ensure unidirectional lasing. The gain media is obtained by using a 16 m erbium-doped fiber operating under bi-directional pump condition. The pump sources are two 980 nm laser diodes coupled to the ring through two 980/1550 WDM couplers. A FOL0906PRO-R17-980 laser diode from FITEL is used for the forward pump source, and a COMSET PM09GL 980 nm laser diode for the backward pump source. An 18 m long dispersion-shifted fiber (DSF) is used for shortening the locked pulse width. A 2 nm bandpass multilayer thin-film optical filter is used to select the operation wavelength of the laser and moderately accommodate the bandwidth of the output pulses. A PC is also employed to maximize the coupling of lightwaves from fiber to the diffused waveguides of the Mach–Zehnder intensity modulator (MZIM). Output pulse trains are extracted via a 90/10 coupler. Mode-locking is obtained by inserting into the ring a 20-GHz 3-dB bandwidth MZIM that periodically modulates the loss of the lightwaves traveling around the ring. The modulator is biased at the quadrature point with a voltage of −1.5 V, and is then driven by superimposing a sinusoidal signal derived from an Anritsu 68347C signal generator.

The output signal is then monitored using an Agilent 86100B wideband oscilloscope with an optical input sampling module, an Ando 6317B optical spectrum analyzer with a resolution of 0.01 nm, and an Agilent E4407B electronic spectrum analyzer. Figure 15.36 shows the picture of the whole system setup for experiment.

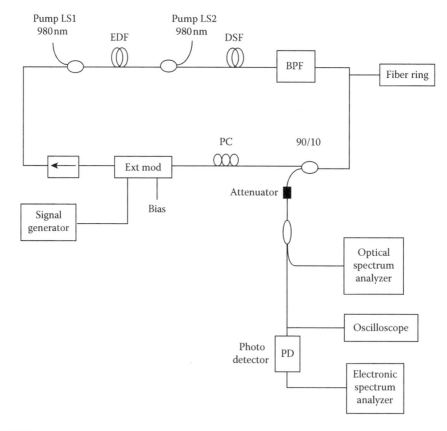

FIGURE 15.35 Active harmonic mode-locked fiber laser experiment setup.

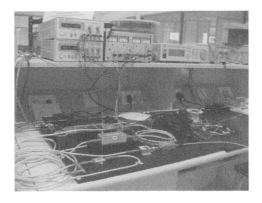

FIGURE 15.36 Active harmonic mode-locked fiber laser.

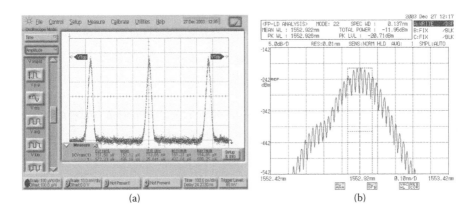

(a)   (b)

FIGURE 15.37 (a) Harmonic mode-locked pulse train with a repetition rate of 2.657 GHz and (b) its optical spectrum.

15.8.2.2 Tunable Wavelength Harmonic Mode-Locked Pulses

The total optical cavity length is 55.4 m, which corresponds to a fundamental frequency f_R of 3.673 MHz (the measurement of f_R is presented in Section 15.8.2.3). When the modulation frequency is set at 2.695038 GHz, or equal to 733 times the f_R, a mode-locking laser pulse is obtained at the output, as shown in Figure 15.37a. The interval between the adjacent pulses is 371 ps, corresponding to a repetition rate of 2.695 GHz. The locked pulses are formed with clear pedestal, with a slightly fluctuated amplitude. The output pulse amplitude is measured, and a peak power is recorded at 9.9 mW.

The pulse width is also measured, and a value of 37 ps FWHM is obtained. However, as this value is taken from the trace of pulse on the oscilloscope, the rise time of the oscilloscope is included in the measured value. Since the rise time of the oscilloscope is 20 ps, which is comparable to the measured value, it cannot be ignored. Besides, the rise time of the photodiode also contributes to the total measured value. Therefore, the actual pulse width is much smaller than the measured value.

Figure 15.37b shows the optical spectrum of the output pulses. The FWHM bandwidth is 0.137 nm with central wavelength at 1552.926 nm. The time bandwidth product (TBP) was determined as

$$TBP = T_{\mathrm{FWHM}} \times BW \tag{15.106}$$

$$TBP = 37 \text{ ps} \times 17.125 \text{ THz} = 0.634$$

The large value obtained here again confirms that the measured value of pulse width is mainly contributed by the rise times of the oscilloscope and the photodiode. The accurate pulse width should be obtained from using an optical auto-correlator.

Alternatively, one can take advantage of the characteristic of the AM MLFL to approximate the pulse width. The pulses generated from AM MLFLs are nearly transform limited [13–16]. This means that the TBP takes the value of 0.44, assuming a Gaussian-shaped pulse. Therefore, the pulse width can be estimated from

$$T_{FWHM} = TBP/BW \tag{15.107}$$

$$T_{FWHM} = 0.44/17.125 \ THz = 26 \ ps$$

It can be seen from Figure 15.37b that the pulse train spectrum has side lobes with a separation of 0.022 nm, which corresponds to a longitudinal mode separation of 2.695 GHz. This spectrum profile is typical for the mode-locked laser, and is usually referred as the *mode-locked structure spectrum* [17,18,23].

The preceding spectrum profile can be explained by taking the Fourier transform of the periodic Gaussian pulse train

$$p(t) = e^{-t^2/2T_0^2} * \mathrm{III}\left(\frac{t}{T}\right) \tag{15.108}$$

where T_0 is the width of the Gaussian pulse, T is the pulse repetition period, and $\mathrm{III}(t/T)$ is the comb function given by

$$\mathrm{III}\left(\frac{t}{T}\right) = \sum_{n=-\infty}^{\infty} \delta(t - nT) \tag{15.109}$$

$$P(\omega) = F(p(t)) = T_0 \sqrt{2\pi} e^{-T_0^2 \omega^2/2} \sum_{n=-\infty}^{\infty} \delta(\omega - n\Omega) \tag{15.110}$$

where $\Omega = 2\pi/T$; and $P(\omega)$ has the structure of a train of Dirac function pulses separated by the repetition frequency with a Gaussian envelope. The laser wavelength is tunable over the whole C-band by tuning the central wavelength of the thin-film bandpass filter inserted in the loop.

The repetition rate of the laser pulses can be increased by increasing the modulation frequency. Figure 15.38 shows the output pulses with the repetition rate of 4 GHz and its optical spectrum.

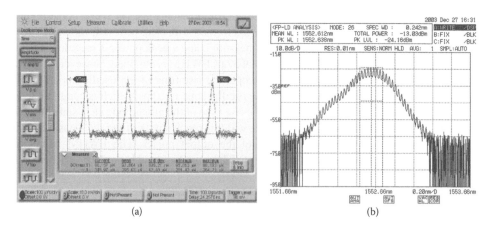

(a)   (b)

FIGURE 15.38   Harmonic mode-locked pulse train with repetition rate of 4 GHz: (a) temporal profile and (b) optical spectrum.

15.8.2.3 Measurement of the Fundamental Frequency

In active harmonic mode-locked fiber laser (HMLFL), the pulses are locked into one of the harmonics of the fundamental frequency f_R by tuning the modulation frequency f_m, so that $f_m = k f_R$. In practice, the laser is mode-locked if f_m is within the locking range around the harmonic frequency—that is, $f_m = k f_R \pm \Delta f$. If f_m falls outside the locking range, the laser becomes unlocked—that is, no pulse trains are observed. If f_m is adjusted until reaching the next harmonic frequency $(k+1)f_R$, the HMLFL is locked again into the next longitudinal resonant mode of the ring. By measuring those locked modulation frequencies, one can determine the fundamental frequency of the ring. The fundamental frequency is determined by

$$(4018635400 - 3981902400)/10 \text{ Hz} \leq f_R \leq (4018636000 - 3981901400)/10 \text{ Hz}$$

$$3673300 \text{ Hz} \leq f_R \leq 3673460 \text{ Hz}$$

$$\text{or } f_R = 3673380 \pm 80 \text{ Hz}$$

15.8.2.4 Effect of the Modulation Frequency

Increasing modulation frequency not only generates higher-repetition-rate mode-locked pulses but also improves other parameters of the pulses such as bandwidth, pulse width, etc.

The solid curve shown in Figure 15.39 indicates the relationship between the pulse width of the generated pulse train and modulation frequency. The pulse width does not change so much when the modulation frequency increase above 2 GHz. This is because the pulse width in this region is small compared to the rise time of the oscilloscope. The actual values should be smaller than the measured values. This is confirmed when examining the equivalent bandwidth of the pulse trains.

Figure 15.40 shows the pulse train bandwidth versus the modulation frequency. The modulation frequency increases as the bandwidth increases. Unlike the behavior of pulse width, the pulse train bandwidth continues to increase even when the modulation frequency is above 2 GHz. This indicates that the actual pulse width should be smaller than its measured value. The pulse width is calculated and plotted as shown by the dash curve in Figure 15.39.

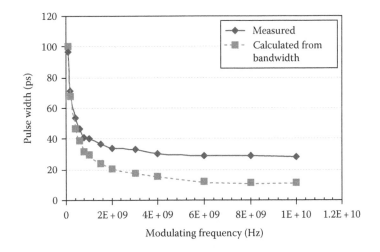

FIGURE 15.39 Pulse width of the HMLFL for various modulation frequencies.

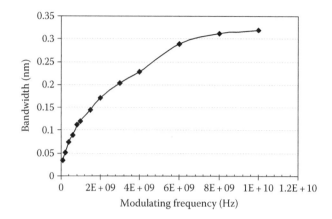

FIGURE 15.40   Bandwidth of the HMLFL for various modulation frequencies.

15.8.2.5   Effect of the Modulation Depth/Index

The same setup shown in Figure 15.35 is used to study the effect of modulation depth on the performance of the mode-locked pulse trains. The settings of the cavity are as following: (1) the total pump power is 274 mW, (2) the modulator is biased at the quadrature point. The modulation depth is varied by changing the signal amplitude (from the signal synthesizer) applied to the MZIM. The pulse width and bandwidth of the output pulse trains are measured for different values of modulation depth.

Figure 15.41 shows the relationship between pulse width and modulation amplitude/power. As the modulation power increases, the pulse width slightly decreases. Therefore, the pulse can be shortened by increasing the modulation power (or the modulation depth). However, the pulse-shortening effect of modulation depth is not as strong as that of modulation frequency. In addition, the modulation depth is limited to a maximum value of 1.

The relationship between pulse width and the modulation depth is verified by examining the effect of modulation power on the bandwidth. When the modulation power increases, the bandwidth decreases, as shown in Figure 15.42. This is consistent with the decrease of pulse width shown in Figure 15.41.

It is found that there is a threshold power of the modulation for locking. When the modulation power is reduced to −4 dBm, corresponding to a modulation depth of 0.08, mode-locking does not occur, and pulse train cannot be obtained.

15.8.2.6   Effect of Fiber Ring Length

The same setup as in Figure 15.35 has been used to study the effect of the fiber ring length on the mode-locked pulse trains. The DSF fiber has been extended from 18 to 118 m. A total pump power of 274 mW is used. The pulse width and bandwidth have been measured and compared with those described in Section 15.8.2.2. Figures 15.43 and 15.44 show the mode-locked pulse temporal train and its optical spectrum for the 18-m DSF ring laser and the 118-m DSF ring laser, respectively.

The pulse trains are very much similar for both cases. Their pulse widths (FWHM) are nearly equal, 33 and 29 ps. However, the pulse optical spectra are quite different, The longer-ring laser has a bandwidth of 3 nm, larger than that of the shorter one, 0.184 nm. The broadening of the bandwidth of the pulse train of the 110 m ring length indicates that the pulse width is shorter than what it is observed by the sampling oscilloscope. The discrepancy here is again due to the effect of the photodiode and the oscilloscope's rise time on the observed pulse.

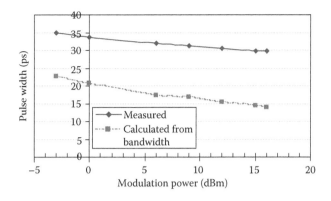

FIGURE 15.41 Pulse width of the HMLFL for various modulation powers.

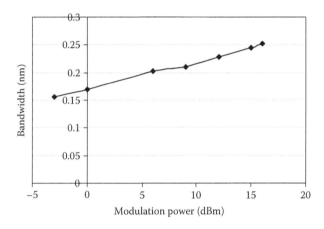

FIGURE 15.42 Bandwidth of the HMLFL for various modulation powers.

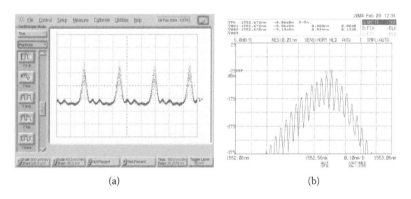

FIGURE 15.43 (a) Locked pulses and (b) its optical spectrum for the 18-m DSF ring laser.

The shortening of the pulse in the longer length ring can be due to the self-phase modulation (SPM) effects or the nonlinear induced phase in the fiber. The high-power pulse traveling in the fiber suffers the nonlinear SPM effect as described in Chapter 2. This happens when the peak of the pulse is increased when the pulse width becomes very short. Under this case the bandwidth of the bound solitons broaden and so the energy distributes to other spectral components and thus the pulse power decreases and the soliton states returns to the stable state again.

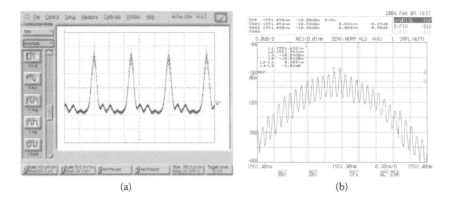

FIGURE 15.44   (a) Locked pulses and (b) its optical spectrum for the 118-m DSF ring laser.

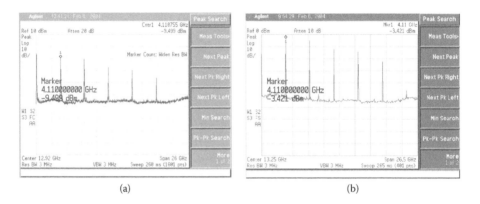

FIGURE 15.45   RF spectrum of HML laser with (a) 18 m DSF and (b) 118 m DSF.

To verify that the pulse shortening is due to the nonlinear effect, the two different-length rings are studied under lower pump power. It is found that when the pump power is reduced to 50 mW, the pulse width and bandwidth of the laser does not change, regardless of the fiber length.

One can estimate the effective nonlinear lengths of the fiber for the pump power of 274 mW and 50 mW. The peak pulse powers are 0.37 and 0.027 W, respectively. The nonlinear lengths can be calculated as [22]:

$$L_{NL}(P_p = 274 \text{ mW}) = \frac{1}{\gamma P_0} = \frac{1}{2 \times 10^{-3} \times 0.37} = 1351(m) \tag{15.111}$$

$$L_{NL}(P_p = 27 \text{ mW}) = \frac{1}{\gamma P_0} = \frac{1}{2 \times 10^{-3} \times 0.027} = 18,518(m) \tag{15.112}$$

It can be seen that the effective nonlinear length of the 118 m long fiber ring is comparable to that of the physical ring length when pumped at a high power. Therefore, the nonlinear effect plays a significant role in this case. When the laser is pumped at a lower power, the effective nonlinear length is much longer than the laser ring length, and hence the SPM effect is minute and cannot be observed. However, one can predict the existence of the SPM effect if the fiber length is increased to be comparable with the nonlinear length.

It is also noticed that the supermode noise was reduced in the case of longer fiber ring. The pulses are more temporally stable and less fluctuating. To verify this, the RF spectrum of the lasers are recorded and compared. It can be seen from Figure 15.45 that the longer length laser has a lower noise floor than the shorter one.

15.8.2.7 Effect of Pump Power

The laser is configured as in Figure 15.35, except that the fiber length is 118 m. Different power levels are pumped into the EDFA to study the laser performance. Figure 15.46 shows the variation of the laser pulse widths under various pump powers. The pulse widths seem to be slightly affected by changes in pump power. However, the effect of pump power is actually stronger, since the pulse widths measured here are not the true pulse widths. They are larger than their true values due to the rise time of the equipment, as discussed earlier. The effect of pump power on the pulse width can be drawn from the relationship between pump power and output pulse bandwidth.

The dependence of the output pulse bandwidth is illustrated in Figure 15.47. As the pump power increases, the bandwidth increases. Since the bandwidth is inversely proportional to the pulse width,

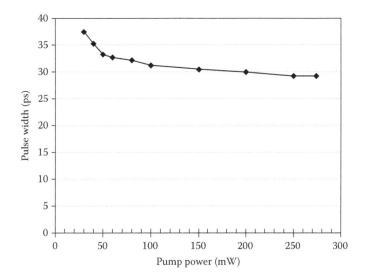

FIGURE 15.46 Pulse width of the HMLFL for various pump powers.

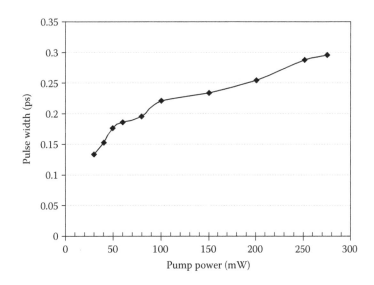

FIGURE 15.47 Change of bandwidth of the HMLFL versus pump power level.

one can infer that the pulse width decreases when the pump power increases. The cause of the decrease in pulse width is attributed to the nonlinear effect. The higher the peak power, the stronger the nonlinear effect, and hence the shorter the pulse width. It is also noticed that the pulse-shortening effect of the pump power is only observed when the fiber length is long, comparable to the effective nonlinear length.

15.9   CONCLUDING REMARKS

We have presented the propagation of solitons through optical fibers through analytical and ISM and numerical approaches. The Lax pairs have enabled the development of eigenvalue solutions for insight observation of the behavior of solitons and interactions of soliton pairs and triple solitons when their relative phase and amplitudes are changed. These dynamic soliton behaviors are then confirmed with experimental demonstration of the formation of solitons and their evolutions in an MLFL.

The time-dependent BPM, the split-step Fourier method, is employed to simulate the propagation of the soliton circulating in an MLFL. Our algorithm has been verified by the analytical approach using the ISM, as described in Section 15.6. As both of these approaches produce consistent results, we further conduct our investigation in the interaction of two and three solitons by using the numerical approach.

As we can conclude, there are three effective methods to overcome the undesirable effects of soliton collapse in our optical soliton transmission systems, which are the method of setting the optimum pulse separation, relative amplitude, and relative phase. The pulse separation method could be used if we are not too concerned about the maximum achievable bit rate. Relative phase method is only useful at 0° and 180°. However, one can use the repulsive nature of the solitons by setting the relative phase to a certain value in order to compensate for the attractive nature of the solitons. However, this method is difficult to implement in practice. Therefore, the relative amplitude method would be the best in overcoming this problem. Only a slight difference in the amplitude of the solitons would prevent the solitons from collapsing.

Thus, we have to implement a device at the transmitting terminal to encode each binary digit with alternate value of amplitudes. This can be carried out by using an external modulator to modulate each pulse with different amplitudes alternately: (1) The simulation assumes a zero-loss fiber. It would be very useful to take into account the losses due to propagation in the optical fiber, and this could be implemented using distributed Raman amplification. (2) Also, the simulation model assumes that the parameters can be controlled exactly. It would be more accurate if the simulation also considers the jittering effects in the phase, amplitudes, as well as the initial separation of the solitons. (3) Finally, the simulation should also consider the effects of the erbium-doped fiber amplifier scheme to boost signals, as well as Raman amplification.

This chapter has also experimentally demonstrated and simulated stable multisoliton bound states that have been generated in an active MLFL under phase matching via optical phase modulation. It is believed that this stable existence of multisoliton bound states is effectively supported by the phase modulation in an anomalous dispersion fiber loop. Simulation results have confirmed the existence of multisoliton bound states in the FM mode-locked fiber laser. The created bound states can easily be harmonic mode-locked to periodically generate multisoliton bound sequence at high repetition rates in this type of fiber laser that is much more prominent than those generated by passive types.

The pulse train of very short pulse width at high repetition rates, up to 10 GHz, generated from an HMLFL, has also been demonstrated. The characteristics of the pulse train such as pulse width and bandwidth have been studied by varying the settings of the cavity. It is found that, as the modulation frequency increases, the pulse becomes shorter, and its corresponding bandwidth increases. The increase of the modulation depth also makes the pulse shorter, but its effect on the pulse is not as strong as modulation frequency. High pump power is also found to help shorten the pulse width in the long cavity laser.

REFERENCES

1. N. Zabusky and M. D. Kruskal, Interaction of solitons in a collisionless plasma and the recurrence of initial states, *Phys. Rev. Lett.*, Vol. 15, p. 240, 1965.
2. G. P. Agrawal, *Nonlinear Fiber Optics*, NY: Academic Press, 1992.
3. V. E. Zaharov and A. B. Shabat, Exact theory of two-dimensional self-focusing and one-dimensional self-modulation of wave in nonlinear media, *Sov. Phys. J. Exp. Theor. Phys.*, Vol. 34, No. 1, pp. 62–69, 1972.
4. R. K. Dodd, J. C. Eilbeck, J. D. Gibbon, and H. C. Morris, *Solitons and Nonlinear Wave Equations*, NY: Academic Press, 1982.
5. M. J. Ablowitz, D. J. Kaup, A. C. Newell, and H. Segur, The inverse scattering transform-Fourier analysis for nonlinear problems, *Studies in Applied Mathematics*, Vol. 53, No. 4, pp. 249–315, 1974.
6. J. Satsuma and N. Yajima, Initial value problems of one-dimensional self-modulation of nonlinear waves in dispersive media, *Supp. Prog. Theor. Phys.*, Vol. 55, pp. 284–306, 1974.
7. D. Anderson and M. Lisak, Bandwidth limits due to mutual pulse interaction in optical soliton communication systems, *Opt. Lett.*, Vol. 10, No. 3, pp. 174–176, 1986.
8. A. Bondeson, M. Lisak, and D. Anderson, Soliton perturbations: A variational principle for soliton parameters, *Phys. Scr.*, Vol. 20, pp. 479–485, 1979.
9. V. I. Karpman and V. V. Solov'ev, A perturbational approach to the two-soliton systems, *Physica*, Vol. 3D, pp. 487–502, 1981.
10. C. Desem and P. L. Chu, Reducing soliton interaction in single-mode optical fibers, *IEE Proc. J.*, Vol. 134, Pt. 3, pp. 145–151, 1987.
11. B. A. Malomed, Bound solitons in coupled nonlinear Schrodinger equation, *J. Phys. Rev. A*, Vol. 45, pp. R8321–R8323, 1991.
12. B. A. Malomed, Bound solitons in the nonlinear Schrodinger-Ginzburg-Landau equation, *J. Phys. Rev. A*, Vol. 44, pp. 6954–6957, 1991.
13. D. Y. Tang, B. Zhao, D. Y. Shen, and C. Lu, Bound-soliton fiber laser, *J. Phys. Rev. A*, Vol. 66, p. 033806, 2002.
14. Y. D. Gong, D. Y. Tang, P. Shum, C. Lu, T. H. Cheng, W. S. Man, and H. Y. Tam, Mechanism of bound soliton pulse formation in a passively mode locked fiber ring laser, *Opt. Eng.*, Vol. 41, No. 11, pp. 2778–2782, 2002.
15. P. Grelu, F. Belhache, and F. Gutty, Relative phase locking of pulses in a passively mode-locked fiber laser, *J. Opt. Soc. Am. B*, Vol. 20, pp. 863–870, 2003.
16. L. M. Zhao, D. Y. Tang, T. H. Cheng, H. Y. Tam, and C. Lu, Bound states of dispersion-managed solitons in a fiber laser at near zero dispersion, *Appl. Opt.*, Vol. 46, pp. 4768–4773, 2007.
17. W. W. Hsiang, C. Y. Lin, and Y. Lai, Stable new bound soliton pairs in a 10 GHz hybrid frequency modulation mode locked Er-fiber laser, *Opt. Lett.*, Vol. 31, pp. 1627–1629, 2006.
18. C. R. Doerr, H. A. Hauss, E. P. Ippen, M. Shirasaki, and K. Tamura, Additive-pulse limiting, *Opt. Lett.*, Vol. 19, pp. 31–33, 1994.
19. R. Davey, N. Langford, and A. Ferguson, Interacting solitons in erbium fiber laser, *Electron. Lett.*, Vol. 27, pp. 1257–1259, 1991.
20. D. Krylov, L. Leng, K. Bergman, J. C. Bronski, and J. N. Kutz, Observation of the breakup of a prechirped N-soliton in an optical fiber, *Opt. Lett.*, Vol. 24, pp. 1191–1193, 1999.
21. J. E. Prilepsky, S. A. Derevyanko, and S. K. Turitsyn, Conversion of a chirped Gaussian pulse to a soliton or a bound multisoliton state in quasi-lossless and lossy optical fiber spans, *J. Opt. Soc. Am. B*, Vol. 24, pp. 1254–1261, 2007.
22. G. P. Agrawal, *Nonlinear Fiber Optics*, NY: Academic Press, 2001.
23. A. P. Agrawal, *Fiber-Optic Communication Systems*, NY: John Wiley & Sons, 1992.
24. J. P. Gordon, Interaction forces among solitons in optical fibers, *Opt. Lett.*, Vol. 8, No. 10, pp. 596–598, 1983.
25. A. Hasegawa, *Optical Solitons in Fibers*, Germany: Springer-Verlag, Berlin, 1989.

16 Higher-Order Spectrum Coherent Receivers

This chapter presents the processing of digital signals using higher-order spectral techniques [1–4] to evaluate the performance of coherent transmission systems, especially the bispectrum method in which the signal distribution in the frequency domain is obtained in a plane. Thus, the cross-correlation as well as the signals themselves can be spatially identified. Thence, we can evaluate the impairments imposed on the transmission systems due to different causes of such degradations. Although equalization has not been described here in this chapter, the techniques for equalization using higher-order spectrum can be found in Refs. [1–3]. The higher order will require additional processors but is compensated by additional dimensions to separate the linear and nonlinear impairments. These issues will be described in the following sections of this chapter.

16.1 BISPECTRUM OPTICAL RECEIVERS AND NONLINEAR PHOTONIC PREPROCESSING

In this section, we present the processing of optical signals before the optoelectronic detection in the optical domain in a nonlinear optical waveguide as an nonlinear (NL) signal processing technique for digital optical receiving system for long-haul optically amplified fiber transmission systems. The algorithm implemented is a high-order spectrum (HOS) technique in which the original signals and two delayed versions are correlated via the four-wave mixing (FWM) or third harmonic conversion process. The optical receivers employing higher-order spectral photonic preprocessors and very-large-scale-integrated (VLSI) electronic systems for the electronic decoding and evaluation of the bit error rate (BER) of the transmission system are presented. A photonicsignal preprocessing system is developed to generate the triple correlation via the third harmonic conversion in a NL optical waveguide. hence the essential part of a triple correlator. The performance of an optical receiver incorporating the HOS processor is given for long-haul phase-modulated fiber transmission.

16.1.1 INTRODUCTORY REMARKS

As stated in previous chapters, tremendous efforts have been taken for reaching higher transmission bit rates and longer haul for optical fiber communication systems. The bit rate can reach several hundreds of Gbps and nearer to the Tbps. In this extremely high-speed operational region, the limits of electronic speed processors have been surpassed, and optical processing is assumed to play an important part of the optical receiving circuitry. Furthermore, novel processing techniques are required in order to minimize the bottlenecks of electronic processing and noises and distortion due to the impairment of the transmission medium, the linear and NL distortion effects.

This chapter deals with the photonic processing of optically modulated signals prior to the electronic receiver for long-haul optically amplified transmission systems. NL optical waveguides in planar or channel structures are studied and employed as a third harmonic converter, so as to generate a triple product of the original optical waves and its two delayed copies. The triple product is then detected by an optoelectronic receiver. Thence, the detected current would be electronically sampled and digitally processed to obtain the bispectrum of the data sequence, and a recovery algorithm is used to recover the data sequence. The generic structures of such high-order spectral optical receivers are shown in Figure 16.1. For the nonlinear photonic processor (Figure 16.1a),

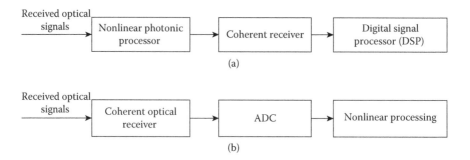

FIGURE 16.1 Generic structure of high-order spectrum optical receiver; (a) photonic preprocessor and (b) nonlinear DSP-based coherent receiver. Note that both in-phase and quadrature components can be processed as combined signals.

the optical signals at the input are delayed and then coupled to a nonlinear photonic device in which the nonlinear conversion process is implemented through the use of third harmonic conversion or degenerate FWM. In the nonlinear digital processor (see Figure 16.1b), the optical signals are detected coherently and then sampled and processed using the nonlinear triple correlation and decoding algorithm.

We propose and simulate this NL photonic preprocessor optical receiver under the MATLAB® and Simulink® platform for differentially coded phase shift keying, the DQPSK modulation scheme.

This section is organized as follows: Section 16.1.3 gives a brief introduction of the triple correlation and bispectrum processing techniques. Section 16.1.4 introduces the simulation platform for the long-haul optically amplified optical fiber communication systems. Section 16.2.2.1 then gives the implementation of the NL optical processor which consists of a nonlinear optical waveguide and an optical receiver associated with a digital signal processing (DSP) sub-systems operating in the electronic domain, so as to recover the data sequence. The performance of the transmission system is given with an evaluation of the BER under linear and NL transmission regimes.

16.1.2 BISPECTRUM

In signal processing, the power spectrum estimation showing the distribution of power in the frequency domain is a useful and popular tool to analyze or characterize a signal or process. However, the phase information between frequency components is suppressed in the power spectrum. Therefore, it is necessarily useful to exploit higher-order spectra known as multidimensional spectra instead of the power spectrum in some cases, especially in nonlinear processes or systems [5]. Different from the power spectrum, the Fourier transform of the autocorrelation, multidimensional spectra are known as Fourier transforms of high-order correlation functions, and hence they provide us not only the magnitude information but also the phase information.

In particular, the two-dimensional spectrum, also called *bispectrum*, is by definition the Fourier transform of the triple correlation or the third-order statistics [2]. For a signal $x(t)$, its triple correlation function C_3 is defined as

$$C_3(\tau_1, \tau_2) = \int x(t)x(t - \tau_1)x(t - \tau_2)\,dt \tag{16.1}$$

where τ_1, τ_2 are the time-delay variables. Thus, the bispectrum can be estimated through the Fourier transform of C_3 as follows

$$B_i(f_1, f_2) \equiv F\{C_3\} = \iint C_3(\tau_1, \tau_2)\exp(-2\pi j(f_1\tau_1 + f_2\tau_2))\,d\tau_1\,d\tau_2 \tag{16.2}$$

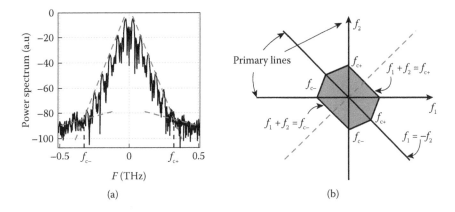

FIGURE 16.2   A description of (a) power spectrum regions and (b) bispectrum regions for explanation.

where $F\{\}$ is the Fourier transform, and f_1, f_2 are the frequency variables. From the definitions of Equations 16.1 and 16.2, both the triple correlation and the bispectrum are represented in a 3D graph with two variables of time and frequency, respectively. Figure 16.2 shows the regions of the power spectrum and bispectrum, respectively, and their relationship. The cutoff frequencies are determined by intersection between the noise and spectral lines of the signal. These frequencies also determine the distinct areas that are basically bounded by a hexagon in a bispectrum. The area inside the hexagon only shows the relationship between frequency components of the signal, whereas the area outside shows the relationship between the signal components and noise. Due to the two-dimensional representation in bispectrum, the variation of the signal and the interaction between signal components can easily be identified.

Because of the unique features of the bispectrum, it is really useful in characterizing the non-Gaussian or nonlinear processes and applicable in various fields such as signal processing [1,2], biomedicine, and image reconstruction. The extension of a number of representation dimensions makes the bispectrum become more easily and significantly a representation of different types of signals and differentiation of various processes, especially nonlinear processes such as doubling and chaos. Hence, the multidimensional spectra technique is proposed as a useful tool to analyze the behaviors of signals generated from these systems.

16.1.3   Bispectrum Coherent Optical Receiver

Figure 16.1 shows the structure of a bispectrum optical receiver in which there are three main sections: an all-optical preprocessor, an optoelectronic detection, and amplification including an analog-to-digital converter (ADC) to generate sampled values of the triple-correlated product. This section is organized as follows. The next subsection gives an introduction to bispectrum and associate noise elimination as well as the benefits of the bispectrum techniques, then the details of the bispectrum processor are given, and then some implementation aspects of the bispectrum processor using VLSI are stated.

16.1.4   Triple Correlation and Bispectra

16.1.4.1   Definition

The power spectrum is the Fourier transform of the autocorrelation of a signal. The bispectrum is the Fourier transform of the triple correlation of a signal. Thus, both the phase and amplitude information of the signals are embedded in the triple-correlated product.

While autocorrelation and its frequency-domain power spectrum do not contain the phase information of a signal, the triple correlation contains both, due to the definition of triple correlation,

$$c(\tau_1, \tau_2) = \int S(t)S(t+\tau_1)S(t+\tau_2)dt \qquad (16.3)$$

where $S(t)$ is the continuous time-domain signal to be recovered, and τ_1, τ_2 are the delay time intervals. For the special cases where $\tau_1 = 0$ or $\tau_2 = 0$, the triple correlation is proportional to the autocorrelation. It means that the amplitude information is also contained in the triple correlation. The benefit of holding phase and amplitude information is that it gives a potential to recover the signal back from its triple correlation. In practice, the delays τ_1 and τ_2 indicate the path difference between the three optical waveguides. These delay times are corresponding to the frequency regions in the spectral domain. Thus, different time intervals would determine the frequency lines in the bispectrum.

16.1.4.2 Gaussian Noise Rejection

Let $S(n)$ be a deterministic sampled signal which the sampled version of the continuous signals $S(t)$. $S(n)$ is corrupted by Gaussian noise w(n), where n is the sampled time index. The observed signal takes the form $Y(n) = S(n) + w(n)$. The polyspectra of any Gaussian process is zero for any order greater than two [2]. The bispectrum is the third-order polyspectrum and offers significant advantage for signal processing over the second-order polyspectrum, commonly known as the power spectrum, which is corrupted by Gaussian noise. Theoretically speaking, the bispectral analysis allows us to extract a non-Gaussian signal from the corrupting effects of Gaussian noise.

Thus, for a signal that has arrived at the optical receiver, the steps to recover the amplitude and phase of the lightwave-modulated signals are as follows:

1. Estimating the bispectrum of $S(n)$ based on observations of $Y(n)$.
2. From the amplitude and phase bispectra, form an estimate of the amplitude and phase distribution in one-dimensional frequency of the Fourier transform of $S(n)$. These form the constituents of the signal $S(n)$ in the frequency domain.
3. Thence, taking the inverse Fourier transform to recover the original signal $S(n)$. This type of receiver can be termed as the *bispectral optical receiver.*

16.1.4.3 Encoding of Phase Information

The bispectra contains almost complete information about the original signal (magnitude and phase). And thus, if the original signal $x(n)$ is real and finite, it can be reconstructed, except for a shift a. Equivalently, the Fourier transform can be determined, except for a linear shift factor of $e^{-j2\pi\omega a}$. By determining two adjacent pulses, any differential phase information will then be readily available [5]. In other words, the bispectra, and hence the triple correlation, contain the phase information of the original signal, allowing it to "pass through" the square law photodiode, which would otherwise destroy this information. The encoded phase information can then be recovered up to a linear phase term, thus necessitating a differential coding scheme.

16.1.4.4 Eliminating Gaussian Noise

For processes that have zero mean and the symmetrical probability density function (PDF), their third-order cumulants are equal to zero. Therefore, in a triple correlation, those symmetrical processes are eliminated. Gaussian noise is assumed to affect the signal quality. Mathematically, the third cumulant is defined as

$$c_3(\tau_1, \tau_2) = m_3(\tau_1, \tau_2) - m_1 \begin{bmatrix} m_2(\tau_1) + m_2(\tau_2) \\ + m_2(\tau_1 - \tau_2)] + 2(m_1)^3 \end{bmatrix} \qquad (16.4)$$

where m_k is the k-th order moment of the signal, especially since it is the mean of the signal. Thus, for the zero mean and symmetrical PDF, its third-order cumulant becomes zero [1,2]. Theoretically, considering the signal as $u(t) = s(t) + n(t)$, where $n(t)$ is an additive Gaussian noise, the triple correlation of $u(t)$ will reject Gaussian noise affecting the $s(t)$.

16.1.5 Transmission and Detection

16.1.5.1 Optical Transmission Route and Simulation Platform

Shown in Figure 16.3 is the schematic of the transmission link over a total length of 700 km, with sections from Melbourne city (in Victoria, Australia) to Gippsland, the inland section in Victoria (city to coastal landing for submarine connection), thence an undersea section of more than 300 km crossing the Bass Strait to George Town of Tasmania (undersea or submarine), and then inland transmission to Hobart of Tasmania (coastal area to city connection). The Gippsland/George Town link is shown in Figure 16.3. Other inland sections in Victoria and Tasmania of Australia are structured with optical fibers and lumped optical amplifiers (Er-doped fiber amplifiers—EDFA). Raman distributed optical amplification (ROA) is used by pump sources located at both ends of the Melbourne-to-Hobart link, including the 300 km undersea section. The undersea section of nearly 300 km long of fibers consisting of only the transmission and dispersion-compensating sections, no active subsystems are included. Only Raman amplification is used with the pump sources injecting the laser beams into both sides of the section. These pump sources are located on the shores of the inland area. This 300 km distance is a fairly long one, and only Raman distributed gain is used. Simulink models of the transmission system including the optical transmitter, the transmission line, and the bispectrum optical receiver are given.

16.1.5.2 FWM and Bispectrum Receiving

We have also integrated the waveguide whose material parameters are real into the MATLAB and Simulink of the transmission system so that the NL parametric conversion system is very close to practice. The spectra of the optical signals before and after this amplification are shown in Figure 16.4, indicating the conversion efficiency. This indicates the performance of the bispectrum optical receiver (Figures 16.5 and 16.6).

16.1.5.3 Performance

We implement the models for both techniques for binary phase shift keying modulation format for serving as a guideline for phase modulation optical transmission systems using nonlinear preprocessing. We note the following:

1. The arbitrary white Gaussian noise (AWGN) block in the Simulink platform can be set in different operating modes. This block then accepts the signal input and assumes the sampling rate of the input signal, then estimates the noise variance based on Gaussian distribution and the specified signal-to-noise ratio (SNR). This is then superimposed on the amplitude of the sampled value. Thus, we believe, at that stage, that the noise is contributed evenly across the entire band of the sampled time (converted to spectral band).

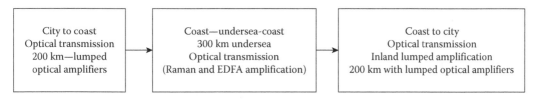

FIGURE 16.3 Schematic of the transmission link including inland and undersea sections between city and coast (inland beach), then undersea and coast (inland) to city.

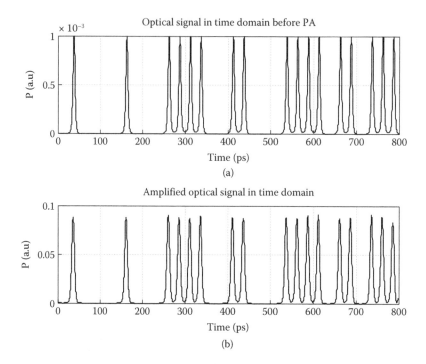

FIGURE 16.4 Time traces of the optical signal: (a) before and (b) after the parametric amplifier.

2. The ideal curve SNR versus the BER plotted in the graph provided is calculated using the commonly used formula in several textbooks on communication theory. This is evaluated based on the geometrical distribution of the phase states, and then the noise distribution over those states. This means that all the modulations and demodulations are assumed to be perfect. However, in digital system simulation, the signals must be sampled. This is even more complicated when a carrier is embedded in the signal, especially when the phase shift keying (PSK) modulation format is used.

3. We thus re-setup the models of (i) AWGN in a complete binary phase shift keying (BPSK) modulation format with both the ideal coherent modulator and demodulator and any necessary filtering required, (ii) AWGN blocks with the coherent modulator and demodulator incorporating the triple correlator and necessary signal processing block. This is done in order to make fair comparison between the two pressing systems.

4. In our former model, the AWGN block was being used incorrectly, in that it was being used in "SNR" mode, which applies the noise power over the entire bandwidth of the channel, which of course is larger than the data bandwidth, meaning that the amount of noise in the data band was a fraction of the total noise applied. We accept that this was an unfair comparison to the theoretical curve that is given against E_b/N_0, as defined in Ref. [1].

5. In the current model, we provide a fair comparison noise that was added to the modulated signal using the AWGN block in the E_b/N_0 mode (E_0 is the energy per bit and N_0 is the noise contained within the bit period), with the "symbol period" set to the carrier period, and, in effect, this set the "carrier-to-noise" ratio, or CNR. Also, the triple correlation receiver was modified a little from the original, namely the addition of the BP filter and some tweaking of the triple correlation delays, which resulted in the BER curve shown in Figure 16.7. Also, an ideal homodyne receiver model was constructed with the noise added and measured in the exact same method as the triple correlation model. This provided a benchmark to compare the triple correlation receiver against.

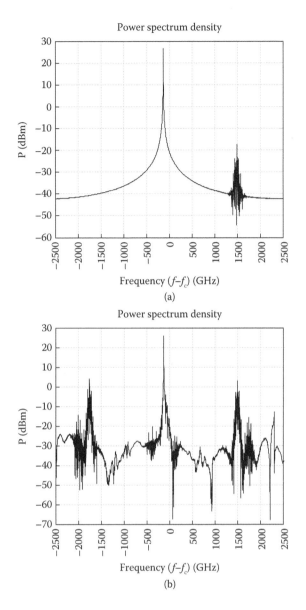

FIGURE 16.5 Corresponding spectra of the optical signal: (a) before and (b) after the parametric amplifier.

6. Furthermore, we can compare the simulated BER values with the theoretical limit set by

$$P_b = \frac{1}{2} \text{erfc} \sqrt{\frac{E_b}{N_0}} \qquad (16.5)$$

by relating the CNR to E_b/N_0 as in the following,

$$\frac{E_b}{N_0} = \text{CNR} \frac{B_W}{f_s} \qquad (16.6)$$

where the channel bandwidth B_W is 1600 Hz, set by the sampling rate, and f_s is the symbol rate—in our case, a symbol is one carrier period (100 Hz), as we are adding noise to

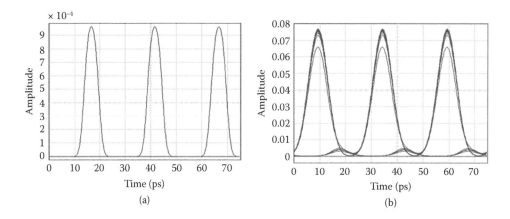

FIGURE 16.6   (a) Input data sequence and (b) detected sequence processed using triple correlation nonlinear photonic processing and recovery scheme bispectrum receiver.

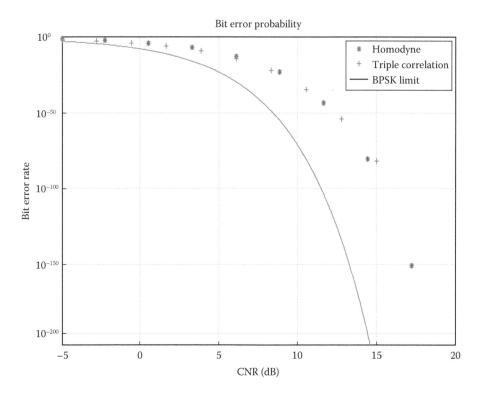

FIGURE 16.7   BER versus carrier/noise ratio in power for nonlinear triple correlation, ideal BPSK under homodyne coherent detection, and ideal BPSK limit. Note that, in practice, the BER of up to 1e−19 would be the lowest value used.

the carrier. These frequencies are set at the normalized level, so as to scale to wherever spectral regions would be of interest. As can be seen in Figure 16.7, the triple correlation receiver matches the performance of the ideal homodyne case and closely approaches the theoretical limit of BPSK (approximately 3 dB at a BER of 10^{-10}). As discussed, the principle benefit from the triple correlation over the ideal homodyne case will be the characterization of the noise of the channel that is achieved by analysis of the regions of symmetry in the 2D bispectrum. Finally, we still expect possible performance improvement when symbol identification is performed directly from the triple correlation matrix, as opposed to the traditional method that involves recovering the pulse shape first. It is not possible at this stage to model the effect of the direct method.

7. Equalization of digital sampled signals can be implemented by employing the algorithms described in Ref. [1].

16.2 NONLINEAR PHOTONIC SIGNAL PROCESSING USING HIGHER-ORDER SPECTRA

16.2.1 INTRODUCTORY REMARKS

With the increasing demand for high capacity, communication networks are facing several challenges, especially in signal processing at the physical layer at ultra-high speeds. When the processing speed is over that of the electronic limit or requires massive parallel and high-speed operations, the processing in the optical domain offers significant advantages. Thus, all-optical signal processing is a promising technology for future optical communication networks. An advanced optical network requires a variety of signal processing functions, including optical regeneration, wavelength conversion, optical switching, and signal monitoring. An attractive way to realize these processing functions in transparent and high-speed modes is to exploit the third-order nonlinearity in optical waveguides, particularly parametric processes.

Nonlinearity is a fundamental property of optical waveguides, including channel, rib-integrated structures, or circular fibers. The origin of nonlinearity comes from the third-order nonlinear polarization in optical transmission media. It is responsible for various phenomena such as self-phase modulation (SPM), cross-phase modulation (XPM), and FWM effects. In these effects, the parametric FWM process is of special interest, because it offers several possibilities for signal processing applications. To implement all-optical signal processing functions, highly nonlinear optical waveguides are required where the field of the guided waves is concentrated in its core region. Hence we must use a guided medium whose nonlinear coefficient is sufficiently high so as to achieve the required energy conversion. Therefore, the highly nonlinear fibers (HNLFs) are commonly employed for this purpose, since the nonlinear coefficient of HNLF is about tenfold higher than that of standard transmission fibers. Indeed, the third-order nonlinearity of conventional fibers is often very small to prevent the degradation of the transmission signal from nonlinear distortions. Recently, nonlinear chalcogenide and tellurite glass waveguides have emerged as a promising device for ultra-high-speed photonic processing. Because of their geometries, these waveguides are called *planar waveguides*. A planar waveguide can confine the lightwaves within an area comparable to the effective wavelength of lighwaves in the medium. Hence, they are very compact for signal processing.

16.2.2 FWM AND PHOTONIC PROCESSING FOR HIGHER-ORDER SPECTRA

16.2.2.1 Bispectral Optical Structures

Figure 16.8 shows the generic and detailed structure of the bispectral optical receiver, respectively, which consists of (1) an all-optical preprocessor front end, followed by (2) a photodetector and electronic amplifier to transfer the detected electronic current to a voltage level appropriate for sampling by an ADC and thus the signals at this stage is in sampled form; (3) the sampled triple correlation product is then transformed to the Fourier domain using the FFT. The product at this stage is the row of the matrix of the bispectral amplitude and phase plane (see Figure 16.10). A number of parallel structures may be required if passive delay paths are used; (4) a recovery algorithm is used to derive the one-dimensional distribution of the amplitude and phase as a function of the frequency, which are the essential parameters required for taking the inverse Fourier transform to recover the time-domain signals.

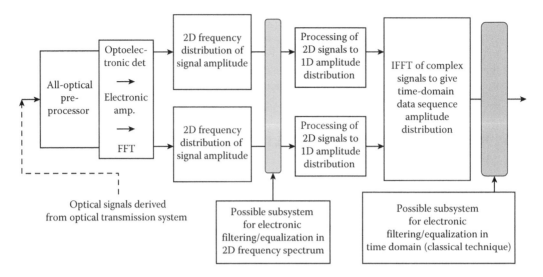

FIGURE 16.8 Generic structure of an optical preprocessing receiver employing bispectrum processing technique.

The physical process of mixing the three waves to generate the fourth wave whose amplitude and phase are proportional to the product of the three input waves is well known in literature of NL optics. This process requires (1) a highly NL medium so as to efficiently convert the energy of the three waves to that of the fourth wave and (2) satisfying the phase-matching conditions of the three input waves of the same frequency (wavelength) to satisfy the conservation of momentum.

16.2.2.2 Phenomena of FWM

The origin of the FWM comes from the parametric processes that lie in the NL responses of bound electrons of a material to applied optical fields. More specifically, the polarization induced in the medium is not linear in the applied field, but contains NL terms whose magnitude is governed by the NL susceptibilities [3,4,6]. The first-, second-, and third-order parametric processes can occur due to these NL susceptibilities [χ^1 χ^2 χ^3]. The coefficient χ^3 is responsible for the FWM that is exploited in this work. Simultaneously with this FWM, there is also a possibility of the generation of third harmonic waves by mixing the three waves and using parametric amplification. The third harmonic generation is normally very small, due to the phase mismatching of the guided wave number (the momentum vector) between the fundamental waves and the third harmonic wave. FWM in guided wave medium such as single-mode optical fibers have been extensively studied due to its efficient mixing to create the fourth wave [4]. The exploitation of the FWM processes in channel optical waveguides has not yet been extensively exploited. In this chapter, we demonstrate theoretical and experimentally the applications of such process in enhancing the sensitivity of the receiver by nonlinear processes.

The three lightwaves are mixed to generate the polarization vector $\vec{P}$ due to the NL third-order susceptibility, given as

$$\vec{P}_{NL} = \varepsilon_0 \chi^{(3)} \vec{E}.\vec{E}.\vec{E} \tag{16.7}$$

where ε_0 is the permittivity in vacuum; $\vec{E}_1, \vec{E}_2, \vec{E}_3$ are the electric field components of the lightwaves; $\vec{E} = \vec{E}_1 + \vec{E}_2 + \vec{E}_3$ is the total field entering the NL waveguide; and $\chi^{(3)}$ is the third-order susceptibility of the NL medium. P_{NL} is the product of the three total optical fields of the three optical waves here that gives the triple product of the waves required for the bispectrum receiver, in which the NL waveguide acts as a multiplier of the three waves that are considered as the pump waves in this section. The mathematical analysis of the coupling equations via the wave equation is complicated, but straightforward. Let ω_1, ω_2, ω_3, and ω_4 be the angular frequencies of the four waves of the FWM process, linearly polarized along the horizontal direction y of the channel waveguides and propagating along the z-direction. The total electric field vector of the four waves is given by

$$\vec{E} = \frac{1}{2}\vec{a}_y \sum_{i=1}^{4} \vec{E}_j e^{j(k_i z - \omega_i t)} + c.c. \tag{16.8}$$

with $\vec{a}_y$ = unit vector along y axis, and $\_c.c.$ = complex conjugate.

The propagation constant can be obtained by $k_i = \dfrac{n_{\mathrm{eff},i}\omega_i}{c}$, where $n_{\mathrm{eff},i}$ is the effective index of the i-th guided waves $E_i(i = 1,...,4)$, which can be either transverse electric (TE) or transverse magnetic (TM) polarized guided mode propagating along the channel NL optical waveguide, and all four waves are assumed to be propagating along the same direction. Substituting Equation 16.8 in Equation 16.7, we obtain

$$\vec{P}_{\mathrm{NL}} = \frac{1}{2}\vec{a}_y \sum_{i=1}^{4} P_i e^{j(k_i z - \omega_i t)} + c.c. \tag{16.9}$$

where $P_i(i = 1,...,4)$ consists of a large number of terms involving the product of three electric fields of the optical-guided waves, for example, the term P_4 can be expressed as

$$P_4 = \frac{3\varepsilon_0}{4}\chi_{xxxx}^{(3)}\left\{ \begin{array}{l} |E_4|^2 E_4 + 2(|E_1|^2 + |E_2|^2 + |E_3|^2)E_4 \\[6pt] + 2E_1 E_2 E_3 e^{j\Phi^+} + 2E_1 E_2 E_3^* e^{j\Phi^-} + c.c. \end{array} \right\} \tag{16.10}$$

with

$$\phi^+ = (k_1 + k_2 + k_3 + k_4)z - (\omega_1 + \omega_2 + \omega_3 + \omega_4)t$$

$$\phi^- = (k_1 + k_2 - k_3 - k_4)z - (\omega_1 + \omega_2 - \omega_3 - \omega_4)t$$

The first four terms of Equation 16.10 represent the SPM and XPM effects, which are dependent on the intensity of the waves. The remaining terms result in FWM. Thus, the question is, which terms are the most effective components that resulted from the parametric mixing process? The effectiveness of the parametric coupling depends on the phase-matching terms governed by ϕ^+ and ϕ^-, or a similar quantity.

It is obvious that significant FWM would occur if the phase matching is satisfied. This requires the matching of both the frequency as well as the wave vectors, as given in Equation 16.10. From Equation 16.10, we can see that the term ϕ^+ corresponds to the case in which three waves are mixed to give the fourth wave whose frequency is three times that of the original wave—this is the third harmonic generation. However, normally the matching of the wave vectors would not be satisfied due to the dispersion effect or the differential mismatching of the wave momentum vectors as guided in a channel optical waveguide. If there is a large mismatching of the momentum vectors of the fundamental and third-order harmonics then only a minute amount of energy would be transferred to the third harmonic components.

The conservation of momentum derived from the wave vectors of the four waves requires that

$$\Delta k = k_1 + k_2 - k_3 - k_4 = \frac{n_{\mathrm{eff},1}\omega_1 + n_{\mathrm{eff},2}\omega_2 - n_{\mathrm{eff},3}\omega_3 - n_{\mathrm{eff},4}\omega_4}{c} = 0 \tag{16.11}$$

The effective refractive indices of the guided modes of the three waves E_1, E_2, and E_3 are the same, as are their frequencies. This condition is automatically satisfied, provided that the NL waveguide is designed such that it supports only a single polarized-mode, TE or TM, and with minimum dispersion difference within the band of the signals.

16.2.3 THIRD-ORDER NONLINEARITY AND PARAMETRIC FWM PROCESS

16.2.3.1 Nonlinear Wave Equation

In optical waveguides, including optical fibers, the third-order nonlinearity is of special importance because it is responsible for all nonlinear effects. The confinement of lightwaves and their propagation in optical waveguides are generally governed by the nonlinear wave equation (NLE), which can be derived from the Maxwell's equations under the coupling of nonlinear polarization. The nonlinear wave propagation in nonlinear waveguide in the time-spatial domain in vector form can be expressed as (see also Chapter 2)

$$\nabla^2 \vec{E} - \frac{1}{c^2}\frac{\partial^2 \vec{E}}{\partial t^2} = \mu_0 \left(\frac{\partial^2 \overrightarrow{P_L}}{\partial t^2} + \frac{\partial^2 \overrightarrow{P_{NL}}}{\partial t^2} \right) \tag{16.12}$$

where $\vec{E}$ is the electric field vector of the lightwave, μ_0 is the vacuum permeability assuming a non-magnetic waveguiding medium, c is the speed of light in vacuum, and $\overrightarrow{P_L}$, $\overrightarrow{P_{NL}}$ are, respectively, the linear and nonlinear polarization vectors, which are formed as

$$\overrightarrow{P_L}(\vec{r},t) = \varepsilon_0 \chi^{(1)} \cdot \vec{E}(\vec{r},t) \tag{16.13}$$

$$\overrightarrow{P_{NL}}(\vec{r},t) = \varepsilon_0 \chi^{(3)} \vdots \vec{E}(\vec{r},t)\vec{E}(\vec{r},t)\vec{E}(\vec{r},t) \tag{16.14}$$

where $\chi^{(3)}$ is the third-order susceptibility. Thus, the linear and nonlinear coupling effects in optical waveguides can be described by Equation 16.12. The second term on the right-hand side is responsible for nonlinear processes including interaction between optical waves through third-order susceptibility.

In most telecommunication applications, only complex envelopes of optical signals is considered in analysis because the bandwidth of the optical signal is much smaller than the optical carrier frequency. To model the evolution of the light propagation in optical waveguides, it requires that Equation 16.12 be further modified and simplified by some assumptions that are valid in most telecommunication applications. Hence, the electrical field $\vec{E}$ can be written as

$$\vec{E}(\vec{r},t) = \frac{1}{2}\hat{x}\left\{ F(x,y)A(z,t)\exp[i(kz - \omega t)] + c.c. \right\} \tag{16.15}$$

where $A(z,t)$ is the slowly varying complex envelope propagating along z in the waveguide, and k is the wave number. After some algebra using the method of separating variables, the following equation for propagation in optical waveguide is obtained [7]

$$\frac{\partial A}{\partial z} + \frac{\alpha}{2}A - i\sum_{n=1}^{\infty} \frac{i^n \beta_n}{n!}\frac{\partial^n A}{\partial t^n} = i\gamma\left(1 + \frac{i}{\omega_0}\frac{\partial}{\partial t}\right) \times A \int_{-\infty}^{\infty} g(t')|A(z,t-t')|^2\, dt' \tag{16.16}$$

where the effect of propagation constant β around ω_0 is Taylor-series expanded, and $g(t)$ is the nonlinear response function including the electronic and nuclear contributions. For the optical pulses wide enough to contain many optical cycles, Equation 16.15 can be simplified as

$$\frac{\partial A}{\partial z} + \frac{\alpha}{2} A + \frac{i\beta_2}{2} \frac{\partial^2 A}{\partial \tau^2} - \frac{\beta_3}{6} \frac{\partial^3 A}{\partial \tau^3} = i\gamma \left[|A|^2 A + \frac{i}{\omega_0} \frac{\partial \left(|A|^2 A \right)}{\partial \tau} - T_R A \frac{\partial \left(|A|^2 \right)}{\partial \tau} \right] \tag{16.17}$$

where a frame of reference moving with the pulse at the group velocity v_g is used by making the transformation $\tau = t - z/v_g \equiv t - \beta_1 z$, and A is the total complex envelope of propagation waves; α, β_k are the linear loss and dispersion coefficients, respectively; $\gamma = \omega_0 n_2 / c A_{eff}$ is the nonlinear coefficient of the guided wave structure; and the first moment of the nonlinear response function is defined as

$$T_R \equiv \int_0^\infty t g(t') dt' \tag{16.18}$$

Equation 16.17 is the basic propagation equation, commonly known as the nonlinear Schrödinger equation (NLSE), which is very useful for investigating the evolution of the amplitude of the optical signal and the phase of the lightwave carrier under the effect of third-order nonlinearity in optical waveguides. The left-hand side (LHS) in Equation 16.17 contains all linear terms, while all nonlinear terms are contained on the right-hand side. In this equation, the first term on the right-hand side is responsible for the intensity-dependent refractive index effects, including FWM.

16.2.3.2 FWM Coupled-Wave Equations

FWM is a parametric process through the third-order susceptibility $\chi^{(3)}$. In the FWM process, the superposition and generation of the propagating of the waves with different amplitudes A_k, frequencies ω_k, and wave numbers k_k through the waveguide can be represented as

$$A = \sum_n A_n e^{[j(k_n z - \omega_n \tau)]} \quad \text{where } n = 1,...,4 \tag{16.19}$$

By ignoring the linear and scattering effects and with the introduction of Equation 16.19 in Equation 16.17, the NLSE can be separated into coupled differential equations, each of which is responsible for one distinct wave in the waveguide

$$\frac{\partial A_1}{\partial z} + \frac{\alpha}{2} A_1 = i\gamma A_1 \left[|A_1|^2 + 2 \sum_{n \neq 1} |A_n|^2 \right] + i\gamma 2 A_3 A_4 A_2^* \exp(-i\Delta k_1 z)$$

$$\frac{\partial A_2}{\partial z} + \frac{\alpha}{2} A_2 = i\gamma A_2 \left[|A_2|^2 + 2 \sum_{n \neq 2} |A_n|^2 \right] + i\gamma 2 A_3 A_4 A_1^* \exp(-i\Delta k_2 z)$$

$$\frac{\partial A_3}{\partial z} + \frac{\alpha}{2} A_3 = i\gamma A_3 \left[|A_3|^2 + 2 \sum_{n \neq 3} |A_n|^2 \right] + i\gamma 2 A_1 A_2 A_4^* \exp(-i\Delta k_3 z) \tag{16.20}$$

$$\frac{\partial A_4}{\partial z} + \frac{\alpha}{2} A_4 = i\gamma A_4 \left[|A_4|^2 + 2 \sum_{n \neq 4} |A_n|^2 \right] + i\gamma 2 A_1 A_2 A_3^* \exp(-i\Delta k_4 z)$$

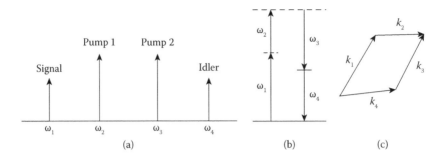

FIGURE 16.9 (a) Position and notation of the distinct waves, (b) diagram of energy conservation, and (c) diagram of momentum conservation in FWM.

where $\Delta k = k_1 + k_2 - k_3 - k_4$ is the wave vector mismatch. The equation system (Equation 16.20) thus describes the interaction between different waves in nonlinear waveguides. The interaction that is represented by the last term in Equation 16.20 can generate new waves. For three waves with different frequencies, a fourth wave can be generated at frequency $\omega_4 = \omega_1 + \omega_2 - \omega_3$. The waves at frequencies ω_1 and ω_2 are called pump waves, whereas the wave at frequency ω_3 is the signal, and the generated wave at ω_4 is called the idler wave, as shown in Figure 16.9a. If all three waves have the same frequency $\omega_1 = \omega_2 = \omega_3$, the interaction is called a degenerate FWM with the new wave at the same frequency. If only two of the three waves are at the same frequency ($\omega_1 = \omega_2 \neq \omega_3$), the process is called *partly degenerate FWM*, which is important for some applications such as the wavelength converter and parametric amplifier.

16.2.3.3 Phase Matching

In parametric nonlinear processes such as FWM, the energy conservation and momentum conservation must be satisfied to obtain a high efficiency of the energy transfer, as shown in Figure 16.9a. The phase-matching condition for the new wave requires

$$\Delta k = k_1 + k_2 - k_3 - k_4 = \frac{1}{c}(n_1\omega_1 + n_2\omega_2 - n_3\omega_3 - n_4\omega_4) = 2\pi\left(\frac{n_1}{\lambda_1} + \frac{n_2}{\lambda_2} - \frac{n_3}{\lambda_3} - \frac{n_4}{\lambda_4}\right) \quad (16.21)$$

During propagation in optical waveguides, the relative phase difference $\theta(z)$ between the four involved waves is determined by

$$\theta(z) = \Delta kz + \phi_1(z) + \phi_2(z) - \phi_3(z) - \phi_4(z) \quad (16.22)$$

where $\phi_k(z)$ relates to the initial phase and the nonlinear phase shift during propagation. An approximation of phase-matching condition can be given as

$$\frac{\partial \theta}{\partial z} \approx \Delta k + \gamma(P_1 + P_2 - P_3 - P_4) = \kappa \quad (16.23)$$

where P_k is the power of the waves, and κ is the phase mismatch parameter. Thus, the FWM process has maximum efficiency for $\kappa = 0$. The mismatch comes from the frequency dependence of the refractive index and the dispersion of optical waveguides. Depending on the dispersion profile of the nonlinear waveguides, it is very important in the selection of pump wavelengths to ensure that the phase mismatch parameter is minimized.

Once the fourth wave is generated, the interaction of the four waves along the section of the waveguide continues to happen, and thus the NL Schrödinger equation must be used to investigate the evolution of the waves.

16.2.3.4 Coupled Equations and Conversion Efficiency

To derive the wave equations to represent the propagation of the three waves to generate the fourth wave, we can resort to Maxwell's equations. It is lengthy to write down all the steps involved in this derivation, so we summarize the standard steps usually employed to derive the wave equations as follows: First, add the NL polarization vector given in Equation 16.7 into the electric field density vector D. Then, taking the curl of the first Maxwell's equation and using the second equation of the four Maxwell's equations and substituting the electric field density vector and using the fourth equation, one would then come up with the vectorial wave equation.

For the FWM process occurring during the interaction of the three waves along the propagation direction of the NL optical channel waveguide, the evolution of the amplitudes, A_1–A_4, of the four waves, E_1–E_4, given in Equation 16.20, is given by (only the A_1 term is given)

$$\frac{dA_1}{dz} = \frac{jn_2\omega_1}{c}\left[\left(\Gamma_{11}|A_1|^2 + 2\sum_{k\neq 1}\Gamma_{1k}|A_k|^2\right)A_1 + 2\Gamma_{1234}A_2^*A_3A_4 e^{j\Delta kz}\right] \tag{16.24}$$

where the wave vector mismatch Δk is given in Equation 16.11, and the * denotes the complex conjugation. Note that the coefficient n_2 in Equation 16.24 is defined as the nonlinear coefficient which is related to the nonlinear susceptibility of the medium. This coefficient can be defined as

$$n_2 = \frac{3}{8n\,\mathrm{Re}(\chi_{xxxx}^3)} \tag{16.25}$$

16.2.4 OPTICAL DOMAIN IMPLEMENTATION

16.2.4.1 Nonlinear Wave Guide

In order to satisfy the condition of FWM and efficient energy transfer between the waves, a guided medium is preferred, thus a rib-waveguide can be employed for guiding. In this case a glass material of AS_2S_3 or TeO_2 whose NL refractive index coefficient is about 100,000 times greater than that of silica is produced for forming the nonlinear waveguide. The three waves are guided in this waveguide structure. Their optical fields are overlapped. The cross-section of the waveguide is in the order of 4×0.4 μm. The waveguide cross-section can be designed such that the dispersion is "flat" over the spectral range of the input waves, ideally from 1520 nm to 1565 nm. This can be done by adjusting the thickness of the rib structure.

The fourth wave generated from the FWM waveguide is then detected by the photodetector, which acts as an integrating device. Thus, the output of this detector is the triple correlation product in the electronic domain that we are looking for.

If equalization or filtering is required, then these functional blocks can be implemented in the bispectral domain as shown in Figure 16.8. Figure 16.10 shows the parallel structures of the bispectral receiver so as to obtain all rows of the bispectral matrix. The components of the structure are almost similar, except the delay time of the optical preprocessor.

In the NL channel waveguide, fabricated using TeO_2 (tellurium oxide) on silica, the interaction of the three waves, one original and two delayed beams, happens via the electronic processes, with highly NL coefficient $[\chi^3]$ converting to the fourth wave.

When the three waves are co-propagating, due to the conservation of the momentum of the interaction and exchange of energy of the four waves the fourth wave is produced that satisfy the FWM phase matching conditions. Indeed, phase matching can also be satisfied by one forward wave and two backward-propagating waves (delayed version of the first wave), leading to almost 100% conversion efficiency to generate the fourth wave.

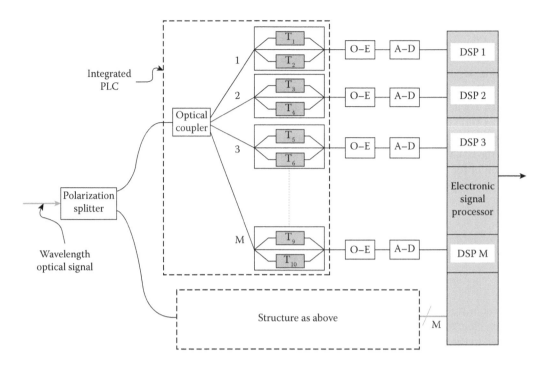

FIGURE 16.10 Parallel structures of photonic preprocessing to generate a triple correlation product in the optical domain.

The interaction of the three waves via the electronic process and the $\chi^{(3)}$ gives rise to the polarization vector P, which in turn couples with the electric field density of the lightwaves and then with the NL Schrödinger wave equation. By solving and modeling this wave equation with the FWM term on the right-hand side of the equation, one can obtain the wave output (the fourth wave) at the end of the NL waveguide section.

16.2.4.2 Third Harmonic Conversion

Third harmonic conversion may happen, but at extremely low efficiencies, at least 1000 times less than that of FWM due to the nonmatching of the effective refractive indices of the guided modes at 1550 nm (fundamental wave) and 517 nm (third harmonic wave).

It is noted that the common term for this process is the matching of the dispersion characteristics, that is, k/omega, where omega is the radial frequency of the waves at 1550 and 1517 nm, versus the thickness of the waveguide.

16.2.4.3 Conservation of Momentum

The conservation of momentum and, thus, phase-matching condition for the FWM are satisfied without much difficulty, as the wavelengths of the three input waves are the same direction. Thus it requires that the generated waves propagating in the guide must satisfy this condition. Thence the optical NL channel waveguide must be designed such that there is no mismatching of the third harmonic conversion and the other forward and backward waves. It is considered that single polarized mode, either TE or TM, will be used to achieve efficient FWM. Thus, the dimension of the channel waveguide would be estimated at about 0.4 μm (height) × 4 μm (width).

16.2.4.4 Estimate of Optical Power Required for FWM

In order to achieve efficient FWM conversion, the NL coefficient $n2$ must be large. This coeffcient is proportional to $\chi^{(3)}$ by a constant $8n/3$, with n as the effective refractive index. This NL coefficient is then multiplied by the intensity of the guided waves to give an estimate of the phase change and, thence, estimation of the efficiency of the FWM. The cross-section of the waveguide can be estimated by using Equation 16.3. Due to high index difference between the waveguide and its cladding, the guided mode is well confined. Hence the effective area of the guided waves is very close to the cross-section area. Thus, an average power of the guided waves would be about 3–5 mW, or about 6 dBm.

The loss of the linear section which is section of multimode interference and delay split is estimated at 3 dB. Thus, the input power of the three waves required for efficient FWM is about 10 dBm (maximum).

16.2.5 Transmission Models and Nonlinear Guided Wave Devices

To model the parametric FWM process between multiwaves, the basic propagation equations described in Section 16.2.3 are used. There are two approaches to simulate the interaction between waves. The first approach, named as the *separating channel technique*, is to use the coupled equations system (Equation 16.20) in which the interactions between different waves are obviously modeled by certain coupling terms in each coupled equation. Thus, each optical wave considered as one separated channel is represented by a phasor. The coupled equations system is then solved to obtain the solutions of the FWM process. The outputs of the nonlinear waveguide are also represented by separated phasors, and hence the desired signal can be extracted without using a filter (Figure 16.11).

The second or alternative approach is to use the propagation equation (Equation 16.17), which allows us to simulate all evolutionary effects of the optical waves in the nonlinear waveguides. In this technique, a total field is used instead of individual waves. The superimposed complex envelope A is represented by only one phasor, which is the summation of individual complex amplitudes of different waves, given as

$$A = \sum_k A_k e^{[j(\omega_k - \omega_0)\tau]} \tag{16.26}$$

where ω_0 is the defined angular central frequency, and A_n, ω_n are the complex envelope and carrier frequency of individual waves, respectively. Hence, various waves at different frequencies are

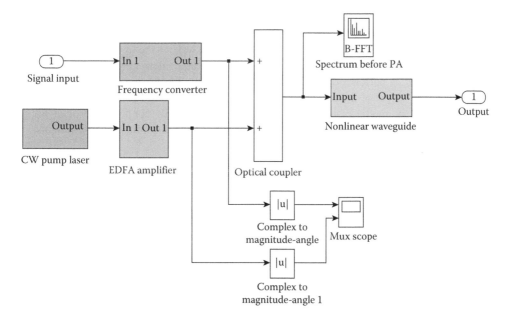

FIGURE 16.11 Typical Simulink® setup of the parametric amplifier using the model of nonlinear waveguide.

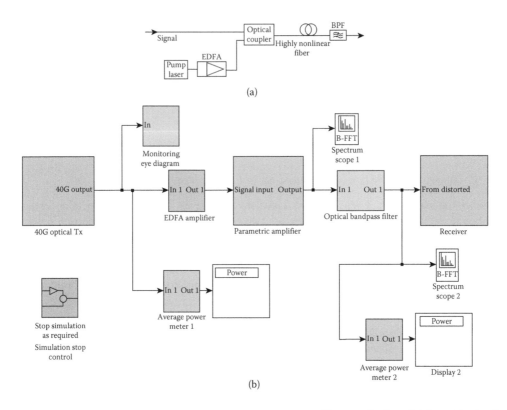

FIGURE 16.12   (a) The typical setup of an optical parametric amplifier and (b) Simulink® model of optical parametric amplifier.

combined into a total signal vector, which facilitates integration of the nonlinear waveguide model into the Simulink platform. Equation 16.17 can also be numerically solved by the split-step Fourier method (SSFM). The Simulink block, representing the nonlinear waveguide, is implemented with an embedded MATLAB program. Because only complex envelopes of the guided waves are considered in the simulation, each of the different optical waves is shifted by a frequency difference between the central frequency and the frequency of the wave to allocate the wave in the frequency band of the total field. Then, the summation of the individual waves, which is equivalent to the combination process at the optical coupler, is performed prior to entering the block of nonlinear waveguide, as depicted in Figure 16.12. The output of the nonlinear waveguide will be selected by an optical bandpass filter (BPF). In this way, the model of nonlinear waveguide can be easily connected to other Simulink blocks that are available in the platform for simulation of optical fiber communication systems.

16.3   SYSTEM APPLICATIONS OF THIRD-ORDER PARAMETRIC NONLINEARITY IN OPTICAL SIGNAL PROCESSING

In this section, a range of signal processing applications are demonstrated through simulations that use the model of nonlinear waveguide to model the wave mixing process.

16.3.1   Parametric Amplifiers

One of the important applications of the $\chi^{(3)}$ nonlinearity is parametric amplification. The optical parametric amplifiers (OPA) offer a wide gain bandwidth, high differential gain, and optional wavelength conversion and operation at any wavelength [7]. These important features of OPA are obtained

TABLE 16.1

Critical Parameters of the Parametric Amplifier in a 40 Gbps System

RZ 40 Gbps Transmitter

λ_s = 1520–1600 nm, λ_0 = 1559 nm

Modulation: RZ-ASK, P_s = 0.01 mW (peak), B_r = 40 Gbps

Parametric Amplifier

Pump source: P_p = 1 W (after EDFA), λ_p = 1560.07 nm

HNLF: L_f = 500 m, D = 0.02 ps/km/nm, S = 0.09 ps/nm²/km, α = 0.5 dB/km, A_{eff} = 12 μm², γ = 13 1/W/km

BPF: $\Delta\lambda_{BPF}$ = 0.64 nm

Receiver

Bandwidth B_e = 28 GHz, i_{eq} = 20 pA/Hz$^{1/2}$, i_d = 10 nA

because the parametric gain process do not rely on energy transitions between energy states, but it is based on highly efficient FWM in which two photons at one or two pump wavelengths interact with a signal photon. The fourth photon, the idler, is formed with a phase such that the phase difference between the pump photons and the signal and idler photons satisfies the phase-matching condition (Equation 16.21). The schematic of the fiber-based parametric amplifier is shown in Figure 16.12a. The parameters of the OPA are given in Table 16.1.

For parametric amplifier using one pump source, from the coupled equations (Equation 16.20) with $A_1 = A_2 = A_p$, $A_3 = A_s$, and $A_4 = A_i$, it is possible to derive three coupled equations for complex field amplitudes of the three waves $A_{p,s,i}$

$$\frac{\partial A_p}{\partial z} = -\frac{\alpha}{2} A_p + i\gamma A_p \left[|A_p|^2 + 2\left(|A_s|^2 + |A_i|^2\right)\right] + i2\gamma A_s A_i A_p^* \exp(-i\Delta kz)$$

$$\frac{\partial A_s}{\partial z} = -\frac{\alpha}{2} A_s + i\gamma A_s \left[|A_s|^2 + 2\left(|A_p|^2 + |A_i|^2\right)\right] + i\gamma A_p^2 A_i^* \exp(-i\Delta kz) \qquad (16.27)$$

$$\frac{\partial A_i}{\partial z} = -\frac{\alpha}{2} A_i + i\gamma A_i \left[|A_i|^2 + 2\left(|A_s|^2 + |A_p|^2\right)\right] + i\gamma A_p^2 A_s^* \exp(-i\Delta kz)$$

The analytical solution of these coupled equations determines the gain of the amplifier [4].

$$G_s(L) = \frac{|A_s(L)|^2}{|A_s(0)|^2} = 1 + \left[\frac{\gamma P_p}{g} \sinh(gL)\right]^2 \qquad (16.28)$$

with L is the length of the highly nonlinear fiber/waveguide, P_p is the pump power, and g is the parametric gain coefficient

$$g^2 = -\Delta k \left(\frac{\Delta k}{4} + \gamma P_p\right) \qquad (16.29)$$

where the phase mismatch Δk can be approximated by extending the propagation constant in a Taylor series around ω_0

$$\Delta k = -\frac{2\pi c}{\lambda_0^2} \frac{dD}{d\lambda} \left(\lambda_p - \lambda_0\right)\left(\lambda_p - \lambda_s\right)^2 \qquad (16.30)$$

Here, $dD/d\lambda$ is the slope of the dispersion factor $D(\lambda)$ evaluated at the zero dispersion of the guided wave component, that is, at the optical wavelength, $\lambda_k = 2\pi c/\omega_k$.

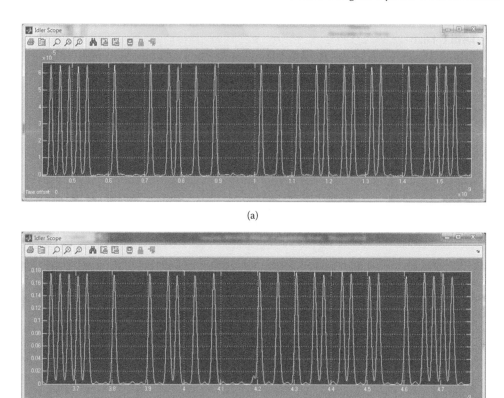

FIGURE 16.13   Time traces of the 40 Gbps signal (a) before and (b) after the parametric amplifier.

Figure 16.13b shows the Simulink setup of the 40 Gbps RZ transmission system using a parametric amplifier. The setup contains a 40 Gbps optical RZ transmitter, an optical receiver for monitoring, a parametric amplifier block, and a BPF that filters the desired signal from the total field output of the amplifier. Details of the parametric amplifier block can be seen in Figure 16.12. The block setup of the parametric amplifier consists of a continuous wave (CW) pump laser source, an optical coupler to combine the signal and the pump, and a highly non-linear fiber block that contains the embedded MATLAB model for nonlinear propagation. The important simulation parameters of the system are listed in Table 16.2.

Figure 16.14 shows the signals before and after the amplifier in the time domain. The time trace indicates the amplitude fluctuation of the amplified signal as a noisy source from a wave mixing process. Their corresponding spectra are shown in Figure 16.15. The noise floor of the output spectrum of the amplifier shows the gain profile of OPA. Simulated dependence of OPA gain on the wavelength difference between the signal and the pump as shown in Figure 16.16, together with theoretical gain using Equation 16.15. The plot shows an agreement between theoretical and simulated results. The peak gain is achieved at phase-matched conditions where the linear phase mismatch is compensated for by the nonlinear phase shift.

16.3.1.1   Wavelength Conversion and Nonlinear Phase Conjugation

Beside the signal amplification in a parametric amplifier, the idler is generated after the wave mixing process. Therefore, this process can also be applied to wavelength conversion. Due to the very fast response of the third-order nonlinearity in optical waveguides, the wavelength conversion based on this effect is transparent to the modulation format and the bit rate of signals. For a flat

TABLE 16.2

Critical Parameters of the Parametric Amplifier in a 40 Gbps Transmission System

RZ 40 Gbps Signal

$\lambda_0 = 1559$ nm, $\lambda_s = \{1531.12, 1537.4, 1543.73, 1550.12\}$ nm, $P_s = 1$ mW (peak), $B_r = 40$ Gbps

Parametric Amplifier

L pump source: $P_p = 100$ mW (after EDFA), $\lambda_p = 1560.07$ nm

HNLF: $L_f = 200$ m, $D = 0.02$ ps/km/nm, $S = 0.03$ ps/nm²/km, $\alpha = 0.5$ dB/km, $A_{eff} = 12$ μm², $\gamma = 13$ 1/W/km

BPF: $\Delta\lambda_{BPF} = 0.64$ nm, $\lambda_i = \{1587.91, 1581.21, 1574.58, 1567.98\}$ nm

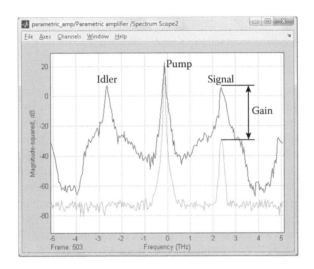

FIGURE 16.14 Optical spectra at the input (light gray-lower) and the output (dark gray-higher) of the OPA.

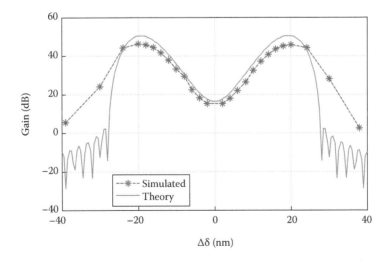

FIGURE 16.15 Calculated and simulated gain of the OPA at $P_p = 30$ dBm.

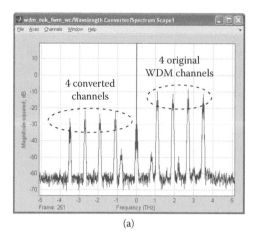
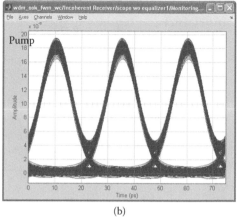

(a)   (b)

FIGURE 16.16 (a) The wavelength conversion of four WDM channels and (b) eye diagram of the converted 40 Gbps signal after BPF.

wideband converter, which is a key device in wavelength-division multiplexing (WDM) networks, a short-length HNLF with a low dispersion slope is required in design. By a suitable selection of the pump wavelength, the wavelength converter can be optimized to obtain a bandwidth of 200 nm. Therefore, the wavelength conversion between bands such as C and L bands can be performed in WDM networks. Figure 16.16 shows an example of the wavelength conversion for four WDM channels at the C band. The important parameters of the wavelength converter are shown in Table 16.2. The WDM signals are converted into the L-band with a conversion efficiency of −12 dB.

Another important application with the same setup is the nonlinear phase conjugation (NPC). A phase-conjugated replica of the signal wave can be generated by the FWM process. From Equation 16.8, the idler wave is approximately given in case of degenerate FWM for simplification: $E_i - A_p^2 A_s^* e^{-j\Delta kz}$ or $E_i - rA_s^* e^{[j(-kz-\omega\tau)]}$ with the signal wave $E_s - A_s e^{[j(kz-\omega\tau)]}$. Thus, the idler field is a complex conjugate of the signal field. In appropriate conditions, optical distortions can be compensated for by using NPC, and optical pulses propagating in the fiber link can be recovered. The basic principle of distortion compensation with NPC refers to spectral inversion. When an optical pulse propagates in an optical fiber, its shape will be spread in time and distorted by the group velocity dispersion. The phase-conjugated replica of the pulse is generated in the middle point of the transmission link by the nonlinear effect. On the contrary, the pulse is spectrally inverted where spectral components in the lower frequency range are shifted to the higher frequency range, and vice versa. If the pulse propagates in the second part of the link in the same manner as in the first part, it is inversely distorted again, which can cancel the distortion in the first part to recover the pulse shape at the end of the transmission link. By using NPC for distortion compensation, a 40–50% increase in transmission distance compared to a conventional transmission link can be obtained. Figure 16.17 shows the setup of a long-haul 40 Gbps transmission system demonstrating the distortion compensation using NPC. The fiber transmission link of the system is divided into two sections by an NPC based on parametric amplifier. Each section consists of five spans with 100 km of standard single-mode fiber (SSMF) in each span. Figure 16.18a shows the eye diagram of the signal after propagating through the first fiber section. After the parametric amplifier at the midpoint of the link, the idler signal, a phase-conjugated replica of the original signal, is filtered for transmission in next section. The signal in the second section suffers the same dispersion as in the first section. At the output of the transmission system, the optical signal is regenerated, as shown in Figure 16.18b. Due to the change in wavelength of the signal in NPC, a tunable dispersion compensator can be required to compensate for the residual dispersion after transmission in real systems. A summary of the parameters of the sub-systems of this transmission system is given in Table 16.3.

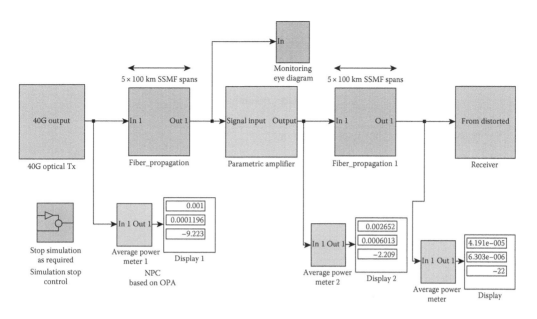

FIGURE 16.17 Simulink® setup of a long-haul 40 Gbps transmission system using NPC for distortion compensation.

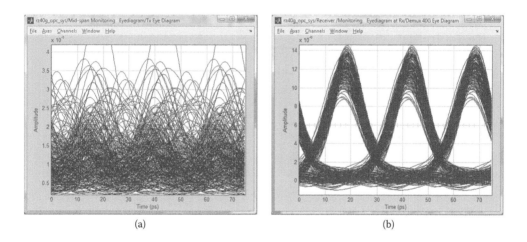

FIGURE 16.18 Eye diagrams of the 40 Gbps signal at the end (a) of the first section and (b) of the transmission link.

16.3.1.2 High-Speed Optical Switching

When the pump is an intensity-modulated signal instead of the CW signal, the gain of the OPA is also modulated due to its exponential dependence on the pump power in a phase-matched condition. The width of gain profile in the time domain is inversely proportional to the product of the gain slope (S_p) or the nonlinear coefficient and the length of the nonlinear waveguide (L). Therefore, an OPA with high gain or large S_pL operates as an optical switch with an ultra-high bandwidth, which is very important in some signal processing applications such as pulse compression or short-pulse generation. A Simulink setup for a 40 GHz short-pulse generator is built with the configuration shown in Figure 16.19. In this setup, the input signal is a CW source with low power, and the pump is amplitude modulated by a Mach–Zehnder intensity modulator (MZIM), which is driven by an RF sinusoidal wave at 40 GHz. The waveform of the modulated pump is shown in Figure 16.20a.

TABLE 16.3
Critical Parameters of the Long-Haul Transmission System Using NPC for Distortion Compensation

RZ 40 Gbps Transmitter

λ_s = 1547 nm, λ_0 = 1559 nm
Modulation: RZ-OOK, P_s = 1 mW (peak), B_r = 40 Gbps

Fiber Transmission Link

SMF: L_{SMF} = 100 km, D_{SMF} = 17 ps/nm/km, α = 0.2 dB/km
EDFA: Gain = 20 dB, NF = 5 dB;
Number of spans: 10 (five in each section), L_{link} = 1000 km

NPC Based on OPA

Pump source: P_p = 1 W (after EDFA), λ_p = 1560.07 nm
HNLF: L_f = 500 m, D = 0.02 ps/km/nm, S = 0.09 ps/nm²/km, α = 0.5 dB/km, A_{eff} = 12 μm², γ = 13 1/W/km
BPF: $\Delta\lambda_{BPF}$ = 0.64 nm

Receiver

Bandwidth B_e = 28 GHz, i_{eq} = 20 pA/Hz$^{1/2}$, i_d = 10 nA

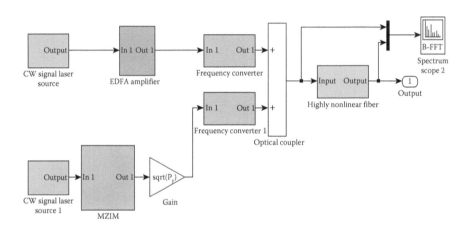

FIGURE 16.19   Simulink® setup of the 40 GHz short-pulse generator.

Important parameters of the FWM-based short-pulse generator are shown in Table 16.4. Figure 16.20b shows the generated short-pulse sequence with a pulse width of 2.6 ps at the signal wavelength after the optical BPF.

Another important application of the optical switch based on the FWM process is the demultiplexer, a key component in ultra-high-speed optical time division multiplexing (OTDM) systems. OTDM is a key technology for Tbps Ethernet transmission that can meet the increasing demand of traffic in future optical networks. A typical scheme of OTDM demultiplexer in which the pump is a mode-locked laser (MLL) to generate short pulses for control is shown in Figure 16.21a. The working principle of the FWM-based demultiplexing is described as follows. The control pulses generated from an MLL at tributary rate are pumped and co-propagated with the OTDM signal through the nonlinear waveguide. The mixing process between the control pulses and the OTDM signal during propagation through the nonlinear waveguide converts the desired tributary channel to a new idler wavelength. Then the demultiplexed signal at the idler wavelength is extracted by a band pass filter before going to a receiver, as shown in Figure 16.21a.

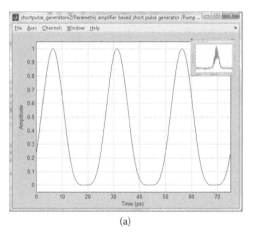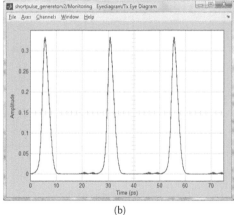

(a)   (b)

FIGURE 16.20 Time traces of (a) the sinusoidal amplitude modulated pump and (b) the generated short-pulse sequence (inset: the pulse spectrum).

TABLE 16.4

Parameters of the 40 GHz Short Pulse Generator

Short-Pulse Generator

Signal: $P_s = 0.7$ mW, $\lambda_s = 1535$ nm, $\lambda_0 = 1559$ nm

Pump source: $P_p = 1$ W (peak), $\lambda_p = 1560.07$ nm, $f_m = 40$ GHz

HNLF: $L_f = 500$ m, $D = 0.02$ ps/km/nm, $S = 0.03$ ps/nm²/km,

$\quad \alpha = 0.5$ dB/km, $A_{eff} = 12$ µm², $\gamma = 13$ 1/W/km

BPF: $\Delta\lambda_{BPF} = 3.2$ nm

However, its stability, especially the walk-off problem, is still a serious obstacle. Recently, planar nonlinear waveguides have emerged as promising devices for ultra-high-speed photonic processing. These nonlinear waveguides offer several advantages, such as no free-carrier absorption, stability at room temperature, no requirement of quasi-phase matching, and the possibility of dispersion engineering. With the same operational principle, planar waveguide-based OTDM demultiplexers are very compact and suitable for photonic integrated solutions. Figure 16.21b shows the Simulink setup of the FWM-based demultiplexer of the on–off keying (OOK) 40 Gbps signal from the 160 Gbps OTDM signal using a highly nonlinear waveguide instead of HNLF. Important parameters of the OTDM system in Table 16.5 are used in the simulation. Figure 16.22a shows the spectrum at the output of the nonlinear waveguide. Then, the demultiplexed signal is extracted by the BPF, as shown in Figure 16.22b. Figure 16.23 shows the time traces of the 160 Gbps OTDM signal, the control signal, and the 40 Gbps demultiplexed signal, respectively. The dark gray dots in Figure 16.23a indicate the timeslots of the desired tributary signal in the OTDM signal. The developed model of OTDM demultiplexer can be applied not only to the conventional OOK format, but also to advanced modulation formats such as DQPSK, which increases the data load of the OTDM system without increase in bandwidth of the signal. By using available blocks developed for the DQPSK system shown Figure 16.21b, a Simulink model of the DQPSK-OTDM system is setup for demonstration. The bit rate of the OTDM system is doubled to 320 Gbps with the same pulse repetition rate. Figure 16.24 shows the simulated performance of the demultiplexer in both 160 Gbps OOK-OTDM and 320 Gbps DQPSK-OTDM systems. The BER curve in case of the DQPSK-OTDM signal shows a low error floor, which may be the result of the influence of nonlinear effects on phase-modulated signals in the waveguide.

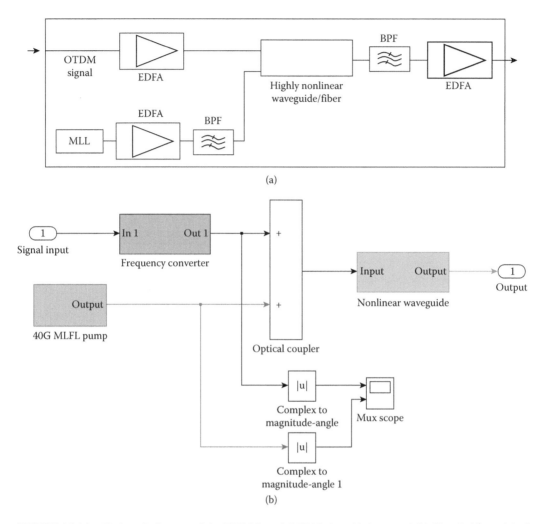

FIGURE 16.21 (a) A typical setup of the FWM-based OTDM demultiplexer and (b) Simulink® model of the OTDM demultiplexer.

TABLE 16.5

Important Parameters of the FWM-Based OTDM Demultiplexer Using a Nonlinear Waveguide

OTDM Transmitter

MLL: $P_0 = 1$ mW, $T_p = 2.5$ ps, $f_m = 40$ GHz

Modulation formats: OOK and DQPSK; OTDM multiplexer: 4×40 Gsymbols/s

FWM-Based Demultiplexer

Pumped control: $P_p = 500$ mW, $T_p = 2.5$ ps, $f_m = 40$ GHz, $\lambda_p = 1556.55$ nm

Input signal: $P_s = 10$ mW (after EDFA), $\lambda_s = 1548.51$ nm

Waveguide: $L_w = 7$ cm, $D_w = 28$ ps/km/nm, $S_w = 0.003$ ps/nm²/km,

   $\alpha = 0.5$ dB/cm, $\gamma = 10^4$ 1/W/km

BPF: $\Delta\lambda_{BPF} = 0.64$ nm

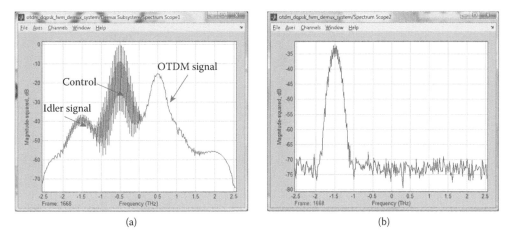

FIGURE 16.22 Spectra at the outputs (a) of nonlinear waveguide and (b) of BPF.

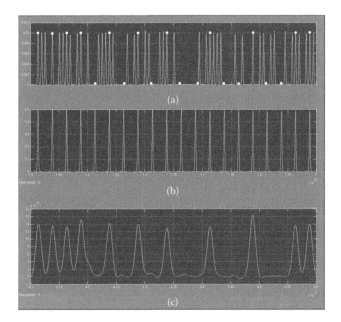

FIGURE 16.23 Time traces of (a) the 160 Gbps OTDM signal, (b) the control signal, and (c) the 40 Gbps demultiplexed signal.

16.3.1.3 Triple Correlation

One of the promising applications exploiting the $\chi^{(3)}$ nonlinearity is the implementation of triple correlation in the optical domain. Triple correlation is a higher-order correlation technique, and its Fourier transform, called bispectrum, is very important in signal processing, especially in signal recovery. The triple correlation of a signal $s(t)$ can be defined as

$$C^3(\tau_1, \tau_2) = \int s(t)s(t-\tau_1)s(t-\tau_2)\,dt \tag{16.31}$$

where τ_1, τ_2 are time-delay variables. Thus, to implement the triple correlation in optical domain, the product of three signals including different delayed versions of the original signal need to be generated

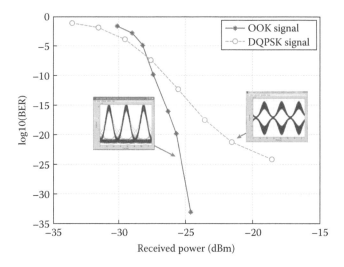

FIGURE 16.24 Simulated performance of the demultiplexed signals for 160 Gbps OOK-OTDM and 320 Gbps DQPSK-OTDM systems (insets: eye diagrams at the receiver).

and then detected by an optical photodiode to perform the integral operation. From the representation of nonlinear polarization vector (see Equation 16.14), this triple product can be generated by the $\chi^{(3)}$ nonlinearity. One way to generate the triple correlation is based on third harmonic generation, where the generated new wave containing the triple-product is at a frequency of three times the original carrier frequency. Thus, if the signal wavelength is in the 1550 nm band, the new wave need to be detected at around 517 nm. The triple optical autocorrelation based on single-stage third harmonic generation has been demonstrated in direct optical pulse shape measurement. However, in this way, it is hard to obtain high efficiency in the wave mixing process due to the difficulty of phase matching between the three signals. Moreover, the triple product wave is in 517 nm, where wideband photodetectors are not available for high-speed communication applications. Therefore, a possible alternative to generate the triple product is based on other nonlinear interactions such as FWM. From Equation 16.20, the fourth wave is proportional to the product of three waves—$A_4 \sim A_1 A_2 A_3^* e^{-j\Delta kz}$. If A_1 and A_2 are the delayed versions of the signal A_3, the mixing of three waves results in the fourth wave A_4, which is obviously the triple product of three signals. As mentioned in Section 16.2, all three waves can take the same frequency; however, these waves should propagate into different directions to possibly distinguish the new generated wave in a diverse propagation direction that requires a strict arrangement of the signals in spatial domain. An alternative way we propose is to convert the three signals into different frequencies (ω_1, ω_2, and ω_3). Then the triple-product wave can be extracted at the frequency $\omega_4 = \omega_1 + \omega_2 - \omega_3$, which is still in the 1550 nm band.

Figure 16.25a shows the Simulink model for the triple correlation based on FWM in a nonlinear waveguide. The structural block consists of two variable delay lines to generate delayed versions of the original signal, as shown in Figure 16.26, and frequency converters to convert the signal into different three waves before combining at the optical coupler to launch into the nonlinear waveguide. Then, the fourth wave signal generated by FWM is extracted by the passband filter. To verify the triple-product based on FWM, another model, shown in Figure 16.25b, to estimate the triple product by using Equation 16.31, is also implemented for comparison. The integration of the generated triple-product signal is then performed at photodetector in the optical receiver to estimate the triple correlation of the signal.

A repetitive signal, which is a dual-pulse sequence with unequal amplitude, is generated for investigation. Important parameters of the setup are shown in Table 16.6. Figure 16.27 shows the waveform of the dual-pulse signal and the spectrum at the output of the nonlinear waveguide. The wavelength spacing between the three waves is unequal to reduce the noise from other

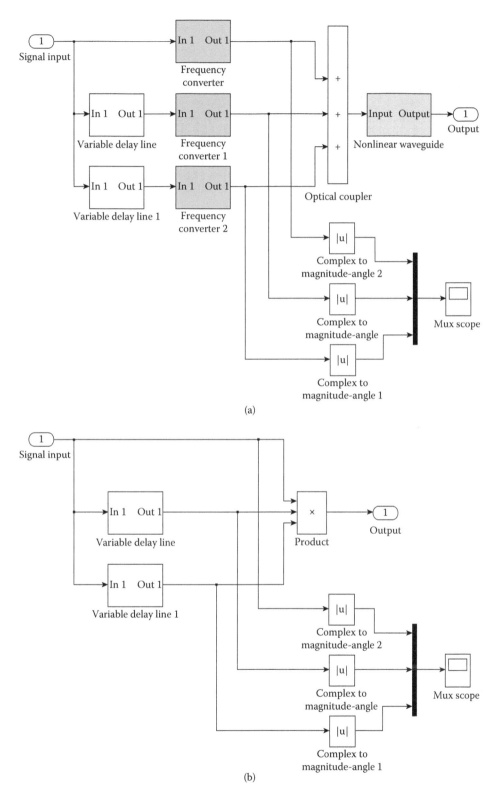

FIGURE 16.25 (a) Simulink® setup of the FWM-based triple-product generation and (b) Simulink setup of the theory-based triple-product generation.

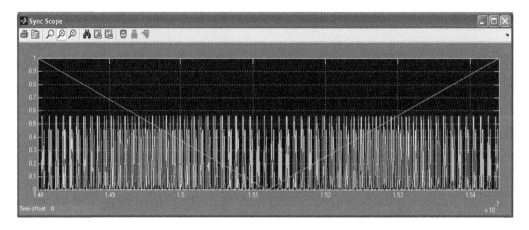

FIGURE 16.26 The variation in time domain of the time delay (gray), the original signal (dark gray), and the delayed signal (light gray).

TABLE 16.6

Important Parameters of the FWM-Based OTDM Demultiplexer Using a Nonlinear Waveguide

Signal Generator

Singl pulse: $P_0 = 100$ mW, $T_p = 2.5$ ps, $f_m = 10$ GHz

Dual pulse: $P_1 = 100$ mW, $P_2 = 2/3\,P_1$, $T_p = 2.5$ ps, $f_m = 10$ GHz

FWM-Based Triple-Product Generator

Original signal: $\lambda_{s1} = 1550$ nm, $\lambda_{s1} = 1552.52$ nm

Delayed τ_1 signal: $\lambda_{s2} = 1552.52$ nm, Delayed τ_2 signal: $\lambda_{s3} = 1554.13$ nm

Waveguide: $L_w = 7$ cm, $D_w = 28$ ps/km/nm, $S_w = 0.003$ ps/nm^2/km,

    $\alpha = 0.5$ dB/cm, $\gamma = 10^4$ 1/W/km

BPF: $\Delta\lambda_{BPF} = 0.64$ nm

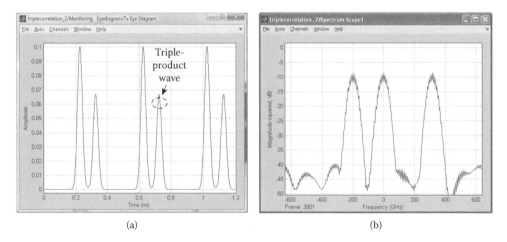

(a)   (b)

FIGURE 16.27 (a) Time trace of the dual-pulse sequence for investigation and (b) spectrum at the output of the nonlinear waveguide.

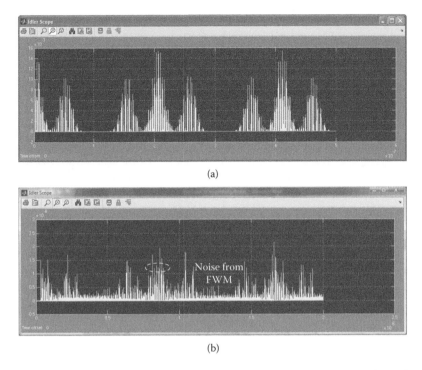

(a)

(b)

FIGURE 16.28   Generated triple-product waves in time domain of the dual-pulse signal based on (a) theory and (b) FWM in nonlinear waveguide.

mixing processes. The triple-product waveforms estimated by theory and FWM process are shown in Table 16.6. In case of the estimation based on FWM, the triple-product signal is contaminated by the noise generated from other mixing processes, as indicated in Table 16.6. Figure 16.28 shows the triple correlations of the signal after processing at the receiver in both cases, based on theory and FWM. The triple correlation is represented by a 3D plot, as displayed in the image. The x- and y- axes of the image represent the time-delay variables (τ_1 and τ_2) in terms of samples with step-size of $T_m/32$, where T_m is the pulse period. The intensity of the triple correlation is represented by colors with scales specified by the color bar. Although the FWM-based triple correlation result is noisy, the triple correlation pattern is still distinguishable, as compared to the theory. Another signal pattern of single pulse that is simpler has also been investigated, as shown in Figures 16.29 and 16.30.

16.3.1.4   Remarks

This section demonstrates the employment of an NL optical waveguide and associated NL effects such as parametric amplification, FWM, and third harmonic generation for the implementation of the triple correlation, and thence the bispectrum creation and signal recovery techniques to reconstruct the data sequence transmitted over a long-haul, optically amplified fiber transmission link.

16.3.2   NONLINEAR PHOTONIC PREPROCESSING IN COHERENT RECEPTION SYSTEMS

This section looks at the uses of nonlinear effects and applications in modern optical communication networks in which 100 Gbps optical Ethernet is expected to be deployed.

The typical performance of a photonic signal preprocessor employing no linear FWM is given, and also that of an advanced processing of such received signals in the electronic domain processed by a digital triple correlation system. At least 10 dB improvement is achieved on the receiver sensitivity.

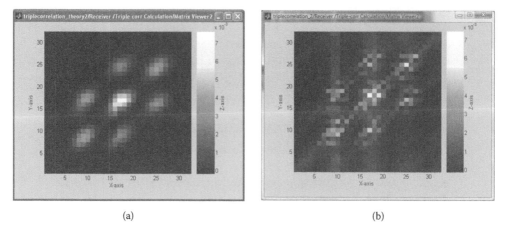

FIGURE 16.29   Triple correlation of the dual-pulse signal based on (a) theoretical estimation and (b) FWM in nonlinear waveguide.

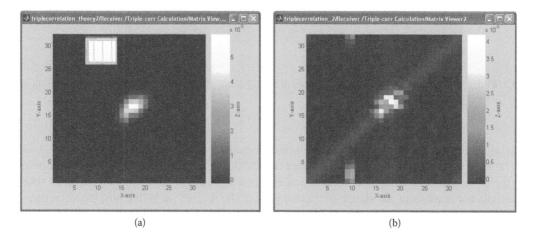

FIGURE 16.30   Triple correlation of the single-pulse signal based on (a) theoretical estimation and (b) FWM in nonlinear waveguide (inset: the single-pulse pattern).

Regarding the nonlinear effects, the nonlinearity of the optical fibers hinders and limits the maximum level of the total average power of all the multiplexed channels for maximizing the transmission distance. These are due to the changes of the refractive index of the guided medium as a function of the intensity of the guided waves. This in turn creates phase changes and hence different group delays, leading to distortion. Furthermore, other associated nonlinear effects such as the four-wave mixing, Raman scattering, Brillouin scattering, intermodulation have also created jittering and distortion of the received pulse sequences after a long transmission distance.

However, we have recently been able to use to our advantage these nonlinear optical effects as a preprocessing element before the optical receiver to improve its sensitivity. A higher-order spectrum technique is employed with the triple correlation implemented in the optical domain via the use of the degenerate FWM effects in a high nonlinear optical waveguide. However, this may add additional optical elements and filtering in the processor and hence complicate the receiver structure. We can overcome this by inserting this nonlinear higher-order spectrum processing sub-system in cascade following the optoelectronic converter and in front of the digital processor (Figure 16.31).

In this section, we illustrate some uses of nonlinear effects and nonlinear processing algorithms for improving the sensitivity of optical receivers employing nonlinear processing at the front end of the photodetector and nonlinear processing algorithm in the electronic domain.

The spectral distribution of the FWM and the simulated spectral conversion can be achieved. There is a degeneracy of the frequencies of the waves, so that efficient conversion can be achieved by satisfying the conservation of momentum. The detected phase states and bispectral properties are depicted in Figure 16.32, in which the phases can be distinguished based on the diagonal spectral lines. Under noisy conditions, these spectral distributions can be observed as depicted in Figure 16.33.

Alternating to the optical processing described earlier, a nonlinear processing technique using HOS technique can be implemented in the electronic domain. This is implemented after the ADC, which samples the incoming electronic signals produced by the coherent optical receiver, as shown in Figure 16.34. The operation of a third-order spectrum analysis is based on the combined interference of three signals (in this case, the complex signals produced at the output of the ADC), two of which are the delayed versions of the original. Thence, the amplitude and phase distribution of the complex signals are obtained in 3D graphs that allow us to determine the signal and noise power, and the phase distribution. These distributions allow us to perform several functions necessary for the evaluation of the performance of optical transmission systems. Simultaneously, the processed signals allow us to monitor the health of the transmission systems—such as the effects due to nonlinear effects, the distortion due to chromatic dispersion of the fiber transmission lines, the noises

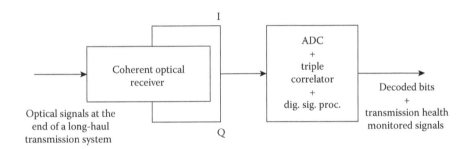

FIGURE 16.31 Schematic diagram of a high-order spectral optical receiver and electronic processing.

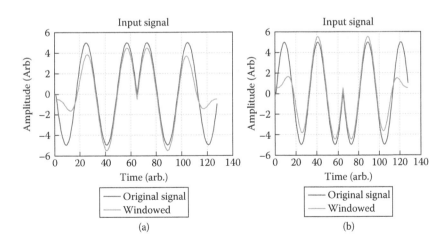

FIGURE 16.32 (a, b) Input waveform with phase changes at the transitions. *(Continued)*

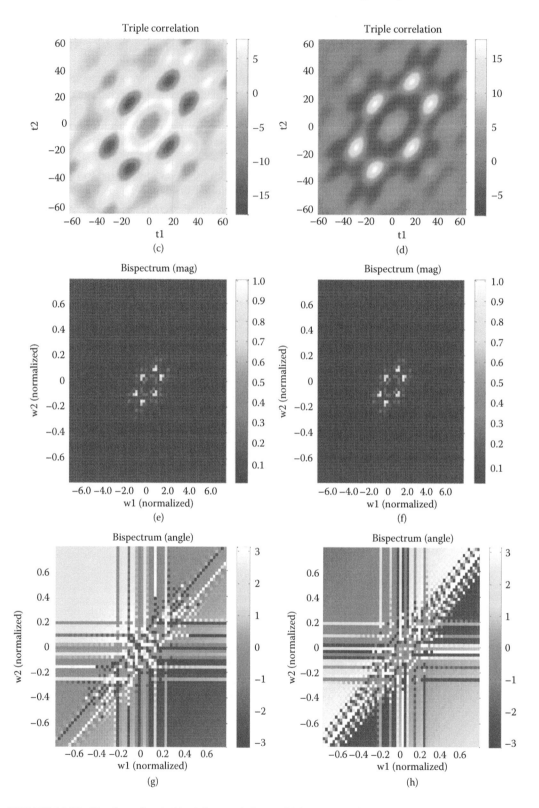

FIGURE 16.32 (Continued) (c–h) triple correlation and bispectrum of both phase and amplitude under off-line DSP processing.

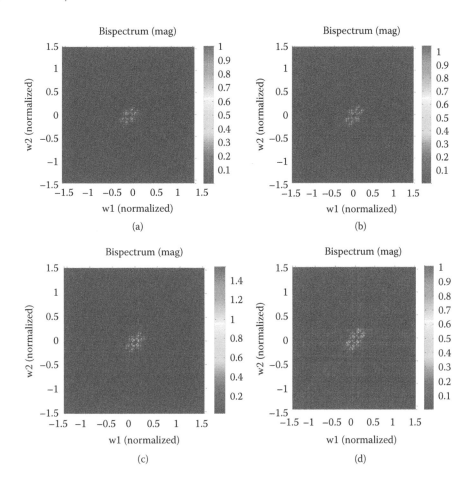

FIGURE 16.33 (a, c) Effect of Gaussian noise on the bispectrum and (b, d) amplitude distribution in 2D phase spectral distribution.

contributed by in-line amplifiers, and so on. A typical curve that compares the performance of this innovative processing with convention detection techniques is shown in Figure 16.34. If a BER of 1e−9 is required at the receivers employing lower order conventional reception, the high-order spectral receiver described here and associate DSP techniques will improve by at least 1000 times. This is equivalent to at least one unit improvement on the quality factor of the eye opening, which is in turn equivalent to about 10 dB in the SNR.

These results are very exciting for network and system operators as significant improvement of the receiver sensitivity can be achieved, and this allows significant flexibility in the operation and management of the transmission systems and networks. Simultaneously, the monitored signals produced by the high-order spectral techniques can be used to determine the distortion and noises of the transmission line, and thus the management of the tuning of the operating parameters of the transmitter, the number of wavelength channels, and the receiver or in-line optical amplifiers.

The effect of additive white Gaussian noise on the bispectrum magnitude is shown in Figure 16.33, and the sequence of figures are as indicated in Figure 16.32. The uncorrupted bispectrum magnitude is shown in Figure 16.32a, while Figure 16.32b through d were generated using SNRs of 10 dB, 3 dB, and 0 dB, respectively. This provides a method of monitoring the integrity of a channel and illustrates another attractive attribute of the bispectrum. It is noted that the bispectrum phase is more sensitive to Gaussian white noise than is the magnitude, and quickly becomes indistinguishable below 6 dB.

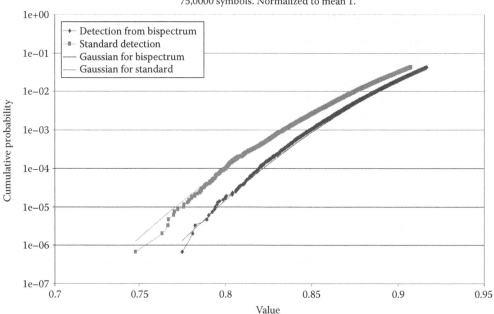

Cumulative probability distribution, with 10% filtered Gaussian noise
for conventional detection and for detection from the bispectrum.
75,0000 symbols. Normalized to mean 1.

FIGURE 16.34 Error estimation of the detection level of a high-order spectrum processor: cumulative probability of error versus normalized amplitude at OSN of 10 dB.

Although the algorithm employing nonlinear processing will involve both hardware and software implementations, from the practical point of view, it needs to deliver the embedded to the market at the right time for systems and networks operating in Tbps speeds. One can thus be facing the following dilemma.

1. The optical preprocessing demands an efficient nonlinear optical waveguide, which must be in the integrated structure whereby efficient coupling and interaction can be achieved. If not, then the gain of about 3 dB in SNR would be defeated by this loss. Furthermore, the integration of the linear optical waveguiding section and a nonlinear optical waveguide is not matched due to differences in the waveguide structures of both regions. For a linear waveguide structure to be efficient for coupling with circular optical fibers, silica on silicon would be best suited due to the small refractive index difference and the technology of burying such waveguides to form an embedded structure whose optical spot size would match with that of a single guided mode fiber. This silica on silicon would not match with an efficient nonlinear waveguide made by As_2S_3 on silicon.

2. On the contrary, if electronic processing is employed, then it requires an ultra-fast ADC, and then fast electronic signal processors. Currently, a 56 GSamples/s ADC is available from Fujitsu, as shown in Figure 16.35. It is noted that the data output samples of the ADC are structured in parallel forms with the referenced clock rate of 1.75 GHz. Thus, all processing of the digital samples must be in parallel form, and, therefore, parallel processing algorithms must be structured in parallel. This is the most challenging problem that we must overcome in the near future.

An application-specific integrated circuit (ASIC) must also be designed for this processor. Hard decisions must also be made in the fitting of such ASICs and associate optical and

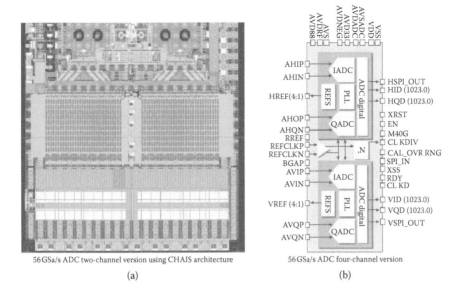

(a) 56 GSa/s ADC two-channel version using CHAIS architecture

(b) 56 GSa/s ADC four-channel version

FIGURE 16.35 Plane view of the Fujitsu ADC operating at 56 GSa/s: (a) integrated chip (b) functional schematic of the ADC.

optoelectronic components into international standard compatible size. Thus, all design and components must meet this requirement.

16.4 CONCLUDING REMARKS

In this section, we have demonstrated a range of signal processing applications exploiting the parametric process in nonlinear waveguides. A brief mathematical description of the parametric process through third-order nonlinearity has been reviewed. A nonlinear waveguide has been proposed to simulate the interaction of multiple waves in optical waveguides, including optical fibers. Based on the developed Simulink modeling platform, a range of signal processing applications exploiting the parametric FWM process has been investigated through simulation. With a CW pump source, applications such as parametric amplifier, wavelength converter, and optical phase conjugator have been implemented for demonstration. Ultra-high-speed optical switching can be implemented by using an intensity-modulated pump to apply in the short-pulse generator and the OTDM demultiplexer. Moreover, the FWM process has been proposed to estimate the triple correlation, which is very important in signal processing. The simulation results showed the possibility of the FWM-based triple correlation using the nonlinear waveguide with different pulse patterns. Although the triple correlation is contaminated by noise from other FWM processes, it is distinguishable. The wavelength positions as well as the power of three delayed signals need to be optimized to obtain the best results.

Furthermore, we have also addressed the important issues of nonlinearity and its uses in optical transmission systems, and the management of networks if the signals that indicate the health of the transmission system are available. There is no doubt that nonlinear phenomena exert distortion impairments on transmitted signals. However these phenomena can also be exploited to improve the transmission quality of the signals as presented in this chapter. This has been briefly described in this chapter using FWM effects and nonlinear signal processing using high-order spectral analysis and processing in the electronic domain. This ultra-high-speed optical preprocessing and/or electronic triple correlation and bispectrum receivers are the first systems using nonlinear processing for 100–400 Gbps and Tbps optical Internet.

The degree of complexity of ADC and DAC, as well as high-speed DSP, will allow higher-complexity processing algorithms embedded for real-time processing. Thus, higher-order spectral processing techniques will be potentially employed for coherent optical communication with higher dimensional processing.

REFERENCES

1. J. M. Mendel, Tutorial on high-order statistic (spectra) in signal processing and system theory: Theoretical results and some applications, *Proc. IEEE*, Vol. 79, No. 3, pp. 278–305, 1991.
2. C. L. Nikias and J. M. Mendel, Signal processing with higher-order spectra, *IEEE Sig. Proc. Mag.*, Vol. 10, No. 3, pp. 10–37, 1993.
3. G. Sundaramoorthy, M. R. Raghuveer, and S. A. Dianat, Bispectral reconstruction of signals in noise: Amplitude reconstruction issues, *IEEE Trans. Acoustics, Speech Sig. Proc.*, Vol. 38, No. 7, pp. 1297–1306, 1990.
4. H. Bartelt, A. W. Lohmann, and B. Wirnitzer, Phase and amplitude recovery from bispectra, *Appl. Opt.*, Vol. 23, pp. 3121–3129, 1984.
5. T. M. Liu, Y. C. Huang, G. W. Chern, K. H. Lin, C. J. Lee, Y. C. Hung, and C. K. Sun, Triple-optical auto-correlation for direct optical pulse-shape measurement, *Appl. Phys. Lett.*, Vol. 81, No. 8, pp. 1402–1404, 2002.
6. D. R. Brillinger. Introduction to polyspectra, *Ann. Math. Stat.*, Vol. 36, pp. 1351–1374, 1965.
7. M. C. Ho, K. Uesaka, M. Marhic, Y. Akasaka, and L. G. Kazovsky, 200-nm-bandwidth fiber optical amplifier combining parametric and Raman gain, *IEEE J. Lightwave Technol.*, Vol. 19, pp. 977–981, 2001.

17 Temporal Lens and Adaptive Electronic/ Photonic Equalization

The duality between the paraxial diffraction of beams in spatial domain and the dispersion of ultra-short pulses in time domain offer significant insights into pulse dynamics when propagating through single-mode optical fibers. A quadratic phase modulation in time (time lens) is the analog of a thin lens in space. Furthermore, an analogy between spatial and temporal imaging can be used to obtain the distortion-less expansion or compression of optical pulse sequences. Temporal imaging systems function toward focusing and defocusing the ultra-short pulses after propagating through a length of single-mode optical fiber. A quadratic phase modulator acts as a time lens via the quadratic phase modulation. Ultra-fast photonic signal processing in optical communication has emerged recently for long-haul optical transmission, and temporal imaging is expected to play a key role in the near future in this emerging technology. The significant application of temporal imaging is the adaptive equalization or, effectively, the refocusing of the broadened pulses by feeding them through an optical modulator that changes the phase of the carriers of the pulses to the opposite of the quadratic phase effects. This type of adaptive equalizer could eliminate any quadratic phase distortion, which normally can be compensated using passive devices such as dispersion-compensating fibers (DCFs), polarization-mode dispersion (PMD) compensator, and eliminators for timing jitter effects.

In this chapter, we demonstrate the propagation and equalization of a single pulse and a sequence of pulses through standard single-mode fiber (SMF). The system transmission performance with and without equalization at 160 Gbps bit rate is given in Figure 17.1. Analytical and simulation results show strong agreement in the effectiveness of this dispersion equalization technique. The 160 Gbps ultra-high-speed transmission system is very sensitive to both higher-order and time-varying dispersions, and it is proven that the equalization has improved the dispersion effects such as the group velocity dispersion (GVD), timing jitter. and polarization-mode dispersion (PMD). Pulse propagation and its equalization are investigated with 120 km SSMF transmission length. Simulation results have shown significant improvement in the bit error rate (BER) under the cases of with and without the equalization.

17.1 INTRODUCTION

When an optical pulse propagates through an SMF, its optical spectrum would travel down the fiber at different speeds, due to material and waveguide dispersion as well as polarization modal delay, and naturally the nonlinear phase effects if the intensity of the pulse is above the SPM threshold level. In case the bit rate reaches 160 Gbps, the compensation and equalization is preferred to be in an active mode, and thus an optical phase modulator operating at a very high speed should be used. An integrated optical phase modulator could change the phase (controlled shift on the phase of a light beam) of the optical signal by applying a driving traveling wave voltage V. When the electric field is generated and applied across an optical channel waveguide formed on an electro-optic substrate, the change in the refractive index would induce a change in the propagation constant of the propagating mode and create a phase shift for all traveling lights through that region [1].

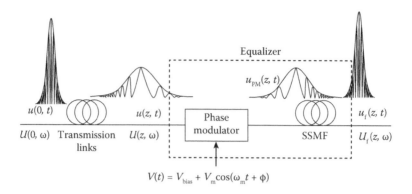

FIGURE 17.1 A schematic of the optical system communication incorporating an equalizer with sinusoidal driving voltage, 40 GHz clock recovered signals for phase modulation.

Figure 17.1 shows a phase shift at the output of the phase modulator. This type of equalization can be implemented adaptively.

Furthermore, it is much more difficult to fully compensate a higher-order dispersion of the third or fourth order in ultra-fast signal processing by using only DCF. Mismatched compensation between SMF and DCF would be accumulated over long distances, causing serious distortion of the received pulses. Therefore, another dispersion compensation scheme that can be implemented at the front end of the receiver in order to fully recover the transmitted signals is reported in this chapter.

Currently, the remarkable progress of integrated photonic technology makes possible the fabrication of active devices such as modulators, integrated transmitters, and so on, which are commercially available for operating speeds of up to several GHz. In particular, ultra-wideband optical phase modulators can offer significant phase modulation over a very short period, thus enabling the alteration of the phase of the lightwave carrier embedded within a very narrow-width pulse sequence. These modulators thus enable the possibility of chirping the carrier frequency, and hence introduce dispersion in the negative or positive sense for equalization of the phase disturbance within an optical pulse due to the quadratic phase effects of single-mode optical fibers. The phase modulation for equalization can be implemented in an adaptive mode.

The propagation of an optical pulse through an optical fiber can be considered as the evolution of a pulse through a quadratic phase medium, that is, the amplitude remains invariant, but its phase, the carrier phase, is altered following a quadratic function of its spectral components. This phenomenon is indeed similar to the diffraction of a lightwave beam through a spatial slit. Alternatively, one could consider the fiber as a temporal lens that defocuses or focuses the guided lightwave beam, depending on the sign of the dispersion factor of the fiber. Therefore, there is duality between the spatial and temporal domains of the propagation of lightwave beams or pulse sequence through a spatial lens or a guided wave optical device. The correction of the blurred optical pulses can be made via the modulation of the phase of the carrier within the pulse period, whether in passive or active and adaptive modes. We demonstrate that equalization can be achieved by using an optical phase modulator, and the performance of this online adaptive equalization is very effective for ultra-high-speed optical transmission systems, especially at 160 Gbps.

The organization of this chapter is as follows. In Section 17.2, space–time duality and its applications as well as principles of temporal imaging are described. The temporal imaging, which is necessary and plays a key role in long-haul-transmission optic fiber, is also discussed. Section 17.3 presents simulation results of the equalization of the impairment due to second-order and third-order dispersion (TOD) of an SMF. Advantages and disadvantages of using sinusoidal and ideal parabolic driving voltage on an optical phase modulator for equalization are also given in detail. Section 17.4 discusses the simulation model by MATLAB® and Simulink® for the transmission, modulation, and equalization techniques of a 160 Gbps transmission system. Section 17.5 outlines the significant results of a single-pulse transmission and equalization of the 160 Gbps transmission system. BER characteristics and eye diagrams are illustrated, together with the transmission performance. Finally, some concluding remarks are stated in Section 17.6.

17.2 SPACE–TIME DUALITY AND EQUALIZATION

The duality between the spatial domain and time domain has been investigated, leading to the formation of temporal imaging, and thus the analogy of a thin lens in spatial domain and a quadratic phase modulation in time domain [2]. Temporal imaging is a technique that enables the expansion or compression of signals in time, and their envelope profiles could be maintained in long-haul transmission [3,4]. Basically, the principle of temporal imaging is based on broadening pulse width due to fiber's dispersion factor and the spreading of a beam due to Fresnel diffraction under far-field consideration. When optical signals propagating through a single-mode optical fiber are transmitted in the air, the pulse width of these signals are broadened due to the optical dispersion or diffraction effects. Consequently, the carrier frequencies of these signals would be chirped up or chirped down, depending on the transmission medium. Thus, the spectra of these waveforms would be broadened or compressed due to the changes in the phase of each frequency component. The Fresnel or Fraunhofer (far-field) diffraction was discussed by Papoulis in the 1960s [5], and signal dispersion is described by Agrawal [6,7]. We can also further clarify in this section.

Furthermore, a key element in a temporal imaging system is a time lens, which is considered a quadratic phase modulation in the time domain. In addition, a dispersive element, such as SSMF, DCF, or fiber Bragg grating (FBG), also performs an equivalent role of diffraction. A quadratic phase modulation could be produced by using several methods, and one of those that is popular for ultra-short-pulse compression or generation is an ultra-high-speed electro-optic phase modulator [8,9]. Assuming that chirped signals are launched into a divergent or convergent lens in the spatial domain or a quadratic phase modulator in the time domain time which would chirp the frequencies of the embedded lightwave carriers in the modulated signal envelope. Each frequency component of a linear chirped pulse generated by a parabolic phase modulator travels at a different speed, due to the GVD in the fiber, and hence the media act as a time lens. Consequently, the frequency components of these signals are reorganized before they are again launched into free air or a dispersive element such as optical fiber. By using a suitable length or dispersion, both before and after a space lens or a time lens, the output pulses could be recovered or even shortened, as compared to the original pulses.

The main purpose of this chapter is to study space–time duality, and how to apply this property to design an adaptive equalization system for ultra-fast optical signal processing and equalization of distorted pulse sequences over an ultra-long-haul and ultra-high-speed optical fiber transmission system. The duality between spatial domain and time domain is proven and summarized. Furthermore, the broadening of a Gaussian pulse in transmission fiber is strongly influenced by the frequency chirping of the modulated carrier.

The duality between light diffraction in spatial domain and narrow-band dispersion in time domain has been studied recently. This duality could be used and applied to expand or compress transmitted pulses in time domain while those pulses' shapes are maintained. This process can be termed as *temporal imaging*. The theory of temporal imaging is based on the duality between par-axial diffraction and narrow-band dispersion. This interesting duality is analyzed and discussed in more detail in Section 17.2.1.1 below. This section gives a review of the analogy between par-axial diffraction of a light beam in spatial domain and the temporal dispersion of a narrow-band pulse in a dielectric medium [2,10,11]. Furthermore, the space–time duality has also led to the employment of quadratic phase modulation devices such as optical adaptive equalizers [12] or electro-optic phase modulators [8] in the time domain, in the analogy of a thin lens in the spatial domain [13], which is equivalent to the phase modulation of the carrier frequency under the pulse envelope in the time domain. In addition, a real-time Fourier transformation that uses a time lens would be considered a temporal equivalence with the spatial Fourier transformation [14]. This equivalence can be used for the implementation of an equalization system, which is introduced in the next section.

17.2.1 SPACE–TIME DUALITY

It is assumed throughout this chapter that the frequency spectrum of the carrier wave is monochromatic. We can then write the evolution of the carriers in the propagation equations without including the frequency term. This section describes the para-axial diffraction and the wave equation representing its dynamic behavior. We then obtain the essential equations for the spatial lens and the time lens for further propagations in an optical link.

17.2.1.1 Paraxial Diffraction

Maxwell's equations would be used to obtain a 3D vector equation. The paraxial form of the Helmholtz equation and its solution are given as follows

$$\nabla^2 E(x,y,z) - j2k \frac{\partial E(x,y,z)}{\partial z} = 0 \tag{17.1}$$

and

$$\frac{\partial^2 E}{\partial x^2} + \frac{\partial^2 E}{\partial y^2} + \frac{\partial^2 E}{\partial z^2} - j2k \frac{\partial E}{\partial z} = 0 \tag{17.2}$$

The propagation direction is in the z-direction and polarization in the transverse plane of x- and y-axes. Owing to the paraxial approximation, the curvature of the field envelope in the direction or propagation (z-direction) is much less than the curvature of the transverse profile.

Therefore, it could be concluded that the term $\left|\frac{\partial^2 E}{\partial z^2}\right|$ is much smaller compared to the terms $\left|\frac{\partial^2 E}{\partial x^2}\right|, \left|\frac{\partial^2 E}{\partial y^2}\right|$, and $\left|2k\frac{\partial E}{\partial z}\right|$, and therefore it might be eliminated. Finally, the electric field propagating down the z-axis can be found as

$$\frac{\partial^2 E}{\partial x^2} + \frac{\partial^2 E}{\partial y^2} - j2k \frac{\partial E}{\partial z} = 0 \tag{17.3}$$

$$\therefore E_z = -\frac{j}{2k}\left(E_{xx} + E_{yy}\right) \tag{17.4}$$

These equations are in parabolic form and similar to the heat diffusion equation. Therefore, the behavior of diffraction and dispersion, which is spreading like a temperature distribution, can be derived by using the solutions of the diffusion problem [2].

17.2.1.2 Governing Nonlinear Schrödinger Equation

An optical signal whose temporal envelope in the z-direction $E(z, t) = u(z, t)e^{j\omega t}$ propagating through an optical fiber can be represented by the nonlinear Schrödinger equation (NLSE)

$$\frac{\partial u}{\partial z} = -\frac{\alpha}{2}u - \frac{j\beta_2}{2}\frac{\partial^2 u}{\partial t^2} + \frac{\beta_3}{6}\frac{\delta^3 u}{\delta t^3} + j\gamma|u|^2 u \tag{17.5}$$

where the amplitude $u = u(z,t)$ is the complex envelope carried by the lightwaves of wavelength λ, along the propagation in z-axis, and t is the time variable. Pulse broadening is a result of the frequency dependence of β. By using Taylor series to expand $\beta(\omega)$ around the carrier frequency ω_c, so that the second, third, and higher-order dispersion can be obtained as $\beta_n = \left.\dfrac{d^n\beta}{d\omega^n}\right|_{\omega=\omega_c}$, $\beta_2 = -D\lambda^2/2\pi c$ is the GVD coefficient and $\beta_3 = \left(\dfrac{2\pi c}{\lambda^2}\right)^{-2}\left[S - \left(\dfrac{4\pi c}{\lambda^3}\right)\beta_3\right]$ the TOD factor, and related to the dispersion slope S. $\gamma = n_2\omega/cA_{eff}$ is the nonlinear coefficient of an optical fiber with an effective area of A_{eff}, corresponding to a mode spot size r_0, where n_2 is the nonlinear refractive index taking a typical value for silica-based glass of $2.6e{-}20$ m²/w. If the losses during transmissions in the optical fiber can be ignored, and a value of β_3 (i.e., about 0.117 ps³/km) is very small compared to a value of β_2 (about 21.6 ps²/km), Equation 17.5 can thus be rewritten in normalized form as

$$\frac{\partial u}{\partial z} = -j\frac{\beta_2}{2}\frac{\partial^2 u}{\partial t^2} + j\gamma|u|^2 u \tag{17.6}$$

17.2.1.3 Diffractive and Dispersive Phases

Different forms of diffusion could be modeled quantitatively using the diffusion equation, which go by different names, depending on the physical situation. Steady-state thermal diffusion is governed by Fourier's law. In all cases of diffusion, the net flux of the transported quantity (atoms, energy, or electrons) equaled a physical property (diffusivity, thermal conductivity, electrical conductivity) multiplied by a gradient (a concentration, thermal, electric field gradient) [2].

The governing equations of the paraxial equation and the nonlinear Schrödinger equation follow the forms

$$E_z = -\frac{j}{2k}\left(E_{xx} + E_{yy}\right) \tag{17.7}$$

$$\frac{\partial u}{\partial z} = -\frac{j\beta_2}{2}\frac{\partial^2 u}{\partial t^2} + j\gamma|u|^2 u \tag{17.8}$$

and the equation of heat diffusion for 2D diffusion is

$$u_t = c(u_{xx} + u_{yy}) \tag{17.9}$$

Paraxial, NLSE, and heat diffusion equations are thus governed by the same parabolic differential equation, and thus they should have similar forms. Let k_x be the Fourier domain variable or the propagation constant in the x direction, and $U(k_x, k_y, 0)$ be the initial Fourier spectrum. By applying the 2D equation of heat diffusion to the two wave equations, solutions for 2D diffraction and wave propagation can be found [2].

$$u(x,y,z) = \frac{1}{(2\pi)^2}\int\int_{-\infty}^{\infty} U(k_x,k_y,0)e^{j(k_x^2+k_y^2)z/2k}e^{j(k_xx+k_yy)}\,dk_x\,dk_y \tag{17.10}$$

$$u(z,t) = \frac{1}{2\pi}\int_{-\infty}^{\infty} U(0,\omega)e^{-j\frac{z}{2}\frac{d^2\beta}{d\omega^2}\omega^2}e^{j\omega t}\,d\omega \tag{17.11}$$

Comparing Equations 17.9 and 17.10, it becomes clear that the envelope of the input pulse is similar to the paraxial diffraction. The terms (k_x, k_y) and x in spatial domain correspond to the terms ω and t in time domain, respectively. In other words, a beam displacement in the spatial domain corresponds to a pulse delay in the time domain [15]. According to Equations 17.9 and 17.10, the diffractive and dispersive phases are given as

$$\phi(t) = \frac{k_x^2 + k_y^2}{2k} z \tag{17.12}$$

$$\phi(\omega) = -\frac{d^2\beta}{2d\omega^2} \omega^2 z \tag{17.13}$$

17.2.1.4 Spatial Lens

A lens exhibits an important property of phase retardation or delay due to the propagation and disturbance of the medium [16]. Furthermore, a lens is said to be thin when a light ray enters at a point on one side of the lens and emerges from the same axial point on the other side of the lens [17]. Therefore, a thin lens delays an incident wavefront by an amount proportional to the thickness of the lens

Recalling the incident monochromatic wavefront $u_i(x,y)$ at plane U_i emerging through the lens would give an output wavefront $u_o(x,y)$ at plane U_o. Thus, the output wavefront could be considered as the product of the input wavefront multiplied by the phase transform function of the lens $(u_o(x,y) = T(x,y) \times u_i(x,y))$, as shown in Figure 17.2a and b.

The effect of a thin lens would be described by a phase function as

$$T(x,y) = \exp\left[jkn\Delta t\right] \exp\left[-jk\frac{x^2 + y^2}{2f}\right] \tag{17.14}$$

where f is the focal length of the lens. The relationship between the focal length, refractive index, two radii of curvatures R1 and R2 of the faces of the lens is calculated as $\frac{1}{f} = (n-1)\left(\frac{1}{R_1} + \frac{1}{R_2}\right)$. Equation 17.13 shows that a thin lens could produce a quadratic phase modulation in real spatial space. In addition the relevant spatial phase transformation of a thin lens can be found as a phase function given by

$$\phi(x,y) = k\frac{x^2 + y^2}{2f} \tag{17.15}$$

17.2.1.5 Time Lens

A time lens is basically just a quadratic phase modulator in time, which can be implemented using a traveling wave electro-optic modulator driven at a microwave frequency ω_m. However, it is difficult to

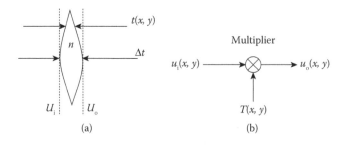

(a) (b)

FIGURE 17.2 (a) The thickness function and (b) analogy system of a lens transformation.

generate a true quadratic phase modulation for an ideal time lens in practice. Therefore, the required modulation function could be approximated by a portion of a sinusoidal phase modulation [15]. By applying a time-dependent sinusoidal wave to the "hot" electrode of a traveling wave electro-optic modulator at a modulation frequency ω_m the applied signal can be written as

$$V(t) = V_o \cos(\omega_m t) \tag{17.16}$$

Then, by using the Taylor series expansion to expand Equation 17.15, the corresponding quadratic approximation around $t = 0$ is

$$V(t) \approx V_o \left(1 - \frac{1}{2} \omega_m^2 t^2 \right) \tag{17.17}$$

By denoting $f_T = \dfrac{\omega_c}{\dfrac{\pi V_c}{V_\pi} \omega_m^2}$ as the equivalent focal time of the time lens, where ω_c is an optical carrier frequency. The approximate quadratic phase modulation of time lens in the time domain would follow a phase function of

$$\phi(t) = \frac{1}{2} \frac{\omega_c}{f_T} t^2 \tag{17.18}$$

17.2.1.6 Temporal Imaging

The space–time duality properties can be summarized as in Table 17.1.

TABLE 17.1
Space–Time Duality

Space–Time Duality	Space	Time (Optical Fiber)
Governing equations	$E_z = -\dfrac{j}{2k}\left(E_{xx} + E_{yy}\right)$	$\dfrac{\partial u}{\partial z} = -\dfrac{j\beta_2}{2}\dfrac{\partial^2 u}{\partial t^2} + j\gamma\lvert u\rvert^2 u$
Propagation variables	z	μ
Transverse variables	x, y	t
Fourier variables	k_x, k_y	ω
Diffractive/dispersive phases	$\phi(t) = \dfrac{k_x^2 + k_y^2}{2k} z$	$\phi(\omega) = -\dfrac{d^2\beta}{2d\omega^2}\omega^2 z$
Lenses/temporal lens phase	$\phi(x, y) = k\dfrac{x^2 + y^2}{2f}$	$\phi(t) = \dfrac{1}{2}\dfrac{\omega_c}{f_T} t^2$
Spatial/temporal focal point	$f = \dfrac{1}{(n-1)\left(\dfrac{1}{R_1} - \dfrac{1}{R_2}\right)}$	$f_T = \dfrac{\omega_c}{\dfrac{\pi V_c}{V_\pi}\omega_m^2}$
Imaging condition	$\dfrac{1}{d_i} + \dfrac{1}{d_o} = \dfrac{1}{f}$	$\dfrac{1}{\mu_i \dfrac{d^2\beta_i}{d\omega^2}} + \dfrac{1}{\mu_o \dfrac{d^2\beta_o}{d\omega^2}} = -\dfrac{\omega_c}{f_\tau}$
Magnification factor M	$-\dfrac{d_o}{d_i}$	$-\dfrac{\mu_o \dfrac{d^2\beta_o}{d\omega^2}}{\mu_i \dfrac{d^2\beta_i}{d\omega^2}}$

The diffraction from an object by a lens produces an image which spatially dependent phase shift [18]. Figure 17.3 shows spatial imaging and temporal imaging systems [2,4]. The dispersion effects from an object pulse to lens and from lens to images are provided by two grating pairs that are installed before and after a quadratic phase modulator [10]. And this quadratic phase modulator acts as a time lens that provides a time-varying phase shift.

The equivalence between the diffraction–dispersion phase functions and space–time lens phase functions, between spatial domain and time domain, leads to the relationship between conventional spatial imaging and temporal imaging configurations, as summarized and described clearly in Figure 17.3a for spatial imaging and Figure 17.3b for temporal imaging. Let an unchirped pulse train enter into temporal and spatial imaging. First, these pulses would be distorted due to diffraction effect in space or dispersion effect in time. At this moment, the phase of each frequency component would be changed in the frequency domain, and therefore a time delay of these frequency components would occur in the time domain. Consequently, the pulse shape of these pulses is blurred or broadened. Next, these pulses are fed into a space lens or time lens. At this time, the carrier frequency is chirped up or down, depending on its applications. Last, this pulse train leaves a space or a time lens, and it is diffracted in the air or dispersed by a grating pair. This diffraction or dispersion is necessary and very important. The reason by all the frequency components of those pulses needed to be reorganized again in order to compress the input pulses due to their different time delays. Moreover, in order to expand or compress a pulse, the magnification parameter M should be modified. This magnification M could be changed by adjusting the object–lens distance,

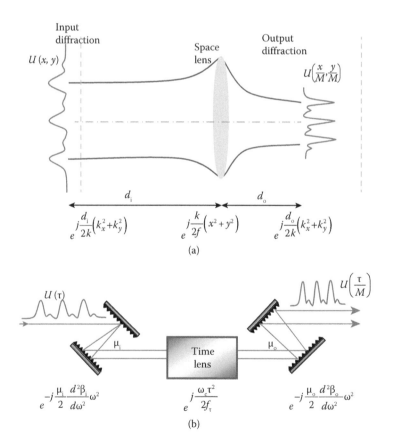

FIGURE 17.3 Configuration for (a) spatial imaging and (b) temporal imaging. (From B. H. Kolner, *IEEE J. Quant. Electron.*, Vol. 30, No. 8, pp. 1951–1963, 1994; C. V. Bennett and B. H. Kolner, *IEEE J. Quant. Electron.*, Vol. 36, No. 6, pp. 649–655, 2000.)

lens–image distance, or lens characteristics in space, or by changing the grating pair separation distance or quadratic phase modulator characteristic in time.

In particular, when these input pulses are compressed, the bit period is shortened. In order to avoid this problem, two thin lenses are used and separated by a mask, which plays the role of a Fourier transform plane. In a practical setup, a pulse train is transmitted to the fist lens, and these pulses are diffracted. A mask that includes an amplitude mask and a phase mask (called *spatial light modulators*), to control both the amplitude and phase of the pulse, could be used to yield any desired pulses. Finally, these pulses are refocused by a second lens. As a result, a received pulse train still keeps its original bit rate, while the pulse width of each individual pulse is reduced.

17.2.1.7 Electro-Optic Phase Modulator as a Time Lens

Electro-optic modulator is an optical device that is used to manipulating either the phase and/or amplitude of the light beam via the change of the refractive index of the medium by applying an electric field, the electro-optic effect. In this section, only phase modulation of the modulated beam is discussed. The electro-optic phase modulator lens has been used very popularly in pulse compression for ultra-short optical pulses, such as a few pico-seconds or a few hundred femto-seconds.

Khayim et al. [8] have introduced the relationship between the applied electric field corresponding to an electro-optic lens as well as the chirped optical carrier frequency, as shown in Figure 17.4 [8]. Furthermore, the focal length of the integrated guided wave electro-optic phase modulator can also be given as

$$f_\tau = -\frac{hd^2}{r_{33}n_e^3 V(t)l_o} \tag{17.19}$$

where mask is the maximum width of the parabolic electrode; h is the spacing between the traveling wave electrodes; l_o is the maximum length of the electrode; r_{33} is the electro-optic coefficient of the $LiNbO_3$ Z-cut X-prop substrate; n_e is the extraordinary index of refraction; and $V(t)$ is the time-variable applied voltage.

Equation 17.18 and Figure 17.4 show that the convex or concave parabolic shape of an applied electric field corresponds to the diverging or converging lens and the down-chirping or up-chirping of the optical carrier frequency, respectively. Depending on specific applications, the carrier frequency would be chirped up or chirped down by adjusting the driving voltage applied to the electro-optic modulator in the time domain, which is equivalent to a converging or diverging lens in the spatial domain.

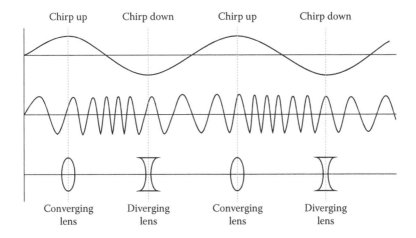

FIGURE 17.4 The relationship between phase modulator electric field, chirped carrier frequency, and corresponding lens. (From T. Khayim et al., *IEEE J. Quant. Electron.*, Vol. 35, No. 10, pp. 1412–1418, 1999.)

17.2.2 Equalization in Transmission System

An ultra-high-speed optical transmission system is considered to be the essential backbone of the next generation of ultra-high-speed networks. When the pulse width reaches less than a few pico-seconds, not only second-order but even higher-order dispersion must be taken into account. On the contrary, third-order and higher-order dispersion effects are not fully compensated by using traditional dispersion-compensating devices such as DCF fibers. Moreover, ultra-fast transmission signals are very sensitive to environmental effects such as temperature, vibration, and so on. Thus, the received signals at receiver end would be distorted dramatically even if only a small outside factor affected the transmission link. Therefore, the tunable adaptive equalizer system is used at the receiver end to recover higher-order dispersion that could not be compensated by the DCF fiber, and to recover distorted signals to improve the BER or reduce the eye-opening penalty (EOP).

The structure of the equalizer system based on spatial and temporal imaging can offer useful and important applications of the time lens optical system, such as compensating GVD, third- and fourth-order dispersion, reducing the timing jitter for a pico-second pulse train, as well as time reversal with magnification [19,20]. Besides these, there are also a number of practical applications of time lens optical systems for temporal magnification and phase reversal optical data that are analogical to spatial diffraction and temporal dispersion [21]. It showed that space–time duality had valuable effects on ultra-fast photonic signal processing, and that its applications could be further explored in the mitigation of nonlinearity impairments. Adaptive dispersion equalization would become a key technique for ultra-high-speed transmission systems, and the temporal imaging theorem could be used to create this adaptive equalizer. Jannson [14] has realized that the spatial Fourier transformations of temporal signals are equivalent to the real-time Fourier transformations in dispersive optical fibers.

Initially, the temporal self-imaging effect in SMF and the time-domain Collett-Wolf equivalence theorem are studied and applied to the transference and propagation of the information contained in the periodic signals in fibers [22,23]. Consequently, this equalizer system model is built and developed based on the implementation of this well-known temporal imaging process. In this case, a single-mode optical fiber acted as free space in the spatial domain, and an adaptive equalizer acted as a time lens in the time domain. Following this quadratic phase modulator is a dispersive element that is used to change the phase of each frequency component in the frequency domain. Thus, different time delays appear between these frequency components, and the final pulse shape would be recovered. A standard SMF fiber link is used as a dispersive element (DCF or FBG could be used instead of using standard SMF fiber). This equalizer system, including quadratic phase modulator and SMF, is used to provide the required quadratic phase modulation [13]. Further investigations, theoretical calculations, and simulation results of transmission systems using adaptive equalizers are discussed in more detail in the later sections.

17.2.2.1 Equalization with Sinusoidal Driven Voltage Phase Modulator

For the optical fiber communication system, the GVD plays a significant role as it distorts signals dramatically, especially at ultra-high speeds. Also, the TOD effects or even the fourth-order dispersion effects also become more significant [24]. A combination of different fibers, such as SMF, reverse dispersion fiber, and DCF, would fully compensate for the second-order dispersion, but residual dispersion effects do exist. For ultra-fast transmitted signals, for example, at 160 Gbps, the system is very sensitive to the environmental effects such as the PMD and fluctuation of the dispersion factor. For example, if there is a small change in temperature, there would be a mismatch in compensation between SMF and DCF. This compensating mismatch would be accumulated over a long-haul transmission, and the received signals would be distorted randomly. Furthermore, for an installed transmission system, the TOD could not be eliminated by using DCF fiber for 160 Gbps transmitted signals. Therefore, an adaptive equalizer can equalize any distortion, adaptively with any variation of this impairment. It could compensate for second-, third-, and even fourth-order dispersion. Moreover, it also eliminated the timing jitter effect, and thus the BER at the receiver.

In order to investigate the effects of using an equalizer, some significant and necessary formulas are briefly given. In applying a sinusoidal driving voltage to phase modulator, a linear chirp would be generated, and hence amplitude distortion. The applied electric field would change the phase of each frequency component in the time domain, and thus the signal spectrum is broadened in the frequency domain. The transmitted signal at the output of the phase modulator is

$$u_{PM}(z,t) = u(z,t)e^{j\frac{1}{2}Kt^2} \tag{17.20}$$

The signal at the input of the receiver end would be the convolution between the signal at the output of the phase modulator and the dispersive device, the SSMF, and is given as

$$u_f(z,t) = u_{PM}(z,t) * e^{-j\frac{t^2}{2D}} \tag{17.21}$$

with $D = \beta_2 L_{PM}$ and $|D| = T_0^2$. Assuming the chirping rate of the phase modulator $K = \dfrac{1}{D}$, the received signal at the output of the equalizer can be obtained as

$$u_f(z,t) = \sqrt{\frac{j}{2\pi D}} A\sqrt{2\pi T_0^2}\, e^{-\frac{T_0^2}{2}\frac{t^2}{D^2}}\, e^{-j\frac{1}{2}Kt^2}\, e^{j\phi\left(\frac{t}{D}\right)} \tag{17.22}$$

In order to recover the initial Gaussian pulse shape, condition $|D| = T_0^2$ should be applied to Equation 17.21 to obtain

$$u_f(z,t) = A\sqrt{\frac{j}{sign(\beta_2)}}\, e^{-\frac{t^2}{2T_0^2}}\, e^{j\left[-\frac{1}{2}Kt^2 + \phi\left(\frac{t}{D}\right)\right]} \tag{17.23}$$

If a sinusoidal driving voltage $V(t) = V_{bias} + V_m \cos(\omega_m t + \phi)$ is applied to the phase modulator to chirp the carrier frequency of signals, then, by using the Taylor series expansion, and by the approximately quadratic driving voltage, the phase of the driving voltage would be

$$\phi(t) = e^{j\left[\frac{\pi}{V_\pi}(V_{bias} + V_m) - \frac{1}{2}\frac{\pi V_m}{V_\pi}\omega_m^2 t^2\right]} \tag{17.24}$$

Comparing Equations 17.22 and 17.23, in order to recover the Gaussian pulse at the end of the system, a chirping rate should be $|K| = \left|\dfrac{\pi V_m}{V_\pi}\right|\omega_m^2$. This depends on the amplitude of the driving voltage and modulation frequency $\omega_m = \dfrac{1}{T_0}\sqrt{\left|\dfrac{V_\pi}{\pi V_m}\right|}$. In a practical system, ω_m could be tuned in order to achieve an almost reconstructed signal. Applying this sinusoidal time-dependent signal to a phase modulator in a 160 Gbps system with a full width at half maximum (FWHM) of an initial un-chirped pulse, $T_{FWHM} = 2$ ps, and the corresponding input pulse half width at $1/e$ intensity point can be estimated as $T_0 = \dfrac{T_{FWHM}}{1.665} \approx 1.2012\,\text{ps}$. Thus, the necessary chirping rate needed to recover the distorted signal is $K = \dfrac{1}{T_0^2} = 6.944 \times 10^{23}\,\text{s}^{-2}$.

Assuming $V_\pi = 3.5V$ and $V_{bias} = V_m = 7V$ so as to shift the sinusoidal wave up by $V_m/2$. Therefore, if a 40 GHz clock signal is used, the chirping rate becomes $K = \left|\dfrac{\pi V_m}{V_\pi}\right|\omega_m^2 = 3.969 \times 10^{23}$. Thus, two 40 GHz phase modulators must be employed to generate the required chirping rate. If the dispersion factor of standard SMF fiber is $D = 17$ ps/(nm.km), then the GVD coefficient would be $\beta_2 = -2.168e-26$ s²/m. Thus, the length of the SMF to be inserted in the equalizer would be calculated as $L_{PM} = \dfrac{T_o^2}{\beta_2} \approx 66.55\,\text{m}$.

17.2.2.2 Equalization with Parabolic Driven Voltage Phase Modulator

The transmission system including a phase equalization subsystem is shown in Figure 17.5.

Because a driving voltage has an ideal parabolic shape, it could create a parabolic phase. Because an SMF transfer function is $H(f) = e^{-j\alpha f^2}$ [25], whose phase variation is parabolic. Therefore, an ideal parabolic driving voltage applied to a phase modulator, could fully equalize the distortion effects on the signals after the transmitting through the fiber link. If a parabolic driving voltage $V(t) = V_{bias} - V_m(\omega_m t)^2$ is applied to the phase modulator instead of a purely sinusoidal driving voltage, then the phase variation at the output of this phase modulator becomes

$$\phi(t) = e^{j\left[\frac{\pi}{V_\pi}V_{bias} - \frac{\pi V_m}{V_\pi}(\omega_m t)^2\right]} \tag{17.25}$$

A needed chirping rate for compensation at this time is $|K| = 2\left|\dfrac{\pi V_m}{V_\pi}\right|f_m^2$, with a modulation frequency $f_m = \dfrac{1}{2\pi T_o}\sqrt{\left|\dfrac{V_\pi}{2\pi V_m}\right|} \approx 40\,\text{GHz}$ for the 160 Gbps bit rate system. The necessary parameters for compensation at this time are listed as: chirping rate $K = 6.944 \times 10^{23}s^{-2}$, $V_\pi = 3.5V$, $V_m = 6V$, $V_{bias} = 2V_m = 12V$, and SMF length in equalizer system is $L_{PM} \approx 66.55m$. The modulation frequency can be estimated as $f_m = \dfrac{1}{2\pi T_o}\sqrt{\left|\dfrac{V_\pi}{2\pi V_m}\right|} \approx 40\,\text{GHz}$.

In summary, if an ideal parabolic driving voltage is assumed, it is possible to achieve full compensation of all the distortions, and only one phase modulator can be used in this case. However, it is very difficult to generate an ideal parabolic driving voltage in practice. The received signals of sinusoidal variation can be acceptable, which could be generated easily. Thus, sinusoidal signal driving-signals applied to the phase modulator are presented in detail by simulation. The outcomes would be compared to the case when employing an ideal parabolic waveform.

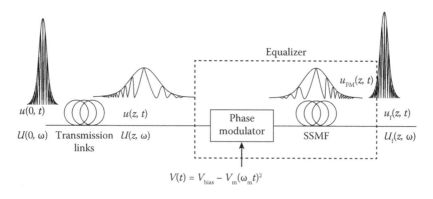

FIGURE 17.5 The overview of the optical system communication by using an equalizer with ideal parabolic driving voltage.

17.3   SIMULATION OF TRANSMISSION AND EQUALIZATION

17.3.1   SINGLE-PULSE TRANSMISSION

Simulation results are carried out for dispersion due to the mismatched lengths between SMF and DCM of the transmission links leading to distortion. Consequently, an adaptive equalizer should be installed at the end of the long-haul transmission link, especially at 160 Gbps, in order to fully compensate for all dispersion effects and timing jitter. Initially, a 2-ps-width single pulse is launched into the optical fiber. The distorted output pulse is monitored after propagating through a transmission link. Then this distorted pulse is led to the equalizer system for recovering its original pulse width and shape.

17.3.1.1   Equalization of Second-Order Dispersion

Figures 17.6 and 17.7 show the equalization of the second-order dispersion effect for a 2 ps Gaussian pulse transmission system.

Let us call T_o a pulse half width at e^{-1} intensity point. By calculating the second-order dispersion length $L_{D2} = \dfrac{T_o^2}{\beta_2} \approx 66.553 \, \text{m}$, and TOD length $L_{D3} = \dfrac{T_o^3}{\beta_3} \approx 17.756 \, \text{km}$ [26], with a group velocity coefficient $\beta_2 = -2.168\text{e} - 26 \, \text{s}^2/\text{m}$, and TOD coefficient $\beta_3 = 9.761\text{e} - 41 \, \text{s}^3/\text{m}$, the output pulse is obviously broadened dramatically if a length of transmission link is greater than L_{D3} in order to investigate a TOD effect. Table 17.2 tabulates the parameters of the equalizer required for the equalization of 160 Gbps transmission due to second- and TOD to demonstrate the effectiveness of the equalizer in the refocusing or reshaping of the transmitted signals.

The simulated results tabulated in Table 17.2 show a strong consistency between estimation and simulation results. Figures 16.6a and 17.7a show the initial pulses before transmission through a 300 m optical fiber length under the two different waveforms of the applied voltage. These pulses would be spreading at the output of the receiver. The equalizer is inserted at the front end of the receiver in order to recover these distorted pulses to its original shape. On the contrary, the pulse shape observed after an equalizer with a sinusoidal driving voltage is not as closed to its original pulse shape with modulator driven by a parabolic driving voltage. The tails of recovered pulse is spreading over and can overlap with other tails from the adjacent pulses. This overlapping contributes to the penalty of the eye diagram. This effect can be observed due to the Taylor approximation, as assumed in Section 17.2. Due to the approximation of the parabolic shape in the region near $t = 0$, the Fourier transformed pulses have prolonged tails at the wings of pulse. This overlapping penalty is considered, discussed, and investigated further in Sections 17.4 and 17.5.

Figures 17.6d, e and 17.7d, e show the evolution of the spectrum of the pulses. The bandwidth of a spectrum before and after being transmitted through an optical fiber is the same, because the phase of each frequency component is changed in the frequency domain. These changes created a time delay of each carrier frequency component in the time domain, and thus the pulse is broadening. And carrier frequency is chirped down while transmitting through the SMF fiber. By using a phase modulator, the carrier frequency under pulse envelope is chirped up. The spectrum is also broadening at this step, because the phase of each frequency component is changed in the time domain, and it leads to a frequency delay in the frequency domain. Therefore, the envelope of these pulses is not changed, while the spectrum is expanded substantially.

Figures 17.6d through f and 17.7d through f plot the phase of the signal at the output of the SMF fiber, at the output of the phase modulator, and at the output of the equalizer. These phase plots indicate that the carrier frequency is chirped down or chirped up. Firstly, when a lightwave-modulated pulse propagating through SMF fiber, the carrier frequency is down chirped, thus carrier phase shape follows a convex parabolic function. However, the applied electric field is approximately concaved parabolic, and the carrier frequency would be chirped up. Only the phases of these frequency components are changed in the time domain, so the pulse shape remains the same. Then, carrier frequency is chirped down again when these pulses are transmitted through SMF in an equalizer.

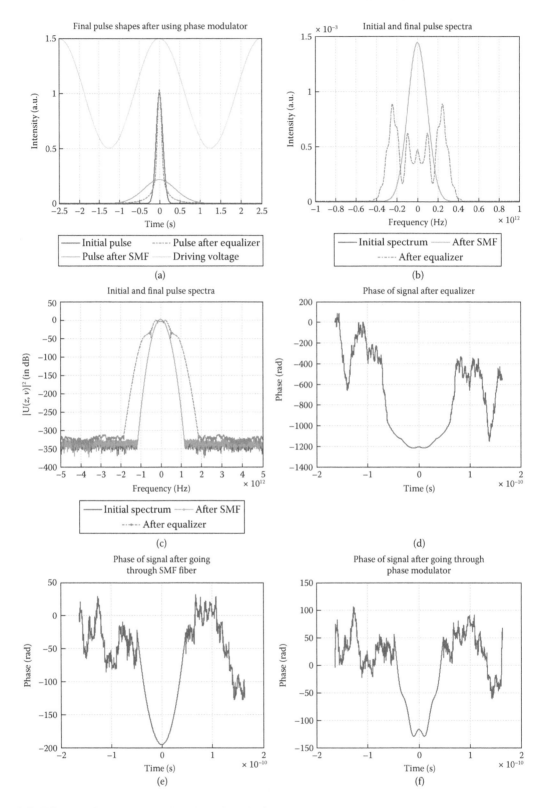

FIGURE 17.6 2 ps Gaussian pulse with a sinusoidal applied driving voltage.

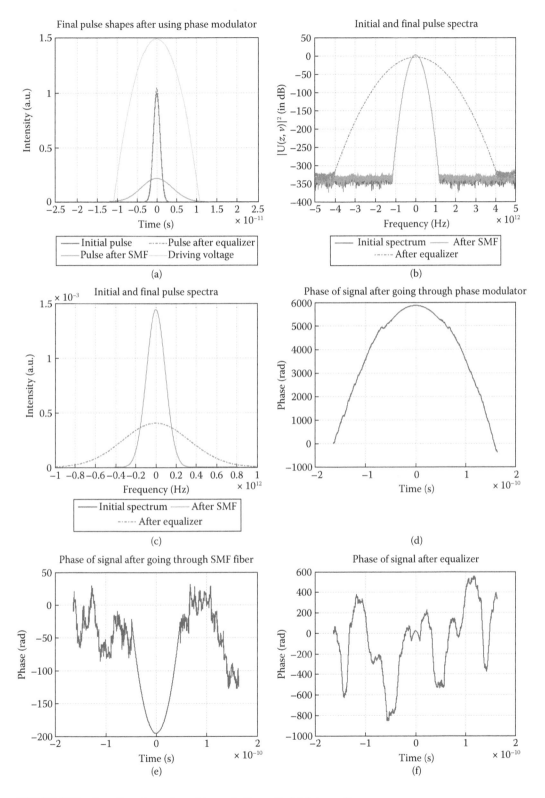

FIGURE 17.7 2 ps Gaussian pulse with parabolic applied driving voltage.

TABLE 17.2

Operation Parameters of the Equalization of the Second-Order Dispersion Effects on the Transmission of a Single Pulse

Parameters	Sinusoidal Driving Voltage	Parabolic Driving Voltage
β_2	−2.168e−26 s²/m	−2.168e−26 s²/m
β_3	0 s³/m	0 s³/m
$L_{\text{Transmission}}$	300 m	300 m
$L_{\text{Equalizer}}$	62 m	62 m
V_π	3.5 V	3.5 V
V_m	7 V	6 V
V_{bias}	7 V	12 V
ϕ	0	0
f_m	40 GHz	40 GHz
Number of PMs	Two phase modulators	One phase modulator

TABLE 17.3

Summary of Parameters for the Equalization of Third-Order Dispersion under Single-Pulse Transmission

Parameters	Sinusoidal Driving Voltage	Parabolic Driving Voltage
β_2	0 s²/m	0 s²/m
β_3	9.761e−41 s³/m	9.761e−41 s³/m
$L_{\text{Transmission}}$	120 km	120 km
$L_{\text{Equalizer}}$	67 m	67 m
V_π	3.5 V	3.5 V
V_m	7 V	6 V
V_{bias}	7 V	12 V
ϕ	$-\dfrac{\pi}{24}$	$-\dfrac{\pi}{24}$
f_m	40 GHz	40 GHz
Number of PMs	Two phase modulators	One phase modulator

During this process, there is a reorganization of each frequency component due to a different time delay of a different frequency component, and the output pulse shape would be recovered. Furthermore, attention should be paid when there is a little concave parabolic shape around the $t = 0$ area. It indicates that the carrier frequency under the pulse envelope is still chirped up. Thus, if the SMF length in the equalizer is increased, the output pulse from the equalizer would be even more compressed than the initial pulse.

In other words, the equalizer is a tunable system that could modify the output pulse width and its shape accordingly for changing some parameters in that equalizer such as voltage amplitude or SMF length.

17.3.1.2   Equalization of TOD

Table 17.3 shows all the necessary parameters for a TOD compensation in the cases where sinusoidal or parabolic driving voltages are used to drive the phase modulator. In this case, an electric field applied to a phase modulator is shifted right by an angle of $\pi/24$, due to the following reasons. First, the oscillated tail of this pulse should be covered by this driving voltage. Second, this distorted pulse is asymmetric, and TOD "pulls" a pulse to the right direction. If the driving voltage is still

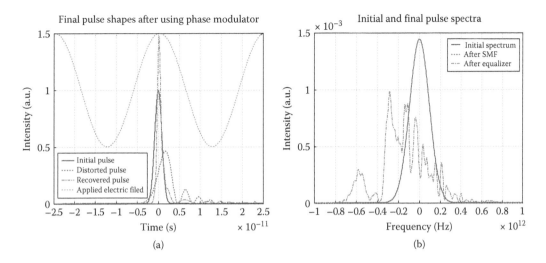

FIGURE 17.8 2 ps Gaussian pulse with sinusoidal applied driving voltage.

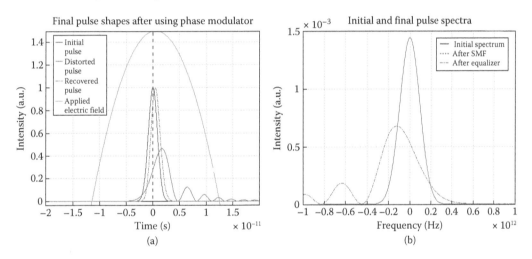

FIGURE 17.9 2 ps Gaussian pulse with parabolic applied driving voltage.

kept in the middle, as in the previous section, the phase of some frequency components might not be changed, or might be changed in the opposite way because the tail of the pulse spreads outside the concave parabolic curve of the driving voltage.

As shown in Figures 17.8a and 17.9a, the output pulse is almost fully recovered according to different voltage waveforms. However, for a transmitted pulse train, if the tail of the individual pulse can be broadened and spreading to neighboring pulses. This is the intersymbol interference and hence difficulty in the recovery of the original pulse states. Figures 17.8b and 17.9b are the spectral profiles corresponding to Figures 17.6a and 17.8a, respectively. It is seen clearly that the distortion in the time domain is converted to that in the frequency domain by Fourier transform, which has been discussed earlier.

17.3.2 Pulse Train Transmission

17.3.2.1 Second-Order Dispersion

Figures 17.10 and 17.11 show the propagating of a pulse train with a bit sequence [1 0 0 1 1 1 0 1], with 2 ps pulse width through 300 and 600 m optical fibers, respectively.

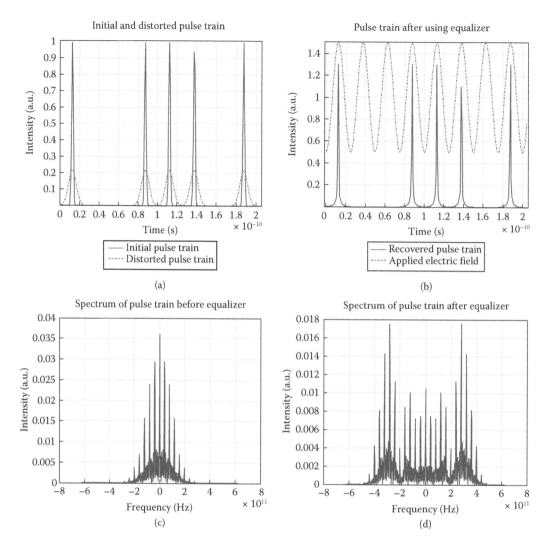

FIGURE 17.10 40 Gbps with 2 ps Gaussian pulse propagating through 300 m optical fiber with the sinusoidal driving voltage.

All other parameters remain the same for these two situations, and they are listed in Table 17.4.

Figures 17.10a and 17.11a show an initial and output pulse train after propagating through 300 and 600 m standard SMF fibers, respectively. The distorted pulse train is recovered and shown in Figures 17.10b and 17.11b. The corresponding spectrum to that pulse train is plotted in Figures 17.10c, d and 17.11c, d. Based on the transmission length of the SMF fiber, there is a huge difference between Figure 17.10 and Figure 17.11, which should be discussed thoroughly.

First of all, when this pulse train is transmitted through a 300 m fiber link, although each pulse in this pulse train is dispersed and broadened, their tails are still not overlapped with each other. Therefore, it is not a very big issue in compensating these dispersive pulses. The spectrum of this signal is also monitored, and it is obviously broadened to satisfy the expectation as explained before. Furthermore, there are several spikes, and the difference between each spike is about 40 GHz, which satisfies the theoretical calculation and expectation.

Since this pulse train is transmitted through 600 m optical fiber, the output pulse train is significantly distorted, and their tails overlap and pulses interfere one another. The overlapping between

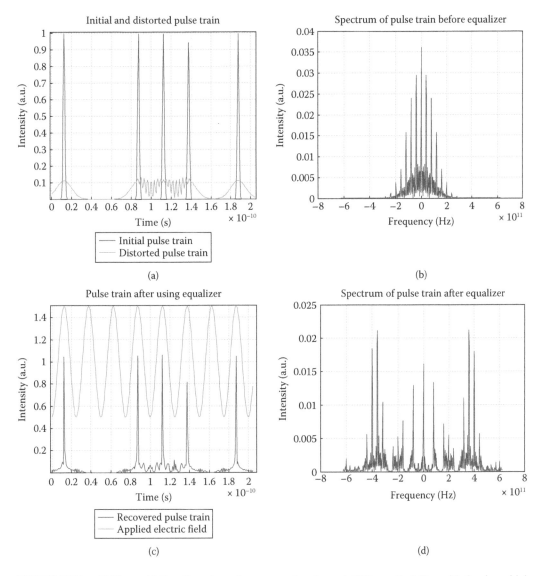

FIGURE 17.11   40 Gbps with 2 ps Gaussian pulse propagating through 600 m optical fiber with the sinusoidal driving voltage.

TABLE 17.4

Parameters for Equalization of Second-Order Elimination under Pulse Train Transmission

	β_2	β_3	$L_{Equalizer}$	V_π	$V_m = V_{bias}$	ϕ	f_m	Number of PMs
Values	−2.168e−26 s²/m	0 s³/m	67 m	3.5 V	7 V	π	40 GHz	2

neighboring pulses would lead to constructive or destructive effects corresponding to the in-phase or out-of-phase status of those carrier frequencies, respectively. Therefore, there is an oscillation when the two tails of two neighboring pulses modulate to each other, as shown in Figure 17.11a. These oscillating tails would contribute to the noise and increase the noise levels of the optical system. Since the output after passing through the equalizer system is plotted in Figure 17.11b, the output pulse train had a significant improvement in dispersion compensation as well as noise reduction.

Although total noise cancellation is not achieved since the driving voltage is just an approximated parabolic waveform, the existing noise level is also reduced significantly, and this noise level, after using an equalizer, is much less than before using it.

17.3.2.2 Equalization of TOD

Figure 17.12a and b shows the initial, distorted, and recovered pulse train after transmitting through a 120 km transmission link, while Figure 17.12c and d shows the spectrum corresponding to that pulse train, respectively. This distorted pulse train, recovered by using an equalizer system and this recovering phenomenon, is discussed clearly in Section 17.3.1.2.

On the contrary, a pulse train is investigated because there are some effects that cannot be obtained or monitored by using a single pulse such as a timing jitter effect. When there are neighboring pulses, two closed pulses will be pulled or pushed from their original position due to reasons such as TOD, nonlinearity effects, and so on. Thus, this effect contributes to the worsening of the BER of transmission systems. One of the most severe influence of this nonlinear effect is the jitter of the pulse

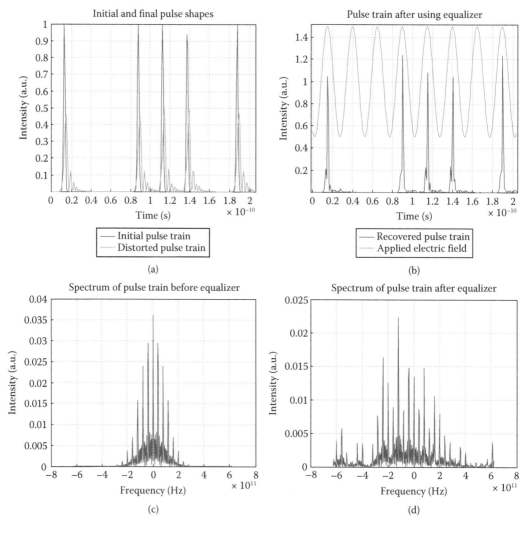

FIGURE 17.12 40 Gbps with 2 ps Gaussian pulse propagating through 120 km optical fiber with the sinusoidal driving voltage.

sequence by its low frequency modulation, this leads to errors in the clock sampling at the receiver and thus the error rate. Therefore, the received signals would not be correct and reliable anymore.

In order to eliminate this problem, an equalizer would be used to push or pull these pulses back to their initial positions as well as remove their distortion. Section 17.3.3 presents further discussions with some simulation results about timing jitter elimination.

17.3.3 Equalization of Timing Jitter and PMD

PMD is a form of modal dispersion where two different polarizations of light in a waveguide, which normally travel at the same speed, travel at different speeds. Thus, it would become broader as the two components disperse along the fiber due to their different group velocities, random imperfections, and asymmetries. In a fiber with constant birefringence, pulse broadening could be estimated from the time delay $\Delta T = \left| \dfrac{L}{V_{gx}} - \dfrac{L}{V_{gy}} \right|$ between two polarization components during propagation of the pulse, where x and y are the two orthogonally polarized modes, V_g is group velocity, and L is a transmission length [27]. There are several PMD compensation schemes using a polarization controller and a phase modulator in the transmitter [9], or using "time lens" [28]. These PMD compensation schemes show that pulses that are placed at different times due to PMD could be shifted into their right positions. Consequently, timing jitter owing to PMD dispersion or other reasons such as nonlinearity effect could also be eliminated. It is pointed out that timing jitter suppression could be applied by using the preceding PMD compensation schemes, as has been proven by Howe and Xu [29] and Jiang [30].

Temporal imaging theorem could be applied in this situation to eliminate timing jitter and PMD. In particular, the phase modulator would apply a very high chirping rate, which could dominate all existing chirping rates owing to PMD, nonlinear effect, and second- and third-order dispersion. Then these pulses with high chirping carrier frequencies are transmitted through the optical fiber link to compress their pulse, recover their original pulse shape, and pull those pulses distorted by timing jitter or PMD back to their original positions.

Figure 17.13 shows timing jitter elimination by using phase modulation with sinusoidal applied voltage, ideal parabolic voltage, and spectral profile corresponding to those recovered pulses. Table 17.5 outlines all parameters of the third order nonlinear coefficients. Table 17.6 indicates all setting necessary parameters in the equalizer system for timing jitter and PMD elimination.

Figure 17.13a shows the timing shift in a range of about ±4 ps added to the input signals with PMD and TOD distortion. The recovered pulse is plotted in Figure 17.13b and c, with sinusoidal applied voltage and ideal parabolic voltage, respectively. The spectral profiles are also plotted in Figure 17.13d and e, corresponding to Figure 17.13b and c, respectively.

When a driving voltage had a sinusoidal waveform, the jittered pulse could not be fully recovered to its original pulse shape. With respect to parabolic modulation, a jittered pulse is fully recovered from TOD and pulled back to its original position. Therefore, it is difficult to fully reduce timing jitter and PMD under sinusoidal modulation, especially when the jittered pulse is broadened within the whole curvature of the modulation. However, under a reasonable jitter or PMD, it is very possible to compensate and pull it back to its original form.

The ideal situation is to use ideal parabolic modulation to fully recover pulse. However, there is a tradeoff between using parabolic and sinusoidal applied voltages. First, it is very hard to generate an ideal parabolic waveform in practice. Second, if it were possible to build a parabolic driving voltage, it would be expensive compared to the very popular sinusoidal waveform. Third, if jittered or PMD pulses are still in a reasonable range, they could still be recovered and pulled back to their initial position with reasonable achieved results. In conclusion, depending on to use either a sinusoidal driving voltage or extensive efforts to generate an ideal or closed to ideal parabolic modulation, the equalization can be approximately implemented. The remaining dispersion or distortion can be subsequently compensated in either the optical or digital electronic domain.

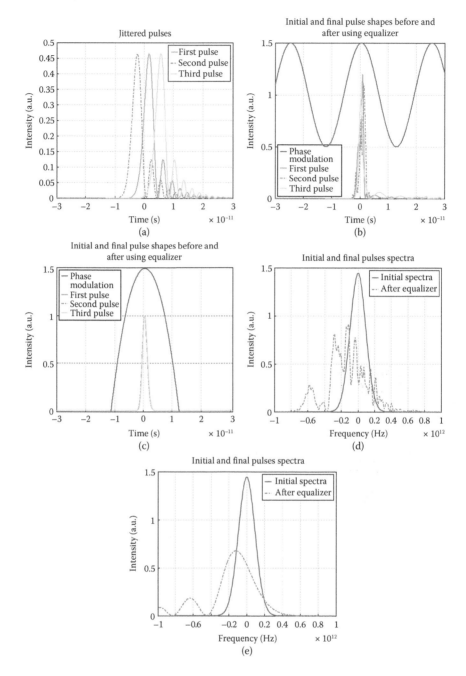

FIGURE 17.13   Investigation of three jittered 2 ps Gaussian pulses.

TABLE 17.5

Parameters for Equalization of Third-Order Elimination under Pulse Train Transmission

	β_2	β_3	$L_{\text{Equalizer}}$	V_π	$V_m = V_{\text{bias}}$	ϕ	f_m	Number of PMs
Values	0 s²/m	9.761e−41 s³/m	80 m	3.5 V	7 V	$\dfrac{5\pi}{6}$	40 GHz	2

TABLE 17.6
Summary of Timing Jitter Elimination Parameters

Parameters	Case: Sinusoidal Driving Voltage	Case: Parabolic Driving Voltage
β_2	0 s²/m	0 s²/m
β_3	9.761e−41 s³/m	9.761e−41 s³/m
$L_{\text{Transmission}}$	120 km	120 km
$L_{\text{Equalizer}}$	80 m	67 m
V_π	3.5 V	3.5 V
V_m	7 V	6 V
V_{bias}	7 V	12 V
ϕ	$-\dfrac{\pi}{24}$	$-\dfrac{\pi}{24}$
f_m	40 GHz	40 GHz
Number of PMs	2 phase modulators	1 phase modulator

17.4 EQUALIZATION IN 160 GBPS TRANSMISSION SYSTEM

17.4.1 SYSTEM OVERVIEW

17.4.1.1 System Configurations

Recently, an SSMF with a GVD value of about 17 ps/(nm · km) has been used widely in long-haul transmission. As discussed earlier, DCF is one of the dispersion management methods to recover signals after a multispan transmission link. However, dispersion slope compensation and wavelength division multiplexed transmission is very hard to be fully compensated, especially for ultra-fast transmitted signals. For example, it would return a significant variation in transmission performance for 160 Gbps, even though there is only a small change in dispersion due to environmental or setting up effects. Therefore, the equalizer system, which includes two main devices—the phase modulator followed by a dispersive element—is used before the receiver. This equalizer is used to compensate not only for GVD but also for third- or fourth-order dispersion [24,31], and also timing jitter [30].

In this study, the equalizer system is used for one span of 120 km transmission fiber (including standard SMF and DCF) to compensate for a GVD of 1.28 ps/nm and a TOD of 1.692 ps/nm². Furthermore, the BER is also calculated when the equalizer is and is not installed, in order to prove the critical improvement when an equalizer is used in the system. In order to investigate the limitation of a 160 Gbps system with and without an equalizer by increasing the number of spans, the transmission length would be increased for each situation. Eye diagrams are also plotted for each situation at a value of receiver power of approximately −32 dBm. From those eye diagrams, significant improvement will be seen for the transmitted system with an equalizer, especially when the number of spans is increased more and more.

17.4.1.2 Experimental Setup

First of all, in order to generate a 160 Gbps optical time division multiplexing (OTDM) signal, a 10 GHz pulse train is created from a mode-locked fiber laser [32], which is modulated at 10 Gbps and then multiplexed optically to 160 Gbps [12,33]. This 160 Gpbs OTDM signal would be transmitted through one span that is a 120 km transmission link, which includes standard SMF and DCF fibers. Afterward, this transmitted signal is demultiplexed to a 10 Gbps signal. Figures 17.14 and 17.15 show the experimental setup with and without an equalizer [12,33].

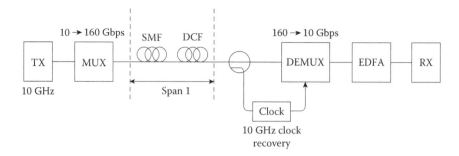

FIGURE 17.14   160 Gbps transmission setup without equalizer.

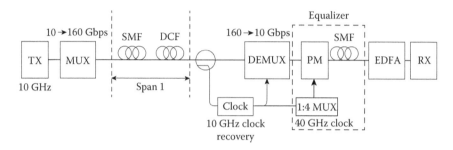

FIGURE 17.15   160 Gbps transmission setup with equalizer. (From T. Hirooka et al., *IEEE Photon. Technol. Lett.*, Vol. 16, No. 10, pp. 2371–2373, 2004; T. Hirooka and M. Nakazawa, *J. Lightwave Technol.*, Vol. 24, No. 7, pp. 2530–2540, 2006.)

The equalizer system includes a phase modulator that is used to modify the chirping factor of carrier frequency, followed by a standard SMF to reshape a received signal, as discussed in the previous section. In addition, a 40 GHz recovery clock is chosen to drive the phase modulator, and this choice is calculated in Section 17.2. Owing to limitation of time, equipment, and experimental knowledge, this experimental setup is applied in a simulation model for further investigation. The Simulink platform is chosen because of its so many advantages. First of all, MATLAB is the standard mathematical tool in academic and R&D laboratories. It is also a very useful and powerful tool for simulation. Furthermore, several existing packages have already been developed for teaching and research purposes, and they are adapted and modified for this simulation purpose. Finally, individual Simulink blocks could be reused and applied for other Simulink models without any conflicts between these models.

Because of the huge advantages of using the Simulink model in MATLAB for running simulations, this platform has been chosen to build a 160 Gbps transmitted signal system with and without an equalizer, based on a real experimental setup. Section 17.4.2 is devoted to introducing and building Simulink models and parameter setting for this study.

17.4.2   Simulation Model Overview

17.4.2.1   System Overview

Modeling of an optical communication system that satisfies some requirements such as simplicity, accuracy in terms of phenomena, and corroborating with experimental systems is very important. Thus, this Simulink model is adapted and modified from some existing Simulink blocks, and they are used, and modified, and thus some new Simulink blocks can be developed. Together, they would create a necessary 160 Gbps system that is built up based on a real experimental setup, which has been introduced briefly in the preceding text (Figure 17.16).

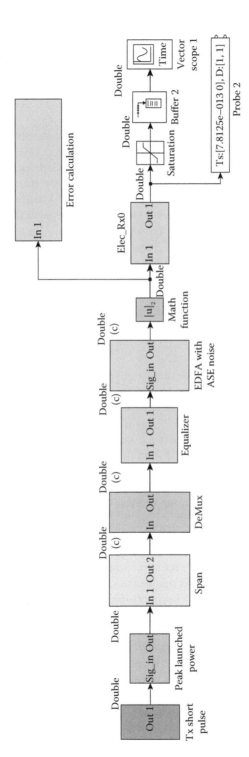

FIGURE 17.16 Overview of Simulink® model for an optical communication system.

The purpose of this Simulink model is to investigate GVD and TOD effects in ultra-fast signal processing. Only an equalization process in linear regime is considered at this stage, and any nonlinearity effects are negligible. Actually, nonlinear effects might change the phase of each frequency component, and it would create a chirp rate on optical carrier frequency. However, the chirp rate that is created from a nonlinear effect is sufficiently small compared to the chirp rate generated from a phase modulator. Thus, it is reasonable to eliminate nonlinearity effect at this stage. This section concentrates on explaining and developing a Simulink model that could be capable of simulating a dispersive compensation before and after using an adaptive equalizer. Some achieved results are also demonstrated in this section.

First of all, a transmitter that could generate 160 Gbps signals transmitted data through fibers at 10 dBm peak power. Although this launching power is high enough to have nonlinearity effects during transmitting signals, it is still ignored, owing to the preceding explanations. Assuming that only one span of transmission fibers of SSMF plus DCF, is employed in the transmission links, is used in these transmission links. When the signal is launched through an SMF fiber, GVD and TOD would affect the propagation signal, and the output signal would be distorted dramatically. Although DCF is used for compensation, the final output signal would not be fully recovered due to other factors such as TOD, PMD, timing jitter, and so on. These problems are also analyzed and discussed in more detail in Sections 17.2 and 17.5, which are devoted to presenting some simulation results about those dispersive effects and their eliminations.

Because the speed of 160 Gbps source is so high that it is currently not yet available, the high-speed OTDM technique would be used to generate this high transmission rate. 160 Gbps or even higher bit rates have already been introduced and generated by using a 10 GHz regenerative mode-locked fiber laser [34–36]. Consequently, this Simulink 160 Gbps laser source is built based on this OTDM technique.

Initially, 16 transmitters of 10 Gbps are used, and these 16 10 Gbps signals are multiplexed by using the OTDM technique. This high-speed signal is transmitted to 120 km (one span), and to several spans of transmission links at later stages. Because the transmitted signals are multiplexed before launching through the optical fiber, they should be demultiplexed back into a 10 Gbps signal. This demultiplexed signal is fed into an equalizer system to recover distorted signals. Erbium-doped fiber is also used to amplify the output signal from the equalizer before receiving this signal at the receiver end. Moreover, an error calculation block is also inserted right before the receiver block in order to count the errors that would be used for calculating BER by the Monte Carlo method. An eye diagram is also plotted as inserted in a pop-up window when the simulation starts running.

17.4.2.2 Transmitter Block

The transmitter is always a necessary device in any communication system. An optical transmitter source is used to launch a laser beam into an optical fiber in a transmission system. This laser beam carries transmitted data that is modulated under different modulation formats such as non-return-to-zero (NRZ), minimum shift keying (MSK), phase shift keying (PSK), and so on, and the return-to-zero (RZ) format is chosen to use in this investigation. The optical transmitter block is shown in Figure 17.1.

A *pulse generator* block is used to generate a 10 GHz square wave with a 2 ps pulse width. The parameters of the pulse generator are set with (1) pulse amplitude = 1; (2) pulse period = 100 ps; and (3) pulse width = 2 ps, that is, about 2% of the pulse period.

Then, this square wave is fed into the MUX block in order to generate 160 GHz square wave transmitted signals. Figure 17.17 shows that a 10 GHz square wave from the pulse generator block is split into 16 input square waves with different time delays by using the *transport delay* block. This time delay could be created by using a microring with different lengths in an experimental setup. Furthermore, this MUX block also included 16 Bernoulli random sources that are considered as 16 different transmitting data signals. These 16 different data signals are multiplied with the square waves mentioned earlier, and these multiplication results are added up to generate 160 Gbps square

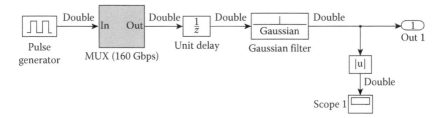

FIGURE 17.17 Transmitter Simulink® model.

wave signals afterward. This output square wave signal has been filtered by using a Gaussian filter to generate a 160 Gbps Gaussian-pulse-shaped signal that is launched into a transmission link under the RZ format at 10 dBm peak launched power. The 160 Gbps Gaussian output multiplexed signal is shown in Figure 17.18.

In order to compare between transmitted and received signals for BER calculation, a "Goto" tag is added to the scope block. In the experimental setup, this multiplexed signal would be demultiplexed into 16 different data packets at the receiver end, and these data packets have been detected at the receiver. When a data packet is extracted and processed in this Simulink model, it is saved via the *Goto* tag and sent to the *error calculation* block for comparison at the receiver end.

17.4.2.3 Transmission Link

The schematic representations, both in subsystem blocks and in the Simulink of a 120 km transmission link, including SMF-DCF fibers, is described in Figures 17.15 and 17.16, respectively. The 160 Gbps RZ transmitted signals of Gaussian shape are launched into a 98 km SSMF at a 10 dBm peak power (Figure 17.19). DCF is used for GVD compensation. Oscilloscopes and spectrum scopes are inserted in front of the SMF, between the SMF and the DCF fiber, and after the DCF fiber in order to observe and measure the pulse broadening as well as compensating factors during a transmission (Figure 17.20). Pulses and spectral profiles are analyzed and discussed in detail in the following section.

17.4.2.4 Demultiplexer

At the receiver end, the 160 Gbps transmitted signals are demultiplexed to a 10 Gbps signal after passing through a demultiplexer block. A 10 GHz clock recovery has been applied to the demultiplexer block. This clock generated a 10 GHz square wave with a 6.25 ps pulse width. The purpose of this clock recovery is illustrated in a simple demultiplexed example in Figure 17.21.

Assuming that 30 Gbps OTDM multiplexed signal to be demultiplexed to 10 Gbps sequence at the receiver end of a transmission system, then the blocks of demultiplexer, clock one and two are used to recover the first, second, and third demultiplexed sequences, respectively, by multiplying this multiplexed signal with each clock recovery. Likewise, synchronization is also an important factor that must be taken into account. Therefore, a unit delay is added in this *DeMux* block for synchronization purposes, as shown in Figure 17.22. The pulse generator, which acts as a clock recovery, is used to generate a 10 GHz square wave with 6.25 ps pulse width. This square wave is delayed by feeding into the unit delay block, and then multiplied with the input signal for the demultiplexing process.

Figure 17.23 demonstrates an example of a successful demultiplexed signal after passing through the *DeMux* block. The higher-position scope (Scope 1) shows the original multiplexed signal, and the demultiplexed output signal is plotted on the lower scope (Scope 3). According to the input signal, the output signal is successfully demultiplexed. In addition, the synchronization issue is even more harder and more significant when TOD and time jittering effects play a critical role during

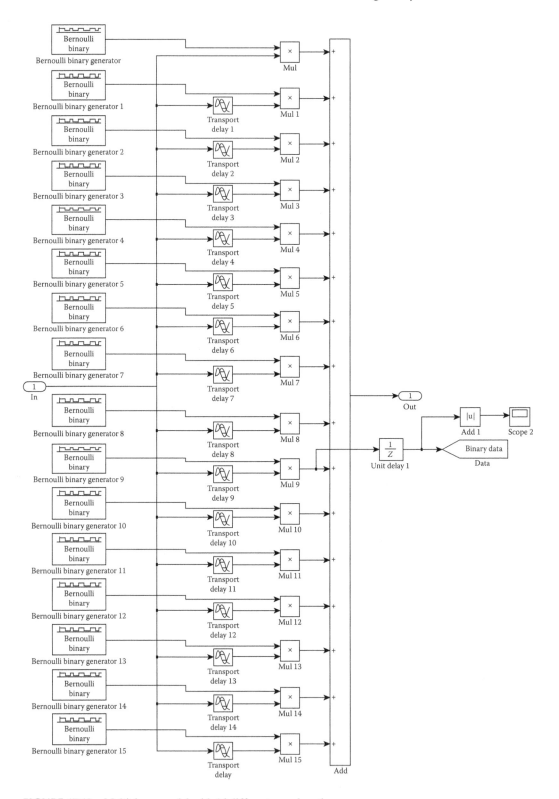

FIGURE 17.18 Multiplexer model with 16 different wavelength sources.

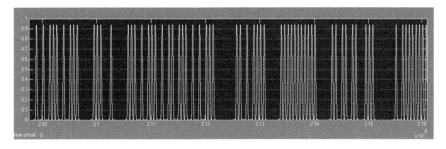

FIGURE 17.19 160 Gbps OTDM output signal.

propagation through the transmission link. Thus, a clock should be synchronized with suitable pulses accurately in order to sample the correct bits.

17.4.2.5 Equalizer System

The equalizer system that has been invented according to the temporal imaging theorem has been introduced and discussed clearly on working algorithms as well as some demonstrated simulation results in the previous sections. This equalizer included a phase modulator, followed by a dispersive element, which is a standard SMF fiber in this Simulink model. Figure 17.24 shows the structure of the subsystem model. The working algorithm of the equalizer system has already been discussed in detail in the previous section. Thus, this section concentrates more on how to build this Simulink model, setting up parameters and some significant results in the sections that follow. The main function of the phase modulator in this model is to create a required chirping-rate-to-carrier frequency of the optical signal. By applying a driving voltage to a phase modulator, the optical signal would be phase modulated because the optical path length is altered by the electric field. Figure 17.24 shows a phase modulator block in more detail, and it reveals the previously explained phase modulation concept.

This *phase modulator* block includes a sinusoidal source to generate a 40 GHz sinusoidal wave to drive a phase modulator. Let V_π be the voltage applied to generate a π phase shift on the optical carrier frequency. Basing on some formulas derived in Section 17.3, a term $\frac{\pi}{V_x}$ is multiplied into that sinusoidal waveform. As with all calculations in Section 17.3, the amplitude of the driving voltage V_m is set to 7 V, which is double the value 3.5 V of V_π, in order to achieve a 2π phase shift on carrier frequency to guarantee the necessary chirp rate. Furthermore, two 40 GHz phase modulators are used in this situation to recover the output signal as a theoretical calculation in Section 17.3 and simulation results in Section 17.4. Thus, the multiplication product is a result of three multiplications between two 40 GHz phase modulators and input signals.

In addition, the synchronization issue is very important, because each output Gaussian pulse of the transmitted data must be fully covered by this 40 GHz sinusoidal driving voltage. It is a necessary process because all optical carrier frequencies need to be chirped correctly by this applied electric field. At this stage, Scope 1 and Scope 2 in Figure 17.25 are used to monitor the applied electric field and input signal, to confirm that the phases of each pulse of signal and sinusoidal waveform is in-phase. Otherwise, the driving voltage should be tuned until an expected position is achieved. This Simulink setup model is also based on the experimental setup described in Section 17.4.1.2. If all the parameters are set correctly, this driving voltage would directly change the phase of each frequency component of the input signal. As a result, the chirping rate of the carrier frequency of this signal had to be changed, and this carrier frequency has been set to be chirped up as the theoretical calculation. These pulses would be compressed and recovered after propagating through an SMF fiber in this model due to reorganization frequency components during the propagation process.

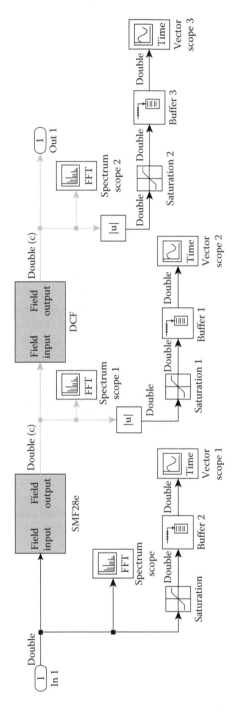

FIGURE 17.20 SMF–DCF fiber in one transmission span.

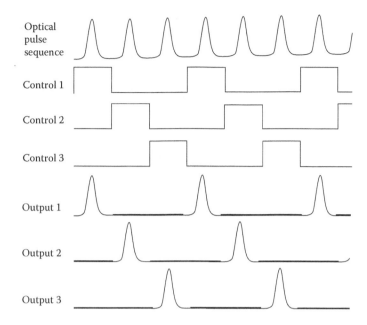

FIGURE 17.21 Optical pulse sequence and control applied signals to demultiplexers and the corresponding outputs.

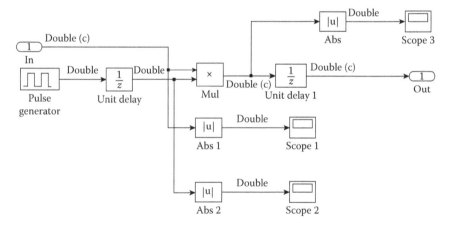

FIGURE 17.22 Demultiplexer block in detail.

17.4.2.6 Errors Calculation

First, this *errors calculation* block is built to calculate the BER based on the Monte Carlo method, which is a widely used class of computational algorithms for simulating the behavior of systems. Furthermore, the Monte Carlo method is employed since it is useful for modeling phenomena with significant uncertainty in inputs. Figure 17.26 shows a BER calculation block using the Monte Carlo method.

The *threshold time decision binary detection* (TTDBD) block is adapted and modified to detect between bit 1 and bit 0 of the received signal. The output from TTDBD is connected to the *error rate calculation* (ERC) block to calculate the BER and the *find delay* block to set the time delay between signals from the transmitter end (from the "Binarydata" tag, which is collected from the transmitter block) and the receiver end to the *integer delay* block. The purpose of the *find delay* block is to synchronize between transmitted and received signals.

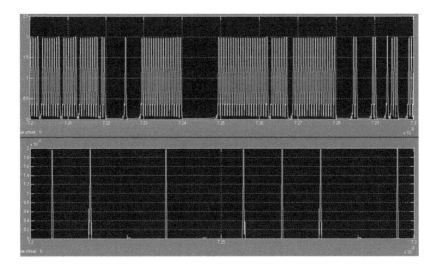

FIGURE 17.23 Demultiplexed signal.

Finally, the computational delay in the ERC block should be set with the same value as the value from *find delay* block. For example, a number "1794," which is returned from *find delay* block, should be set for the *integer delay* block. As a result, the ERC block should also ignore the first 1794 samples at the beginning of the comparison in this block. It follows that this number should obviously be set for a computation delay parameter. In addition, the correlation window length (samples) parameter of the *find delay* block should be set sufficiently large so that the computed delay eventually stabilizes at a constant value. However, there is a tradeoff between the reliability of the computed delay and the processing time to compute the delay. Thus, a reasonable value for the correlation window length should also be taken into account.

17.4.3 SIMULATION RESULTS

17.4.3.1 Single-Pulse Transmission

17.4.3.1.1 Equalization of Distortion Due to GVD

The Simulink model for this GVD elimination is modified a bit to cope with the purpose of this section. The transmitter is used temporarily to transmit a single pulse, and thus the multiplexer and demultiplexer blocks are taken out. Furthermore, only 300 m of SMF fiber is used in this transmission link, and DCF is not necessarily used in this section.

With respect to the pulse shape of a single pulse of the transmitter output from the oscilloscope, the full width at the half maximum of this single pulse is measured to be approximately 2 ps (Figure 17.27a). After propagating through 300 m of standard SMF fiber, this pulse is broadened to about 9 ps (Figure 17.27b). This achieved result is consistent with the simulation result in Section 17.3 and the theoretical calculation. The final recovered pulse shape is plotted in Figure 17.27c with a pulse width value of 2 ps, but it had a longer tail as compared to the original pulse shape.

In addition, the spectral profiles of a single pulse before and after using the equalizer are shown in Figure 17.27d and e. The spectrum of a lightwave-modulated single pulse before and after propagating through a SMF fiber remain the same, because only the phases of the spectral components are changed. Thus these phase changes superimpose different time delays on each individual spectral component due to the wavelength dependent of the effective indexes of the guided lightwaves. Moreover, these delays are a reason for pulse broadening. Compared to the spectra after the equalizer that is broadened dramatically, it indicates that the optical carrier frequency is chirped, and it is actually up-chirped, which is consistent with theoretical calculations and expectations.

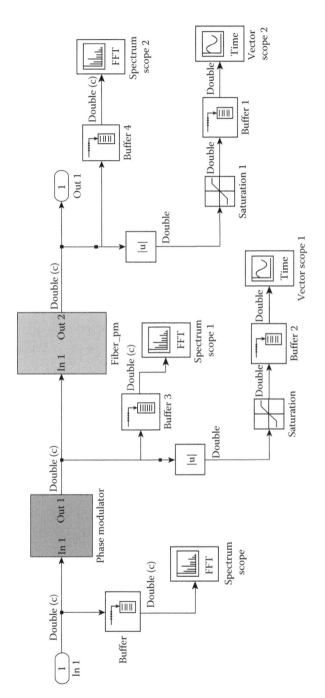

FIGURE 17.24 Equalizer system including phase modulator followed by an SMF.

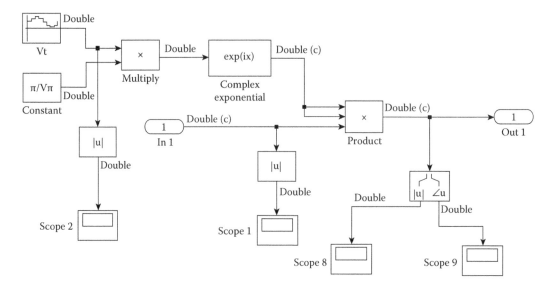

FIGURE 17.25 Phase modulator block in detail.

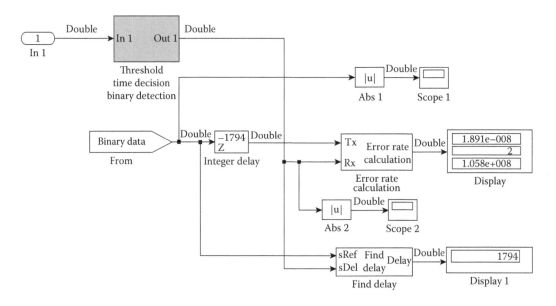

FIGURE 17.26 The BER calculation block.

Finally, all simulation results for a transmitted single pulse have matched with the obtained results from Section 17.4 and the expectations. Therefore, an equalizer could eliminate the GVD effect during transmission.

17.4.3.1.2 TOD Elimination

TOD plays a significant role in ultra-high-speed signals in the transmission system. From theoretical calculation and simulation results from Sections 17.2 and 17.3, respectively, sinusoidal modulation would not have fully recovered the distorted signals due to their asymmetric TOD. On the contrary, the oscillated tail of the recovered signal is a reasonable reduction, and thus the noise level created from this tail does not have a big effect on the transmission system, and its contribution to BER is acceptable. Figure 17.28a through c shows an initial single pulse before being transmitted through

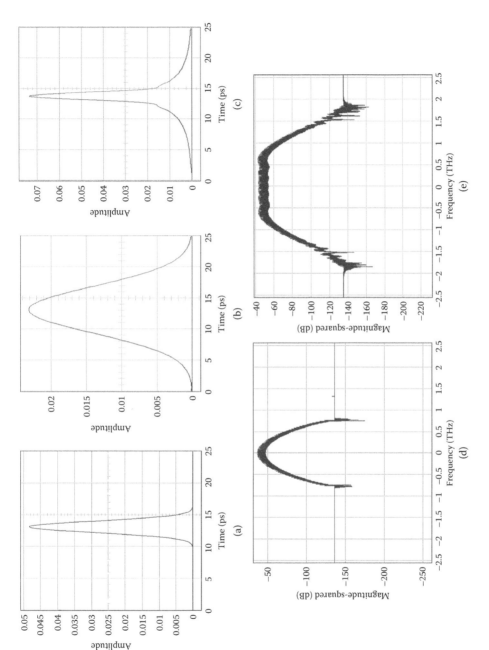

FIGURE 17.27 GVD effects under single-pulse transmission: (a) at a transmitter; (b) after 300 m SMF; (c) after equalizer; (d) and (e) spectral profile of a single pulse before and after the equalizer, respectively.

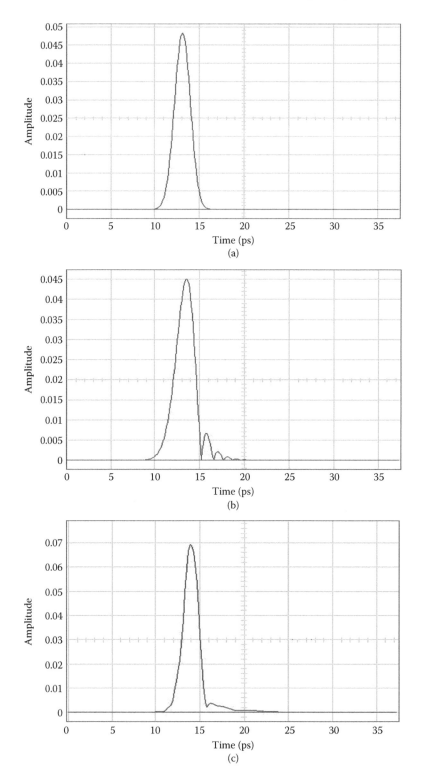

FIGURE 17.28 TOD elimination for a single-pulse transmission (a) at a transmitter; (b) after 120 km SMF; (c) after the equalizer.   *(Continued)*

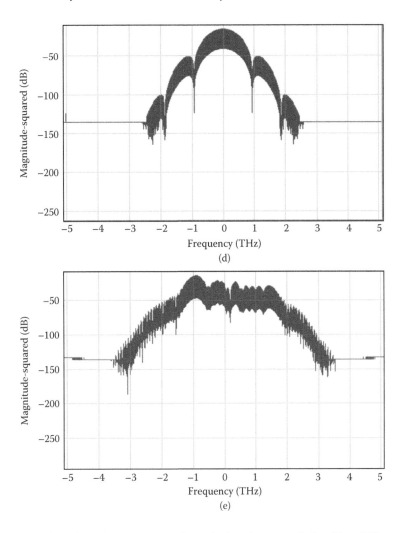

FIGURE 17.28 (Continued) TOD elimination for a single-pulse transmission (d) and (e) spectral profiles before and after the equalizer, respectively.

the optical fiber, a distorted pulse that is affected critically by TOD after propagating through a 120 km transmission link, and recovered pulse after the equalizer, respectively. Figure 17.28d and e plots the spectra of this single pulse before and after the equalizer. It shows that the carrier frequency is chirped at a very high chirping rate, which dominates all other effects such as nonlinearity of the second-order, third-order, jitter, or PMD effects.

17.4.3.2 160 Gbps Transmission and Equalization

Ultra-high-speed communication systems have been becoming more and more popular because of the increase in demand for high-capacity optical networks. Furthermore, transmission distance for this ultra-fast transmitted signal is improving and developing. In order to achieve longer transmission links with high-speed transmission, different modulation formats such as differential PSK is applied [37]. For the OTDM system, use of the on–off keying format has been extended up to approximately 600 km transmission distance. The limit of the OTDM technique is that, due to its sensitivity properties, even small perturbations in the optical fibers such as GVD, TOD, PMD, and timing jitter could distort signals dramatically [38,39]. In this section, some significant results are given for a 120 km transmission link with BER characteristics. The transmission length is a parameter for the evaluation of the equalized system, and the eye diagram is illustrated for the received power of about −32 dBm.

17.4.3.2.1   GVD Equalization of 120 km SSMF Transmission

Simulation results for GVD equalization are shown in Table 17.7 for three situations: back-to-back (transmitter and receiver are connected directly together without a transmission link), without equalizer, and with equalizer. These data are then plotted in Figure 17.29 with a GVD value of 1.28 ps/nm. The TOD, nonlinear effect, and timing jitter are set to zero in order to investigate only the GVD effect.

In this simulation setup, a 97.84 km standard SMF fiber with dispersion parameter $D = 17$ ps/(nm · km) is used as the transmission link. In order to have a second-order dispersion compensation, a 22.16 km DCF fiber length is used, with dispersion parameter $D = -75$ ps/(nm · km). Therefore, a GVD value of 1.28 ps/nm is the mismatched dispersion.

TABLE 17.7

BER Characteristic for GVD Equalization of 160 Gbps System for Back-to-Back, with and without Equalizer

Back-to-Back		Without Equalizer		With Equalizer	
Received Power (dBm)	log(BER)	Received Power (dBm)	log(BER)	Received Power (dBm)	log(BER)
−37	−4.521	−36.375	−4.453	−36.745	−4.495
−36	−5.647	−35.193	−5.521	−35.498	−5.821
−35	−6.727	−34.152	−6.470	−34.437	−7.006
−34	−7.853	−33.012	−7.455	−33.414	−8.054
−33	−8.972	−31.983	−8.196	−32.356	−9.017
−32	−9.987				

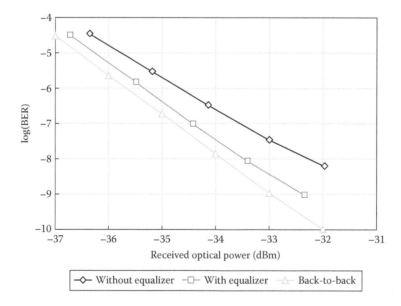

FIGURE 17.29   BER versus receiver sensitivity of 10 Gbps demultiplexing signal under GVD linear impairment under the case of back-to-back "triangle."; transmission with "square"; and without equalizer "diamond."

With a low value of received power, there is an improvement of about 1/2 dBm between with and without an equalizer at a log(BER) value of −5. However, this difference is increased when the received power is increased. For example, at a log(BER) value of −8, this improvement is greater than 1 dBm. This is further improved as a function of the receiver sensitivity. Thus, using an equalizer could improve long-haul transmission performance, especially at a higher received power (or at a higher signal-to-noise ratio).

The back-to-back BER characteristic is also plotted in Figure 17.29. Although the improvement of the EOP for using the equalizer is still approximately half a dBm worse than the back-to-back system, it still shows a better improvement when the equalizer is used. This small residual penalty is due to the approximation of using a sinusoidal waveform instead of an ideal parabolic waveform. If the received power is increased further, there would be a significant BER improvement, with saturation at around −32. Thus, the error floor would appear following a diamond-trend at the higher received power value.

Figure 17.30 shows eye diagrams corresponding to the individual curve of the BER transmission performance. For example, Figure 17.30a1 shows an eye diagram in the case before demultiplexing, and Figure 17.30a2 and a3 shows eye diagrams after demultiplexing at the received power values of −36.375 and −36.745 dBm under without and with equalizer, respectively. Before demultiplexing, the input signals of 160 Gbps are observed as depicted in Figure 17.30(a1). The middle and right-hand-side eye diagrams (a2) and (a3) are the demultiplexed 10 Gbps signals at the receiver end under without and with equalizer, respectively.

Through the eye diagrams obtained for different received power levels, it is observed that higher the received power, the more the opening of the eye. The FWHM of the distorted eye is about 3 ps, while the FWHM of the recovered eye is only approximately 2 ps at about −32 dBm received power. Thus, there is a significant improvement when the equalizer is inserted, especially in long-haul transmission under accumulation of mismatched dispersion.

17.4.3.2.2   TOD Equalization of 120 km SSMF Transmission

For TOD elimination case, a second-order dispersion is assumed to be fully compensated by the DCF fiber. A 104.7 km standard SMF and a 15.3 km DCF with dispersion slope values of $S = 0.06$ ps/(nm^2 . km) and 15.3 ps/(nm^2 . km), respectively, are used in each span of the transmission system. Thus, the dispersion slope TOD is calculated with a value of about 1.692 ps/nm^2.

Table 17.8 and Figure 17.31 show the simulation results, the BER characteristics of the TOD equalization scheme. Although the TOD effect for 120 km is not very significant when compared to the GVD effect, as discussed in the previous section, the equalizer still shows its usefulness in improving transmission performance. However, this BER characteristic when using the equalizer could still not get closer to the back-to-back line (triangle line). The error floor is possible due to the diamond line for nonequalization at a higher received power, while a square-line for using the equalizer system still drops steeply. Consequently, the improvement between using and not using an equalizer would be much greater than 1 dBm (as at about log(BER) = −8 in Figure 17.31).

In conclusion, an equalizer offers significant improvement of the transmission performance in long-haul optical transmissions. Furthermore, when the signal-to-noise ratio is improved, there is a significant improvement between using and not using an equalizer system. Some eye diagrams are also plotted in Figure 17.32.

Figure 17.32a1, a2, and a3 shows eye diagrams for 160 Gbps signals before being demultiplexed, 10 Gbps signals after being demultiplexed not using equalizer, and 10 Gbps signals after being demultiplexed using equalizer, respectively, and, from (a) to (e), showed eye diagrams for each received power values from simulation. There is an improvement when a higher power is received at the receiver end. Furthermore, with the same received power value, eye diagrams in the TOD case are a bit bigger than the eye diagrams in the GVD case. Therefore, a received BER from simulation also showed lower BER values in the TOD case for a case using an equalizer.

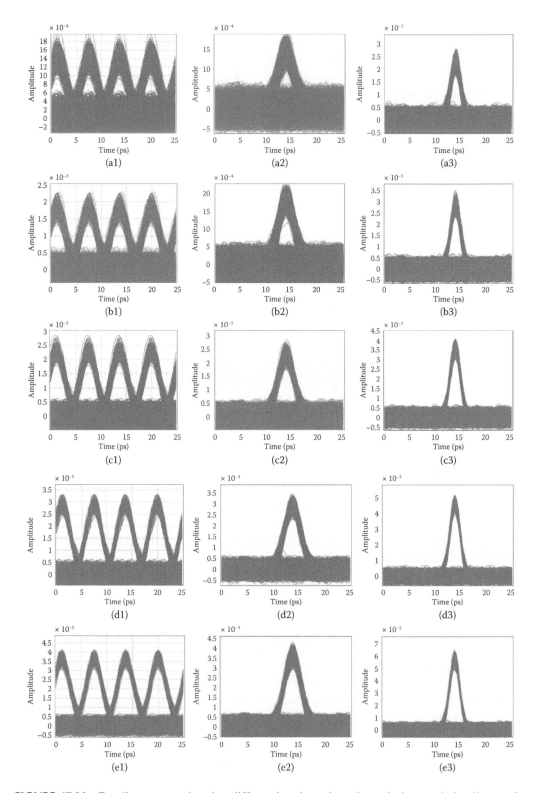

FIGURE 17.30 Eye diagrams monitored at different locations along the optical transmission lines under without and with GVD effects and with different received power for three cases (from left to right): before demultiplexing, after demultiplexing without and with equalizer.

TABLE 17.8

BER Characteristic Summary of TOD Elimination for 160 Gbps System for Back-to-Back, with and without Equalizer

Back-to-Back		Without Equalizer		With Equalizer	
Received Power (dBm)	log (BER)	Received Power (dBm)	log (BER)	Received Power (dBm)	log (BER)
−37	−4.521	−36.366	−4.673	−36.703	−4.526
−36	−5.647	−35.375	−5.521	−35.616	−5.708
−35	−6.727	−34.363	−6.570	−34.597	−6.820
−34	−7.853	−33.202	−7.489	−33.468	−7.984
−33	−8.972	−32.134	−8.296	−32.413	−9.056
−32	−9.987				

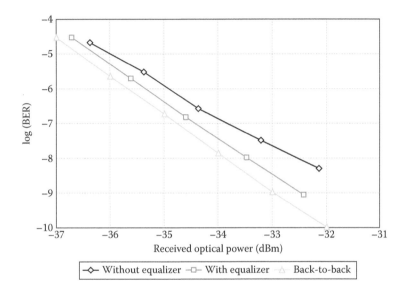

FIGURE 17.31 BER versus receiver sensitivity of 10 Gbps demultiplexing signal in TOD elimination under the case of back-to-back "triangle." and transmission with "square" and without equalizer "diamond."

Although the BER of the system remained pretty much the same as the GVD elimination due to the approximation of sinusoidal applied voltage, it still confirmed its usefulness in compensating for second-order and third-order effects (might also be timing jitter and PMD during transmission) through Figures 17.30 and 17.32. Furthermore, the FWHM of the distorted eye is about 2.8 ps, while the FWHM of the recovered eye is approximately 2.2 ps at the received power value of about −32 dBm.

17.4.3.2.3   Equalization of TOD with Variable Fiber Lengths

The TOD effects critically impose the distortion of the pulse transmission over the long-haul transmission link. Unlike the effect of the GVD that could be fully compensated for by using the DCF fiber, it is very difficult to compensate for TOD fully by using existing DCFs. Therefore, there is always distortion due to the mismatch between SMFs and DCFs. To compensate for the TOD, the equalizer offers some additional equalization potential that may be necessary at the end of the transmission system.

The five pairs of eye diagrams in Figure 17.33 show the recovered signals after they are transmitted through 240 km (two spans), 480 km (four spans), 720 km (six spans), 960 km (eight spans), and 1200 km (10 spans) of optical links, respectively, at a received power value of approximately −32 dBm. All set parameters for the equalizer system remain the same and are inserted at the end of transmission links. Erbium-doped fiber with a gain value of 30 dB and a 5 dB

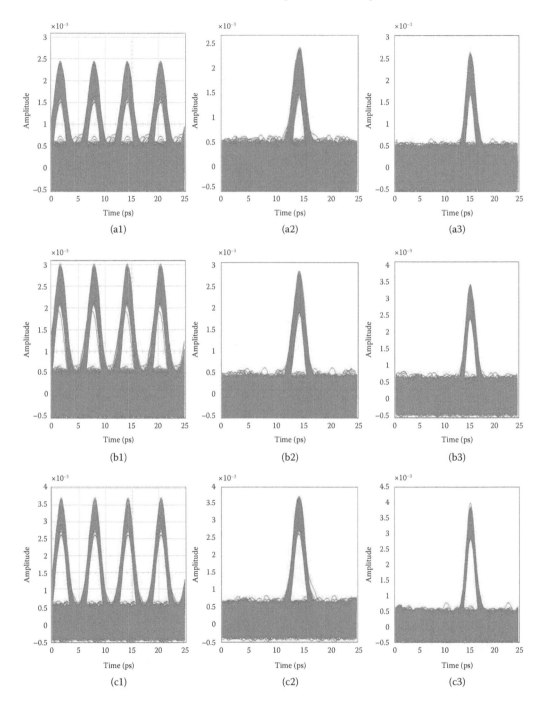

FIGURE 17.32 Eye diagrams for TOD with ranges of received power for three cases (from left to right): before demultiplexing, after demultiplexing without and with equalization. *(Continued)*

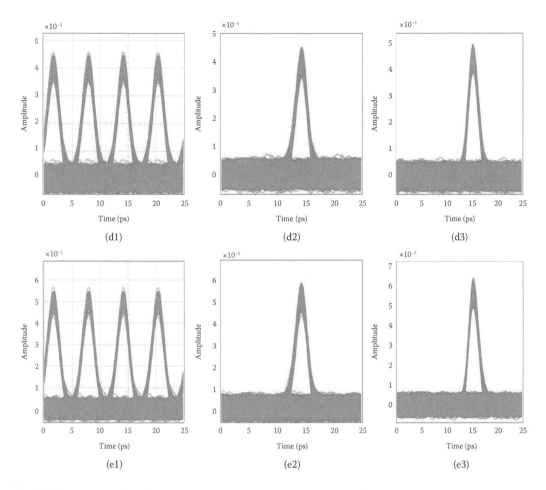

FIGURE 17.32 (Continued)   Eye diagrams for TOD with ranges of received power for three cases (from left to right): before demultiplexing, after demultiplexing without and with equalization.

noise figure are used between each 120 km transmission span to compensate for the optical losses of transmitted signals due to attenuation. Figure 17.33a1 through a5 indicate that TOD is accumulated over long transmission lengths, and the received signals would be distorted rapidly. For two spans of the transmission link, the received signals have not been distorted too much, and the FWHM of an eye is measured at about 2.6 ps. When a transmission link is increased up to four spans (480 km), a third-order dispersive effect is seen clearly. However, the eye is still widely opening, and the FWHM is about 3 ps at this time. On the contrary, when there are more than six spans, greater than 720 km in fiber length, the received signals are distorted dramatically, especially at their tails. These long oscillation tails would substantially affect the neighboring pulses during transmission.

Figure 17.33b1 through b5 shows the TOD equalization. The eye diagrams of five situations from two spans to ten spans are also distorted, since more than six spans are used in the transmission link, but these distortions are not too much, and the eye windows of these eye diagrams are still opening wide enough to have a reasonable BER. Furthermore, the FWHM of an eye when two spans are used is about 2.2 ps, while this FWHM is approximately 3 ps after transmitting through ten spans (1200 km fiber length). The received results after using an equalizer are reasonable, and there is a huge improvement as compared to signals before using the equalizer.

Further investments might be done by running simulations to count the number of errors and by calculating the BER of this transmitted system for variable transmission lengths afterward.

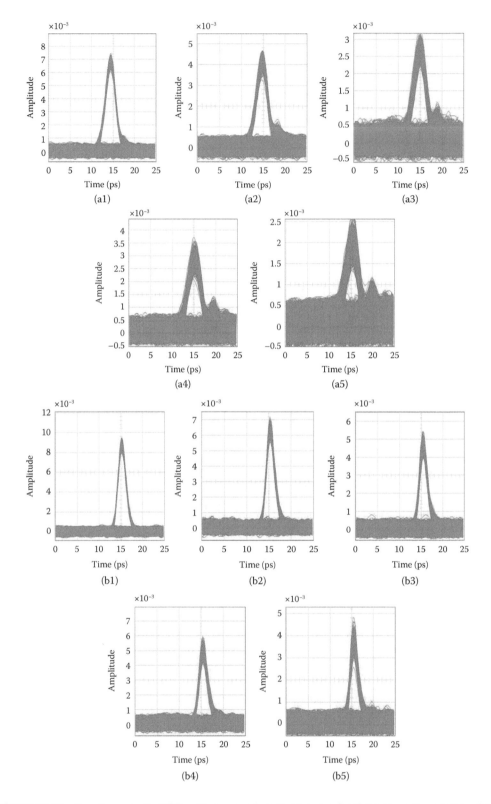

FIGURE 17.33 Eye diagrams for TOD with ranges of received power for three cases (from left to right): before demultiplexing, after demultiplexing without and with equalizer.

This would give correct and exact estimations and calculations on how much the received signals could be improved by using the equalizer.

17.5 CONCLUDING REMARKS

Photonic signal processing based on space–time duality is very important in ultra-high-speed transmission systems. In this chapter, a foundation for the principles of temporal imaging and space–time duality have been studied and proven. From the equivalence between paraxial diffraction in space and narrow-band dispersion in time has been used to construct on analogous approximations to the relative bandwidths of spatial and temporal Fourier spectra. By using the diffraction–dispersion analogy, a quadratic phase modulation acts as a time lens, and the focal length of spatial and time lens have already been derived. The duality properties between spatial and time domains has led to temporal imaging, which would be very useful in future works, especially for designing adaptive equalizers by using phase modulation to expand or compress pulses in ultra-high-speed communication systems. Analyzing the impulse response of the temporal imaging system reveals the resolution that is related to the Fourier transform of the aperture function, and thus it proves the equivalence of the time-domain interpretation of the Fraunhofer diffraction pattern [2]. It is a very important result that solidifies the nature of the space–time duality.

With respect to photonic signal processing, a single-pulse transmission has been studied by propagating through SSMF with GVD and TOD effects without an initial chirp factor. The knowledge from the simulation results for the single pulse is then used for the design of the dispersion-compensating scheme that could eliminate not only second-order dispersion but also TOD, PMD, and timing jitter.

The space–time duality and temporal imaging are studied and then applied in the design of an adaptive equalizer system that includes a phase modulator followed by a dispersive element that could recover the initial pulse shape from distorted pulse. An efficient adaptive equalizer acts as a time lens in the time domain for optical pulse width compression by using a phase modulator, and a dispersive element is proposed, analyzed, and proven by simulation results. By using the equalizer system, the GVD effect is fully compensated. Furthermore, TOD, PMD, and timing jitter are also reduced significantly in ultra-high-speed transmission systems using a parabolic sinusoidal voltage waveform in the driving of the phase modulator. It has also been shown that an ideal parabolic phase modulation could fully recover the original pulse.

Furthermore, this adaptive equalizer has also been applied in ultra-high-speed transmission and equalization at 160 Gbps. Higher bit rates such as 320 Gbps or even higher might be possible, and would be investigated at a later stage. An evaluation of the improvement between with and without using an equalizer is also made. By monitoring the eye diagrams at the receiver end of this system, an eye of signals without an equalizer is distorted significantly, and it would return a much higher BER during transmission. On the contrary, when an adaptive equalizer is inserted, there is a critical improvement in BER as well as a wide opening eye in an eye diagram. The transmission length is increased up to 1200 km in order to investigate its limitation. We have shown the agreement between theoretical and simulation results of the performance of the equalization technique for a 160 Gbps optical transmission system under the influence of GVD, TOD, PMD impairment, and the elimination of the timing jitter, not only with a single pulse but also with 160 Gbps random sequence data transmission. The nonlinear phase shift and noises in the system might challenge the system stability.

REFERENCES

1. G. Lifante, *Integrated Photonics: Fundamentals*, West Sussex: John Wiley, 2003.
2. B. H. Kolner, Space-time duality and the theory of temporal imaging, *IEEE J. Quant. Electron.*, Vol. 30, No. 8, pp. 1951–1963, 1994.

3. C. V. Bennett and B. H. Kolner, Principles of parametric temporal imaging—Part I: System configurations, *IEEE J. Quant. Electron.*, Vol. 36, No. 4, pp. 430–437, 2000.

4. C. V. Bennett and B. H. Kolner, Principles of parametric temporal imaging—Part II: System performance, *IEEE J. Quant. Electron.*, Vol. 36, No. 6, pp. 649–655, 2000.

5. A. Papoulis, *Systems and Transforms with Applications in Optics*, New York: McGraw-Hill, 1968.

6. A. R. Adams and Y. Suematsu, *Handbook of Semiconductor Lasers and Photonic Integrated Circuits*, New York: Chapman and Hall, 1994.

7. G. P. Agrawal, *Nonlinear Fiber Optics*, 3rd ed., San Diego, CA: Academic Press, 2001.

8. T. Khayim, M. Yamauchi, D. S. Kim, and T. Kobayashi, Femtosecond optical pulse generation from a CW laser using an electro-optic phase modulator featuring lens modulation, *IEEE J. Quant. Electron.*, Vol. 35, No. 10, pp. 1412–1418, 1999.

9. L. S. Yan, Q. Yu, T. Luo, and X. S. Yao, Compensation of higher order polarization-mode dispersion using phase modulation and polarization control in the transmitter, *IEEE Photon. Technol. Lett.*, Vol. 14, No. 6, pp. 858–1012, 2002.

10. E. B. Treacy, Optical pulse compression with diffraction gratings, *IEEE J. Quant. Electron.*, Vol. QE-5, pp. 454–458, 1969.

11. A. Papoulis, Pulse compression, fiber communication, and diffraction: A unified approach, *J. Opt. Soc. Am. A,* Vol. 11, No. 1, pp. 3–13, 1994.

12. T. Hirooka, M. Nakazawa, F. Futami, and S. Watanabe, A new adaptive equalization scheme for a 160-Gbps transmitted signal using time-domain optical Fourier transformation, *IEEE Photon. Technol. Lett.*, Vol. 16, No. 10, pp. 2371–2373, 2004.

13. J. Azana and M. A. Muriel, Real-time optical spectrum analysis based on the time-space duality in chirped fiber gratings, *IEEE J. Quant. Electron.*, Vol. 36, No. 5, pp. 517–526, 2000.

14. T. Jannson, Real-time Fourier transformation in dispersive optical fibers, *Opt. Lett.*, Vol. 8, No. 4, pp. 232–234, 1983.

15. A. A. Godil, B. A. Auld, and D. M. Bloom, Picosecond time-lenses, *IEEE J. Quant. Electron.*, Vol. 30, No. 3, pp. 827–837, 1994.

16. F. T. S. Yu and I. C. Khoo, *Principles of Optical Engineering*, New York: John Wiley, 1990.

17. J. W. Goodman, *Introduction to Fourier Optic*, New York: McGraw Hill, 1968.

18. M. T. Kauffman, A. A. Godil, B. A. Auld, W. C. Banyai, and D. M. Bloom, Applications of time lens optical systems, *Electron. Lett.*, Vol. 29, No. 3, pp. 268–269, 1993.

19. B. H. Kolner and M. Nazarathy, Temporal imaging with a time lens, *Opt. Lett.*, Vol. 14, No. 12, pp. 630–632, 1989.

20. B. H. Kolner and M. Nazarathy, Temporal imaging with a time lens: Erratum, *Opt. Lett.*, Vol. 15, No. 11, p. 655, 1990.

21. C. V. Bennet, R. P. Scott, and B. H. Kolner, Temporal magnification and reversal of 100 Gbps optical data with an un-conversion time micro-scope, *Appl. Phys. Lett.*, Vol. 65, No. 20, pp. 2513–2515, 1994.

22. T. Jannson and J. Jannson, Temporal self-imaging effect in single-mode fibers, *J. Opt. Soc. Am.*, Vol. 71, No. 11, pp. 1373–1376, 1981.

23. B. E. A. Saleh and M. I. Irshid, Collett-Wolf equivalence theorem and propagation of a pulse in a signal mode optical fiber, *Opt. Lett.*, Vol. 7, No. 7, pp. 342–343, 1982.

24. T. Yamamoto and M. Nakazawa, Third- and fourth-order active dispersion compensation with a phase modulator in a terabit-per-second optical time-division multiplexed transmission, *Opt. Lett.*, Vol. 11, No. 9, pp. 647–649, 2001.

25. A. F. Elrefaie, R. E. Wagner, D. A. Atlas, and D. G. Daut, Chromatic dispersion limitations in coherent lightwave transmission systems, *J. Lightwave Technol.*, Vol. 6, No. 5, pp. 704–709, 1988.

26. X. Li, X. Chen, and M. Qasmi, A broad-band digital filtering approach for time-domain simulation of pulse propagation in optical fiber, *J. Lightwave Technol.*, Vol. 23, No. 2, pp. 864–875, 2005.

27. G. P. Agrawal, *Fiber-Optic Communication Systems*, 3rd ed., New York: John Wiley, 2002.

28. M. Romagnoli, P. Franco, R. Corsini, A. Schiffini, and M. Midrio, Time-domain Fourier optics for polarization-mode dispersion, *Opt. Lett.*, Vol. 24, No. 17, pp. 1197–1199, 1999.

29. J. V. Howe and C. Xu, Ultrafast optical signal processing based upon space-time dualities, *J. Lightwave Technol.*, Vol. 24, No. 7, pp. 2649–1662, 2006.

30. L. A. Jiang, M. E. Grein, H. A. Haus, E. P. Ippen, and H. Yokoyama, Timing Jitter eater for optical pulse trains, *Opt. Lett.*, Vol. 28, pp. 78–80, 2003.

31. E. Hellstrom, H. Sunnerud, M. Westlund, and M. Karlsson, Third-order dispersion compensation using a phase modulator, *J. Lightwave Technol.*, Vol. 21, No. 5, pp. 1188–1197, 2003.

32. M. Nakazawa, E. Yoshida, and Y. Kimura, Ultrastable harmonically and regeneratively modelocked polarization-maintaining erbium fiber ring laser, *Electron. Lett.*, Vol. 30, pp. 1603–1604, 1994.
33. T. Hirooka and M. Nakazawa, Optical adaptive equalization of high-speed signals using time-domain optical Fourier transformation, *J. Lightwave Technol.*, Vol. 24, No. 7, pp. 2530–2540, 2006.
34. M. Nakazawa, T. Yamamoto, and K. R. Tamura, 1.28 Tbits/s–70 km OTDM transmission using third- and fourth-order simultaneous dispersion compensation with a phase modulator, *Electron. Lett.*, Vol. 36, No. 24, pp. 2027–2029, 2000.
35. S. Kawanishi, H. Takara, T. Morioka, O. Kamatani, K. Takiguchi, T. Kitoh, and M. Saruwatari, Single channel 400 Gbit/s time-division-multiplexed transmission of 0.98 ps pulses over 40 km employing dispersion slope compensation, *Electron. Lett.*, Vol. 32, pp. 916–918, 1996.
36. S. Diez, R. Ludwig, and H. G. Wener, All-optical switch for TDM and WDM/TDM systems demonstrated in a 640 Gbit/s demultiplexing experiment, *Electron. Lett.*, Vol. 34, pp. 803–805, 1998.
37. M. Daikoku, T. Miyazaki, I. Morita, H. Tanaka, F. Kubota, and M. Suzuki, 160 Gbit/s-based field transmission experiments with single polarization RZ-DPSK signals and simple PMD compensator, in *European Conference on Optical Communication, 2005 (ECOC 2005)*, Glasgow, Scotland, September 2005, paper We2.2.1.
38. G. Lehmann, W. Schairer, H. Rohde, E. Sikora, Y. R. Zhou, A. Lord, D. Payne, et al., 160 Gbit/s OTDM transmission field trial over 550 km of legacy SSMF, in *European Conference on Optical Communication 2004. (ECOC 2004)*, Stockholm, Sweden, September 2004, paper We1.5.2.
39. T. Hirooka, K. I. Hagiuda, T. Kumakura, K. Osawa, and M. Nakazawa, 160-Gbps–600-km OTDM transmission using time-domain optical Fourier transformation, *IEEE Photon. Technol. Lett.*, Vol. 18, No. 24, pp. 2647–2649, 2006.

18 Comparison of Modulation Formats for Digital Optical Communications

In this chapter, the transmission performances of modulation formats, including discrete phase, continuous phase, amplitude shift keying, partial response duobinary, and multicarrier orthogonal modulation are compared. First, a summary of amplitude shift keying (ASK), differential phase shift keying (DPSK), and differential quadrature phase shift keying (DQPSK) of various FWHM pulse widths is presented. Continuous phase frequency shift keying (CPFSK) and minimum shift keying (MSK) formats are contrasted with discrete phase shift keying and then the duobinary. The potential multicarrier format orthogonal frequency division multiplexing (OFDM) is then contrasted to all those modulation formats and, finally, the specific roles of electrical equalization in coherent and incoherent detection systems is given.

18.1 IDENTIFICATION OF MODULATION FEATURES FOR COMBATING IMPAIRMENT EFFECTS

A sequence of information is used to modulate the lightwave for transmission over the optical fiber, the guided optical medium that exerts variable delay difference of lightwaves of the two sidebands of the passband signals, leading to linear chromatic dispersion. Furthermore, the random variation of the core geometry and that of the physical stresses along the length of the fiber links, generating the asymmetric changes of the propagation constants of the two polarized modes of the linearly polarized mode guided in the fiber, thus the polarization-mode dispersion (PMD) is created and hence lowering the corner frequency of the passband of the fiber. The total average intensity of the multiplexed optical channels would also create the nonlinear phase distortion effects on the modulated pulse sequences. Thus the nonlinear threshold power level would be the limit level of the lightwave average power.

Modulation formats would be the best way to reduce these linear and nonlinear impairments. In order to identify the roles of the digital signals and modulation techniques as well as the pulse formats for combating these undesirable effects, the spectral properties of the passband signals can be identified under the amplitude, phase, or frequency modulation.

One can identify the following features of modulation formats.

18.1.1 Binary Digital Optical Signals

The amplitude modulation at high speeds exhibits the rise and fall of the intensity or optical field at the edges of the bits, these rising and falling edges act like the sampling of a continuous wave and thus the peaking features of the spectrum of the passband signals. These peaks would contribute to the total average signal power and are hence limited by the nonlinear SPM effects. These peaks would be about 6 dB higher in power than those under phase modulation.

Phase shift keying is a discrete phase switching at the transition of the phase of the carrier at the boundary of the bit period. The amplitude of phase shift keying is constant. This constant envelope gives a smoother spectral property. However there are still spectral peaks due to the sampling of the switching as similar to the case with the ASK signals. This leads to higher energy concentration in the spectrum but with some peaking. Depending on the phase shift, the roll-off of the spectra is sharp or slow, and the suppression of the side lobes of the spectrum.

Continuous phase shift keying or continuous phase frequency shift keying offers a smooth spectrum similar to discrete phase shift keying—sharp roll-off and high suppression of the side lobes of the spectrum. The detection of the logical bits can be either phase detection with balanced receiver or frequency discrimination using two narrow band optical filters with an appropriate optical delay line.

Vestigial and single-sideband modulation formats offer a narrower band, as compared with the DSB method, of the signals for transmission to combat linear impairments. However, a partial response modulation technique such as duobinary offers truly single-sideband signals with alternating phase shift to effectively combat the linear dispersion impairments and nonlinear effects. Furthermore, it can be detected directly and uses standard optical receivers.

All modulation techniques can also be applied in optical coherent detection systems to achieve high improvement of SNR, especially when integrated with the estimation of the phase of the demodulated signals using digital signal processors.

18.1.2 M-ARY DIGITAL OPTICAL SIGNALS

QPSK and DQPSK are typical examples of the enhancement of transmission capacity by using distribution of signals over two $\pi/2$ phase shifts of the two DPSK constellations. M-ary signals can be formed by employing both amplitude levels and phase states such as ADPSK, M-ary STAR-QAM. Provided that the detection of optical signals does not suffer losses, the M-ary modulation signals exhibit an equivalent lower symbol rate for very speed signals (e.g., 100 Gbps). A major problem of the M-ary signals is the transition edges of the signals between the levels that create narrower eye opening in the time domain, the temporal eye.

18.1.3 MULTI-SUBCARRIER DIGITAL OPTICAL SIGNALS

OFDM offers the distribution of signal energy into subcarriers within the signal band. These sub-signals are combined and demodulated after propagation. Thus, a high-frequency signal can be lowered into several low-frequency signals to combat linear and nonlinear impairments. They are then optically modulated for transmission using digital modulation techniques, for example, DPSK, DQPSK, and so on discussed in the preceding text. This modulation technique also takes advantage of the high-speed signal processing systems and thus offers much flexibility in the tuning of the signals to the transmission media.

18.1.4 MODULATION FORMATS AND ELECTRONIC EQUALIZATION

In Chapter 11, it is identified that the optical carrier contributes to the linear and nonlinear dispersion of the channel on the amplitude or phase modulation. Under discrete or continuous phase modulation, the carrier is embedded in the signal band and contributes insignificantly to the linear and nonlinear distortion of the data pulse sequence. The electronic equalization including linear and nonlinear equalizers perform well when the passband signals exhibit single sideband, as their time-domain signals are linear with respect to the frequency term, and thus the linear equalizer such as FFE performs best. It is thus expected that VSB also behaves well under electronic equalization. Furthermore, the MSK under the detection of optical filtering offers significant improvements in the resilience to chromatic dispersion (CD), and this modulation format would also be most suitable under linear equalization.

Thus, for long-haul optically amplified transmission systems, it is expected that research works would be concentrating in the demonstration of digital phase modulation in long-haul filtering-cascade optically amplified transmission systems with the integration of electronic equalization. The baud rate can reach 50 GBauds and the integrate bit rates can be in the multi-Tbps.

The transmission distance will be limited by the nonlinearity and hence to a distance about 1500 km for 400 GBauds and 3500 km for 100 GBaud Tbps rates.

Under the intense interests in the delivery of 10 Gbps Ethernet and thus 100 Gbps Ethernet, at this speed the electronic processing of the optical receivers and electronic equalization and optical modulators would not work very well. To solve this bottleneck, multilevel modulation formats are expected to contribute due to their effectiveness in lowering the transmission symbol rate down to matured electronic processing speeds. However, multilevel signaling would suffer a reduction on the sensitivity due to the nonlinear limit in the power level that causes difficulties in the electronic equalization.

18.2   AMPLITUDE, PHASE, AND FREQUENCY MODULATION FORMATS IN DISPERSION-COMPENSATING SPAN TRANSMISSION SYSTEMS

18.2.1   ASK—DPSK AND DPSK—DQPSK UNDER SELF-HOMODYNE RECEPTION

18.2.1.1   Dispersion Sensitivity of Different Modulation Formats of ASK and DPSK

ASK and DPSK modulation formats with pulse shapes of NRZ, RZ, and CSRZ at 40 Gbps are measured in an experimental test bed, as shown in Figure 18.1, with the channels launched into an SSMF at power level below the nonlinear threshold. The output of the transmitter is then fed into an optical filter of bandwidth of 1.2 nm. Used "clock," which is derived directly from the PRBS generator, is fed into the error analyzer for non-DPSK signals. The optoelectronic receiver is a 45 GHz 3 dB bandwidth receiver. The PRBS bit pattern generator and error analyzer are SHF AG Product [1]. Optical amplifiers are included so as to boost the signal power to sufficient average power to meet the demand of the sensitivity of the receiver. An optical attenuator is used to vary the power into the receiver, and a 10 dB coupler is used to tap the signal in front of the receiver and feed into an optical power meter for measuring the sensitivity in dBm. The optical transmitter is set up so that the biasing of the data modulator can be tuned to different biasing points, so that the RZ and CSRZ formats can be generated. We have measured the penalty of the eye monitored by an Agilent sampling oscilloscope for different ASK and DPSK modulation formats.

18.2.2   NRZ-ASK AND NRZ-DPSK UNDER SELF-HOMODYNE RECEPTION

Figure 18.2 shows the receiver sensitivity of the ASK and DPSK modulation with the format NRZ. It is observed that the DPSK signals behave more linearly than that of ASK. The carrier residual power of the ASK modulation has contributed to the nonlinear behavior due to the shift of the carrier, as pointed out in Chapter 11, and it is expected that the DPSK formats would offer slightly better receiver sensitivity when the dispersion length is not sufficiently long. At 4 km, the distortion for 40 Gbps becomes serious, and it is observed that DPSK has a 1 dB better sensitivity at 1e–11 BER as compared with that of NRZ-ASK, as shown in Figure 18.2a and b (dark gray square).

Figure 18.3b and c shows the BER versus receiver sensitivity under the uses of one single filter and two filters. This is typical when the optical signals are transmitted with multiplexer and demultiplexer

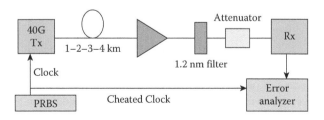

FIGURE 18.1   Experimental test bed for ASK and DPSK modulation formats. "Cheated Clock" is a clock derived from the PRBS pattern generator.

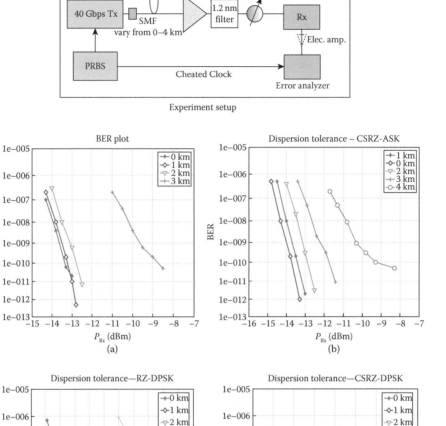

FIGURE 18.2 Experimental BER versus receiver sensitivity of (a) RZ-ASK, (b) CSRZ-ASK, (c) RZ-DPSK, (d) CSRZ-DPSK under residual dispersion of SSMF of 1, 2, 3, and 4 km. Due to 10 dB coupler, the Rx power level should be read with an addition –10 dB, that is, the label is –15, –14, ..., –10, and –9 dBm.

in an optical transmission system. The DPSK formats behave more resiliently to the dispersion with very compatible receiver sensitivity. The setup of the experiment is shown in Figure 18.3a.

The effects of cascading of optical filters can be observed from the plot of the measurements of the BER versus the receiver sensitivity of 40 Gbps CSRZ-DPSK modulation format. The multiplexing and demultiplexing by cascading two NEL 100 GHz AWGs of 0.5 nm 3 dB bandwidth and a tunable optical filter of 1.2 nm passband, no fiber is used and thus only two AWGs offering narrow passband to pass 40 G CSRZ-DPSK sufficiently as observed in Figure 18.3. The transmission of

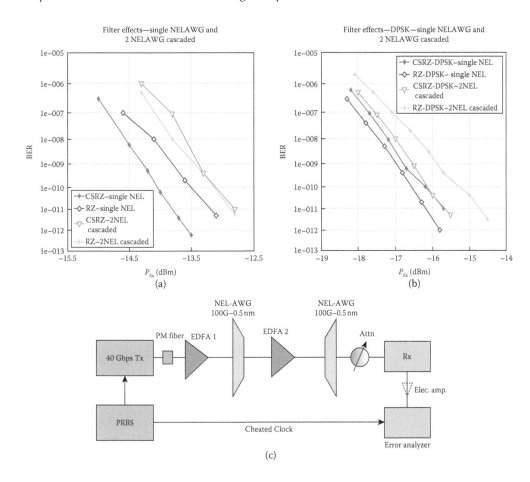

FIGURE 18.3 Experimental BER versus receiver sensitivity of (a) ASK and (b) DPSK with formats CSRZ and RZ with single filter, and duo-filters with (c) the experimental setup.

40 Gbps over 10 Gbps optical transmission link is also important for upgrading the transmission rate of one or a few wavelength channels to 40 Gbps. With 100 GHz spacing and 0.5 nm passband filters in the network, the BER has been measured versus the receiver sensitivity of 40 Gbps, and 10 Gbps of adjacent wavelength channels indicate that the CSRZ-DPSK behaves very well and suffers no penalty, as shown in Figure 18.56. We thus expect that optical MSK channels would also perform better under conventional 10 Gbps transmission systems.

18.2.3 RZ-ASK AND RZ-DPSK UNDER SELF-HOMODYNE RECEPTION

The transmission performance of the RZ formats, shown in Figure 18.4 for ASK and DPSK modulation, is also measured under the same experimental platform. It is observed that improvement in receiver sensitivity of about 2, 2, and 4 dB for 1, 2, and 3 km of SSMF transmission, respectively, can be achieved for a BER of 1e−11; and 1 dB at 4 km of SSMF at BER of 1e−6. The RZ format would exhibit a switching or sampling of the waveform, and thus spikes separated at 40 GHz is expected in the DPSK spectrum but at the carrier frequency. At short lengths below 3 km, the contribution of the nonlinear phase distortion is not high enough, and hence the improvement of the receiver sensitivity can be achieved, while at 4 km SSMGF of 40 Gbps, the nonlinear distortion takes over the linear distortion, and hence not much improvement in the eye diagram can be observed.

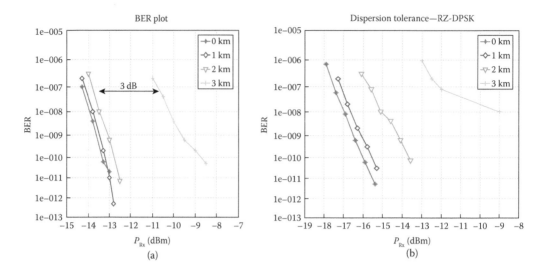

FIGURE 18.4   BER versus receiver sensitivity of (a) RZ-ASK and (b) RZ-DPSK under dispersion of SSMF of 1, 2, 3, and 4 km. Due to the 10 dB coupler, the Rx power level should be read with an additional −10 dB, that is, the label is −15, −14,...,−10 and −9 dBm.

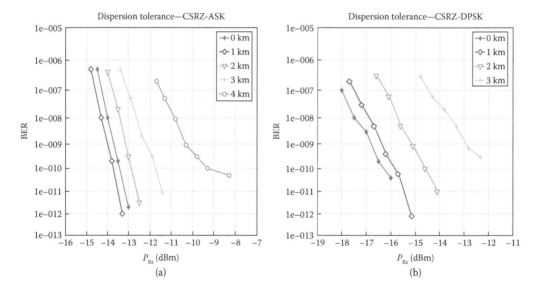

FIGURE 18.5   BER versus receiver sensitivity of (a) CSRZ-ASK and (b) CSRZ-DPSK under dispersion of SSMF of 1, 2, 3, and 4 km. Due to the 10 dB coupler, the Rx power level should be read with an additional −10 dB—that is, the label is −15, −14,...,−10, and −9 dBm.

18.2.4   CSRZ-ASK AND CSRZ-DPSK UNDER SELF-HOMODYNE RECEPTION

As shown in Figure 18.5, a ~3–4 dB improvement of CSRZ-DPSK over CSRZ-ASK formats is achieved. It is understood that a complete removal of the carrier would offer no linear distortion but only linear type, but much lower than that suffered by the CSRZ-ASK. Thus, we observe a better improvement with CSRZ-DPSK than with CSRAZ-DPSK.

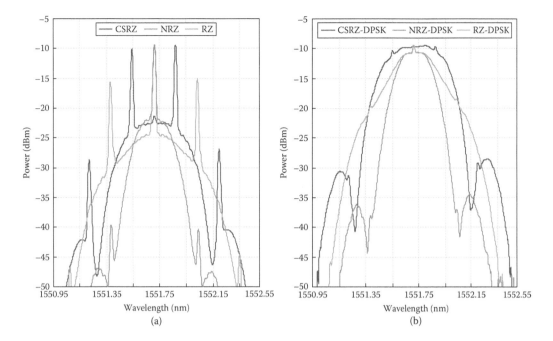

FIGURE 18.6   Optical spectra of (a) ASK and (b) DPSK modulation with different modulation formats.

18.2.5   ASK AND DPSK SPECTRA

Shown in Figure 18.6 are the spectra of different formats on ASK and DPSK modulation, and the spikes due to the on–off of the ASK indicate nonlinear distortion in the waveform under high dispersion effects than for DPSK modulation, whose time-domain waveform has a constant envelope. The energy concentration of the signals over the spectrum of DPSK modulation allows better performance than ASK. Furthermore, the linear equalization works for the modulations formats with discrete phase shift keying using noncoherent detection and naturally balanced receiver for an additional 3 dB improvement.

18.2.6   ASK AND DPSK UNDER SELF-HOMODYNE RECEPTION IN LONG-HAUL TRANSMISSION

Figure 18.7 shows the receiver sensitivity versus the BER for different ASK and DPSK formats over 320 km SSMF with dispersion compensation and eight channels of 100 GHz spacing with the matched filter at the center of 192.33 GHz. Further improvement of the receiver sensitivity can be expected for 40 Gbps DQPSK, in which two bits per symbol is used, and the effective bandwidth is about half of that of RZ-DPSK. Polarization multiplexing can also be used to reduce the symbol rate and hence achieve better improvement due to lower dispersion and quantum shot noises. Furthermore, the reduction of the symbol rate allows the operation of bit rate reaching 107 Gbps for 10 Gbps Ethernet. At this speed, the processing of electronic circuit speed has reached its limit, and thus the reduction of the symbol rate is essential.

18.3   NONLINEAR EFFECTS IN ASK AND DPSK UNDER SELF-HOMODYNE RECEPTION IN LONG-HAUL TRANSMISSION

Nonlinear effects are investigated for ASK and DPSK modulation formats. The measurement on the transmission systems is set up as shown in Figure 18.1, but with a 50 km SSMF spool and dispersion

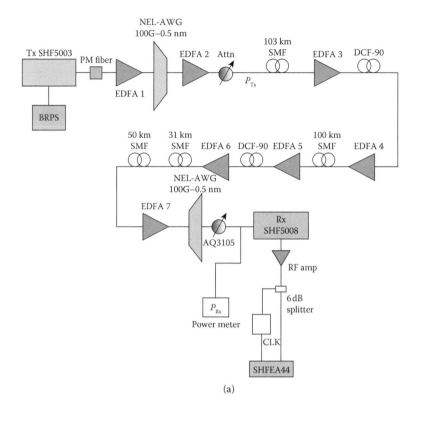

(a)

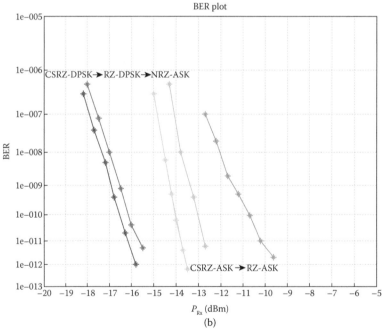

(b)

FIGURE 18.7 BER versus receiver sensitivity of 230 km transmission with optical amplifiers and dispersion-compensating modules, 40 Gbps bit rate, and DWDM eight channels of 100 GHz spacing. Only one optical filter NEL array waveguide demultiplexer is used: (a) experimental setup (b) BER of ASK and DPSK modulation schemes versus received power.

compensators and optical amplifiers. The photonic components are set up with the following conditions.

1. Laser source: tunable laser set at 1551.72 nm for centered wavelength (lambda 5) and 1548.51 nm for lambda 1 to 1560 nm for lambda 16.
2. Optical filters: muxes NEL-AWG = frequency spacing 100 G 3 dB BW of 0.5 nm; circulating property, so spectrum would appear cyclic (note the spectrum); make note of the channel input and output; for example, input at port 5 of lambda 5, then output at port 8. Input lambda 1 at port 1, then output at port 8 (lambda 1).
3. Optical transmitters can be set at CSRZ-DPSK or RZ-DPSK.
4. Clock recovery using 6 dB splitter with hardware not derived from the PRBS BPG (bit pattern generator).
5. Tuning of MZ phase decoder at the Rx is necessary when lambda channels are changed.
6. EDFA 1 driven at 178 mW pump power (saturation mode) and EDFA2 driven at 197 mW (saturation mode).
7. The attenuator is used to vary the optical power input at the receiver.

It is noted that DCF has lower nonlinear power threshold. Hence, it is expected that nonlinearities start their effects at lower launched power as compared to 50 km SMF-DCF.

The measurement of the receiver sensitivity is conducted with (1) tuning the wavelength of the laser source: lambda 1 = 1548.51 nm, lambda 5 = 1551.72 nm and lambda 16 = 1560.61 nm (according to ITU grid standard); this aims to show different channels multiplexed or demultiplexed through the multiplexer or demultiplexer may suffer different dispersion amount. Thus it is important to conduct some distortion compensation at the receiving end to achieve error-free performance; (2) the mod formats implemented are RZ-DPSK and CSRZ-DPSK; (3) power launched into the Sumitomo DCFM was measured to be 7.1 dBm; (4) 40 GHz clock recovery unit was utilized; (5) use two NELAWG mux/demux filters of 100 GHz spacing with a 3 dB bandwidth 0.45 nm, → similar to the currently deployed real 100 GHz spacing system; (6) DCF module for dispersion compensation is Sumitomo DCFM (−850 ps/nm @ 1550 nm); (7) power used in the BER plot is the reading power at the power meter. The received power can easily be calculated by: received power = reading power + 10 dB (1:10 coupler) − 0.7 (insertion loss of the coupler).

The following gives the transmission performance of the transmission system for ASK and DPSK formats.

18.3.1 Performance of DWDM RZ-DPSK and CSRZ-DPSK

At first, the BER under low launched average power is measured versus the receiver sensitivity for different wavelength channels. Three channels are selected, with wavelengths of the center C-band region and two others at the extremes, the shortest and longest ones. This is to ensure that the error-free detection can be achieved for all channels of the system. The BER versus the receiver sensitivity is shown in Figure 18.8.

18.3.2 Nonlinear Effects on CSRZ-DPSK and RZ-DPSK

The receiver sensitivity is plotted against the BER for different launched power for RZ-DPSK and CSRZ-DPSK formats, setup as shown in Figure 18.9a through c. It is observed that the CSRZ is more resilient to the nonlinear effects at a launched power of 11 dBm, while the RZ-DPSK suffers at this power level. This can be due to the suppression of the carrier that would in turn lead to lowering the total average power lower contained in the RZ-DPSK format.

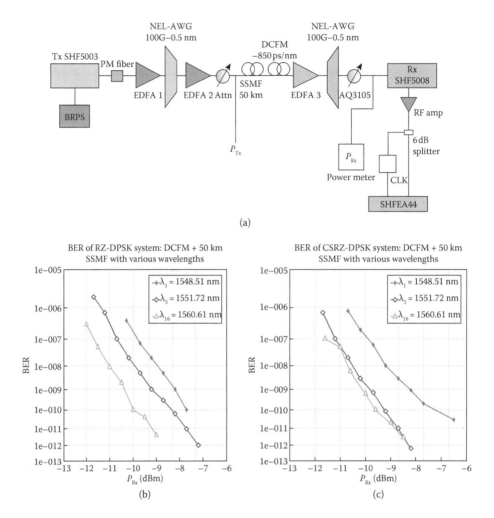

FIGURE 18.8 BER versus receiver sensitivity with different wavelength channels as a parameter of 50 km transmission with optical amplifiers and dispersion-compensating modules, 40 Gbps bit rate for (a) experimental setup, (b) RZ-DPSK, and (c) CSRZ-DPSK modulation formats. Note that the channels are center and lowest and longest region.

18.3.3 Nonlinear Effects on CSRZ-ASK and RZ-ASK

The ASK modulation with formats RZ and CSRZ does not suffer nonlinear phase distortion due to the SPM up to 11 dBm of launched power, but the slope of the BER curve indicates that it suffers from nonlinear phase distortion due to second harmonics, as explained earlier. The CSRZ-ASK offers at least 1 dB in power better than that of the RZ-ASK, while the BER slope is nearly the same though, as shown in Figure 18.10. The experimental setup is shown in Figure 18.10 for contrasting ASK and DPSK modulation formats. This is, as explained in Chapter 11, owing to the second harmonic distortion of the modulation. When the average power is increased to 13 dBm, the distortion is quite severe, and error free in detection could not be achieved.

18.3.4 Continuous Phase versus Discrete Phase Shift Keying under Self-Homodyne Reception

The continuous phase modulation offers much narrower bandwidth in the passband spectrum as compared to the discrete phase shift keying modulation formats, due to the continuity of the

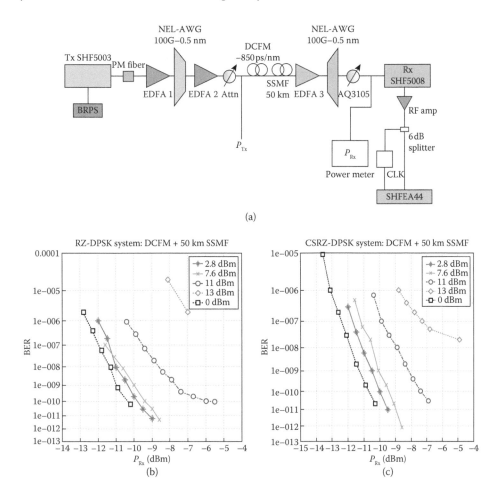

FIGURE 18.9 BER versus receiver sensitivity with launched power as a parameter of 50 km transmission with optical amplifiers and dispersion-compensating modules, 40 Gbps bit rate for (b) RZ-DPSK and (c) CSRZ-DPSK modulation formats. Experiment setup shown in (a) similar to Figure 18.8a, but controlling the launched power.

carrier phase at the boundary of the bit period. That means there is chirp modulating of the carrier from one-bit period to the other, depending on the differential bits of the input data sequence. Furthermore, the carrier is embedded in the signal band observed as the peak at the center of the spectrum of continuous phase shift keying formats.

When the separation between the frequencies of the continuous phase modulation equals a quarter of the bit rate, then orthogonality is achieved. This orthogonality allows improvement in the detection of the MSK signals either by balanced phase detection or by frequency discrimination. If there is a photonic processing system that can distinguish between the orthogonal channels, then the MSK signals would offer much ease of implementation in the photonic domain.

18.3.5 MULTI-SUBCARRIER VERSUS SINGLE/DUAL CARRIER MODULATION UNDER SELF-HOMODYNE RECEPTION

The lowering of the bit rate can be significantly reduced with multicarriers, especially when they are orthogonal and generated by the electronic processor of IFFT and FFT in which the nature of orthogonality can be provided via the discrete Fourier transform, the OFDM. Combining with

DQPSK, the scheme offers better resilience to nonlinearity and lowering of dispersion effects. However, the high ratio of the peak power and average power makes this scheme unattractive. Furthermore, the FWM effect may be very serious among subcarriers, as they are very close to each other, and phase matching can easily occur.

18.3.6 MULTILEVEL VERSUS BINARY OR I–Q MODULATION UNDER SELF-HOMODYNE RECEPTION

16 STAR-ASK and MADPSK multilevel formats offer the lowering of symbol rates, and polarization multiplexing will allow the symbol rate to reduce the bit rate of 100 Gbps to around 12.5 Gbps, making the use of digital electronic signal processing attractive for both coherent and incoherent

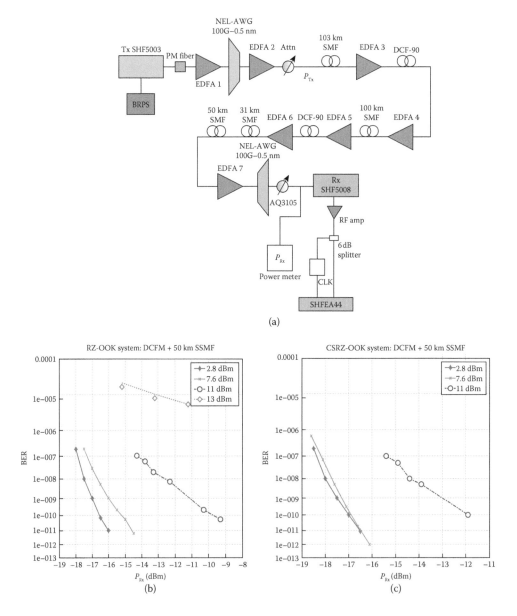

FIGURE 18.10 BER versus receiver sensitivity with launched power as a parameter of 50 km and 230 km transmission with optical amplifiers and dispersion-compensating modules, 40 Gbps bit rate for (b) RZ-ASK, (c) CSRZ-ASK. The transmission experiment setup is shown in (a). *(Continued)*

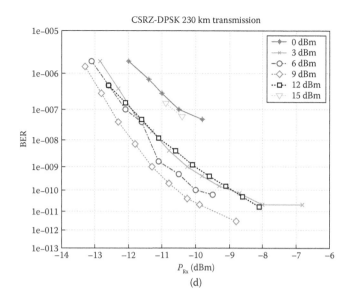

FIGURE 18.10 (Continued)   (d) CSRZ-DPSK modulation formats.

transmission and detection. The pattern dependency is one of the major problems of the multilevel amplitude and phase modulation.

In the following, the multilevel formats of 16 STAR-QAM and 16-ADPSK are characterized and compared with binary DPSK formats:

- *Implementation of transmitter and receiver*: 16 ADPSK can be implemented by using one dual-drive MZIM which is more complex as compared to a single-drive modulator type. The dual-drive type requires more complex and high speed microwave drivers. Then, a second MZIM is used to provide the amplitude modulation. A synchronization of the bit pattern in these two modulators is essential.
- *Receiver sensitivity*: The theoretical estimation that a 1 dB loss may be suffered from the same bit rate (symbol rate) of ASK modulation format is due to the pattern dependency between the amplitude levels.
- *Robustness toward nonlinear fiber effects*: Due to the multilevel of the amplitudes, it may suffer nonlinearity of the fiber than conventional NRZ-ASK of the same symbol rate.
- *Dispersion tolerance*: Multilevel is expected to offer better dispersion tolerance than its DPSK binary counterparts.
- *Spectral efficiency*: The spectral efficiency of multilevel would be higher than the binary DPSK of the same symbol rate due to its narrower passband property.

Coherent detection of 16 STAR-QAM may offer much better improvement, in the order of 10–20 dB, if electronic signal processing can be used for the phase estimation of the detected signals.

18.3.7   SINGLE-SIDEBAND AND PARTIAL RESPONSE MODULATION UNDER SELF-HOMODYNE RECEPTION

In Chapter 11, we have addressed the impact of the carrier on the fundamental and higher-order harmonic distortion in the nature of the modulation schemes, ASK, DPSK, and the SSB, as well as the duobinary formats. SSB and duobinary suffer the least nonlinear second harmonic distortion.

Furthermore, the duobinary modulation offers a true single sideband with reduction of the carrier plus its ability for direct detection without resorting to the balanced receiver structure. Thus, these types of modulation formats would be considered to be the best binary modulation formats as they permit the reduction of linear CD effects and the electronic dispersion compensation at the receiver and shared pre distortion and post compensation, as described in Chapter 11.

18.4   100 G AND TBPS HOMODYNE RECEPTION TRANSMISSION SYSTEMS

18.4.1   GENERATION OF MULTI-SUBCARRIERS

Table 18.1 tabulates the parameters for the generation of multi-subcarriers for modulation to generate Tbps channels employing the recirculating frequency shifting (RCFS) technique and nonlinear driving of I–Q modulators to generate subharmonic subcarriers. These subcarriers are then fed into an I–Q optical modulator, described in Chapter 3, to generate multichannels whose bit rate is determined by the electrical broadband signals applied to the I–Q modulator. Thus, using PDM-QPSK with a baud rate of 25 G, the aggregate bit rate reaches 100 G or, with FEC, 28 GB, leading to 112 Gbps effective rate.

18.4.2   NYQUIST SIGNAL GENERATION USING DAC BY EQUALIZATION IN FREQUENCY DOMAIN

Single-channel 1 Tbps transmission is impossible to realize using only one carrier or subcarrier and modulation. If we keep the non-FEC baud rate close to 25 GBaud, we need PDM-4096QAM modulation scheme to achieve 1 Tbps per single lightwave carrier. This is impossible under the constraint of maximum laser amplitude power available to date. Furthermore, as the OSNR requirement increases exponentially to the modulation format level, it is impossible to reach the transmission distance with limited laser power and the threshold of the nonlinear level of the transmission fiber. Even PDM-16QAM needs, theoretically, 7 dB more OSNR than PDM-QPSK to reach the same distance. To solve this problem, one can either increase the baud rate or employ more subcarriers, that is, a superchannel with multiple subcarriers. Increasing the baud rate leads to higher demand on the bandwidth for O/E components, thus challenging the technology advances and cost. Hence, employing high-spectral-efficient superchannel with subchannel bandwidth of 50 GHz or below seems to be the most favorable choice.

Considering the optical transport network (OTN) framer and the overhead rate with the use of FEC of either 7% or 20%, the total bit rate can increase to 28 G from 25 GBaud per channel. So with PDM-QPSK (polarization division multiplexing – quadrature phase shift keying) and with 10 channels this leads to 1.12 Tbps. If PDM-16QAM is used then the aggregate bit rate for 10 channels is 2.24 Tbps. Figure 18.11 shows the architecture for Tbps or superchannel, which is created by modulating a set of subcarriers generated by a comb generator. For the sake of simplicity, all subcarriers can be modulated by one I/Q optical modulator with polarization multiplexing components. A probe channel can be generated separately and injected into the spectral window of a channel, which is filtered out. At the receiver end, the transmitter channels are demultiplexed into individual ones whose polarized components are separated. Their I and Q components are also separated via a π/2 hybrid coupler and then detected by the PDP (photodetector pair), electronic amplification and then sampled by an analog-to-digital converter (ADC) to convert to the digital domain signals for further processing in the DSP sub-systems.

For multiple subcarrier transmission, we would need multicarrier modulation techniques and corresponding multicarrier demodulation and coherent detection. Currently, there are a few methods and apparatus to generate multicarriers (see more details later in Section 18.4.5.1). The commonly used one is by using a single ECL laser source and a recirculating shift loop and an MZM by an RF signal. The frequency of the RF signal is the spacing of the subchannels. Tx-DSP and DAC (see Figure 18.12) are needed for shaping the signal pulse, for example, Nyquist pulse shaping, and pulse shaping for compensating nonideal O/E components transfer function and CD precompensation, and so on.

TABLE 18.1
Tbps Superchannel Transmission System

Parameter	Superchannel RCFS Comb Generator	Superchannel Nonlinear Comb Generator	Some Specs	Remarks
Technique				
Bit rate	1,2, ..., N Tbps (whole C-band)	1 Tbps, 2,..., N Tbps	~1.28 Tbps @ 28–32 GB	20% OH for OTN, FEC
Number of external cavity lasers (ECLs)	1	$N \times 2$		
Nyquist roll-off α	0.1 or less	0.1 or less		DAC pre-equalization required
Baud rate (GBauds)	28–32	28–32	28, 30, or 31.5 GBaud	Pending on FEC coding allowance
Transmission distance	2500	2500	1200 (16 span) ~2000 km (25 spans) 2500 km (30 spans) 500 km	20% FEC required for long-haul application Metro application
Modulation format	QPSK/16QAM	QPSK/16QAM	Multicarrier Nyquist-WDM PDM-DQPSK/QAM Multicarrier Nyquist-WDM PDM-16QAM	For long haul For long haul For metro
Channel spacing			4×50 GHz 2×50 GHz	For long haul
Launch power	<<0 dBm if 20 Tbps is used		~ −3–1 dBm lower if $N > 2$	Depending on QPSK/16QAM and long haul/metro can be different
B2B ROSNR @ 2e−2 (BOL) (dB)	14.5	14.5	15 dB for DQPSK 22 dB for 16QAM	1 dB hardware penalty 1 dB narrow filtering penalty
Fiber type	SSMF G.652 (or G.655)	SSMF G.652 (or 655)	G.652 SSMF	
Span loss	22	22	22 dB (80 km)	
Amplifier	EDFA ($G > 22$ dB); $NF < 5$ dB		EDFA(OAU or OBU)	
BER	2e−3	2e−3	Pre-FEC 2e−2 (20%) or 1e−3 classic FEC (7%)	
CD penalty (dB)			0 dB @ +/−3000 ps/nm <0.3 dB @ +/−30,000 ps/nm;	16.8 ps/nm/km and 0.092 ps/(nm².km)
PMD penalty (DGD)			0.5 dB @ 75 ps, 2.5 symbol periods	
SOP rotation speed	10 MHz	10 MHz	10 MHz	
Filters cascaded penalty			<1 dB @ 12 pcs wavelength selective switch (WSS)	
Driver linearity	Required	Required	THD <3%	16QAM even more strict

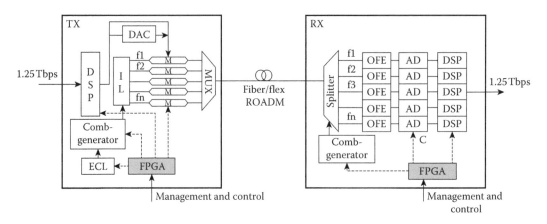

FIGURE 18.11   Architecture of a superchannel 1 Tbps transmission platform.

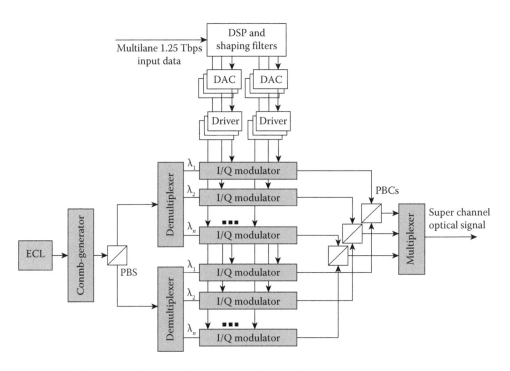

FIGURE 18.12   General architecture of a superchannel transmitter.

Interleaver can be used to separate the subcarriers for individual subchannel modulation. At the output, all the signals of the subchannels are multiplexed together to form a superchannel 1.25 Tbps optical signal.

The principal question is how we can create superchannels under laboratory conditions to achieve the proof of concept of the Tbps transmission, that is, how to generate superchannels, transmission, and detection/demodulation at the receiving end. Thus, one of the main principles of this chapter is to describe techniques employed to generate superchannels and to inject the probe channel to evaluate the performance of the transmission systems for Tbps superchannel.

In Nyquist-WDM, the goal is to place different optical subchannels as close as possible to their baud rate, the Nyquist rate, but minimizing channel crosstalk through the formation of approximately close to rectangular spectrum shape of each channel by either optical filtering or electrical filtering and equalization.

The rectangular spectrum has a $\sin c$, that is, $\left(\dfrac{\sin x}{x}\right)$ time-domain impulse response. At the sampling instant $t = kT$ ($k = 1,2,...,N$ is nonzero integer), its amplitudes reach zero. This implies that, at the ideal sampling instants, the ISI from neighboring symbols is negligible, or free of intersymbol interference (ISI). Such Nyquist pulse and its spectrum for either a single channel or multiple channels is described in Chapter 3. Note that the maximum of the next pulse raise is the minimum of the previous impulse of the consecutive Nyquist channel.

Considering one subchannel carrying 25 GBaud PDM-DQPSK signal, the resulting capacity is 100 Gbps for a subchannel, and hence, to reach 1 Tbps, ten subchannels would be required. To increase the spectral efficiency, the bandwidth of these ten subchannels must be packed densely together. The most likely technique for packing the channel as close as possible in the frequency with minimum ISI is the Nyquist pulse shaping, which is described later in this section. Thus, the name Nyquist-WDM system is coined. However, in practice, such "brick wall"-like spectrum is impossible to obtain, and hence a nonideal solution for non-ISI pulse shape should be found, the raised-cosine pulse with some roll-off property condition though to be met.

The raised-cosine filter is an implementation of a low-pass Nyquist filter, that is, one that has the property of vestigial symmetry. This means that its spectrum exhibits odd symmetry about $\dfrac{1}{2T_s}$, where T_s is the symbol period. Its frequency-domain representation is "brick-wall-like" function. The frequency response is characterized by two values: β, the *roll-off factor*, and T_s, the reciprocal of the symbol rate in Sym/s; that is, $\dfrac{1}{2T_s}$ is the half bandwidth of the filter. The impulse response of such a filter can be obtained by analytically taking the inverse Fourier transformation of the frequency spectrum, in terms of the normalized $\sin c$ function, as

$$h(t) = \sin c\left(\frac{t}{T_s}\right)\frac{\cos\left(\dfrac{\pi\beta t}{T_s}\right)}{1-\left(2\dfrac{\pi\beta t}{T_s}\right)^2} \tag{18.1}$$

where the roll-off factor, β, is a measure of the *excess bandwidth* of the filter; that is, the bandwidth occupied beyond the Nyquist bandwidth as from the amplitude at 1/2T. Figure 18.13 depicts the frequency spectra of raised-cosine pulse with various roll-off factors. Their corresponding time-domain pulse shapes are given in Figure 18.14.

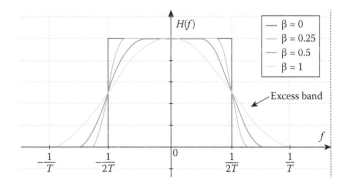

FIGURE 18.13   Frequency response of raised-cosine filter with various values of the roll-off factor β.

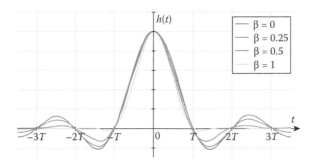

FIGURE 18.14 Impulse response of raised-cosine filter with the roll-off factor β as a parameter.

When used to filter a symbol stream, a Nyquist filter has the property of eliminating ISI, as its impulse response is zero at all nT (where n is an integer), except when $n = 0$. Therefore, if the transmitted waveform is correctly sampled at the receiver, the original symbol values can be recovered completely. However, in many practical communications systems, a matched filter is used at the receiver, so as to minimize the effects of noises. For zero ISI, the net response of the product of the transmitting and receiving filters must equate to $H(f)$, and thus we can write

$$H_R(f)H_T(f) = H(f)$$ (18.2)

Or, alternatively, we can rewrite that

$$\left|H_R(f)\right| = \left|H_T(f)\right| = \sqrt{\left|H(f)\right|}$$ (18.3)

The filters that can satisfy the conditions of Equation 18.3 are the root-raised-cosine filters. The main problem with root-raised-cosine filters is that they occupy a larger frequency band than that of the Nyquist *sin c* pulse sequence. Thus, for the transmission system, we can split the overall raised-cosine filter with a root-raised-cosine filter at both the transmitting and receiving ends, provided the system is linear. This linearity is to be specified accordingly. An optical fiber transmission system can be considered to be linear if the total power of all channels is under the nonlinear SPM threshold limit. When it is over this threshold, a weakly linear approximation can be used.

The design of a Nyquist filter influences the performance of the overall transmission system. Oversampling factor, selection of roll-off factor for different modulation formats, and FIR Nyquist filter design are the key parameters to be determined, as shown in Figure 18.15. If taking into account the transfer functions of the overall transmission channel, including fiber, WSS, and the cascade of the transfer functions of all O/E components, the total channel transfer function is more Gaussian like. To compensate for this effect in the Tx-DSP, one would thus need a special Nyquist filter to achieve the overall frequency response equivalent to that of the rectangular or raised cosine with a roll-off factor as described in Chapter 3.

18.4.3 FUNCTION MODULES OF A NYQUIST-WDM SYSTEM

A generic schematic of the functional modules of the Nyquist-WDM system is described in Figure 18.2. Two special features should be taken into account. At the Tx side, pulse shaping is so defined that the overall transfer function from driver to the ADC on the Rx side should ideally be rectangular or of NRZ form. The implementation can be either by using a look-up-table or a Nyquist low-pass filter in the frequency domain or time domain. Alternatively, the optical spectrum of the modulated lightwaves at the output of the optical modulator can be obtained via the port of

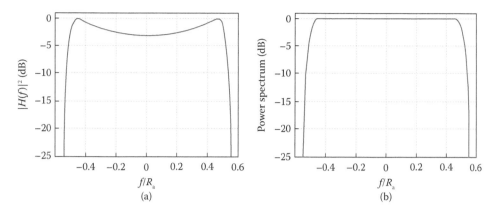

FIGURE 18.15 (a) Desired Nyquist filter for spectral equalization and (b) output spectrum of the Nyquist-filtered QPSK signal.

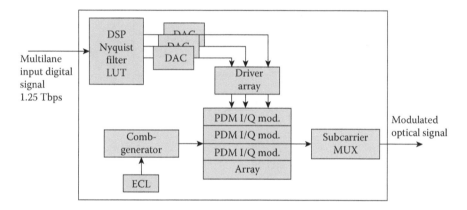

FIGURE 18.16 Transmitter architecture of a Nyquist-WDM system with electrical Nyquist filter.

the optical spectrum analyzer (OSA). This is then equalized to achieve a "brick wall"-like spectrum. Thus, the required spectrum on the DAC to achieve this equalization is known and then used to modify the DAC, driving output voltage levels at (H_I^+, H_I^-) *and* (H_Q^+, H_Q^-) and (V_I^+, V_I^-) *and* (V_Q^+, V_Q^-), where the equalized optical spectra can be obtained for the two polarized modes of the linearly polarized mode to be launched into the SSMF optically amplified transmission lines.

For each subchannel, 4 × DACs are needed to convert the discrete digital signal to analog signal (X_I, X_Q, Y_I, Y_Q) (see Figure 18.16). In order to match the voltage amplitude (power) requirement of the I–Q modulator, 4 × RF broadband drivers are required to provide appropriate RF signal amplitude for driving the MZM so as to obtain the phase difference between the in-phase and quadrature-phase constellation points. At the output of the transmitter, all subchannels are multiplexed together to form a superchannel.

There is another way to generate the Nyquist signal, which is suitable when Tx-DSP is not available. The Nyquist spectrum is generated by an optical rectangular (or raised-cosine roll-off) filter, as shown in Figure 18.17. However, in practice, it is very difficult to achieve a sharp filter roll-off without much phase fluctuation at the edges of the filter.

At the receiver, multi carrier parallel demodulation is used, as shown in Figure 18.18. At first, the superchannel will be separated into individual subchannels. Each subchannel will be demodulated like a single-channel reception: OFE down-converts the optical spectrum to baseband (homodyne coherent demodulation), and the four-lane analog signal will be converted into a digital

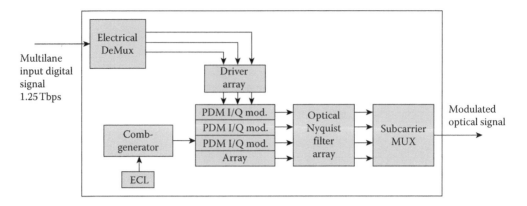

FIGURE 18.17   Schematic of the Nyquist optical transmitter architecture with optical Nyquist filter.

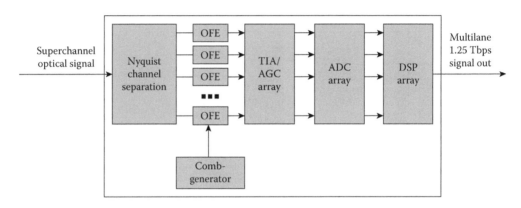

FIGURE 18.18   Receiver architecture of the Nyquist-WDM system.

signal using 4 × ADC. The digital signal will be processed by ASIC modules CD compensation, timing recovery, MIMO FIR filter, carrier recovery, and so on.

18.4.4   DSP Architecture

Single subchannel DSP can be either using the traditional FDEQ + TDEQ architecture or pure FDEQ architecture, as shown in Figure 18.19.

18.4.5   Key Hardware Subsystems

18.4.5.1   Recirculating Frequency Shifting

The main modules of a multicarrier generator (comb generator) are illustrated as follows (Figure 18.20):

1. A continuous-wave (CW) ECL or multilaser bank of ECLs whose linewidth is sufficiently narrow, possibly in the order of <100 KHz, can be employed as the original lightwave carrier/carriers.
2. An MZM for pulse shaping (CSRZ), which will make the output spectrum of the super-channel flat.

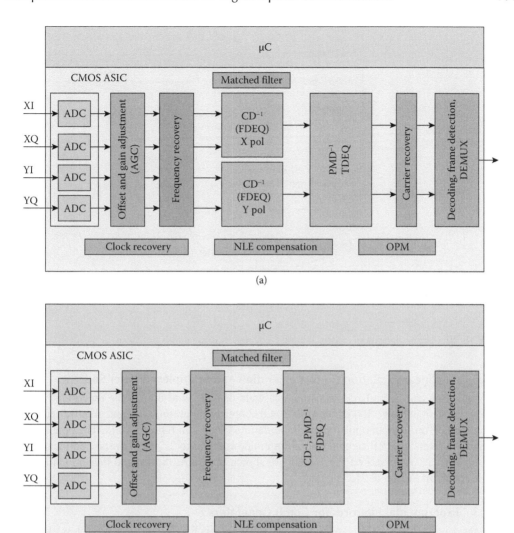

FIGURE 18.19 (a) DSP architecture, FDEQ + TDEQ, and (b) pure FDEQ.

3. An MZM used as a phase modulator, the phase control signal coming from an RF signal generator.
4. RF generator, its frequency is the spacing of the subchannels.
5. Sinusoidal signal generator.
6. 90° phase shifter, which makes the spectrum the only single-sideband.

Under the condition that the modulator, an I–Q modulator, can be driven such that the amplitude swings to $2V_\pi$, the generation of first- and second-order frequency shifting components can be formed. We operate the modulator in the region such that no suppression of the primary carrier can be achieved, and thus five subcarriers can be generated from one main carrier. Thus, using two main carriers, we can generate ten subcarriers, and hence the modulation of these subcarriers can form 10×100 Gbps or 1 Tbps superchannels.

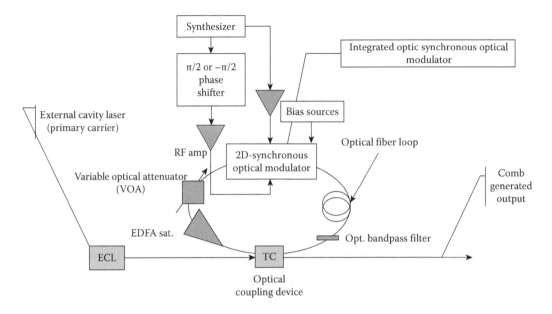

FIGURE 18.20 Block diagram of a recirculating frequency shifting comb generator.

18.4.5.2 Nonlinear Excitation Comb Generation and Multiplexed Laser Sources

Comb generation can be realized by using a cascade of CSRZs to double the repetition rate of a low-RF-driving signal (e.g., 10 GHz to give 20 GHz optical repetition rate optical pulse sequence), and then feeding through an optical phase modulator as shown in Figure 18.21a. The optical phase modulator is driven with RF signals whose amplitude is equal to $2V_\pi$, so that the nonlinear region can be exploited. The generated subcarriers are shown in Figure 18.21b, and their modulation by 28 GB is shown in Figure 18.21c.

18.4.5.3 Experimental Platform for Comb Generators

The general path flow of the experiment platform for an RCFS comb generator is shown in Figure 18.22a. There are two separate paths that need to be noted (in order of importance).

1. The main structure is the optical fiber or an integrated optic loop in which besides the optical path interconnecting all optical components, the shifting of the frequency of the lightwaves is implemented by the optical modulation via the optical modulator. The output is obtained via the coupling of the optical coupler.
2. The optical coupler is used for both injecting the lightwaves from a source operating in CW mode, and tapping the frequency-shifted lightwaves out to the output port.
3. The optical modulator incorporated in the loop performs the frequency shifting. Single-sideband operating is used by RF phase delay by $\pi/2$ with respect to each other, hence the Hibert transformation and the suppression of the right or left sideband accordingly, depending on the relative phase shift between the in-phase and quadrature-phase electrodes of the I–Q modulator.
4. The insert of Figure 18.22 is the integrated optic plan view of the I–Q modulator waveguide and electrodes. There are two child interferometers with overlaid electrodes that perform the suppression of the carrier and generate the sidebands under the biasing and driving condition of the applied RF waves (see Figure 18.22). Note that the electrical phase shifters are inserted in both arms of the RF paths before applying to the electrodes to create the $\pi/2$ phase shift, and thence suppression of one sideband.

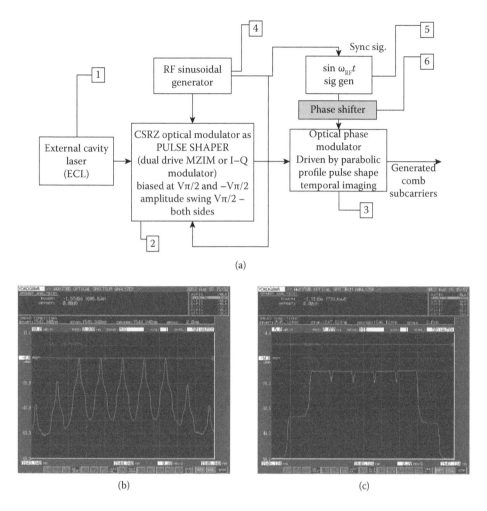

FIGURE 18.21 Comb generation using nonlinear driving conditions on MZIM I/Q modulator (a) comb line, (b) spectrum of nonlinear comb generator 1 × 5 subcarriers at 30 GHz spacing, and (c) modulated subchannels.

5. In addition, there is a parent interferometric optic structure in which both child interferometers are positioned. A pair of electrodes is also employed on one branch of this interferometer, so that phase shifting between the emerging lightwaves from the child interferometers can be altered with respect to each other.

6. A variable optical attenuator is also used to ensure that the total loop gain is less than unity, so that lasing effects would not occur. This is also used to adjust the gain equalization due to different gain factors.

7. A sinusoidal RF wave generator is also required. However, in practice, such an oscillator is simple, provided that the frequency is known.

8. It is noted that the output optical port of the comb generator must be connected with an angled PM fiber connector, so that there is no feedback of optical comb waves back to the recirculating loop. If not, backscattering would then create interference effects, and noise disturbance on modulated signals would be observed, hence much higher BER.

9. Figure 18.22b and c show the spectrum of generate subcarriers and their modulated spectra and (c) spectrum of the extracted modulated sub-carrier spectrum from the superchannel.

10. Figure 18.23 shows the pictorial setup of hardware system for the generation of the super channels.

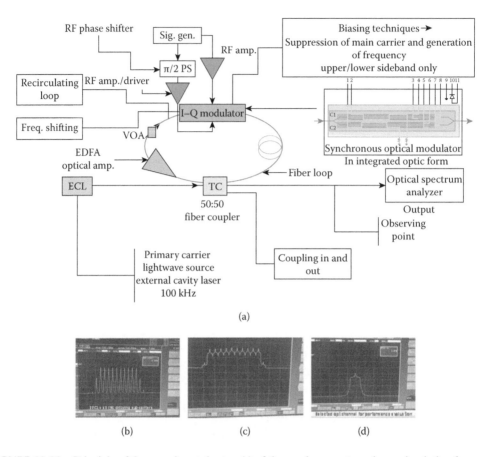

(a)

FIGURE 18.22 Principle of the experimental setup (a) of the comb generator using recirculating frequency shifting; insert is the structure of an integrated I–Q modulator; (b) subcarrier's spectrum; (c) modulated spectrum of subcarriers; and (d) extracted single-channel modulated spectrum.

FIGURE 18.23 Image of setup of a comb generator structure including optical modulator and RF phase shifters, two branches to RF port inputs.

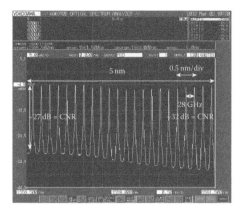

FIGURE 18.24 Spectrum of generated multicarriers.

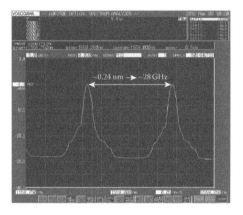

FIGURE 18.25 Measured subchannel spacing between two adjacent channels.

Referring to Figures 18.24 and 18.25, we can observe that:

1. The spectra of the generated subcarriers are equalized over a wide region of 5 nm, and the carrier noise ratio (CNR) reaches 27~32 dB.
2. The spacing between two adjacent subcarriers can be varied by tuning the excitation RF frequency, shown as 28 GHz spacing.

18.4.6 NON-DCF 1 TBPS AND 2 TBPS SUPERCHANNEL TRANSMISSION PERFORMANCE

18.4.6.1 Transmission Platform

The schematic structure of the transmission system is shown in Figure 18.26, consisting of an optically amplified multi-SSMF-span fiber transmission incorporating variable optical attenuators, and optical transmitters and coherent optical receiver with off-line DSP. Optical bandpass filter (OBPF) is employed using either the D40 demux or Yenista sharp roll-off (500 dB/nm) to extract the subchannel of the superchannel whose performance is to be measured, normally in terms of BER versus OSNR. The multispan optically amplified fiber transmission line consists of two main parts: the fiber spools and banks of optical amplifiers and VOA as shown in Figure 18.26 and Figure 18.27a, respectively. The launched power to the fiber of the first span can be adjusted by using the variable optical attenuator (VOA) and booster amplifiers located at the input of into each span so that the EDFA of the spans operate at saturation level. Thus, in order to achieve the

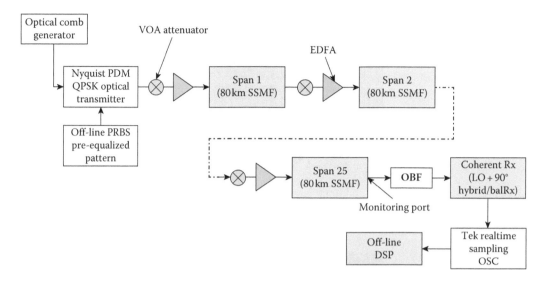

FIGURE 18.26 Schematic structure of the Nyquist-QPSK PDM optically amplified multispan transmission line.

same launched power into each span, the remote control is employed to adjust these input launched power span by span, especially when the launched power is varied to obtain different OSNRs. Back-to-back (B2B) is measured by inserting noises at different levels to the signal power monitoring port. A D40 demultiplexer is also available and incorporated in the 19″ rack, so that a subchannel of the superchannel can be extracted for measuring the transmission or B2B performance.

The transmitter employs Nyquist-QPSK as described in the previous section, with variable channel spacing and baud rate and optimization of the DAC. The optical receiver is a coherent type with an external local oscillator (LO) whose wavelength is tuned to the right wavelength location of the subchannel whose performance is to be measured. In practice, this LO must be remotely/automatically locked to the wavelength of the subchannel [2].

18.4.6.2 Performance

18.4.6.2.1 Tbps Transmission using Three Subchannel Transmission Test

We considered a test of transmission of three channels only for transmission over 2000 km optically amplified multispan to ensure that the signal quality can be achieved using Nyquist-QPSK with interference due to linear crosstalk due to overlapping of subchannels, as they are packed so close to each other using the Nyquist criteria. Therefore, the three sub-channels can be generated by splitting the ECL source into three branches of equal power. One optical path is fed though a CSRZ optical modulator driven with a sinusoidal signal source of frequency equal to the subchannel spacing, and the other branch would then be modulated with a Nyquist-QPSK format combined with the two CSRZ carriers, which are also modulated by Nyquist-QPSK but with a different random pattern. Thus, we do have three subchannels that are decorrelated, and thus the BER versus OSNR can be obtained to ensure that the effects of overlapping can be justified. Thus, the schematic of this arrangement is shown in Figure 18.28. The transmission performance measured with BER against launched power and subchannel spacing of 28 and 30 GHz with 28 GBauds Nyquist-QPSK transmitted over a 1600 km line is shown in Figure 18.30.

The BER is optimum at the launched power of −1 to 0 dBm for both probe channel and the side channels. The probe channel is at the canter of the three channels. This proves that the Nyquist pulse shaping can offer a performance close to that of a single-channel transmission over 1600 km optically amplified multispan and nondispersion-compensating fiber line. These performance results

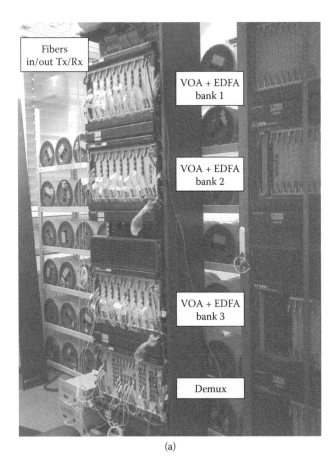

(a)

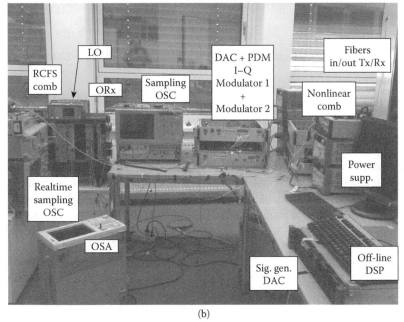

(b)

FIGURE 18.27 (a) Fiber lines running from optically amplified multifiber span transmission line to (b) transmitter and receiver platform.

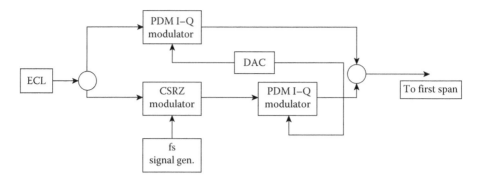

FIGURE 18.28 Schematic of generation of three subcarrier channel transmitters for testing transmission in the initial phase for a Tbps transmission system.

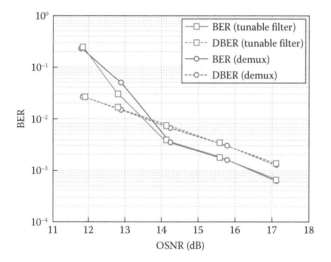

FIGURE 18.29 Performance of channel 9 of the superchannel transmission over 2000 km with D40 demux or optical bandpass Yenistar filter.

give much confidence in the pulse shaping, and for proceeding to superchannel of 1 Tbps and 2 Tbps over a single dual-polarized mono-mode fiber.

Furthermore, we do also use optical filter at the receiver, that is, before feeding into the hybrid coupler, the optical filter is employed to extract the probe channel. The transmission performance for a five-channel superchannel is shown in Figure 18.29; three channels are shown in Figure 18.30 and Figure 18.31; and over 1600 km multispan non-DCF SSMF 80 km optical amplified spans link is shown in Figure 18.32. In general, the sharp roll-off filter would offer 1 dB improvement over the wavelength demultiplexer, which commonly has a parabolic roll-off filtering characteristic.

18.4.6.2.2   1 Tbps, 2 Tbps, or N Tbps Transmission

We can now extend the transmission of Nyquist-pulse-shaped subchannels with the total capacity of all channels of 1 Tbps and thence 2 Tbps. For 1 Tbps, we use nine subchannels plus-one probe channel, which are all modulated and pulse shaped, satisfying Nyquist criteria with raised-cosine filter and pre-equalization. Thus, we have ten channels generated either using RCFS or nonlinear comb generation. Then one channel is suppressed by using band stop filter of a WSS, and then combined with another Nyquist-shaped QPSK channel, independently driven by a decorrelated sequence generated from DAC ports. This inserted channel acts as the probe channel, and hence

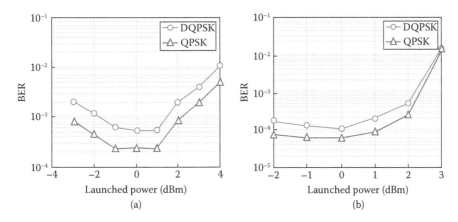

FIGURE 18.30 BER versus launched power (dBm) using three channels with central channel as probe channel over a transmission distance of 1600 km non-DCF, 20 optically amplified 80 km SSMF spans with VOA inserted in front of EDFA; 28 GBaud PDM-Nyquist-QPSK (a) single channel and (b) three channels.

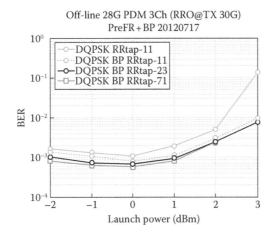

FIGURE 18.31 1600 km transmission Nyquist-QPSK processed with FIR 11 to 23 taps: BER versus OSNR. Number of subchannels is three, generated and modulated as shown in Figure 18.30.

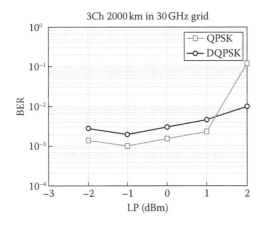

FIGURE 18.32 Three-channel test with central channel as probe over 2000 km transmission by simulation; 31.8 GHz channel spacing, 28 GBaud QPSK with processing QPSK and DPQSK (to optimize for cycle slippage).

one can choose to measure the performance of transmission for any subchannel of the superchannel as required, instead of all independently driven subchannels due to the limitation of independent sequence from one DAC subsystem. Figure 18.33 shows the generic scheme for generating such a probe channel within the superchannel. We can select whichever channel spectral location to insert the probe channel by tuning the source wavelength and the WSS, depending on whether the RCFS or nonlinear comb generator is employed. Figure 18.34 shows the B2B performance of ten subchannels (2 × 5 by nonlinear driving modulator, described earlier) under Nyquist pulse shaping and PDM-QPSK modulation format. Individual in-phase and quadrature-phase components are also processed to see the effects of one component on other channels. Thus, the total BER of QPSK is the total sum of all these individual components. DQPSK processing offers 1 dB better in OSNR for the same BER and a BER of 1e−3, achieved for an OSNR of 15.2 and 16 dB, respectively, for QPSK and DQPSK, respectively. The main reasons for evaluating all individual in-phase and quadrature-received signals are to ensure that the DAC-generated signals at the ports $(V_I^+, V_Q^+)(H_I^+, H_Q^+)\_and\_(V_I^-, V_Q^-)(H_I^-, H_Q^-)$ enforce no penalties on the coherent detected signals. With this scheme for the generation of 1 and 2 Tbps, it is not difficult to extend to N *Tpbs*, provided that the RCFS comb generator is used, or one has to employ $\frac{N}{5}$ ECLs if nonlinear Comb generation technique is resorted to. The principal problems that we would face in this case would be the nonlinear distortion effects that are described in the following section.

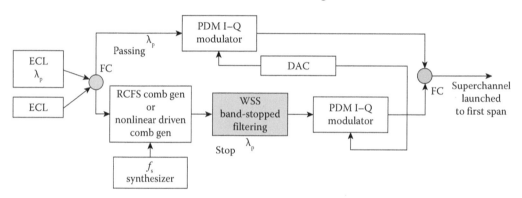

FIGURE 18.33 Generation of ($N - 1$) subcarrier dummy channels and plus-one probe channel for Tbps superchannel transmission systems.

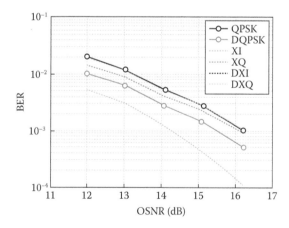

FIGURE 18.34 Ten subchannels of 112 Gbps 28 GBauds Nyquist-QPSK superchannel generation and transmission using nonlinear driven comb generator. 28 GBauds, 30 G grid subchannel spacing, ten subchannels with one probe channel. In the key box: D = differential; XQ, XI is x-values of the Q and I components.

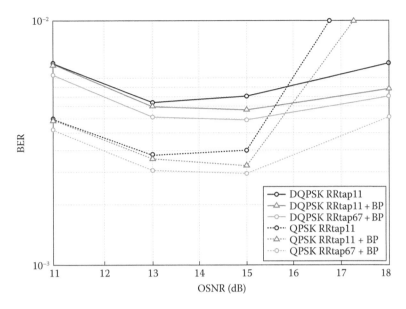

FIGURE 18.35 BER versus launched power (−2 dBm to 1.5 dBm with 0.5 dBm per division in horizontal scale) 28 GBauds Nyquist-QPSK with 30 GHz spacing with and without back propagation additional processing and 11 tap FIR and 2000 km transmission non-DCF fiber multispan line probe channel selected using D40 demux.

Using a nonlinear comb generator, ten subchannels are generated, and hence one channel is eliminated by passing through the WSS set by band-stopped filtering. These subcarriers are then modulated and combined with the probe channel, and then amplified and launched into the first span. The B2B performance is shown in Figure 18.34. The channel spacing is 30 GHz with 28 GBauds as the symbol rate, the BER obtained is 1e−3 at 14.5 dB. The transmission performance of the ten subchannel superchannel over 200 km of optically amplified non-DCF SSMF multispan transmission line is shown in Figures 18.34 and 18.35, in which B2B with four in-phase and quadrature components of the PDM-28 G Nyquist-QPSK are analyzed with their BER plotted against OSNR at the initial phase prior to transmission and then over the complete link. The attenuation per span is 22 dB, and the EDFA stages are optimized so that the launched power into each span is kept the same at each span. The optimum launched power is −1–0 dBm with a BER of 2e−3 with 11 taps FIR filter and back propagation to moderately compensate for fiber nonlinearity (Figures 18.36 through 18.38).

The back propagation is conducted by propagating the received sampled signals at the receiver, then converted into the optical-domain level and the propagation though span by span, with the nonlinear coefficient equal and in opposite sign with those of the SSMF. The back propagation distance per span is about 22 km as the effective length of the fiber under nonlinear SPM effects.

18.4.6.2.3 Tbps Transmission Incorporating FEC at Coherent DSP Receiver

When the superchannel is employed with FEC at the transmitter and decoder at the receiver, the improvement in the BER with respect to the OSR can be achieved as shown in Figure 18.39. The FEC is incorporated in the software loaded to the DAC, hence the name *soft FEC*.

Figure 18.40 shows the FEC improvement on BER versus OSNR for Nyquist-QPSK with BCJR and without FEC incorporated and loaded in the DAC and decoded also in the DSP at the receiver, permitting the erroneous channels becoming error free.

When the baud rate is increased to 32 GB, the PDM Nyquist-shaping QPSK sequence, the transmission and reception under 64 GSa/s ADC are achieved as shown in Figure 18.41. Similar performance as at 28 GB can be observed under precompensation and noncompensation. This compensation is necessary as the bandwidth of the DAC is limited to only 13.7 GHz. Similarly, the

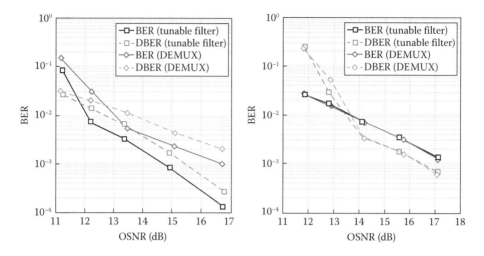

FIGURE 18.36 Transmission performance of channel 9 and channel 20 of the >2 Tbps superchannel with channel spacing of 30 GHz and 28 GBauds Nyquist-QPSK roll-off factor 0.1 using D40 demux and Yenistar sharp roll-off filter (500 dB/nm).

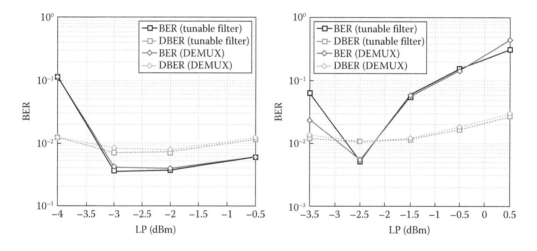

FIGURE 18.37 BER versus launched power of superchannel; probe channels are channel numbers 9 and 20; transmission distance of 2000 km multispan optically amplified non-DCF line. LP = launched power.

performance under soft FEC, shown in Figures 18.42 and 18.43, can be achieved for 2000 km transmission of non-DCF 80 km SSMF span link.

18.4.6.2.4 224 Gbps 16QAM Tbps Transmission

In this subsection, we will describe the experimental setup that we have employed for the generation of a 224 Gbps POLMUX-RZ-16QAM signal. In order to overcome the problem of unavailability of high-speed DACs, higher modulation QAM, the 16QAM, is employed. The constellation and eye diagram can be observed as shown in Figure 18.44. The generation of such 16QAM can be either from a DAC or from two ports of 28 G bit pattern generator (BPG) where one port is 3 dB attenuated with respect to the other, and then combined via an RF combiner. Tunable delay must be done to synchronize the two ports due to different propagation electrical paths, as shown in Figure 18.45. Due to the different amplitudes, we can see the three levels in the eye diagram. The spectrum of a single channel

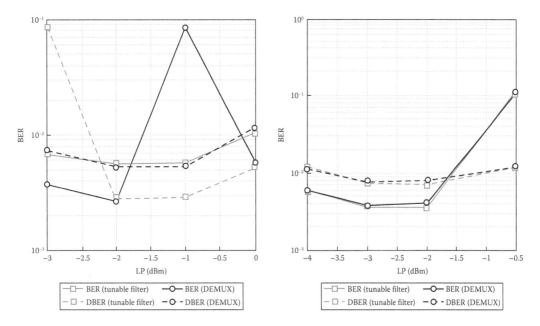

FIGURE 18.38 Five channels nonlinear comb generator subcarrier transmission over 2000 km: BER versus OSNR using D40 demux or sharp roll-off optical filter (Yenistar). LP = launched power.

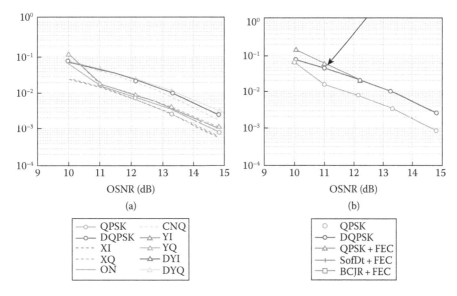

FIGURE 18.39 Soft differential FEC processing for Nyquist-QPSK, BER versus OSNR. 28 GBauds with individual components of QPSK, that is, real and imaginary parts (in-phase and quadrature-phase components), B2B setup (a) individual I, Q components, (b) combined with differential and nondifferential QPSK with FEC. Legend: XI, YI = x-value and y-values of inphase component; DYI, DYQ = differential y-values of I, Q components. Other defined in text.

and multichannel (superchannel) by modulating a comb generator as described in this section is shown in Figure 18.46a and b, respectively. The measured B2B of BER versus OSNR for 224 Gbps is shown in Figure 18.47a. Comparing with QPSK, we can see that, for 16QAM, the required OSNR would be about 5–6 dB higher. This will affect the nonlinear limit and the launched power into the fiber transmission spans. Transmission over distances of more than 650 km can be obtained as shown

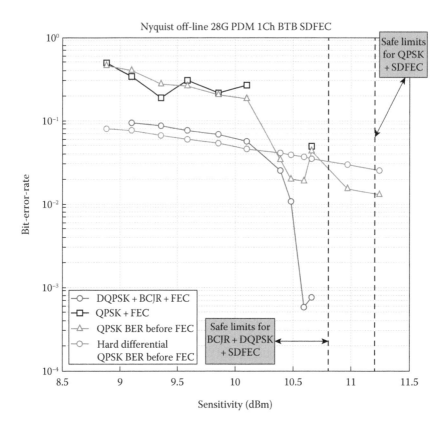

FIGURE 18.40 FEC improvement BER versus OSNR for Nyquist-QPSK with BCJR and without FEC.

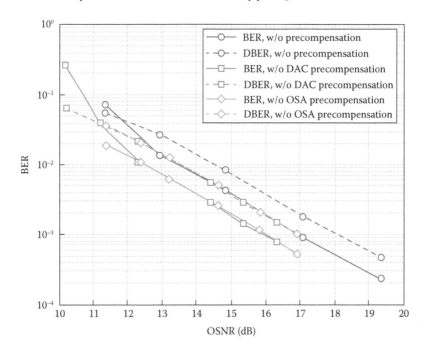

FIGURE 18.41 32 GBauds PDM Nyquist-QPSK transmission, BER versus OSNR with and without electronic compensation in DAC, B2B scenario.

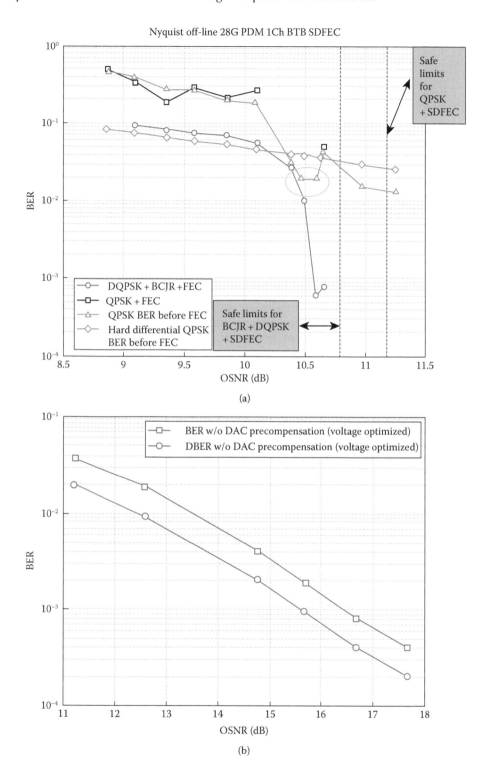

FIGURE 18.42 Performance of transmission systems under (a) Nyquist 28G PDM QPSK 1-Ch Back-to-Back with SDFEC and (b) BER without DAC precompensation with optimized voltage level and DBER without DAC precompensation with optimized applied voltage. Baud rate of 28 G and 32 GBauds, back-to-back scenario, modulation format Nyquist-QPSK.

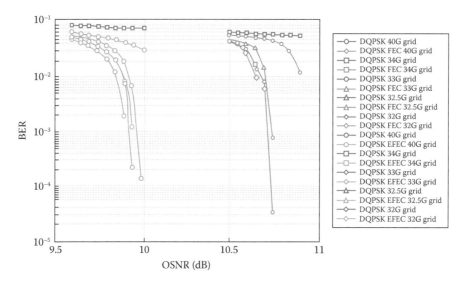

FIGURE 18.43   32 GB Nyquist-QPSK under a roll-off factor of 0.1 with sub-channel spacing as parameter. Transmission distance >2000 km non-DCF optically amplified SSMF spans under scenarios of with and without FEC 20%, far left and far right of the graph, respectively.

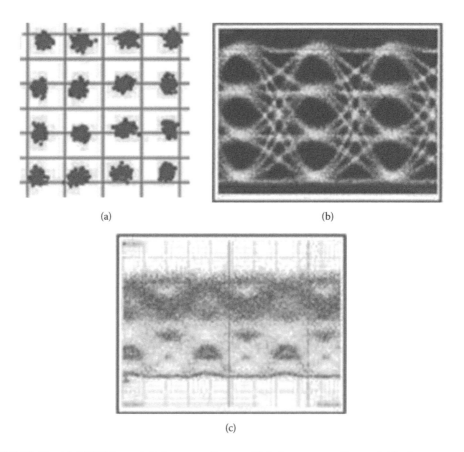

FIGURE 18.44   (a) 16QAM constellation; eye diagram (b) before propagation and (c) after propagation, respectively, with baud rate of 28 GB.

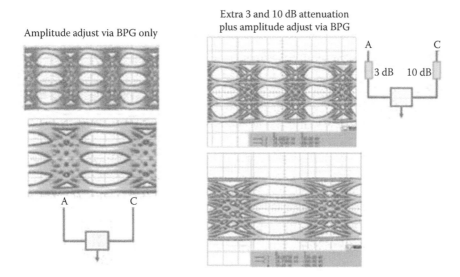

Amplitude adjust via BPG only

Extra 3 and 10 dB attenuation
plus amplitude adjust via BPG

FIGURE 18.45 Circuit diagram and eye diagrams generated from SHF BPG. A and C are two output ports of the SHF 12103A unit.

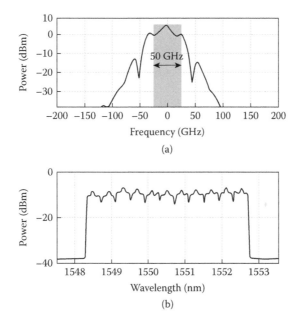

FIGURE 18.46 Spectrum of 16QAM at 28 GB (a) single channel and (b) 11 channels as a superchannel.

in Figure 18.47b. Indeed, we have conducted the field trial for 1 T (100 G ×10 superchannels) and 2 T with 224 G × 10 subcarriers, and the transmission distances demonstrated with FEC of 1e−3 limit are 3500 km and 1470 km with optimization of optical amplifiers in the link.

18.4.6.3   Coding Gain of FEC and Transmission Simulation

Simulations are conducted to estimate the gain and BER against OSNR with error coding and non-coding so as to assist with the experimental performance. The simulation proves the coding gain for Nyquist-QPSK of 0.1 roll-off factor, as shown in Figure 18.48. Under coding, we can see that the OSNR requirements for 1e−3 may be reduced down to 11 dB. Thus, an extra margin of 4 dB

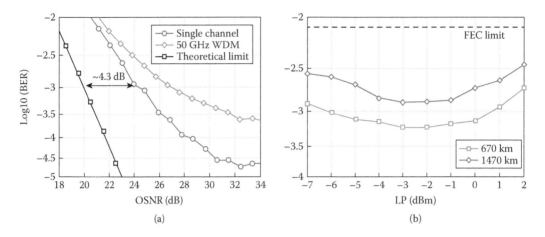

FIGURE 18.47 Measured BER versus OSNR of (a) B2B and (b) after multispan link of non-DCF optical amplifiers and 80 km SSMF span length transmission over distances as noted. The launched power is set at −1 dBm.

can be gained to allow the extension of the transmission reach of Nyquist-QPSK, an extra 500 km. Figure 18.48a and b shows the effects of 30 and 28.5 GHz subchannel spacing on the BER of the Nyquist-QPSK with LDPC and significant improvement of FEC on their performance. Similarly, Figure 18.48c and d is related to 33 and 32.5 subchannel spacing with 32 GBauds. Figure 18.48e and f then prove the improvement when BCJR additional coding is superimposed to obtain further coding gain. Figure 18.48g displays all gain curves into one graph, and Figure 18.48h and i shows the spectra of subchannels, the odd channels only.

18.4.6.4 MIMO Filtering Process to Extend Transmission Reach

The filter structure incorporated in the DSP processing is shown in Figure 18.49. Due to whitening noise effects in the optical amplification stages in the link, the root-raised-cosine filter is used at the receiver for match filtering. That is, the complete filtering process in the link forms a complete Nyquist filter, satisfying the Nyquist-shaping criteria. Following the Rx imperfections, compensation, and CD compensation, a complex butterfly FIR structure is applied to compensate for PMD. Each of the four complex FIR filters is realized by a butterfly structure of corresponding real FIR filters. The recursive constant modulus algorithm (CMA) LMS algorithm continuously updates the filter taps, which guarantee the initial convergence and tracking of time-variant channel distortions. In the steady state, the complex butterfly structure is a digital, real representation of the inverse impulse response determined by the tap coefficients.

The output signals from the FIR equalization stage (x' and y') at time k are related to the input signal vectors (x and y) containing samples $k-L+1$ to k by

$$x'(k) = \mathbf{h}_{xx} \cdot \mathbf{x}(k) + \mathbf{h}_{yx} \cdot \mathbf{y}(k)$$
$$y'(k) = \mathbf{h}_{yy} \cdot \mathbf{y}(k) + \mathbf{h}_{xy} \cdot \mathbf{x}(k)$$

(18.4)

where $\mathbf{h}_{xx}, \mathbf{h}_{xy}, \mathbf{h}_{yy}, \mathbf{h}_{yx}$ are the $T/2$-spaced tap vectors (T is symbol period) for the FIR filter, and the dot "." denotes the vector dot product. The length L of the tap vector is equal to the impulse response of the distorted transmission medium to be compensated. Initial equalizer acquisition is performed on the first several thousand symbols, depending on the "learning/training" process. These symbols are subsequently discarded in the error counting. The equalizer tap vectors are then updated continuously throughout the processing of the data set in order to track channel changes.

The MIMO filter length in the commercial 100 G receivers without tailoring for Nyquist transmission is usually set between 7 and 11 as a trade-off between complexity and requirements since

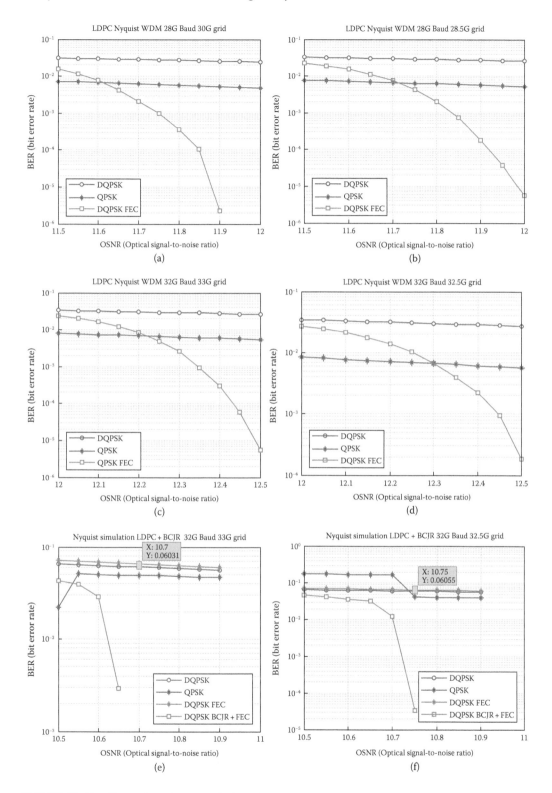

FIGURE 18.48 Simulation of Nyquist-QPSK with and without error coding gain: (a, b) LDPC coding and gain at 28 GHz grid; (c, d) LDPC Nyquist-QPSK with 33 GHz grid; (e, f) BCJR + LDPC and 3 GHz grid.

(Continued)

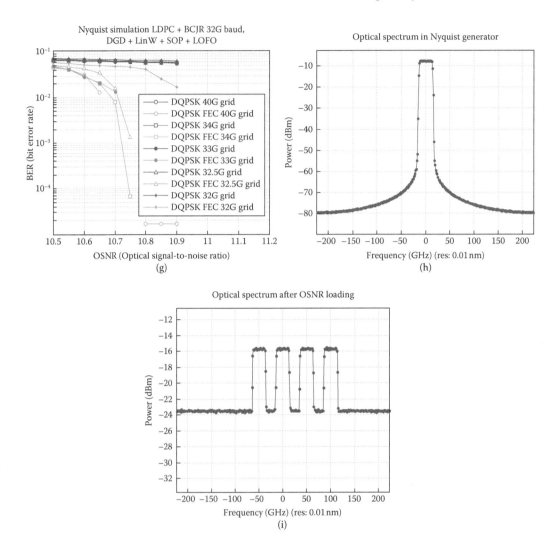

FIGURE 18.48 (Continued) Simulation of Nyquist-QPSK with and without error coding gain; (g) Nyquist-QPSK with and without coding gain with grid frequency spacing as parameters (32–40 GHz) as summary of (a–f); (h, i) spectra of channels before and after ASE noise loading.

the measured mean DGD in real long-haul optical links is around 25 ps. Further increasing the FIR complexity would enhance the gain in case of a longer pulse response to be employed due to the limited transfer function of the transmission system. This is verified in our experimental platform with the BER against OSNR, with the launched power and number of taps as depicted in Figure 18.50a through d. It is noted that a tap number higher than 9 does not offer any gain improvement in performance, as depicted in Figure 18.50. The tap length of the MIMO filter can be reasonably short for experimentation under the following scenarios (a) BTB performance, L stands for the filter tap length, (b) 1500 km transmission line, (c) 2000 km transmission line, and (d) MIMO filter with tap length of 23 convergence results. $\mathbf{h}_{xx}$ for the in-phase real part of $\mathbf{h}_{xx}$, $\mathbf{h}_{xx}$ for the quadrature imaginary part of $\mathbf{h}_{xx}$.

In our Nyquist experimental platform, the pulse sequence is shaped by a Nyquist square root filter (RRC) of a roll-off factor of 0.1 generated by a time-domain FIR filter with 65 taps. Simulation results do not show any penalty caused by our FIR tap settings. However, hardware imperfections likely require filter pulse response of more FIR taps to achieve the best performance. Furthermore, the convergence of the CMA algorithm would fail for FIR with a longer tap length, as shown in Figure 18.51.

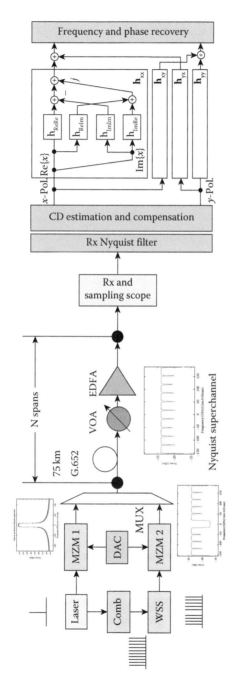

FIGURE 18.49 Nyquist superchannel experimental setup and receiver structure; each of four complex MIMO filters consists of four real FIR filters.

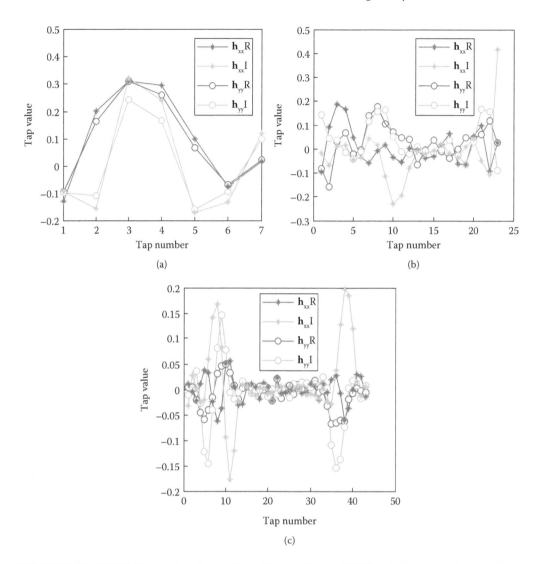

FIGURE 18.50 MIMO filter tap length extending effects in Nyquist-WDM optical transmission experimental results; (a) tap length 7, (b) tap length 23, (c) tap length 43, $\mathbf{h}_{xx}R$ for the real part of $\mathbf{h}_{xx}$, and $\mathbf{h}_{xx}I$ for the imaginary part of $\mathbf{h}_{xx}$.

With seven taps, we are able to acquire the channel with the FIR filter and no convergence with 23 and 43. Therefore, the conventional CMA method is used and the performance is shown in Figure 18.51a. The coarse step applies fast learning via larger values of a weighting factor μ_1 and a forgetting factor α_1. The fine step uses smaller values of these two parameters and improves the final performance.

To solve the convergence problem, we perform the acquisition procedure in two steps, with smaller and larger FIR filter lengths. First, a shorter FIR filter with a smaller number of taps L_1, for example, less than 9, and larger μ_1 and α_1 can be used in the initial step to find the main values of the starting taps. After the preconvergence phase, the filter is extended with the reiterative tap values obtained in the first procedure, while the extended taps are set to null in association with smaller values of μ_2 and α_2. This method ensures the filter convergence (Figure 18.52).

Using this new algorithm, the B2B performance can be improved by up to 0.7 dB at a BER of 10^{-3} with a tap extension from 7 to 23, as shown in Figure 18.51a. The performance of the new algorithm is further verified over the transmission in 1500 and 2000 km links. The tap extension from

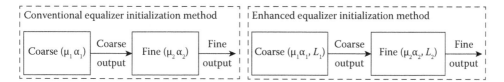

FIGURE 18.51 Conventional and enhanced MIMO filter tap convergence algorithm.

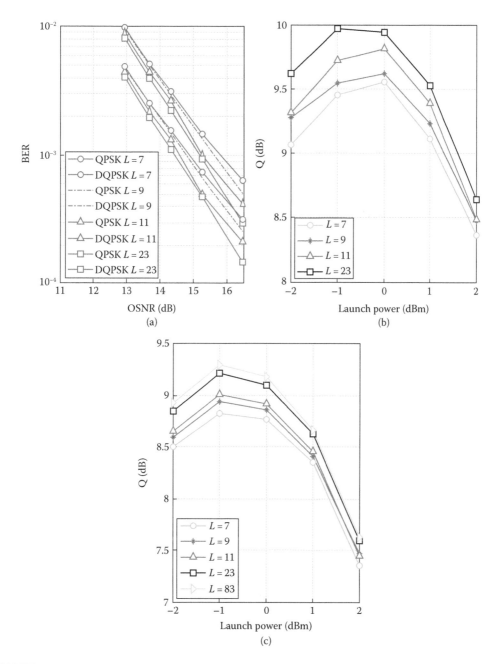

FIGURE 18.52 Enhanced MIMO filter tap length extending performance (BER versus OSNR) in Nyquist-WDM optical transmission experiments: (a) BTB, (b) 1500 km transmission results, and (c) 2000 km transmission.

7 to 23 enables a Q gain of almost 0.5 dB at two optimum launch power of -1 and 0 dBm. To test the MIMO filter performance improvement by extending the filter length, we checked the filter of length 83. The improvement could only be about 0.1 dB, indicating that further increasing the complexity brings only a negligible gain. Thus, a filter length of 23 is the FIR filter limit.

18.4.7 MULTICARRIER SCHEME COMPARISON

A preliminary comparison and analysis of Nyquist-QPSK transmission scheme for superchannel Tbps transmission over long-haul and expected metro networks is shown in Table 18.2.

Hardware complexity: A comb generator is required at the transmitter side of all schemes, which can be either using one ECL, then employing an RCFS technique to generate N subcarriers of locked phase and frequency, or a multiplied factor of five subcarriers per ECL using nonlinear driving method. Furthermore, a set of parallel PDM I/Q modulators and DAC with eight ports of positive and complementary signals, an optical subchannel demux, and a mux would be necessary for generating independent sets of in-phase and quadrature signals for modulating the PDM I–Q modulators. At the receiver side, a standard coherent receiving system can be used with the insertion of a sharp optical filter at the front end of the optical hybrid coupler, which is required to separate the subchannels. This may require a set of optical filters, one for each subchannel. Thus, a technical solution should be provided to avoid this complex and expensive solution. This is the set back of the Nyquist-WDM technique.

DSP complexity: For Nyquist-WDM, because the Nyquist filtering results in ISI, at the receiver side, one needs sequence estimation algorithms such as MAP and MLSE. For Nyquist-QPSK and different comb-generated subcarriers and subchannels, FIR with lower number of taps would be suitable, so that the complexity can be acceptable.

ECLs: For RCFS comb generation, there needs to be only one ECL source, but additional modulator and optic components are required, while for nonlinear driving comb generators, the number of ECLs would be increased accordingly. To cover the whole C-band with superchannels of 30 GHz spacing, there would be around 12 ECLs. This number may be high, especially when they are to be packaged in the same line card.

TABLE 18.2
Comparisons of Tbps Transmission Schemes

	Nyquist-WDM	CO-OFDM	eOFDM
Hardware complexity	Tx: similar	Tx: similar	Tx: similar
	Rx: multi-OFE	Rx: need OFFT	Rx: multi-OFE
DSP complexity	More complex, possibly due to compensation— but no higher degree of complexity, possibly more processing time required for CPU	Normal	Normal
SE (theoretically)	Similar	Similar	Similar
Tx-DSP + DAC	Essential	Possibly not	Essential
ADC sampling rate	$1.2 \times$ Baud	$2 \times$ Baud	$2 \times$ Bandwidth
Bandwidth requirement on O/E components	Depending on subchannel spacing	Depending on subchannel spacing	Depending on subchannel spacing
Special requirements	DSP for sequence estimation (MAP, MLSE)	Orthogonal channel separation	Cyclic prefix and guard band for OFDM symbols
Flexibility	Medium	Medium	High

Spectral efficiency: All Nyquist pulse shaping schemes can enhance the SE by a factor of about 2; the overlapping between subchannels would create minimum crosstalk if the roll-off factor is less than 0.1; and the third-order harmonics of the subchannels and comb-generated subcarriers is more than 30 dB below the primary carrier.

Tx-DSP and DAC: Ideally, CO-OFDM may not need DSP, but for compensation of components and transmission impairments, it is preferred to use Tx-DSP. The other two schemes must use Tx-DSP and DAC to shape the pulse or generate the designed signal.

ADC sampling rate: This depends on channel spacing, ideally by a factor of 2x of the single-sided signal bandwidth. 56–64 GS/s ADC and DAC are available from Fujitsu, which allow the generation of random sequence for 28–32 Gbps.

Bandwidth requirements on O/E components: Similar to the sampling rate, Nyquist-WDM needs the lowest bandwidth, assuming the same baud rate. The other two schemes are similar.

Flexibility: All three schemes have flexibility in the number of subchannels, modulation formats, and bandwidth of subchannels, while eOFDM can adjust some parameters in the electrical domain by means of Tx-DSP; it has more flexibility than the other two schemes.

18.5 MODULATION FORMATS AND ALL-OPTICAL NETWORKING

18.5.1 Advanced Modulation Formats in Long-Haul Transmission Systems

Over the years, especially in the last ten years, research works have been ongoing both in theory and simulation and experiments under various modulation formats. It is noted that, along the transmission systems, optical add/drop multiplexers (OADMs) are commonly inserted in order to inject new channels or dropping selected channels to the location. The channels can then be fed into optical cross-connects (OXCs) for routing to further transmission lines. Thus, we expect that the optical signals must be resilient to optical filtering cascaded along the transmission path. Optical filtering can thus assist the electronic filtering of the optical equalization at the receiver or as predistortion at the transmitter. The design of optical filtering can be integrated with these electronic processing subsystems.

Coherent transmission and detection techniques have become a reality with the availability of narrow band lasers for use as LOs. The increase of high bit rate also reduces the demand of the LO laser linewidth. Furthermore, the processing of mixed signal detection and electronic processing for phase estimation of the received electronic signals will support the implementation and installation of coherent transmission in all-optical networks and systems.

18.5.2 Advanced Modulation Formats in All-Optical Networks

Future transparent networks would expect to be all-optical, and thus be spectrally efficient through the transmission systems, OADMs, and OXCs for optical routing. In such networks, the advanced modulation formats that offer narrow optical passband would minimize the effects of tight filtering at OADM and OXC sites. Formats, especially phase shift keying or continuous phase—such as SSB and VSB, MSK, RZ-DPSK and RZ-DQPSK, and duobinary (as an exception)—are shown to offer excellent transmission performance and adaptivity to the electronic dispersion compensation placed at the receiver (see Chapter 11). Optical prefiltering does generate the formats with a truly single-sideband such as SSB, VSB, and duobinary signals. In optical networks, the cascading of several of these optical subsystems is necessary, and modulation format signals must be resilient to optical filter cascading [3,4]. The complexity of network structures in next-generation optical networks requires flexibility and adaptability of optical signals and spectral efficiency as well as ease of electronic equalization. Advanced modulation formats could offer to resolve a number of challenging issues that are to be integrated in the design and arrangement of OADMs and OXCs in future networks. Monitoring of modulation formats, wavelength channels, and eye distortion for adaptive equalization are also important for all-optical networking. Figures 18.53 and 18.54 shows the BER vriation as a function of the receiver sensitivity.

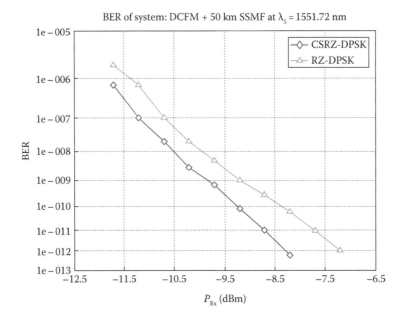

FIGURE 18.53 Measured BER versus receiver sensitivity for two formats CSRZ-DPSK and RZ-DPSK at wavelength 1551.72 nm.

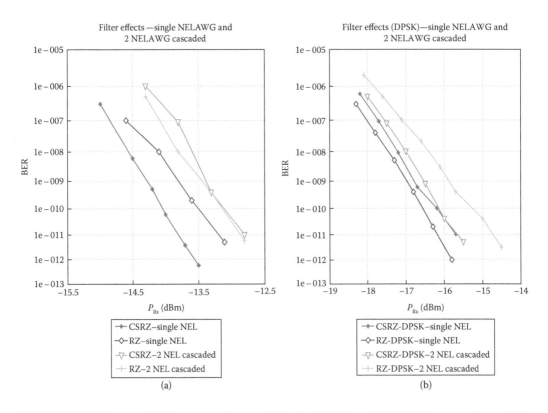

FIGURE 18.54 Measured BER versus receiver sensitivity for (a) ASK and (b) DPSK formats under multiple filtering effects—typical in all-optical networking.

The effects of filter cascading can be observed from the measure of the BER versus the receiver sensitivity of 40 Gbps CSRZ-DPSK modulation under the multiplexing and demultiplexing by using two NEL 100 GHz AWGs. One AWG has a 3 dB bandwidth of 0.5 nm and the other of 1.2 nm passband and tunable. The cascaded AWGs offer narrow passband to sufficiently pass the 40 G CSRZ-DPSK as observed in Figure 18.55. The transmission of 40 Gbps over 10 Gbps optical transmission systems is important for upgrading the transmission rate of one or few wavelength channels to 40 Gbps. With 100 GHz spacing and 0.5 nm passband filters in the network, the BER has been measured versus the receiver sensitivity of 40 Gbps, and 10 Gbps of adjacent wavelength channels indicate that the CSRZ-DPSK behaves very well and suffers no penalty, as shown in Figure 18.56. We thus expect that the optical MSK channel would also perform better under conventional 10 Gbps transmission systems.

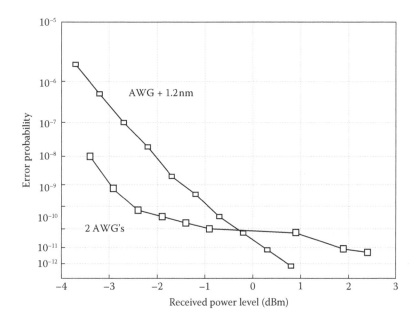

FIGURE 18.55 Measured BER versus receiver sensitivity for CSRZ-DPSK under the cascading of two optical filters (AWGS) and one AWG. Horizontal axis must add −10 dB.

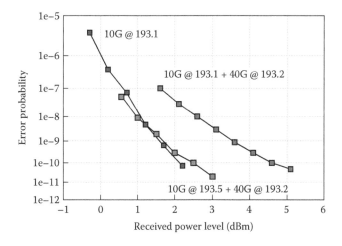

FIGURE 18.56 Measured BER versus receiver sensitivity for 40 Gbps CSRZ-DPSK and 10 Gbps under normal 10 Gbps optical transmission setup.

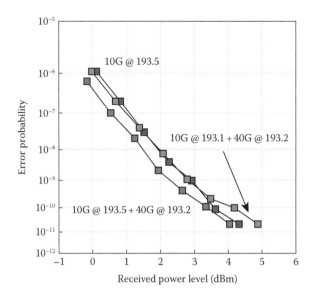

FIGURE 18.57   320 km transmission 40 G impact on 10 G channel: probability of error versus receiver sensitivity (dBm)—effects of 40 G (CSRZ-DPSK) with 10 G (NRZ-ASK) channel simultaneously transmitted for NRZ-ASK and CSRZ-DPSK formats—lighter gray dots for 1.2 nm thin film filter and dots for 0.5 nm AWG filter (demux with 100 GHz spacing).

18.5.3   HYBRID 40 GBPS OVER 10 GBPS OPTICAL NETWORKS: 328 KM SSMF + DCF FOR 320 KM TX—IMPACT OF ADJACENT 10 G/40 G CHANNELS

An experimental test bed is set up for a 320 km transmission performance with a 10 G NRZ-ASK and a CSRZ-DPSK 40 G channel to test adjacent and nonadjacent channel performance with a 100 GHz AWG mux and a 1.2 nm tunable filter at the input of the Rx. Significant penalty when adjacent 40 G channel is switched on, owing to the narrow width of the tunable filter. Non-adjacent channel too far off to have any impact, as observed from the results shown in Figures 18.55 through 18.57.

Result shows that there is no appreciable impact on 10 G signal of an adjacent 40 G signal. For the case that two 100 GHz AWGs are used as the mux and Rx filter for CSRZ-DPSK transmission over 320 km, we now measure the performance of the 40 G stream transmitted simultaneously with adjacent and nonadjacent 10 G channel. Note the problem with using two AWGs at 40 Gbps as there is significant error floor. However, we can see that 10 G adjacent or nonadjacent has no impact on 40 G signal, as expected. The small improvement with the 40 G channel is probably due to less noise from EDFA.

18.6   ULTRA-FAST OPTICAL NETWORKS

Future large-scale deployment of broadband connection in the access area will require transport and add or drop of high-speed-capacity information with bit rates reaching 1600 Gbps or even higher to 170 Gbps, including forward error coding.

Optical time division multiplexing (OTDM) should be employed to multiplex real-time-domain signal channels to a higher rate, for example, 4 × 40 Gbps to 160 Gbps. In the previous chapters of this book, most of our studies are based on the base rates of 40 or 10 Gbps, with different modulation formats of amplitude, phase, and frequency of the lightwave carriers. This section addresses a number of technological issues that would enable the transmission of ultra-high-speed 160 Gbps signals over these systems and networks.

160 Gbps is normally composed of time-interleaved RZ-coded lower bit rate, generally at 10 or 40 Gbps, usually generated by mode-locked fiber lasers [5]. The transmission of these

ultra-high-bit-rate signals requires the precise compensation of the dispersion effects using optical phase modulation, as addressed in Chapter 10.

Furthermore, to recover the lower-rate channels, time division demultiplexing and adding or dropping other channels is required. This enables the insertion and extraction of lower-bit-rate channels. In order to achieve the insertion and extraction, two principal photonic devices are required—a photonic clock recovery operating at the lower rate of the channel, and an ultra-fast gating device, usually an electroabsorption modulator (EAM), operating in a short time window within which other pulse sequences would be isolated, so that no interference occurs. A contrast ratio of at least 20 dB is necessary.

Photonic ultra-fast gates can be based on a number of operating principles, such as phase shifting in interferometric-type gate/switch, frequency shifting such as parametric frequency conversion or four-wave mixing, or gain/absorption variation by electrical or optical excitation. The gating function can be provided by EAM [6], SOA—semiconductor-amplifier-based ultra-fast nonlinear interferometers [7,8], SOA-based interferometers, and Sagnac interferometers incorporating highly nonlinear fifers [9,10]. Optical or electrical clock pulses are required, depending on the operating domain of the gate.

Wavelength conversion for the ultra-fast routing of these OTDM channels is also required in order to avoid the congestion at nodes in the networks, as shown in Figure 18.58. The EAM seems

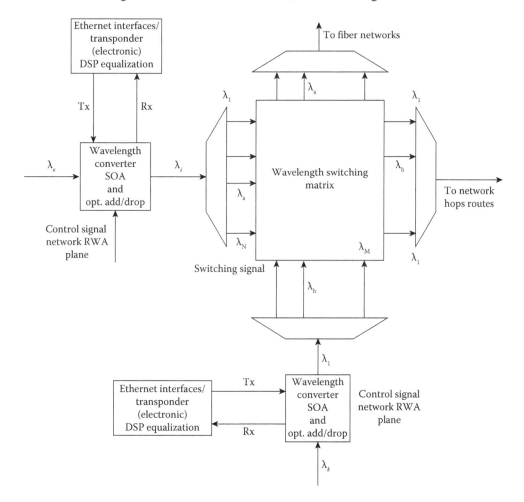

FIGURE 18.58 Detailed structure of an optical node for ultra-high-speed optical Ethernet. Wavelength conversion and routing by an SOA and demultiplexer and switching matrix for routing a specific wavelength to a designated output port for further routing.

to be the simplest photonic gate for such a network node. The design and demonstration of such network nodes are critical for the future deployment of ultra-fast optical networking.

18.7 CONCLUDING REMARKS

Optical networks have rapidly evolved in the past decade to cope with the dramatic increase in data traffic, mostly driven by IP transport and Internet applications. Whereas the pace of large fiber infrastructure building has slowed down, transport technologies went through several transformations in all areas of the networks, from the core to the metro to the access with bit rate per wavelength subcarrier reaching 100 Gbps and even higher. At the turn of the century, most optical networks were carrying a mix of 2.5 and 10 G line traffics, while the investment in newer technologies (mostly 40 G) had started, and fell soon after as a result of the subsequent economic downturn (the dot-com collapse). With the first transmission research on 40 G starting in 1996, the actual deployments only started a decade later, around 2006. Aside from the economy, this long lag was due to an unstable ecosystem: fragmented supply chains, numerous technological steps, missing standardization, discrete designs, and so on. With the massive 40 G and currently 100 G deployment that have been taking place in the past ten years, this ecosystem is stabilizing only now, with the availability of MSA modules and focus on two-line transmission technologies: DPSK and coherently demodulated polarization-multiplexed QPSK.

This later technology, accelerated by remarkable advances in DSP at 64 GSa/s allowing the implementation of powerful coherent receivers and by a highly focused effort in standardization bodies, has paved the way for the advent of 100 G solutions by quickly forming stable 100 G eco-networks. The lag between the first 100 G research results and actual network implementation now seems to be less than half that of 40 G: from 2007 to 2014. Recently, research at rates higher than 100 G has rapidly shown the suitability of complex, multilevel modulation formats with polarization multiplexing and strong DSP to rates from 200 Gbps to 1 Tbps per channel, hence flexible spectral bandwidth and flexible grid are required. This may point to an even more rapid implementation of such solutions into optical transport platforms. However, some fundamental limitations may limit this pace, and trade-offs between spectral density, fiber capacity, transparent reach, implementation complexity, power consumption, space, and operational flexibility will define the boundaries for future optical communication networks.

Optical fiber technology tackles the vast and growing field of high-capacity optical transport, at channel rates of 100 G and beyond. The chapters of this book are provided to focus on both theoretical and practical approaches by global telecommunication operators; content providers; system, subsystem, and component vendors; fiber manufacturers; instrumentation vendors; and academic institutions, thus representing the entire field of active research and development in next-generation OTNs. These chapters address the latest findings and issues in DSP-based homodyne reception for 100 Gbps and higher bit rate optical transport systems. Aspects of design, implementation, capacity limits, performance, modulation formats, components, power, space, and network architecture are among the covered topics. The chapters presenting an overview of the subject as well as the current and projected state of the industry are followed by those providing detailed reviews of specific high-rate transponder technologies, for example, line-side optics, client interfaces, forward error correction, and DSP. Spatial multiplexing technology is also covered in the chapter addressing optical fibers as a promising alternate means to achieve high-capacity transport. Furthermore, specific enabling technologies for next-generation high-capacity optical networks, that is, optimized optical line fibers and wavelength-selective switches, are described.

In particular, this chapter addresses the contrast of different modulation format transmission systems under self-coherent and coherent reception incorporating ADC-DSP and pulse-shaping DAC-based optical transmitters. There is no doubt that the self-coherent and direct detection optical systems will find significant deployment in metro and access networks due to their low costs, whereas coherent reception will find extensive deployment in long-haul core networks. Both types of reception techniques for optical transmission systems are presented in this chapter.

With the explosion on Cloud radio access networking (C-RAN) technology and massive MIMO wireless systems and the data centers, the capacity of front and backhaul OTNs will reach Petabps levels in the near future, with optical routers servicing the OTNs in which routing in the optical domain will be implemented to reduce the power consumption, simplifying network implementation as well as adding speed. The network infrastructures are still well behind the advances made in data center technologies, and advances in research are expected to address these issues as well as Silicon-integrated photonics to provide high-speed transponders and optical routers based on switching matrices and tunable spectral widths.

REFERENCES

1. A. G. SHF, Einsteinstr, Berlin, Germany, available at shf.de.
2. L. N. Binh, Optical PLL super combed carrier cloning: Circuit and methodology for locking superchannel coherent receivers, private communications.
3. G. Raybon, Performance of advanced modulation format in optically routed networks, in *Proceeding of the Optical Fiber Conference 2006*, paper OThR1, OFC 2006, Anaheim CA, 2006.
4. P. J. Winzer, G. Raybon, M. Duelk, "107-Gb/s optical ETDM transmitter for 100G Ethernet transport", *ECOC*, Vol. 6, pp. 1–6, 2005.
5. H. Q. Lam, P. Shum, L. N. Binh, and Y. D. Gong, "Mode-locked fiber lasers," Polarization switching in an active harmonic mode locked fiber laser, in *Conference on Optical Internet COIN-ACOFT*, Melbourne, June 24–26, 2007.
6. D. Rafique, J. Zhao, A. D. Ellis, Digital back-propagation for spectrally efficient WDM 112 Gbit/s PM m-ary QAM transmission, *Optics Express* Vol. 19, No. 6, pp. 5219–5224, 2011.
7. C. Schubert, C. Schmidt, S. Ferber, R. Ludwig, H. G. Weber, "Error-free all-optical add-drop multiplexing at 160 Gbit/s", *Elect. Lett.*, Vol. 39, No.19, pp. 1074–1076, 2003.
8. J. P. Turkiewicz, H. Rohde, W. Schairer, G. Lehmann, E. Tangdiongga, G.D. Khoe, and H. de Waardt, "All-optical OTDM Add-Drop Node at 16×10 Gbit/s in between two Fibre Links of 150 km", *Proc. ECOC'03,* paper PDP - Th4.4.5, pp. 84–86, 2003.
9. M. Heid, S. L. Jansen, E. Meissner, W. Vogt, and H. Melchior, "160-Gbit/s demultiplexing to baserates of 10 and 40 Gbit/s with a monolithically integrated SOA-Mach-Zehnder interferometer," in 28th European Conference on Optical Communication (ECOC '02), Copenhagen, Vol. 3, pp. 1–2, 2002.
10. T. Morioka, "Extreme OTDM transmission issue: Beyond 100 Gbit/s," Proc. 26th European Conf on POptical Communications, 2004 (ECOC'04), Stockholm, Thl.2, pp. 234–236, 2004.

Annex 1: Technical Data of Single-Mode Optical Fibers

A1.1 STANDARD SINGLE-MODE OPTICAL FIBERS

Corning® Single-Mode Optical Fiber

The Standard For Performance

Corning® SMF-28™ single-mode optical fiber has set the standard for value and performance for telephony, cable television, submarine, and utility network applications. Widely used in the transmission of voice, data, and/or video services, SMF-28 fiber is manufactured to the most demanding specifications in the industry. SMF-28 fiber meets or exceeds ITU-T Recommendation G.652, TIA/EIA-492CAAA, IEC Publication 60793-2 and GR-20-CORE requirements.

Taking advantage of today's high-capacity, low-cost transmission components developed for the 1310 nm window, SMF-28 fiber features low dispersion and is optimized for use in the 1310 nm wavelength region. SMF-28 fiber also can be used effectively with TDM and WDM systems operating in the 1550 nm wavelength region.

Features And Benefits

- Versatility in 1310 nm and 1550 nm applications.

- Outstanding geometrical properties for low splice loss and high splice yields.

- OVD manufacturing reliability and product consistency.

- Optimized for use in loose tube, ribbon, and other common cable designs.

The Sales Leader

Corning SMF-28 fiber is the world's best selling fiber. In 2000, SMF-28 fiber was deployed in over 45 countries around the world. All types of network providers count on this fiber to support network expansion into the 21st Century.

Protection And Versatility

SMF-28 fiber is protected for long-term performance and reliability by the CPC™ coating system. Corning's enhanced, dual acrylate CPC coatings provide excellent fiber protection and are easy to work with. CPC coatings are designed to be mechanically stripped and have an outside diameter of 245 μm. They are optimized for use in many single- and multi-fiber cable designs including loose tube, ribbon, slotted core, and tight buffer cables.

Patented Quality Process

SMF-28 fiber is manufactured using the Outside Vapor Deposition (OVD) process, which produces a totally synthetic ultra-pure fiber. As a result, Corning SMF-28 fiber has consistent geometric properties, high strength, and low attenuation. Corning SMF-28 fiber can be counted on to deliver excellent performance and high reliability, reel after reel. Measurement methods comply with ITU recommendations G.650, IEC 60793-1, and Bellcore GR-20-CORE.

Optical Specifications

Attenuation

Standard Attenuation Cells

Wavelength	Attenuation Cells (dB/km)	
(nm)	Premium *	Standard
1310	≤0.35	≤0.40
1550	≤0.25	≤0.30

* Lower attenuation available in limited quantities.

Point Discontinuity

No point discontinuity greater than 0.10 dB at either 1310 nm or 1550 nm.

Attenuation at the Water Peak

The attenuation at 1383 ± 3 nm shall not exceed 2.1 dB/km.

Attenuation vs. Wavelength

Range (nm)	Ref. λ (nm)	Max. α Difference (dB/km)
1285 - 1330	1310	0.05
1525 - 1575	1550	0.05

The attenuation in a given wavelength range does not exceed the attenuation of the reference wavelength (λ) by more than the value α.

Attenuation with Bending

Mandrel Diameter (mm)	Number of Turns	Wavelength (nm)	Induced Attenuation* (dB)
32	1	1550	≤0.50
50	100	1310	≤0.05
50	100	1550	≤0.10

*The induced attenuation due to fiber wrapped around a mandrel of a specified diameter.

Cable Cutoff Wavelength (λ_{ccf})
$\lambda_{ccf} \leq 1260$ nm

Mode-Field Diameter
9.2 ± 0.4 μm at 1310 nm
10.4 ± 0.8 μm at 1550 nm

Dispersion
Zero Dispersion Wavelength (λ_0):
1302 nm ≤ λ_0 ≤ 1322 nm

Zero Dispersion Slope (S_0):
≤ 0.092 ps/(nm²·km)

$$\text{Dispersion} = D(\lambda) := \frac{S_0}{4}\left[\lambda - \frac{\lambda_0^4}{\lambda^3}\right] \text{ps/(nm·km)},$$
$$\text{for } 1200 \text{ nm} \leq \lambda \leq 1600 \text{ nm}$$
$$\lambda = \text{Operating Wavelength}$$

Polarization Mode Dispersion

Fiber Polarization Mode Dispersion (PMD)

	Value (ps/√km)
PMD Link Value	≤ 0.1*
Maximum Individual Fiber	≤ 0.2

* Complies with IEC SC 86A/WG1, Method 1, September 1997.

The PMD link value is a term used to describe the PMD of concatenated lengths of fiber (also known as the link quadrature average). This value is used to determine a statistical upper limit for system PMD performance.

Individual PMD values may change when cabled. Corning's fiber specification supports network design requirements for a 0.5 ps/√km maximum PMD.

Environmental Specifications

Environmental Test Condition	Induced Attenuation (dB/km)	
	1310 nm	1550 nm
Temperature Dependence -60°C to +85°C*	≤0.05	≤0.05
Temperature-Humidity Cycling -10°C to +85°C*, up to 98% RH	≤0.05	≤0.05
Water Immersion, 23°± 2°C*	≤0.05	≤0.05
Heat Aging, 85° ± 2°C*	≤0.05	≤0.05

*Reference temperature = +23°C

Operating Temperature Range

-60°C to +85°C

Dimensional Specifications

Standard Length (km/reel): 2.2 - 50.4*
* Longer spliced lengths available at a premium.

Glass Geometry

Fiber Curl: ≥ 4.0 m radius of curvature
Cladding Diameter: 125.0 ± 1.0 μm
Core-Clad Concentricity: ≤ 0.5 μm
Cladding Non-Circularity: ≤ 1.0%

Defined as: $\left[1 - \frac{\text{Min. Cladding Diameter}}{\text{Max. Cladding Diameter}}\right] \times 100$

Coating Geometry

Coating Diameter: 245 ± 5 μm
Coating-Cladding Concentricity: <12 μm

Mechanical Specifications

Proof Test

The entire fiber length is subjected to a tensile proof stress ≥ 100 kpsi (0.7 GN/m²)*.
* Higher proof test levels available at a premium.

Performance Characterizations
Characterized parameters are typical values.

Core Diameter: 8.2 μm

Numerical Aperture: 0.14

NA is measured at the one percent power level of a one-dimensional far-field scan at 1310 nm.

Zero Dispersion Wavelength (λ_0): 1313 nm

Zero Dispersion Slope (S_0): 0.086 ps /(nm²·km)

Refractive Index Difference: 0.36%

Effective Group Index of Refraction, (N_{eff} @ nominal MFD):

1.4677 at 1310 nm
1.4682 at 1550 nm

Fatigue Resistance Parameter (n_d): 20

Coating Strip Force:

Dry: 0.6 lbs. (3N)
Wet, 14-day room temperature: 0.6 lbs. (3N)

Rayleigh Backscatter Coefficient (for 1 ns pulse width):

1310 nm: -77 dB
1550 nm: -82 dB

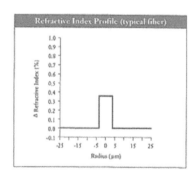

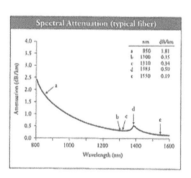

A1.2   ENHANCED STANDARD SINGLE-MODE OPTICAL FIBERS

Corning® SMF-28e™ Optical Fiber
Product Information

PI1344
Issued: April 2001
ISO 9001 Registered

Corning® Single-Mode Optical Fiber

Introducing SMF-28e™ Optical Fiber

Corning® SMF-28e™ single-mode optical fiber continues Corning's long history and performance as a premium fiber supplier. While allowing low water peak attenuation for system installers seeking a low water peak fiber, it meets all requirements for standard single-mode fiber.

Reduced Attenuation

SMF-28e fiber is designed with low water peak attenuation and complies with the requirements of newly adopted U.S. standard, TIA/EIA 492-CAAB, and the international standard, ITU G.652.C. These requirements define standard single-mode fibers for use across a broad wavelength range including the Extended Band (1360 nm-1460 nm).

SMF-28e fiber provides superior attenuation performance throughout the 1260 nm to 1625 nm wavelength range, including a specified attenuation at 1383 nm $\leq$ 0.31 dB/km.

The Impact of Low Water Peak

Typical optical fiber displays an attenuation increase at or about 1383 nm. This "water peak" region (1375-1400 nm) is where light is strongly absorbed by hydroxyl (-OH) ions that are present in the glass core leading to increased attenuation over this wavelength range. SMF-28e fiber has had these ions removed during manufacture, thus reducing the attenuation and leading to a smooth curve across the Extended Band.

SMF-28e fiber boasts superior low water peak performance, providing excellent first-day attenuation as well as unparalleled lifetime performance.

Features And Benefits

- Versatility in the 1310 nm, 1383 nm and 1550 nm windows.
- Low attenuation throughout the Extended Band.
- Characterized out to 1625 nm.
- Longer transmission distance.
- Enables emerging technologies like CWDM and SOA.
- More options for network management.
- Lengths up to 50.4 km/spool.
- Protected by CPC™ coating system.

Optical Specifications

Premium Attenuation

Wavelength (nm)	Attenuation (dB/km)
1310	≤ 0.35
1383	≤ 0.31*
1550	≤ 0.25

*Attenuation increases due to hydrogen aging at this wavelength will be ≤0.01 dB/km and evaluated in accordance with the IEC 60793-2 test procedure.

Point Discontinuity

No point discontinuity greater than 0.10 dB at either 1310 nm or 1550 nm.

Attenuation vs. Wavelength

Range (nm)	Ref. λ (nm)	Max. α Difference (dB/km)
1285 - 1330	1310	0.05
1525 - 1575	1550	0.05

The attenuation in a given wavelength range does not exceed the attenuation of the reference wavelength (λ) by more than the value α.

Attenuation with Bending

Mandrel Diameter (mm)	Number of Turns	Wavelength (nm)	Induced Attenuation* (dB)
32	1	1550	≤0.50
50	100	1310	≤0.05
50	100	1550	≤0.10
75	100	1625	≤0.50

*The induced attenuation due to fiber wrapped around a mandrel of a specified diameter.

Cable Cutoff Wavelength (λ_{ccf})

$\lambda_{ccf} \leq 1260$ nm

Mode-Field Diameter

9.2 ± 0.4 µm at 1310 nm
10.4 ± 0.8 µm at 1550 nm

Dispersion

Zero Dispersion Wavelength (λ_0):
1302 nm $\leq \lambda_0 \leq$ 1322 nm

Zero Dispersion Slope (S_0):
≤ 0.092 ps/(nm²·km)

$$\text{Dispersion} = D(\lambda) \approx \frac{S_0}{4}\left[\lambda - \frac{\lambda_0^4}{\lambda^3}\right] \text{ ps/(nm·km)},$$
$$\text{for } 1200 \text{ nm} \leq \lambda \leq 1625 \text{ nm}$$

λ = Operating Wavelength

Polarization Mode Dispersion

Fiber Polarization Mode Dispersion (PMD)

	Value (ps/√km)
PMD Link Value	≤0.1*
Maximum Individual Fiber	≤0.2

* Complies with IEC SC 86A/WGI, Method 1, September 1997.

The PMD link value is a term used to describe the PMD of concatenated lengths of fiber (also known as the link quadrature average). This value is used to determine a statistical upper limit for system PMD performance.

Individual PMD values may change when cabled. Corning's fiber specification supports network design requirements for a 0.5 ps/√km maximum PMD.

Environmental Specifications

Environmental Test Condition	Induced Attenuation (dB/km)	
	1310 nm	1550 nm
Temperature Dependence -60°C to +85°C*	≤0.05	≤0.05
Temperature Humidity Cycling -10°C to +85°C* up to 98% RH	≤0.05	≤0.05
Water Immersion, 23°± 2°C	≤0.05	≤0.05
Heat Aging, 85°± 2°C*	≤0.05	≤0.05

*Reference temperature = +23°C

Operating Temperature Range

-60°C to +85°C

Dimensional Specifications

Standard Length (km/spool): 2.2 - 50.4

Glass Geometry

Fiber Curl: ≥ 4.0 m radius of curvature
Cladding Diameter: 125.0 ± 1.0 µm
Core-Clad Concentricity: ≤ 0.5 µm
Cladding Non-Circularity: ≤1.0%

Defined as: $\left[1 - \dfrac{\text{Min. Cladding Diameter}}{\text{Max. Cladding Diameter}} \right] \times 100$

Coating Geometry

Coating Diameter: 245 ± 5 µm
Coating-Cladding Concentricity: < 12 µm

Mechanical Specifications

Proof Test

The entire fiber length is subjected to a tensile stress ≥100 kpsi (0.7 GN/m²)*.

*Higher proof test levels available at a premium.

Performance Characterizations

Characterized parameters are typical values.

Core Diameter: 8.2 µm

Numerical Aperture: 0.14

NA is measured at the one percent power level of a one-dimensional far-field scan at 1310 nm.

Zero Dispersion Wavelength (λ_0): 1313 nm

Zero Dispersion Slope (S_0): 0.086 ps/(nm²•km)

Refractive Index Difference: 0.36%

Effective Group Index of Refraction (N_{eff}):

1.4677 at 1310 nm
1.4682 at 1550 nm

Fatigue Resistance Parameter (n_d): 20

Coating Strip Force:

Dry: 0.6 lbs. (3N)
Wet, 14-day room temperature: 0.6 lbs. (3N)

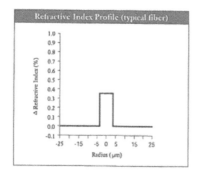

Refractive Index Profile (typical fiber)

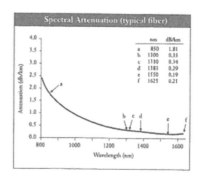

Spectral Attenuation (typical fiber)

nm	dB/km
a 850	1.81
b 1300	0.35
c 1310	0.34
d 1383	0.29
e 1550	0.19
f 1625	0.21

A1.3   LARGE EFFECTIVE AREA FIBER

Corning® LEAF® Optical Fiber
Product Information

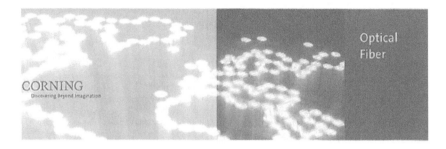

PI1107
Issued: May 2001
Supercedes: April 2001
ISO 9001 Registered

A Powerful Network Needs:
Backbone by LEAF Fiber.

With the ever-accelerating race for bandwidth, network designers are challenged to build a network for the present that will also maximize future technologies. Deploy the fiber that revolutionized network technology and gives you room to move. Break the bandwidth barrier with a fiber so technologically advanced it gives you the optical backbone you need for today's and tomorrow's networks – Corning® LEAF® optical fiber.

Find out what the world's most powerful networks have in common: Backbone by LEAF fiber.

The Large Effective Area Advantage

LEAF fiber's large effective area (A_{eff}) offers higher power-handling capability, improved optical signal-to-noise ratio, longer amplifier spacing, and maximum dense wavelength division multiplexing (DWDM) channel plan flexibility compared with other non-zero dispersion-shifted fibers (NZ-DSFs). Fiber with a large A_{eff} also provides a critical performance advantage – the ability to uniformly reduce all non-linear effects (Figure 1). Non-linear effects represent the greatest performance limitations in today's multi-channel DWDM systems.

The Next Generation

In addition to outperforming other NZ-DSFs in the conventional band (C-Band: 1530-1565 nm), LEAF fiber facilitates the next technological development in fiber-optic networks -- the migration to the long band (L-Band: 1565-1625 nm). In both C-Band and L-Band operation, LEAF fiber has demonstrated greater ability to handle more channels by reducing non-linear effects such as four-wave mixing, self-phase modulation and cross-phase modulation in multi-channel DWDM transmission.

Reduce Network Costs

With its increased optical reach advantage, LEAF fiber requires fewer amplifiers and regenerators, and therefore provides immediate and long-term cost savings. LEAF fiber is also compatible with installed base fibers and photonic components. In fact, LEAF fiber's slightly larger mode-field diameter improves its splicing performance, especially when connecting to standard single-mode fiber such as Corning® SMF-28™ fiber. And, as with all Corning optical fiber, LEAF fiber's geometry package is the best in the industry. With LEAF fiber, it is easy and economical to increase the information-carrying capacity of your network.

Figure 1

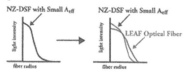

LEAF fiber's larger A$_{eff}$ increases the area where the light can propagate, thereby reducing non-linear effects.

Fiber For Today & Tomorrow

While LEAF fiber is exceptionally suited to operate with already-installed 2.5 Gbps systems, it is techno-economically optimized for today's high-channel-count 10 Gbps systems, and provides the ability to upgrade in the future to tomorrow's high bit systems. Additionally, LEAF fiber's unparalleled specifications on polarization mode dispersion (PMD) allow fiber installed today to operate at data rates higher than 10 Gbps. The combination of LEAF fiber's large A$_{eff}$ and its demonstrated Raman upgradeability allows transmission engineers to design and build networks advantaged over other fiber plants. As the world's most advanced NZ-DSF, LEAF fiber is ready for future technology when your network is.

LEAF Fiber – All About Value

With LEAF fiber's proven large A$_{eff}$ advantage, the industry's best geometry package, and inherent future-proof design, LEAF fiber continues to be the fiber of choice for today's high-capacity and tomorrow's all-optical networks. Network providers on the cutting edge have embraced large A$_{eff}$ technology as the fiber "backbone" for high-data-rate networks now and in the future.

Technology Awards

Corning Incorporated has received multiple industry awards for its patented LEAF optical fiber. Independent panels of experts have chosen LEAF fiber based on its technical merits for the following awards:

> "Annual Technology Award" from Fiberoptic Product News

> "Commercial Technology Achievement Award for Fiber-Optics" from Laser Focus World Magazine

> "Circle of Excellence Award" from Photonics Spectra Magazine

> "R&D 100 Award" from R&D Magazine

Coating

Corning fiber is protected for long-term performance and reliability by the CPC™ coating system. Corning's enhanced, dual acrylate CPC coatings provide excellent fiber protection and are easy to work with. CPC coatings are designed to be mechanically stripped and have an outside diameter of 245 µm. CPC coatings are optimized for use in many single- and multi-fiber cable designs, including loose tube, ribbon, slotted core and tight buffer cables.

Optical Specifications

Attenuation

≤ 0.25 dB/km at 1550 nm

≤ 0.25 dB/km at 1625 nm

> No point discontinuity greater than 0.10 dB at 1550 nm

> Attenuation at 1383 ± 3 nm shall not exceed 1.0 dB/km

Attenuation vs Wavelength		
Range (nm)	Ref. λ (nm)	Max Increase α (dB/km)
1525–1575	1550	0.05
1625	1550	0.05

The attenuation in a given wavelength range does not exceed the attenuation of the reference wavelength (λ) by more than the value α. In all cases, a maximum attenuation of ≤ 0.25 dB/km applies at 1550 nm and 1625 nm.

Attenuation With Bending			
Mandrel Diameter (mm)	Number of Turns	Wavelength (nm)	Induced Attenuation (dB)
32	1	1550 & 1625	≤0.50
75	100	1550 & 1625	≤0.05

The induced attenuation due to fiber wrapped around a mandrel of a specified diameter.

Mode-Field Diameter

9.20 to 10.00 μm at 1550 nm

Dispersion

Total Dispersion: 2.0 to 6.0 psec/(nm•km) over the range 1530 to 1565 nm

4.5 to 11.2 psec/(nm•km) over the range of 1565 to 1625 nm

Fiber Polarization Mode Dispersion (PMD)	
	Value (ps/√km)
PMD Link Design Value	≤0.04*
Maximum Individual Fiber	≤0.1

Complies with IEC SC 86A/WG1, Method 1, September 1997 (n=24, Q=0.1%)

The PMD link design value is a term used to describe the PMD of concatenated lengths of fiber (also known as PMDQ). This value represents a statistical upper limit for total link PMD.

PMD values may change when fiber is cabled. Corning's fiber specification supports emerging network design requirements for high-data-rate systems operating at 10 Gbps (TDM) rates and higher.

Environmental Specifications

Environmental Test Condition	Induced Attenuation (dB/km) 1550 nm
Temperature Dependence -60°C to +85°C*	≤ 0.05
Temperature – Humidity Cycling -10°C to +85°C* and up to 98% RH	≤ 0.05
Water Immersion, 23°C	≤ 0.05
Heat Aging, 85°C*	≤ 0.05

Operating Temperature Range: -60°C to +85°C
Reference Temperature = +23°C

Dimensional Specifications

Standard Length (km/reel) 4.4 - 25.2*
*Longer spliced lengths available at a premium.

Glass Geometry

Fiber Curl: ≥ 4.0 m radius of curvature
Cladding Diameter: 125.0 ± 1.0 μm
Core/Clad Concentricity: ≤ 0.5 μm
Cladding Non-Circularity: ≤ 1.0%

Defined as:

$$\left[1 - \frac{\text{Min. Cladding Diameter}}{\text{Max. Cladding Diameter}} \right] \times 100$$

Coating Geometry

Coating Diameter: 245 ± 5 μm
Coating/Cladding Concentricity: < 12 μm

Mechanical Specifications

Proof Test
The entire length of fiber is subjected to a tensile proof stress ≥ 100 kpsi (0.7 GN/m²)*
*Higher proof test available at a premium.

Performance Characterizations

Characterized parameters are typical values.

Effective Area (A_{eff})
72 μm²
Effective Group Index of Refraction (N_{eff})
1.469 at 1550 nm
Fatigue Resistance Parameter (n_d)
20
Coating Strip Force
Dry, 2.8 N (0.6 lbs)
Wet, 14 days room temperature: 2.7 N (0.6 lbs)

Consistency with Global Standards

The values in this product information sheet demonstrate Corning® LEAF® fiber's conformity with ITU-T Recommendation G.655, IEC 60793-2 for B4 class fibers and Belcore/Telcordia GR-20-CORE.

Dispersion Calculation

$$\text{Dispersion} = D\,(\lambda) = \left(\frac{D(1565\text{ nm}) - D(1530\text{ nm})}{35} \cdot (\lambda - 1565) \right) + D(1565\text{ nm})$$

λ = Operating wavelength up to 1565

$$\text{Dispersion} = D\,(\lambda) = \left(\frac{D(1625\text{ nm}) - D(1565\text{ nm})}{60} \cdot (\lambda - 1625) \right) + D(1625\text{ nm})$$

λ = Operating wavelength from 1565–1625

Special selections of LEAF fiber attributes are available upon request.

Annex 2: Coherent Balanced Receiver and Method for Noise Suppression

It has been shown that a balanced optical receiver can suppress the excess noise intensity generated from the local oscillator [1–3]. In a balanced receiver, an optical coupler is employed to mix a weak optical signal with a local oscillator field of average powers of E_S^2 and E_L^2, respectively. Figure A2.1 shows the generic diagram of a balanced optical receiver under a coherent detection scheme in which two photodetectors and amplifiers are operating in a push–pull mode.

Usually, the magnitude of the optical field of the local oscillator E_L is much greater than E_s, the signal field. The fields at the output of a 2×2 coupler can thus be written as

$$
\begin{bmatrix} \tilde{E}_1 \\ \tilde{E}_2 \end{bmatrix} = \begin{bmatrix} \sqrt{1-k} & \sqrt{k}\,e^{j\pi/2} \\ \sqrt{k}\,e^{j\pi/2} & \sqrt{1-k} \end{bmatrix} \begin{bmatrix} \tilde{E}_S \\ \tilde{E}_L \end{bmatrix}
\tag{A2.1}
$$

where E_s is the field of the received optical signal, E_L is the field of the local oscillator (possibly a distributed feedback laser), and k is the intensity coupling coefficient of the coupler. For a 3-dB coupler, we have $k = 0.5$. The E_s and E_L fields can then be written as

$$
E_S = \sqrt{2}\,\tilde{E}_S\,e^{j(\omega_1 t + \phi_1)}
$$
$$
E_L = \sqrt{2}\,\tilde{E}_L\,e^{j(\omega_2 t + \phi_2)}
\tag{A2.2}
$$

where $\tilde{E}_S, \tilde{E}_L$ are the magnitude of the optical fields of the signal and the local oscillator laser, respectively. The electronic current generated from the PDs corresponding to the average optical power and the signals are given by

$$
i_1(t) = \frac{\Re}{2}|E_1|^2 = \frac{\Re_1}{2}\left\{ \begin{matrix} \frac{1}{2}\left(E_S^2 + E_L^2\right) + \\ E_S E_L \cos\left[(\omega_1 + \omega_2)t + \phi_1 + \phi_2 - \frac{\pi}{2}\right] \end{matrix} \right\} + N_1(t)
\tag{A2.3}
$$

$$
i_2(t) = \frac{\Re}{2}|E_2|^2 = \frac{\Re_2}{2}\left\{ \begin{matrix} \frac{1}{2}\left(E_S^2 + E_L^2\right) \\ + E_S E_L \cos\left[\begin{matrix}(\omega_1 + \omega_2)t \\ + \phi_1 + \phi_2 - \frac{\pi}{2} \end{matrix}\right] \end{matrix} \right\} + N_2(t)
\tag{A2.4}
$$

where ϕ_1, ϕ_2 and ω_1, ω_2 are the phase and frequency of the signal and local oscillator, respectively; $N_1(t)$ and $N_2(t)$ are the noises resulting from the photodetection process; and $\Re_1$, $\Re_2$ are the

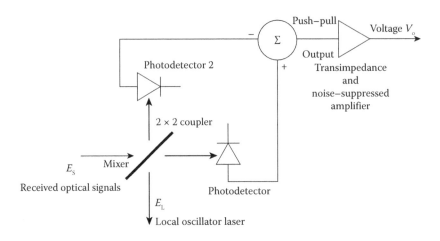

FIGURE A2.1   Generic block diagram of a coherent balanced detector optical receiver.

responsivity of the PDs 1 and 2, respectively. It is assumed that the two photodiodes have the same quantum efficiency, and thence their responsivity. Thus, we have the total current of a back-to-back PD pair, as referred to the input of the electronic amplified, as

$$
i_{\text{Teq}}(t) = \left\{
\begin{array}{l}
I_{\text{dc1}} - I_{\text{dc2}} \\[2ex]
+ (\Re_1 + \Re_2) E_S E_L \cos \left[
\begin{array}{l}
(\omega_1 - \omega_2)t \\[1ex]
+ \phi_1 - \phi_2 - \dfrac{\pi}{2}
\end{array}
\right]
\end{array}
\right\} + N_1(t) + N_2(t)
\qquad (A2.5)
$$

where the first two terms I_{dc1} and I_{dc2} are the detector currents generated by the detector pair due to the reception of the power of the local oscillator, which is a continuous wave source, and thus these terms appear as the DC constant currents that are normally termed as *shot noises* due to the optical power of the local oscillator. The third term is the beating between the local oscillator and the signal carrier, the signal envelope. The noise currents $N_{\text{1eq}}(t)$ and $N_{\text{2eq}}(t)$ are seen as the equivalence at the input of the electronic amplifier from the noise processes that are generated by the PDs and electronic amplifiers. Noise components are the shot noise due to bias currents, the quantum shot noise that is dependent on the strength of the signal and the local oscillator, the thermal noise due to the input resistance of the amplifier, and the equivalent noise current referred to the input of the electronic amplifier contributed by all the noise sources at the output port of the electronic amplifier. We denote $N_{\text{1eq}}(t)$ as the quantum shot noise generated due to the average current produced by the PD pair in reception of the average optical power of the signal sequence.

The difference of the produced electronic currents can be derived using the following techniques: (1) the generated electronic currents of the PDs can be coupled through a 180° microwave coupler and then fed into an electronic amplifier; (2) the currents are fed into a differential electronic amplifier; or (3) a balanced PD pair connected back-to-back and the fed to a small-signal electronic amplifier [4]. The first two techniques require stringent components and are normally not preferred, as contrasted by the high performance of the balanced receiver structure of (3).

Regarding the electronic amplifier and electronic preamplifier, a transimpedance configuration is selected due to its wideband and high dynamics. However, it suffers high noises due to the equivalent input impedance of the shunt feedback impedance, normally around a few hundred ohms. In the next section, the theoretical analyses of noises of the optical balanced receiver is described.

A2.1 ANALYTICAL NOISE EXPRESSIONS

In the electronic preamplifier, the selection of the transistor as the first-stage amplifying device is very critical. Either an FET or a BJT could be used. However, the BJT is preferred for wideband applications due to its robustness to noises. The disadvantages of using the BJT as compared with the FET are due to its small base resistance, which leads to high thermal noise. However, for shunt feedback amplifier, the resistance of the Millers' equivalent resistance as referred to the input of the amplifier is much smaller, and thus dominates over that of the base resistance. The advantage of the BJT over the FET is that its small-signal gain follows an exponential trend with respect to the small variation of the driving current derived from the photodetection, as compared with parabolic for FET. The FET may also offer high input impedance between the gate to source of the input port, but may not offer much improvement in a feedback amplification configuration in term of noises. This section focuses on the use and design of BJT multistage shunt feedback electronic amplifiers.

Noises of electronic amplifying devices can be represented by superimposing all the noise current generators to the small-signal equivalent circuit. These noise generators represent the noises introduced into the circuit by physical sources/processes at different nodes. Each noise generator can be expressed by a noise spectral density current square or power, as shown in Figure A2.2. This figure gives a general model of small-signal equivalent circuit, including noise current sources of any

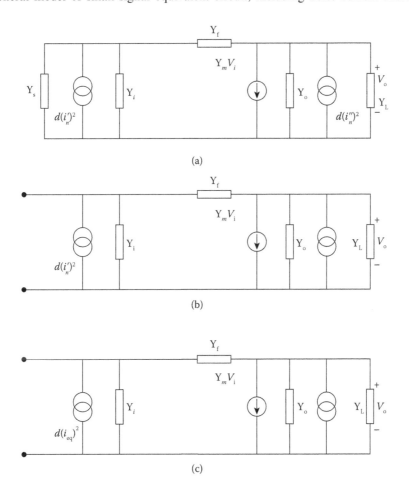

FIGURE A2.2 Small equivalent circuits including noise sources: (a) Y-parameter model representing the ideal current model and all current noise sources at the input and output ports, (b) with noise sources at input and output ports, and (c) with a total equivalent noise source at the input.

transistor, which can be represented by the transfer $[y]$ matrix parameters. Indeed, the contribution of the noise sources of the first stage of the electronic preamplifier is dominant, plus that of the shunt feedback resistance.

The current generator $\dfrac{di_n'^2}{df}$ is the total equivalent noise generator contributed by all the noise sources to node 1, including the thermal noises of all resistors connected to the node. Similarly, for the noise source, $\dfrac{di_n''^2}{df}$ is referred to the output node.

A2.2 NOISE GENERATORS

Electrical shot noises are generated by the random generation of streams of electrons (current). In optical detection, shot noises are generated by (1) biasing currents in electronic devices and (2) photo currents generated by the photodiode.

A biasing current I generates a shot noise power spectral density S_I, given by

$$S_I = \frac{d(i_I^2)}{df} = 2qI \quad A^2/Hz \tag{A2.6}$$

with q as the electronic charge. The quantum shot noise $< i_s^2 >$ generated by the PD by an optical signal with an average optical power Pin is given by

$$S_Q = \frac{d\left\langle i_s^2 \right\rangle}{df} = 2q\left\langle i_s^2 \right\rangle \quad A^2/Hz \tag{A2.7}$$

If the PD is an APD type, then the noise spectral density is given by

$$S_Q = \frac{d\left\langle i_s^2 \right\rangle}{df} = 2q\left\langle i_s^2 \right\rangle\left\langle G_n^2 \right\rangle \quad A^2/Hz \tag{A2.8}$$

It is noted here again that the dark currents generated by the PD must be included with the total equivalent noise current at the input after it is evaluated. These currents are generated even in the absence of the optical signal. These dark currents can be eliminated by cooling the PD to at least below the temperature of liquid nitrogen (77°K).

At a certain temperature, the conductivity of a conductor varies randomly. The random movement of electrons generates a fluctuating current, even in the absence of an applied voltage. The thermal noise spectral density of a resistor R is given by

$$S_R = \frac{d\left(i_R^2\right)}{df} = \frac{4k_B T}{R} \quad \text{in } A^2/Hz \tag{A2.9}$$

where K_B is the Boltzmann constant, T is the absolute temperature (in °K), R is the resistance in ohms, and i_R denotes the noise current due to resistor R.

A2.3 EQUIVALENT INPUT NOISE CURRENT

Our goal is to obtain an analytical expression of the noise spectral density equivalent to a source looking into the electronic amplifier, including the quantum shot noises of the PD. A general method for deriving the equivalent noise current at the input is given by representing the electronic device by a Y-equivalent linear network, as shown in Figure A2.3. The two current noise sources, $di_n'^2$ and $di_n''^2$, represent the summation of all noise currents at the input and output of the Y-network. This can be transformed into a Y-network circuit with the noise current referred to the input as follows:

The output voltages V_0 of Figure A2.2a can be written as

$$V_0 = \frac{i_N'(Y_f - Y_m) + i_N''(Y_i + Y_f)}{Y_f(Y_m + Y_i + Y_0 + Y_L) + Y_i(Y_0 + Y_L)} \tag{A2.10}$$

and for Figure A2.2b

$$V_0 = \frac{(i_N')_{eq}(Y_f - Y_m)}{Y_f(Y_m + Y_i + Y_o + Y_L) + Y_i(Y_o + Y_L)} \tag{A2.11}$$

Thus, comparing the preceding two equations, we can deduce the equivalent noise current at the input of the detector as

$$i_{Neq} = i_N' + i_N'' \frac{Y_i + Y_f}{Y_f - Y_m} \tag{A2.12}$$

Then, reverting to mean square generators for a noise source, we have

$$d(i_{Neq})^2 = d(i_N')^2 + d(i_N'')^2 \left| \frac{Y_i + Y_f}{Y_f - Y_m} \right|^2 \tag{A2.13}$$

It is therefore expected that, if the Y-matrix of the front-end section of the amplifier is known, the equivalent noise at the input of the amplifier can be obtained by using Equation A2.13.

We propose a three-stage electronic preamplifier in AC configuration, as shown in Figure A2.3. The details of the design of this amplifier are given in the next section. The small-signal and associated noise sources of this amplifier are given in Section A2.2. As can be seen, this general configuration is a forward path of the series and shut stages which reduce the interaction between stages due to the impedance matching levels [5]. Shunt-shunt feedback is placed around the forward path, and hence stable transfer function is the transfer impedance that is important for transferring the generated electronic photodetected current to the voltage output for further amplification and data recovery.

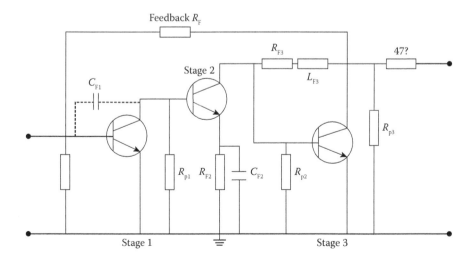

FIGURE A2.3 AC circuit model of a three-stage-feedback electronic preamplifier.

For a given source, the input noise current power of a BJT front end can be found by

$$i_{\text{Neq}}^2 = \int_0^B d(i_{\text{Neq}})^2 = a + \frac{b}{r_E} + c r_E \qquad (A2.14)$$

where B is the bandwidth of the electronic preamplifier and r_E is the emitter resistance of the front-end transistor of the preamplifier. The parameters a, b, and c are dependent on the circuit elements and amplifier bandwidth. Hence, an optimum value of r_E can be found, and an optimum biasing current can be set for the collector current of the BJT, such that i_{Neq}^2 is at its minimum, as

$$r_{\text{Eopt}} = \sqrt{\frac{b}{c}} \quad \text{hence} \quad \rightarrow i_{\text{Neq}}^2\Big|_{r_E = r_{\text{Eopt}}} = a + 2\sqrt{bc} \qquad (A2.15)$$

If two types of BJT are considered, Phillips BFR90A and BFT24, then a good approximation of the equivalent noise power can be found as

$$a = \frac{8\pi B^3}{3}\left\{r_B C_s^2 + (C_s + C_f + C_{tE})\tau_T\right\}$$

$$b = \frac{B}{\beta_N} \quad \text{and} \qquad (A2.16)$$

$$c = \frac{4\pi B^3}{3}(C_s + C_f + C_{tE})^2$$

The theoretical estimation of the parameters of the transistors can be derived, from the measured scattering parameters as given by the manufacturer [6], as

For   transistor   type $BFR90A$

$r_{\text{Eopt}} = 59\,\Omega$ for $I_{\text{Eopt}} = 0.44\,\text{mA}$

$I_{\text{eq}}^2 = 7.3 \times 10^{-16}\,A^2$ hence $\dfrac{I_{\text{eq}}^2}{B} = 4.9 \times 10^{-24}\,A^2/\text{Hz}$ (A2.17)

$I_{\text{eq}} = 27\,\text{nA}$ thus $\dfrac{I_{\text{eq}}}{\sqrt{B}} = 2.21\,\text{pA}/\sqrt{\text{Hz}}$

For   transistor   type $BFT24$

$r_{\text{Eopt}} = 104\,\Omega$ for $I_{\text{Eopt}} = 0.24\,\text{mA}$

$I_{\text{eq}}^2 = 79.2 \times 10^{-16}\,A^2$ hence $\dfrac{I_{\text{eq}}^2}{B} = 6.1 \times 10^{-24}\,A^2/\text{Hz}$ (A2.18)

$I_{\text{eq}} = 30.2\,\text{nA}$ thus $\dfrac{I_{\text{eq}}}{\sqrt{B}} = 2.47\,\text{pA}/\sqrt{\text{Hz}}$

Note that the equivalent noise current depends largely on some not-well-defined values such as the capacitance, transit times, base-spreading resistance, and short-circuit current gain β_N. The term $\dfrac{I_{\text{eq}}}{\sqrt{B}}$ is usually specified as the equivalent noise spectral density referred to the input of the electronic preamplifier.

A2.4   POLE-ZERO PATTERN AND DYNAMICS

An AC model of a three-stage electronic preamplifier is shown in Figure A2.3, and the design circuit is shown in Figure A2.4. As briefly mentioned earlier, there are three stages with feedback impedance from the output to the input. The subscripts of the resistors, capacitors, and inductors indicate the order of the stages. The first stage is a special structure of shunt feedback amplification in which the shunt resistance is increased to infinity. The shunt resistance is in the order of hundreds of ohms for the required bandwidth, thus contributing to the noise of the amplifier, which is not acceptable. The shunt resistance increases the pole of this stage and approaches the origin. The magnitude of this pole is reduced by the same amount as that of the forward path gain. Thus, the poles of the close loop amplifier remain virtually unchanged. As R_F is increased, the pole p_1 decreases, but G_1 is increased. Hence, $\dfrac{G_1 p_1}{s - p_1}$ remains constant. Thus, the position of the root locus is almost unchanged.

A compensating technique for reducing the bandwidth of the amplifier is to add capacitance across the base-collector of the first stage. This may be necessary if oscillation occurs, due to the phase shift becoming unacceptable at $GH = 1$, where G is the open-loop gain, and H is the feedback transfer function.

The second stage is a series feedback stage with feedback peaking. The capacitance C_{F2} is chosen such that, at high frequencies, it begins to bypass the feedback resistor. Thence, the feedback admittance partially compensates for the normal high-frequency drop in gain associated with the base and stray capacitances. Also, if the capacitance is chosen such that $R_{F2}C_{F2} = \tau_r$, the transfer admittance and input impedance become single pole, which would be desirable [7]. The first and second stages are direct coupled (Figure A2.5).

The third stage uses the inductive peaking technique. For a shunt stage with a resistive load, the forward path gain has only one pole, and hence there is only one real pole in the closed loop response. A complex pole pair can be obtained by placing a zero in Z_{F2}, and hence a pole in the feedback transfer function $H(\omega)$. Figure A2.6a through c shows the high-frequency singularity pattern of individual stages. Figure A2.6d shows the root locus diagram. It can be calculated that the poles take up the positions when the loop gain GH is 220.

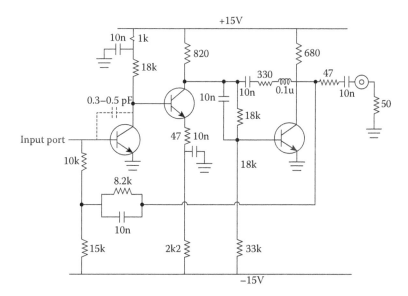

FIGURE A2.4   Design circuit of the electronic preamplifier for the balanced optical amplifier.

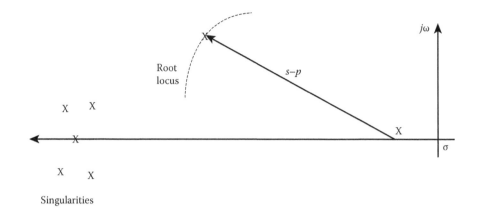

Singularities

FIGURE A2.5 Effect of first-stage pole on root locus of a shunt feedback electronic amplifier. The root locus is given by $1 = (GH)_{\text{mid-band}} \sum \dfrac{s - z_i}{z_i} \sum \dfrac{p_i}{s - p_i}$.

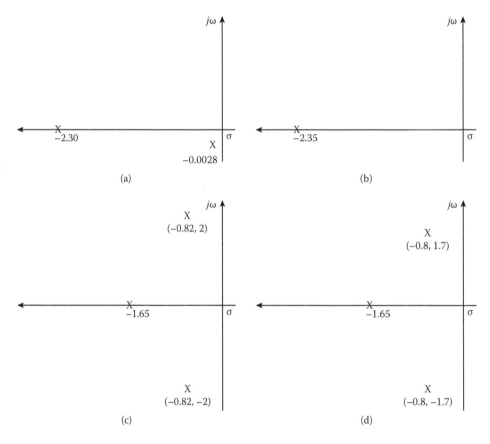

FIGURE A2.6 High-frequency root locus diagram: (a) singularity pattern for first stage, (b) second stage, (c) third stage, and (d) root locus for close loop pattern for GH = 220.

In the low-frequency region, the loop gain has five poles and three zeroes on the negative axis, and two zeroes at the origin. The largest low-frequency pole is set at approximately the input coupling capacitor, and the input resistance leading to a low-frequency cutoff of around tens of kHz.

The singularities of individual stages can be approximated by

Stage 1: The dominant pole at

$$p = -\frac{1}{\beta_N R_{L1} C_{F1}}$$ (A2.19)

and the other pole located at

$$p = -\frac{C_{F1}}{C_{F1} + C_{L1}} \frac{1}{\tau_T + C_{F1}/g_m}$$ (A2.20)

and the midband gain = $\beta_N R_{L1}$ with the load of stage 1 of R_{L1}

Stage 2: The dominant pole at

$$p = -\frac{1}{r_B G_{T2} \tau_T} \text{ ; and the midband gain} = G_{T2} = 1/R_{F2}$$ (A2.21)

Stage 3: Complex pole pair at

$$|p|^2 = -\frac{R_L}{\tau_T^2 R_{F3}} ; \sigma = -\frac{(R_L + R_{F3})}{2\tau_F R_{F3}} ; \tau_F = \frac{L_{F3}}{R_{F3}} ; \text{ and a zero at } -\frac{1}{\tau_F} ; \text{ with a gain} = -R_{F3}$$ (A2.22)

The feedback configuration of the circuit of the transimpedance electronic preamplifier has a 10 or 15 K resistor whose noises would contribute to the total noise current of the amplifier. The first and second stages are direct coupled, hence eliminating the level shifter and a significant amount of stray capacitance. The peaking capacitor required for the second stage is on the order of 0.5 pF, and hence no discrete component can be used. This may not require any component, as the stay capacitance may suffice.

The equivalent noise current at the input of the transimpedance amplifier (TIA) is approximately proportional to the square of the capacitance at the base of the transistor. Therefore, minimization of the stray capacitance at this point must be conducted by shortening the connection lead as much as possible. As a guide, the critical points of the circuit lies on the stray capacitances which must be minimized, and the grounding for smallest capacitances is necessary. For instance, at the base of the output stage, the capacitance to the ground is more tolerable at this point (and therefore at the collector to the ground of the second transistor), than say the collector of the last stage, which should ideally have no capacitance to ground.

An acceptable step response can be achieved by manipulating the values of changing C_{F1}, R_{F3}, L_{F3}. Since C_{F1} contributes to the capacitance at the base of the first stage, C_{F1} is minimized. The final value of C_{F1} is 0.5 pF (about half twist of two wires, R_{F1} is 330 ohms, and L_{F1} is about 0.1 μF (2 cm of wire). The amplifier is sensitive to parasitic capacitance between the feedback resistor and the first two stages. Thus, a grounded shield is placed between the 10 K resistor and the first and second stages.

The expressions for the singularities for the first two stages are similar, as described earlier. The base-collector capacitance of the third stage affected the position of the poles considerably. The poles of the singularities of this stage are as follows: the midband transresistance is $-R_{F3}$, with $zero @ z = \frac{1}{\tau_F}$, and a complex pole pair is given by

$$|p_p|^2 = \frac{1}{L_F C_F + R_F C_F \tau_T + \tau_T \tau_F \frac{R_F}{R_L}} ; \sigma = \frac{1}{2}|p_p|^2 \left(R_F C_F + \tau_T \left(1 + \frac{R_F}{R_L} \right) \right)$$ (A2.23)

In addition, a large pole is given approximately at

$$p = -\frac{1}{L_F C_F \tau_T \left| p_p \right|^2} \tag{A2.24}$$

Shown in Figure A2.7 are the open-loop singularities in (a), the root locus diagram in (b), and the close loop in (c). The two large poles can be ignored without any significant difference. Similarly, for the root locus diagram, the movement of the large poles is negligible, the pole and zero pair can

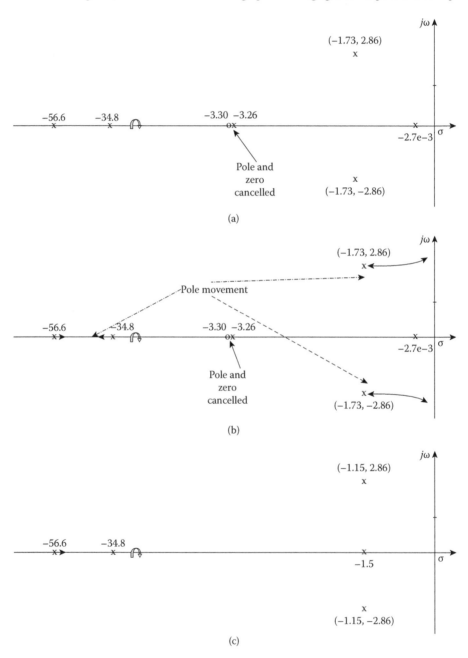

FIGURE A2.7   Pole-zero pattern of the transimpedance amplifier given in Figure A2.4 (a) original pattern, (b) movements of poles and zeros due to feedback, and (c) final stable pattern.

be ignored, and the movement of the remaining three poles can be calculated by considering just these three poles as the open-loop singularities.

A2.5   RESPONSE AND NOISE MEASUREMENTS

A2.5.1   RISE TIME AND 3 dB BANDWIDTH

Based on the pole-zero patterns given in the previous section, the step responses can be estimated and contrasted with the measured curve, as shown in Figure A2.8. The experimental setup for the rise time measurement is shown in Figure A2.9, in the electrical domain without using a photodetector. An artificial current source is implemented using a series resistor with minimum stray capacitance. This testing in the electrical domain is preferred over the optical technique as the rise time of an electrical time-domain reflectometer (TDR) is sufficiently short, so that it does not influence the measurement of the rise time. If the optical method is used, the principal problem is that we must be able to modulate the intensity of the lightwaves with a very sharp step function.

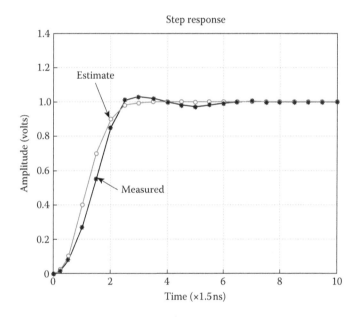

FIGURE A2.8   Step response of the amplifier: o = estimated, * = measured.

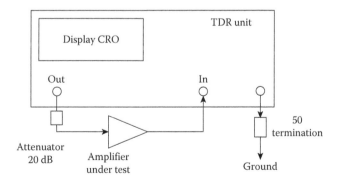

FIGURE A2.9   Experimental setup for measurement of the rise time in the electrical domain using a time-domain reflectometer (TDR).

This is not possible when the Mach–Zehnder intensity modulator (MZIM) is used, as its transfer characteristics follow a square of a cosine function. A very short pulse sequence laser such as the mode-locked fiber laser can be used. However, this optical high speed pulse is not employed in this work because the bandwidth of our amplifier is not that very wide, and the TDR measurement is sufficient to give us reasonable measurement value of the expected rise time.

The passband of the TIA is also confirmed with the measurement of the frequency response by using a scattering parameter test set HP 85046A. The scattering parameter S_{21} frequency response is measured, and the 3 dB passband is confirmed at 190 MHz.

A2.5.2 Noise Measurement and Suppression

Two pig-tailed PDs are used and mounted with back-to-back configuration at the input of the transistor of stage 1. A spectrum analyzer is used to measure the noise of the amplifier shown in Figure A2.10. The background noise of the analyzer was measured to be −88 dBm. The expected noise referred to the input is 9.87×10^{-24} A²/Hz. When the amplifier is connected to the spectrum analyzer, this noise level is increased to −85 dBm, which indicates that the noise at the output of the amplifier is around −88 dBm.

Since this power is measured into 50 Ω, using a spectral bandwidth of 300 kHz, the input current noise can be estimated as

$$\int di_{\text{Neq}}^2 = \frac{v_N^2}{R_T^2} = \frac{10^{-11} \times 50}{(5k)^2} = 2 \times 10^{-17.5}\,\text{A}^2 \rightarrow \frac{di_{\text{Neq}}^2}{df} = \frac{2e^{-17.5}\text{A}^2}{3 \times 10^5} = 1.06 \times 10^{-23}\,\text{A}^2/\text{Hz} \qquad (\text{A2.25})$$

Thus, the measured noise is very close to the expected value.

A2.5.3 Requirement for Quantum Limit

The noise required for near-quantum-limit operation can be estimated. The total shot noise referred to the input is

$$i_{\text{T–shot}}^2 = (\Re_1 + \Re_2)|E_L|^2 \qquad (\text{A2.26})$$

where $\Re_{1,2}$ is the responsivity of the photodetectors 1 and 2, respectively. The total excess noise referred to the input is

$$i_{\text{NT–shot}}^2 = (\Re_1 + \Re_2)|E_L|^2 ; i_{\text{N–excess}}^2 = 2q\gamma(\omega)(\Re_1 + \Re_2)^2|E_L|^4 \qquad (\text{A2.27})$$

The excess noise from each detector is correlated, so γ is a function of frequency, typically between 10^4 and 10^{10} A⁻¹. A receiver would operate within 1 dB of the quantum limit if the shot

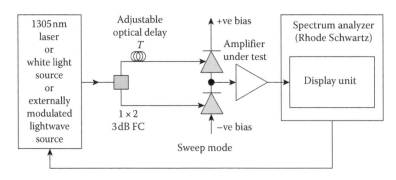

FIGURE A2.10 Experimental setup to measure excess noise impact and cancellation of a balanced receiver. FC = fiber coupler.

noise is about 6 dB above the excess noise and the amplifier noise. The amplifier noise power spectral density is 6.09×10^{-14} A^2/Hz. Assuming that a complete cancellation of the excess noise can be achieved, the local oscillator required power is given as

$$i^2_{\text{NT-shot}} > 4i^2_{\text{NTeq}} \rightarrow \quad |E_\text{L}|^2 > 0.24\,\text{mW} \approx -6.2\,\text{dBm} \tag{A2.28}$$

Since there is bound to be excess noise due to imperfection of the balancing of the two detectors, the power of the local oscillator is expected to be slightly greater than this value.

The output voltage of the amplifier can be monitored so noises contributed by different processes can be measured. When the detectors are not illuminated, we have

$$v_0^2 = Z_\text{T}^2 i^2_{\text{Neq}} \tag{A2.29}$$

where Z_T is the transimpedance, and v_o is the output voltage of the amplifier. It is known that incoherent white light is free from excess intensity noise, so when the detectors are illuminated with white light sources, we have

$$v_0^2 = Z_\text{T}^2 \left(i^2_{\text{Neq}} + i^2_{\text{N-shot}} \right) \tag{A2.30}$$

When the local oscillator is turned on illuminating the detector pair, then

$$v_0^2 = Z_\text{T}^2 \left(i^2_{\text{Neq}} + i^2_{\text{N-shot}} + i^2_{\text{N-excess}} \right) \tag{A2.31}$$

A2.5.4   EXCESS NOISE CANCELLATION TECHNIQUE

The presence of uncancelled excess noise can be shown by observing the output voltage of the electronic preamplifier using a spectrum analyzer. First, the detectors are illuminated with incoherent white light, and the output noise is measured. Then the detectors are illuminated with the laser local oscillator at a power level equal to that of the white light source, and thus the shot noise would be at the same magnitude. Thus, the increase in the output noise will be due to the uncancelled intensity noise. This, however, does not demonstrate the cancellation of the local oscillator intensity noise.

A well-known method for this cancellation is to bias the local oscillator just above its threshold [3]. This causes the relaxation resonance frequency to move over the passband. Thus, when the outputs of the single or balanced receivers are compared, the resonance peak is observed for the former case, and not in the latter case in which the peak is suppressed. Since the current is only in the μA scale in each detector for a local oscillator operating in the region near the threshold, the dominant noise is that of the electronic amplifier. This method is thus not suitable here. However, the relaxation resonance frequency is typically between 1 and 10 GHz. A better method can be developed by modulating the local oscillator with a sinusoidal source and by measuring the difference between single and dual photodetection receivers. The modulation can be implemented at either one of the frequencies to observe the cancellation of the noise peak power, or by sweeping the local oscillator over the bandwidth of the amplifier, so that the cancellation of the amplifier can be measured. For fair comparison, the local oscillator power would be twice that of the single-detector case, so the same received power of both cases can be almost identical, and hence the signal-to-noise ratio (SNR) at the output of the amplifier can be derived and compared.

A2.5.5   EXCESS NOISE MEASUREMENT

An indication of excess noise and its cancellation within the 0–200 MHz frequency range by the experimental setup can be observed as shown in Figure A2.10. A laser of 1305 nm wavelength with an output power of 2dBm is employed as the local oscillator, so that the dominant noise is that of the electronic transimpedance preamplifier, as explained earlier. The local oscillator is intensity modulated by the sweep generator from the spectrum analyzer of 3 GHz frequency range (Rhode Schwartz model). Noise suppression pattern of the same trend can be observed with suppression about every 28 MHz interval over 200 MHz, the entire range of the TIA.

Figure A2.11 shows the noise spectral density for the case of single and dual detectors. Note that the optical paths have different lengths. This path difference is used for cancellation of the excess noise, as described in Ref. [5,8]. It can be shown that the optical delay line leads to a cancellation of noise following a relationship of $\{1-\cos(2\pi fT)\}$, with T as the delay time of the optical path of one of the fibers from the fiber coupler to the detector. Hence, the maxima of the cancellation occurs at $f = m/T$, where m is an integer. From Figure A2.10, we can deduce the optical path length as $f = 30$ MHz or $T = 31.7$ ns, and thus $d = cT/n_{eff} = 6.8$ m with n_{eff} (~1.51 @ 1550 nm) is the effective refractive index of the guided mode in the fiber. The total length of the fiber delay is equivalent to about 6.95 m, in which the effective refractive index of the propagation mode is estimated to be 1.482 at the measured wavelength. Thus, there is a small discrepancy, which accounts for the uncertainty of the exact value of the effective refractive index of the fiber.

Using this method, Abbas et al. [1] achieved a cancellation of 3.5 dB, while 8 dB is obtained in our setup. This discrepancy can be accounted for by a mismatch combination of the photodetector pair which can be tuned to be identical to improve the sensitivity of the receiver.

In our initial measurements, we have observed some discrepancies, which are due to (1) unmatched properties of the photodetection of the balanced detector pair; and (2) the delay path has three extra optical connections, and, therefore, more loss, so that even when there is no delay, the excess noise would not be completely cancelled. Thus, at some stage, the longer path was implemented with a tunable optical delay path so as to match the delay path to the null frequency of the RF wave to achieve cancellation of the excess noise. This noise cancellation can lead to an improvement of 16 dB in the optical signal-to-noise ratio (OSNR) due to the dual-detector configuration, and thus an improvement of near 100 km length of standard single-mode fiber.

The noise cancellation is periodic, following a relationship with respect to the delay time T imposed by the optical path of the interferometer T as $(1-\cos 2\pi fT)$. The optical delay path difference can be minimized by adjusting the optical delay line which can be implemented by a fiber path coupled out of a fiber end to open air path and coupled back to another fiber path via a pair of Selfoc lenses. The air path can be changed to adjust the delay time.

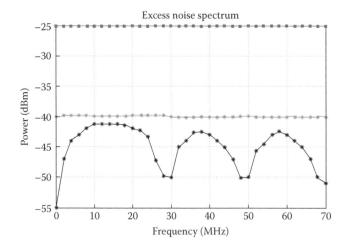

FIGURE A2.11   Excess noise for single-detector and dual detectors with noise cancellation using delay and filter at the input port of the preamplifier.

It is noted that the local oscillator light source can be directly modulated (DD) with a sweep sinusoidal signal derived from the electrical spectrum analyzer (ESA). This light source has also been externally modulated using an MZIM of bandwidth much larger than that of the electronic amplifier (about 25 GHz). The excess noise characteristics obtained using both types of sources are almost identical. Furthermore, with the excess noise reduction of 8 dB, and the signal power for the balanced receiver operating in a push–pull mode, the received electrical current would be double that of a single-detector receiver. That means a gain of 3 dB in the SNR by this balanced receiver.

A2.6 REMARKS

In this annex, we consider the design and implementation of a balanced receiver using a dual-detector configuration and a BJT front-end preamplifier. In view of the recent growing interest in coherent optical transmission systems with electronic digital signal processing [5,9,10], a noise-suppressed balanced receiver operating in the multi-GHz bandwidth range would be essential. Noise cancellation using optical delay line in one of the detection paths leads to suppression of excess noise in the receiver.

In this section, we demonstrate, as an example, by design analysis and implementation, a discrete wideband optical amplifier with a bandwidth of around 190 MHz and a total input noise equivalent spectral density of $10^{-23}\,A^2/Hz$. This agrees well with the predicted value using a noise model analysis. It is shown that, with sensible construction of the discrete amplifier, minimizing the effects of stay capacitances and appropriate application of compensation techniques at a bandwidth of 190 MHz can be easily obtained for transistors with a few GHz transition frequencies. Further, by using an optimum emitter current for the first stage and minimizing the biasing and feedback resistance and capacitance at the input, a low-noise amplifier with an equivalent noise current of about 1.0 pA/$\sqrt{Hz}$ is achieved. Furthermore, the excess noise cancellation property of the receiver is found to give a maximum SNR of 8 dB with a matching of the two photodetectors.

Although the bandwidth of the electronic amplifier as reported here, is only 190 MHz, which is about 1% of the 100 Gbps target bit rate for modern optical Ethernet, the bandwidth of our amplifier has reached its optimum value. The amplifier is constructed using discrete transistor stages whose transition frequency is only 5 GHz. The amplifier configuration reported here can be scaled up to the multi-GHz region for a 100 Gbps receiver using an integrated electronic amplification device without much difficulty. We thus believe that the design procedures for determining the pole and zero patterns on the s-plane and their dynamics for stability consideration are essential. Therefore the designed circuit can be adopted for wideband systems by either integrated Si-Ge microelectronic technology can use. When the amplifier is implemented in the multi-GHz range, then the microwave design technique employing the noise figure method as described in Chapter 5 may be most useful when incorporating the design methodology reported in this section.

A2.7 NOISE EQUATIONS

Referring to the small-signal and noise model given in Figures A2.12 and A2.13, the spectral density of noises due to the collector bias current I_c, small-signal transconductance g_m, and base conductance g_B generated in a BJT can be expressed as

$$di_1^2 = 4k_B T g_B df; di_2^2 = 2q I_c df + 2k_B T g_m df$$

$$di_3^2 = 2q I_B df + 2k_B T (1-\alpha_N) g_m df \quad \text{and} \tag{A2.32}$$

$$di_4^2 = \sum \text{shot noise of diodes and thermal noise of bias in } g_B \text{ resistors}$$

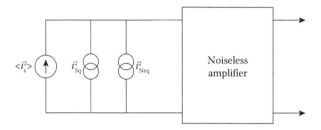

FIGURE A2.12   Equivalent noise current at the input and noiseless amplifier model; i_{Sq}^2 is the quantum shot noise, which is signal dependent, and i_{Neq}^2 is the total equivalent noise current referred to the input of the electronic amplifier.

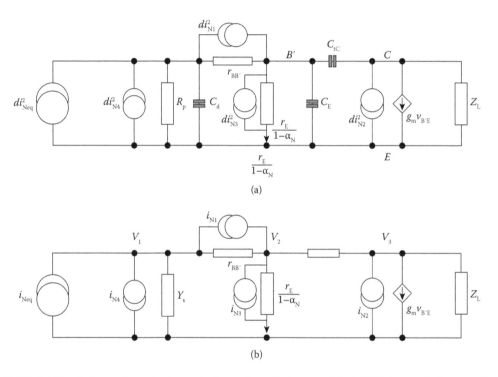

FIGURE A2.13   (a) Approximated noise equivalent and small-signal model of a BJT front end; (b) generalized noise and small-signal model circuit. Note that $r_B = r_{sd} + r_{BB'}$ and $C_d = C_p + C_i$ with r_{BB} is the base resistance, r_{sd} is the diode resistance, C_d is the photodiode capacitance, and C_i is the input capacitance.

From nodal analysis of the small-signal equivalent circuit given in Figure A2.13, we can obtain the relationship

$$\begin{bmatrix} Y_s + g_B & -g_B & 0 \\ -g_B & g_B + y_l + y_f & -y_f \\ 0 & g_m - y_f & y_f + y_2 \end{bmatrix} \begin{bmatrix} V_1 \\ V_2 \\ V_3 \end{bmatrix} = \begin{bmatrix} i_{eq} - i_{N1} \\ i_{N1} + i_{N3} \\ i_{N2} \end{bmatrix} \qquad (A2.33)$$

Hence, V_3, V_2, V_1, can be found by using Euler's rule for the matrix relationship

$$V_3 = \frac{\Delta_{13}}{\Delta}(i_{eq} - i_{N1}) + \frac{\Delta_{23}}{\Delta}(i_{N1} + i_{N3}) + \frac{\Delta_{33}}{\Delta}(i_{N2}) \qquad (A2.34)$$

Hence, the noise currents as referred to the input are

$$d(i_1'^2) = \left| \frac{Y_s}{g_B} \right|^2 dI_1^2 = \omega_2 C_s^2 r_B 4kT df \tag{A2.35}$$

$$d(i_2'^2) = \left| (Y + y_1 + y_f) + \frac{Y_s}{g_B}(y_1 + y_f) \right|^2 \left| \frac{1}{y_f - g_m} \right|^2 = \left| \frac{1}{y_f - g_m} \right|^2 dI_2^2$$

$$= \left(\frac{1}{\beta_N r_E^2} + \omega^2 \left(C_0^2 - \frac{2C_B r_B}{\beta_N r_E} \right) + \omega^4 2C_s r_B C_B^2 \right) 2kT r_E df \tag{A2.36}$$

and

$$d(i_3'^2) = \left| \frac{Y_s}{g_B} + 1 \right|^2 dI_3^2 = \left(\omega \ C_s^2 r_B^2 + 1 \right) \frac{2kT}{\beta r_E} df \tag{A2.37}$$

In which we have assumed that

$$\omega C_f \ll g_m$$

$$C_0 = C_s + C_E + C_f + \frac{C_s r_s}{\beta r_E} \approx C_s + C_{tE} + C_f + \frac{\tau_t}{r_E} = C_x + \frac{\tau_t}{r_E} \tag{A2.38}$$

with $\quad C_x = C_E + C_f; \quad C_E = C_{tE} + \dfrac{\tau_t}{r_E}$

The total noise spectral density referred to the input of the amplifier as shown in Figures A2.12 and A2.13 are thus given as

$$\frac{di_{Neq}^2}{df^2} = \frac{di_{1eq}^2}{df^2} + \frac{di_{2eq}^2}{df^2} + \frac{di_{3eq}^2}{df^2} = = 2k_B T \left[\frac{1}{\beta r_E} \frac{\beta+1}{\beta} + \omega^2 \left(\frac{2}{r_B} + \frac{1}{\beta r_E} \right) + C_0 r_E - \frac{2C_s r_B C_x}{\beta} + \omega^2 C_s^2 r_B^2 C_x^2 r_E \right] \tag{A2.39}$$

Thence, the total noise power referred to the input is given as

$$i_{Neq}^2 = \int_0^B di_{Neq}^2 = a + \frac{b}{r_E} + c r_E; \text{ with}$$

$$a \approx 2k_B T \left[\frac{8\pi^2 B^3}{3} \left(C_s r_B + C_x \tau_t + \frac{C_s^2 r_B^2}{\beta} \right) + \frac{(2\pi)^4}{5} B^5 C_s^2 r_B^2 2C_y \tau_t \right]$$

$$b \approx 2k_B T \left[\frac{B}{\beta} + \frac{4\pi^2 B^3}{3} \left(\frac{C_s r_B^2}{\beta} + \tau_t^2 \right) + \frac{16\pi^4 B^5}{5} C_s^2 r_B^2 \tau_t^2 \right] \tag{A2.40}$$

$$c \approx 2k_B T \left[\frac{4\pi^2 B^3 C_x^2}{3} + \frac{16\pi^4 B^5}{5} C_s^2 r_B^2 C_y^2 \right] \text{ with } \quad C_y = C_{tE} + \frac{\tau_t}{r_E}$$

where the first term in each coefficient a, b, and c is dominant for a bandwidth of 190 MHz of the two transistor types Phillip *BFR90A* and *BFT24* under consideration.

If the dependence of C_E on I_E is ignored, then Equation A2.40 becomes

$$i_{Neq}^2 = 2k_B T \left[\frac{1}{r_E} \frac{B}{\beta} + r_E \left(\frac{4\pi^2 B^3}{3} C_0^2 \right) + \frac{16\pi^4 B^5}{5} C_s^2 r_B^2 C_x^2 + \frac{8\pi^2 B^3 C_s^2 r_B}{3} \right] \tag{A2.41}$$

For the transistor BFR90A, the optimum emitter resistance is $r_{Eopt} = 49 \, \Omega$ and the total equivalent noise power referred to the input is $4.3 \times 10^{-16} \, A^2$, which is moderately high.

Clearly, the noise power is a cubic dependence on the bandwidth of the amplifier. This is thus very critical for an ultra-wideband optical receiver. Thus, it is very important to suppress the excess noise of the optical receiver due to the quantum shot noise of the local oscillator.

REFERENCES

1. G. L. Abbas, V. Chan, and T. K. Yee, A dual-detector optical heterodyne receiver for local oscillator noise suppression, *IEEE J. Lightwave Tech.*, Vol. LT_3, No. 5, pp. 1110–1122, 1985.
2. B. Kasper, C. Burns, J. Talman, and K. Hall, Balanced dual detector receiver for optical heterodyne communication at Gb/s rate, *Elect. Lett.*, Vol. 22, No. 8, pp. 413–414, 1986.
3. S. Alexander, Design of wide-band optical heterodyne balanced mixer receivers, *IEEE J. Lightwave Tech.*, Vol. LT-5, No. 4, pp. 523–537, 1987.
4. L. N. Binh, *Digital Optical Communications*, Boca Raton, FL: CRC Press, 2008, Chapter 4.
5. R. Sterlin, R. Battiig, P. Henchoz, and H. Weber, Excess noise suppression in a fiber optic balanced heterodyne detection system, *Optical and Quantum Electronics*, vol. 18, pp. 445–454, 1986.
6. Phillips Handbook of Semiconductor, Vol. S10, 1987.
7. E. M. Cherry and D. E. Hooper, *Amplifying and Low Pass Amplifier Design*, New York: John Wiley, 1968, Chapters 4 and 8.
8. J. Kahn and E. Ip, Principles of digital coherent receivers for optical communications, *in Proceedings of OFC 2009*, paper OTuG5, San Diego, CA, USA, March 2009.
9. C. Zhang, Y. Mori, K. Igarashi, K. Katoh, and K. Kikuchi, Demodulation of 1.28-Tbit/s polarization-multiplexed 16-QAM signals on a single carrier with digital coherent receiver, in *Proceedings of OFC 2009*, Paper OTuG3, San Diego, CA, USA, March 2009.
10. L. N. Binh, *Digital Optical Communications*, Boca Raton, FL: CRC Press, 2008, Chapter 11.

Annex 3: RMS Definition and Power Measurement

A3.1 DEFINITIONS AND MATHEMATICAL REPRESENTATION

In mathematics, the root mean square (abbreviated as "RMS" or "rms"), also known as the *quadratic mean*, is a statistical measure of the magnitude of a varying quantity. It is especially useful when the variates are positive and negative—for example, sinusoids. RMS is used in various fields, including electrical engineering and optical technology, and especially in the measurement of the optical power for evaluating the optical signal-to-noise ratio (OSNR) in optical transmission systems.

It can be calculated for a series of discrete values or for a continuously varying function. Its name comes from its definition as the square root of the mean of the squares of the values. It is a special case of the generalized mean, with the exponent $p = 2$.

The RMS value of a set of values (or a continuous-time waveform) is the square root of the arithmetic mean (average) of the squares of the original values (or the square of the function that defines the continuous waveform).

In the case of a set of n values $\{x_1, x_2, \ldots, x_n\}$, the RMS value can be mathematically written as

$$x_{\mathrm{rms}} = \sqrt{\frac{1}{n}\left(x_1^2 + x_2^2 + \cdots + x_n^2\right)} \tag{A3.1}$$

The corresponding formula for a continuous function (or waveform) $f(t)$, defined over the interval $T_1 \leq t \leq T_2$, is

$$f_{\mathrm{rms}} = \sqrt{\frac{1}{T_2 - T_1} \int_{T_1}^{T_2} \left[f(t)\right]^2 dt} \tag{A3.2}$$

and the RMS for a function over all time is

$$f_{\mathrm{rms}} = \lim_{T \to \infty} \sqrt{\frac{1}{T} \int_{0}^{T} \left[f(t)\right]^2 dt} \tag{A3.3}$$

Thus, the RMS over all time of a periodic function is equal to the RMS of one period of the function. The RMS value of a continuous function or signal can be approximated by taking the RMS of a series of equally spaced samples. In addition, the RMS value of various waveforms can also be determined without calculus. In the case of the RMS statistic of a random process, the expected value is used instead of the mean, and the square of the RMS of the random noise term is the sum of the square of the means plus the standard deviation of the random function, for example, Gaussian noise.

$$x_{\mathrm{rms}}^2 = \bar{x}^2 + \sigma_x^2 \tag{A3.4}$$

Thus, the RMS value is always greater than the mean value.

A3.2   RMS OF COMMON FUNCTIONS

The mathematical representations of a sinusoidal function and for a number of common functions are tabulated in Table A3.1, and illustrated in Figures A3.1 and A3.2, respectively. For optical waves, the term V can be replaced with the electric field of the optical waves. Thus, a measurement of the optical power is indeed the measurement of the square of the RMS value of the optical waves.

TABLE A3.1
Waveform and RMS Values

Waveform	Equation	RMS
DC, constant	$y = a$	a
Sine wave	$y = a\sin(2\pi ft)$	$\dfrac{a}{\sqrt{2}}$
Square wave	$y = \begin{cases} a & \{ft\} < 0.5 \\ -a & \{ft\} > 0.5 \end{cases}$	a
DC-shifted square wave	$y = \begin{cases} a + DC & \{ft\} < 0.5 \\ -a + DC & \{ft\} > 0.5 \end{cases}$	$\sqrt{a^2 + DC^2}$
Modified square wave	$y = \begin{cases} 0 & \{ft\} < 0.25 \\ a & 0.25 < \{ft\} < 0.5 \\ 0 & 0.5 < \{ft\} < 0.75 \\ -a & \{ft\} > 0.75 \end{cases}$	$\dfrac{a}{\sqrt{2}}$
Triangle wave	$y = \left\|2a\{ft\} - a\right\|$	$\dfrac{a}{\sqrt{3}}$
Sawtooth wave	$y = 2a\{ft\} - a$	$\dfrac{a}{\sqrt{3}}$
Pulse sequence	$y = \begin{cases} a & \{ft\} < D \\ 0 & \{ft\} > D \end{cases}$	$a\sqrt{D}$

Notes: t is time; f is frequency; a is amplitude (peak value); D is the duty cycle or the percentage (%) spent high of the period $(1/f)$; $\{r\}$ is the fractional part of r.

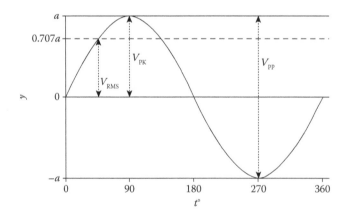

FIGURE A3.1   Generic feature of the RMS value of a periodic sinus function versus the time. V can be replaced by the field $\underset{\sim}{E}$ of the optical waves.

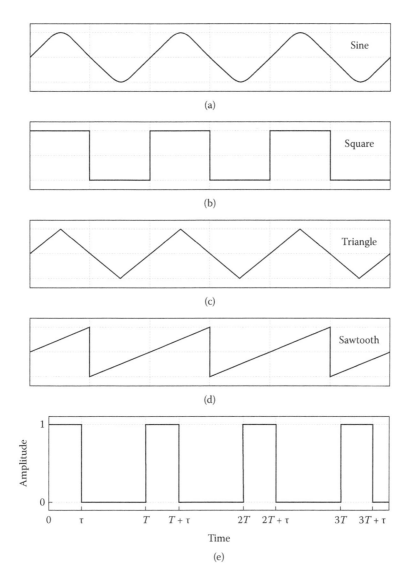

FIGURE A3.2 Typical periodic waveforms (a) sinusoidal, (b) square, (c) triangular, (d) sawtooth, and (e) pulse sequence with duty cycle of $\tau/T = D$.

Hence, in the case that the optical waves are modulated by a sequence of the square wave sequence of modulated lightwaves, then the RMS value of this modulated sequence equals the product of the RMS value of the sinusoidal wave function of the optical carrier (assumed normalized peak value) and the RMS value of the square function, thus equal to $\frac{1}{\sqrt{2}} a \sqrt{D}$, where a is the amplitude of the pulse sequence and D is the duty cycle of the pulse sequence, which can be derived without difficulty from the return-to-zero or non-return-to-zero pulse shaping.

Annex 4: Power Budget

A4.1 OVERVIEW

Power budget is the counterpart of the rise time budget to ensure that the signals entering the reception subsystem has sufficient energy for detection that satisfies a certain criteria for error-free reception. The two scenarios of distinct reception sensitivities of a communication receiver are when FEC is employed or not. If no FEC is employed, then for a Gaussian PDF, the bit error rate (BER) is required to reach 10^{-9}, or the amplitude ratio of the average levels between the "1" and "0" must be at least six times the total standard deviation of the noise levels of the "1" and "0." With FEC then, this BER can be lowered to about 10^{-3} or thereabouts, depending on the error coding method.

Thus, the following steps must be employed to determine the power budget:

1. Identifying the modulation format of the signals, and hence the Euclidean distance between the states. Thence, the required ratio between the energy level and the noise power for a specified BER can be determined.
2. Identifying the probability density function of the transmission and detection processes, and hence the noises. Thence, identifying the noise sources contributed to the modulated and transmitted signals or signal levels.
3. Obtain the attenuation factor, for example, dB/km in fiber, and any noise sources superimposing on the signals during transmission, for example, the amplification spontaneous emission (ASE) noises of in-line optical amplifiers.
4. Thence, determine the required receiver sensitivity of the reception systems, which is defined as the minimum power level required of the signals to be available at the input of the receiver at a specified BER.
5. Work back, using a linear log scale diagram to estimate the required signal power to be launched at the output of the transmitter in order to achieve this reception-receiving sensitivity.

The preceding steps can be done in our heads without resorting to computing devices, not even a simple calculator. This can be done by converting all power and attenuation in log scale or in the dB level, and thence all estimations are just additions and subtractions.

A4.2 POWER BUDGET ESTIMATION EXAMPLE

As an example, consider the transmission system given in Figure A4.1. The modulation is assumed to be binary OOK, the transmission link can be a multispan SSMF optically amplified type, and the reception system is based on direct detection technique with a PD and followed by TIA and AGC, if required.

Under the binary OOK modulation scheme, we can estimate that the required signal-to-noise ratio (SNR) is 6 for a BER of 1e−9. If there is any uncertainty in the transmission system, then a BER = 1e−12 can be used, leading to an SNR of 7 (the Q factor in linear scale) being required. Figure A4.2 depicts the required SNR for BER values over ten decades. That is the Q factor increase by one unit required for a decade of BER for a Gaussian probability density function (PDF) detection process. Thus, corresponding to a Q value of 6 and 7, the OSNR can be determined to be $10\,Log_{10}\,6{\sim}15\,dB$.

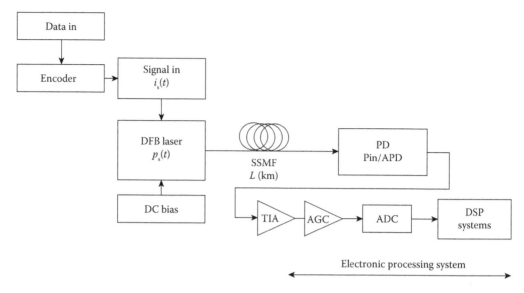

FIGURE A4.1   Intensity modulation direct detection transmission scheme.

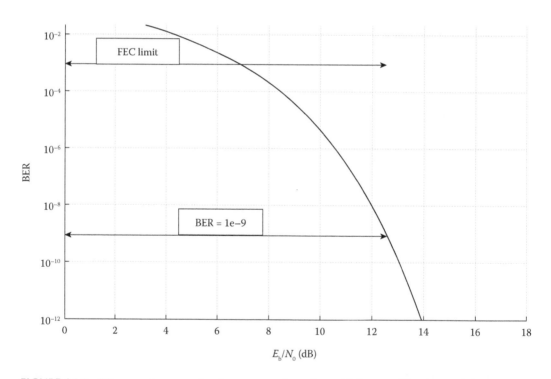

FIGURE A4.2   Bit error rate versus signal energy to noise ratio per bit for on–off keying modulation.

Therefore, for an optical receiver, how do we determine the noise power at its input port, that is, at the front of the photodetector? This means we have to estimate all noise-equivalent power to the input port of the receiver, or alternatively we find all equivalent noise currents referred to the input port of the receiver as shown in Figure A4.3, in which we can identify: (a) $<i_s^2>$ represents average input current as converted by the photodetection process of the input signal; (b) total equivalent

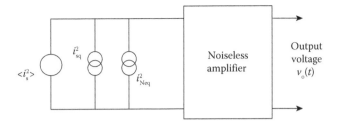

FIGURE A4.3 Equivalent current model referred to the input of the optical receiver, average signal current, and equivalent noise current of the electronic preamplifier as seen from its input port.

noise current i_{Neq}^2 as referred to the input of the electronic preamplifier followed the PD*; and (c) total noise equivalent, i_{sq}^2, generated during the opto-electronic conversion such as the quantum shot noise by the signal input and the dark current noise of the PD.

It is noted that, under direct detection, the amplifier noise current normally dominates over the quantum shot noises, while this is opposite in the case of coherent detection, in which the quantum shot noises due to the high average power of the coherent sources can dominate.

Now, reverting back to the optical fiber transmission link, and for direct modulation the laser can be biased at current a current $I_b = 5I$th with a magnitude of 4.5Ith then an average optical power at the output of the DFB can reach 10 dBm. However, for the "0," we would have to drive with a magnitude of about 1.5Ith. Hence, we expect noisy signals at the base level, a "high" level, and we expect a distinction ratio of more than 20 dB. Under this driving condition, let us assume that the average power of 0 dBm can be obtained at the output of the DFB laser.

The attenuation and power requirements for each subsystem of the link can now be estimated as follows:

1. The fiber loss for 80 and 100 km SSMF would be about 20–25 dB.
2. Using the pin or APD (avalanched gain of 10) photodetector and a differential trans-impedance amplifier (TIA) (Inphi model 3250A) with a noise spectral density referred to

   the input of $\left(\dfrac{di_N^2}{df}\right)^{1/2} = 20 - 40\text{pA}/\sqrt{Hz}$, we would be −7 dBm and −17 dBm, respectively,

   for a required OSNR of 15 dB, and a BER of 1e−9 to 1e−12 (δ or Q factor of 6 or 7 under Gaussian PDF).
3. This indicates that an APD is preferred as the PD, so that we can avoid the use of an optical amplifier. The projection variation of transmission distance with respect to the launched power at the output of the transmitter is shown in Figure A4.4.
4. With the p-i-n PD, we expect to transmit 25 G NRZ channel over only 40 km with a launched power of 4 dBm. However, with the APD of a gain of 10, the transmission distance limited by the power budget can reach over 80 km, on condition that the pulse sequence is equalized at the DSP-based receiver, or an SSB modulation scheme is employed, such as the duobinary modulation format.

---

* Commercial amplifiers normally specify the noise characteristics as the total equivalent noise spectral density referred to the input in units of $A/\sqrt{Hz}$. This parameter is normally measured by connecting the output of the amplifier to a spectrum analyzer with no signals input. The output is integrated over the bandwidth of the amplifier to obtain the total noise voltage at the output. In case of a TIA, this output voltage noise is then transferred back to the input by a simple division. Hence, we can get the total equivalent noise current referred to the input. This noise current referred to the input can be converted to equivalent optical power by using the responsivity of the photodetector. From this value and the OSNR, we can justify whether the transmission link would operate to give the required BER.

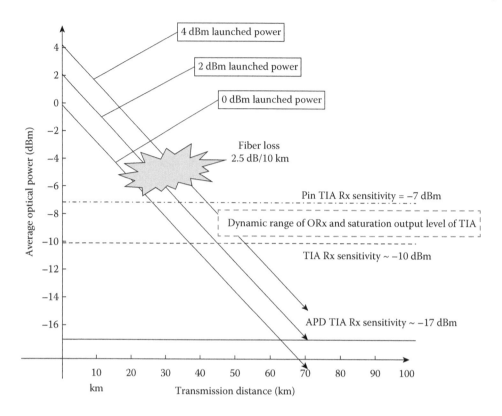

FIGURE A4.4   Power budget with an average launched power of 0, 2, and 4 dBm with receiver sensitivity of −7 dBm and −17 dBm, and a fiber loss of 2.5 dB/10 km.

A4.3   SNR AND OPTICAL SNR

Referring to Figure A4.3, the circuit represents a noisy amplifying circuit by the total noises as referred to the input cascade with a noiseless amplifying devices. The photodetected signal current is presented by a pure current source connected parallel to the noise sources. It must be noted that there are also other noise sources superimposing on the signal current, which are commonly known as the signal-dependent *quantum shot noise*. This noise is dependent on the signal average current whose spectral density is given by

$$S_{NQ} = 2q\langle i_S \rangle \text{ in } A^2/Hz$$

$$\rightarrow i_{NQ} \text{ } by\_integrating\_over\_BW \& sqrt$$

(A4.1)

Furthermore, there are noise currents which are represented by dark current of the PD i_{Nd}, that is, the randomly generated electric current in the semiconductor junction at temperatures above the absolute reference temperature even when the PD is not receiving any optical signals. Thus the total noise current referred to the input of an optical receiver be written as the sum of these noise currents, by

$$i_{NT} = i_{NA} + i_{NQ} + i_{Nd}$$

(A4.2)

Thus, the SNR, in terms of power, can be formed as

$$SNR = \frac{\langle i_S^2 \rangle}{i_{NT}^2}$$

(A4.3)

The BER can be estimated for ASK binary signals under Gaussian noise assumptions as

$$\text{BER} = \frac{1}{2}\text{ercf}\left(\frac{Q}{\sqrt{2}}\right) \text{ where } Q = \frac{I_1 - I_0}{\sigma_1 - \sigma_0} \tag{A4.4}$$

where I_1, I_0 are the average currents detected by the PD for "1" and "0," respectively, and σ_1, σ_0 are the standard deviations of the noise levels due to "1" and "0," respectively. If the modulation format is QAM, then the errors of detection can be considered as the summation of the two orthogonal "binary" signals, the complex and real components. The BER can be estimated for both directions, as given in Equation A4.4.

Thus, we can see that, by estimating the total current noise referred to the input of the electronic preamplifier, normally a TIA, and then the dark current and the signal-dependent signal current noise with a typical average optical power at the input, we can obtain a rough estimate of the SNR and the optical OSNR. Note that the estimation of the noises should be considered for the dominating noise term.

Naturally, the electronic current at the output of the preamplifier of the receiver would normally not be sufficient for the ADC and signal processor, so a voltage gain amplifier with or without an automatic gain control stage (see Figure A4.1) would commonly be employed to boost up the signal levels. The noise current referred to the input of this voltage gain stage of the optical receiver is normally much smaller than the total noise current referred to the input of the preamplifier, so it can be ignored.

Other modulations can be derived using the principles of superposition of probability of errors contributed by the considered constellation point by the neighboring points.

A4.4   TRANSIMPEDANCE AMPLIFIER: DIFFERENTIAL AND NONDIFFERENTIAL TYPES

Transimpedance amplifier (TIA) is the critical amplifying device which is connected in cascade with a high-speed photodetector (PD) to provide the pre amplification of the current produced in the PD to a voltage output whose levels may a level appropriate for analog-to-digital conversion, so that digitized signals can then be processed in the digital domain by a digital signal processor (DSP). The total equivalent noises referred to the input port of this amplifier are very critical on the impact of the SNR of the optical receiver. The method for calculations of noise processes and noise currents, as well as the total equivalent noises referred to the input, are given in Chapter 4.

Traditionally, single-input or nondifferential-input TIA has been extensively exploited over the years [1]. However, currently, differential TIAs (DTIAs) offer wider bandwidths and high differential transimpedance gains of around 3000–5000 Ω over the 30–40 GHz bandwidth by using Si-Ge technology, whose transition frequency f_T can reach 280 GHz [2,3]. However, these differential TIAs [4] are limited in their dynamic range, and an automatic gain control stage would normally be required. A typical setup of photodetector pair and a differential TIA is shown in Figure A4.5. Typical circuit diagrams of the DTIA employing Si-Ge and InP are shown in Figure A4.6, in which there are the main differential long-tail pair, followed by two common collector stages connected to the differential output of this long-tail pair. The outputs of these stages are then fed back by a shunt feedback to obtain the highest transfer impedance and wideband property. The outputs are fed back to the differential input. These differential ports are then fed into a further differential voltage gain stage (without any feedback stage).

The DTIA can offer a transfer impedance of 4000–5000 Ω. Thus, with an optical signal average power input of 0 dBm (or 1 mW), and a responsivity of the photodetector of 0.9~1.0 A/W, a current of 1 mA is produced and fed into the differential TIA. This in turn produces a differential voltage at the output of 4–5 V signals. Unfortunately, this level will saturate the outputs of the DTIA. So, normally, an automatic gain control stage would be employed to increase the dynamic range of

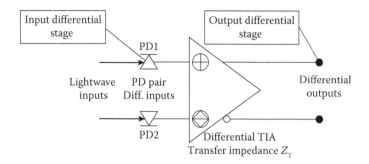

FIGURE A4.5 Schematic of a balanced optoelectronic receiver using a differential TIA.

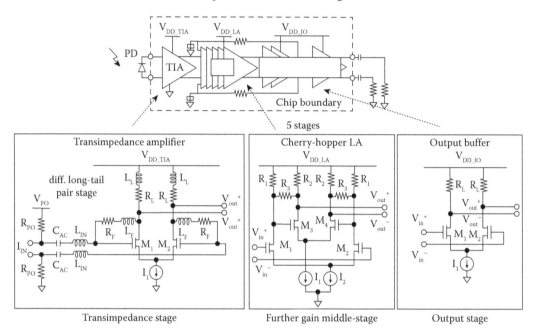

FIGURE A4.6 Typical circuit topologies of three stage differential TIA producible by SiGe and InP technology.

the optical receiver. The operational parameters of the differential TIA can be specified together with its total input-referred noise spectral density noise current $25\,\mathrm{pA}/\sqrt{\mathrm{Hz}}$ on average to a bandwidth of 40 GHz.

REFERENCES

1. T. V. Muoi, Receiver design for high speed optical fiber systems, *IEEE J. Lightwave Technol.*, pp. 243–266, 1984.
2. P. Kempf and M. Racanelli, Silicon germanium BiCMOS technology, *IEEE GaAs IC Symp. Dig.*, pp. 3–6, 2002.
3. B. Jagannathan, M. Khater, F. Pagette, J.-S. Rieh, D. Angell, H. Chen, J. Florkey, et al., Self-aligned SiGe NPN transistors with 285 GHz f_{max} 207 GHz f_T in a manufacturable technology, *IEEE Electron. Device Lett.*, Vol. 23, pp. 258–260, 2002.
4. J. S. Weiner, A. Leven, V. Houtsma, Y. Baeyens, Y.-K. Chen, P. Paschke, Y. Yang, et al., SiGe differential trans-impedance amplifier with 50-GHz bandwidth, *IEEE J. Solid-State Circuits*, Vol. 38, No. 9, p. 1512, 2003.

Annex 5: Modeling of Digital Photonic Transmission Systems

A5.1 OVERVIEW

A digital photonic transmission system can be divided into three main subsystems: optical transmitter, optical fiber channel, and optical receiver, as demonstrated in Figure A5.1.

The first key subsystem is the optical transmitter. The main function of an optical transmitter is to generate lightwaves carrying a particular modulation format. Modulation formats are classified into three groups depending on whether the amplitude, phase, or frequency component of the lightwave carrier is modulated. In modern photonic communications, the modulation process is implemented by using external optical modulators that can be categorized into phase and intensity modulators. The advantages of external data modulation over direct data modulation on semiconductor lasers and operational descriptions of external optical modulators are presented in the next section.

The second subsystem is the optical fiber channel. Classification of optical fibers is based on their dispersion characteristics, and some notable types are standard single-mode fiber (SSMF), commonly known as ITU-G.652 type, and nonzero dispersion-shifted fiber (NZ-DSF), ITU-G.655 type. SSMF has a chromatic dispersion (CD) factor of about ±17 ps/(nm.km) at 1550 nm wavelength, compared to small values of ±(2–6) ps/(nm.km) for NZ-DSF. The dispersion induced from these optical transmission fibers is compensated for by the dispersion-compensating fiber (DCF), which has negative CD factors. In addition, recent progress in fiber design has also introduced several new types of fibers, such as dispersion-flattened fibers (DFF) and Corning Vascade fibers [1,2].

Optical fibers consist of several impairments that cause severe degradations to the system performance. These impairments are grouped into fiber dispersions and fiber nonlinearities. Fiber dispersion includes the second-order CD dispersion, third-order dispersion slope, and polarization-mode dispersion (PMD). On the contrary, fiber nonlinearities, which are power-dependent impairments, contain a number of effects, including intrachannel self-phase modulation (SPM), interchannel cross-phase modulation (XPM), and four-wave mixing (FWM). Dispersion and nonlinearity impairments are embedded into the nonlinear Schrödinger equation (NLSE), which governs signal propagation along the optical fiber. The most popular method to solve NLSE numerically is the symmetric split-step Fourier method (SSFM) [3]. This method, however, encounters a number of issues, such as long computation time and artificial errors caused by the windowing effect of fast Fourier transform (FFT) and inverse FFT (IFFT) operations. This annex provides, in Section A5.4, detailed descriptions of the fiber impairments, symmetric SSFM, and techniques to overcome the modeling limiting factors.

Optical signals are attenuated when propagating along the optical fiber channel, thus necessitating signal amplification. This amplification is carried out in photonic domain by using Erbium-doped fiber amplifiers (EDFAs). Figure A5.1 illustrates the conventional configuration in which DCFs are normally accompanied by two EDFAs. The first EDFA compensates for the attenuation of the preceding SSMF span, while the other EDFA boosts optical intensity to a designated level before launching into the next transmission span. From the system point of view, there are two key parameters modeling an EDFA: amplified spontaneous emission (ASE) noise and noise figure (NF). These parameters are formulated in Section A5.5. Unless specifically stated, it is assumed in this thesis that optical amplifiers are operating in their saturation modes. Figure A5.1 also shows optical filters whose bandwidths have significant impact on the system performance. The modeling of these optical filters as well as electrical filters is presented in Section A5.6.

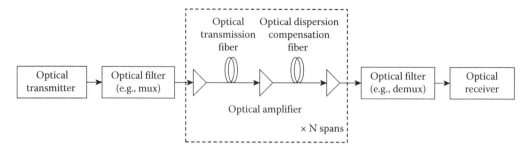

FIGURE A5.1 Generalized diagram of optical transmission systems.

The last key subsystem is the direct detection optical receiver, which can be classified into two types: single-photodiode receivers and Mach–Zehnder delay interferometer (MZDI)-balanced receivers for incoherent detection of optical OOK and DPSK signals, respectively [4,5]. The performance of these receivers is influenced by several noise sources whose formulations are provided in Section A5.7.

Evaluation of the system performance is critical in fiber optic communications for both modeling/simulation and practical experiments. The performance of optical transmission systems can be characterized by several measures. A quick measure is based on simple metrics such as eye opening (EO) and eye opening penalty (EOP) of detected signals. However, the most popular measure is bit error rate (BER) as a function of either optical-signal-to-noise ratios (OSNRs), average received powers, or average input powers. Moreover, power or OSNR penalties can be inferred from the obtained BER curves, at a particular BER level. Conventional methods for calculating BER to evaluate the system performance are presented in Section A5.8, while Section 2.9 provides thorough descriptions of statistical methods for novel applications in fiber optic communications. Finally, Section 2.10 discusses the advantages of the developed MATLAB® and Simulink® modeling platform before a summary of this annex is provided.

A5.2 OPTICAL TRANSMITTER

An optical transmitter normally consists of a narrow-linewidth laser source, external optical modulators, a bit pattern data generator, and, optionally, an electrical precoder or an electrical shaping filter (see Figure A5.2).

The narrow-linewidth laser source is normally a distributed feedback (DFB) laser [6,7] with generated wavelengths complying to the ITU-Grid standard. The laser can be biased at a constant current to provide continuous-wave (CW) lightwaves that pass through an external optical modulator for the data modulation process.

A5.2.1 BACKGROUND OF EXTERNAL OPTICAL MODULATORS

In the 1980s and early 1990s, direct modulation of semiconductor lasers was the main modulation technique. However, this technique faces several limiting factors [8,9]:

1. Direct modulation induces unwanted chirps, resulting in signal spectral broadening, causing severe dispersion penalties.
2. Directly modulated optical signals experience fluctuation in the intensity; this arises from the relative intensity noise (RIN) of the semiconductor laser.
3. Laser phase noise induced from nonzero linewidth of laser sources also limits the application of the direct modulation technique in high-speed transmission systems.

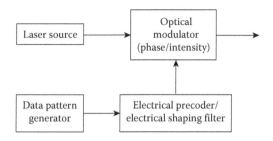

FIGURE A5.2   Generic block diagram of an optical transmitter.

External data modulation has thus been the preferred technique over direct modulation as it minimizes the preceding problems for digital photonic transmission systems. External data modulation can be implemented using either electroabsorption modulators (EAMs) or electro-optic modulators (EOMs). EOMs have been preferable due to the advantages of electro-optic materials such as the linear response characteristic, high extinction ratio, and, particularly, the capability to control either phase, frequency, or amplitude of the lightwave carrier [10,11]. The operation of EOM is based on principles of Pockels electro-optic effects of solid-state, polymeric, or semiconductor materials [8,10]. Over the years, waveguides of EOMs are mainly integrated on lithium niobate (LiNbO$_3$) material. It has prominent properties such as high electro-optic coefficients, low attenuation, and the possibility of generating chirp-free signals [10,12]. LiNbO$_3$ EOMs have been developed since the early 1980s, but were not popular until the advent of EDFA in the late 1980s [10,13]. They were employed in coherent optical communications to mitigate the problems of broad linewidth and RIN of the laser source in direct modulation, as mentioned earlier. The knowledge of using these external modulators has recently been revisited for the generation of advanced modulation formats in incoherent transmission systems. The data modulation is conducted in photonic domain by using either the optical phase or intensity modulator.

A5.2.2   OPTICAL PHASE MODULATOR

An electro-optic phase modulator (EOPM) employs a single electrode, as shown in Figure A5.3. Its operation is based on the Pockels electro-optic effect, that is, when a driving voltage is applied onto the electrode, the refractive index (RI) of electro-optic materials changes accordingly, thus inducing delays to lightwaves. Since delays correspond to phase changes, EOPM is able to manipulate the phase of the lightwave carrier.

The induced phase variation corresponding to a particular radio frequency (RF) electrical driving voltage $V(t)$ is given as

$$\phi(t) = \pi \frac{\left(V(t) + V_{\text{bias}}\right)}{V_\pi} \tag{A5.1}$$

where V_π is the driving voltage required to create a π phase shift on the lightwave carrier and has typical values within the range 3–6 V [12,14]; $V(t)$ is a time-varying driving signal voltage, and V_{bias} is a DC bias voltage. The complex-envelop representation of the optical field E_o at the output of EOPM is expressed as

$$E_o(t) = E_i(t)e^{j\phi(t)} \tag{A5.2}$$

EOPMs operating at high speed, for example, 40 Gbps, and using resonant-type electrodes have been recently reported [15,16]. In addition, due to the linear response characteristics of electro-optical

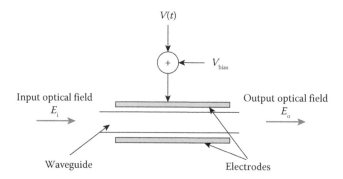

FIGURE A5.3   Electro-optic phase modulator (EOPM).

materials (reflected in Equation A5.1), EOPM also has the capability to carry out the frequency modulation by manipulating the slope of the time-varying RF driving signal.

A5.2.3   OPTICAL INTENSITY MODULATOR

An intensity modulator comprises two EOPMs in a parallel structure to form a Mach–Zehnder Interferometer (MZI), which is commonly known as Mach–Zehnder intensity modulator (MZIM) (see Figure A5.4).

Lightwave intensity is split equally when entering the two arms of MZIM. Each arm of the MZIM is actually an EOPM that can conduct the phase modulation on the optical carrier. At the output of MZIM, optical fields from two arms are coupled and interfered with each other either constructively or destructively. This enables the "on–off" modulation of the lightwave intensity. Figure A5.5 shows a sample of an MZIM packaged in our lab at Monash University and its modulation transfer characteristic [17]. This MZIM has a 3 dB bandwidth of up to 26 GHz.

MZIM can be classified into single-drive and dual-drive types, which are described in the following subsections.

A5.2.3.1   Single-Drive MZIM

A single-drive MZIM has only a single driving voltage applied to either arm of the MZIM. For instance, it is assumed that there is no voltage driving onto arm 1, while a voltage $V(t)$ is applied on arm 2 (refer to Figure A5.4). The transmitted optical field $E_o(t)$ at the output of a single-drive MZIM is a function of $V(t)$ and a bias DC voltage V_{bias}. Written in the low-pass equivalent format, the expression of $E_o(t)$ is given by

$$E_o(t) = \frac{E_i(t)}{2}\left[1+e^{j\pi\frac{(V(t)+V_{bias})}{V_\pi}}\right] = E_i \cos\left[\frac{\pi}{2}\frac{(V(t)+V_{bias})}{V_\pi}\right]e^{-j\left[\frac{\pi}{2}\frac{(V(t)+V_{bias})}{V_\pi}\right]} \quad (A5.3)$$

The phase term in Equation A5.3 implies the existence of the phase modulation of the optical carrier, that is, the chirping effect. Thus, by using a single-drive MZIM, the generated optical signals are theoretically not chirp-free, particularly in the case of using a z-cut LiNbO$_3$ MZIM that has an asymmetrical structure of field distributions [12]. However, an x-cut MZIM can provide a modest amount of chirping, thanks to its symmetrical distribution of the electrical fields [12]. In practice, a small amount of chirping might be useful for transmission [8].

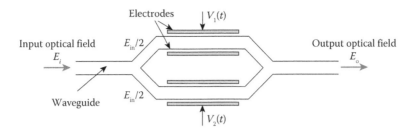

FIGURE A5.4 Optical intensity modulator with a Mach–Zehnder interferometric structure.

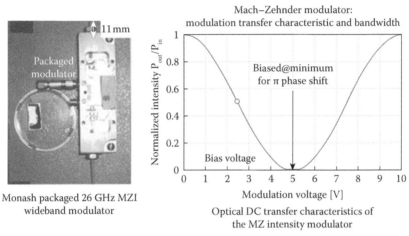

FIGURE A5.5 An MZIM packaged at Monash University and its operating the transfer curve.

A5.2.3.2 Dual-Drive MZIM

A dual-drive MZIM has a push–pull arrangement for dual-driving voltages (i.e., $V_1(t)$ and $V_2(t)$ are inverse to each other; $V_2(t) = -V_1(t)$), hence the chirping effect in the modulation of the light-wave carrier can be completely eliminated. The transmitted optical field $E_o(t)$ in Equation A5.3 can be rewritten as

$$E_o(t) = \frac{E_i(t)}{2}\left[e^{j\pi\frac{(V(t)+V_{bias})}{V_\pi}} + e^{j\pi\frac{-(V(t)+V_{bias})}{V_\pi}} \right] = E_i \cos\left[\frac{\pi}{2}\frac{(V(t)+V_{bias})}{V_\pi} \right] \tag{A5.4}$$

In Equation A5.4, the phase term no longer exists, indicating that the chirping effect is totally eliminated.

A5.3 IMPAIRMENTS OF OPTICAL FIBER

A5.3.1 CHROMATIC DISPERSION

This section presents key properties of the CD impairment in a single-mode fiber. The initial point when mentioning fiber CD is the expansion of the mode propagation constant or "wave number" parameter β using the Taylor series [8]

$$\beta(\omega) = \frac{\omega n(\omega)}{c} = \beta_0 + \beta_1\Delta\omega + \frac{1}{2}\beta_2\Delta\omega^2 + \frac{1}{6}\beta_3\Delta\omega^3 + \ldots + \frac{1}{n!}\beta_n\Delta\omega^n \tag{A5.5}$$

where ω is the angular optical frequency and $n(\omega)$ is the fiber RI. The parameters β_n represents the n-th derivative of β, and their meanings are described in the following

1. β_0 involves the phase velocity, v_p, of the optical carrier ω_0, and v_p is defined as

$$v_p = \frac{\omega_0}{\beta_0} = \frac{c}{n(\omega_0)} \tag{A5.6}$$

2. β_1 determines the group velocity v_g that is related to β of the guided mode by

$$v_g = \frac{1}{\beta_1} = \left(\left. \frac{d\beta}{d\omega} \right|_{\omega=\omega_0} \right)^{-1} \tag{A5.7}$$

3. β_2 is the derivative of the group velocity v_g with respect to frequency, and, hence, β_2 clearly shows the frequency dependence of the group velocity. This means that different frequency components of an optical pulse propagate along the optical fiber at different velocities, thus leading to the spreading of the pulse, that is, the dispersion. The parameter β_2 is commonly known as group velocity dispersion (GVD). The optical fiber exhibits normal dispersions for $\beta_2 > 0$ or anomalous dispersions for $\beta_2 < 0$. A pulse having the spectral width of $\Delta\omega$ and traveling through a length L of fiber is broadened by an amount of time ΔT, given by $\Delta T = \beta_2 L \Delta\omega$ [3]. In practice, a more common factor to represent the fiber CD of a single-mode optical fiber is the dispersion factor D with the unit of ps/(nm.km). D is closely related to GVD β_2 by

$$D = -\left(\frac{2\pi c}{\lambda^2} \right)\beta_2 \tag{A5.8}$$

where λ is the operating wavelength.

4. β_3 is the derivative of β_2, and it contributes to the dispersion slope $S(\lambda)$ as follows

$$S = \frac{dD}{d\lambda} = \left(\frac{2\pi c}{\lambda^2} \right)\beta_3 + \left(\frac{4\pi c}{\lambda^3} \right)\beta_2 \tag{A5.9}$$

A5.3.1.1 Chromatic Dispersion as a Total of Material Dispersion and Waveguide Dispersion

From the view of fiber design [8,18], D is a sum of material dispersion (D_M) and waveguide dispersion (D_W)

$$D = -\left(\frac{2\pi c}{\lambda^2} \right)\beta_2 \equiv D_M + D_W \tag{A5.10}$$

The following equations describe how D_M and D_W are obtained. In this case, a step-index optical fiber with a core radius a is considered, and the RIs of the core and cladding of the SSMF are denoted as n_1 and n_2, respectively. The significant transverse propagation constants of guided lightwaves u and v in the core and cladding regions are formulated as

$$u = a\sqrt{k^2 n_1^2 - \beta^2} \tag{A5.11}$$

$$v = a\sqrt{\beta^2 - k^2 n_2^2} \tag{A5.12}$$

where $k^2 n_1^2$ and $k^2 n_2^2$ are the plane-wave propagation constants in the core and cladding, respectively. The guided wave number β is calculated as

$$\beta = \sqrt{k^2 (b(n_1^2 - n_2^2) + n_2^2)} \qquad (A5.13)$$

where b is the normalized propagation constant whose values for guided modes fall within the range of [0,1], and b is calculated as

$$b = \frac{\dfrac{\beta}{k} - n_2}{n_1 - n_2} \qquad (A5.14)$$

The normalized frequency V is expressed as

$$V = ak\sqrt{n_1^2 - n_2^2} \qquad (A5.15)$$

The waveguide dispersion D_W can be calculated in the following equation [18–21]

$$D_W = -\left(\frac{n_1 - n_2}{c\lambda}\right) V \frac{d^2(Vb)}{dV^2} \qquad (A5.16)$$

where $Vd^2(Vb)/dV^2$ is defined as the normalized waveguide dispersion parameter. An effective approximation based on polynomial interpolation has been developed to calculate the waveguide dispersion parameter of a multicladding DCF [18].

The material dispersion of an optical fiber is due to the wavelength dependence of the RI of the core and the cladding. The RI $n(\lambda)$ is estimated by Sellmeier equation [21]

$$n^2(\lambda) = 1 + \sum_{i=1}^{M} \frac{B_i \lambda^2}{\left(\lambda^2 - \lambda_i^2\right)} \qquad (A5.17)$$

where λ_i indicates the i-th resonance wavelength, B_i is its corresponding oscillator strength, and n stands for n_1 or n_2 for either the core or cladding RIs, respectively. The material dispersion factor D_M is then obtained by

$$D_M = -\frac{\lambda}{c}\left(\frac{d^2 n(\lambda)}{d\lambda^2}\right) \qquad (A5.18)$$

where c is the light velocity in vacuum. For pure silica and over the wavelength of 1.25–1.66 μm, D_M can also be approximated by an empirical relation [19,21]

$$D_M = 122\left(1 - \frac{\lambda_{ZD}}{\lambda}\right) \qquad (A5.19)$$

where λ_{ZD} is the zero material dispersion wavelength, which is defined as the wavelength at $D_M(\lambda) = 0$—for instance, $\lambda_{ZD} = 1.276$ μm for pure silica. λ_{ZD} can vary according to various doping concentrations in the core and cladding of different materials such as Germanium (Ge) or Fluorine (F).

With the demand of reducing effects of fiber CD, several types of fibers including dispersion-shifted fiber (DSF) and NZ-DSF were proposed. The latter type requires a nonzero local dispersion value in order to avoid the phase matching between the wavelengths in a DWDM transmission system, that is, to avoid the FWM effect. Values of D for various types of fibers within C-band

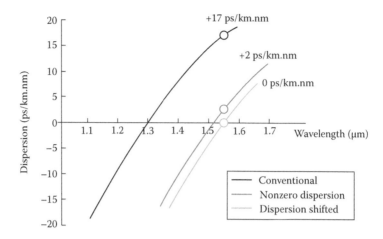

FIGURE A5.6 Typical values of fiber CD for different types of fiber.

wavelengths of the ITU-Grid standard are demonstrated in Figure A5.6, and the circle values are located at 1550 nm wavelength.

The optical transmission medium also involves DCF and several other types of fibers such as DFF and Corning Vascade fibers. DCF is usually used in-line with SSMF in a dispersion-managed optical system, so that fiber CD is fully compensated within a span. The dispersion factor D of a DCF has negative values. On the contrary, DFF is manufactured for the specific purpose of flattening dispersion factors over a wide range of wavelengths in order to reduce fiber CH effects. However, these wavelengths travel at nearly the same velocities and thus the phase-matching conditions are easily satisfied. This results in FWM effects with the introduction of a ghost pulse. However, DFF has an important application in parametric amplification utilizing the FWM nonlinearity [3]. Furthermore, Corning Vascade fibers are designed for ultra-long-haul and transoceanic optical transmission systems, and they provide a complete built-in CD compensation.

The design of DCF and DFF involves a multicladding/core structure, and is more complicated (see Figure A5.7) as compared to the simple step-index profile of SSMF [18].

A5.3.1.2 Dispersion Length

An important parameter to govern the effects of fiber CD on optical pulses is the dispersion length L_D. This length corresponds to the distance at which a pulse has broadened by one-bit interval [22,23]. For high-capacity and long-haul transmission employing external modulation, L_D can be estimated as [23]

$$L_D = \frac{10^5}{D.R^2} \tag{A5.20}$$

where R is the bit rate (Gbps), D is in ps/(nm.km), and L_D is in kilometer.

Equation A5.20 provides a reasonable approximation, even though the accurate computation of L_D depends on a number of factors: the modulation format, pulse shaping, and the optical receiver performance. It is found that this transmission limit is inversely proportional to the square of the bit rate. Thus, for 10 Gbps OC-192 OOK systems ($D = \pm 17$ ps/(nm.km)), the dispersion length L_D has a value of approximately 60 km SSMF. This SSMF length corresponds to a residual dispersion of about ± 1000 ps/nm. In the case of 40 Gbps OC-768 systems, L_D is about 4 km, or equivalent to ± 60 ps/nm.

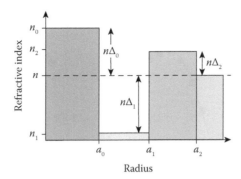

FIGURE A5.7   Index profile of a triple cladding fiber for a design of DCF or DFF.

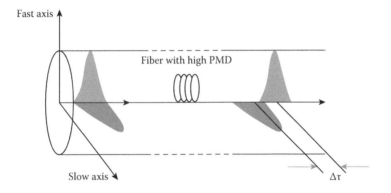

FIGURE A5.8   Delay caused by DGD of two PSPs along the fiber propagation.

A5.3.2   Polarization-Mode Dispersion

PMD represents another cause of pulse spreading and relates to the differential group delay (DGD) between two orthogonal principal states of polarizations (PSPs) of the propagating optical field, as illustrated in Figure A5.8.

Fiber PMD is caused by either the asymmetry of the fiber core or the deformation of optical fibers. These are the consequences of defects in the manufacturing process, the external stress to the fiber, the aging problem, or the variation of temperature over time.

The delay between two PSPs is normally negligibly small at 10 Gbps. However, at high bit rates and in ultra-long-haul transmission, PMD severely degrades the system performance [24–27]. The instantaneous value of DGD ($\Delta\tau$) varies along the fiber and follows a Maxwellian distribution [25,28,29] (see Figure A5.9).

The Maxwellian distribution is governed by the following expression

$$f(\Delta\tau) = \frac{32(\Delta\tau)^2}{\pi^2 \langle \Delta\tau \rangle^3} \exp\left\{ -\frac{4(\Delta\tau)^2}{\pi \langle \Delta\tau \rangle^2} \right\} \quad \Delta\tau \geq 0 \qquad (A5.21)$$

The mean DGD value $\langle \Delta\tau \rangle$ is commonly termed *fiber PMD* and provided in the fiber specifications. The following expression gives an estimate of the maximum transmission limit L_{max} due to the PMD effect [30]

$$L_{\max} = \frac{0.02}{\langle \Delta\tau \rangle^2 \cdot R^2} \qquad (A5.22)$$

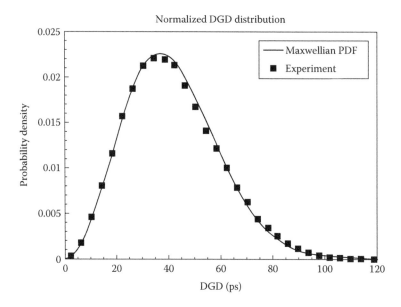

FIGURE A5.9   Maxwellian distribution of PMD random process.

where R is the bit rate. Based on Equation A5.22, L_{max} for both old fiber vintages and contemporary fibers are obtained as follows

- $\langle \Delta \tau \rangle = 1$ ps/km (old fiber vintages)
  - $R = 40$ Gbps; $L_{max} = 12.5$ km
  - $R = 10$ Gbps; $L_{max} = 200$ km
- $\langle \Delta \tau \rangle = 0.1$ ps/km (contemporary fiber for modern optical systems)
  - $R = 40$ Gbps; $L_{max} = 1250$ km
  - $R = 10$ Gbps; $L_{max} = 20,000$ km

A5.3.3 FIBER NONLINEARITY

Fiber RI is dependent on both operating wavelengths and lightwave intensity. This intensity-dependent phenomenon is known as the Kerr effect and is the cause of fiber nonlinear effects. The power dependence of RI is expressed as [3]

$$n' = n + \bar{n}_2 (P/A_{eff}) \qquad (A5.23)$$

where P is the average optical intensity inside the fiber, $\bar{n}_2$ is the fiber nonlinear coefficient, and A_{eff} is the effective area of the fiber.

Fiber nonlinear effects include intrachannel SPM, interchannel XPM, FWM, stimulated Raman scattering (SRS), and stimulated Brillouin scattering (SBS). SRS and SBS are not the main degrading factors, as their effects are only getting noticeably large with very high optical power. On the contrary, FWM severely degrades the performance of an optical system with the generation of ghost pulses only if the phases of optical signals are matched with each other. However, with high local dispersions such as in SSMF, effects of FWM become negligible [3,31]. In terms of XPM, its effects can be considered to be negligible in a DWDM system in the following scenarios [32–37]: (1) highly locally dispersive system and (2) large channel spacing. However, XPM should be taken into account for optical transmission systems deploying NZ-DSF fiber where local dispersion values are small. Thus, SPM is usually the dominant nonlinear effect for systems employing transmission fiber with high local dispersions, for example, SSMF and DCF. The effect of SPM is normally coupled with the nonlinear phase shift ϕ_{NL}, defined as [3]

$$\phi_{NL} = \int_0^L \gamma P(z)\, dz = \gamma L_{eff} P$$

$$\gamma = \omega_c \bar{n}_2 / (A_{eff} c)$$

$$L_{eff} = (1 - e^{-\alpha L}) / \alpha$$

(A5.24)

where ω_c is the lightwave carrier, L_{eff} is the effective transmission length, and α is the fiber attenuation factor that normally has a value of 0.17–0.2 dB/km for the current 1550 nm window of operating wavelengths. The temporal variation of the nonlinear phase f_{NL} results in the generation of new spectral components far apart from the lightwave carrier ω_c, indicating the broadening of the signal spectrum. This spectral broadening $\delta\omega$ can be obtained from the time dependence of the nonlinear phase shift, as follows

$$\delta\omega = -\frac{\partial \phi_{NL}}{\partial T} = -\gamma \frac{\partial P}{\partial T} L_{eff}$$

(A5.25)

Equation A5.25 indicates that $\delta\omega$ is proportional to the time derivative of the average signal power P. In addition, the generation of new spectral components occur mainly at the rising and falling edges of optical pulses, that is, the amount of generated chirps are larger for an increased steepness of the pulse edges.

A5.4 MODELING OF FIBER PROPAGATION

A5.4.1 SYMMETRICAL SPLIT-STEP FOURIER METHOD (SSFM)

The evolution of slow varying complex envelopes $A(z,t)$ of optical pulses along a single-mode optical fiber is governed by NLSE [3]

$$\frac{\partial A(z,t)}{\partial z} + \frac{\alpha}{2} A(z,t) + \beta_1 \frac{\partial A(z,t)}{\partial t} + \frac{j}{2}\beta_2 \frac{\partial^2 A(z,t)}{\partial t^2} - \frac{1}{6}\beta_3 \frac{\partial^3 A(z,t)}{\partial t^3} = -j\gamma |A(z,t)|^2 A(z,t)$$

(A5.26)

where z is the spatial longitudinal coordinate, α accounts for fiber attenuation, β_1 indicates DGD, β_2 and β_3 represent second- and third-order factors of fiber CD, and γ is the nonlinear coefficient. In a single-channel transmission, Equation A5.26 includes the following effects: fiber attenuation, fiber CD and PMD, dispersion slope, and SPM nonlinearity. Fluctuation of optical intensity caused by the Gordon–Mollenauer effect [38] is also included in this equation.

The solution of NLSE and hence the modeling of pulse propagation along a single-mode optical fiber is solved numerically by using SSFM [3]. In SSFM, fiber length is divided into a large number of small segments, δz. In practice, fiber dispersion and nonlinearity are mutually interactive at any distance along the fiber. However, these mutual effects are small within δz, and thus the effects of fiber dispersion and fiber nonlinearity over δz are assumed to be statistically independent of each other. As a result, SSFM can separately define two operators: (1) the linear operator that involves fiber attenuation and fiber dispersion effects and (2) the nonlinearity operator that takes into account fiber nonlinearities. These linear and nonlinear operators are formulated as follows

$$\hat{D} = -\frac{i\beta_2}{2}\frac{\partial^2}{\partial T^2} + \frac{\beta_3}{6}\frac{\partial^3}{\partial T^3} - \frac{\alpha}{2}$$

$$\hat{N} = i\gamma \mid A \mid^2$$

(A5.27)

where A replaces $A(z,t)$ for simpler notation and $T = t - z/v_g$ is the reference time frame moving at the group velocity. Equation A5.27 can be rewritten in a shorter form, given by

$$\frac{\partial A}{\partial z} = (\hat{D} + \hat{N})A$$

(A5.28)

and the complex amplitudes of optical pulses propagating from z to $z + \delta z$ are calculated using the following approximation

$$A(z+h,T) \approx \exp(h\hat{D})\exp(h\hat{N})A(z,T)$$

(A5.29)

Equation A5.29 is accurate to the second-order of the step size δz [3]. The accuracy of SSFM can be improved by including the effect of fiber nonlinearity in the middle of the segment rather than at the segment boundary (see Figure A5.10). This modified SSFM is known as the symmetric SSFM.

Equation A5.29 is now modified as

$$A(z+\delta z,T) \approx \exp\left(\frac{\delta z}{2}\hat{D}\right)\exp\left(\int_z^{z+\delta z}\hat{N}(z')dz'\right)\exp\left(\frac{\delta z}{2}\hat{D}\right)A(z,T)$$

(A5.30)

This method is accurate to the third-order of the step size δz. In symmetric SSFM, the optical pulse propagates along a fiber segment δz in two stages. First, the optical pulse propagates through the linear operator that has a step of $\delta z/2$, which takes into account fiber attenuation and dispersion effects. Then, the fiber nonlinearity is calculated in the middle of the segment. After that, the pulse propagates through the second half of the linear operator. The process continues repetitively in consecutive segments of size δz until the end of the fiber. It should be highlighted that the linear operator is computed in the frequency domain, while the nonlinear operator is calculated in the time domain.

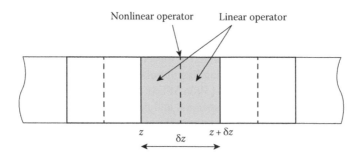

FIGURE A5.10   Schematic illustration of symmetric SSFM.

A5.4.1.1 Modeling of PMD

First-order PMD can be implemented by modeling the optical fiber as two separate paths representing the propagation of two PSPs. Symmetrical SSFM is carried out on each transmission path before the outputs of these two paths are superimposed to calculate the output optical field. The transfer function to represent the first-order PMD is given by

$$H(f) = H^+(f) + H^-(f) \qquad (A5.31)$$

where $H^+(f) = \sqrt{k}\,\exp\left[j2\pi f\left(-\dfrac{\Delta\tau}{2}\right)\right]$ and $H^-(f) = \sqrt{k}\,\exp\left[j2\pi f\left(-\dfrac{\Delta\tau}{2}\right)\right]$, in which k is the power splitting ratio, and $k = 1/2$ when using a 3 dB or 50:50 optical coupler/splitter, and $\Delta\tau$ is the instantaneous DGD value following a Maxwell distribution (refer to Equation A5.31).

A5.4.2 Optimization of Symmetrical SSFM

A5.4.2.1 Optimization of Computational Time

A huge amount of time is spent in symmetric SSFM for FFT and IFFT operations, particularly when fiber nonlinear effects are involved. In practice, when optical pulses propagate toward the end of a fiber span, the pulse intensity has been greatly attenuated due to fiber attenuation. As a result, fiber nonlinear effects are getting negligible for the rest of that fiber span, and, hence, the transmission is operating in a linear domain in this range. In this research, a technique to configure symmetric SSFM is proposed in order to reduce the computational time. If the peak power of an optical pulse is lower than the nonlinear threshold of the transmission fiber, for example, around -4 dBm, symmetrical SSFM is switched to a linear-mode operation. This linear mode involves only fiber dispersions and fiber attenuation, and its low-pass equivalent transfer function for the optical fiber is

$$H(\varpi) = \exp\left\{-j\left[(1/2)\beta_2\varpi^2 + (1/6)\beta_3\varpi^3\right]\right\} \qquad (A5.32)$$

If β_3 is not considered in this fiber transfer function, which is normally the case due to its negligible effects on 40 Gbps and lower-bit-rate transmission systems, the preceding transfer function has a parabolic phase profile [39,40].

A5.4.2.2 Mitigation of Windowing Effect and Waveform Discontinuity

In symmetric SSFM, mathematical operations of FFT and IFFT play very significant roles. However, due to the finite window length required for FFT and IFFT operations, these operations normally introduce overshooting at two boundary regions of the FFT window, commonly known as the windowing effect of FFT. In addition, since the FFT operation is a block-based process, there exists the issue of waveform discontinuity, that is, the right-most sample of the current output block does not start at the same position of the left-most sample of the previous output block. The windowing effect and the waveform discontinuity problems are resolved with the following solutions (see Figure A5.11).

1. The actual window length for FFT/IFFT operations consists of two blocks of samples (2N sample length). The output, however, is a truncated version, with the length of one block (N samples) and output samples taken in the middle of the two input blocks.
2. The next FFT window overlaps the previous one by one block of N sample.

A5.5 OPTICAL AMPLIFIER

In the modeling of an EDFA, the two key parameters are ASE noises and NF.

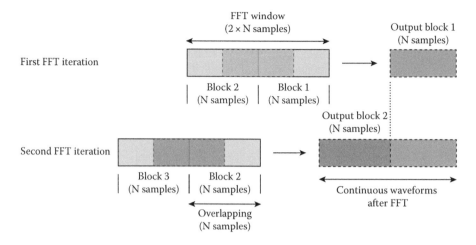

FIGURE A5.11 Proposed solution for mitigating the windowing effect and waveform discontinuity caused by FFT/IFFT operations.

A5.5.1 ASE NOISES

The following formulation accounts for the average power of ASE noises (N_{ASE}) [8]

$$N_{ASE} = mn_{sp}hf(G - 1)B_o \tag{A5.33}$$

where f is the optical operating frequency, G is the amplifier optical gain, n_{sp} is the spontaneous emission factor, h is Planck's constant, m is the number of polarization modes (usually $m = 2$ for systems without using polarization multiplexing), and B_o is the bandwidth of an optical filter.

A5.5.2 NOISE FIGURE (NF)

NF of an EDFA is defined as the ratio between the output OSNR and the OSNR at the input of the EDFA, given by

$$NF = \frac{OSNR_{in}}{OSNR_{out}} \approx 2n_{sp} \text{ for G} \gg 1 \tag{A5.34}$$

Assuming that EDFAs are operating in the saturation regions, and by substituting NF from Equation A5.33 in Equation A5.34, the N_{ASE} can be related to NF as follows

$$N_{ASE} = (NF \cdot G - 1)hfB_o \tag{A5.35}$$

A5.6 OPTICAL AND ELECTRICAL FILTERS

In this research, filtering of noise-corrupted optical signals is conducted with a Gaussian-type optical filter whose impulse response is governed by the following expression

$$h_{Gauss}(t) = \frac{1}{\sqrt{2\pi}\xi} e^{\left(\frac{-t^2}{2\xi^2}\right)} \tag{A5.36}$$

where $\xi = \dfrac{\sqrt{\ln(2)}}{2\pi BT}$, in which B is the one-sided 3 dB bandwidth of the Gaussian filter and T is the bit period. Variations of these two parameters are reflected by the change of the BT product.

The modeling of an electrical filter can also be implemented with a Gaussian filter having a similar impulse response to Equation A5.35. Alternatively, a fifth-order Bessel filter is used that can be easily designed using the filter design toolbox in MATLAB. The MATLAB pseudo-codes for designing a fifth-order Bessel filter are shown as follows

$[b,a]$ = besself(5th order, 2*pi*BT/sampling_factor)
$[bz,az]$ = impinvar(b,a,1);
$[hf\ t1]$ = impz(bz,az,2*delay* sampling_factor + 1, sampling_factor)

where BT product is defined similarly to that of the Gaussian filter.

A5.7 OPTICAL RECEIVER

The original message is recovered in the electrical domain, and thus a conversion of lightwaves into electrical signals is required. In optical communications, this process is widely implemented with a positive-intrinsic-negative (PIN) photodiode for either coherent or incoherent detections. The first type requires a local oscillator to coherently convert modulated lightwaves from optical frequency range down to intermediate frequency (IF) range. On the contrary, the incoherent detection that has been preferable since the past decade is based on the square law envelope detection of optical signals. The incoherent detection, however, still requires the recovery of the clock timing. In the modeling of digital photonic transmission systems in this research, the ideal clock timing is assumed. Another key parameter of an optical receiver is the responsivity parameter of the PIN photodiode, which is a measure of the efficiency of the photonic-electronic conversion. Photodiodes with high responsivity of around 0.8–0.9 A/W are commercially available. In this research, it is assumed that the responsivity is equal to 1 A/W. Moreover, the induced electrical current is usually amplified with a transimpedance amplifier (TIA) and then passed through an electrical filter. At this stage, electrical eye diagrams are observed, and the sampling process of electrically filtered received signals is carried out for recovering the data information.

Received signals are corrupted by noise from several sources. They include the shot noise (σ_{shot}^2), the electronic noise (σ_{elec}^2), the dark current noise (σ_{dark}^2), and the interactions between signals and ASE noises ($\sigma_{signal,ASE}^2$), as well as between ASE noises themselves ($\sigma_{ASE,ASE}^2$). The summation of these noise sources reflects the total receiver noise.

$$\sigma_{total}^2 = \sigma_{shot}^2 + \sigma_{elec}^2 + \sigma_{dark}^2 + \sigma_{signal,ASE}^2 + \sigma_{ASE,ASE}^2 \qquad (A5.37)$$

These noise sources are modeled with normal distributions whose variances represent the noise power, and they are described as follows:

1. *Shot noise* (σ_{shot}^2) is caused by the intrinsic opto-electronic phenomenon of the semiconductor photodiode. A random number of electron-hole pairs is generated with the receipt of photons, causing the randomness of the induced photo-current. The shot noise is given by

$$\sigma_{shot}^2 = 2 \cdot q \langle i_s \rangle B_e \qquad (A5.38)$$

where B_e is the 3 dB bandwidth of the electrical filter and $<i_s>$ is the average signal current (with the unit of A/Hz).

2. *Electronic noise* source σ_{elec}^2 is injected from the TIA. It is modeled by an equivalent noise current i_{Neq} over the bandwidth B_e of the electrical filter. The unit of i_{Neq} is $A/\sqrt{Hz}$ and the value of σ_{elec}^2 is obtained from

$$\sigma_{elec}^2 = (i_{Neq})^2 B_e \tag{A5.39}$$

3. *Dark current* i_{dark} is normally specified for a photodiode and has the unit of A/Hz. Hence, the noise power σ_{dark}^2 is calculated as

$$\sigma_{dark}^2 = 2 \cdot q i_{dark} B_e \tag{A5.40}$$

4. The variances of amplitude fluctuations due to the beating of signal and ASE noises and between ASE noises itself are given in the following expressions

$$\sigma_{signal,ASE}^2 = 4 \cdot i_S i_N \frac{B_e}{B_{opt}} \tag{A5.41}$$

$$\sigma_{ASE-ASE}^2 = i_N^2 \frac{B_e}{B_{opt}^2} (2 \cdot B_{opt} - B_e) \tag{A5.42}$$

where B_{opt} is the 3 dB bandwidth of the optical filter and i_N is the noise-induced photocurrent. In practice, the value of $\sigma_{ASE,ASE}^2$ is normally negligible compared to the value of $\sigma_{signal,ASE}^2$ and can be ignored without affecting the performance of the receiver. In addition, in an optically preamplified receiver, that is, the optical signal is amplified at a stage before the photo-detector, the $\sigma_{signal,ASE}^2$ is the dominant factor compared to other noise sources.

A5.8   PERFORMANCE EVALUATION

The performance of an optical transmission system can be evaluated by using conventional techniques such as the Monte Carlo method and the single Gaussian distribution method. However, these conventional techniques have several limiting factors. The main limitation in the Monte Carlo method is the large amount of time needed for a simulation experiment, whereas the single Gaussian method does not take into account distortions caused by the dynamic effects of optical fibers. To cope with these issues, two statistical methods are presented in this research for novel applications in optical fiber communication, and they offer flexible and fast processing methods to obtain the BER performance. Several main characteristics of these methods are as follows

1. The first method implements the expected maximization (EM) theorem, in which the probability distribution function (PDF) of the received electrical signals is estimated as a mixture of multiple Gaussian distributions (MGDs). Although the application of this method in optical communications has recently been reported [41,42], the guidelines for optimizing the accuracy of this method are yet to be presented.
2. The second method, which is based on the generalized extreme values (GEV) theorem, is the generalized Pareto distribution (GPD) method. This method predicts the probability of the occurrence of extreme values that occur within the long tail of the signal PDF. Although the GPD method is popularly used in several fields such as finance [43], meteorology [44], and climate forecasting [45]; it has not yet been applied in the field of optical communications.

BER values calculated from all of these evaluation methods are normally plotted as a function of OSNR, which is discussed in the next subsection.

A5.8.1 OSNR

OSNR is a metric for the quality assessment of received signals that are corrupted by ASE noises of EDFAs. OSNR is defined as the ratio of the average optical signal power to the average optical noise power. For a single EDFA with the output power P_{out} and the noise power N_{ASE}, OSNR is computed as [8]

$$\text{OSNR} = \frac{P_{out}}{N_{ASE}} = \frac{P_{out}}{(NF \cdot G - 1)hf\Delta f} \tag{A5.43}$$

where NF is the noise figure, G is the amplifier gain, hf is the photon energy, and Δf is the optical measurement bandwidth. When addressing an OSNR value, it is important to define an optical reference bandwidth for the calculation of OSNR. A bandwidth Δf of 12.5 GHz (or $\Delta\lambda = 0.1$ nm) is the typical reference bandwidth for calculating OSNR values.

A5.8.1.1 OSNR Penalty

OSNR penalty is obtained from the BER curves and determined at a particular BER. A value of the OSNR penalty is obtained by comparing the values of OSNR before and after the change of the parameters that are under test, as given by

$$\text{OSNR\_Penalty} = 10\log\left(\frac{\text{OSNR}_{before}}{\text{OSNR}_{after}}\right) \tag{A5.44}$$

A5.8.2 Eye Opening

OSNR is a time-averaged indicator for the ratio of average optical signal power over average optical noise power. Hence, OSNR is used most effectively when noise is the main degrading factor to the system performance. However, the OSNR metric is getting less accurate the system degradation is mainly due to waveform distortions. These waveform distortions can originate either from the ISI problem caused by fiber CD and PMD, from fiber nonlinearities, or from the effects of narrowband optical/electrical filtering. In contrast, Waveform distortions are taken into account by using the eye opening (EO) metric. The EO is determined from the difference between the "mark" and "space" levels. In addition, EOP is the penalty of an EO when comparing to a reference EO. This reference EO is usually obtained from a back-to-back configuration when signal waveforms are not distorted at all. The EOP is normally in log scale (dB) and given by

$$\text{EOP} = 10\log\left(\frac{\text{EO}_{ref}}{\text{EO}_{received}}\right) \tag{A5.45}$$

EO and EOP metrics are useful for noise-free systems as they provide a good measure for pulse distortions. If noise is present, calculations of EO and EOP become less precise. In addition, the accuracy of EO and EOP calculations relies on the sampling instance. The detected pulses are usually sampled at the middle of the eye diagrams where the EO is maximum.

A5.8.3 Conventional Evaluation Methods

A5.8.3.1 Monte Carlo Method

The BER in simulation experiments are computed as the ratio between the number of error occurrences (N_{error}) and the total number of transmitted data bits (N_{tot})

$$\text{BER} = \frac{N_{error}}{N_{trans}} \tag{A5.46}$$

The Monte Carlo method provides a precise BER calculation, as it takes into account effects arising from all the fiber impairment. However, the Monte Carlo method requires a huge number of transmitted information bits to obtain low values of BER, thus leading to excessive computational time. A BER of 1e−9, which is considered as "error free" in fiber optic communications, requires at least 1e−10 transmitted bits. Furthermore, a BER of up to 1e−12 is now becoming more common for modern high-capacity and high-speed digital photonic transmission systems. These huge numbers are not feasible for simulation experiments conducted using personal computers, as done presently. In addition, time-consuming operations such as FFT/IFFT while implementing the symmetrical SSFM also significantly increase the computational time.

However, the Monte Carlo method is still used effectively in simulation experiments because commercial optical transmission systems usually employ forward error correction (FEC) coding schemes to significantly enhance system performance. Pre-FEC BERs can thus be as low as 2e−3 (the FEC limit), provided that no sign of long-burst errors is observed [23]. Figure A5.12 demonstrates significant improvements in system performance by using FEC schemes: 7% single-stage and 23% concatenated Reed-Solomon code [23].

With the current processing speed of contemporary personal computers, a BER of up to 1e−6 is of interest for simulation experiments. BERs calculated from the Monte Carlo method are usually taken as benchmarks for BER values computed by other methods.

A5.8.3.2 Single Gaussian Statistical Method

This method implements a statistical process to calculate BER. It should be noted that signals are normally in voltage unit since photodiode-induced currents are amplified by a transimpedance electrical amplifier which then transfers the current into voltage. A particular voltage serves as a reference for distinguishing "0" and "1" levels, known as the threshold voltage (V_{th}). In addition, the received signals are sampled at a particular instance. As a result, based on this V_{th}, normalized histograms of received signals for "0" and "1" are obtained, thus leading to the achievement of the PDF. The PDF normally complies to a distribution such as Gaussian, Chi-square, and so on. The BER is calculated as [46]

$$BER = P("1")P("0"|"1")+P("0")P("1"|"0") \tag{A5.47}$$

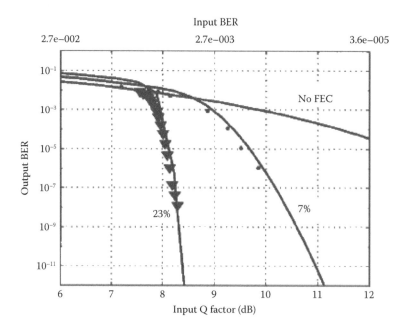

FIGURE A5.12 Demonstration of effectiveness of FEC schemes in improving system performance.

where:

P("1") is the probability that a bit "1" is sent.

P("0"|"1") is the probability of error due to receiving "0" where actually a "1" is sent.

P("0") is the probability that a "0" is sent.

P("1"|"0") is the probability of error due to receiving "1" where actually a "0" is sent.

In the case of binary digital transmission, the probability of transmitting a "0" or "1" is equal, that is, P("1") = P("0") = 1/2. P("0"|"1") and P("1"|"0") are calculated from integrating the overlap region of the PDF that exceeds the threshold voltage.

A popular assumption is that the PDF of received electrical signals follow a Gaussian distribution. This enables the fast calculation for BER values by using complementary error functions (ERFC) [46]

$$\text{BER} = \frac{1}{2}\left[\text{erfc}\left(\frac{|\mu_1 - V_{\text{th}}|}{\sqrt{2}\sigma_1}\right) + \text{erfc}\left(\frac{|\mu_0 - V_{\text{th}}|}{\sqrt{2}\sigma_0}\right)\right] \qquad (A5.48)$$

where μ_0, μ_1, and σ_0, σ_1 are means and variances of the Gaussian PDFs of "0" and "1" received signals, respectively.

Apart from BER, the quality (Q) factor is another common metric to assess system performance. Q values are calculated from μ_0, μ_1, and σ_1, σ_1 as follows [46]

$$Q = \frac{\mu_1 - \mu_0}{\sigma_1 - \sigma_0} \qquad (A5.49)$$

and Q factors are either in linear scale or in dB scale. The BER can be obtained from the Q factor by [46]

$$\text{BER} = \frac{1}{2}\text{erfc}\left(\frac{Q}{\sqrt{2}}\right) \qquad (A5.50)$$

A5.8.3.2.1 *Improving Accuracy of Histograms*

The PDFs of "0" and "1" received signals are determined from normalized histograms. The estimation of these normalized histograms thus considerably affects the BER's accuracy. Thus, a proper estimation for histogram value is very important. A histogram is normally divided into a number of bins whose binwidths are the same with a sufficiently large number of transmitted bits (N_0). A good estimate for the width (W_{bin}) of each equally spaced histogram bin is given by $W_{\text{bin}} = \sqrt{N_0}$ [47].

A5.8.4 Novel Statistical Methods

The single Gaussian distribution (the Q factor) method considers only the effects of noise corruption on the detected signals while ignoring waveform distortions caused by fiber dispersion and fiber nonlinear effects. These distortions result in multipeak and non-Gaussian PDFs (see Figure A5.13). Such PDFs cannot be correctly estimated by the conventional single Gaussian distribution method.

The preceding issue is resolved with the following two statistical methods: the MGD method based on the EMax (EM) theorem, and the GPD method based on the GEV theorem. The implementation of these methods is conducted in MATLAB. It should be noted that the conventional single Gaussian distribution method is a particular case of the MGD method.

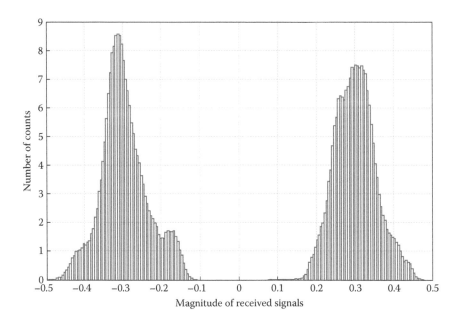

FIGURE A5.13   Demonstration of multipeak and non-Gaussian distributions of received signals.

A5.8.4.1   Multi-Gaussian Distributions Method

One of the most popular applications of the EM theorem is to obtain parameters of mixed probability densities. This theorem is based on the fact that most of deterministic distributions can be considered as a superposition of multiple distributions. The probability distribution function $p(x|\Theta)$ for a set of received data can be expressed as the mixture of M different distributions [48]

$$p(x|\Theta) = \sum_{i=1}^{M} w_i p_i(x|\theta_i); \quad \Theta = (w_1,...,w_M,\theta_1,...,\theta_M) \tag{A5.51}$$

where $p_i(x|\theta_i)$ represents the PDF of each distribution in the mixture, and each PDF has a weight w_i such that $\sum_{i=1}^{M} w_i = 1$. This weight indicates the probability of each PDF. When adopted for optical communications, the EM algorithm is implemented with a mixture of MGDs. A critical stage in the multi-Gaussian distributions (MGD) method that affects the accuracy of BER calculations is the estimation of the number of Gaussian distributions for use in the mixture.

A5.8.4.1.1   Selection of Number of Gaussian Distributions in MGD Method

The number of Gaussian distributions to be used is estimated by the number of peak and valley pairs in the first and second derivatives of the original data set. This is illustrated in Figure A5.14 (courtesy of Ref. [49]). This figure is based on the "Heming Lake Pike" example [50,51]. In this example, the data of five age-groups give the lengths of 523 pikes, and they were sampled in 1965 from Heming Lake, Manitoba, Canada. The components are heavily overlapped, and the resultant PDF is obtained with a mixture of five Gaussian distributions (refer to the top figure of Figure A5.14). The number of Gaussian distributions is then estimated from the number of peak and valley pairs in the first- and second-derivative curves of the original data set. As seen from Figure A5.14, the first derivative of the mixed PDF clearly shows four pairs of peaks (mid gray dots) and valleys (blue dots), suggesting that there should be at least four Gaussian distributions contributing to the original PDF. However, by taking the second derivative, it is realized that there is actually up to five contributed Gaussian distributions.

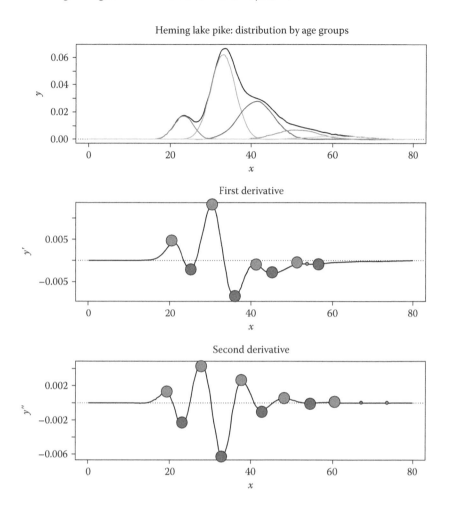

FIGURE A5.14 Process of estimating the number of Gaussian PDFs for MGD fitting based on the number of pairs of inflection peaks and valleys in the first- and second-derivative curves.

A5.8.4.1.2 Steps for Implementing MGD Method to Obtain BERs

1. Obtaining the PDF by the normalized histogram of received electrical signals.
2. Estimating the number of Gaussian distributions (N_{Gaus}) used for fitting the PDF of the original data set. This selection is based on the guidelines explained earlier.
3. Implementing the EM algorithm in MATLAB with the mixture of N_{Gaus} Gaussian distributions to calculate values of mean, variance, and weight for each distribution.
4. Calculating BERs by integrating the tails of these Gaussian distributions when these tails cross the threshold value.

A5.8.4.2 Generalized Pareto Distribution (GPD) Method

Generalized Pareto distribution (GPD) is a subset of the GEV theorem, which consists of GEV and GPD distributions. Both methods can be used to determine the probabilities of extreme events occurring in the tails of the data PDF. However, the key difference between these two methods is that GEV requires the whole signal PDF, whereas GPD only needs the tail regions of the signal PDF. There is only one report on the application of GEV theorem in fiber optic communications [52]. However, this study looks at GEV distribution for the OOK optical system and only involves noise effects while neglecting the effects of fiber impairments. Moreover, similar to the single Gaussian distribution, GEV distribution fails to precisely estimate the multipeak PDF of the received signals, which are mainly caused by waveform distortions.

When nonlinearity is the dominant shortcoming in the performance of optical transmission systems, sampled received signals usually introduce a long-tailed PDF. This differs from the Gaussian PDF, which has slow roll-off tails. As a result, the conventional BER based on the assumption of Gaussian PDF is no longer valid, and often underestimates the BER.

A wide range of analytical techniques have recently been studied for optical communications, such as importance sampling, and multicanonical and covariance matrix methods [53–56]. Although these techniques provide precise BERs, they are quite complicated. In contrast, the GPD method has been widely used in various fields [43,44], and it has become available in recent MATLAB versions (since MATLAB version 7.1). Thus, GPD provides the potential of a fast and convenient method for evaluating the system performance in both practical and simulation scenarios.

The PDF for the GPD fitting function is defined as follows [57]

$$y = f(x|k,\sigma,\theta) = \left(\frac{1}{\sigma}\right)\left(1+k\frac{(x-\theta)}{\sigma}\right)^{-1-\frac{1}{k}} \text{ for } \theta < x \text{ when } k > 0 \text{ or for } \theta < x < \frac{-\sigma}{k} \ k < 0 \quad \text{(A5.52)}$$

where k is the shape parameter ($k \neq 0$), σ is the scale parameter, and θ is the threshold parameter. The key constraints for the Equation A5.52 are described in the following:

1. When $k > 0$ and $\theta < x$: there is no upper bound for x.
2. When $k < 0$ and $\theta < x < -\sigma/k$: zero probability for the case $x > -\sigma/k$.
3. When $k = 0$, Equation A5.52 turns to

$$y = f(x|0,\sigma,\theta) = \left(\frac{1}{\sigma}\right)e^{-\frac{(x-\theta)}{\sigma}} \text{ for } \theta < x \quad \text{(A5.53)}$$

4. When $k = 0$ and $\theta = 0$, GPD is equivalent to the exponential distribution.
5. When $k > 0$ and $\theta = \sigma$, GPD is equivalent to the Pareto distribution.

Accordingly, the GPD method has three basic classes of the underlying distributions:

1. Distributions whose tails decrease exponentially, such as the normal distribution, have shape parameters equal to zero.
2. Distributions with tails decreasing as a polynomial, such as Student's t distribution, lead to a positive shape parameter.
3. Distributions having finite tails, such as the beta distribution, have negative shape parameters.

The first step and also the most critical step affecting the accuracy of BER calculations is to find the precise threshold for the GPD fitting function.

A5.8.4.2.1 Selection of GPD Threshold

The threshold value (V_{GPD}) used for the GPD fitting function indicates the start of the generalized Pareto distribution. There have been several suggested guidelines to aid the selection of the GPD threshold [58–60]. However, they are either too complicated or not applicable in optical communications.

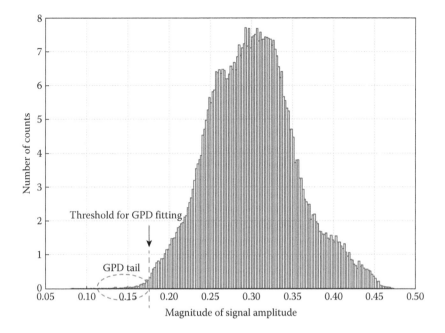

FIGURE A5.15   Selection of threshold for GPD fitting.

In this research, simple guidelines to determine the GPD threshold value are proposed. These guidelines are based on the observation that extreme values in the long tail region normally comply with a slow exponential slope compared to a faster decaying slope when they are close to the peak of the distribution. The interception region of these two slopes gives a good estimated region for the GPD threshold value (see Figure A5.15).

The accuracy in the selection of V_{GPD} is evaluated by using the cumulative density function (CDF) and the Quantile–Quantile plot (Q–Q plot). If there is a high correlation between the distribution of the tail of the original data set and the implemented GPD function, there should be a good fit between the empirical CDF of the original data set and the GPD-fitted CDF (see Figure A5.16). Furthermore, this high correlation is also reflected in a linear trend of the Q–Q plot, as observed in Figure A5.17.

Figures A5.18 and A5.19 demonstrate the inaccuracy of the GPD method caused by an improper selection of V_{GPD}. Figure A5.18 shows the discrepancy between the GPD-fitted CDF and the empirical CDF of data, while Figure A5.19 clearly displays a nonlinear trend of the Q–Q plot instead of a linear trend.

A5.9   VALIDATION OF MGD AND GPD METHODS

The validation of MGD and GPD methods is conducted in an optical DPSK transmission system over 880 km SSMF dispersion-managed optical link (see Figure A5.20).

Each span consists of 100 km SSMF and 10 km of DCF whose dispersion values are +17 ps/(nm·km) and −170 ps/(nm.km) at 1550 nm wavelength, respectively. Both CD and the dispersion slopes are fully compensated, that is, zero residual dispersion. The average optical input power (P_{in}) into each span is set for the system to operate in the nonlinear transmission region. The attenuation of each span is fully compensated for by using EDFA1 and EDFA2, which have optical gains of 10 and 19 dB, respectively.

The nonlinear SPM effect is the main factor degrading system performance. This is of interest since fiber SPM creates a nondeterministic long-tailed region in the obtained PDFs of the received

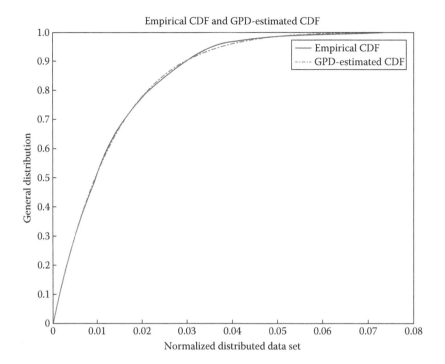

FIGURE A5.16   Demonstration of high correlation between GPD-fitted and empirical CDF.

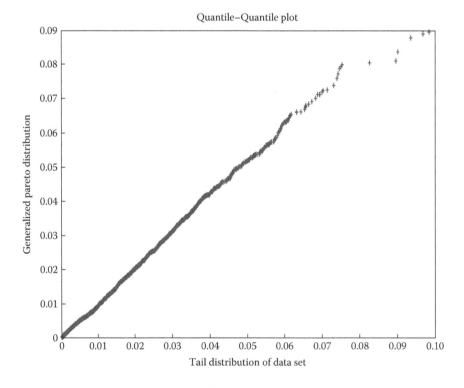

FIGURE A5.17   Q–Q plot of a high correlation GPD-fitting function.

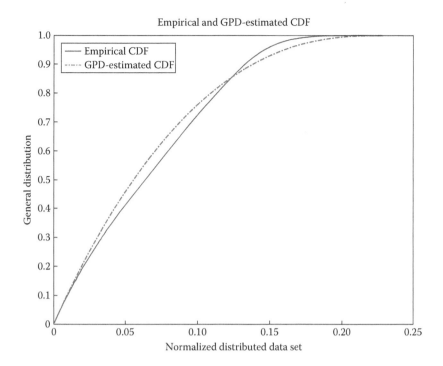

FIGURE A5.18 Demonstration of incompliance between the empirical CDF and the GPD-fitted CDF due to improper selection of V_{GPD}.

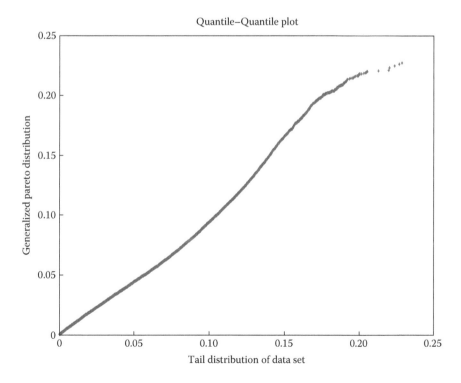

FIGURE A5.19 Nonlinear trend in the Q–Q plot caused by an improper selection of V_{GPD}.

electrical signals. The BER results calculated from the novel statistical methods are compared to those from the Monte Carlo and the semi-analytical methods. The semi-analytical method to obtain the BER for optical DPSK transmission systems has been reported previously [1–3], and is expressed in the following

$$\text{BER} = \frac{1}{2} - \frac{\rho e^{-\rho}}{2} \sum_{k=0}^{\infty} \frac{(-1)^k}{2k+1} \left[I_k \left(\frac{\rho}{2} \right) + I_{k+1} \left(\frac{\rho}{2} \right) \right]^2 e^{-\frac{1}{2}(2k+1)^2 \sigma_{\text{NLP}}^2} \tag{A5.54}$$

where ρ is the obtained OSNR, σ_{NLP}^2 is the variance of nonlinear phase noise, and I_k is the first-kind modified Bessel function. In order to calculate the BERs of optical DPSK transmission systems involving the nonlinear phase noise, Equation A5.1 requires the estimation of OSNR and the variance of the nonlinear phase noise distribution. These parameters are obtained from MATLAB by processing the stored values obtained in the sampling process. Figure A5.21a and b illustrates the fitting curves implemented in the MGD method for the PDFs of bit "0" and bit "1" received signals ($P_{\text{in}} = 10$ dBm).

The selection of the optimal threshold V_{GPD} for the GPD fitting follows the guideline as described in the previous section. BERs for the cases of 10 and 11 dBm fiber input powers obtained from the MGD and GPD statistical methods are compared to the Monte Carlo and semi-analytical methods in Table A5.1.

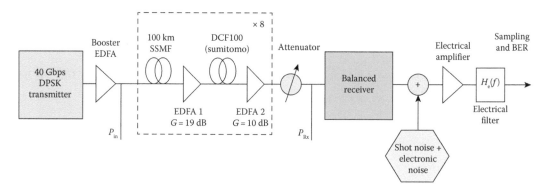

FIGURE A5.20 Simulation setup for validation of novel statistical methods.

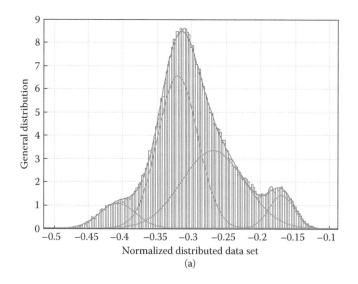

FIGURE A5.21 Demonstration of MGD-fitting curves for received signals of (a) bit "0". *(Continued)*

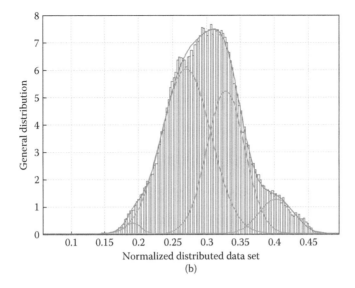

FIGURE A5.21 (Continued)   Demonstration of MGD-fitting curves for received signals of (b) bit "1".

TABLE A5.1

Comparison of BER Values Obtained by Novel Statistical Methods to Monte Carlo and Semi-Analytical Methods

| Input Power | Evaluation Methods | | | |
	Monte Carlo	Semi-Analytical	MGD	GPD
10 dBm	1.7e−5	2.58e−5	5.3e−6	3.56e−4
11 dBm	NA	1.7e−8	2.58e−9	4.28e−8

Table A5.1 shows the adequate accuracy of the proposed novel statistical methods. The discrepancies of the BER values compared to the Monte Carlo and semi-analytical BER are within one decade. Therefore, these methods offer fast processing techniques while maintaining the accuracy of the obtained BER within the acceptable limit.

A5.10   MATLAB® AND SIMULINK® MODELING PLATFORM

A simulation package has been developed on the MATLAB® and Simulink® platform for modeling advanced digital optical transmission systems. The modeling platform developed in this research mainly aims to investigate and verify the benefits and shortcomings of the advanced modulation formats used in fiber optic communications. Thus, single-channel optical transmission systems are of main interest in this research. An earlier version of this MATLAB and Simulink simulation package has been used by Lockheed Martin Corporation, USA. The MATLAB and Simulink modeling platform has several advantages, as follows:

1. The simulator provides toolboxes and block-sets for setting up complicated transmission configurations. In addition, the initialization process for all key parameters of subsystem components can be automatically conducted at the start of any simulation. Furthermore, the initialization file is written in a separate MATLAB file; hence, the simulation parameters can be modified easily.

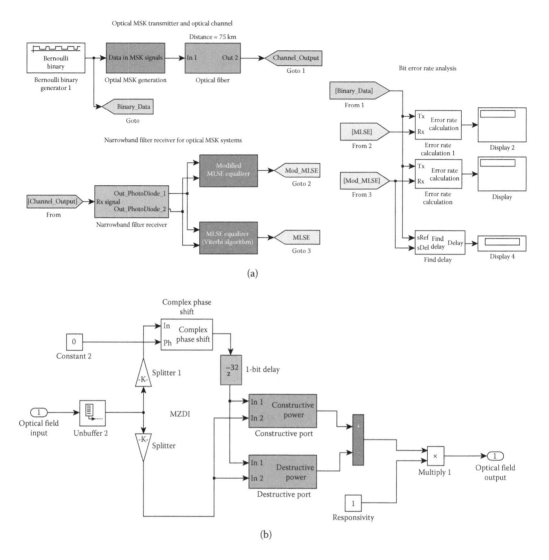

FIGURE A5.22 Demonstration of Simulink® simulation test-bed for (a) OFDR-based MSK transmission systems and MLSE equalizers and (b) balanced receiver integrated with MZDI.

2. Signal monitoring can be carried out easily on any point along the propagation path in a simulation with the simple plug-and-see monitoring scopes provided by Simulink block-sets.
3. Numerical data can be easily stored for postprocessing in MATLAB. This offers a complete package for generating and processing numerical data for the achievement of BER.

A typical modeling platform under MATLAB and Simulink is shown in Figure A5.22.

A5.11 SUMMARY

This annex presented in detail the modeling of critical subsystem components of a digital photonic transmission system, starting from the optical transmitter, onto the properties of the optical fiber and to the optical receiver. Detailed description about the fiber impairments including CD, PMD, and Kerr-effect nonlinearities were provided. Moreover, the symmetrical SSFM was optimized to reduce the computational time and to mitigate artificial errors induced from the windowing effect of FFT/IFFT. The modeling of ASE noises as well as the receiver noise was also discussed.

Apart from the conventional evaluation methods such as Monte Carlo and single Gaussian distribution methods, two statistic methods have been adopted for optical communications. These are the MGD method implementing the EM theorem and the GPD method based on the GEV theorem. The guidelines for optimizing the accuracy of these two methods were provided. These methods offer fast processing techniques to obtain BER values. Finally, a MATLAB and Simulink modeling platform has been developed for advanced digital photonic transmission systems. This platform takes advantage of the user-friendly MATLAB and Simulink, thus making it easy for further development.

REFERENCES

1. S. Ten—Corning Inc, Advanced fibers for submarine and long-Haul applications, in *Proceedings of IEEE LEOS'04,* paper WJ2, 2004.
2. S. Ten—Corning Inc, Advanced fibers for submarine networks, in *Proceedings of SubOptic'07* (invited paper), Maryland, 2007.
3. G. P. Agrawal, *Nonlinear Fiber Optics*, 3rd ed., San Diego: Academic Press, 2001.
4. K. P. Ho, *Phase-Modulated Optical Communication Systems*, New York: Springer, 2005.
5. A. H. Gnauck and P. J. Winzer, Optical phase-shift-keyed transmission, *IEEE J. Lightwave Technol.,* Vol. 23, No. 1, pp. 115–130, 2005.
6. H. Ghafouri-Shiraz, *Distributed Feedback Laser Diodes*, New York: Wiley, 1995.
7. G. Morthier and P. Vankwikelberge, *Handbook of Distributed Feedback Laser Diodes*, Norwood, MA: Artech House, 1995.
8. G. P. Agrawal, *Fiber-Optic Communication Systems*, 3rd ed., New York: Wiley, 2002.
9. M. Lax, Rate equations and amplitude noise, *J. Quant. Electron.,* Vol. 3, No. 37, pp. 37–46, 1967.
10. I. P. Kaminow and T. Li, *Optical Fiber Communications*, Vol. IVA, Academic Press–Elsevier Science (USA), 2002, chap. 6.
11. T. Kawanishi, T. Sakamoto, and M. Izutsu, High-speed control of lightwave amplitude, phase, and frequency by use of electrooptic effect, *IEEE J. Sel. Topics Quant. Electron.,* Vol. 13, No. 1, pp. 79–91, 2007.
12. G. P. Agrawal, *Fiber-Optic Communication Systems*, 3rd ed., New York: Wiley, 2002, chap. 3.
13. E. L. Wooten, K. M. Kissa, A. Yi-Yan, E. J. Murphy, D. A. Lafaw, and P. F. Hallemeier, A review of lithium niobate modulators for fiber-optic communications systems, *IEEE Journal of Selected Topics in Quantum Electronics*, Vol. 6, No. 1, pp. 69–82, 2000.
14. Operating Manual, DPSK Optical Transmitter—SHF 5003, SHF Communication Technologies AG—Germany.
15. T. Kawanishi, S. Shinada, T. Sakamoto, S. Oikawa, K. Yoshiara, and M. Izutsu, Reciprocating optical modulator with resonant modulating electrode, *Electron. Lett.,* Vol. 41, No. 5, pp. 271–272, 2005.
16. R. Krahenbuhl, J. H. Cole, R. P. Moeller, and M. M. Howerton, High-speed optical modulator in LiNbO$_3$ with cascaded resonant-type electrodes, *IEEE J. Lightwave Technol.,* Vol. 24, No. 5, pp. 2184–2189, 2006.
17. L. N. Binh, Tutorial Part I on optical systems design, in *Proceedings of ICOCN 2002,* Singapore, November 2002.
18. L. N. Binh, T. L. Huynh, K. Y. Chin, and D. Sharma, Design of dispersion flattened and compensating fibers for dispersion-managed optical communications systems, *Int. J. Wireless Opt. Commun.,* Vol. 2, No. 1, pp. 63–82, 2004.
19. G. P. Agrawal, *Fiber-Optic Communications Systems*, 3rd ed., John Wiley & Sons, New York, 2001.
20. J. B. Jeunhomme, *Single Mode Fibre Optics, Principles and Applications*, 2nd ed., Marcel Dekker Pub., New York, 1990.
21. J. A. Buck, *Fundamentals of Optical Fibers*, New York: Wiley, 1995.
22. G. P. Agrawal, *Fiber-Optic Communication Systems*, 3rd ed., New York: Wiley, 2001.
23. I. P. Kaminow and T. Li, *Optical Fiber Communications*, Vol. IVB, Academic Press, San Diego, CA, USA, 2002, chap. 5.
24. J. P. Gordon and H. Kogelnik, PMD fundamentals: Polarization mode dispersion in optical fibers, *PNAS,* Vol. 97, No. 9, pp. 4541–4550, 2000.
25. Corning. Inc, An Introduction to the Fundamentals of PMD in Fibers, White Paper, July 2006.

26. A. Galtarossa and L. Palmieri, Relationship between pulse broadening due to polarisation mode dispersion and differential group delay in long singlemode fiber, *Electron. Lett.,* Vol. 34, No. 5, pp. 492–493, 1998.

27. J. M. Fini and H. A. Haus, Accumulation of polarization-mode dispersion in cascades of compensated optical fibers, *IEEE Photon. Technol. Lett.,* Vol. 13, No. 2, pp. 124–126, 2001.

28. A. Carena, V. Curri, R. Gaudino, P. Poggiolini, and S. Benedetto, A time-domain optical transmission system simulation package accounting for nonlinear and polarization-related effects in fiber, *IEEE J. Sel. Areas Commun.,* Vol. 15, No. 4, pp. 751–765, 1997.

29. S. A. Jacobs, J. J. Refi, and R. E. Fangmann, Statistical estimation of PMD coefficients for system design, *Electron. Lett.,* Vol. 33, No. 7, pp. 619–621, 1997.

30. I. Kaminow and T. Koch, *Optical Fiber Communications IIIA,* San Diego, CA: Academic Press, 1997.

31. J. Leibrich and W. Rosenkranz, Efficient numerical simulation of multichannel WDM transmission systems limited by XPM, *IEEE Photon. Technol. Lett.,* Vol. 15, No. 3, pp. 395–397, 2003.

32. D. Marcuse, A. R. Chraplyvy, and R. W. Tkach, Dependence of cross-phase modulation on channel number in fiber WDM systems, *IEEE J. Lightwave Technol.,* Vol. 12, No. 5, pp. 885–890, 1994.

33. T. Mizuochi, K. Ishida, T. Kobayashi, J. Abe, K. Kinjo, K. Motoshima, and K. Kasahara, A comparative study of DPSK and OOK WDM transmission over transoceanic distances and their performance degradations due to nonlinear phase noise, *IEEE J. Lightwave Technol.,* Vol. 21, No. 9, pp. 1933–1943, 2003.

34. K. Hoon, Cross-phase-modulation-induced nonlinear phase noise in WDM direct-detection DPSK systems, *IEEE J. Lightwave Technol.,* Vol. 21, No. 8, pp. 1770–1774, 2003.

35. S. Bigo, G. Bellotti, and M. W. Chbat, Investigation of cross-phase modulation limitation over various types of fiber infrastructures, *IEEE Photon. Technol. Lett.,* Vol. 11, No. 5, pp. 605–607, 1999.

36. C. Furst, J. P. Elbers, C. Scheerer, and C. Glingener, Limitations of dispersion-managed DWDM systems due to cross-phase modulation, in *Proceedings of Annual Meeting LEOS'00,* Vol. 1, Rio Grande, 2000, pp. 23–24.

37. H. J. Thiele, R. I. Killey, and P. Bayvel, Influence of transmission distance on XPM-induced intensity distortion in dispersion-managed, amplified fibre links, *Electron. Lett.,* Vol. 35, No. 5, pp. 408–409, 1999.

38. J. P. Gordon and L. F. Mollenauer, Phase noise in photonic communications systems using linear amplifiers, *Opt. Lett.,* Vol. 15, No. 23, pp. 1351–1353, 1990.

39. A. F. Elrefaie and R. E. Wagner, Chromatic dispersion limitations for FSK and DPSK systems with direct detection receivers, *IEEE Photon. Technol. Lett.,* Vol. 3, No. 1, pp. 71–73, 1991.

40. A. F. Elrefaie, R. E. Wagner, D. A. Atlas, and A. D. Daut, Chromatic dispersion limitation in coherent lightwave systems, *IEEE J. Lightwave Technol.,* Vol. 6, No. 5, pp. 704–710, 1988.

41. D. Ye and W. D. Zhong, Improved BER monitoring based on amplitude histogram and multi-Gaussian curve fitting, *J. Opt. Networking,* Vol. 6, No. 6, pp. 584–598, 2007.

42. L. Ding, W.-D. Zhong, C. Lu, and Y. Wang, New bit-error-rate monitoring technique based on histograms and curve fitting, *Opt. Express,* Vol. 12, No. 11, pp. 2507–2511, 2004.

43. E. J. Bomhoff, *Financial Forecasting for Business and Economics,* Spiral ed., San Diego, CA: Academic Press, 1995.

44. B. B. Brabson and J. P. Palutikof, Test of the generalized Pareto distribution for predicting extreme wind speed, *J. Appl. Meteorol.,* Vol. 39, pp. 1627–1640, 2000.

45. T. Schneider, Analysis of incomplete climate data: Estimation of mean values and covariance matrices and imputation of missing values, *J. Clim.,* Vol. 14, pp. 853–871, 2000.

46. J. G. Proakis, *Digital Communications,* 4th ed., New York: McGraw-Hill, 2001.

47. W. H. Tranter, K. S. Shanmugam, T. S. Rappaport, and K. L. Kosbar, *Principles of Communication Systems Simulation with Wireless Applications,* NJ: Prentice Hall, 2004.

48. A. P. Dempster, N. M. Laird, and D. B. Rubin, Maximum-likelihood from the incomplete data via the EM algorithm, *J. Roy. Stat. Soc.,* Vol. 39, pp. 1–38, 1977.

49. E. F. Glynn, *Mixtures of Gaussians,* Stowers Institute for Medical Research, Research notes, Retrieved http://research.stowers-institute.org/mcm/efg/R/Statistics/MixturesOfDistributions/index.htm, from November 2014.

50. P. D. M. Macdonald, Analysis of length-frequency distributions, in *Age and Growth of Fish,* Ames, IA: Iowa State University Press, 1987, pp. 371–384.

51. P. D. M. Macdonald and T. J. Pitcher, Age-groups from size-frequency data: a versatile and efficient method of analysing distribution mixtures, *J. Fish. Res. Board Can.* Vol. 36, pp. 987–1001, 1979.

52. Y. Kopsinis, J. Thompson, and B. Mulgrew, Performance evaluation of optical communication systems using extreme value theory, in *Proceedings of IEE Seminar on Optical Fibre Communications and Electronic Signal Processing,* London, UK, December 2005.

53. Y. Yadin, M. Shtaif, and M. Orenstein, Bit-error rate of optical DPSK in fiber systems by multicanonical Monte Carlo Simulations, *IEEE Photon. Technol. Lett.,* Vol. 17, No. 6, pp. 1355–1357, 2005.

54. W. Pellegrini, J. Zweck, C. R. Menyuk, and R. Holzlohner, Computation of bit error ratios for a dense WDM system using the noise covariance matrix and multicanonical Monte Carlo methods, *IEEE Photon. Technol. Lett.,* Vol. 17, No. 8, pp. 1644–1646, 2005.

55. D. Yevick, Multicanonical communication system modeling-application to PMD statistics, *IEEE Photon. Technol. Lett.,* Vol. 14, No. 11, pp. 1512–1514, 2002.

56. N. B. Mandayam and B. Aazhang, Importance sampling for analysis of direct detection optical communication systems, *IEEE Trans. Commun.,* Vol. 43, No. 234, pp. 229–239, 1995.

57. MATLAB Helpdesk, Statistical Toolbox, Generalized Pareto Distribution, http://www.mathworks.com/access/helpdesk/help/toolbox/stats.

58. R. L. Smith, *Handbook of Applicable Mathematics; Extreme Value Theory*, Wiley, New York, 1989.

59. A. C. Davidson, Models for exceedances over high thresholds, *J. Roy. Stat. Soc.,* Vol. B52, pp. 393–442, 1990.

60. R. L. Smith and I. Weissman, Estimating the extremal index, *J. Roy. Stat. Soc.,* Vol. B56, pp. 515–528, 1994.

Index